Grundlehren der mathematischen Wissenschaften 317

A Series of Comprehensive Studies in Mathematics

R. Tyrrell Rockafellar • Roger J-B Wets

Variational Analysis

with figures drawn by Maria Wets

Springer

R. Tyrrell Rockafellar
Department of Mathematics
University of Washington
Seattle, WA 98195-4350
USA
rtr@math.washington.edu

Roger J-B Wets
Department of Mathematics
University of California at Davis
One Shields Ave.
Davis, CA 95616
USA
rjbwets@ucdavis.edu

ISSN 0072-7830
ISBN 978-3-540-62772-2 e-ISBN 978-3-642-02431-3
DOI 10.1007/978-3-642-02431-3
Springer Dordrecht Heidelberg London New York

Library of Congress Control Number: 2009929711

Mathematics Subject Classification (2000): 47H05, 49J40, 49J45, 49J52, 49K40, 49N15, 52A50, 52A41, 54B20, 54C60, 54C65, 90C31

Cover design: WMXDesign GmbH, Heidelberg

Printed on acid-free paper

Springer is part of Springer Science+Business Media (www.springer.com)

PREFACE

In this book we aim to present, in a unified framework, a broad spectrum of mathematical theory that has grown in connection with the study of problems of optimization, equilibrium, control, and stability of linear and nonlinear systems. The title *Variational Analysis* reflects this breadth.

For a long time, 'variational' problems have been identified mostly with the 'calculus of variations'. In that venerable subject, built around the minimization of integral functionals, constraints were relatively simple and much of the focus was on infinite-dimensional function spaces. A major theme was the exploration of variations around a point, within the bounds imposed by the constraints, in order to help characterize solutions and portray them in terms of 'variational principles'. Notions of perturbation, approximation and even generalized differentiability were extensively investigated. Variational theory progressed also to the study of so-called stationary points, critical points, and other indications of singularity that a point might have relative to its neighbors, especially in association with existence theorems for differential equations.

With the advent of computers, there has been a tremendous expansion of interest in new problem formulations that similarly demand such modes of analysis but are far from being covered by classical concepts, not to speak of classical results. For those problems, finite-dimensional spaces of arbitrary dimensionality are important alongside of function spaces, and theoretical concerns go hand in hand with the practical ones of mathematical modeling and the design of numerical procedures.

It is time to free the term 'variational' from the limitations of its past and to use it to encompass this now much larger area of modern mathematics. We see 'variations' as referring not only to movement away from a given point along rays or curves, and to the geometry of tangent and normal cones associated with that, but also to the forms of perturbation and approximation that are describable by set convergence, set-valued mappings and the like. Subgradients and subderivatives of functions, convex and nonconvex, are crucial in analyzing such 'variations', as are the manifestations of Lipschitzian continuity that serve to quantify rates of change.

Our goal is to provide a systematic exposition of this broader subject as a coherent branch of analysis that, in addition to being powerful for the problems that have motivated it so far, can take its place now as a mathematical discipline ready for new applications.

Rather than detailing all the different approaches that researchers have been occupied with over the years in the search for the right ideas, we seek to reduce the general theory to its key ingredients as now understood, so as to make it accessible to a much wider circle of potential users. But within that consolidation, we furnish a thorough and tightly coordinated exposition of facts and concepts.

Several books have already dealt with major components of the subject. Some have concentrated on convexity and kindred developments in realms of nonconvexity. Others have concentrated on tangent vectors and subderivatives more or less to the exclusion of normal vectors and subgradients, or vice versa, or have focused on topological questions without getting into generalized differentiability. Here, by contrast, we cover set convergence and set-valued mappings to a degree previously unavailable and integrate those notions with both sides of variational geometry and subdifferential calculus. We furnish a needed update in a field that has undergone many changes, even in outlook. In addition, we include topics such as maximal monotone mappings, generalized second derivatives, and measurable selections and integrands, which have not in the past received close attention in a text of this scope. (For lack of space, we say little about the general theory of critical points, although we see that as a close neighbor to variational analysis.)

Many parts of this book contain material that is new not only in its manner of presentation but also in research. Each chapter provides motivations at the beginning and throughout, and each concludes with extensive notes which furnish credits and references together with historical perspective on how the ideas gradually took shape. These notes also explain the reasons for some of the decisions about notation and terminology that we felt were expedient in streamlining the subject so as to prepare it for wider use.

Because of the large volume of material and the challenge of unifying it properly, we had to draw the line somewhere. We chose to keep to finite-dimensional spaces so as not to cloud the picture with the many complications that a treatment of infinite-dimensional spaces would bring. Another reason for this choice was the fact that many of the concepts have multiple interpretations in the infinite-dimensional context, and more time may still be needed for them to be sorted out. Significant progress continues, but even in finite-dimensional spaces it is only now that the full picture is emerging with clarity. The abundance of applications in finite-dimensional spaces makes it desirable to have an exposition that lays out the most effective patterns in that domain, even if, in some respects, such patterns are not able go further without modification.

We envision that this book will be useful to graduate students, researchers and practitioners in a range of mathematical sciences, including some front-line areas of engineering and statistics that draw on optimization. We have aimed at making available a handy reference for numerous facts and ideas that cannot be found elsewhere except in technical papers, where the lack of a coordinated terminology and notation is currently a formidable barrier. At the same time, we have attempted to write this book so that it is helpful to readers who want to learn the field, or various aspects of it, step by step. We have provided many figures and examples, along with exercises accompanied by guides.

We have divided each chapter into a main part followed by sections marked by *, so as to signal to the reader a stage at which it would be reasonable, in a first run, to skip ahead to the next chapter. The results placed in the * sections are often important as well as necessary for the completeness of the theory, but they can suitably be addressed at a later time, once other developments begin to draw on them.

For updates and errata, see http://math.ucdavis.edu/~rjbw.

Acknowledgment. We are grateful for all the assistance we have received in the course of this project. The figures were computer-drawn and fine-tuned by Maria Wets, who also in numerous other ways generously gave technical and logistical support.

For the first printing, help with references was provided by Alexander Ioffe, Boris Mordukhovich, and René Poliquin, in particular. Lisa Korf was extraordinarily diligent in reading parts of the manuscript for possible glitches. Useful feedback came not only from these individuals but many others, including Amir Abdessamad, Hedy Attouch, Jean-Pierre Aubin, Gerald Beer, Michael Casey, Xiaopeng M. Dong, Asen Dontchev, Hélène Frankowska, Grant Galbraith, Rafal Goebel, René Henrion, Alejandro Jofré, Claude Lemaréchal, Adam Levy, Teck Lim, Roberto Lucchetti, Juan Enrique Martinez-Legaz, Madhu Nayakkankuppam, Vicente Novo, Georg Pflug, Werner Römisch, Chengwu Shao, Thomas Strömberg, and Kathleen Wets. The chapters on set convergence and epi-convergence benefited from the scrutiny of a seminar group consisting of Gül Gürkan, Douglas Lepro, Yonca Özge, and Stephen Robinson. Conversations we had over the years with our students and colleagues contributed significantly to the final form of the book as well. Grants from the National Science Foundation were essential in sustaining the long effort.

The changes in this third printing mainly concern various typographical, corrections, and reference omissions, which came to light in the first and second printing. Many of these reached our notice through our own re-reading and that of our students, as well as the individuals already mentioned. Really major input, however, arrived from Shu Lu and Michel Valadier, and above all from Lionel Thibault. He carefully went through almost every detail, detecting numerous places where adjustments were needed or desirable. We are extremely indebted for all these valuable contributions.

CONTENTS

Chapter 1. Max and Min **1**

A. Penalties and Constraints 2
B. Epigraphs and Semicontinuity 7
C. Attainment of a Minimum 11
D. Continuity, Closure and Growth 13
E. Extended Arithmetic 15
F. Parametric Dependence 16
G. Moreau Envelopes 20
H. Epi-Addition and Epi-Multiplication 23
I.* Auxiliary Facts and Principles 28
Commentary 34

Chapter 2. Convexity **38**

A. Convex Sets and Functions 38
B. Level Sets and Intersections 42
C. Derivative Tests 45
D. Convexity in Operations 49
E. Convex Hulls 53
F. Closures and Continuity 57
G.* Separation 62
H.* Relative Interiors 64
I.* Piecewise Linear Functions 67
J.* Other Examples 71
Commentary 74

Chapter 3. Cones and Cosmic Closure **77**

A. Direction Points 77
B. Horizon Cones 80
C. Horizon Functions 86
D. Coercivity Properties 90
E.* Cones and Orderings 95
F.* Cosmic Convexity 97
G.* Positive Hulls 99
Commentary 105

Chapter 4. Set Convergence **108**

A. Inner and Outer Limits 109
B. Painlevé-Kuratowski Convergence 111
C. Pompeiu-Hausdorff Distance 117
D. Cones and Convex Sets 118
E. Compactness Properties 120
F. Horizon Limits 122
G.* Continuity of Operations 125
H.* Quantification of Convergence 131
I.* Hyperspace Metrics 138
Commentary 144

Chapter 5. Set-Valued Mappings **148**

A. Domains, Ranges and Inverses 149
B. Continuity and Semicontinuity 152

C. Local Boundedness 157
D. Total Continuity 164
E. Pointwise and Graphical Convergence 166
F. Equicontinuity of Sequences 173
G. Continuous and Uniform Convergence 175
H.* Metric Descriptions of Convergence 181
I.* Operations on Mappings 183
J.* Generic Continuity and Selections 187
Commentary 192

Chapter 6. Variational Geometry **196**

A. Tangent Cones 196
B. Normal Cones and Clarke Regularity 199
C. Smooth Manifolds and Convex Sets 202
D. Optimality and Lagrange Multipliers 205
E. Proximal Normals and Polarity 212
F. Tangent-Normal Relations 217
G.* Recession Properties 222
H.* Irregularity and Convexification 225
I.* Other Formulas 227
Commentary 232

Chapter 7. Epigraphical Limits **238**

A. Pointwise Convergence 239
B. Epi-Convergence 240
C. Continuous and Uniform Convergence 250
D. Generalized Differentiability 255
E. Convergence in Minimization 262
F. Epi-Continuity of Function-Valued Mappings 269
G.* Continuity of Operations 275
H.* Total Epi-Convergence 278
I.* Epi-Distances 282
J.* Solution Estimates 286
Commentary 292

Chapter 8. Subderivatives and Subgradients **298**

A. Subderivatives of Functions 299
B. Subgradients of Functions 300
C. Convexity and Optimality 308
D. Regular Subderivatives 311
E. Support Functions and Subdifferential Duality 317
F. Calmness 322
G. Graphical Differentiation of Mappings 324
H.* Proto-Differentiability and Graphical Regularity 329
I.* Proximal Subgradients 333
J.* Other Results 336
Commentary 343

Chapter 9. Lipschitzian Properties **349**

A. Single-Valued Mappings 349
B. Estimates of the Lipschitz Modulus 354
C. Subdifferential Characterizations 358
D. Derivative Mappings and Their Norms 364
E. Lipschitzian Concepts for Set-Valued Mappings 368

F. Aubin Property and Mordukhovich Criterion 376
G. Metric Regularity and Openness 386
H.* Semiderivatives and Strict Graphical Derivatives 390
I.* Other Properties 399
J.* Rademacher's Theorem and Consequences 403
K.* Mollifiers and Extremals 408
Commentary 415

Chapter 10. Subdifferential Calculus **421**

A. Optimality and Normals to Level Sets 421
B. Basic Chain Rule 426
C. Parametric Optimality 432
D. Rescaling 438
E. Piecewise Linear-Quadratic Functions 440
F. Amenable Sets and Functions 442
G. Semiderivatives and Subsmoothness 446
H.* Coderivative Calculus 452
I.* Extensions 458
Commentary 469

Chapter 11. Dualization **473**

A. Legendre-Fenchel Transform 473
B. Special Cases of Conjugacy 476
C. The Role of Differentiability 480
D. Piecewise Linear-Quadratic Functions 484
E. Polar Sets and Gauges 490
F. Dual Operations 493
G. Duality in Convergence 500
H. Dual Problems of Optimization 502
I. Lagrangian Functions 508
J.* Minimax Problems 514
K.* Augmented Lagrangians and Nonconvex Duality 518
L.* Generalized Conjugacy 525
Commentary 529

Chapter 12. Monotone Mappings **533**

A. Monotonicity Tests and Maximality 533
B. Minty Parameterization 537
C. Connections with Convex Functions 542
D. Graphical Convergence 551
E. Domains and Ranges 553
F.* Preservation of Maximality 556
G.* Monotone Variational Inequalities 559
H.* Strong Monotonicity and Strong Convexity 562
I.* Continuity and Differentiability 567
Commentary 575

Chapter 13. Second-Order Theory **579**

A. Second-Order Differentiability 579
B. Second Subderivatives 582
C. Calculus Rules 591
D. Convex Functions and Duality 603
E. Second-Order Optimality 606
F. Prox-Regularity 609

G. Subgradient Proto-Differentiability 618
H. Subgradient Coderivatives and Perturbation 622
I.* Further Derivative Properties 625
J.* Parabolic Subderivatives 633
Commentary 638

Chapter 14. Measurability **642**

A. Measurable Mappings and Selections 643
B. Preservation of Measurability 651
C. Limit Operations 655
D. Normal Integrands 660
E. Operations on Integrands 669
F. Integral Functionals 675
Commentary 679

References **684**

Index of Statements **710**

Index of Notation **725**

Index of Topics **726**

1. Max and Min

Questions about the maximum or minimum of a function f relative to a set C are fundamental in variational analysis. For problems in n real variables, the elements of C are vectors $x = (x_1, \ldots, x_n) \in \mathbb{R}^n$. In any application, f or C are likely to have special structure that needs to be addressed, but we begin here with concepts associated with maximization and minimization as general operations.

It's convenient for many purposes to consider functions f that are allowed to be *extended-real-valued*, i.e., to take values in $\overline{\mathbb{R}} = [-\infty, \infty]$ instead of just $\mathbb{R} = (-\infty, \infty)$. In the extended real line $\overline{\mathbb{R}}$, which has all the properties of a compact interval, every subset $R \subset \overline{\mathbb{R}}$ has a *supremum* (least upper bound), which is denoted by $\sup R$, and likewise an *infimum* (greatest lower bound), $\inf R$, either of which could be infinite. (Caution: The case of $R = \emptyset$ is anomalous, in that $\inf \emptyset = \infty$ but $\sup \emptyset = -\infty$, so that $\inf \emptyset > \sup \emptyset$!) Custom allows us to write $\min R$ in place of $\inf R$ if the greatest lower bound $\inf R$ actually belongs to the set R. Likewise, we have the option of writing $\max R$ in place of $\sup R$ if the value $\sup R$ is in R.

For $\inf R$ and $\sup R$ in the case of the set $R = \{f(x) \mid x \in C\} \subset \overline{\mathbb{R}}$, we introduce the notation

$$\inf_C f := \inf_{x \in C} f(x) := \inf \{f(x) \mid x \in C\},$$
$$\sup_C f := \sup_{x \in C} f(x) := \sup \{f(x) \mid x \in C\}.$$

(The symbol ':=' means that the expression on the left is *defined* as equal to the expression on the right. On occasion we'll use '=:' as the statement or reminder of a definition that goes instead from right to left.) When desirable for emphasis, we permit ourselves to write $\min_C f$ in place of $\inf_C f$ when $\inf_C f$ is one of the values in the set $\{f(x) \mid x \in C\}$ and likewise $\max_C f$ in place of $\sup_C f$ when $\sup_C f$ belongs to $\{f(x) \mid x \in C\}$.

Corresponding to this, but with a subtle difference dictated by the interpretations that will be given to ∞ and $-\infty$, we introduce notation also for the sets of points x where the minimum or maximum of f over C is regarded as being attained:

$$\begin{aligned}
\operatorname{argmin}_C f &:= \operatorname*{argmin}_{x\in C} f(x)\\
&:= \begin{cases} \{x \in C \mid f(x) = \inf_C f\} & \text{if } \inf_C f \neq \infty,\\ \emptyset & \text{if } \inf_C f = \infty,\end{cases}\\
\operatorname{argmax}_C f &:= \operatorname*{argmax}_{x\in C} f(x)\\
&:= \begin{cases} \{x \in C \mid f(x) = \sup_C f\} & \text{if } \sup_C f \neq -\infty,\\ \emptyset & \text{if } \sup_C f = -\infty.\end{cases}
\end{aligned}$$

Note that we don't regard the minimum as being attained at any $x \in C$ when $f \equiv \infty$ on C, even though we may write $\min_C f = \infty$ in that case, nor do we regard the maximum as being attained at any $x \in C$ when $f \equiv -\infty$ on C. The reasons for these exceptions will be explained shortly. Quite apart from whether $\inf_C f < \infty$ or $\sup_C f > -\infty$, the sets $\operatorname{argmin}_C f$ and $\operatorname{argmax}_C f$ could be empty in the absence of appropriate conditions of continuity, boundedness or growth. A simple and versatile statement of such conditions will be devised in this chapter.

The roles of ∞ and $-\infty$ deserve close attention here. Let's look specifically at minimizing f over C. If there is a point $x \in C$ where $f(x) = -\infty$, we know at once that x furnishes the minimum. Points $x \in C$ where $f(x) = \infty$, on the other hand, have virtually the opposite significance. They aren't even worth contemplating as candidates for furnishing the minimum, unless f has ∞ as its value everywhere on C, a case that can be set aside as expressing a form of degeneracy—which we underline by defining $\operatorname{argmin}_C f$ to be empty then. In effect, the side condition $f(x) < \infty$ is considered to be *implicit* in minimizing $f(x)$ over $x \in C$. Everything of interest is the same as if we were minimizing over $C' := \{x \in C \mid f(x) < \infty\}$ instead of C.

A. Penalties and Constraints

This gives birth to an important idea in the context of C being a subset of $\mathbb{R}^n$. Perhaps f is merely real-valued on C, but whether this is true or not, we can transform the problem of minimizing f over C into one of *minimizing f over all of $\mathbb{R}^n$* just by defining (or as the case may be, redefining) $f(x)$ to be ∞ for all the points $x \in \mathbb{R}^n$ such that $x \notin C$. This helps in thinking abstractly about minimization and in achieving a single framework for the development of properties and results.

1.1 Example (equality and inequality constraints). *A set $C \subset \mathbb{R}^n$ may be specified as consisting of the vectors $x = (x_1, \ldots, x_n)$ such that*

$$x \in X \ \text{ and } \begin{cases} f_i(x) \le 0 & \text{for } i \in I_1,\\ f_i(x) = 0 & \text{for } i \in I_2,\end{cases}$$

where X is some subset of $\mathbb{R}^n$ and I_1 and I_2 are index sets for families of functions $f_i : \mathbb{R}^n \to \mathbb{R}$ called constraint functions. The conditions $f_i(x) \le 0$

are inequality constraints on x*, while those of form* $f_i(x) = 0$ *are equality constraints; the condition* $x \in X$ *(where in particular* X *could be all of* $\mathbb{R}^n$*) is an abstract or geometric constraint.*

A problem of minimizing a function $f_0 : \mathbb{R}^n \to \mathbb{R}$ *subject to all of these constraints can be identified with the problem of minimizing the function* $f : \mathbb{R}^n \to \overline{\mathbb{R}}$ *defined by taking* $f(x) = f_0(x)$ *when* x *satisfies the constraints but* $f(x) = \infty$ *otherwise. The possibility of having* $\inf f = \infty$ *corresponds then to the possibility that* $C = \emptyset$*, i.e., that the constraints may be inconsistent.*

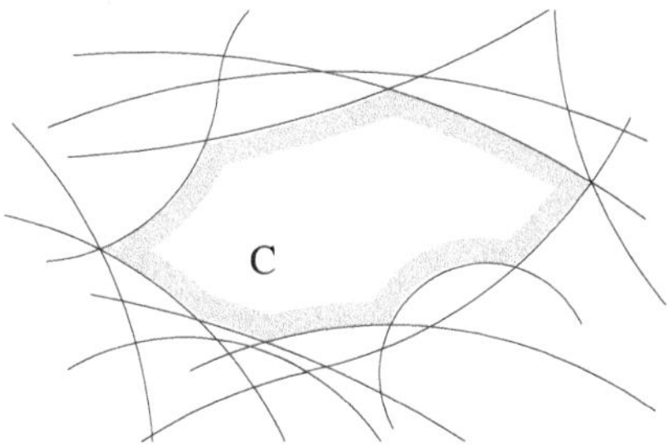

Fig. 1–1. A set defined by inequality constraints.

Constraints can also have the form $f_i(x) \le c_i$, $f_i(x) = c_i$ or $f_i(x) \ge c_i$ for values $c_i \in \mathbb{R}$, but this doesn't add real generality because f_i can always be replaced by $f_i - c_i$ or $c_i - f_i$. Strict inequalities are rarely seen in constraints, however, since they could threaten the attainment of a maximum or minimum.

An abstract constraint $x \in X$ is often convenient in representing conditions of a more complicated or open-ended nature, to be worked out later, but also for conditions too simple to be worth introducing constraint functions for, such as upper or lower bounds on the variables x_j as components of x.

1.2 Example (box constraints). *A set* $X \subset \mathbb{R}^n$ *is called a box if it is a product* $X_1 \times \cdots \times X_n$ *of closed intervals* X_j *of* $\mathbb{R}$*, not necessarily bounded. The condition* $x \in X$*, a box constraint on* $x = (x_1, \ldots, x_n)$*, then restricts each variable* x_j *to* X_j*. For instance, the nonnegative orthant*

$$\mathbb{R}^n_+ := \{x = (x_1, \ldots, x_n) \mid x_j \ge 0 \text{ for all } j\} = [0, \infty)^n$$

is a box in $\mathbb{R}^n$*; the constraint* $x \in \mathbb{R}^n_+$ *restricts all variables to be nonnegative. With* $X = \mathbb{R}^s_+ \times \mathbb{R}^{n-s} = [0, \infty)^s \times (-\infty, \infty)^{n-s}$*, only the first* s *variables* x_j *would have to be nonnegative. In other cases, useful for technical reasons, some intervals* X_j *could have the degenerate form* $[c_j, c_j]$*, which would force* $x_j = c_j$*.*

Constraints refer to the structure of the set over which the minimization or maximization should effectively take place, and in the approach of identifying a problem with a function $f : \mathbb{R}^n \to \overline{\mathbb{R}}$ they enter the specification of f. But the structure of the function being minimized or maximized can be affected by constraint representations in other ways as well.

1.3 Example (penalties and barriers). *Instead of imposing a direct constraint $f_i(x) \leq 0$ or $f_i(x) = 0$, one could add a term $\theta_i(f_i(x))$ to the function being minimized, where $\theta_i : I\!R \to \overline{I\!R}$ has $\theta_i(t) = 0$ for $t \leq 0$ but is positive for values of f_i that violate the condition in question. Then θ_i is a penalty function associated with f_i. A problem of minimizing $f_0(x)$ over $x \in X$ subject to constraints on $f_1(x), \ldots, f_m(x)$, might in this way be replaced by:*

$$\text{minimize } f_0(x) + \theta_1(f_1(x)) + \cdots + \theta_m(f_m(x)) \text{ subject to } x \in X.$$

A related idea in lieu of $f_i(x) \leq 0$ is adding a term $\theta_i(f_i(x))$ where θ_i is a barrier function: $\theta_i(t) = \infty$ for $t \geq 0$, and $\theta_i(t) \to \infty$ as $t \nearrow 0$.

Fig. 1–2. (a) A penalty function with rewards. (b) A barrier function.

As a penalty substitute for a constraint $f_i(x) \leq 0$, for instance, a term $\theta_i(f_i(x))$ with $\theta_i(t) = \lambda t_+$, where $t_+ := \max\{0, t\}$ and $\lambda > 0$, would give so-called *linear penalties*, while $\theta_i(t) = \frac{1}{2}\lambda t_+^2$ would give *quadratic penalties*. The penalty function

$$\theta_i(t) = \begin{cases} 0 & \text{if } t \leq 0, \\ \infty & \text{if } t > 0, \end{cases}$$

would enforce $f_i(x) \leq 0$ by triggering an infinite cost for any transgression. This is the limiting case of linear or quadratic penalties when $\lambda \to \infty$. Penalty functions can also involve negative values (*rewards*) when a constraint $f_i(x) \leq 0$ is satisfied with room to spare, cf. Figure 1–2(a); the same for barrier functions, cf. Fig 1–2(b). Common barrier functions for $f_i(x) \leq 0$ are, for any $\varepsilon > 0$,

$$\theta_i(t) = \begin{cases} \varepsilon/|t| & \text{when } t < 0, \\ \infty & \text{when } t \geq 0, \end{cases} \quad \text{or} \quad \theta_i(t) = \begin{cases} -\varepsilon \log|t| & \text{when } t < 0, \\ \infty & \text{when } t \geq 0. \end{cases}$$

These examples underscore the useful range of possibilities opened up by admitting extended-real-valued functions. They also show that properties like differentiability which are routinely assumed in classical analysis can't be counted on in variational analysis. A function of the composite kind in 1.3 can fail to be differentiable regardless of the degree of differentiability of the f_i's because of kinks and jumps induced by the θ_i's, which may be essential to the problem model being used.

Everything said about minimization can be translated into the language of maximization, with $-\infty$ taking the part of ∞. Such symmetry is reassuring, but it must be understood that a basic asymmetry is implicit too in the approach we're taking. In passing from the minimization of a given function over C to the minimization of a corresponding function over $\mathbb{R}^n$, we've resorted to an extension by the value ∞, but in the case of maximization it would be $-\infty$. The extended function would then be different, and so would be the properties we'd like it to have. In effect we're abandoning any predisposition toward having a theory that treats maximization and minimization together on an equal footing. In the assumptions eventually imposed to identify the classes of functions most suitable for applying these operations, we mark out separate territories for each.

In actual practice there's rarely a need to consider both minimization and maximization simultaneously for a single combination of a function f and a set C, so this approach causes no discomfort. Rather than spend too many words on parallel statements, we adopt minimization as the vehicle of exposition and mention maximization only from time to time, taking for granted that the reader will generally understand the accommodations needed in that direction. We thereby enter a pattern of working mainly with extended-real-valued functions on $\mathbb{R}^n$ and treating them in a one-sided manner where ∞ has a qualitatively different role from that of $-\infty$ in our formulas, and where the terminology and notation reflect this bias.

Starting off now on this path, we introduce for $f : \mathbb{R}^n \to \overline{\mathbb{R}}$ the set

$$\operatorname{dom} f := \{x \in \mathbb{R}^n \mid f(x) < \infty\},$$

called the *effective domain* of f, and write

$$\begin{aligned} \inf f &:= \inf_x f(x) := \inf_{x \in \mathbb{R}^n} f(x) = \inf_{x \in \operatorname{dom} f} f(x), \\ \operatorname{argmin} f &:= \operatorname{argmin}_x f(x) := \operatorname*{argmin}_{x \in \mathbb{R}^n} f(x) = \operatorname*{argmin}_{x \in \operatorname{dom} f} f(x). \end{aligned}$$

We call f a *proper* function if $f(x) < \infty$ for at least one $x \in \mathbb{R}^n$, and $f(x) > -\infty$ for all $x \in \mathbb{R}^n$, or in other words, if $\operatorname{dom} f$ is a nonempty set on which f is finite; otherwise it is *improper*. The proper functions $f : \mathbb{R}^n \to \overline{\mathbb{R}}$ are thus the ones obtained by taking a nonempty set $C \subset \mathbb{R}^n$ and a function $f : C \to \mathbb{R}$, and putting $f(x) = \infty$ for all $x \notin C$. All other kinds of functions $f : \mathbb{R}^n \to \overline{\mathbb{R}}$ are termed *improper* in this context. While proper functions are our central concern, improper functions may arise indirectly and can't always be excluded from consideration.

The developments so far can be summarized as follows in the language of optimization.

1.4 Example (principle of abstract minimization). *Problems of minimizing a finite function over some subset of $\mathbb{R}^n$ correspond one-to-one with problems of minimizing over all of $\mathbb{R}^n$ a function $f : \mathbb{R}^n \to \overline{\mathbb{R}}$, under the identifications:*

$$\begin{aligned} \operatorname{dom} f &= \textit{ set of feasible solutions,} \\ \operatorname{argmin} f &= \textit{ set of optimal solutions,} \\ \inf f &= \textit{ optimal value.} \end{aligned}$$

The convention that $\operatorname{argmin} f = \emptyset$ when $f \equiv \infty$ ensures that a problem is not regarded as having an optimal solution if it doesn't even have a feasible solution. A lack of feasible solutions is signaled by the optimal value being ∞.

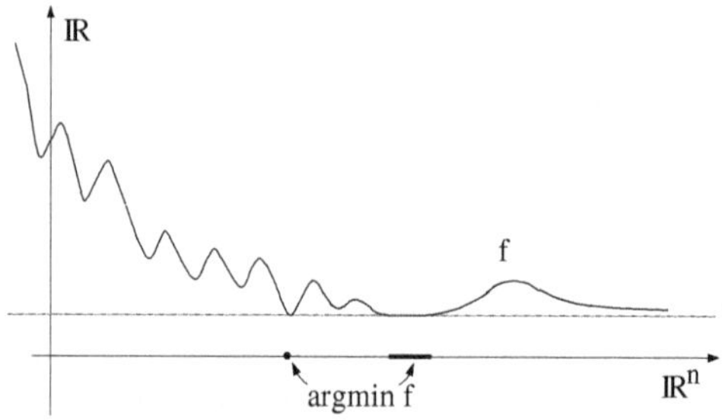

Fig. 1–3. Local and global optimality in a difficult yet classical case.

It should be emphasized here that the notation $\operatorname{argmin} f$ refers to points $\bar{x}$ giving a *global* minimum of f. A *local* minimum occurs at $\bar{x}$ if $f(\bar{x}) < \infty$ and $f(x) \geq f(\bar{x})$ for all $x \in V$, where

$$V \in \mathcal{N}(\bar{x}) := \text{ the collection of all neighborhoods of } \bar{x}.$$

Then $\bar{x}$ is a *locally* optimal solution to the problem of minimizing f. By a *neighborhood* of x one means any set having x in its interior, for example a closed ball

$$I\!B(x, \lambda) := \{x' \mid d(x, x') \leq \lambda\},$$

where we use the notation

$$\begin{aligned} &d(x, x') := |x - x'| \quad \text{(Euclidean distance)}, \quad \text{with} \\ &|x| := |(x_1, \ldots, x_n)| = \sqrt{x_1^2 + \cdots + x_n^2} \quad \text{(Euclidean norm)}. \end{aligned}$$

A point $\bar{x}$ giving a local minimum of f can also be viewed as giving the global minimum in an auxiliary problem in which the function agrees with f on some neighborhood of $\bar{x}$ but takes the value ∞ elsewhere, so the study of local optimality can to a large extent be subsumed into the study of global optimality.

An extremely useful type of function in the framework we're adopting is the *indicator* function δ_C of a set $C \subset I\!R^n$, which is defined by

$$\delta_C(x) = 0 \text{ if } x \in C, \qquad \delta_C(x) = \infty \text{ if } x \notin C.$$

The indicator functions on $I\!R^n$ are characterized as a class by taking on no value other than 0 or ∞. The constant function 0 is the indicator of $C = I\!R^n$, while the constant function ∞ is the indicator of $C = \emptyset$. Obviously $\operatorname{dom} \delta_C = C$, and δ_C is proper if and only if C is nonempty.

There's no question of our wanting to minimize a function like δ_C, but indicators nonetheless are important in problems of minimization. To take a simple but revealing case, suppose we're given a finite-valued function $f_0 : \mathbb{R}^n \to \mathbb{R}$ and wish to minimize it over a set $C \subset \mathbb{R}^n$. This is equivalent, as already seen, to minimizing a certain extended-real-valued function f over all of $\mathbb{R}^n$, namely the one defined by $f(x) = f_0(x)$ for $x \in C$ but $f(x) = \infty$ for $x \notin C$. The new observation is that $f = f_0 + \delta_C$. The constraint $x \in C$ can thus be enforced by adding its indicator to the function being minimized. Similarly, the condition that $\bar{x}$ be locally optimal in minimizing f can be expressed by $\bar{x} \in \operatorname{argmin} \big(f + \delta_V\big)$ for some $V \in \mathcal{N}(\bar{x})$.

By identifying each set C with its indicator δ_C, we can pass between facts about subsets of $\mathbb{R}^n$ and facts about extended-real-valued functions on $\mathbb{R}^n$. This helps to cross-fertilize between geometrical and analytical concepts. Further assistance comes from identifying functions on $\mathbb{R}^n$ with certain subsets of $\mathbb{R}^{n+1}$, as we explain next.

B. Epigraphs and Semicontinuity

Ideas of geometry have traditionally been brought to bear in the study of functions and mappings by applying them to graphs. In variational analysis, graphs continue to serve this purpose for vector-valued functions, but extended-real-valued functions require a different perspective. The graph of such a function on $\mathbb{R}^n$ would generally be a subset of $\mathbb{R}^n \times \overline{\mathbb{R}}$ rather than $\mathbb{R}^{n+1}$, and this wouldn't be convenient because $\mathbb{R}^n \times \overline{\mathbb{R}}$ isn't a vector space. Anyway, even if extended values weren't an issue, the geometry of graphs wouldn't convey the properties that turn out to be crucial for our purposes. Graphs have to be replaced by 'epigraphs'.

For $f : \mathbb{R}^n \to \overline{\mathbb{R}}$, the *epigraph* of f is the set

$$\operatorname{epi} f := \big\{(x, \alpha) \in \mathbb{R}^n \times \mathbb{R} \,\big|\, \alpha \geq f(x)\big\}$$

(see Figure 1–4). The epigraph thus consists of all the points of $\mathbb{R}^{n+1}$ lying on or above the graph of f. (Note that $\operatorname{epi} f$ is truly a subset of $\mathbb{R}^{n+1}$, not just of $\mathbb{R}^n \times \overline{\mathbb{R}}$.) The image of $\operatorname{epi} f$ under the projection $(x, \alpha) \mapsto x$ is $\operatorname{dom} f$. The points x where $f(x) = \infty$ are the ones such that the vertical line $(x, \mathbb{R}) := \{x\} \times \mathbb{R}$ misses $\operatorname{epi} f$, whereas the points where $f(x) = -\infty$ are the ones such that this line is entirely included in $\operatorname{epi} f$.

What distinguishes the class of subsets of $\mathbb{R}^{n+1}$ that are the epigraphs of the extended-real-valued functions on $\mathbb{R}^n$? Clearly E belongs to this 'epigraphical' class if and only if it intersects every vertical line $(x, \mathbb{R})$ in a closed interval which, unless empty, is unbounded above. The associated function in that case is proper if and only if E includes no entire vertical line, and $E \neq \emptyset$.

Every property of f has its counterpart in a property of $\operatorname{epi} f$, because the correspondence between functions and epigraphs is one-to-one. Many properties also relate very naturally to the various level sets of f. In general, we'll find it useful to have the notation

$$\begin{aligned}
\operatorname{lev}_{\leq\alpha} f &:= \{x \in \mathbb{R}^n \mid f(x) \leq \alpha\},\\
\operatorname{lev}_{<\alpha} f &:= \{x \in \mathbb{R}^n \mid f(x) < \alpha\},\\
\operatorname{lev}_{=\alpha} f &:= \{x \in \mathbb{R}^n \mid f(x) = \alpha\},\\
\operatorname{lev}_{>\alpha} f &:= \{x \in \mathbb{R}^n \mid f(x) > \alpha\},\\
\operatorname{lev}_{\geq\alpha} f &:= \{x \in \mathbb{R}^n \mid f(x) \geq \alpha\}.
\end{aligned}$$

The most important of these in the context of minimization are the lower level sets $\operatorname{lev}_{\leq\alpha} f$. For α finite, they correspond to the 'horizontal cross sections' of $\operatorname{epi} f$. For $\alpha = \inf f$, one has $\operatorname{lev}_{\leq\alpha} f = \operatorname{lev}_{=\alpha} f = \operatorname{argmin} f$.

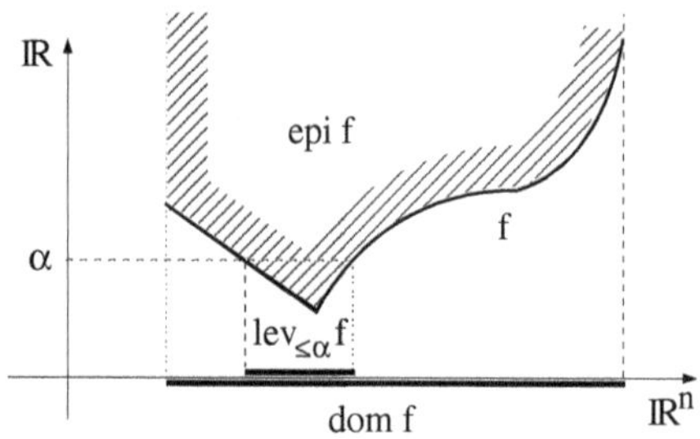

Fig. 1–4. Epigraph and effective domain of an extended-real-valued function.

We're ready now to answer a basic question about a function $f : \mathbb{R}^n \to \overline{\mathbb{R}}$. What property of f translates into the sets $\operatorname{lev}_{\leq\alpha} f$ all being closed? The answer depends on a one-sided concept of limit.

1.5 Definition (lower limits and lower semicontinuity). *The lower limit of a function $f : \mathbb{R}^n \to \overline{\mathbb{R}}$ at $\bar{x}$ is the value in $\overline{\mathbb{R}}$ defined by*

$$\begin{aligned}
\liminf_{x\to\bar{x}} f(x) :&= \lim_{\delta\searrow 0}\left[\inf_{x\in B(\bar{x},\delta)} f(x)\right]\\
&= \sup_{\delta>0}\left[\inf_{x\in B(\bar{x},\delta)} f(x)\right] = \sup_{V\in\mathcal{N}(\bar{x})}\left[\inf_{x\in V} f(x)\right].
\end{aligned} \tag{1(1)}$$

The function $f : \mathbb{R}^n \to \overline{\mathbb{R}}$ is lower semicontinuous (lsc) at $\bar{x}$ if

$$\liminf_{x\to\bar{x}} f(x) \geq f(\bar{x}), \quad \textit{or equivalently} \quad \liminf_{x\to\bar{x}} f(x) = f(\bar{x}), \tag{1(2)}$$

and lower semicontinuous on $\mathbb{R}^n$ if this holds for every $\bar{x} \in \mathbb{R}^n$.

The two versions in 1(2) agree because $\inf\{f(x) \mid x \in \mathbb{B}(\bar{x},\delta)\} \leq f(\bar{x})$ for all $\delta > 0$. For this reason too,

$$\liminf_{x \to \bar{x}} f(x) \leq f(\bar{x}) \text{ always.} \tag{1(3)}$$

In replacing the limit as $\delta \searrow 0$ by the supremum over $\delta > 0$ in 1(1) we appeal to the general fact that

$$\inf_{x \in X_1} f(x) \leq \inf_{x \in X_2} f(x) \text{ when } X_1 \supset X_2.$$

1.6 Theorem (characterization of lower semicontinuity). *The following properties of a function $f : \mathbb{R}^n \to \overline{\mathbb{R}}$ are equivalent:*

(a) *f is lower semicontinuous on $\mathbb{R}^n$;*

(b) *the epigraph set $\operatorname{epi} f$ is closed in $\mathbb{R}^n \times \mathbb{R}$;*

(c) *the level sets of type $\operatorname{lev}_{\leq \alpha} f$ are all closed in $\mathbb{R}^n$.*

These equivalences will be established after some preliminaries. An example of a function on $\mathbb{R}$ that happens to be lower semicontinuous at every point but two is displayed in Figure 1–5. Notice how the defect is associated with the failure of the epigraph to include all of its boundary.

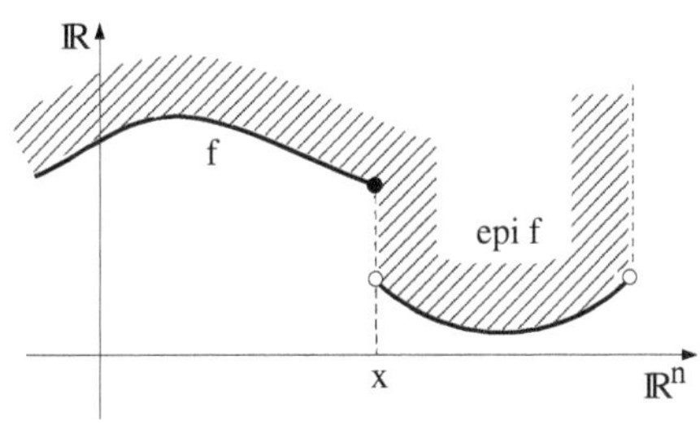

Fig. 1–5. An example where lower semicontinuity fails.

In the proof of Theorem 1.6 and throughout the book, we use sequence notation in which the running index is always superscript ν (Greek 'nu'). We symbolize the natural numbers by $\mathbb{N}$, so that $\nu \in \mathbb{N}$ means $\nu = 1, 2, \ldots$. The notation $x^\nu \to x$, or $x = \lim_\nu x^\nu$, refers then to a sequence $\{x^\nu\}_{\nu \in \mathbb{N}}$ in $\mathbb{R}^n$ that converges to x, i.e., has $|x^\nu - x| \to 0$ as $\nu \to \infty$. We speak of x as a *cluster point* of x^ν as $\nu \to \infty$ if, instead of necessarily claiming $x^\nu \to x$, we wish merely to assert that some subsequence converges to x. (Every bounded sequence in $\mathbb{R}^n$ has at least one cluster point. A sequence in $\mathbb{R}^n$ converges to x if and only if it is bounded and has x as its only cluster point.)

1.7 Lemma (characterization of lower limits).

$$\liminf_{x \to \bar{x}} f(x) = \min\{\alpha \in \overline{\mathbb{R}} \mid \exists x^\nu \to \bar{x} \text{ with } f(x^\nu) \to \alpha\}.$$

(Here the constant sequence $x^\nu \equiv \bar{x}$ is admitted and yields $\alpha = f(\bar{x})$.)

Proof. By the conventions explained at the outset of this chapter, the 'min' in place of 'inf' means that the value on the left is not only the greatest lower bound in $\overline{\mathbb{R}}$ of the possible limit values α described in the set on the right, it's actually attained as the limit corresponding to some sequence $x^\nu \to \bar{x}$. Let $\bar{\alpha} = \liminf_{x\to\bar{x}} f(x)$. We first suppose that $x^\nu \to \bar{x}$ with $f(x^\nu) \to \alpha$ and show this implies $\alpha \geq \bar{\alpha}$. For any $\delta > 0$ we eventually have x^ν in the ball $\mathbb{B}(\bar{x},\delta)$ and therefore $f(x^\nu) \geq \inf\{f(x) \mid x \in \mathbb{B}(\bar{x},\delta)\}$. Taking the limit in ν with δ fixed, we get $\alpha \geq \inf\{f(x) \mid x \in \mathbb{B}(\bar{x},\delta)\}$ for arbitrary $\delta > 0$, hence $\alpha \geq \bar{\alpha}$.

Next we must demonstrate the existence of $x^\nu \to \bar{x}$ such that actually $f(x^\nu) \to \bar{\alpha}$. Let $\bar{\alpha}^\nu = \inf\{f(x) \mid x \in \mathbb{B}(\bar{x},\delta^\nu)\}$ for a sequence of values $\delta^\nu \searrow 0$. The definition of the lower limit $\bar{\alpha}$ assures us that $\bar{\alpha}^\nu \to \bar{\alpha}$. For each ν it is possible to find $x^\nu \in \mathbb{B}(\bar{x},\delta^\nu)$ for which $f(x^\nu)$ is as near as we please to $\bar{\alpha}^\nu$, say in the interval $[\bar{\alpha}^\nu, \alpha^\nu]$, where α^ν is chosen to satisfy $\alpha^\nu > \bar{\alpha}^\nu$ and $\alpha^\nu \to \bar{\alpha}$. (If $\bar{\alpha} = \infty$, we get $f(x^\nu) = \bar{\alpha}^\nu = \infty$ automatically.) Then obviously $x^\nu \to x$ and $f(x^\nu)$ has the same limit as $\bar{\alpha}^\nu$, namely $\bar{\alpha}$. □

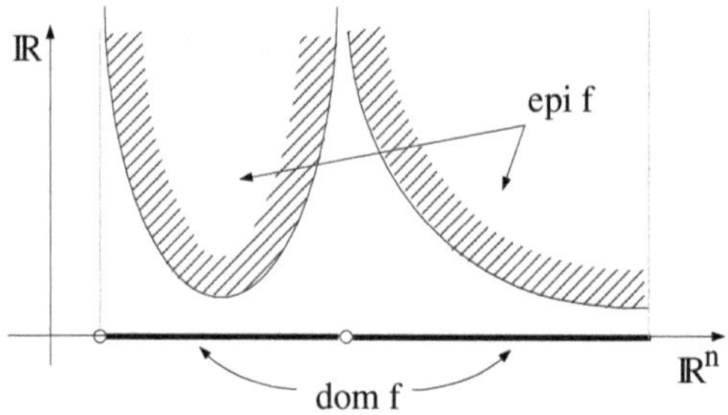

Fig. 1–6. An lsc function with effective domain not closed or connected.

Proof of 1.6. (a)⇒(b). Suppose $(x^\nu,\alpha^\nu) \in \operatorname{epi} f$ and $(x^\nu,\alpha^\nu) \to (\bar{x},\alpha)$ with α finite. We have $x^\nu \to \bar{x}$ and $\alpha^\nu \to \alpha$ with $\alpha^\nu \geq f(x^\nu)$ and must show that $\alpha \geq f(\bar{x})$, so that $(\bar{x},\alpha) \in \operatorname{epi} f$. The sequence $\{f(x^\nu)\}$ has at least one cluster point $\beta \in \overline{\mathbb{R}}$. We can suppose (through replacing the sequence $\{(x^\nu,\alpha^\nu)\}_{\nu\in\mathbb{N}}$ by a subsequence if necessary) that $f(x^\nu) \to \beta$. In this case $\alpha \geq \beta$, but also $\beta \geq \liminf_{x\to\bar{x}} f(x)$ by Lemma 1.7. Then $\alpha \geq f(\bar{x})$ by our assumption of lower semicontinuity.

(b)⇒(c). When $\operatorname{epi} f$ is closed, so too is the intersection $[\operatorname{epi} f] \cap (\mathbb{R}^n, \alpha)$ for each $\alpha \in \mathbb{R}$. This intersection in $\mathbb{R}^n \times \mathbb{R}$ corresponds geometrically to the set $\operatorname{lev}_{\leq\alpha} f$ in $\mathbb{R}^n$, which therefore is closed. The set $\operatorname{lev}_{\leq -\infty} f = \operatorname{lev}_{=-\infty} f$, being the intersection of these closed sets as α ranges over $\mathbb{R}$, is closed also, whereas $\operatorname{lev}_{\leq\infty} f$ is just the whole space $\mathbb{R}^n$.

(c)⇒(a). Fix any $\bar{x}$ and let $\bar{\alpha} = \liminf_{x\to\bar{x}} f(x)$. To establish that f is lsc at $\bar{x}$, it will suffice to show $f(\bar{x}) \leq \bar{\alpha}$, since the opposite inequality is automatic. The case of $\bar{\alpha} = \infty$ is trivial, so suppose $\bar{\alpha} < \infty$. Consider a sequence $x^\nu \to \bar{x}$ with $f(x^\nu) \to \bar{\alpha}$, as guaranteed by Lemma 1.7. For any $\alpha > \bar{\alpha}$ it will eventually be true that $f(x^\nu) \le \alpha$, or in other words, that x^ν belongs to $\operatorname{lev}_{\le\alpha} f$. Since $x^\nu \to \bar{x}$, this level set, which by assumption is closed, must contain $\bar{x}$. Thus we have $f(\bar{x}) \le \alpha$ for every $\alpha > \bar{\alpha}$. Obviously, then, $f(\bar{x}) \le \bar{\alpha}$. □

When Theorem 1.6 is applied to indicator functions, it reduces to the fact that δ_C is lsc if and only if the set C is closed. The lower semicontinuity of a general function $f : \mathbb{R}^n \to \overline{\mathbb{R}}$ doesn't require $\operatorname{dom} f$ to be closed, however, even when $\operatorname{dom} f$ happens to be bounded. Figure 1–6 illustrates this.

C. Attainment of a Minimum

Another question can now be addressed. What conditions on a function $f : \mathbb{R}^n \to \overline{\mathbb{R}}$ ensure that f attains its minimum over $\mathbb{R}^n$ at some x, i.e., that the set $\operatorname{argmin} f$ is nonempty? The issue is central because of the wide spectrum of minimization problems that can be put into this simple-looking form.

A fact customarily cited is this: *a continuous function on a compact set attains its minimum.* It also, of course, attains its maximum; this assertion is symmetric with respect to max and min. A more flexible approach is desirable, however. We don't always wish to single out a compact set, and constraints might not even be present. The very distinction between constrained and unconstrained minimization is suppressed in working with the principle of abstract minimization in 1.4, not to mention problem formulations involving penalty expressions as in 1.3. It's all just a matter of whether the function f being minimized takes on the value ∞ in some regions or not. Another feature is that the functions we want to deal with may be far from continuous. The one in Figure 1–6 is a case in point, but that function f does attain its minimum. A property that's crucial in this regard is the following.

1.8 Definition (level boundedness). *A function $f : \mathbb{R}^n \to \overline{\mathbb{R}}$ is (lower) level-bounded if for every $\alpha \in \mathbb{R}$ the set $\operatorname{lev}_{\le\alpha} f$ is bounded (possibly empty).*

Note that only finite values of α are considered in this definition. The level boundedness property corresponds to having $f(x) \to \infty$ as $|x| \to \infty$.

1.9 Theorem (attainment of a minimum). *Suppose $f : \mathbb{R}^n \to \overline{\mathbb{R}}$ is lower semicontinuous, level-bounded and proper. Then the value $\inf f$ is finite and the set $\operatorname{argmin} f$ is nonempty and compact.*

Proof. Let $\bar{\alpha} = \inf f$; because f is proper, $\bar{\alpha} < \infty$. For $\alpha \in (\bar{\alpha}, \infty)$, the set $\operatorname{lev}_{\le\alpha} f$ is nonempty; it's closed because f is lsc (cf. 1.6) and bounded because f is level-bounded. The sets $\operatorname{lev}_{\le\alpha} f$ for $\alpha \in (\bar{\alpha}, \infty)$ are therefore compact and nested: $\operatorname{lev}_{\le\alpha} f \subset \operatorname{lev}_{\le\beta} f$ when $\alpha < \beta$. The intersection of this family of sets, which is $\operatorname{lev}_{\le\bar{\alpha}} f = \operatorname{argmin} f$, is therefore nonempty and compact. Since f doesn't have the value $-\infty$ anywhere, we conclude also that $\bar{\alpha}$ is finite. Under these circumstances, $\inf f$ can be written as $\min f$. □

1.10 Corollary (lower bounds). *If $f : \mathbb{R}^n \to \overline{\mathbb{R}}$ is lsc and proper, then it is bounded from below (finitely) on each bounded subset of $\mathbb{R}^n$ and in fact attains a minimum relative to any compact subset of $\mathbb{R}^n$ that meets* $\operatorname{dom} f$.

Proof. For any bounded set $B \subset \mathbb{R}^n$ apply the theorem to the function g defined by $g(x) = f(x)$ when $x \in \operatorname{cl} B$ but $g(x) = \infty$ when $x \notin \operatorname{cl} B$. The case where $g \equiv \infty$ can be dealt with as a triviality, while in all other cases g is lsc, level-bounded and proper. □

The conclusion of Theorem 1.9 would hold with level boundedness replaced by the weaker assumption that, for some $\alpha \in \mathbb{R}$, the set $\operatorname{lev}_{\leq \alpha} f$ is bounded and nonempty; this is easily gleaned from the proof. But level boundedness is more convenient to work with in applications, and it's typically present anyway in situations where the attainment of a minimum is sought.

The crucial ingredient in Theorem 1.9 is the fact that when f is both lsc and level-bounded it is *inf-compact*, which means that the sets $\operatorname{lev}_{\leq \alpha} f$ for $\alpha \in \mathbb{R}$ are all compact. This property is very flexible in providing a criterion for the existence of optimal solutions, and it can be applied to a variety of problems, with or without constraints.

1.11 Example (level boundedness relative to constraints). *For a problem of minimizing a continuous function $f_0 : \mathbb{R}^n \to \mathbb{R}$ over a nonempty, closed set $C \subset \mathbb{R}^n$, if all sets of the form*

$$\{x \in C \mid f_0(x) \leq \alpha\} \quad \text{for } \alpha \in \mathbb{R}$$

are bounded, then the minimum of f_0 over C is finite and attained on a nonempty, compact subset of C.

This criterion is fulfilled in particular if C is bounded or if f_0 is level bounded, with the latter condition covering even the case of unconstrained minimization, where $C = \mathbb{R}^n$.

Detail. The problem corresponds to minimizing $f = f_0 + \delta_C$ over $\mathbb{R}^n$. Here f is proper because $C \neq \emptyset$, and it's lsc by 1.6 because its level sets of the form $C \cap \{x \mid f_0(x) \leq \alpha\}$ for $\alpha < \infty$ are closed—by virtue of the closedness of C and the continuity of f_0. In assuming these sets are also bounded, we get the desired conclusions from 1.9. □

An illustration of existence in the pattern of Example 1.11 with C not necessarily bounded but f_0 inf-compact is furnished by $f_0(x) = |x|$. The minimization problem consists then of finding the point or points of C nearest to the origin of $\mathbb{R}^n$. Theorem 1.9 is also applicable, of course, to minimization problems that do not fit the pattern of 1.11 at all. For instance, in minimizing the function in Figure 1–6 one isn't simply minimizing a continuous function relative to a closed set, but the conditions in 1.9 are satisfied and a minimizing point exists. This is the kind of situation encountered in general when dealing with barrier functions, for instance.

D. Continuity, Closure and Growth

These results for minimization can be extended in evident ways to maximization. Instead of the lower limit of f at $\bar{x}$, the required concept in dealing with maximization is that of the *upper limit*

$$\begin{aligned}\limsup_{x\to\bar{x}} f(x) :&= \lim_{\delta\searrow 0}\left[\sup_{x\in\mathbf{B}(\bar{x},\delta)} f(x)\right]\\ &= \inf_{\delta>0}\left[\sup_{x\in\mathbf{B}(\bar{x},\delta)} f(x)\right] = \inf_{V\in\mathcal{N}(\bar{x})}\left[\sup_{x\in V} f(x)\right].\end{aligned} \qquad 1(4)$$

The function f is *upper semicontinuous* (usc) at $\bar{x}$ if this value equals $f(\bar{x})$. Upper semicontinuity at every point of $\mathbb{R}^n$ corresponds geometrically to the closedness of the *hypograph* of f, which is the set

$$\operatorname{hypo} f := \left\{(x,\alpha)\in\mathbb{R}^n\times\mathbb{R}\,\middle|\,\alpha\le f(x)\right\}, \qquad 1(5)$$

and the closedness of the upper level sets $\operatorname{lev}_{\ge\alpha} f$. Corresponding to the lower limit formula in Lemma 1.7, there's the upper limit formula

$$\limsup_{x\to\bar{x}} f(x) = \max\left\{\alpha\in\overline{\mathbb{R}}\,\middle|\,\exists\, x^\nu\to\bar{x} \text{ with } f(x^\nu)\to\alpha\right\}.$$

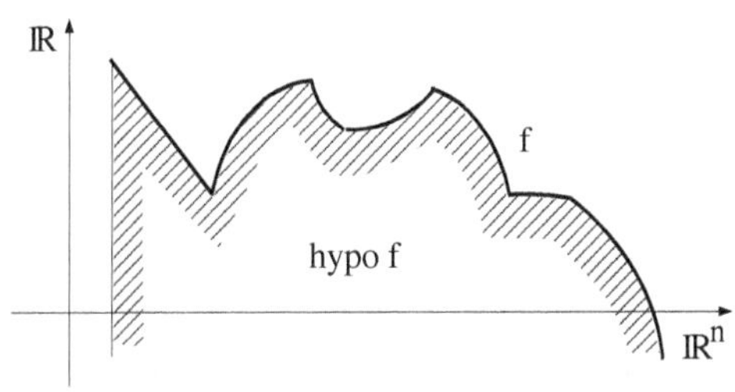

Fig. 1–7. The hypograph of a function.

Of course, f is regarded as *continuous* if $x\to\bar{x}$ implies $f(x)\to f(\bar{x})$, with the obvious interpretation being made when $f(\bar{x})=\infty$ or $f(\bar{x})=-\infty$.

1.12 Exercise (continuity of functions). *A function $f:\mathbb{R}^n\to\overline{\mathbb{R}}$ is continuous if and only if it is both lower semicontinuous and upper semicontinuous:*

$$\lim_{x\to\bar{x}} f(x) = f(\bar{x}) \iff \liminf_{x\to\bar{x}} f(x) = \limsup_{x\to\bar{x}} f(x).$$

Upper and lower limits of the values of f also serve to describe the closure and interior of $\operatorname{epi} f$. In stating the facts in this connection, we use the notation that

$$\begin{aligned}
\operatorname{cl} C &= \text{ closure of } C = \{x \mid \forall V \in \mathcal{N}(x),\, V \cap C \neq \emptyset\},\\
\operatorname{int} C &= \text{ interior of } C = \{x \mid \exists V \in \mathcal{N}(x),\, V \subset C\},\\
\operatorname{bdry} C &= \text{ boundary of } C = \operatorname{cl} C \setminus \operatorname{int} C \quad \text{(set difference)}.
\end{aligned}$$

1.13 Exercise (closures and interiors of epigraphs). *For an arbitrary function $f : \mathbb{R}^n \to \overline{\mathbb{R}}$ and a pair of elements $\bar{x} \in \mathbb{R}^n$ and $\bar{\alpha} \in \mathbb{R}$, one has*

(a) $(\bar{x}, \bar{\alpha}) \in \operatorname{cl}(\operatorname{epi} f)$ *if and only if* $\bar{\alpha} \geq \liminf_{x \to \bar{x}} f(x)$,

(b) $(\bar{x}, \bar{\alpha}) \in \operatorname{int}(\operatorname{epi} f)$ *if and only if* $\bar{\alpha} > \limsup_{x \to \bar{x}} f(x)$,

(c) $(\bar{x}, \bar{\alpha}) \notin \operatorname{cl}(\operatorname{epi} f)$ *if and only if* $(\bar{x}, \bar{\alpha}) \in \operatorname{int}(\operatorname{hypo} f)$,

(d) $(\bar{x}, \bar{\alpha}) \notin \operatorname{int}(\operatorname{epi} f)$ *if and only if* $(\bar{x}, \bar{\alpha}) \in \operatorname{cl}(\operatorname{hypo} f)$.

Semicontinuity properties of a function $f : \mathbb{R}^n \to \overline{\mathbb{R}}$ are 'constructive' to a certain extent. If f is not lower semicontinuous, its epigraph is not closed (cf. 1.6), but the set $E := \operatorname{cl}(\operatorname{epi} f)$ is not only closed, it's the epigraph of another function. This function is lsc and is the greatest (the highest) of all the lsc functions g such that $g \leq f$. It is called *lsc regularization*, or more simply, the *lower closure* of f, and is denoted by $\operatorname{cl} f$; thus

$$\operatorname{epi}(\operatorname{cl} f) := \operatorname{cl}(\operatorname{epi} f). \tag{1(6)}$$

The direct formula for $\operatorname{cl} f$ in terms of f is seen from 1.13(a) to be

$$(\operatorname{cl} f)(x) = \liminf_{x' \to x} f(x'). \tag{1(7)}$$

To understand this further, the reader may try the operation out on the function in Figure 1–5. Of course, $\operatorname{cl} f \leq f$ always.

The *usc regularization* or *upper closure* of f is analogously defined in terms of closing $\operatorname{hypo} f$, which amounts to taking the upper limit of f at every point x. (With $\operatorname{cl} f$ denoting the lower closure, $-\operatorname{cl}(-f)$ is the upper closure.) Although lower and upper semicontinuity can separately be arranged in this manner, f obviously can't be redefined to be continuous at x unless the two regularizations happen to agree at x.

Lower and upper limits of f *at infinity* instead of at a point $\bar{x}$ are also of interest, especially in connection with various growth properties of f in the large. They are defined by

$$\liminf_{|x| \to \infty} f(x) := \lim_{r \nearrow \infty} \inf_{|x| \geq r} f(x), \qquad \limsup_{|x| \to \infty} f(x) := \lim_{r \nearrow \infty} \sup_{|x| \geq r} f(x). \tag{1(8)}$$

1.14 Exercise (growth properties). *For a function $f : \mathbb{R}^n \to \overline{\mathbb{R}}$ and exponent $p \in (0, \infty)$, if f is lsc and $f > -\infty$ one has*

$$\liminf_{|x| \to \infty} \frac{f(x)}{|x|^p} = \sup\Big\{\gamma \in \mathbb{R} \;\Big|\; \exists \beta \in \mathbb{R} \text{ with } f(x) \geq \gamma |x|^p + \beta \text{ for all } x\Big\},$$

whereas if f is usc and $f < \infty$ one has

$$\limsup_{|x|\to\infty} \frac{f(x)}{|x|^p} = \inf\left\{\gamma \in I\!R \,\middle|\, \exists \beta \in I\!R \text{ with } f(x) \leq \gamma|x|^p + \beta \text{ for all } x\right\}.$$

Guide. For the first equation, denote the 'lim inf' by $\bar{\gamma}$ and the set on the right by Γ. Begin by showing that for any $\gamma \in \Gamma$ one has $\gamma \leq \bar{\gamma}$. Next argue that for finite $\gamma < \bar{\gamma}$ the inequality $f(x) \geq \gamma|x|^p$ will hold for all x outside a certain bounded set B. Then, by appealing to 1.10 on B, demonstrate that by subtracting off a sufficiently large constant from $\gamma|x|^p$ an inequality can be made to hold on all of $I\!R^n$ that corresponds to γ being in Γ. □

E. Extended Arithmetic

In applying the results so far to particular functions f, one has to be able to verify the needed semicontinuity. As with continuity, it's helpful to know how semicontinuity behaves relative to the operations often used in constructing a function from others, and criteria for this will be developed shortly. A question that sometimes must be settled first is the very meaning of such operations for functions having infinite values. Expressions like $f_1(x) + f_2(x)$ and $\lambda f(x)$ have to be interpreted properly in cases involving ∞ and $-\infty$.

Although the arithmetic of $I\!R$ doesn't carry over to $\overline{I\!R}$ without certain deficiencies, many rules extend in an obvious manner. For instance, $\infty + \alpha$ should be regarded as ∞ for any real α. The only combinations raising controversy are $0{\cdot}\infty$ and $\infty - \infty$. It's expedient to set

$$0{\cdot}\infty = 0 = 0{\cdot}(-\infty),$$

but there's no single, symmetric way of handling $\infty - \infty$. Because we orient toward minimization, the convention we'll generally use is *inf-addition*:

$$\infty + (-\infty) = (-\infty) + \infty = \infty.$$

(The opposite convention in $\overline{I\!R}$ is *sup-addition*; we won't invoke it without explicit warning.) Extended arithmetic then obeys the associative, commutative and distributive laws of ordinary arithmetic with one crucial exception:

$$\lambda{\cdot}(\infty - \infty) \neq (\lambda{\cdot}\infty - \lambda{\cdot}\infty) \text{ when } \lambda < 0.$$

With a little experience, it's as easy to cope with this as it is to keep on the lookout for implicit division by 0 in algebraic formulas. Since $\alpha - \alpha \neq 0$ when $\alpha = \infty$ or $\alpha = -\infty$, one must in particular refrain from canceling from both sides of an equation a term that might be ∞ or $-\infty$.

Lower and upper limits for functions are closely related to the concept of the lower and upper limit of a sequence of numbers $\alpha^\nu \in \overline{I\!R}$, defined by

$$\liminf_{\nu\to\infty} \alpha^\nu := \lim_{\nu\to\infty} \left[\inf_{\kappa\geq\nu} \alpha^\kappa\right], \qquad \limsup_{\nu\to\infty} \alpha^\nu := \lim_{\nu\to\infty} \left[\sup_{\kappa\geq\nu} \alpha^\kappa\right]. \tag{1(9)}$$

1.15 Exercise (lower and upper limits of sequences). *The cluster points of any sequence* $\{\alpha^\nu\}_{\nu\in\mathbb{N}}$ *in* $\overline{\mathbb{R}}$ *form a closed set of numbers in* $\overline{\mathbb{R}}$, *of which the lowest is* $\liminf_\nu \alpha^\nu$ *and the highest is* $\limsup_\nu \alpha^\nu$. *Thus, at least one subsequence of* $\{\alpha^\nu\}_{\nu\in\mathbb{N}}$ *converges to* $\liminf_\nu \alpha^\nu$, *and at least one converges to* $\limsup_\nu \alpha^\nu$.

In applying 'lim inf' and 'lim sup' to sums and scalar multiples of sequences of numbers there's an important caveat: *the rules for* $\infty - \infty$ *and* $0{\cdot}\infty$ *aren't necessarily preserved under limits*:

- $\alpha^\nu \to \alpha$ and $\beta^\nu \to \beta \;\;\not\Rightarrow\;\; \alpha^\nu + \beta^\nu \to \alpha + \beta$ when $\alpha + \beta = \infty - \infty$;
- $\alpha^\nu \to \alpha$ and $\beta^\nu \to \beta \;\;\not\Rightarrow\;\; \alpha^\nu{\cdot}\beta^\nu \to \alpha{\cdot}\beta$ when $\alpha{\cdot}\beta = 0{\cdot}(\pm\infty)$.

Either of the sequences $\{\alpha^\nu\}$ or $\{\beta^\nu\}$ could overpower the other, so limits involving $\infty - \infty$ or $0{\cdot}\infty$ may be 'indeterminate'.

F. Parametric Dependence

The themes of extended-real-valued representation, semicontinuity and level boundedness pervade the parametric study of problems of minimization as well. From 1.4, a minimization problem in n variables can be specified by a single function on $\mathbb{R}^n$, as long as infinite values are admitted. Therefore, a problem in n variables that depends on m parameters can be specified by a single function $f : \mathbb{R}^n \times \mathbb{R}^m \to \overline{\mathbb{R}}$: for each vector $u = (u_1, \ldots, u_m)$ there is the problem of minimizing $f(x,u)$ with respect to $x = (x_1, \ldots, x_n)$. No loss of generality is involved in having u range over all of $\mathbb{R}^m$, since applications where u lies naturally in some subset U of $\mathbb{R}^m$ can be handled by defining $f(x,u) = \infty$ for $u \notin U$.

Important issues are the behavior with respect to u of the optimal value and optimal solutions of this problem in x. A parametric extension of the level boundedness concept in 1.8 will help in the analysis.

1.16 Definition (uniform level boundedness). *A function* $f : \mathbb{R}^n \times \mathbb{R}^m \to \overline{\mathbb{R}}$ *with values* $f(x,u)$ *is level-bounded in* x *locally uniformly in* u *if for each* $\bar{u} \in \mathbb{R}^m$ *and* $\alpha \in \mathbb{R}$ *there is a neighborhood* $V \in \mathcal{N}(\bar{u})$ *along with a bounded set* $B \subset \mathbb{R}^n$ *such that* $\{x \mid f(x,u) \le \alpha\} \subset B$ *for all* $u \in V$; *or equivalently, there is a neighborhood* $V \in \mathcal{N}(\bar{u})$ *such that the set* $\{(x,u) \mid u \in V,\ f(x,u) \le \alpha\}$ *is bounded in* $\mathbb{R}^n \times \mathbb{R}^m$.

1.17 Theorem (parametric minimization). *Consider*

$$p(u) := \inf_x f(x,u), \qquad P(u) := \operatorname{argmin}_x f(x,u),$$

in the case of a proper, lsc function $f : \mathbb{R}^n \times \mathbb{R}^m \to \overline{\mathbb{R}}$ *such that* $f(x,u)$ *is level-bounded in* x *locally uniformly in* u.

(a) *The function* p *is proper and lsc on* $\mathbb{R}^m$, *and for each* $u \in \operatorname{dom} p$ *the set* $P(u)$ *is nonempty and compact, whereas* $P(u) = \emptyset$ *when* $u \notin \operatorname{dom} p$.

(b) *If* $x^\nu \in P(u^\nu)$, *and if* $u^\nu \to \bar{u} \in \operatorname{dom} p$ *in such a way that* $p(u^\nu) \to p(\bar{u})$ *(as when* p *is continuous at* $\bar{u}$ *relative to a set* U *containing* $\bar{u}$ *and* u^ν*), then the sequence* $\{x^\nu\}_{\nu\in\mathbb{N}}$ *is bounded, and all its cluster points lie in* $P(\bar{u})$.

(c) *For* p *to be continuous at a point* $\bar{u}$ *relative to a set* U *containing* $\bar{u}$, *a sufficient condition is the existence of some* $\bar{x} \in P(\bar{u})$ *such that* $f(\bar{x}, u)$ *is continuous in* u *at* $\bar{u}$ *relative to* U.

Proof. For each $u \in \mathbb{R}^m$ let $f_u(x) = f(x,u)$. As a function on $\mathbb{R}^n$, either $f_u \equiv \infty$ or f_u is proper, lsc and level-bounded, so that Theorem 1.9 is applicable to the minimization of f_u. The first case, which corresponds to $p(u) = \infty$, can't hold for every u, because $f \not\equiv \infty$. Therefore $\operatorname{dom} p \neq \emptyset$, and for each $u \in \operatorname{dom} p$ the value $p(u) = \inf f_u$ is finite and the set $P(u) = \operatorname{argmin} f_u$ is nonempty and compact. In particular, $p(u) \leq \alpha$ if and only if there is an x with $f(x,u) \leq \alpha$. Hence for $V \subset \mathbb{R}^m$ we have

$$(\operatorname{lev}_{\leq\alpha} p) \cap V = \left[\text{ image of } (\operatorname{lev}_{\leq\alpha} f) \cap (\mathbb{R}^n \times V) \text{ under } (x,u) \mapsto u\,\right].$$

Since the image of a compact set under a continuous mapping is compact, we see that $(\operatorname{lev}_{\leq\alpha} p) \cap V$ is closed whenever V is such that $(\operatorname{lev}_{\leq\alpha} f) \cap (\mathbb{R}^n \times V)$ is closed and bounded. From the uniform level boundedness assumption, any $\bar{u} \in \mathbb{R}^m$ has a neighborhood V such that $(\operatorname{lev}_{\leq\alpha} f) \cap (\mathbb{R}^n \times V)$ is bounded; replacing V by a smaller, closed neighborhood of $\bar{u}$ if necessary, we can get $(\operatorname{lev}_{\leq\alpha} f) \cap (\mathbb{R}^n \times V)$ also to be closed, because f is lsc. Thus, each $\bar{u} \in \mathbb{R}^m$ has a neighborhood whose intersection with $\operatorname{lev}_{\leq\alpha} p$ is closed, hence $\operatorname{lev}_{\leq\alpha} p$ itself is closed. Then p is lsc by 1.6(c). This proves (a).

In (b) we have for any $\alpha > p(\bar{u})$ that eventually $\alpha > p(u^\nu) = f(x^\nu, u^\nu)$. Again taking V to be a closed neighborhood of $\bar{u}$ as in Definition 1.16, we see that for all ν sufficiently large the pair (x^ν, u^ν) lies in the compact set $(\operatorname{lev}_{\leq\alpha} f) \cap (\mathbb{R}^n \times V)$. The sequence $\{x^\nu\}_{\nu\in\mathbb{N}}$ is therefore bounded, and for any cluster point $\bar{x}$ we have $(\bar{x}, \bar{u}) \in \operatorname{lev}_{\leq\alpha} f$. This being true for arbitrary $\alpha > p(\bar{u})$, we see that $f(\bar{x}, \bar{u}) \leq p(\bar{u})$, which means that $\bar{x} \in P(\bar{u})$.

In (c) we have $p(u) \leq f(\bar{x}, u)$ for all u and $p(\bar{u}) = f(\bar{x}, \bar{u})$. The upper semicontinuity of $f(\bar{x}, \cdot)$ at $\bar{u}$ relative to any set U containing $\bar{u}$ therefore implies the upper semicontinuity of p at $\bar{u}$. Inasmuch as p is already known to be lsc at $\bar{u}$, we can conclude in this case that p is continuous at $\bar{u}$ relative to U. □

A simple example of how $p(u) = \inf_x f(x,u)$ can fail to be lsc is furnished by $f(x,u) = e^{xu}$ on $\mathbb{R}^1 \times \mathbb{R}^1$. This has $p(u) = 0$ for all $u \neq 0$, but $p(0) = 1$; the set $P(u) = \operatorname{argmin}_x f(x,u)$ is empty for all $u \neq 0$, but $P(0) = (-\infty, \infty)$. Switching to $f(x,u) = |2e^{xu} - 1|$, we get the same function p and the same set $P(0)$, but $P(u) \neq \emptyset$ for $u \neq 0$, with $P(u)$ consisting of a single point. In this case like the previous one, Theorem 1.17 isn't applicable; actually, $f(x,u)$ isn't level-bounded in x for any u. A more subtle example is obtained with $f(x,u) = \min\{\,|x - u^{-1}|, 1 + |x|\,\}$ when $u \neq 0$, but $f(x,u) = 1 + |x|$ when $u = 0$. This is continuous in (x,u) and level-bounded in x for each u, but not locally uniformly in u. Once more, $p(u) = 0$ for $u \neq 0$ but $p(0) = 1$; on the other hand, $P(u) = \{1/u\}$ for $u \neq 0$, but $P(0) = \{0\}$.

The important question of when actually $p(u^\nu) \to p(\bar{u})$ for a sequence $u^\nu \to \bar{u}$, as called for in 1.17(b), goes far beyond the simple sufficient condition offered in 1.17(c). A fully satisfying answer will have to await the theory of 'epi-convergence' in Chapter 7.

Parametric minimization has a simple geometric interpretation.

1.18 Proposition (epigraphical projection). *Suppose $p(u) = \inf_x f(x,u)$ for a function $f : I\!R^n \times I\!R^m \to \overline{I\!R}$, and let E be the image of $\operatorname{epi} f$ under the projection $(x,u,\alpha) \mapsto (u,\alpha)$. If for each $u \in \operatorname{dom} p$ the set $P(u) = \operatorname{argmin}_x f(x,u)$ is attained, then $\operatorname{epi} p = E$. In general, $\operatorname{epi} p$ is the set obtained by adjoining to E any lower boundary points that might be missing, i.e., by closing the intersection of E with each vertical line in $I\!R^m \times I\!R$.*

Proof. This is clear from the projection argument given for Theorem 1.17. □

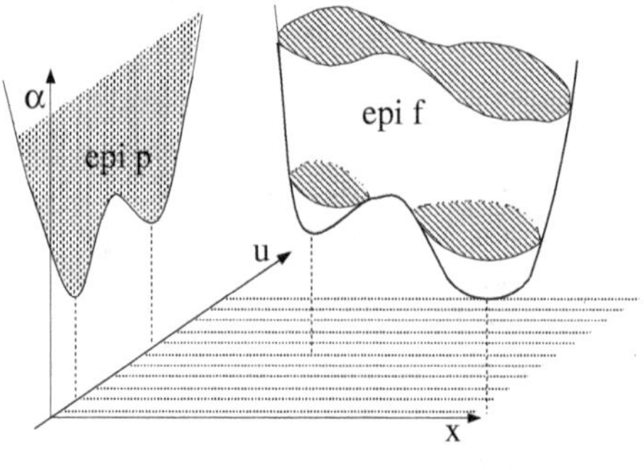

Fig. 1–8. Epigraphical projection in parametric minimization.

1.19 Exercise (constraints in parametric minimization). *For each u in a closed set $U \subset I\!R^m$ let $p(u)$ denote the optimal value and $P(u)$ the optimal solution set in the problem*

$$\text{minimize } f_0(x,u) \text{ over all } x \in X \text{ satisfying } f_i(x,u) \begin{cases} \leq 0 & \text{for } i \in I_1, \\ = 0 & \text{for } i \in I_2, \end{cases}$$

for a closed set $X \subset I\!R^n$ and continuous functions $f_i : X \times U \mapsto I\!R$ (for $i \in \{0\} \cup I_1 \cup I_2$). Suppose that for each $\bar{u} \in U$, $\varepsilon > 0$ and $\alpha \in I\!R$ the set of pairs $(x,u) \in X \times U$ satisfying $|u - \bar{u}| \leq \varepsilon$ and $f_0(x,u) \leq \alpha$, along with all the constraints indexed by I_1 and I_2, is bounded in $I\!R^n \times I\!R^m$.

Then p is lsc on U, and for every $u \in U$ with $p(u) < \infty$ the set $P(u)$ is nonempty and compact. If only f_0 depends on u, and the constraints are satisfied by at least one x, then $\operatorname{dom} p = U$, and p is continuous relative to U. In that case, whenever $x^\nu \in P(u^\nu)$ with $u^\nu \to \bar{u}$ in U, all the cluster points of the sequence $\{x^\nu\}_{\nu \in I\!N}$ are in $P(\bar{u})$.

Guide. This is obtained from Theorem 1.17 by taking $f(x,u) = f_0(x,u)$ when

$(x,u) \in X \times U$ and all the constraints are satisfied, but $f(x,u) = \infty$ otherwise. Then $p(u)$ is assigned the value ∞ when $u \notin U$. □

1.20 Example (distance functions and projections). *For any nonempty, closed set $C \subset \mathbb{R}^n$, the distance $d_C(x)$ of a point x from C depends continuously on x, while the projection $P_C(x)$, consisting of the points of C nearest to x is nonempty and compact. Whenever $w^\nu \in P_C(x^\nu)$ and $x^\nu \to \bar{x}$, the sequence $\{w^\nu\}_{\nu \in \mathbb{N}}$ is bounded and all its cluster points lie in $P_C(\bar{x})$.*

Detail. Taking $f(w,x) = |w - x| + \delta_C(w)$, we get $d_C(x) = \inf_w f(w,x)$ and $P_C(x) = \operatorname{argmin}_w f(w,x)$, and we can then apply Theorem 1.17. □

1.21 Example (convergence of penalty methods). *Suppose a problem of type*

$$\text{minimize } f(x) \text{ over all } x \in \mathbb{R}^n \text{ satisfying } F(x) \in D,$$

with proper, lsc $f : \mathbb{R}^n \to \overline{\mathbb{R}}$, continuous $F : \mathbb{R}^n \to \mathbb{R}^m$, and closed $D \subset \mathbb{R}^m$, is approximated through some penalty scheme by a problem of type

$$\text{minimize } f(x) + \theta\big(F(x), r\big) \text{ over all } x \in \mathbb{R}^n$$

with parameter $r \in (0,\infty)$, where the function $\theta : \mathbb{R}^m \times (0,\infty) \to \overline{\mathbb{R}}$ is lsc with $-\infty < \theta(u,r) \nearrow \delta_D(u)$ as $r \to \infty$. Assume that for some $\bar{r} \in (0,\infty)$ sufficiently high the level sets of the function $x \mapsto f(x) + \theta\big(F(x),\bar{r}\big)$ are bounded, and consider any sequence of parameter values $r^\nu \geq \bar{r}$ with $r^\nu \to \infty$. Then:

(a) *The optimal value in the approximate problem for r^ν converges to the optimal value in the true problem.*

(b) *Any sequence $\{x^\nu\}_{\nu \in \mathbb{N}}$ chosen with x^ν an optimal solution to the approximate problem for r^ν is a bounded sequence such that every cluster point $\bar{x}$ is an optimal solution to the true problem.*

Detail. Set $\bar{s} = 1/\bar{r}$ and define $g(x,s) := f(x) + \tilde{\theta}\big(F(x),s\big)$ on $\mathbb{R}^n \times \mathbb{R}$ for

$$\tilde{\theta}(u,s) := \begin{cases} \theta(u,1/s) & \text{when } s > 0 \text{ and } s \leq \bar{s}, \\ \delta_D(u) & \text{when } s = 0, \\ \infty & \text{when } s < 0 \text{ or } s > \bar{s}. \end{cases}$$

Identifying the given problem with that of minimizing $g(x,0)$ in $x \in \mathbb{R}^n$, and identifying the approximate problem for parameter $r \in [\bar{r},\infty)$ with that of minimizing $g(x,s)$ in $x \in \mathbb{R}^n$ for $s = 1/r$, we aim at applying Theorem 1.17 to the 'inf' $p(s)$ and the 'argmin' $P(s)$.

The function $\tilde{\theta}$ on $\mathbb{R}^m \times \mathbb{R}$ is proper (because $D \neq \emptyset$), and furthermore it's lsc. The latter is evident at all points where $s \neq 0$, while at points where $s = 0$ it follows from having $\tilde{\theta}(u,s) \nearrow \tilde{\theta}(u,0)$ as $s \searrow 0$, since then for any $\alpha \in \mathbb{R}$ the set $\operatorname{lev}_{\leq\alpha} \tilde{\theta}(\cdot,s)$ in $\mathbb{R}^m$ decreases as s decreases in $(0,\bar{s}]$, the intersection over all $s > 0$ being $\operatorname{lev}_{\leq\alpha} \tilde{\theta}(\cdot,0)$. The assumptions that f is lsc, F is continuous and D is closed, ensure through this that g is lsc, and proper as well unless $g \equiv \infty$, in which event everything would trivialize.

The choice of $\bar{s}$, along with the monotonicity of $\tilde{\theta}(u,s)$ in $s \in (0,\bar{s}]$, ensures that g is level bounded in x locally uniformly in s, and that p is nonincreasing on $[0,\bar{s}]$. From 1.17(a) we have then that $p(s) \to p(0)$ as $s \searrow 0$. In terms of $s^\nu := 1/r^\nu$ this gives claim (a), and from 1.17(b) we then have claim (b). □

Facts about barrier methods of constraint approximation (cf. 1.3) can likewise be deduced from the properties of parametric optimization in 1.17.

G. Moreau Envelopes

These properties support a method of approximating general functions in terms of certain 'envelope functions'.

1.22 Definition (Moreau envelopes and proximal mappings). *For a proper, lsc function $f : \mathbb{R}^n \to \overline{\mathbb{R}}$ and parameter value $\lambda > 0$, the Moreau envelope function $e_\lambda f$ and proximal mapping $P_\lambda f$ are defined by*

$$e_\lambda f(x) := \inf_w \Big\{ f(w) + \frac{1}{2\lambda}|w - x|^2 \Big\} \le f(x),$$
$$P_\lambda f(x) := \operatorname{argmin}_w \Big\{ f(w) + \frac{1}{2\lambda}|w - x|^2 \Big\}.$$

Here we speak of $P_\lambda f$ as a mapping in the 'set-valued' sense that will later be developed in Chapter 5. Note that if f is an indicator function δ_C, then $P_\lambda f$ coincides with the projection mapping P_C in 1.20, while $e_\lambda f$ is $(1/2\lambda)d_C^2$ for the distance function d_C in 1.20.

In general, $e_\lambda f$ approximates f from below in the manner depicted in Figure 1–9. For smaller and smaller λ, it's easy to believe that, $e_\lambda f$ approximates f better and better, and indeed, $1/\lambda$ can be interpreted as a penalty parameter for a putative constraint $w - x = 0$ in the minimization defining $e_\lambda f(x)$. We'll apply Theorem 1.17 in order to draw exact conclusions about this behavior, which has very useful implications in variational analysis. First, though, we need an associated definition.

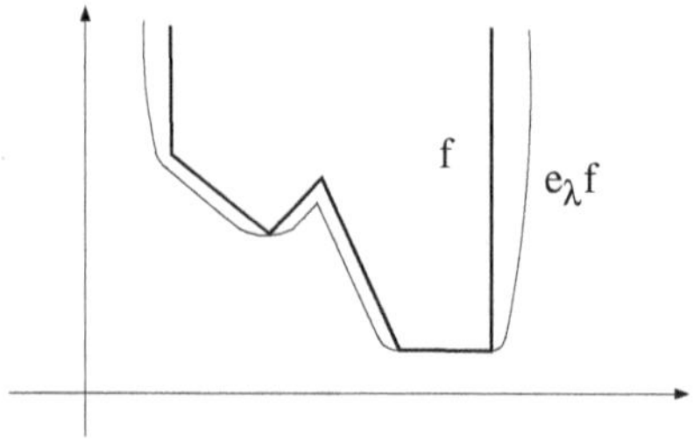

Fig. 1–9. Approximation by a Moreau envelope function.

1.23 Definition (prox-boundedness). *A function $f : \mathbb{R}^n \to \overline{\mathbb{R}}$ is prox-bounded if there exists $\lambda > 0$ such that $e_\lambda f(x) > -\infty$ for some $x \in \mathbb{R}^n$. The supremum of the set of all such λ is the threshold λ_f of prox-boundedness for f.*

1.24 Exercise (characteristics of prox-boundedness). *For a proper, lsc function $f : \mathbb{R}^n \to \overline{\mathbb{R}}$, the following properties are equivalent:*

(a) *f is prox-bounded;*

(b) *f majorizes a quadratic function (i.e., $f \geq q$ for a polynomial function q of degree two or less);*

(c) *for some $r \in \mathbb{R}$, $f + \frac{1}{2}r|\cdot|^2$ is bounded from below on $\mathbb{R}^n$;*

(d) $\liminf_{|x|\to\infty} f(x)/|x|^2 > -\infty$.

Indeed, if r_f is the infimum of all r for which (c) holds, the limit in (d) is $-\frac{1}{2}r_f$ and the proximal threshold for f is $\lambda_f = 1/\max\{0, r_f\}$ (with '$1/0 = \infty$').

In particular, if f is bounded from below, i.e., has $\inf f > -\infty$, then f is prox-bounded with threshold $\lambda_f = \infty$.

Guide. Utilize 1.14 in connection with (d). Establish the sufficiency of (b) by arguing that this condition implies (d). □

1.25 Theorem (proximal behavior). *Let $f : \mathbb{R}^n \to \overline{\mathbb{R}}$ be proper, lsc, and prox-bounded with threshold $\lambda_f > 0$. Then for every $\lambda \in (0, \lambda_f)$ the set $P_\lambda f(x)$ is nonempty and compact, whereas the value $e_\lambda f(x)$ is finite and depends continuously on (λ, x), with*

$$e_\lambda f(x) \nearrow f(x) \text{ for all } x \text{ as } \lambda \searrow 0.$$

In fact, $e_{\lambda^\nu} f(x^\nu) \to f(\bar{x})$ whenever $x^\nu \to \bar{x}$ while $\lambda^\nu \searrow 0$ in $(0, \lambda_f)$ in such a way that the sequence $\{|x^\nu - \bar{x}|/\lambda^\nu\}_{\nu \in \mathbb{N}}$ is bounded.

Furthermore, if $w^\nu \in P_{\lambda^\nu} f(x^\nu)$, $x^\nu \to \bar{x}$ and $\lambda^\nu \to \lambda \in (0, \lambda_f)$, then the sequence $\{w^\nu\}_{\nu \in \mathbb{N}}$ is bounded and all its cluster points lie in $P_\lambda f(\bar{x})$.

Proof. Fixing any $\lambda_0 \in (0, \lambda_f)$, we apply 1.17 to the problem of minimizing $h(w, x, \lambda)$ in w, where $h(w, x, \lambda) = f(w) + h_0(w, x, \lambda)$ with

$$h_0(w, x, \lambda) := \begin{cases} (1/2\lambda)|w - x|^2 & \text{when } \lambda \in (0, \lambda_0], \\ 0 & \text{when } \lambda = 0 \text{ and } w = x, \\ \infty & \text{otherwise.} \end{cases}$$

Here h_0 is lsc, in fact continuous when $\lambda > 0$ and also on sets of the form $\{(w, x, \lambda) \mid |w - x| \leq \mu\lambda,\ 0 \leq \lambda \leq \lambda_0\}$ for any $\mu > 0$. Hence h is lsc and proper. We verify next that $h(w, x, \lambda)$ is level-bounded in w locally uniformly in (x, λ). If not, we could have where $h(w^\nu, x^\nu, \lambda^\nu) \leq \bar{\alpha} < \infty$ with $(x^\nu, \lambda^\nu) \to (\bar{x}, \lambda)$ but $|w^\nu| \to \infty$. Then $w^\nu \neq x^\nu$ (at least for large ν), so $\lambda^\nu \in (0, \lambda_0]$ and $f(w^\nu) + (1/2\lambda^\nu)|w^\nu - x^\nu|^2 \leq \bar{\alpha}$. The choice of λ_0 ensures through the definition of λ_f the existence of $\lambda_1 > \lambda_0$ and $\beta \in \mathbb{R}$ such that $f(w) \geq -(1/2\lambda_1)|w|^2 + \beta$. Then $-(1/2\lambda_1)|w^\nu|^2 + (1/2\lambda_0)|w^\nu - x^\nu|^2 \leq \bar{\alpha} - \beta$. In dividing this by $|w^\nu|^2$ and taking the limit as $\nu \to \infty$, we get $-(1/2\lambda_1) + (1/2\lambda_0) \leq 0$, a contradiction.

The conclusions come now from Theorem 1.17, with the continuity properties of h_0 being used in 1.17(c) to see that $h(\bar{x},\cdot,\cdot)$ is continuous relative to sets containing the sequences $\{(x^\nu,\lambda^\nu)\}$ that come up. □

The fact that $e_\lambda f$ is a finite, continuous function, whereas f itself may merely be lower semicontinuous and extended-real-valued, shows that approximation by Moreau envelopes has a natural effect of 'regularizing' a function f. This hints at some of the uses of such approximation which will later be prominent in many phases of theory.

In the concept of Moreau envelopes, minimization is used as a means of *defining* one function in terms of another, just like composition or integration can be used in classical analysis. Minimization and maximization have this role also in defining for any family of functions $\{f_i\}_{i\in I}$ from $\mathbb{R}^n$ to $\overline{\mathbb{R}}$ another such function, called the *pointwise supremum*, by

$$\big(\sup_{i\in I} f_i\big)(x) := \sup_{i\in I} f_i(x), \tag{1(10)}$$

as well as a function, called the *pointwise infimum* of the family, by

$$\big(\inf_{i\in I} f_i\big)(x) := \inf_{i\in I} f_i(x). \tag{1(11)}$$

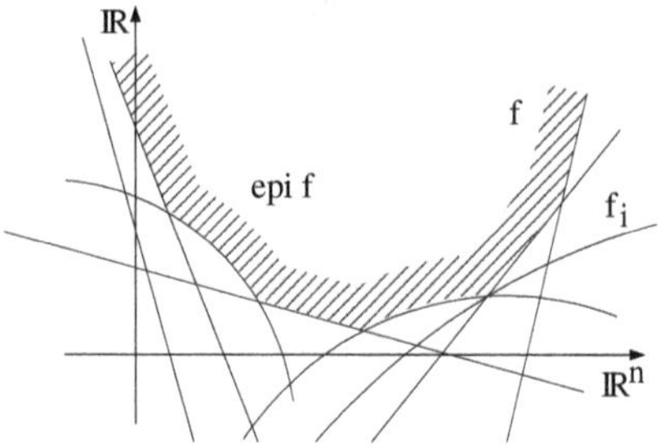

Fig. 1–10. Pointwise max operation: intersection of epigraphs.

Geometrically, these functions have a nice interpretation, cf. Figures 1–10 and 1–11. The epigraph of the pointwise supremum is the intersection of the sets $\operatorname{epi} f_i$, whereas the epigraph of the pointwise infimum is the 'vertical closure' of the union of the sets $\operatorname{epi} f_i$ (vertical closure in $\mathbb{R}^n \times \mathbb{R}$ being the operation that closes a set's intersection with each vertical line; cf. 1.18).

1.26 Proposition (semicontinuity under pointwise max and min).

(a) $\sup_{i\in I} f_i$ *is lsc if each* f_i *is lsc;*

(b) $\inf_{i\in I} f_i$ *is lsc if each* f_i *is lsc and the index set* I *is finite;*

(c) $\sup_{i\in I} f_i$ *and* $\inf_{i\in I} f_i$ *are both continuous if each* f_i *is continuous and the index set* I *is finite.*

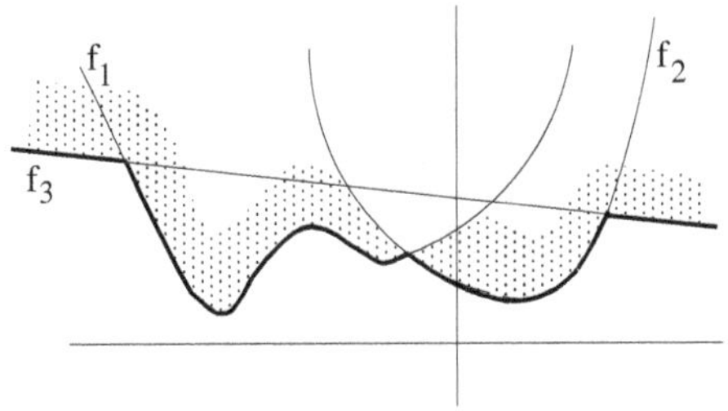

Fig. 1–11. Pointwise min operation: union of epigraphs.

Proof. In (a) and (b) we apply the epigraphical criterion in 1.6; the intersection of closed sets is closed, as is the union if only finitely many sets are involved. We get (c) from (b) and its usc counterpart, using 1.12. □

The pointwise min operation is parametric minimization with the index i as the parameter, and this is a way of finding criteria for lower semicontinuity beyond the one in 1.26(b). In fact, to say that $p(u) = \inf_x f(x,u)$ for $f : \mathbb{R}^n \times \mathbb{R}^m \to \overline{\mathbb{R}}$ is to say that p is the pointwise infimum of the family of functions $f_x(u) := f(x,u)$ indexed by $x \in \mathbb{R}^n$.

Variational analysis is heavily influenced by the fact that, in general, operations like pointwise maximization and minimization *can fail to preserve smoothness*. A function is said to be *smooth* if it is continuously differentiable, i.e., of class $\mathcal{C}^1$; otherwise it is *nonsmooth*. (It is twice smooth if of class $\mathcal{C}^2$, which means that all first and second partial derivatives exist and are continuous, and so forth.) Regardless of the degree of smoothness of a collection of the f_i's, the functions $\sup_{i\in I} f_i$ and $\inf_{i\in I} f_i$ aren't likely to be smooth, cf. Figures 1–10 and 1–11. Therefore, they typically fall outside the domain of classical differential analysis, as do the functions in optimization that arise through penalty expressions. The search for ways around this obstacle has led to concepts of one-sided derivatives and 'subgradients' that support a robust *nonsmooth analysis*, which we'll start to look at in Chapter 8. Despite this tendency of maximization and minimization to destroy smoothness, the process of forming Moreau envelopes will often be found to create smoothness where none was present initially. (This will be seen for instance in 2.26.)

H. Epi-Addition and Epi-Multiplication

Especially interesting as an operation based on minimization, for this and other reasons, is *epi-addition*, also called *inf-convolution*. For functions $f_1 : \mathbb{R}^n \to \overline{\mathbb{R}}$ and $f_2 : \mathbb{R}^n \to \overline{\mathbb{R}}$, the *epi-sum* is the function $f_1 \# f_2 : \mathbb{R}^n \to \overline{\mathbb{R}}$ defined by

$$\begin{aligned}(f_1 \# f_2)(x) &:= \inf_{x_1+x_2=x} \big\{ f_1(x_1) + f_2(x_2) \big\} \\ &= \inf_w \big\{ f_1(x-w) + f_2(w) \big\} = \inf_w \big\{ f_1(w) + f_2(x-w) \big\}.\end{aligned} \qquad 1(12)$$

(Here the inf-addition rule $\infty - \infty = \infty$ is to be used in case of conflicting infinities.) For instance, epi-addition is the operation behind 1.20 and 1.22:

$$d_C = \delta_C \# |\cdot|, \qquad e_\lambda f = f \# \frac{1}{2\lambda}|\cdot|^2. \tag{1(13)}$$

1.27 Proposition (properties of epi-addition). *Let f_1 and f_2 be lsc and proper on $\mathbb{R}^n$, and suppose for each bounded set $B \subset \mathbb{R}^n$ and $\alpha \in \mathbb{R}$ that the set*

$$\left\{(x_1, x_2) \in \mathbb{R}^n \times \mathbb{R}^n \,\middle|\, f_1(x_1) + f_2(x_2) \leq \alpha,\ x_1 + x_2 \in B\right\}$$

is bounded (as is true in particular if one of the functions is level-bounded while the other is bounded below). Then $f_1 \# f_2$ is lsc and proper. Furthermore, $f_1 \# f_2$ is continuous at any point $\bar{x}$ where its value is finite and expressible as $f_1(\bar{x}_1) + f_2(\bar{x}_2)$ with $\bar{x}_1 + \bar{x}_2 = \bar{x}$ such that either f_1 is continuous at $\bar{x}_1$ or f_2 is continuous at $\bar{x}_2$.

Proof. This follows from Theorem 1.17 in the case of minimizing $f(w, x) = f_1(x - w) + f_2(w)$ in w with x as parameter. The symmetry between the roles of f_1 and f_2 yields the symmetric form of the continuity assertion. □

Epi-addition is commutative and associative; the formula in the case of more than two functions works out to

$$(f_1 \# f_2 \# \cdots \# f_r)(x) = \inf_{x_1 + x_2 + \cdots + x_r = x} \left\{f_1(x_1) + f_2(x_2) + \cdots + f_r(x_r)\right\}.$$

One has $f \# \delta_{\{0\}} = f$ for all f, where $\delta_{\{0\}}$ is of course the indicator of the singleton set $\{0\}$. A companion operation is *epi-multiplication* by scalars $\lambda \geq 0$; the epi-multiple $\lambda \star f$ is defined by

$$\begin{aligned} (\lambda \star f)(x) &:= \lambda f(\lambda^{-1} x) \ \text{ for } \lambda > 0 \\ (0 \star f)(x) &:= \begin{cases} 0 & \text{if } x = 0,\ f \not\equiv \infty, \\ \infty & \text{otherwise.} \end{cases} \end{aligned} \tag{1(14)}$$

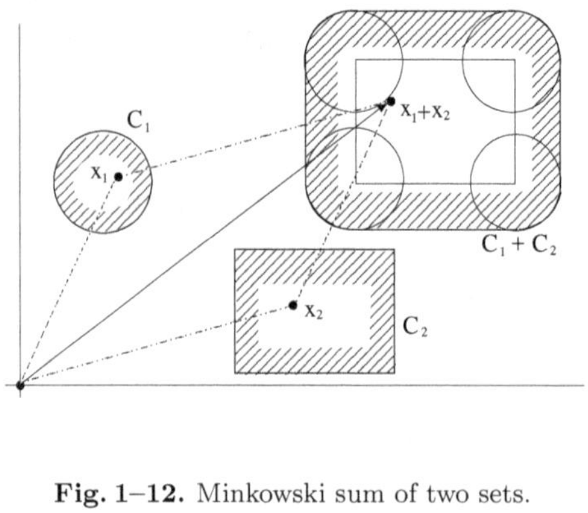

Fig. 1–12. Minkowski sum of two sets.

The names of these operations come from a connection with *set algebra*. The *translate* of a set C by a vector a is $C + a := \{x + a \mid x \in C\}$. General Minkowski scalar multiples and sums of sets in $\mathbb{R}^n$ are defined by

$$\begin{aligned} \lambda C &:= \{\lambda x \mid x \in C\}, \qquad C/\tau = \tau^{-1}C, \qquad -C = (-1)C, \\ C_1 + C_2 &:= \{x_1 + x_2 \mid x_1 \in C_1,\ x_2 \in C_2\}, \\ C_1 - C_2 &:= \{x_1 - x_2 \mid x_1 \in C_1,\ x_2 \in C_2\}. \end{aligned}$$

In general, $C_1 + C_2$ can be interpreted as the union of all the translates $C_1 + x_2$ of C_1 obtained through vectors $x_2 \in C_2$, but it is also the union of all the translates $C_2 + x_1$ of C_2 for $x_1 \in C_1$. Minkowski sums and scalar multiples are particularly useful in working with neighborhoods. For instance, one has

$$\mathbb{B}(x, \varepsilon) = x + \varepsilon\mathbb{B} \text{ for } \mathbb{B} := \mathbb{B}(0,1) \text{ (closed unit ball).} \qquad 1(15)$$

Similarly, for any set $C \subset \mathbb{R}^n$ and $\varepsilon > 0$ the 'fattened' set $C + \varepsilon\mathbb{B}$ consists of all points x lying in a closed ball of radius ε around some point of C, cf. Figure 1–13. Note that $C_1 + C_2$ is empty if either C_1 or C_2 is empty.

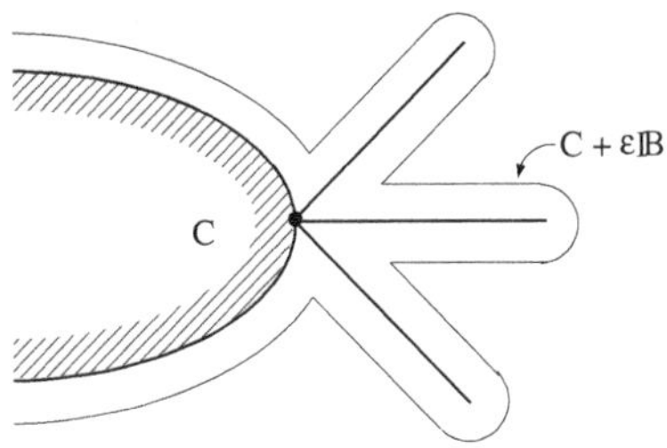

Fig. 1–13. A fattened set.

1.28 Exercise (sums and multiples of epigraphs).

(a) *For functions f_1 and f_2 on $\mathbb{R}^n$, the epi-sum $f_1 \# f_2$ satisfies*

$$\operatorname{epi}(f_1 \# f_2) = \operatorname{epi} f_1 + \operatorname{epi} f_2$$

as long as the infimum defining $(f_1 \# f_2)(x)$ is attained when finite. Regardless of such attainment, one always has

$$\{(x, \alpha) \mid (f_1 \# f_2)(x) < \alpha\} = \{(x_1, \alpha_1) \mid f_1(x_1) < \alpha_1\} + \{(x_2, \alpha_2) \mid f_2(x_2) < \alpha_2\}.$$

(b) *For a function f and a scalar $\lambda > 0$, the epi-multiple $\lambda \star f$ satisfies*

$$\operatorname{epi}(\lambda \star f) = \lambda(\operatorname{epi} f).$$

Guide. Caution is needed in (a) because the functions f_i can take on ∞ and $-\infty$ as values, whereas elements (x_i, α_i) of $\operatorname{epi} f_i$ can only have α_i finite. □

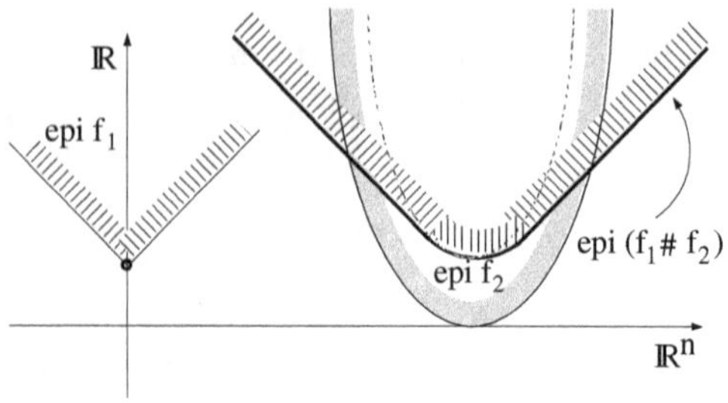

Fig. 1–14. Epi-addition of two functions: geometric interpretation.

For $\lambda = 0$, $\lambda(\text{epi} f)$ would no longer be an epigraph in general: $0 \,\text{epi} f = \{(0,0)\}$ when $f \not\equiv \infty$, while $0 \,\text{epi} f = \emptyset$ when $f \equiv \infty$. The definition of $0{\star}f$ in 1(14) fits naturally with this and, through the epigraphical interpretations in 1.28, supports the rules

$$\lambda{\star}(f_1 \# f_2) = (\lambda{\star}f_1) \# (\lambda{\star}f_2) \text{ for } \lambda \geq 0,$$
$$\lambda_1{\star}(\lambda_2{\star}f) = (\lambda_1\lambda_2){\star}f \text{ for } \lambda_1 \geq 0,\ \lambda_2 \geq 0.$$

In the case of *convex* functions in Chapter 2 it will turn out that another form of distributive law holds as well, cf. 2.24(c). For such functions a fundamental duality will be revealed in Chapter 11 between epi-addition and epi-multiplication on the one hand and ordinary addition and scalar multiplication on the other hand.

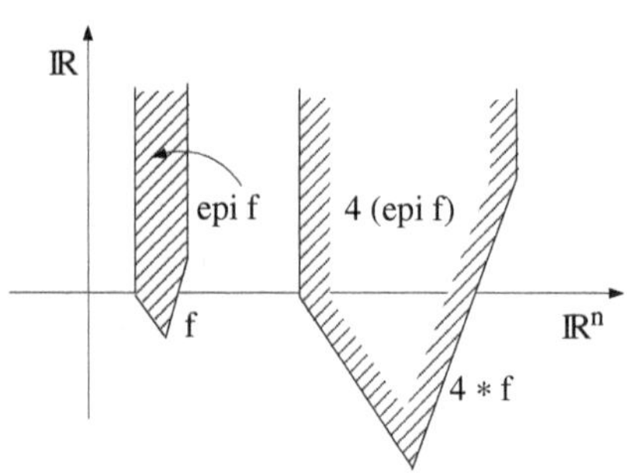

Fig. 1–15. Epi-multiplication of a function: geometric interpretation.

1.29 Exercise (calculus of indicators, distances, and envelopes).

(a) $\delta_{C+D} = \delta_C \# \delta_D$, *whereas* $\delta_{\lambda C} = \lambda{\star}\delta_C$ *for* $\lambda > 0$,

(b) $d_{C+D} = d_C \# d_D$, *whereas* $d_{\lambda C} = \lambda{\star}d_C$ *for* $\lambda > 0$,

(c) $e_{\lambda_1+\lambda_2}f = e_{\lambda_1}(e_{\lambda_2}f)$ *for* $\lambda_1 > 0,\ \lambda_2 > 0$.

Guide. In part (b), verify that the function $h(x) = |\cdot|$ satisfies $h \# h = h$ and $\lambda{\star}h = h$ for $\lambda > 0$. Demonstrate in the first equation that both sides equal

$\delta_C \# \delta_D \# h$ and in the second that both sides equal $(\lambda\star\delta_C)\# h$. In part (c), use the associative law to reduce the issue to whether $(1/2\lambda_1)|\cdot|^2 \# (1/2\lambda_2)|\cdot|^2 = (1/2\lambda)|\cdot|^2$ for $\lambda = \lambda_1 + \lambda_2$, and verify this by direct calculation. □

1.30 Example (vector-max and log-exponential functions). *The functions*

$$\operatorname{vecmax}(x) := \max\{x_1, \ldots, x_n\}, \qquad \operatorname{logexp}(x) := \log\big(e^{x_1} + \cdots + e^{x_n}\big),$$

for $x = (x_1, \ldots, x_n) \in \mathbb{R}^n$ *are related by the estimate*

$$\varepsilon\star\operatorname{logexp} - \varepsilon \log n \;\leq\; \operatorname{vecmax} \;\leq\; \varepsilon\star\operatorname{logexp} \quad \textit{for any } \varepsilon > 0. \tag{1(16)}$$

Thus, for any smooth functions $f_1, \ldots, f_m$ *on* $\mathbb{R}^n$, *viewed as the components of a mapping* $F : \mathbb{R}^n \to \mathbb{R}^m$, *the generally nonsmooth function*

$$f(x) = \operatorname{vecmax}\big(F(x)\big) = \max\big\{f_1(x), \ldots, f_m(x)\big\}$$

can be approximated uniformly, to any accuracy, by a smooth function

$$f_\varepsilon(x) = (\varepsilon\star\operatorname{logexp})\big(F(x)\big) = \varepsilon \log\big(e^{f_1(x)/\varepsilon} + \cdots + e^{f_m(x)/\varepsilon}\big).$$

Detail. With $\mu = \operatorname{vecmax}(x)$, we have $e^{\mu/\varepsilon} \leq \sum_{j=1}^n e^{x_j/\varepsilon} \leq n e^{\mu/\varepsilon}$. In taking logarithms and multiplying by ε, we get the estimates in 1(16). From those inequalities the function $\varepsilon\star\operatorname{logexp}$ converges uniformly to the function vecmax as $\varepsilon \searrow 0$. The epigraphical significance of this will emerge in Chapter 3. □

Another operation through which new functions are constructed by minimization is *epi-composition*. This combines a function $f : \mathbb{R}^n \to \overline{\mathbb{R}}$ with a mapping $F : \mathbb{R}^n \to \mathbb{R}^m$ to get a function $Ff : \mathbb{R}^m \to \overline{\mathbb{R}}$ by

$$(Ff)(u) := \inf\big\{f(x) \,\big|\, F(x) = u\big\}. \tag{1(17)}$$

The name comes from the fact that the epigraph of Ff is essentially obtained by 'composing' f with the epigraphical mapping $(x, \alpha) \mapsto \big(F(x), \alpha\big)$ associated with F. Here is a precise statement which exhibits parallels with 1.28(a).

1.31 Exercise (images of epigraphs). *For a function* $f : \mathbb{R}^n \to \overline{\mathbb{R}}$ *and a mapping* $F : \mathbb{R}^n \to \mathbb{R}^m$, *the epi-composite function* $Ff : \mathbb{R}^m \to \overline{\mathbb{R}}$ *has*

$$\operatorname{epi} Ff \;=\; \big\{(F(x), \alpha) \,\big|\, (x, \alpha) \in \operatorname{epi} f\big\}$$

as long as the infimum defining $(Ff)(u)$ *is attained when it is finite. Regardless of such attainment, one always has*

$$\big\{(u, \alpha) \,\big|\, (Ff)(u) < \alpha\big\} \;=\; \big\{(F(x), \alpha) \,\big|\, f(x) < \alpha\big\}.$$

Epi-composition as defined by 1(17) clearly concerns the special case of parametric minimization in which the parameter vector u gives the right side of a system of equality constraints, but its usefulness as a concept goes beyond what this simple description might suggest.

1.32 Proposition (properties of epi-composition). *Let $f : I\!R^n \to \overline{I\!R}$ be proper and lsc, and let $F : I\!R^n \to I\!R^m$ be continuous. Suppose for each $\alpha \in I\!R$ and $u \in I\!R^m$ that the set $\{x \mid f(x) \leq \alpha,\ F(x) \in I\!B(u,\varepsilon)\}$ is bounded for some $\varepsilon > 0$. Then the epi-composite function $Ff : I\!R^m \to \overline{I\!R}$ is proper and lsc, and the infimum defining $(Ff)(u)$ is attained for each u for which it is finite.*

Proof. Let $\bar{f}(x,u) = f(x)$ when $u = F(x)$ but $\bar{f}(x,u) = \infty$ when $u \neq F(x)$. Then $(Ff)(u) = \inf_x \bar{f}(x,u)$. Applying 1.17, we immediately find justification for all the claims. □

I.* Auxiliary Facts and Principles

A sidelight on lower and upper semicontinuity is provided by their localization. A set $C \subset I\!R^n$ is said to be *locally closed* at a point $\bar{x}$ (not necessarily in C) if $C \cap V$ is closed for some closed neighborhood $V \in \mathcal{N}(\bar{x})$. A test for whether C is (globally) closed is whether C is locally closed at every $\bar{x}$.

1.33 Definition (local semicontinuity). *A function $f : I\!R^n \to \overline{I\!R}$ is locally lower semicontinuous at $\bar{x}$, a point where $f(\bar{x})$ is finite, if there is an $\varepsilon > 0$ such that all sets of the form $\{x \in I\!B(\bar{x},\varepsilon) \mid f(x) \leq \alpha\}$ with $\alpha \leq f(\bar{x}) + \varepsilon$ are closed. The definition of f being locally upper semicontinuous at $\bar{x}$ is parallel.*

1.34 Exercise (consequences of local semicontinuity).

(a) *f is locally lsc at $\bar{x}$, a point where $f(\bar{x})$ is finite, if and only if $\operatorname{epi} f$ is locally closed at $(\bar{x}, f(\bar{x}))$.*

(b) *f is finite and continuous on some neighborhood of $\bar{x}$ if and only if $f(\bar{x})$ is finite and f is both locally lsc and locally usc at $\bar{x}$.*

Several rules of calculation are often useful in dealing with maximization or minimization. Let's look first at one that's reminiscent of standard rules for interchanging the order of summation or integration.

1.35 Proposition (interchange of order of minimization). *For $f : I\!R^n \times I\!R^m \to \overline{I\!R}$ one has in terms of $p(u) := \inf_x f(x,u)$ and $q(x) := \inf_u f(x,u)$ that*

$$\begin{aligned} \inf_{x,u} f(x,u) &= \inf_u p(u) = \inf_x q(x), \\ \operatorname{argmin}_{x,u} f(x,u) &= \Big\{ (\bar{x},\bar{u}) \,\Big|\, \bar{u} \in \operatorname{argmin}_u p(u),\ \bar{x} \in \operatorname{argmin}_x f(x,\bar{u}) \Big\} \\ &= \Big\{ (\bar{x},\bar{u}) \,\Big|\, \bar{x} \in \operatorname{argmin}_x q(x),\ \bar{u} \in \operatorname{argmin}_u f(\bar{x},u) \Big\}. \end{aligned}$$

Proof. Elementary. □

Note here that $f(x,u)$ could have the form $f_0(x,u) + \delta_C(x,u)$ for a function f_0 on $I\!R^n \times I\!R^m$ and a set $C \subset I\!R^n \times I\!R^m$, where C could in turn be specified by a system of constraints $f_i(x,u) \leq 0$ or $f_i(x,u) = 0$. Then in minimizing relative to x to get $p(u)$ one is in effect minimizing $f_0(x,u)$ subject to these constraints as parameterized by u.

1.36 Exercise (max and min of a sum). *For any extended-real-valued functions f_1 and f_2, one has (under the convention $\infty - \infty = \infty$) the estimates*

$$\inf_{x \in X} \{f_1(x) + f_2(x)\} \ge \inf_{x \in X} f_1(x) + \inf_{x \in X} f_2(x),$$

$$\sup_{x \in X} \{f_1(x) + f_2(x)\} \le \sup_{x \in X} f_1(x) + \sup_{x \in X} f_2(x),$$

as well as the rule

$$\inf_{(x_1,x_2) \in X_1 \times X_2} \Big\{ f_1(x_1) + f_2(x_2) \Big\} = \inf_{x_1 \in X_1} f_1(x_1) + \inf_{x_2 \in X_2} f_2(x_2).$$

1.37 Exercise (scalar multiples versus max and min). *For any extended-real-valued function f one has (under the convention $0 \cdot \infty = 0$) that*

$$\inf_{x \in X} \lambda f(x) = \lambda \inf_{x \in X} f(x), \qquad \sup_{x \in X} \lambda f(x) = \lambda \sup_{x \in X} f(x) \qquad \text{when } \lambda \ge 0,$$

while on the other hand

$$\inf_{x \in X} f(x) = - \sup_{x \in X} \{ -f(x) \}, \qquad \sup_{x \in X} f(x) = - \inf_{x \in X} \{ -f(x) \}.$$

1.38 Proposition (lower limits of sums). *For arbitrary extended-real-valued functions one has*

$$\liminf_{x \to \bar{x}} \big[f_1(x) + f_2(x) \big] \ge \liminf_{x \to \bar{x}} f_1(x) + \liminf_{x \to \bar{x}} f_2(x)$$

if the sum on the right is not $\infty - \infty$. On the other hand, one always has

$$\liminf_{x \to \bar{x}} \lambda f(x) = \lambda \liminf_{x \to \bar{x}} f(x) \quad \text{when } \lambda \ge 0.$$

Proof. From the definition of lower limits in 1.5 we calculate

$$\begin{aligned}\liminf_{x \to \bar{x}} \big[f_1(x) + f_2(x) \big] &= \lim_{\delta \searrow 0} \inf_{x \in \boldsymbol{B}(\bar{x},\delta)} \big[f_1(x) + f_2(x) \big] \\ &\ge \lim_{\delta \searrow 0} \Big[\inf_{x \in \boldsymbol{B}(\bar{x},\delta)} f_1(x) + \inf_{x \in \boldsymbol{B}(\bar{x},\delta)} f_2(x) \Big]\end{aligned}$$

by the first rule in 1.36 and then continue with

$$\ge \lim_{\delta \searrow 0} \inf_{x \in \boldsymbol{B}(\bar{x},\delta)} f_1(x) + \lim_{\delta \searrow 0} \inf_{x \in \boldsymbol{B}(\bar{x},\delta)} f_2(x) = \liminf_{x \to \bar{x}} f_1(x) + \liminf_{x \to \bar{x}} f_2(x)$$

as long as this sum is not $\infty - \infty$. The second formula in the proposition is easily verified from the scalar multiplication rule in 1.37 when $\lambda > 0$. When $\lambda = 0$, it's trivial because both sides of the equation have to be 0. □

Since lower semicontinuity is basic to the geometry of epigraphs, there's a need for conditions that can facilitate the verification of lower semicontinuity in circumstances where a function is not just 'abstract' but expressed through operations performed on other functions. Some criteria have already been given in 1.17, 1.26 and 1.27, and here are others.

1.39 Proposition (semicontinuity of sums, scalar multiples).

(a) *$f_1 + f_2$ is lsc if f_1 and f_2 are lsc and proper.*

(b) *λf is lsc if f is lsc and $\lambda \geq 0$.*

Proof. Both facts follow from 1.38. □

The properness assumption in 1.39(a) ensures that $f_1(x) + f_2(x)$ isn't $\infty - \infty$, as required in applying 1.38. For $f_1 + f_2$ to inherit properness, we'd have to know that $\operatorname{dom} f_1 \cap \operatorname{dom} f_2$ isn't empty.

1.40 Exercise (semicontinuity under composition).

(a) *If $f(x) = g\big(F(x)\big)$ with $F : \mathbb{R}^n \to \mathbb{R}^m$ continuous and $g : \mathbb{R}^m \to \overline{\mathbb{R}}$ lsc, then f is lsc.*

(b) *If $f(x) = \theta\big(g(x)\big)$ with $g : \mathbb{R}^n \to \overline{\mathbb{R}}$ lsc, $\theta : \mathbb{R} \to \overline{\mathbb{R}}$ lsc and nondecreasing, and with θ extended to infinite arguments by $\theta(\infty) = \sup\theta$ and $\theta(-\infty) = \inf\theta$, then f is lsc.*

1.41 Exercise (level-boundedness of sums and epi-sums). *For functions f_1 and f_2 on $\mathbb{R}^n$ that are bounded from below,*

(a) *$f_1 + f_2$ is level-bounded if either f_1 or f_2 is level-bounded,*

(b) *$f_1 \# f_2$ is level-bounded if both f_1 and f_2 are level-bounded.*

An extension of Theorem 1.9 serves in the study of computational methods. A *minimizing sequence* for a function $f : \mathbb{R}^n \to \overline{\mathbb{R}}$ is a sequence $\{x^\nu\}_{\nu \in \mathbb{N}}$ such that $f(x^\nu) \to \inf f$.

1.42 Theorem (minimizing sequences). *Let $f : \mathbb{R}^n \to \overline{\mathbb{R}}$ be lsc, level-bounded and proper. Then every minimizing sequence $\{x^\nu\}$ for f is bounded and has all of its cluster points in* $\operatorname{argmin} f$. *If f attains its minimum at a unique point $\bar{x}$, then necessarily $x^\nu \to \bar{x}$.*

Proof. Let $\{x^\nu\}$ be a minimizing sequence and let $\bar{\alpha} = \inf f$, which is finite by 1.9; then $f(x^\nu) \to \bar{\alpha}$. For any $\alpha \in (\bar{\alpha}, \infty)$, the point x^ν eventually lies in $\operatorname{lev}_{\leq\alpha} f$, which on the basis of our assumptions is compact. The sequence $\{x^\nu\}$ is thus bounded and has all of its cluster points in $\operatorname{lev}_{\leq\alpha} f$. Since this is true for arbitrary $\alpha \in (\bar{\alpha}, \infty)$, such cluster points all belong in fact to the set $\operatorname{lev}_{\leq\bar{\alpha}} f$, which is the same as $\operatorname{argmin} f$. □

The concept of ε-*optimal* solutions to the problem of minimizing f on $\mathbb{R}^n$ is natural in a context of optimizing sequences and provides a further complement to Theorem 1.9. Such points form the set

$$\varepsilon\text{–}\operatorname{argmin}_x f(x) := \big\{ x \,\big|\, f(x) \leq \inf f + \varepsilon \big\}. \tag{1(18)}$$

According to the next result, they can be approximated by nearby points that are true optimal solutions to a slightly perturbed problem of minimization.

1.43 Proposition (variational principle; Ekeland). *Let $f : \mathbb{R}^n \to \overline{\mathbb{R}}$ be lsc with $\inf f$ finite, and let $\bar{x} \in \varepsilon$–$\operatorname{argmin} f$, where $\varepsilon > 0$. Then for any $\delta > 0$ there exists a point*

$$\tilde{x} \in \mathbb{B}(\bar{x}, \varepsilon/\delta) \quad \textit{with} \quad f(\tilde{x}) \leq f(\bar{x}), \quad \operatorname{argmin}_x \{f(x) + \delta|x - \tilde{x}|\} = \{\tilde{x}\}.$$

Proof. Let $\bar{\alpha} = \inf f$ and $\bar{f}(x) = f(x) + \delta|x - \bar{x}|$. The function $\bar{f}$ is lsc, proper and level-bounded: $\operatorname{lev}_{\leq\alpha} \bar{f} \subset \{x \mid \bar{\alpha} + \delta|x - \bar{x}| \leq \alpha\} = \mathbb{B}(\bar{x}, (\alpha - \bar{\alpha})/\delta)$. The set $C := \operatorname{argmin} \bar{f}$ is therefore nonempty and compact (by 1.9). Then the function $\tilde{f} := f + \delta_C$ is lsc, proper and level-bounded, so $\operatorname{argmin} \tilde{f}$ is nonempty (by 1.9). Let $\tilde{x} \in \operatorname{argmin} \tilde{f}$; this means that $\tilde{x}$ is a point minimizing f over $\operatorname{argmin} \bar{f}$. For all x belonging to $\operatorname{argmin} f$ we have $f(\tilde{x}) \leq f(x)$, while for all x not belonging to it we have $\bar{f}(\tilde{x}) < \bar{f}(x)$, which can be written as $f(\tilde{x}) < f(x) + \delta|x - \bar{x}| - \delta|\tilde{x} - \bar{x}|$, where $|x - \bar{x}| - |\tilde{x} - \bar{x}| \leq |x - \tilde{x}|$. It follows that $f(\tilde{x}) < f(x) + \delta|x - \tilde{x}|$ for all $x \neq \tilde{x}$, so that the set $\operatorname{argmin}_x \{f(x) + \delta|x - \tilde{x}|\}$ consists of the single point $\tilde{x}$. Moreover, $f(\tilde{x}) \leq f(\bar{x}) - \delta|\tilde{x} - \bar{x}|$, where $f(\bar{x}) \leq \bar{\alpha} + \varepsilon$, so $\tilde{x} \in \operatorname{lev}_{\leq\alpha} f$ for $\alpha = \bar{\alpha} + \varepsilon$. Hence $\tilde{x} \in \mathbb{B}(\bar{x}, \varepsilon/\delta)$. □

The following example shows further how the pointwise supremum operation, as a way of generating new functions from given ones, has interesting theoretical uses. It also reveals another role of the envelope functions in 1.22. Here a function f is regularized from below by a sort of maximal quadratic interpolation based on specifying a 'curvature parameter' $\lambda > 0$. The interpolated function $h_\lambda f$ that is obtained can serve as an approximation to f whose properties in some respects are superior to those of f, yet are different from those of the envelope $e_\lambda f$.

1.44 Example (proximal hulls). *For a function $f : \mathbb{R}^n \to \overline{\mathbb{R}}$ and any $\lambda > 0$, the λ-proximal hull of f is the function $h_\lambda f : \mathbb{R}^n \to \overline{\mathbb{R}}$ defined as the pointwise supremum of the collection of all the quadratic functions of the elementary form $x \mapsto \alpha - (1/2\lambda)|x - w|^2$ that are majorized by f. This function $h_\lambda f$, satisfying $e_\lambda f \leq h_\lambda f \leq f$, is related to the envelope $e_\lambda f$ by the formulas*

$$h_\lambda f(x) = \sup_{w \in \mathbb{R}^n} \Big\{ e_\lambda f(w) - \frac{1}{2\lambda}|x - w|^2 \Big\}, \quad \text{so } h_\lambda f = -e_\lambda[-e_\lambda f],$$

$$e_\lambda f(x) = \inf_{w \in \mathbb{R}^n} \Big\{ h_\lambda f(w) + \frac{1}{2\lambda}|x - w|^2 \Big\}, \quad \text{so } e_\lambda f = e_\lambda[h_\lambda f].$$

It is lsc, and it is proper as long as f is proper and prox-bounded with threshold $\lambda_f > \lambda$, in which case one has

$$h_\lambda f(x) \nearrow f(x) \ \ \textit{for all } x \ \textit{as } \lambda \searrow 0.$$

At all points x that are λ-proximal for f, in the sense that $f(x)$ is finite and $x \in P_\lambda f(w)$ for some w, one actually has $h_\lambda f(x) = f(x)$. Thus, $h_\lambda f$ always agrees with f on $\operatorname{rge} P_\lambda f$. When $h_\lambda f$ agrees with f everywhere, f is called a λ-proximal function.

Detail. In the notation $j_\lambda(u) = (1/2\lambda)|u|^2$, let $q_{\lambda,w,\alpha}(x) = \alpha - j_\lambda(x-w)$, with the pair $(w,\alpha) \in \mathbb{R}^n \times \mathbb{R}$ as a parameter element. By definition, $h_\lambda f$ is the pointwise supremum of the collection of all the functions $q_{\lambda,w,\alpha}$ majorized by f. These functions are continuous, so $h_\lambda f$ is lsc by 1.26(a). We have $q_{\lambda,w,\alpha} \leq f$ if and only if $\alpha \leq f(x) + j_\lambda(x-w)$ for all x, which in the envelope notation of 1.22 means that $\alpha \leq e_\lambda f(w)$. Therefore, $h_\lambda f$ can just as well be viewed as the pointwise supremum of the collection $\{g_w\}_{w\in\mathbb{R}^n}$, where $g_w(x) := e_\lambda f(w) - j_\lambda(x-w)$. This is what the displayed formula for $h_\lambda f(x)$ says.

The observation that $e_\lambda f$ is determined by the collection of all quadratics $q_{\lambda,w,\alpha} \leq f$, with these being the same as the collection of all $q_{\lambda,w,\alpha} \leq h_\lambda f$, reveals that $e_\lambda f = e_\lambda[h_\lambda f]$. The displayed formula for $e_\lambda f$ in terms of $h_\lambda f$ is therefore valid as well.

When $x \in P_\lambda(w)$, we have $f(x) + j_\lambda(x-w) \leq f(x') + j_\lambda(x'-w)$ for all x'. As long as $f(x)$ is finite, this comes out as saying that $f \geq q_{\lambda,w,\alpha}$ for $\alpha := f(x) + j_\lambda(x-w)$, with these two functions agreeing at x itself. Then obviously $h_\lambda f(x) = f(x)$.

In the case of $f \equiv \infty$ we have $h_\lambda f \equiv \infty$ for all λ. On the other hand, if f isn't prox-bounded, we have $h_\lambda f \equiv -\infty$ for all λ. Otherwise $h_\lambda f \not\equiv \infty$, yet $h_\lambda f \geq e_\lambda f$ (as seen from taking $w = x$ in the displayed formula). Since $e_\lambda f$ is finite for $\lambda \in (0, \lambda_f)$, $h_\lambda f$ must be proper for such λ.

The inequality $h_\lambda f \leq h_{\lambda'} f$ holds when $\lambda' \leq \lambda$, for the following reason. A second-order Taylor expansion of $q_{\lambda,w,\alpha}$ at any point $\bar{x}$ shows there's a function $q_{\lambda',w',\alpha'} \leq q_{\lambda,w,\alpha}$ with $q_{\lambda',w',\alpha'}(\bar{x}) = q_{\lambda,w,\alpha}(\bar{x})$. Hence the sup of all $q_{\lambda',w',\alpha'} \leq f$ is at least as high everywhere as the sup of all $q_{\lambda,w,\alpha} \leq f$. Thus, $h_\lambda f(x)$ increases, if anything, as $\lambda \searrow 0$. Because $e_\lambda f \leq h_\lambda f \leq f$, and $e_\lambda f(x) \nearrow f(x)$ as $\lambda \searrow 0$ by 1.25 when f is prox-bounded and lsc, we must likewise have $h_\lambda f(x) \nearrow f(x)$ as $\lambda \searrow 0$ in these circumstances. □

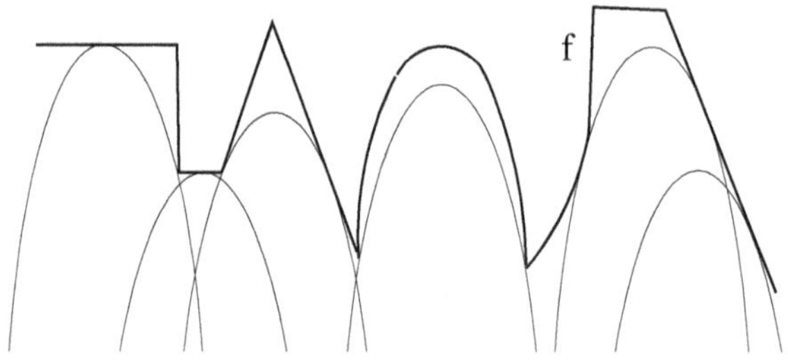

Fig. 1–16. Interpolation of a function by a proximal hull.

It's intriguing that each of the functions $e_\lambda f$ and $h_\lambda f$ in Example 1.44 completely embodies all the information necessary to derive the other. This is a relationship of 'duality', and many more examples of such a phenomenon will appear as we go along, especially in Chapters 6, 8, and 11. (More will be seen about this specific instance of duality in 11.63.)

1.45 Exercise (proximal inequalities). *For functions f_1 and f_2 on $\mathbb{R}^n$ and any $\lambda > 0$, one has*

$$e_\lambda f_1 \leq e_\lambda f_2 \iff h_\lambda f_1 \leq h_\lambda f_2.$$

Thus in particular, f_1 and f_2 have the same Moreau λ-envelope if and only if they have the same λ-proximal hull.

Guide. Get this from the description of these envelopes and proximal hulls in terms of quadratics majorized by f_1 and f_2. (See 1.44 and its justification.) □

Also of interest in the rich picture of Moreau envelopes and proximal hulls are the following functions, which illustrate further the potential role of max and min in generating approximations of a given function. Their special 'subsmoothness' properties will be documented in 12.62.

1.46 Example (double envelopes). *For a proper, lsc function $f : \mathbb{R}^n \to \overline{\mathbb{R}}$ that is prox-bounded with threshold λ_f, and parameter values $0 < \mu < \lambda < \lambda_f$, the Lasry-Lions double envelope $e_{\lambda,\mu} f$ is defined by*

$$e_{\lambda,\mu} f(x) := \sup_w \Big\{ e_\lambda f(w) - \frac{1}{2\mu}|w - x|^2 \Big\}, \quad \text{so } e_{\lambda,\mu} f = -e_\mu[-e_\lambda f].$$

This function is bracketed by $e_\lambda f \leq e_{\lambda,\mu} f \leq e_{\lambda-\mu} f \leq f$ and satisfies

$$e_{\lambda,\mu} f = e_{\lambda-\mu}[h_\lambda f] = h_\mu[e_{\lambda-\mu} f]. \qquad 1(19)$$

Also, $\inf e_{\lambda,\mu} f = \inf e_\lambda f = \inf f$ *and* $\operatorname{argmin} e_{\lambda,\mu} f = \operatorname{argmin} e_\lambda f = \operatorname{argmin} f$.

Detail. From $e_\lambda f = f \# j_\lambda$, with $j_\lambda(w) = (1/2\lambda)|w|^2$, one easily gets $\inf e_\lambda f = \inf f$ and $\operatorname{argmin} e_\lambda f = \operatorname{argmin} f$. The agreement of these with $\inf e_{\lambda,\mu} f$ and $\operatorname{argmin} e_{\lambda,\mu} f$ will follow from proving that $e_{\lambda,\mu} f$ lies between $e_\lambda f$ and f.

It's elementary from the definition of Moreau envelopes that $e_{\lambda-\mu} f \leq f$. Likewise $e_\mu[-e_\lambda f] \leq -e_\lambda f$, hence $e_\lambda f \leq e_{\lambda,\mu} f$. The first equation of 1(19) will similarly yield $e_{\lambda,\mu} f \leq e_{\lambda-\mu} f$, inasmuch as $h_\lambda f \leq f$, so we concentrate henceforth on 1(19). The formulas in 1.44 and 1.29(c) give $e_{\lambda,\mu} f = -e_\mu[-e_\lambda f] = -e_\mu[-e_\mu[e_{\lambda-\mu} f]] = h_\mu[e_{\lambda-\mu} f]$. Because $e_\lambda f = e_\lambda[h_\lambda f]$, this implies also that $e_{\lambda,\mu} f = e_{\lambda,\mu}[h_\lambda f] = h_\mu[e_{\lambda-\mu}[h_\lambda f]]$.

All that remains is verifying for $g := e_{\lambda-\mu}[h_\lambda f]$ that $h_\mu g = g$, or equivalently through the description of proximal hulls in 1.44, that g is the pointwise supremum of all the functions of the form $q_{\mu,w,\alpha}(x) = \alpha - j_\mu(x - w)$ that it majorizes. It suffices to demonstrate for arbitrary $\bar{x}$ the existence of some $q_{\mu,\bar{w},\bar{\alpha}} \leq g$ such that $q_{\mu,\bar{w},\bar{\alpha}}(\bar{x}) = g(\bar{x})$. Because we're dealing with a class of functions f that's invariant under translations, we can focus on $\bar{x} = 0$.

Let $\tilde{w} \in P_{\lambda-\mu}[h_\lambda f](0)$, which is nonempty by 1.25, since $\lambda - \mu < \lambda < \lambda_f$ (this being also the proximal threshold for $h_\lambda f$). Then $k(w) \geq k(\tilde{w}) = g(0)$ for $k := h_\lambda f + j_{\lambda-\mu}$. Next, observe from $h_\lambda f(w) = \sup_z \{ e_\lambda f(z) - j_\lambda(w - z) \}$ (cf. 1.44) and the quadratic expansion

$$j_\lambda(a + \tau\omega) - j_\lambda(a) = \tau\big[j_\lambda(a + \omega) - j_\lambda(a)\big] - \tau(1 - \tau) j_\lambda(\omega), \qquad 1(20)$$

as applied to $a = \tilde{w} - z$, that $h_\lambda f(\tilde{w} + \tau\omega)$ can be estimated for $\tau \in (0,1)$ by

$$\begin{aligned} &\sup_z\{e_\lambda f(z) - j_\lambda(\tilde{w} + \tau w - z)\} \\ &\quad = \sup_z\{e_\lambda f(z) - (1-\tau)j_\lambda(\tilde{w} - z) - \tau j_\lambda(\tilde{w} - z + \omega) - \tau(1-\tau)j_\lambda(\omega)\} \\ &\quad \leq (1-\tau)\sup_z\{e_\lambda f(z) - j_\lambda(\tilde{w} - z)\} + \tau \sup_z\{e_\lambda f(z) - j_\lambda(\tilde{w} + \omega - z)\} \\ &\qquad - \tau(1-\tau)j_\lambda(\omega) \;=\; (1-\tau)h_\lambda f(\tilde{w}) + \tau h_\lambda f(\tilde{w} + \omega) - \tau(1-\tau)j_\lambda(\omega). \end{aligned}$$

Thus, for such τ we have the 'upper expansion'

$$h_\lambda f(\tilde{w} + \tau\omega) - h_\lambda f(\tilde{w}) \;\leq\; \tau\big[h_\lambda f(\tilde{w} + \omega) - h_\lambda f(\tilde{w})\big] - \tau(1-\tau)j_\lambda(\omega),$$

which can be added to the expansion in 1(20) of $j_{\lambda-\mu}$ instead of j_λ, this time at $a = \tilde{w}$, to get for all $\tau \in (0,1)$ that

$$k(\tilde{w} + \tau\omega) - k(\tilde{w}) \leq \tau\big[k(\tilde{w} + \omega) - k(\tilde{w})\big] + \tau(1-\tau)\big[j_\lambda(\omega) - j_{\lambda-\mu}(\omega)\big].$$

Since $k(\tilde{w}+\tau\omega)-k(\tilde{w}) \geq 0$, we must have $k(\tilde{w}+\omega)-k(\tilde{w})+j_\lambda(\omega)-j_{\lambda-\mu}(\omega) \geq 0$ and, in substituting $w = \tilde{w} + \omega$, can write this as

$$h_\lambda f(w) \;\geq\; g(0) - j_{\lambda-\mu}(w) + j_{\lambda-\mu}(w - \tilde{w}) - j_\lambda(w - \tilde{w}) \;\text{ for all } w. \qquad 1(21)$$

The quadratic function of w on the right side of 1(21) has its second-order part coming only from the j_λ term, due to cancellation in the $j_{\lambda-\mu}$ terms, so it must be of the form $q_{\lambda,\bar{w},\bar{\alpha}}(w)$ for some $\bar{w}$ and $\bar{\alpha}$. From $h_\lambda f \geq q_{\lambda,\bar{w},\bar{\alpha}}$ we get $e_{\lambda-\mu}[h_\lambda f] \geq e_{\lambda-\mu}[q_{\lambda,\bar{w},\bar{\alpha}}]$. But the latter is $q_{\mu,\bar{w},\bar{\alpha}}$; this follows from the rule that $[-j_\lambda]\# j_{\lambda-\mu} = -j_\mu$, which can be checked by simple calculation. Hence $g(x) \geq q_{\mu,\bar{w},\bar{\alpha}}(x)$ for all x. On the other hand,

$$\begin{aligned} q_{\mu,\bar{w},\bar{\alpha}}(0) &= e_{\lambda-\mu}[q_{\lambda,\bar{w},\bar{\alpha}}](0) \;=\; \inf_w\{q_{\lambda,\bar{w},\bar{\alpha}}(w) + j_{\lambda-\mu}(w)\} \\ &= \inf_w\{g(0) + j_{\lambda-\mu}(w - \tilde{w}) - j_\lambda(w - \tilde{w})\} \;=\; g(0), \end{aligned}$$

because $j_{\lambda-\mu} \geq j_\lambda$ when $0 < \mu < \lambda$. Thus, the quadratic function $q_{\lambda,\bar{w},\bar{\alpha}}$ fits the prescription. □

Commentary

The key notion that a problem of minimizing f over a subset C of a given space can be represented by defining f to have the value ∞ outside of C, and that lower semicontinuity is then the essential property to look for, originated independently with Moreau [1962], [1963a], [1963b], and the dissertation of Rockafellar [1963]. Both authors were influenced by unpublished but widely circulated lecture notes of Fenchel [1951], which for a long time were a prime source for the theory of convex functions.

Fenchel's tactic was to deal with pairs (C, f) where C is a nonempty subset of $\mathbb{R}^n$ and f is a real-valued function on C, and to call such a pair 'closed' if the set

$$\{(x,\alpha) \in C \times \mathbb{R} \,|\, f(x) \leq \alpha\}$$

is closed in $\mathbb{R}^n \times \mathbb{R}$. Moreau and Rockafellar observed that in extending f beyond C with ∞, not only could notation be simplified, but Fenchel's closedness property would reduce to the lower semicontinuity of f, a concept already well understood in analysis, and the set on which Fenchel focused his attention as a substitute for the graph of f over C could be studied then as the epigraph of f over $\mathbb{R}^n$. At the same time, Fenchel's operation of closing a pair (C, f) could be implemented simply by taking lower limits of the extended function f, so as to obtain the function we've denoted by $\operatorname{cl} f$. (It must be mentioned that our general usage of 'closure' and '$\operatorname{cl} f$' departs from the traditions of convex analysis in the context of convex functions f, where a slightly different operation, 'biconjugate closure', has been expressed in this manner. In passing to the vast territory beyond convex analysis, it's better to reserve a different symbol, $\operatorname{cl}^* f$, for the latter operation in its relatively limited range of interest. For a convex function f, $\operatorname{cl} f$ and $\operatorname{cl}^* f$ agree in every case except the very special one where $\operatorname{cl} f$ is identically $-\infty$ on its effective domain (cf. 2.5), this domain D being nonempty and yet not the whole space; then $\operatorname{cl}^* f$ differ for points $x \notin D$, where $(\operatorname{cl} f)(x) = \infty$ but $(\operatorname{cl}^* f)(x) := -\infty$.)

Moreau and Rockafellar recognized also that, in the extended-real-valued framework, 'indicator' functions δ_C could assume an operational character like that of characteristic functions in integration. For Fenchel, the corresponding object associated with a set C would have been a pair $(C, 0)$. This didn't convey the fertile idea that constraints can be imposed by adding an indicator to a given function, or that an indicator function was an extreme case of a penalty or barrier function.

Penalty and barrier functions have some earlier history in association with numerical methods, but in optimization they were popularized by Fiacco and McCormick [1968]. The idea that a sequence of such functions, with parameter values increasingly severe, can be viewed as *converging to an indicator*—in the geometric sense of converging epigraphs (i.e., epi-convergence, which will be laid out in Chapter 7)—originated with Attouch and Wets [1981]. The parametric approach in 1.21 by way of 1.17 is new.

The study of how optimal values and optimal solutions might depend on the parameters in a given problem is an important topic with a large literature. The representation of the main issues as concerned with minimizing a single extended-real-valued function $f(x, u)$ on $\mathbb{R}^n \times \mathbb{R}^m$ with respect to x, and looking then to the behavior with respect to u of the infimum $p(u)$ and argmin set $P(u)$, offers a great conceptual simplification in comparison to what is often seen in this subject, and at the same time it supports a broader range of applications. Such an approach was first adopted full scale by Rockafellar [1968a], [1970a], in work with optimization problems of convex type and their possible modes of dualization. The geometric interpretation of $\operatorname{epi} p$ as the projection of $\operatorname{epi} f$ (cf. 1.18 and Fig. 1–8) was basic to this.

The fundamental result of parametric optimization in Theorem 1.17, in which properties of $p(u)$ and $P(u)$ are derived from a uniform level-boundedness assumption on $f(x, u)$, goes back to Wets [1974]. He employed a form of 'inf-compactness', a concept developed by Moreau [1963c] for the sake of obtaining the attainment of a minimum as in Theorem 1.9 and other uses. For problems in $\mathbb{R}^n$, we have found it convenient instead to keep the boundedness aspect of inf-compactness separate from the lower semicontinuity aspect, hence the introduction of the term 'level-boundedness'. Of course in infinite-dimensional spaces the boundedness and closedness of a level set wouldn't guarantee its compactness.

Questions of the existence of solutions and how they depend on a problem's parameters have long been recognized as crucial to applications of mathematics, not only in optimization, and treated under the heading of 'well-posedness'. In the past, that term was usually taken to refer to the existence *and uniqueness* of a solution and its *continuous* behavior in response to data perturbations, as well as accompanying robustness properties in the convergence of sequences of approximate solutions. Studies in the calculus of variations, optimal control and numerical methods of minimization, however, have shown that uniqueness and continuity are often too restrictive to be adopted as the standards of nicety; see Dontchev and Zolezzi [1993] for a broad exposition of the subject and its classical roots. It's increasingly apparent that forms of *semicontinuity* in a problem's data elements and solution mappings, along with potential multivaluedness in the latter, are the practical concepts. Semicontinuity of multivalued mappings will be taken up in Chapter 5, while their generalized differentiation, which likewise is central to well-posedness as now understood, will be studied in Chapter 8 and beyond.

Semicontinuity of functions was itself introduced by Baire in his 1899 thesis; see Baire [1905] for more on early developments, for instance the fact that a lower semicontinuous function on a compact set attains its minimum (cf. Theorem 1.9), and the fact that the pointwise limit of an increasing sequence of continuous functions is lower semicontinuous. Interestingly, although Baire never published anything about parametric optimization, he attributes to that topic his earliest inspiration for studying separately the two inequalities which Cauchy had earlier distilled as the essence of continuity. In an historical note (Baire [1927]) he says that in late 1896 he considered (as translated to our setting) a function $f(x,u)$ of two real variables on a product of closed, bounded intervals X and U, assuming continuity in x and u separately, but not jointly. He looked at $p(u) = \min_{x \in X} f(x,u)$ and observed it to be upper semicontinuous with respect to $u \in U$, even if not necessarily continuous. (A similar observation is the basis for part (c) of our Theorem 1.17.) Baire explains this in order to make clear that he arrived at semicontinuity not out of a desire to generalize, but because he was led to it naturally by his investigations. The same could now be said for numerous other 'one-sided' concepts that have come to be very important in variational analysis.

Epi-addition originated with Fenchel [1951] in the theory of convex functions, but Fenchel's version was slightly different: he automatically replaced $f_1 \# f_2$ by its lsc hull, $\mathrm{cl}(f_1 \# f_2)$. Also, Fenchel kept to a format in which a pair (C_1, f_1) was combined with a pair (C_2, f_2) to get a new pair (C, f), which was cumbersome and did not suggest the ripe possibilities. But Fenchel did emphasize what we now see as addition of epigraphs, and he recognized that such addition was dual to ordinary addition of functions with respect to the kind of dualization of convex functions that we'll take up in Chapter 11. For the very special case of finite, continuous functions on $[0, \infty)$, a related operation was developed in a series of papers by Bellman and Karush [1961], [1962a], [1962b], [1963]. They called this the 'maximum transform'. Although unaware of Fenchel's results, they too recognized the duality of this operation in a certain sense with that of ordinary function addition. Moreau [1963b] and Rockafellar [1963] changed the framework to that of extended-real-valued functions, with Rockafellar emphasizing convex functions on $\mathbb{R}^n$ but Moreau considering also nonconvex functions and infinite-dimensional spaces.

The term 'inf-convolution' is due to Moreau, who brought to light an analogy with integral convolution. Our alternative term 'epi-addition' is aimed at better reflecting Fenchel's geometric motivation and the duality with addition, as well as providing a parallel opportunity for introducing the term 'epi-multiplication' for an operation that has a closely associated history, but which hitherto has been nameless. The new symbols '$\#$' for epi-addition and '$\star$' for epi-multiplication are designed as reminders of the ties to '$+$' and '$\cdot$'.

The results about epi-addition in 1.27 show the power of placing this operation in the general framework of parametric optimization. In contrast, Moreau [1963b] only treats the case where one of the functions f_1 or f_2 is inf-compact, while the other is bounded below. For other work on epi-addition see Moreau [1970] and especially Strömberg [1994]. Applications of this operation, although not yet formulated as such, can be detected as far back as the 1950s in research on solving Hamilton-Jacobi equations when $n = 1$; cf. Lax [1957].

Set addition and scalar multiplication have long been a mainstay in functional analysis, but have their roots in the geometric theories of Minkowski [1910], [1911]. Extended arithmetic, with its necessary conventions for handling $\pm\infty$, was first explored by Moreau [1963c]. The operation of epi-composition was developed by Rockafellar [1963], [1970a].

The notion of a proximal mapping, due to Moreau [1962], [1965], wasn't treated by him in the parametric form we feature here, but only for $\lambda = 1$ in our notation. He concentrated on convex functions f and the ways that proximal mappings generalize projection operators. Although he investigated many properties of the associated case of the epi-addition operation, which yields what we have called an 'envelope' function, he didn't treat such functions as providing an approximation or regularization of f. That idea, tied to the increasingly close relationship between f and its λ-envelope as $\lambda \searrow 0$, stems from subsequent work of Attouch [1977] in the convex case and Attouch and Wets [1983a] for arbitrary functions. Nonetheless, it has seemed appropriate to attach Moreau's name to such envelope functions, since it's he who initiated the whole study of epi-addition with a squared norm and brought the rich consequences to everyone's attention. The term 'Moreau-Yosida regularization', which has sometimes been used in referring to envelopes, doesn't seem appropriate in our setting because, at best, it makes sense only for the case of convex functions f (through a connection with the subgradient mappings of such functions, cf. 10.2).

The double envelopes in 1.46 were introduced by Lasry and Lions [1986] as approximants that can be better behaved than the Moreau envelopes; for more on this, see also Attouch and Azé [1993]. The surprising 'interchange' formulas for these double envelopes in 1(19) were discovered by Strömberg [1996].

The variational principle of Ekeland [1974] in Proposition 1.43 is especially useful in infinite-dimensional Banach spaces and general metric spaces, where (with a broader statement and a different proof relying on completeness instead of compactness) it provides an important handle on the existence of 'approximate solutions' to various problems. Another powerful variational principle which relies on the norm-squared instead of the norm of a Banach space, has been developed by Borwein and Preiss [1987].

2. Convexity

The concept of convexity has far-reaching consequences in variational analysis. In the study of maximization and minimization, the division between problems of convex or nonconvex type is as significant as the division in other areas of mathematics between problems of linear or nonlinear type. Furthermore, convexity can often be introduced or utilized in a local sense and in this way serves many theoretical purposes.

A. Convex Sets and Functions

For any two different points x_0 and x_1 in $\mathbb{R}^n$ and parameter value $\tau \in \mathbb{R}$ the point

$$x_\tau := x_0 + \tau(x_1 - x_0) = (1-\tau)x_0 + \tau x_1 \qquad 2(1)$$

lies on the line through x_0 and x_1. The entire line is traced as τ goes from $-\infty$ to ∞, with $\tau = 0$ giving x_0 and $\tau = 1$ giving x_1. The portion of the line corresponding to $0 \leq \tau \leq 1$ is called the *closed line segment* joining x_0 and x_1, denoted by $[x_0, x_1]$. This terminology is also used if x_0 and x_1 coincide, although the line segment reduces then to a single point.

2.1 Definition (convex sets and convex functions).

(a) *A subset C of $\mathbb{R}^n$ is convex if it includes for every pair of points the line segment that joins them, or in other words, if for every choice of $x_0 \in C$ and $x_1 \in C$ one has $[x_0, x_1] \subset C$:*

$$(1-\tau)x_0 + \tau x_1 \in C \quad \text{for all } \tau \in (0,1). \qquad 2(2)$$

(b) *A function f on a convex set C is convex relative to C if for every choice of $x_0 \in C$ and $x_1 \in C$ one has*

$$f\big((1-\tau)x_0 + \tau x_1\big) \leq (1-\tau)f(x_0) + \tau f(x_1) \quad \text{for all } \tau \in (0,1), \qquad 2(3)$$

and f is strictly convex relative to C if this inequality is strict for points $x_0 \neq x_1$ with $f(x_0)$ and $f(x_1)$ finite.

In plain English, the word 'convex' suggests a bulging appearance, but the convexity of a set in the mathematical sense of 2.1(a) indicates rather the absence of 'dents', 'holes', or 'disconnectedness'. A set like a cube or floating disk in $\mathbb{R}^3$, or even a general line or plane, is convex despite aspects of flatness. Note also that the definition doesn't require C to contain two different points, or even a point at all: the empty set is convex, and so is every singleton set $C = \{x\}$. At the other extreme, $\mathbb{R}^n$ is itself a convex set.

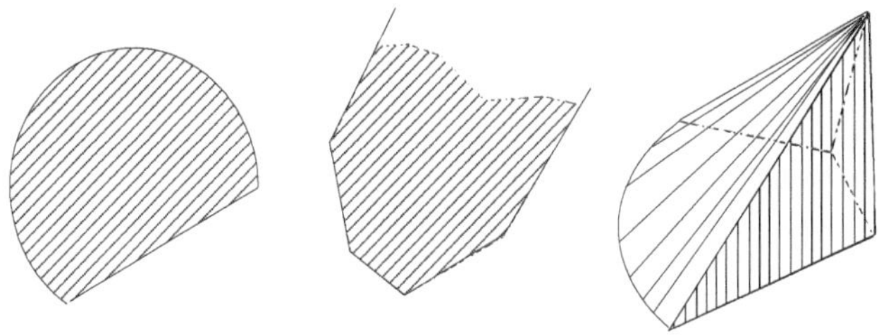

Fig. 2–1. Examples of closed, convex sets, the middle one unbounded.

Many connections between convex sets and convex functions will soon be apparent, and the two concepts are therefore best treated in tandem. In both 2.1(a) and 2.1(b) the τ interval (0,1) could be replaced by [0,1] without really changing anything. Extended arithmetic as explained in Chapter 1 is to be used in handling infinite values of f; in particular the inf-addition convention $\infty - \infty = \infty$ is to be invoked in 2(3) when needed. Alternatively, the convexity condition 2(3) can be recast as the condition that

$$\begin{aligned} &f(x_0) < \alpha_0 < \infty, \quad f(x_1) < \alpha_1 < \infty, \quad \tau \in (0,1) \\ &\quad \Longrightarrow \quad f((1-\tau)x_0 + \tau x_1) \leq (1-\tau)\alpha_0 + \tau\alpha_1. \end{aligned} \tag{2(4)}$$

An extended-real-valued function f is said to be *concave* when $-f$ is convex; similarly, f is strictly concave if $-f$ is strictly convex. Concavity thus corresponds to the opposite direction of inequality in 2(3) (with the sup-addition convention $\infty - \infty = -\infty$).

The geometric picture of convexity, indispensable as it is to intuition, is only one motivation for this concept. Equally compelling is an association with 'mixtures', or 'weighted averages'. A *convex combination* of elements $x_0, x_1, \ldots, x_p$ of $\mathbb{R}^n$ is a linear combination $\sum_{i=0}^p \lambda_i x_i$ in which the coefficients λ_i are nonnegative and satisfy $\sum_{i=0}^p \lambda_i = 1$. In the case of just two elements a convex combination can equally well be expressed in the form $(1-\tau)x_0 + \tau x_1$ with $\tau \in [0,1]$ that we've seen so far. Besides having an interpretation as weights in many applications, the coefficients λ_i in a general convex combination can arise as probabilities. For a discrete random vector variable that can take on the value x_i with probability λ_i, the 'expected value' is the vector $\sum_{i=0}^p \lambda_i x_i$.

2.2 Theorem (convex combinations and Jensen's inequality).

(a) *A set C is convex if and only if C contains all convex combinations of its elements.*

(b) *A function f is convex relative to a convex set C if and only if for every choice of points $x_0, x_1, \ldots, x_p$ in C one has*

$$f\Big(\sum_{i=0}^{p}\lambda_i x_i\Big) \leq \sum_{i=0}^{p}\lambda_i f(x_i) \ \textit{when}\ \lambda_i \geq 0,\ \sum_{i=0}^{p}\lambda_i = 1. \qquad 2(5)$$

Proof. In (a), the convexity of C means by definition that C contains all convex combinations of two of its elements at a time. The 'if' assertion is therefore trivial, so we concentrate on 'only if'. Consider a convex combination $x = \lambda_0 x_0 + \cdots + \lambda_p x_p$ of elements x_i of a convex set C in the case where $p > 1$, we aim at showing that $x \in C$. Without loss of generality we can assume that $0 < \lambda_i < 1$ for all i, because otherwise the assertion is trivial or can be reduced notationally to this case by dropping elements with coefficient 0. Writing

$$x = (1-\lambda_p)\sum_{i=0}^{p-1}\lambda_i' x_i + \lambda_p x_p \ \text{ with } \ \lambda_i' = \frac{\lambda_i}{1-\lambda_p},$$

where $0 < \lambda_i' < 1$ and $\sum_{i=0}^{p-1}\lambda_i' = 1$, we see that $x \in C$ if the convex combination $x' = \sum_{i=0}^{p-1}\lambda_i' x_i$ is in C. The same representation can now be applied to x' if necessary to show that it lies on the line segment joining x_{p-1} with some convex combination of still fewer elements of C, and so forth. Eventually one gets down to combinations of only two elements at a time, which do belong to C. A closely parallel argument establishes (b). □

Any convex function f on a convex set $C \subset \mathbb{R}^n$ can be identified with a convex function on all of $\mathbb{R}^n$ by defining $f(x) = \infty$ for all $x \notin C$. Convexity is thereby preserved, because the inequality 2(3) holds trivially when $f(x_0)$ or $f(x_1)$ is ∞. For most purposes, the study of convex functions can thereby be reduced to the framework of Chapter 1 in which functions are everywhere defined but extended-real-valued.

2.3 Exercise (effective domains of convex functions). *For any convex function $f : \mathbb{R}^n \to \overline{\mathbb{R}}$, $\operatorname{dom} f$ is a convex set with respect to which f is convex. The proper convex functions on $\mathbb{R}^n$ are thus the functions obtained by taking a finite, convex function on a nonempty, convex set $C \subset \mathbb{R}^n$ and giving it the value ∞ everywhere outside of C.*

The indicator δ_C of a set $C \subset \mathbb{R}^n$ is convex if and only if C is convex. In this sense, convex sets in $\mathbb{R}^n$ correspond one-to-one with special convex functions on $\mathbb{R}^n$. On the other hand, convex functions on $\mathbb{R}^n$ correspond one-to-one with special convex sets in $\mathbb{R}^{n+1}$, their epigraphs.

2.4 Proposition (convexity of epigraphs). *A function $f : \mathbb{R}^n \to \overline{\mathbb{R}}$ is convex if and only if its epigraph set $\operatorname{epi} f$ is convex in $\mathbb{R}^n \times \mathbb{R}$, or equivalently, its strict epigraph set $\{(x,\alpha) \mid f(x) < \alpha < \infty\}$ is convex.*

Proof. The convexity of $\operatorname{epi} f$ means that whenever $(x_0, \alpha_0) \in \operatorname{epi} f$ and $(x_1, \alpha_1) \in \operatorname{epi} f$ and $\tau \in (0,1)$, the point $(x_\tau, \alpha_\tau) := (1-\tau)(x_0,\alpha_0) + \tau(x_1,\alpha_1)$ belongs to $\operatorname{epi} f$. This is the same as saying that whenever $f(x_0) \leq \alpha_0 \in \mathbb{R}$ and $f(x_1) \leq \alpha_1 \in \mathbb{R}$, one has $f(x_\tau) \leq \alpha_\tau$. The latter is equivalent to the convexity inequality 2(3) or its variant 2(4). The 'strict' version follows similarly. □

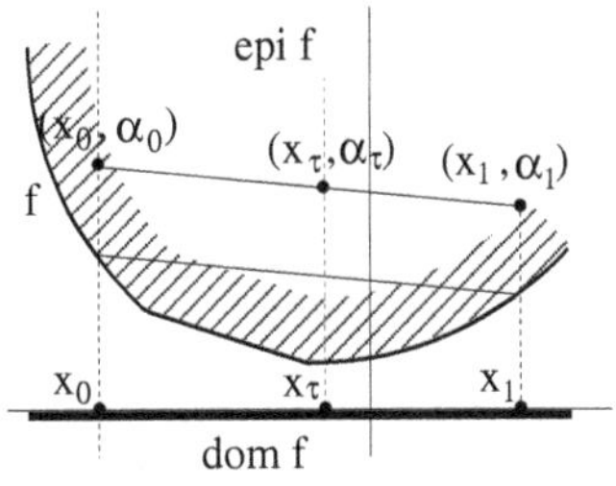

Fig. 2–2. Convexity of epigraphs and effective domains.

For concave functions on $\mathbb{R}^n$ it is the hypograph rather than the epigraph that is a convex subset of $\mathbb{R}^n \times \mathbb{R}$.

2.5 Exercise (improper convex functions). *An improper convex function f must have $f(x) = -\infty$ for all $x \in \operatorname{int}(\operatorname{dom} f)$. If such a function is lsc, it can only have infinite values: there must be a closed, convex set D such that $f(x) = -\infty$ for $x \in D$ but $f(x) = \infty$ for $x \notin D$.*

Guide. Argue first from the definition of convexity that if $f(x_0) = -\infty$ and $f(x_1) < \infty$, then $f(x_\tau) = -\infty$ at all intermediate points x_τ as in 2(1). □

Improper convex functions are of interest mainly as possible by-products of various constructions. An example of an improper convex function having finite as well as infinite values (it isn't lsc) is

$$f(x) = \begin{cases} -\infty & \text{for } x \in (0,\infty) \\ 0 & \text{for } x = 0 \\ \infty & \text{for } x \in (-\infty,0). \end{cases} \tag{2(6)}$$

The chief significance of convexity and strict convexity in optimization derives from the following facts, which use the terminology in 1.4 and its sequel.

2.6 Theorem (characteristics of convex optimization). *In a problem of minimizing a convex function f over $\mathbb{R}^n$ (where f may be extended-real-valued), every locally optimal solution is globally optimal, and the set of all such optimal solutions (if any), namely* $\operatorname{argmin} f$, *is convex.*

Furthermore, if f is strictly convex and proper, there can be at most one optimal solution: the set $\operatorname{argmin} f$, *if nonempty, must be a singleton.*

Proof. If x_0 and x_1 belong to $\operatorname{argmin} f$, or in other words, $f(x_0) = \inf f$ and $f(x_1) = \inf f$ with $\inf f < \infty$, we have for $\tau \in (0,1)$ through the convexity inequality 2(3) that the point x_τ in 2(1) satisfies

$$f(x_\tau) \leq (1-\tau)\inf f + \tau \inf f = \inf f,$$

where strict inequality is impossible. Hence $x_\tau \in \operatorname{argmin} f$, and $\operatorname{argmin} f$ is convex. When f is strictly convex and proper, this shows that x_0 and x_1 can't be different; then $\operatorname{argmin} f$ can't contain more than one point.

In a larger picture, if x_0 and x_1 are any points of $\operatorname{dom} f$ with $f(x_0) > f(x_1)$, it's impossible for x_0 to furnish a local minimum of f because every neighborhood of x_0 contains points x_τ with $\tau \in (0,1)$, and such points satisfy $f(x_\tau) \leq (1-\tau)f(x_0) + \tau f(x_1) < f(x_0)$. Thus, there can't be any locally optimal solutions outside of $\operatorname{argmin} f$, where global optimality reigns. □

The uniqueness criterion provided by Theorem 2.6 for optimal solutions is *virtually the only one that can be checked in advance*, without somehow going through a process that would individually determine all the optimal solutions to a problem, if any.

B. Level Sets and Intersections

For any optimization problem of convex type in the sense of Theorem 2.6 the set of feasible solutions is convex, as seen from 2.3. This set may arise from inequality constraints, and here the convexity of constraint functions is vital.

2.7 Proposition (convexity of level sets). *For a convex function $f : \mathbb{R}^n \to \overline{\mathbb{R}}$ all the level sets of type $\operatorname{lev}_{\leq\alpha} f$ and $\operatorname{lev}_{<\alpha} f$ are convex.*

Proof. This follows right from the convexity inequality 2(3). □

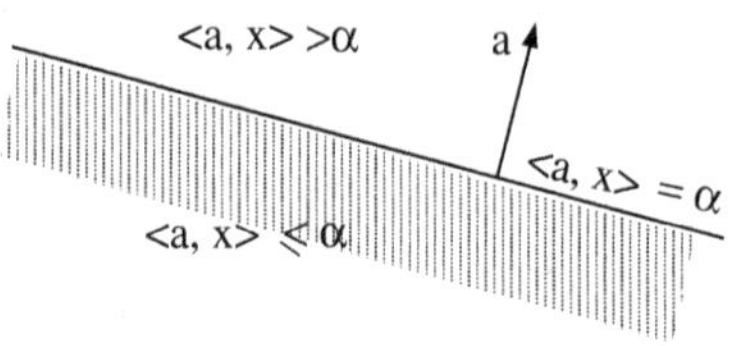

Fig. 2–3. A hyperplane and its associated half-spaces.

Level sets of the type $\operatorname{lev}_{\geq\alpha} f$ and $\operatorname{lev}_{>\alpha} f$ are convex when, instead, f is concave. Those of the type $\operatorname{lev}_{=\alpha} f$ are convex when f is both convex and concave at the same time, and in this respect the following class of functions is important. Here we denote by $\langle x, y\rangle$ the canonical inner product in $\mathbb{R}^n$:

$$\langle x, y\rangle = x_1 y_1 + \cdots + x_n y_n \text{ for } x = (x_1,\ldots,x_n),\ y = (y_1,\ldots,y_n).$$

2.8 Example (affine functions, half-spaces and hyperplanes). *A function f on $\mathbb{R}^n$ is said to be affine if it differs from a linear function by only a constant:*

$$f(x) = \langle a, x\rangle + \beta \text{ for some } a \in \mathbb{R}^n \text{ and } \beta \in \mathbb{R}.$$

Any affine function is both convex and concave. As level sets of affine functions, all sets of the form $\{x \,|\, \langle a,x\rangle \leq \alpha\}$ and $\{x \,|\, \langle a,x\rangle \geq \alpha\}$, as well as all those of the form $\{x \,|\, \langle a,x\rangle < \alpha\}$ and $\{x \,|\, \langle a,x\rangle > \alpha\}$, are convex in $\mathbb{R}^n$, and so too are all those of the form $\{x \,|\, \langle a,x\rangle = \alpha\}$. For $a \neq 0$ and α finite, the sets in *the last category are the hyperplanes in* $\mathbb{R}^n$, *while the others are the closed half-spaces and open half-spaces associated with such hyperplanes.*

Affine functions are the only *finite* functions on $\mathbb{R}^n$ that are both convex and concave, but other functions with infinite values can have this property, not just the constant functions ∞ and $-\infty$ but examples such as 2(6).

A set defined by several equations or inequalities is the intersection of the sets defined by the individual equations or inequalities, so it's useful to know that convexity of sets is preserved under taking intersections.

2.9 Proposition (intersection, pointwise supremum and pointwise limits).

(a) $\bigcap_{i\in I} C_i$ *is convex if each set* C_i *is convex.*

(b) $\sup_{i\in I} f_i$ *is convex if each function* f_i *is convex.*

(c) $\sup_{i\in I} f_i$ *is strictly convex if each* f_i *is strictly convex and* I *is finite.*

(d) f *is convex if* $f(x) = \limsup_\nu f^\nu(x)$ *for all* x *and each* f^ν *is convex.*

Proof. These assertions follow at once from Definition 2.1. Note that (b) is the epigraphical counterpart to (a): taking the pointwise supremum of a collection of functions corresponds to taking the intersection of their epigraphs. □

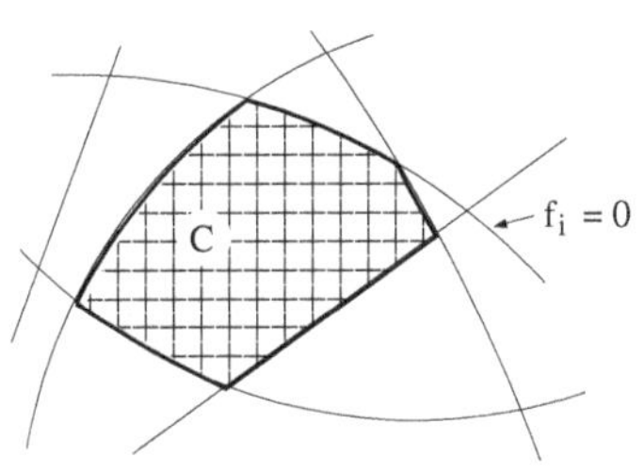

Fig. 2–4. A feasible set defined by convex inequalities.

As an illustration of the intersection principle in 2.9(a), any set $C \subset \mathbb{R}^n$ consisting as in Example 1.1 of the points satisfying a constraint system

$$x \in X \ \text{ and } \ \begin{cases} f_i(x) \leq 0 & \text{for } i \in I_1, \\ f_i(x) = 0 & \text{for } i \in I_2, \end{cases}$$

is convex if the set $X \subset \mathbb{R}^n$ is convex and the functions f_i are convex for $i \in I_1$ but affine for $i \in I_2$. Such sets are common in convex optimization.

2.10 Example (polyhedral sets and affine sets). *A set* $C \subset \mathbb{R}^n$ *is said to be a polyhedral set if it can be expressed as the intersection of a finite family of closed half-spaces or hyperplanes, or equivalently, can be specified by finitely many linear constraints, i.e., constraints* $f_i(x) \leq 0$ *or* $f_i(x) = 0$ *where* f_i *is affine. It is called an affine set if it can be expressed as the intersection of hyperplanes alone, i.e., in terms only of constraints* $f_i(x) = 0$ *with* f_i *affine.*

Affine sets are in particular polyhedral, while polyhedral sets are in particular closed, convex sets. The empty set and the whole space are affine.

Detail. The empty set is the intersection of two parallel hyperplanes, whereas $\mathbb{R}^n$ is the intersection of the 'empty collection' of hyperplanes in $\mathbb{R}^n$. Thus, the empty set and the whole space are affine sets, hence polyhedral. Note that since every hyperplane is the intersection of two opposing closed half-spaces, hyperplanes are superfluous in the definition of a polyhedral set.

The alternative descriptions of polyhedral and affine sets in terms of linear constraints are based on 2.8. For an affine function f_i that happens to be a constant function, a constraint $f_i(x) \leq 0$ gives either the empty set or the whole space, and similarly for a constraint $f_i(x) = 0$. Such possible degeneracy in a system of linear constraints therefore doesn't affect the geometric description of the set of points satisfying the system as being polyhedral or affine. □

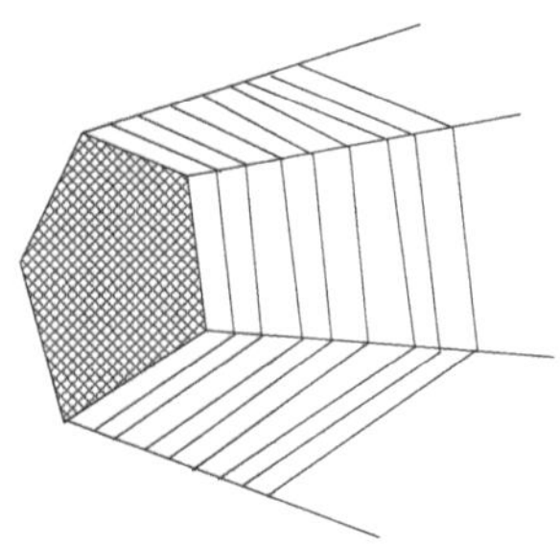

Fig. 2–5. A polyhedral set.

2.11 Exercise (characterization of affine sets). *For a nonempty set $C \subset \mathbb{R}^n$ the following properties are equivalent:*

(a) *C is an affine set;*

(b) *C is a translate $M + p$ of a linear subspace M of $\mathbb{R}^n$ by a vector p;*

(c) *C has the form $\{x \mid Ax = b\}$ for some $A \in \mathbb{R}^{m \times n}$ and $b \in \mathbb{R}^m$.*

(d) *C contains for each pair of distinct points the entire line through them: if $x_0 \in C$ and $x_1 \in C$ then $(1-\tau)x_0 + \tau x_1 \in C$ for all $\tau \in (-\infty, \infty)$.*

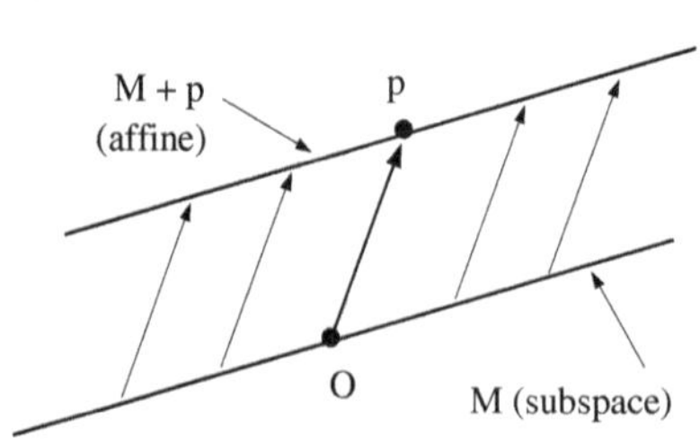

Fig. 2–6. Affine sets as translates of subspaces.

Guide. Argue that an affine set containing 0 must be a linear subspace. A subspace M of dimension $n-m$ can be represented as the set of vectors orthogonal to certain vectors $a_1, \ldots, a_m$. Likewise reduce the analysis of (d) to the case where $0 \in C$. □

Taking the pointwise supremum of a family of functions is just one of many convexity-preserving operations. Others will be described shortly, but we first expand the criteria for verifying convexity directly. For differentiable functions, conditions on first and second derivatives serve this purpose.

C. Derivative Tests

We begin the study of such conditions with functions of a single real variable. This approach is expedient because convexity is essentially a one-dimensional property; behavior with respect to line segments is all that counts. For instance, a set C is convex if and only if its intersection with every line is convex. By the same token, *a function f is convex if and only if it's convex relative to every line.* Many of the properties of convex functions on $\mathbb{R}^n$ can thus be derived from an analysis of the simpler case where $n=1$.

2.12 Lemma (slope inequality). *A real-valued function f on an interval $C \subset \mathbb{R}$ is convex on C if and only if for arbitrary points $x_0 < y < x_1$ in C one has*

$$\frac{f(y)-f(x_0)}{y-x_0} \leq \frac{f(x_1)-f(x_0)}{x_1-x_0} \leq \frac{f(x_1)-f(y)}{x_1-y}. \qquad 2(7)$$

Then for any $x \in C$ the difference quotient $\Delta_x(y) := \big[f(y)-f(x)\big]/(y-x)$ is a nondecreasing function of $y \in C \setminus \{x\}$, i.e., one has $\Delta_x(y_0) \leq \Delta_x(y_1)$ for all choices of y_0 and y_1 not equal to x with $y_0 < y_1$.

Similarly, strict convexity is characterized by strict inequalities between the difference quotients, and then $\Delta_x(y)$ is an increasing function of $y \in C \setminus \{x\}$.

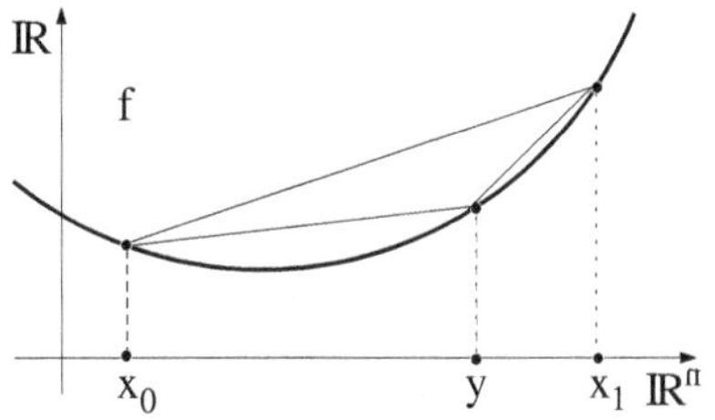

Fig. 2–7. Slope inequality.

Proof. The convexity of f is equivalent to the condition that

$$f(y) \leq \frac{x_1-y}{x_1-x_0} f(x_0) + \frac{y-x_0}{x_1-x_0} f(x_1) \text{ when } x_0 < y < x_1 \text{ in } C, \qquad 2(8)$$

since this is 2(3) when y is x_τ for $\tau = (y - x_0)/(x_1 - x_0)$. The first inequality is what one gets by subtracting $f(x_0)$ from both sides in 2(8), whereas the second corresponds to subtracting $f(x_1)$. The case of strict convexity is parallel. □

2.13 Theorem (one-dimensional derivative tests). *For a differentiable function f on an open interval $O \subset I\!R$, each of the following conditions is both necessary and sufficient for f to be convex on O:*

(a) *f' is nondecreasing on O, i.e., $f'(x_0) \leq f'(x_1)$ when $x_0 < x_1$ in O;*
(b) *$f(y) \geq f(x) + f'(x)(y - x)$ for all x and y in O;*
(c) *$f''(x) \geq 0$ for all x in O (assuming twice differentiability).*

Similarly each of the following conditions is both necessary and sufficient for f to be strictly convex on O:

(a′) *f' is increasing on O: $f'(x_0) < f'(x_1)$ when $x_0 < x_1$ in O.*
(b′) *$f(y) > f(x) + f'(x)(y - x)$ for all x and y in O with $y \neq x$.*

A sufficient (but not necessary) condition for strict convexity is:

(c′) *$f''(x) > 0$ for all x in O (assuming twice differentiability).*

Proof. The equivalence between (a) and (c) when f is twice differentiable is well known from elementary calculus, and the same is true of the implication from (c′) to (a′). We'll show now that [convexity] ⇒ (a) ⇒ (b) ⇒ [convexity]. If f is convex, we have

$$f'(x_0) \leq \frac{f(x_1) - f(x_0)}{x_1 - x_0} = \frac{f(x_0) - f(x_1)}{x_0 - x_1} \leq f'(x_1) \text{ when } x_0 < x_1 \text{ in } O$$

from the monotonicity of difference quotients in Lemma 2.12, and this gives (a). On the other hand, if (a) holds we have for any $y \in O$ that the function $g_y(x) := f(x) - f(y) - f'(y)(x - y)$ has $g_y'(x) \geq 0$ for all $x \in (y, \infty) \cap O$ but $g_y'(x) \leq 0$ for all $x \in (-\infty, y) \cap O$. Then g_y is nondecreasing to the right of y but nonincreasing to the left, and it therefore attains its global minimum over O at y. This means that (b) holds. Starting now from (b), consider the family of affine functions $l_y(x) = f(y) + f'(y)(x - y)$ indexed by $y \in O$. We have $f(x) = \max_{y \in O} l_y(x)$ for all $x \in O$, so f is convex on O by 2.9(b).

In parallel fashion we see that [strict convexity] ⇒ (a′) ⇒ (b′). To establish that (b′) implies not just convexity but strict convexity, consider $x_0 < x_1$ in O and an intermediate point x_τ as in 2(1). For the affine function $l(x) = f(x_\tau) + f'(x_\tau)(x - x_\tau)$ we have $f(x_0) > l(x_0)$ and $f(x_1) > l(x_1)$, but $f(x_\tau) = l(x_\tau) = (1 - \tau)l(x_0) + \tau l(x_1)$. Therefore, $f(x_\tau) < (1 - \tau)f(x_0) + \tau f(x_1)$. □

Here are some examples of functions of a single variable whose convexity or strict convexity can be established by the criteria in Theorem 2.13, in particular by the second-derivative tests:

- $f(x) = ax^2 + bx + c$ on $(-\infty, \infty)$ when $a \geq 0$; strictly convex when $a > 0$.
- $f(x) = e^{ax}$ on $(-\infty, \infty)$; strictly convex when $a \neq 0$.
- $f(x) = x^r$ on $(0, \infty)$ when $r \geq 1$; strictly convex when $r > 1$.
- $f(x) = -x^r$ on $(0, \infty)$ when $0 \leq r \leq 1$; strictly convex when $0 < r < 1$.

- $f(x) = x^{-r}$ on $(0,\infty)$ when $r > 0$; strictly convex in fact on this interval.
- $f(x) = -\log x$ on $(0,\infty)$; strictly convex in fact on this interval.

The case of $f(x) = x^4$ on $(-\infty,\infty)$ furnishes a counterexample to the common misconception that positivity of the second derivative in 2.13(c$'$) is not only sufficient but necessary for strict convexity: this function is strictly convex relative to the entire real line despite having $f''(0) = 0$. The strict convexity can be verified by applying the condition to f on the intervals $(-\infty,0)$ and $(0,\infty)$ separately and then invoking the continuity of f at 0. In general, it can be seen that 2.13(c$'$) remains a sufficient condition for the strict convexity of a twice differentiable function when relaxed to allow $f''(x)$ to vanish at finitely many points x (or even on a subset of O having Lebesgue measure zero).

To extend the criteria in 2.13 to functions of $x = (x_1,\ldots,x_n)$, we must call for appropriate conditions on gradient vectors and Hessian matrices. For a differentiable function f, the *gradient* vector and *Hessian* matrix at x are

$$\nabla f(x) := \left[\frac{\partial f}{\partial x_j}(x)\right]_{j=1}^{n}, \qquad \nabla^2 f(x) := \left[\frac{\partial^2 f}{\partial x_i \partial x_j}(x)\right]_{i,j=1}^{n,n}.$$

Recall that a matrix $A \in \mathbb{R}^{n\times n}$ is called *positive-semidefinite* if $\langle z, Az\rangle \geq 0$ for all z, and *positive-definite* if $\langle z, Az\rangle > 0$ for all $z \neq 0$. This terminology applies even if A is not symmetric, but of course $\langle z, Az\rangle$ depends only on the *symmetric part of* A, i.e., the matrix $\frac{1}{2}(A + A^*)$, where A^* denotes the transpose of A. In terms of the components a_{ij} of A and z_j of z, one has

$$\langle z, Az\rangle = \sum_{i=1}^{n}\sum_{j=1}^{n} a_{ij} z_i z_j.$$

2.14 Theorem (higher-dimensional derivative tests). *For a differentiable function f on an open convex set $O \subset \mathbb{R}^n$, each of the following conditions is both necessary and sufficient for f to be convex on O:*

(a) $\langle x_1 - x_0, \nabla f(x_1) - \nabla f(x_0)\rangle \geq 0$ *for all x_0 and x_1 in O;*
(b) $f(y) \geq f(x) + \langle \nabla f(x), y - x\rangle$ *for all x and y in O;*
(c) $\nabla^2 f(x)$ *is positive-semidefinite for all x in O (f twice differentiable).*

For strict convexity, a necessary and sufficient condition is (a) holding with strict inequality when $x_0 \neq x_1$, or (b) holding with strict inequality when $x \neq y$. A condition that is sufficient for strict convexity (but not necessary) is the positive definiteness of the Hessian matrix in (c) for all x in O.

Proof. As already noted, f is convex on O if and only if it is convex on every line segment in O. This is equivalent to the property that for every choice of $y \in O$ and $z \in \mathbb{R}^n$ the function $g(t) = f(y + tz)$ is convex on any open interval of t values for which $y + tz \in O$. Here $g'(t) = \big\langle z, \nabla f(y + tz)\big\rangle$ and $g''(t) = \big\langle z, \nabla^2 f(y + tz)z\big\rangle$. The asserted conditions for convexity and strict convexity are equivalent to requiring in each case that the corresponding condition in 2.13 hold for all such functions g. □

Twice differentiable concave or strictly concave functions can similarly be characterized in terms of Hessian matrices that are negative-semidefinite or negative-definite.

2.15 Example (quadratic functions). *A function f on $\mathbb{R}^n$ is quadratic if it's expressible as $f(x) = \frac{1}{2}\langle x, Ax\rangle + \langle a, x\rangle + \text{const.}$, where the matrix $A \in \mathbb{R}^{n\times n}$ is symmetric. Then $\nabla f(x) = Ax + a$ and $\nabla^2 f(x) \equiv A$, so f is convex if and only if A is positive-semidefinite. Moreover, a function f of this type is strictly convex if and only if A is positive-definite.*

Detail. Note that the positive definiteness of the Hessian is being asserted as necessary for the strict convexity of a quadratic function, even though it was only listed in Theorem 2.14 merely as sufficient in general. The reason is that if A is positive-semidefinite, but not positive-definite, there's a vector $z \neq 0$ such that $\langle z, Az\rangle = 0$, and then along the line through the origin in the direction of z it's impossible for f to be strictly convex. □

Because level sets of convex functions are convex by 2.7, it follows from Example 2.15 that every set of the form

$$C = \left\{ x \,\middle|\, \tfrac{1}{2}\langle x, Ax\rangle + \langle a, x\rangle \leq \alpha \right\} \text{ with } A \text{ positive-semidefinite}$$

is convex. This class of sets includes all closed Euclidean balls as well as general ellipsoids, paraboloids, and 'cylinders' with ellipsoidal or paraboloidal base.

Algebraic functions of the form described in 2.15 are often called *quadratic*, but that name seems to carry the risk of suggesting sometimes that $A \neq 0$ is assumed, or that second-order terms—only—are allowed. Sometimes, we'll refer to them as *linear-quadratic* as a way of emphasizing the full generality.

2.16 Example (convexity of vector-max and log-exponential). *In terms of $x = (x_1, \ldots, x_n)$, the functions*

$$\text{vecmax}(x) := \max\{x_1, \ldots, x_n\}, \qquad \text{logexp}(x) := \log\left(e^{x_1} + \cdots + e^{x_n}\right),$$

are convex on $\mathbb{R}^n$ but not strictly convex.

Detail. Convexity of $f = \text{logexp}$ is established via 2.14(c) by calculating in terms of $\sigma(x) = \sum_{j=1}^n e^{x_j}$ that

$$\begin{aligned}\left\langle z, \nabla^2 f(x) z\right\rangle &= \frac{1}{\sigma(x)} \sum_{j=1}^n e^{x_j} z_j^2 - \frac{1}{\sigma(x)^2} \sum_{j=1}^n \sum_{i=1}^n e^{(x_i+x_j)} z_i z_j \\ &= \frac{1}{2\sigma(x)^2} \sum_{i=1}^n \sum_{j=1}^n e^{(x_i+x_j)} (z_i - z_j)^2 \geq 0.\end{aligned}$$

Strict convexity fails because $f(x + t\mathbb{1}) = f(x) + t$ for $\mathbb{1} = (1, 1, \ldots, 1)$. As $f = \text{vecmax}$ is the pointwise max of the n linear functions $x \mapsto x_j$, it's convex by 2.9(c). It isn't strictly convex, because $f(\lambda x) = \lambda f(x)$ for $\lambda \geq 0$. □

2.17 Example (norms). *By definition, a norm on* $\mathbb{R}^n$ *is a real-valued function* $h(x) = \|x\|$ *such that*

$$\|\lambda x\| = |\lambda|\|x\|, \quad \|x+y\| \leq \|x\| + \|y\|, \quad \|x\| > 0 \text{ for } x \neq 0.$$

Any such a function h *is convex, but not strictly convex. The corresponding balls* $\{x \mid \|x - x_0\| \leq \rho\}$ *and* $\{x \mid \|x - x_0\| < \rho\}$ *are convex sets. Beyond the Euclidean norm* $|x| = (\sum_{j=1}^n |x_j|^2)^{1/2}$ *these properties hold for the* l_p *norms*

$$\|x\|_p := \Big(\sum\nolimits_{j=1}^n |x_j|^p\Big)^{1/p} \text{ for } 1 \leq p < \infty, \qquad \|x\|_\infty := \max_{j=1,\dots,n} |x_j|. \qquad 2(9)$$

Detail. Any norm satisfies the convexity inequality 2(3), but the strict version fails when $x_0 = 0$, $x_1 \neq 0$. The associated balls are convex as translates of level sets of a convex function. For any $p \in [1,\infty]$ the function $h(x) = \|x\|_p$ obviously fulfills the first and third conditions for a norm. In light of the first condition, the second can be written as $h\big(\frac{1}{2}x + \frac{1}{2}y\big) \leq \frac{1}{2}h(x) + \frac{1}{2}h(y)$, which will follow from verifying that h is convex, or equivalently that $\operatorname{epi} h$ is convex. We have $\operatorname{epi} h = \{\lambda(x,1) \mid x \in B,\ \lambda \geq 0\}$, where $B = \{x \mid \|x\|_p \leq 1\}$. The convexity of $\operatorname{epi} h$ can easily be derived from this formula once it is known that the set B is convex. For $p = \infty$, B is a box, while for $p \in [1,\infty)$ it is $\operatorname{lev}_{\leq 1} g$ for the convex function $g(x) = \sum_{j=1}^n |x_j|^p$, hence it is convex in that case too. □

D. Convexity in Operations

Many important convex functions lack differentiability. Norms can't ever be differentiable at the origin. The vector-max function in 2.16 fails to be differentiable at $\bar{x} = (\bar{x}_1, \dots, \bar{x}_n)$ if two coordinates $\bar{x}_j$ and $\bar{x}_k$ tie for the max. (At such a point the function has 'kinks' along the lines parallel to the x_j-axis and the x_k axis.) Although derivative tests can't be used to establish the convexity of functions like these, other criteria can fill the need, for instance the fact in 2.9 that a pointwise supremum of convex functions is convex. (A pointwise infimum of convex functions is convex only in cases like a decreasing sequence of convex functions or a 'convexly parameterized' family as will be treated in 2.22(a).)

2.18 Exercise (addition and scalar multiplication). *For convex functions* $f_i : \mathbb{R}^n \to \overline{\mathbb{R}}$ *and real coefficients* $\lambda_i \geq 0$, *the function* $\sum_{i=1}^m \lambda_i f_i$ *is convex. It is strictly convex if for at least one index* i *with* $\lambda_i > 0$, f_i *is strictly convex.*

Guide. Work from Definition 2.1. □

2.19 Exercise (set products and separable functions).

(a) *If* $C = C_1 \times \cdots \times C_m$ *where each* C_i *is convex in* $\mathbb{R}^{n_i}$, *then* C *is convex in* $\mathbb{R}^{n_1} \times \cdots \times \mathbb{R}^{n_m}$. *In particular, any box in* $\mathbb{R}^n$ *is a closed, convex set.*

(b) *If* $f(x) = f_1(x_1)+\cdots+f_m(x_m)$ *for* $x = (x_1, \ldots, x_m)$ *in* $I\!R^{n_1}\times\cdots\times I\!R^{n_m}$, *where each* f_i *is convex, then* f *is convex. If each* f_i *is strictly convex, then* f *is strictly convex.*

2.20 Exercise (convexity in composition).

(a) *If* $f(x) = g(Ax+a)$ *for a convex function* $g : I\!R^m \to \overline{I\!R}$ *and some choice of* $A \in I\!R^{m\times n}$ *and* $a \in I\!R^m$, *then* f *is convex.*

(b) *If* $f(x) = \theta\big(g(x)\big)$ *for a convex function* $g : I\!R^n \to \overline{I\!R}$ *and a nondecreasing convex function* $\theta : I\!R \to \overline{I\!R}$, *the convention being used that* $\theta(\infty) = \infty$ *and* $\theta(-\infty) = \inf\theta$, *then* f *is convex. Furthermore,* f *is strictly convex in the case where* g *is strictly convex and* θ *is increasing.*

(c) *Suppose* $f(x) = g\big(F(x)\big)$ *for a convex function* $g : I\!R^m \to \overline{I\!R}$ *and a mapping* $F : I\!R^n \to I\!R^m$, *where* $F(x) = \big(f_1(x), \ldots, f_m(x)\big)$ *with* f_i *convex for* $i = 1, \ldots, s$ *and* f_i *affine for* $i = s+1, \ldots, m$. *Suppose that* $g(u_1, \ldots, u_m)$ *is nondecreasing in* u_i *for* $i = 1, \ldots, s$. *Then* f *is convex.*

An example of a function whose convexity follows from 2.20(a) is $f(x) = \|Ax - b\|$ for any matrix A, vector b, and any norm $\|\cdot\|$. Some elementary examples of convex and strictly convex functions constructed as in 2.20(b) are:

- $f(x) = e^{g(x)}$ is convex when g is convex, and strictly convex when g is strictly convex.
- $f(x) = -\log|g(x)|$ when $g(x) < 0$, $f(x) = \infty$ when $g(x) \geq 0$, is convex when g is convex, and strictly convex when g is strictly convex.
- $f(x) = g(x)^2$ when $g(x) \geq 0$, $f(x) = 0$ when $g(x) < 0$, is convex when g is convex.

As an example of the higher-dimensional composition in 2.20(c), a vector of convex functions $f_1, \ldots, f_m$ can be composed with vecmax or logexp, cf. 2.16. This confirms that convexity is preserved when a nonsmooth function, given as the pointwise maximum of finitely many smooth functions, is approximated by a smooth function as in Example 1.30. Other examples in the mode of 2.20(c) are obtained with $g(z)$ of the form

$$|z|_+ := \max\{z, 0\} \ \text{ for } \ z \in I\!R$$

in 2.20(b), or more generally

$$|z|_+ := \sqrt{|z_1|_+^2 + \cdots + |z_m|_+^2} \ \text{ for } \ z = (z_1, \ldots, z_m) \in I\!R^m \qquad 2(10)$$

in 2.20(c). This way we get the convexity of a function of the form $f(x) = \big|(f_1(x), \ldots, f_m(x))\big|_+$ when f_i is convex. Then too, for instance, the pth power of such a function is convex for any $p \in [1, \infty)$, as seen from further composition with the nondecreasing, convex function $\theta : t \mapsto |t|_+^p$.

2.21 Proposition (images under linear mappings). *If* $L : I\!R^n \to I\!R^m$ *is linear, then* $L(C)$ *is convex in* $I\!R^m$ *for every convex set* $C \subset I\!R^n$, *while* $L^{-1}(D)$ *is convex in* $I\!R^n$ *for every convex set* $D \subset I\!R^m$.

Proof. The first assertion is obtained from the definition of the convexity of C and the linearity of L. Specifically, if $u = (1-\tau)u_0 + \tau u_1$ for points $u_0 = L(x_0)$ and $u_1 = L(x_1)$ with x_0 and x_1 in C, then $u = L(x)$ for the point $x = (1-\tau)x_0 + \tau x_1$ in C. The second assertion can be proved quite as easily, but it may also be identified as a specialization of the composition rule in 2.20(a) to the case of g being the indicator δ_C. □

Projections onto subspaces are examples of mappings L to which 2.21 can be applied. If D is a subset of $\mathbb{R}^n \times \mathbb{R}^d$ and C consists of the vectors $x \in \mathbb{R}^n$ for which there's a vector $w \in \mathbb{R}^d$ with $(x, w) \in D$, then C is the image of D under the mapping $(x, w) \mapsto x$, so it follows that C is convex when D is convex.

2.22 Proposition (convexity in inf-projection and epi-composition).

(a) *If* $p(u) = \inf_x f(x, u)$ *for a convex function* f *on* $\mathbb{R}^n \times \mathbb{R}^m$, *then* p *is convex on* $\mathbb{R}^m$. *Also, the set* $P(u) = \operatorname{argmin}_x f(x, u)$ *is convex for each* u.

(b) *For any convex function* $f : \mathbb{R}^n \to \overline{\mathbb{R}}$ *and matrix* $A \in \mathbb{R}^{m \times n}$ *the function* $Af : \mathbb{R}^m \to \overline{\mathbb{R}}$ *defined by* $(Af)(u) := \inf\{f(x) \mid Ax = u\}$ *is convex.*

Proof. In (a), the set $D = \{(u, \alpha) \mid p(u) < \alpha < \infty\}$ is the image of the set $E = \{(x, u, \alpha) \mid f(x, u) < \alpha < \infty\}$ under the linear mapping $(x, u, \alpha) \mapsto (u, \alpha)$. The convexity of E through 2.4 implies that of D through 2.21, and this ensures that p is convex. The convexity of $P(u)$ follows from 2.6.

Likewise in (b), the set $D' = \{(u, \alpha) \mid (Af)(u) < \alpha < \infty\}$ is the image of $E' = \{(x, \alpha) \mid f(x) < \alpha < \infty\}$ under the linear transformation $(x, \alpha) \mapsto (Ax, \alpha)$. Again, the claimed convexity is justified through 2.4 and 2.21. □

The notation Af in 2.22(b) is interchangeable with Lf for the mapping $L : x \mapsto Ax$, which fits with the general definition of epi-composition in 1(17).

2.23 Proposition (convexity in set algebra).

(a) $C_1 + C_2$ *is convex when* C_1 *and* C_2 *are convex.*

(b) λC *is convex for every* $\lambda \in \mathbb{R}$ *when* C *is convex.*

(c) *When* C *is convex, one has* $(\lambda_1 + \lambda_2)C = \lambda_1 C + \lambda_2 C$ *for* $\lambda_1, \lambda_2 \geq 0$.

Proof. Properties (a) and (b) are immediate from the definitions. In (c) it suffices through rescaling to deal with the case where $\lambda_1 + \lambda_2 = 1$. The equation is then merely a restatement of the definition of the convexity of C. □

As an application of 2.23, the 'fattened' set $C + \varepsilon\mathbb{B}$ is convex whenever C is convex; here $\mathbb{B}$ is the closed, unit Euclidean ball, cf. 1(15).

2.24 Exercise (epi-sums and epi-multiples).

(a) $f_1 \# f_2$ *is convex when* f_1 *and* f_2 *are convex.*

(b) $\lambda \star f$ *is convex for every* $\lambda > 0$ *when* f *is convex.*

(c) *When* f *is convex, one has* $(\lambda_1 + \lambda_2) \star f = (\lambda_1 \star f) \# (\lambda_2 \star f)$ *for* $\lambda_1, \lambda_2 \geq 0$.

Guide. Get these conclusions from 2.23 through the geometry in 1.28. □

2.25 Example (distance functions and projections). *For a convex set* C, *the distance function* d_C *is convex, and as long as* C *is closed and nonempty, the* *projection mapping P_C is single-valued, i.e., for each point $x \in \mathbb{R}^n$ there is a unique point of C nearest to x. Moreover P_C is continuous.*

Detail. The distance function (from 1.20) arises through epi-addition, cf. 1(13), so its convexity is a consequence of 2.24. The elements of $P_C(x)$ minimize $|w-x|$ over all $w \in C$, but they can also be regarded as minimizing $|w-x|^2$ over all $w \in C$. Since the function $w \mapsto |w-x|^2$ is strictly convex (via 2.15), there can't be more than one such element (see 2.6), but on the other hand there's at least one (by 1.20). Hence $P_C(x)$ is a singleton. The cluster point property in 1.20 ensures then that, as a single-valued mapping, P_C is continuous. □

For a convex function f, the Moreau envelopes and proximal mappings defined in 1.22 have remarkable properties.

2.26 Theorem (proximal mappings and envelopes under convexity). *Let $f : \mathbb{R}^n \to \overline{\mathbb{R}}$ be lsc, proper, and convex. Then f is prox-bounded with threshold ∞, and the following properties hold for every $\lambda > 0$.*

(a) *The proximal mapping $P_\lambda f$ is single-valued and continuous. In fact $P_\lambda f(x) \to P_{\bar\lambda} f(\bar x)$ whenever $(\lambda, x) \to (\bar\lambda, \bar x)$ with $\bar\lambda > 0$.*

(b) *The envelope function $e_\lambda f$ is convex and continuously differentiable, the gradient being*

$$\nabla e_\lambda f(x) = \frac{1}{\lambda}\big[x - P_\lambda f(x)\big].$$

Proof. From Definition 1.22 we have $e_\lambda f(x) := \inf_w g_\lambda(x,w)$ and $P_\lambda f(x) := \operatorname{argmin}_w g_\lambda(x,w)$ for the function $g_\lambda(x,w) := f(w) + (1/2\lambda)|w-x|^2$, which our assumptions imply to be lsc, proper and convex in (x,w), even strictly convex in w. If the threshold for f is ∞ as claimed, then $e_\lambda f$ and $P_\lambda f$ have all the properties in 1.25, and in addition $e_\lambda f$ is convex by 2.22(a), while $P_\lambda f$ is single-valued by 2.6. This gives everything in (a) and (b) except for the differentiability in (b), which will need a supplementary argument. Before proceeding with that argument, we verify the prox-boundedness.

In order to show that the threshold for f is ∞, it suffices to show for arbitrary $\lambda > 0$ that $e_\lambda f(0) > -\infty$, which can be accomplished through Theorem 1.9 by demonstrating the boundedness of the level sets of $g_\lambda(0,\cdot)$. If the latter property were absent, there would exist $\alpha \in \mathbb{R}$ and points x^ν with $f(x^\nu) + (1/2\lambda)|x^\nu|^2 \le \alpha$ such that $1 < |x^\nu| \to \infty$. Fix any x_0 with $f(x_0) < \infty$. Then in terms of $\tau^\nu = 1/|x^\nu| \in (0,1)$ and $\bar x^\nu := (1-\tau^\nu)x_0 + \tau^\nu x^\nu$ we have $\tau^\nu \to 0$ and

$$\begin{aligned} f(\bar x^\nu) &\le (1-\tau^\nu) f(x_0) + \tau^\nu f(x^\nu) \\ &\le (1-\tau^\nu) f(x_0) + \tau^\nu \alpha - (1/2\lambda)|x^\nu| \to -\infty. \end{aligned}$$

The sequence of points $\bar x^\nu$ is bounded, so this is incompatible with f being proper and lsc (cf. 1.10). The contradiction proves the claim.

The differentiability claim in (b) is next. Continuous differentiability will follow from the formula for the gradient mapping, once that is established, since $P_\lambda f$ is already known to be a continuous, single-valued mapping. Consider any point $\bar{x}$, and let $\bar{w} = P_\lambda f(\bar{x})$ and $\bar{v} = (\bar{x} - \bar{w})/\lambda$. Our task is to show that $e_\lambda f$ is differentiable at $\bar{x}$ with $\nabla e_\lambda f(\bar{x}) = \bar{v}$, or equivalently in terms of $h(u) := e_\lambda f(\bar{x}+u) - e_\lambda f(\bar{x}) - \langle \bar{v}, u\rangle$ that h is differentiable at 0 with $\nabla h(0) = 0$. We have $e_\lambda f(\bar{x}) = f(\bar{w}) + (1/2\lambda)|\bar{w} - \bar{x}|^2$, whereas $e_\lambda f(\bar{x} + u) \leq f(\bar{w}) + (1/2\lambda)|\bar{w} - (\bar{x} + u)|^2$, so that

$$h(u) \leq \frac{1}{2\lambda}|\bar{w} - (\bar{x} + u)|^2 - \frac{1}{2\lambda}|\bar{w} - \bar{x}|^2 - \frac{1}{\lambda}\langle \bar{x} - \bar{w}, u\rangle = \frac{1}{2\lambda}|u|^2.$$

But h inherits the convexity of $e_\lambda f$ and therefore has $\frac{1}{2}h(u) + \frac{1}{2}h(-u) \geq h\big(\frac{1}{2}u + \frac{1}{2}(-u)\big) = h(0) = 0$, so from the inequality just obtained we also get

$$h(u) \geq -h(-u) \geq -\frac{1}{2\lambda}|-u|^2 = -\frac{1}{2\lambda}|u|^2.$$

Thus we have $\big|h(u)\big| \leq (1/2\lambda)|u|^2$ for all u, and this obviously yields the desired differentiability property. □

Especially interesting in Theorem 2.26 is the fact that regardless of any nonsmoothness of f, the envelope approximations $e_\lambda f$ are always smooth.

E. Convex Hulls

Nonconvex sets can be 'convexified'. For $C \subset I\!R^n$, the *convex hull* of C, denoted by $\operatorname{con} C$, is the smallest convex set that includes C. Obviously $\operatorname{con} C$ is the intersection of all the convex sets $D \supset C$, this intersection being a convex set by 2.9(a). (At least one such set D always exists—the whole space.)

2.27 Theorem (convex hulls from convex combinations). *For a set $C \subset I\!R^n$, $\operatorname{con} C$ consists of all the convex combinations of elements of C:*

$$\operatorname{con} C = \Big\{ \sum\nolimits_{i=0}^{p} \lambda_i x_i \,\Big|\, x_i \in C,\ \lambda_i \geq 0,\ \sum\nolimits_{i=0}^{p} \lambda_i = 1,\ p \geq 0 \Big\}.$$

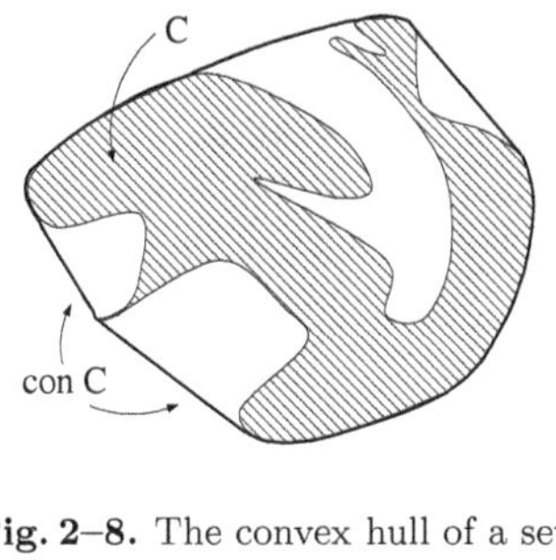

Fig. 2–8. The convex hull of a set.

Proof. Let D be the set of all convex combinations of elements of C. We have $D \subset \operatorname{con} C$ by 2.2, because $\operatorname{con} C$ is a convex set that includes C. But

D is itself convex: if x is a convex combination of $x_0, \ldots, x_p$ in C and x' is a convex combination of $x'_0, \ldots, x'_{p'}$ in C, then for any $\tau \in (0,1)$, the vector $(1-\tau)x + \tau x'$ is a convex combination of $x_0, \ldots, x_p$ and $x'_0, \ldots, x'_{p'}$ together. Hence $\operatorname{con} C = D$. □

Sets expressible as the convex hull of a *finite* subset of $\mathbb{R}^n$ are especially important. When $C = \{a_0, a_1, \ldots, a_p\}$, the formula in 2.27 simplifies: $\operatorname{con} C$ consists of all convex combinations $\lambda_0 a_0 + \lambda_1 a_1 + \cdots + \lambda_p a_p$. If the points $a_0, a_1, \ldots, a_p$ are *affinely independent*, the set $\operatorname{con}\{a_0, a_1, \ldots, a_p\}$ is called a *p-simplex* with these points as its *vertices*. The affine independence condition means that the only choice of coefficients $\lambda_i \in (-\infty, \infty)$ with $\lambda_0 a_0 + \lambda_1 a_1 + \cdots + \lambda_p a_p = 0$ and $\lambda_0 + \lambda_1 + \cdots + \lambda_p = 0$ is $\lambda_i = 0$ for all i; this holds if and only if the vectors $a_i - a_0$ for $i = 1, \ldots, p$ are linearly independent. A 0-simplex is a point, whereas a 1-simplex is a closed line segment joining a pair of distinct points. A 2-simplex is a triangle, and a 3-simplex a tetrahedron.

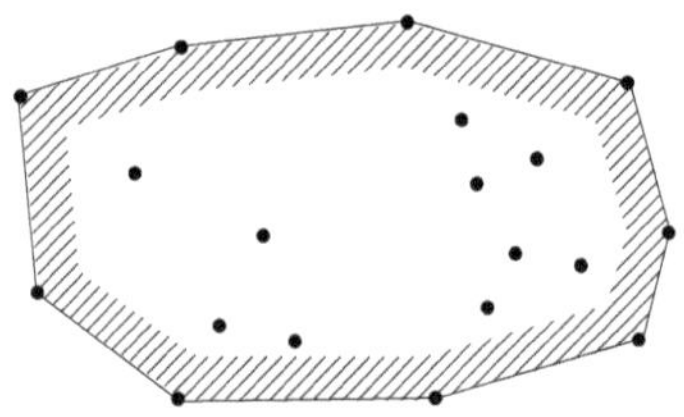

Fig. 2–9. The convex hull of finitely many points.

Simplices are often useful in technical arguments about convexity. The following are some of the facts commonly employed.

2.28 Exercise (simplex technology).

(a) *Every simplex $S = \operatorname{con}\{a_0, a_1, \ldots, a_p\}$ is a polyhedral set, in particular closed and convex.*

(b) *When $a_0, \ldots, a_n$ are affinely independent in $\mathbb{R}^n$, every $x \in \mathbb{R}^n$ has a unique expression in barycentric coordinates: $x = \sum_{i=0}^n \lambda_i a_i$ with $\sum_{i=0}^n \lambda_i = 1$.*

(c) *The expression of each point of a p-simplex $S = \operatorname{con}\{a_0, a_1, \ldots, a_p\}$ as a convex combination $\sum_{i=0}^p \lambda_i a_i$ is unique: there is a one-to-one correspondence, continuous in both directions, between the points of S and the vectors $(\lambda_0, \lambda_1, \ldots, \lambda_p) \in \mathbb{R}^{p+1}$ such that $\lambda_i \geq 0$ and $\sum_{i=0}^p \lambda_i = 1$.*

(d) *Every n-simplex $S = \operatorname{con}\{a_0, a_1, \ldots, a_n\}$ in $\mathbb{R}^n$ has nonempty interior, and $x \in \operatorname{int} S$ if and only if $x = \sum_{i=0}^n \lambda_i a_i$ with $\lambda_i > 0$ and $\sum_{i=0}^n \lambda_i = 1$.*

(e) *Simplices can serve as neighborhoods: for every point $\bar{x} \in \mathbb{R}^n$ and neighborhood $V \in \mathcal{N}(\bar{x})$ there is an n-simplex $S \subset V$ with $\bar{x} \in \operatorname{int} S$.*

(f) *For an n-simplex $S = \operatorname{con}\{a_0, a_1, \ldots, a_n\}$ in $\mathbb{R}^n$ and sequences $a_i^\nu \to a_i$, the set $S^\nu = \operatorname{con}\{a_0^\nu, a_1^\nu, \ldots, a_n^\nu\}$ is an n-simplex once ν is sufficiently large. Furthermore, if $x^\nu \to x$ with barycentric representations $x^\nu = \sum_{i=0}^n \lambda_i^\nu a_i^\nu$ and $x = \sum_{i=0}^n \lambda_i a_i$, where $\sum_{i=0}^n \lambda_i^\nu = 1$ and $\sum_{i=0}^n \lambda_i = 1$, then $\lambda_i^\nu \to \lambda_i$.*

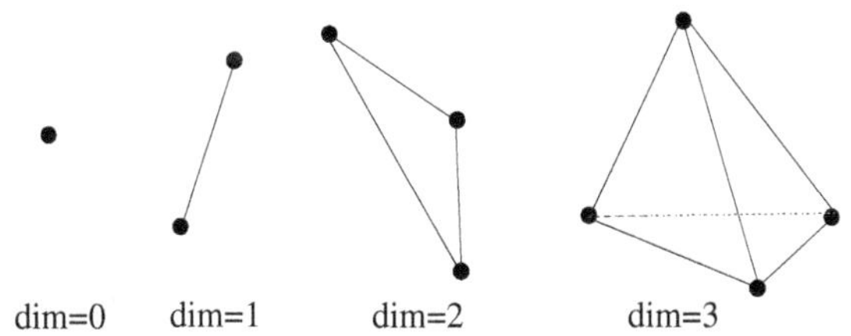

Fig. 2–10. Simplices of dimensions 0, 1, 2 and 3.

Guide. In (a)-(d) simplify by translating the points to make $a_0 = 0$; in (a) and (c) augment $a_1, \ldots, a_p$ by other vectors (if $p < n$) in order to have a basis for $\mathbb{R}^n$. Consider the linear transformation that maps a_i to $e_i = (0, \ldots, 1, 0, \ldots, 0)$ (where the 1 in e_i appears in ith position). For (e), the case of $\bar{x} = 0$ and convex neighborhoods $V = \mathbb{B}(0, \varepsilon)$ is enough. Taking any n-simplex S with $0 \in \operatorname{int} S$, show there's a $\delta > 0$ such that $\delta S \subset \mathbb{B}(0, \varepsilon)$. In (f) let A be the matrix having the vectors $a_i - a_0$ as its columns; similarly, A^ν. Identify the simplex assumption on S with the nonsingularity of A and argue (by way of determinants for instance) that then A^ν must eventually be nonsingular. For the last part, note that $x = Az + a_0$ for $z = (\lambda_1, \ldots, \lambda_n)$ and similarly $x^\nu = A^\nu z^\nu + a_0^\nu$ for ν sufficiently large. Establish that the convergence of A^ν to A entails the convergence of the inverse matrices. □

In the statement of Theorem 2.27, the integer p isn't fixed and varies over all possible choices of a convex combination. This is sometimes inconvenient, and it's valuable then to know that a fixed choice will suffice.

2.29 Theorem (convex hulls from simplices; Carathéodory). *For a set $C \neq \emptyset$ in $\mathbb{R}^n$, every point of $\operatorname{con} C$ belongs to some simplex with vertices in C and thus can be expressed as a convex combination of $n+1$ points of C (not necessarily different). For every point in $\operatorname{bdry} C$, the boundary of C, n points suffice. When C is connected, then every point of $\operatorname{con} C$ can be expressend as the combination of no more than n points of C.*

Proof. First we show that $\operatorname{con} C$ is the union of all simplices formed from points of C: each $x \in \operatorname{con} C$ can be expressed not only as a convex combination $\sum_{i=0}^p \lambda_i x_i$ with $x_i \in C$, but as one in which $x_0, x_1, \ldots, x_p$ are affinely independent. For this it suffices to show that when the convex combination is chosen with p minimal (which implies $\lambda_i > 0$ for all i), the points x_i can't be affinely dependent. If they were, we would have coefficients μ_i, at least one of them positive, such that $\sum_{i=0}^p \mu_i x_i = 0$ and $\sum_{i=0}^p \mu_i = 0$. Then there is a largest $\tau > 0$ such that $\tau \mu_i \leq \lambda_i$ for all i, and by setting $\lambda_i' = \lambda_i - \tau \mu_i$ we would get a representation $x = \sum_{i=0}^p \lambda_i' x_i$ in which $\sum_{i=0}^p \lambda_i' = 1$, $\lambda_i' \geq 0$, and actually $\lambda_i' = 0$ for some i. This would contradict the minimality of p.

Of course when $x_0, x_1, \ldots, x_p$ are affinely independent, we have $p \leq n$. If $p < n$ we can choose additional points $x_{p+1}, \ldots, x_n$ arbitrarily from C and expand the representation $x = \sum_{i=0}^p \lambda_i x_i$ to $x = \sum_{i=0}^n \lambda_i x_i$ by taking $\lambda_i = 0$ for $i = p+1, \ldots, n$. This proves the first assertion of the theorem.

By passing to a lower dimensional space if necessary, one can suppose that $\operatorname{con} C$ is n-dimensional. Then $\operatorname{con} C$ is the union of all n-simplices whose vertices are in C. A point in $\operatorname{bdry} C$, while belonging to some such n-simplex, can't be in its interior and therefore must be on boundary of that n-simplex. But the boundary of an n-simplex is a union of $(n-1)$-simplices whose vertices are in C, i.e., every point in $\operatorname{bdry} C$ can be obtained as the convex combination of no more than n point in C.

Finally, let's show that if some point $\bar{x}$ in $\operatorname{con} C$, but not in C, has a minimal representation $\sum_{i=0}^{n} \bar{\lambda}_i x_i$ involving exactly $n+1$ points $x_i \in C$, then C can't be connected. By a translation if necessary, we can simplify to the case where $\bar{x} = 0$; then $\sum_{i=0}^{n} \bar{\lambda}_i x_i = 0$ with $0 < \bar{\lambda}_i < 1$, $\sum_{i=0}^{n} \bar{\lambda}_i = 1$, and $x_0, x_1, \ldots, x_n$ affinely independent. In this situation the vectors $x_1, \ldots, x_n$ are linearly independent, for if not we would have an expression $\sum_{i=1}^{n} \mu_i x_i = 0$ with $\sum_{i=1}^{n} \mu_i = 0$ (but not all coefficients 0), or one with $\sum_{i=1}^{n} \mu_i = 1$; the first case is precluded by $x_0, x_1, \ldots, x_n$ being affinely independent, while the second is impossible by the uniqueness of the coefficients $\bar{\lambda}_i$, cf. 2.28(b). Thus the vectors $x_1, \ldots, x_n$ form a basis for $\mathbb{R}^n$, and every $x \in \mathbb{R}^n$ can be expressed uniquely as a linear combination $\sum_{i=1}^{n} \alpha_i x_i$, where the correspondence $x \leftrightarrow (\alpha_1, \ldots, \alpha_n)$ is continuous in both directions. In particular, $x_0 \leftrightarrow (-\bar{\lambda}_1/\bar{\lambda}_0, \ldots, -\bar{\lambda}_n/\bar{\lambda}_0)$.

Let D be the set of all $x \in \mathbb{R}^n$ with $\alpha_i \leq 0$ for $i = 1, \ldots, n$; this is a closed set whose interior consists of all x with $\alpha_i < 0$ for $i = 1, \ldots, n$. We have $x_0 \in \operatorname{int} D$ but $x_i \notin D$ for all $i \neq 0$. Thus, the open sets $\operatorname{int} D$ and $\mathbb{R}^n \setminus D$ both meet C. If C is connected, it can't lie entirely in the union of these disjoint sets and must meet the boundary of D. There must be a point $x_0' \in C$ having an expression $x_0' = \sum_{i=1}^{n} \alpha_i x_i$ with $\alpha_i \leq 0$ for $i = 1, \ldots, n$ but also $\alpha_i = 0$ for some i, which we can take to be $i = n$. Then $x_0' + \sum_{i=1}^{n-1} |\alpha_i| x_i = 0$, and in dividing through by $1 + \sum_{i=1}^{n-1} |\alpha_i|$ we obtain a representation of 0 as a convex combination of only n points of C, which is contrary to the minimality that was assumed. □

2.30 Corollary (compactness of convex hulls). *For any compact set $C \subset \mathbb{R}^n$, $\operatorname{con} C$ is compact. In particular, the convex hull of a finite set of points is compact; thus, simplices are compact.*

Proof. Let $D \subset (\mathbb{R}^n)^{n+1} \times \mathbb{R}^{n+1}$ consist of all $w = (x_0, \ldots, x_n, \lambda_0, \ldots, \lambda_n)$ with $x_i \in C$, $\lambda_i \geq 0$, $\sum_{i=0}^{n} \lambda_i = 1$. From the theorem, $\operatorname{con} C$ is the image of D under the mapping $F : w \mapsto \sum_{i=0}^{n} \lambda_i x_i$. The image of a compact set under a continuous mapping is compact. □

For a function $f : \mathbb{R}^n \to \overline{\mathbb{R}}$, there is likewise a notion of convex hull: $\operatorname{con} f$ is the greatest convex function majorized by f. (Recall that a function $g : \mathbb{R}^n \to \overline{\mathbb{R}}$ is *majorized* by a function $h : \mathbb{R}^n \to \overline{\mathbb{R}}$ if $g(x) \leq h(x)$ for all x. With the opposite inequality, g is *minorized* by h.) Clearly, $\operatorname{con} f$ is the pointwise supremum of all the convex functions $g \leq f$, this pointwise supremum being a convex function by 2.9(b); the constant function $-\infty$ always serves as such a function g. Alternatively, $\operatorname{con} f$ is the function obtained by taking epi(con f) to be the epigraphical closure of con(epi f). The condition $f = \operatorname{con} f$ means that f is convex. For $f = \delta_C$ one has $\operatorname{con} f = \delta_D$, where $D = \operatorname{con} C$.

2.31 Proposition (convexification of a function). *For* $f : \mathbb{R}^n \to \overline{\mathbb{R}}$,

$$(\operatorname{con} f)(x) = \inf\Big\{\sum\nolimits_{i=0}^{n} \lambda_i f(x_i) \,\Big|\, \sum\nolimits_{i=0}^{n} \lambda_i x_i = x,\ \lambda_i \geq 0,\ \sum\nolimits_{i=0}^{n} \lambda_i = 1\Big\}.$$

Proof. We apply Theorem 2.29 to epi f in $\mathbb{R}^{n+1}$: every point of con(epi f) is a convex combination of at most $n+2$ points of epi f. Actually, at most $n+1$ points x_i at a time are needed in determining the values of con f, since a point $(\bar{x}, \bar{\alpha})$ of con(epi f) not representable by fewer than $n+2$ points of epi f would lie in the interior of some $n+1$-simplex S in con(epi f). The vertical line through $(\bar{x}, \bar{\alpha})$ would meet the boundary of S in a point $(\bar{x}, \tilde{\alpha})$ with $\bar{\alpha} > \tilde{\alpha}$, and such a boundary point is representable by $n+1$ of the points in question, cf. 2.28(d). Thus, (con f)(x) is the infimum of all numbers α such that there exist $n+1$ points $(x_i, \alpha_i) \in \operatorname{epi} f$ and scalars $\lambda_i \geq 0$, with $\sum_{i=0}^{n} \lambda_i (x_i, \alpha_i) = (x, \alpha)$, $\sum_{i=0}^{n} \lambda_i = 1$. This description translates to the formula claimed. □

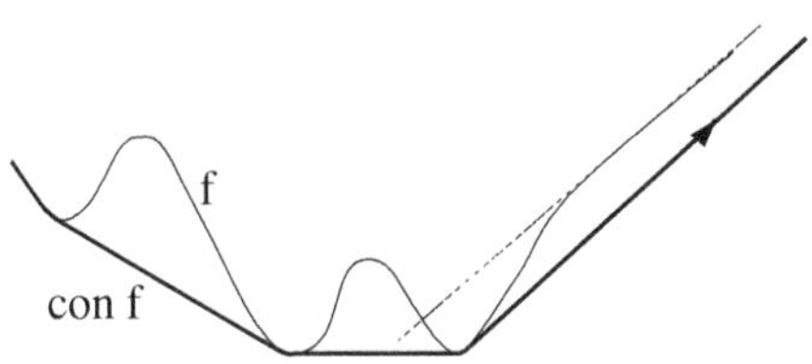

Fig. 2–11. The convex hull of a function.

F. Closures and Continuity

Next we consider the relation of convexity to some topological concepts like closures and interiors of sets and semicontinuity of functions.

2.32 Proposition (convexity and properness of closures). *For a convex set* $C \subset \mathbb{R}^n$, cl C *is convex. Likewise, for a convex function* $f : \mathbb{R}^n \to \overline{\mathbb{R}}$, cl f *is convex. Moreover* cl f *is proper if and only if* f *is proper.*

Proof. First, for sequences of points x_0^ν and x_1^ν in C converging to x_0 and x_1 in cl C, and for any $\tau \in (0,1)$, the points $x_\tau^\nu = (1-\tau)x_0^\nu + \tau x_1^\nu \in C$, converge to $x_\tau = (1-\tau)x_0 + \tau x_1$, which therefore lies in cl C. This proves the convexity of cl C. In the case of a function f, the epigraph of the function cl f (the lsc regularization of f introduced in 1(6)-1(7)) is the closure of the epigraph of f, which is convex, and therefore cl f is convex as well, cf. 2.4.

If f is improper, so is cl f by 2.5. Conversely, if cl f is improper it is $-\infty$ on the convex set $D = \operatorname{dom}(\operatorname{cl} f) = \operatorname{cl}(\operatorname{dom} f)$ by 2.5. Consider then a simplex $S = \operatorname{con}\{a_0, a_1, \ldots, a_p\}$ of maximal dimension in D. By a translation if necessary, we can suppose $a_0 = 0$, so that the vectors $a_1, \ldots, a_p$ are linearly independent. The p-dimensional subspace M generated by these vectors must include D, for if not we could find an additional vector $a_{p+1} \in D$ linearly independent of the others, and this would contradict the maximality of p. Our analysis can therefore be reduced to M, which under a change of coordinates can be identified with $I\!R^p$. To keep notation simple, we can just as well assume that $M = I\!R^n$, $p = n$. Then S has nonempty interior; let $x \in \operatorname{int} S$, so that $x = \sum_{i=0}^n \lambda_i a_i$ with $\lambda_i > 0$, $\sum_{i=0}^n \lambda_i = 1$, cf. 2.28(d). For each i we have $(\operatorname{cl} f)(a_i) = -\infty$, so there is a sequence $a_i^\nu \to a_i$ with $f(a_i^\nu) \to -\infty$. Then for ν sufficiently large we have representations $x = \sum_{i=0}^n \lambda_i^\nu a_i^\nu$ with $\lambda_i^\nu > 0$, $\sum_{i=0}^n \lambda_i^\nu = 1$, cf. 2.28(f). From Jensen's inequality 2.2(b), we obtain $f(x) \le \sum_{i=0}^n \lambda_i^\nu f(a_i^\nu) \le \max\{f(a_0^\nu), f(a_1^\nu), \ldots, f(a_n^\nu)\} \to -\infty$. Therefore, $f(x) = -\infty$, and f is improper. □

The most important topological consequences of convexity can be traced to a simple fact about line segments which relates the closure $\operatorname{cl} C$ to the interior $\operatorname{int} C$ of a convex set C, when the interior is nonempty.

2.33 Theorem (line segment principle). *A convex set C has $\operatorname{int} C \neq \emptyset$ if and only if $\operatorname{int}(\operatorname{cl} C) \neq \emptyset$. In that case, whenever $x_0 \in \operatorname{int} C$ and $x_1 \in \operatorname{cl} C$, one has $(1-\tau)x_0 + \tau x_1 \in \operatorname{int} C$ for all $\tau \in (0,1)$. Thus, $\operatorname{int} C$ is convex. Moreover,*

$$\operatorname{cl} C = \operatorname{cl}(\operatorname{int} C), \qquad \operatorname{int} C = \operatorname{int}(\operatorname{cl} C).$$

Proof. Leaving the assertions about $\operatorname{int}(\operatorname{cl} C)$ to the end, start just with the assumption that $\operatorname{int} C \neq \emptyset$. Choose $\varepsilon_0 > 0$ small enough that the ball $I\!B(x_0, \varepsilon_0)$ is included in C. Writing $I\!B(x_0, \varepsilon_0) = x_0 + \varepsilon_0 I\!B$ for $I\!B = I\!B(0,1)$, note that our assumption $x_1 \in \operatorname{cl} C$ implies $x_1 \in C + \varepsilon_1 I\!B$ for all $\varepsilon_1 > 0$. For arbitrary fixed $\tau \in (0,1)$, it's necessary to show that the point $x_\tau = (1-\tau)x_0 + \tau x_1$ belongs to $\operatorname{int} C$. For this it suffices to demonstrate that $x_\tau + \varepsilon_\tau I\!B \subset C$ for some $\varepsilon_\tau > 0$. We do so for $\varepsilon_\tau := (1-\tau)\varepsilon_0 - \tau\varepsilon_1$, with ε_1 fixed at any positive value small enough that $\varepsilon_\tau > 0$ (cf. Figure 2–12), by calculating

$$\begin{aligned} x_\tau + \varepsilon_\tau I\!B &= (1-\tau)x_0 + \tau x_1 + \varepsilon_\tau I\!B \subset (1-\tau)x_0 + \tau(C + \varepsilon_1 I\!B) + \varepsilon_\tau I\!B \\ &= (1-\tau)x_0 + (\tau\varepsilon_1 + \varepsilon_\tau)I\!B + \tau C = (1-\tau)(x_0 + \varepsilon_0 I\!B) + \tau C \\ &\subset (1-\tau)C + \tau C = C, \end{aligned}$$

where the convexity of C and $I\!B$ has been used in invoking 2.23(c).

As a special case of the argument so far, if $x_1 \in \operatorname{int} C$ we get $x_\tau \in \operatorname{int} C$; thus, $\operatorname{int} C$ is convex. Also, any point of $\operatorname{cl} C$ can be approached along a line segment by points of $\operatorname{int} C$, so $\operatorname{cl} C = \operatorname{cl}(\operatorname{int} C)$.

It's always true, on the other hand, that $\operatorname{int} C \subset \operatorname{int}(\operatorname{cl} C)$, so the nonemptiness of $\operatorname{int} C$ implies that of $\operatorname{int}(\operatorname{cl} C)$. To complete the proof of the theorem, no longer assuming outright that $\operatorname{int} C \neq \emptyset$, we suppose $\bar{x} \in \operatorname{int}(\operatorname{cl} C)$ and aim at showing $\bar{x} \in \operatorname{int} C$.

By 2.28(e), some simplex $S = \operatorname{con}\{a_0, a_1, \ldots, a_n\} \subset \operatorname{cl} C$ has $\bar{x} \in \operatorname{int} S$.

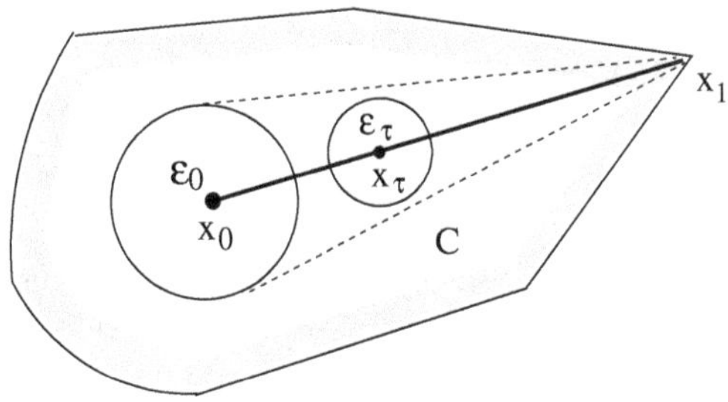

Fig. 2–12. Line segment principle for convex sets.

Then $\bar{x} = \sum_{i=0}^{n} \lambda_i a_i$ with $\sum_{i=0}^{n} \lambda_i = 1$, $\lambda_i > 0$, cf. 2.28(d). Consider in C sequences $a_i^\nu \to a_i$, and let $S^\nu = \text{con}\{a_0^\nu, a_1^\nu, \ldots, a_n^\nu\} \subset C$. For large ν, S^ν is an n-simplex too, and $\bar{x} = \sum_{i=0}^{n} \lambda_i^\nu a_i^\nu$ with $\sum_{i=0}^{n} \lambda_i^\nu = 1$ and $\lambda_i^\nu \to \lambda_i$; see 2.28(f). Eventually $\lambda_i^\nu > 0$, and then $\bar{x} \in \text{int}\, S^\nu$ by 2.28(d), so $\bar{x} \in \text{int}\, C$. □

2.34 Proposition (interiors of epigraphs and level sets). *For f convex on $I\!R^n$,*

$$\text{int}(\text{epi}\, f) = \{(x, \alpha) \in I\!R^n \times I\!R \mid x \in \text{int}(\text{dom}\, f),\ f(x) < \alpha\},$$
$$\text{int}(\text{lev}_{\leq \alpha} f) = \{x \in \text{int}(\text{dom}\, f) \mid f(x) < \alpha\} \quad \textit{for } \alpha \in (\inf f, \infty).$$

Proof. Obviously, if $(\bar{x}, \bar{\alpha}) \in \text{int}(\text{epi}\, f)$ there's a ball around $(\bar{x}, \bar{\alpha})$ within $\text{epi}\, f$, so that $\bar{x} \in \text{int}(\text{dom}\, f)$ and $f(\bar{x}) < \bar{\alpha}$.

On the other hand, if the latter properties hold there's a simplex $S = \text{con}\{a_0, a_1, \ldots, a_n\}$ with $\bar{x} \in \text{int}\, S \subset \text{dom}\, f$, cf. 2.28(e). Each $x \in S$ is a convex combination $\sum_{i=0}^{n} \lambda_i a_i$ and satisfies $f(x) \leq \sum_{i=0}^{n} \lambda_i f(a_i)$ by Jensen's inequality in 2.2(b) and therefore also satisfies $f(x) \leq \max\{f(a_0), \ldots, f(a_n)\}$. For $\tilde{\alpha} := \max\{f(a_0), \ldots, f(a_n)\}$ the open set $\text{int}\, S \times (\tilde{\alpha}, \infty)$ lies then within $\text{epi}\, f$. The vertical line through $(\bar{x}, \bar{\alpha})$ thus contains a point $(\bar{x}, \alpha_0) \in \text{int}(\text{epi}\, f)$ with $\alpha_0 > \bar{\alpha}$, but it also contains a point $(\bar{x}, \alpha_1) \in \text{epi}\, f$ with $\alpha_1 < \bar{\alpha}$, inasmuch as $f(\bar{x}) < \bar{\alpha}$. The set $\text{epi}\, f$ is convex because f is convex, so by the line segment principle in 2.33 all the points between $(\bar{x}, \alpha_0)$ and $(\bar{x}, \alpha_1)$ must belong to $\text{int}(\text{epi}\, f)$. This applies to $(\bar{x}, \bar{\alpha})$ in particular.

Consider now a level set $\text{lev}_{\leq \bar{\alpha}} f$. If $\bar{x} \in \text{int}(\text{dom}\, f)$ and $f(\bar{x}) < \bar{\alpha}$, then some ball around $(\bar{x}, \bar{\alpha})$ lies in $\text{epi}\, f$, so $f(x) \leq \bar{\alpha}$ for all x in some neighborhood of $\bar{x}$. Then $\bar{x} \in \text{int}(\text{lev}_{\leq \bar{\alpha}} f)$. Conversely, if $\bar{x} \in \text{int}(\text{lev}_{\leq \bar{\alpha}} f)$ and $\inf f < \bar{\alpha} < \infty$, we must have $\bar{x} \in \text{int}(\text{dom}\, f)$, but also there's a point $x_0 \in \text{dom}\, f$ with $f(x_0) < \bar{\alpha}$. For $\varepsilon > 0$ sufficiently small the point $x_1 = \bar{x} + \varepsilon(\bar{x} - x_0)$ still belongs to $\text{lev}_{\leq \bar{\alpha}} f$. Then $\bar{x} = (1 - \tau)x_0 + \tau x_1$ for $\tau = 1/(1 + \varepsilon)$, and we get $f(\bar{x}) \leq (1 - \tau) f(x_0) + \tau f(x_1) < \bar{\alpha}$, which is the desired inequality. □

2.35 Theorem (continuity properties of convex functions). *A convex function $f : I\!R^n \to \overline{I\!R}$ is continuous on $\text{int}(\text{dom}\, f)$ and therefore agrees with $\text{cl}\, f$ on $\text{int}(\text{dom}\, f)$, this set being the same as $\text{int}(\text{dom}(\text{cl}\, f))$. (Also, f agrees with $\text{cl}\, f$ as having the value ∞ outside of $\text{dom}(\text{cl}\, f)$, hence in particular outside of $\text{cl}(\text{dom}\, f)$.) Moreover,*

$$(\mathrm{cl}\, f)(x) \;=\; \lim_{\tau \nearrow 1} f\big((1-\tau)x_0 + \tau x\big) \;\; \textit{for all } x \textit{ if } x_0 \in \operatorname{int}(\operatorname{dom} f). \qquad 2(11)$$

If f is lsc, it must in addition be continuous relative to the convex hull of any finite subset of $\operatorname{dom} f$, in particular any line segment in $\operatorname{dom} f$.

Proof. The closure formula will be proved first. We know from the basic expression for $\mathrm{cl}\, f$ in 1(7) that $(\mathrm{cl}\, f)(x) \le \liminf_{\tau \nearrow 1} f((1-\tau)x_0+\tau x)$, so it will suffice to show that if $(\mathrm{cl}\, f)(x) \le \alpha \in I\!R$ then $\limsup_{\tau \nearrow 1} f((1-\tau)x_0+\tau x) \le \alpha$. The assumption on α means that $(x, \alpha) \in \mathrm{cl}(\operatorname{epi} f)$, cf. 1(6). On the other hand, for any real number $\alpha_0 > f(x_0)$ we have $(x_0, \alpha_0) \in \operatorname{int}(\operatorname{epi} f)$ by 2.34. Then by the line segment principle in Theorem 2.33, as applied to the convex set $\operatorname{epi} f$, the points $(1-\tau)(x_0, \alpha_0) + \tau(x, \alpha)$ for $\tau \in (0,1)$ belong to $\operatorname{int}(\operatorname{epi} f)$. In particular, then, $f((1-\tau)x_0 + \tau x) < (1-\tau)\alpha_0 + \tau\alpha$ for $\tau \in (0,1)$. Taking the upper limit on both sides as $\tau \nearrow 1$, we get the inequality needed.

When the closure formula is applied with $x = x_0$ it yields the fact that $\mathrm{cl}\, f$ agrees with f on $\operatorname{int}(\operatorname{dom} f)$. Hence f is lsc on $\operatorname{int}(\operatorname{dom} f)$. But f is usc there by 1.13(b) in combination with the characterization of $\operatorname{int}(\operatorname{epi} f)$ in 2.34. It follows that f is continuous on $\operatorname{int}(\operatorname{dom} f)$. We have $\operatorname{int}(\operatorname{dom} f) = \operatorname{int}(\operatorname{dom}(\mathrm{cl}\, f))$ by 2.33, because $\operatorname{dom} f \subset \operatorname{dom}(\mathrm{cl}\, f) \subset \mathrm{cl}(\operatorname{dom} f)$. The same inclusions also yield $\mathrm{cl}(\operatorname{dom} f) = \mathrm{cl}(\operatorname{dom}(\mathrm{cl}\, f))$.

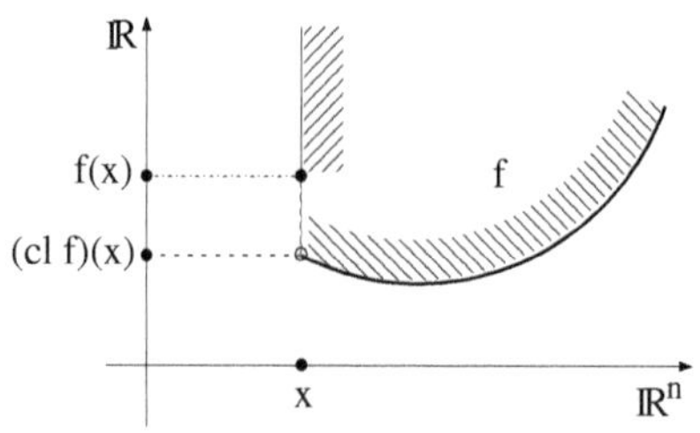

Fig. 2–13. Closure operation on a convex function.

Consider now a finite set $C \subset \operatorname{dom} f$. Under the assumption that f is lsc, we wish to show that f is continuous relative to $\operatorname{con} C$. But $\operatorname{con} C$ is the union of the simplices generated by the points of C, as shown in the first part of the proof of Theorem 2.29, and there are only finitely many of these. If a function is continuous on a finite family of sets, it is also continuous on their union. It suffices therefore to show that f is continuous relative to any simplex $S = \operatorname{con}\{a_0, a_1, \ldots, a_p\} \subset \operatorname{dom} f$. For simplicity of notation we can translate so that $0 \in S$ and investigate continuity at 0; we have a unique representation of 0 as a convex combination $\sum_{i=0}^{p} \bar\lambda_i a_i$.

We argue next that S is the union of the finitely many other simplices having 0 as one vertex and certain a_i's as the other vertices. This will enable us to reduce further to the study of continuity relative to such a simplex.

Consider any point $\tilde x \neq 0$ in S and represent it as a convex combination $\sum_{i=0}^{p} \tilde\lambda_i a_i$. The points $x_\tau = (1-\tau)0 + \tau\tilde x$ on the line through 0 and $\tilde x$ can't all be in S, which is compact (by 2.30), so there must be a highest τ with $x_\tau \in S$. For this we have $\tau \geq 1$ and $x_\tau = \sum_{i=0}^p \mu_i a_i$ for $\mu_i = (1-\tau)\bar{\lambda}_i + \tau\tilde{\lambda}_i$ with $\sum_{i=0}^p \mu_i = 1$, where $\mu_i \geq 0$ for all i but $\mu_i = 0$ for at least one i such that $\bar{\lambda}_i > \tilde{\lambda}_i$ (or τ wouldn't be the highest). We can suppose $\mu_0 = 0$; then in particular $\bar{\lambda}_0 > 0$. Since $\tilde{x}$ lies on the line segment joining 0 with x_τ, it belongs to $\operatorname{con}\{0, a_1, \ldots, a_p\}$. This is a simplex, for if not the vectors $a_1, \ldots, a_p$ would be linearly dependent: we would have $\sum_{i=1}^p \eta_i a_i = 0$ for certain coefficients η_i, not all 0. It's impossible that $\sum_{i=1}^p \eta_i = 0$, because $\{a_0, a_1, \ldots, a_p\}$ are affinely independent, so if this were the case we could, by rescaling, arrange that $\sum_{i=1}^p \eta_i = 1$. But then in defining $\eta_0 = 0$ we would be able to conclude from the affine independence that $\eta_i = \bar{\lambda}_i$ for $i = 0, \ldots, p$, because $0 = \sum_{i=0}^p (\eta_i - \bar{\lambda}_i) a_i$ with $\sum_{i=0}^p (\eta_i - \bar{\lambda}_i) = 0$. This would contradict $\bar{\lambda}_0 > 0$.

We have gotten to where a simplex $S_0 = \operatorname{con}\{0, a_1, \ldots, a_p\}$ lies in $\operatorname{dom} f$ and we need to prove that f is not just lsc relative to S_0 at 0, as assumed, but also usc. Any point of S_0 has a unique expression $\sum_{i=1}^p \lambda_i a_i$ with $\lambda_i \geq 0$ and $\sum_{i=1}^p \lambda_i \leq 1$, and as the point approaches 0 these coefficients vanish, cf. 2.28(c). The corresponding value of f is bounded above by $\lambda_0 f(0) + \sum_{i=1}^p \lambda_i f(a_i)$ through Jensen's inequality 2.2(b), where $\lambda_0 = 1 - \sum_{i=1}^p \lambda_i$, and in the limit this bound is $f(0)$. Thus, $\limsup_{x \to 0} f(x) \leq f(0)$ for $x \in S_0$. □

2.36 Corollary (finite convex functions). *A finite, convex function f on an open, convex set $O \neq \emptyset$ in $\mathbb{R}^n$ is continuous on O. Such a function has a unique extension to a proper, lsc, convex function f on $\mathbb{R}^n$ with* $\operatorname{dom} f \subset \operatorname{cl} O$.

Proof. Apply the theorem to the convex function g that agrees with f on O but takes on ∞ everywhere else. Then $\operatorname{int}(\operatorname{dom} g) = O$. □

Finite convex functions will be seen in 9.14 to have the even stronger property of Lipschitz continuity locally.

2.37 Corollary (convex functions of a single real variable). *Any lsc, convex function $f : \mathbb{R} \to \overline{\mathbb{R}}$ is continuous with respect to* $\operatorname{cl}(\operatorname{dom} f)$.

Proof. This is clear from 2(11) when $\operatorname{int}(\operatorname{dom} f) \neq \emptyset$; otherwise it's trivial. □

Not everything about the continuity properties of convex functions is good news. The following example provides clear insight into what can go wrong.

2.38 Example (discontinuity and unboundedness). *On $\mathbb{R}^2$, the function*

$$f(x_1, x_2) = \begin{cases} x_1^2/2x_2 & \text{if } x_2 > 0, \\ 0 & \text{if } x_1 = 0 \text{ and } x_2 = 0, \\ \infty & \text{otherwise,} \end{cases}$$

is lsc, proper, convex and positively homogeneous. Nonetheless, f fails to be continuous relative to the compact, convex set $C = \{(x_1, x_2) \mid x_1^4 \leq x_2 \leq 1\}$ in $\operatorname{dom} f$, despite f being continuous relative to every line segment in C. In fact f is not even bounded above on C.

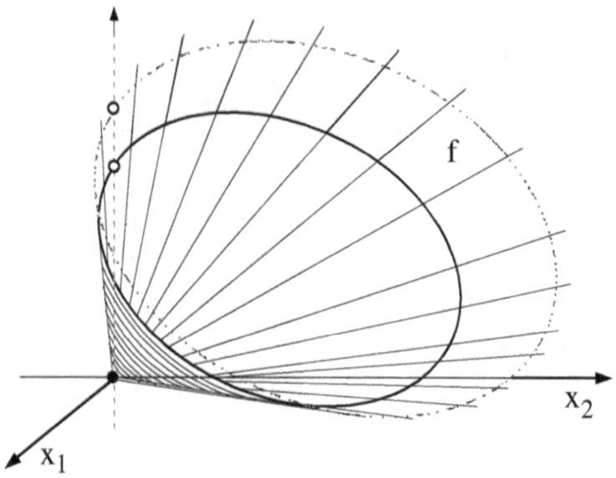

Fig. 2–14. Example of discontinuity.

Detail. The convexity of f on the open half-plane $H = \{(x_1, x_2) \mid x_2 > 0\}$ can be verified by the second-derivative test in 2.14, and the convexity relative to all of $\mathbb{R}^n$ follows then via the extension procedure in 2.36. The epigraph of f is actually a circular cone whose axis is the ray through $(0,1,1)$ and whose boundary contains the rays through $(0,1,0)$ and $(0,0,1)$, see Figure 2–14. The unboundedness and lack of continuity of f relative to H are seen from its behavior along the boundary of C at 0, as indicated from the piling up of the level sets of this function as shown in Figure 2–15. □

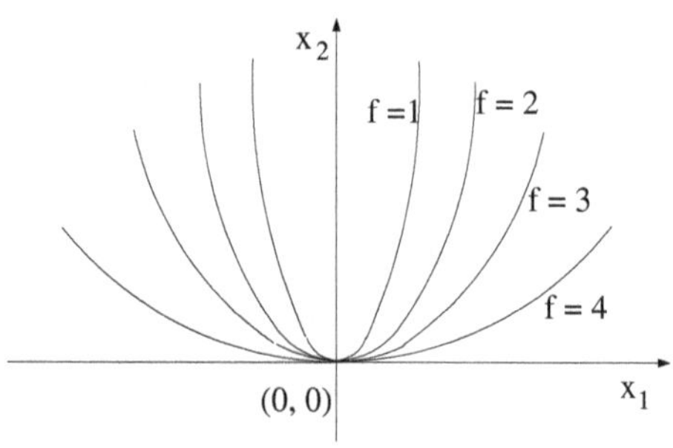

Fig. 2–15. Nest of level sets illustrating discontinuity.

G*. Separation

Properties of closures and interiors of convex sets lead also to a famous principle of separation. A hyperplane $\{x \mid \langle a, x\rangle = \alpha\}$, where $a \neq 0$ and $\alpha \in \mathbb{R}$, is said to *separate* two sets C_1 and C_2 in $\mathbb{R}^n$ if C_1 is included in one of the corresponding closed half-spaces $\{x \mid \langle a, x\rangle \leq \alpha\}$ or $\{x \mid \langle a, x\rangle \geq \alpha\}$, while C_2 is included in the other. The separation is said to be *proper* if the hyperplane itself doesn't actually include *both* C_1 and C_2. As a related twist in wording, we say that C_1 and C_2 *can, or cannot, be separated* according to whether a separating hyperplane exists; similarly for whether they can, or cannot, be separated *properly*. Additional variants are *strict* separation, where the two sets lie in complementary *open* half-spaces, and *strong* separation, where they lie in different half-spaces $\{x \mid \langle a, x\rangle \leq \alpha_1\}$ and $\{x \mid \langle a, x\rangle \geq \alpha_2\}$ with $\alpha_1 < \alpha_2$. Strong separation implies strict separation, but not conversely.

2.39 Theorem (separation). *Two nonempty, convex sets C_1 and C_2 in $\mathbb{R}^n$ can be separated by a hyperplane if and only if $0 \notin \operatorname{int}(C_1 - C_2)$. The separation must be proper if also $\operatorname{int}(C_1 - C_2) \neq \emptyset$. Both conditions certainly hold when $\operatorname{int} C_1 \neq \emptyset$ but $C_2 \cap \operatorname{int} C_1 = \emptyset$, or when $\operatorname{int} C_2 \neq \emptyset$ but $C_1 \cap \operatorname{int} C_2 = \emptyset$.*

Strong separation is possible if and only if $0 \notin \operatorname{cl}(C_1 - C_2)$. This is ensured in particular when $C_1 \cap C_2 = \emptyset$ with both sets closed and one of them bounded.

Proof. In the case of a separating hyperplane $\{x \mid \langle a, x\rangle = \alpha\}$ with $C_1 \subset \{x \mid \langle a, x\rangle \leq \alpha\}$ and $C_2 \subset \{x \mid \langle a, x\rangle \geq \alpha\}$, we have $0 \geq \langle a,\, x_1 - x_2\rangle$ for all $x_1 - x_2 \in C_1 - C_2$, in which case obviously $0 \notin \operatorname{int}(C_1 - C_2)$. This condition is therefore necessary for separation. Moreover if the separation weren't proper, we would have $0 = \langle a,\, x_1 - x_2\rangle$ for all $x_1 - x_2 \in C_1 - C_2$, and this isn't possible if $\operatorname{int}(C_1 - C_2) \neq \emptyset$. In the case where $\operatorname{int} C_1 \neq \emptyset$ but $C_2 \cap \operatorname{int} C_1 = \emptyset$, we have $0 \notin (\operatorname{int} C_1) - C_2$ where the set $(\operatorname{int} C_1) - C_2$ is convex (by 2.23 and the convexity of interiors in 2.33) as well as open (since it is the union of translates $(\operatorname{int} C_1) - x_2$, each of which is an open set). Then $(\operatorname{int} C_1) - C_2 \subset C_1 - C_2 \subset \operatorname{cl}\big[(\operatorname{int} C_1) - C_2\big]$ (through 2.33), so actually $\operatorname{int}(C_1 - C_2) = (\operatorname{int} C_1) - C_2$ (once more through the relations in 2.33). In this case, therefore, we have $0 \notin \operatorname{int}(C_1 - C_2) \neq \emptyset$. Similarly, this holds when $\operatorname{int} C_2 \neq \emptyset$ but $C_1 \cap \operatorname{int} C_2 = \emptyset$.

The sufficiency of the condition $0 \notin \operatorname{int}(C_1 - C_2)$ for the existence of a separating hyperplane is all that remains to be established. Note that because $C_1 - C_2$ is convex (cf. 2.23), so is $C := \operatorname{cl}(C_1 - C_2)$. We need only produce a vector $a \neq 0$ such that $\langle a, x\rangle \leq 0$ for all $x \in C$, for then we'll have $\langle a, x_1\rangle \leq \langle a, x_2\rangle$ for all $x_1 \in C_1$ and $x_2 \in C_2$, and separation will be achieved with any α in the nonempty interval between $\sup_{x_1 \in C_1}\langle a, x_1\rangle$ and $\inf_{x_2 \in C_2}\langle a, x_2\rangle$.

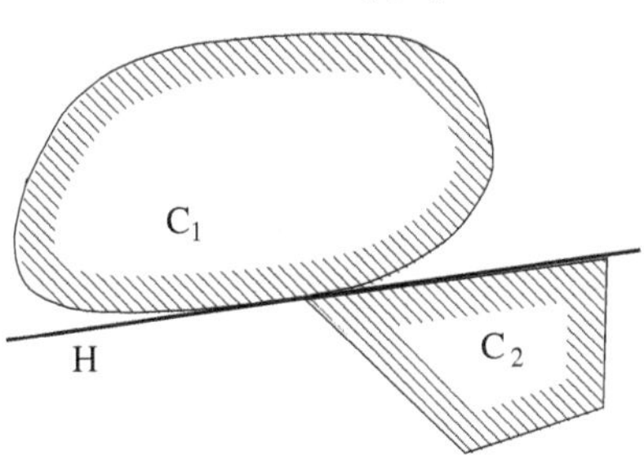

Fig. 2–16. Separation of convex sets.

Let $\bar{x}$ denote the unique point of C nearest to 0 (cf. 2.25). For any $x \in C$ the segment $[\bar{x}, x]$ lies in C, so the function $g(\tau) = \frac{1}{2}|(1-\tau)\bar{x} + \tau x|^2$ satisfies

$g(\tau) \geq g(0)$ for $\tau \in (0,1)$. Hence $0 \leq g'(0) = \langle \bar{x}, x - \bar{x} \rangle$. If $0 \notin C$, so that $\bar{x} \neq 0$, we can take $a = -\bar{x}$ and be done. A minor elaboration of this argument confirms that *strong* separation is possible *if and only if* $0 \notin C$. Certainly we have $0 \notin C$ in particular when $C_1 \cap C_2 = \emptyset$ and $C_1 - C_2$ is closed, i.e., $C_1 - C_2 = C$; it's easy to see that $C_1 - C_2$ is closed if both C_1 and C_2 are closed and one is actually compact.

When $0 \in C$, so that $\bar{x} = 0$, we need to work harder to establish the sufficiency of the condition $0 \notin \operatorname{int}(C_1 - C_2)$ for the existence of a separating hyperplane. From $C = \operatorname{cl}(C_1 - C_2)$ and $0 \notin \operatorname{int}(C_1 - C_2)$, we have $0 \notin \operatorname{int} C$ by Theorem 2.33. Then there is a sequence $x^\nu \to 0$ with $x^\nu \notin C$. Let $\bar{x}^\nu$ be the unique point of C nearest to x^ν; then $\bar{x}^\nu \neq x^\nu$, $\bar{x}^\nu \to 0$. Applying to x^ν and $\bar{x}^\nu$ the same argument we earlier applied to 0 and its projection on C when this wasn't necessarily 0 itself, we verify that $\langle x - \bar{x}^\nu, x^\nu - \bar{x}^\nu \rangle \leq 0$ for every $x \in C$. Let a be any cluster point of the sequence of vectors $a^\nu := (x^\nu - \bar{x}^\nu)/|x^\nu - \bar{x}^\nu|$, which have $|a^\nu| = 1$ and $\langle x - \bar{x}^\nu, a^\nu \rangle \leq 0$ for every $x \in C$. Then $|a| = 1$, so $a \neq 0$, and in the limit we have $\langle x, a \rangle \leq 0$ for every $x \in C$, as required. □

H.* Relative Interiors

Every nonempty affine set has a well determined *dimension*, which is the dimension of the linear subspace of which it is a translate, cf. 2.11. Singletons are 0-dimensional, lines are 1-dimensional, and so on. The hyperplanes in $\mathbb{R}^n$ are the affine sets that are $(n-1)$-dimensional.

For any convex set C in $\mathbb{R}^n$, the *affine hull* of C is the smallest affine set that includes C (it's the intersection of all the affine sets that include C). The interior of C relative to its affine hull is the *relative interior* of C, denoted by $\operatorname{rint} C$. This coincides with the true interior when the affine hull is all of $\mathbb{R}^n$, but is able to serve as a robust substitute for $\operatorname{int} C$ when $\operatorname{int} C = \emptyset$.

2.40 Proposition (relative interiors of convex sets). *For $C \subset \mathbb{R}^n$ nonempty and convex, the set $\operatorname{rint} C$ is nonempty and convex with $\operatorname{cl}(\operatorname{rint} C) = \operatorname{cl} C$ and $\operatorname{rint}(\operatorname{cl} C) = \operatorname{rint} C$. If $x_0 \in \operatorname{rint} C$ and $x_1 \in \operatorname{cl} C$, then $\operatorname{rint}[x_0, x_1] \subset \operatorname{rint} C$.*

Proof. Through a translation if necessary, we can suppose that $0 \in C$. We can suppose further that C contains more than just 0, because otherwise the assertion is trivial. Let $a_1, \ldots, a_p$ be a set of linearly independent vectors chosen from C with p as high as possible, and let M be the p-dimensional subspace of $\mathbb{R}^n$ generated by these vectors. Then M has to be the affine hull of C, because the affine hull has to be an affine set containing M, and yet there can't be any point of C outside of M or the maximality of p would be contradicted. The p-simplex $\operatorname{con}\{0, a_1, \ldots, a_p\}$ in C has nonempty interior relative to M (which can be identified with $\mathbb{R}^p$ through a change of coordinates), so $\operatorname{rint} C \neq \emptyset$. The relations between $\operatorname{rint} C$ and $\operatorname{cl} C$ follow then from the ones in 2.33. □

2.41 Exercise (relative interior criterion). *For a convex set $C \subset \mathbb{R}^n$, one has $x \in \operatorname{rint} C$ if and only if $x \in C$ and, for every $x_0 \neq x$ in C, there exists $x_1 \in C$ such that $x \in \operatorname{rint}[x_0, x_1]$.*

Guide. Reduce to the case where C is n-dimensional. □

The properties in Proposition 2.40 are crucial in building up a calculus of closures and relative interiors of convex sets.

2.42 Proposition (relative interiors of intersections). *For a family of convex sets $C_i \subset \mathbb{R}^n$ indexed by $i \in I$ and such that $\bigcap_{i\in I} \operatorname{rint} C_i \neq \emptyset$, one has $\operatorname{cl} \bigcap_{i\in I} C_i = \bigcap_{i\in I} \operatorname{cl} C_i$. If I is finite, then also $\operatorname{rint} \bigcap_{i\in I} C_i = \bigcap_{i\in I} \operatorname{rint} C_i$.*

Proof. On general grounds, $\operatorname{cl} \bigcap_{i\in I} C_i \subset \bigcap_{i\in I} \operatorname{cl} C_i$. For the reverse consider any $x_0 \in \bigcap_{i\in I} \operatorname{rint} C_i$ and $x_1 \in \bigcap_{i\in I} \operatorname{cl} C_i$. We have $[x_0, x_1) \subset \bigcap_{i\in I} \operatorname{rint} C_i \subset \bigcap_{i\in I} C_i$ by 2.40, so $x_1 \in \operatorname{cl} \bigcap_{i\in I} C_i$. This argument has the by-product that $\operatorname{cl} \bigcap_{i\in I} C_i = \operatorname{cl} \bigcap_{i\in I} \operatorname{rint} C_i$. In taking the relative interior on both sides of this equation, we obtain via 2.40 that $\operatorname{rint} \bigcap_{i\in I} C_i = \operatorname{rint} \bigcap_{i\in I} \operatorname{rint} C_i$. The fact that $\operatorname{rint} \bigcap_{i\in I} \operatorname{rint} C_i = \bigcap_{i\in I} \operatorname{rint} C_i$ when I is finite is clear from 2.41. □

2.43 Proposition (relative interiors in product spaces). *For a convex set G in $\mathbb{R}^n \times \mathbb{R}^m$, let X be the image of G under the projection $(x, u) \mapsto x$, and for each $x \in X$ let $S(x) = \{u \mid (x, u) \in G\}$. Then X and $S(x)$ are convex, and*

$$(x, u) \in \operatorname{rint} G \iff x \in \operatorname{rint} X \;\; \textit{and} \;\; u \in \operatorname{rint} S(x).$$

Proof. We have X convex by 2.21 and $S(x)$ convex by 2.9(a). We begin by supposing that $(x, u) \in \operatorname{rint} G$ and proving $x \in \operatorname{rint} X$ and $u \in \operatorname{rint} S(x)$. Certainly $x \in X$ and $u \in S(x)$. For any $x_0 \in X$ there exists $u_0 \in S(x_0)$, and we have $(x_0, u_0) \in G$. If $x_0 \neq x$, then $(x_0, u_0) \neq (x, u)$ and we can obtain from the criterion in 2.41 as applied to $\operatorname{rint} G$ the existence of $(x_1, u_1) \neq (x, u)$ in G such that $(x, u) \in \operatorname{rint}\big[(x_0, u_0), (x_1, u_1)\big]$. Then $x_1 \in X$ and $x \in \operatorname{rint}[x_0, x_1]$. This proves by 2.41 (as applied to X) that $x \in \operatorname{rint} X$. At the same time, from the fact that $u \in S(x)$ we argue via 2.41 (as applied to G) that if $u_0' \in S(x)$ and $u_0' \neq u$, there exists $(x_1', u_1') \in G$ for which $(x, u) \in \operatorname{rint}\big[(x, u_0'), (x_1', u_1')\big]$. Then necessarily $x_1' = x$, and we conclude $u \in \operatorname{rint}[u_0', u_1']$, thereby establishing by criterion 2.41 (as applied to $S(x)$) that $u \in \operatorname{rint} S(x)$.

We tackle the converse now by assuming $x \in \operatorname{rint} X$ and $u \in \operatorname{rint} S(x)$ (so in particular $(x, u) \in G$) and aiming to prove $(x, u) \in \operatorname{rint} G$. Relying yet again on 2.41, we suppose $(x_0, u_0) \in G$ with $(x_0, u_0) \neq (x, u)$ and reduce the argument to showing that for some choice of $(x_1, u_1) \in G$ the pair (x, u) lies in $\operatorname{rint}\big[(x_0, u_0), (x_1, u_1)\big]$. If $x_0 = x$, this follows immediately from criterion 2.41 (as applied to $S(x)$ with $x_1 = x$). We assume therefore that $x_0 \neq x$. Since $x \in \operatorname{rint} X$, there then exists by 2.41 a vector $\bar{x}_1 \in X$, $\bar{x}_1 \neq x_0$, with $x \in \operatorname{rint}[x_0, \bar{x}_1]$, i.e., $x = (1 - \tau)x_0 + \tau\bar{x}_1$ for a certain $\tau \in (0, 1)$. We have $(\bar{x}_1, \bar{u}_1) \in G$ for some $\bar{u}_1$, and consequently $(1 - \tau)(x_0, u_0) + \tau(\bar{x}_1, \bar{u}_1) \in G$. If the point $u_0' := (1 - \tau)u_0 + \tau\bar{u}_1$ coincides with u, we get the desired sort of representation of (x, u) as a relative interior point of the line segment $\big[(x_0, u_0), (\bar{x}_1, \bar{u}_1)\big]$, whose

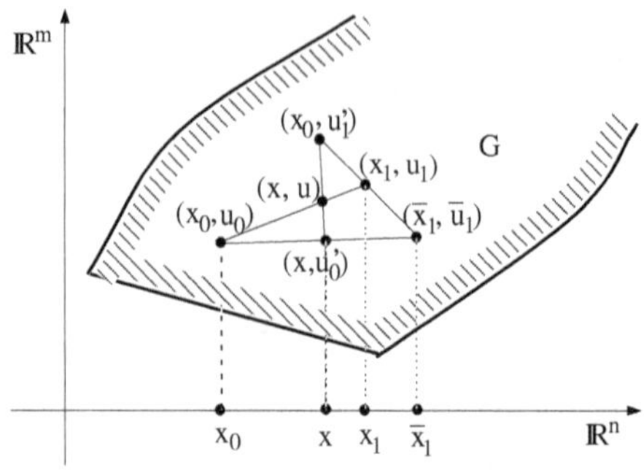

Fig. 2–17. Relative interior argument in a product space.

endpoints lie in G. If not, u_0' is a point of $S(x)$ different from u and, because $u \in \operatorname{rint} S(x)$, we get from 2.41 the existence of some $u_1' \in S(x)$, $u_1' \neq u_0'$, with $u \in \operatorname{rint}[u_0', u_1']$, i.e., $u = (1-\mu)u_0' + \mu u_1'$ for some $\mu \in (0,1)$. This yields

$$\begin{aligned}(x,u) &= (1-\mu)(x,u_0') + \mu(x,u_1')\\ &= (1-\mu)\big[(1-\tau)(x_0,u_0) + \tau(\bar{x}_1,\bar{u}_1)\big] + \mu(x,u_1')\\ &= \tau_0(x_0,u_0) + \tau_1(\bar{x}_1,\bar{u}_1) + \tau_2(x,u_1'),\\ &\qquad\qquad \text{where } 0 < \tau_i < 1,\ \tau_0+\tau_1+\tau_2 = 1.\end{aligned}$$

We can write this as $(x,u) = (1-\tau')(x_0,u_0) + \tau'(x_1,u_1)$ with $\tau' := \tau_1 + \tau_2 = 1 - \tau_0 \in (0,1)$ and $(x_1,u_1) := [\tau_1/(\tau_1+\tau_2)](\bar{x}_1,\bar{u}_1) + [\tau_2/(\tau_1+\tau_2)](x,u_1') \in G$. Here $(x_1,u_1) \neq (x_0,u_0)$, because otherwise the definition of x_1 would imply $x = \bar{x}_1$ in contradiction to our knowledge that $x \in \operatorname{rint}[x_0,\bar{x}_1]$. Thus, $(x,u) \in \operatorname{rint}\big[(x_0,u_0),(x_1,u_1)\big]$, and we can conclude that $(x,u) \in \operatorname{rint} G$. □

2.44 Proposition (relative interiors of set images). *For linear $L : I\!R^n \to I\!R^m$,*

(a) $\operatorname{rint} L(C) = L(\operatorname{rint} C)$ *for any convex set $C \subset I\!R^n$,*

(b) $\operatorname{rint} L^{-1}(D) = L^{-1}(\operatorname{rint} D)$ *for any convex set $D \subset I\!R^m$ such that* $\operatorname{rint} D$ *meets the range of L, and in this case also* $\operatorname{cl} L^{-1}(D) = L^{-1}(\operatorname{cl} D)$.

Proof. First we prove the relative interior equation in (b). In $I\!R^n \times I\!R^m$ let $G = \big\{(x,u) \,\big|\, u = L(x) \in D\big\} = M \cap (I\!R^n \times D)$, where M is the graph of L, a certain linear subspace of $I\!R^n \times I\!R^m$. We'll apply 2.43 to G, whose projection on $I\!R^n$ is $X = L^{-1}(D)$. Our assumption that $\operatorname{rint} D$ meets the range of L means $M \cap \operatorname{rint}[I\!R^n \times D] \neq \emptyset$, where $M = \operatorname{rint} M$ because M is an affine set. Then $\operatorname{rint} G = M \cap \operatorname{rint}(I\!R^n \times D)$ by 2.42; thus, (x,u) belongs to $\operatorname{rint} G$ if and only if $u = L(x)$ and $u \in \operatorname{rint} D$. But by 2.43 (x,u) belongs to $\operatorname{rint} G$ if and only if $x \in \operatorname{rint} X$ and $u \in \operatorname{rint}\{L(x)\} = \{L(x)\}$. We conclude that $x \in \operatorname{rint} L^{-1}(D)$ if and only if $L(x) \in \operatorname{rint} D$, as claimed in (b). The closure assertion follows similarly from the fact that $\operatorname{cl} G = M \cap \operatorname{cl}(I\!R^n \times D)$ by 2.42.

The argument for (a) just reverses the roles of x and u. We take $G = \big\{(x,u) \,\big|\, x \in C,\ u = L(x)\big\} = M \cap (C \times I\!R^m)$ and consider the projection U of G in $I\!R^m$, once again applying 2.42 and 2.43. □

2.45 Exercise (relative interiors in set algebra).

(a) *If* $C = C_1 \times C_2$ *for convex sets* $C_i \subset \mathbb{R}^{n_i}$, *then* $\operatorname{rint} C = \operatorname{rint} C_1 \times \operatorname{rint} C_2$.

(b) *For convex sets* C_1 *and* C_2 *in* $\mathbb{R}^n$, $\operatorname{rint}(C_1 + C_2) = \operatorname{rint} C_1 + \operatorname{rint} C_2$.

(c) *For a convex set* C *and any scalar* $\lambda \in \mathbb{R}$, $\operatorname{rint}(\lambda C) = \lambda(\operatorname{rint} C)$.

(d) *For convex sets* C_1 *and* C_2 *in* $\mathbb{R}^n$, *the condition* $0 \in \operatorname{int}(C_1 - C_2)$ *holds if and only if* $0 \in \operatorname{rint}(C_1 - C_2)$ *but there is no hyperplane* $H \supset C_1 \cup C_2$. *This is true in particular if either* $C_1 \cap \operatorname{int} C_2 \neq \emptyset$ *or* $C_2 \cap \operatorname{int} C_1 \neq \emptyset$.

(e) *Convex sets* $C_1 \neq \emptyset$ *and* $C_2 \neq \emptyset$ *in* $\mathbb{R}^n$ *can be separated properly if and only if* $0 \notin \operatorname{rint}(C_1 - C_2)$. *This condition is equivalent to* $\operatorname{rint} C_1 \cap \operatorname{rint} C_2 = \emptyset$.

Guide. Get (a) from 2.41 or 2.43. Get (b) and (c) from 2.44(a) through special choices of a linear transformation. In (d) verify that $C_1 \cup C_2$ lies in a hyperplane if and only if $C_1 - C_2$ lies in a hyperplane. Observe too that when $C_1 \cap \operatorname{int} C_2 \neq \emptyset$ the set $C_1 - \operatorname{int} C_2$ is open and contains 0. For (e), work with Theorem 2.39. Get the equivalent statement of the relative interior condition out of the algebra in (b) and (c). □

2.46 Exercise (closures of functions).

(a) *If* f_1 *and* f_2 *are convex on* $\mathbb{R}^n$ *with* $\operatorname{rint}(\operatorname{dom} f_1) = \operatorname{rint}(\operatorname{dom} f_2)$, *and on this common set* f_1 *and* f_2 *agree, then* $\operatorname{cl} f_1 = \operatorname{cl} f_2$.

(b) *If* f_1 *and* f_2 *are convex on* $\mathbb{R}^n$ *with* $\operatorname{rint}(\operatorname{dom} f_1) \cap \operatorname{rint}(\operatorname{dom} f_2) \neq \emptyset$, *then* $\operatorname{cl}(f_1 + f_2) = \operatorname{cl} f_1 + \operatorname{cl} f_2$.

(c) *If* $f = g \circ L$ *for a linear mapping* $L : \mathbb{R}^n \to \mathbb{R}^m$ *and a convex function* $g : \mathbb{R}^m \to \overline{\mathbb{R}}$ *such that* $\operatorname{rint}(\operatorname{dom} g)$ *meets the range of* L, *then* $\operatorname{rint}(\operatorname{dom} f) = L^{-1}\big(\operatorname{rint}(\operatorname{dom} g)\big)$ *and* $\operatorname{cl} f = (\operatorname{cl} g) \circ L$.

Guide. Adapt Theorem 2.35 to relative interiors. In (b) use 2.42. In (c) argue that $\operatorname{epi} f = L_0^{-1}(\operatorname{epi} g)$ with $L_0 : (x, \alpha) \mapsto (L(x), \alpha)$, and apply 2.44(b). □

I.* Piecewise Linear Functions

Polyhedral sets are important not only in the geometric analysis of systems of linear constraints, as in 2.10, but also in piecewise linearity.

2.47 Definition (piecewise linearity).

(a) *A mapping* $F : D \to \mathbb{R}^m$ *for a set* $D \subset \mathbb{R}^n$ *is piecewise linear on* D *if* D *can be represented as the union of finitely many polyhedral sets, relative to each of which* $F(x)$ *is given by an expression of the form* $Ax + a$ *for some matrix* $A \in \mathbb{R}^{m \times n}$ *and vector* $a \in \mathbb{R}^m$.

(b) *A function* $f : \mathbb{R}^n \to \overline{\mathbb{R}}$ *is piecewise linear if it is piecewise linear on* $D = \operatorname{dom} f$ *as a mapping into* $\mathbb{R}$.

Piecewise linear functions and mappings should perhaps be called 'piecewise affine', but the popular term is retained here. Definition 2.47 imposes no conditions on the extent to which the polyhedral sets whose union is D may overlap or be redundant, although if such a representation exists a refinement with better properties may well be available; cf. 2.50 below. Of course, on the intersection of any two sets in a representation the two formulas must agree, since both give $f(x)$, or as the case may be, $F(x)$. This, along with the finiteness of the collection of sets, ensures that the function or mapping in question must be continuous relative to D. Also, D must be closed, since polyhedral sets are closed, and the union of finitely many closed sets is closed.

2.48 Exercise (graphs of piecewise linear mappings). *For $F : D \to I\!R^m$ to be piecewise linear relative to D, it is necessary and sufficient that its graph, the set $G = \{(x,u) \mid x \in D,\ u = F(x)\}$, be expressible as the union of a finite collection of polyhedral sets in $I\!R^n \times I\!R^m$.*

Here we're interested primarily in what piecewise linearity means for convex functions. An example of a convex, piecewise linear function is the vector-max function in 1.30 and 2.16. This function f on $I\!R^n$ has the linear formula $f(x_1,\ldots,x_n) = x_k$ on $C_k := \{(x_1,\ldots,x_n) \mid x_j - x_k \le 0 \text{ for } j = 1,\ldots,n\}$, and the union of these polyhedral sets C_k is the set $\operatorname{dom} f = I\!R^n$.

The graph of any piecewise linear function $f : I\!R^n \to \overline{I\!R}$ must have a representation like that in 2.48 relative to $\operatorname{dom} f$. In the convex case, however, a more convenient representation appears in terms of $\operatorname{epi} f$.

2.49 Theorem (convex piecewise linear functions). *A proper function f is both convex and piecewise linear if and only if $\operatorname{epi} f$ is polyhedral.*

In general for a function $f : I\!R^n \to \overline{I\!R}$, the set $\operatorname{epi} f \subset I\!R^{n+1}$ is polyhedral if and only if f has a representation of the form

$$f(x) = \begin{cases} \max\{l_1(x),\ldots,l_p(x)\} & \text{when } x \in D, \\ \infty & \text{when } x \notin D, \end{cases}$$

where D is a polyhedral set in $I\!R^n$ and the functions l_i are affine on $I\!R^n$; here $p = 0$ is admitted and interpreted as giving $f(x) = -\infty$ when $x \in D$. Any function f of this type is convex and lsc, in particular.

Proof. The case of $f \equiv \infty$ being trivial, we can suppose $\operatorname{dom} f$ and $\operatorname{epi} f$ to be nonempty. Of course when $\operatorname{epi} f$ is polyhedral, $\operatorname{epi} f$ is in particular convex and closed, so that f is convex and lsc (cf. 2.4 and 1.6).

To say that $\operatorname{epi} f$ is polyhedral is to say that $\operatorname{epi} f$ can be expressed as the set of points $(x,\alpha) \in I\!R^n \times I\!R$ satisfying a finite system of linear inequalities (cf. 2.10 and the details after it); we can write this system as

$$\gamma_i \ge \langle (c_i,\eta_i),(x,\alpha)\rangle = \langle c_i, x\rangle + \eta_i\alpha \ \text{ for } i = 1,\ldots,m$$

for certain vectors $c_i \in I\!R^n$ and scalars $\eta_i \in I\!R$ and $\gamma_i \in I\!R$. Because $\operatorname{epi} f \ne \emptyset$, and $(x,\alpha') \in \operatorname{epi} f$ whenever $(x,\alpha) \in \operatorname{epi} f$ and $\alpha' \ge \alpha$, it must be true that $\eta_i \le 0$ for all i. Arrange the indices so that $\eta_i < 0$ for $i = 1,\ldots,p$ but $\eta_i = 0$ for $i = p+1,\ldots,m$. (Possibly $p = 0$, i.e., $\eta_i = 0$ for all i.) Taking $b_i = c_i/|\eta_i|$ and $\beta_i = \gamma_i/|\eta_i|$ for $i = 1,\ldots p$, we get $\operatorname{epi} f$ expressed as the set of points

$(x, \alpha) \in \mathbb{R}^n \times \mathbb{R}$ satisfying

$$\langle b_i, x\rangle - \beta_i \leq \alpha \text{ for } i = 1, \ldots, p, \quad \langle c_i, x\rangle \leq \gamma_i \text{ for } i = p+1, \ldots, m.$$

This gives a representation of f of the kind specified in the theorem, with D the polyhedral set in $\mathbb{R}^n$ defined by the constraints $\langle c_i, x\rangle \leq \gamma_i$ for $i = p+1, \ldots, m$ and l_i the affine function with formula $\langle b_i, x\rangle - \beta_i$ for $i = 1, \ldots, p$.

Under this kind of representation, if f is proper (i.e., $p \neq 0$), $f(x)$ is given by $\langle b_k, x\rangle - \beta_k$ on the set C_k consisting of the points $x \in D$ such that

$$\langle b_i, x\rangle - \beta_i \leq \langle b_k, x\rangle - \beta_k \text{ for } i = 1 \ldots, p, \ i \neq k.$$

Clearly C_k is polyhedral and $\bigcup_{k=1}^p C_k = D$, so f is piecewise linear, cf. 2.47.

Suppose now on the other hand that f is proper, convex and piecewise linear. Then $\operatorname{dom} f = \bigcup_{k=1}^r C_k$ with C_k polyhedral, and for each k there exist $a_k \in \mathbb{R}^n$ and $\alpha_k \in \mathbb{R}$ such that $f(x) = \langle a_k, x\rangle - \alpha_k$ for all $x \in C_k$. Let

$$E_k := \big\{(x, \alpha) \,\big|\, x \in C_k, \ \langle a_k, x\rangle - \alpha_k \leq \alpha < \infty\big\}.$$

Then $\operatorname{epi} f = \bigcup_{k=1}^r E_k$. Each set E_k is polyhedral, since an expression for C_k by a system of linear constraints in $\mathbb{R}^n$ readily extends to one for E_k in $\mathbb{R}^{n+1}$. The union of the polyhedral sets E_k is the convex set $\operatorname{epi} f$. The fact that $\operatorname{epi} f$ must then itself be polyhedral is asserted by the next lemma, which we state separately for its independent interest. □

2.50 Lemma (polyhedral convexity of unions). *If a convex set C is the union of a finite collection of polyhedral sets C_k, it must itself be polyhedral.*

Moreover if $\operatorname{int} C \neq \emptyset$, the sets C_k with $\operatorname{int} C_k = \emptyset$ are superfluous in the representation. In fact C can then be given a refined expression as the union of a finite collection of polyhedral sets $\{D_j\}_{j \in J}$ such that

(a) *each set D_j is included in one of the sets C_k,*

(b) $\operatorname{int} D_j \neq \emptyset$, *so* $D_j = \operatorname{cl}(\operatorname{int} D_j)$,

(c) $\operatorname{int} D_{j_1} \cap \operatorname{int} D_{j_2} = \emptyset$ *when* $j_1 \neq j_2$.

Proof. Let's represent the sets C_k for $k = 1, \ldots, r$ in terms of a single family of affine functions $l_i(x) = \langle a_i, x\rangle - \alpha_i$ indexed by $i = 1, \ldots, m$: for each k there is a subset I_k of $\{1, \ldots, m\}$ such that $C_k = \big\{x \,\big|\, l_i(x) \leq 0 \text{ for all } i \in I_k\big\}$. Let I denote the set of indices $i \in \{1, \ldots, m\}$ such that $l_i(x) \leq 0$ for all $x \in C$. We'll prove that

$$C = \big\{x \,\big|\, l_i(x) \leq 0 \text{ for all } i \in I\big\}.$$

Trivially the set on the right includes C, so our task is to demonstrate that the inclusion cannot be strict. Suppose $\bar{x}$ belongs to the set on the right but not to C. We'll argue this to a contradiction.

For each index $k \in \{1, \ldots, r\}$ let C'_k be the set of points $x \in C$ such that the line segment $[x, \bar{x}]$ meets C_k, this set being closed because C_k is closed. For each $x \in C$, a set which itself is closed (because it is the union of finitely many closed sets C_k), let $K(x)$ denote the set of indices k such that $x \in C'_k$.

Select a point $\hat{x} \in C$ for which the number of indices in $K(\hat{x})$ is as low as possible. Any point x in $C \cap [\hat{x}, \bar{x}]$ has $K(x) \subset K(\hat{x})$ by the definition of $K(x)$ and consequently $K(x) = K(\hat{x})$ by the minimality of $K(\hat{x})$. We can therefore suppose (by moving to the 'last' point that still belongs to C, if necessary) that the segment $[\hat{x}, \bar{x}]$ touches C only at $\hat{x}$.

For any k, the set of $x \in C$ with $k \notin K(x)$ is complementary to C'_k and therefore open relative to C. Hence the set of $x \in C$ satisfying $K(x) \subset K(\hat{x})$ is open relative to C. The minimality of $K(\hat{x})$ forbids strict inclusion, so there has to be a neighborhood V of $\hat{x}$ such that $K(x) = K(\hat{x})$ for all $x \in C \cap V$.

Take any index $k_0 \in K(\hat{x})$. Because $\hat{x}$ is the only point of $[\hat{x}, \bar{x}]$ in C, it must also be the only point of $[\hat{x}, \bar{x}]$ in C_{k_0}, and accordingly there must be an index $i_0 \in I_{k_0}$ such that $l_{i_0}(\hat{x}) = 0 < l_{i_0}(\bar{x})$. For each $x \in C \cap V$ the line segment $[x, \bar{x}]$ meets C_{k_0} and thus contains a point x' satisfying $l_{i_0}(x') \leq 0$ as well as the point $\bar{x}$ satisfying $l_{i_0}(\bar{x}) > 0$, so necessarily $l_{i_0}(x) \leq 0$ (because l_{i_0} is affine). More generally then, for any $x \neq \hat{x}$ in C (but not necessarily in V) the line segment $[x, \hat{x}]$, which lies entirely in C by convexity, contains a point $x^* \neq \hat{x}$ in V, therefore satisfying $l_{i_0}(x^*) \leq 0$. Since $l_{i_0}(\hat{x}) = 0$, this implies $l_{i_0}(x) \leq 0$ (again because l_{i_0} is affine). Thus, i_0 is an index with the property that $C \subset \{x \mid l_{i_0}(x) \leq 0\}$, or in other words, $i_0 \in I$. But since $l_{i_0}(\bar{x}) > 0$ this contradicts the choice of $\bar{x}$ as satisfying $l_i(\bar{x}) \leq 0$ for every $i \in I$.

In passing now to the case where $\operatorname{int} C \neq \emptyset$, there's no loss of generality in supposing that none of the affine functions l_i is a constant function. Then

$$\operatorname{int} C_k = \{x \mid l_i(x) < 0 \text{ for } i \in I_k\}, \qquad \operatorname{int} C = \{x \mid l_i(x) < 0 \text{ for } i \in I\}.$$

Let K_0 denote the set of indices k such that $\operatorname{int} C_k \neq \emptyset$, and K_1 the set of indices k such that $\operatorname{int} C_k = \emptyset$. We wish to show next that $C = \bigcup_{k \in K_0} C_k$. It will be enough to show that $\operatorname{int} C \subset \bigcup_{k \in K_0} C_k$, since $C \supset \bigcup_{k \in K_0} C_k$ and $C = \operatorname{cl}(\operatorname{int} C)$ (cf. 2.33). If $\operatorname{int} C \not\subset \bigcup_{k \in K_0} C_k$, the open set $\operatorname{int} C \setminus \bigcup_{k \in K_0} C_k \neq \emptyset$ would be contained in $\bigcup_{k \in K_1} C_k$, the union of the sets with empty interior. This is impossible for the following reason. If the latter union had nonempty interior, there would be a minimal index set $K \subset K_1$ such that $\operatorname{int} \bigcup_{k \in K} C_k \neq \emptyset$. Then K would have to be more than a singleton, and for any $k^* \in K$ the open set $\left[\operatorname{int} \bigcup_{k \in K} C_k\right] \setminus \bigcup_{k \in K \setminus \{k^*\}} C_k$ would be nonempty and included in C_{k^*}, in contradiction to $\operatorname{int} C_{k^*} = \emptyset$.

To construct a refined representation meeting conditions (a), (b), and (c), consider as index elements j the various partitions of $\{1, \ldots, m\}$; each partition j can be identified with a pair (I_+, I_-), where I_+ and I_- are disjoint subsets of $\{1, \ldots, m\}$ whose union is all of $\{1, \ldots, m\}$. Associate with such each partition j the set D_j consisting of all the points x (if any) such that

$$l_i(x) \leq 0 \text{ for all } i \in I_-, \qquad l_i(x) \geq 0 \text{ for all } i \in I_+.$$

Let J_0 be the index set consisting of the partitions $j = (I_+, I_-)$ such that $I_- \supset I_k$ for some k, so that $D_j \subset C_k$. Since every $x \in C$ belongs to at least one C_k, it also belongs to a set D_j for at least one $j \in J_0$. Thus, C is the union of the sets D_j for $j \in J_0$. Hence also, by the argument given earlier, it is the union of the ones with $\operatorname{int} D_j \neq \emptyset$, the partitions j with this property comprising a possibly smaller index set J within J_0. For each $j \in J$ the nonempty set $\operatorname{int} D_j$ consists of the points x satisfying

$$l_i(x) < 0 \text{ for all } i \in I_-, \qquad l_i(x) > 0 \text{ for all } i \in I_+.$$

For any two different partitions j_1 and j_2 in J there must be some $i^* \in \{1, \ldots, m\}$ such that one of the sets D_{j_1} and D_{j_2} is contained in the open half-space $\{x \mid l_{i^*}(x) < 0\}$, while the other is contained in the open half-space $\{x \mid l_{i^*}(x) > 0\}$. Hence $\operatorname{int} D_{j_1} \cap \operatorname{int} D_{j_2} = \emptyset$ when $j_1 \neq j_2$. The collection $\{D_j\}_{j \in J}$ therefore meets all the stipulations. □

J.* Other Examples

For any convex function f, the sets $\operatorname{lev}_{\leq \alpha} f$ are convex, as seen in 2.7. But a nonconvex function can have that property as well; e.g. $f(x) = \sqrt{|x|}$.

2.51 Exercise (functions with convex level sets). *For a differentiable function $f : \mathbb{R}^n \to \mathbb{R}$, the sets $\operatorname{lev}_{\leq \alpha} f$ are convex if and only if*

$$\langle \nabla f(x_0), x_0 - x_1 \rangle \geq 0 \text{ whenever } f(x_1) \leq f(x_0).$$

Furthermore, if such a function f is twice differentiable at $\bar{x}$ with $\nabla f(\bar{x}) = 0$, then $\nabla^2 f(\bar{x})$ must be positive-semidefinite.

The convexity of the log-exponential function in 2.16 is basic to the treatment of a much larger class of functions often occurring in applications, especially in problems of engineering design. These functions are described next.

2.52 Example (posynomials). *A posynomial in the variables $y_1, \ldots, y_n$ is an expression of the form*

$$g(y_1, \ldots, y_n) = \sum\nolimits_{i=1}^{r} c_i y_1^{a_{i1}} y_2^{a_{i2}} \cdots y_n^{a_{in}}$$

where (1) only positive values of the variables y_j are admitted, (2) all the coefficients c_i are positive, but (3) the exponents a_{ij} can be arbitrary real numbers. Such a function may be far from convex, yet convexity can be achieved through a logarithmic change of variables. Setting $b_i = \log c_i$ and

$$f(x_1, \ldots, x_n) = \log g(y_1, \ldots, y_n) \text{ for } x_j = \log y_j,$$

one obtains a convex function $f(x) = \operatorname{logexp}(Ax + b)$ defined for all $x \in \mathbb{R}^n$, where b is the vector in $\mathbb{R}^r$ with components b_i, and A is the matrix in $\mathbb{R}^{r \times n}$ with components a_{ij}. Still more generally, any function

$$g(y) = g_1(y)^{\lambda_1} \cdots g_p(y)^{\lambda_p} \text{ with } g_k \text{ posynomial, } \lambda_k > 0,$$

can be converted by the logarithmic change of variables into a convex function of form $f(x) = \sum_{k=1}^{p} \lambda_k \operatorname{logexp}(A_k x + b_k)$.

Detail. The convexity of f follows from that of logexp (in 2.16) and the composition rule in 2.20(a). The integer r is called the *rank* of the posynomial. When $r = 1$, f is merely an affine function on $\mathbb{R}^n$. □

The wide scope of example 2.52 is readily appreciated when one considers for instance that the following expression is a posynomial:

$$g(y_1, y_2, y_3) = 5y_1^7/y_3^4 + \sqrt{y_2 y_3} = 5y_1^7 y_2^0 y_3^{-4} + y_1^0 y_2^{1/2} y_3^{1/2} \text{ for } y_i > 0.$$

The next example illustrates the approach one can take in using the derivative conditions in 2.14 to verify the convexity of a function defined on more than just an open set.

2.53 Example (weighted geometric means). *For any choice of weights $\lambda_j > 0$ satisfying $\lambda_1 + \cdots + \lambda_n \leq 1$ (for instance $\lambda_j = 1/n$ for all j), one gets a proper, lsc, convex function f on $\mathbb{R}^n$ by defining*

$$f(x) = \begin{cases} -x_1^{\lambda_1} x_2^{\lambda_2} \cdots x_n^{\lambda_n} & \text{for } x = (x_1, \ldots, x_n) \text{ with } x_j \geq 0, \\ \infty & \text{otherwise.} \end{cases}$$

Detail. Clearly f is finite and continuous relative to $\operatorname{dom} f$, which is nonempty, closed, and convex. Hence f is proper and lsc. On $\operatorname{int}(\operatorname{dom} f)$, which consists of the vectors $x = (x_1, \ldots, x_n)$ with $x_k > 0$, the convexity of f can be verified through condition 2.14(c) and the calculation that

$$\langle z, \nabla^2 f(x) z \rangle = |f(x)| \left[\sum\nolimits_{j=1}^{n} \lambda_j (z_j/x_j)^2 - \left(\sum\nolimits_{j=1}^{n} \lambda_j (z_j/x_j) \right)^2 \right];$$

this expression is nonnegative by Jensen's inequality 2.2(b) as applied to the function $\theta(t) = t^2$. (Specifically, $\theta\big(\sum_{j=0}^{n} \lambda_j t_j\big) \leq \sum_{j=0}^{n} \lambda_j \theta(t_j)$ in the case of $t_j = z_j/x_j$ for $j = 1, \ldots, n$, $t_0 = 0$, $\lambda_0 = 1 - \sum_{j=1}^{n} \lambda_j$.) The convexity of f relative to all of $\operatorname{dom} f$ rather than merely $\operatorname{int}(\operatorname{dom} f)$, is obtained then by taking limits in the convexity inequality. □

A number of interesting convexity properties hold for spaces of matrices. The space $\mathbb{R}^{n \times n}$ of all square matrices of order n is conveniently treated in terms of the inner product

$$\langle A, B \rangle := \sum_{i,j=1}^{n,n} a_{ij} b_{ji} = \operatorname{tr} AB, \tag{2(12)}$$

where $\operatorname{tr} C$ denotes the *trace* of a matrix $C \in \mathbb{R}^{n \times n}$, which is the sum of the diagonal elements of C. (The rule that $\operatorname{tr} AB = \operatorname{tr} BA$ is obvious from this inner product interpretation.) Especially important is

$$\mathbb{R}^{n \times n}_{\text{sym}} := \text{ space of symmetric real matrices of order } n, \tag{2(13)}$$

which is a linear subspace of $\mathbb{R}^{n\times n}$ having dimension $n(n+1)/2$. This can be treated as a Euclidean vector space in its own right relative to the trace inner product 2(12). In $\mathbb{R}^{n\times n}_{\rm sym}$, the positive-semidefinite matrices form a closed, convex set whose interior consists of the positive-definite matrices.

2.54 Exercise (eigenvalue functions). *For each matrix $A \in \mathbb{R}^{n\times n}_{\rm sym}$ denote by $\operatorname{eig} A = (\lambda_1, \ldots, \lambda_n)$ the vector of eigenvalues of A in descending order (with eigenvalues repeated according to their multiplicity). Then for $k = 1, \ldots, n$ one has the convexity on $\mathbb{R}^{n\times n}_{\rm sym}$ of the function*

$$\Lambda_k(A) := \text{ sum of the first } k \text{ components of } \operatorname{eig} A.$$

Guide. Show that $\Lambda_k(A) = \max_{P\in\mathcal{P}_k} \operatorname{tr}[PAP]$, where $\mathcal{P}_k$ is the set of all matrices $P \in \mathbb{R}^{n\times n}_{\rm sym}$ such that P has rank k and $P^2 = P$. (These matrices are the orthogonal projections of $\mathbb{R}^n$ onto its linear subspaces of dimension k.) Argue in terms of diagonalization. Note that $\operatorname{tr}[PAP] = \langle A, P\rangle$. □

2.55 Exercise (matrix inversion as a gradient mapping). *On $\mathbb{R}^{n\times n}_{\rm sym}$, the function*

$$j(A) := \begin{cases} \log(\det A) & \text{if } A \text{ is positive-definite,} \\ -\infty & \text{if } A \text{ is not positive-definite,} \end{cases}$$

is concave and usc, and, where finite, differentiable with gradient $\nabla j(A) = A^{-1}$.

Guide. Verify first that for any symmetric, positive-definite matrix C one has $\operatorname{tr} C - \log(\det C) \geq n$, with strict inequality unless $C = I$; for this consider a diagonalization of C. Next look at arbitrary symmetric, positive-definite matrices A and B, and by taking $C = B^{1/2}AB^{1/2}$ establish that

$$\log(\det A) + \log(\det B) \leq \langle A, B\rangle - n$$

with equality holding if and only if $B = A^{-1}$. Show that in fact

$$j(A) = \inf_{B\in\mathbb{R}^{n\times n}_{\rm sym}} \big\{\langle A, B\rangle - j(B) - n\big\} \text{ for all } A \in \mathbb{R}^{n\times n}_{\rm sym},$$

where the infimum is attained if and only if A is positive-definite and $B = A^{-1}$. Using the differentiability of j at such A, along with the inequality $j(A') \leq \langle A', A^{-1}\rangle - j(A^{-1}) - n$ for all A', deduce then that $\nabla j(A) = A^{-1}$. □

The inversion mapping for symmetric matrices has still other remarkable properties with respect to convexity. To describe them, we'll use for matrices A, $B \in \mathbb{R}^{n\times n}_{\rm sym}$ the notation

$$A \succeq B \iff A - B \text{ is positive-semidefinite.} \qquad 2(14)$$

2.56 Proposition (convexity property of matrix inversion). *Let P be the open, convex subset of $\mathbb{R}^{n\times n}_{\rm sym}$ consisting of the positive-definite matrices, and let $J : P \to P$ be the inverse-matrix mapping: $J(A) = A^{-1}$. Then J is convex with respect to $\succeq$ in the sense that*

$$J\big((1-\lambda)A_0 + \lambda A_1\big) \preceq (1-\lambda)J(A_0) + \lambda J(A_1)$$

for all A_0, $A_1 \in P$ and $\lambda \in (0,1)$. In addition, J is order-inverting:

$$A_0 \preceq A_1 \iff A_0^{-1} \succeq A_1^{-1}.$$

Proof. We know from linear algebra that any two positive-definite matrices A_0 and A_1 in $\mathbb{R}^{n\times n}_{\text{sym}}$ can be diagonalized simultaneously through some choice of a basis for $\mathbb{R}^n$. We can therefore suppose without loss of generality that A_0 and A_1 are diagonal, in which case the assertions reduce to properties that are to hold component by component along the diagonal, i.e. in terms of the eigenvalues of the two matrices. We come down then to the one-dimensional case: for the mapping $J:(0,\infty)\to(0,\infty)$ defined by $J(\alpha)=\alpha^{-1}$, is J convex and order-inverting? The answer is evidently yes. □

2.57 Exercise (convex functions of positive-definite matrices). *On the convex subset P of $\mathbb{R}^{n\times n}_{\text{sym}}$ consisting of the positive-definite matrices, the relations*

$$f(A) = g(A^{-1}), \qquad g(B) = f(B^{-1}),$$

give a one-to-one correspondence between the convex functions $f: P \to \overline{\mathbb{R}}$ that are nondecreasing (with $f(A_0) \le f(A_1)$ if $A_0 \preceq A_1$) and the convex functions $g: P \to \overline{\mathbb{R}}$ that are nonincreasing (with $g(B_0) \ge g(B_1)$ if $B_0 \preceq B_1$). In particular, for any subset S of $\mathbb{R}^{n\times n}_{\text{sym}}$ the function

$$g(B) = \sup_{C\in S} \operatorname{tr}\big[CB^{-1}C^*\big] = \sup_{C\in S} \big\langle C^*C, B^{-1}\big\rangle \qquad 2(15)$$

is convex on P and nonincreasing.

Guide. This relies on the properties of the inversion mapping J in 2.56, composing it with f and g in analogy with 2.20(c). The example at the end corresponds to $f(A) = \sup_{C\in S} \operatorname{tr}[CAC^*]$. □

Functions of the kind in 2(15) are important in theoretical statistics. As a special case of 2.57, we see from 2.54 that the function $g(B) = \Lambda_k(B^{-1})$, giving the sum of the first k eigenvalues of B^{-1}, is convex and nonincreasing on the set of positive-definite matrices $B \in \mathbb{R}^{n\times n}_{\text{sym}}$. This corresponds to taking S to be the set of all projection matrices of rank k.

Commentary

The role of convexity in inspiring many of the fundamental strategies and ideas of variational analysis, such as the emphasis on extended-real-valued functions and the geometry of their epigraphs, has been described at the end of Chapter 1. The lecture notes of Fenchel [1951] were instrumental, as were the cited works of Moreau and Rockafellar that followed.

Extensive references to early developments in convex analysis and their historical background can be found in the book of Rockafellar [1970a]. The two-volume opus of Hiriart-Urruty and Lemaréchal [1993] provides a more recent perspective on the subject and its relevance to optimization. The book of Ekeland and Temam [1974] has long served as a source for extensions to infinite-dimensional spaces and applications to variational problems related to partial differential equations, as have the lecture notes of Moreau [1967].

Geometry in its basic sense provided much of the original motivation for studying convex sets, as evidenced by the classical works of Minkowski [1910], [1911]. From another angle, however, convexity has long been seen as vital for understanding the topologies of infinite-dimensional vector spaces, again in part through the contributions of Minkowski in associating gauge functions and norms with convex sets and in developing polar duality, such as will be discussed in Chapter 11 (see 11.19). Oddly, though, convex functions received relatively little attention in that context, apart from the case of norms. For many years, results pertaining to the differentiability properties of convex functions, for instance, were developed and presented almost exclusively in the case of norms and seen mainly as an adjunct to the study of Banach space geometry instead of being viewed more broadly, which would easily have been possible. Little was made of the characterization of the convexity of a function by the convexity of its epigraph, which provides such a strong bridge between convex geometry and analysis. Perhaps this was because the geometry of the time was focused on bounded sets, whereas epigraphs are inherently unbounded.

An example of this conceptual limitation is furnished by the famous Hahn-Banach theorem, which in its traditional formulation says that if a linear function on a linear subspace is bounded from above by some multiple of a given norm, it can be extended to the whole space while preserving that bound. Actually, this is nothing more than a very special case of the separation theorem for convex sets (cf. Theorem 2.39 in the finite-dimensional setting) as applied to the epigraph of the norm function. For applications nowadays, separation properties are much more important than extensions of linear functions. Yet, textbooks on functional analysis in linear spaces often persist in giving only the traditional version of the Hahn-Banach theorem rather than its more potent geometric counterpart.

The earliest version of the separation theorem in 2.39, due to Minkowski [1911], concerned a pair of disjoint convex sets, one of which is bounded. The general version in 2.45(e), using relative interiors, was developed by Fenchel [1951]. The equivalence of Fenchel's condition with $0 \notin \operatorname{rint}(C_1 - C_2)$ in 2.45(d) was first utilized in Rockafellar [1970a]. The related condition $0 \notin \operatorname{int}(C_1 - C_2)$ has ties especially to infinite-dimensional theory; see Rockafellar [1974a]. For further refinements in the subject of separation, see Klee [1968].

The usefulness of the classical concept of 'relative interior' for convex sets is one of the basic features distinguishing finite-dimensional from infinite-dimensional convexity. The key is the fact that every nonempty convex set in $\mathbb{R}^n$ lies in the closure of its relative interior (cf. 2.40). The algebra of the 'rint' operation was developed by Rockafellar [1970a].

The upper bound of $n+1$ points in Carathéodory's theorem 2.29 for sets $C \subset \mathbb{R}^n$ is primarily useful for compactness arguments which require that only a fixed number of points come into play at any one time. For the fact that n points suffice when the set C is connected, see Bonnesen and Fenchel [1934]. Actually this result holds as long as C is the union of at most n connected sets; cf. Hanner and Rådström [1951].

The results in 2.26(b) about the Moreau envelopes of convex functions and the expression of proximal mappings as gradients come essentially from Moreau [1965].

In speaking of polyhedral sets, we make a distinction between such sets and 'polyhedra' in the sense developed in topology in reference to certain unions of simplices. A polyhedral set is a *convex* polyhedron. Accordingly, many authors use 'polyhedral convex set' for this concept, but 'polyhedral set' is simpler and has the advantage that, as an independent term not employed elsewhere, it can be defined directly as in 2.10 with respect to systems of linear constraints. In contrast, one isn't really authorized to speak of such a set as a convex polyhedron without first demonstrating that it fits the topological definition of a polyhedron, which requires the serious machinery of the Minkowski-Weyl theorem—see 3.52 and 3.54. On the other hand, with polyhedral sets defined simply and directly, one can equally well speak of general polyhedra as *piecewise polyhedral* sets.

Convex functions with polyhedral epigraphs were studied by Rockafellar [1970a], but their characterization in 2.49 in terms of the definition of piecewise linearity in 2.47 has not previously been made explicit.

Functions f for which the sets $\mathrm{lev}_{\leq\alpha} f$ are convex, as in 2.51, are called *quasiconvex*. This usage dates back many decades in game theory and economics, but an entirely different meaning for 'quasiconvexity' prevails in the branch of variational calculus related to partial differential equations. That meaning, due to Morrey [1952], refers instead to possibly nonconvex functions f on $\mathbb{R}^n$ that, as 'integrands', yield convex integral functionals on certain Sobolev spaces.

The 'posynomials' in 2.52 were introduced to optimization theory by Duffin, Peterson and Zener [1967] for the promotion of various applications in engineering. The conversion of posynomials to convex functions through a logarithmic change of variables (and the convexity of the log-exponential function in 2.16) was subsequently demonstrated by Rockafellar [1970a].

The class of all convex functions $f(A)$ on $\mathbb{R}^{n\times n}_{\mathrm{sym}}$ that, like the functions $\Lambda_k(A)$ in 2.54, depend only on the eigenvalues of A was characterized by Davis [1957]. Additional results in this direction may be found in Lewis [1996]. The log-determinant function in 2.55, which falls into this category as well, has recently gained significance in the development of interior-point methods for solving optimization problems that concern matrices. For a survey of that subject see Boyd and Vandenberghe [1995]; for eigenvalue optimization more generally see Lewis and Overton [1996].

3. Cones and Cosmic Closure

An important advantage that the extended real line $\overline{\mathbb{R}}$ has over the real line $\mathbb{R}$ is *compactness*: every sequence of elements has a convergent subsequence. This property is achieved by adjoining to $\mathbb{R}$ the special elements ∞ and $-\infty$, which can act as limits for unbounded sequences under special rules. An analogous compactification is possible for $\mathbb{R}^n$. It serves in characterizing basic 'growth' properties that sets and functions may have in the large.

Every vector $x \neq 0$ in $\mathbb{R}^n$ has both *magnitude* and *direction*. The magnitude of x is $|x|$, which can be manipulated in familiar ways. The direction of x has often been underplayed as a mathematical entity, but our interest now lies in a rigorous treatment where directions are viewed as 'points at infinity' to be adjoined to ordinary space.

A. Direction Points

Operationally we wish to think of the direction of x, denoted by $\operatorname{dir} x$, as an attribute associated with x under the rule that $\operatorname{dir} x = \operatorname{dir} x'$ means $x' = \lambda x \neq 0$ for some $\lambda > 0$. The zero vector is regarded as having no direction; $\operatorname{dir} 0$ is undefined. There is a one-to-one correspondence, then, between the various directions of vectors $x \neq 0$ in $\mathbb{R}^n$ and the *rays* in $\mathbb{R}^n$, a ray being a closed half-line emanating from the origin. Every direction can thus be *represented* uniquely by a ray, but it will be preferable to think of directions themselves as abstract points, designated as *direction points*, which lie beyond $\mathbb{R}^n$ and form a set called the *horizon* of $\mathbb{R}^n$, denoted by $\operatorname{hzn} \mathbb{R}^n$.

For $n = 1$ there are only two direction points, symbolized by ∞ and $-\infty$; it's in adding these direction points to $\mathbb{R}$ that $\overline{\mathbb{R}}$ is obtained. We follow the same procedure now in higher dimensions by adding all the direction points in $\operatorname{hzn} \mathbb{R}^n$ to $\mathbb{R}^n$ to form the extended space

$$\operatorname{csm} \mathbb{R}^n := \mathbb{R}^n \cup \operatorname{hzn} \mathbb{R}^n,$$

which will be called the *cosmic closure* of $\mathbb{R}^n$, or *n-dimensional cosmic space*. Our aim is to supply this extended space with geometry and topology so that it can serve as a useful companion to $\mathbb{R}^n$ in questions of analysis, just as $\overline{\mathbb{R}}$ serves alongside of $\mathbb{R}$. (Caution: while $\operatorname{csm} \mathbb{R}^1$ can be identified with $\overline{\mathbb{R}}$, $\operatorname{csm} \mathbb{R}^n$ isn't the product space $\overline{\mathbb{R}}^n$.)

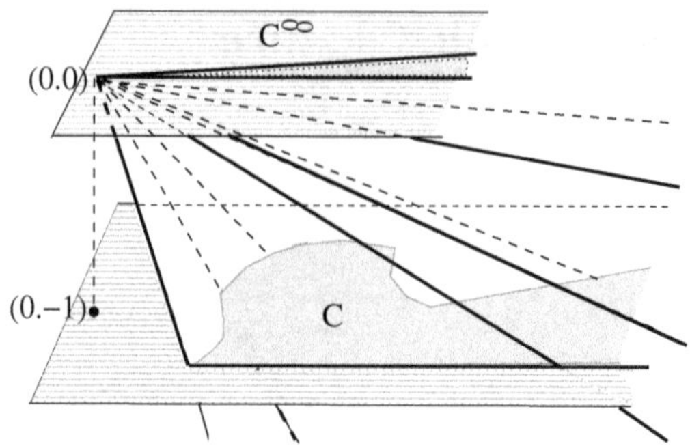

Fig. 3–1. The ray space model for n-dimensional cosmic space.

There are several ways of interpreting the space $\operatorname{csm} \mathbb{R}^n$ geometrically, all of them leading to the same mathematical structure but having different advantages. The simplest is the *celestial model*, in which $\mathbb{R}^n$ is imagined as shrunk down to the interior of the n-dimensional ball $\mathbb{B}$ through the mapping $x \mapsto x/(1+|x|)$, and $\operatorname{hzn} \mathbb{R}^n$ is identified with the surface of this ball. For $n = 3$, this brings to mind the picture of the universe as bounded by a 'celestial sphere'. For $n = 2$, we get a Flatland universe comprised of a closed disk, the edge of which corresponds to the horizon set.

A second approach to setting up the structure of analysis in n-dimensional cosmic space utilizes the *ray space model.* This refers to the representation of $\operatorname{csm} \mathbb{R}^n$ in $\mathbb{R}^n \times \mathbb{R}$ that's illustrated in Figure 3–1. Each $x \in \mathbb{R}^n$ corresponds uniquely to a downward sloping ray in $\mathbb{R}^n \times \mathbb{R}$, namely the one passing through $(x, -1)$. The points in $\operatorname{hzn} \mathbb{R}^n$, on the other hand, correspond uniquely to the 'horizontal' rays in $\mathbb{R}^n \times \mathbb{R}$, which are the ones lying in the hyperplane $(\mathbb{R}^n, 0)$; for the meaning to attach to the set C^∞ see Definition 3.3. This model, although less intuitive than the celestial model and requiring an extra dimension, is superior for many purposes, such as extensions of convexity, and it will turn out also to be the most direct in applications involving generalized differentiation as will be met in Chapter 8. (Such applications, by the way, dictate the choice of downward instead of upward sloping rays; see Theorem 8.9 and the comment after its proof.)

A third approach, which fits between the other two, uses the fact that each ray in the ray space model in $\mathbb{R}^n \times \mathbb{R}$, whether associated with an ordinary point of $\mathbb{R}^n$ or a direction point in $\operatorname{hzn} \mathbb{R}^n$, pierces the closed unit hemisphere

$$H_n := \{(x, \beta) \in \mathbb{R}^n \times \mathbb{R} \,|\, \beta \le 0,\ |x|^2 + \beta^2 = 1\} \qquad 3(1)$$

in a unique point. Thus, H_n furnishes an alternative model of $\operatorname{csm} \mathbb{R}^n$ in which the rim of H_n represents the horizon of $\mathbb{R}^n$. This will be called the *hemispherical model* of $\operatorname{csm} \mathbb{R}^n$.

This framework leads very naturally to an extended form of convergence. In expressing it we'll systematically rely on the notation:

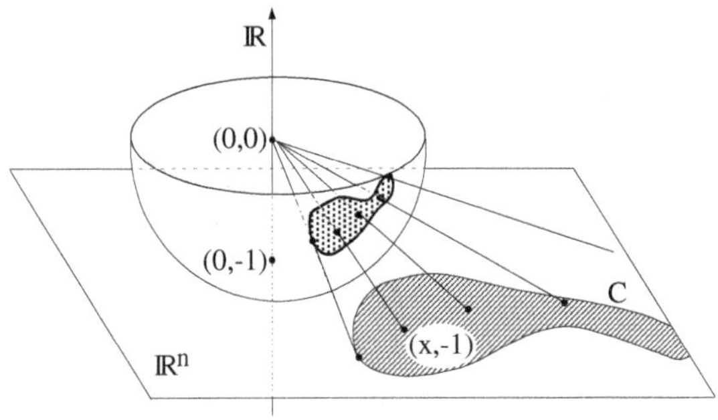

Fig. 3–2. The hemispherical model for n-dimensional cosmic space.

$$\lambda^\nu \searrow 0 \quad \Longleftrightarrow \quad \lambda^\nu \to 0 \text{ with } \lambda^\nu > 0.$$

3.1 Definition (cosmic convergence to direction points). *A sequence of points $x^\nu \in \mathbb{R}^n$ converges to a direction point* $\operatorname{dir} x \in \operatorname{hzn} \mathbb{R}^n$*, written $x^\nu \to \operatorname{dir} x$ (where $x \neq 0$), if $\lambda^\nu x^\nu \to x$ for some choice of $\lambda^\nu \searrow 0$. Likewise, a sequence of direction points $\operatorname{dir} x^\nu \in \operatorname{hzn} \mathbb{R}^n$ converges to a direction point $\operatorname{dir} x \in \operatorname{hzn} \mathbb{R}^n$, written $\operatorname{dir} x^\nu \to \operatorname{dir} x$, if $\lambda^\nu x^\nu \to x$ for some choice of scalars $\lambda^\nu > 0$.*

A mixed sequence of ordinary points and direction points converges to $\operatorname{dir} x$ if every subsequence consisting of ordinary points converges to $\operatorname{dir} x$, and the same holds for every subsequence consisting of direction points.

This definition isn't sensitive to the particular nonzero vectors chosen in representing given direction points in the 'dir' notation. Clearly, a sequence of points in $\operatorname{csm} \mathbb{R}^n$ converges in the extended sense if and only if the corresponding sequence of points in H_n converges in the ordinary sense.

3.2 Theorem (compactness of cosmic space). *Every sequence of points in $\operatorname{csm} \mathbb{R}^n$ (whether ordinary points, direction points or some mixture) has a convergent subsequence (in the cosmic sense). In this setting, the bounded sequences in $\mathbb{R}^n$ are characterized as the sequences of ordinary points such that no cluster point is a direction point.*

Proof. This is immediate from the interpretation in the hemispherical model, since H_n is compact in $\mathbb{R}^n \times \mathbb{R}$. □

Obviously $\operatorname{hzn} \mathbb{R}^n$ is itself compact as a subset of $\operatorname{csm} \mathbb{R}^n$, whereas $\mathbb{R}^n$, its complement, is open as a subset of $\operatorname{csm} \mathbb{R}^n$.

Cosmic convergence can be quantified through the introduction of a metric, which can be done in several ways. One possibility is to start from the Poincaré metric, taking the distance between two points x and y in $\mathbb{R}^n$ to be the Euclidean distance between $x/(1+|x|)$ and $y/(1+|y|)$. Then $\operatorname{csm} \mathbb{R}^n$ can be identified with the completion of $\mathbb{R}^n$. The extension of Poincaré distance to this completion furnishes a metric on $\operatorname{csm} \mathbb{R}^n$. Another possibility is to measure distances between points of $\operatorname{csm} \mathbb{R}^n$ in terms of the distances between the corresponding points of H_n in the hemispherical model, either geodesically or in the Euclidean sense in $\mathbb{R}^{n+1}$. A different tactic is better, however. The cosmic metric $d_{\rm csm}$ that will be adopted in Chapter 4 (see 4.48) is based instead on the ray space model, which is the real workhorse for analysis in $\operatorname{csm} \mathbb{R}^n$.

For a set $C \subset \mathbb{R}^n$ we must distinguish between $\operatorname{csm} C$, the *cosmic closure* of C in which direction points in $\operatorname{csm} \mathbb{R}^n$ are allowed as possible limits, and $\operatorname{cl} C$, the ordinary closure of C in $\mathbb{R}^n$. The collection of all direction point limits will be called the *horizon* of C and denoted by $\operatorname{hzn} C$. Thus,

$$\operatorname{hzn} C = \operatorname{csm} C \cap \operatorname{hzn} \mathbb{R}^n, \qquad \operatorname{csm} C = \operatorname{cl} C \cup \operatorname{hzn} C. \tag{3(2)}$$

The set $\operatorname{hzn} C$, closed within $\operatorname{hzn} \mathbb{R}^n$, furnishes a precise description of any unboundedness exhibited by C, and it will be quite important in what follows.

However, rather than working directly with $\operatorname{hzn} C$ and other subsets of $\operatorname{hzn} \mathbb{R}^n$ in developing specific conditions, we'll generally find it more expedient to work with representations of such sets in terms of rays in $\mathbb{R}^n$. The reason is that such representations lend themselves better to calculations and, when applied later to functions, yield properties more attuned to 'analysis'.

B. Horizon Cones

A set $K \subset \mathbb{R}^n$ is called a *cone* if $0 \in K$ and $\lambda x \in K$ for all $x \in K$ and $\lambda > 0$. Aside from the *zero cone* $\{0\}$, the cones K in $\mathbb{R}^n$ are characterized as the sets expressible as nonempty unions of rays. The set of direction points represented by the rays in K will be denoted by $\operatorname{dir} K$. There is thus a one-to-one correspondence between cones in $\mathbb{R}^n$ and subsets of the horizon of $\mathbb{R}^n$,

$$K \ \text{(cone in } \mathbb{R}^n) \quad \longleftrightarrow \quad \operatorname{dir} K \ \text{(direction set in } \operatorname{hzn} \mathbb{R}^n),$$

where the cone is closed in $\mathbb{R}^n$ if and only if the associated direction set is closed in $\operatorname{hzn} \mathbb{R}^n$. Cones K are said in this manner to *represent* sets of direction points. The zero cone corresponds to the empty set of direction points, while the *full cone* $K = \mathbb{R}^n$ gives the entire horizon.

A general subset of $\operatorname{csm} \mathbb{R}^n$ can be expressed in a unique way as

$$C \cup \operatorname{dir} K$$

for some set $C \subset \mathbb{R}^n$ and some cone $K \subset \mathbb{R}^n$; C gives the *ordinary* part of the set and K the *horizon* part. The corresponding cone in the ray space model in Figure 3–1 is then $\{(\lambda x, -\lambda) \mid x \in C, \lambda > 0\} \cup \{(x, 0) \mid x \in K\}$. We'll typically express properties of $C \cup \operatorname{dir} K$ in terms of properties of the pair C and K. Results in which such a pair of sets appears, with K a cone, will thus be one of the main vehicles for statements about analysis in $\operatorname{csm} \mathbb{R}^n$.

In dealing with possibly unbounded subsets C of $\mathbb{R}^n$, the cone K that represents $\operatorname{hzn} C$ interests us especially.

3.3 Definition (horizon cones). *For a set $C \subset \mathbb{R}^n$, the horizon cone is the closed cone $C^\infty \subset \mathbb{R}^n$ representing the direction set* $\operatorname{hzn} C$, *so that*

$$\operatorname{hzn} C = \operatorname{dir} C^\infty, \qquad \operatorname{csm} C = \operatorname{cl} C \cup \operatorname{dir} C^\infty,$$

this cone being given therefore by

$$C^\infty = \begin{cases} \{x \mid \exists\, x^\nu \in C,\ \lambda^\nu \searrow 0,\ \textit{with}\ \lambda^\nu x^\nu \to x\} & \textit{when } C \neq \emptyset, \\ \{0\} & \textit{when } C = \emptyset. \end{cases}$$

Note that $(\operatorname{cl} C)^\infty = C^\infty$ always. If C happens itself to be a cone, C^∞ coincides with $\operatorname{cl} C$. The rule that $C^\infty = \{0\}$ when $C = \emptyset$ isn't merely a convention, but is dictated by the principle of having $\operatorname{hzn} C = \operatorname{dir} C^\infty$ and $\operatorname{csm} C = \operatorname{cl} C \cup \operatorname{dir} C^\infty$, since the empty set of direction points is represented by the cone $K = \{0\}$.

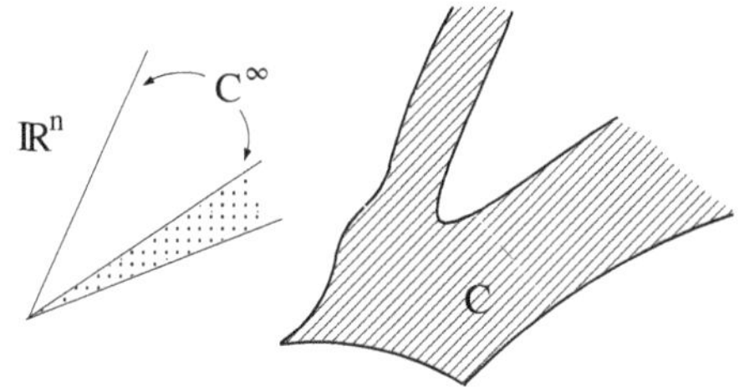

Fig. 3–3. The horizon cone associated with an unbounded set.

3.4 Exercise (cosmic closedness). *A subset of* $\operatorname{csm} \mathbb{R}^n$, *written as* $C \cup \operatorname{dir} K$ *for a set $C \subset \mathbb{R}^n$ and a cone $K \subset \mathbb{R}^n$, is closed in the cosmic sense if and only if C and K are closed in $\mathbb{R}^n$ and $C^\infty \subset K$. In general, its cosmic closure is*

$$\operatorname{csm}(C \cup \operatorname{dir} K) = \operatorname{cl} C \cup \operatorname{dir}(C^\infty \cup \operatorname{cl} K).$$

3.5 Theorem (horizon criterion for boundedness). *A set $C \subset \mathbb{R}^n$ is bounded if and only if its horizon cone is just the zero cone: $C^\infty = \{0\}$.*

Proof. A set is unbounded if and only if it contains an unbounded sequence. Equivalently by the facts in 3.2, a set is bounded if and only if $\operatorname{hzn} C = \emptyset$. Since C^∞ is the cone representing $\operatorname{hzn} C$, this means that $C^\infty = \{0\}$. □

Of course, the equivalence in Theorem 3.5 is also evident directly from the limit formula for C^∞ in 3.3. The proof of Theorem 3.5 reminds us, however, of the underlying principles. The formula just serves as an operational expression of the correspondence between horizon cones and horizon sets; the cone that stands for the empty set of directions points is the null cone.

A calculus of horizon cones will soon be developed, and this will make it possible to apply the criterion in Theorem 3.5 to many situations. A powerful tool will be the special horizon properties of convex sets.

3.6 Theorem (horizon cones of convex sets). *For a convex set C the horizon cone C^∞ is convex, and for any point $\bar{x} \in C$ it consists of the vectors w such that $\bar{x} + \tau w \in \operatorname{cl} C$ for all $\tau \geq 0$.*

In particular, when C is closed it is bounded unless it actually includes a half-line $\{\bar{x} + \tau w \mid \tau \geq 0\}$ from some point $\bar{x}$, and it must then also include all parallel half-lines $\{x + \tau w \mid \tau \geq 0\}$ emanating from other points $x \in C$.

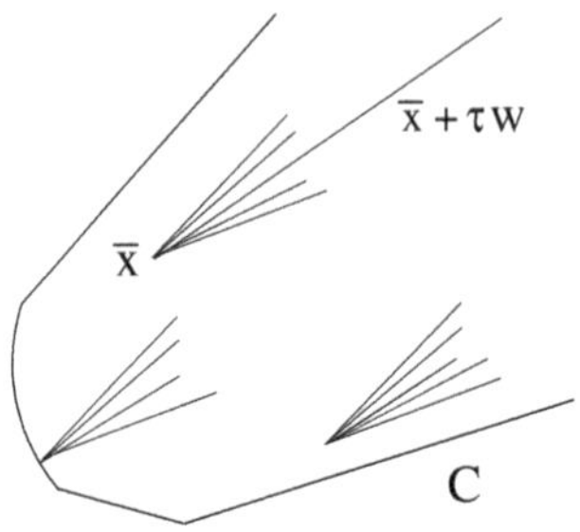

Fig. 3–4. Half-line characterization of horizon cones of convex sets.

Proof. There is nothing to prove when $C = \emptyset$, so suppose $C \neq \emptyset$ and fix any point $\bar{x} \in C$. If w has the property that $\bar{x} + \tau w \in \operatorname{cl} C$ for all $\tau \geq 0$, we can find points $x^\nu \in C$ with $\left|(\bar{x} + \nu w) - x^\nu\right| \leq 1/\nu$ for all $\nu \in I\!N$. Then for $\lambda^\nu = 1/\nu$ we have $\lambda^\nu x^\nu \to w$ and can conclude that $w \in C^\infty$. On the other hand, starting from the assumption that $w \in C^\infty$ and taking any $\tau \in (0, \infty)$, we know that $\tau w \in C^\infty$ (because C^∞ is a cone) and hence that there are sequences of points $x^\nu \in C$ and scalars $\lambda^\nu \in (0,1)$ with $\lambda^\nu \searrow 0$ and $\lambda^\nu x^\nu \to \tau w$. Then the points $(1 - \lambda^\nu)\bar{x} + \lambda^\nu x^\nu$ belong to C by convexity and converge to $\bar{x} + \tau w$, which therefore belongs to $\operatorname{cl} C$.

This establishes the criterion for membership in C^∞. The assertion about boundedness is then obvious from Theorem 3.5. The convexity of C^∞ now follows through the convexity of $\operatorname{cl} C$ in 2.32: if both $\bar{x} + \tau w_0$ and $\bar{x} + \tau w_1$ belong to $\operatorname{cl} C$ for all $\tau > 0$ then for any $\theta \in (0,1)$ the point

$$(1-\theta)[\bar{x} + \tau w_0] + \theta[\bar{x} + \tau w_1] = \bar{x} + \tau\big[(1-\theta)w_0 + \theta w_1\big]$$

likewise belongs to $\operatorname{cl} C$ for all $\tau > 0$, so $(1-\theta)w_0 + \theta w_1$ belongs to C^∞. □

The properties in Theorem 3.6 and Figure 3–4 associate the horizon cone of a convex set very closely with its global 'recession cone', as will be seen in 6.34. Local and global recession cones for convex and nonconvex sets will be defined in 6.33.

3.7 Exercise (convex cones). *For $K \subset I\!R^n$ the following are equivalent:*

(a) *K is a convex cone.*

(b) *K is a cone such that $K + K \subset K$.*

(c) *K is nonempty and contains* $\sum_{i=1}^{p} \lambda_i x_i$ *whenever* $x_i \in K$ *and* $\lambda_i \geq 0$.

Examples of convex cones include all linear subspaces of $\mathbb{R}^n$, in particular the zero cone $K = \{0\}$ and the full cone $K = \mathbb{R}^n$. Half-spaces $\{x \mid \langle a, x\rangle \leq 0\}$ and $\{x \mid \langle a, x\rangle \geq 0\}$ are also in this class, as is the nonnegative orthant $\mathbb{R}^n_+$ in Example 1.2.

3.8 Proposition (convex cones versus subspaces). *For a convex cone* K *in* $\mathbb{R}^n$, *the set* $M = K \cap (-K)$ *is the largest of the linear subspaces* M *such that* $M \subset K$, *while* $M = K - K$ *is the smallest one such that* $M \supset K$. *For* K *itself to be a linear subspace, it is necessary and sufficient that* $K = -K$.

Proof. A set K is a convex cone if and only if it contains the origin and is closed under the operations of addition and nonnegative scalar multiplication, cf. 3.7(c). The only thing lacking for a convex cone to be a linear subspace is the possibility of multiplying by -1. This justifies the last statement of 3.8. It's elementary that the sets $K \cap (-K)$ and $K - K$ are themselves closed under all the operations in question and are, respectively, the largest such set within K and the smallest such set that includes K. □

Some rules will now be developed for determining horizon cones and using them to ascertain the closedness of sets constructed by various operations.

3.9 Proposition (intersections and unions). *For any collection of sets* $C_i \subset \mathbb{R}^n$ *for* $i \in I$, *an arbitrary index set, one has*

$$\Big[\bigcap\nolimits_{i\in I} C_i\Big]^\infty \subset \bigcap\nolimits_{i\in I} C_i^\infty, \qquad \Big[\bigcup\nolimits_{i\in I} C_i\Big]^\infty \supset \bigcup\nolimits_{i\in I} C_i^\infty.$$

The first inclusion holds as an equation for closed, convex sets C_i *having nonempty intersection. The second holds as an equation when* I *is finite.*

Proof. The general facts follow directly from the definition of the horizon cones in question. The special result in the convex case is seen from the characterization in 3.6. □

3.10 Theorem (images under linear mappings). *For a linear mapping* $L : \mathbb{R}^n \to \mathbb{R}^m$ *and a closed set* $C \subset \mathbb{R}^n$, *a sufficient condition for the set* $L(C)$ *to be closed is that* $L^{-1}(0) \cap C^\infty = \{0\}$. *Under this condition one has* $L(C^\infty) = L(C)^\infty$, *although in general only* $L(C^\infty) \subset L(C)^\infty$.

Proof. The condition will be shown first to imply that L carries unbounded sequences in C into unbounded sequences in $\mathbb{R}^m$. If this were not true, there would exist by 3.2 a sequence of points $x^\nu \in C$ converging to a point of hzn $\mathbb{R}^n$, but such that the sequence of images $u^\nu = L(x^\nu)$ is bounded. We would then have scalars $\lambda^\nu \searrow 0$ with $\lambda^\nu x^\nu$ converging to a vector $x \neq 0$. By definition we would have $x \in C^\infty$, but also (because linear mappings are continuous)

$$L(x) = \lim_{\nu\to\infty} L(\lambda^\nu x^\nu) = \lim_{\nu\to\infty} \lambda^\nu u^\nu = 0,$$

which the condition prohibits.

Assuming henceforth that the condition holds, we prove now that $L(C)$ is a closed set. Suppose $u^\nu \to u$ with $u^\nu \in L(C)$, or in other words, $u^\nu = L(x^\nu)$ for some $x^\nu \in C$. The sequence $\{L(x^\nu)\}_{\nu \in \mathbb{N}}$ is bounded, so the sequence $\{x^\nu\}_{\nu \in \mathbb{N}}$ must be bounded as well, for the reasons just given, and it therefore has a cluster point x. We have $x \in C$ (because C is closed) and $u = L(x)$ (again because linear mappings are continuous), and therefore $u \in L(C)$.

The next task is to show that $L(C^\infty) \subset L(C)^\infty$. This is trivial when $C = \emptyset$, so we assume $C \neq \emptyset$. Any $x \in C^\infty$ is in this case the limit of $\lambda^\nu x^\nu$ for some choice of $x^\nu \in C$ and $\lambda^\nu \searrow 0$. The points $u^\nu = L(x^\nu)$ in $L(C)$ then give $\lambda^\nu u^\nu \to L(x)$, showing that $L(x) \in L(C)^\infty$.

For the inclusion $L(C^\infty) \supset L(C)^\infty$, we consider any $u \in L(C)^\infty$, writing it as $\lim_\nu \lambda^\nu u^\nu$ with $u^\nu \in L(C)$ and $\lambda^\nu \searrow 0$. For any choice of $x^\nu \in C$ with $L(x^\nu) = u^\nu$, we have $L(\lambda^\nu x^\nu) \to u$. The boundedness of the sequence $\{L(\lambda^\nu x^\nu)\}$ implies the boundedness of the sequence $\{\lambda^\nu x^\nu\}$, which therefore must have a cluster point x. Then $x \in C^\infty$ and $L(x) = u$, and we conclude that $u \in L(C^\infty)$. □

The hypothesis in Theorem 3.10 is crucial to the conclusions: counterexamples are encountered even when L is a linear projection mapping. Suppose for instance that L is the linear mapping from $\mathbb{R}^2$ into itself given by $L(x_1, x_2) = (x_1, 0)$. If C is the hyperbola defined by the equation $x_1 x_2 = 1$, the cone C^∞ is the union of the x_1-axis and the x_2-axis and doesn't satisfy the condition assumed in 3.10: the x_2-axis gives vectors $x \neq 0$ in C^∞ that project onto 0. The image $L(C)$ in this case consists of the x_1-axis with the origin deleted; it is not closed. For a different example, where only the inclusion $L(C^\infty) \subset L(C)^\infty$ holds, consider the same mapping L but let C be the parabola defined by $x_2 = x_1^2$. Then C^∞ is the nonnegative x_2-axis, and $L(C^\infty) = \{0\}$. But $L(C)$ is the x_1-axis (hence a closed set), and $L(C)^\infty$ is the x_1-axis too.

3.11 Exercise (products of sets). *For sets $C_i \subset \mathbb{R}^{n_i}$, $i = 1, \ldots, m$, one has*

$$(C_1 \times \cdots \times C_m)^\infty \subset C_1^\infty \times \cdots \times C_m^\infty.$$

If every C_i is convex and nonempty, this holds as an equation. It also holds as an equation when no more than one of the nonempty sets C_i is unbounded.

Guide. Derive the inclusion from the definition of the horizon cone. In the convex case, use the characterization provided in 3.6. □

The equality in 3.11 can fail in some situations where convexity is absent. An example with $m = 2$ is obtained by taking $C_1 = C_2 = \{2^k \mid k \in \mathbb{N}\}$ in $\mathbb{R}^1$. One has $C_1^\infty = C_2^\infty = \mathbb{R}_+$, but $(C_1 \times C_2)^\infty \neq \mathbb{R}_+ \times \mathbb{R}_+$. This is because a vector (a, b) belongs to $(C_1 \times C_2)^\infty$ if and only if one has $\lambda^\nu(2^{k^\nu}, 2^{l^\nu}) \to (a, b)$ for some sequence $\lambda^\nu \searrow 0$ and exponents k^ν and l^ν; for $a > 0$ and $b > 0$ this is realizable if and only if a/b is an *integral power of* 2 (the exponent being positive, negative or zero).

3.12 Exercise (sums of sets). *For closed sets $C_i \subset \mathbb{R}^n$, $i = 1, \ldots, m$, a sufficient condition for $C_1 + \cdots + C_m$ to be closed is the nonexistence of vectors $x_i \in C_i^\infty$*

such that $x_1 + \cdots + x_m = 0$, *except with* $x_i = 0$ *for all* i. *Then also*

$$(C_1 + \cdots + C_m)^\infty \subset C_1^\infty + \cdots + C_m^\infty,$$

where the inclusion holds as an equation if the sets $\emptyset \neq C_i$ *are convex, or if at most one of them is unbounded.*

In particular, if C *and* B *are closed sets with* $\emptyset \neq B$ *bounded, then* $C + B$ *is closed with* $(C + B)^\infty = C^\infty$.

Guide. Apply the mapping $L : (x_1, \ldots, x_m) \mapsto x_1 + \cdots + x_m$ to the product set in 3.11, using 3.10. □

The possibility that $C_1 + C_2$ might not be closed, even though C_1 and C_2 themselves are closed, is demonstrated in Figure 3–5. This can't happen if one of the two sets is also bounded, as the last part of 3.12 makes clear.

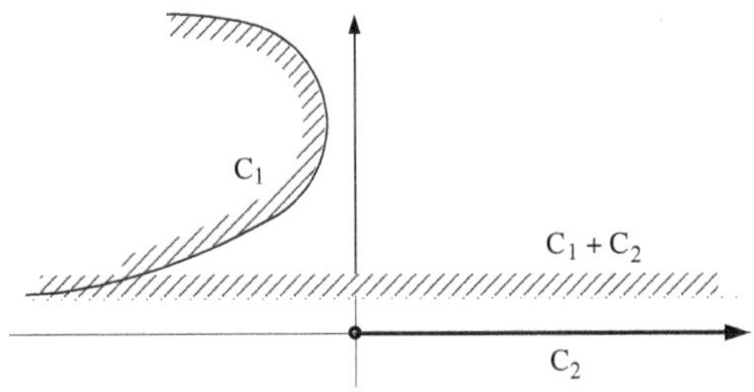

Fig. 3–5. Closed, convex sets whose sum is not closed.

This result can be applied to the convex hull operation, which for cones takes a special form. The following concept of 'pointedness' will be helpful in determining when the convex hull of a closed cone is closed.

3.13 Definition (pointed cones). *A cone* $K \subset \mathbb{R}^n$ *is pointed if the equation* $x_1 + \cdots + x_p = 0$ *is impossible with* $x_i \in K$ *unless* $x_i = 0$ *for all* i.

3.14 Proposition (pointedness of convex cones). *A convex cone* $K \subset \mathbb{R}^n$ *is pointed if and only if* $K \cap -K = \{0\}$.

Proof. This is immediate from Definition 3.13 and the fact that a convex cone contains any sum of its elements (cf. 3.7). □

3.15 Theorem (convex hulls of cones). *For a cone* $K \subset \mathbb{R}^n$, *a vector* x *belongs to* $\operatorname{con} K$ *if and only if it can be expressed as a sum* $x_1 + \cdots + x_p$ *with* $x_i \in K$. *When* $x \neq 0$, *the vectors* x_i *can be taken to be linearly independent (so* $p \leq n$*).*

In particular, $\operatorname{con} K = K + \cdots + K$ *(*n *terms). Moreover, if* K *is closed and pointed, so too is* $\operatorname{con} K$.

Proof. From 2.27, $\operatorname{con} K$ consists of all convex combinations $\lambda_1 x_1 + \cdots + \lambda_p x_p$ of elements $x_i \in K$. But K contains all nonnegative multiples of its elements, so the coefficients λ_i are superfluous; $\operatorname{con} K$ consists of all sums $x_1 + \cdots + x_p$

with $x_i \in K$. Since K is a union of rays emanating from the origin, it is a connected subset of $\mathbb{R}^n$, and therefore by 2.29, one can restrict p to be n.

On the other hand, if x has an expression $x_1 + \cdots + x_p$ with $x_i \in K$ and p minimal, the vectors x_i must be linearly independent, for otherwise there would exist coefficients μ_i, at least one positive, with $\mu_1 x_1 + \cdots + \mu_p x_p = 0$. In this case, supposing without loss of generality that the largest of the positive components is μ_p, and $\mu_p = 1$, we would get $x = x'_1 + \cdots + x'_{p-1}$ for $x'_i = (1 - \mu_i)x_i \in K$, and this would contradict the minimality of p.

The fact that $\operatorname{con} K$ consists of all sums $x_1 + \cdots + x_n$ with $x_i \in K$ means that $\operatorname{con} K$ is the image of K^n (product of n copies of K) under the linear transformation $L : (x_1, \ldots, x_n) \mapsto x_1 + \cdots + x_n$. According to Theorem 3.10, this image is a closed set if K^n is closed (true when K is closed) and $L^{-1}(0)$ meets the horizon cone of K^n, which is K^n itself (because K^n is a cone), only at 0. The latter condition is precisely the condition that K be pointed. Hence $\operatorname{con} K$ is closed when K is closed and pointed.

Moreover, $\operatorname{con} K \cap [-\operatorname{con} K] = \{0\}$ in this case, for otherwise some nonzero sum $x_1 + \cdots + x_p$ of elements of K would equal $-[x'_1 + \cdots + x'_q]$ for certain other elements of K, and then $x_1 + \cdots + x_p + x'_1 + \cdots + x'_q = 0$ for a combination of elements of K which aren't all 0, in contradiction to the pointedness of K. We conclude from 3.14 that $\operatorname{con} K$ is pointed when K is closed and pointed. □

The criterion in 3.10 for closed images under linear mappings has the following extension, which however says nothing about the horizon cone of the image. Other extensions will be developed in Chapter 5 (cf. 5.25, 5.26).

3.16 Exercise (images under nonlinear mappings). *Consider a continuous mapping $F : \mathbb{R}^n \to \mathbb{R}^m$ and a closed set $C \subset \mathbb{R}^n$. A sufficient condition for $F(C)$ to be closed in $\mathbb{R}^m$ is the nonexistence of an unbounded sequence $\{x^\nu\}_{\nu\in\mathbb{N}}$ with $\{F(x^\nu)\}_{\nu\in\mathbb{N}}$ bounded, such that $x^\nu \to \operatorname{dir} x$ for a vector $x \neq 0$ in C^∞.*

Guide. Pattern the argument after the proof of Theorem 3.10. □

The condition in 3.16 holds in particular when $|F(x)| \to \infty$ as $|x| \to \infty$, or if $C^\infty = \{0\}$, the latter case being classical through 3.5.

C. Horizon Functions

Our next goal is the application of these 'cosmic' ideas to the understanding of the behavior-in-the-large of functions $f : \mathbb{R}^n \to \overline{\mathbb{R}}$. We look at the direction points in the cosmic closure of $\operatorname{epi} f$, which form a subset of $\operatorname{hzn} \mathbb{R}^{n+1}$ represented by a certain closed cone in $\mathbb{R}^{n+1}$, namely the horizon cone $(\operatorname{epi} f)^\infty$. A crucial observation is that—as long as $f \not\equiv \infty$—*this horizon cone is itself an epigraph*: not only is it a cone, it has the property that if a vector (x, β) belongs to $(\operatorname{epi} f)^\infty$, then for all $\beta' \in (\beta, \infty)$ one also has $(x, \beta') \in (\operatorname{epi} f)^\infty$. This is evident from Definition 3.3 as applied to the set $\operatorname{epi} f$, and it leads to the following concept.

3.17 Definition (horizon functions). *For any function $f : \mathbb{R}^n \to \overline{\mathbb{R}}$, the associated horizon function is the function $f^\infty : \mathbb{R}^n \to \overline{\mathbb{R}}$ specified by*

$$\operatorname{epi} f^\infty = (\operatorname{epi} f)^\infty \;\text{ if }\; f \not\equiv \infty, \qquad f^\infty = \delta_{\{0\}} \;\text{ if }\; f \equiv \infty.$$

The exceptional case in the definition arises because the function $f \equiv \infty$ has $\operatorname{epi} f = \emptyset$, so $(\operatorname{epi} f)^\infty = \{(0,0)\}$. The set $(\operatorname{epi} f)^\infty$ is not an epigraph then, since it doesn't contain along with $(0,0)$ all the points $(0,\beta)$ with $0 < \beta < \infty$. When such points are added, the set becomes the epigraph of $\delta_{\{0\}}$.

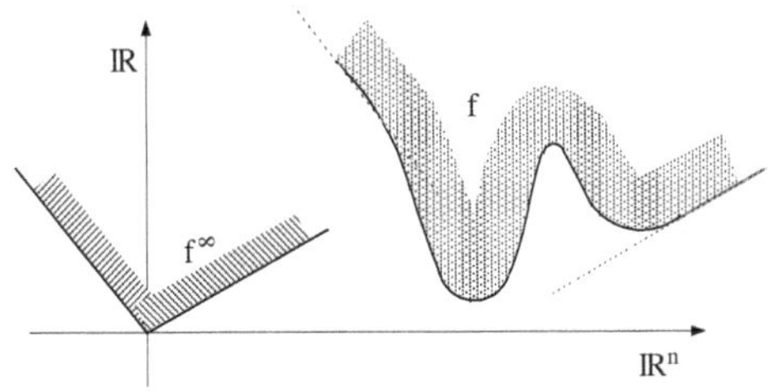

Fig. 3–6. An example of a horizon function.

To gain an appreciation of horizon functions and their properties, we must make concrete what it means for the epigraph of a function to be a cone.

3.18 Definition (positive homogeneity and sublinearity). *A function $h : \mathbb{R}^n \to \overline{\mathbb{R}}$ is positively homogeneous if $0 \in \operatorname{dom} h$ and $h(\lambda x) = \lambda h(x)$ for all x and all $\lambda > 0$. It is sublinear if in addition*

$$h(x + x') \leq h(x) + h(x') \;\text{ for all }\; x \;\text{ and }\; x'.$$

Norms and linear functions are examples of positively homogeneous functions that are actually sublinear. An indicator function δ_C is positively homogeneous if and only if C is a cone.

3.19 Exercise (epigraphs of positively homogeneous functions). *A function h is positively homogeneous if and only if $\operatorname{epi} h$ is a cone. Then either $h(0) = 0$ or $h(0) = -\infty$. When h is lsc with $h(0) = 0$, it must be proper.*

Sublinearity of h is equivalent to h being both positively homogeneous and convex. It holds if and only if $\operatorname{epi} h$ is a convex cone.

For every function f one has $(\operatorname{cl} f)^\infty = f^\infty$. If f is positively homogeneous, one has $f^\infty = \operatorname{cl} f$. These relations just specialize to the set $E = \operatorname{epi} f$ the fact that $(\operatorname{cl} E)^\infty = E^\infty$, and if E is a cone, also $E^\infty = \operatorname{cl} E$.

3.20 Exercise (sublinearity versus linearity). *For a proper, convex, positively homogeneous function $h : \mathbb{R}^n \to \overline{\mathbb{R}}$, the set*

$$\{x \mid h(-x) = -h(x)\}$$

is a linear subspace of $I\!R^n$ relative to which h is linear. It is all of $I\!R^n$ if and only if h is a linear function on $I\!R^n$.

Guide. Apply 3.8 to epi h. □

3.21 Theorem (properties of horizon functions). *For any $f : I\!R^n \to \overline{I\!R}$, the function f^∞ is lsc and positively homogeneous and, if $f \not\equiv \infty$, is given by*

$$f^\infty(w) = \liminf_{\substack{\lambda \searrow 0 \\ x \to w}} (\lambda\star f)(x) := \lim_{\delta \searrow 0} \inf_{\substack{\lambda \in (0,\delta) \\ x \in \boldsymbol{B}(w,\delta)}} \lambda f(\lambda^{-1}x). \qquad 3(3)$$

When f is convex, f^∞ is sublinear (in particular convex), and if f is also lsc and proper one has for any $\bar{x} \in \operatorname{dom} f$ that

$$f^\infty(w) = \lim_{\tau \to \infty} \frac{f(\bar{x} + \tau w) - f(\bar{x})}{\tau} = \sup_{\tau \in (0,\infty)} \frac{f(\bar{x} + \tau w) - f(\bar{x})}{\tau}. \qquad 3(4)$$

Proof. Trivially f^∞ is lsc and positively homogeneous when it is $\delta_{\{0\}}$, so suppose $f \not\equiv \infty$. Then the epigraph of f^∞ is the closed set $(\operatorname{epi} f)^\infty$ by definition, and this implies f^∞ is lsc (cf. 1.6). Since $(\operatorname{epi} f)^\infty$ is a cone, f^∞ is positively homogeneous (cf. 3.19). We have $f^\infty(w) = \inf\{\beta \mid (w,\beta) \in (\operatorname{epi} f)^\infty\}$ for any $w \in I\!R^n$. The formula for $(\operatorname{epi} f)^\infty$ stemming from Definition 3.3 tells us then that $f^\infty(w)$ is the lowest value of β such that for some choice of w^ν and λ^ν with $\lambda^\nu \searrow 0$ and $\lambda^\nu w^\nu \to w$, one can find $\alpha^\nu \geq f(w^\nu)$ with $\lambda^\nu \alpha^\nu \to \beta$. Equivalently, $f^\infty(w)$ is the lowest value of β such that for some choice of $x^\nu \to w$ and $\lambda^\nu \searrow 0$, one has $\lambda^\nu f(x^\nu/\lambda^\nu) \to \beta$. Through the sequential characterization of lower limits in Lemma 1.7 this reduces to saying that $f^\infty(w)$ is given by 3(3).

The special formula 3(4) in the convex case follows from the characterization of C^∞ in 3.6 as applied to the closed, convex set $C = \operatorname{epi} f$ at $(\bar{x}, f(\bar{x}))$. The fact that the limit in τ agrees with the supremum in τ comes from the monotonicity of difference quotients of convex functions, cf. 2.12. □

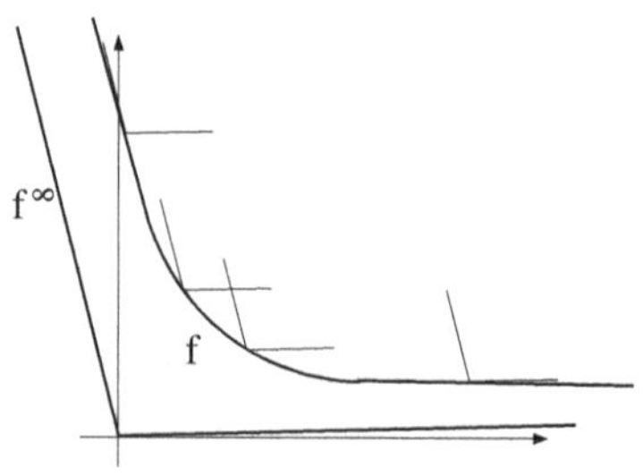

Fig. 3–7. Horizon properties of a convex function.

The illustration in Figure 3–6 underscores the geometric ideas behind horizon functions. The special properties in the case of convex functions are brought out in Figure 3–7. An interesting example in the convex case is furnished by the vector-max and log-exponential functions in 1.30 and 2.16, where

$$(\text{logexp})^{\infty} = \text{vecmax} \qquad 3(5)$$

by formula 3(3) or 3(4) and the estimates in 1(16). For convex quadratic functions, we have

$$f(x) = \tfrac{1}{2}\langle x, Ax\rangle + \langle b, x\rangle + c \quad \Longrightarrow \quad f^{\infty}(x) = \langle b, x\rangle + \delta(x|Ax = 0).$$

This is seen through either 3(3) or 3(4). For a nonconvex quadratic function f, given by the same formula but with A not positive-semidefinite (cf. 2.15), only 3(3) is applicable; then $f^{\infty}(x) = -\infty$ for vectors x with $\langle x, Ax\rangle \leq 0$, but $f^{\infty}(x) = \infty$ for x with $\langle x, Ax\rangle > 0$. This demonstrates that the horizon function of a proper—even finite—function f can be improper if the growth properties of f so dictate.

For a set $C \subset \mathbb{R}^n$ one has $\delta_C^{\infty} = \delta_{C^{\infty}}$. The function $f(x) = |x - a| + \alpha$ yields $f^{\infty} = |\cdot|$, while $f(x) = |x - a|^2 + \alpha$ yields $f^{\infty} = \delta_{\{0\}}$. In general,

$$g^{\infty} = f^{\infty} \text{ when } g(x) = f(x - a) + \alpha. \qquad 3(6)$$

Since $\operatorname{epi} g$ in this formula is simply the translate of $\operatorname{epi} f$ by the vector (a, α), the equation is just an analytical expression of the geometric fact that the set of direction points in the cosmic closure of a set is unaffected when the set is translated. Indeed, 3(6) reduces to this fact when applied to $f = \delta_C$ and $\alpha = 0$.

3.22 Corollary (monotonicity on lines). *If a proper, lsc, convex function f on $\mathbb{R}^n$ is bounded from above on a half-line $\{\bar{x} + \tau w \mid \tau \geq 0\}$, then for every $x \in \mathbb{R}^n$ the function $\tau \mapsto f(x + \tau w)$ is nonincreasing:*

$$f(x + \tau w) \geq f(x + \tau' w) \text{ when } -\infty < \tau < \tau' < \infty.$$

In fact, this has to hold if for any sequence $\{x^{\nu}\}_{\nu \in \mathbb{N}}$ converging to dir w *such that the sequence $\{f(x^{\nu})\}_{\nu \in \mathbb{N}}$ is bounded from above.*

Proof. A vector w as described is one such that $(w, 0) \in (\operatorname{epi} f)^{\infty}$, i.e., $f^{\infty}(w) \leq 0$. The claim is justified therefore by formula 3(4). □

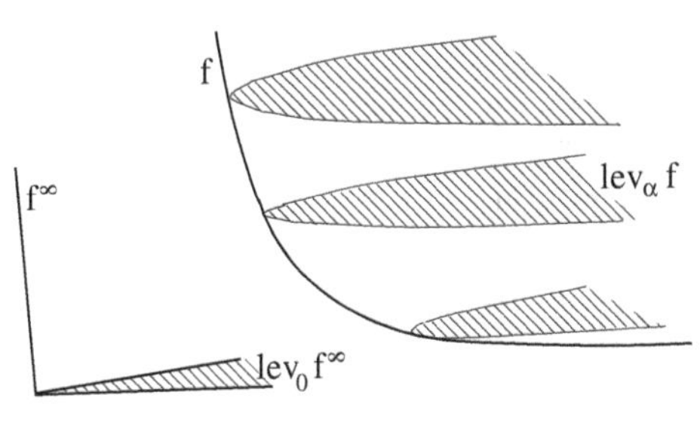

Fig. 3–8. Epigraphical interpretation of the horizon cone to a level set.

Through horizon functions, the results about horizon cones of sets can be sharpened to take advantage of constraint representations.

3.23 Proposition (horizon cone of a level set). *For any function* $f : \mathbb{R}^n \to \overline{\mathbb{R}}$ *and any* $\alpha \in \mathbb{R}$, *one has* $(\mathrm{lev}_{\leq \alpha} f)^\infty \subset \mathrm{lev}_{\leq 0} f^\infty$, *i.e.,*

$$\{x \mid f(x) \leq \alpha\}^\infty \subset \{x \mid f^\infty(x) \leq 0\}.$$

This is an equation when f *is convex, lsc and proper, and* $\mathrm{lev}_{\leq\alpha} f \neq \emptyset$. *Thus, for such a function* f, *if some set* $\mathrm{lev}_{\leq\alpha} f$ *is both nonempty and bounded, for instance the level set* $\mathrm{argmin}\, f$, *then* f *must be level-bounded.*

Proof. The inclusion is trivial if $\mathrm{lev}_{\leq\alpha} f$ is empty, because the horizon cone of this set is then $\{0\}$ by definition, whereas $f^\infty(0) \leq 0$ always because f^∞ is positively homogeneous (cf. 3.21). Suppose therefore that $\mathrm{lev}_{\leq\alpha} f \neq \emptyset$ and consider any $x \in (\mathrm{lev}_{\leq\alpha} f)^\infty$. We must show $f^\infty(x) \leq 0$, and for this only the case of $x \neq 0$ requires argument. There exist x^ν in $\mathrm{lev}_{\leq\alpha} f$ and $\lambda^\nu \searrow 0$ with $\lambda^\nu x^\nu \to x$. Let $w^\nu = \lambda^\nu x^\nu$. Then $\lambda^\nu f(w^\nu/\lambda^\nu) = \lambda^\nu f(x^\nu) \leq \lambda^\nu \alpha \to 0$ with $w^\nu \to x$. This implies by formula 3(3) that $f^\infty(x) \leq 0$.

When f is convex, lsc and proper, the property in 3(4), which holds for arbitrary $\bar{x} \in \mathrm{dom}\, f$, yields the further conclusions. □

3.24 Exercise (horizon cones from constraints). *Let*

$$C = \{x \in X \mid f_i(x) \leq 0 \ \text{ for } \ i \in I\}$$

for a nonempty set $X \subset \mathbb{R}^n$ *and finite functions* f_i *on* $\mathbb{R}^n$. *Then*

$$C^\infty \subset \{x \in X^\infty \mid f_i^\infty(x) \leq 0 \ \text{ for } \ i \in I\}.$$

The inclusion holds as an equation when C *is nonempty,* X *is closed and convex, and each function* f_i *is convex.*

Guide. Combine 3.23 with 3.9, invoking 2.36. □

D. Coercivity Properties

Growth properties of a function f involving an exponent $p \in (0, \infty)$ have already been met with in Chapter 1 (cf. 1.14). The case of $p = 1$, which is the most important, is closely tied to properties of the horizon function f^∞.

3.25 Definition (coercivity properties). *A function* $f : \mathbb{R}^n \to \overline{\mathbb{R}}$ *is level-coercive if it is bounded below on bounded sets and satisfies*

$$\liminf_{|x|\to\infty} \frac{f(x)}{|x|} > 0,$$

whereas it is coercive if it is bounded below on bounded sets and

$$\liminf_{|x|\to\infty} \frac{f(x)}{|x|} = \infty.$$

It is counter-coercive if instead

$$\liminf_{|x|\to\infty} \frac{f(x)}{|x|} = -\infty.$$

For example, if $f(x) = |x|^p$ for an exponent $p \in (1, \infty)$, f is coercive, but for $p = 1$ it's merely level-coercive. For $p \in (0, 1)$, f isn't even level-coercive, although it's level-bounded. These cases are shown in Figure 3–9. Likewise, the function $f(x) = \sqrt{1 + |x|^2}$ is level-coercive but not coercive.

Note that properties of $\operatorname{dom} f$ can have a major effect: if f is lsc and proper (hence bounded below on bounded sets by 1.10), and $\operatorname{dom} f$ is bounded, then f is coercive. Thus, the function f defined by $f(x) = 1/(1 - |x|)$ when $|x| < 1$, $f(x) = \infty$ when $|x| \geq 1$, is coercive. An indicator function δ_C is coercive if and only if C is bounded.

Counter-coercivity is exhibited by the function $f(x) = -|x|^p$ when $p \in (1, \infty)$, but not when $p \in (0, 1)$.

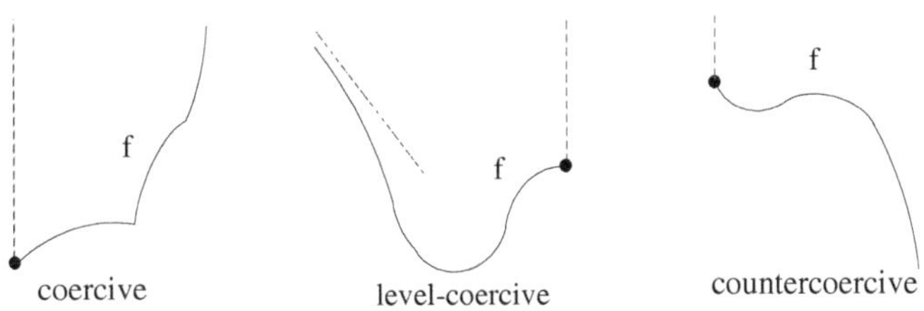

Fig. 3–9. Coercivity examples.

3.26 Theorem (horizon functions and coercivity). *Let $f : \mathbb{R}^n \to \overline{\mathbb{R}}$ be lsc and proper. Then*

$$\liminf_{|x|\to\infty} \frac{f(x)}{|x|} \;=\; \inf_{|x|=1} f^\infty(x), \tag{3(7)}$$

and in consequence

(a) *f is level-coercive if and only if $f^\infty(x) > 0$ for all $x \neq 0$; this means for some $\gamma \in (0, \infty)$ there exists $\beta \in (-\infty, \infty)$ with $f(x) \geq \gamma|x| + \beta$ for all x;*

(b) *f is coercive if and only if $f^\infty(x) = \infty$ for all $x \neq 0$; this means for every $\gamma \in (0, \infty)$ there exists $\beta \in (-\infty, \infty)$ with $f(x) \geq \gamma|x| + \beta$ for all x;*

(c) *f is counter-coercive if and only if $f^\infty(x) = -\infty$ for some $x \neq 0$, or equivalently, $f^\infty(0) = -\infty$; this means that for no $\gamma \in (-\infty, \infty)$ does there exist $\beta \in (-\infty, \infty)$ with $f(x) \geq \gamma|x| + \beta$ for all x.*

Proof. The sequential characterization of lower limits in Lemma 1.7 adapts to the kind of limit on the left in 3(7) as well as to the one giving $f^\infty(x)$ in 3(3). The right side of 3(7) can in this way be expressed as

$$\begin{aligned}
&\inf\{\gamma \mid \exists\, x,\ \lambda^\nu \searrow 0,\ x^\nu,\ \alpha^\nu \geq f(x^\nu) \text{ with } |x| = 1,\ \lambda^\nu\alpha^\nu \to \gamma,\ \lambda^\nu x^\nu \to x\}\\
&\quad = \inf\{\gamma \mid \exists\, \lambda^\nu \searrow 0,\ x^\nu,\ \alpha^\nu \geq f(x^\nu) \text{ with } \lambda^\nu\alpha^\nu \to \gamma,\ |\lambda^\nu x^\nu| \to 1\}\\
&\quad = \inf\{\gamma \mid \exists\, \lambda^\nu \searrow 0,\ x^\nu \text{ with } \lambda^\nu f(x^\nu) \to \gamma,\ \lambda^\nu |x^\nu| \to 1\}\\
&\quad = \inf\{\gamma \mid \exists\, x^\nu \text{ with } |x^\nu| \to \infty,\ f(x^\nu)/|x^\nu| \to \gamma\},
\end{aligned}$$

where we have arrived at an expression representing the left side of 3(7). Therefore, 3(7) is true. From 1.14 we know that the left side of 3(7) is the supremum of the set of all $\gamma \in I\!R$ for which there exists $\beta \in I\!R$ with $f(x) \geq \gamma|x| + \beta$ for all x. Everything in (a), (b) and (c) then follows, except for the claim in (c) that the conditions there are also equivalent to having $f^\infty(0) = -\infty$. If $f^\infty(x) = -\infty$ for some $x \neq 0$, then $f^\infty(0) = -\infty$ because f^∞ is lsc and positively homogeneous (by 3.21). On the other hand, if $f^\infty(0) = -\infty$ it's clear from 3(3) that there can't exist γ and β in $I\!R$ with $f(x) \geq \gamma|x| + \beta$ for all x. □

3.27 Corollary (coercivity and convexity). *For any proper, lsc function f on $I\!R^n$, level coercivity implies level boundedness. When f is convex the two properties are equivalent. No proper, lsc, convex function is counter-coercive.*

Proof. The first assertion combines 3.26(a) with 3.23. The second combines 3.26(c) with the fact that by formula 3(4) in Theorem 3.21 every proper, lsc, convex function f has $f^\infty(0) = 0$. □

3.28 Example (coercivity and prox-boundedness). *If a proper, lsc function f on $I\!R^n$ is not counter-coercive, it is prox-bounded with threshold $\lambda_f = \infty$.*

Detail. The hypothesis ensures through 3.26(c) the existence of values $\gamma \in I\!R$ and $\beta \in I\!R$ such that $f(x) \geq \gamma|x| + \beta$. Then $\liminf_{|x|\to\infty} f(x)/|x|^2 \geq 0$, hence by 1.24 the threshold of prox-boundedness for f is ∞. □

Coercivity properties are especially useful because they can readily be verified for functions obtained through various constructions or 'perturbations' of other functions. The following are some elementary rules.

3.29 Exercise (horizon functions in addition). *Let f_1 and f_2 be lsc and proper on $I\!R^n$, and suppose that neither is counter-coercive. Then*

$$(f_1 + f_2)^\infty \geq f_1^\infty + f_2^\infty,$$

where the inequality becomes an equation when both functions are convex and $\operatorname{dom} f_1 \cap \operatorname{dom} f_2 \neq \emptyset$. Thus,

(a) *if either f_1 or f_2 is level-coercive while the other is bounded from below (as is true if it too is level-coercive), then $f_1 + f_2$ is level-coercive;*

(b) *if either f_1 or f_2 is coercive, then $f_1 + f_2$ is coercive.*

Guide. Use formula 3(3) while being mindful of 1.36, but in the convex case use formula 3(4). For the coercivity conclusions apply Theorem 3.26. □

A counterexample to equality always holding in 3.29 in the absence of convexity, even when the domain condition is satisfied, can be obtained from the counterexample after 3.11, concerning the horizon cones of product sets, by considering indicators of such sets.

3.30 Proposition (horizon functions in pointwise max and min). *For a collection* $\{f_i\}_{i\in I}$ *of functions* $f_i : \mathbb{R}^n \to \overline{\mathbb{R}}$, *one has*

$$\big(\sup_{i\in I} f_i\big)^\infty \geq \sup_{i\in I} f_i^\infty, \qquad \big(\inf_{i\in I} f_i\big)^\infty \leq \inf_{i\in I} f_i^\infty.$$

The first inequality is an equation when $\sup_{i\in I} f_i \not\equiv \infty$ *and the functions are convex, lsc and proper. The second inequality is an equation whenever the index set is finite.*

Proof. This applies 3.9 to the epigraphs in question. □

3.31 Theorem (coercivity in parametric minimization). *For a proper, lsc function* $f : \mathbb{R}^n \times \mathbb{R}^m \to \overline{\mathbb{R}}$, *a sufficient condition for* $f(x,u)$ *to be level-bounded in* x *locally uniformly in* u, *which is also necessary when* f *is convex, is*

$$f^\infty(x,0) > 0 \ \text{ for all } \ x \neq 0. \tag{3(8)}$$

If this is fulfilled, the function $p(u) := \inf_x f(x,u)$ *has*

$$p^\infty(u) = \inf_x f^\infty(x,u), \ \text{ attained when finite.} \tag{3(9)}$$

Thus, p *is level-coercive if* f *is level-coercive, and* p *is coercive if* f *is coercive.*

Proof. The question of whether $f(x,u)$ is level-bounded in x locally uniformly in u revolves by definition around whether, for balls $\mathbb{B}(\bar{u},\varepsilon)$ and scalars $\alpha \in \mathbb{R}$, sets of the form $C = \big(\mathbb{R}^n \times \mathbb{B}(\bar{u},\varepsilon)\big) \cap \operatorname{lev}_{\leq\alpha} f$ are bounded when nonempty. Boundedness corresponds by Theorem 3.5 to the horizon cone of such a nonempty set C being just the zero cone. The horizon cone can be calculated from 3.9 and 3.23:

$$\begin{aligned} C^\infty &\subset \big(\mathbb{R}^n \times \mathbb{B}(\bar{u},\varepsilon)\big)^\infty \cap \big(\operatorname{lev}_{\leq\alpha} f\big)^\infty \\ &\subset \big(\mathbb{R}^n \times \{0\}\big) \cap \big\{(x,u) \,\big|\, f^\infty(x,u) \leq 0\big\} = \big\{(x,0) \,\big|\, f^\infty(x,0) \leq 0\big\}, \end{aligned}$$

where equality holds throughout when f is convex. Clearly, then, condition 3(8) is always sufficient, and if f is convex it is both necessary and sufficient. The level boundedness property of f ensures through 1.17 and 1.18 that p is a proper, lsc function on $\mathbb{R}^m$ whose epigraph is the image of that of f under the linear mapping $L : (x,u,\alpha) \mapsto (u,\alpha)$. We have $L^{-1}(0,0) \cap (\operatorname{epi} f)^\infty = (0,0,0)$ under 3(8), because $(\operatorname{epi} f)^\infty = \operatorname{epi} f^\infty$ by definition. This ensures through Theorem 3.10 that $L\big((\operatorname{epi} f)^\infty\big) = L(\operatorname{epi} f)^\infty$, which is the assertion of 3(9) (cf. the general principle in 1.18 again). □

3.32 Corollary (boundedness in convex parametric minimization). *Let*

$$p(u) = \inf_x f(x,u), \qquad P(u) = \operatorname{argmin}_x f(x,u),$$

for a proper, lsc, convex function $f : \mathbb{R}^n \times \mathbb{R}^m \to \overline{\mathbb{R}}$, *and suppose that for* *some $\bar{u}$ the set $P(\bar{u})$ is nonempty and bounded. Then $P(u)$ is bounded for all u, and p is a proper, lsc, convex function with horizon function given by 3(9).*

Proof. Since $P(\bar{u})$ is one of the level sets of $g(x) = f(x, \bar{u})$, the hypothesis implies through 3.23 that $g^\infty(x) > 0$ for all $x \neq 0$. But $g^\infty(x) = f^\infty(x, 0)$ by formula 3(4) in Theorem 3.21. Therefore, assumption 3(8) of Theorem 3.31 is satisfied in this case. We see also that all functions $f(\cdot, u)$ on $\mathbb{R}^n$ have $f^\infty(\cdot, u)$ as their horizon function, so all are level-bounded. In particular, all sets $P(u)$ must be bounded. □

3.33 Corollary (coercivity in epi-addition).

(a) *Suppose that $f = f_1 \# f_2$ for proper, lsc functions f_1 and f_2 on $\mathbb{R}^n$ such that $f_1^\infty(-w) + f_2^\infty(w) > 0$ for all $w \neq 0$. Then f is a proper, lsc function and the infimum in its definition is attained when finite. Moreover $f^\infty \geq f_1^\infty \# f_2^\infty$. When f_1 and f_2 are convex, this holds as an equation.*

(b) *Suppose that $f = f_1 \# f_2$ for proper, lsc functions f_1 and f_2 on $\mathbb{R}^n$ such that f_2 is coercive and f_1 is not counter-coercive. Then f is a proper, lsc function and the infimum in its definition is attained when finite. Moreover $f^\infty \geq f_1^\infty$. When f_1 and f_2 are convex, this holds as an equation.*

Proof. We have $f(x) = \inf_w g(w,x)$ for $g(w,x) = f_1(x-w) + f_2(w)$. Theorem 3.31 can be applied using the estimate $g^\infty(w,x) \geq f_1^\infty(w-x) + f_2^\infty(w)$, which comes from 3.29 and holds as an equation when f_1 and f_2 are convex. This yields (a). Then (b) is the special case where f_2^∞ has the property in 3.26(b) but f_1^∞ avoids the property in 3.26(c). □

3.34 Theorem (cancellation in epi-addition). *If $f_1 \# g = f_2 \# g$ for proper, lsc, convex functions f_1, f_2 and g such that g is coercive, then $f_1 = f_2$.*

Proof. The hypothesis implies through 3.33(b) that $f_i \# g$ is finite (since f_i is not counter-coercive, cf. 3.27). Also, $f_i \# g$ is convex (cf. 2.24), hence for all $\lambda > 0$ the Moreau envelope $e_\lambda(f_i \# g)$ is finite, convex and differentiable (cf. 2.26). In terms of $j_\lambda(x) := (1/2\lambda)|x|^2$ we have

$$e_\lambda(f_i \# g) = (f_i \# g) \# j_\lambda = (f_i \# j_\lambda) \# g = e_\lambda f_i \# g.$$

Then for any choice of $v \in \mathbb{R}^n$ we have

$$\begin{aligned}
\inf_x \Big\{ [e_\lambda(f_i \# g)](x) - \langle v, x\rangle \Big\} &= \inf_x \Big\{ [e_\lambda f_i \# g](x) - \langle v, x\rangle \Big\} \\
&= \inf_x \Big(\inf_{w+w'=x} \big\{ e_\lambda f_i(w) + g(w') \big\} - \langle v, x\rangle \Big) \\
&= \inf_{w,w'} \big\{ e_\lambda f_i(w) + g(w') - \langle v, w+w'\rangle \big\} \\
&= \inf_w \big\{ e_\lambda f_i(w) - \langle v, w\rangle \big\} + \inf_{w'} \big\{ g(w') - \langle v, w'\rangle \big\},
\end{aligned}$$

where the last infimum has a finite value because of the coercivity of g. Our assumption that $f_1 \# g = f_2 \# g$ implies that $e_\lambda f_1 \# g = e_\lambda f_2 \# g$ and consequently through this calculation that

$$\inf_w \{e_\lambda f_1(w) - \langle v, w\rangle\} = \inf_w \{e_\lambda f_2(w) - \langle v, w\rangle\} \text{ for all } v. \tag{3(10)}$$

If the targeted conclusion $f_1 = f_2$ were false, we would have to have $e_\lambda f_1 \neq e_\lambda f_2$ for all $\lambda > 0$ sufficiently small, cf. 1.25. Then for any such λ and any x where these Moreau envelopes disagree, say $e_\lambda f_1(x) > e_\lambda f_2(x)$, the vector $v = \nabla e_\lambda f_1(x)$ would satisfy $e_\lambda f_1(w) \geq e_\lambda f_1(x) + \langle v, w - x\rangle$ for all w, so that we would have

$$\begin{aligned}\inf_w \{e_\lambda f_1(w) - \langle v, w\rangle\} &\geq e_\lambda f_1(x) - \langle v, x\rangle \\ &> e_\lambda f_2(x) - \langle v, x\rangle \geq \inf_w \{e_\lambda f_2(w) - \langle v, w\rangle\}.\end{aligned}$$

The conflict between this strict inequality and the general equation in 3(10) shows that necessarily $f_1 = f_2$. □

3.35 Corollary (cancellation in set addition). *If $C_1 + B = C_2 + B$ for nonempty, closed, convex sets C_1, C_2 and B such that B is bounded, then $C_1 = C_2$.*

Proof. This specializes Theorem 3.34 to $f_i = \delta_{C_i}$ and $g = \delta_B$. □

Alternative proofs of 3.34 and 3.35 can readily be based on the duality that will be developed in Chapter 11.

3.36 Corollary (functions determined by their Moreau envelopes). *If f_1, f_2, are proper, lsc, convex functions with $e_\lambda f_1 = e_\lambda f_2$ for some $\lambda > 0$, then $f_1 = f_2$.*

Proof. This case of Theorem 3.34, for $g = (1/2\lambda)|\cdot|^2$, was a stepping stone in its proof. □

3.37 Corollary (functions determined by their proximal mappings). *If f_1, f_2, are proper, lsc, convex functions such that $P_\lambda f_1 = P_\lambda f_2$ for some $\lambda > 0$, then $f_1 = f_2 + \text{const.}$*

Proof. From the gradient formula in Theorem 2.26, it's clear that $P_\lambda f_1 = P_\lambda f_2$ if and only if $e_\lambda f_1 = e_\lambda f_2 + \text{const.} = e_\lambda(f_2 + \text{const.})$. The preceding corollary then comes into play. □

E.* Cones and Orderings

Besides their importance in connection with growth properties and unboundedness, cones are useful also in many other ways beyond representing sets of direction points.

3.38 Proposition (vector inequalities). *For an arbitrary closed, convex cone $K \subset \mathbb{R}^n$, define the inequality $x \geq_K y$ for vectors x and y in $\mathbb{R}^n$ to mean that $x - y \in K$. The partial ordering $\geq_K$ then satisfies:*

(a) *$x \geq_K x$ for all x;*

(b) *$x \geq_K y$ implies $-y \geq_K -x$;*

(c) $x \geq_K y$ *implies* $\lambda x \geq_K \lambda y$ *for all* $\lambda \geq 0$;
(d) $x \geq_K y$ *and* $x' \geq_K y'$ *imply* $x + x' \geq_K y + y'$;
(e) $x^\nu \geq_K y^\nu$, $x^\nu \to x$, $y^\nu \to y$, *imply* $x \geq_K y$.

Conversely, if a relation '$\geq$' for vectors has these properties, it must be of the form $\geq_K$ *for a closed, convex cone* K.

To have the additional property that $x = y$ *when both* $x \geq_K y$ *and* $y \geq_K x$, *it is necessary and sufficient to have* K *be pointed.*

Proof. The first part is immediate from the definition of $\geq_K$. In the converse part, one notes from (d) that $x \geq y$ must be equivalent to $x - y \geq 0$, so the ordering must be of type $\geq_K$ for $K = \{x \mid x \geq 0\}$. The various conditions, taken with $y = 0$ in (c), and $y = y' = 0$ in(d), and $y^\nu = 0$ in (e), then force K to be a closed, convex cone. The role of pointedness is seen from 3.14. □

The standard case for vector inequalities in $\mathbb{R}^n$ is the one where K is the nonnegative orthant $\mathbb{R}^n_+$ in 1.2, which in particular is a pointed, closed, convex cone. For this case the customary notation is simply $x \geq y$:

$$(x_1, \ldots, x_n) \geq (y_1, \ldots, y_n) \iff x_j \geq y_j \text{ for } j = 1, \ldots, n.$$

Such notation is often convenient in representing function inequalities as well, for instance, a system $f_i(x) \leq 0$ for $i = 1, \ldots, m$ can be written as $F(x) \leq 0$ for the mapping $F : x \mapsto (f_1(x), \ldots, f_m(x))$.

Another partial ordering that fits the general pattern in 3.38 is the matrix ordering described at the end of Chapter 2.

3.39 Example (matrix inequalities). *In* $\mathbb{R}^{n\times n}_{\text{sym}}$, *the space of symmetric real matrices of order* n, *the partial ordering* $A \succeq B$ *is the one associated with the pointed, closed, convex cone consisting of the positive-semidefinite matrices, and it therefore obeys the rules:*

(a) $A \succeq A$ *for all* A;
(b) $A \succeq B$ *implies* $-B \succeq -A$;
(c) $A \succeq B$ *implies* $\lambda A \succeq \lambda B$ *for all* $\lambda \geq 0$;
(d) $A \succeq B$ *and* $A' \succeq B'$ *imply* $A + A' \succeq B + B'$;
(e) $A^\nu \succeq B^\nu$, $A^\nu \to A$, $B^\nu \to B$, *imply* $A \succeq B$.
(f) $A \succeq B$ *and* $B \succeq A$ *imply* $A = B$.

Detail. The set $K \subset \mathbb{R}^{n\times n}_{\text{sym}}$ consisting of the positive-semidefinite matrices is defined by the system of inequalities $0 \leq l_x(A) =: \langle x, Ax\rangle$ indexed by the vectors $x \in \mathbb{R}^n$. Each function l_x is linear on $\mathbb{R}^{n\times n}_{\text{sym}}$, and the set of solutions to a system of linear inequalities is always a closed, convex cone. The reason K is pointed is that the eigenvalues of a symmetric, positive-semidefinite matrix A are nonnegative. If $-A$ is positive-semidefinite too, the eigenvalues must all be 0, so A has to be the zero matrix. □

The class of cones is preserved under a number of common operations, and so too is the class of positively homogeneous functions.

3.40 Exercise (operations on cones and positively homogeneous functions).

(a) $\bigcap_{i \in I} K_i$ *and* $\bigcup_{i \in I} K_i$ *are cones when each* K_i *is a cone.*

(b) $K_1 + K_2$ *is a cone when* K_1 *and* K_2 *are cones.*

(c) $L(K)$ *is a cone when* K *is a cone and* L *is linear.*

(d) $L^{-1}(K)$ *is a cone when* K *is a cone and* L *is linear.*

(e) $\sup_{i \in I} h_i$ *and* $\inf_{i \in I} h_i$ *are positively homogeneous functions when each* h_i *is positively homogeneous.*

(f) $h_1 + h_2$ *and* $h_1 \# h_2$ *are positively homogeneous functions when* h_1 *and* h_2 *are positively homogeneous.*

(g) λh *is positively homogeneous when* h *is positively homogeneous,* $\lambda \geq 0$.

(h) $h \circ L$ *is positively homogeneous when* h *is positively homogeneous and the mapping* L *is linear.*

In (e), (f), (g) *and* (h), *the assertions remain true when positive homogeneity is replaced by sublinearity, except for the inf case in (e).*

F.* Cosmic Convexity

Convexity can be introduced in cosmic space, and through that concept the special role of horizon cones of convex sets can be better understood.

3.41 Definition (convexity in cosmic space). *A general subset of* $\operatorname{csm} \mathbb{R}^n$, *written as* $C \cup \operatorname{dir} K$ *for a set* $C \subset \mathbb{R}^n$ *and a cone* $K \subset \mathbb{R}^n$, *is said to be convex if* C *and* K *are convex and* $C + K \subset C$.

In the case of a subset $C \cup \operatorname{dir} K$ of $\operatorname{csm} \mathbb{R}^n$ that actually lies in $\mathbb{R}^n$ (because $K = \{0\}$), this extended definition of convexity agrees with the one at the beginning of Chapter 2. In general, the condition $C + K \subset C$ means that for every point $\bar{x} \in C$ and every direction point $\operatorname{dir} w \in \operatorname{dir} K$, the half-line $\{\bar{x} + \tau w \mid \tau \geq 0\}$ is included in C, which of course is the property developed in Theorem 3.6 when C happens to be a closed, convex set and $K = C^\infty$. This property can be interpreted as generalizing to $C \cup \operatorname{dir} K$ the line segment criterion for convexity: half-lines are viewed as infinite line segments joining ordinary points with direction points. The idea is depicted in Figure 3–10. A similar interpretation can be made of the convexity condition on K in Definition 3.41 in terms of 'horizon line segments' being included in $C \cup \operatorname{dir} K$.

3.42 Exercise (cone characterization of cosmic convexity). *A set in* $\operatorname{csm} \mathbb{R}^n$ *is convex if and only if the corresponding cone in the ray space model is convex.*

Guide. The cone corresponding to $C \cup \operatorname{dir} K$ in the ray space model of $\operatorname{csm} \mathbb{R}^n$ consists of the vectors $\lambda(x, -1)$ with $\lambda > 0$, $x \in C$, along with the vectors $(x, 0)$ with $x \in K$. Apply to this cone the convexity criterion in 3.7(b). □

The convexity of a subset $C \cup \operatorname{dir} K$ of $\operatorname{csm} \mathbb{R}^n$ is equivalent also to the convexity of the corresponding subset of H_n in the hemispherical model, when

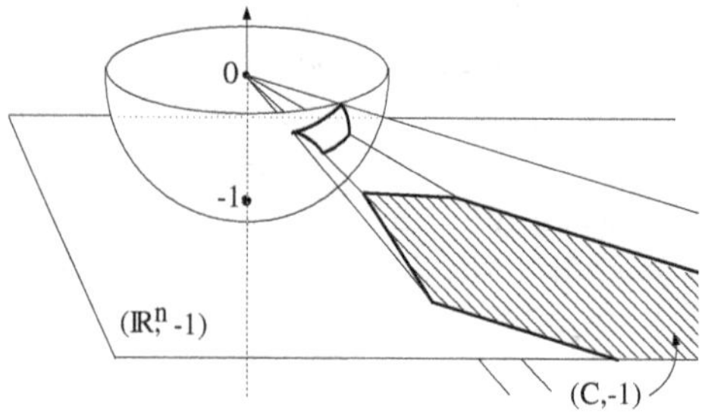

Fig. 3–10. Cosmic convexity.

such convexity is taken in the sense of *geodesic* segments joining pairs of points of H_n that are not *antipodal*, i.e., not just opposite each other on the rim of H_n. For antipodal pairs of points a geodesic link is not well defined, nor is one needed for the sake of cosmic convexity. Indeed, a subset of the horizon of $\mathbb{R}^n$ consisting of just two points $\operatorname{dir} x$ and $\operatorname{dir}(-x)$ is convex by Definition 3.41 and the criterion in 3.42, since the corresponding cone in the ray space model is the line $\{\lambda(x,0) \mid -\infty < \lambda < \infty\}$ (a convex subset of $\mathbb{R}^{n+1}$).

3.43 Exercise (extended line segment principle). *For a convex set in* $\operatorname{csm}\mathbb{R}^n$*, written as* $C \cup \operatorname{dir} K$ *for a set* $C \subset \mathbb{R}^n$ *and a cone* $K \subset \mathbb{R}^n$*, one has for any point* $\bar{x} \in \operatorname{int} C$ *and vector* $w \in K$ *that* $\bar{x} + \tau w \in \operatorname{int} C$ *for all* $\tau \in (0, \infty)$.

Guide. Deduce this from Theorem 2.33 as applied to the cone representing $C \cup \operatorname{dir} K$ in the ray space model. □

Convex hulls can be investigated in the cosmic framework too. For a set $E \subset \operatorname{csm}\mathbb{R}^n$, $\operatorname{con} E$ is defined of course to be the smallest convex set in $\operatorname{csm}\mathbb{R}^n$ that includes E. When E happens not to contain any direction points, i.e., E is merely a subset of $\mathbb{R}^n$, $\operatorname{con} E$ is the convex hull studied earlier.

3.44 Exercise (cosmic convex hulls). *For a general subset of* $\operatorname{csm}\mathbb{R}^n$*, written as* $C \cup \operatorname{dir} K$ *for a set* $C \subset \mathbb{R}^n$ *and a cone* $K \subset \mathbb{R}^n$*, one has*

$$\operatorname{con}(C \cup \operatorname{dir} K) = (\operatorname{con} C + \operatorname{con} K) \cup \operatorname{dir}(\operatorname{con} K).$$

Guide. Work with the cone in $\mathbb{R}^{n+1}$ corresponding to $C \cup \operatorname{dir} K$ in the ray space model for $\operatorname{csm}\mathbb{R}^n$. The convex hull of this cone corresponds to the cosmic convex hull of $C \cup \operatorname{dir} K$. □

3.45 Proposition (extended expression of convex hulls). *If* $D = \operatorname{con} C + \operatorname{con} K$ *for a pointed, closed cone* K *and a nonempty, closed set* $C \subset \mathbb{R}^n$ *with* $C^\infty \subset K$*, then* D *is closed and is the smallest closed onvex set that includes* C *and whose horizon cone includes* K*. In fact* $D^\infty = \operatorname{con} K$.

Proof. This can be interpreted through 3.4 as concerning a closed subset $C \cup \operatorname{dir} K$ of $\operatorname{csm}\mathbb{R}^n$. The corresponding cone in the ray space model of $\operatorname{csm}\mathbb{R}^n$ is not only closed but pointed, because K is pointed, so its convex hull is closed by 3.15. The latter cone corresponds to $(\operatorname{con} C + \operatorname{con} K) \cup \operatorname{dir}(\operatorname{con} K)$ by 3.44. Hence this subset of $\operatorname{csm} \mathbb{R}^n$ is closed, and the claimed properties follow. □

In the language of subsets of $\operatorname{csm} \mathbb{R}^n$, this theorem can be summarized simply by saying that if $C \cup \operatorname{dir} K$ is cosmically closed and K is pointed, then $\operatorname{con}(C \cup \operatorname{dir} K)$ is cosmically closed as well. The case of $K = \{0\}$, where there aren't any direction points in $E = C \cup \operatorname{dir} K$, is that of the convex hull of a compact set, already treated in 2.30.

3.46 Corollary (closures of convex hulls). *Let $C \subset \mathbb{R}^n$ be a closed set such that C^∞ is pointed. Then*

$$\operatorname{cl}(\operatorname{con} C) = \operatorname{con} C + \operatorname{con} C^\infty, \qquad (\operatorname{cl} \operatorname{con} C)^\infty = \operatorname{con} C^\infty.$$

Proof. Take $K = C^\infty$ in 3.45. □

3.47 Corollary (convex hulls of coercive functions). *Let $f : \mathbb{R}^n \to \overline{\mathbb{R}}$ be proper, lsc and coercive. Then $\operatorname{con} f$ is proper, lsc and coercive, and for each x in the set $\operatorname{dom}(\operatorname{con} f) = \operatorname{con}(\operatorname{dom} f)$ the infimum in the formula for $(\operatorname{con} f)(x)$ in 2.31 is attained.*

Proof. Here we apply 3.46 to $\operatorname{epi} f$, which is a closed set having $\operatorname{epi} f^\infty$ as its horizon cone. By the coercivity assumption this horizon cone consists just of the nonnegative vertical axis in $\mathbb{R}^n \times \mathbb{R}$, cf. 3.26(b). □

G.* Positive Hulls

Another operation of interest is that of forming the smallest cone containing a set $C \subset \mathbb{R}^n$. This cone, called the *positive hull* of C, has the formula

$$\operatorname{pos} C = \{0\} \cup \{\lambda x \mid x \in C, \lambda > 0\}, \tag{3(11)}$$

see Figure 3–11. (If $C = \emptyset$, one has $\operatorname{pos} C = \{0\}$, but if $C \neq \emptyset$, one has $\operatorname{pos} C = \{\lambda x \mid x \in C, \lambda \geq 0\}$.) Clearly, C is itself a cone if and only if $C = \operatorname{pos} C$.

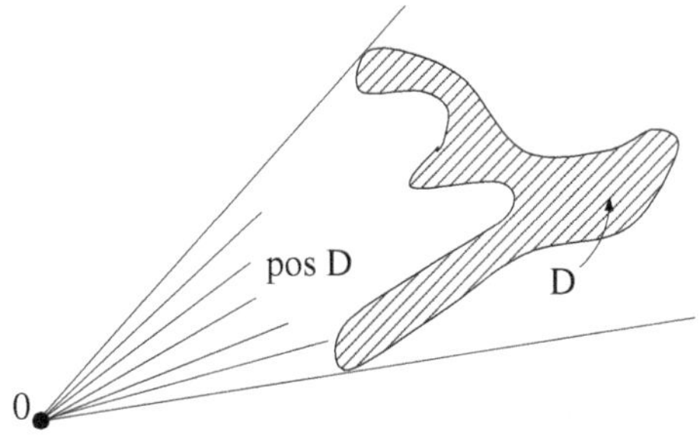

Fig. 3–11. The positive hull of a set.

Similarly, one can form the greatest positively homogeneous function majorized by a given function $f : \mathbb{R}^n \to \overline{\mathbb{R}}$. This function is called the *positive hull* of f. It has the formula

$$(\operatorname{pos} f)(x) = \inf \{\alpha \mid (x, \alpha) \in \operatorname{pos}(\operatorname{epi} f)\}, \qquad 3(12)$$

because its epigraph must be the smallest cone of epigraphical type in $\mathbb{R}^n \times \mathbb{R}$ containing $\operatorname{epi} f$, cf. 3.19. Using the fact that the set $\lambda \operatorname{epi} f$ for $\lambda > 0$ is the epigraph of the function $\lambda \star f : x \mapsto \lambda f(\lambda^{-1}x)$, one can write the formula in question as

$$(\operatorname{pos} f)(x) = \inf_{\lambda \geq 0} (\lambda \star f)(x) \ \text{ when } f \not\equiv \infty. \qquad 3(13)$$

3.48 Exercise (closures of positive hulls).

(a) *Let $C \subset \mathbb{R}^n$ be closed with $0 \notin C$. Then $\operatorname{cl}(\operatorname{pos} C) = (\operatorname{pos} C) \cup C^\infty$. If C is bounded, then $\operatorname{pos} C$ is closed.*

(b) *Let $f : \mathbb{R}^n \to \overline{\mathbb{R}}$ be lsc with $f \not\equiv \infty$ and $f(0) > 0$. Then $\operatorname{cl}(\operatorname{pos} f) = \min\{\operatorname{pos} f, f^\infty\}$. If in addition $\operatorname{pos} f \leq f^\infty$, as is true in particular when f is coercive or $\operatorname{dom} f$ is all of $\mathbb{R}^n$, then $\operatorname{pos} f$ is lsc.*

Guide. In (a) consider the cone that corresponds to C in the ray space model, observing that $\operatorname{pos} C$ is the image of this cone under the projection mapping $L : (x, \beta) \mapsto x$. Apply Theorem 3.10 to L and the closure of this cone. In (b) apply (a) epigraphically. □

3.49 Exercise (convexity of positive hulls).

(a) *For a convex set C, the cone $\operatorname{pos} C$ is convex.*

(b) *For a convex function f, the positively homogeneous function $\operatorname{pos} f$ is convex, hence sublinear.*

(c) *For a proper, convex function f on $\mathbb{R}^n$, the function*

$$h(\lambda, x) = \begin{cases} \lambda f(\lambda^{-1}x) & \text{when } \lambda > 0, \\ 0 & \text{when } \lambda = 0 \text{ and } x = 0, \\ \infty & \text{otherwise,} \end{cases}$$

is proper and convex with respect to $(\lambda, x) \in \mathbb{R} \times \mathbb{R}^n$, in fact sublinear in these variables. The lower closure of h, likewise sublinear but also lsc, is expressed in terms of the lower closure of f by

$$(\operatorname{cl} h)(\lambda, x) = \begin{cases} \lambda (\operatorname{cl} f)(\lambda^{-1}x) & \text{when } \lambda > 0, \\ f^\infty(x) & \text{when } \lambda = 0, \\ \infty & \text{otherwise.} \end{cases}$$

Guide. Derive (b) from (a) from epigraphs. In (c), determine that $h = \operatorname{pos} g$ for the function $g(\lambda, x) = f(x)$ when $\lambda = 1$, but $g(\lambda, x) = \infty$ when $\lambda \neq 1$. In taking lower closures, apply 3.48(b) to g. □

The joint convexity in 3.49(c) in the variables λ and x is surprising in many situations and might be hard to recognize without this insight. For instance, for any positive-definite, symmetric matrix $A \in \mathbb{R}^{n\times n}$ the expression

$$h(\lambda, x) = \frac{1}{2\lambda}\langle x, Ax\rangle$$

is convex as a function of $(\lambda, x) \in (0,\infty) \times \mathbb{R}^n$. This is the case of $f(x) = \frac{1}{2}\langle x, Ax\rangle$. The function in Example 2.38 arises in this way.

3.50 Example (gauge functions). *Let $C \subset \mathbb{R}^n$ be closed and convex with $0 \in C$. The gauge of C is the function $\gamma_C : \mathbb{R}^n \to \overline{\mathbb{R}}$ defined by*

$$\gamma_C(x) := \inf\{\lambda \geq 0 \,|\, x \in \lambda C\}, \quad \text{so } \gamma_C = \operatorname{pos}(\delta_C + 1).$$

This function is nonnegative, lsc and sublinear (hence convex) with level sets

$$C = \{x \,|\, \gamma_C(x) \leq 1\}, \quad C^\infty = \{x \,|\, \gamma_C(x) = 0\}, \quad \operatorname{pos} C = \{x \,|\, \gamma_C(x) < \infty\}.$$

It is actually a norm, having C as its unit ball, if and only if C is bounded with nonempty interior and is symmetric: $-C = C$.

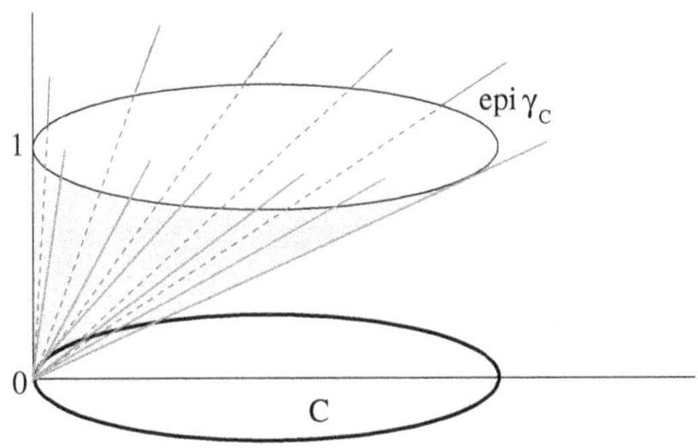

Fig. 3–12. The gauge function of a set as a positive hull.

Detail. The fact that C is convex with $0 \in C$ ensures having $\lambda C \subset \lambda' C$ when $\lambda < \lambda'$, and $C^\infty = \bigcup_{\lambda>0} \lambda C$ (cf. 3.6). From C being closed, we deduce that $\operatorname{lev}_{\leq\lambda} \gamma_C = \lambda C$ for all $\lambda \in (0,\infty)$, whereas $\operatorname{lev}_{\leq 0} \gamma_C = C^\infty$. Hence γ_C is lsc with $\operatorname{dom} \gamma_C = \operatorname{pos} C$. Because $\gamma_C = \operatorname{pos} f$ for $f = \delta_C + 1$, which is convex, γ_C is sublinear by 3.49(b).

Boundedness of C, which is equivalent by 3.5 to $C^\infty = \{0\}$, corresponds to the property that $\gamma_C(x) = 0$ only for $x = 0$. Symmetry of C corresponds to having $\gamma_C(-x) = \gamma_C(x)$. The only additional property required for γ_C to be a norm (as described in 2.17) is finiteness, which means that for every $x \neq 0$ there exists $\lambda \in \mathbb{R}_+$ with $x \in \lambda C$, or in other words, $\operatorname{pos} C = \mathbb{R}^n$. Obviously this is true when $0 \in \operatorname{int} C$.

Conversely, if $\operatorname{pos} C = \mathbb{R}^n$ consider any simplex neighborhood $V = \operatorname{con}\{a_0, \ldots, a_n\}$ of 0 (cf. 2.28), and for each a_i select $\lambda_i > 0$ such that $a_i \in \lambda_i C$.

Let ε be the lowest of the values λ_i^{-1}. Then $\varepsilon a_i \in C$ for all i, so that the set $\operatorname{con}\{\varepsilon a_0, \ldots, \varepsilon a_n\} = \varepsilon V$ is (by the convexity of C) a neighborhood of 0 within C; thus $0 \in \operatorname{int} C$. When C is symmetric, 0 has to belong to its interior if that's nonempty, as is clear from the line segment principle in 2.33 (because 0 is an intermediate point on the line segment joining any point $x \in \operatorname{int} C$ with the point $-x$, also belonging to $\operatorname{int} C$). □

3.51 Exercise (entropy functions). *Under the convention that* $0 \log 0 = 0$, *the function* $g : \mathbb{R}^n \to \overline{\mathbb{R}}$ *defined for* $y = (y_1, \ldots, y_n)$ *by*

$$g(y) = \begin{cases} \sum_{j=1}^n y_j \log y_j & \text{when } y_j \geq 0,\ \sum_{j=1}^n y_j = 1, \\ \infty & \text{otherwise,} \end{cases}$$

is proper, lsc and convex. Furthermore, the function $h : \mathbb{R}^n \to \overline{\mathbb{R}}$ *defined by*

$$h(y) = \begin{cases} \sum_{j=1}^n y_j \log y_j - (\sum_{j=1}^n y_j) \log(\sum_{j=1}^n y_j) & \text{when } y_j \geq 0, \\ \infty & \text{otherwise,} \end{cases}$$

is proper, lsc and sublinear.

Guide. To get the convexity of g, write $g(y)$ as $\sum_{j=1}^n \theta(y_j) + \delta_{\{0\}}(1 - \sum_{j=1}^n y_j)$ where $\theta(t)$ has the value $t \log t$ for $t > 0$, 0 for $t = 0$, and ∞ for $t < 0$. Then to get the properties of h show that $h = \operatorname{pos} g$. □

The function g in 3.51 will be seen in 11.12 to be dual to the log-exponential function in the sense of the Legendre-Fenchel transform.

The theory of positive hulls and convex hulls has important implications for the study of polyhedral sets and convex, piecewise linear functions.

3.52 Theorem (generation of polyhedral cones; Minkowski-Weyl). *A cone* K *is polyhedral if and only if it can be expressed as* $\operatorname{con}(\operatorname{pos}\{b_1, \ldots, b_r\})$ *for some finite collection of vectors* $b_1, \ldots, b_r$.

Proof. Suppose first that $K = \operatorname{con}\big(\operatorname{pos}\{b_1, \ldots, b_r\}\big)$. By 3.15, K is the union of $\{0\}$ and the finitely many cones $K_J = \operatorname{con}\big(\operatorname{pos}\{b_j \mid j \in J\}\big)$ corresponding to index sets $J \subset \{1, \ldots, r\}$ such that the vectors b_j for $j \in J$ are linearly independent. Each of the convex cones K_J is closed, even polyhedral; this is obvious from the coordinate representation of K_J relative to a basis for $\mathbb{R}^n$ that includes $\{b_j \,|\, j \in J\}$. As a convex set expressible as the union of a finite collection of polyhedral sets, K itself is polyhedral by Lemma 2.50.

Suppose now instead that K is polyhedral. Because K is a cone, any closed half-space $\{x \,|\, \langle a, x\rangle \leq \alpha\}$ that includes K must have $\alpha \geq 0$, and then the half-space $\{x \,|\, \langle a, x\rangle \leq 0\}$ includes K as well. Therefore, K must actually have a representation of the form

$$K = \{x \,|\, \langle a_i, x\rangle \leq 0 \text{ for } i = 1, \ldots, m\}. \tag{3(14)}$$

In particular, $K^\infty = K$. For the subspace $M = K \cap (-K)$ (cf. 3.8) we have $K + M = K$ by 3.6, and therefore in terms of the orthogonal complement

$M^\perp := \{w \,|\, \langle x, w\rangle = 0$ for all $x \in M\}$ actually that

$$K = K' + M \text{ for } K' := K \cap M^\perp.$$

Here K' is a closed, convex cone too, and K' is pointed, because $K' \cap (-K') \subset M \cap M^\perp = \{0\}$ (cf. 3.14). Moreover K' is polyhedral: for any basis $w_1, \ldots, w_d$ of M, K' consists of the vectors x satisfying not only $\langle a_i, x\rangle \le 0$ for $i = 1, \ldots, m$ in 3(14), but also $\langle \pm w_k, x\rangle \le 0$ for $k = 1, \ldots, d$. If we can find a representation $K' = \operatorname{con}\big(\operatorname{pos}\{b_1, \ldots, b_r\}\big)$, we will have from $K = K' + M$ a corresponding representation $K = \operatorname{con}\big(\operatorname{pos}\{b_1, \ldots, b_r, \pm w_1, \ldots, \pm w_d\}\big)$.

This argument shows there's no loss of generality in focusing on polyhedral cones that are pointed. Thus, we may assume that $K \cap (-K) = \{0\}$, but also, to avoid triviality, that $K \neq \{0\}$. For each index set $I \subset \{1, \ldots, m\}$ in 3(14) let K_I denote the polyhedral cone consisting of the vectors $x \in K$ such that $\langle a_i, x\rangle = 0$ for all $i \in I$. Obviously, the union of all the cones K_I is K. Let K_0 be the union of all the cones K_I that happen to be single rays, i.e., to have dimension 1. We'll demonstrate that if K_I has dimension greater than 1, then any nonzero vector in K_I belongs to a sum $K_{I'} + K_{I''}$ for index sets I' and I'' properly larger than I. This will establish, through repeated application, that every nonzero vector in K can be represented as a sum of vectors belonging to cones K_I of dimension 1, or in other words, as a sum of vectors in K_0. We'll know then from 3.15 that $K = \operatorname{con} K_0$; in particular $K_0 \neq \emptyset$. Because K_0 is the union of finitely many rays we'll have $K_0 = \operatorname{pos}\{b_1, \ldots, b_r\}$ for certain vectors b_j, and the desired representation of K will be achieved.

Suppose therefore that $0 \neq \bar{x} \in K_I$ and $\dim K_I > 1$. Then there's a vector $\tilde{x} \neq 0$ in K_I that isn't just a scalar multiple of $\bar{x}$. Let

$$\overline{T} = \{\tau \in \mathbb{R}_+ \,|\, \bar{x} - \tau\tilde{x} \in K\}, \qquad \widetilde{T} = \{\tau \in \mathbb{R}_+ \,|\, \tilde{x} - \tau\bar{x} \in K\}.$$

It's clear that $\overline{T}$ is a closed interval containing 0, and the same for $\widetilde{T}$. On the other hand, $\overline{T}$ can't be unbounded , for if it contained a sequence $0 < \tau^\nu \nearrow \infty$ we would have $(1/\tau^\nu)[\bar{x} - \tau^\nu\tilde{x}] \in K$ for all ν and consequently in the limit $-\tilde{x} \in K$, in contradiction to our knowledge that $\tilde{x} \in K$ and $K \cap (-K) = \{0\}$. Therefore, $\overline{T}$ has a highest element $\tau' > 0$. The vector $x' = \bar{x} - \tau'\tilde{x}$ must then be such that the index set $I' := \{i \,|\, \langle a_i, x'\rangle = 0\}$ is properly larger than I. Likewise, $\widetilde{T}$ has a highest element $\tau'' > 0$, and the vector $x'' = \tilde{x} - \tau''\bar{x}$ must be such that the index set $I'' := \{i \,|\, \langle a_i, x''\rangle = 0\}$ is properly larger than I. We note that $(1/\tau'')\tilde{x} - \bar{x} \in K$, so $1/\tau'' \le \tau'$ by the definition of τ'. It's impossible that $1/\tau'' = \tau'$, because then $(1/\tau'')\tilde{x} - \bar{x}$ would be the vector $-x'$, and we would have both x' and $-x'$ in K with $x' \neq 0$ (because $\tilde{x}$ and $\bar{x}$ aren't multiples of each other), contrary to $K \cap (-K) = \{0\}$. Hence $\tau'\tau'' < 1$. Let $\bar{x}' = [1/(1 - \tau'\tau'')]x'$ and $\bar{x}'' = [\tau'/(1 - \tau'\tau'')]x''$. Then $\bar{x}' \in K_{I'}$, $\bar{x}'' \in K_{I''}$,

$$\bar{x}' + \bar{x}'' = \frac{1}{1 - \tau'\tau''}\big(\bar{x} - \tau'\tilde{x}\big) + \frac{\tau'}{1 - \tau'\tau''}\big(\tilde{x} - \tau''\bar{x}\big) = \bar{x}.$$

Thus, we have a representation $\bar{x} \in K_{I'} + K_{I''}$ of the kind sought. □

3.53 Corollary (polyhedral sets as convex hulls). *A set C is polyhedral if and only if it can be represented as the convex hull of at most finitely many ordinary points and finitely many direction points, i.e., in the form*

$$C = \operatorname{con}\{a_1, \ldots, a_m\} + \operatorname{con}\big(\operatorname{pos}\{a_{m+1}, \ldots, a_r\}\big).$$

Proof. A nonempty polyhedral set C specified by a finite system of inequalities $\langle c_j, x\rangle \leq \gamma_j$ has $\operatorname{csm} C$ corresponding in the ray space model to the cone given by the inequalities $\big\langle (c_j, \gamma_j), (x, \beta)\big\rangle \leq 0$ and $\big\langle (0, -1), (x, \beta)\big\rangle \leq 0$, and conversely such cones correspond to polyhedral sets C. The result is obtained by applying Theorem 3.52 to these cones in $\mathbb{R}^{n+1}$. The rays that generate them can be normalized to the two types $(a_i, -1)$ and $(a_i, 0)$. □

3.54 Exercise (generation of convex, piecewise linear functions). *A function $f : \mathbb{R}^n \to \overline{\mathbb{R}}$ has $\operatorname{epi} f$ polyhedral if and only if, for some collection of vectors a_i and scalars c_i, f can be expressed in the form*

$$f(x) = \begin{cases} \text{infimum of } \; t_1 c_1 + \cdots + t_m c_m + t_{m+1} c_{m+1} + \cdots + t_r c_r \\ \text{subject to } \; t_1 a_1 + \cdots + t_m a_m + t_{m+1} a_{m+1} + \cdots + t_r a_r = x \\ \text{with } \; t_i \geq 0 \text{ for } i = 1, \ldots, r, \; \sum_{i=1}^m t_i = 1, \end{cases}$$

where the infimum is attained when it is not infinite. Thus, among functions having such a representation, those for which the infimum for at least one x is finite are the proper, convex, piecewise linear functions f with $\operatorname{dom} f \neq \emptyset$.

Guide. Use 2.49 and 3.53 to identify the class of functions in question with the ones whose epigraph is the extended convex hull of finitely many ordinary points in $\mathbb{R}^{n+1}$ and direction points in $\operatorname{hzn} \mathbb{R}^{n+1}$. Work out the meaning of such a representation in terms of a formula for $f(x)$. Derive attainment of the infimum from the closedness of the epigraph. □

3.55 Proposition (polyhedral and piecewise linear operations).

(a) *If C is polyhedral, then C^∞ is polyhedral. If C_1 and C_2 are polyhedral, then so too are $C_1 \cap C_2$ and $C_1 + C_2$. Furthermore, $L(C)$ and $L^{-1}(D)$ are polyhedral when C and D are polyhedral and L is linear.*

(b) *If f is proper, convex and piecewise linear, f^∞ has these properties as well. If f_1 and f_2 are convex and piecewise linear, then so too are $\max\{f_1, f_2\}$, $f_1 + f_2$ and $f_1 \# f_2$, when proper. Also, $f \circ L$ is convex and piecewise linear when f is convex and piecewise linear and the mapping L is linear. Finally, if $p(u) = \inf_x f(x, u)$ for a function f that is convex and piecewise linear, then p is convex and piecewise linear unless it has no values other than ∞ and $-\infty$.*

Proof. The polyhedral convexity of C^∞ is seen from the special case of 3.24 where the functions f_i are affine (or from the proof of 3.53). That of $C_1 \cap C_2$ and $L^{-1}(D)$ follows from considering C_1, C_2 and D as intersections of finite collections of closed half-spaces as in the definition of polyhedral sets in 2.10.

For C_1+C_2 and $L(C)$ one appeals instead to representations as extended convex hulls of finitely many elements in the mode of 3.53.

The convex piecewise linearity of f^∞ follows from that of f because this property corresponds to a polyhedral epigraph; cf. 2.49 and the definition of f^∞ in 3.17. The convex piecewise linearity of $\max\{f_1, f_2\}$, $f_1 + f_2$, and $f \circ L$ is evident from the definition of piecewise linearity and the fact that these operations preserve convexity. For $f_1 \# f_2$ and p one can appeal to the epigraphical interpretations in 1.18 and 1.28, invoking representations as in 3.54. □

Commentary

Many compactifications of $\mathbb{R}^n$ have been proposed and put to use. The simplest is the one-point compactification, in which a single abstract element is added to represent 'infinity', but there is also the well known Stone-Čech compactification (making every bounded continuous function on $\mathbb{R}^n$ have a continuous extension to the larger space) and the compactification of n-dimensional projective geometry, in which a new point is added for each family of parallel lines in $\mathbb{R}^n$. The latter is closest in spirit to the 'cosmic' compactification developed here, but it's also quite different because it doesn't distinguish between directions that are opposite to each other.

For all the importance of 'directions' in analysis, it's surprising that the notion has been so lacking in mathematical formalization, aside from the one-dimensional case served by adjoining ∞ and $-\infty$ to $\mathbb{R}$. Most often, authors have been content with identifying 'directions' with vectors of length one, which works to a degree but falls short of full potential because of the easy confusion of such vectors with ones in an ordinary role, and the lack of a single space in which ordinary points and direction points can be contemplated together. The portrayal of directions as abstract points corresponding to equivalence classes of half-lines under parallelism, or in other words as corresponding one-to-one with rays emanating from the origin, was offered in Rockafellar [1970a], but only algebraic issues connected with convexity were then pursued. A similar idea can be glimpsed in the geometric thinking of Bouligand [1932a] much earlier, but on an informal basis only. Not until here has the concept been developed fully as a topological compactification with all its ramifications, although a precursor was the paper of Rockafellar and Wets [1992].

Interest in cosmic compactification ideas has been driven especially by applications to the behavior of the 'min' operation under the epigraphical convergence of functions studied in Chapter 7 (and the underpinnings of this theory in Chapter 5), and by the need for a proper understanding of the 'horizon subgradients' that are crucial in the subdifferential analysis of Chapters 8–10.

Many of the properties of the cosmic closure of $\mathbb{R}^n$ have been implicit in other work, of course, especially with regard to convexity. What we have called the 'horizon cone' of a set was introduced as the 'asymptotic cone' by Steinitz [1913], [1914], [1916]. For a *convex* set C that's *closed*, it's the same as the *recession cone* of C defined in Rockafellar [1970a]. We have preferred the term 'horizon' here to 'asymptotic' for several reasons. It better expresses the underlying geometry and motivation for the compactification: horizon cones represent sets of horizon points lying in the horizon of

$\mathbb{R}^n$. It better lends itself to broader usage in association with functions and mappings, as well as special kinds of limits in the theory of set convergence in Chapter 4 for which the term 'asymptotic' would be awkward. Also, by being less encumbered than 'asymptotic' by other meanings, it helps to make clear how cosmic compactification ideas enter into variational analysis.

The term 'recession cone' was temporarily used by Rockafellar [1981b] in the present sense of 'horizon cone' for nonconvex as well as convex sets, but we reserve that now for a different concept which better extends the meaning of 'recession' beyond the territory of convexity; see 6.33–6.34.

The horizon properties of unbounded convex sets were already well understood by Steinitz, who can be credited with the facts in Theorem 3.6, in particular. The corresponding theory of horizon cones was developed further by Stoker [1940]. Such cones were used by Choquet [1962] in expressing the closures of convex hulls of unions of convex sets and for closure results in convex analysis more generally by Rockafellar [1970a] (§8). Some facts, such as the horizon cone criterion for the closedness of the sum of two sets (specializing the criterion for an arbitrary number of sets in 3.12) were obtained earlier by Debreu [1959] even for nonconvex sets. The possibility of inequality in the horizon cone formula for product sets in 3.11 (as demonstrated by the example following that result) hasn't previously been noted.

The application of horizon cone theory to epigraphs to generate growth properties of functions carries forward to the nonconvex case a theme of Rockafellar [1963], [1966b], [1970a], for convex functions. Growth properties of nonconvex functions, even on infinite-dimensional spaces, have been analyzed in this manner by Baiocchi, Buttazzo, Gastaldi and Tomarelli [1988], Zălinescu [1989], and Auslender [1996] for the purpose of understanding the existence of optimal solutions and the convergence of methods for finding them.

The coercivity result in 3.26 is new, at least in such detail. We have tried here to straighten out the terminology of 'coercivity' so as to avoid some of the conflicts and ambiguities that have arisen in what this means for functions $f : \mathbb{R}^n \to \overline{\mathbb{R}}$. For convex functions f, coercivity was called 'co-finiteness' in Rockafellar [1970a] because of its tie to the finiteness of the conjugate convex function (see 11.8(d)), but that term isn't very apt for a general treatment of nonconvex functions. Coercivity concepts have an important role in nonlinear analysis not just for functions f into $\mathbb{R}$ or $\overline{\mathbb{R}}$ but also vector-valued mappings from one linear space into another; cf. Brezis [1968], [1972], Browder [1968a], [1968b], and Rockafellar [1970c].

The application of coercivity to parametric minimization in Theorem 3.31 appears for the convex case in Rockafellar [1970a] along with the facts in 3.32. A generalization to the nonconvex case was presented by Auslender [1996].

The sublinearity fact in 3.20 goes back to the theory of convex bodies (cf. Bonnesen and Fenchel [1934]), to which it is related through the notion of 'support function' that will be explained in Chapter 8.

The cancellation rule in 3.35 for sums of convex sets is due to Rådström [1952] for compact C_1 and C_2. The general case with possibly unbounded sets and the epi-addition version in 3.34 for $f_1 \# g = f_2 \# g$ don't seem to have been observed before, although Zagrodny [1994] has demonstrated the latter for g strictly convex and explored the matter further in an infinite-dimensional setting. These cancellation results will later be easy consequences of the duality correspondence between convex sets and their 'support functions' (in Chapter 8) and the Legendre-Fenchel transform for convex functions in (Chapter 11), which convert addition of convex sets and epi-addition of convex functions into ordinary addition of functions where cancellation becomes trivial (when the function being canceled is finite). Rådström approached the matter abstractly, but the interpretation via support functions was made soon after by Hörmander [1954].

For convex cones, pointedness has customarily been defined in terms of the property in 3.14. The extension of pointedness to general cones through Definition 3.13, which gives the same concept when the cone is convex, was proposed by Rockafellar [1981b] for the express purpose of dealing with convex hulls of unbounded sets. But that paper attended mostly to other issues and didn't openly develop results corresponding to 3.15, 3.45 and 3.46, although these were to a certain degree implicit.

The concept of convex hulls of sets consisting of both ordinary points and horizon points, or in other words, in the setting of cosmic convexity as in 3.44, stems from Rockafellar [1970a]. That book also made the first applications of convex hull theory to functions, including the dual representation for 'polyhedral' functions in 3.53. The basic Minkowski-Weyl result in 3.52, which effectively furnishes a convex hull representation of generalized type for any polyhedral convex set, comes from Weyl [1935], but the proof given here is new.

The importance of convex cones in setting up general vector inequalities (cf. 3.38) that might be used in describing constraints has long been recognized in optimization theory, e.g. Duffin [1956]. Matrix inequalities such as in 3.39 were first utilized for this purpose by Bellman and Fan [1963] and are now popular in the subject called 'positive-definite programming'; cf. Boyd and Vanderberghe [1995].

The entropy function $g(y)$ described in 3.51 is fundamental to Boltzman-Shannon entropy in statistical mechanics. The related function $h(y)$ in 3.51 turns out to be the key to duality theory in log-exponential programming; see Duffin, Peterson and Zener [1967] and Rockafellar [1970a] (§30).

4. Set Convergence

The precise meaning of such basic concepts in analysis as differentiation, integration and approximation is dictated by the choice of a notion of limit for sequences of functions. In the past, pointwise limits have received most of the attention. Whether 'uniform' or invoked in an 'almost everywhere' sense, they underlie the standard definitions of derivatives and integrals as well as the very meaning of a series expansion. In variational analysis, however, pointwise limits are inadequate for such mathematical purposes. A different approach to convergence is required in which, on the geometric level, limits of sequences of sets have the leading role.

Motivation for the development of this geometric approach has come from optimization, stochastic processes, control systems and many other subjects. When a problem of optimization is approximated by a simpler problem, or a sequence of such problems, for instance, it's of practical interest to know what might be expected of the behavior of the associated sets of feasible or optimal solutions. How close will they be to those for the given problem? Related challenges arise in approximating functions that may be extended-real-valued and mappings that may be set-valued. The limiting behavior of a sequence of such functions and mappings, possibly discontinuous and not having the same effective domains, can't be well understood in a framework of pointwise convergence. And this fundamentally affects the question of how 'differentiation' might be extended to meet the demands of variational analysis, since that's inevitably tied to ideas of local approximation.

The theory of set convergence will provide ways of approximating set-valued mappings through convergence of graphs (Chapter 5) and extended-real-valued functions through convergence of epigraphs (Chapter 7). It will lead to tangent and normal cones to general sets (Chapter 6) and to subderivatives and subgradients of nonsmooth functions (Chapter 8 and beyond).

When should a sequence of sets C^ν in $I\!R^n$ be said converge to another such set C? For operational reasons in handling statements about sequences, it will be convenient to work with the following collections of subsets of $I\!N$:

$$\begin{aligned}\mathcal{N}_\infty &:= \{ N \subset I\!N \mid I\!N \setminus N \text{ finite} \} \\ &= \{ \text{ subsequences of } I\!N \text{ containing all } \nu \text{ beyond some } \bar\nu \}, \\ \mathcal{N}_\infty^\# &:= \{ N \subset I\!N \mid N \text{ infinite} \} = \{ \text{ all subsequences of } I\!N \}.\end{aligned}$$

The set $\mathcal{N}_\infty$ gives the 'filter' of neighborhoods of ∞ that are implicit in the notation $\nu \to \infty$, and $\mathcal{N}_\infty^\#$ is its associated 'grill'. Obviously, $\mathcal{N}_\infty \subset \mathcal{N}_\infty^\#$. The subsequences of a sequence $\{x^\nu\}_{\nu\in I\!N}$ have the form $\{x^\nu\}_{\nu\in N}$ with $N \in \mathcal{N}_\infty^\#$, while those that are 'tails' of $\{x^\nu\}_{\nu\in I\!N}$ have this form with $N \in \mathcal{N}_\infty$.

Subsequences will pervade much of what we do, so an initial investment in this notation will pay off in simpler formulas and even in bringing analogies to light that might otherwise be missed. We write $\lim_\nu$, $\lim_{\nu\to\infty}$ or $\lim_{\nu\in I\!N}$ when $\nu \to \infty$ as usual in $I\!N$, but $\lim_{\nu\in N}$ or $\lim_{\nu \underset{N}{\to} \infty}$ in the case of convergence of a subsequence designated by an index set N in $\mathcal{N}_\infty^\#$ or $\mathcal{N}_\infty$. The relations

$$\begin{aligned}\mathcal{N}_\infty^\# &= \Big\{N \subset I\!N \;\Big|\; \forall N' \in \mathcal{N}_\infty,\ N \cap N' \neq \emptyset\Big\}\\ \mathcal{N}_\infty &= \Big\{N \subset I\!N \;\Big|\; \forall N' \in \mathcal{N}_\infty^\#,\ N \cap N' \neq \emptyset\Big\}\end{aligned} \qquad 4(1)$$

express a natural duality between $\mathcal{N}_\infty$ and $\mathcal{N}_\infty^\#$. An appeal to this duality can be helpful in arguments involving limit operations.

A. Inner and Outer Limits

The issue of whether a sequence of subsets of $I\!R^n$ has a limit can best be approached through the study of two 'semilimits' which always exist.

4.1 Definition (inner and outer limits). *For a sequence $\{C^\nu\}_{\nu\in I\!N}$ of subsets of $I\!R^n$, the outer limit is the set*

$$\begin{aligned}\limsup_{\nu\to\infty} C^\nu :&= \Big\{x \;\Big|\; \exists N \in \mathcal{N}_\infty^\#,\ \exists x^\nu \in C^\nu\ (\nu \in N)\ \text{with}\ x^\nu \underset{N}{\to} x\Big\}\\ &= \Big\{x \;\Big|\; \forall V \in \mathcal{N}(x),\ \exists N \in \mathcal{N}_\infty^\#,\ \forall \nu \in N\colon\ C^\nu \cap V \neq \emptyset\Big\},\end{aligned}$$

while the inner limit is the set

$$\begin{aligned}\liminf_{\nu\to\infty} C^\nu :&= \Big\{x \;\Big|\; \exists N \in \mathcal{N}_\infty,\ \exists x^\nu \in C^\nu\ (\nu \in N)\ \text{with}\ x^\nu \underset{N}{\to} x\Big\}\\ &= \Big\{x \;\Big|\; \forall V \in \mathcal{N}(x),\ \exists N \in \mathcal{N}_\infty,\ \forall \nu \in N\colon\ C^\nu \cap V \neq \emptyset\Big\}.\end{aligned}$$

The limit of the sequence exists if the outer and inner limit sets are equal:

$$\lim_{\nu\to\infty} C^\nu := \limsup_{\nu\to\infty} C^\nu = \liminf_{\nu\to\infty} C^\nu.$$

The inner and outer limits of a sequence $\{C^\nu\}_{\nu\in I\!N}$ always exist, although the limit itself might not, but they could sometimes be $\emptyset$. When $C^\nu \neq \emptyset$ for all ν, the set $\liminf_\nu C^\nu$ consists of all possible limit points of sequences $\{x^\nu\}_{\nu\in I\!N}$ with $x^\nu \in C^\nu$ for all ν, whereas $\limsup_\nu C^\nu$ consists of all possible cluster points of such sequences. In any case, it's clear from the inclusion $\mathcal{N}_\infty \subset \mathcal{N}_\infty^\#$ that $\liminf_\nu C^\nu \subset \limsup_\nu C^\nu$ always.

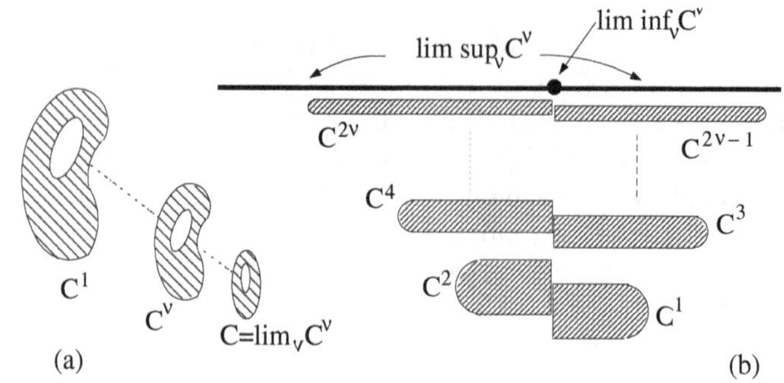

Fig. 4–1. Limit concepts. (a) A sequence of sets that converges to a limit set. (b) Inner and outer limits for a nonconvergent sequence of sets.

Without loss of generality, the neighborhoods V in Definition 4.1 can be taken to be of the form $\mathbb{B}(x,\varepsilon)$. Because the condition $\mathbb{B}(x,\varepsilon)\cap C^\nu\neq\emptyset$ is equivalent to $x\in C^\nu+\varepsilon\mathbb{B}$, the formulas can then be written just as well as

$$\begin{aligned}\liminf_{\nu\to\infty} C^\nu &= \Big\{x \;\Big|\; \forall\varepsilon>0,\ \exists N\in\mathcal{N}_\infty,\ \forall\nu\in N\colon\ x\in C^\nu+\varepsilon\mathbb{B}\Big\},\\ \limsup_{\nu\to\infty} C^\nu &= \Big\{x \;\Big|\; \forall\varepsilon>0,\ \exists N\in\mathcal{N}_\infty^\#,\ \forall\nu\in N\colon\ x\in C^\nu+\varepsilon\mathbb{B}\Big\}.\end{aligned} \qquad 4(2)$$

Inner and outer limits can also be expressed in terms of distance functions or operations of intersection and union. Recall that the distance of a point x from a set C is denoted by $d_C(x)$ (cf. 1.20), or alternatively by $d(x,C)$ when that happens to be more convenient. For $C=\emptyset$, we have $d(x,C)\equiv\infty$. Aside from that case, not only is the function d_C finite everywhere and continuous (as asserted in 1.20), it satisfies

$$d_C(x') \;\le\; d_C(x)+|x'-x| \;\text{ for all } x' \text{ and } x. \qquad 4(3)$$

This is evident from the fact that $|y-x'|\le|y-x|+|x'-x|$ for all $y\in C$.

4.2 Exercise (characterizations of set limits).

(a)
$$\liminf_{\nu\to\infty} C^\nu = \Big\{x \;\Big|\; \limsup_{\nu\to\infty} d(x,C^\nu)=0\Big\},$$
$$\limsup_{\nu\to\infty} C^\nu = \Big\{x \;\Big|\; \liminf_{\nu\to\infty} d(x,C^\nu)=0\Big\},$$

(b)
$$\liminf_{\nu\to\infty} C^\nu = \bigcap_{N\in\mathcal{N}_\infty^\#}\operatorname{cl}\bigcup_{\nu\in N} C^\nu, \qquad \limsup_{\nu\to\infty} C^\nu = \bigcap_{N\in\mathcal{N}_\infty}\operatorname{cl}\bigcup_{\nu\in N} C^\nu,$$

(c)
$$\liminf_{\nu\to\infty} C^\nu = \bigcap_{\varepsilon>0}\Big[\bigcup\nolimits_{\nu=1}^\infty \bigcap\nolimits_{\kappa=\nu}^\infty (C^\kappa+\varepsilon\mathbb{B})\Big].$$

B. Painlevé-Kuratowski Convergence

When $\lim_\nu C^\nu$ exists in the sense of Definition 4.1 and equals C, the sequence $\{C^\nu\}_{\nu\in\mathbb{N}}$ is said to *converge* to C, written

$$C^\nu \to C.$$

Set convergence in this sense is known more specifically as *Painlevé-Kuratowski* convergence. It is the basic notion for most purposes, although variants involving the cosmic closure of $\mathbb{R}^n$ will be developed for some applications. A competing notion of convergence with respect to 'Pompeiu-Hausdorff distance' will be explained later as amounting to the same thing for *bounded* sequences of sets (all in some bounded region of $\mathbb{R}^n$), but otherwise being inappropriate in its meaning and unworkable as a tool for dealing with sets like cones, epigraphs, and the graphs of mappings (see 4.13). Some preliminary examples of set limits are:

- A sequence of balls $\mathbb{B}(x^\nu, \rho^\nu)$ converges to the ball $\mathbb{B}(x, \rho)$ when $x^\nu \to x$ and $\rho^\nu \to \rho$. When $\rho^\nu \to \infty$, these balls converge to $\mathbb{R}^n$ while their complements converge to $\emptyset$.
- Consider a sequence that alternates between two different closed sets D_1 and D_2 in $\mathbb{R}^n$, with $C^\nu = D_1$ when ν is odd but $C^\nu = D_2$ when ν is even. Such a sequence fails to converge. Its inner limit is $D_1 \cap D_2$, whereas its outer limit is $D_1 \cup D_2$.
- For a set $D \subset \mathbb{R}^n$ with $\operatorname{cl} D = \mathbb{R}^n$ but $D \neq \mathbb{R}^n$ (for instance D could be the set of vectors whose coordinates are all rational), the constant sequence $C^\nu \equiv D$ converges to $C = \mathbb{R}^n$, not to D.

4.3 Exercise (limits of monotone and sandwiched sequences).
(a) $\lim_\nu C^\nu = \operatorname{cl} \bigcup_{\nu\in\mathbb{N}} C^\nu$ *whenever* $C^\nu \nearrow$, *meaning* $C^\nu \subset C^{\nu+1} \subset \cdots$;
(b) $\lim_\nu C^\nu = \bigcap_{\nu\in\mathbb{N}} \operatorname{cl} C^\nu$ *whenever* $C^\nu \searrow$, *meaning* $C^\nu \supset C^{\nu+1} \supset \cdots$;
(c) $C^\nu \to C$ *whenever* $C_1^\nu \subset C^\nu \subset C_2^\nu$ *with* $C_1^\nu \to C$ *and* $C_2^\nu \to C$.

4.4 Proposition (closedness of limits). *For any sequence of sets $C^\nu \subset \mathbb{R}^n$, both the inner limit set $\liminf_\nu C^\nu$ and the outer limit set $\limsup_\nu C^\nu$ are closed. Furthermore, they depend only on the closures $\operatorname{cl} C^\nu$, in the sense that*

$$\operatorname{cl} C^\nu = \operatorname{cl} D^\nu \quad \Longrightarrow \quad \begin{cases} \liminf_\nu C^\nu = \liminf_\nu D^\nu \\ \limsup_\nu C^\nu = \limsup_\nu D^\nu. \end{cases}$$

Thus, whenever $\lim_\nu C^\nu$ exists, it is closed. (If $C^\nu \equiv C$, then $\lim_\nu C^\nu = \operatorname{cl} C$.)

Proof. This is obvious from the intersection formulas in 4.2(b). □

The closure facts in Proposition 4.4 identify the natural setting for the study of set convergence as the space of all *closed* subsets of $\mathbb{R}^n$. Limit sets of all types are closed, and in passing to the limit of a sequence of sets it's only the closures of the sets that matter. The consideration of sets that aren't closed is really unnecessary in the context of convergence, and indeed, such sets can raise awkward issues, as shown by the last of the examples before 4.3. On the other hand, from the standpoint of exposition a systematic focus only on closed sets could be burdensome and might even seem like a restriction; an extra assumption would have to be added to the statement of every result. We continue therefore with general sets, as long as it's expedient to do so.

The next theorem and its corollary provide the major criteria for checking set convergence.

4.5 Theorem (hit-and-miss criteria). *For C^ν, $C \subset \mathbb{R}^n$ with C closed, one has*

(a) $C \subset \liminf_\nu C^\nu$ *if and only if for every open set $O \subset \mathbb{R}^n$ with $C \cap O \neq \emptyset$ there exists $N \in \mathcal{N}_\infty$ such that $C^\nu \cap O \neq \emptyset$ for all $\nu \in N$;*

(b) $C \supset \limsup_\nu C^\nu$ *if and only if for every compact set $B \subset \mathbb{R}^n$ with $C \cap B = \emptyset$ there exists $N \in \mathcal{N}_\infty$ such that $C^\nu \cap B = \emptyset$ for all $\nu \in N$;*

(a′) $C \subset \liminf_\nu C^\nu$ *if and only if whenever $C \cap \operatorname{int} \mathbb{B}(x, \rho) \neq \emptyset$ for a ball $\mathbb{B}(x, \rho)$, there exists $N \in \mathcal{N}_\infty$ such that $C^\nu \cap \operatorname{int} \mathbb{B}(x, \rho) \neq \emptyset$ for all $\nu \in N$;*

(b′) $C \supset \limsup_\nu C^\nu$ *if and only if, whenever $C \cap \mathbb{B}(x, \rho) = \emptyset$ for a ball $\mathbb{B}(x, \rho)$, there exists $N \in \mathcal{N}_\infty$ such that $C^\nu \cap \mathbb{B}(x, \rho) = \emptyset$ for all $\nu \in N$.*

(c) *It suffices in (a′) and (b′) to consider the countable collection of all balls $\mathbb{B}(x, \rho)$ such that ρ and the coordinates of x are rational numbers.*

Proof. In (a), it's evident from Definition 4.1 that '$\Rightarrow$' holds. The condition in the second half of (a) obviously implies, in turn, the condition in the second half of (a′). By demonstrating that the special version of the latter in (c) (for rational balls only) guarantees $C \subset \liminf_\nu C^\nu$, we'll establish the equivalences in both (a) and (a′).

Consider any $x \in C$ and rational $\varepsilon > 0$. There is a rational point $x' \in \operatorname{int} \mathbb{B}(x, \varepsilon/2)$. For such a point x' we have $C \cap \operatorname{int} \mathbb{B}(x', \varepsilon/2) \neq \emptyset$, so by assumption there exists $N \in \mathcal{N}_\infty$ with $C^\nu \cap \operatorname{int} \mathbb{B}(x', \varepsilon/2) \neq \emptyset$ for all $\nu \in N$. Then in particular $x' \in C^\nu + (\varepsilon/2)\mathbb{B}$, so that $x \in C^\nu + (\varepsilon/2)\mathbb{B} + (\varepsilon/2)\mathbb{B} = C^\nu + \varepsilon\mathbb{B}$ for all $\nu \in N$. Thus, x satisfies the defining condition in 4.1 for membership in $\liminf_\nu C^\nu$.

Likewise, '$\Rightarrow$' holds in (b) on the basis of Definition 4.1, while the condition in the second half of (b) implies in turn the condition in the second half of (b′) and then its rational version in (c). We have to argue from the latter back to the property that $C \supset \limsup_\nu C^\nu$. Thus, in assuming the rational version of the condition in the second half of (b′) and considering an arbitrary point $x \notin C$, we need to demonstrate that x fails to belong to $\limsup_\nu C^\nu$.

Because C is closed, there's a rational $\varepsilon > 0$ such that $C \cap \mathbb{B}(x, 2\varepsilon) = \emptyset$. A rational point x' can be selected from $\operatorname{int} \mathbb{B}(x, \varepsilon)$, and we then have $x \in \operatorname{int} \mathbb{B}(x', \varepsilon)$ and $C \cap \mathbb{B}(x', \varepsilon) = \emptyset$. By assumption, there must exist $N \in \mathcal{N}_\infty$ such that $C^\nu \cap \mathbb{B}(x', \varepsilon) = \emptyset$ for all $\nu \in N$. Since $x \in \operatorname{int} \mathbb{B}(x', \varepsilon)$, it's impossible in this case for x to belong to $\limsup_\nu C^\nu$, as seen from Definition 4.1. □

4.6 Exercise (index criterion for convergence). *The following is sufficient for $\lim_\nu C^\nu$ to exist: whenever the index set $N = \{\nu \mid C^\nu \cap O \neq \emptyset\}$ for an open set O belongs to $\mathcal{N}_\infty^\#$, it actually belongs to $\mathcal{N}_\infty$.*

Guide. Argue from Theorem 4.5. □

The following consequence of Theorem 4.5 provides a bridge to the quantification of set convergence through distance functions.

4.7 Corollary (pointwise convergence of distance functions). *For sets C^ν and C in $I\!R^n$ with C closed, one has $C^\nu \to C$ if and only if $d(x, C^\nu) \to d(x, C)$ for every $x \in I\!R^n$. In fact*

(a) *$C \subset \liminf_\nu C^\nu$ if and only if $d(x, C) \geq \limsup_\nu d(x, C^\nu)$ for all x,*

(b) *$C \supset \limsup_\nu C^\nu$ if and only if $d(x, C) \leq \liminf_\nu d(x, C^\nu)$ for all x.*

Proof. Nothing is lost by assuming C^ν to be closed. For any closed set C',

$$\begin{aligned} d(x, C') < \alpha &\iff C' \cap \operatorname{int} I\!B(x, \alpha) \neq \emptyset, \\ d(x, C') > \beta &\iff C' \cap I\!B(x, \beta) = \emptyset, \end{aligned}$$

so (a) and (b) are just reformulations of 4.5(a′) and 4.5(b′). □

In applying Corollary 4.7 to the case where actually $C = \liminf_\nu C^\nu$ or $C = \limsup_\nu C^\nu$, one obtains a distance function equation in the second case, but not in the first, as noted next.

4.8 Exercise (equality in distance limits). *For a sequence of sets C^ν in $I\!R^n$ one always has $\liminf_\nu d(x, C^\nu) = d(x, \limsup_\nu C^\nu)$, but in general only $\limsup_\nu d(x, C^\nu) \leq d(x, \liminf_\nu C^\nu)$.*

Guide. The specialization of Corollary 4.7(a) and (b) to set limit equalities doesn't directly furnish equalities for the limits of the distance functions, just special inequalities. In the case of $\limsup_\nu C^\nu$, however, one can argue that when $\alpha > d(x, \limsup_\nu C^\nu)$ there must be a sequence $\{x^\nu\}_{\nu \in N}$ with $N \in \mathcal{N}_\infty^\#$, $x^\nu \in C^\nu$ for $\nu \in N$, such that $|x - x^\nu| < \alpha$. This leads to the opposite inequality.

An example showing that the inequality in the case of $\liminf_\nu C^\nu$ can't always be strengthened to an equality is generated by taking $C^\nu = (I\!R, 0) \subset I\!R^2$ when ν is even, $C^\nu = (0, I\!R)$ when ν is odd. □

Set convergence can be described in terms of 'gap' measurements, too. The *gap distance* between two nonempty sets C and D in $I\!R^n$ is

$$\begin{aligned} \operatorname{gap}(C, D) :&= \inf\{|x - y| \mid x \in C,\, y \in D\} \\ &= \inf_z \{d(z, C) + d(z, D)\}. \end{aligned} \qquad 4(4)$$

Observe that $\operatorname{gap}(C, \{x\}) = d(x, C)$ for any singleton $\{x\}$; more generally one has $\operatorname{gap}(C, I\!B(x, \rho)) = \max[d(x, C) - \rho, 0]$ for any $\rho \in (0, \infty)$. It follows then from 4.7 that a sequence of sets C^ν converges to a closed set $C \neq \emptyset$ if and only if $\operatorname{gap}(C^\nu, I\!B(x, \rho)) \to \operatorname{gap}(C, I\!B(x, \rho))$ for all $x \in I\!R^n$ and $\rho > 0$.

Not only distances but also projections (as in 1.20) can be used in characterizing set convergence.

4.9 Proposition (set convergence through projections). *For nonempty, closed sets C^ν and C in $\mathbb{R}^n$, one has $C^\nu \to C$ if and only if* $\limsup_\nu d(0, C^\nu) < \infty$ *and the projection mappings P_{C^ν} have the property that*

$$\operatorname{limsup}_\nu P_{C^\nu}(x) \subset P_C(x) \;\text{ for all } x.$$

When the sets C^ν and C are also convex, one simply has that $C^\nu \to C$ if and only if $P_{C^\nu}(x) \to P_C(x)$ for all x.

Proof. Necessity in general: From $C^\nu \to C$ we have $d(x, C^\nu) \to d(x, C) < \infty$ for all x by 4.7 (in particular $x = 0$, so that $\limsup_\nu d(0, C^\nu) < \infty$). Consider any $\bar{x} \in \limsup_\nu P_{C^\nu}(x)$. There's an index set $N \in \mathcal{N}_\infty^\#$ such that $\bar{x} = \lim_{\nu \in N} \bar{x}^\nu$ for points $\bar{x}^\nu \in C^\nu$ with $|\bar{x}^\nu - x| = d(x, C^\nu)$. Taking limits in this equation we get $|\bar{x} - x| = d(x, C)$, but also $\bar{x} \in C$, so that $\bar{x} \in P_C(x)$.

Sufficiency in general: Consider any $x \in \mathbb{R}^n$. It suffices by 4.7 to verify that $d(x, C^\nu) \to d(x, C)$. The sets $P_{C^\nu}(x)$ are nonempty by 1.20, so for each ν we can choose some $\bar{x}^\nu \in P_{C^\nu}(x)$, i.e., $\bar{x} \in C^\nu$ with $|\bar{x}^\nu - x| = d(x, C^\nu)$. These distances form a bounded sequence, because $d(x, C^\nu) \le d(0, C^\nu) + |x|$ and $\limsup_\nu d(0, C^\nu) < \infty$. Any cluster point of $\{d(x, C^\nu)\}_{\nu \in \mathbb{N}}$ must therefore be of the form $|\bar{x} - x|$ for some cluster point $\bar{x}$ of $\{\bar{x}^\nu\}_{\nu \in \mathbb{N}}$. But such a cluster point $\bar{x}$ belongs by assumption to $P_C(x)$ and therefore has $|\bar{x} - x| = d(x, C)$. Hence the unique cluster point of the bounded sequence $\{d(x, C^\nu)\}_{\nu \in \mathbb{N}}$ is $d(x, C)$, and the desired conclusion is at hand.

The simplified characterization in the convex case comes from the fact that the projections $P_{C^\nu}(x)$ and $P_C(x)$ are singletons then; cf. 2.25. Of course $d(0, C^\nu) \to d(0, C)$ when $P_{C^\nu}(0) \to P_C(0)$. □

The characterization of set convergence in Proposition 4.9 will be extended in 5.35 to the 'graphical convergence' of the associated projection mappings.

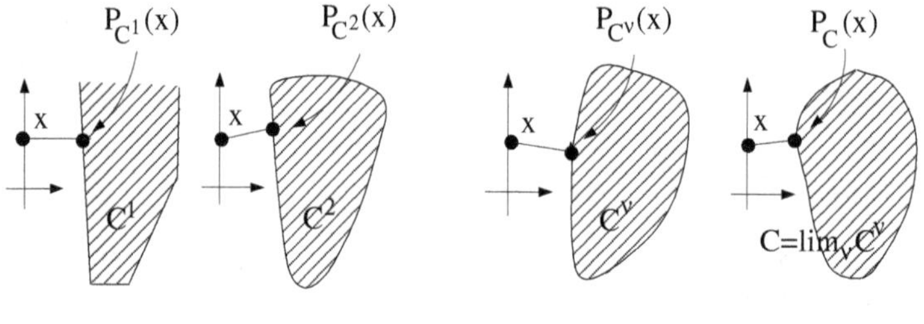

Fig. 4–2. Projections onto converging convex sets.

The next result furnishes clear geometric insight into the 'closeness' relationships between sets that are the hallmark of set convergence.

4.10 Theorem (uniformity of approximation in set convergence). *For subsets C^ν, $C \subset \mathbb{R}^n$ with C closed, one has*

(a) *$C \subset \liminf_\nu C^\nu$ if and only if for every $\rho > 0$ and $\varepsilon > 0$ there is an index set $N \in \mathcal{N}_\infty$ with $C \cap \rho\mathbb{B} \subset C^\nu + \varepsilon\mathbb{B}$ for all $\nu \in N$;*

(b) $C \supset \limsup_\nu C^\nu$ *if and only if for every* $\rho > 0$ *and* $\varepsilon > 0$ *there is an index set* $N \in \mathcal{N}_\infty$ *with* $C^\nu \cap \rho \mathbb{B} \subset C + \varepsilon \mathbb{B}$ *for all* $\nu \in N$.

Thus, $C = \lim_\nu C^\nu$ *if and only if for every* $\rho > 0$ *and* $\varepsilon > 0$ *there is an index set* $N \in \mathcal{N}_\infty$ *such that both inclusions hold. The following variants of the characterizations in (a) and (b) are valid as well:*

(a′) $C \subset \liminf_\nu C^\nu$ *if and only if for every* $\bar{x} \in \mathbb{R}^n$, $\rho > 0$ *and* $\varepsilon > 0$, *there is an index set* $N \in \mathcal{N}_\infty$ *with* $C \cap \mathbb{B}(\bar{x}, \rho) \subset C^\nu + \varepsilon \mathbb{B}$ *for all* $\nu \in N$;

(b′) $C \supset \limsup_\nu C^\nu$ *if and only if for every* $\bar{x} \in \mathbb{R}^n$, $\rho > 0$ *and* $\varepsilon > 0$, *there is an index set* $N \in \mathcal{N}_\infty$ *with* $C^\nu \cap \mathbb{B}(\bar{x}, \rho) \subset C + \varepsilon \mathbb{B}$ *for all* $\nu \in N$.

In all of these characterizations, it suffices that the inclusions be satisfied for all ρ *large enough (i.e. for every* $\rho \geq \bar{\rho}$ *for some* $\bar{\rho} \geq 0$*). In addition,* ρ *and* ε *can be restricted to be rational.*

Proof. Sufficiency in (a): Suppose the inclusion in (a) holds for all $\rho \geq \bar{\rho}$ for some $\bar{\rho} \geq 0$. Consider any $x \in C$ and let $\rho \geq \max\{\bar{\rho}, |x|\}$. Then $x \in C \cap \rho \mathbb{B}$, so for any $\varepsilon > 0$ there is an index set $N \in \mathcal{N}_\infty$ with $x \in C^\nu + \varepsilon \mathbb{B}$ for all $\nu \in N$. This means that $x \in \liminf_\nu C^\nu$.

Necessity in (a): Suppose to the contrary that one can find $\rho > 0$, $\varepsilon > 0$ and $N \in \mathcal{N}_\infty^\#$ such that points $x^\nu \in [C \cap \rho \mathbb{B}] \setminus [C^\nu + 2\varepsilon \mathbb{B}]$ exist for $\nu \in N$ with $x^\nu \underset{N}{\rightarrow} \bar{x}$. When $\nu \in N$ is large enough, one has

$$2\varepsilon \leq d(x^\nu, C^\nu) \leq d(\bar{x}, C^\nu) + |\bar{x} - x^\nu| \leq d(\bar{x}, C^\nu) + \varepsilon.$$

Letting $\nu \underset{N}{\rightarrow} \infty$ and taking Corollary 4.7(a) into account, one gets $2\varepsilon \leq \limsup_\nu d(\bar{x}, C^\nu) + \varepsilon \leq d(\bar{x}, C) + \varepsilon$, where $d(\bar{x}, C) = 0$ because $\bar{x} \in C \cap \rho \mathbb{B}$. Then $2\varepsilon \leq \varepsilon$, an impossibility.

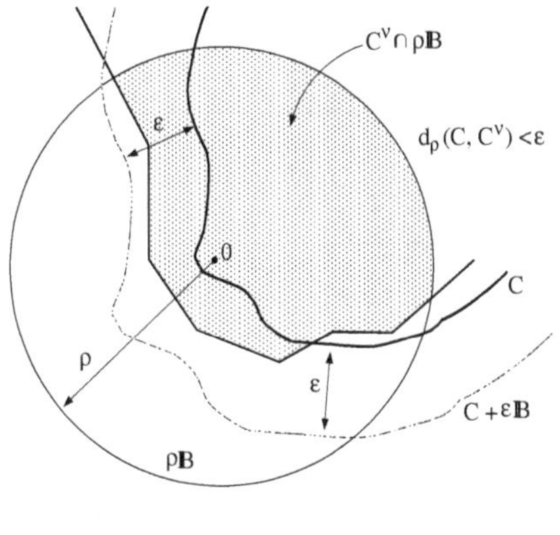

Fig. 4–3. Closeness between sets in convergence to a limit.

Sufficiency in (b): Let $x^\nu \underset{N}{\rightarrow} \bar{x}$ for some $N \in \mathcal{N}_\infty^\#$ with $x^\nu \in C^\nu$ for all $\nu \in N$. We think of $\bar{x}$ as an arbitrary point in $\limsup_\nu C^\nu$. If the inclusion in (b) is satisfied for all $\rho \geq \bar{\rho}$ for some $\bar{\rho} \geq 0$, it follows that for all $\rho > \max\{\bar{\rho}, |\bar{x}|\}$ and $\varepsilon > 0$, $\bar{x} \in C + \varepsilon \mathbb{B}$, i.e., $\bar{x}$ also belongs to C.

Necessity in (b): Suppose to the contrary that one can find $\rho > 0$, $\varepsilon > 0$ and $N \in \mathcal{N}_\infty^\#$ such that for all $\nu \in N$, there exists $x^\nu \in \big[C^\nu \cap \rho\mathbb{B}\big] \setminus \big[C + \varepsilon\mathbb{B}\big]$. Let $\bar{x}$ be a cluster point of the x^ν. Then $\bar{x} \in \limsup_\nu C^\nu$ and $\bar{x} \notin C + \mathrm{int}(\varepsilon\mathbb{B})$, so $\limsup_\nu C^\nu$ can't be included in C.

Justification of (a′) and (b′): These are obtained simply by replacing the ball $\rho\mathbb{B}$ in the proof of (a) and (b) by $\mathbb{B}(\bar{x}, \rho)$ with $\bar{x}$ chosen arbitrarily.

Restriction of ρ and ε to being rational is possible with impunity because any real number can be bracketed arbitrarily closely by rational numbers, and this guarantees that every ball can itself be bracketed arbitrarily closely by balls having rational radius. □

4.11 Corollary (escape to the horizon). *The condition $C^\nu \to \emptyset$ (or equivalently, $\limsup_\nu C^\nu = \emptyset$) holds for a sequence $\{C^\nu\}_{\nu\in\mathbb{N}}$ in $\mathbb{R}^n$ if and only if for every $\rho > 0$ there is an index set $N \in \mathcal{N}_\infty$ such that $C^\nu \cap \rho\mathbb{B} = \emptyset$ for all $\nu \in N$ (or, in other words, $d_{C^\nu}(0) \to \infty$). Further, for closed sets C^ν and C one has*

(a) $C \subset \liminf_\nu C^\nu$ *if and only if for all* $\varepsilon > 0$, $C \setminus (C^\nu + \varepsilon\mathbb{B}) \to \emptyset$;

(b) $C \supset \limsup_\nu C^\nu$ *if and only if for all* $\varepsilon > 0$, $C^\nu \setminus (C + \varepsilon\mathbb{B}) \to \emptyset$.

It suffices in these characterizations that the inclusions be satisfied for all ρ larger than some $\bar{\rho}$; in addition, ρ and ε can be restricted to be rational.

Proof. The first criterion is obtained by taking $C = \emptyset$ in 4.10(b). Assertions (a) and (b) reformulate 4.10(a) and 4.10(b) to take advantage of this view in terms of the convergence of 'excesses'. □

Clearly both 4.10 and 4.11 could be stated equivalently with an arbitrary bounded set B replacing the arbitrarily large ball $\rho\mathbb{B}$.

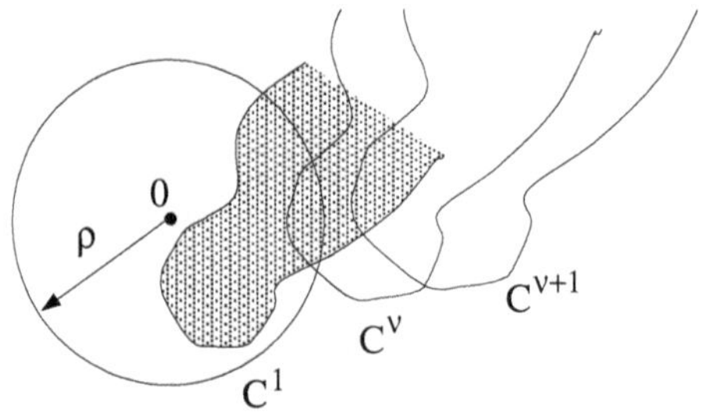

Fig. 4–4. A sequence of sets escaping to the horizon.

In the situation described in Corollary 4.11, the terminology that the sequence $\{C^\nu\}_{\nu\in\mathbb{N}}$ *escapes to the horizon* is appropriate—not only because the sequence eventually departs from any bounded region of $\mathbb{R}^n$, but also in light of the cosmic closure theory in Chapter 3. Although no ordinary point occurs as a limit of a subsequence of points $x^\nu \in C^\nu$, and this is the meaning of $C^\nu \to \emptyset$, various direction points in $\mathrm{hzn}\,\mathbb{R}^n$ may show up as limits in the cosmic sense. Horizon limit concepts will be essential later in understanding the behavior of set convergence with respect to operations like set addition.

Corollary 4.11 can also be exploited to obtain the boundedness of the sets C^ν of a sequence converging to a limit set C that is bounded, provided that the sets C^ν are connected.

4.12 Corollary (limits of connected sets). *Let $C^\nu \subset I\!R^n$ be connected with $\limsup_\nu C^\nu$ bounded and no subsequence escaping to the horizon. Then there is a bounded set $B \subset I\!R^n$ such that $C^\nu \subset B$ for all ν in some $N \in \mathcal{N}_\infty$.*

Proof. Let $C = \limsup_\nu C^\nu$ and $B = C + \varepsilon I\!B$ for some $\varepsilon > 0$, these sets being bounded. Let $B' = \rho I\!B$ for ρ big enough that $B \subset \operatorname{int} B'$. From 4.11(b) we have $C^\nu \setminus B \to \emptyset$. Then $(C^\nu \setminus B) \cap B' = \emptyset$ eventually by 4.11, but $C^\nu \setminus B \neq C^\nu$ eventually as well, since no subsequence of $\{C^\nu\}$ escapes to the horizon. Hence for all ν in some $N \in \mathcal{N}_\infty$ we have $C^\nu = (C^\nu \cap B) \cup (C^\nu \setminus B')$ with $C^\nu \cap B \neq \emptyset$. But C^ν is connected and $B \subset \operatorname{int} B'$, so $C^\nu \setminus B' = \emptyset$, i.e., $C^\nu \subset B$. □

C. Pompeiu-Hausdorff Distance

Closely related to ordinary—Painlevé-Kuratowski—convergence $C^\nu \to C$, but in some important respects distinct from it, is convergence with respect to Pompeiu-Hausdorff distance.

4.13 Example (Pompeiu-Hausdorff distance). *For $C, D \subset I\!R^n$ closed and nonempty, the Pompeiu-Hausdorff distance between C and D is the quantity*

$$d\!l_\infty(C, D) := \sup_{x \in I\!R^n} \big| d_C(x) - d_D(x) \big|,$$

where the supremum could equally be taken just over $C \cup D$, yielding the alternative formula

$$d\!l_\infty(C, D) = \inf\Big\{ \eta \geq 0 \,\Big|\, C \subset D + \eta I\!B,\ D \subset C + \eta I\!B \Big\}. \qquad 4(5)$$

A sequence $\{C^\nu\}_{\nu \in I\!N}$ is said to converge with respect to Pompeiu-Hausdorff distance to C when $d\!l_\infty(C^\nu, C) \to 0$ (these sets being closed and nonempty).

This property entails ordinary set convergence $C^\nu \to C$ and is equivalent to it when there is a bounded set $X \subset I\!R^n$ such that C^ν, $C \subset X$. But convergence with respect to Pompeiu-Hausdorff distance is not equivalent to ordinary set convergence without this boundedness restriction. Indeed, it is possible to have $C^\nu \to C$ with $d\!l_\infty(C^\nu, C) \equiv \infty$. Even for compact sets C^ν and C, it is possible to have $C^\nu \to C$ while $d\!l_\infty(C^\nu, C) \to \infty$.

Detail. Since C and D are closed, the expression on the right side of 4(5) is the same as the infimum of all $\eta \geq 0$ such that $d_D(x) \leq \eta$ for all $x \in C$ and $d_C(x) \leq \eta$ for all $x \in D$. Thus it is the same as the value obtained when the supremum defining $d\!l_\infty(C, D)$ is restricted $x \in C \cup D$. This value can't be greater than $d\!l_\infty(C, D)$, but it can't be less either, for the following reason. If $d_D \leq \eta$ on C, we have for any $x \in I\!R^n$ and $x' \in C$ that $d_D(x) \leq |x - x'| + d_D(x') \leq |x - x'| + \eta$, and consequently (in taking the infimum over $x' \in C$) that $d_D(x) \leq d_C(x) + \eta$. Likewise, if $d_C \leq \eta$ on D we have $d_C(x) \leq d_D(x) + \eta$ for all $x \in \mathbb{R}^n$. Then $|d_C(x) - d_D(x)| \leq \eta$ for all $x \in \mathbb{R}^n$.

The implication from Pompeiu-Hausdorff convergence to ordinary set convergence is clear from 4.10, as is the equivalence between the two notions under the boundedness restriction. An example where $C^\nu \to C$ but $d\!l_\infty(C^\nu, C) \equiv \infty$ is seen by taking the sets C^ν to be rays that rotate to the ray C. An example of compact sets $C^\nu \to C$ with $d\!l_\infty(C^\nu, C) \to \infty$ is obtained by taking $C^\nu = \{a^\nu, b^\nu\}$ with $a^\nu \to a$ but $|b^\nu| \to \infty$, and letting $C = \{a\}$. □

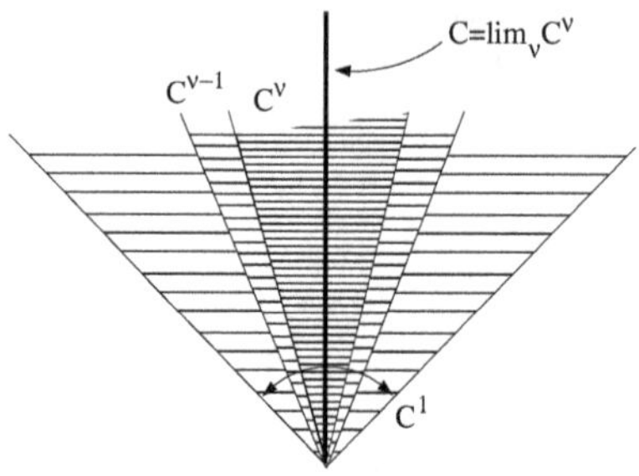

Fig. 4–5. A converging sequence $C^\nu \to C$ with Pompeiu-Hausdorff distance always ∞.

The notation $d\!l_\infty(C, D)$ for Pompeiu-Hausdorff distance conforms to a pattern that will come into view later when expressions $d\!l_\rho(C, D)$ and $\hat{d\!l}_\rho(C, D)$, depending on a parameter $\rho \geq 0$, are introduced to quantify ordinary set convergence as reflected by the estimates in 4.10; see 4(11). Note that because $d\!l_\infty(C, D)$ can have the value ∞, *Pompeiu-Hausdorff distance doesn't furnish a metric* for the space of nonempty, closed subsets of $\mathbb{R}^n$, although it does so when restricted to the subsets of a bounded set $X \subset \mathbb{R}^n$. (This isn't just a peculiarity of $\mathbb{R}^n$ but would arise for the space of closed subsets of any metric space in which distances can be unbounded, as can be seen from the initial formula for $d\!l_\infty(C, D)$.)

Anyway, the convergence shortcomings in Example 4.13 make clear that Pompeiu-Hausdorff distance is unsuitable for analyzing sequences of unbounded sets or even unbounded sequences of bounded sets, except perhaps in very special circumstances. A distance expression $d\!l(C, D)$ that does fully furnish a metric for set convergence will be provided later, in 4(12).

D. Cones and Convex Sets

For special classes of sets, such as cones and convex sets, special convergence properties are available.

4.14 Exercise (limits of cones). *For a sequence of cones K^ν in $\mathbb{R}^n$, the inner and outer limits, as well as the limit if it exists, are cones. If $K^\nu \neq \{0\}$ for all ν, or just for all ν in some $N \in \mathcal{N}_\infty^\#$, then* $\limsup_\nu K^\nu \neq \{0\}$.

In the presence of convexity, set convergence displays an 'internal' uniformity property of approximation which complements the one in 4.10.

4.15 Proposition (limits of convex sets). *For a sequence $\{C^\nu\}_{\nu \in \mathbb{N}}$ of convex subsets of $\mathbb{R}^n$, $\liminf_\nu C^\nu$ is convex, and so too, when it exists, is $\lim_\nu C^\nu$. (But $\limsup_\nu C^\nu$ need not be convex in general.)*

Moreover, if $C = \liminf_\nu C^\nu$ and $\operatorname{int} C \neq \emptyset$, for any compact set $B \subset \operatorname{int} C$ there exists an index set $N \in \mathcal{N}_\infty$ such that $B \subset \operatorname{int} C^\nu$ for all $\nu \in N$.

Proof. Let $C = \liminf_\nu C^\nu$. The convexity of C is elementary: if x_0 and x_1 belong to C, we can find for all ν in some set $N \in \mathcal{N}_\infty$ points x_0^ν and x_1^ν in C^ν such that $x_0^\nu \underset{N}{\to} x_0$ and $x_1^\nu \underset{N}{\to} x_1$. Then for arbitrary $\lambda \in [0,1]$ we have for $x_\lambda^\nu := (1-\lambda)x_0^\nu + \lambda x_1^\nu$ and $x_\lambda := (1-\lambda)x_0 + \lambda x_1$ that $x_\lambda^\nu \underset{N}{\to} x_\lambda$, so $x_\lambda \in C$.

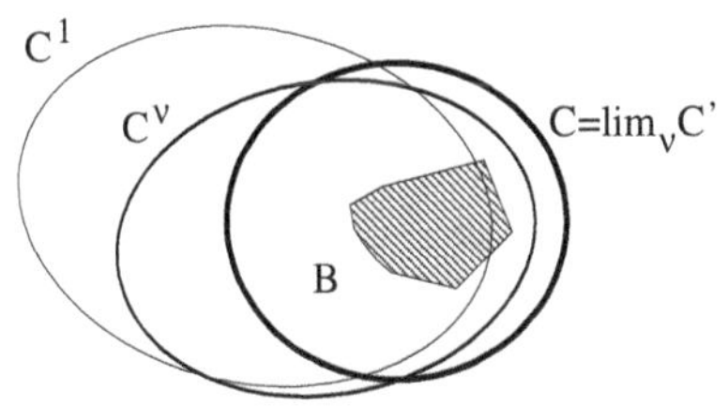

Fig. 4–6. Internal approximation in the convergence of convex sets.

Suppose B is a compact subset of $\operatorname{int} C$. For small enough $\varepsilon > 0$ we have $B + 2\varepsilon\mathbb{B} \subset C$, as seen from the fact that the distance function associated with the complement of C is positive on B and hence by its continuity (in 1.20) has a positive minimum on B. Choose ρ large enough so that $B + 2\varepsilon\mathbb{B} \subset \rho\mathbb{B}$ (i.e., $\rho \geq 2\varepsilon + \max_{x\in B}|x|$), and apply 4.10(a): there exists $N \in \mathcal{N}_\infty$ such that

$$B + 2\varepsilon\mathbb{B} \subset C \cap \rho\mathbb{B} \subset C^\nu + \varepsilon\mathbb{B} \text{ for all } \nu \in N.$$

From the cancellation law 3.35 we get $B + \varepsilon\mathbb{B} \subset C^\nu$ for all $\nu \in N$. □

Simple examples show that $\limsup_\nu C^\nu$ need not be convex when every C^ν is convex. For instance, let D_1 and D_2 be any two closed, convex sets in $\mathbb{R}^n$, and let $C^\nu = D_1$ for ν odd but $C^\nu = D_2$ for ν even. Then $\limsup_\nu C^\nu$ is the set $D_1 \cup D_2$, which may well be nonconvex.

The internal approximation property in Proposition 4.15 holds in particular whenever a sequence of convex sets C^ν converges to a set C with $\operatorname{int} C \neq \emptyset$. This agreeable property, illustrated in Figure 4–6, can't be expected for sequences of nonconvex sets, even when they are closed and the limit C happens anyway to be convex. The kind of difficulty that may arise is shown in Figure

4–7. The sets C^ν can converge to C even though they are riddled with holes, as long as the holes get finer and finer and thus vanish in the limit. Of course, for sets C^ν that aren't closed there's also the sort of example described just before 4.3.

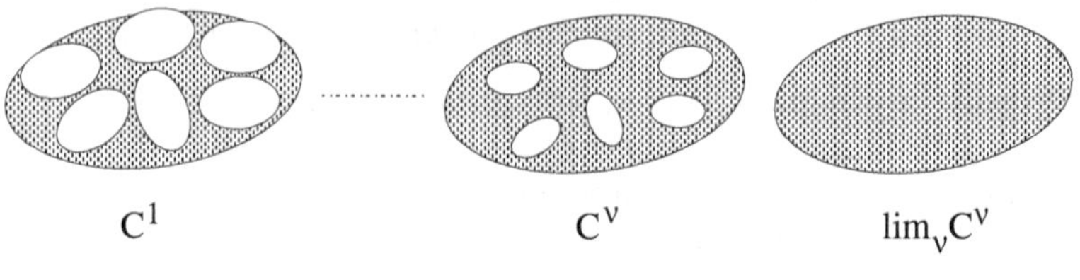

Fig. 4–7. Nonconvex sets converging to a convex set despite holes.

In the case of convex sets, the convergence criterion of Theorem 4.10 can be cast in a simpler form involving the Pompeiu-Hausdorff distance between truncations.

4.16 Exercise (convergence of convex sets through truncations). *For convex sets C^ν, closed and nonempty, one has $C^\nu \to C$ if and only if there exists $\rho_0 \geq 0$ such that, for all $\rho \geq \rho_0$, the truncations $C^\nu \cap \rho I\!B$ converge to $C \cap \rho I\!B$ with respect to Pompeiu-Hausdorff distance, i.e., $d\!l_\infty(C^\nu \cap \rho I\!B, C \cap \rho I\!B) \to 0$. (Without convexity, the 'only if' part fails.)*

For another special result, recall that a set $D \subset I\!R^n$ is *star-shaped (with respect to q)* if there's a point $q \in D$ such that the line segment $[q, x]$ lies in D for each $x \in D$.

4.17 Exercise (limits of star-shaped sets). *If a bounded sequence of star-shaped sets C^ν converges to a set C, then C must be star-shaped. (Without the boundedness, this can fail.)*

Guide. Show that if $\bar{q}$ is a cluster point of a sequence $\{q^\nu\}_{\nu \in I\!N}$ such that C^ν is star-shaped at q^ν, then C is star-shaped at $\bar{q}$. For a counterexample to C being star-shaped in the absence of the boundedness assumption on the sequence of sets C^ν, consider in $I\!R^2$ the sets $C^\nu = \big[(-1,0),(1,\nu)\big] \cup \big[(1,0),(-1,\nu)\big]$ for $\nu = 1, 2, \ldots$. □

E. Compactness Properties

A remarkable feature of set convergence is the existence of convergent subsequences for any sequence $\{C^\nu\}_{\nu \in I\!N}$ of subsets of $I\!R^n$.

4.18 Theorem (extraction of convergent subsequences). *Every sequence of nonempty sets C^ν in $I\!R^n$ either escapes to the horizon or has a subsequence converging to a nonempty set C in $I\!R^n$.*

Proof. If the sequence doesn't escape to the horizon, there's a point $\hat{x}$ in its outer limit. Then for ν in some index set $N^0 \in \mathcal{N}_\infty^\#$ one can find points $x^\nu \in C^\nu$ such that $x^\nu \underset{N^0}{\to} \hat{x}$. Consider the countable collection of open balls in 4.5(c)(a′), writing it as a sequence $\{O^\mu\}_{\mu\in\mathbb{N}}$. Construct a nest $N^0 \supset N^1 \supset N^2 \supset \ldots$ of index sets $N^\mu \in \mathcal{N}_\infty^\#$ by defining

$$N^\mu = \begin{cases} \{\nu \in N^{\mu-1} \,|\, C^\nu \cap O^\mu \neq \emptyset\} & \text{if this set of indices is infinite,} \\ \{\nu \in N^{\mu-1} \,|\, C^\nu \cap O^\mu = \emptyset\} & \text{otherwise.} \end{cases}$$

Finally, put together an index set N by taking the first index in N to be the first index in N^0 and, step by step, letting the μth element of N be the first index in N^μ larger than all indices previously added to N. Then $N \in \mathcal{N}_\infty^\#$, and for each μ either $C^\nu \cap O^\mu \neq \emptyset$ for all but finitely many $\nu \in N$ or, quite the opposite, $C^\nu \cap O^\mu = \emptyset$ for all but finitely many $\nu \in N$. Let $C = \limsup_{\nu\in N} C^\nu$. The set C contains $\hat{x}$, hence is nonempty. For each of the balls O^μ meeting C, it can't be true that $C^\nu \cap O^\mu = \emptyset$ for all but finitely many ν, so such balls O^μ must be in the other category in the construction scheme: we must have $C^\nu \cap O^\mu \neq \emptyset$ for all but finitely many $\nu \in N$. Therefore $C \subset \liminf_{\nu\in N} C^\nu$ by 4.5(c)(a′), so actually $C^\nu \underset{N}{\to} C$. □

The compactness property in Theorem 4.18 yields yet another way of looking at inner and outer limits.

4.19 Proposition (cluster description of inner and outer limits). *For any sequence of sets C^ν in $\mathbb{R}^n$, let $\mathcal{L}$ be the collection of all sets C that are cluster points in the sense of set convergence, i.e., such that $C^\nu \underset{N}{\to} C$ for some index set $N \in \mathcal{N}_\infty^\#$. Then*

$$\liminf_{\nu\to\infty} C^\nu = \bigcap\nolimits_{C\in\mathcal{L}} C, \qquad \limsup_{\nu\to\infty} C^\nu = \bigcup\nolimits_{C\in\mathcal{L}} C.$$

Proof. Any $x \in \liminf_\nu C^\nu$ is the limit of a sequence $\{x^\nu\}_{\nu\in N_x}$ selected with $x^\nu \in C^\nu$ and $N_x \in \mathcal{N}_\infty$. It's then also the limit of any subsequence $\{x^\nu\}_{\nu\in N}$ with $N \subset N_x$ and $N \in \mathcal{N}_\infty^\#$, so it belongs to the set limit of the corresponding subsequence $\{C^\nu\}_{\nu\in N}$ if that happens to exist. The fact that some subsequence of $\{C^\nu\}_{\nu\in\mathbb{N}}$ does converge is provided by Theorem 4.18. Therefore, $\liminf_\nu C^\nu \subset \bigcap_{C\in\mathcal{L}} C$.

The opposite inclusion for the lim inf comes from the fact that if x doesn't belong to $\liminf_\nu C^\nu$ there must be an index set $N_x \in \mathcal{N}_\infty^\#$ such that every C^ν for $\nu \in N_x$ has empty intersection with a certain neighborhood V of x. By selecting a subsequence $\{C^\nu\}_{\nu\in N}$ with $N \subset N_x$, $N \in \mathcal{N}_\infty^\#$, such that $C^\nu \underset{N}{\to} C$, which is again guaranteed possible by 4.18, we get $C \in \mathcal{L}$ but $x \notin C$.

In the case of the lim sup the inclusion $\supset$ is obvious. On the other hand, any point x belonging to $\limsup_\nu C^\nu$ is the limit of a sequence $\{x^\nu\}_{\nu\in N_x}$ chosen with $x^\nu \in C^\nu$, $N_x \in \mathcal{N}_\infty^\#$. By 4.18 there's then a subsequence $\{C^\nu\}_{\nu\in N}$ with $N \subset N_x$, $N \in \mathcal{N}_\infty^\#$, which converges to a set C. In particular we have $x^\nu \underset{N}{\to} x$, so $x \in C$. This gives the opposite inclusion. □

Observe that in the union describing the outer limit in 4.19 there is no need to apply the closure operation. Despite the possibility of an infinite collection $\mathcal{L}$ being involved, this union is always closed.

F. Horizon Limits

Set limits can equally well be developed in the context of the cosmic space $\operatorname{csm}\mathbb{R}^n$ introduced in Chapter 3, and this is useful for a number of purposes, such as gaining insight later into what happens to a convergent sequence of sets when various operations are performed on it. The idea is very simple: in the formulas in Definition 4.1 consider not only ordinary sequences $x^\nu \underset{N}{\to} x$ in $\mathbb{R}^n$ but also sequences that may converge in the extended sense of Definition 3.1 to a point $\operatorname{dir} x \in \operatorname{hzn}\mathbb{R}^n$; such sequences may consist of ordinary points, direction points or a mixture. In accordance with the unique representation of any subset of $\operatorname{csm}\mathbb{R}^n$ as $C \cup \operatorname{dir} K$ with C a subset of $\mathbb{R}^n$ and K a cone in $\mathbb{R}^n$, we may express convergence in this *cosmic* sense by

$$C^\nu \cup \operatorname{dir} K^\nu \overset{c}{\to} C \cup \operatorname{dir} K, \qquad \text{or} \qquad C \cup \operatorname{dir} K = \text{c-lim}_\nu \big[C^\nu \cup \operatorname{dir} K^\nu\big].$$

It's clear from the foundations of Chapter 3 that such convergence in $\operatorname{csm}\mathbb{R}^n$ is equivalent to ordinary convergence of the corresponding sets in H_n, the hemispherical model for $\operatorname{csm}\mathbb{R}^n$, or for that matter, ordinary convergence of the corresponding cones in $\mathbb{R}^{n+1}$ in the ray space model for $\operatorname{csm}\mathbb{R}^n$. Of course, cosmic outer and inner limits can be considered along with cosmic limits.

To make this concept easier to work with, it helps to introduce as the *horizon outer limit* and the *horizon inner limit* of a sequence of sets $C^\nu \subset \mathbb{R}^n$ the cones in $\mathbb{R}^n$ representing the sets of direction points in $\operatorname{hzn}\mathbb{R}^n$ that belong, respectively, to the cosmic outer limit and the cosmic inner limit of this sequence, namely

$$\begin{aligned} \limsup_\nu{}^\infty C^\nu &:= \{0\} \cup \Big\{x \,\Big|\, \exists N \in \mathcal{N}_\infty^\#,\ x^\nu \in C^\nu,\ \lambda^\nu \searrow 0,\ \lambda^\nu x^\nu \underset{N}{\to} x\Big\}, \\ \liminf_\nu{}^\infty C^\nu &:= \{0\} \cup \Big\{x \,\Big|\, \exists N \in \mathcal{N}_\infty,\ x^\nu \in C^\nu,\ \lambda^\nu \searrow 0,\ \lambda^\nu x^\nu \underset{N}{\to} x\Big\}. \end{aligned} \tag{4(6)}$$

(In these formulas the union with $\{0\}$ is superfluous when $C^\nu \neq \emptyset$, but it's needed for instance to make the limits come out as $\{0\}$ when $C^\nu \equiv \emptyset$.) We say that the *horizon limit* of the sets C^ν exists when these are equal:

$$\lim_\nu{}^\infty C^\nu = K \quad \iff \quad \limsup_\nu{}^\infty C^\nu = K = \liminf_\nu{}^\infty C^\nu.$$

4.20 Exercise (cosmic limits through horizon limits). *For any sequence of sets $C^\nu \subset \mathbb{R}^n$, the cones $\limsup_\nu^\infty C^\nu$ and $\liminf_\nu^\infty C^\nu$ are closed. The cosmic outer limit of a sequence of sets $C^\nu \cup \operatorname{dir} K^\nu$ (for cones $K^\nu \subset \mathbb{R}^n$) is*

$$\big(\limsup_\nu C^\nu\big) \cup \operatorname{dir}\big(\limsup_\nu{}^\infty C^\nu \cup \limsup_\nu K^\nu\big),$$

whereas the cosmic inner limit is

$$\left(\operatorname{liminf}_\nu C^\nu\right) \cup \operatorname{dir}\left(\liminf_\nu^\infty (C^\nu \cup K^\nu)\right).$$

Thus, $C^\nu \cup \operatorname{dir} K^\nu \xrightarrow{c} C \cup \operatorname{dir} K$ *if and only if* $C^\nu \to C$ *and*

$$\operatorname{limsup}_\nu^\infty C^\nu \cup \operatorname{limsup}_\nu K^\nu \ \subset\ K\ \subset\ \liminf_\nu^\infty C^\nu \cup K^\nu.$$

Guide. Rely on the geometric principles in the foregoing discussion. □

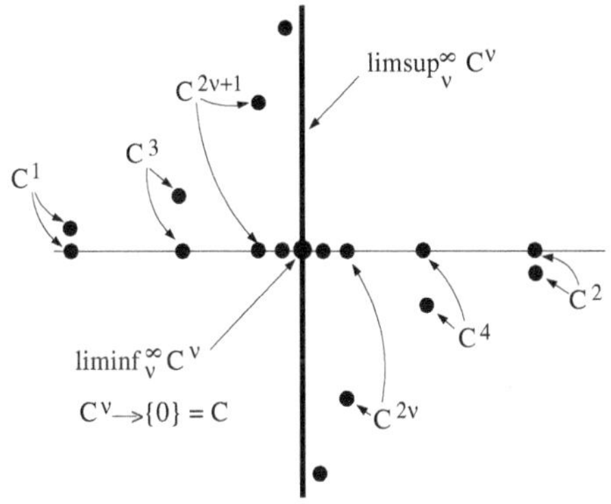

Fig. 4–8. Horizon outer and inner limits.

4.21 Exercise (properties of horizon limits). *For any sequence of sets* $C^\nu \subset \mathbb{R}^n$*, the horizon limit sets* $\liminf_\nu^\infty C^\nu$ *and* $\limsup_\nu^\infty C^\nu$*, as well as* $\lim_\nu^\infty C^\nu$ *when it exists, are closed cones which depend only on the sequence* $\{\operatorname{cl} C^\nu\}_{\nu\in\mathbb{N}}$ *and have the following properties:*

(a) $\liminf_\nu^\infty C^\nu \subset \limsup_\nu^\infty C^\nu$,

(b) $\liminf_\nu [C^\nu]^\infty \subset \liminf_\nu^\infty C^\nu$ *and* $\limsup_\nu [C^\nu]^\infty \subset \limsup_\nu^\infty C^\nu$,

(c) $\liminf_\nu^\infty C^\nu \supset C^\infty$ *when* $\liminf_\nu C^\nu \supset C$,

(d) $\lim_\nu^\infty C^\nu = C^\infty$ *when* $C^\nu \equiv C$,

(e) $\operatorname{liminf}_\nu^\infty C^\nu = \bigcap_{N\in\mathcal{N}_\infty^\#} \Big[\bigcup_{\nu\in N} C^\nu\Big]^\infty, \quad \operatorname{limsup}_\nu^\infty C^\nu = \bigcap_{N\in\mathcal{N}_\infty} \Big[\bigcup_{\nu\in N} C^\nu\Big]^\infty.$

Guide. Utilize 4.20 and, for (e), also 4.2(b) as applied cosmically. □

4.22 Example (eventually bounded sequences). *A sequence of sets* $C^\nu \subset \mathbb{R}^n$ *has the property that* $\limsup_\nu^\infty C^\nu = \{0\}$ *if and only if it is eventually bounded in the sense that for some index set* $N \in \mathcal{N}_\infty$ *the set* $\bigcup_{\nu\in N} C^\nu$ *is bounded.*

Our main interest for now with cosmic convergence ideas lies in applying them in the context of sequences of sets in $\mathbb{R}^n$ itself.

4.23 Definition (total set convergence). *A sequence of sets $C^\nu \subset \mathbb{R}^n$ is said to converge totally to a closed set $C \subset \mathbb{R}^n$, written $C^\nu \xrightarrow{t} C$, if $\operatorname{csm} C^\nu \xrightarrow{c} \operatorname{csm} C$, or equivalently $C^\nu \xrightarrow{c} \operatorname{csm} C$, in the context of the cosmic space $\operatorname{csm} \mathbb{R}^n$.*

The equivalence in this definition holds because 'csm' is a closure operation; the principle in 4.4 can be applied in the hemispherical model for $\operatorname{csm} \mathbb{R}^n$. Figure 4–8 supplies an example where a sequence of sets converges, but not totally. Total set convergence $C^\nu \xrightarrow{t} C$ automatically entails ordinary set convergence $C^\nu \to C$, and indeed the relationship between these two concepts can be characterized as follows.

4.24 Proposition (horizon criterion for total convergence). *For sets C^ν and C in $\mathbb{R}^n$, one has*

$$C^\nu \xrightarrow{t} C \qquad \Longleftrightarrow \qquad \lim_\nu C^\nu = C, \quad \operatorname{limsup}_\nu^\infty C^\nu \subset C^\infty,$$

in which case actually $\lim_\nu^\infty C^\nu = C^\infty$.

Proof. Either way, we have $C^\nu \to C$. Hence C is closed, so $\operatorname{csm} C = C \cup \operatorname{dir} C^\infty$ by 3.4. The cosmic outer and inner limits of C^ν are the same as those of $\operatorname{csm} C^\nu$, because limits of sets aren't affected when closures are taken (by 4.4—as applied cosmically). We now invoke 4.20 with $K^\nu = \{0\}$, $K = C^\infty$, and are done. □

Total convergence $C^\nu \xrightarrow{t} C$, by making demands on the behavior of unbounded sequences of selected points x^ν, imposes a requirement on how the sets converge 'in the large', in contrast to ordinary convergence $C^\nu \to C$, which is local in character and at best refers to uniformities relative to bounded regions as in 4.10. For this reason total convergence is important in situations where the remote parts of a set can have far-reaching influence on the outcome of a construction or operation. In such situations ordinary convergence is often too feeble to ensure the desired properties of limits. Fortunately, some of the most common cases encountered in dealing with sequences of sets are ones in which total convergence is an automatic consequence of ordinary convergence.

4.25 Theorem (automatic cases of total convergence). *In each of the following cases, ordinary convergence $C^\nu \to C \neq \emptyset$ entails total convergence $C^\nu \xrightarrow{t} C$:*

(a) *C^ν is convex for all ν;*

(b) *C^ν is a cone for all ν;*

(c) *$C^\nu \subset C^{\nu+1}$ for all ν;*

(d) *$C^\nu \subset B$ for all ν, where B is bounded;*

(e) *C^ν converges to C with respect to Pompeiu-Hausdorff distance.*

Proof. In each case we work from the assumption that $C^\nu \to C$ and show that the additional condition is enough to guarantee that $\limsup_\nu^\infty C^\nu \subset C^\infty$, so that actually $C^\nu \xrightarrow{t} C$. We treat the cases in reverse order.

Case (e) entails that $C^\nu \subset C + \varepsilon^\nu \mathbb{B}$ for a sequence $\varepsilon^\nu \searrow 0$. Then $\limsup_\nu^\infty C^\nu \subset \limsup_\nu^\infty (C + \varepsilon^\nu \mathbb{B}) = C^\infty$. Case (d) has both $\limsup_\nu^\infty C^\nu = \{0\}$ and $C^\infty = \{0\}$. In (c) the relation $C = \operatorname{cl} \bigcup_\nu C^\nu$ (cf. 4.3(a)) implies that any sequence of points $x^\nu \in C^\nu$ lies in C. The inclusion $\limsup_\nu^\infty C^\nu \subset C^\infty$ is then immediate from the definition of the two sets. In (b) it is elementary that $\limsup_\nu^\infty C^\nu = \limsup_\nu C^\nu$ and that C is a cone as well, hence $C^\infty = C$. But $\limsup_\nu C^\nu = C$ in consequence of our assumption that $C^\nu \to C$.

For case (a) consider $w \in \limsup_\nu^\infty C^\nu$. We must show that $w \in C^\infty$, and for that it suffices to show for $\bar{x} \in C$ that $C \supset \{\bar{x}+\tau w \mid \tau \geq 0\}$ (cf. 3.6). For any $\tau \geq 0$ the vector τw belongs to $\limsup_\nu^\infty C^\nu$ (because this set is a cone), so there exist $N \in \mathcal{N}_\infty^\#$, $\lambda^\nu \searrow 0$, and $x^\nu \in C^\nu$ such that $\lambda^\nu x^\nu \underset{N}{\to} \tau w$. For the index set N in this condition we have $w \in \liminf_{\nu \in N}^\infty C^\nu$ and $C = \liminf_{\nu \in N} C^\nu$. Hence for any $\bar{x} \in C$ there is a sequence $\bar{x}^\nu \underset{N}{\to} \bar{x}$ with $\bar{x}^\nu \in C^\nu$. Convexity of C^ν ensures that when ν is large enough so that $\lambda^\nu \leq 1$, one has $(1-\lambda^\nu)\bar{x}^\nu + \lambda^\nu x^\nu \in C^\nu$; these points converge to $\bar{x} + \tau w \in \liminf_{\nu \in N} C^\nu = C$. Thus, $\bar{x} + \tau w \in C$ for arbitrary $\tau \geq 0$, as required. □

The criterion in 4.25(d), meaning that sequence $\{C^\nu\}_{\nu \in \mathbb{N}}$ is bounded, can be broadened slightly: eventual boundedness as defined in 4.22 is enough.

G.* Continuity of Operations

With these cosmic notions at our disposal along with the basic ones of set convergence, we turn to questions of continuity of operations. If a set is produced by operations performed on other sets, will an approximation of it be produced when the same operations are performed on approximations to these other sets? Often we'll see that approximations in the sense of total convergence rather than ordinary convergence are needed in order to get good answers. Sometimes conditions of convexity must be imposed.

4.26 Theorem (convergence of images). *For sets $C^\nu \subset \mathbb{R}^n$ and a continuous mapping $F : \mathbb{R}^n \to \mathbb{R}^m$, one always has*

$$F\big(\liminf_\nu C^\nu\big) \subset \liminf_\nu F(C^\nu), \qquad F\big(\limsup_\nu C^\nu\big) \subset \limsup_\nu F(C^\nu).$$

The second of these inclusions is an equality if $K \cap \limsup_\nu^\infty C^\nu = \{0\}$ for the cone $K \subset \mathbb{R}^n$ consisting of the origin and all vectors $x \neq 0$ giving directions $\operatorname{dir} x$ that are limits of unbounded sequences on which F is bounded. Under this condition, therefore,

$$C^\nu \to C \quad \Longrightarrow \quad F(C^\nu) \to F(C).$$

In particular, the latter holds when the sequence $\{C^\nu\}_{\nu \in \mathbb{N}}$ is eventually bounded, or when F has the property that $\big|F(x)\big| \to \infty$ as $|x| \to \infty$.

Proof. The two general inclusions are elementary consequences of the definitions of inner and outer limits. Assuming the additional condition, consider now a point $u \in \limsup_\nu F(C^\nu)$: for some index set $N_0 \in \mathcal{N}_\infty^\#$ we have $u = \lim_{\nu \in N_0} F(x^\nu)$ with $x^\nu \in C^\nu$. The sequence $\{x^\nu\}_{\nu \in N_0}$ must be bounded, for if not there would be an index set $N \subset N_0$, $N \in \mathcal{N}_\infty^\#$, such that $x^\nu \underset{N}{\to} \operatorname{dir} x$

for some $x \neq 0$. Then $x \in \limsup_\nu^\infty C^\nu$ by definition, and yet $\{F(x^\nu)\}_{\nu\in N}$ is bounded, because $F(x^\nu) \underset{N}{\to} u$; this has been excluded. The boundedness of $\{x^\nu\}_{\nu\in N_0}$ implies the existence of a cluster point x, belonging by definition to $\limsup_\nu C^\nu$, and because F is continuous, we have $F(x) = u$. Thus, $u \in F(\limsup_\nu C^\nu)$, and equality in the outer limit inclusion is established. With this equality we get, in the case of $C^\nu \to C$, that

$$\operatorname{limsup}_\nu F(C^\nu) = F(C) \subset \operatorname{liminf}_\nu F(C^\nu),$$

hence $\lim_\nu F(C^\nu) = F(C)$. The condition $K \cap \limsup_\nu^\infty C^\nu = \{0\}$ is satisfied trivially if the sequence of sets C^ν is eventually bounded (cf. 4.22) or if $K = \{0\}$. Certainly $K = \{0\}$ if $|F(x)| \to \infty$ as $|x| \to \infty$. □

In the case of an eventually bounded sequence $\{C^\nu\}_{\nu\in I\!N}$ as in 4.22, the sequence $\{F(C^\nu)\}_{\nu\in I\!N}$ is eventually bounded as well (because F is bounded on bounded regions of $I\!R^n$), and the conclusion can be written in the form $C^\nu \xrightarrow{t} C \implies F(C^\nu) \xrightarrow{t} F(C)$, cf. 4.25. Other cases where total convergence is preserved can be identified as follows.

4.27 Theorem (total convergence of linear images). *For a linear mapping $L : I\!R^n \to I\!R^m$, if $C^\nu \xrightarrow{t} C$ and $L^{-1}(0) \cap C^\infty = \{0\}$, then $L(C^\nu) \xrightarrow{t} L(C)$.*

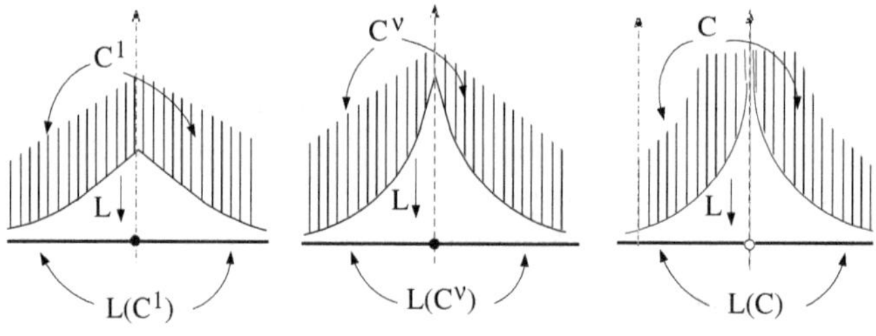

Fig. 4–9. Converging sets without convergence of their projections.

Proof. From $C^\nu \xrightarrow{t} C$ we have $\limsup_\nu^\infty C^\nu \subset C^\infty$ (by 4.24). On the other hand, the cone K in 4.26 is $L^{-1}(0)$ in the case of linear $F = L$. The condition $L^{-1}(0) \cap C^\infty = \{0\}$ therefore implies by 4.26 that $L(C^\nu) \to L(C)$. It also implies by 3.10 that $L(C^\infty) \subset L(C)^\infty$, so in order to verify that actually $L(C^\nu) \xrightarrow{t} L(C)$, it will suffice (again by 4.24) to show that $\limsup_\nu^\infty L(C^\nu) \subset L(C^\infty)$.

Suppose $u \in \limsup_\nu^\infty L(C^\nu)$. For indices ν in some set $N_0 \in \mathcal{N}_\infty^\#$ there exist $x^\nu \in C^\nu$ and $\lambda^\nu \searrow 0$ with $\lambda^\nu L(x^\nu) \underset{N_0}{\to} u$. Then $L(\lambda^\nu x^\nu) \underset{N_0}{\to} u$, because L is linear. If the sequence $\{\lambda^\nu x^\nu\}_{\nu\in N_0}$ were unbounded, it would have a cluster point of the form $\operatorname{dir} x$ for some $x \neq 0$, and then $\mu^\nu \lambda^\nu x^\nu \underset{N}{\to} x$ for some index set $N \subset N_0$, $N \in \mathcal{N}_\infty^\#$, and choice of scalars $\mu^\nu \searrow 0$. Then $x \in \limsup_\nu^\infty C^\nu \subset C^\infty$, yet also $L(x) = \lim_{\nu\in N} L(\mu^\nu\lambda^\nu x^\nu) = \lim_{\nu\in N} \mu^\nu L(\lambda^\nu x^\nu) = 0$, because $L(\lambda^\nu x^\nu) \underset{N}{\to} u$. This is impossible because $L^{-1}(0) \cap C^\infty = \{0\}$. Hence the sequence $\{\lambda^\nu x^\nu\}_{\nu\in N_0}$ is bounded and has a cluster point x; we have $\lambda^\nu x^\nu \underset{N}{\to} x$

for some index set $N \subset N_0$, $N \in \mathcal{N}_\infty^\#$. Then $x \in \limsup_\nu^\infty C^\nu \subset C^\infty$ and $u = \lim_{\nu \in N} L(\lambda^\nu x^\nu) = L(x)$. Hence, $u \in L(C^\infty)$. □

4.28 Example (convergence of projections of convex sets). *Let M be a linear subspace of $\mathbb{R}^n$, and let P_M be the projection mapping onto M. For convex sets $C^\nu \subset \mathbb{R}^n$, if $C^\nu \to C \neq \emptyset$ and $M^\perp \cap C^\infty = \{0\}$, then $P_M(C^\nu) \to P_M(C)$.*

Detail. This applies Theorem 4.27. The mapping P_M is linear with $P_M^{-1}(0) = M^\perp$. For convex sets, convergence and total convergence coincide, cf. 4.26. □

The need for the condition $M^\perp \cap C^\infty = \{0\}$ in this example (and for the condition on C^∞ more generally in Theorem 4.27) is illustrated in $\mathbb{R}^2$ by

$$C^\nu = \{(x_1, x_2) \,|\, x_2 \geq x_1^{-1},\ x_1 \geq \nu^{-1}\}, \qquad C = \{(x_1, x_2) \,|\, x_2 \geq x_1^{-1},\ x_1 > 0\},$$

with M taken to be the x_1-axis. Then $M^\perp \cap C^\infty = \{0\} \times \mathbb{R}_+$, and $C^\nu \to C$ but $P_M(C^\nu) = [\nu^{-1}, \infty) \not\to P_M(C) = (0, \infty)$.

4.29 Exercise (convergence of products and sums).

(a) *If $C_i^\nu \to C_i$ in $\mathbb{R}^{n_i}$, $i = 1, \ldots, m$, then $C_1^\nu \times \cdots \times C_m^\nu \to C_1 \times \cdots \times C_m$. If actually $C_i^\nu \xrightarrow{\text{t}} C_i$ and $C_1^\infty \times \cdots \times C_m^\infty = (C_1 \times \cdots \times C_m)^\infty$, one has*

$$C_1^\nu \times \cdots \times C_m^\nu \xrightarrow{\text{t}} C_1 \times \cdots \times C_m.$$

(b) *If $\liminf_\nu C_i^\nu \supset C_i$ for all i, $\liminf_\nu (C_1^\nu + \cdots + C_m^\nu) \supset C_1 + \cdots + C_m$.*

(c) *If $C_1^\nu \xrightarrow{\text{t}} C_1$, $C_2^\nu \xrightarrow{\text{t}} C_2$ and $C_1^\infty \cap (-C_2^\infty) = \{0\}$, then $C_1^\nu + C_2^\nu \to C_1 + C_2$.*

(d) *If $C_i^\nu \xrightarrow{\text{t}} C_i \subset \mathbb{R}^n$, $i = 1, \ldots, m$, $C_1^\infty \times \cdots \times C_m^\infty = (C_1 \times \cdots \times C_m)^\infty$, and the only way to choose vectors $x_i \in C_i^\infty$ satisfying $x_1 + \cdots + x_m = 0$ is to take $x_i = 0$ for all i, then*

$$C_1^\nu + \cdots + C_m^\nu \xrightarrow{\text{t}} C_1 + \cdots + C_m.$$

Guide. Derive (a) and (b) directly. To get the total convergence claim in (d), combine the total convergence fact in (a) with 4.27, taking L to be the linear mapping $(x_1, \ldots, x_m) \mapsto x_1 + \cdots + x_m$.

To obtain the inclusion $\limsup_\nu (C_1^\nu + C_2^\nu) \subset C_1 + C_2$ in (c), rely on the condition $C_1^\infty \cap (-C_2^\infty) = \{0\}$, and on $\limsup_\nu^\infty C_i^\nu = C_i^\infty$ for $i = 1, 2$, to show that given any sequence $x^\nu \to \bar{x}$, the sequence of sets $\{C_1^\nu \cap (x^\nu - C_2^\nu), \nu \in \mathbb{N}\}$ is eventually bounded. Now let $x^\nu \underset{N}{\to} \bar{x}$ be such that $N \in \mathcal{N}_\infty^\#$ and for all $\nu \in N$, $x^\nu \in C_1^\nu + C_2^\nu$. In particular, this means that $\bar{x} \in \limsup_\nu (C_1^\nu + C_2^\nu)$, and that there exist $N_0 \in \mathcal{N}_\infty^\#, N_0 \subset N$, $u^\nu \underset{N_0}{\to} \bar{u}$ with $u^\nu \in C_1^\nu \cap (x^\nu - C_2^\nu)$ for $\nu \in N_0$. There only remains to observe that $\bar{u} \in C_1$, and $(\bar{x} - \bar{u}) \in C_2$. □

The condition $C_1^\infty \times \cdots \times C_m^\infty = (C_1 \times \cdots \times C_m)^\infty$ called for here is satisfied in particular when the sets C_i are nonempty convex sets, or when no more than one of them is unbounded, cf. 3.11. An example of how total convergence of products can fail without this condition is provided by $C_1^\nu \times C_2^\nu$ with $C_1^\nu = C_2^\nu = \{2, 2^2, \ldots, 2^\nu\} \cup [2^{\nu+1}, \infty)$. To see that the assumptions in part (c) do not yield total convergence of the sums, let $C_1 = C_1^\nu = C \times \{0\}$ and

$C_2 = C_2^\nu = \{0\} \times C$ with $C = \{2^k \mid k \in I\!N\}$. Note that $C_1^\infty + C_2^\infty = I\!R^2_+$ but $\limsup_\nu^\infty (C_1^\nu + C_2^\nu) = (C_1 + C_2)^\infty \neq I\!R^2_+$, cf. the example after Exercise 3.11.

4.30 Proposition (convergence of convex hulls).

(a) *If $K^\nu \to K$ for cones $K, K^\nu \subset I\!R^n$ such that K is pointed, then $\operatorname{con} K^\nu \to \operatorname{con} K$.*

(b) *If $C^\nu \to C$ for sets C^ν that are all contained in some bounded region of $I\!R^n$, then $\operatorname{con} C^\nu \to \operatorname{con} C$.*

(c) *If $C^\nu \xrightarrow{t} C$ for sets $C, C^\nu \subset I\!R^n$ with $C \neq \emptyset$ and C^∞ pointed, then $\operatorname{con} C^\nu \xrightarrow{t} \operatorname{cl}(\operatorname{con} C) = \operatorname{con} C + \operatorname{con} C^\infty$.*

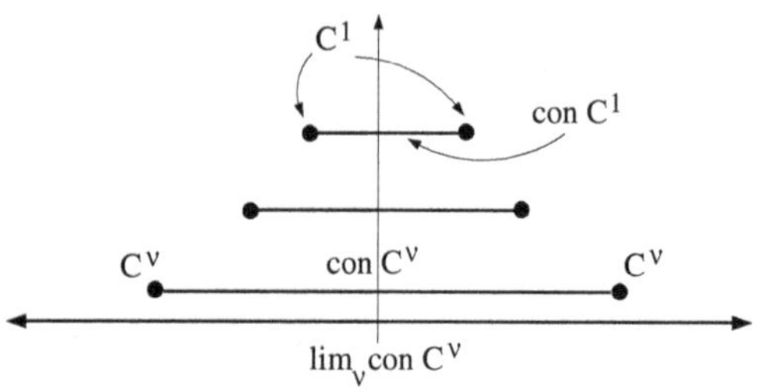

Fig. 4–10. Subsets C^ν of $I\!R^2$ with $C^\nu \to \emptyset$ while $\operatorname{con} C^\nu \to I\!R \times \{0\}$.

Proof. The statement in (a) applies 4.29(d) in the context of $\operatorname{con} K$ being closed and given by the formula $\operatorname{con} K = K + \cdots + K$ (n terms), cf. 3.15. Here $[K \times \cdots \times K]^\infty = K \times \cdots \times K = K^\infty \times \cdots \times K^\infty$. Pointedness of K ensures by definition that the only way to get $x_1 + \cdots x_n = 0$ with $x_i \in K$ is to take $x_i = 0$ for all i. Note that K^ν must then be pointed as well, for all ν in some index set $N \in \mathcal{N}_\infty$, so that $\operatorname{con} K^\nu$ is also pointed for such ν.

Next we address the statement in (c). Define $K \subset I\!R^{n+1}$ by

$$K := \{\lambda(x, -1) \mid x \in C,\ \lambda > 0\} \cup \{(x, 0) \mid x \in C^\infty\},$$

this being the closure of the cone that represents C in the ray space model for $\operatorname{csm} I\!R^n$ in Chapter 3. Similarly define K^ν for C^ν. To say that $C^\nu \xrightarrow{t} C$ is to say that $K^\nu \to K$. The assumption that C^∞ is pointed guarantees that K is pointed. Then $\operatorname{con} K^\nu \to \operatorname{con} K$ by (a). Furthermore, as noted above, $\operatorname{con} K^\nu$ like $\operatorname{con} K$ must be closed for all ν sufficiently large. Then $\operatorname{con} K^\nu$ and $\operatorname{con} K$ are the cones in the ray space model that correspond to $\operatorname{csm}(\operatorname{con} C^\nu)$ and $\operatorname{csm}(\operatorname{con} C)$. Hence $\operatorname{cl}(\operatorname{con} C^\nu) \xrightarrow{t} \operatorname{cl}(\operatorname{con} C) = \operatorname{con} C + \operatorname{con} C^\infty$.

Finally, we note that (b) is a case of (c) where $C^\infty = \{0\}$. □

4.31 Exercise (convergence of unions). *For $C_i^\nu \subset I\!R^n$, $i = 1, \ldots, m$, one has*

$$C_i^\nu \to C_i \quad \Longrightarrow \quad \bigcup_{i=1}^m C_i^\nu \to \bigcup_{i=1}^m C_i,$$

$$C_i^\nu \xrightarrow{t} C_i \quad \Longrightarrow \quad \bigcup_{i=1}^m C_i^\nu \xrightarrow{t} \bigcup_{i=1}^m C_i.$$

Guide. Rely on the definitions and, in the case of total convergence, 4.24. □

In contrast to the convergence of unions, the question of convergence of intersections is troublesome and can't be answered without making serious restrictions. The obvious elementary rule for outer limits, that

$$\limsup_\nu \bigcap\nolimits_{i=1}^{m} C_i^\nu \subset \bigcap\nolimits_{i=1}^{m} \limsup_\nu C_i^\nu, \tag{4(7)}$$

isn't reflected in anything so simple for inner limits of sets C_i^ν in general.

For pairs of *convex* sets, however, a positive result about convergence of intersections can be obtained under a mild assumption involving the degree of overlap of the limit sets. Recall from Chapter 2 that to say two convex sets C_1 and C_2 in $I\!R^n$ *can't be separated* (even improperly) is to say there is no hyperplane H such that C_1 lies in one of the closed half-spaces associated with H while C_2 lies in the other. This property, already characterized through Theorem 2.39 (see also 2.45), will be the key to a powerful fact: for convex sets $C_1^\nu \to C_1$ and $C_2^\nu \to C_2$, we'll show that if C_1 and C_2 can't be separated, then $C_1^\nu \cap C_2^\nu \to C_1 \cap C_2$.

We'll establish this in the next theorem by embedding it in a broader statement about convergence of solutions to constraint systems of the form

$$x \in X \;\text{ and }\; F(x) \in D, \;\text{ where }\; X \subset I\!R^n,\; D \subset I\!R^m,\; F : I\!R^n \to I\!R^m,$$

which in particular could specialize the constraint system in Example 1.1 through interpretation of the components of $F(x) = \big(f_1(x), \ldots, f_m(x)\big)$ as constraint functions. When $m = n$ and F is the identity mapping, the solution set is $X \cap D$, so the study of approximations to this kind of constraint system will cover convergence of set intersections as a particular case.

A background fact in this direction, not requiring convexity, is that for a continuous mapping $F : I\!R^n \to I\!R^m$ and sets $D^\nu \subset I\!R^m$ one always has

$$\begin{aligned} \liminf_\nu F^{-1}(D^\nu) &\subset F^{-1}\big(\liminf_\nu D^\nu\big), \\ \limsup_\nu F^{-1}(D^\nu) &\subset F^{-1}\big(\limsup_\nu D^\nu\big). \end{aligned} \tag{4(8)}$$

Another general fact, allowing also for the approximation of F, is that

$$\begin{aligned} &\limsup_\nu \big\{x^\nu \in X^\nu \bigm| F^\nu(x^\nu) \in D^\nu\big\} \subset \big\{x \in X \bigm| F(x) \in D\big\} \;\text{ when} \\ &\limsup_\nu X^\nu \subset X, \quad \limsup_\nu D^\nu \subset D, \;\text{ and }\; F^\nu(x^\nu) \to F(x),\, \forall\, x^\nu \to x. \end{aligned} \tag{4(9)}$$

To draw a sharper conclusion than 4(9), involving actual convergence of the set of solutions under the assumption that $X^\nu \to X$ and $D^\nu \to D$, specialization to the case where all the sets are convex and the mappings are linear is essential. For linear mappings L^ν and L from $I\!R^n$ to $I\!R^m$, pointwise convergence $L^\nu \to L$ is equivalent to convergence $A^\nu \to A$ of the associated matrices in $I\!R^{n\times m}$, where $A^\nu \to A$ means that each component of A^ν converges to the corresponding component of A. (For more about the convergence of matrices see 9.3 and the comments that follow.) From this matrix characterization it's evident that pointwise convergence $L^\nu \to L$ of linear mappings automatically entails having $L^\nu(x^\nu) \to L(x)$ whenever $x^\nu \to x$.

4.32 Theorem (convergence of solutions to convex systems). *Let*

$$C^\nu = \{x \in X^\nu \mid L^\nu(x) \in D^\nu\}, \qquad C = \{x \in X \mid L(x) \in D\},$$

for linear mappings $L^\nu, L : \mathbb{R}^n \to \mathbb{R}^m$ and convex sets $X^\nu, X \subset \mathbb{R}^n$ and $D^\nu, D \subset \mathbb{R}^m$, such that $L(X)$ cannot be separated from D. If $L^\nu \to L$, $\liminf_\nu X^\nu \supset X$ and $\liminf_\nu D^\nu \supset D$, then $\liminf_\nu C^\nu \supset C$. Indeed,

$$L^\nu \to L,\ X^\nu \to X,\ D^\nu \to D \quad \Longrightarrow \quad C^\nu \to C.$$

The following are special cases.

(a) *For linear mappings $L^\nu \to L$ and convex sets $D^\nu \to D$, if D and $\operatorname{rge} L$ cannot be separated, then $(L^\nu)^{-1}(D^\nu) \to L^{-1}(D)$.*

(b) *For matrices $A^\nu \to A$ in $\mathbb{R}^{m\times n}$ and vectors $b^\nu \to b$ in $\mathbb{R}^m$, if A has full rank m, then $\{x \mid A^\nu x = b^\nu\} \to \{x \mid Ax = b\}$.*

(c) *For convex sets C_1^ν and C_2^ν in $\mathbb{R}^n$, the inclusion $\liminf_\nu (C_1^\nu \cap C_2^\nu) \supset \liminf_\nu C_1^\nu \cap \liminf_\nu C_2^\nu$ holds if the convex sets $\liminf_\nu C_1^\nu$ and $\liminf_\nu C_2^\nu$ cannot be separated. Indeed,*

$$C_1^\nu \to C_1,\ C_2^\nu \to C_2 \quad \Longrightarrow \quad C_1^\nu \cap C_2^\nu \to C_1 \cap C_2$$

as long as C_1 and C_2 cannot be separated.

Proof. From 4(9) we have $C \supset \limsup_\nu C^\nu$ when $\limsup_\nu X^\nu \subset X$ and $\limsup_\nu D^\nu \subset D$, so we concentrate on showing that $C \subset \liminf_\nu C^\nu$ when $\liminf_\nu X^\nu \supset X$ and $\liminf_\nu D^\nu \supset D$. Let $\bar{x} \in C$; we must produce $x^\nu \in C^\nu$ with $x^\nu \to \bar{x}$. We have $\bar{x} \in X$ and for $\bar{u} := L(\bar{x})$ also $\bar{u} \in D$. Hence there exist $\bar{x}^\nu \in X^\nu$ with $\bar{x}^\nu \to \bar{x}$ and $\bar{u}^\nu \in D^\nu$ with $\bar{u}^\nu \to \bar{u}$. For $\bar{z}^\nu := L^\nu(\bar{x}^\nu) - \bar{u}^\nu$ we have $\bar{z}^\nu \to 0$.

The nonseparation assumption is equivalent by Theorem 2.39 to having $0 \in \operatorname{int}\big(L(X) - D\big)$. Then there's a simplex neighborhood S of 0 in $L(X) - D$, cf. 2.28(e); $S = \operatorname{con}\{z_0, z_1, \ldots, z_m\}$ with $z_i = L(x_i) - u_i$ for certain vectors $x_i \in X$ and $u_i \in D$. Accordingly by 2.28(d) there's a representation

$$0 = \sum\nolimits_{i=0}^m \lambda_i z_i \ \text{ with } \lambda_i > 0,\ \sum\nolimits_{i=0}^m \lambda_i = 1.$$

Since $X^\nu \to X$ and $D^\nu \to D$, we can find $x_i^\nu \in X^\nu$ and $u_i^\nu \in D^\nu$ with $x_i^\nu \to x_i$ and $u_i^\nu \to u_i$. Then for $z_i^\nu = L^\nu(x_i^\nu) - u_i^\nu$ we have $z_i^\nu \to z_i$, so by 2.28(f) there exists for ν sufficiently large a representation

$$0 = \sum\nolimits_{i=0}^m \lambda_i^\nu z_i^\nu = \sum\nolimits_{i=0}^m \lambda_i^\nu \big(L^\nu(x_i^\nu) - u_i^\nu\big) \ \text{ with } \lambda_i^\nu > 0,\ \sum\nolimits_{i=0}^m \lambda_i^\nu = 1,$$

where $\lambda_i^\nu \to \lambda_i$. At the same time, since $\bar{z}^\nu \to 0$, there exists by 2.28(f) for ν sufficiently large a representation

$$\bar{z}^\nu = \sum_{i=0}^m \bar{\lambda}_i^\nu z_i^\nu = \sum_{i=0}^m \bar{\lambda}_i^\nu \big(L^\nu(x_i^\nu) - u_i^\nu\big) \text{ with } \bar{\lambda}_i^\nu > 0, \ \sum_{i=0}^m \bar{\lambda}_i^\nu = 1,$$

where $\bar{\lambda}_i^\nu \to \lambda_i$. With $\theta^\nu = \min\{1, \lambda_0^\nu/\bar{\lambda}_0^\nu, \ldots, \lambda_m^\nu/\bar{\lambda}_m^\nu\}$ and for ν large enough, $0 \leq \theta^\nu \leq 1$ and $\theta^\nu \to 1$. Then for $\mu_i^\nu = \lambda_i^\nu - \theta^\nu \bar{\lambda}_i^\nu$ we have $0 \leq \mu_i^\nu \to 0$ and $\sum_{i=0}^m \mu_i^\nu + \theta^\nu = 1$, so that the vectors $x^\nu := \sum_{i=0}^m \mu_i^\nu x_i^\nu + \theta^\nu \bar{x}^\nu$ and $u^\nu := \sum_{i=0}^m \mu_i^\nu u_i^\nu + \theta^\nu \bar{u}^\nu$ belong to X^ν and D^ν by convexity and converge to $\bar{x}$ and $\bar{u}$. We have $0 = L^\nu(x^\nu) - u^\nu$, so $x^\nu \in C^\nu$ as desired.

For the special case in (a) take $X^\nu = \mathbb{R}^n$. For (b), set $D^\nu = \{b^\nu\}$ in (a), $L^\nu(x) = A^\nu x$, $L(x) = Ax$. For (c), let $X^\nu = C_1^\nu$, $D^\nu = C_2^\nu$, and $L^\nu = I$. □

The convergence in Theorem 4.32 is of course total, because the sets are convex; cf. 4.25. The proof of the result reveals a broader truth for situations where it's not necessarily true that $X^\nu \to X$ and $D^\nu \to D$:

$$\operatorname{liminf}_\nu X^\nu \supset X, \quad \operatorname{liminf}_\nu D^\nu \supset D \qquad \Longrightarrow \qquad \operatorname{liminf}_\nu C^\nu \supset C$$

whenever X and D are convex sets such that $L(X)$ and D can't be separated.

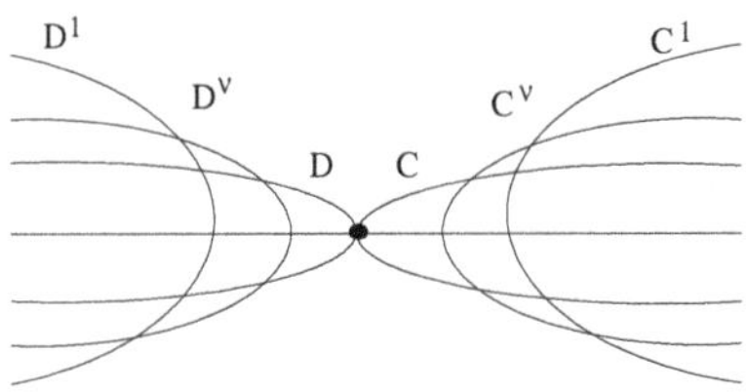

Fig. 4–11. An example where $C^\nu \to C$ and $D^\nu \to D$ but $C^\nu \cap D^\nu \not\to C \cap D$.

The pairwise intersection result in 4.32(c) can be generalized as follows to multiple intersections.

4.33 Exercise (convergence of convex intersections). *For sequences of convex sets $C_i^\nu \to C_i$ in $\mathbb{R}^n$ one has*

$$C_1^\nu \cap \cdots \cap C_q^\nu \ \to \ C_1 \cap \cdots \cap C_q$$

if none of the limit sets C_i can be separated from the intersection $\bigcap_{k=1,\, k\neq i}^q C_k$ of the others.

Guide. Apply 4.32 to $X^\nu = \mathbb{R}^n$, $D^\nu = C_1^\nu \times \cdots \times C_q^\nu$, $L^\nu(x) = (x, \ldots, x)$. □

H.* Quantification of Convergence

The properties of distance functions in 4.7 provide the springboard to a description of set convergence in terms of a metric on a space whose elements are sets. As background for this development, we need to strengthen the assertion in 4.7 about the convergence of distance functions, namely from pointwise convergence to uniform convergence on bounded sets. The following relations will be utilized.

4.34 Lemma (distance function relations). *Let C_1 and C_2 be closed subsets of $\mathbb{R}^n$. Let $\varepsilon > 0$, $\rho > 0$, $\rho' \geq 2\rho + d_{C_1}(0)$. Then*

$$\begin{array}{llcl}
\text{(a)} & C_1 \cap \rho\mathbb{B} \subset C_2 + \varepsilon\mathbb{B} & \Longleftarrow & d_{C_2} \leq d_{C_1} + \varepsilon \text{ on } \rho\mathbb{B}, \\
\text{(b)} & d_{C_2} \leq d_{C_1} + \varepsilon \text{ on } \rho\mathbb{B} & \Longleftarrow & C_1 \cap \rho'\mathbb{B} \subset C_2 + \varepsilon\mathbb{B}, \\
\text{(c)} & d_{C_2} \leq d_{C_1} + \varepsilon \text{ on } \mathbb{R}^n & \Longleftarrow & C_1 \subset C_2 + \varepsilon\mathbb{B}, \\
\text{(d)} & d_{C_2} \geq d_{C_1} \text{ on } \rho\mathbb{B} & \Longleftarrow & 2\rho + d_{C_1}(0) \leq d_{C_2}(0).
\end{array}$$

If C_1 is convex, 2ρ can be replaced by ρ in the inequality imposed on ρ'. If also $0 \in C_1$, then ρ' can be replaced simply by ρ in (b), so that the implications in (a) and (b) combine to give

$$C_1 \cap \rho\mathbb{B} \subset C_2 + \varepsilon\mathbb{B} \iff d_{C_2} \leq d_{C_1} + \varepsilon \text{ on } \rho\mathbb{B}.$$

This equivalence holds also, even without convexity, when C_1 is a cone.

Proof. Suppose $C_1 \neq \emptyset$, since everything is trivial otherwise. If $d_{C_2} \leq d_{C_1} + \varepsilon$ on $\rho\mathbb{B}$, we have for every $x \in C_1 \cap \rho\mathbb{B}$ that $d_{C_2}(x) \leq \varepsilon$ (because $d_{C_1}(x) = 0$). As C_2 is closed, this means $C_1 \cap \rho\mathbb{B} \subset C_2 + \varepsilon\mathbb{B}$, which gives (a). For (b) and (c), note that for any x and any set D satisfying $D \subset C_2 + \varepsilon\mathbb{B}$, we have

$$\begin{aligned}
d(x, D) \geq d(x, C_2 + \varepsilon\mathbb{B}) &= \inf\left\{ |(y + \varepsilon z) - x| \,\middle|\, y \in C_2,\, z \in \mathbb{B} \right\} \\
&\geq \inf\left\{ |y - x| - \varepsilon|z| \,\middle|\, y \in C_2,\, z \in \mathbb{B} \right\} = d(x, C_2) - \varepsilon,
\end{aligned}$$

so that $d_{C_2} \leq d_D + \varepsilon$ on $\mathbb{R}^n$. With $D = C_1$ we get (c). Taking $D = C_1 \cap \rho'\mathbb{B}$ we can obtain (b) by verifying that $d(x, C_1 \cap \rho'\mathbb{B}) = d(x, C_1)$ when $x \in \rho\mathbb{B}$ and $\rho' \geq 2\rho + d_{C_1}(0)$. (When C_1 is a cone, it's enough to have $\rho' \geq \rho$, because the projection of any $x \in \rho\mathbb{B}$ on any ray in C_1 lies then in $\rho'\mathbb{B}$; hence the special assertion at the end of the lemma.)

To proceed with the verification, suppose $|x| \leq \rho$ and consider any $x_1 \in P_{C_1}(x)$ (this projection being nonempty by 1.20, since C_1 is closed). It will suffice to demonstrate that $x_1 \in \rho'\mathbb{B}$ when ρ' satisfies the inequality specified. We have $|x_1| \leq |x| + |x_1 - x|$ with $|x_1 - x| = d(x, C_1) \leq d(x, 0) + d(0, C_1)$, so $|x_1| \leq 2|x| + d(0, C_1) \leq 2\rho + d(0, C_1) \leq \rho'$, as required. In the special case where C_1 is convex, more can be gleaned by considering also the point $x_0 \in C_1$ with $|x_0| = d_{C_1}(0)$. For any $\tau \in (0,1)$ the point $x_\tau = (1-\tau)x_0 + \tau x_1$ lies in C_1 by convexity, so that

$$0 \leq |x_\tau|^2 - |x_0|^2 = 2\tau\langle x_0,\, x_1 - x_0\rangle + \tau^2|x_1 - x_0|^2.$$

On dividing by τ and taking the limit as $\tau \searrow 0$ we see that $\langle x_0,\, x_1 - x_0\rangle \geq 0$. Likewise, from $x_\tau - x = (x_1 - x) - (1-\tau)(x_1 - x_0)$ we obtain

$$0 \le |x_\tau - x|^2 - |x_1 - x|^2 = -2(1-\tau)\langle x_1 - x,\, x_1 - x_0\rangle + (1-\tau)^2|x_1 - x_0|^2,$$

from which it follows on dividing by $1-\tau$ and taking the limit as $\tau \nearrow 1$ that $\langle x - x_1,\, x_1 - x_0\rangle \ge 0$. In combination with the fact that $\langle x_0,\, x_1 - x_0\rangle \ge 0$, we get $\langle x - x_1 + x_0,\, x_1 - x_0\rangle \ge 0$. This gives $|x_1 - x_0|^2 \le \langle x,\, x_1 - x_0\rangle \le |x||x_1 - x_0|$, hence $|x_1 - x_0| \le |x| \le \rho$. Then $|x_1| \le |x_1 - x_0| + |x_0| \le \rho + d_{C_1}(0)$, so that we only need to have $\rho' \ge \rho + d_{C_1}(0)$ in order to conclude that $x_1 \in \rho' I\!B$.

The implication in (d) comes simply from observing that for all $x \in \rho I\!B$ one has $d(x, C_2) \ge d(0, C_2) - d(x, 0) \ge d(0, C_2) - \rho$, while similarly $d(x, C_1) \le d(x, 0) + d(0, C_1) \le \rho + d(0, C_1)$. Thus, $d_{C_2} \ge d_{C_1}$ on $\rho I\!B$ when $\rho + d(0, C_1) \le d(0, C_2) - \rho$, which is true when $d(0, C_2) \ge 2\rho + d(0, C_1)$. □

The relations in Lemma 4.34 give a special role to the origin, but the generalization to balls centered at any point $\bar{x}$ is immediate through the device of applying this lemma to the translates $C_1 - \bar{x}$ and $C_2 - \bar{x}$ instead C_1 and C_2. The effect of this is to replace 0 by $\bar{x}$ and the balls $\rho I\!B$ and $\rho' I\!B$ by $I\!B(\bar{x}, \rho)$ and $I\!B(\bar{x}, \rho')$ in the lemma's statement.

4.35 Theorem (uniformity in convergence of distance functions). *For subsets C^ν and C of $I\!R^n$ with C closed and nonempty, one has $C^\nu \to C$ if and only if the distance functions d_{C^ν} converge uniformly to d_C on all bounded sets $B \subset I\!R^n$. In more detail with focus on sets $B = \rho I\!B$,*

(a) *$C \subset \liminf_\nu C^\nu$ if and only if there exists for each $\rho > 0$ and $\varepsilon > 0$ an index set $N \in \mathcal{N}_\infty$ with $d(x, C^\nu) \le d(x, C) + \varepsilon$ for all $x \in \rho I\!B$ when $\nu \in N$;*

(b) *$C \supset \limsup_\nu C^\nu$ if and only if there exists for each $\rho > 0$ and $\varepsilon > 0$ an index set $N \in \mathcal{N}_\infty$ with $d(x, C^\nu) \ge d(x, C) - \varepsilon$ for all $x \in \rho I\!B$ when $\nu \in N$.*

Proof. Corollary 4.7 already tells us that $C^\nu \to C$ if and only if $d_{C^\nu}(x) \to d_C(x)$ for all $x \in I\!R^n$. Uniform convergence over all bounded sets is an automatic consequence of convergence at each point when a sequence of functions is equicontinuous at each point. This is just what we have here. According to 4(3) the inequality $d_C(x_1) \le d_C(x_2) + |x_1 - x_2|$ holds for all pairs of points x_1 and x_2, and likewise for each function d_{C^ν}. Thus, we can concentrate on the conditions in (a) and (b) in verifying the equivalence with the asserted set inclusions.

We rely now on Lemma 4.34 and the geometric characterization of set convergence in 4.10. Taking $C_1 = C$ and $C_2 = C^\nu$ in 4.34(a), we see that if $d_{C^\nu} \le d_C + \varepsilon$ on $\rho I\!B$, then $C \cap \rho I\!B \subset C^\nu + \varepsilon I\!B$. The uniformity condition in (a) thus implies the condition in 4.10(a). Conversely, if the condition in 4.10(a) holds, we can take for any ρ the value $\rho' = 2\rho + d_C(0)$ and, by applying 4.10(a) to ρ', obtain through 4.34(b) that $d_{C^\nu} \le d_C + \varepsilon$ on $\rho I\!B$. Then the uniformity condition in (a) is satisfied.

The argument for the equivalence between the uniformity condition in (b) and the condition in 4.10(b) is almost identical, with $C_1 = C^\nu$ and $C_2 = C$ in 4.34. The twist is that we must, in the converse part, choose ρ' so as to have $\rho' \ge 2\rho + d_{C^\nu}(0)$ for all $\nu \in N$ such that $d_{C^\nu}(0) < 2\rho + d_C(0)$. The value $\rho' = 4\rho + d_C(0)$ definitely suffices. □

Theorem 4.35 is valid in the case of $C = \emptyset$ as well, provided only that the right interpretations are given to the uniform convergence conditions to accommodate the fact that $d(x, C) \equiv \infty$ in that case. The main adjustment is that in (b) one must consider, instead of arbitrarily small ε, an arbitrarily high real number α and look to having $d(x, C^\nu) \geq \alpha$ for all $x \in \rho I\!B$ when $\nu \in N$. (The assertions in (a) are trivial when $C = \emptyset$.) The appropriate extension of uniform convergence notions to sequences of general extended-real-valued functions will be seen in Chapter 7 (starting with 7.12).

In 4.13 it was observed that the Pompeiu-Hausdorff distance $d\!l_\infty$ fails to provide a quantification of set convergence except under boundedness restrictions. A quantification in terms of a metric is possible nevertheless, as will soon be seen. An intermediate quantification of set convergence, which we'll pass through first and which is valuable for its own sake, involves not a single metric, but a family of pseudo-metrics. A *pseudo-metric*, we recall, satisfies the criteria for a metric (nonnegative real values, symmetry in the two arguments, and the triangle inequality), except that the distance between two different elements might in some cases be zero.

Because set convergence doesn't distinguish between a set and its closure (cf. 4.4), a full metric space interpretation of set convergence isn't possible without restriction to sets that are closed. In the notation

$$\begin{aligned} \text{sets}(I\!R^n) &:= \text{ the space of all subsets of } I\!R^n, \\ \text{cl-sets}(I\!R^n) &:= \text{ the space of all closed subsets of } I\!R^n, \\ \text{cl-sets}_{\neq\emptyset}(I\!R^n) &:= \text{ the space of all nonempty, closed subsets of } I\!R^n, \end{aligned} \qquad 4(10)$$

it's therefore cl-sets$(I\!R^n)$ rather than sets$(I\!R^n)$ that we'll be turning to in this context, or actually cl-sets$_{\neq\emptyset}(I\!R^n)$ in keeping with our pattern of treating sequences $C^\nu \to \emptyset$ alternatively as escaping to the horizon in the sense of 4.11.

Two basic measures of distance between sets will be utilized in tandem. We define for every choice of the parameter $\rho \in I\!R_+ = [0, \infty)$ and pair of *nonempty* sets C and D the values

$$\begin{aligned} d\!l_\rho(C, D) &:= \max_{|x| \leq \rho} \Big| d_C(x) - d_D(x) \Big|, \\ \hat{d\!l}_\rho(C, D) &:= \inf\Big\{ \eta \geq 0 \Big| \; C \cap \rho I\!B \subset D + \eta I\!B, \;\; D \cap \rho I\!B \subset C + \eta I\!B \Big\}, \end{aligned} \qquad 4(11)$$

where in particular $d\!l_0(C, D) = |d_C(0) - d_D(0)|$. Clearly, $\hat{d\!l}_\rho$ relates to the uniform approximation property in 4.10, whereas $d\!l_\rho$ relates to the one in 4.35, and they take off in different ways from the equivalent formulas for $d\!l_\infty(C, D)$ in 4.13. We refer to $d\!l_\rho(C, D)$ as the *ρ-distance* between C and D, although it is really just a pseudo-distance; the quantity $\hat{d\!l}_\rho(C, D)$ serves to provide effective estimates for the ρ-distance (cf. 4.37(a) below). We don't insist on applying these expressions only to closed sets, but the main interest lies in thinking of $d\!l_\rho$ and $\hat{d\!l}_\rho$ as functions on cl-sets$_{\neq\emptyset}(I\!R^n) \times$ cl-sets$_{\neq\emptyset}(I\!R^n)$ with values in $I\!R_+$.

4.36 Theorem (quantification of set convergence). *For each $\rho \geq 0$, $d\!l_\rho$ is a pseudo-metric on the space* cl-sets$_{\neq\emptyset}(\mathbb{R}^n)$, *but $\hat{d\!l}_\rho$ is not. Both families $\{d\!l_\rho\}_{\rho\geq 0}$ and $\{\hat{d\!l}_\rho\}_{\rho\geq 0}$ characterize set convergence: for any $\bar{\rho} \in \mathbb{R}_+$, one has*

$$\begin{aligned} C^\nu \to C \quad &\Longleftrightarrow \quad d\!l_\rho(C^\nu, C) \to 0 \ \text{ for all } \ \rho \geq \bar{\rho} \\ &\Longleftrightarrow \quad \hat{d\!l}_\rho(C^\nu, C) \to 0 \ \text{ for all } \ \rho \geq \bar{\rho}. \end{aligned}$$

Proof. Theorem 4.35 gives us the characterization of set convergence in terms of $d\!l_\rho$, while 4.10 gives it to us for $\hat{d\!l}_\rho$. For $d\!l_\rho$, the pseudo-metric properties of nonnegativity $d\!l_\rho(C_1, C_2) \in \mathbb{R}_+$, symmetry $d\!l_\rho(C_1, C_2) = d\!l_\rho(C_2, C_1)$, and the triangle inequality $d\!l_\rho(C_1, C_2) \leq d\!l_\rho(C_1, C) + d\!l_\rho(C, C_2)$, are obvious from the definition 4(11) and the inequality

$$|d_{C_1}(x) - d_{C_2}(x)| \ \leq \ |d_{C_1}(x) - d_C(x)| + |d_C(x) - d_{C_2}(x)|.$$

The triangle inequality can fail for $\hat{d\!l}_\rho$: take $C_1 = \{1\} \subset \mathbb{R}$, $C_2 = \{-1\}$, $C = \{-6/5, 6/5\}$ and $\rho = 1$. Thus $\hat{d\!l}_\rho$ isn't a pseudo-metric. □

The distance expressions in 4(11) and Theorem 4.35 utilize origin-centered balls $\rho\mathbb{B}$, but there's really nothing special about the origin in this. The balls $\mathbb{B}(\bar{x}, \rho)$ centered at any point $\bar{x}$ could play the same role. More generally one could work just as well with the collection of all nonempty, bounded sets $B \subset \mathbb{R}^n$, defining $d\!l_B$ and $\hat{d\!l}_B$ in the manner of 4(11) through replacement $\rho\mathbb{B}$ by B. The end results would essentially be the same, but freed of a seeming dependence on the origin. For simplicity, though, $d\!l_\rho$ and $\hat{d\!l}_\rho$ suffice.

The example in the proof of 4.36, showing that $\hat{d\!l}_\rho$ doesn't satisfy the triangle inequality, can be supplemented by the following example, which leads to further insights. Fix any $\rho > 0$ and two different vectors a_1 and a_2 with $|a_1| = 1 = |a_2|$. Choose any sequence $\rho^\nu \searrow \rho$ and define

$$C_1^\nu = \{0, \rho^\nu a_1\}, \quad C_2^\nu = \{0, \rho^\nu a_2\}, \quad C_1 = \{0, \rho a_1\}, \quad C_2 = \{0, \rho a_2\},$$

noting that $C_1^\nu \to C_1$ and $C_2^\nu \to C_2$. As a matter of fact, $\hat{d\!l}_\rho(C_1^\nu, C_1) = \hat{d\!l}_\rho(C_2^\nu, C_2) = \rho^\nu - \rho \to 0$. But $\hat{d\!l}_\rho(C_1^\nu, C_2^\nu) = 0$ for all ν, while $\hat{d\!l}_\rho(C_1, C_2) = \rho \min\big\{|a_1|, |a_2|, |a_1 - a_2|\big\} > 0$. Hence for large ν one has

$$\hat{d\!l}_\rho(C_1, C_2) > \hat{d\!l}_\rho(C_1, C_1^\nu) + \hat{d\!l}_\rho(C_1^\nu, C_2^\nu) + \hat{d\!l}_\rho(C_2^\nu, C_2),$$

which would be impossible if $\hat{d\!l}_\rho$ enjoyed the triangle inequality.

This example shows how it's possible to have sequences $C_1^\nu \to C_1$ and $C_2^\nu \to C_2$ in cl-sets$_{\neq\emptyset}(\mathbb{R}^n)$ such that $\hat{d\!l}_\rho(C_1^\nu, C_2^\nu) \not\to \hat{d\!l}_\rho(C_1, C_2)$. Of course $d\!l_\rho(C_1^\nu, C_2^\nu) \to d\!l_\rho(C_1, C_2)$, because this is a consequence of the pseudo-metric property of $d\!l_\rho$ along with the fact that $d\!l_\rho(C_1^\nu, C_1) \to 0$ and $d\!l_\rho(C_2^\nu, C_2) \to 0$:

$$\begin{aligned} d\!l_\rho(C_1, C_2) &\leq d\!l_\rho(C_1, C_1^\nu) + d\!l_\rho(C_1^\nu, C_2^\nu) + d\!l_\rho(C_2^\nu, C_2), \\ d\!l_\rho(C_1^\nu, C_2^\nu) &\leq d\!l_\rho(C_1^\nu, C_1) + d\!l_\rho(C_1, C_2) + d\!l_\rho(C_2, C_2^\nu). \end{aligned}$$

In short, $d\!l_\rho$ is better behaved than $\hat{d\!l}_\rho$, therefore more convenient for a number of technical purposes. This doesn't mean, however, that $\hat{d\!l}_\rho$ may be ignored. The important virtue of the $\hat{d\!l}_\rho$ family of distance expressions is their direct tie to the set inclusions in 4.10, which are the solid basis for most geometric thinking about set convergence. Luckily it's easy to work with the two families side by side, making use of the following properties.

4.37 Proposition (distance estimates). *The distance expressions $d\!l_\rho(C_1, C_2)$ and $\hat{d\!l}_\rho(C_1, C_2)$ (for nonempty, closed sets C_1 and C_2 in $I\!R^n$) are nondecreasing functions of ρ on $I\!R_+$, and $d\!l_\rho(C_1, C_2)$ depends continuously on ρ. One has*

(a) $\quad \hat{d\!l}_\rho(C_1, C_2) \leq d\!l_\rho(C_1, C_2) \leq \hat{d\!l}_{\rho'}(C_1, C_2)$

$\qquad\qquad$ *for* $\rho' \geq 2\rho + \max\{d_{C_1}(0), d_{C_2}(0)\}$,

(b) $\quad \hat{d\!l}_\rho(C_1, C_2) = d\!l_\rho(C_1, C_2) = d\!l_{\rho_0}(C_1, C_2)$

$\qquad\qquad$ *for* $\rho \geq \rho_0$ *if* $C_1 \cup C_2 \subset \rho_0 I\!B$,

(c) $\quad d\!l_\rho(C_1, C_2) \leq \max\Big\{d_{C_1}(0), d_{C_2}(0)\Big\} + \rho$,

(d) $\quad \big|d\!l_\rho(C_1, C_2) - d\!l_{\rho_0}(C_1, C_2)\big| \leq 2|\rho - \rho_0|$ *for any* $\rho_0 \geq 0$.

If C_1 and C_2 are convex, 2ρ can be replaced by ρ in (a). If they also contain 0, then ρ' can be taken to be ρ in (a), so that

$$\hat{d\!l}_\rho(C_1, C_2) = d\!l_\rho(C_1, C_2) \text{ for all } \rho \geq 0.$$

Proof. The monotonicity of $d\!l_\rho(C_1, C_2)$ and $\hat{d\!l}_\rho(C_1, C_2)$ in ρ is evident from the formulas in 4(11). To verify the continuity of $d\!l_\rho(C_1, C_2)$ with respect to ρ we argue from the fact that $\big|d_{C_1}(x') - d_{C_1}(x)\big| \leq |x' - x|$, cf. 4(3). We obtain for the function $\varphi(x) := \big|d_{C_1}(x) - d_{C_2}(x)\big|$ that

$$\begin{aligned}\big|\varphi(x') - \varphi(x)\big| \;&\leq\; \big|[d_{C_1}(x') - d_{C_2}(x')] - [d_{C_1}(x) - d_{C_2}(x)]\big| \\ &\leq\; \big|d_{C_1}(x') - d_{C_1}(x)\big| + \big|d_{C_2}(x') - d_{C_2}(x)\big| \;\leq\; 2|x' - x|.\end{aligned}$$

Since for any $x' in \rho I\!B$ there exists $x \in \rho_0 I\!B$ with $|x' - x| \leq |\rho - \rho_0|$, this gives us in 4(11) that, for any $\rho_0 \geq 0$,

$$d\!l_\rho(C_1, C_2) = \max_{|x'| \leq \rho} \varphi(x') \leq \max_{|x| \leq \rho_0} \varphi(x) + 2|\rho - \rho_0| \;=\; d\!l_{\rho_0}(C_1, C_2) + 2|\rho - \rho_0|,$$

so we have not just continuity but also the stronger property claimed in (d).

The inequalities in (a) are immediate from the implications in 4.34(a)(b). The special feature in the convex case is covered by the last statement of 4.34. The equalities in (b) come from the fact that when both C_1 and C_2 are within $\rho_0 I\!B$, the inclusions $C_1 \cap \rho I\!B \subset C_2 + \eta I\!B$ and $C_2 \cap \rho I\!B \subset C_1 + \eta I\!B$ are equivalent for all $\rho \geq \rho_0$ to $C_1 \subset C_2 + \eta I\!B$ and $C_2 \subset C_1 + \eta I\!B$, which by 4.34(c) imply $|d_{C_1} - d_{C_2}| \leq \eta$ everywhere. We get (c) from observing that every $x \in \rho I\!B$ has by the triangle inequality both $d_{C_1}(x) \leq d_{C_1}(0) + \rho$ and $d_{C_2}(x) \leq d_{C_2}(0) + \rho$, and therefore $\left|d_{C_1}(x) - d_{C_2}(x)\right| \leq \max\{d_{C_1}(0), d_{C_2}(0)\} + \rho$. □

Note that 4.37(a) yields for $\hat{d\!l}_\rho$ a triangle inequality of sorts: for all $\rho \geq 0$, $\hat{d\!l}_\rho(C_1, C_2) \leq \hat{d\!l}_{\rho'}(C_1, C) + \hat{d\!l}_{\rho'}(C, C_2)$ for $\rho' = 2\rho + \max\{d_{C_1}(0), d_{C_2}(0), d_C(0)\}$.

The estimates in 4.37 could be extended to the corresponding distance expressions based on balls $\mathbb{B}(\bar{x}, \rho)$ instead of balls $\rho\mathbb{B} = \mathbb{B}(0, \rho)$, but rather than pursuing the details of this we can merely think of applying the estimates as they stand to the translates $C_1 - \bar{x}$ and $C_2 - \bar{x}$ in place of C_1 and C_2.

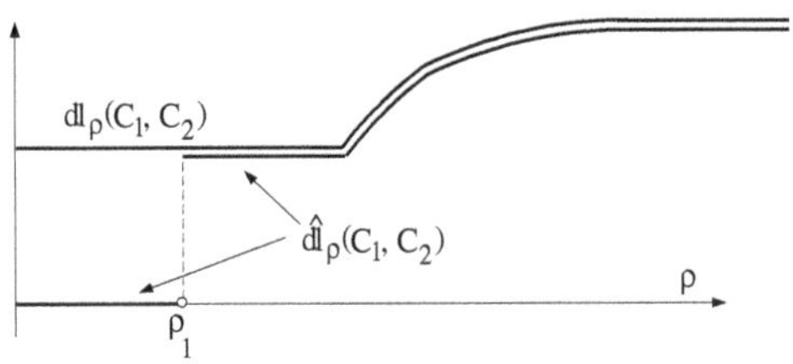

Fig. 4–12. Set distance expressions as functions of ρ when C_1 and C_2 are bounded.

4.38 Corollary (Pompeiu-Hausdorff distance as a limit). *When $\rho \to \infty$, both $d\!l_\rho(C, D)$ and $\hat{d\!l}_\rho(C, D)$ tend to the Pompeiu-Hausdorff distance $d\!l_\infty(C, D)$:*

$$d\!l_\infty(C, D) = \lim_{\rho\to\infty} d\!l_\rho(C, D) = \lim_{\rho\to\infty} \hat{d\!l}_\rho(C, D).$$

Proof. This is evident from 4.13 and the inequalities in 4.37(a). □

The properties of Pompeiu-Hausdorff distance in 4.38 are supplemented in the case of convex sets by the following fact about truncations, which quantifies the convergence result in 4.16.

4.39 Proposition (distance between convex truncations). *For nonempty, closed, convex sets C_1 and C_2, let $\rho_0 := \max\{d_{C_1}(0), d_{C_2}(0)\}$. Then*

$$\hat{d\!l}_\rho(C_1, C_2) \leq d\!l_\infty\big(C_1 \cap \rho\mathbb{B},\, C_2 \cap \rho\mathbb{B}\big) \leq 4\hat{d\!l}_\rho(C_1, C_2) \quad \textit{for } \rho > 2\rho_0.$$

Proof. The first inequality is obvious because $C_2 \cap \rho\mathbb{B} \subset C_1 \cap \rho\mathbb{B} + \varepsilon\mathbb{B}$ implies $C_2 \cap \rho\mathbb{B} \subset C_1 + \varepsilon\mathbb{B}$. To get the second inequality it's enough to show that if $\rho > 2\rho_0$ and $C_2 \cap \rho\mathbb{B} \subset C_1 + \varepsilon\mathbb{B}$, then $C_2 \cap \rho\mathbb{B} \subset C_1 \cap \rho\mathbb{B} + 4\varepsilon\mathbb{B}$.

Take arbitrary $x_2 \in C_2 \cap \rho\mathbb{B}$. On the basis of our assumptions there exists $x_1 \in C_1$ with $|x_2 - x_1| \leq \varepsilon$ along with $x_0 \in C_1$ with $|x_0| \leq \rho_0$. For $\tau \in [0, 1]$ and $x_\tau := (1 - \tau)x_0 + \tau x_1$, a point lying in C_1 by convexity, we estimate

$$|x_\tau| \leq (1 - \tau)|x_0| + \tau|x_1| \leq (1 - \tau)\rho_0 + \tau(|x_2| + \varepsilon) \leq \rho_0 + \tau[\rho - \rho_0 + \varepsilon].$$

Thus, $x_\tau \in C_1 \cap \rho\mathbb{B}$ when $\rho_0 + \tau[\rho - \rho_0 + \varepsilon] \leq \rho$, which is equivalent to $\tau \in [0, \bar{\tau}]$ for $\bar{\tau} := (\rho - \rho_0)/(\rho - \rho_0 + \varepsilon)$. Let $\bar{x}_1 := x_{\bar{\tau}}$. We have

$$\begin{aligned} d(x_2, C_1 \cap \rho I\!B) &\leq |x_2 - \bar{x}_1| \leq |x_2 - x_1| + |x_1 - \bar{x}_1| \\ &\leq \varepsilon + (1 - \bar{\tau}(|x_0| + |x_1|) \leq \varepsilon + (1 - \bar{\tau})(\rho_0 + \rho + \varepsilon) \\ &= 2\varepsilon \left[1 + \frac{\rho_0}{\rho - \rho_0 + \varepsilon}\right] \leq 2\varepsilon \left[1 + \frac{\rho_0}{\rho - \rho_0}\right] \leq 4\varepsilon. \end{aligned}$$

This being true for arbitrary $x_2 \in C_2 \cap \rho I\!B$, the claimed inclusion is valid. □

A sequence $\{C^\nu\}_{\nu \in I\!N}$ is called *(equi-) bounded* if there's a bounded set B such that $C^\nu \subset B$ for all ν. Otherwise it's *unbounded*, but it's *eventually* bounded as long as there's an index set $N \in \mathcal{N}_\infty$ such that the tail subsequence $\{C^\nu\}_{\nu \in N}$ is bounded. Thus, not only must a bounded sequence consist of bounded sets, the boundedness must be 'uniform'. For any set $X \subset I\!R^n$ we'll use the notation

$$\begin{aligned} \text{cl-sets}(X) &:= \text{ the space of all closed subsets of } X, \\ \text{cl-sets}_{\neq\emptyset}(X) &:= \text{ the space of all nonempty, closed subsets of } X. \end{aligned}$$

The following facts amplify the assertions in 4.13.

4.40 Exercise (properties of Pompeiu-Hausdorff distance).

(a) *For a bounded sequence $\{C^\nu\}_{\nu \in I\!N}$ in* cl-sets$_{\neq\emptyset}(I\!R^n)$ *and a closed set C, one has $C^\nu \to C$ if and only if $d\!l_\infty(C^\nu, C) \to 0$.*

(b) *For an unbounded sequence $\{C^\nu\}_{\nu \in I\!N}$ in* cl-sets$_{\neq\emptyset}(I\!R^n)$*, it is impossible to have $d\!l_\infty(C^\nu, C) \to 0$ without having $(C^\nu)^\infty = C^\infty$ for all ν in some index set $N \in \mathcal{N}_\infty$. Indeed, $d\!l_\infty(C_1, C_2) = \infty$ when $C_1^\infty \neq C_2^\infty$.*

(c) *Relative to* cl-sets$_{\neq\emptyset}(X)$ *for any nonempty, bounded subset X of $I\!R^n$, the distance $d\!l_\infty(C_1, C_2)$ gives a metric.*

Guide. In (a), choose ρ large enough that $\bigcup_{\nu \in I\!N} C^\nu \subset \rho I\!B$ and utilize the last part of 4.37(b). Derive (b) by arguing that the inclusion $C_1 \subset C_2 + \eta I\!B$ implies through 3.12 that $C_1^\infty \subset C_2^\infty$. Part (c) follows from 4.38 and the observation that sets $C_1, C_2 \in$ cl-sets(X) are distinct if and only if $d\!l_\infty(C_1, C_2) \neq 0$. □

I* Hyperspace Metrics

To obtain a metric that fully characterizes convergence in cl-sets$_{\neq\emptyset}(I\!R^n)$, we have to look elsewhere than the Pompeiu-Hausdorff distance, which, as just confirmed, only works for the subspaces cl-sets$_{\neq\emptyset}(X)$ of cl-sets$_{\neq\emptyset}(I\!R^n)$ that correspond to bounded sets $X \subset I\!R^n$. Such a metric can be derived in many ways from the family of pseudo-metrics $d\!l_\rho$, but a convenient expression that eventually will be seen to enjoy some especially attractive properties is

$$d\!l(C, D) := \int_0^\infty d\!l_\rho(C, D) e^{-\rho} d\rho. \tag{4(12)}$$

This will be called the *(integrated) set distance* between C and D. Note that

$$d\!\!l(C,D) \le d\!\!l_\infty(C,D), \tag{4(13)}$$

because $d\!\!l_\rho(C,D) \le d\!\!l_\infty(C,D)$ for all ρ, while $\int_0^\infty e^{-\rho}d\rho = 1$.

4.41 Lemma (estimates for the integrated set distance). *For any nonempty, closed subsets C_1 and C_2 of $I\!R^n$ and any $\rho \in I\!R_+$, one has*

(a) $d\!\!l(C_1,C_2) \ge (1-e^{-\rho})\big|d_{C_1}(0) - d_{C_2}(0)\big| + e^{-\rho}d\!\!l_\rho(C_1,C_2)$,

(b) $d\!\!l(C_1,C_2) \le (1-e^{-\rho})d\!\!l_\rho(C_1,C_2) + e^{-\rho}\big(\max\{d_{C_1}(0), d_{C_2}(0)\} + \rho + 1\big)$,

(c) $\big|d_{C_1}(0) - d_{C_2}(0)\big| \le d\!\!l(C_1,C_2) \le \max\{d_{C_1}(0), d_{C_2}(0)\} + 1$.

Proof. We write

$$d\!\!l(C_1,C_2) = \int_0^\rho d\!\!l_\tau(C_1,C_2)e^{-\tau}d\tau + \int_\rho^\infty d\!\!l_\tau(C_1,C_2)e^{-\tau}d\tau$$

and note from the monotonicity of $d\!\!l_\rho(C_1,C_2)$ in ρ (cf. 4.37) that

$$\begin{aligned}
d\!\!l_0(C_1,C_2)\int_0^\rho e^{-\tau}d\tau &\le \int_0^\rho d\!\!l_\tau(C_1,C_2)e^{-\tau}d\tau \le d\!\!l_\rho(C_1,C_2)\int_0^\rho e^{-\tau}d\tau,\\
d\!\!l_\rho(C_1,C_2)\int_\rho^\infty e^{-\tau}d\tau &\le \int_\rho^\infty d\!\!l_\tau(C_1,C_2)e^{-\tau}d\tau\\
&\le \int_\rho^\infty \big[\max\{d_{C_1}(0), d_{C_2}(0)\} + \tau\big]e^{-\tau}d\tau,
\end{aligned}$$

where the last inequality comes from 4.37(c). The lower estimates calculate out to the inequality in (a), and the upper estimates to the one in (b).

In (c), the inequality on the left comes from the limit as $\rho \nearrow \infty$ in (a), where the term $e^{-\rho}d\!\!l_\rho(C_1,C_2)$ tends to 0 because of 4.37(c). The inequality on the right comes from taking $\rho = 0$ in (b). □

4.42 Theorem (metric description of set convergence). *The expression $d\!\!l$ gives a metric on* cl-sets$_{\neq\emptyset}(I\!R^n)$ *which characterizes ordinary set convergence:*

$$C^\nu \to C \iff d\!\!l(C^\nu, C) \to 0.$$

Furthermore, $\big($cl-sets$_{\neq\emptyset}(I\!R^n), d\!\!l\big)$ *is a complete metric space in which a sequence $\{C^\nu\}_{\nu\in I\!N}$ escapes to the horizon if and only if for some set C in this space (and then for every C) one has $d\!\!l(C^\nu, C) \to \infty$.*

Proof. We get $d\!\!l(C_1,C_2) \ge 0$, $d\!\!l(C_1,C_2) = d\!\!l(C_2,C_1)$, and the triangle inequality $d\!\!l(C_1,C_2) \le d\!\!l(C_1,C) + d\!\!l(C,C_2)$ from the corresponding properties of the pseudo-metrics $d\!\!l_\rho(C_1,C_2)$. The estimate in 4.41(c) gives us $d\!\!l(C_1,C_2) < \infty$. Since for closed sets C_1 and C_2 the distance functions d_{C_1} and d_{C_2} are continuous and vanish only on these sets, respectively, we have $\big|d_{C_1}(x) - d_{C_2}(x)\big|$ positive on some open set unless $C_1 = C_2$. Thus $d\!\!l(C_1,C_2) > 0$ unless $C_1 = C_2$. This proves that $d\!\!l$ is a metric.

It's clear from the estimates in 4.41(a) and (b) that $d\!\!l(C^\nu, C) \to 0$ if and only if $d\!\!l_\rho(C^\nu, C) \to 0$ for every $\rho \ge 0$. In view of Theorem 4.36, we know therefore that the metric $d\!l$ on cl-sets$_{\neq\emptyset}(I\!R^n)$ characterizes set convergence.

From 4.42, a sequence $\{C^\nu\}$ in cl-sets$_{\neq\emptyset}(I\!R^n)$ escapes to the horizon if and only if it eventually misses every ball $\rho I\!B$, or equivalently, has $d_{C^\nu}(0) \to \infty$. Since by taking $\rho = 0$ in the inequalities in 4.41(a)(b) one has

$$\big|d_{C^\nu}(0) - d_C(0)\big| \;\leq\; d\!l(C^\nu, C) \;\leq\; \max\big\{d_{C^\nu}(0), d_C(0)\big\} + 1,$$

the sequence escapes to the horizon if and only if $d\!l(C^\nu, C) \to \infty$ for every C, or for just one C. Because a Cauchy sequence $\{C^\nu\}_{\nu\in N}$ in particular has the property that, for any fixed ν_0, the distance sequence $\{d\!l(C^\nu, C^{\nu_0})\}_{\nu\in N}$ is bounded, it can't have any subsequence escaping to the horizon. Hence by the compactness property in 4.18, every Cauchy sequence in cl-sets$_{\neq\emptyset}(I\!R^n)$ has a subsequence converging to an element of cl-sets$_{\neq\emptyset}(I\!R^n)$, this element necessarily then being the actual limit of the sequence. Therefore, the metric space $\big(\text{cl-sets}_{\neq\emptyset}(I\!R^n), d\!l\big)$ is complete. □

4.43 Corollary (local compactness in metric spaces of sets). *The metric space $\big(\text{cl-sets}_{\neq\emptyset}(I\!R^n), d\!l\big)$ has the property that for every one of its elements C_0 and every $r > 0$ the ball $\big\{C \,\big|\, d\!l(C, C_0) \leq r\big\}$ is compact.*

Proof. This is a consequence of the compactness in 4.18 and the criterion in 4.42 for escape to the horizon. □

4.44 Example (distances between cones). *For closed cones K_1, $K_2 \subset I\!R^n$, the Pompeiu-Hausdorff distance is always $d\!l_\infty(K_1, K_2) = \infty$ unless $K_1 = K_2$, whereas the integrated set distance is the ρ-distance for $\rho = 1$: one has*

$$d\!l(K_1, K_2) = d\!l_1(K_1, K_2) = \hat{d\!l}_1(K_1, K_2) = d\!l_\infty\big(K_1 \cap I\!B, K_2 \cap I\!B\big) \leq 1,$$

and on the other hand

$$d\!l_\rho(K_1, K_2) = \hat{d\!l}_\rho(K_1, K_2) = \rho d\!l(K_1, K_2) \;\; \text{for all} \;\; \rho \geq 0.$$

In the metric space $\big(\text{cl-sets}_{\neq\emptyset}(I\!R^n), d\!l\big)$, the set of all closed cones is compact.

Detail. If $K_1 \not\subset K_2$, there must be a ray $R \subset K_1$ such that $R \not\subset K_2$. Then $R \not\subset K_2 + \eta I\!B$ for all $\eta \in I\!R_+$, so that $d(K_1, K_2) = \infty$. Likewise this has to be true when $K_2 \not\subset K_1$. Thus, $d\!l_\infty(K_1, K_2) = \infty$ unless $K_1 = K_2$.

Because the equivalence at the end of Lemma 4.34 always holds for cones, regardless of convexity, we have $\hat{d\!l}_\rho(K_1, K_2) = d\!l_\rho(K_1, K_2)$ for all $\rho \geq 0$. It's clear also in the cone case that $d\!l_\rho(K_1, K_2) = \rho d\!l_1(K_1, K_2)$. Then $d\!l(K_1, K_2) = d\!l_1(K_1, K_1)$ by definition 4(12), inasmuch as $\int_0^\infty \rho e^{-\rho} d\rho = 1$.

We have $\hat{d\!l}_1(K_1, K_2) \leq 1$ because $K_1 \cap I\!B \subset K_2 + I\!B$ and $K_2 \cap I\!B \subset K_1 + I\!B$ when K_1 and K_2 contain 0. The equation $\hat{d\!l}_1(K_1, K_2) = d\!l_\infty\big(K_1 \cap I\!B, K_2 \cap I\!B\big)$ holds on the basis of the definitions of these quantities, since

$$(K + \eta I\!B) \cap I\!B \subset [K \cap I\!B] + \eta I\!B \;\; \text{for any closed cone} \;\; K.$$

The latter comes from the observation that the projection of any point of $I\!B$ on a ray R belongs to $R \cap \mathbb{B}$, which implies that any point of $\mathbb{B}$ within distance η of K is also within distance η of $K \cap \mathbb{B}$. □

To round out the discussion we record an approximation fact and use it to establish the separability of our metric space of sets.

4.45 Proposition (separability and approximation by finite sets). *Every closed set C in $\mathbb{R}^n$ can be expressed as the limit of a sequence of sets C^ν, each of which consists of just finitely many points.*

In fact, the points can be chosen to have rational coordinates. Thus, the countable collection consisting of all finite sets of rational points in $\mathbb{R}^n$ is dense in the metric space $\big(\text{cl-sets}_{\neq\emptyset}(\mathbb{R}^n), d\!\!l\big)$, which therefore is separable.

Proof. Because the collection of subsets of $\mathbb{R}^n$ consisting of all finite sets whose points have rational coordinates is a countable collection, it can be indexed as a single sequence $\{C^\nu\}_{\nu\in\mathbb{N}}$. We need only show for an arbitrary set C in $\text{cl-sets}_{\neq\emptyset}(\mathbb{R}^n)$ that C is a cluster point of this sequence.

For each $\mu \in \mathbb{N}$ the set $C + \mu^{-1}\mathbb{B}$ is closed through the fact that $\mathbb{B}$ is compact (cf. 3.12). Thus $C + \mu^{-1}\mathbb{B}$, like C, belongs to $\text{cl-sets}_{\neq\emptyset}(\mathbb{R}^n)$. The nested sequence $\{C + \mu^{-1}\mathbb{B}\}_{\mu\in\mathbb{N}}$ is decreasing and has C as its intersection, so $C + \mu^{-1}\mathbb{B} \to C$ (cf. 4.3(b)). It will suffice therefore to show for arbitrary $\mu \in \mathbb{N}$ that $C + \mu^{-1}\mathbb{B}$ is a cluster point of $\{C^\nu\}_{\nu\in\mathbb{N}}$, since by diagonalization the same will then be true for C itself.

To get an index set $N^\mu \in \mathcal{N}_\infty^\#$ such that $C^\nu \underset{N^\mu}{\to} C + \mu^{-1}\mathbb{B}$, we can simply choose N^μ to consist of the indices ν such that $C^\nu \subset C + \mu^{-1}\mathbb{B}$. This choice trivially ensures that $\limsup_{\nu\in N^\mu} C^\nu \subset C + \mu^{-1}\mathbb{B}$, but also, because points with rational coordinates are dense in $C + \mu^{-1}\mathbb{B}$ (this set being the closure of the union of all open balls of radius μ^{-1} centered at points of C), it ensures that $\liminf_{\nu\in N^\mu} C^\nu \supset C + \mu^{-1}\mathbb{B}$. Hence we do in this way have $\lim_{\nu\in N^\mu} C^\nu = C + \mu^{-1}\mathbb{B}$, as required. □

Cosmic set convergence can likewise be quantified in terms of a metric. This can be accomplished through the identification of subsets of $\operatorname{csm}\mathbb{R}^n$ with cones in $\mathbb{R}^{n+1}$ in the ray space model for $\operatorname{csm}\mathbb{R}^n$. Distances between such cones can be measured with the set metric $d\!\!l$ already developed.

When a subset of $\operatorname{csm}\mathbb{R}^n$ is designated by $C \cup \operatorname{dir} K$ (with $C, K \subset \mathbb{R}^n$, K a cone), the corresponding cone in the ray space model is

$$\operatorname{pos}(C,-1) \cup (K,0) = \big\{\lambda(x,-1) \,\big|\, \lambda \geq 0,\ x \in C\big\} \cup \big\{(x,0) \,\big|\, x \in K\big\}.$$

Accordingly, we define the *cosmic* set metric $d\!\!l_{\text{csm}}$ by

$$\begin{aligned} &d\!\!l_{\text{csm}}\big(C_1 \cup \operatorname{dir} K_1,\ C_2 \cup \operatorname{dir} K_2\big) \\ &\qquad := d\!\!l\big(\operatorname{pos}(C_1,-1) \cup (K_1,0),\ \operatorname{pos}(C_2,-1) \cup (K_2,0)\big) \end{aligned} \tag{4(14)}$$

for subsets $C_1 \cup \operatorname{dir} K_1$ and $C_2 \cup \operatorname{dir} K_2$ of $\operatorname{csm}\mathbb{R}^n$. Through the formulas in 4.44 for $d\!\!l$ as applied to cones, this *cosmic distance* between $C_1 \cup \operatorname{dir} K_1$ and $C_2 \cup \operatorname{dir} K_2$ can alternatively be expressed in other ways. For instance, it's the Pompeiu-Hausdorff distance $d\!l_\infty(B_1, B_2)$ between the subsets B_1 and B_2 of $I\!R^{n+1}$ obtained by intersecting the unit ball of $I\!R^{n+1}$ with the cones $\text{pos}(C_1, -1) \cup (K_1, 0)$ and $\text{pos}(C_2, -1) \cup (K_2, 0)$. No matter how the distance is expressed, we get at once a characterization of set convergence in $\text{csm}\, I\!R^n$.

4.46 Theorem (metric description of cosmic set convergence). *On the space* cl-sets(csm $I\!R^n$), $d\!l_{\text{csm}}$ *is a metric that characterizes cosmic set convergence:*

$$C^\nu \cup \text{dir}\, K^\nu \xrightarrow{c} C \cup \text{dir}\, K \iff d\!l_{\text{csm}}\big(C^\nu \cup \text{dir}\, K^\nu,\ C \cup \text{dir}\, K\big) \to 0.$$

The metric space $\big(\text{cl-sets}(\text{csm}\, I\!R^n), d\!l_{\text{csm}}\big)$ *is separable and compact.*

Proof. Because cosmic set convergence is equivalent by definition to the ordinary set convergence of the corresponding cones in the ray space model, this is immediate from Theorem 4.36 as invoked for cones in $I\!R^{n+1}$. The separability and compactness are seen from 4.45 and 4.18. □

A restriction to nonempty subsets of $\text{csm}\, I\!R^n$ isn't needed, because the definition of $d\!l_{\text{csm}}$ in 4(14) uses $d\!l$ only for cones, and cones always contain the origin. But of course, in the topology of cosmic set convergence, the empty set is an isolated element; it isn't the limit of any sequence of nonempty sets, since any sequence of points in $\text{csm}\, I\!R^n$ has a cluster point (cf. 3.2).

The metric $d\!l_{\text{csm}}$ can be applied in particular to subsets C_1 and C_2 of $I\!R^n$ (with $K_1 = K_2 = \{0\}$, $\text{dir}\, K_1 = \text{dir}\, K_2 = \emptyset$). One has

$$\begin{aligned} d\!l_{\text{csm}}(C_1, C_2) &= d\!l\big(\text{pos}(C_1, -1), \text{pos}(C_2, -1)\big) \\ &= d\!l\big(\text{cl}\,\text{pos}(C_1, -1), \text{cl}\,\text{pos}(C_2, -1)\big) \\ &= d\!l_{\text{csm}}\big(\text{cl}\, C_1 \cup \text{dir}\, C_1^\infty,\ \text{cl}\, C_2 \cup \text{dir}\, C_2^\infty\big), \end{aligned} \qquad 4(15)$$

since the closure of the cone representing a set $C \subset I\!R^n$ in the ray space model for $\text{csm}\, I\!R^n$ is, by definition, the cone representing $\text{cl}\, C \cup \text{dir}\, C^\infty$ (see 3.4). This yields a characterization of total set convergence in $I\!R^n$.

4.47 Corollary (metric description of total set convergence). *On the space* cl-sets($I\!R^n$), $d\!l_{\text{csm}}$ *is a metric that characterizes total set convergence:*

$$C^\nu \xrightarrow{t} C \iff d\!l_{\text{csm}}(C^\nu, C) \to 0.$$

The metric space $\big(\text{cl-sets}(I\!R^n), d\!l_{\text{csm}}\big)$ *is locally compact and separable, and its completion is* $\big(\text{cl-sets}(\text{csm}\, I\!R^n), d\!l_{\text{csm}}\big)$*; it forms an open set within the latter.*

Proof. All this is clear from Theorem 4.46 and the definition of total convergence in 4.23. The claims of local compactness and openness rest on the fact that, with respect to cosmic convergence, the space cl-sets(hzn $I\!R^n$) is a closed subset of cl-sets(csm $I\!R^n$). (Sequences of direction points can converge only to direction points.) □

The cosmic set metric $d\!l_{\text{csm}}$ can be interpreted as arising from a special non-Euclidean metric for measuring distances between points of $I\!R^n$ itself. This comes out as follows.

4.48 Exercise (cosmic metric properties). *Let $\theta\big((x,\alpha),(y,\beta)\big)$ denote the angle between two nonzero vectors (x,α) and (y,β) in $\mathbb{R}^n\times\mathbb{R}$, and let*

$$s\big((x,\alpha),(y,\beta)\big) := \begin{cases} \sin\theta\big((x,\alpha),(y,\beta)\big) & \text{if } \theta\big((x,\alpha),(y,\beta)\big) < \pi/2,\\ 1 & \text{if } \theta\big((x,\alpha),(y,\beta)\big) \geq \pi/2. \end{cases}$$

Define

$$d_{\rm csm}(x,y) := s\big((x,-1),(y,-1)\big) \ \text{ for } x,\, y\in\mathbb{R}^n. \tag{4(16)}$$

Then $d_{\rm csm}$ is a metric on $\mathbb{R}^n$ such that the completion of the space $\big(\mathbb{R}^n, d_{\rm csm}\big)$ is the space $\big(\operatorname{csm}\mathbb{R}^n, d_{\rm csm}\big)$ with $d_{\rm csm}$ extended by

$$\begin{aligned} d_{\rm csm}(x,\operatorname{dir} y) &= s\big((x,-1),(y,0)\big),\\ d_{\rm csm}(\operatorname{dir} x,\operatorname{dir} y) &= s\big((x,0),(y,0)\big). \end{aligned} \tag{4(17)}$$

This point metric $d_{\rm csm}$ on $\operatorname{csm}\mathbb{R}^n$ is compatible with the set metric $\mathbb{d}_{\rm csm}$ on $\text{cl-sets}(\operatorname{csm}\mathbb{R}^n)$ in the sense that

$$\begin{aligned} d_{\rm csm}(x,y) &= \mathbb{d}_{\rm csm}\big(\{x\},\{y\}\big),\\ d_{\rm csm}(x,\operatorname{dir} y) &= \mathbb{d}_{\rm csm}\big(\{x\},\{\operatorname{dir} y\}\big),\\ d_{\rm csm}(\operatorname{dir} x,\operatorname{dir} y) &= \mathbb{d}_{\rm csm}\big(\{\operatorname{dir} x\},\{\operatorname{dir} y\}\big), \end{aligned} \tag{4(18)}$$

while on the other hand, $\mathbb{d}_{\rm csm}$ on $\text{cl-sets}_{\neq\emptyset}(\operatorname{csm}\mathbb{R}^n)$ is the Pompeiu-Hausdorff distance generated by $d_{\rm csm}$: in denoting by $d_{\rm csm}(x, C\cup\operatorname{dir} K)$ the distance with respect to $d_{\rm csm}$ from a point $x\in\mathbb{R}^n$ to a subset $C\cup\operatorname{dir} K$ of $\operatorname{csm}\mathbb{R}^n$, one has

$$\begin{aligned} &\mathbb{d}_{\rm csm}\big(C_1\cup\operatorname{dir} K_1,\, C_2\cup\operatorname{dir} K_2\big)\\ &\qquad = \sup_{x\in\mathbb{R}^n} \big| d_{\rm csm}(x, C_1\cup\operatorname{dir} K_1) - d_{\rm csm}(x, C_2\cup\operatorname{dir} K_2)\big|. \end{aligned} \tag{4(19)}$$

Guide. Start by going backwards from the formulas in 4(18); in other words, begin with the fact that, for points $x,y\in\mathbb{R}^n$, $\mathbb{d}_{\rm csm}(\{x\},\{y\})$ is by definition 4(14) equal to $\mathbb{d}(\operatorname{pos}(x,-1),\operatorname{pos}(y,-1))$. Argue through 4.44 that this has the value $\mathbb{d}_\infty(B_x,B_y)$, where $B_x=\mathbb{B}\cap\operatorname{pos}(x,-1)$ and $B_y=\mathbb{B}\cap\operatorname{pos}(y,-1)$, which calculates out to $s((x,-1),(y,-1))$. That provides the foundation for getting everything else. The supremum in 4(19) is adequately taken over $\mathbb{R}^n$ instead of over general elements of $\operatorname{csm}\mathbb{R}^n$ because the expression being maximized has a unique continuous extension to $\operatorname{csm}\mathbb{R}^n$. □

Commentary

Up to the mid 1970s, the study of set convergence was carried out almost exclusively by topologists. A key article by Michael [1951] and the book by Nadler [1978] highlight their concerns: the description and analysis of a topology compatible with set convergence on the *hyperspace* cl-sets($\mathbb{R}^n$), or more generally on cl-sets(X) where X is an arbitrary topological space. Ever-expanding applications nowadays to optimization problems, random sets, economics, and other topics, have rekindled the interest in the subject and have refocused the research; the book by Beer [1993] and the survey articles of Sonntag and Zălinescu [1991] and Lucchetti and Torre [1994] reflect this shift in emphasis.

The concepts of inner and outer limits for a sequence of sets are due to the French mathematician-politician Painlevé, who introduced them in 1902 in his lectures on analysis at the University of Paris; set convergence was defined as the equality of these two limits. Hausdorff [1927] and Kuratowski [1933] popularized such convergence by including it in their books, and that's how Kuratowski's name ended up to be associated with it.

In calling the two kinds of set limits 'inner' and 'outer', we depart from the terms 'lower' and 'upper', which until now have commonly been used. Our terminology carries over in Chapter 5 to 'inner semicontinuity' and 'outer semicontinuity' of set-valued mappings, in contrast to 'lower semicontinuity' and 'upper semicontinuity'. The reasons why we have felt the need for such a change are twofold. 'Inner' and 'outer' are geometrically more accurate and reflect the nature of the concepts, whereas 'lower' and 'upper', words suggesting possible spatial relationships other than inclusion, can be misleading. More importantly, however, this switch helps later in getting around a serious difficulty with what 'upper semicontinuity' has come to mean in the literature of set-valued mappings. That term is now incompatible with the notion of continuity naturally demanded in the many applications where boundedness of the sets and sequences of interest isn't assured. We are obliged to abandon 'upper semicontinuity' and use different words for the concept that we require. 'Outer semicontinuity' fits, and as long as we are passing from upper to outer it makes sense to pass at the same time from lower to inner, although that wouldn't strictly be necessary. This will be explained more fully in Chapter 5; see the Commentary for that chapter and the discussion in the text itself around Figure 5–7.

Probably because set convergence hasn't been covered in the standard texts on topology and hasn't therefore achieved wide familiarity, despite having been on the scene for a long time, researchers have often turned to convergence with respect to the Pompeiu-Hausdorff distance as a substitute, even when that might be inappropriate. In terms of the *excess* of a set C over another set D given by $e(C,D) := \sup\{d(x,D) \mid x \in C\}$, Pompeiu [1905], a student of Painlevé, defined the distance between C and D, when they are nonempty, by $e(C,D) + e(D,C)$. Hausdorff [1927] converted this to $\max\{e(C,D), e(D,C)\}$, a distance expression inducing the same convergence. Our equivalent way of defining this distance in 4.13 has the advantage of pointing to the modifications that quantify set convergence in general.

The hit-and-miss criteria in Theorem 4.5 originated with the description by Fell [1962] of the hyperspace topology associated with set convergence. Wijsman [1966] and Holmes [1966] were responsible for the characterization of set convergence as pointwise convergence of distance functions (cf. Corollary 4.7); the extension to gap functions (cf. 4(4)) comes from Beer and Lucchetti [1993]. The observation that *con-*

vex sets converge if and only if their projection mappings converge pointwise is due to Sonntag [1976] (see also Attouch [1984]), but the statement in 4.9 about the convergence of the projections of arbitrary sequences of sets is new. The approximation results involving ε-fattening of sets (in Theorem 4.10 and Corollary 4.11) were first formulated by Salinetti and Wets [1981]. In an arbitrary metric space, each of these characterizations leads to a different notion of convergence that has abundantly been explored in the literature; see e.g. Constantini, Levi and Zieminska [1993], Beer [1993] and Sonntag and Zălinescu [1991].

Still other convergence notions for sets in $I\!R^n$ aren't equivalent to Painlevé-Kuratowski convergence but have significance for certain applications. One of these, like convergence with respect to the Pompeiu-Hausdorff distance, is more restrictive: C^ν converges to C *in the sense of Fisher* [1981] when $C \subset \liminf C^\nu$ and $e(C^\nu, C) \to 0$ (the latter referring to the 'excess' defined above). Another notion is convergence *in the sense of Vietoris* [1921] when $C = \liminf_\nu C^\nu$ and for every closed set $F \subset I\!R^n$ with $C \cap F = \emptyset$ there exists $N \in \mathcal{N}_\infty$ such that $C^\nu \cap F = \emptyset$ for all $\nu \in N$, i.e., when in the hit-and-miss criterion in Theorem 4.5 about missing compact sets has been switched to missing closed sets. Other such notions aren't comparable to Painlevé-Kuratowski convergence at all. An example is 'rough' convergence, which was introduced to analyze the convergence of probability measures (Lucchetti, Salinetti and Wets [1994]) and of packings and tilings (Wicks [1994]); see Lucchetti, Torre and Wets [1993]. A sequence of sets C^ν converges *roughly* to C when $C = \limsup_\nu C^\nu$ and $\operatorname{cl} D \supset \limsup_\nu D^\nu$, where $D = I\!R^n \setminus C$ and $D^\nu = I\!R^n \setminus C^\nu$.

The observation in 4.3 about monotone sequences can be found in Mosco [1969], at least for convex sets. The criterion in Corollary 4.12 for the connectedness of a limit set can be traced back to Janiszewski, as recorded by Choquet [1947]. The convexity of the inner limit associated with a collection of convex sets (Proposition 4.15) is part of the folklore; the generalization in 4.17 to star-shaped sets is due to Beer and Klee [1987]. The use of truncations to quantify the convergence of convex sets, as in 4.16, comes from Salinetti and Wets [1979]. The assertion in 4.15, that a compact set contained in the interior of the inner limit of a sequence must also be contained in the interior of the approaching sets, can essentially be found in Robert [1974], but the proof furnished here is new.

Theorem 4.18 on the compactness of the hyperspace cl-sets($I\!R^n$) has a long history starting with Zoretti [1909], another student of Painlevé, who proved a version of this fact for a bounded sequence of continua in the plane. Developments culminated in the late 1920s in the version presented here, with proofs provided variously by Zarankiewicz [1927], Hausdorff [1927], Lubben [1928] and R.L. Moore (1925, unpublished). The cluster description of inner and outer limits (in Proposition 4.19) can be found in Choquet [1947]. For a sequence of nonempty, convex sets, one can combine 4.15 with 4.18 to obtain the 'selection theorem' of Blaschke [1914]: in $I\!R^n$, any bounded sequence $\{C^\nu\}_{\nu \in I\!N}$ of such sets has a subsequence converging to some nonempty, compact, convex set C.

Attempts at ascertaining when set convergence is preserved under various operations (as in 4.27 and 4.30) have furnished the chief motivation for our introduction of horizon and cosmic limits with their associated cosmic metric. Partial results along these lines were reported in Rockafellar and Wets [1992], but the full properties of such limits are brought out here for the first time along with the fundamental role they play in variational analysis. The concept of 'total' convergence and its quantification in 4.46 are new as well. The horizon limit formula in 4.21(e) was discovered by M. Dong (unpublished).

For sequences of *convex* sets, McLinden and Bergstrom [1981] obtained forerunners of Theorem 4.27 and Example 4.28. Our introduction of total convergence bears fruit in enabling us to go beyond convexity in this context. The convergence criteria for the sums of sets in 4.29 are new, as are the total convergence aspects of 4.30 and 4.31. New too is Theorem 4.32 about the convergence of convex systems, at least in the way it's formulated here, but it could essentially be derived from results in McLinden and Bergstrom [1981] about sequences of convex functions; see also Azé and Penot [1990].

The metrizability of cl-sets($\mathbb{R}^n$) in the topology of set convergence has long been known on the general principles of metric space theory. Here we have specifically introduced the metric $d\!l$ on cl-sets$_{\neq\emptyset}(\mathbb{R}^n)$ in 4.42, constructing it from the pseudo-metrics $d\!l_\rho$. One of the benefits of this choice, through the cone properties in 4.44 (newly reported here), is the ease with which we are able to go on to provide metric characterizations of cosmic set convergence and total set convergence in 4.46 and 4.47.

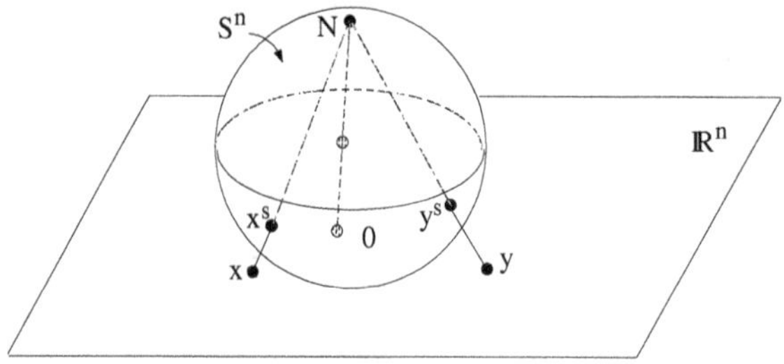

Fig. 4–13. Stereographic images of x and y in $\mathbb{R}^n$ on a sphere in $\mathbb{R}^{n+1}$.

An alternative way of quantifying ordinary set convergence would be to use the *stereographic Pompeiu-Hausdorff metric* mentioned in Rockafellar and Wets [1984]. It relies on measuring the distance between two sets by means of the Pompeiu-Hausdorff distance, but bases the latter not on the Euclidean distance $d(x,y) = |x-y|$ between points x and y in $\mathbb{R}^n$ but on their stereographic distance $d^s(x,y) = |x^s - y^s|$, where x^s and y^s are the stereographic images of x and y on S^n, the n-dimensional sphere of radius 1 in $\mathbb{R}^{n+1}$ with center at $(0,\ldots,0,1)$, as indicated in Figure 4–13 in terms of the 'north pole' N of S^n. (When determining the distance between two subsets of $\mathbb{R}^n$ in this manner, N is automatically added to the corresponding image subsets of S^n; thus in particular, the empty set corresponds to $\{N\}$, and its stereographic distance to other sets is finite.) This metric isn't convenient operationally, however, and by its association with the one-point compactification of $\mathbb{R}^n$ it deviates from the cosmic framework we are keen on maintaining.

The use of $\hat{d\!l}_\rho$ to measure the distance between sets was proposed by Walkup and Wets [1967] for convex cones and by Mosco [1969] for convex sets in general. However, it was not until Attouch and Wets [1991] that these distance expressions where studied in depth, and not merely for convex sets. The pseudo-metrics $d\!l_\rho$ come from the specialization to indicator functions of pseudo-metrics introduced in Attouch and Wets [1986] to measure the distance between (arbitrary) functions. The investigation of the relationship between $\hat{d\!l}_\rho$ and $d\!l_\rho$ and of the 'uniformities' that they induce was carried out in Attouch, Lucchetti and Wets [1991].

The inequalities in Lemma 4.34 are slightly sharper than the ones in those papers, and the same holds for Proposition 4.37. The much smaller value for ρ' in the convex case in 4.37(a) is recorded here for the first time. The particular expression used in 4(12) for the metric $d\!l$ on cl-sets$_{\neq\emptyset}(I\!R^n)$ is new, as are the resulting inequalities in Lemma 4.41. So too is the recognition that Pompeiu-Hausdorff distance is the common limit of $\hat{d\!l}_\rho$ and $d\!l_\rho$. The distance estimate for truncated convex sets in 4.39 comes from a lemma of Loewen and Rockafellar [1994] sharpening an earlier one of Clarke [1983].

The local compactness of the metric space (cl-sets$_{\neq\emptyset}(I\!R^n), d\!l$) in 4.43 is mentioned in Fell [1962]. The separability property in 4.45 can be traced back to Kuratowski [1933]. The 'constructive' proof of it provided here is inspired by a related result for set-valued mappings of Salinetti and Wets [1981].

5. Set-Valued Mappings

The concept of a variable set is of great importance. Abstractly we can think of two spaces X and U and the assignment to each $x \in X$ of a set $S(x) \subset U$, i.e., an element of the space

$$\text{sets}(U) := \text{ collection of all subsets of } U.$$

It is natural to speak then of a *set-valued mapping* S, but in so doing we must be careful to interpret the terminology in the manner most favorable for our purposes. A variable set can often be viewed with advantage as generalizing a variable point, especially in situations where the set might often have only one element. It is preferable therefore to identify as the *graph* of S a subset of $X \times U$, namely

$$\operatorname{gph} S := \{(x, u) \mid u \in S(x)\},$$

rather than a subset of $X \times \text{sets}(U)$ as might literally seem to be dictated by the words 'set-valued mapping'. To emphasize this we write $S : X \rightrightarrows U$ instead of $S : X \to \text{sets}(U)$. The emphasis could further be conveyed when deemed necessary by speaking of S as a *multifunction* or *correspondence*, but the simpler language of set-valuedness will be used in what follows.

Obviously S is fully described by $\operatorname{gph} S$, and every set $G \subset X \times U$ is the graph of a uniquely determined set-valued mapping $S : X \rightrightarrows U$:

$$S(x) = \{u \mid (x, u) \in G\}, \qquad G = \operatorname{gph} S.$$

It will be useful to work with this pairing as an extension of the familiar one between functions from X to U and their graphs in $X \times U$, and even to think of functions as special cases of set-valued mappings. In effect, a mapping $S : X \to U$ and the associated mapping from X to singletons $\{u\}$ in $\text{sets}(U)$ are to be regarded as the same mathematical object seen from two different angles.

More than one interpretation will therefore be possible in strict terms when we speak of a set-valued mapping, but the appropriate meaning will always be clear from the context. In general we allow ourselves to refer to $S : X \rightrightarrows U$ just as a *mapping* and say that S is *empty-valued*, *single-valued* or *multivalued* at x according to whether $S(x)$ is the empty set, a singleton, or a set containing more than one element. The case of S being single-valued everywhere on X is specified by the notation $S : X \to U$. We say S is *compact-valued* or *convex-*

valued when $S(x)$ is compact or convex, and so forth. (Although 'mapping' will have wider reference, 'function' will always entail single-valuedness.)

A. Domains, Ranges and Inverses

In line with this broad view, the *domain* and *range* of $S : X \rightrightarrows U$ are taken to be the sets

$$\operatorname{dom} S := \{x \mid S(x) \neq \emptyset\}, \qquad \operatorname{rge} S := \{u \mid \exists x \text{ with } u \in S(x)\},$$

which are the images of $\operatorname{gph} S$ under the projections $(x,u) \mapsto x$ and $(x,u) \mapsto u$, see Figure 5–1. The *inverse* mapping $S^{-1} : U \rightrightarrows X$ is defined by $S^{-1}(u) := \{x \mid u \in S(x)\}$; obviously $(S^{-1})^{-1} = S$. The *image* of a set C under S is

$$S(C) := \bigcup\nolimits_{x \in C} S(x) = \{u \mid S^{-1}(u) \cap C \neq \emptyset\},$$

while the *inverse image* of a set D is

$$S^{-1}(D) := \bigcup\nolimits_{u \in D} S^{-1}(u) = \{x \mid S(x) \cap D \neq \emptyset\}.$$

Note that $\operatorname{dom} S^{-1} = \operatorname{rge} S = S(X)$, whereas $\operatorname{rge} S^{-1} = \operatorname{dom} S = S^{-1}(U)$.

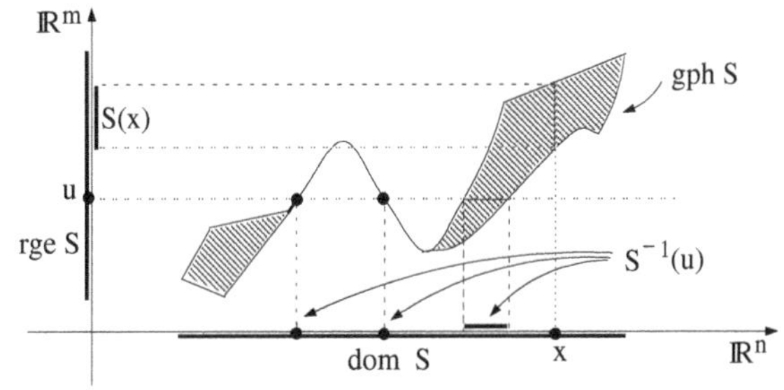

Fig. 5–1. Notational scheme for set-valued mappings.

For the most part we'll be concerned with the case where X and U are subsets of finite-dimensional real vector spaces, say $\mathbb{R}^n$ and $\mathbb{R}^m$. Notation can then be streamlined very conveniently because any mapping $S : X \rightrightarrows U$ is at the same time a mapping $S : \mathbb{R}^n \rightrightarrows \mathbb{R}^m$. One merely has $S(x) = \emptyset$ for $x \notin X$. Note that because we follow the pattern of identifying S with a subset of $X \times U$ as its graph, this is not actually a matter of *extending* S from X to the rest of $\mathbb{R}^n$, since $\operatorname{gph} S$ is unaffected. We merely choose to regard this graph set as lying in $\mathbb{R}^n \times \mathbb{R}^m$ rather than just $X \times U$. Images and inverse images under S stay the same, as do the sets $\operatorname{dom} S$ and $\operatorname{rge} S$, but now also

$$\operatorname{dom} S = S^{-1}(\mathbb{R}^m), \qquad \operatorname{rge} S = S(\mathbb{R}^n).$$

To illustrate some of these ideas, a single-valued mapping $F : X \to \mathbb{R}^m$ given on a set $X \subset \mathbb{R}^n$ can be treated in terms of $S : \mathbb{R}^n \rightrightarrows \mathbb{R}^m$ defined by $S(x) = \{F(x)\}$ when $x \in X$, but $S(x) = \emptyset$ when $x \notin X$. Then $\operatorname{dom} S = X$. Although S isn't multivalued anywhere, the notation $S : \mathbb{R}^n \to \mathbb{R}^m$ wouldn't be correct, because S isn't single-valued except on X; as an alternative to introducing S one could think of F itself as a mapping $\mathbb{R}^n \rightrightarrows \mathbb{R}^m$ with $\operatorname{dom} F = X$. Anyway, the inverse mapping S^{-1}, which is identical to F^{-1}, may well be multivalued. The range of this inverse is X.

Important examples of set-valued mappings that we've already been dealing with, in addition to the inverses of single-valued mappings, are the projection mappings P_C onto sets $C \subset \mathbb{R}^n$ and the proximal mappings $P_\lambda f$ associated with functions $f : \mathbb{R}^n \to \overline{\mathbb{R}}$; cf. 1.20 and 1.22. The study of how the feasible set and optimal set in a problem of optimization can depend on the problem's parameters leads to such mappings too.

5.1 Example (constraint systems). *Consider a set $X \subset \mathbb{R}^n$ and a mapping $F : X \to \mathbb{R}^m$, $F(x) = \big(f_1(x), \ldots, f_m(x)\big)$. For each $u = (u_1, \ldots, u_m) \in \mathbb{R}^m$ as a parameter vector, $F^{-1}(u)$ is the set of all solutions $x = (x_1, \ldots, x_n)$ to the equation system*

$$f_i(x_1, \ldots, x_n) = u_i \ \text{ for } i = 1, \ldots, m \ \text{ with } (x_1, \ldots, x_n) = x \in X.$$

For a box $D = D_1 \times \cdots \times D_m$ in $\mathbb{R}^m$, the set $F^{-1}(D)$ consists of all vectors x satisfying the constraint system

$$f_i(x_1, \ldots, x_n) \in D_i \ \text{ for } i = 1, \ldots, m \ \text{ with } (x_1, \ldots, x_n) = x \in X.$$

When $m = n$ in this example, the number of equations matches the number of unknowns, and hopes rise that F^{-1} might be single-valued on its effective domain, $F(X)$. Beyond the issue of single-valuedness it may be important to understand 'continuity' properties in the dependence of $F^{-1}(u)$ on u. This may be a concern even when $m \neq n$ and the study of solutions lies fully in the context of a set-valued mapping $F^{-1} : \mathbb{R}^m \rightrightarrows \mathbb{R}^n$.

5.2 Example (generalized equations and implicit mappings). *Beyond an equation system written as $F(x) = \bar{u}$ for a single-valued mapping F, one may wish to solve a problem of the type*

$$\text{determine } \bar{x} \text{ such that } S(\bar{x}) \ni \bar{u}$$

for some kind of mapping $S : \mathbb{R}^n \rightrightarrows \mathbb{R}^m$. The desired solution set is then $S^{-1}(\bar{u})$. Perturbations could be studied in terms of replacing $\bar{u}$ by u, with emphasis on the behavior of $S^{-1}(u)$ when u is near $\bar{u}$. Or, parameter vectors u could be introduced more broadly. Starting from $S : \mathbb{R}^n \times \mathbb{R}^m \rightrightarrows \mathbb{R}^p$, the analysis could target properties of the mapping $T : \mathbb{R}^m \rightrightarrows \mathbb{R}^n$ defined by

$$T(w) = \big\{x \,\big|\, S(x, w) \ni \bar{u}\big\}$$

and thus concern set-valued analogs of the implicit mapping theorem rather than of the inverse mapping theorem. As a special case, one could have, for mappings $F : I\!R^n \times I\!R^d \to I\!R^m$ *and* $N : I\!R^n \rightrightarrows I\!R^m$,

$$T(w) = \{x \mid -F(x,w) \in N(x)\}.$$

Generalized equations like $S(\bar{x}) \ni 0$ or $-F(\bar{x}) \in N(\bar{x})$, the latter corresponding to $S(x) = F(x) + N(x)$, will later be seen to arise from characterizations of optimality in problems of constrained minimization (see 6.13 and 10.1). The implicit mapping T in Example 5.2 suggests the convenience of being able to treat single-valuedness, multivaluedness and empty-valuedness as properties that can temporarily be left in the background, if desired, without holding up the study of other features like continuity.

Set-valued mappings can be combined in a number of ways to get new mappings. Addition and scalar multiplication are defined by

$$(S_1 + S_2)(x) := S_1(x) + S_2(x), \qquad (\lambda S)(x) = \lambda S(x),$$

where the right sides use Minkowski addition and scalar multiplication of sets as introduced in Chapter 1; similarly for $S_1 - S_2$ and $-S$. Composition of $S : I\!R^n \rightrightarrows I\!R^m$ with $T : I\!R^m \rightrightarrows I\!R^p$ is defined by

$$(T \circ S)(x) := T(S(x)) = \bigcup_{u \in S(x)} T(u) = \{w \mid S(x) \cap T^{-1}(w) \neq \emptyset\}$$

to get a mapping $T \circ S : I\!R^n \rightrightarrows I\!R^p$. In the case of single-valued S and T, this reduces to the usual notion of composition. Clearly $(T \circ S)^{-1} = S^{-1} \circ T^{-1}$.

Many problems of applied mathematics can be solved by computing a fixed point of a possibly multivalued mapping from $I\!R^n$ into itself, and this generates interest in continuity and convergence properties of such mappings.

5.3 Example (algorithmic mappings and fixed points). *For a mapping* $S : I\!R^n \rightrightarrows I\!R^n$, *a fixed point is a point* $\bar{x}$ *such that* $\bar{x} \in S(\bar{x})$. *An approach to finding such a point* $\bar{x}$ *is to generate a sequence* $\{x^\nu\}_{\nu \in I\!N}$ *from a starting point* x^0 *by the rule* $x^\nu \in S(x^{\nu-1})$. *This implies that*

$$x^1 \in S(x^0), \quad x^2 \in (S \circ S)(x^0), \quad \ldots, \quad x^\nu \in (S \circ \cdots \circ S)(x^0).$$

More generally, a numerical procedure may be built out of a sequence of algorithmic mappings $T^\nu : I\!R^n \rightrightarrows I\!R^n$ *through the rule* $x^\nu \in T^\nu(x^{\nu-1})$, *where* T^ν *is some kind of approximation to* S *at* $x^{\nu-1}$.

In the framework of Example 5.3, not only are the continuity notions important, but also the ways in which a sequence of set-valued mappings might be said to converge to another set-valued mapping. Questions of local approximation, perhaps through some form of generalized differentiation, also pose a challenge. A central aim of the theory of set-valued mappings is to provide the concepts and results that support these needs of analysis. Continuity will be studied here and generalized differentiability in Chapter 8.

B. Continuity and Semicontinuity

Continuity properties of mappings $S : I\!R^n \rightrightarrows I\!R^m$ can be developed in terms of outer and inner limits like those in 4.1:

$$\begin{aligned}
\limsup_{x \to \bar{x}} S(x) : &= \bigcup_{x^\nu \to \bar{x}} \limsup_{\nu \to \infty} S(x^\nu) \\
&= \Big\{ u \,\Big|\, \exists\, x^\nu \to \bar{x},\ \exists\, u^\nu \to u \text{ with } u^\nu \in S(x^\nu) \Big\}, \\
\liminf_{x \to \bar{x}} S(x) : &= \bigcap_{x^\nu \to \bar{x}} \liminf_{\nu \to \infty} S(x^\nu) \\
&= \Big\{ u \,\Big|\, \forall\, x^\nu \to \bar{x},\ \exists\, N \in \mathcal{N}_\infty,\ u^\nu \underset{N}{\to} u \text{ with } u^\nu \in S(x^\nu) \Big\}.
\end{aligned} \qquad 5(1)$$

5.4 Definition (continuity and semicontinuity). *A set-valued mapping $S : I\!R^n \rightrightarrows I\!R^m$ is outer semicontinuous (osc) at $\bar{x}$ if*

$$\limsup_{x \to \bar{x}} S(x) \subset S(\bar{x}),$$

or equivalently $\limsup_{x \to \bar{x}} S(x) = S(\bar{x})$, *but inner semicontinuous (isc) at $\bar{x}$ if*

$$\liminf_{x \to \bar{x}} S(x) \supset S(\bar{x}),$$

or equivalently when S is closed-valued, $\liminf_{x \to \bar{x}} S(x) = S(\bar{x})$. *It is called continuous at $\bar{x}$ if both conditions hold, i.e., if $S(x) \to S(\bar{x})$ as $x \to \bar{x}$.*

These terms are invoked relative to X, a subset of $I\!R^n$ containing $\bar{x}$, when the properties hold in restriction to convergence $x \to \bar{x}$ with $x \in X$ (in which case the sequences $x^\nu \to x$ in the limit formulas are required to lie in X).

The equivalences follow from the fact that the constant sequence $x^\nu \equiv \bar{x}$ is among those considered in 5(1). For this reason $S(\bar{x})$ must be a closed set when S is outer semicontinuous at $\bar{x}$, whether in the main sense or merely relative to some subset X. Note further that when S is inner semicontinuous at a point $\bar{x} \in \operatorname{dom} S$ relative to X, there must be a neighborhood $V \in \mathcal{N}(\bar{x})$ such that $X \cap V \subset \operatorname{dom} S$. When $X = I\!R^n$ this requires $\bar{x} \in \operatorname{int}(\operatorname{dom} S)$.

Clearly, a single-valued mapping $F : X \to I\!R^m$ is continuous in the usual sense at $\bar{x}$ relative to X if and only if it is continuous at $\bar{x}$ relative to X as a mapping $I\!R^n \rightrightarrows I\!R^m$ handled under Definition 5.4.

5.5 Example (profile mappings). *For a function $f : I\!R^n \to \overline{I\!R}$, the epigraphical profile mapping $E_f : I\!R^n \rightrightarrows I\!R^1$, defined by $E_f(x) = \{\alpha \in I\!R \mid \alpha \geq f(x)\}$, has* $\operatorname{gph} E_f = \operatorname{epi} f$, $\operatorname{dom} E_f = \operatorname{dom} f$, *and* $E_f^{-1}(\alpha) = \operatorname{lev}_{\leq \alpha} f$.

Furthermore, E_f is osc at $\bar{x}$ if and only if f is lsc at $\bar{x}$, whereas it is isc at $\bar{x}$ if and only if f is usc at $\bar{x}$. Thus too, E_f is continuous at $\bar{x}$ if and only if f is continuous at $\bar{x}$. On the other hand, the level-set mapping $\alpha \mapsto \operatorname{lev}_{\leq \alpha} f$ is osc everywhere if and only if f is lsc everywhere.

Analogous properties hold for the hypographical profile mapping $H_f : \mathbb{R}^n \rightrightarrows \mathbb{R}^1$ *with* $H_f(x) = \{\alpha \in \mathbb{R} \mid \alpha \le f(x)\}$, $\operatorname{gph} H_f = \operatorname{hypo} f$.

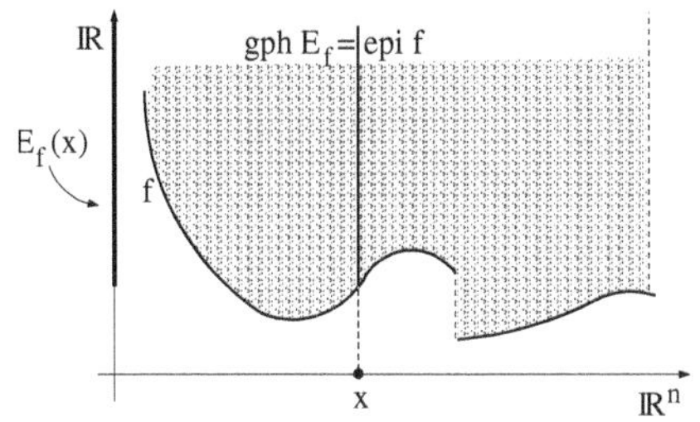

Fig. 5–2. The epigraphical profile mapping associated with a function f.

As a further illustration of the meaning of Definition 5.4, Figure 5–3(a) displays a mapping that fails to be inner semicontinuous at x despite being outer semicontinuous at x and in fact continuous at every $x' \neq x$. In Figure 5–3(b) the mapping S is isc at x but fails to be osc at that point.

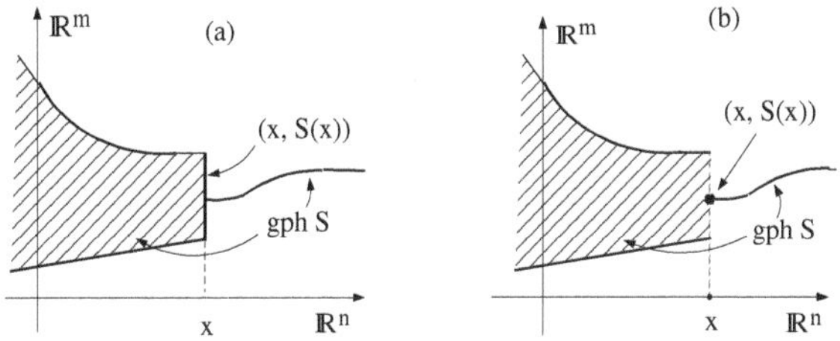

Fig. 5–3. (a) An osc mapping that fails to be isc at x. (b) An isc mapping.

5.6 Exercise (criteria for semicontinuity at a point). *Consider a mapping* $S : \mathbb{R}^n \rightrightarrows \mathbb{R}^m$, *a set* $X \subset \mathbb{R}^n$, *and any point* $\bar{x} \in X$.

(a) *S is osc at $\bar{x}$ relative to X if and only if for every $u \notin S(\bar{x})$ there are neighborhoods $W \in \mathcal{N}(u)$ and $V \in \mathcal{N}(\bar{x})$ such that $X \cap V \cap S^{-1}(W) = \emptyset$.*

(b) *S is isc at $\bar{x}$ relative to X if and only if for every $u \in S(\bar{x})$ and $W \in \mathcal{N}(u)$ there is a neighborhood $V \in \mathcal{N}(\bar{x})$ such that $X \cap V \subset S^{-1}(W)$.*

(c) *S is osc at $\bar{x}$ relative to X if and only if $x^\nu \in X$, $x^\nu \to \bar{x}$ and $S(x^\nu) \to D$ imply $D \subset S(\bar{x})$.*

(d) *S is isc at $\bar{x}$ relative to X if and only if $x^\nu \in X$, $x^\nu \to \bar{x}$ and $S(x^\nu) \to D$ imply $D \supset S(\bar{x})$.*

Guide. For (a) and (b) use the hit-and-miss criteria for set convergence in Theorem 4.5. For (c) and (d) appeal to the cluster description of inner and outer limits in Proposition 4.19. □

5.7 Theorem (characterizations of semicontinuity). *For $S : I\!R^n \rightrightarrows I\!R^m$,*

(a) *S is osc (everywhere) if and only if* $\operatorname{gph} S$ *is closed in $I\!R^n \times I\!R^m$; moreover, S is osc if and only if S^{-1} is osc;*

(b) *when S closed-valued, S is osc relative to a set $X \subset I\!R^n$ if and only if $S^{-1}(B)$ is closed relative to X for every compact set $B \subset I\!R^m$;*

(c) *S is isc relative to a set $X \subset I\!R^n$ if and only if $S^{-1}(O)$ is open relative to X for every open set $O \subset I\!R^m$.*

Proof. The claims in (a) are obvious from the definition of outer semicontinuity. Every sequence in a compact set $B \subset I\!R^m$ has a convergent subsequence, and on the other hand, a set consisting of a convergent sequence and its limit is a compact set. Therefore, the image condition in (b) is equivalent to the condition that whenever $u^\nu \to \bar{u}$, $x^\nu \in S^{-1}(u^\nu)$ and $x^\nu \to \bar{x}$ with $x^\nu \in X$ and $\bar{x} \in X$, one has $\bar{x} \in S^{-1}(\bar{u})$. Since $x^\nu \in S^{-1}(u^\nu)$ is the same as $u^\nu \in S(x^\nu)$, this is precisely the condition for S to be osc relative to X.

Failure of the condition in (c) means the existence of an open set O and a sequence $x^\nu \to \bar{x}$ in X such that $\bar{x} \in S^{-1}(O)$ but $x^\nu \notin S^{-1}(O)$; in other words, $S(\bar{x}) \cap O \neq \emptyset$ yet $S(x^\nu) \cap O = \emptyset$ for all ν. This property says that $\liminf_\nu S(x^\nu) \not\supset S(\bar{x})$. Thus, the condition in (c) fails if and only if S fails to be isc relative to X at some point $\bar{x} \in X$. □

Although $\operatorname{gph} S$ is closed when S is osc, and S is *closed-valued* in that case (the sets $S(x)$ are all closed), the sets $\operatorname{dom} S$ and $\operatorname{rge} S$ don't have to be closed then. For example, the set-valued mapping $S : I\!R^1 \rightrightarrows I\!R^1$ defined by

$$\operatorname{gph} S = \big\{ (x,u) \in I\!R^2 \,\big|\, x \neq 0,\ u \geq 1/x^2 \big\}$$

is osc but has $\operatorname{dom} S = I\!R^1 \setminus \{0\}$ and $\operatorname{rge} S = (0,\infty)$, neither of which is closed.

5.8 Example (feasible-set mappings). *Suppose $T : I\!R^d \rightrightarrows I\!R^n$ is defined relative to a set $W \subset I\!R^d$ by $T(w) = \emptyset$ for $w \notin W$, but otherwise*

$$T(w) = \big\{ x \in X \,\big|\, f_i(x,w) \leq 0 \text{ for } i \in I_1 \text{ and } f_i(x,w) = 0 \text{ for } i \in I_2 \big\},$$

where X is a closed subset of $I\!R^n$ and each f_i is a continuous real-valued function on $X \times W$. (Here W could be all of $I\!R^d$, and X all of $I\!R^n$.)

If W is closed, then T is osc. Even if W is not closed, T is osc at any point $\bar{w} \in \operatorname{int} W$. Moreover $\operatorname{dom} T$ is the set of vectors $w \in W$ for which the constraints in x that define $T(w)$ are consistent.

Detail. When W is closed, $\operatorname{gph} T$ is closed—because it is the intersection of $X \times W$ with the various sets $\big\{ (x,w) \,\big|\, f_i(x,w) \leq 0 \big\}$ for $i \in I_1$ and $\big\{ (x,w) \,\big|\, f_i(x,w) = 0 \big\}$ for $i \in I_2$, which are closed by the continuity of the f_i's. This is why T is osc then. The assertion when W isn't itself closed comes from replacing W by a closed neighborhood of $\bar{w}$ within W. □

It's useful to observe that outer semicontinuity is a constructive property in the sense that it can be created by passing, if necessary, from a given mapping $S : I\!R^n \rightrightarrows I\!R^m$ to the mapping $\operatorname{cl} S : I\!R^n \rightrightarrows I\!R^m$ defined by

$$\operatorname{gph}(\operatorname{cl} S) := \operatorname{cl}(\operatorname{gph} S). \tag{5(2)}$$

This mapping is called the *closure* of S, or its *osc hull*, and it has the formula

$$(\operatorname{cl} S)(x) = \limsup_{x' \to x} S(x') \quad \text{for all } x. \tag{5(3)}$$

Inner semicontinuity of a given mapping is typically harder to verify than outer semicontinuity, and it isn't constructive in such an easy sense. (The mapping defined as in 5(3), but with liminf on the right, isn't necessarily isc.) But an effective calculus of 'strict' continuity, a Lipschitzian property of mappings introduced in Chapter 9 which entails continuity and in particular inner semicontinuity, will be developed in Chapter 10. For now, we content ourselves with special criteria for inner semicontinuity that depend on convexity.

As already noted, a mapping $S : \mathbb{R}^n \rightrightarrows \mathbb{R}^m$ is called *convex-valued* when $S(x)$ is a convex set for all x. A stronger property, implying this, is of interest as well: S is called *graph-convex* when the set $\operatorname{gph} S$ is convex in $\mathbb{R}^n \times \mathbb{R}^m$, cf. Figure 5–4, in which case $\operatorname{dom} S$ and $\operatorname{rge} S$ are convex too. Graph-convexity of S is equivalent to having

$$S\big((1-\tau)x_0 + \tau x_1\big) \supset (1-\tau)S(x_0) + \tau S(x_1) \quad \text{for } \tau \in (0,1). \tag{5(4)}$$

5.9 Theorem (inner semicontinuity from convexity). *Consider a mapping $S : \mathbb{R}^n \rightrightarrows \mathbb{R}^m$ and a point $\bar{x} \in \mathbb{R}^n$.*

(a) *If S is convex-valued and $\operatorname{int} S(\bar{x}) \neq \emptyset$, then a necessary and sufficient condition for S to be isc relative to $\operatorname{dom} S$ at $\bar{x}$ is that for all $u \in \operatorname{int} S(\bar{x})$ there exists $W \in \mathcal{N}(\bar{x}, u)$ such that $W \cap (\operatorname{dom} S \times \mathbb{R}^m) \subset \operatorname{gph} S$; in particular, S is isc at $\bar{x}$ if and only if $(\bar{x}, u) \in \operatorname{int}(\operatorname{gph} S)$ for every $u \in \operatorname{int} S(\bar{x})$.*

(b) *If S is graph-convex and $\bar{x} \in \operatorname{int}(\operatorname{dom} S)$, then S is isc at $\bar{x}$.*

(c) *If S is isc at $\bar{x}$, then so is the convex hull mapping $T : x \mapsto \operatorname{con} S(x)$.*

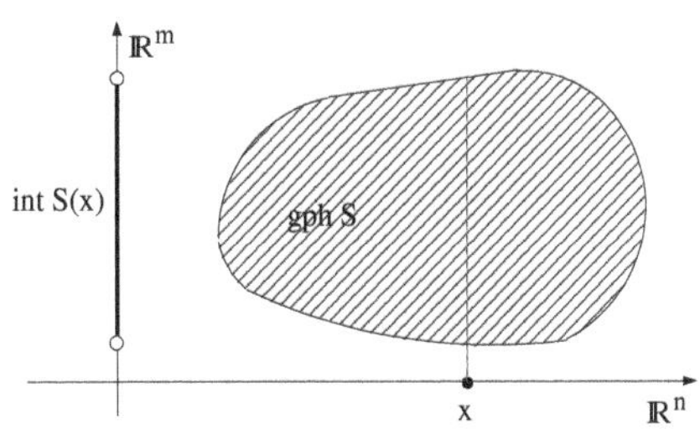

Fig. 5–4. Inner semicontinuity through graph-convexity.

Proof. In (a), the assumption that there exists $(V \times U) \in \mathcal{N}(\bar{x}) \times \mathcal{N}(\bar{u})$ such that $(V \cap \operatorname{dom} S) \times U \subset \operatorname{gph} S$ implies that every sequence $\{x^\nu\}_{\nu \in \mathbb{N}} \subset \operatorname{dom} S$ converging to $\bar{x}$ will eventually enter and stay in $V \cap \operatorname{dom} S$, and for those x^ν one can find $u^\nu \in U \subset S(x^\nu)$ converging to $\bar{u}$. This clearly yields the inner semicontinuity of S relative to $\operatorname{dom} S$. For the converse, suppose that S is isc relative to $\operatorname{dom} S$ at $\bar{x}$ and let B be a compact neighborhood of $\bar{u}$ within $\operatorname{int} S(\bar{x})$. For any sequence $x^\nu \to \bar{x}$ in $\operatorname{dom} S$, we eventually have $B \subset \operatorname{int} S(x^\nu)$ by 4.15. It follows that for some neighborhood $V \in \mathcal{N}(\bar{x})$, one has $B \subset \operatorname{int} S(x)$ for all $x \in V \cap \operatorname{dom} S$. Then $V \times B$ is a neighborhood of $(\bar{x}, \bar{u})$ such that $(V \times B \cap \operatorname{dom} S \times I\!R^n) \subset \operatorname{gph} S$.

In (b), denote the convex set $\operatorname{gph} S$ by G and the linear mapping $(x, u) \mapsto x$ by L. To say that $S(x)$ depends inner semicontinuously on x is to say that $L^{-1}(x) \cap G$ has this property. But that's true by 4.32(c) at any point $\bar{x} \in \operatorname{int}(\operatorname{dom} S)$, because the convex sets $\{\bar{x}\}$ and $L(G) = \operatorname{dom} S$ can't be separated.

If $\bar{u} \in T(\bar{x}) = \operatorname{con} S(\bar{x})$, then $\bar{u} = \sum_{k=0}^m \lambda^k u^k$ with $\lambda^k \geq 0$, $\sum_{k=0}^m \lambda^k = 1$, and $u^k \in S(\bar{x})$ (2.27). Inner semicontinuity of S means that when $x^\nu \to \bar{x}$, one can find $u^{k\nu} \to u^k$ such that $u^{k\nu} \in S(x^\nu)$. The sequence $u^\nu = \sum_{k=0}^m \lambda^k u^{k\nu}$ converges to $\bar{u}$ and for all ν, $u^\nu \in S(x^\nu)$. This implies that $T(\bar{x}) \subset \liminf_\nu T(x^\nu)$ for all $x^\nu \to x$, i.e., T is isc at $\bar{x}$ as claimed in (c). □

Automatic continuity properties of graph-convex mappings that go farther than the one in 5.9(b) will be developed in Chapter 9 (see 9.33, 9.34, 9.35).

5.10 Example (parameterized convex constraints). *Suppose*

$$T(w) = \{x \mid f_i(x, w) \leq 0 \ \text{ for } \ i = 1, \ldots, m\}$$

for finite, continuous functions f_i on $I\!R^n \times I\!R^d$ such that $f_i(x, w)$ is convex in x for each w. If for $\bar{w}$ there is a point $\bar{x}$ such that $f_i(\bar{x}, \bar{w}) < 0$ for $i = 1, \ldots, m$, then T is continuous not only at $\bar{w}$ but at every w in some neighborhood of $\bar{w}$.

Detail. Let $f(x, w) = \max\{f_1(x, w), \ldots, f_m(x, w)\}$. Then f is continuous in (x, w) (by 1.26(c)) and convex in x (by 2.9(b)), with $\operatorname{lev}_{\leq 0} f = \operatorname{gph} T$. The level set $\operatorname{lev}_{\leq 0} f$ is closed in $I\!R^n \times I\!R^d$, hence T is osc by 5.7(a). For each w, $T(w)$ is the level set $\operatorname{lev}_{\leq 0} f(\cdot, w)$ in $I\!R^n$, which is convex (by 2.7). For $\bar{x}$ and $\bar{w}$ as postulated we have $f(\bar{x}, \bar{w}) < 0$, and then by continuity $f(\bar{x}, w) < 0$ for all w in some open set O containing $\bar{w}$. According to 2.34 this implies

$$\operatorname{int} T(w) = \{x \mid f(x, w) < 0\} \neq \emptyset \ \text{ for all } \ w \in O.$$

For any $\tilde{w} \in O$ and any $\tilde{x} \in \operatorname{int} T(\tilde{w})$, the continuity of f and the fact that $f(\tilde{x}, \tilde{w}) < 0$ yield a neighborhood $W \subset O \times \operatorname{int} T(\tilde{w})$ of $(\tilde{x}, \tilde{w})$ which is contained in $\operatorname{gph} T$ and such that $f < 0$ on W. Then certainly $\tilde{x}$ belongs to the inner limit of $T(w)$ as $w \to \tilde{w}$. This inner limit, which is a closed set, therefore includes $\operatorname{int} T(\tilde{w})$, so it also includes $\operatorname{cl}\big(\operatorname{int} T(\tilde{w})\big)$, which is $T(\tilde{w})$ by Theorem 2.33 because $T(\tilde{w})$ is a closed, convex set and $\operatorname{int} T(\tilde{w}) \neq \emptyset$. This tells us that T is isc at $\tilde{w}$. □

5.11 Proposition (continuity of distances). *For a closed-valued mapping $S : I\!R^n \rightrightarrows I\!R^m$ and a point $\bar{x}$ in a set $X \subset I\!R^n$,*

(a) *S is osc at $\bar{x}$ relative to X if and only if for every $u \in I\!R^m$ the function $x \mapsto d\big(u, S(x)\big)$ is lsc at $\bar{x}$ relative to X.*

(b) *S is isc at $\bar{x}$ relative to X if and only if for every $u \in I\!R^m$ the function $u \mapsto d\big(u, S(x)\big)$ is usc at $\bar{x}$ relative to X.*

(c) *S is continuous at $\bar{x}$ relative to X if and only if for every $u \in I\!R^m$ the function $x \mapsto d\big(u, S(x)\big)$ is continuous at $\bar{x}$ relative to X.*

Proof. This is immediate from the criteria for set convergence in 4.7. □

5.12 Proposition (uniformity of approximation in semicontinuity). *For a closed-valued mapping $S : I\!R^n \rightrightarrows I\!R^m$ and a point $\bar{x}$ in a set $X \subset I\!R^n$:*

(a) *S is osc at $\bar{x}$ relative to X if and only if for every $\rho > 0$ and $\varepsilon > 0$ there is a neighborhood $V \in \mathcal{N}(\bar{x})$ such that*

$$S(x) \cap \rho I\!B \subset S(\bar{x}) + \varepsilon I\!B \ \text{ for all } \ x \in X \cap V.$$

(b) *S is isc at $\bar{x}$ relative to X if and only if for every $\rho > 0$ and $\varepsilon > 0$ there is a neighborhood $V \in \mathcal{N}(\bar{x})$ such that*

$$S(\bar{x}) \cap \rho I\!B \subset S(x) + \varepsilon I\!B \ \text{ for all } \ x \in X \cap V.$$

Proof. This just adapts of 4.10 to the language of set-valued mappings. □

5.13 Exercise (uniform continuity). *Consider a closed-valued mapping $S : I\!R^n \rightrightarrows I\!R^m$ and a compact set $X \subset \operatorname{dom} S$. If S is continuous relative to X, then for any $\rho > 0$ and $\varepsilon > 0$ there exists $\delta > 0$ such that*

$$S(x') \cap \rho I\!B \subset S(x) + \varepsilon I\!B \ \text{ for all } \ x', x \in X \ \text{ with } \ |x' - x| \leq \delta.$$

Guide. Apply 5.12(a) and 5.12(b) simultaneously at each point $\bar{x}$ of X to get an open neighborhood of that point for which both of the inclusions in 5.12 hold simultaneously. Invoking compactness, cover B by finitely many such neighborhoods. Choose δ small enough that for every closed ball $I\!B(\tilde{x}, \delta)$ with $\tilde{x} \in X$, the set $I\!B(\tilde{x}, \delta) \cap X$ must lie within at least one of these covering neighborhoods. □

Both 5.12 and 5.13 obviously continue to hold when the balls $\rho I\!B$ are replaced by the collection of all bounded sets $B \subset I\!R^m$. Likewise, the balls $\varepsilon I\!B$ can be replaced by the collection of all neighborhoods U of the origin in $I\!R^m$.

C. Local Boundedness

A continuous single-valued mapping carries bounded sets into bounded sets. In understanding the connections between continuity and boundedness properties of potentially multivalued mappings, the following concept is the key.

5.14 Definition (local boundedness). *A mapping $S : I\!R^n \rightrightarrows I\!R^m$ is locally bounded at a point $\bar{x} \in I\!R^n$ if for some neighborhood $V \in \mathcal{N}(\bar{x})$ the set*

$S(V) \subset \mathbb{R}^m$ is bounded. It is called locally bounded on $\mathbb{R}^n$ if this holds at every $\bar{x} \in \mathbb{R}^n$. It is bounded on $\mathbb{R}^n$ if rge S *is a bounded subset of $\mathbb{R}^m$.*

Local boundedness at a point $\bar{x}$ requires $S(\bar{x})$ to be a bounded set, but more, namely that for all x in some neighborhood V of $\bar{x}$, the sets $S(x)$ all lie within a single bounded set B, cf. Figure 5–5. This property can equally well be stated as follows: there exist $\delta > 0$ and $\rho > 0$ such that if $|x - \bar{x}| \leq \delta$ and $u \in S(x)$, then $|u| \leq \rho$. In the notation for closed Euclidean balls this means that $S(\mathbb{B}(\bar{x}, \delta)) \subset \mathbb{B}(0, \rho)$.

In particular, S is locally bounded at any point $\bar{x} \notin \mathrm{cl}(\mathrm{dom}\, S)$. This fits trivially with the definition, because for such a point $\bar{x}$ there is a neighborhood $V \in \mathcal{N}(\bar{x})$ that misses $\mathrm{dom}\, S$ and therefore has $S(V) = \emptyset$. (The empty set is, of course, regarded as bounded.)

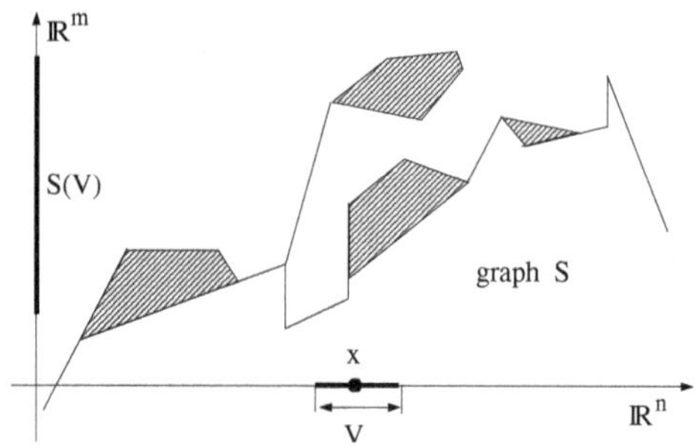

Fig. 5–5. Local boundedness.

5.15 Proposition (boundedness of images). *A mapping $S : \mathbb{R}^n \rightrightarrows \mathbb{R}^m$ is locally bounded if and only if $S(B)$ is bounded for every bounded set B. This is equivalent to the property that whenever $u^\nu \in S(x^\nu)$ and the sequence $\{x^\nu\}_{\nu \in \mathbb{N}}$ is bounded, then the sequence $\{u^\nu\}_{\nu \in \mathbb{N}}$ is bounded.*

Proof. The condition is sufficient, since for any point x we can select a bounded neighborhood $V \in \mathcal{N}(x)$ and conclude that $S(V)$ is bounded. To show that it is necessary, consider any bounded set $B \subset \mathbb{R}^n$, and for each $x \in \mathrm{cl}\, B$ use the local boundedness to select an open $V_x \in \mathcal{N}(x)$ such that $S(V_x)$ is bounded. The set $\mathrm{cl}\, B$ is compact, so it is covered by a finite collection of the sets V_x, say $V_{x_1}, \ldots, V_{x_k}$. Denote the union of these by V. Then $S(B) \subset S(V) = S(V_{x_1}) \cup \ldots \cup S(V_{x_k})$, where the set on the right, being the union of a finite collection of bounded sets, is bounded. Thus $S(B)$ is bounded. The sequential version of the condition is apparent. □

5.16 Exercise (local boundedness of inverses). *The inverse S^{-1} of a mapping $S : \mathbb{R}^n \rightrightarrows \mathbb{R}^m$ is locally bounded if and only if*

$$|x^\nu| \to \infty,\ u^\nu \in S(x^\nu) \quad \Longrightarrow \quad |u^\nu| \to \infty.$$

It should be remembered in 5.16 that the condition holds—vacuously—when there are *no* sequences $\{x^\nu\}_{\nu \in \mathbb{N}}$ and $\{u^\nu\}_{\nu \in \mathbb{N}}$ with $u^\nu \in S(x^\nu)$ such that $|x^\nu| \to \infty$. This is the case where the set $\operatorname{dom} S = \operatorname{rge} S^{-1}$ is bounded, so S^{-1} is a bounded mapping, not just locally bounded.

5.17 Example (level boundedness as local boundedness).

(a) *For a function* $f : \mathbb{R}^n \to \overline{\mathbb{R}}$, *the mapping* $\alpha \mapsto \operatorname{lev}_{\leq\alpha} f$ *is locally bounded if and only if* f *is level-bounded.*

(b) *For a function* $f : \mathbb{R}^n \times \mathbb{R}^m \to \overline{\mathbb{R}}$, *one has* $f(x,u)$ *level-bounded in* x *locally uniformly in* u *if and only if, for each* $\alpha \in \mathbb{R}$, *the mapping* $u \mapsto \{x \mid f(x,u) \leq \alpha\}$ *is locally bounded. The mapping* $(u,\alpha) \mapsto \{x \mid f(x,u) \leq \alpha\}$ *is then locally bounded too.*

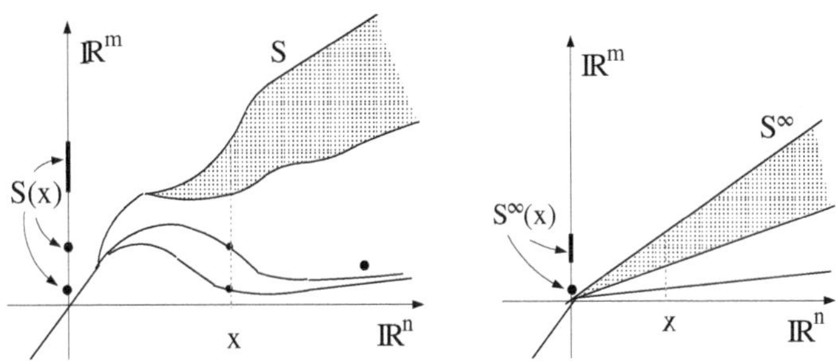

Fig. 5–6. A mapping S and the associated horizon mapping S^∞.

A useful criterion for the local boundedness of $S : \mathbb{R}^n \rightrightarrows \mathbb{R}^m$ can be stated in terms of the *horizon mapping* $S^\infty : \mathbb{R}^n \rightrightarrows \mathbb{R}^m$, which is specified by

$$\operatorname{gph} S^\infty := (\operatorname{gph} S)^\infty. \tag{5(5)}$$

This graphical definition means that

$$S^\infty(x) = \{u = \lim_\nu \lambda^\nu u^\nu \mid u^\nu \in S(x^\nu),\ \lambda^\nu x^\nu \to x,\ \lambda^\nu \searrow 0\}. \tag{5(6)}$$

Note that since the graph of S^∞ is a closed cone in $\mathbb{R}^n \times \mathbb{R}^m$, S^∞ is osc with $0 \in S^\infty(0)$. Also, $S^\infty(\lambda x) = \lambda S^\infty(x)$ for $\lambda > 0$; in other words, S^∞ is positively homogeneous. If S is graph-convex, so is S^∞. Clearly $(S^{-1})^\infty = (S^\infty)^{-1}$.

When S is a linear mapping L, one has $S^\infty = L$ because the set $\operatorname{gph} S = \operatorname{gph} L$ is a linear subspace of $\mathbb{R}^n \times \mathbb{R}^m$ and thus is its own horizon cone. An affine mapping $S(x) = L(x) + b$ has $S^\infty = L$.

5.18 Theorem (horizon criterion for local boundedness). *A sufficient condition for a mapping* $S : \mathbb{R}^n \rightrightarrows \mathbb{R}^m$ *to be locally bounded is* $S^\infty(0) = \{0\}$, *and then the horizon mapping* S^∞ *is locally bounded as well.*

When S *is graph-convex and osc, this criterion is fulfilled if there exists a point* $\bar{x}$ *such that* $S(\bar{x})$ *is nonempty and bounded.*

Proof. If S is not locally bounded at a point $\bar{x}$, there exist by 5.15 sequences $x^\nu \to \bar{x}$ and $u^\nu \in S(x^\nu)$ with $0 < |u^\nu| \to \infty$. Then the sequence of points

(x^ν, u^ν) in $\operatorname{gph} S$ is unbounded, but for $\lambda^\nu = 1/|u^\nu|$ the sequence of points $\lambda^\nu(x^\nu, u^\nu)$ is bounded and has a cluster point $(0, u)$ with $|u| = 1$. Then $(0, u) \in (\operatorname{gph} S)^\infty$ and $0 \neq u \in S^\infty(0)$, so the condition is violated.

Because $(\operatorname{gph} S)^\infty$ is a closed cone, we have $((\operatorname{gph} S)^\infty)^\infty = (\operatorname{gph} S)^\infty$ and consequently $(S^\infty)^\infty = S^\infty$. Thus, when $S^\infty(0) = \{0\}$, the mapping $T = S^\infty$ has $T^\infty(0) = \{0\}$ and, by the argument already given, is locally bounded.

When S is graph-convex and osc, $\operatorname{gph} S$ is convex and closed. Then for any pair $(\bar{x}, \bar{u}) \in \operatorname{gph} S$, a vector $u \in S^\infty(0)$ if and only if $(\bar{x}, \bar{u}) + \tau(0, u) \in \operatorname{gph} S$ for all $\tau \geq 0$, or in other words, $\bar{u} + \tau u \in S(\bar{x})$ for all $\tau \geq 0$ (cf. 3.6). This can only hold for $u = 0$ when $S(\bar{x})$ is bounded. □

Theorem 5.18 opens the way to establishing the local boundedness of a mapping S by applying to $\operatorname{gph} S$ the horizon cone formulas in Chapter 3. The fact that the condition isn't necessary for S to be locally bounded, merely sufficient, is seen from examples like $S : I\!R \to I\!R$ with $S(u) = u^2$. This mapping is locally bounded, yet $S^\infty(0) = [0, \infty)$. The condition in 5.18 is helpful anyway because of the convenient calculus that can be built around it.

5.19 Theorem (outer semicontinuity under local boundedness). *Suppose that $S : I\!R^n \rightrightarrows I\!R^m$ is locally bounded at the point $\bar{x}$. Then the following condition is equivalent to S being osc at $\bar{x}$: the set $S(\bar{x})$ is closed, and for every open set $O \supset S(\bar{x})$ there is a neighborhood $V \in \mathcal{N}(\bar{x})$ such that $S(V) \subset O$.*

Proof. Suppose S is osc at $\bar{x}$; then $S(\bar{x})$ is closed. Consider any open set $O \supset S(\bar{x})$. If there were no $V \in \mathcal{N}(\bar{x})$ such that $S(V) \subset O$, a sequence $x^\nu \to \bar{x}$ would exist with $S(x^\nu) \not\subset O$. Then for each ν we could choose $u^\nu \in S(x^\nu) \setminus O$ and get a sequence that's bounded, by virtue of the local boundedness assumption on S, cf. 5.14, yet lies entirely in the complement of O, which is closed. This sequence would have a cluster point $\bar{u}$, likewise in the complement of O and not, therefore, in $S(\bar{x})$. Thus, for some index set $N \in \mathcal{N}_\infty^\#$ we would have $x^\nu \underset{N}{\to} \bar{x}$, $u^\nu \underset{N}{\to} \bar{u}$, $u^\nu \in S(x^\nu)$ but $\bar{u} \notin S(\bar{x})$, in contradiction of the supposed outer semicontinuity of S at $\bar{x}$. Hence the condition is necessary.

For the sufficiency, assume that the proposed condition holds. Consider arbitrary sequences $x^\nu \to \bar{x}$ and $u^\nu \to \bar{u}$ with $u^\nu \in S(x^\nu)$. We must verify that $\bar{u} \in S(\bar{x})$. If this were not the case, then, since $S(\bar{x})$ is closed, there would be a closed ball $I\!B(\bar{u}, \delta)$ having empty intersection with $S(\bar{x})$. Let O be the complement of this ball. Then O is an open set which includes $S(\bar{x})$ and is such that eventually u^ν lies outside of O, in contradiction to the assumption that $S(x^\nu) \subset O$ once x^ν comes within a certain neighborhood $V \in \mathcal{N}(\bar{x})$. □

5.20 Corollary (continuity of single-valued mappings). *For any single-valued mapping $F : I\!R^n \to I\!R^m$, viewed as a special case of a set-valued mapping, the following properties are equivalent:*

(a) *F is continuous at $\bar{x}$;*

(b) *F is osc at $\bar{x}$ and locally bounded at $\bar{x}$;*

(c) *F is isc at $\bar{x}$.*

A single-valued mapping $F : I\!R^n \to I\!R^m$ that isn't continuous at a point $\bar{x}$ can be osc at $\bar{x}$ as a set-valued mapping without being locally bounded there. This is illustrated by the case of $F : I\!R^1 \to I\!R^1$ defined by $F(0) = 0$ but $F(x) = 1/x$ for $x \neq 0$. Local boundedness fails at $\bar{x} = 0$.

The neighborhood property in 5.19 need not hold for a set-valued mapping S for which $S(x)$ may be unbounded, even if S is continuous. That's just as well, because the property is incompatible with intuitive notions of an unbounded set varying continuously with parameters, despite its formal appeal. This is evident in Figure 5–7, which depicts a ray rotating at a uniform rate about the origin in $I\!R^2$. The ray is $S(t)$, and we consider what happens to it as $t \to t_0$. A particular choice of open set $O \supset S(t_0)$ is indicated. No matter how near t is to t_0, as long as $t \neq t_0$, the set $S(t)$ is never included in O. Thus, if we were to demand that the neighborhood property in 5.19 be fulfilled as part of a general definition of continuity for set-valued mappings, unbounded as well as bounded, we would be unable to say that the rotating ray moves 'continuously', which would be a highly unsatisfactory state of affairs. The ray does rotate continuously in the sense of Definition 5.4.

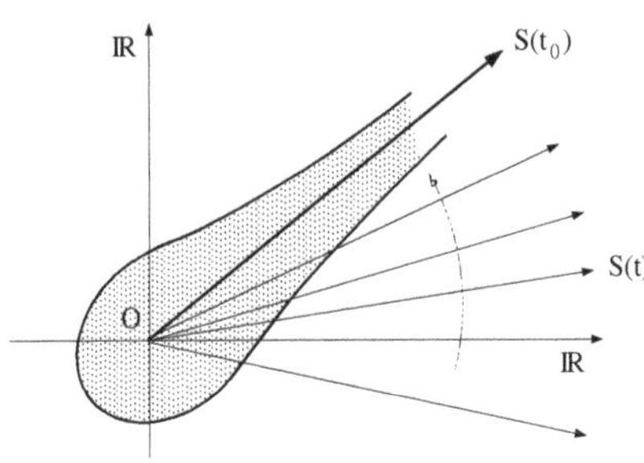

Fig. 5–7. A rotating ray.

The appropriately weaker version of the neighborhood property in 5.19 that corresponds precisely to outer semicontinuity is easily obtained through consideration of *truncations* of S, i.e., mappings of the form

$$S_{\cap B} : x \mapsto S(x) \cap B.$$

Clearly S is osc if and only if all its truncations $S_{\cap B}$ for compact sets B are osc; hence S is osc if and only if all such truncations (which themselves are locally bounded mappings) satisfy the neighborhood property in question.

Theorem 5.19 helps to clarify the extent to which continuity can be characterized using the Pompeiu-Hausdorff distance $d\!l_\infty(C, D)$ between sets C and D, as defined in 4.13. The fact that a *bounded* sequence of nonempty, closed sets C^ν converges to a nonempty, closed set C if and only if $d\!l_\infty(C^\nu, C) \to 0$ (cf. 4.40) gives the following.

5.21 Corollary (continuity of locally bounded mappings). *Let* $S : I\!R^n \rightrightarrows I\!R^m$

be locally bounded at $\bar{x}$ with $S(x)$ nonempty and closed for all x in some neighborhood of $\bar{x}$. Then S is continuous at $\bar{x}$ if and only if the Pompeiu-Hausdorff distance $d\!l_\infty\big(S(x), S(\bar{x})\big)$ tends to 0 as $x \to \bar{x}$.

When local boundedness is absent, Pompeiu-Hausdorff distance is no guide to continuity, since one can have $S(x) \to S(\bar{x})$ while $d\!l_\infty\big(S(x), S(\bar{x})\big) \to \infty$, even in some cases where $S(x)$ is nonempty and compact for all x; cf. 4.13. To get a distance description of continuity under all circumstances, one has to appeal to a metric such as $d\!l\big(S(x), S(\bar{x})\big)$; cf. 4.36. Alternatively, one can work with the family of pseudo-metrics $d\!l_\rho$ and their estimates $\hat{d\!l}_\rho$ as in 4(11); cf. 4.35.

Local boundedness of a mapping S, like outer semicontinuity, can be considered *relative to a set* X by replacing the neighborhood V in Definition 5.14 by $V \cap X$. Such local boundedness at a point $\bar{x} \in X$ is nothing more than the ordinary local boundedness of the mapping $S|_X$ that 'restricts' S to X,

$$S|_X(x) := \begin{cases} S(x) & \text{if } x \in X, \\ \emptyset & \text{if } x \notin X, \end{cases}$$

which can also be viewed as an *inverse truncation*: $S|_X = \big(S^{-1}_{\cap X}\big)^{-1}$. All the results that have been stated about local boundedness, in particular 5.19 and 5.20, carry over to such relativization in the obvious manner, through application to $S|_X$ in place of S. This is useful in situations like the following.

5.22 Example (optimal-set mappings). *Suppose $P(u) := \operatorname{argmin}_x f(x,u)$ for a proper, lsc function $f : I\!R^n \times I\!R^m \to \overline{I\!R}$ with $f(x,u)$ level-bounded in x locally uniformly in u. Let $p(u) := \inf_x f(x,u)$, and consider a set $U \subset \operatorname{dom} p$.*

The mapping $P : I\!R^m \rightrightarrows I\!R^n$ is locally bounded relative to U if p is locally bounded from above relative to U. It is osc relative to U in addition, if p is actually continuous relative to U.

In particular, P is continuous relative to U at any point $\bar{u} \in U$ where it is single-valued and p is continuous relative to U.

Detail. This restates part of 1.17 in the terminology now available. □

5.23 Example (proximal mappings and projections).

(a) *For any nonempty set $C \subset I\!R^n$, the projection mapping P_C is everywhere osc and locally bounded.*

(b) *For any proper, lsc function $f : I\!R^n \to \overline{I\!R}$ that is proximally bounded with threshold λ_f, and any $\lambda \in (0, \lambda_f)$, the proximal mapping $P_\lambda f$ is everywhere osc and locally bounded.*

Detail. This specializes 5.22 to the mappings in 1.20 and 1.22; cf. 1.25. □

5.24 Exercise (continuity of perturbed mappings). *Suppose $S = S_0 + T$ for mappings $S_0, T : I\!R^n \rightrightarrows I\!R^m$, and let $\bar{x}$ be a point where T is continuous and locally bounded. If S_0 is osc, isc, or continuous at $\bar{x}$, then that property holds also for S at $\bar{x}$.*

In particular, if $S : \mathbb{R}^n \times \mathbb{R}^m$ *has the form* $S(x) = C + F(x)$ *for a closed set* $C \subset \mathbb{R}^m$ *and a continuous mapping* $F : \mathbb{R}^n \to \mathbb{R}^m$, *then* S *is continuous.*

The outer semicontinuity of a mapping $S : \mathbb{R}^n \rightrightarrows \mathbb{R}^m$ doesn't entail necessarily that $S(C)$ and $S^{-1}(D)$ are closed sets when C and D are closed, as we've already seen in the case of $S(\mathbb{R}^n)$ and $S^{-1}(\mathbb{R}^m)$. Other assumptions have to be added for such conclusions, and again local boundedness has a role.

5.25 Theorem (closedness of images). *Let* $S : \mathbb{R}^n \rightrightarrows \mathbb{R}^m$ *be osc,* C *a closed subset of* $\mathbb{R}^n$, *and* D *a closed subset of* $\mathbb{R}^m$. *Then,*

(a) $S(C)$ *is closed if* C *is compact or* S^{-1} *is locally bounded (implying that* $\operatorname{rge} S$ *is closed).*

(b) $S^{-1}(D)$ *is closed if* D *is compact or* S *is locally bounded (implying that* $\operatorname{dom} S$ *is closed).*

Proof. In (b), the first assertion applies 5.7(b) with $X = \mathbb{R}^n$. For the second assertion, suppose S is locally bounded as well as osc, and consider any closed set $D \subset \mathbb{R}^m$; we wish to verify that $S^{-1}(D)$ is closed. It's enough to demonstrate that $S^{-1}(D)$ is locally closed everywhere, or equivalently, that $S^{-1}(D) \cap B$ is closed for every compact set $B \subset \mathbb{R}^n$. But $S^{-1}(D) \cap B$ is the image of $(\operatorname{gph} S) \cap (B \times D)$ under the projection $(x, u) \mapsto x$. The set $(\operatorname{gph} S) \cap (B \times D)$ is closed because $\operatorname{gph} S$, B and D are all closed, and it is bounded because it's included in $B \times S(B)$, which is bounded (by 5.15); hence it is compact. The image of a compact set under a continuous mapping is compact, so we conclude that $S^{-1}(D) \cap B$ is compact, hence closed, as required.

Now (a) follows by symmetry, since the outer semicontinuity of S is equivalent to that of S^{-1}, cf. 5.7(a). □

5.26 Exercise (horizon criterion for a closed image). *Let* $S : \mathbb{R}^n \rightrightarrows \mathbb{R}^m$ *be osc.*

(a) $S(C)$ *is closed when* C *is closed and* $(S^\infty)^{-1}(0) \cap C^\infty = \{0\}$ *(as is true if* $(S^\infty)^{-1}(0) = \{0\}$ *or if* $C^\infty = \{0\}$*). Then* $S(C)^\infty \subset S^\infty(C^\infty)$.

(b) $S^{-1}(D)$ *is closed when* D *is closed and* $S^\infty(0) \cap D^\infty = \{0\}$ *(as is true if* $S^\infty(0) = \{0\}$ *or if* $D^\infty = \{0\}$*). Then* $S^{-1}(D)^\infty \subset (S^\infty)^{-1}(D^\infty)$.

Guide. By 5.7(a), it suffices to verify (b). The closedness of $S^{-1}(D)$ can be established by showing that the mapping $S_D : u \mapsto S(u) \cap D$ is locally bounded when the horizon assumption is fulfilled. This yields $S_D^\infty(0) = \{0\}$. □

A counterexample to the inclusion $S(C)^\infty \subset S^\infty(C^\infty)$ holding without some extra assumption is furnished by the single-valued mapping $S : \mathbb{R}^1 \to \mathbb{R}^1$ with $S(u) = \sqrt{|u|}$ and the set $C = [0, \infty)$. A case in which $S(C)^\infty \subset S^\infty(C^\infty)$ but $S(C)^\infty \neq S^\infty(C^\infty)$ is encountered in the mapping $S : \mathbb{R}^1 \to \mathbb{R}^1$ with $S(u) = u \sin u$ and the set $C = \{0, \pm\pi, \pm 2\pi, \dots\}$.

D. Total Continuity

The notion of continuity in Definition 5.4 for set-valued mappings S is powerful and apt for many purposes, and it makes topological sense in corresponding to the continuity of the single-valued mapping which assigns to each point x a set $S(x)$ as an element of a space of sets, this 'hyperspace' being endowed with the topology induced by Painlevé-Kuratowski convergence (as developed near the end of Chapter 4). Yet this notion of continuity has shortcomings as well. Figure 5–8 depicts a mapping $S : \mathbb{R}^1 \rightrightarrows \mathbb{R}^1$ which is continuous at 0 according to 5.4, despite the appearance of a feature looking very much like a 'discontinuity'. Here $S(x) = \{1 + x^{-1}, 1\}$ for $x \neq 0$, while $S(0) = \{1\}$, so that $S(x)$ does tend to $S(0)$ as x tends to 0, just as continuity requires. Note that S isn't locally bounded at 0.

This example clashes with our geometric intuition because convergence of unbounded sequences to direction points isn't taken into account by Definition 5.4. If we think of the sets $S(x)$ as lying in the one-dimensional cosmic space $\operatorname{csm} \mathbb{R}^1$, identified with $\overline{\mathbb{R}}$ in letting $\infty \leftrightarrow \operatorname{dir} 1$ and $-\infty \leftrightarrow \operatorname{dir}(-1)$, we see that $S(x) \xrightarrow{c} \{1, \infty\}$ as x tends to 0 from the right, whereas $S(x) \xrightarrow{c} \{1, -\infty\}$ as x tends to 0 from the left. These cosmic limits differ from each other and from $S(0) = \{1\}$, and that's the source of our feeling of 'discontinuity', even though $S(x) \to S(0)$ from both sides in the Painlevé-Kuratowski context. Intuitively, therefore, we find it hard in some situations to accept the version of continuity dictated by ordinary set convergence and may wish to have at our disposal an alternative version that utilizes cosmic limits.

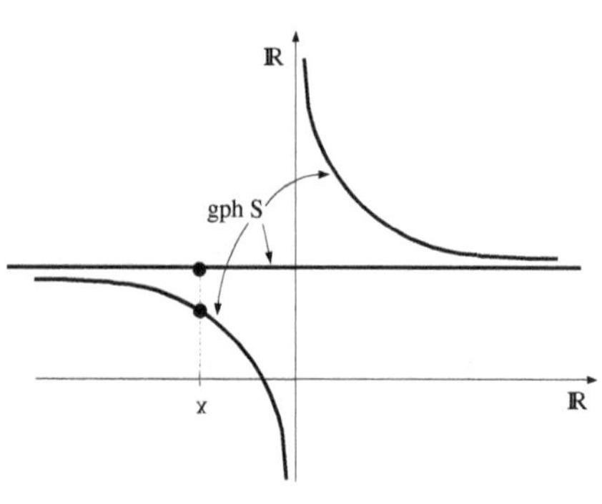

Fig. 5–8. An everywhere continuous set-valued mapping: $S(x) \to S(0)$ as $x \to 0$.

Such thinking leads us to introduce continuity concepts that appeal to the framework of mappings $S : \mathbb{R}^n \rightrightarrows \operatorname{csm} \mathbb{R}^m$ in order to quantify the behavior of unbounded sequences of elements $u^\nu \in S(x^\nu)$ in terms of convergence to points of $\operatorname{hzn} \mathbb{R}^m$. Hardly any additional effort is needed, beyond a straightforward translation of the conditions in Definition 5.4 along the same lines that were followed for the corresponding extension of set convergence in Chapter 4.

Thus, *cosmic continuity at* $\bar{x} \in \mathbb{R}^n$ for a mapping $S : \mathbb{R}^n \rightrightarrows \operatorname{csm} \mathbb{R}^m$ is taken to mean that $S(x^\nu) \xrightarrow{c} S(\bar{x})$ whenever $x^\nu \to \bar{x}$; similarly for cosmic outer and inner semicontinuity. Cosmic outer semicontinuity at every point $\bar{x}$ corresponds to $\operatorname{gph} S$ being closed as a subset of $\mathbb{R}^n \times \operatorname{csm} \mathbb{R}^m$. The definition of 'subgradients' in Chapter 8 will draw on the notion of cosmic outer semicontinuity (cf. 8.7), and the following fact will then be helpful. Here we employ horizon limits in the same extended sense that was introduced in 5(1) for ordinary set limits.

5.27 Proposition (cosmic semicontinuity). *A mapping* $S : \mathbb{R}^n \rightrightarrows \operatorname{csm} \mathbb{R}^m$, *written as* $S(x) = C(x) \cup \operatorname{dir} K(x)$ *with* $C(x)$ *a set in* $\mathbb{R}^m$ *and* $K(x)$ *a cone in* $\mathbb{R}^m$, *is cosmically outer semicontinuous at* $\bar{x}$ *if and only if*

$$C(\bar{x}) \supset \limsup_{x \to \bar{x}} C(x), \qquad K(\bar{x}) \supset \limsup_{x \to \bar{x}}{}^{\infty} C(x) \cup \limsup_{x \to \bar{x}} K(x),$$

whereas the corresponding condition for cosmic inner semicontinuity is

$$C(\bar{x}) \subset \liminf_{x \to \bar{x}} C(x), \qquad K(\bar{x}) \subset \liminf_{x \to \bar{x}}{}^{\infty} C(x) \cup \liminf_{x \to \bar{x}} K(x).$$

Proof. This is obvious from 4.20. □

Note that the mapping in Figure 5–8 isn't cosmically osc at 0, much less cosmically continuous there, although it's continuous at 0 in the ordinary, non-cosmic sense, as already observed. This highlights very well the distinction between ordinary and cosmic continuity and the reasons why, in some situations, it might be desirable to insist on the latter. Actually, for most purposes it's not necessary to pass to the full cosmic setting, because the concept of total convergence introduced in 4.23 can serve the same ends in $\mathbb{R}^n$ itself.

5.28 Definition (total continuity). *A mapping* $S : \mathbb{R}^n \rightrightarrows \mathbb{R}^m$ *is said to be totally continuous at* $\bar{x}$ *if* $S(x) \xrightarrow{t} S(\bar{x})$ *whenever* $x \to \bar{x}$, *or equivalently if*

$$\lim_{x \to \bar{x}} S(x) = S(\bar{x}), \qquad \lim_{x \to \bar{x}}{}^{\infty} S(x) = S(\bar{x})^{\infty}.$$

It is totally outer semicontinuous at $\bar{x}$ *if at least*

$$\limsup_{x \to \bar{x}} S(x) \subset S(\bar{x}), \qquad \limsup_{x \to \bar{x}}{}^{\infty} S(x) \subset S(\bar{x})^{\infty}.$$

Total inner semicontinuity at a point $\bar{x} \in \operatorname{dom} S$ could likewise be defined as meaning that

$$\liminf_{x \to \bar{x}} S(x) \supset S(\bar{x}), \qquad \liminf_{x \to \bar{x}}{}^{\infty} S(x) \supset S(\bar{x})^{\infty},$$

but this would be superfluous because the first of these inclusions always entails the second (cf. 4.21(c)). Total inner semicontinuity therefore wouldn't be any stronger than ordinary inner semicontinuity.

5.29 Proposition (criteria for total continuity). *A mapping* $S : \mathbb{R}^n \rightrightarrows \mathbb{R}^m$ *is totally continuous at* $\bar{x}$ *if and only if it is continuous at* $\bar{x}$ *and has*

$$\limsup_{x \to \bar{x}}{}^{\infty} S(x) \subset S(\bar{x})^{\infty}.$$

Total continuity at $\bar{x}$ is automatic from continuity when S is convex-valued or cone-valued on a neighborhood of $\bar{x}$, or when S is locally bounded at $\bar{x}$.

Proof. These facts are obvious from 4.24 and 4.25. □

5.30 Exercise (images of converging sets). *Consider a mapping $S : \mathbb{R}^n \rightrightarrows \mathbb{R}^m$ and sets $C^\nu \subset \mathbb{R}^n$.*

(a) *If S is isc, one has $\liminf_\nu S(C^\nu) \supset S(\liminf_\nu C^\nu)$.*

(b) *If S is osc, one has $\limsup_\nu S(C^\nu) \subset S(\limsup_\nu C^\nu)$ provided that S^{-1} is locally bounded, or alternatively that $(S^\infty)^{-1}(0) \cap \limsup_\nu^\infty C^\nu = \{0\}$.*

(c) *If S is continuous, one has $S(C^\nu) \to S(C)$ whenever $C^\nu \to C$ and S^{-1} is locally bounded, or alternatively, whenever $C^\nu \xrightarrow{t} C$ and $(S^\infty)^{-1}(0) \cap C^\infty = \{0\}$.*

(d) *If S is totally continuous, one has $S(C^\nu) \xrightarrow{t} S(C)$ whenever $C^\nu \xrightarrow{t} C$, $(S^\infty)^{-1}(0) \cap C^\infty = \{0\}$ and $(S^\infty)(C^\infty) \subset S(C)^\infty$.*

Guide. Extend the argument for 4.26, also utilizing ideas in 5.26. □

Some results about the preservation of continuity when mappings are added or composed together will be provided in 5.51 and 5.52.

E. Pointwise and Graphical Convergence

The 'convergence' of a sequence of mappings can have a number of meanings. Let's start with pointwise convergence.

5.31 Definition (pointwise limits of mappings). *For a sequence of mappings $S^\nu : \mathbb{R}^n \rightrightarrows \mathbb{R}^m$, the pointwise outer limit and the pointwise inner limit are the mappings $\text{p-}\limsup_\nu S^\nu$ and $\text{p-}\liminf_\nu S^\nu$ defined at each point x by*

$$\big(\text{p-}\limsup_\nu S^\nu\big)(x) := \limsup_\nu S^\nu(x),$$
$$\big(\text{p-}\liminf_\nu S^\nu\big)(x) := \liminf_\nu S^\nu(x).$$

When the pointwise outer and inner limits agree, the pointwise limit $\text{p-}\lim_\nu S^\nu$ is said to exist; thus, $S = \text{p-}\lim_\nu S^\nu$ if and only if $S \supset \text{p-}\limsup_\nu S^\nu$ and $S \subset \text{p-}\liminf_\nu S^\nu$. In this case the notation $S^\nu \xrightarrow{p} S$ is also used, and the mappings S^ν are said to converge pointwise to S. Thus,

$$S^\nu \xrightarrow{p} S \iff S^\nu(x) \to S(x) \text{ for all } x.$$

Obviously $\text{p-}\limsup_\nu S^\nu \supset \text{p-}\liminf_\nu S^\nu$ always. Here we use '$\subset$' and '$\supset$' in the sense of the natural ordering among mappings $\mathbb{R}^n \rightrightarrows \mathbb{R}^m$:

$$S_1 \subset S_2 \text{ when } \operatorname{gph} S_1 \subset \operatorname{gph} S_2, \qquad S_1 \supset S_2 \text{ when } \operatorname{gph} S_1 \supset \operatorname{gph} S_2.$$

Definition 5.31 focuses on whole mappings, but it's convenient sometimes to say S^ν *converges pointwise to S at a point $\bar{x}$* when $S^\nu(\bar{x}) \to S(\bar{x})$.

Pointwise convergence of mappings $S^\nu : \mathbb{R}^n \rightrightarrows \mathbb{R}^m$ is an attractive notion because it reduces in the single-valued case to something very familiar. It has many important applications, but it only provides one part of the essential picture of what convergence can signify. For many purposes it's necessary instead, or as well, to consider a different kind of convergence, obtained by applying the theory of set convergence to the sets $\operatorname{gph} S^\nu$ in $\mathbb{R}^n \times \mathbb{R}^m$.

5.32 Definition (graphical limits of mappings). *For a sequence of mappings $S^\nu : \mathbb{R}^n \rightrightarrows \mathbb{R}^m$, the graphical outer limit, denoted by $\text{g-lim sup}_\nu S^\nu$, is the mapping having as its graph the set $\limsup_\nu (\operatorname{gph} S^\nu)$:*

$$\begin{aligned} \operatorname{gph}\big(\text{g-lim sup}_\nu S^\nu\big) &= \limsup_\nu \big(\operatorname{gph} S^\nu\big), \\ \big(\text{g-lim sup}_\nu S^\nu\big)(x) &= \big\{u \,\big|\, \exists N \in \mathcal{N}_\infty^\#,\ x^\nu \underset{N}{\to} x,\ u^\nu \underset{N}{\to} u,\ u^\nu \in S^\nu(x^\nu)\big\}. \end{aligned}$$

The graphical inner limit, denoted by $\text{g-lim inf}_\nu S^\nu$, is the mapping having as its graph the set $\liminf_\nu (\operatorname{gph} S^\nu)$:

$$\begin{aligned} \operatorname{gph}\big(\text{g-lim inf}_\nu S^\nu\big) &= \liminf_\nu \big(\operatorname{gph} S^\nu\big), \\ \big(\text{g-lim inf}_\nu S^\nu\big)(x) &= \big\{u \,\big|\, \exists N \in \mathcal{N}_\infty,\ x^\nu \underset{N}{\to} x,\ u^\nu \underset{N}{\to} u,\ u^\nu \in S^\nu(x^\nu)\big\}. \end{aligned}$$

If these outer and inner limits agree, the graphical limit $\text{g-lim}_\nu S^\nu$ exists; thus, $S = \text{g-lim}_\nu S^\nu$ if and only if $S \supset \text{g-lim sup}_\nu S^\nu$ and $S \subset \text{g-lim inf}_\nu S^\nu$. In this case the notation $S^\nu \xrightarrow{g} S$ is also used, and the mappings S^ν are said to converge graphically to S. Thus,

$$S^\nu \xrightarrow{g} S \quad \Longleftrightarrow \quad \operatorname{gph} S^\nu \to \operatorname{gph} S.$$

The mappings $\text{g-lim sup}_\nu S^\nu$ and $\text{g-lim inf}_\nu S^\nu$ are always osc, and so too is $\text{g-lim}_\nu S^\nu$ when it exists. This is evident from the fact that their graphs, as certain set limits, are closed (by 4.4). These limit mappings need not be isc, however, even when every S^ν is isc. But $\text{g-lim inf}_\nu S^\nu$ is graph-convex when every S^ν is graph-convex, and then $\text{g-lim inf}_\nu S^\nu$ is isc on the interior of $\operatorname{dom}\big(\text{g-lim inf}_\nu S^\nu\big)$ by 5.9(b).

5.33 Proposition (graphical limit formulas at a point). *For any sequence of mappings $S^\nu : \mathbb{R}^n \rightrightarrows \mathbb{R}^m$ one has*

$$\begin{aligned} \big(\text{g-liminf}_\nu S^\nu\big)(x) &= \bigcup_{\{x^\nu \to x\}} \liminf_{\nu\to\infty} S^\nu(x^\nu) = \lim_{\delta \searrow 0} \Big[\liminf_{\nu\to\infty} S^\nu(x + \delta \mathbb{B})\Big], \\ \big(\text{g-limsup}_\nu S^\nu\big)(x) &= \bigcup_{\{x^\nu \to x\}} \limsup_{\nu\to\infty} S^\nu(x^\nu) = \lim_{\delta \searrow 0} \Big[\limsup_{\nu\to\infty} S^\nu(x + \delta \mathbb{B})\Big], \end{aligned}$$

where the unions are taken over all sequences $x^\nu \to x$. Thus, S^ν converges graphically to S if and only if, at each point $\bar{x} \in \mathbb{R}^n$, one has

$$\bigcup_{\{x^\nu \to \bar{x}\}} \limsup_{\nu\to\infty} S^\nu(x^\nu) \subset S(\bar{x}) \subset \bigcup_{\{x^\nu \to \bar{x}\}} \liminf_{\nu\to\infty} S^\nu(x^\nu). \qquad 5(7)$$

Proof. Let $\overline{S} = \text{g-limsup}_\nu S^\nu$ and $\underline{S} = \text{g-liminf}_\nu S^\nu$. The expressions of $\overline{S}(x)$ and $\underline{S}(x)$ as unions over all sequences $x^\nu \to x$ merely restate the formulas in Definition 5.32.

Because $\operatorname{gph} \overline{S} = \limsup_\nu \operatorname{gph} S^\nu$, we also have through 4.1 that $u \in \overline{S}(x)$ if and only if, for all $\varepsilon > 0$ and $\delta > 0$, there exists $N \in \mathcal{N}_\infty^\#$ such that $\mathbb{B}(x,\delta) \times \mathbb{B}(u,\varepsilon)$ meets $\operatorname{gph} S^\nu$ for all $\nu \in N$, or in other words

$$(u + \varepsilon\mathbb{B}) \cap S^\nu(x + \delta\mathbb{B}) \neq \emptyset \text{ for all } \nu \in N. \tag{5(8)}$$

This means that $u + \varepsilon\mathbb{B}$ meets $\limsup_\nu S^\nu(x + \delta\mathbb{B})$ for all $\varepsilon > 0$ and $\delta > 0$. Since this outer limit set decreases if anything as δ decreases, we conclude that $u \in \overline{S}(x)$ if and only if $u \in \lim_{\delta \searrow 0} \limsup_\nu S^\nu(x + \delta\mathbb{B})$.

Similarly, because $\operatorname{gph} \underline{S} = \liminf_\nu \operatorname{gph} S^\nu$ we obtain through 4.1 that $u \in \underline{S}(x)$ if and only if, for all $\varepsilon > 0$ and $\delta > 0$, there is an index set $N \in \mathcal{N}_\infty$ (instead of $\mathcal{N}_\infty^\#$) such that 5(8) holds. Then $u + \varepsilon\mathbb{B}$ meets $\liminf_\nu S^\nu(x + \delta\mathbb{B})$ for all $\varepsilon > 0$ and $\delta > 0$, which is equivalent to the condition that $u \in \lim_{\delta \searrow 0} \liminf_\nu S^\nu(x + \delta\mathbb{B})$. □

The characterization of graphical convergence in 5.33 prompts us to define *graphical convergence of S^ν to S at a point $\bar{x}$* as meaning that 5(7) holds at $\bar{x}$. In this sense, S^ν converges graphically to S if and only if it does so at every point. More generally, we say that S^ν converges graphically to S *relative to a set X* if 5(7), with x^ν constrained to X, holds for every $\bar{x} \in X$. For closed X, this is the same as saying that the restrictions $S^\nu|_X$ converge graphically to the restriction $S|_X$ (this being the mapping that agrees with S on X but is empty-valued everywhere else).

The expression for graphical outer limits in 5.33 says that

$$\big(\text{g-lim sup}_\nu S^\nu\big)(\bar{x}) = \limsup_{\substack{\nu\to\infty \\ x\to\bar{x}}} S^\nu(x) = \limsup_{\substack{\nu\to\infty \\ \delta \searrow 0}} S^\nu(\bar{x} + \delta\mathbb{B}). \tag{5(9)}$$

The expression for graphical inner limits doesn't have such a simple bivariate interpretation, however. For instance, graphical convergence at $\bar{x}$ is generally *weaker* than the assertion that $S^\nu(\bar{x} + \delta\mathbb{B}) \to S(\bar{x})$ as $\nu \to \infty$ and $\delta \searrow 0$. The relationship between this property and graphical convergence of S^ν to S will be clarified later through the discussion of 'continuous convergence' of mappings (see 5.43 and 5.44).

5.34 Exercise (uniformity in graphical convergence). *Let $S, S^\nu : \mathbb{R}^n \rightrightarrows \mathbb{R}^m$.*

(a) *Suppose the mapping S is closed-valued. Then $\text{g-lim}_\nu S^\nu = S$ if and only if, for every $\varepsilon > 0$ and $\rho > 0$, there exists $N \in \mathcal{N}_\infty$ such that*

$$\left.\begin{aligned} S(x) \cap \rho\mathbb{B} &\subset S^\nu(\mathbb{B}(x,\varepsilon)) + \varepsilon\mathbb{B} \\ S^\nu(x) \cap \rho\mathbb{B} &\subset S(\mathbb{B}(x,\varepsilon)) + \varepsilon\mathbb{B} \end{aligned}\right\} \text{ when } |x| \le \rho \text{ and } \nu \in N.$$

(b) *Suppose the mappings S^ν are connected-valued (e.g., convex-valued), $S = \text{g-lim sup}_\nu S^\nu$, $S(\bar{x})$ is bounded and $\limsup_{x\to\bar{x},\nu\to\infty} d(0, S^\nu(x)) < \infty$; this* *last condition is certainly satisfied when $S = g\lim_\nu S^\nu$ and $S(\bar{x}) \neq \emptyset$. Then, there exist $N \in \mathcal{N}_\infty$, $V \in \mathcal{N}(\bar{x})$ and a bounded set B such that*

$$S(x) \subset B \ \text{ and } \ S^\nu(x) \subset B \ \text{ for all } x \in V,\ \nu \in N.$$

Guide. Derive (a) from the uniformity of approximation in set convergence in 4.10. Note incidentally that the balls $\rho\mathbb{B}$ could be replaced by arbitrary bounded sets $B \subset \mathbb{R}^m$, while the balls $\varepsilon\mathbb{B}$ could be replaced by arbitrary neighborhoods U of the origin in $\mathbb{R}^m$.

For (b) rely on the formula for the graphical outer limit in 5(9) and then appeal to 4.12. □

5.35 Example (graphical convergence of projection mappings). *For closed sets $C^\nu, C \subset \mathbb{R}^n$, one has $P_{C^\nu} \xrightarrow{g} P_C$ if and only if $C^\nu \to C$.*

Detail. The case where $C = \emptyset$, $\operatorname{gph} P_C = \emptyset$, is trivial, so attention can be concentrated on the case where C is nonempty and $\operatorname{dom} P_C = \mathbb{R}^n$; then either side of the proposed equivalence entails the nonemptiness of C^ν for all ν sufficiently large. We may as well assume $C^\nu \neq \emptyset$ for all ν.

Invoking the characterization of set convergence in 4.9, we see, with a minor extension of the argument developed there, that the convergence of C^ν to C is equivalent to having

$$\begin{cases} \limsup_\nu d(0, C^\nu) < \infty, \\ \limsup_\nu P_{C^\nu}(x^\nu) \subset P_C(x) \ \text{ when } x^\nu \to x, \end{cases} \tag{5(12)}$$

the latter being the same as $\text{g-}\limsup_\nu P_{C^\nu} \subset P_C$. From knowing not only that $\text{g-}\limsup_\nu P_{C^\nu} \subset P_C$ but also $\text{g-}\liminf_\nu P_{C^\nu} \supset P_C$, we readily conclude 5(12), since the convergence of points $(x^\nu, \bar{x}^\nu) \in \operatorname{gph} P_{C^\nu}$ to $(x, \bar{x}) \in \operatorname{gph} P_C$ implies that $d(0, C^\nu) \leq |\bar{x}^\nu| \to |\bar{x}| < \infty$.

Conversely, suppose 5(12) holds and consider any x_0 and $\bar{x}_0 \in P_C(x_0)$. We must verify that $(x_0, \bar{x}_0) \in \liminf_\nu \operatorname{gph} P_{C^\nu}$. The fact that $\bar{x}_0 \in P_C(x_0)$ implies for arbitrary $\varepsilon \in (0,1)$ that for $x_\varepsilon = (1-\varepsilon)x_0 + \varepsilon\bar{x}_0$ that $P_C(x_\varepsilon) = \{\bar{x}_0\}$. For each ν, choose any $x^\nu \in P_{C^\nu}(x_\varepsilon)$. On the basis of 5(12), the sequence $\{x^\nu\}_{\nu\in\mathbb{N}}$ is bounded and has all its cluster points in $P_C(x_\varepsilon)$; hence it has $\bar{x}_0$ as its only cluster point. Thus the points $(x_\varepsilon, x^\nu) \in \operatorname{gph} P_{C^\nu}$ converge to $(x_\varepsilon, \bar{x}_0) \in \operatorname{gph} P_C$, so that $(x_\varepsilon, \bar{x}_0) \in \liminf_\nu \operatorname{gph} P_{C^\nu}$. This being true for arbitrary $\varepsilon > 0$, we must have $(x_0, \bar{x}_0) \in \liminf_\nu \operatorname{gph} P_{C^\nu}$. □

An important property of graphical limits, which isn't enjoyed by pointwise limits, is their stability under taking inverses. The geometry of the definition yields at once that

$$S^\nu \xrightarrow{g} S \iff (S^\nu)^{-1} \xrightarrow{g} S^{-1}, \tag{5(10)}$$

and similarly for outer and inner limits.

Another special feature of graphical limits is the possibility always of selecting graphically convergent subsequences. To state this property, we use the terminology that a sequence of mappings $S^\nu : \mathbb{R}^n \rightrightarrows \mathbb{R}^m$ *escapes to the horizon* if for every choice of bounded sets $C \subset \mathbb{R}^n$ and $D \subset \mathbb{R}^m$, an index set $N \in \mathcal{N}_\infty$ exists such that

$$S^\nu(x) \cap D = \emptyset \text{ for all } \nu \in N \text{ when } x \in C.$$

5.36 Theorem (extraction of graphically convergent subsequences). *A sequence of mappings $S^\nu : \mathbb{R}^n \rightrightarrows \mathbb{R}^m$ either escapes to the horizon or has a subsequence converging graphically to a mapping $S : \mathbb{R}^n \rightrightarrows \mathbb{R}^m$ with $\operatorname{dom} S \neq \emptyset$.*

Proof. This is evident from 4.18, the corresponding compactness result for set convergence, as applied to the sequence of sets $\operatorname{gph} S^\nu$ in $\mathbb{R}^n \times \mathbb{R}^m$. □

Graphical convergence doesn't generally imply pointwise convergence, and pointwise convergence doesn't generally imply graphical convergence. A sequence of mappings S^ν can even be such that both its graphical limit and its pointwise limit exist, but the two are different! An example of this phenomenon is displayed in Figure 5–9. Nevertheless, certain basic relations between graphical and pointwise convergence follow from 5.33 and the definitions:

$$\text{p-liminf}_\nu S^\nu \quad \subset \quad \left\{ \begin{array}{c} \text{g-liminf}_\nu S^\nu \\ \text{p-limsup}_\nu S^\nu \end{array} \right\} \quad \subset \quad \text{g-limsup}_\nu S^\nu. \qquad 5(11)$$

In particular, it's always true that $\text{p-lim}_\nu S^\nu \subset \text{g-lim}_\nu S^\nu$ when both limits exist.

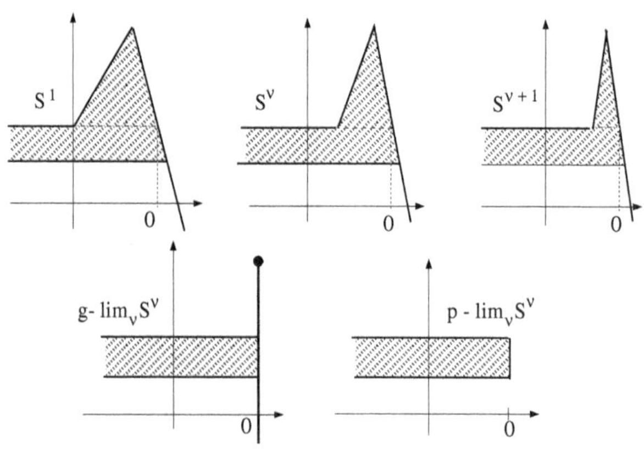

Fig. 5–9. Possible distinctness of pointwise and graphical limits.

Later we'll arrive at a complete answer to the question of what circumstances induce graphical convergence and pointwise convergence to be the same, cf. Theorem 5.40. More important for now is the question of what features of graphical convergence make it interesting in circumstances where it *doesn't* agree with pointwise convergence, or where pointwise limits might not even exist. One such feature, certainly, is the compactness property in 5.36. Others are suggested by the ability of graphical convergence to make sense of situations where a sequence of single-valued mappings might appropriately have a multivalued limit.

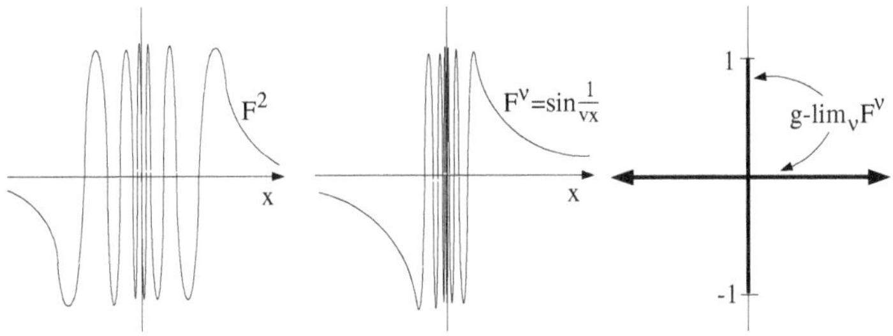

Fig. 5–10. Convergence of rapidly oscillating functions.

For example, in models of phenomena with rapidly oscillating states the graphical limit of a sequence of single-valued mappings may well be multivalued at certain points which represent instability or turbulence. A suggestive illustration is furnished in Figure 5–10, where $F^\nu(x) = \sin(1/\nu x)$. In this case $(\text{g-lim}_\nu F^\nu)(0) = [-1,1]$, while $(\text{g-lim}_\nu F^\nu)(x) = \{0\}$ for all $x \neq 0$. For $F^\nu(x) = \sin(\nu x)$ instead, $(\text{g-lim}_\nu F^\nu)(x) = [-1,1]$ for all x.

Probability theory provides other insights into the potential advantages of working with graphical convergence. Some of these come up in the analysis of convergence of distribution functions on $I\!R$, these being nondecreasing functions $F : I\!R \to I\!R$ such that $F(x) \to 0$ as $x \searrow -\infty$, and $F(x) \to 1$ as $x \to \infty$. Traditionally such functions are normalized by taking them to be continuous on the right, and the analysis proceeds through attempts to rely on pointwise convergence in special, restricted ways. But distribution functions have limits everywhere from the left as well as from the right, and instead of normalizing them through right continuity, or for that matter left continuity, it's possible to identify them with the special mappings $S : I\!R \rightrightarrows I\!R$ obtained by inserting a vertical interval to fill in the graph whenever there would otherwise be a gap due to a jump.

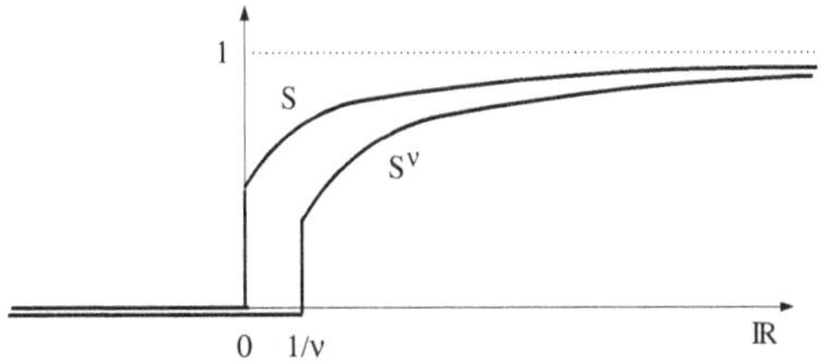

Fig. 5–11. Convergence of probability distribution functions with jumps.

Then, instead of pointwise convergence properties of F^ν to F, one can turn to graphical convergence of S^ν to S as the key. This is demonstrated in Figure 5–11 for the distribution functions

$$F^\nu(x) = \begin{cases} \left(\frac{1}{2} - \frac{1}{\nu}\right) + \left(\frac{1}{2} + \frac{1}{\nu}\right)\left(1 - e^{-(x-\nu^{-1})}\right) & \text{for } x \geq \nu^{-1}, \\ 0 & \text{for } x < \nu^{-1}, \end{cases}$$

where the corresponding S^ν is single-valued at points $x \neq \nu^{-1}$, agreeing there with $F^\nu(x)$, but $S^\nu(x) = \big[0, F(x)\big]$ at $x = \nu^{-1}$. The bounded, osc mappings S^ν converge graphically to the mapping $S : \mathbb{R} \rightrightarrows \mathbb{R}$ having the single value $1 - \frac{1}{2}e^{-x}$ when $x > 0$ and the single value 0 when $x < 0$, but $S(0) = \big[0, \frac{1}{2}\big]$. This mapping S corresponds to a distribution function F with a jump at 0.

The ways graphical convergence might come up in nonclassical approaches to differentiation, which will be explored in depth later, can be appreciated from the case of the convex functions $f^\nu : \mathbb{R} \to \mathbb{R}$ defined by

$$f^\nu(x) = \begin{cases} (\nu/2)x^2 & \text{for } x \in [-\nu^{-1}, \nu^{-1}], \\ |x| - (1/2\nu) & \text{otherwise.} \end{cases}$$

These converge pointwise to the convex function $f(x) = |x|$. The derivative functions

$$\left(f^\nu\right)'(x) = \begin{cases} -1 & \text{for } x < -\nu^{-1}, \\ \nu x & \text{for } -\nu^{-1} \leq x \leq \nu^{-1}, \\ 1 & \text{for } x > \nu^{-1} \end{cases}$$

have both a pointwise limit and a graphical limit; both have the single value 1 on the positive axis and the single value -1 on the negative axis, but the pointwise limit has the value 0 at the origin, whereas the graphical limit has the interval $[-1, 1]$ there. The graphical limit makes more sense than the pointwise limit as a possible candidate for a generalized kind of derivative for the limit function f, and indeed this will eventually fit the pattern adopted.

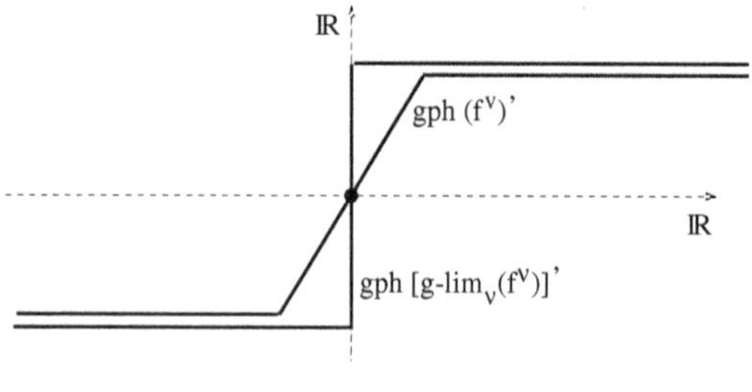

Fig. 5–12. Graphical convergence of derivatives of convex functions.

The critical role played by graphical convergence stems in large part from the following theorem, which virtually stands as a characterization of the concept. This role has often been obscured in the past by ad hoc efforts aimed at skirting issues of multivaluedness.

5.37 Theorem (approximation of generalized equations). *For closed-valued mappings $S, S^\nu : \mathbb{R}^n \rightrightarrows \mathbb{R}^m$ and vectors $\bar{u}, \bar{u}^\nu \in \mathbb{R}^m$, consider the generalized equation $S^\nu(x) \ni \bar{u}^\nu$ as an approximation to the generalized equation $S(x) \ni \bar{u}$, the respective solution sets being $(S^\nu)^{-1}(\bar{u}^\nu)$ and $S^{-1}(\bar{u})$, with the elements of the former referred to as approximate solutions and the elements of the latter as true solutions.*

(a) *As long as $\text{g-lim sup}_\nu S^\nu \subset S$, one has for every choice of $\bar{u}^\nu \to \bar{u}$ that $\limsup_\nu (S^\nu)^{-1}(\bar{u}^\nu) \subset S^{-1}(\bar{u})$. Thus, any cluster point of a sequence of approximate solutions is a true solution.*

(b) *If $\text{g-lim inf}_\nu S^\nu \supset S$, one has $S^{-1}(\bar{u}) \subset \bigcap_{\varepsilon>0} \liminf_\nu (S^\nu)^{-1}\big(\mathbb{B}(\bar{u},\varepsilon)\big)$. In this case, therefore, every true solution is the limit of approximate solutions corresponding to some choice of $\bar{u}^\nu \to \bar{u}$.*

(c) *When $S^\nu \xrightarrow{g} S$, both conclusions hold.*

Proof. In (a) the definition of the graphical outer limit is applied directly. In (b) the representation of the inner limit in 5.33 is applied to the mappings $(S^\nu)^{-1}$ and S^{-1}. □

For the wealth of applications covered by this theorem, see Example 5.2 and the surrounding discussion. In particular the condition $S(x) \ni b$ could be an equation $F(x) = b$ for a mapping $F : \mathbb{R}^n \to \mathbb{R}^m$.

F. Equicontinuity of Sequences

In order to understand graphical convergence further, not only in its relation to pointwise convergence but other convergence concepts for set-valued mappings, properties of equicontinuity will be helpful.

5.38 Definition (equicontinuity properties). *A sequence of set-valued mappings $S^\nu : \mathbb{R}^n \rightrightarrows \mathbb{R}^m$ is equi-osc at $\bar{x}$ relative to X (a set containing $\bar{x}$) if for every $\varepsilon > 0$ and $\rho > 0$ there exists $V \in \mathcal{N}(\bar{x})$ such that*

$$S^\nu(x) \cap \rho\mathbb{B} \subset S^\nu(\bar{x}) + \varepsilon\mathbb{B} \ \text{ for all } \nu \in \mathbb{N} \ \text{ when } x \in V \cap X.$$

It is asymptotically equi-osc if, instead of necessarily for all $\nu \in \mathbb{N}$, this holds for all ν in an index set $N \in \mathcal{N}_\infty$ which, like V, can depend on ε and ρ.

On the other hand, a sequence is equi-isc at $\bar{x}$ relative to X if for every $\varepsilon > 0$ and $\rho > 0$ there exists $V \in \mathcal{N}(\bar{x})$ such that

$$S^\nu(\bar{x}) \cap \rho\mathbb{B} \subset S^\nu(x) + \varepsilon\mathbb{B} \ \text{ for all } \nu \in \mathbb{N} \ \text{ when } x \in V \cap X.$$

It is asymptotically equi-isc if, instead of necessarily for all $\nu \in \mathbb{N}$, this holds for all ν in an index set $N \in \mathcal{N}_\infty$ which, like V, can depend on ε and ρ.

Finally, a sequence of set-valued mappings $S^\nu : \mathbb{R}^n \rightrightarrows \mathbb{R}^m$ is equicontinuous at $\bar{x}$ relative to X if it is both equi-isc and equi-osc at $\bar{x}$ relative to

X. Likewise, it is asymptotically equicontinuous when it is both asymptotically equi-isc and asymptotically equi-osc.

Any sequence of mappings that is equicontinuous at $\bar{x}$ is in particular asymptotically equicontinuous at $\bar{x}$, of course. But equicontinuity is a considerably more restrictive property. In particular it entails all of the mappings being continuous, whereas a sequence can be asymptotically equicontinuous at $\bar{x}$ without any of the mappings themselves being continuous there. For instance, the sequence of mappings $S^\nu : I\!R \to I\!R$ defined by

$$S^\nu(x) = \begin{cases} \{1/\nu\} & \text{if } x > 0, \\ [-1/\nu, 1/\nu] & \text{if } x = 0, \\ \{-1/\nu\} & \text{if } x < 0, \end{cases}$$

has this character at $\bar{x} = 0$. The same can be said about sequences that are asymptotically equi-osc or asymptotically equi-isc.

5.39 Exercise (equicontinuity of single-valued mappings). *A sequence of continuous single-valued mappings $F^\nu : I\!R^n \to I\!R^m$ is equicontinuous at $\bar{x}$ if and only if it is equi-osc at $\bar{x}$. Moreover, this means that for all $\varepsilon > 0$ and $\rho > 0$ there exist $V \in \mathcal{N}(\bar{x})$ and $N \in \mathcal{N}_\infty$ such that*

$$\left|F^\nu(x) - F^\nu(\bar{x})\right| \le \varepsilon \;\text{ for all } \nu \in N \;\text{ and } x \in V \text{ having } \left|F^\nu(x)\right| \le \rho.$$

The mention of ρ can be dropped when the sequence $\{F^\nu\}_{\nu \in I\!N}$ is eventually locally bounded at $\bar{x}$, in the sense that there exist $V \in \mathcal{N}(\bar{x})$, $N \in \mathcal{N}_\infty$ and a bounded set B such that $F^\nu(x) \in B$ for all $x \in V$ when $\nu \in N$.

Similarly, a sequence of single-valued mappings is asymptotically equicontinuous at $\bar{x}$ if and only if it is asymptotically equi-osc at $\bar{x}$.

The traditional definition of equicontinuity in the case of single-valued mappings doesn't involve ρ along with ε, as in 5.39, but traditional applications are limited anyway to collections of mappings that are uniformly bounded, where the ρ feature makes no difference. Thus, 5.39 identifies the correct extension that equicontinuity should have for collections that aren't uniformly bounded. This view is obviously important in maintaining a theory of continuity of set-valued mappings that specializes appropriately in the single-valued case, because statements in which a bounded set $\rho I\!B$ enters in the image space are essential in set convergence.

5.40 Theorem (graphical versus pointwise convergence). *If a sequence of closed-valued mappings $S^\nu : I\!R^n \rightrightarrows I\!R^m$ is asymptotically equi-osc at $\bar{x}$, then*

$$(\text{g-liminf}_\nu\, S^\nu)(\bar{x}) = (\text{p-liminf}_\nu\, S^\nu)(\bar{x}),$$
$$(\text{g-limsup}_\nu\, S^\nu)(\bar{x}) = (\text{p-limsup}_\nu\, S^\nu)(\bar{x}).$$

Thus in particular, if the sequence is asymptotically equi-osc everywhere, one has $S^\nu \xrightarrow{g} S$ if and only if $S^\nu \xrightarrow{p} S$.

More generally for a set $X \subset \mathbb{R}^n$ and a point $\bar{x} \in X$, any two of the following conditions implies the third:

(a) *the sequence is asymptotically equi-osc at $\bar{x}$ relative to X;*

(b) *S^ν converges graphically to S at $\bar{x}$ relative to X;*

(c) *S^ν converges pointwise to S at $\bar{x}$ relative to X.*

Proof. To obtain the first equation it will suffice in view of 5(11) to show that $G := (\text{g-liminf}_\nu S^\nu)(\bar{x}) \subset (\text{p-liminf}_\nu S^\nu)(\bar{x})$. For any $\bar{u} \in G$ there's a sequence $(x^\nu, u^\nu) \to (\bar{x}, \bar{u})$ with $u^\nu \in S(x^\nu)$. Take $\rho > |u^\nu|$ for all ν and consider any $\varepsilon > 0$. Equi-outer semicontinuity at $\bar{x}$ implies the existence of $V \in \mathcal{N}(\bar{x})$ and $N_0 \in \mathcal{N}_\infty$ such that $S^\nu(x^\nu) \cap \rho\mathbb{B} \subset S^\nu(\bar{x}) + \varepsilon\mathbb{B}$ when $x^\nu \in V$ and $\nu \in N_0$, or equivalently (since $x^\nu \to \bar{x}$), for all $\nu \in N \subset N_0$ for some index set $N \in \mathcal{N}_\infty$ such that $x^\nu \in V$ when $\nu \in N$. This means that $u^\nu \in S^\nu(\bar{x}) + \varepsilon\mathbb{B}$ for all $\nu \in N$. Since such a $N \in \mathcal{N}_\infty$ exists for every $\varepsilon > 0$, we may conclude, from the inner limit formula in 4(2), that $\bar{u} \in \liminf_\nu S^\nu(\bar{x})$.

The proof of the second equation is identical, except N is taken to belong to $\mathcal{N}_\infty^\#$ rather than $\mathcal{N}_\infty$.

For the rest, we can redefine $S^\nu(x)$ to be empty outside of X if necessary in order to reduce without loss of generality to the case of $X = \mathbb{R}^n$. Then the implication (a)+(b) $\Rightarrow$ (c) and the implication (a)+(c) $\Rightarrow$ (b) both follow directly from the identities just established. There remains only to show that (b)+(c) $\Rightarrow$ (a). Suppose this isn't true, i.e., that despite both (b) and (c) holding the sequence fails to be asymptotically equi-osc at $\bar{x}$: there exist $\varepsilon > 0$, $\rho > 0$, $N \in \mathcal{N}_\infty^\#$ and $x^\nu \underset{N}{\to} \bar{x}$ such that

$$S^\nu(x^\nu) \cap \rho\mathbb{B} \not\subset S^\nu(\bar{x}) + \varepsilon\mathbb{B} \text{ when } \nu \in N.$$

In this case, for each $\nu \in N$ we can choose $u^\nu \in S^\nu(x^\nu)$ with $|u^\nu| \leq \rho$ but $u^\nu \notin S^\nu(\bar{x}) + \varepsilon\mathbb{B}$. The sequence $\{u^\nu\}_{\nu \in N}$ then has a cluster point $\bar{u}$, which by virtue of (b) must belong to $S(\bar{x})$. Yet $u^\nu + \varepsilon\mathbb{B}$ doesn't meet $S^\nu(\bar{x})$, which converges to $S(\bar{x})$ under (c). Hence $\bar{u} \notin S(\bar{x})$, a contradiction. □

G. Continuous and Uniform Convergence

Two other concepts of convergence, which are closely related, will further light up the picture of graphical convergence.

5.41 Definition (continuous and uniform limits of mappings). *A sequence of mappings $S^\nu : \mathbb{R}^n \rightrightarrows \mathbb{R}^m$ is said to converge continuously to a mapping S at $\bar{x}$ if $S^\nu(x^\nu) \to S(\bar{x})$ for all sequences $x^\nu \to \bar{x}$. If this holds at all $\bar{x} \in \mathbb{R}^n$, the sequence S^ν converges continuously to S. It does so relative to a set $X \subset \mathbb{R}^n$ if this holds at all $\bar{x} \in X$ when $x^\nu \in X$.*

The mappings S^ν converge uniformly to S on a subset X if for every $\varepsilon > 0$ and $\rho > 0$ there exists $N \in \mathcal{N}_\infty$ such that

$$\left.\begin{array}{l} S^\nu(x) \cap \rho I\!B \subset S(x) + \varepsilon I\!B \\ S(x) \cap \rho I\!B \subset S^\nu(x) + \varepsilon I\!B \end{array}\right\} \text{ for all } x \in X \text{ when } \nu \in N.$$

The inclusion property for uniform convergence should be compared to the one automatically present by virtue of Theorem 4.10 when S^ν converges to S pointwise at an individual point $\bar{x}$: for every $\varepsilon > 0$ and $\rho > 0$ there exists $N \in \mathcal{N}_\infty$ such that

$$\left.\begin{array}{l} S^\nu(\bar{x}) \cap \rho I\!B \subset S(\bar{x}) + \varepsilon I\!B \\ S(\bar{x}) \cap \rho I\!B \subset S^\nu(\bar{x}) + \varepsilon I\!B \end{array}\right\} \text{ when } \nu \in N.$$

By the same token, continuous convergence of S^ν to S at $\bar{x}$ can be identified with the condition that for every $\varepsilon > 0$ and $\rho > 0$ there exists $N \in \mathcal{N}_\infty$ along with a neighborhood $V \in \mathcal{N}(\bar{x})$ such that

$$\left.\begin{array}{l} S^\nu(x) \cap \rho I\!B \subset S(\bar{x}) + \varepsilon I\!B \\ S(\bar{x}) \cap \rho I\!B \subset S^\nu(x) + \varepsilon I\!B \end{array}\right\} \text{ for all } x \in V \text{ when } \nu \in N.$$

For continuous convergence relative to X, x must of course be restricted to X.

Another way of looking at these notions is through distance functions.

5.42 Exercise (distance function descriptions of convergence). *For mappings $S, S^\nu : I\!R^n \rightrightarrows I\!R^m$ with S closed-valued, one has*

(a) *S^ν converges pointwise to S at a point $\bar{x}$ if and only if, for each u, $d(u, S^\nu(\bar{x})) \to d(u, S(\bar{x}))$ as $\nu \to \infty$;*

(b) *S^ν converges continuously to S at a point $\bar{x}$ if and only if, for each u, $d(u, S^\nu(x)) \to d(u, S(\bar{x}))$ as $\nu \to \infty$ and $x \to \bar{x}$;*

(c) *S^ν converges uniformly to S on a set X if and only if, for each $u \in I\!R^m$ and $\eta > 0$ the sequence of functions $h^\nu(x) = \min\{d(u, S^\nu(x)), \eta\}$ converges uniformly on X to $h(x) = \min\{d(u, S(x)), \eta\}$.*

Guide. The relationship between set convergence and pointwise convergence of distance functions in 4.7 easily yields (a), (b) and the necessity of the function convergence in (c). For the sufficiency in (c) one has to appeal also to the continuity of distance functions (in 1.20). The role of η is to handle the possibility of $S^\nu(x)$ or $S(x)$ being empty. □

It's easy to extend 5.42(b) to continuous convergence relative to a set X: simply restrict x to X in taking limits $x \to \bar{x}$.

Beyond the characterizations of convergence in 5.42, there are others that can be based on the distance expressions $d\!l_\rho$ and $\hat{d\!l}_\rho$ introduced in 4(11). Indeed, the pair of inclusions used above in defining uniform convergence can be written as $\hat{d\!l}_\rho(S^\nu(x), S(x)) \leq \varepsilon$, whereas the ones used in defining continuous convergence correspond to having $\hat{d\!l}_\rho(S^\nu(x), S(\bar{x})) \leq \varepsilon$. Here $d\!l_\rho$ could substitute for $\hat{d\!l}_\rho$ by virtue of the inequalities in 4.55(a), which bracket these expressions relative to each other as ρ varies.

Continuous convergence is the 'pointwise localization of uniform convergence'. This is made precise in the next theorem, which extends to the context of set-valued mappings some facts that are well known for functions.

5.43 Theorem (continuous versus uniform convergence). *For mappings $S, S^\nu : I\!R^n \rightrightarrows I\!R^m$ and a set $X \subset I\!R^n$, the following conditions are equivalent:*

(a) *S^ν converges continuously to S relative to X;*

(b) *S^ν converges uniformly to S on all compact subsets of X, and S is continuous relative to X. Here the continuity of S is automatic from the uniform convergence if each S^ν is continuous relative to X and each $x \in X$ has a compact neighborhood relative to X.*

In general, whenever S^ν converges continuously to S relative to a set X at $\bar{x} \in X$ and at the same time converges pointwise to S on a neighborhood of $\bar{x}$ relative to X, S must be continuous at $\bar{x}$ relative to X.

Proof. Through 5.42, the entire argument can be translated to the setting of a bounded sequence of functions h^ν converging in one way or another to a function h. Specifically, uniform convergence of S^ν to S on X is characterized in 5.42(c) as meaning that for each $u \in I\!R^m$ and $\rho \in I\!R_+$ the sequence of functions h^ν defined in 5.42(c) converges uniformly on X to the function h defined in 5.42(c). On the other hand, continuous convergence of S^ν to S at $\bar{x}$ relative to X is equivalent through 5.42(b) to the property that, again for each $u \in I\!R^m$ and $\rho \in I\!R_+$, h^ν converges continuously to the function h at $\bar{x}$ relative to X. Next, S is continuous relative to X at $\bar{x}$ if and only if, for all $u \in I\!R^m$ and $\rho \in I\!R_+$, the function h in 5.42(c) is continuous at $\bar{x}$ relative to X. Finally, S^ν converges pointwise to S on a set $V \cap X$ if and only if, in the same context, h^ν converges pointwise to h on $V \cap X$.

Proceeding to the task as translated, we consider $h, h^\nu : X \to [0, \rho]$. Suppose h^ν converges continuously to h at $\bar{x}$ relative to X at $\bar{x}$ and at the same time pointwise on $V \cap X$ for a neighborhood $V \in \mathcal{N}(\bar{x})$. We'll prove that h must be continuous at $\bar{x}$. Consider any $\varepsilon > 0$. By continuous convergence there exist $V' \in \mathcal{N}(\bar{x})$ and $N \in \mathcal{N}_\infty$ such that $\left|h^\nu(x) - h(\bar{x})\right| \leq \varepsilon$ when $x \in V'$ and $\nu \in N$. Then because $h^\nu(x) \to h(x)$ on $V \cap X$ we have $\left|h(x) - h(\bar{x})\right| \leq \varepsilon$ for $x \in V' \cap V \cap X$ and $\nu \in N$. Since $V' \cap V \in \mathcal{N}(\bar{x})$, this establishes the claim.

Suppose now that h^ν converges continuously to h relative to X. In particular, h^ν converges pointwise to h on X, so by the argument just given, h must be continuous relative to X. Let B be any compact subset of X. We'll verify that h^ν converges uniformly to h on B. If not,

$$\exists\, \varepsilon > 0, N_0 \in \mathcal{N}_\infty^\# \text{ such that } \left\{x \in B \,\middle|\, |h^\nu(x) - h(x)| \geq \varepsilon\right\} \neq \emptyset \text{ for all } \nu \in N_0.$$

Choosing for each $\nu \in N_0$ an element x^ν of this set we would get a sequence in B, which by compactness would have a cluster point $\bar{x} \in B$. There would be an index set $N \in \mathcal{N}_\infty^\#$ within N_0 such that $x^\nu \underset{N}{\to} \bar{x}$. Then $h^\nu(x^\nu) \underset{N}{\to} h(\bar{x})$ by the assumed continuous convergence, while also $h(x^\nu) \underset{N}{\to} h(\bar{x})$ by the continuity of h, in contradiction to x^ν having been selected with $\left|h^\nu(x^\nu) - h(x^\nu)\right| \geq \varepsilon$.

Conversely, suppose now instead that h is continuous relative to X and that h^ν converges to h uniformly on every compact subset B of X. We'll prove that h^ν converges continuously to h relative to X. Consider any point $\bar{x} \in X$ and any sequence $x^\nu \to \bar{x}$ in X. Let B be the compact subset of X consisting of $\bar{x}$ and all the points x^ν. Consider any $\varepsilon > 0$. By the continuity of h relative to X there exists $V \in \mathcal{N}(\bar{x})$ such that

$$\big|h(x) - h(\bar{x})\big| \leq \varepsilon/2 \text{ for all } x \in V \cap B.$$

Next, from the uniform convergence, there exists $N \in \mathcal{N}_\infty$ such that

$$\big|h^\nu(x) - h(x)\big| \leq \frac{\varepsilon}{2} \text{ for all } x \in B \text{ when } \nu \in N.$$

In combination we obtain $\big|h^\nu(x) - h(\bar{x})\big| \leq \varepsilon$ for all $x \in V \cap B$ when $\nu \in N$, which signals continuous convergence on B, hence $h^\nu(x^\nu) \to h(\bar{x})$.

Finally, suppose that the functions h^ν are continuous and that they converge uniformly to h on all compact subsets of X, and consider any $\bar{x} \in X$. We need to demonstrate that h is continuous at $\bar{x}$ relative to X, provided that $\bar{x}$ has a compact neighborhood B relative to X. On such a neighborhood B, the functions h^ν converge uniformly to h: for any $\varepsilon > 0$ there's an index $\bar{\nu} \in I\!N$ such that $\big|h^{\bar{\nu}}(x) - h(x)\big| \leq \varepsilon/3$ for all $x \in B$. But also by the continuity of $h^{\bar{\nu}}$ relative to X there's a $\delta > 0$ such that $\big|h^{\bar{\nu}}(x) - h^{\bar{\nu}}(\bar{x})\big| \leq \varepsilon/3$ for all $x \in B$ with $|x - \bar{x}| \leq \delta$. We can choose δ small enough that all the points $x \in X$ with $|x - \bar{x}| \leq \delta$ belong to B. For such points x we then have

$$\begin{aligned}\big|h(x) - h(\bar{x})\big| &\leq \big|h(x) - h^{\bar{\nu}}(x)\big| + \big|h^{\bar{\nu}}(x) - h^{\bar{\nu}}(\bar{x})\big| + \big|h^{\bar{\nu}}(\bar{x}) - h(\bar{x})\big| \\ &\leq (\varepsilon/3) + (\varepsilon/3) + (\varepsilon/3) = \varepsilon.\end{aligned}$$

Since we were able to get this for any $\varepsilon > 0$ by taking $\delta > 0$ sufficiently small, we conclude that h is continuous at $\bar{x}$ relative to X. □

5.44 Theorem (graphical versus continuous convergence). *For mappings $S, S^\nu : I\!R^n \rightrightarrows I\!R^m$ and a set $X \subset I\!R^n$, the following properties at $\bar{x} \in X$ are equivalent:*

(a) *S^ν converges continuously to S at $\bar{x}$ relative to X;*

(b) *S^ν converges graphically to S at $\bar{x}$ relative to X, and the sequence is asymptotically equicontinuous at $\bar{x}$ relative to X.*

Proof. Suppose first that (a) holds. Obviously this condition implies that S^ν converges both pointwise and graphically to S at $\bar{x}$ relative to X, and therefore by 5.40 that the sequence is asymptotically equi-osc relative to X. We must show that the sequence is also asymptotically equi-isc relative to X. If not, there would exist $\varepsilon > 0$, $\rho > 0$, $N \in \mathcal{N}_\infty^\#$ and $x^\nu \underset{N}{\to} \bar{x}$ in X such that

$$S^\nu(\bar{x}) \cap \rho I\!B \not\subset S^\nu(x^\nu) + \varepsilon I\!B \text{ when } \nu \in N.$$

It would be possible then to choose for each $\nu \in N$ an element $u^\nu \in S^\nu(\bar{x})$ with $|u^\nu| \leq \rho$ and $u^\nu \notin S^\nu(x^\nu) + \varepsilon I\!B$. Such a sequence would have a cluster point $\bar{u}$, necessarily in $S(\bar{x})$ because $S^\nu(\bar{x}) \to S(\bar{x})$, yet this is impossible when $u^\nu + \varepsilon\mathbb{B}$ doesn't meet $S^\nu(x^\nu)$, which also converges to $S(\bar{x})$. The contradiction establishes the required equicontinuity.

Now suppose that (b) holds. Consider any sequence $x^\nu \to \bar{x}$ in X. Graphical convergence yields that $\limsup_\nu S^\nu(x^\nu) \subset S(\bar{x})$, so the burden of our effort is to show for arbitrary $\bar{u} \in S(\bar{x})$ that $\bar{u} \in \liminf_\nu S^\nu(x^\nu)$. Because the sequence of mappings is asymptotically equi-osc, we have $S^\nu(\bar{x}) \to S(\bar{x})$ by 5.40, so for indices ν in some set $N_0 \in \mathcal{N}_\infty$ we can find $u^\nu \in S^\nu(\bar{x})$ with $u^\nu \underset{N}{\to} \bar{u}$. Take ρ large enough that $u^\nu \in \rho\mathbb{B}$ for all $\nu \in N$. Let $\varepsilon > 0$. Because the sequence is asymptotically equi-isc, there exist $V \in \mathcal{N}(\bar{x})$ and $N_1 \in \mathcal{N}_\infty$ with the property that

$$S^\nu(\bar{x}) \cap \rho\mathbb{B} \subset S^\nu(x) + \varepsilon\mathbb{B} \text{ for all } x \in V \text{ when } \nu \in N_1.$$

Then for some $N \in \mathcal{N}_\infty$ with $N \subset N_0 \cap N_1$ we have $u^\nu \in S^\nu(x^\nu) + \varepsilon\mathbb{B}$ for all $\nu \in N$. We have shown that for arbitrary $\varepsilon > 0$ there exists $N \in \mathcal{N}_\infty$ with this property, and therefore through 4(2) that $\bar{u} \in \liminf_\nu S^\nu(x^\nu)$. □

5.45 Corollary (graphical convergence of single-valued mappings). *For single-valued mappings $F, F^\nu : \mathbb{R}^n \to \mathbb{R}^m$, the following conditions are equivalent:*

(a) *F^ν converges continuously to F at $\bar{x}$;*

(b) *F^ν converges graphically to F at $\bar{x}$, and the sequence is eventually locally bounded at $\bar{x}$, i.e., there exist $V \in \mathcal{N}(\bar{x})$, $N \in \mathcal{N}_\infty$ and a bounded set B such that $F^\nu(x) \in B$ for all $x \in V$ when $\nu \in N$.*

Proof. It's evident from 5.44 and the definition of continuous convergence that (a) implies (b). On the other hand, the local boundedness in (b) implies that every sequence $F^\nu(x^\nu)$ with $x^\nu \to \bar{x}$ is bounded, while the graphical convergence in (b) along with the single-valuedness of F ensures that the only possible cluster point of such a sequence is $F(\bar{x})$. Hence in (b), $F^\nu(x^\nu) \to F(\bar{x})$ whenever $x^\nu \to \bar{x}$. □

5.46 Proposition (graphical convergence from uniform convergence). *Consider $S^\nu, S : \mathbb{R}^n \rightrightarrows \mathbb{R}^m$ and a set $X \subset \mathbb{R}^n$.*

(a) *If the mappings S^ν are osc relative to X and converge uniformly to S on X, then S is osc relative to X and $S^\nu \xrightarrow{g} S$ relative to X.*

(b) *If the mappings S^ν are continuous relative to X and converge uniformly to S on all compact subsets of X, and if each point of X has a compact neighborhood relative to X (as is true when X is closed or open), then S^ν converges graphically to S relative to X. In particular, this holds when S^ν and S are single-valued on X.*

Proof. In (a), let's begin by verifying that S is osc relative to X. Suppose $\lim_{\kappa\to\infty}(x^\kappa, u^\kappa) = (\bar{x}, \bar{u})$ with $x^\kappa, \bar{x} \in X$ and $u^\kappa \in S(x^\kappa)$. We have to show that $\bar{u} \in S(\bar{x})$. Pick ρ large enough that $u^\kappa, \bar{u} \in \operatorname{int} \rho\mathbb{B}$ and fix $\varepsilon > 0$. The uniform convergence of S^ν to S on X gives us the existence of $N \in \mathcal{N}_\infty$ such that (through the second inclusion in Definition 5.41):

$$u^\kappa \in S(x^\kappa) \cap \rho I\!B \subset S^\nu(x^\kappa) + \varepsilon I\!B \text{ for all } \kappa \in I\!N \text{ when } \nu \in N.$$

Taking the limsup with respect to κ for each fixed $\nu \in N$, we get

$$\bar{u} \in \operatorname{limsup}_\kappa \left[S^\nu(x^\kappa) + \varepsilon I\!B\right] = \left[\operatorname{limsup}_\kappa S^\nu(x^\kappa)\right] + \varepsilon I\!B \subset S^\nu(\bar{x}) + \varepsilon I\!B;$$

here the equality follows from $\varepsilon I\!B$ being compact, while the inclusion follows from S^ν being osc relative to X. Because $|\bar{u}| < \rho$, this allows us to conclude that $d(\bar{u}, S(\bar{x})) \leq 2\varepsilon$, since by the first inclusion in Definition 5.41 we necessarily have, for ν sufficiently large, that $S^\nu(\bar{x}) \cap \rho I\!B \subset S(\bar{x}) + \varepsilon I\!B$. This implies that $\bar{u} \in S(\bar{x})$, inasmuch as $\varepsilon > 0$ can be chosen arbitrarily small.

To get $S^\nu \xrightarrow{g} S$ relative to X, we must show for $G := (X \times I\!R^m) \cap \operatorname{gph} S$ and $G^\nu := (X \times I\!R^m) \cap \operatorname{gph} S^\nu$ that

$$(X \times I\!R^m) \cap \operatorname{limsup}_\nu G^\nu \subset G \subset \operatorname{liminf}_\nu G^\nu. \tag{5(13)}$$

For the first inclusion, suppose $(x^\nu, u^\nu) \to (\bar{x}, \bar{u})$ with $\bar{x} \in X$ and $(x^\nu, u^\nu) \in G^\nu$ for all ν in some $N_0 \in \mathcal{N}_\infty$. Choose ρ large enough that $u^\nu, \bar{u} \in \operatorname{int} \rho I\!B$. The uniform convergence of S^ν to S on X yields for any $\varepsilon > 0$ the existence of $N \in \mathcal{N}_\infty$ such that $N \subset N_0$ and (through the first inclusion in Definition 5.41):

$$u^\nu \in S^\nu(x^\nu) \cap \rho I\!B \subset S(x^\nu) + \varepsilon I\!B \text{ for all } \nu \in N.$$

Since S is osc relative to X, as just demonstrated, we obtain on taking limsup with respect to ν that $\bar{u} \in S(\bar{x}) + \varepsilon I\!B$, where again $\limsup_\nu \left[S(x^\nu) + \varepsilon I\!B\right] = \left[\limsup_\nu S(x^\nu)\right] + \varepsilon I\!B$. The choice of $\varepsilon > 0$ being arbitrary, we get $\bar{u} \in S(\bar{x})$, hence $(\bar{x}, \bar{u}) \in G$. Thus, the first inclusion in 5(13) is correct.

For the second inclusion in 5(13), consider any $(\bar{u}, \bar{x}) \in G$ and any $\rho > |\bar{u}|$. The uniform convergence provides for any $\kappa \in I\!N$ a set $N_\kappa \in \mathcal{N}_\infty$ such that (through the second inclusion in Definition 5.41):

$$\bar{u} \in S(\bar{x}) \cap \rho I\!B \subset S^\nu(\bar{x}) + (1/\kappa) I\!B \text{ for all } \nu \in N_\kappa.$$

Then $d(\bar{u}, S^\nu(\bar{x})) \leq 1/\kappa$ for $\nu \in N_\kappa$, hence $\limsup_\nu d(\bar{u}, S^\nu(\bar{x})) = 0$. It follows that we can find $u^\nu \in S^\nu(\bar{x})$ with $u^\nu \to \bar{u}$. Setting $x^\nu = \bar{x}$ for all $\nu \in I\!N$, we obtain $(x^\nu, u^\nu) \in G^\nu$ with $(x^\nu, u^\nu) \to (\bar{x}, \bar{u})$, as required.

Part (b) is obtained immediately from the combination of Theorem 5.44 with Theorem 5.43. □

The next two convergence theorems extend well known facts about single-valued mappings to the framework of set-valued mappings.

5.47 Theorem (Arzelà-Ascoli, set-valued version). *If a sequence of mappings $S^\nu : I\!R^n \rightrightarrows I\!R^m$ is asymptotically equicontinuous relative to a set X, it admits a subsequence converging uniformly on all compact subsets of X to a mapping $S : I\!R^n \rightrightarrows I\!R^m$ that is continuous relative to X.*

Proof. This combines 5.44 with the compactness property in 5.36 and the characterization of continuous convergence in 5.43. □

In the following we write $S^\nu(x) \nearrow S(x)$ to mean that $S^\nu(x) \to S(x)$ with $S^\nu(x) \subset S^{\nu+1}(x) \subset \cdots$, and similarly we write $S^\nu(x) \searrow S(x)$ when the opposite inclusions hold; see 4.3 for such monotone set convergence.

5.48 Theorem (uniform convergence of monotone sequences). *Consider set-valued mappings $S, S^\nu : \mathbb{R}^n \rightrightarrows \mathbb{R}^m$ and a set $X \subset \operatorname{dom} S$. Suppose either*

(a) *$S^\nu(x) \nearrow S(x)$ for all $x \in X$, with S osc and S^ν isc relative to X, or*

(b) *$S^\nu(x) \searrow S(x)$ for all $x \in X$, with S isc and S^ν osc relative to X.*

Then S must actually be continuous relative to X, and the mappings S^ν must converge uniformly to S on every compact subset of X.

Proof. It suffices by Theorem 5.43 to demonstrate that, under either of these hypotheses, S^ν converges continuously to S relative to X. For each $x \in X$, $S(x)$ has to be closed, and there's no loss of generality in assuming $S^\nu(x)$ to be closed too. The distance characterizations of semicontinuity in 5.11 and continuous convergence in 5.42(b) then provide a bridge to a simpler context.

For (a), fix any $u \in \mathbb{R}^m$ and let $d^\nu(x) = d\big(u, S^\nu(x)\big)$ and $d(x) = d\big(u, S(x)\big)$. By assumption we have $d^\nu(x) \searrow d(x)$ for all $x \in X$, with d lsc and d^ν usc relative to X. We must show that $d^\nu(x^\nu) \to d(\bar{x})$ whenever $x^\nu \to \bar{x}$ in X. For any $\varepsilon > 0$ there's an index ν_0 such that $d^\nu(\bar{x}) \le d(\bar{x}) + \varepsilon$ when $\nu \ge \nu_0$. Further, there's a neighborhood V of $\bar{x}$ relative to X such that $d(x) \ge d(\bar{x}) - \varepsilon$ and $d^{\nu_0}(x) \le d^{\nu_0}(\bar{x}) + \varepsilon$ when $x \in V$. Next, there's an index $\nu_1 \ge \nu_0$ such that $x^\nu \in V$ when $\nu \ge \nu_1$. Putting these properties together, we see for $\nu \ge \nu_1$ that $d^\nu(x^\nu) \ge d(x^\nu) \ge d(\bar{x}) - \varepsilon$, and on the other hand, $d^\nu(x^\nu) \le d^{\nu_0}(x^\nu) \le d^{\nu_0}(\bar{x}) + \varepsilon \le d(\bar{x}) + 2\varepsilon$. Hence $\big|d^\nu(x^\nu) - d(\bar{x})\big| \le 2\varepsilon$ when $\nu \ge \nu_1$.

For (b), the argument is the same but the inequalities are reversed because, instead, $d^\nu(x) \nearrow d(x)$ for $x \in X$, with d usc and d^ν lsc relative to X. □

When the mappings S and S^ν are assumed to be continuous relative to X in Theorem 5.48, cases (a) and (b) coalesce, and the conclusion that S^ν converges uniformly to S on compact subsets of X is analogous to Dini's theorem on the convergence of monotone sequences of continuous real-valued functions.

H.* Metric Descriptions of Convergence

Continuous convergence and uniform convergence of set-valued mappings can also be characterized with the metric for set convergence that was developed in the last part of Chapter 4.

5.49 Proposition (metric version of continuous and uniform convergence).

(a) *A sequence of mappings $S^\nu : \mathbb{R}^n \rightrightarrows \mathbb{R}^m$ converges continuously at $\bar{x}$ to a mapping S if and only if, for all $\varepsilon > 0$, there exist $V \in \mathcal{N}(\bar{x})$ and $N \in \mathcal{N}_\infty$ such that, for all $x \in V$ and $\nu \in N$, one has $d\!\!l\big(S^\nu(x), S(\bar{x})\big) \le \varepsilon$.*

(b) *A sequence of mappings $S^\nu : \mathbb{R}^n \rightrightarrows \mathbb{R}^m$ converges uniformly on a set $X \subset \mathbb{R}^n$ to a mapping S if and only if, for all $\varepsilon > 0$, there exist $N \in \mathcal{N}_\infty$ such that, for all $\nu \in N$ and $x \in X$, one has $d\!\!l\big(S^\nu(x), S(x)\big) \le \varepsilon$.*

Proof. We confine ourselves to proving (b); the proof of (a) proceeds on the same lines. We appeal to estimates in terms of $\hat{d\!l}_\rho$ and what they say for $d\!l_\rho$. From the definition of $\hat{d\!l}_\rho$, it's immediate that the sequence $\{S^\nu\}_{\nu\in\mathbb{N}}$ converges uniformly to S on X if and only if

$$\forall\,\varepsilon>0,\,\forall\,\rho\geq 0,\,\exists\,N\in\mathcal{N}_\infty:\quad \hat{d\!l}_\rho\big(S^\nu(x),S(x)\big)\leq\varepsilon\quad\forall\,x\in X,\,\forall\,\nu\in N.$$

The relationship between $\hat{d\!l}_\rho$ and $d\!l_\rho$ recorded in 4.37(a) allows us to rewrite this condition as

$$\forall\,\varepsilon>0,\,\forall\,\rho\geq 0,\,\exists\,N\in\mathcal{N}_\infty:\quad d\!l_\rho\big(S^\nu(x),S(x)\big)\leq\varepsilon\quad\forall\,x\in X,\,\forall\,\nu\in N.$$

It remains to be shown that this latter condition is equivalent to

$$\forall\,\varepsilon>0,\,\exists\,N\in\mathcal{N}_\infty:\quad d\!l\big(S^\nu(x),S(x)\big)\leq\varepsilon\quad\forall\,x\in X,\,\forall\,\nu\in N.$$

Let's begin with '$\Leftarrow$'; suppose it's false. Then for some $x\in X$, $\varepsilon>0$, $\rho\geq 0$ and $N\in\mathcal{N}_\infty^\#$, we have $d\!l_\rho\big(S^\nu(x),S(x)\big)>\varepsilon$ for all $\nu\in N$. The inequality in 4.41(a) implies then that $d\!l\big(S^\nu(x),S(x)\big)>\varepsilon'=e^{-\rho}\varepsilon$ for all $\nu\in N$, thus invalidating the assumption that for ε' there is $N'\in\mathcal{N}_\infty$ such that $d\!l\big(S^\nu(x),S(x)\big)\leq\varepsilon'$ for all $\nu\in N'$.

For the '$\Rightarrow$' part we likewise proceed by contradiction. Suppose that for some $x\in X$, there are $\varepsilon>0$ and $N\in\mathcal{N}_\infty^\#$ such that for all $\nu\in N$, $d\!l\big(S^\nu(x),S(x)\big)>\varepsilon$. The inequality in 4.41(b) for $d\!l$ implies that, for any $\rho\in\mathbb{R}_+$ and $\nu\in N$, we have

$$(1-e^{-\rho})d\!l_\rho\big(S^\nu(x),S(x)\big)+e^{-\rho}(\beta^\nu+\rho+1)\geq d\!l\big(S^\nu(x),S(x)\big)>\varepsilon,$$

where $\beta^\nu=\max\big\{d\big(0,S^\nu(x)\big),d\big(0,S(x)\big)\big\}$. By assumption, $d\!l_0\big(S^\nu(x),S(x)\big)$ is arbitrarily small for ν large enough, say less than ε; without loss of generality, this may be taken to be the case for all $\nu\in N$. Then, with $\beta=d\big(0,S(x)\big)$, since $\beta^\nu\leq\beta+\varepsilon$ for any $\rho\in\mathbb{R}_+$ and all $\nu\in N$, we have

$$d\!l_\rho\big(S^\nu(x),S(x)\big)>\varepsilon_\rho:=\frac{\varepsilon-e^{-\rho}(\beta+\varepsilon+\rho+1)}{1-e^{-\rho}}.$$

In fixing $\rho>0$ arbitrarily and taking $\varepsilon'=\varepsilon_\rho$, it follows that there couldn't be an index set $N'\in\mathcal{N}_\infty$ such that $d\!l_\rho\big(S^\nu(x),S(x)\big)\leq\varepsilon'$ for all $\nu\in N'$. $\square$

Graphical convergence of mappings can be quantified as well. Such convergence is identified with set convergence of graphs, so one can rely on a metric for set convergence in the space $\text{cl-sets}_{\neq\emptyset}(\mathbb{R}^n\times\mathbb{R}^m)$ to secure a metric for graph convergence in the space

$$\text{osc-maps}_{\neq\emptyset}(\mathbb{R}^n,\mathbb{R}^m):=\big\{S:\mathbb{R}^n\rightrightarrows\mathbb{R}^m\;\big|\;S\text{ osc},\;\operatorname{dom}S\neq\emptyset\big\}.$$

It's enough to introduce for any two mappings $S,T\in\text{osc-maps}_{\neq\emptyset}(\mathbb{R}^n,\mathbb{R}^m)$ the *graph distance*

$$d\!\!l(T,S) := d\!\!l(\operatorname{gph} S, \operatorname{gph} T)$$

along with, for $\rho \geq 0$, the expressions

$$d\!\!l_\rho(T,S) := d\!\!l_\rho(\operatorname{gph} S, \operatorname{gph} T), \qquad \hat{d\!\!l}_\rho(T,S) := \hat{d\!\!l}_\rho(\operatorname{gph} S, \operatorname{gph} T).$$

All the estimates and relationships involving $d\!\!l$, $d\!\!l_\rho$ and $\hat{d\!\!l}_\rho$ in the context of set convergence in 4.37–4.45 carry over immediately then to the context of graphical convergence.

5.50 Theorem (quantification of graphical convergence). *The graph distance $d\!\!l$ is a metric on the space* osc-maps$_{\neq\emptyset}(I\!R^n, I\!R^m)$, *whereas $d\!\!l_\rho$ is a pseudo-metric for each $\rho \geq 0$ (but $\hat{d\!\!l}_\rho$ is not). They all characterize graphical convergence:*

$$\begin{aligned} S^\nu \to S \quad &\iff \quad d\!\!l(S^\nu, S) \to 0 \\ &\iff \quad d\!\!l_\rho(S^\nu, S) \to 0 \ \textit{ for all } \rho \textit{ greater than some } \bar{\rho} \\ &\iff \quad \hat{d\!\!l}_\rho(S^\nu, S) \to 0 \ \textit{ for all } \rho \textit{ greater than some } \bar{\rho}. \end{aligned}$$

Moreover, $\big($osc-maps$_{\neq\emptyset}(I\!R^n, I\!R^m), d\!\!l\big)$ *is a separable, complete metric space that is locally compact.*

Proof. This merely translates 4.36, 4.42, 4.43 and 4.45 to the case where the sets involved are the graphs of osc mappings. □

In the quantification of set distances in Chapter 4, it was observed that the expressions could also be utilized without requiring sets to be closed. Similarly here, one can invoke $d\!\!l_\rho(S,T)$ and $\hat{d\!\!l}_\rho(S,T)$ even when S and T aren't osc, but of course these values are the same as $d\!\!l_\rho(\operatorname{cl} S, \operatorname{cl} T)$ and $\hat{d\!\!l}_\rho(\operatorname{cl} S, \operatorname{cl} T)$. In this sense we can speak of the graph distance between any two elements of

$$\operatorname{maps}(I\!R^n, I\!R^m) := \ \text{the space of all } S : I\!R^n \rightrightarrows I\!R^m.$$

For this larger space, however, $d\!\!l$ is not a metric but just a pseudo-metric.

Graph-distances can be used to obtain estimates between the solutions of generalized equations like those in Theorem 5.37, although we won't take this up here. For more about solution estimates in that setting, see Theorem 9.43 and its sequel.

I.* Operations on Mappings

Various operations can be used in constructing set-valued mappings, and questions arise about the extent to which these operations preserve properties of continuity and local boundedness. We now look into such matters systematically. First are some results about sums of mappings, which augment the ones in 5.24.

5.51 Proposition (addition of set-valued mappings). *Let $S_1, S_2 : I\!R^n \rightrightarrows I\!R^m$.*

(a) *$S_1 + S_2$ is locally bounded if both S_1 and S_2 are locally bounded.*

(b) *$S_1 + S_2$ is osc if the mappings S_1 and S_2 are osc and the mapping $(x, y) \mapsto S_1(x) \cap [y - S_2(x)]$ is locally bounded, the latter being true certainly if either S_1 or S_2 is locally bounded.*

(c) *$S_1 + S_2$ is totally continuous at $\bar{x}$ if S_1 and S_2 are totally continuous at $\bar{x}$ with $[S_1(\bar{x}) \times S_2(\bar{x})]^\infty = S_1(\bar{x})^\infty \times S_2(\bar{x})^\infty$, and $S_1(\bar{x})^\infty \cap [-S_2(\bar{x})^\infty] = \{0\}$. Both of these conditions are satisfied if either S_1 or S_2 is locally bounded at $\bar{x}$, whereas the first is satisfied if $S_1(\bar{x})$ and $S_2(\bar{x})$ are convex sets.*

Proof. In (a), the local boundedness of S_1 and S_2 is equivalent to having $S_1(B)$ and $S_2(B)$ bounded whenever B is bounded (by 5.15). The elementary fact that $(S_1 + S_2)(B) \subset S_1(B) + S_2(B)$ then gives us $(S_1 + S_2)(B)$ bounded whenever B is bounded.

In (b), the mapping $T : (x, y) \mapsto S_1(x) \cap [y - S_2(x)]$ has $\operatorname{dom} T = \operatorname{gph}(S_1 + S_2)$. This mapping is osc: its graph is the intersection of the set $\{(x, y, u_1) \mid (x, u_1) \in \operatorname{gph} S_1\}$ with the set $\{(x, y, y - u_2) \mid (x, u_2) \in \operatorname{gph} S_2\}$, both being closed since S_1 and S_2 are osc. When T is also locally bounded, $\operatorname{dom} T$ has to be closed, and this means that $S_1 + S_2$ is osc.

Similar reasoning establishes that the mapping T in the argument for (a) is locally bounded when either S_1 or S_2 is locally bounded, because $T(B, B') \subset S_1(B) \cap (B' - S_2(B))$ for any bounded sets $B \subset I\!R^n$ and $B' \subset I\!R^m$.

In (c), we simply invoke the result in 4.29 on convergence of sums of sets, but do so in the context of total continuity in 5.28. Then we make use of the facts in 3.12 about the closure of a sum of sets. □

5.52 Proposition (composition of set-valued mappings). *Consider the mapping $T \circ S : I\!R^n \rightrightarrows I\!R^p$ for $S : I\!R^n \rightrightarrows I\!R^m$ and $T : I\!R^m \rightrightarrows I\!R^p$.*

(a) *$T \circ S$ is locally bounded if both S and T are locally bounded.*

(b) *$T \circ S$ is osc if S and T are osc and the mapping $(x, w) \mapsto S(x) \cap T^{-1}(w)$ is locally bounded, as is true when either S or T^{-1} is locally bounded.*

(c) *$T \circ S$ is continuous if S and T are continuous and S is locally bounded.*

(d) *$T \circ S$ is continuous at $\bar{x}$ if S is continuous at $\bar{x}$ and T is continuous with T^{-1} locally bounded, or alternatively, if S it totally continuous at $\bar{x}$ and T is continuous with $(T^\infty)^{-1}(0) \cap S(\bar{x})^\infty = \{0\}$.*

(e) *$T \circ S$ is totally continuous at $\bar{x}$ if S is totally continuous at $\bar{x}$, T is totally continuous, $(T^\infty)^{-1}(0) \cap S(\bar{x})^\infty = \{0\}$, and $(T^\infty)(S(\bar{x})^\infty) \subset T(S(\bar{x}))^\infty$.*

Proof. For (a), we can rely on the criterion in 5.15: If S and T are both locally bounded, we know for every bounded set $B \subset I\!R^n$ that $S(B)$ is bounded in $I\!R^m$ and therefore that $T(S(B)) = (T \circ S)(B)$ is bounded in $I\!R^p$.

For (b), denote the set-valued mapping $(x, w) \mapsto S(x) \cap T^{-1}(w)$ by R. The graph of R is closed, because it consists of all (x, w, u) such that (x, u, w) belongs to $[(\operatorname{gph} S) \times I\!R^p] \cap [I\!R^n \times \operatorname{gph} T]$; the latter is closed by the outer semicontinuity of S and T. Thus, R is osc. Under the assumption that R is locally bounded as well, $\operatorname{dom} R$ is closed by 5.25(a). But $\operatorname{dom} R = \operatorname{gph}(T \circ S)$. Therefore, $T \circ S$ is osc under this assumption.

Assertions (d) and (e) are immediate from 5.30(c) and (d) and the definitions of continuity and total continuity, while (c) is a special case of the second version of (d) applied at every point $\bar{x}$. □

As a special case in 5.52, either S or T might be single-valued, of course. Continuity in the usual sense could then be substituted for continuity or outer semicontinuity as a set-valued mapping, cf. 5.20.

Other results can similarly be derived from the ones in Chapter 4 on the calculus of convergent sets. For example, if $S(x) = S_1(x) \cap S_2(x)$ for continuous, convex-valued mappings S_1 and S_2, and if $\bar{x}$ is a point where $S_1(\bar{x})$ and $S_2(\bar{x})$ can't be separated, then S is continuous at $\bar{x}$. This is clear from 4.32(c).

In a similar vein are results about the preservation of continuity of set-valued mappings under various forms of convergence of mappings.

Let's now turn to the convergence of images and the preservation of graphical convergence under various operations. The conditions we'll need involve restrictions on how sequences of points and sets can escape to the horizon, and consequently they will draw on the concept of total set convergence (cf. Definition 4.23). By *total graph convergence* $S^\nu \xrightarrow{t} S$ one means that $\operatorname{gph} S^\nu \xrightarrow{t} \operatorname{gph} S$. By appealing to horizon limits and to S^∞, the horizon mapping associated with S, this mode of convergence is supplied with a characterization that is well suited to various manipulations. Namely, it follows from the description of total set convergence in 4.24 that

$$S^\nu \xrightarrow{t} S \quad \Longleftrightarrow \quad S^\nu \xrightarrow{g} S, \quad \operatorname{limsup}_\nu^\infty \operatorname{gph} S^\nu \subset \operatorname{gph} S^\infty. \tag{5(14)}$$

From the criteria collected in Proposition 4.25, one has that $S^\nu \xrightarrow{g} S$ entails the stronger property $S^\nu \xrightarrow{t} S$ whenever the graphs $\operatorname{gph} S^\nu$ are convex sets or cones, or are nondecreasing or uniformly bounded. Moreover, again from the developments in Chapter 4, one has

$$S_1^\nu \xrightarrow{t} S_1,\ S_2^\nu \xrightarrow{t} S_2 \quad \Longrightarrow \quad S_1^\nu \cup S_2^\nu \xrightarrow{t} S_1 \cup S_2,$$

and provided that $(S_1 \times S_2)^\infty = S_1^\infty \times S_2^\infty$,

$$S_1^\nu \xrightarrow{t} S_1,\ S_2^\nu \xrightarrow{t} S_2 \quad \Longrightarrow \quad S_1^\nu \times S_2^\nu \xrightarrow{t} S_1 \times S_2.$$

It should be noted that total convergence $S^\nu \xrightarrow{t} S$ is not ensured by S^ν converging continuously to S. A counterexample is provided by the mappings $S^\nu : \mathbb{R}^1 \rightrightarrows \mathbb{R}^1$ defined by $S^\nu(x) = \{0, \nu\}$ when $x = 1/\nu$ but $S^\nu(x) = \{0\}$ for all other x. Taking $S(x) = \{0\}$ for all x we get $S^\nu(x^\nu) \to S(x)$ whenever $x^\nu \to x$, yet it's not true that $\operatorname{gph} S^\nu \xrightarrow{t} \operatorname{gph} S$.

A series of criteria (4.21–4.28) have already been provided for the preservation of set limits under certain mappings (linear, addition, etc.). We now consider the images of sets under an arbitrary mapping $S : \mathbb{R}^n \rightrightarrows \mathbb{R}^m$ as well as the images of sets under converging sequences of mappings.

5.53 Theorem (images of converging mappings). *Let $S, S^\nu : I\!R^n \rightrightarrows I\!R^d$ be mappings and C, C^ν subsets of $I\!R^n$.*

(a) $\liminf_\nu S^\nu(C^\nu) \supset S(C)$ *whenever* $\liminf_\nu C^\nu \supset C$ *and* S^ν *converges continuously to* S.

(b) $\limsup_\nu S^\nu(C^\nu) \subset S(C)$ *whenever* $\limsup_\nu C^\nu \subset C$, $\limsup_\nu^\infty C^\nu \subset C^\infty$, $S^\nu \xrightarrow{t} S$, *and* $(S^\infty)^{-1}(0) \cap C^\infty = \{0\}$.

(c) *Further,* $\limsup_\nu^\infty S^\nu(C^\nu) \subset S(C)^\infty$ *if in addition,* $(S^\infty)(C^\infty) \subset S(C)^\infty$ *and* $\limsup_\nu^\infty \operatorname{gph} S^\nu \subset \operatorname{gph} S^\infty$.

In particular, $S^\nu(C^\nu) \to S(C)$ *when* S^ν *converges continuously to* S, $C^\nu \xrightarrow{t} C$ *and* $(S^\infty)^{-1}(0) \cap C^\infty = \{0\}$, *or when* S^ν *converges both continuously and totally to* S, $C^\nu \to C$ *and* $(S^\infty)^{-1}(0) = \{0\}$.

Proof. Statement (a) is a direct consequence of the definitions of the inner limit (in 4.1) and continuous convergence; recall from 5.44 that continuous convergence of S^ν to S implies the continuity of S.

For (b) we need to show that $u \in S(C)$ when $u^\nu \underset{N}{\to} u$ for some $N \in \mathcal{N}_\infty^\#$ with $u^\nu \in S^\nu(C^\nu)$ for all $\nu \in N$. Pick $x^\nu \in (S^\nu)^{-1}(u^\nu) \cap C^\nu$. If the sequence $\{x^\nu\}_{\nu \in N}$ clusters at a point x, then $x \in C$ (because $C \supset \limsup_\nu C^\nu$), and since $S^\nu \xrightarrow{t} S$ one has $u \in S(x) \subset S(C)$. Otherwise, the x^ν cluster at a point in the horizon of $I\!R^n$, say $\operatorname{dir} x$ (with $x \neq 0$). Since $(x^\nu, u^\nu) \in \operatorname{gph} S^\nu$, this would imply that $(x, 0) \in \limsup_\nu^\infty \operatorname{gph} S^\nu \subset \operatorname{gph} S^\infty$, i.e., $x \in (S^\infty)^{-1}(0)$, but the assumption $(S^\infty)^{-1}(0) \cup C^\infty = \{0\}$ excludes such a possibility.

For (c) one shows that $\limsup_\nu^\infty S^\nu(C^\nu) \subset S(C)^\infty$ when $\limsup_\nu^\infty \operatorname{gph} S^\nu \subset \operatorname{gph} S^\infty$, i.e., $u \in S(C)^\infty$ whenever $u^\nu \underset{N}{\to} \operatorname{dir} u$ for some index set $N \in \mathcal{N}_\infty^\#$ with $u^\nu \in S^\nu(C^\nu)$ for $\nu \in N$. Pick $x^\nu \in (S^\nu)^{-1}(u^\nu) \cap C^\nu$. If the points x^ν cluster at $x \in I\!R^n$, then $x \in C$ (which includes $\limsup C^\nu$) and since by assumption $\limsup_\nu^\infty \operatorname{gph} S^\nu \subset \operatorname{gph} S^\infty$, it follows that $u \in S^\infty(0) \subset S^\infty(C^\infty) \subset S(C)^\infty$. Otherwise, there exist $N_0 \subset N$, $N_0 \in \mathcal{N}_\infty^\#$, $x \neq 0$ such that $x^\nu \underset{N_0}{\to} \operatorname{dir} x$; note that then $x \in C^\infty \supset \limsup_\nu^\infty C^\nu$. Since for $\nu \in N_0$, $(x^\nu, u^\nu) \subset \operatorname{gph} S^\nu$ and $|u^\nu| \nearrow \infty$, $|x^\nu| \nearrow \infty$, there exists $\lambda^\nu \searrow 0$, $\nu \in N_0$ such that $\lambda^\nu(x^\nu, u^\nu)$ clusters at a point of the type $(\alpha x, \beta u) \neq (0, 0)$ with $\alpha \geq 0$, $\beta \geq 0$. If $\beta = 0$, then $0 \neq x \in (S^\infty)^{-1}(0) \cap C^\infty$ and that is ruled out by the assumption that $(S^\infty)^{-1}(0) \cap C^\infty = \{0\}$. Thus $\beta > 0$, and $u \in (S^\infty)(\alpha\beta^{-1}x) \subset (S^\infty)(C^\infty) \subset S(C)^\infty$ where the second inclusion comes from the last assumption in (c).

The two remaining statements are just a rephrasing of the consequences of (a) and (b) when limits or total limits exists, making use of the properties of the inner and outer horizon limits recorded in 4.20. □

5.54 Exercise (convergence of positive hulls). *Let $C^\nu, C \subset I\!R^n$ be such that $C^\nu \xrightarrow{t} C$ with C compact, and $0 \notin C$. Then $\operatorname{pos} C^\nu \xrightarrow{t} \operatorname{pos} C$.*

Guide. Let V be an open neighborhood of 0 such that $V \cap C = \emptyset$. Let S be the mapping define by $S(x) := \operatorname{pos}\{x\}$ on its domain $I\!R^n \setminus V$. Observe that S is continuous relative to $\operatorname{dom} S$, and that's enough to guarantee that $S(C^\nu) \to S(C)$ under the conditions $C^\nu \xrightarrow{t} C$ and $C^\infty = \{0\}$. Finally, use the fact that the sets $\operatorname{pos} C^\nu$ and $\operatorname{pos} C$ are cones in order to pass from ordinary convergence to total convergence by the criterion in 4.25(b). □

J.* Generic Continuity and Selections

Semicontinuous mappings are 'mostly' continuous. The next theorem makes this precise. It employs the following terminology. A subset A of a set $X \subset I\!R^n$ is called *nowhere dense* in X if no point $x \in X \cap \operatorname{cl} A$ has a neighborhood $V \in \mathcal{N}(x)$ with $X \cap V \subset \operatorname{cl} A$, or in other words if

$$\operatorname{int}_X(\operatorname{cl}_X A) = \emptyset, \quad \text{where} \quad \begin{cases} \operatorname{cl}_X \text{ denotes closure relative to } X, \\ \operatorname{int}_X \text{ denotes interior relative to } X. \end{cases}$$

It's called *meager* in X if it's the union of countably many sets that are nowhere dense in X. Any subset of a meager set is itself meager, inasmuch as any subset of a nowhere dense set is nowhere dense. Elementary examples of sets A that are nowhere dense in X are sets the form $A = Y \setminus (\operatorname{int}_X Y)$ for $Y \subset X$ with $Y = \operatorname{cl}_X Y$, or of the form $A = (\operatorname{cl}_X Y) \setminus Y$ for $Y \subset X$ with $Y = \operatorname{int}_X Y$. It's known that when X is closed in $I\!R^n$ (e.g., when $X = I\!R^n$ itself), or for that matter when X is open in $I\!R^n$, every meager subset A of X has dense complement: $\operatorname{cl}[X \setminus A] \supset X$.

5.55 Theorem (generic continuity from semicontinuity). *For $S : I\!R^n \rightrightarrows I\!R^m$ closed-valued, if S is osc relative to $X \subset I\!R^n$, or isc relative to X, then the set of points $x \in X$ where S fails to be continuous relative to X is meager in X.*

Proof. Let $\mathcal{B}$ be the collection of all rational closed balls in $I\!R^m$ (i.e., balls $B = I\!B(u, \rho)$ for which ρ and the coordinates of u are rational). This is a countable collection which suffices in generating neighborhoods in $I\!R^m$: for every $\bar{u} \in I\!R^m$ and $W \in \mathcal{N}(\bar{u})$ there exists $B \in \mathcal{B}$ with $u \in \operatorname{int} B$ and $B \subset W$.

Let's first deal with the case where S is osc relative to X. Let X_0 consist of the points of X where S fails to be isc relative to X. We must demonstrate that X_0 can be covered by a union of countably many nowhere dense subsets of X. To know that S is isc relative to X at a point $\bar{x} \in X$, it suffices by 5.6(b) to know that whenever $\bar{u} \in S(\bar{x})$ and $B \in \mathcal{N}(\bar{u}) \cap \mathcal{B}$, there's a neighborhood $V \in \mathcal{N}(\bar{x})$ with $X \cap V \subset S^{-1}(B)$. In other words, the points $\bar{x} \in X \setminus X_0$ are characterized by the property that

$$\forall B \in \mathcal{B}: \qquad \bar{x} \in S^{-1}(\operatorname{int} B) \implies \bar{x} \in \operatorname{int}_X \big[S^{-1}(B) \cap X\big].$$

Thus, each point of X_0 belongs to $[S^{-1}(\operatorname{int} B) \cap X] \setminus \operatorname{int}_X [S^{-1}(B) \cap X]$ for some $B \in \mathcal{B}$. Therefore

$$X_0 \subset \bigcup_{B \in \mathcal{B}} \{Y_B \setminus (\operatorname{int}_X Y_B)\} \text{ with } Y_B := S^{-1}(B) \cap X.$$

Each of the sets $Y_B \setminus (\operatorname{int}_X Y_B)$ in this countable union is nowhere dense in X, because Y_B is closed relative to X by 5.7(b).

We turn now to the claim under the alternative assumption that S is isc relative to X. This time, let X_0 consist of the points of X where S is osc relative to X. Again the task is to demonstrate that X_0 can be covered by a union of countably many nowhere dense subsets of X. For any $\bar{x} \in X$ and $\bar{u} \notin S(\bar{x})$, there's a neighborhood $B \in \mathcal{N}(\bar{u}) \cap \mathcal{B}$ such that $B \cap S(\bar{x}) = \emptyset$; here we utilize the assumption that S is closed-valued at $\bar{x}$. Hence in view of 5.6(a), for S to be osc relative to X at such a point $\bar{x}$, it's necessary and sufficient that, whenever $B \in \mathcal{B}$ and $S(\bar{x}) \cap B = \emptyset$, there should exist $V \in \mathcal{N}(\bar{x})$ yielding $S(x) \cap \operatorname{int} B = \emptyset$ for all $x \in X \cap V$. Equivalently, the points $\bar{x} \in X \setminus X_0$ are characterized by the property that

$$\forall B \in \mathcal{B}: \qquad \bar{x} \notin S^{-1}(B) \implies \bar{x} \notin \operatorname{cl}_X\big[S^{-1}(\operatorname{int} B) \cap X\big].$$

Thus, each point of X_0 belongs to $\operatorname{cl}_X[S^{-1}(\operatorname{int} B) \cap X] \setminus [S^{-1}(B) \cap X]$ for some $B \in \mathcal{B}$. Therefore

$$X_0 \subset \bigcup_{B \in \mathcal{B}} \big\{(\operatorname{cl}_X Y_B) \setminus Y_B\big\} \text{ with } Y_B := S^{-1}(\operatorname{int} B) \cap X.$$

Each of the sets $(\operatorname{cl}_X Y_B) \setminus Y_B$ in this countable union is nowhere dense in X, because Y_B is open relative to X by 5.7(c). □

5.56 Corollary (generic continuity of extended-real-valued functions). *If a function $f : \mathbb{R}^n \to \overline{\mathbb{R}}$ is lsc relative to X, or usc relative to X, then the set of points $x \in X$ where f fails to be continuous relative to X is meager in X.*

Proof. This applies Theorem 5.55 to the profile mappings in 5.5. □

The continuity properties we've been studying for set-valued mappings have an interesting connection with selection properties. A *selection* of $S : \mathbb{R}^n \rightrightarrows \mathbb{R}^m$ is a single-valued mapping $s : \operatorname{dom} S \to \mathbb{R}^m$ such that $s(x) \in S(x)$ for each $x \in \operatorname{dom} S$. It's important for various purposes to know the circumstances under which there must exist selections s that are continuous relative to $\operatorname{dom} S$.

Set-valued mappings that are merely osc can't be expected to admit continuous selections, but if a mapping $S : \mathbb{R}^n \rightrightarrows \mathbb{R}^m$ is isc at $\bar{x}$ relative to $\operatorname{dom} S$, there exists for each $\bar{u} \in S(\bar{x})$ a selection $s : \operatorname{dom} S \to \mathbb{R}^m$ such that s is continuous at $\bar{x}$ relative to $\operatorname{dom} S$ and $s(\bar{x}) = \bar{u}$. This follows from the observation through 5.11(b) that $\limsup_{x \to \bar{x}} d\big(\bar{u}, S(x)\big) \leq d\big(\bar{u}, S(\bar{x})\big) = 0$; for each $x \in \operatorname{dom} S$ one can choose $s(x)$ to be any point u of $S(x)$ with $|u - \bar{u}| \leq d\big(\bar{u}, S(x)\big) + |x - \bar{x}|$. However, it doesn't follow from S being isc on a neighborhood of $\bar{x}$ that a selection s can be found that is continuous on a neighborhood of $\bar{x}$. Something more must usually be demanded of S besides a continuity property in order to get selections that are continuous at more than just a single point. The 'something' is convex-valuedness.

The picture is much simpler when the mapping is continuous rather than just isc, so we look at that case first.

5.57 Example (projections as continuous selections). *Let $S : I\!R^n \rightrightarrows I\!R^m$ be continuous relative to* $\operatorname{dom} S$ *and convex-valued. For each $u \in I\!R^m$ define $s_u : I\!R^n \to I\!R^m$ by taking $s_u(x)$ to be the projection $P_{S(x)}(u)$ of u on $S(x)$. Then s_u is a selection of S that is continuous relative to* $\operatorname{dom} S$.

Furthermore, for any choice of $U \subset I\!R^m$ such that $\operatorname{cl} U \supset \operatorname{rge} S$ the family of continuous selections $\{s_u\}_{u \in U}$ fully determines S, in the sense that

$$S(x) = \operatorname{cl}\left\{s_u(x) \,\middle|\, u \in U\right\} \quad \textit{for every } x \in \operatorname{dom} S.$$

Detail. The continuity of S relative to $\operatorname{dom} S$ ensures that S is closed-valued. Thus $S(x)$ is a nonempty, closed, convex set for each $x \in \operatorname{dom} S$, and the projection mapping $P_{S(x)}$ is accordingly single-valued and continuous (cf. 2.25); in fact $P_{S(x)}(u)$ depends continuously on x for each u (cf. 4.9). In other words, $s_u(x)$ is a well defined, uniquely determined element of $S(x)$, and the mapping s_u is continuous. If $\operatorname{cl} U \supset \operatorname{rge} S$, there exists for each $x \in \operatorname{dom} S$ and $\bar{u} \in S(x)$ a sequence of points $u^\nu \in U$ with $u^\nu \to \bar{u}$, and for this we have $s_{u^\nu}(x) \to s_{\bar{u}}(x)$ by the continuity of $P_{S(x)}(u)$ with respect to u. This yields the closure formula claimed for $S(x)$. □

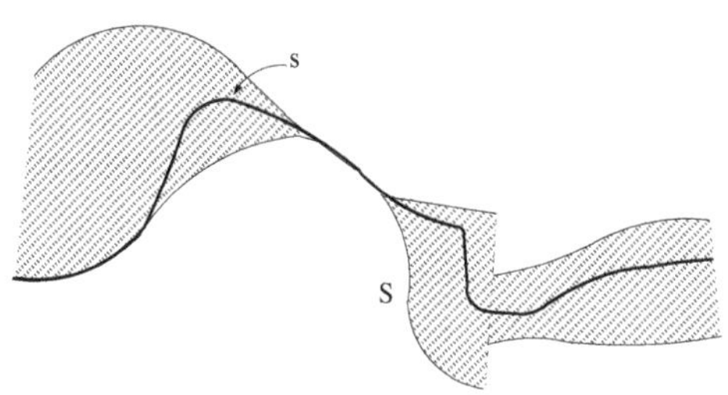

Fig. 5–13. A continuous selection s from an isc mapping S.

Let's note in 5.57 that U could be taken to be all of $I\!R^m$, and then actually $S(x) = \{s_u(x) \mid u \in U\}$, since $s_u(x) = u$ when $u \in S(x)$. On the other hand, U could be taken to be any countable, dense subset of $I\!R^n$ (such as the set of all vectors with rational coordinates), and we would then have a *countable* family of continuous selections that fully determines S.

The main theorem on continuous selections asserts the existence of such a countable family even when the continuity of S is relaxed to inner semicontinuity, as long as $\operatorname{dom} S$ exhibits σ-compactness. Recall that a set X is *σ-compact* if it can be expressed as a countable union of compact sets, and note that in $I\!R^n$ all closed sets and all open sets are σ-compact.

5.58 Theorem (Michael representations of isc mappings). *Suppose the mapping $S : I\!R^n \rightrightarrows I\!R^m$ is isc relative to* $\operatorname{dom} S$ *as well as closed-convex-valued, and that* *dom S is σ-compact. Then S has a continuous selection and in fact admits a Michael representation: there exists a countable collection* $\{s_i\}_{i\in I}$ *of selections* s_i *for S that are continuous relative to* dom S *and such that*

$$S(x) = \text{cl}\,\big\{s_i(x)\,\big|\, i \in I\big\} \quad \textit{for every } x \in \text{dom}\, S.$$

Proof. *Part 1.* Let's begin by showing that *if X is a compact set such that* $\text{int}\, S(x) \neq \emptyset$ *for all* $x \in X$*, then there is a continuous mapping* $s : X \to \mathbb{R}^m$ *with* $s(x) \in \text{int}\, S(x)$ *for all* $x \in X$. For each $x \in X$ choose any $u_x \in \text{int}\, S(x)$. By Theorem 5.9(a) there's an open neighborhood V_x of x such that $u_x \in \text{int}\, S(x')$ for all $x' \in V_x \cap \text{dom}\, S$. The family $\{V_x\}_{x\in X}$ is an open covering of X, so by the compactness of X, one can extract a finite subcovering, say $\{V_i\}_{i\in I}$ with I finite. Associated with each V_i is a point u_i such that

$$u_i \in \text{int}\, S(x) \quad \text{ for all } \quad x \in V_i \cap \text{dom}\, S.$$

Define on X the functions $\theta_i(x) := \min\big\{1, d(x, X \setminus V_i)\big\}$, which are continuous (since distance functions are continuous, see 1.20). Let $\theta(x) := \sum_{i\in I} \theta_i(x) > 0$ and $\lambda_i(x) := \theta_i(x)/\theta(x)$, noting that these expressions are continuous relative to $x \in X$ with $\lambda_i(x) \geq 0$ and $\sum_{i\in I} \lambda_i(x) = 1$. Then take

$$s(x) := \sum_{i\in I} \lambda_i(x) u_i \quad \text{ for all } \quad x \in X.$$

Since $\lambda_i(x) = 0$ when $x \notin V_i$, only points $u_i \in \text{int}\, S(x)$ play a role in the sum defining $s(x)$. Hence $s(x) \in \text{int}\, S(x)$ by the convexity of $\text{int}\, S(x)$ (which comes from 2.33).

Part 2. Next we argue that *for any compact set* $X \subset \text{dom}\, S$ *and any* $\eta > 0$*, it's possible to find a continuous mapping* $s : X \to \mathbb{R}^m$ *with* $d(s(x), S(x)) < \eta$ *for all* $x \in \text{dom}\, S$. We apply Part 1 to the mapping $S_\eta : X \rightrightarrows \mathbb{R}^m$ defined by $S_\eta(x) := S(x) + \eta\mathbb{B}$. To see that S_η fits the assumptions there, note that it's closed-convex-valued (by 3.12 and 2.23) with $\text{int}\, S_\eta(x) \neq \emptyset$ for all $x \in X$ (by 2.45(b)). Moreover S_η is isc relative to X by virtue of 5.24.

Part 3. We pass now to the higher challenge of showing that *for any compact set* $X \subset \text{dom}\, S$ *there's a continuous mapping* $s : X \to \mathbb{R}^m$ *with* $s(x) \in S(x)$ *for all* $x \in X$. From Part 2 with $\eta = 1$, we first get a continuous mapping s^0 with $d(s^0(x), S(x)) < 1$ for all $x \in X$. This initializes the following 'algorithm'. Having obtained a continuous mapping s^ν with $d(s^\nu(x), S(x)) < 2^{-\nu}$ for all $x \in X$, form the convex-valued mapping $S^\nu : X \rightrightarrows \mathbb{R}^m$ by

$$S^\nu(x) := S(x) \cap \big[s^\nu(x) + 2^{-\nu}\mathbb{B}\big]$$

and observe that it is isc relative to $\text{dom}\, S^\nu = X$ by 4.32 (and 5.24). Applying Part 2, get a continuous mapping $s^{\nu+1} : X \to \mathbb{R}^m$ with the property that $d(s^{\nu+1}(x), S^\nu(x)) < 2^{-(\nu+1)}$ for all $x \in X$. Then in particular

$$d(s^{\nu+1}(x), S(x)) < 2^{-(\nu+1)} \quad \text{ for all } \quad x \in X,$$

but also $d(s^{\nu+1}(x), s^\nu(x)) < 2^{-(\nu+1)} + 2^{-\nu} < 2^{-(\nu-1)}$. By induction, one gets

$$d(s^{\nu+\kappa}(x), s^\nu(x)) < 2^{-(\nu+\kappa-2)} + \cdots + 2^{-(\nu+1)} + 2^{-\nu} + 2^{-(\nu-1)} < 2^{-(\nu-2)},$$

so that for each $x \in X$ the sequence $\{s^\nu(x)\}_{\nu\in I\!N}$ has the Cauchy property. For each x the limit of this sequence exists; denote it by $s(x)$. It follows from $d(s^{\nu+1}(x), S(x)) < 2^{-(\nu+1)}$ that $s(x) \in S(x)$. Since the convergence of the functions s^ν is uniform on X, the limit function s is also continuous (cf. 5.43).

Part 4. To pass from the case in Part 3 where X is compact to the case where X is merely σ-compact, so that we can take $X = \operatorname{dom} S$, we note that the argument in Part 1 yields in the more general setting a countable (instead of finite) index set I giving a *locally finite* covering of X by open sets V_i: each x belongs to only finitely many of these sets V_i, so that in the sums defining $\lambda_i(x)$ and $s(x)$ only finitely many nonzero terms are involved.

Part 5. So far we have established the existence of at least one continuous selection s for S, but we must go on now to the existence of a Michael representation for S. Let $\mathbb{Q}$ stand for the rational numbers and $\mathbb{Q}_+$ for the nonnegative rational numbers. Let

$$Q := \big\{(u,\rho) \in \mathbb{Q}^n \times \mathbb{Q}_+ \,\big|\, (\operatorname{rge} S) \cap \operatorname{int} I\!B(u,\rho) \neq \emptyset\big\}.$$

For any (u,ρ) the set $S^{-1}(\operatorname{int} I\!B(u,\rho))$ is open relative to $\operatorname{dom} S$ (cf. 5.7(c)), hence σ-compact because $\operatorname{dom} S$ is σ-compact. For each $(u,\rho) \in Q$, the mapping $S_{u,\rho} : I\!R^n \rightrightarrows I\!R^m$ with $S_{u,\rho}(x) := \operatorname{cl}\big[S(x) \cap \operatorname{int} I\!B(u,\rho)\big]$, has $\operatorname{dom} S_{u,\rho} = S^{-1}(\operatorname{int} I\!B(u,\rho))$, and it is isc relative to this set (by 4.32(c)). Also, $S_{u,\rho}$ is convex-valued (by 2.9, 2.40). Hence by Part 4 it has a continuous selection $s_{u,\rho} : S^{-1}(\operatorname{int} I\!B(u,\rho)) \mapsto I\!R^m$.

Let $\mathbb{Z}$ denote the integers. For each $k \in I\!N$ and $(u,\rho) \in Q$ let $C^k_{u,\rho}$ be the union of all the balls of the form

$$B \subset S^{-1}(\operatorname{int} I\!B(u,\rho)) \text{ with } B = I\!B\Big(\frac{1}{k}x, \Big(\frac{1}{2k} - \frac{1}{(2k)^2}\Big)\Big) \text{ for some } x \in \mathbb{Z}^n.$$

Observe that $C^k_{u,\rho}$ is closed and $\bigcup_{k=1}^\infty C^k_{u,\rho} = S^{-1}(\operatorname{int} I\!B(u,\rho))$. For each $k \in I\!N$ and $(u,\rho) \in Q$ the mapping $S^k_{u,\rho} : I\!R^n \rightrightarrows I\!R^m$ defined by

$$S^k_{u,\rho} = \begin{cases} \{s_{u,\rho}(x)\} & \text{if } x \in C^k_{u,\rho}, \\ S(x) & \text{otherwise,} \end{cases}$$

is isc relative to the set $\operatorname{dom} S^k_{u,\rho} = \operatorname{dom} S$ (because the set $\operatorname{dom} S \setminus C^k_{u,\rho}$ is open relative to $\operatorname{dom} S$). Also, $S^k_{u,\rho}$ is convex-valued. Again by Part 4, we get the existence of a continuous selection: $s^k_{u,\rho} : \operatorname{dom} S \to I\!R^m$. In particular, $s^k_{u,\rho}$ is a continuous selection for S itself.

This countable collection of continuous selections $\{s^k_{u,\rho}\}_{(u,\rho)\in Q,\, k\in I\!N}$ furnishes a Michael representation of S. To establish this, we have to show that

$$\operatorname{cl}\big\{s^k_{u,\rho}(x) \,\big|\, (u,\rho) \in Q,\ k \in I\!N\big\} = S(x).$$

This will follow from the construction of the selections. It suffices to demonstrate that for any given $\varepsilon > 0$ and $\bar{u} \in S(\bar{x})$ there exists (u, ρ) with $|\bar{u} - s^k_{u,\rho}(\bar{x})| < \varepsilon$. But one can always find $u \in \mathbb{Q}^n$ and $\rho \in \mathbb{Q}_+$ with $\rho < \varepsilon$ such that $\bar{u} \in S(\bar{x}) \cap \operatorname{int} \mathbb{B}(u, \rho)$ and then pick k sufficiently large that $\bar{x}$ belongs to $C^k_{u,\rho}$. In that case $s^k_{u,\rho}(\bar{x})$ has the desired property. □

5.59 Corollary (extensions of continuous selections). *Suppose $S : \mathbb{R}^n \rightrightarrows \mathbb{R}^m$ is isc relative to* $\operatorname{dom} S$ *as well as convex-valued, and that* $\operatorname{dom} S$ *is σ-compact. Let $\bar{s}$ be a continuous selection for S relative to a closed set $X \subset \operatorname{dom} S$ (i.e., $\bar{s}(x) \in S(x)$ for $x \in X$, and $\bar{s}$ is continuous relative to X). Then there exists a continuous selection s for S that agrees with $\bar{s}$ on X (i.e., $s(x) \in S(x)$ for $x \in \operatorname{dom} S$ with $s(x) = \bar{s}(x)$ if $x \in X$, and s is continuous relative to* $\operatorname{dom} S$*).*

Proof. Define $\overline{S}(x) = \{\bar{s}(x)\}$ for $x \in X$ but $\overline{S}(x) = S(x)$ otherwise. Then $\operatorname{dom} \overline{S} = \operatorname{dom} S$, and the requirements are met for applying Theorem 5.58 to $\overline{S}$. Any continuous selection s for $\overline{S}$ has the properties demanded. □

Commentary

Vasilesco [1925], a student of Lebesgue, initiated the study of the topological properties of set-valued mappings, but he only looked at the special case of mappings $S : \mathbb{R} \rightrightarrows \mathbb{R}$ with bounded values. His definitions of continuity and semicontinuity were based on expressions akin to what we now call Pompeiu-Hausdorff distance. For his continuous mappings he obtained results on continuous extensions, a theorem on approximate covering by piecewise linear functions, an implicit function theorem and, by relying on what he termed quasi-uniform continuity, a generalization of Arzelá's theorem on the continuity of the pointwise limit of a sequence of continuous mappings. It was not until the early 1930s that notions of continuity and semicontinuity were systematically investigated for mappings $S : \mathbb{R}^n \rightrightarrows \mathbb{R}^m$, or more generally from one metric space into another. The way was led by Bouligand [1932a], [1933], Kuratowski [1932], [1933], and Blanc [1933].

Advances in the 1940s and 1950s paralleled those in the theory of set convergence because of the close relationship to limits of sequences of sets. This period also produced some fundamental results that make essential use of semicontinuity at more than just one point. In this category are the fixed point theorem of Kakutani [1941], the genericity theorem of Kuratowski [1932] and Fort [1951], and the selection theorem of Michael [1956]. The book by Berge [1959] was instrumental in disseminating the theory to a wide range of potential users. More recently the books of Klein and Thompson [1984] and Aubin and Frankowska [1990] have helped in furthering access.

With burgeoning applications in optimal control theory (Filippov [1959]), statistics (Kudo [1954], Richter [1963]), and mathematical economics (Debreu [1967], Hildenbrand [1974]), the 1960s saw the development of a measurability and integration theory for set-valued mappings (for expositions see Castaing and Valadier [1977] and our Chapter 14), which in turn stimulated additional interest in continuity and semicontinuity in such mappings as a special case. The theory of maximal monotone operators (Minty [1962], Brezis [1973], Browder [1976]), manifested in particular by subgradient mappings associated with convex functions (Rockafellar [1970d]), as will be taken up in Chapter 12, focused further attention on multivaluedness and also the concept of local boundedness (Rockafellar [1969b]). Strong incentive came too from the study of differential inclusions (Ważewski [1961a] [1962], Filippov [1967], Olech [1968], Aubin and Cellina [1984], Aubin [1991]).

Convergence theory for set-valued mappings originated in the 1970s, the motivation arising mostly from approximation questions in dynamical systems and partial differential equations (Brezis [1973], Sbordone [1974], Spagnolo [1976], Attouch [1977], De Giorgi [1977]), stochastic optimization (Salinetti and Wets [1981]), and classical approximation theory (Deutsch and Kenderov [1983], Sendov [1990]).

The terminology of 'inner' and 'outer' semicontinuity, instead of 'lower' and 'upper', has been forced on us by the fact that the prevailing definition of 'upper semicontinuity' in the literature is out of step with developments in set convergence and the scope of applications that must be handled, now that mappings S with unbounded range and even unbounded value sets $S(x)$ are so important. In the way the subject has widely come to be understood, S is *upper semicontinuous* at $\bar{x}$ if $S(\bar{x})$ is closed and for each open set O with $S(\bar{x}) \subset O$ the set $\{ x \mid S(x) \subset O \}$ is a neighborhood of $\bar{x}$. On the other hand, S is *lower semicontinuous* at $\bar{x}$ if $S(\bar{x})$ is closed and for each open set O with $S(\bar{x}) \cap O \neq \emptyset$ the set $\{ x \mid S(x) \cap O \neq \emptyset \}$ is a neighborhood of $\bar{x}$. These definitions have the mathematically appealing feature that upper semicontinuity everywhere corresponds to $S^{-1}(C)$ being closed whenever C is closed, whereas lower semicontinuity everywhere corresponds to $S^{-1}(O)$ being open whenever O is open. Lower semicontinuity agrees with our inner semicontinuity (cf. 5.7(c)), but upper semicontinuity differs from our outer semicontinuity (cf. 5.7(b)) and is seriously troublesome in its narrowness.

When the framework is one of mappings into compact spaces, as authors primarily concerned with abstract topology have especially found attractive, upper semicontinuity is indeed equivalent to outer semicontinuity (cf. Theorem 5.19), but beyond that, in situations where S isn't locally bounded, it's easy to find examples where S fails to meet the test of upper semicontinuity at $\bar{x}$ even though $S(\bar{x}) = \limsup_{x \to \bar{x}} S(x)$. In consequence, if one goes on to define continuity as the combination of upper and lower semicontinuity, one ends up with a notion that doesn't correspond to having $S(x) \to S(\bar{x})$ as $x \to \bar{x}$, and thus is at odds with what continuity really ought to mean in many applications, as we have explained in the text around Figure 5–7. In applications to functional analysis the mismatch can be even more bizarre; for instance in an infinite-dimensional Hilbert space the mapping that associates with each point x the closed ball of radius 1 around x fails to be upper semicontinuous even though it's 'Lipschitz continuous' (in the usual sense of that term—see 9.26); see cf. p. 28 of Yuan [1999].

The literature is replete with ad hoc remedies for this difficulty, which often confuse and mislead the hurried, and sometimes the not-so-hurried, reader. For example, what we call outer semicontinuous mappings were called 'closed' by Berge [1959], but upper semicontinuous by Choquet [1969], whereas those that Berge refers to as upper semicontinuous were called upper semicontinuous 'in the strong sense' by Choquet. Although closedness is descriptive as a global property of a graph, it doesn't work very well as a term for signaling that $S(\bar{x}) = \limsup_{x \to \bar{x}} S(x)$ at an individual point $\bar{x}$. More recent authors, such as Klein and Thompson [1984], and Aubin and Frankowska [1990], have reflected the terminology of Berge. Choquet's usage, however, is that of Bouligand, who in his pioneering efforts did define upper semicontinuity instead by set convergence and specifically thought of it in that way when dealing with unboundedness; cf. Bouligand [1935, p. 12], for instance. Yet in contrast to Bouligand, Choquet didn't take 'continuity' to be the combination of Bouligand's upper semicontinuity with lower semicontinuity, but that of upper semicontinuity 'in the strong sense' with lower semicontinuity. This idea of 'continuity' does have topological content—it corresponds to the topology developed for general spaces of sets by Vietoris (see the Commentary for Chapter 4)—but, again, it's based on pure-mathematical considerations which have somehow gotten divorced from the kinds of examples that should have served as a guide.

Despite the historical justification, the tide can no longer be turned in the meaning of 'upper semicontinuity', yet the concept of 'continuity' is too crucial for applications to be left in the poorly usable form that rests on such an unfortunately restrictive property. We adopt the position that continuity of S at $\bar{x}$ must be identified with having both $S(\bar{x}) = \limsup_{x \to \bar{x}} S(x)$ and $S(\bar{x}) = \liminf_{x \to \bar{x}} S(x)$. To maintain this while trying to keep conflicts with other terminology at bay, we speak of the two equations in this formulation as signaling outer semicontinuity and inner semicontinuity. There is the side benefit then too from the fact that 'outer' and 'inner' better convey the attendant geometry; cf. Figure 5–3. For clarity in contrasting this osc+isc notion of continuity in the Bouligand sense with the restrictive usc+lsc notion, it's appropriate to refer to the latter as *Vietoris-Berge continuity.*

The diverse descriptions of semicontinuity listed between 5.7 and 5.12 are essentially classical, except for those in Propositions 5.11 and 5.12, which follow directly from related characterizations of set convergence. Theorem 5.7(a), that closed graphs characterize osc mappings, appears in Choquet [1969]. Theorem 5.7(c), that a mapping is isc if and only if the inverse images of open sets are open, is already in Kuratowski [1932]. Dantzig, Folkman and Shapiro [1967] and Walkup and Wets [1968] in the case of linear constraints, and Walkup and Wets [1969b] and Evans and Gould [1970] in the case of nonlinear constraints, were probably among the first to rely on the properties of feasible-set mappings (Example 5.8) to obtain stability results for the solutions of optimization problems; cf. also Hogan [1973]. The fact that certain convexity properties yield inner semicontinuity comes from Rockafellar [1971b] in the case of convex-valued mappings (Theorem 5.9(a)). The fact that inner semicontinuity is preserved when taking convex hulls can be traced to Michael [1951].

Proposition 5.15 and Theorem 5.18 are new, as is the concept of an horizon mapping 5(6). The characterization of outer semicontinuity under local boundedness in Theorem 5.19 was of course the basis for the definition of upper semicontinuity in the period when the study of set-valued mappings was confined to mappings into compact spaces. A substantial part of the statements in Example 5.22 can be found in Berge [1959]. Theorem 5.25 is new, at least in this formulation. All the results 5.27–5.30 dealing with cosmic or total (semi)continuity appear here for the first time.

The systematic study of pointwise convergence of set-valued mappings was initiated in the context of measurable mappings, Salinetti and Wets [1981]. The origin of graphical convergence, a more important convergence concept for variational analysis, is more diffuse. One could trace it to a definition for the convergence of elliptic differential operators in terms of their resolvants (Kato [1966]). But it was not until the work of Spagnolo [1976], Attouch [1977], De Giorgi [1977] and Moreau [1978] that its pivotal importance was fully recognized. The expressions for graphical convergence at a point in Proposition 5.33 are new, as are those in 5.34(a); the uniformity result in 5.34(b) for connected-valued mappings stems from Bagh and Wets [1996]. Theorem 5.37 has not been stated explicitly in the literature, but, except possibly for 5.37(b), has been part of the folklore.

The concept of equi-outer semicontinuity, as well as the results that follow, up to the generalized Arzelá-Ascoli Theorem 5.47, were developed in the process of writing this book; extensions of these results to mappings defined on a topological space and whose values are subsets of an arbitrary metric space appear in Bagh and Wets [1996]. Dolecki [1982] introduced a notion related to equi-osc, which he called *quasi equi-semicontinuity.* His definition, however, works well only when dealing with collections of mappings that all have their ranges contained within a fixed bounded set. Kowalczyk [1994] defines equi-isc, which he calls *lower equicontinuity* and introduces a notion related to equi-osc and, even more closely to Dolecki's quasi equicontinuity, which he calls *upper equicontinuity.* It refers to upper semicontinuity, as defined above, which lead to Vietoris-Berge continuity. Equicontinuity in Kowalczyk's sense is the combination of equi-isc and upper equicontinuity.

Hahn [1932] came up with the notion of continuous convergence for real-valued functions. Del Prete, Dolecki and Lignola [1986] proposed a definition of continuous convergence for set-valued mappings, but in the context of mappings that are continuous in the Vietoris-Berge sense, discussed above, and this resulted in a more restrictive condition. The definition of uniform convergence in 5.41 is equivalent to that in Salinetti and Wets [1981]. The fact in 5.48 that a decreasing sequence of mappings converges uniformly on compact sets is an extension of a result of Del Prete and Lignola [1983].

The use of set distances between graphs as a measure of the distance between mappings goes back to Attouch and Wets [1991]. Theorem 5.50 renders the implications more explicit.

The results in 5.51–5.52 about the preservation of semicontinuity under various operations are new, as are those about the images of converging mappings in 5.53–5.54.

The generic continuity result in Theorem 5.55 is new in its applicability to arbitrary semicontinuous mappings S from a set $X \subset \mathbb{R}^n$ into $\mathbb{R}^m$. Kuratowski [1932] and Fort [1951] established such properties for mappings into compact spaces, but invoking those results in this context would require assuming that S is locally bounded.

Theorem 5.58, on the existence of Michael representations (dense covering of isc mappings by continuous selections), is due to Michael [1956], [1959]; our proof is essentially the one found in Aubin and Cellina [1984]. Example 5.57, finding selections by projection, is already mentioned in Ekeland and Valadier [1971]. For osc mappings S, Olech [1968] and Cellina [1969] were the first to obtain approximate continuous selections. Olech was concerned with a continuous selection that is close in a pointwise sense, whereas Cellina focused on a selection that is close in the graphical sense, of which the more inclusive version that follows is due to Beer [1983]: Let $S : \mathbb{R}^n \rightrightarrows \mathbb{R}^m$ be osc, convex-valued and with $\operatorname{rge} S$ bounded. Then for all $\varepsilon > 0$ there exists a continuous function $s : \operatorname{dom} S \to \mathbb{R}^m$ such that $d(\operatorname{gph} s, \operatorname{gph} S) < \varepsilon$.

6. Variational Geometry

In the study of 'variations', constraints can present a major complication. Before the effects of variations can be ascertained it may be necessary to determine the directions in which something can be varied at all. This may be difficult, whether the variations are aimed at tests of optimality or stability, or arise in trying to understand the consequences of perturbations in the data parameters on which a mathematical model might depend.

In maximizing or minimizing a function over a set $C \subset \mathbb{R}^n$, for instance, properties of the boundary of C can be crucial in characterizing a solution. When C is specified by a system of constraints such as inequalities, however, the boundary may have all kinds of curvilinear facets, edges and corners. Standard methods of geometric analysis can't cope with such a lack of smoothness except in simple cases where the pieces making up the boundary of C are neatly laid out and can be dealt with one by one.

An approach to geometry is needed through which the main variational properties of a set C can be identified, characterized and placed in a coordinated framework despite the possibility of boundary complications. Such an approach can be worked out in terms of associating with each point of C certain *cones* of tangent vectors and normal vectors, which generalize the tangent and normal *subspaces* in classical differential geometry. Several such cones come into play, but they fit into a tight pattern which eventually emerges in Figure 6–17. The central developments in this chapter, and the basic choices made in terminology and notation, are aimed at spotlighting this pattern and making it easy to appreciate and remember.

A. Tangent Cones

A primitive notion of variation at a point $\bar{x} \in \mathbb{R}^n$ is that of taking a vector $w \neq 0$ and replacing $\bar{x}$ by $\bar{x} + \tau w$ for small values of τ. Directional derivatives are often defined relative to such variations, for example. When constraints are present, however, straight-line variations of this sort might not be permitted. It may be hard even to know which 'curves', if any, might serve as feasible paths of variation away from $\bar{x}$. But sequences that converge to $\bar{x}$ without violating the constraints can be viewed as representing modes of variation in reverse, and the concept of direction can still then be utilized.

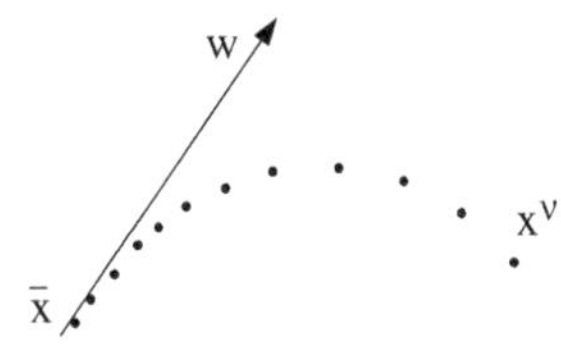

Fig. 6–1. Convergence to a point from a particular direction.

A sequence $x^\nu \to \bar{x}$ is said to *converge from the direction* $\operatorname{dir} w$ if for some sequence of scalars $\tau^\nu \searrow 0$ the vectors $[x^\nu - \bar{x}]/\tau^\nu$ converge to w. Here the direction $\operatorname{dir} w$ of a vector $w \neq 0$ has the formal meaning assigned at the beginning of Chapter 3 and is independent of the scaling of w. An idea closely related to such directional convergence is that of w being the right derivative

$$\xi'_+(0) := \lim_{\tau \searrow 0} \frac{\xi(\tau) - \xi(0)}{\tau}$$

of a vector-valued function $\xi : [0, \varepsilon] \to \mathbb{R}^n$ with $\xi(0) = \bar{x}$. In this case w is the limit of $[\xi(\tau^\nu) - \bar{x}]/\tau^\nu$ for *every* choice of a sequence $\tau^\nu \searrow 0$ in $[0, \varepsilon]$.

In considering these notions relative to a set C, it will be useful to have the notation

$$x^\nu \underset{C}{\to} \bar{x} \quad \Longleftrightarrow \quad x^\nu \to \bar{x} \text{ with } x^\nu \in C. \qquad 6(1)$$

6.1 Definition (tangent vectors and geometric derivability). *A vector $w \in \mathbb{R}^n$ is tangent to a set $C \subset \mathbb{R}^n$ at a point $\bar{x} \in C$, written $w \in T_C(\bar{x})$, if*

$$[x^\nu - \bar{x}]/\tau^\nu \to w \ \text{ for some } x^\nu \underset{C}{\to} \bar{x},\ \tau^\nu \searrow 0, \qquad 6(2)$$

or in other words if $\operatorname{dir} w$ *is a direction from which some sequence in C converges to $\bar{x}$, or if $w = 0$. Such a tangent vector w is derivable if there actually exists $\xi : [0, \varepsilon] \to C$ with $\varepsilon > 0$, $\xi(0) = \bar{x}$ and $\xi'_+(0) = w$. The set C is geometrically derivable at $\bar{x}$ if every tangent vector w to C at $\bar{x}$ is derivable.*

The tangent vectors to a relatively nice set C at a point $\bar{x}$ are illustrated in Figure 6–2. These vectors form a cone, namely the one representing the subset of $\operatorname{hzn} \mathbb{R}^n$ that consists of the directions from which sequences in C can converge to $\bar{x}$. In this example C is geometrically derivable at $\bar{x}$. But tangent vectors aren't always derivable, even though the condition involving a function $\xi : [0, \varepsilon] \to C$ places no assumptions of continuity, not to speak of differentiability, on ξ except at 0. An example of this will be furnished shortly.

6.2 Proposition (tangent cone properties). *At any point $\bar{x}$ of a set $C \subset \mathbb{R}^n$, the set $T_C(\bar{x})$ of all tangent vectors is a closed cone expressible as an outer limit:*

$$T_C(\bar{x}) = \limsup_{\tau \searrow 0} \tau^{-1}(C - \bar{x}). \qquad 6(3)$$

The subset of $T_C(\bar{x})$ consisting of the derivable tangent vectors is given by the corresponding inner limit (i.e., with lim inf *in place of* lim sup*) and is a closed* *cone as well. Thus, C is geometrically derivable at $\bar{x}$ if and only if the sets $[C - \bar{x}]/\tau$ actually converge as $\tau \searrow 0$, so that the formula for $T_C(\bar{x})$ can be written as a full limit.*

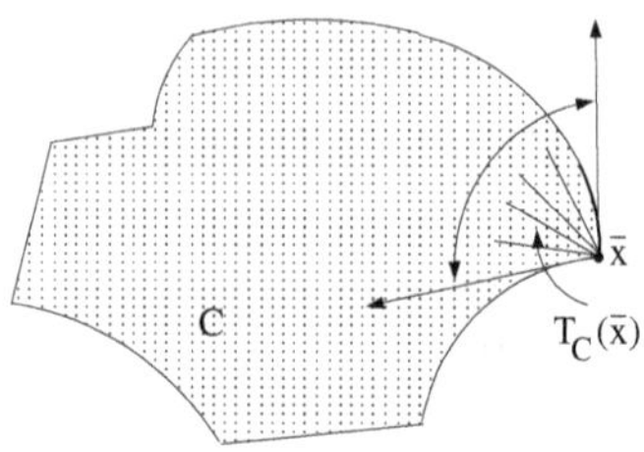

Fig. 6–2. A tangent cone.

We'll refer to $T_C(\bar{x})$ as the *tangent cone* to C at $\bar{x}$. The set of derivable tangent vectors forms the *derivable cone* to C at $\bar{x}$, but we'll have less need to consider it independently and therefore won't introduce notation for it here.

The geometric derivability of C at $\bar{x}$ is, by 6.2, a property of local approximation. Think of $\bar{x} + (C - \bar{x})/\tau$ as the image of C under the one-to-one transformation $L_\tau : x \mapsto \bar{x} + \tau^{-1}(x - \bar{x})$, which amounts to a global magnification around $\bar{x}$ by the factor τ^{-1}. As $\tau \searrow 0$ the scale blows up, but the progressively magnified images $L_\tau(C) = \bar{x} + (C - \bar{x})/\tau$ may anyway converge to something. If so, that limit set must be $\bar{x} + T_C(\bar{x})$, and C is then geometrically derivable at $\bar{x}$, cf. Figure 6–3.

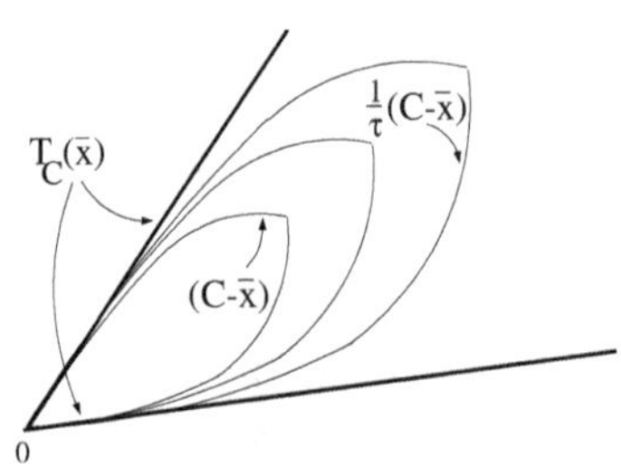

Fig. 6–3. Geometric derivability: local approximation through infinite magnification.

The convergence theory in Chapter 4 can be applied to glean a great amount of information about this mode of tangential approximation. For example, the uniformity result in 4.10 tells us that when C is geometrically derivable at $\bar{x}$ there exists for every $\rho > 0$ and $\varepsilon > 0$ a $\bar{\tau} > 0$ such that, for all $\tau \in (0, \bar{\tau})$:

$$\tau^{-1}(C - \bar{x}) \cap \rho I\!B \subset T_C(\bar{x}) + \varepsilon I\!B, \qquad T_C(\bar{x}) \cap \rho I\!B \subset \tau^{-1}(C - \bar{x}) + \varepsilon I\!B.$$

Here the first inclusion would hold on the basis of the definition of $T_C(\bar{x})$ as an outer limit, but the second inclusion is crucial to geometric derivability as combining the outer limit with the corresponding inner limit.

It's easy to obtain examples from this of how a set C can fail to be geometrically derivable. Such an example is shown in Figure 6–4, where C is the closed subset of $\mathbb{R}^2$ consisting of $\bar{x} = (0,0)$ and all the points $x = (x_1, x_2)$ with $x_1 \neq 0$ and $x_2 = x_1 \sin(\log x_1)$. The tangent cone $T_C(\bar{x})$ clearly consists of all $w = (w_1, w_2)$ with $|w_2| \leq |w_1|$, but the derivable cone consists of just $w = (0,0)$; no nonzero tangent vector w is a derivable tangent vector. This is true because, no matter what the 'scale of magnification', the set $[C - \bar{x}]/\tau = \tau^{-1}C$ will always have the same wavy appearance and therefore can't ever satisfy the conditions of uniform approximation just given.

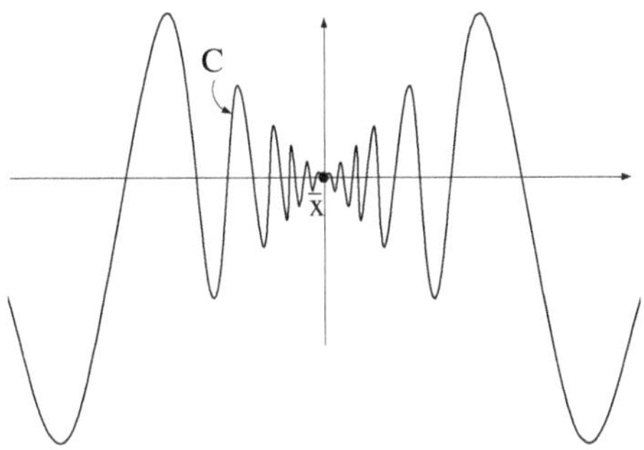

Fig. 6–4. An example where geometric derivability fails.

B. Normal Cones and Clarke Regularity

A natural counterpart to 'tangency' is 'normality', which we develop next. Following traditional patterns, we'll denote by $o(|x - \bar{x}|)$ for $x \in C$ a term with the property that $o(|x - \bar{x}|)/|x - \bar{x}| \to 0$ when $x \xrightarrow[C]{} \bar{x}$ with $x \neq \bar{x}$.

6.3 Definition (normal vectors). *Let $C \subset \mathbb{R}^n$ and $\bar{x} \in C$. A vector v is normal to C at $\bar{x}$ in the regular sense, or a regular normal, written $v \in \widehat{N}_C(\bar{x})$, if*

$$\langle v, x - \bar{x} \rangle \leq o(|x - \bar{x}|) \text{ for } x \in C. \tag{6(4)}$$

It is normal to C at $\bar{x}$ in the general sense, or simply a normal vector, written $v \in N_C(\bar{x})$, if there are sequences $x^\nu \xrightarrow[C]{} \bar{x}$ and $v^\nu \to v$ with $v^\nu \in \widehat{N}_C(x^\nu)$.

6.4 Definition (Clarke regularity of sets). *A set $C \subset \mathbb{R}^n$ is regular at one of its points $\bar{x}$ in the sense of Clarke if it is locally closed at $\bar{x}$ and every normal vector to C at $\bar{x}$ is a regular normal vector, i.e., $N_C(\bar{x}) = \widehat{N}_C(\bar{x})$.*

Note that normal vectors can be of any length, and indeed the zero vector is technically regarded as a regular normal to C at every point $\bar{x} \in C$. The 'o' inequality in 6(4) means that

$$\limsup_{\substack{x \xrightarrow[C]{} \bar{x} \\ x \neq \bar{x}}} \frac{\langle v, x - \bar{x} \rangle}{|x - \bar{x}|} \leq 0, \qquad 6(5)$$

since this is equivalent to asserting that $\max\{0, \langle v, x - \bar{x}\rangle\}$ is a term $o(|x - \bar{x}|)$.

6.5 Proposition (normal cone properties). *At any point $\bar{x}$ of a set $C \subset \mathbb{R}^n$, the set $N_C(\bar{x})$ of all normal vectors is a closed cone, and so too is the set $\widehat{N}_C(\bar{x})$ of all regular normal vectors, which in addition is convex and characterized by*

$$v \in \widehat{N}_C(\bar{x}) \iff \langle v, w \rangle \leq 0 \text{ for all } w \in T_C(\bar{x}). \qquad 6(6)$$

Furthermore,

$$N_C(\bar{x}) = \limsup_{x \xrightarrow[C]{} \bar{x}} \widehat{N}_C(x) \supset \widehat{N}_C(\bar{x}). \qquad 6(7)$$

Proof. The fact that $N_C(\bar{x})$ and $\widehat{N}_C(\bar{x})$ contain 0 has already been noted. Obviously, when either set contains v it also contains λv for any $\lambda \geq 0$. Thus, both sets are cones. The closedness of $N_C(\bar{x})$ is immediate from 6(7), which merely restates the definition of $N_C(\bar{x})$. The closedness and convexity of $\widehat{N}_C(\bar{x})$ will follow from establishing 6(6), since that relation expresses $\widehat{N}_C(\bar{x})$ as the intersection of a family of closed half-spaces $\{v \mid \langle v, w \rangle \leq 0\}$.

First we'll verify the implication '⇒' in 6(6). Consider any $v \in \widehat{N}_C(\bar{x})$ and $w \in T_C(\bar{x})$. By the definition of tangency there exist sequences $x^\nu \xrightarrow[C]{} \bar{x}$ and $\tau^\nu \searrow 0$ such that the vectors $w^\nu = [x^\nu - \bar{x}]/\tau^\nu$ converge to w. Because v satisfies 6(4) we have $\langle v, w^\nu \rangle \leq o(|\tau^\nu w^\nu|)/\tau^\nu \to 0$ and consequently $\langle v, w \rangle \leq 0$.

For the implication '⇐' in 6(6), suppose $v \notin \widehat{N}_C(\bar{x})$, so that 6(4) doesn't hold. From the equivalence of 6(4) with 6(5), there must be a sequence $x^\nu \xrightarrow[C]{} \bar{x}$ with $x^\nu \neq \bar{x}$ such that

$$\liminf_{\nu \to \infty} \frac{\langle v, x^\nu - \bar{x} \rangle}{|x^\nu - \bar{x}|} > 0.$$

Let $w^\nu = [x^\nu - \bar{x}]/|x^\nu - \bar{x}|$ so that $\liminf_\nu \langle v, w^\nu \rangle > 0$ with $|w^\nu| = 1$. Passing to a subsequence if necessary, we can suppose that w^ν converges to a vector w. Then $\langle v, w \rangle > 0$, but also $w \in T_C(\bar{x})$ because w is the limit of $[x^\nu - \bar{x}]/\tau^\nu$ with $\tau^\nu = |x^\nu - \bar{x}| \searrow 0$. Thus, the condition on the right of 6(6) fails for v. □

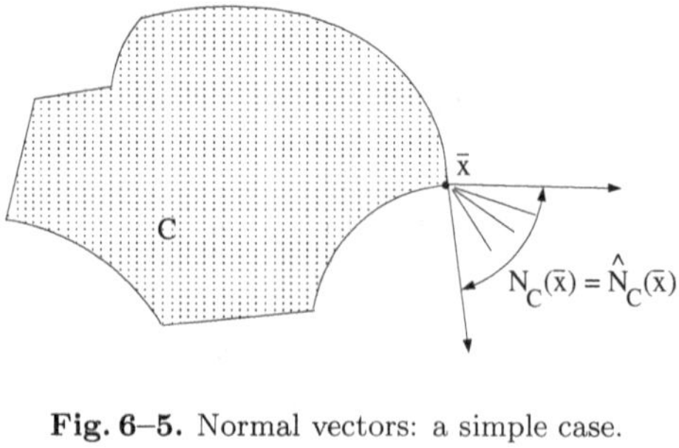

Fig. 6–5. Normal vectors: a simple case.

We call $N_C(\bar{x})$ the *normal cone* to C at $\bar{x}$ and $\widehat{N}_C(\bar{x})$ the cone of regular normals, or the *regular normal cone*.

Some possibilities are shown in Figure 6–5, where as in Figure 6–2 the cone associated with a point $\bar{x}$ is displayed under the 'floating vector' convention as emanating from $\bar{x}$ instead of from the origin, as would be required if their elements were being interpreted as 'position vectors'. When $\bar{x}$ is any point on a curved boundary of the set C, the two cones $N_C(\bar{x})$ and $\widehat{N}_C(\bar{x})$ reduce to a ray which corresponds to the outward normal direction indicated classically. When $\bar{x}$ is an outward corner point at the bottom of C, the two cones still coincide but constitute more than just a ray; a multiplicity of normal directions is present. Interior points $\bar{x}$ of C have $N_C(\bar{x}) = \widehat{N}_C(\bar{x}) = \{0\}$; there aren't any normal directions at such points.

At all points of the set C in Figure 6–5 considered so far, C is regular, so only one cone is exhibited. But C isn't regular at the 'inward corner point' of C. There $\widehat{N}_C(\bar{x}) = \{0\}$, while $N_C(\bar{x})$ is comprised of two rays. This phenomenon is shown in more detail in Figure 6–6, again at an 'inward corner point' $\bar{x}$ where two solitary rays appear. The directions of these two rays arise as limits in hzn $\mathbb{R}^n$ of the regular normal directions at neighboring boundary points, and the angle between them depends on the angle at which the boundary segments meet. In the extreme case of an inward cusp, the two rays would have opposite directions and the normal cone would be a full line.

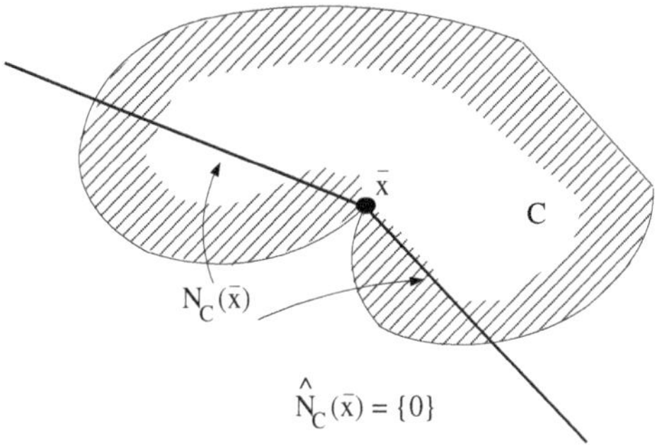

Fig. 6–6. An absence of Clarke regularity.

Nonetheless, in all these cases C is geometrically derivable at $\bar{x}$ despite the absence of regularity. Derivability therefore isn't equivalent to regularity, but we'll see later (in 6.30) that it's a sure consequence of such regularity.

The picture in Figure 6–6 illustrates how normal vectors in the general sense can, in peculiar situations for certain sets C, actually point *into* a part of C. This possibility causes some linguistic discomfort over 'normality', but the cone of such limiting normal vectors comes to dominate technically in formulas and proofs, so there are compelling advantages in reserving the simplest name and notation for it. The limit process in Definition 6.3 is essential, for instance, in achieving closedness properties like the following. Many key results would fail if we tried to make do with regular normal vectors alone.

6.6 Proposition (limits of normal vectors). *If $x^\nu \xrightarrow[C]{} \bar{x}$, $v^\nu \in N_C(x^\nu)$ and $v^\nu \to v$, then $v \in N_C(\bar{x})$. In other words, the set-valued mapping $N_C : x \mapsto N_C(x)$ is outer semicontinuous at $\bar{x}$ relative to C.*

Proof. The set $\{(x,v) \mid v \in N_C(x)\}$ is by definition the closure in $C \times \mathbb{R}^n$ of $\{(x,v) \mid v \in \widehat{N}_C(x)\}$. Hence it is closed relative to $C \times \mathbb{R}^n$. □

In harmony with the general theory of set-valued mappings, it's convenient to think of N_C and $\widehat{N}_C$ not just as mappings on C but of type $\mathbb{R}^n \rightrightarrows \mathbb{R}^n$ with

$$N_C(\bar{x}) = \widehat{N}_C(\bar{x}) := \emptyset \text{ when } \bar{x} \notin C. \tag{6(8)}$$

Then $C = \operatorname{dom} N_C = \operatorname{dom} \widehat{N}_C$, and one has

$$N_C(\bar{x}) = \limsup_{x \to \bar{x}} \widehat{N}_C(x) \text{ for all } \bar{x} \in \mathbb{R}^n \text{ when } C \text{ is closed.}$$

Equivalently $\operatorname{gph} N_C = \operatorname{cl}(\operatorname{gph} \widehat{N}_C)$ in $\mathbb{R}^n \times \mathbb{R}^n$, or $N_C = \operatorname{cl} \widehat{N}_C$ in the sense of the closure or osc hull of a mapping (cf. 5(2)), as long as C is closed.

C. Smooth Manifolds and Convex Sets

The concepts of normal vector and tangent vector are of course invariant under changes of coordinates, whether linear or nonlinear. The effects of such a change can be expressed by way of a smooth (i.e., continuously differentiable) mapping F and its *Jacobian* at a point $\bar{x}$, which we denote by $\nabla F(\bar{x})$. In terms of $F(x) = (f_1(x), \ldots, f_m(x))$ for $x = (x_1, \ldots, x_n)$ the Jacobian is the matrix

$$\nabla F(x) := \left[\frac{\partial f_i}{\partial x_j}(x) \right]_{i,j=1}^{m,n} \in \mathbb{R}^{m \times n}.$$

The expansion $F(x) = F(\bar{x}) + \nabla F(\bar{x})(x - \bar{x}) + o(|x - \bar{x}|)$ summarizes the m coordinate expansions $f_i(x) = f_i(\bar{x}) + \langle \nabla f_i(\bar{x}), x - \bar{x} \rangle + o(|x - \bar{x}|)$. The transpose matrix $\nabla F(\bar{x})^*$ has the gradients $\nabla f_i(\bar{x})$ as its columns, so that

$$\nabla F(\bar{x})^* y = y_1 \nabla f_1(\bar{x}) + \cdots + y_m \nabla f_m(\bar{x}) \text{ for } y = (y_1, \ldots, y_m). \tag{6(9)}$$

6.7 Exercise (change of coordinates). *Let $C = F^{-1}(D) \subset \mathbb{R}^n$ for a smooth mapping $F : \mathbb{R}^n \to \mathbb{R}^m$ and a set $D \subset \mathbb{R}^m$, and suppose $\nabla F(\bar{x})$ has full rank m at a point $\bar{x} \in C$ with image $\bar{u} = F(\bar{x}) \in D$. Then*

$$\begin{aligned} T_C(\bar{x}) &= \{ w \mid \nabla F(\bar{x}) w \in T_D(\bar{u}) \}, \\ N_C(\bar{x}) &= \{ \nabla F(\bar{x})^* y \mid y \in N_D(\bar{u}) \}, \\ \widehat{N}_C(\bar{x}) &= \{ \nabla F(\bar{x})^* y \mid y \in \widehat{N}_D(\bar{u}) \}. \end{aligned}$$

Guide. Consider first the case where $m = n$. Use the inverse mapping theorem to show that only a smooth change of local coordinates is involved, and this doesn't affect the normals and tangents in question except for their coordinate expression. Next tackle the case of $m < n$ by choosing $a_1, \ldots, a_{n-m}$ to be a basis for the $(n-m)$-dimensional subspace $\{w \in \mathbb{R}^n \mid \nabla F(\bar{x})w = 0\}$, and defining $F_0 : \mathbb{R}^n \to \mathbb{R}^m \times \mathbb{R}^{n-m}$ by $F_0(x) := \big(F(x), \langle a_1, x\rangle, \ldots, \langle a_{n-m}, x\rangle\big)$. Then $C = F_0^{-1}(D_0)$ for $D_0 = D \times \mathbb{R}^{n-m}$. The $n \times n$ matrix $\nabla F_0(\bar{x})$ has rank n, so the earlier argument can be applied. □

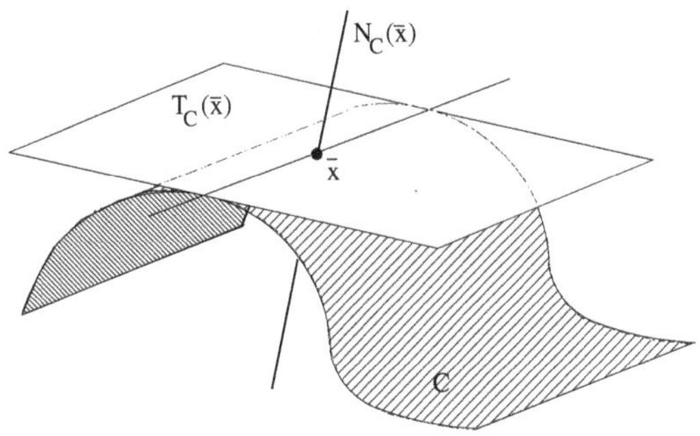

Fig. 6–7. Tangent and normal subspaces for a smooth manifold.

On the basis of this fact it's easy to verify that the general tangent and normal cones defined here reduce in the case of a smooth manifold in $\mathbb{R}^n$ to the tangent and normal spaces associated with such a manifold in classical differential geometry.

6.8 Example (tangents and normals to smooth manifolds). *Let C be a d-dimensional smooth manifold in $\mathbb{R}^n$ around the point $\bar{x} \in C$, in the sense that C can be represented relative to an open neighborhood $O \in \mathcal{N}(\bar{x})$ as the set of solutions to $F(x) = 0$, where $F : O \to \mathbb{R}^m$ is a smooth (i.e., $\mathcal{C}^1$) mapping with $\nabla F(\bar{x})$ of full rank m, where $m = n - d$. Then C is regular at $\bar{x}$ as well as geometrically derivable at $\bar{x}$, and the tangent and normal cones to C at $\bar{x}$ are linear subspaces orthogonally complementary to each other, namely*

$$\begin{aligned} T_C(\bar{x}) &= \{w \in \mathbb{R}^n \mid \nabla F(\bar{x})w = 0\}, \\ N_C(\bar{x}) &= \{v = \nabla F(\bar{x})^* y \mid y \in \mathbb{R}^m\}. \end{aligned}$$

Detail. Locally we have $C = F^{-1}(0)$, so the formulas and the regularity follow from 6.7 with $D = \{0\}$. □

Another important case illustrating both Clarke regularity and geometric derivability is that of a convex subset of $\mathbb{R}^n$, which we take up next.

6.9 Theorem (tangents and normals to convex sets). *A convex set $C \subset \mathbb{R}^n$ is geometrically derivable at any point $\bar{x} \in C$, with*

$$N_C(\bar{x}) = \widehat{N}_C(\bar{x}) = \left\{ v \,\middle|\, \langle v,\, x - \bar{x} \rangle \le 0 \ \text{ for all } x \in C \right\},$$
$$T_C(\bar{x}) = \operatorname{cl} \left\{ w \,\middle|\, \exists \lambda > 0 \ \text{ with } \bar{x} + \lambda w \in C \right\},$$
$$\operatorname{int} T_C(\bar{x}) = \left\{ w \,\middle|\, \exists \lambda > 0 \ \text{ with } \bar{x} + \lambda w \in \operatorname{int} C \right\}.$$

Furthermore, C is regular at $\bar{x}$ as long as C is locally closed at $\bar{x}$.

Proof. Let $K = \{w \mid \exists \lambda > 0 \text{ with } \bar{x} + \lambda w \in C\}$. Because C includes for any of its points the entire line segment joining that point with $\bar{x}$, the vectors in K are precisely the ones expressible as positive scalar multiples of vectors $x - \bar{x}$ for $x \in C$. Thus, they are derivable tangent vectors. Not only do we have $K \subset T_C(\bar{x})$ but also $T_C(\bar{x}) \subset \operatorname{cl} K$ by Definition 6.1, so $T_C(\bar{x}) = \operatorname{cl} K$ and C is geometrically derivable at $\bar{x}$. The relationship between $T_C(\bar{x})$ and $\widehat{N}_C(\bar{x})$ in 6.5 then gives us $\widehat{N}_C(\bar{x}) = \{v \mid \langle v, w \rangle \le 0 \text{ for all } w \in K\}$, and this is the same as the formula for $\widehat{N}_C(\bar{x})$ claimed in the theorem.

Consider now a vector $v \in N_C(\bar{x})$. Fix any $x \in C$. By Definition 6.3 there are sequences $v^\nu \to v$ and $x^\nu \underset{C}{\to} \bar{x}$ with $v^\nu \in \widehat{N}_C(x^\nu)$. According to the formula just established we have $\langle v^\nu,\, x - x^\nu \rangle \le 0$, hence in the limit, $\langle v,\, x - \bar{x} \rangle \le 0$. This is true for arbitrary $x \in C$, so it follows that $x \in \widehat{N}_C(\bar{x})$. The inclusion $N_C(\bar{x}) \supset \widehat{N}_C(\bar{x})$ holds always, so we have $N_C(\bar{x}) = \widehat{N}_C(\bar{x})$.

To deal with $\operatorname{int} T_C(\bar{x})$, let $K_0 = \{w \mid \exists \lambda > 0 \text{ with } \bar{x} + \lambda w \in \operatorname{int} C\}$. Obviously K_0 is a open subset of K, but also $K \subset \operatorname{cl} K_0$ when $K_0 \ne \emptyset$, because $C \subset \operatorname{cl}(\operatorname{int} C)$ when $\operatorname{int} C \ne \emptyset$, hence $K_0 = \operatorname{int} K = \operatorname{int}(\operatorname{cl} K)$; cf. 2.33. Since $\operatorname{cl} K = T_C(\bar{x})$, we conclude that $K_0 = \operatorname{int} T_C(\bar{x})$. □

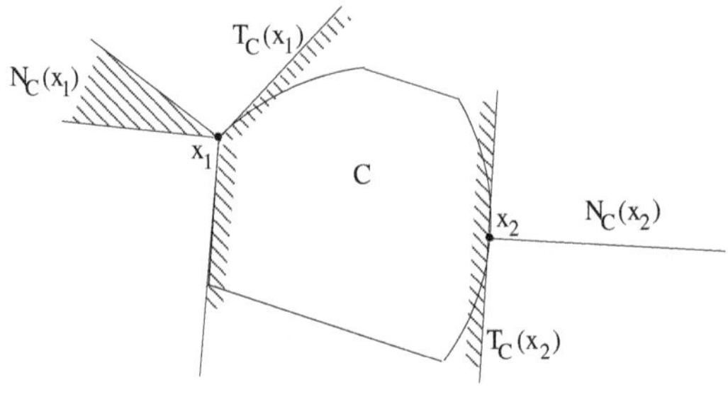

Fig. 6–8. Variational geometry of a convex set.

The normal cone formula in 6.9 relates to the notion of a *supporting half-space* to a convex set C at a point $\bar{x} \in C$, this being a closed half-space $H \supset C$ having $\bar{x}$ on its boundary. The formula says that $N_C(\bar{x})$ consists of (0 and) all the vectors $v \ne 0$ normal to such half-spaces.

6.10 Example (tangents and normals to boxes). *Suppose $C = C_1 \times \cdots \times C_n$, where each C_j is a closed interval in $I\!R$ (not necessarily bounded, perhaps just consisting of a single number). Then C is regular and geometrically derivable at every one of its points $\bar{x} = (\bar{x}_1, \ldots, \bar{x}_n)$. Its tangent cones have the form*

$$T_C(\bar{x}) = T_{C_1}(\bar{x}_1) \times \cdots \times T_{C_n}(\bar{x}_n), \quad \textit{where}$$

$$T_{C_j}(\bar{x}_j) = \begin{cases} (-\infty, 0] & \text{if } \bar{x}_j \text{ is (only) the right endpoint of } C_j, \\ [0, \infty) & \text{if } \bar{x}_j \text{ is (only) the left endpoint of } C_j, \\ (-\infty, \infty) & \text{if } \bar{x}_j \text{ is an interior point of } C_j, \\ [0, 0] & \text{if } C_j \text{ is a one-point interval,} \end{cases}$$

while its normal cones have the form

$$N_C(\bar{x}) = N_{C_1}(\bar{x}_1) \times \cdots \times N_{C_n}(\bar{x}_n), \quad \textit{where}$$

$$N_{C_j}(\bar{x}_j) = \begin{cases} [0, \infty) & \text{if } \bar{x}_j \text{ is (only) the right endpoint of } C_j, \\ (-\infty, 0] & \text{if } \bar{x}_j \text{ is (only) the left endpoint of } C_j, \\ [0, 0] & \text{if } \bar{x}_j \text{ is an interior point of } C_j, \\ (-\infty, \infty) & \text{if } C_j \text{ is a one-point interval.} \end{cases}$$

Detail. In particular, C is a closed convex set. The formulas in Theorem 6.9 relative to a tangent vector $w = (w_1, \ldots, w_m)$ or a normal vector $v = (v_1, \ldots, v_n)$ translate directly into the indicated requirements on the signs of the components w_j and v_j. □

D. Optimality and Lagrange Multipliers

The nonzero *regular* normals to any set C at one of its points $\bar{x}$ can always be interpreted as the normals to the 'curvilinear supporting half-spaces' to C at $\bar{x}$. This description is provided by the following characterization of regular normals. It echoes the geometry in the convex case, where 'linear' supporting half-spaces were seen to suffice.

6.11 Theorem (gradient characterization of regular normals). *A vector v is a regular normal to C at $\bar{x}$ if and only if there is a function h that achieves a local maximum relative to C at $\bar{x}$ and is differentiable there with $\nabla h(\bar{x}) = v$. In fact h can be taken to be smooth on $\mathbb{R}^n$ and such that its global maximum relative to C is achieved uniquely at $\bar{x}$.*

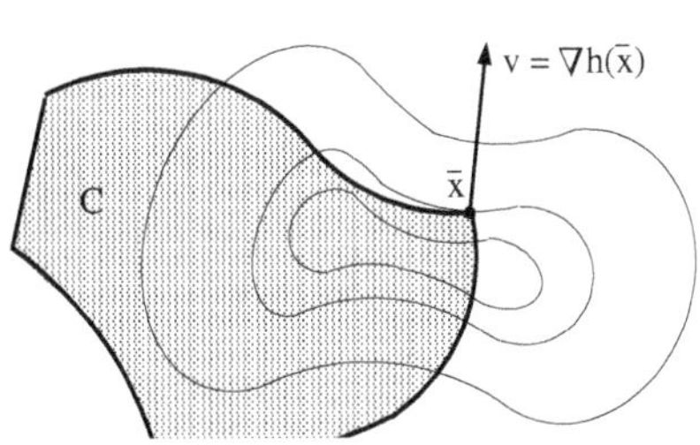

Fig. 6–9. Regular normals as gradients.

Proof. Sufficiency: If h has a local maximum on C at $\bar{x}$ and is differentiable there with $\nabla h(\bar{x}) = v$, we have $h(\bar{x}) \geq h(x) = h(\bar{x}) + \langle v, x - \bar{x} \rangle + o(|x - \bar{x}|)$ locally for $x \in C$. Then $\langle v, x - \bar{x} \rangle + o(|x - \bar{x}|) \leq 0$ for $x \in C$, so that $v \in \widehat{N}_C(\bar{x})$.

Necessity: If $v \in \widehat{N}_C(\bar{x})$, the expression

$$\theta_0(r) := \sup\{\langle v, x - \bar{x}\rangle \mid x \in C, |x - \bar{x}| \le r\} \le r|v|$$

is nondecreasing on $[0, \infty)$ with $0 = \theta_0(0) \le \theta_0(r) \le o(r)$. The function $h_0(x) = \langle v, x - \bar{x}\rangle - \theta_0(|x - \bar{x}|)$ is therefore differentiable at $\bar{x}$ with $\nabla h_0(\bar{x}) = v$ and $h_0(x) \le 0 = h_0(\bar{x})$ for $x \in C$. It has its global maximum over C at $\bar{x}$.

Although h_0 is known only to be differentiable at $\bar{x}$, there exists, as will be demonstrated next, an everywhere continuously differentiable function h with $\nabla h(\bar{x}) = \nabla h_0(\bar{x}) = v$ and $h(\bar{x}) = h_0(\bar{x})$, but $h(x) < h_0(x)$ for all $x \ne \bar{x}$ in C, so that h achieves its maximum over C uniquely at $\bar{x}$. This will finish the proof. We'll produce h in the form $h(x) := \langle v, x - \bar{x}\rangle - \theta(|x - \bar{x}|)$ for a suitable function θ on $[0, \infty)$. It will suffice to construct θ in such a way that $\theta(0) = 0$, $\theta(r) > \theta_0(r)$ for $r > 0$, and θ is continuously differentiable on $(0, \infty)$ with $\theta'(r) \to 0$ as $r \searrow 0$ and $\theta(r)/r \to 0$ as $r \searrow 0$. Since $\theta(0) = 0$, the latter means that at 0 the right derivative of θ exists and equals 0; then certainly $\theta(r) \to 0$ as $r \searrow 0$.

As a first step, define θ_1 by $\theta_1(r) := (1/r)\int_r^{2r} \theta_0(s)ds$ for $r > 0$, $\theta_1(0) = 0$. The integral is well defined despite the possible discontinuities of θ_0, because θ_0 is nondecreasing, a property implying further that θ_0 has right and left limits $\theta_0(r+)$ and $\theta_0(r-)$ at any $r \in (0, \infty)$. The integrand in the definition of $\theta_1(r)$ is bounded below on $(r, 2r)$ by $\theta_0(r+)$ and above by $\theta_0(2r-)$, so we have $\theta_0(r+) \le \theta_1(r) \le \theta_0(2r-)$ for all $r \in (0, \infty)$. Also, $\theta_1(r) = (1/r)[\varphi(2r) - \varphi(r)]$ for the function $\varphi(r) := \int_0^r \theta_0(s)ds$, which is continuous on $(0, \infty)$ with right derivative $\varphi'_+(r) = \theta_0(r+)$ and left derivative $\varphi'_-(r) = \theta_0(r-)$. Hence θ_1 is continuous on $(0, \infty)$ with right derivative $(1/r)[2\theta_0(2r+) - \theta_0(r+) - \theta_1(r)]$ and left derivative $(1/r)[2\theta_0(2r-) - \theta_0(r-) - \theta_1(r)]$, both of which are nonnegative (because $\theta_0(r+) \le \theta_1(r) \le \theta_0(2r-)$ and θ_0 is nondecreasing); consequently θ_1 is nondecreasing. These derivatives and $\theta_1(r)$ itself approach 0 as $r \searrow 0$. Because $\theta_0(r+)/r \le \theta_1(r)/r \le \theta_0(2r-)/r$, we have $\theta_1(r)/r \to 0$ as $r \searrow 0$. Thus, θ_1 has the crucial properties of θ_0 but in addition is continuous on $[0, \infty)$ with left and right derivatives, which agree at points r such that θ_0 is continuous at both r and $2r$, and which are themselves continuous at such points r.

Next define $\theta_2(r) := (1/r)\int_r^{2r} \theta_1(s)ds$ for $r > 0$, $\theta_2(0) = 0$. We have $\theta_2 \ge \theta_1$, hence $\theta_2 \ge \theta_0$. By the reasoning just given, θ_2 inherits from θ_1 the crucial properties of θ_0, but in addition, because of the continuity of θ_1, θ_2 is continuously differentiable on $(0, \infty)$ with $\theta_2'(r) \to 0$ as $r \searrow 0$. Finally, take $\theta(r) = \theta_2(r) + r^2$. This function meets all requirements. □

Although a general normal vector $v \in N_C(\bar{x})$ that isn't regular can't always be described in the manner of Theorem 6.11 and Figure 6–9 as the gradient of a function maximized relative to C at $\bar{x}$, it can be viewed as arising from a sequence of 'nearby' situations of such character, since by definition it's a limit of regular normals, each corresponding to a maximum of some smooth function. Obviously out of such considerations, the limits in Definition 6.3 are indispensable in building up a theory that will eventually be able to take on the issues of what happens when optimization problems are perturbed. This view of what normality signifies is the source of many applications.

The next theorem provides a fundamental connection between variational geometry and optimality conditions more generally.

6.12 Theorem (basic first-order conditions for optimality). *Consider a problem of minimizing a differentiable function f_0 over a set $C \subset \mathbb{R}^n$. A necessary condition for $\bar{x}$ to be locally optimal is*

$$\big\langle \nabla f_0(\bar{x}), w\big\rangle \geq 0 \text{ for all } w \in T_C(\bar{x}), \tag{6(10)}$$

which is the same as $-\nabla f_0(\bar{x}) \in \widehat{N}_C(\bar{x})$ and implies

$$-\nabla f_0(\bar{x}) \in N_C(\bar{x}), \quad \text{or } \nabla f_0(\bar{x}) + N_C(\bar{x}) \ni 0. \tag{6(11)}$$

When C is convex, these tangent and normal cone conditions are equivalent and can be written also in the form

$$\big\langle \nabla f_0(\bar{x}), x - \bar{x}\big\rangle \geq 0 \text{ for all } x \in C, \tag{6(12)}$$

which means that the linearized function $l(x) := f_0(\bar{x}) + \langle \nabla f_0(\bar{x}), x - \bar{x}\rangle$ achieves its minimum over C at $\bar{x}$. When f_0 too is convex, the equivalent conditions are sufficient for $\bar{x}$ to be globally optimal.

Proof. Local optimality of $\bar{x}$ means that $f_0(x) - f_0(\bar{x}) \geq 0$ for all x in a neighborhood of $\bar{x}$ in C. But $f_0(x) - f_0(\bar{x}) = \langle \nabla f_0(\bar{x}), x - \bar{x}\rangle + o\big(|x - \bar{x}|\big)$. Therefore, $\langle -\nabla f_0(\bar{x}), x - \bar{x}\rangle \leq o\big(|x - \bar{x}|\big)$ for $x \in C$, which by definition is the condition $-\nabla f_0(\bar{x}) \in \widehat{N}_C(\bar{x})$. This condition is equivalent by 6.5 to 6(10) and implies 6(11) because $N_C(\bar{x}) \supset \widehat{N}_C(\bar{x})$. In the convex case, 6(10) and 6(11) are equivalent to 6(12) through Theorem 6.9. The sufficiency for global optimality comes then from the inequality $f_0(x) \geq f_0(\bar{x}) + \langle \nabla f_0(\bar{x}), x - \bar{x}\rangle$ in 2.14. □

The normal cone condition 6(11) will later be seen to be equivalent to the tangent cone condition 6(10) not just in the convex case but whenever C is regular at $\bar{x}$ (cf. 6.29). It's typically the most versatile first-order expression of optimality because of its easy modes of specialization and the powerful calculus that can be built around it. When $\bar{x} \in \operatorname{int} C$, it reduces to *Fermat's rule*, the classical requirement that $\nabla f_0(\bar{x}) = 0$, inasmuch as $N_C(\bar{x}) = \{0\}$. In general, it reflects the nature of the boundary of C near $\bar{x}$, as it should.

For a smooth manifold as in 6.8 and Figure 6–7, we see from 6(11) that $\nabla f_0(\bar{x})$ must satisfy an orthogonality condition at any point where f_0 has a local minimum. For a convex set C, the necessary condition takes the form of requiring $-\nabla f_0(\bar{x})$ to be normal to a supporting half-space to C at $\bar{x}$, as observed through the equivalent statement 6(12). For a box as in 6.10, it comes down to *sign restrictions* on the components $(\partial f_0/\partial x_j)(\bar{x})$ of $\nabla f_0(\bar{x})$.

The first-order conditions in Theorem 6.12 fit into a broader picture in which the gradient mapping ∇f_0 is replaced by any mapping $F : C \to \mathbb{R}^n$.

This format has rich applications beyond the characterization of a minimum relative to C, for instance in descriptions of 'equilibrium'.

6.13 Example (variational inequalities and complementarity). *For any set $C \subset \mathbb{R}^n$ and any mapping $F : C \to \mathbb{R}^n$, the relation*

$$F(\bar{x}) + N_C(\bar{x}) \ni 0$$

is the variational condition for C and F, the vector $\bar{x} \in C$ being a solution. When C is convex, it is also called the variational inequality for C and F, because it can be written equivalently in the form

$$\bar{x} \in C, \qquad \big\langle F(\bar{x}), x - \bar{x} \big\rangle \geq 0 \ \text{ for all } x \in C,$$

and interpreted as saying that the linear function $l(x) = \langle F(\bar{x}), x \rangle$ achieves its minimum over C at $\bar{x}$. In the special case where $C = \mathbb{R}^n_+$ it is known as the complementarity condition for the mapping F, because it comes out as

$$\begin{aligned} &\bar{x}_j \geq 0, \quad \bar{v}_j \geq 0, \quad \bar{x}_j \bar{v}_j = 0 \ \text{ for } j = 1, \ldots n, \ \text{ where} \\ &\bar{v} = F(\bar{x}), \quad \bar{x} = (\bar{x}_1, \ldots, \bar{x}_n), \quad \bar{v} = (\bar{v}_1, \ldots, \bar{v}_n), \end{aligned}$$

which can be summarized vectorially by the notation $0 \leq \bar{x} \perp F(\bar{x}) \geq 0$.

Detail. The reduction in the convex case is seen from 6.9. When $C = \mathbb{R}^n_+$ the condition that $\bar{v} + N_C(\bar{x}) \ni 0$ says that $-\bar{v}_j \in N_{\mathbb{R}_+}(\bar{x}_j)$ for $j = 1, \ldots n$, cf. 6.10. This requires $\bar{v}_j = 0$ when $\bar{x}_j > 0$ but merely $\bar{v}_j \geq 0$ when $\bar{x}_j = 0$. □

For $F = \nabla f_0$ the complementarity condition in 6.13 is the first-order optimality condition of Theorem 6.12 for the minimization of f_0 over $\mathbb{R}^n_+$.

Theorem 6.12 leads to a broad theory of Lagrange multipliers. The most classical case of Lagrange multipliers, for minimization subject to smooth equality constraints only, is already at hand in the combination of condition 6(11) with the formula for $N_C(\bar{x})$ in Example 6.8, where C is specified by the constraints $f_i(x) = 0$, $i = 1, \ldots, m$. If the gradients of the constraint functions are linearly independent at $\bar{x}$, a local minimum of f_0 at $\bar{x}$ relative to these constraints entails the existence of a vector $\bar{y} = (\bar{y}_1, \ldots, \bar{y}_m) \in \mathbb{R}^m$ such that

$$\nabla f_0(\bar{x}) + \bar{y}_1 \nabla f_1(\bar{x}) + \cdots + \bar{y}_m \nabla f_m(\bar{x}) \ = \ 0,$$

cf. 6(9). The next theorem extends the normal cone formula in Example 6.8 to constraint systems that are far more general, and in so doing it gives rise to much wider results involving multipliers $\bar{y}_i$.

6.14 Theorem (normal cones to sets with constraint structure). *Let*

$$C = \big\{ x \in X \,\big|\, F(x) \in D \big\}$$

for closed sets $X \subset \mathbb{R}^n$ and $D \subset \mathbb{R}^m$ and a $\mathcal{C}^1$ mapping $F : \mathbb{R}^n \to \mathbb{R}^m$, written componentwise as $F(x) = \big(f_1(x), \ldots, f_m(x)\big)$. At any $\bar{x} \in C$ one has

$$\widehat{N}_C(\bar{x}) \supset \left\{ \sum_{i=1}^m y_i \nabla f_i(\bar{x}) + z \,\middle|\, y \in \widehat{N}_D\big(F(\bar{x})\big),\ z \in \widehat{N}_X(\bar{x}) \right\},$$

where $y = (y_1, \ldots, y_m)$. *On the other hand, one has*

$$N_C(\bar{x}) \subset \left\{ \sum_{i=1}^m y_i \nabla f_i(\bar{x}) + z \,\middle|\, y \in N_D\big(F(\bar{x})\big),\ z \in N_X(\bar{x}) \right\}$$

at any $\bar{x} \in C$ *satisfying the constraint qualification that*

$$\begin{cases} \text{the only vector } y \in N_D\big(F(\bar{x})\big) \text{ for which} \\ -\sum_{i=1}^m y_i \nabla f_i(\bar{x}) \in N_X(\bar{x}) \text{ is } y = (0, \ldots, 0). \end{cases}$$

If in addition to this constraint qualification the set X *is regular at* $\bar{x}$ *and* D *is regular at* $F(\bar{x})$, *then* C *is regular at* $\bar{x}$ *and*

$$N_C(\bar{x}) = \left\{ \sum_{i=1}^m y_i \nabla f_i(\bar{x}) + z \,\middle|\, y \in N_D\big(F(\bar{x})\big),\ z \in N_X(\bar{x}) \right\}.$$

Proof. For simplicity we can assume that X is compact, and hence that C is compact, since the analysis is local and wouldn't be affected if X were replaced by its intersection with some ball $\mathbb{B}(\bar{x}, \delta)$. Likewise we can assume D is compact, since nothing would be lost by intersecting D with a ball $\mathbb{B}(F(\bar{x}), \varepsilon)$ and taking δ small enough that $|F(x) - F(\bar{x})| < \varepsilon$ when $|x - \bar{x}| \le \delta$.

We'll first verify the inclusion for $\widehat{N}_C(\bar{x})$. The notation in 6(9) will be convenient. Suppose $v = \nabla F(\bar{x})^* y + z$ with $y \in \widehat{N}_D(\bar{x})$ and $z \in \widehat{N}_X(\bar{x})$. We have $\big\langle y,\, F(x) - F(\bar{x}) \big\rangle \le o\big(F(x) - F(\bar{x})\big)$ when $F(x) \in D$, where

$$F(x) - F(\bar{x}) = \nabla F(\bar{x})(x - \bar{x}) + o\big(|x - \bar{x}|\big).$$

Therefore $\big\langle y, \nabla F(\bar{x})(x - \bar{x}) \big\rangle \le o\big(|x - \bar{x}|\big)$ when $F(x) \in D$, the inner product on the left being the same as $\langle \nabla F(\bar{x})^* y,\, x - \bar{x} \rangle$, i.e., $\langle v - z,\, x - \bar{x} \rangle$. At the same time we have $\langle z,\, x - \bar{x} \rangle \le o\big(|x - \bar{x}|\big)$ when $x \in X$, so we get $\langle v,\, x - \bar{x} \rangle \le o\big(|x - \bar{x}|\big)$ for $x \in C$ and conclude that $v \in \widehat{N}_C(\bar{x})$.

For the sake of deriving the inclusion for $N_C(\bar{x})$, assume from now on that the constraint qualification holds at $\bar{x}$. It must also hold at all points $x \in C$ in some neighborhood of $\bar{x}$ relative to C, for otherwise we could contradict it by considering a sequence $x^\nu \underset{C}{\to} \bar{x}$ that yields $-\nabla F(x^\nu)^* y^\nu \in N_X(x^\nu)$ for nonzero vectors $y^\nu \in N_D(F(x^\nu))$; such a sequence can be normalized to $|y^\nu| = 1$, and then by selecting any cluster point y we would get $-\nabla F(\bar{x})^* y \in N_X(\bar{x})$ through 6.6, yet $|y| = 1$.

Next, as a transitory step, we demonstrate that the inclusion claimed for $N_C(\bar{x})$ holds for $\widehat{N}_C(\bar{x})$. Let $v \in \widehat{N}_C(\bar{x})$. By Theorem 6.11 there's a smooth function h on $\mathbb{R}^n$ such that $\operatorname{argmax}_C h = \{\bar{x}\}$, $\nabla h(\bar{x}) = v$. We take any sequence of values $\tau^\nu \searrow 0$ and analyze for each ν the problem of minimizing over $X \times D$ the $\mathcal{C}^1$ function

$$\varphi^\nu(x, u) := -h(x) + \frac{1}{2\tau^\nu} \big| F(x) - u \big|^2.$$

Through our arrangement that X and D are compact, the minimum is attained at some point (x^ν, u^ν) (not necessarily unique). Moreover $(x^\nu, u^\nu) \to (\bar{x}, F(\bar{x}))$; cf. 1.21. The optimality condition in Theorem 6.12 gives us

$$-\nabla_x \varphi^\nu(x^\nu, u^\nu) =: z^\nu \in N_X(x^\nu), \qquad -\nabla_u \varphi^\nu(x^\nu, u^\nu) =: y^\nu \in N_D(u^\nu),$$

inasmuch as $\operatorname{argmin}_{x\in X} \varphi^\nu(x, u^\nu) = \{x^\nu\}$ and $\operatorname{argmin}_{u\in D} \varphi^\nu(x^\nu, u) = \{u^\nu\}$. Differentiating φ^ν in u, we see that $y^\nu = [F(x^\nu) - u^\nu]/\tau^\nu$. Differentiating φ^ν next in x, we get

$$z^\nu = \nabla h(x^\nu) - \nabla F(x^\nu)^* y^\nu \ \text{ with } \ \nabla F(x^\nu) \to \nabla F(\bar{x}),\ \nabla h(x^\nu) \to v.$$

By passing to subsequences, we can reduce to having the sequence of vectors $y^\nu \in N_D(u^\nu)$ either convergent to some y or such that $\lambda^\nu y^\nu \to y \neq 0$ for a choice of scalars $\lambda^\nu \searrow 0$. In both cases we have $y \in N_D\big(F(\bar{x})\big)$ by 6.6, because $N_D(u^\nu)$ is a cone and $u^\nu \to F(\bar{x})$.

If $y^\nu \to y$, we have at the same time that $z^\nu \to z := v - \nabla F(\bar{x})^* y$ with $z \in N_X(\bar{x})$, again by virtue of 6.6. This yields the desired representation $v = \nabla F(\bar{x})^* y + z$. On the other hand, if $\lambda^\nu y^\nu \to y \neq 0$, $\lambda^\nu \searrow 0$, we obtain from $\lambda^\nu z^\nu = \lambda^\nu \nabla h(x^\nu) - \nabla F(x^\nu)^* \lambda^\nu y^\nu$ that $\lambda^\nu z^\nu \to z := -\nabla F(\bar{x})^* y$, $z \in N_X(\bar{x})$, which produces a representation $0 = \nabla F(\bar{x})^* y + z$ of the sort forbidden by the constraint qualification. Therefore, only the first case is viable.

This demonstrates that $\widehat{N}_C(\bar{x}) \subset S(\bar{x})$, where we now denote by $S(x)$ for any $x \in C$ the set of all vectors of the form $\nabla F(x)^* y + z$ with $y \in N_D(F(x))$ and $z \in N_X(x)$. The argument has depended on our assumption that the constraint qualification is satisfied at $\bar{x}$, but we've observed that it's satisfied then for all x in a neighborhood of $\bar{x}$ relative to C. Hence we've actually proved that $\widehat{N}_C(x) \subset S(x)$ for all x in such a neighborhood. In order to get $N_C(\bar{x}) \subset S(\bar{x})$, we can therefore use the fact that $N_C(\bar{x}) = \limsup_{x\to\bar{x}} \widehat{N}_C(x)$ to reduce the task to verifying that S is osc at $\bar{x}$ relative to C.

Let $x^\nu \xrightarrow[C]{} \bar{x}$ and $v^\nu \to v$ with $v^\nu \in S(x^\nu)$, so that $v^\nu = \nabla F(x^\nu)^* y^\nu + z^\nu$ with $y^\nu \in N_D(F(x^\nu))$ and $z^\nu \in N_X(x^\nu)$. We can revert once more to two cases: either $(y^\nu, z^\nu) \to (y, z)$ or $\lambda^\nu (y^\nu, z^\nu) \to (y, z) \neq (0,0)$ for some sequence $\lambda^\nu \searrow 0$. In the first case we obtain in the limit that $v = \nabla F(\bar{x})^* y + z$ with $y \in N_D(F(\bar{x}))$ and $z \in N_X(\bar{x})$ by 6.6, hence $v \in S(\bar{x})$ as desired. But the second case is impossible, because it would give us $\lambda^\nu v^\nu = \nabla F(x^\nu)^* \lambda^\nu y^\nu + \lambda^\nu z^\nu$ and in the limit $0 = \nabla F(\bar{x})^* y + z$ in contradiction to the constraint qualification. This confirms that $N_C(\bar{x}) \subset S(\bar{x})$.

All that remains is the theorem's assertion when X is regular at $\bar{x}$ and D is regular at $F(\bar{x})$. Then $\widehat{N}_D\big(F(\bar{x})\big) = N_D\big(F(\bar{x})\big)$ and $\widehat{N}_X(\bar{x}) = N_X(\bar{x})$, so the inclusions already developed for $\widehat{N}_C(\bar{x})$ and $N_C(\bar{x})$, along with the general inclusion $\widehat{N}_C(\bar{x}) \subset N_C(\bar{x})$, imply that these cones coincide and equal $S(\bar{x})$. Since C is closed (because X and D are closed and F is continuous), this tells us also that C is regular at $\bar{x}$. □

A result for tangent cones, parallel to Theorem 6.14, will emerge in 6.31.

It's notable that the normal vector representation for smooth manifolds in Example 6.8 is established independently by 6.14 as the case of $D = \{0\}$. This is interesting technically because it sidesteps the usual appeal to the classical inverse mapping theorem through a change of coordinates as in 6.7, which was the justification of Example 6.8 that we resorted to earlier.

The case of Theorem 6.14 where C is defined by inequalities $f_i(x) \leq 0$ for $i = 1, \ldots, m$ corresponds to $D = I\!R^n_-$ and is shown in Figure 6–10. Then $N_C(\bar{x})$ is the convex cone generated by the gradients $\nabla f_i(\bar{x})$ of the constraints that are active at $\bar{x}$.

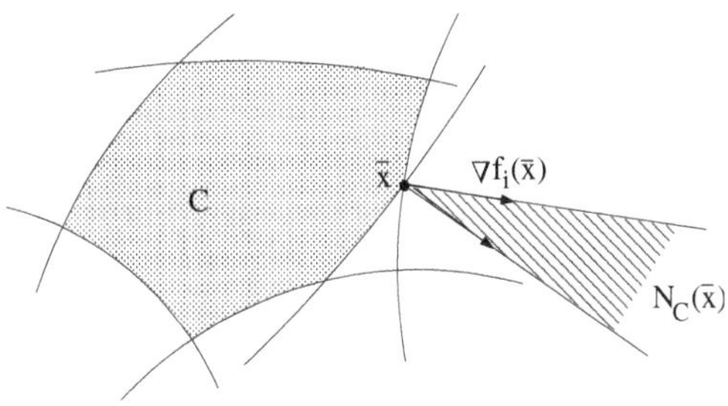

Fig. 6–10. Normals to a set defined by inequality constraints.

This idea carries forward to situations where f_i is constrained to lie in a closed interval D_i, which may place an upper bound, a lower bound, or both on $f_i(x)$, or represent an equality constraint when D_i is a one-point interval. Then D is a general box, and when the normal vector representation formula in 6.10 is combined with the optimality condition 6(11) we obtain a powerful Lagrange multiplier rule.

6.15 Corollary (Lagrange multipliers). *Consider the problem*

$$\text{minimize } f_0(x) \text{ subject to } x \in X \text{ and } f_i(x) \in D_i \text{ for } i = 1, \ldots, m,$$

where X is a closed set in $I\!R^n$, D_i is a closed interval in $I\!R$, and the functions f_i are of class $\mathcal{C}^1$. Let $\bar{x}$ be locally optimal, and suppose the following constraint qualification holds at $\bar{x}$: no vector $y = (y_1, \ldots, y_m) \neq (0, \ldots, 0)$ satisfies

$$-[y_1 \nabla f_1(\bar{x}) + \cdots + y_m \nabla f_m(\bar{x})] \in N_X(\bar{x})$$

and meets the sign restrictions that

$$\begin{cases} y_i = 0 & \text{if } f_i(\bar{x}) \text{ lies in the interior of } D_i, \\ y_i \geq 0 & \text{if } f_i(\bar{x}) \text{ is (only) the right endpoint of } D_i, \\ y_i \leq 0 & \text{if } f_i(\bar{x}) \text{ is (only) the left endpoint of } D_i, \\ y_i \text{ free} & \text{if } D_i \text{ is a one-point interval.} \end{cases}$$

Then there is a vector $\bar{y} = (\bar{y}_1, \ldots, \bar{y}_m)$ meeting these sign restrictions with

$$-[\nabla f_0(\bar{x}) + \bar{y}_1 \nabla f_1(\bar{x}) + \cdots + \bar{y}_m \nabla f(\bar{x})] \in N_X(\bar{x}).$$

Proof. This is the case of 6.14 with $D = D_1 \times \cdots \times D_m$, as applied to 6(11). For normal vectors to boxes, see 6.10. □

The four cases of sign restriction in this Lagrange multiplier rule correspond to the constraint $f_i(x) \in D_i$ being inactive, upper active, lower active, or doubly active—an equality constraint. When $X = I\!R^n$, the cone $N_X(\bar{x})$ reduces to $\{0\}$ and the gradient relations turn into simple equations. When X is a box, on the other hand, these relations take the form of sign restrictions on the partial derivatives of these gradient combinations, cf. 6.10.

E. Proximal Normals and Polarity

The basic optimality condition in Theorem 6.12 is useful not only as a foundation for such multiplier rules but for theoretical purposes. This is illustrated in the following analysis of a special kind of normal vector.

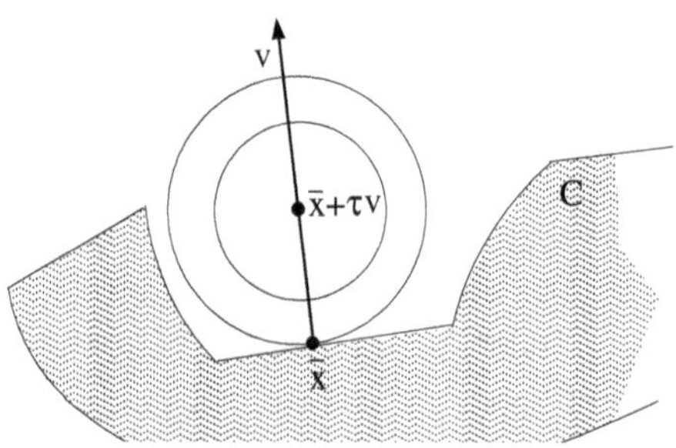

Fig. 6–11. Proximal normals from nearest-point projections.

6.16 Example (proximal normals). *Consider a set $C \subset I\!R^n$ and its projection mapping P_C (which assigns to each $x \in I\!R^n$ the point, or points, of C nearest to x). For any $x \in I\!R^n$,*

$$\bar{x} \in P_C(x) \quad \Longrightarrow \quad x - \bar{x} \in \widehat{N}_C(\bar{x}), \ \text{ so } \lambda(x - \bar{x}) \in \widehat{N}_C(\bar{x}) \ \text{ for all } \lambda \geq 0.$$

Any such vector $v = \lambda(x - \bar{x})$ is called a proximal normal to C at $\bar{x}$. The proximal normals to C at $\bar{x}$ are thus the vectors v such that $\bar{x} \in P_C(\bar{x} + \tau v)$ for some $\tau > 0$. Then actually $P_C(\bar{x} + \tau' v) = \{\bar{x}\}$ for every $\tau' \in (0, \tau)$.

Detail. For any point $\tilde{x} \in I\!R^n$, we have $P_C(\tilde{x}) = \operatorname{argmin}_{x \in C} f_0(x)$ with $f_0(x) = \frac{1}{2}|x - \tilde{x}|^2$, $\nabla f_0(x) = x - \tilde{x}$. Hence by 6.12, $\bar{x} \in P_C(\tilde{x})$ implies $\tilde{x} - \bar{x} \in \widehat{N}_C(\bar{x})$. Here we apply these relationships to cases where $\tilde{x} = \bar{x} + \tau v$, see Figure 6–11. It's elementary from the triangle inequality that when $\bar{x}$ is one of the points of C nearest to $\tilde{x}$, then for all intermediate points x on the line segment joining $\tilde{x}$ with $\bar{x}$, the *unique* nearest point of C to x is $\bar{x}$. □

In essence, a vector $v \neq 0$ is a proximal normal to C at $\bar{x}$ when v points from $\bar{x}$ toward the center of a closed ball that touches C only at $\bar{x}$. This condition is more restrictive than one characterizing regular normals in Theorem 6.11 and amounts to the existence of $\varepsilon > 0$ such that

$$\langle v, x - \bar{x} \rangle \leq \varepsilon |x - \bar{x}|^2 \text{ for all } x \in C.$$

6.17 Proposition (proximality of normals to convex sets). *For a convex set C, every normal vector is a proximal normal vector. The normal cone mapping N_C and the projection mapping P_C are thus related by*

$$N_C = P_C^{-1} - I, \qquad P_C = (I + N_C)^{-1}.$$

Proof. For any vector $v \in N_C(\bar{x})$, the convex function $f(x) = \frac{1}{2}\left|x - (\bar{x} + v)\right|^2$ has gradient $\nabla f(\bar{x}) = -v$ and thus satisfies the first-order optimality condition $-\nabla f(\bar{x}) \in N_C(\bar{x})$. When C is convex, this condition is not only necessary by 6.12 for $\bar{x}$ to minimize f over C but sufficient, so that we have $v \in N_C(\bar{x})$ if and only if $\bar{x} \in P_C(\bar{x} + v)$, in fact $P_C(\bar{x} + v) = \{\bar{x}\}$ because f is strictly convex. Hence every normal is a proximal normal, and the graph of P_C consists of the pairs (z, x) such that $z - x \in N_C(x)$, or equivalently $z \in (I + N_C)(x)$. This means that $P_C = (I + N_C)^{-1}$, or equivalently $N_C = P_C^{-1} - I$. □

For a nonconvex set C, there can be regular normals that aren't proximal normals, even when C is defined by smooth inequalities. This is illustrated by

$$C = \left\{x = (x_1, x_2) \in \mathbb{R}^2 \,\middle|\, x_2 \geq x_1^{3/5},\, x_2 \geq 0\right\},$$

where the vector $v = (1, 0)$ is a regular normal vector at $\bar{x} = (0, 0)$ but no point of $\left\{\bar{x} + \tau v \,\middle|\, \tau > 0\right\}$ projects onto $\bar{x}$, cf. Figure 6–12(a).

The proximal normals at $\bar{x}$ always form a cone, and this cone is convex; these facts are evident from the description of proximal normals just provided. But in contrast to the cone of regular normals the cone of proximal normals needn't be closed—as Figure 6–12(a) likewise makes clear. Nor is it true that the closure of the cone of proximal normals always equals the cone of regular normals, as seen from the similar example in Figure 6–12(b), where only the zero vector is a proximal normal at $\bar{x}$.

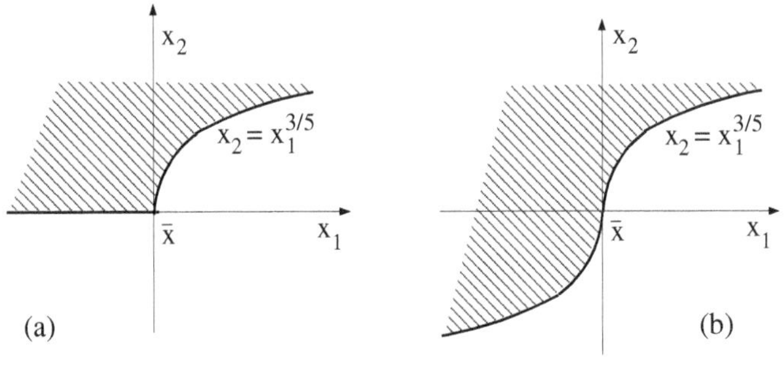

Fig. 6–12. Regular normals versus proximal normals.

6.18 Exercise (approximation of normals). *Let C be a closed subset of $\mathbb{R}^n$, and let $\bar{x} \in C$ and $\bar{v} \in N_C(\bar{x})$.*

(a) *For any $\varepsilon > 0$ there exist $x \in \mathbb{B}(\bar{x}, \varepsilon) \cap C$ and $v \in \mathbb{B}(\bar{v}, \varepsilon) \cap N_C(x)$ such that v is a proximal normal to C at x.*

(b) *For any sequence of closed sets $C^\nu \neq \emptyset$ such that $\limsup_\nu C^\nu = C$ there is a subsequence $\{C^\nu\}_{\nu \in N}$ (the index set $N \in \mathcal{N}_\infty^\#$ being identifiable with $\mathbb{N}$ itself when actually $C^\nu \to C$) along with points $x^\nu \in C^\nu$ and proximal normals $v^\nu \in N_{C^\nu}(x^\nu)$ such that $x^\nu \underset{N}{\to} \bar{x}$ and $v^\nu \underset{N}{\to} \bar{v}$. Thus in particular, in terms of graphical convergence of normal cone mappings, one has*

$$N_C \subset \text{g-lim sup}_\nu N_{C^\nu}.$$

Guide. In (a), argue from Definition 6.3 that it suffices to treat the case where $\bar{v}$ is a regular normal with $|\bar{v}| = 1$. For a sequence of values $\varepsilon^\nu \searrow 0$ consider the points $\tilde{x}^\nu := \bar{x} + \varepsilon^\nu \bar{v}$ and show that their projections $P_C(\tilde{x}^\nu)$ yield points $x^\nu \to \bar{x}$ with proximal normals $v^\nu \to \bar{v}$.

In (b) it suffices through a diagonalization argument based on (a) to consider the case where $\bar{v}$ is a proximal normal to C at $\bar{x}$: for some $\tau > 0$ one has $\bar{x} \in P_C(\bar{x} + \tau\bar{v})$. By virtue of 4.19 there's an index set $N \in \mathcal{N}_\infty^\#$ such that $\bar{x} \in \lim_{\nu \in N} C^\nu =: D$. Argue that $\bar{x} \in P_D(\bar{x} + \tau\bar{v})$ and invoke the fact in 5.35 that correspondingly $\text{g-lim}_{\nu \in N} P_{C^\nu} = P_D$ in order to generate the required sequences of elements x^ν and v^ν. □

An example where the graphical convergence inclusion in 6.18(b) is strict is furnished in $\mathbb{R}^2$ by taking C to be the horizontal axis and C^ν to be the graph of $x_2 = \nu^{-1}\sin(\nu x_1)$. At each point $\bar{x} = (\bar{x}_1, 0)$ of C, the cone $N_C(\bar{x})$ is $\{0\} \times \mathbb{R}$, whereas the cone $\left[\text{g-lim sup}_\nu N_{C^\nu}\right](\bar{x})$ is all of $\mathbb{R}^2$. Here, by the way, every normal to C^ν or C is a proximal normal.

6.19 Exercise (characterization of boundary points). *For a closed set $C \subset \mathbb{R}^n$, a point $\bar{x} \in C$ is a boundary point if and only if there is a vector $v \neq 0$ in $N_C(\bar{x})$. On the other hand, $\bar{x} \in \operatorname{int} C$ if and only if $N_C(\bar{x})$ is just the zero cone. For nonclosed C, one has $N_C(\bar{x}) \subset N_{\operatorname{cl} C}(\bar{x})$.*

Guide. Argue that a boundary point $\bar{x}$ can be approached by a sequence of points $\tilde{x}^\nu \notin C$, and the projections of those points on C yield proximal normals of length 1 at points arbitrarily close to $\bar{x}$. □

This characterization of boundary and interior points has an important consequence for closed convex sets, whose nonzero normals correspond to supporting half-spaces.

6.20 Theorem (envelope representation of convex sets). *A nonempty, closed, convex set in $\mathbb{R}^n$ is the intersection of its supporting half-spaces. Thus, a set C is closed and convex if and only if C is the intersection of a collection of closed half-spaces, or equivalently, the set of solutions to some system of linear inequality constraints. Such a set is regular at every point. (Here $\mathbb{R}^n$ is the intersection of the empty collection of closed half-spaces.)*

For a closed, convex cone, all supporting half-spaces have the origin as a boundary point. On the other hand the intersection of any collection of closed half-spaces having the origin as a boundary point is a closed, convex cone.

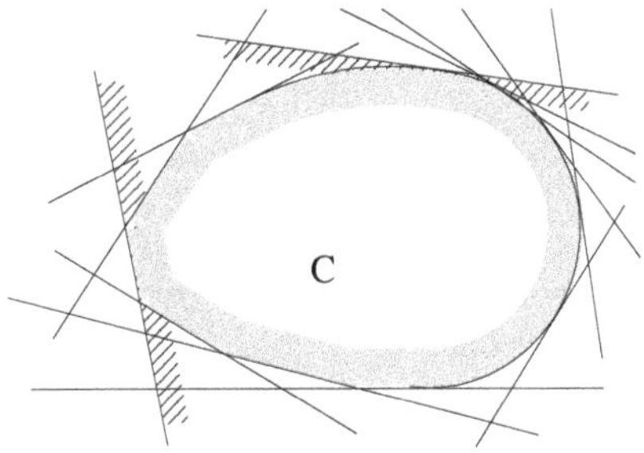

Fig. 6–13. A closed, convex set as the intersection of its supporting half-spaces.

Proof. If C is the intersection of a collection of closed half-spaces, each of those being itself a closed, convex set expressible by a single linear inequality, then certainly C is closed and convex. On the other hand, if C is a closed, convex set, and $\tilde{x}$ is any point not in C, any point $\bar{x} \in P_C(\tilde{x})$ (at least one of which exists, cf. 1.20) gives a proximal normal $v = \tilde{x} - \bar{x} \neq 0$ to C at $\bar{x}$ as in 6.16. The half-space $H = \{x \mid \langle v, x - \bar{x}\rangle \leq 0\}$ then supports C at $\bar{x}$ (cf. 6.9), but it doesn't contain $\tilde{x}$, because $\langle v, \tilde{x} - \bar{x}\rangle = |v|^2 > 0$.

When C is a cone, any half-space $\{x \mid \langle a, x\rangle \leq \alpha\} \supset C$ must have $\alpha \geq 0$ (because $0 \in C$), and then also $\{x \mid \langle a, x\rangle \leq 0\} \supset C$ (because otherwise for some $x \in C$ and large $\lambda > 0$ we would have $\langle a, \lambda x\rangle > \alpha$ despite $\lambda x \in C$). □

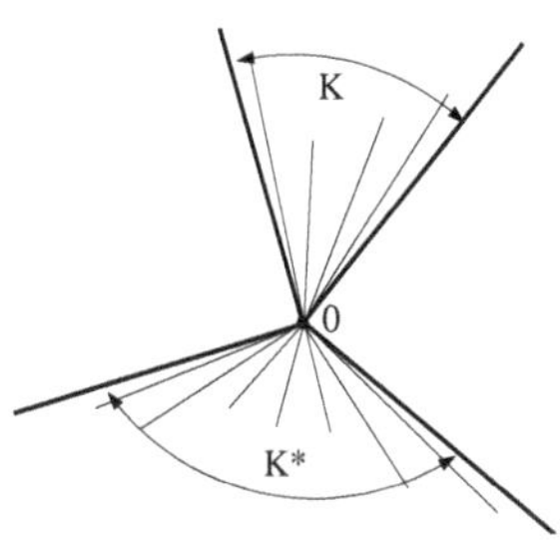

Fig. 6–14. Convex cones polar to each other.

The envelope representation of convex cones in Theorem 6.20 expresses a form of duality which will be important in understanding the relationships between tangent vectors and normal vectors, especially in connection with regularity. For any cone $K \subset \mathbb{R}^n$, the *polar* of K is defined to be the cone

$$K^* := \{v \mid \langle v, w\rangle \leq 0 \text{ for all } w \in K\}. \qquad 6(14)$$

The *bipolar* is the cone $K^{**} = (K^*)^*$. Whenever $K_1 \subset K_2$, one has $K_1^* \supset K_2^*$ and $K_1^{**} \subset K_2^{**}$.

6.21 Corollary (polarity correspondence). *For a cone $K \subset \mathbb{R}^n$, the polar cone K^* is closed and convex, and $K^{**} = \operatorname{cl}(\operatorname{con} K)$. Thus, in the class of all closed, convex cones the correspondence $K \leftrightarrow K^*$ is one-to-one, with $K^{**} = K$.*

Proof. By definition, K^* is the intersection of a collection of closed half-spaces H having $0 \in \operatorname{bdry} H$, namely all those of the form $H = \{v \mid \langle v, w\rangle \leq 0\}$ with $w \in K$. Hence it is a closed, convex cone. But K^{**} is the intersection of all the half-spaces $\{w \mid \langle v, w\rangle \leq 0\}$ that include K, or equivalently, include $\operatorname{cl}(\operatorname{con} K)$. This intersection equals $\operatorname{cl}(\operatorname{con} K)$ by Theorem 6.20. □

6.22 Exercise (pointedness and polarity). *A convex cone K has nonempty interior if and only if its polar cone K^* is pointed. In fact*

$$w \in \operatorname{int} K \quad \Longleftrightarrow \quad \langle v, w\rangle < 0 \ \textit{ for all nonzero } \ v \in K^*.$$

Here if $K = K_0^*$ for some closed cone K_0, not necessarily convex, one can replace K^* by K_0 in each instance.

Guide. Argue that a convex set containing 0 has empty interior if and only if it lies in a hyperplane through the origin (e.g., utilize the existence of simplex neighborhoods, cf. 2.28(e)). Argue next that a closed, convex cone fails to be pointed if and only if it includes some line through the origin (cf. 3.7 and 3.13). Then use the fact that at boundary points of $\operatorname{cl} K$ a supporting hyperplane exists (cf. 2.33—or 6.19, 6.9). For a closed cone K_0 such that $K^* = \operatorname{cl}(\operatorname{con} K_0)$, the pointedness of K^* is equivalent to that of K_0 (cf. 3.15). □

The polar of the nonnegative orthant $\mathbb{R}^n_+$ is the nonpositive orthant $\mathbb{R}^n_-$, and vice versa. The zero cone $\{0\}$ and the full cone $\mathbb{R}^n$ likewise furnish an example of cones that are polar to each other. The polar of a ray $\{\tau w \mid \tau > 0\}$, where $w \neq 0$, is the half-space $\{v \mid \langle v, w\rangle \leq 0\}$.

6.23 Example (orthogonal subspaces). *Orthogonality of subspaces is a special case of polarity of cones: for a linear subspace M of $\mathbb{R}^n$, one has*

$$M^* = M^\perp = \{v \mid \langle v, w\rangle = 0 \ \textit{ for all } \ w \in M\}, \qquad M^{**} = M^{\perp\perp} = M.$$

According to this, the relationship in 6.8 and Figure 6–7 between tangents and normals to a smooth manifold fits the context of cones that are polar to each other. But so too does the relationship in 6.9 and Figure 6–8 between tangents and normals to a convex set.

6.24 Example (polarity of normals and tangents to convex sets). *For any convex set $C \subset \mathbb{R}^n$ (closed or not), and any point $\bar{x} \in C$, the cones $N_C(\bar{x})$ and $T_C(\bar{x})$ are polar to each other. Moreover, $N_C(\bar{x})$ is pointed if and only if $\operatorname{int} C \neq \emptyset$.*

Detail. This is seen from 6.9. □

In particular from 6.24, the polar relationship holds between $N_C(\bar{x})$ and $T_C(\bar{x})$ when C is a box. Then the cones in question are themselves boxes (typically unbounded) as described in 6.10.

F. Tangent-Normal Relations

How far does polarity go in general in describing the connections between normal vectors and tangent vectors? In the pursuit of this question, another concept of tangent vector, more special in character, will be valuable for its remarkable properties and eventually its close tie to Clarke regularity.

6.25 Definition (regular tangent vectors). *A vector $w \in \mathbb{R}^n$ is tangent to C at a point $\bar{x} \in C$ in the regular sense, or a regular tangent vector, indicated by $w \in \widehat{T}_C(\bar{x})$, if for every sequence $\tau^\nu \searrow 0$ and every sequence $\bar{x}^\nu \underset{C}{\to} \bar{x}$ there is a sequence $x^\nu \underset{C}{\to} \bar{x}$ with $(x^\nu - \bar{x}^\nu)/\tau^\nu \to w$. In other words,*

$$\widehat{T}_C(\bar{x}) := \liminf_{\substack{x \underset{C}{\to} \bar{x} \\ \tau \searrow 0}} \frac{C - x}{\tau}. \tag{6(15)}$$

Often $\widehat{T}_C(\bar{x})$ coincides with the cone $T_C(\bar{x})$ of tangent vectors in the general sense of Definition 6.1, but not always. Insights are provided by Figure 6–15. The two kinds of tangent vector are the same at all points of the set C in that figure except at the 'inward corner'. There the regular tangent vectors form a convex cone smaller than the general tangent cone, which isn't convex.

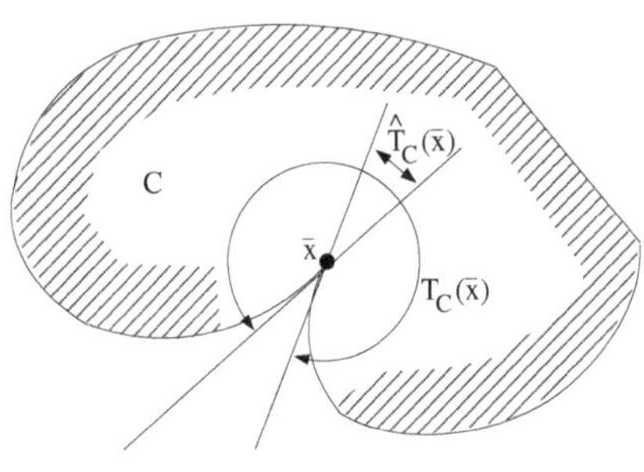

Fig. 6–15. Tangent cones in the regular and general sense.

The parallel between this discrepancy in Figure 6–15 and the one in Figure 6–6 for normal cones to the same set is no accident. It will emerge in 6.29 that when C is locally closed at $\bar{x}$, not only does regularity correspond to every normal vector at $\bar{x}$ being a regular normal vector, but equally well to every tangent vector there being a regular tangent vector. This, of course, is the ultimate reason for calling this type of tangent vector 'regular'.

6.26 Theorem (regular tangent cone properties). *For $C \subset \mathbb{R}^n$ and $\bar{x} \in C$, every regular tangent vector $w \in \widehat{T}_C(\bar{x})$ is in particular a derivable tangent vector, and $\widehat{T}_C(\bar{x})$ is a closed, convex cone with $\widehat{T}_C(\bar{x}) \subset T_C(\bar{x})$.*

When C is locally closed at $\bar{x}$, one has $w \in \widehat{T}_C(\bar{x})$ if and only if, for every sequence $\bar{x}^\nu \underset{C}{\to} \bar{x}$, there are vectors $w^\nu \in T_C(\bar{x}^\nu)$ such that $w^\nu \to w$. Thus,

$$\widehat{T}_C(\bar{x}) = \liminf_{x \underset{C}{\to} \bar{x}} T_C(x). \tag{6(16)}$$

Proof. From the definition it's clear that $\widehat{T}_C(\bar{x})$ contains 0 and for any w all positive multiples λw. Hence $\widehat{T}_C(\bar{x})$ is a cone. As an inner limit, it's closed. Every $w \in \widehat{T}_C(\bar{x})$ can in particular be expressed as a limit of the form in Definition 6.25 with $\bar{x}^\nu \equiv \bar{x}$ for any sequence of values $\tau^\nu \searrow 0$, so it's a derivable tangent vector (cf. Definition 6.1). Thus also, $\widehat{T}_C(\bar{x}) \subset T_C(\bar{x})$ (cf. 6.2).

Inasmuch as $\widehat{T}_C(\bar{x})$ is a cone, we can establish its convexity by demonstrating for arbitrary w_0 and w_1 in $\widehat{T}_C(\bar{x})$ that $w_0 + w_1 \in \widehat{T}_C(\bar{x})$ (cf. 3.7). Consider sequences $\bar{x}^\nu \underset{C}{\to} \bar{x}$ and $\tau^\nu \searrow 0$. To prove that $w_0 + w_1 \in \widehat{T}_C(\bar{x})$ we need to show there is a sequence $x^\nu \underset{C}{\to} \bar{x}$ such that $(x^\nu - \bar{x}^\nu)/\tau^\nu \to w_0 + w_1$. We know there exist, by the assumption that $w_0 \in \widehat{T}_C(\bar{x})$, points $\tilde{x}^\nu \underset{C}{\to} \bar{x}$ with $(\tilde{x}^\nu - \bar{x}^\nu)/\tau^\nu \to w_0$. Then, since $w_1 \in \widehat{T}_C(\bar{x})$, there exist points $x^\nu \underset{C}{\to} \bar{x}$ with $(x^\nu - \tilde{x}^\nu)/\tau^\nu \to w_1$. It follows that $(x^\nu - \bar{x}^\nu)/\tau^\nu \to w_0 + w_1$, as desired.

Suppose now that C is locally closed at $\bar{x}$. Replacing C by $C \cap V$ for a closed neighborhood $V \in \mathcal{N}(\bar{x})$, we can reduce the verification of 6(16) to the case where C is closed. Let $K(\bar{x})$ stand for the set given by the 'lim inf' on the right side of 6(16). Our goal is to demonstrate that $w \notin K(\bar{x})$ if and only if $w \notin \widehat{T}_C(\bar{x})$. The definition of $K(\bar{x})$ means that

$$w \notin K(\bar{x}) \iff \exists \varepsilon > 0,\ \tilde{x}^\nu \underset{C}{\to} \bar{x},\ \text{ with } d\big(w, T_C(\tilde{x}^\nu)\big) \geq \varepsilon, \tag{6(17)}$$

while the limit formula 6(15) for $\widehat{T}_C(\bar{x})$ says that

$$w \notin \widehat{T}_C(\bar{x}) \iff \begin{cases} \exists \varepsilon > 0,\ \bar{x}^\nu \underset{C}{\to} \bar{x},\ \bar{\tau}^\nu \searrow 0 \\ \text{with } \big(\bar{x}^\nu + \bar{\tau}^\nu I\!B(w,\varepsilon)\big) \cap C = \emptyset. \end{cases} \tag{6(18)}$$

If $w \notin K(\bar{x})$, we get from 6(17) the existence of $\tilde{\varepsilon} > 0$ and points $\tilde{x}^\nu \underset{C}{\to} \bar{x}$ with $I\!B(w, \tilde{\varepsilon}) \cap T_C(\tilde{x}^\nu) = \emptyset$. Then for some $\tilde{\tau}^\nu > 0$ we have $I\!B(w,\tilde{\varepsilon}) \cap \tau^{-1}(C - \tilde{x}^\nu) = \emptyset$ for all $\tau \in (0, \tilde{\tau}^\nu]$, which means $\big(\tilde{x}^\nu + \tau I\!B(w,\tilde{\varepsilon})\big) \cap C = \emptyset$. Selecting $\tau^\nu \in (0, \tilde{\tau}^\nu]$ in such a way that $\tau^\nu \searrow 0$, we conclude through 6(18) that $w \notin \widehat{T}_C(\bar{x})$.

If we start instead by supposing $w \notin \widehat{T}_C(\bar{x})$, we have $\bar{x}^\nu$ and $\bar{\tau}^\nu$ as in 6(18). The task is to demonstrate from this the existence of a sequence of points $\tilde{x}^\nu$ as in 6(17). For this purpose it will suffice to prove the following fact in simpler notation: under the assumption that $\hat{x}$ is a point of the closed set C such that for some $\varepsilon > 0$ and $\bar{\tau} > 0$ the ball $\hat{x} + \bar{\tau} I\!B(w, \varepsilon)$ doesn't meet C, there exists $\tilde{x} \in C \cap I\!B\big(\hat{x}, \bar{\tau}(|w| + \varepsilon)\big)$ such that $d\big(w, T_C(\tilde{x})\big) \geq \varepsilon$.

The set of all $\tau \in [0, \bar{\tau}]$ such that the ball $I\!B(\hat{x} + \tau w, \tau\varepsilon) = \hat{x} + \tau I\!B(w, \varepsilon)$ meets C is closed, and by assumption it does not contain $\bar{\tau}$. Let $\hat{\tau}$ be the highest number it does contain; then $0 \leq \hat{\tau} < \bar{\tau}$. The set

$$D := \hat{x} + [\hat{\tau}, \bar{\tau}] I\!B(w, \varepsilon) = \bigcup \big\{ \hat{x} + \tau(w + \varepsilon I\!B) \,\big|\, \tau \in [\hat{\tau}, \bar{\tau}] \big\}$$

(see Figure 6–16) then meets C, although $\operatorname{int} D$ doesn't; D is compact and convex in $\mathbb{R}^n$. Let $\tilde{x} \in C \cap D$ and $\tilde{\tau} = \bar{\tau} - \hat{\tau} > 0$. Suppose therefore that $\tilde{\tau} > 0$. From the definition of D, $|\tilde{x} - \hat{x}| \le \tilde{\tau}(|w| + \varepsilon)$. For arbitrary $\tilde{\varepsilon} \in (0, \varepsilon)$ we prove now that $\big(\tilde{x} + (0, \tilde{\tau}]\mathbb{B}(w, \tilde{\varepsilon})\big) \cap C = \emptyset$, so that $\mathbb{B}(w, \tilde{\varepsilon}) \cap \tau^{-1}(C - \tilde{x}) = \emptyset$ for all $\tau \in (0, \tilde{\tau}]$. This will establish that $d\big(w, T_C(\tilde{x})\big) \ge \varepsilon$.

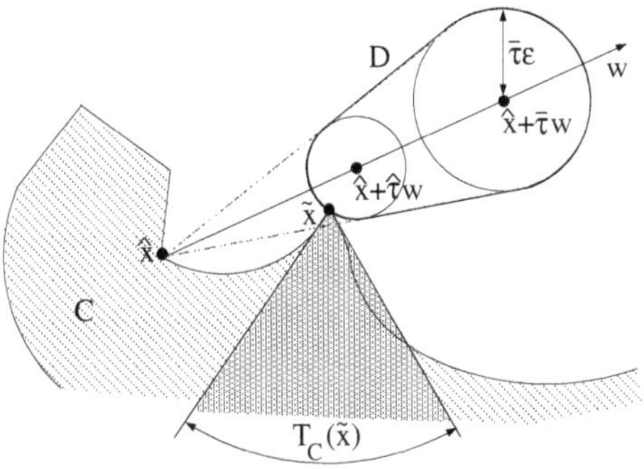

Fig. 6–16. Perturbation argument.

By its selection, $\tilde{x}$ belongs to the ball $\hat{x} + \hat{\tau}\mathbb{B}(w, \varepsilon)$ and consequently $\hat{\tau}\varepsilon \ge |\tilde{x} - (\hat{x} + \hat{\tau}w)| = |(\tilde{x} + \tilde{\tau}w) - (\hat{x} + \bar{\tau}w)|$. The ball of radius $\bar{\tau}\varepsilon - \hat{\tau}\varepsilon = \tilde{\tau}\varepsilon$ around $\tilde{x} + \tilde{\tau}w$ lies therefore in the ball $(\hat{x} + \bar{\tau}w) + \bar{\tau}\varepsilon\mathbb{B} = \hat{x} + \bar{\tau}\mathbb{B}(w, \varepsilon)$ within D. The ball $(\tilde{x} + \tilde{\tau}w) + \tilde{\tau}\tilde{\varepsilon}\mathbb{B} = \tilde{x} + \tilde{\tau}\mathbb{B}(w, \tilde{\varepsilon})$ lies accordingly in $\operatorname{int} D$. By the convexity of D, so then do all the line segments joining points of this ball with $\tilde{x}$, except for $\tilde{x}$ itself (cf. 2.33). Thus, $\big(\tilde{x} + (0, \tilde{\tau}]\mathbb{B}(w, \tilde{\varepsilon})\big) \cap C = \emptyset$. □

It might be imagined from 6.26 that the mapping $\widehat{T}_C : x \mapsto \widehat{T}_C(x)$ can be counted on to be isc relative to C. But that may fail to be true. For example, if C is the subset of $\mathbb{R}^3$ formed by the union of the graph of $x_3 = x_1 x_2$ with that of $x_3 = -x_1 x_2$, the cone $\widehat{T}_C(\bar{x})$ at any point $\bar{x}$ on the x_1-axis or the x_2-axis, except at the origin, is a line, but at the origin it's a plane.

The regular tangent cone at the inner corner point in Figure 6–15 is the polar of the normal cone at that point in Figure 6–6. We'll show that this relationship always holds when C is locally closed at the point in question. The estimate in part (b) of the next proposition will be crucial in this.

6.27 Proposition (normals to tangent cones). *Consider a set $C \subset \mathbb{R}^n$ and a point $\bar{x} \in C$ where C is locally closed. For the cone $T = T_C(\bar{x})$, one has*

(a) $N_T(0) = \bigcup_{w \in T} N_T(w) \subset N_C(\bar{x})$;

(b) *for any vector $w \notin T$ there is a vector $\bar{v} \in N_C(\bar{x})$ with $|\bar{v}| = 1$ such that $d_T(w) = \langle \bar{v}, w \rangle$; thus in particular,*

$$\min_{v \in N_C(\bar{x}) \cap \mathbf{B}} \langle v, w \rangle \le d_T(w) \le \max_{v \in N_C(\bar{x}) \cap \mathbf{B}} \langle v, w \rangle \quad \textit{for all } w.$$

Proof. In (a), note first that because T is a cone one has $N_T(w) = N_T(\lambda w)$ for all $\lambda > 0$. This implies that $N_T(w) \subset N_T(0)$, since $\limsup_\nu N_T(w^\nu) \subset N_T(0)$ when $w^\nu \underset{T}{\to} 0$. Thus, the equation for $N_T(0)$ is correct.

Consider now any vector $v \in N_T(0)$. Since $T = \limsup_{\tau \searrow 0}[C - \bar{x}]/\tau$, there exists on the basis of 6.18(b) a sequence $\tau^\nu \searrow 0$ along with points $w^\nu \in T^\nu := [C - \bar{x}]/\tau^\nu$ and vectors $v^\nu \in N_{T^\nu}(w^\nu)$ such that $w^\nu \to 0$ and $v^\nu \to v$. But $N_{T^\nu}(w^\nu) = N_C(x^\nu)$ for $x^\nu = \bar{x} + \tau^\nu w^\nu \to \bar{x}$. Hence actually $v \in N_C(\bar{x})$ by the limit property in 6.6. This proves (a).

In (b), choose $\bar{w}$ in the projection $P_T(w)$. Then $|w - \bar{w}| = d_T(w) > 0$, while $w - \bar{w}$ is a proximal normal to T at $\bar{w}$; in particular, $w - \bar{w} \in N_T(\bar{w})$. Because $\tau\bar{w} \in T$ for all $\tau \geq 0$, the minimum of $\varphi(\tau) := |w - \tau\bar{w}|^2$ over such τ is attained at $\tau = 1$, so $0 = \varphi'(0) = -2\langle w - \bar{w}, \bar{w}\rangle$. Let $v := (w - \bar{w})/|w - \bar{w}|$. Then $|v| = 1$ and $\langle v, \bar{w}\rangle = 0$, so that $\langle v, w\rangle = \langle v, w - \bar{w}\rangle = |w - \bar{w}| = d_T(w)$. Furthermore $v \in N_T(\bar{w})$, which implies by (a) that $v \in N_C(\bar{x})$. This establishes the first assertion of (b) and shows that the double inequality holds when $w \notin T$. It holds trivially though when $w \in T$, since $N_C(\bar{x}) \cap \mathbb{B}$ contains $v = 0$. □

6.28 Theorem (tangent-normal polarity). *For $C \subset \mathbb{R}^n$ and $\bar{x} \in C$,*

(a) $\widehat{N}_C(\bar{x}) = T_C(\bar{x})^*$ *always,*

(b) $\widehat{T}_C(\bar{x}) = N_C(\bar{x})^*$ *as long as C is locally closed at $\bar{x}$.*

Proof. The first polarity relation merely restates 6(6). For the second polarity relation under the assumption of local closedness, we begin by considering any $w \in \widehat{T}_C(\bar{x})$ and $v \in N_C(\bar{x})$. The definition of $N_C(\bar{x})$ gives us sequences $\bar{x}^\nu \underset{C}{\to} \bar{x}$ and $v^\nu \to v$ with $v^\nu \in \widehat{N}_C(\bar{x}^\nu)$. According to 6(16) we can then find a sequence $w^\nu \to w$ with $w^\nu \in T_C(\bar{x}^\nu)$. We have $\langle v^\nu, w^\nu\rangle \leq 0$ by the first polarity relation, so in the limit we get $\langle v, w\rangle \leq 0$. Thus, every $w \in \widehat{T}_C(\bar{x})$ satisfies $\langle v, w\rangle \leq 0$ for all $v \in N_C(\bar{x})$, and the inclusion $\widehat{T}_C(\bar{x}) \subset N_C(\bar{x})^*$ is valid.

Suppose now that $w \notin \widehat{T}_C(\bar{x})$. We must show that also $w \notin N_C(\bar{x})^*$, or in other words, that for some $v \in N_C(\bar{x})$ one has $\langle v, w\rangle > 0$. The condition $w \notin \widehat{T}_C(\bar{x})$ is equivalent through 6(16) to the existence of $\bar{x}^\nu \underset{C}{\to} \bar{x}$ and $\varepsilon > 0$ such that $d(w, T_C(\bar{x}^\nu)) \geq \varepsilon$. The tangent-normal relation in 6.27(b) suffices to finish the proof, because it can be applied to $\bar{x}^\nu$ (where C is locally closed as well, once ν is sufficiently high) and then yields vectors $v^\nu \in N_C(\bar{x}^\nu) \cap \mathbb{B}$ with $\langle v^\nu, w\rangle \geq \varepsilon$. Any cluster point v of such a sequence belongs to $N_C(\bar{x})$ by 6.6 and satisfies $\langle v, w\rangle \geq \varepsilon$. □

6.29 Corollary (characterizations of Clarke regularity). *At any $\bar{x} \in C$ where C is locally closed, the following are equivalent and mean that C is regular at $\bar{x}$:*

(a) $N_C(\bar{x}) = \widehat{N}_C(\bar{x})$, *i.e., all normal vectors at $\bar{x}$ are regular;*

(b) $T_C(\bar{x}) = \widehat{T}_C(\bar{x})$, *i.e., all tangent vectors at $\bar{x}$ are regular;*

(c) $N_C(\bar{x}) = \{v \mid \langle v, w\rangle \leq 0 \text{ for all } w \in T_C(\bar{x})\} = T_C(\bar{x})^*$,

(d) $T_C(\bar{x}) = \{w \mid \langle v, w\rangle \leq 0 \text{ for all } v \in N_C(\bar{x})\} = N_C(\bar{x})^*$,

(e) $\langle v, w\rangle \leq 0$ *for all* $w \in T_C(\bar{x})$ *and* $v \in N_C(\bar{x})$,

(f) *the mapping $\widehat{N}_C$ is osc at $\bar{x}$ relative to C,*

(g) *the mapping T_C is isc at $\bar{x}$ relative to C.*

Proof. Property (a) is what we have defined regularity to be (in the presence of local closedness). Theorem 6.28, along with the basic inclusions $\widehat{N}_C(\bar{x}) \subset$

$N_C(\bar{x})$ and $\widehat{T}_C(\bar{x}) \subset T_C(\bar{x})$ in 6.5 and 6.26, yields at once the equivalence of (b), (c), (d), and (e). Here we make use of the basic facts about polarity in 6.21. The equivalence of (f) with (a) holds by the limit formula in 6.5, while that of (g) with (b) holds by limit formula in 6.26. □

6.30 Corollary (consequences of Clarke regularity). *If C is regular at $\bar{x}$, the cones $T_C(\bar{x})$ and $N_C(\bar{x})$ are convex and polar to each other. Furthermore, C is geometrically derivable at $\bar{x}$.*

Proof. The convexity comes from the polarity relations, cf. 6.21, while the geometric derivability comes from 6.29(b) and the fact that regular tangent vectors are derivable, cf. 6.26. □

The 'inward corner point' in Figure 6–15 illustrates how a set can be geometrically derivable at a point without being regular there, and how the tangent cone in that case need not be convex.

The basic relationships between tangents and normals that have been established so far in the case of C locally closed at $\bar{x}$ are shown in Figure 6–17. When C is regular at $\bar{x}$, the left and right sides of the diagram join up.

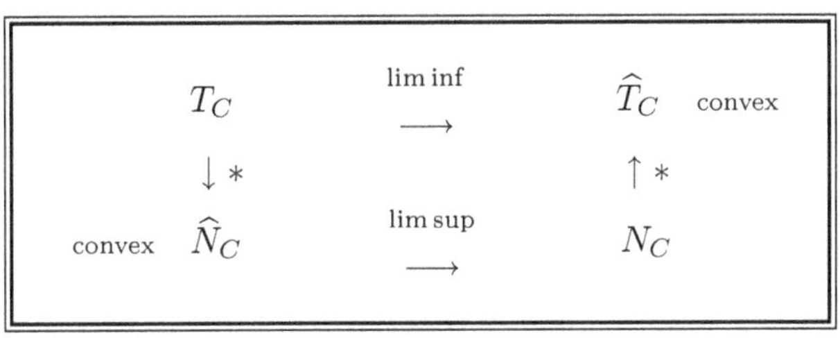

Fig. 6–17. Diagram of tangent and normal cone relationships for closed sets.

The polarity results enable us to state a tangent cone counterpart to the normal cone result in Theorem 6.14.

6.31 Theorem (tangent cones to sets with constraint structure). *Let*

$$C = \{x \in X \mid F(x) \in D\}$$

for closed sets $X \subset \mathbb{R}^n$ and $D \subset \mathbb{R}^m$ and a $\mathcal{C}^1$ mapping $F : \mathbb{R}^n \to \mathbb{R}^m$. At any $\bar{x} \in C$ one has

$$T_C(\bar{x}) \subset \Big\{ w \in T_X(\bar{x}) \,\Big|\, \nabla F(\bar{x}) w \in T_D\big(F(\bar{x})\big) \Big\}.$$

If the constraint qualification in Theorem 6.14 is satisfied at $\bar{x}$, one also has

$$\widehat{T}_C(\bar{x}) \supset \Big\{ w \in \widehat{T}_X(\bar{x}) \,\Big|\, \nabla F(\bar{x}) w \in \widehat{T}_D\big(F(\bar{x})\big) \Big\}.$$

If in addition X is regular at $\bar{x}$ and D is regular at $F(\bar{x})$, then

$$T_C(\bar{x}) = \widehat{T}_C(\bar{x}) = \big\{ w \in T_X(\bar{x}) \,\big|\, \nabla F(\bar{x})w \in T_D\big(F(\bar{x})\big) \big\}.$$

Proof. The first inclusion is elementary from the definitions, while the second follows from the normal cone inclusion in Theorem 6.14 by the polarity relation in 6.28(b). In the regular case the relations give equations, as seen in 6.29. □

A regular tangent counterpart to the tangent cone formula within 6.7 can also be stated.

6.32 Exercise (regular tangents under a change of coordinates). *Let $C = F^{-1}(D) \subset I\!R^n$ for a smooth mapping $F : I\!R^n \to I\!R^m$ and a set $D \subset I\!R^m$. Let $\bar{x} \in C$, $\bar{u} = F(\bar{x}) \in D$, and suppose $\nabla F(\bar{x})$ has full rank m. Then*

$$\widehat{T}_C(\bar{x}) = \big\{ w \,\big|\, \nabla F(\bar{x})w \in \widehat{T}_D(\bar{u}) \big\}.$$

Guide. See Exercise 6.7. □

G*. Recession Properties

Exploration of the following auxiliary notion related to tangency will bring to light other important connections between tangents and normals.

6.33 Definition (recession vectors). *A vector w is a local recession vector for a set $C \subset I\!R^n$ at a point $\bar{x} \in C$, written $w \in R_C(\bar{x})$, if for some neighborhood $V \in \mathcal{N}(\bar{x})$ and $\varepsilon > 0$ one has $x + \tau w \in C$ for all $x \in C \cap V$ and $\tau \in [0, \varepsilon]$. It is a global recession vector for C if actually $x + \tau w \in C$ for all $x \in C$ and $\tau \geq 0$.*

Roughly, a local recession vector $w \neq 0$ for C at $\bar{x}$ has the property that if C is translated by a sufficiently small amount in the direction of w, it locally 'recedes within itself' around $\bar{x}$. A global recession vector w translates C into itself entirely: $C + \tau w \subset C$ for all $\tau \geq 0$.

6.34 Exercise (recession cones and convexity).

(a) *The set $R_C(\bar{x})$ of all local recession vectors to C at $\bar{x}$ is a convex cone lying within $\widehat{T}_C(\bar{x})$. In fact, when $W = \operatorname{con}\{w_1, \ldots, w_r\}$ with $w_k \in R_C(\bar{x})$ there exist $V \in \mathcal{N}(\bar{x})$ and $\varepsilon > 0$ such that*

$$x + \tau W \subset C \text{ for all } x \in C \cap V \text{ and } \tau \in [0, \varepsilon].$$

(b) *If C is closed, the set of global recession vectors to C is $\bigcap_{x \in C} R_C(x)$, and this cone is not only convex but closed.*

(c) *If C is convex, the global recession cone $\bigcap_{x \in C} R_C(x)$ lies in C^∞, and when C is also closed this inclusion is an equation.*

Guide. In (a), rely on the definition of $R_C(\bar{x})$. In (b), verify one direction by arguing that if a half-line $\big\{\bar{x} + \tau w \,\big|\, \tau \geq 0\big\}$ from $\bar{x} \in C$ doesn't fully lie in C, it must emerge at a point $\tilde{x}$, and then $w \notin R_C(\tilde{x})$. In (c), draw on 3.6. □

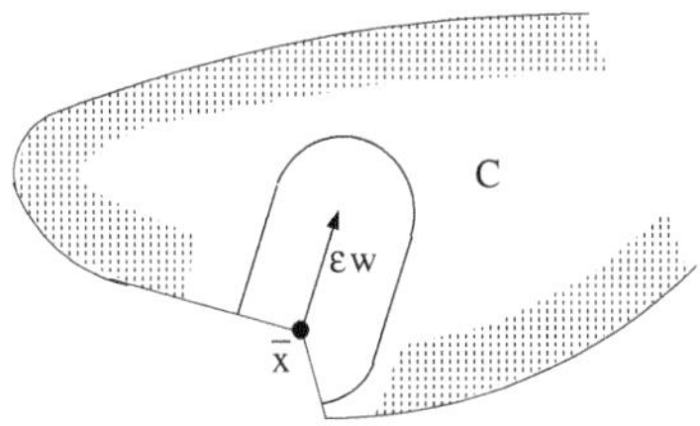

Fig. 6–18. A local recession vector.

6.35 Proposition (recession vectors from tangents and normals). *Let $\bar{x}$ be a point of $C \subset \mathbb{R}^n$ where C is locally closed. Then $w \in R_C(\bar{x})$ if and only if there is an open neighborhood $V \in \mathcal{N}(\bar{x})$ for which one of the following conditions holds, all these conditions being equivalent to each other:*

(a) $w \in T_C(x)$ *for all* $x \in C \cap V$;

(b) $w \in \widehat{T}_C(x)$ *for all* $x \in C \cap V$;

(c) $\langle v, w\rangle \leq 0$ *for all* $v \in N_C(x)$ *when* $x \in C \cap V$;

(d) $\langle v, w\rangle \leq 0$ *for all* $v \in \widehat{N}_C(x)$ *when* $x \in C \cap V$.

Proof. If $w \in R_C(\bar{x})$ we have from Definition 6.33 that (a) holds for an open neighborhood V. Theorem 6.26 ensures the equivalence of (a) with (b) for any open set V, since $\widehat{T}_C(x) \subset T_C(x)$, while the equivalence of (b) with (c) is clear from 6.28. The equivalence of (c) with (d) is immediate from the fact that $N_C(\bar{x})$ is the outer limit of $\widehat{N}_C(\bar{x})$ as $x \underset{C}{\to} \bar{x}$, cf. 6.5. To complete the claims of equivalence, we suppose $w \notin R_C(\bar{x})$ and prove that (c) doesn't hold for any $V \in \mathcal{N}(\bar{x})$. Replacing C by a local portion at $\bar{x}$ if necessary, we can assume in this that C is closed. Also, there is no loss of generality in taking $|w| = 1$.

Because $w \notin R_C(\bar{x})$, we know that for every neighborhood V of $\bar{x}$ and every $\varepsilon > 0$ we can find a point $\hat{x} \in C \cap V$ and a value $\tau \in (0, \varepsilon]$ such that $\hat{x} + \tau w \notin C$. The closed set $C \cap \{\hat{x} + \tau w \mid 0 \leq \tau \leq \varepsilon\}$ then contains a point $\tilde{x}$ with the property that $\tilde{x} + \tau w \notin C$ for all τ in some interval $(0, \delta)$. Thus, there are points $\tilde{x} \in C$ arbitrarily near to $\bar{x}$ that have this property. It will suffice to show that for such a point $\tilde{x}$ there is an arbitrarily close point $\tilde{x}'$ of C for which a vector $v \in N_C(\tilde{x}')$ satisfies $\langle v, w\rangle > 0$.

Let A be the matrix of the projection mapping from $\mathbb{R}^n$ onto the $(n-1)$-dimensional subspace orthogonal to w; thus, $Ax = x - \langle x, w\rangle w$. The line $L := \{\tilde{x} + \tau w \mid -\infty < \tau < \infty\}$ consists of the points x satisfying $A(x - \tilde{x}) = 0$, the corresponding τ coordinate of such a point being $l(x) := \langle x - \tilde{x}, w\rangle$. Let B be a compact neighborhood of $\tilde{x}$ with the property that the points $\tilde{x} + \tau w$ belonging to B all have $\tau < \delta$. The assumption that the points $\tilde{x} + \tau w$ with $\tau \in (0, \delta)$ all lie outside of C translates under this notation into the assertion that the problem

$$\text{minimize } f(x) := -l(x) + \delta_{C \cap B}(x) \text{ subject to } A(x - \tilde{x}) = 0$$

has $x = \tilde{x}$ as its unique optimal solution. For a sequence of positive penalty coefficients $r^\nu \to \infty$ consider the relaxed problems

$$\text{minimize } f^\nu(x) := f(x) + \tfrac{1}{2}r^\nu|A(x-\tilde{x})|^2 \text{ over } x \in I\!R^n.$$

Each of these problems has a least one optimal solution x^ν, because f is lsc with $\operatorname{dom} f = C \cap B$, which is compact. We have $x^\nu \to \tilde{x}$ by 1.21. Eventually $x^\nu \in \operatorname{int} B$, and once this occurs x^ν is a point at which the function $h^\nu(x) = l(x) - \frac{1}{2}r^\nu|A(x-\tilde{x})|^2$ achieves a local maximum relative to C. Then $\nabla h^\nu(x^\nu) \in \widehat{N}_C(x^\nu)$ by 6.12. We need only verify that the vector $v^\nu := \nabla h^\nu(x^\nu)$, which in particular lies in $N_C(x^\nu)$, satisfies $\langle v^\nu, w\rangle > 0$. We calculate $v^\nu = \nabla l(x^\nu) - r^\nu A^* A(x^\nu - \tilde{x})$, with $\nabla l(x^\nu) = w$ and $A^*A = A^2 = A$ because A is the matrix of an orthogonal projection (hence self-adjoint and idempotent). Thus $v^\nu = w - A(x^\nu - \tilde{x})$, where $A(x^\nu - \tilde{x})$ belongs to the subspace orthogonal to w, and therefore $\langle v^\nu, w\rangle = \langle w, w\rangle = 1$. □

6.36 Theorem (interior tangents and recession). *At a point $\bar{x} \in C$ where C is locally closed, the following conditions on a vector $\bar{w}$ are equivalent:*

(a) $\bar{w} \in \operatorname{int} \widehat{T}_C(\bar{x})$,

(b) $\bar{w} \in \operatorname{int} R_C(\bar{x})$,

(c) *there is a neighborhood $W \in \mathcal{N}(\bar{w})$ along with $V \in \mathcal{N}(\bar{x})$ and $\varepsilon > 0$ such that $x + \tau W \subset C$ for all $x \in C \cap V$ and $\tau \in [0, \varepsilon]$,*

(d) *there exists $V \in \mathcal{N}(\bar{x})$ such that $\langle v, \bar{w}\rangle < 0$ for all $v \in N_C(x)$ with $|v| = 1$ when $x \in C \cap V$,*

(e) *$\langle v, \bar{w}\rangle < 0$ for all $v \in N_C(\bar{x})$ with $|v| = 1$.*

Such a $\bar{w}$ exists if and only if $N_C(\bar{x})$ is pointed, and then $\widehat{T}_C(\bar{x}) = \operatorname{cl} R_C(\bar{x})$.

Proof. We have (c) $\Leftrightarrow$ (b) by 6.34(a), since every neighborhood of $\bar{w}$ includes a neighborhood of the form $W = \operatorname{con}\{w_1, \ldots, w_r\}$, cf. 2.28(e). Obviously (b) $\Rightarrow$ (a) because $R_C(\bar{x}) \subset \widehat{T}_C(\bar{x})$. On the other hand, (a) $\Leftrightarrow$ (e) by 6.22, where the existence of such a vector $\bar{w}$ is identified also with the pointedness of $N_C(\bar{x})$. We have (e) $\Leftrightarrow$ (d) by the outer semicontinuity of the mapping N_C in 6.6. Hence through the characterization in 6.35(c), (e) implies in particular that $\bar{w} \in R_C(\bar{x})$. But the set of vectors $\bar{w}$ satisfying (e) for fixed $\bar{x}$ is open. Thus, (e) actually implies (b), and the pattern of equivalences is complete.

A convex set is the closure of its interior when that's nonempty, so from $\operatorname{int} R_C(\bar{x}) = \operatorname{int} \widehat{T}_C(\bar{x}) \neq \emptyset$ we get $\operatorname{cl} R_C(\bar{x}) = \operatorname{cl} \widehat{T}_C(\bar{x}) = \widehat{T}_C(\bar{x})$. □

Along with other uses, Theorem 6.36 can assist in determining the regular tangent cone $\widehat{T}_C(\bar{x})$ geometrically. Often the local recession cone $R_C(\bar{x})$ is evident from the geometry of C around $\bar{x}$ and has nonempty interior, and in that event $\widehat{T}_C(\bar{x})$ is simply the closure of $R_C(\bar{x})$. In Figure 6–15, for instance, it's easy to see what $R_C(\bar{x})$ is at every point $\bar{x} \in C$, and always $\operatorname{int} R_C(\bar{x}) \neq \emptyset$.

Theorem 6.36 also reveals that the set limits involved in forming tangent cones have an internal uniform approximation property like the one for convex set limits in 4.15, even though the difference quotients in these limits aren't necessarily convex sets.

6.37 Corollary (internal approximation in tangent limits). *Let C be locally closed at $\bar{x} \in C$, and let B be any compact subset of $\operatorname{int} \widehat{T}_C(\bar{x})$ (or of $\operatorname{int} T_C(\bar{x})$ if C is regular at $\bar{x}$). Then there exist $\varepsilon > 0$ and $V \in \mathcal{N}(\bar{x})$ such that*

$$\tau^{-1}(C - x) \supset B \text{ for all } \tau \in (0, \varepsilon) \text{ when } x \in C \cap V,$$

and hence in particular $T_C(x) \supset B$ for all $x \in C \cap V$.

Proof. Each point $\bar{w}$ of B enjoys the property in 6.36(c) for some $W \in \mathcal{N}(\bar{w})$, $V \in \mathcal{N}(\bar{x})$ and $\varepsilon > 0$. Since B is compact, it can be covered by finitely many such neighborhoods W_k with associated V_k and ε_k, $k = 1, \ldots, m$. The desired property holds then for $V = V_1 \cap \cdots \cap V_m$ and $\varepsilon = \min\{\varepsilon_1, \ldots, \varepsilon_m\}$. □

H.* Irregularity and Convexification

The picture of normals and tangents in the case of sets that aren't Clarke regular can be completed through convexification. The *convexified* normal and tangent cones to C at $\bar{x}$ are defined by

$$\overline{N}_C(\bar{x}) = \operatorname{cl} \operatorname{con} N_C(\bar{x}), \qquad \overline{T}_C(\bar{x}) = \operatorname{cl} \operatorname{con} T_C(\bar{x}). \tag{6(19)}$$

6.38 Exercise (convexified normal and tangent cones). *Let C be a closed subset of $\mathbb{R}^n$, and let $\bar{x} \in C$.*

(a) *The cones $\overline{N}_C(\bar{x})$ and $\widehat{T}_C(\bar{x})$ are polar to each other:*

$$\overline{N}_C(\bar{x}) = \widehat{T}_C(\bar{x})^*, \qquad \widehat{T}_C(\bar{x}) = \overline{N}_C(\bar{x})^*.$$

When $N_C(\bar{x})$ is pointed, or equivalently $\widehat{T}_C(\bar{x})$ has nonempty interior, a vector v belongs to $\overline{N}_C(\bar{x})$ if and only if it can be expressed as a sum $v_1 + \cdots + v_r$ of vectors $v_k \in N_C(\bar{x})$, with $r \leq n$.

(b) *The cones $\overline{T}_C(\bar{x})$ and $\widehat{N}_C(\bar{x})$ are polar to each other:*

$$\overline{T}_C(\bar{x}) = \widehat{N}_C(\bar{x})^*, \qquad \widehat{N}_C(\bar{x}) = \overline{T}_C(\bar{x})^*.$$

When $T_C(\bar{x})$ is pointed, or equivalently $\widehat{N}_C(\bar{x})$ has nonempty interior, a vector w belongs to $\overline{T}_C(\bar{x})$ if and only if it can be expressed as a sum $w_1 + \cdots + w_r$ of vectors $w_k \in T_C(\bar{x})$, with $r \leq n$.

Guide. Combine the relations in 6.26 with the facts about polarity in 6.21, 6.24, and convex hulls of cones 3.15. □

The fundamental constraint qualification in terms of normal cones has a counterpart in terms of tangent cones, which however is a more stringent assumption in the absence of Clarke regularity.

6.39 Exercise (constraint qualification in tangent cone form). *Let $\bar{x} \in C = \{x \in X \mid F(x) \in D\}$ for a $\mathcal{C}^1$ mapping $F : \mathbb{R}^n \to \mathbb{R}^m$ and closed sets $X \subset \mathbb{R}^n$ and $D \subset \mathbb{R}^m$. For $\bar{x}$ to satisfy the constraint qualification of 6.14, namely that*

$$\begin{cases} \textit{the only vector } y \in N_D\big(F(\bar{x})\big) \textit{ for which} \\ -\sum_{i=1}^{m} y_i \nabla f_i(\bar{x}) \in N_X(\bar{x}) \textit{ is } y = (0,\ldots,0), \end{cases}$$

any of the following conditions, which are equivalent to each other, is sufficient; they are necessary as well when X is regular at $\bar{x}$ and D is regular at $F(\bar{x})$:

(a) $\widehat{T}_D\big(F(\bar{x})\big) - \nabla F(\bar{x})\widehat{T}_X(\bar{x}) = I\!R^m$;

(b) *the only vector $y \perp \widehat{T}_D\big(F(\bar{x})\big)$ with $\nabla F(\bar{x})^* y \perp \widehat{T}_X(\bar{x})$ is $y = 0$, but there is a vector $\tilde{w} \in \operatorname{rint} \widehat{T}_X(\bar{x})$ with $\nabla F(\bar{x})\tilde{w} \in \operatorname{rint} \widehat{T}_D\big(F(\bar{x})\big)$;*

(c) *for the convexified normal cones $\overline{N}_X(\bar{x})$ and $\overline{N}_D\big(F(\bar{x})\big)$,*

$$\begin{cases} \textit{the only vector } y \in \overline{N}_D\big(F(\bar{x})\big) \textit{ for which} \\ -\sum_{i=1}^{m} y_i \nabla f_i(\bar{x}) \in \overline{N}_X(\bar{x}) \textit{ is } y = (0,\ldots,0). \end{cases}$$

Guide. First verify that (c) is sufficient for the constraint qualification, and that it is necessary when the regularity properties hold. Next, in order to obtain the equivalence of (c) with (a), let $K = \widehat{T}_D(F(\bar{x})) - \nabla F(\bar{x})\widehat{T}_X(\bar{x})$. Show that K^* consists of the vectors $y \in \widehat{T}_D(F(\bar{x}))^*$ such that $-\nabla F(\bar{x})^* y \in \widehat{T}_X(\bar{x})^*$. Apply 6.38, noting along the way that if a convex set has $I\!R^m$ as its closure, it must in fact be $I\!R^m$ (cf. 2.33). For the equivalence of (a) with (b), argue that $K = I\!R^m$ if and only if $0 \in \operatorname{rint} K$ but there's no vector $y \neq 0$ such that $y \perp K$. Use 2.44 and 2.45 to see that $\operatorname{rint} K = \operatorname{rint} \widehat{T}_D(F(\bar{x})) - \nabla F(\bar{x}) \operatorname{rint} \widehat{T}_X(\bar{x})$. □

An example where the normal cone constraint qualification in 6.14 is satisfied, but the tangent cone constraint qualification in 6.39(a) [or (b)] isn't, is obtained with $F(x) \equiv x$ when X is the heart-shaped set in Figures 6–6 and 6–15, $\bar{x}$ is the inward corner point of X, and D is the closed half-plane consisting of the horizontal line through $\bar{x}$ and all points below it. The normal cone version is generally more robust, therefore, than the tangent cone version.

This example also illustrates, by the way, the value of working mainly with the possibly nonconvex normal cone $N_C(\bar{x})$ in Definition 6.3 instead of the convexified cone $\overline{N}_C(\bar{x})$. While convexification may be desirable or essential for some applications, it should usually be invoked only in the final stages of a development, or some degree of robustness may be lost.

When $X = I\!R^n$ in 6.39, while D is regular at $\bar{x}$ (as when D is convex), the tangent cone constraint qualification in version (b) comes out simply as requiring the existence of a vector $\tilde{w}$ with $\nabla F(\bar{x})\tilde{w} \in \operatorname{rint} T_D\big(F(\bar{x})\big)$, yet the nonexistence of a nonzero vector $y \perp T_D\big(F(\bar{x})\big)$ with $\nabla F(\bar{x})^* y = 0$. The classical case is the following one.

6.40 Example (Mangasarian-Fromovitz constraint qualification). *For smooth functions $f_i : I\!R^n \to I\!R$, $i = 1,\ldots,m$, let*

$$C = \big\{ x \in I\!R^n \bigm| f_i(x) \leq 0 \textit{ for } i \in [1,s],\ f_i(x) = 0 \textit{ for } i \in [s+1,m] \big\}$$

and view this as $C = \{x \in X \mid F(x) \in D\}$ *with* $X = I\!R^n$, $F = (f_1, \ldots, f_m)$, *and* $D = \{(u_1, \ldots, u_m) \mid u_i \le 0 \text{ for } i \in [1,s],\ u_i = 0 \text{ for } i \in [s+1, m]\}$.

The constraint qualifications in 6.14 and 6.39(a)(b)(c) are equivalent then to the stipulation that the gradients $\nabla f_i(\bar{x})$ *for* $i \in [s+1, m]$ *are linearly independent and there exists* $\tilde{w}$ *satisfying*

$$\begin{cases} \langle \nabla f_i(\bar{x}), \tilde{w} \rangle < 0 & \text{for all indices } i \in [1,s] \text{ with } f_i(\bar{x}) = 0, \\ \langle \nabla f_i(\bar{x}), \tilde{w} \rangle = 0 & \text{for all indices } i \in [s+1, r]. \end{cases}$$

Detail. This is based on the foregoing remarks. The vector $\nabla F(\bar{x})\tilde{w}$ has components $\langle \nabla f_i(\bar{x}), \tilde{w} \rangle$. On the other hand, $T_D(F(\bar{x}))$ consists of the vectors $(u_1, \ldots, u_m)$ such that $u_i \le 0$ for $i \in [1,s]$ with $f_i(\bar{x}) = 0$, but $u_i = 0$ for $i \in [s+1, m]$, and $\operatorname{rint} T_D(F(\bar{x}))$ is the same except with strict inequalities. □

I.* Other Formulas

In addition to the formulas for normal cones in 6.14 and the ones for tangent cones in 6.31 (see also 6.7 and 6.32), there are other rules that can sometimes be used to determine normals and tangents.

6.41 Proposition (tangents and normals to product sets). *With* $I\!R^n$ *expressed as* $I\!R^{n_1} \times \cdots \times I\!R^{n_m}$, *write* $x \in I\!R^n$ *as* $(x_1, \ldots, x_m)$ *with components* $x_i \in I\!R^{n_i}$. *If* $C = C_1 \times \cdots \times C_m$ *for closed sets* $C_i \in I\!R^{n_i}$, *then at any* $\bar{x} = (\bar{x}_1, \ldots, \bar{x}_m)$ *with* $\bar{x}_i \in C_i$ *one has*

$$\begin{aligned} N_C(\bar{x}) &= N_{C_1}(\bar{x}_1) \times \cdots \times N_{C_m}(\bar{x}_m), \\ \widehat{N}_C(\bar{x}) &= \widehat{N}_{C_1}(\bar{x}_1) \times \cdots \times \widehat{N}_{C_m}(\bar{x}_m), \\ T_C(\bar{x}) &\subset T_{C_1}(\bar{x}_1) \times \cdots \times T_{C_m}(\bar{x}_m), \\ \widehat{T}_C(\bar{x}) &= \widehat{T}_{C_1}(\bar{x}_1) \times \cdots \times \widehat{T}_{C_m}(\bar{x}_m). \end{aligned}$$

Furthermore, C *is regular at* $\bar{x}$ *if and only if each* C_i *is regular at* $\bar{x}_i$. *In the regular case the inclusion for* $T_C(\bar{x})$ *becomes an equation like the others.*

Proof. The formula for $\widehat{N}_C(\bar{x})$ is immediate from the description of regular normals in 6.11. Noting that $x^\nu \underset{C}{\to} \bar{x}$ if and only if $x_i^\nu \underset{C}{\to} \bar{x}_i$ for each i, we obtain the formula for $N_C(\bar{x})$ as well. The truth of the regularity assertion is then obvious. The inclusion for $T_C(\bar{x})$ is an elementary consequence of Definition 6.1, while the equation for $\widehat{T}_C(\bar{x})$ follows by the second polarity relation in 6.28 from the equation for $N_C(\bar{x})$. Because $T_C(\bar{x}) = \widehat{T}_C(\bar{x})$ when C is regular at $\bar{x}$, the inclusion for $T_C(\bar{x})$ is an equation in that case. □

An example where the inclusion for $T_C(\bar{x})$ in 6.41 is strict is encountered for $C = C_1 \times C_2 \subset I\!R \times I\!R$ with $C_1 = C_2 := \{0\} \cup \{2^{-k} \mid k \in I\!N\}$. At $\bar{x} = (\bar{x}_1, \bar{x}_2) = (0,0)$ every vector $w = (w_1, w_2)$ with positive coordinates belongs to $T_{C_1}(\bar{x}_1) \times T_{C_2}(\bar{x}_2)$, but such a vector belongs to $T_C(\bar{x})$ if and only if there is a choice of $\tau^\nu \searrow 0$ and exponents k^ν and l^ν with $(2^{-k^\nu}, 2^{-l^\nu})/\tau^\nu) \to (w_1, w_2)$. That can't hold unless w_1/w_2 happens to be an integral power of 2. Of course, the polar of the formula for $\widehat{N}_C(\bar{x})$ in 6.41 provides anyway that

$$\operatorname{con} T_C(\bar{x}) = \operatorname{con} T_{C_1}(\bar{x}_1) \times \cdots \times \operatorname{con} T_{C_m}(\bar{x}_m).$$

6.42 Theorem (tangents and normals to intersections). *Let $C = C_1 \cap \cdots \cap C_m$ for closed sets $C_i \subset \mathbb{R}^n$, and let $\bar{x} \in C$. Then*

$$\begin{aligned} T_C(\bar{x}) &\subset T_{C_1}(\bar{x}) \cap \cdots \cap T_{C_m}(\bar{x}), \\ \widehat{N}_C(\bar{x}) &\supset \widehat{N}_{C_1}(\bar{x}) + \cdots + \widehat{N}_{C_m}(\bar{x}). \end{aligned}$$

Under the condition that the only combination of vectors $v_i \in N_{C_i}(\bar{x})$ with $v_1 + \cdots + v_m = 0$ is $v_i = 0$ for all i (this being satisfied for $m = 2$ when C_1 and C_2 are convex and cannot be separated), one also has

$$\begin{aligned} \widehat{T}_C(\bar{x}) &\supset \widehat{T}_{C_1}(\bar{x}) \cap \cdots \cap \widehat{T}_{C_m}(\bar{x}), \\ N_C(\bar{x}) &\subset N_{C_1}(\bar{x}) + \cdots + N_{C_m}(\bar{x}). \end{aligned}$$

If in addition every C_i is regular at $\bar{x}$, then C is regular at $\bar{x}$ and

$$\begin{aligned} T_C(\bar{x}) &= T_{C_1}(\bar{x}) \cap \cdots \cap T_{C_m}(\bar{x}), \\ N_C(\bar{x}) &= N_{C_1}(\bar{x}) + \cdots + N_{C_m}(\bar{x}). \end{aligned}$$

Proof. This applies Theorems 6.14 and 6.31 to $D := C_1 \times \cdots \times C_m \subset (\mathbb{R}^n)^m$ and the mapping $F : x \mapsto (x, \ldots, x) \in (\mathbb{R}^n)^m$ with $X = \mathbb{R}^n$. Normals and tangents to D are calculated from 6.41. □

6.43 Theorem (tangents and normals to image sets). *Let $D = F(C)$ for a closed set $C \subset \mathbb{R}^n$ and a smooth mapping $F : \mathbb{R}^n \to \mathbb{R}^m$. At any $\bar{u} \in D$ one has*

$$\begin{aligned} T_D(\bar{u}) &\supset \bigcup_{\bar{x} \in F^{-1}(\bar{u}) \cap C} \left\{ \nabla F(\bar{x}) w \,\middle|\, w \in T_C(\bar{x}) \right\}, \\ \widehat{N}_D(\bar{u}) &\subset \bigcap_{\bar{x} \in F^{-1}(\bar{u}) \cap C} \left\{ y \,\middle|\, \nabla F(\bar{x})^* y \in \widehat{N}_C(\bar{x}) \right\}. \end{aligned}$$

If there exists $U \in \mathcal{N}(\bar{u})$ such that $F^{-1}(U) \cap C$ is bounded, then also

$$\begin{aligned} \widehat{T}_D(\bar{u}) &\supset \bigcap_{\bar{x} \in F^{-1}(\bar{u}) \cap C} \left\{ \nabla F(\bar{x}) w \,\middle|\, w \in \widehat{T}_C(\bar{x}) \right\}, \\ N_D(\bar{u}) &\subset \bigcup_{\bar{x} \in F^{-1}(\bar{u}) \cap C} \left\{ y \,\middle|\, \nabla F(\bar{x})^* y \in N_C(\bar{x}) \right\}, \end{aligned}$$

where the union including $N_D(\bar{u})$ is itself a closed cone. When C is convex and F is linear with $F(x) = Ax$ for a matrix $A \in \mathbb{R}^{m \times n}$, one simply has

$$\left.\begin{array}{rcl} T_D(\bar u) &=& \widehat T_D(\bar u) = \mathrm{cl}\left\{Aw \,\middle|\, w\in T_C(\bar x)\right\}\\ N_D(\bar u) &=& \widehat N_D(\bar u) = \left\{y \,\middle|\, A^*y\in N_C(\bar x)\right\}\end{array}\right\} \text{ for every } \bar x\in F^{-1}(\bar u)\cap C.$$

Proof. If $y\in \widehat N_D(\bar u)$, there exists by 6.11 a smooth function h around $\bar u$ such that h has a local maximum relative to D at $\bar u$, and $\nabla h(\bar u)=y$. Then for any point $\bar x\in F^{-1}(\bar u)\cap C$ the function $k(x)=h\big(F(x)\big)$ has a local maximum relative to C at $\bar x$. Since $\nabla k(\bar x)=\nabla F(\bar x)^*y$, we deduce from 6.11 that $\nabla F(\bar x)^*y\in \widehat N_C(\bar x)$. This proves the first of the normal cone inclusions. The second then follows from the limit definition of $N_D(\bar u)$ and the closedness property in 6.6. The boundedness condition ensures in this that when $F(x^\nu)=u^\nu\to\bar u$ with $x^\nu\in C$, the sequence $\{x^\nu\}_{\nu\in\mathbb{N}}$ is bounded, thus having a cluster point $\bar x\in C$. The same argument supports the assertion that the cone on the right in this second inclusion is closed. Note that it also ensures D is locally closed at $\bar u$.

The first of the tangent cone inclusions stems from the recollection that whenever $\bar x\in C$ and $w\in T_C(\bar x)$ there are sequences $x^\nu \underset{C}{\to} \bar x$ and $\tau^\nu\searrow 0$ such that $(x^\nu-\bar x)/\tau^\nu\to w$. In terms of $F(\bar x)=\bar u$ and $F(x^\nu)=u^\nu$ in D, we then have $(u^\nu-\bar u)/\tau^\nu\to\nabla F(\bar x)w$, hence $\nabla F(\bar x)w\in T_D(\bar u)$.

For the second of the tangent cone inclusions, consider any vector $\bar z$ belonging to all the cones $\left\{\nabla F(\bar x)w \,\middle|\, w\in \widehat T_C(\bar x)\right\}$ for $\bar x\in F^{-1}(\bar u)\cap C$. We must show that $\bar z\in \widehat T_D(\bar u)$. Because D is locally closed at $\bar u$, we know from formula 6(16) of Theorem 6.26 that $\widehat T_D(\bar u)$ is the inner limit of $T_D(u)$ as $u\underset{D}{\to}\bar u$. Thus it will suffice to consider an arbitrary sequence $u^\nu\underset{D}{\to}\bar u$ and demonstrate the existence of $z^\nu\in T_D(u^\nu)$ such that $z^\nu\to\bar z$.

Because $u^\nu\in D$, there exist points $x^\nu\in F^{-1}(u^\nu)\cap C$. Our boundedness condition makes the sequence $\{x^\nu\}$ have a cluster point $\bar x$; we can assume simply that $x^\nu\to\bar x$. Then $\bar x\in F^{-1}(\bar u)\cap C$ and it follows by assumption that $\bar z\in\left\{\nabla F(\bar x)w \,\middle|\, w\in\widehat T_C(\bar x)\right\}$. Thus, there exists $\bar w\in\widehat T_C(\bar x)$ with $\nabla F(\bar x)\bar w=\bar z$. Then by the inner limit formula 6(16) of Theorem 6.26 for $\widehat T_C(\bar x)$, there are vectors $w^\nu\in T_C(x^\nu)$ such that $w^\nu\to\bar w$. Let $z^\nu=\nabla F(x^\nu)w^\nu$. We have $z^\nu\to\bar z$ by the continuity of ∇F. Moreover the first of the tangent cone inclusions, as invoked for $T_D(u^\nu)$, gives us $z^\nu\in T_D(u^\nu)$ as required.

When C is convex and $F(x)=Ax$, so that D is convex, the condition $z\in N_D(\bar u)$ translates through the characterization of normals to convex sets in 6.9 into saying, for any choice of $\bar x\in C$ with $A\bar x=\bar u$, that $\langle z, Ax-A\bar x\rangle\le 0$ for all $x\in C$. This is identical to having $\langle A^*z, x-\bar x\rangle\le 0$ for all $x\in C$, which by 6.9 again means that $A^*z\in N_C(\bar x)$.

The tangent cone formula in the convex case can be derived similarly from characterization of tangent cones to convex sets in 6.9. For any $\bar x\in C$ with $A\bar x=\bar u$, a vector z has the property that $\bar u+\varepsilon z\in D$ if and only if there is a point $x\in C$ with the property that $z=(Ax-A\bar x)/\varepsilon$. This corresponds to having $z=Aw$ for a vector w such that $\bar x+\varepsilon w\in C$. □

Formulas for tangent and normal cones to an *inverse* image set $C=F^{-1}(D)$, in the case of a closed set $D\subset\mathbb{R}^m$ and a smooth mapping

$F : I\!R^n \to I\!R^m$, are of course furnished already by 6.7 and 6.32 and more importantly by Theorems 6.14 and 6.31 in the case of $X = I\!R^n$. In comparing such formulas with the ones in for the 'forward' images in Theorem 6.43, a significant difference is seen. Theorems 6.14 and 6.31 bring in a constraint qualification to produce a pair of inclusions going in the *opposite* direction from the elementary ones first obtained, whereas Theorem 6.43 features a second pair of inclusions having the *same* direction as the first pair. The earlier results achieve calculus rules in equation form for the broad class of regular sets, but 6.43 only offers such rules for convex sets, and in quite a special way, useful as it may be. These two different patterns of formulas will be evident also in the calculus of subgradients and subderivatives in Chapter 10.

6.44 Exercise (tangents and normals under set addition). *Let $C = C_1 + \cdots + C_m$ for closed sets $C_i \subset I\!R^n$. Then at any point $\bar{x} \in C$ one has*

$$T_C(\bar{x}) \supset \bigcup_{\substack{\bar{x}_1+\cdots+\bar{x}_m=\bar{x}\\ \bar{x}_i\in C_i}} \Big[T_{C_1}(\bar{x}_i) + \cdots + T_{C_m}(\bar{x}_m)\Big],$$

$$\widehat{N}_C(\bar{x}) \subset \bigcap_{\substack{\bar{x}_1+\cdots+\bar{x}_m=\bar{x}\\ \bar{x}_i\in C_i}} \Big[\widehat{N}_{C_1}(\bar{x}_1) \cap \cdots \cap \widehat{N}_{C_m}(\bar{x}_m)\Big].$$

If $\bar{x}$ has a neighborhood V such that the set of $(x_1, \ldots, x_m) \in C_1 \times \cdots \times C_m$ with $x_1 + \cdots + x_m \in V$ is bounded, then

$$\widehat{T}_C(\bar{x}) \supset \bigcap_{\substack{\bar{x}_1+\cdots+\bar{x}_m=\bar{x}\\ \bar{x}_i\in C_i}} \Big[\widehat{T}_{C_1}(\bar{x}_i) + \cdots + \widehat{T}_{C_m}(\bar{x}_m)\Big],$$

$$N_C(\bar{x}) \subset \bigcup_{\substack{\bar{x}_1+\cdots+\bar{x}_m=\bar{x}\\ \bar{x}_i\in C_i}} \Big[N_{C_1}(\bar{x}_1) \cap \cdots \cap N_{C_m}(\bar{x}_m)\Big].$$

If each C_i is convex, then, for any choice of $\bar{x}_i \in C_i$ with $\bar{x}_1 + \cdots + \bar{x}_m = \bar{x}$,

$$T_C(\bar{x}) = \operatorname{cl}\big[T_{C_1}(\bar{x}_i) + \cdots + T_{C_m}(\bar{x}_m)\big],$$
$$N_C(\bar{x}) = N_{C_1}(\bar{x}_1) \cap \cdots \cap N_{C_m}(\bar{x}_m).$$

Guide. Here $C = F\big(C_1 \times \cdots \times C_m\big)$ for $F(x_1, \ldots, x_m) = x_1 + \cdots + x_m$. Apply Theorem 6.43. □

The variational geometry of polyhedral sets has special characteristics worth noting.

6.45 Lemma (polars of polyhedral cones; Farkas). *The polar of a cone of form $K = \big\{x \,\big|\, \langle a_i, x\rangle \le 0 \text{ for } i = 1, \ldots, m\big\}$ is $K^* = \operatorname{con}\big(\operatorname{pos}\{a_1, \ldots, a_m\}\big)$, the cone consisting of all linear combinations $y_1 a_1 + \cdots + y_m a_m$ with $y_i \ge 0$.*

Proof. There's no loss of generality in supposing that $a_i \ne 0$ in $I\!R^n$. Let $J = \operatorname{con}\big(\operatorname{pos}\{a_1, \ldots, a_m\}\big)$. The cone J is polyhedral by 3.52, in particular closed. It's elementary that $J^* = K$, hence $K^* = J^{**} = J$ by 6.20. □

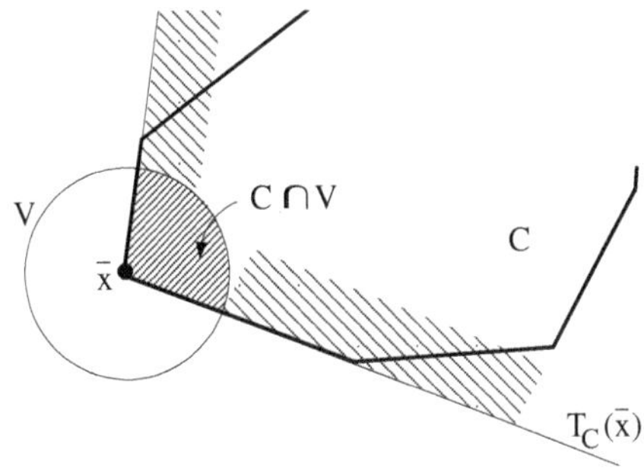

Fig. 6–19. Tangent cones to a polyhedral set.

6.46 Theorem (tangents and normals to polyhedral sets). *For a polyhedral set $C \subset \mathbb{R}^n$ and any point $\bar{x} \in C$, the cones $T_C(\bar{x})$ and $N_C(\bar{x})$ are polyhedral. Indeed, relative to any representation*

$$C = \{x \mid \langle a_i, x\rangle \le \alpha_i \text{ for } i = 1, \dots, m\}$$

and the active index set $I(\bar{x}) := \{i \mid \langle a_i, \bar{x}\rangle = \alpha_i\}$, one has

$$\begin{aligned} T_C(\bar{x}) &= \{w \mid \langle a_i, w\rangle \le 0 \text{ for all } i \in I(\bar{x})\}, \\ N_C(\bar{x}) &= \{y_1 a_1 + \cdots + y_m a_m \mid y_i \ge 0 \text{ for } i \in I(\bar{x}),\ y_i = 0 \text{ for } i \notin I(\bar{x})\}. \end{aligned}$$

Proof. Since tangents and normals are 'local', we can suppose that all the constraints are active at $\bar{x}$. Then $T_C(\bar{x})$ consists of the vectors w such that $\langle a_i, w\rangle \le 0$ for $i = 1, \dots, m$, so $T_C(\bar{x})$ is polyhedral. Since $N_C(\bar{x})$ is the polar of $T_C(\bar{x})$ by 6.24, it has the description in 6.45 and is polyhedral by 3.52. □

6.47 Exercise (exactness of tangent approximations). *For a polyhedral set C and any $\bar{x} \in C$, there exists $V \in \mathcal{N}(\bar{x})$ with $C \cap V = [\bar{x} + T_C(\bar{x})] \cap V$.*

Guide. Use the formula for $T_C(\bar{x})$ that comes out of a description of C by finitely many linear constraints. □

6.48 Exercise (separation of convex cones). *Let $K_1, K_2 \subset \mathbb{R}^n$ be convex cones.*

(a) *K_1 and K_2 can be separated if and only if $K_1 - K_2 \ne \mathbb{R}^n$. This is equivalent to having $K_1^* \cap [-K_2^*] \ne \{0\}$.*

(b) *K_1 and K_2 can be separated properly in particular if $K_1 - K_2 \ne \mathbb{R}^n$ and $\operatorname{int}[K_1 - K_2] \ne \emptyset$. This is equivalent to having not only $K_1^* \cap [-K_2^*] \ne \{0\}$ but also $K_1^* \cap [-K_2^*]$ pointed.*

(c) *When K_1 and K_2 are closed, they can be separated almost strictly, in the sense of $K_1 \setminus \{0\}$ and $K_2 \setminus \{0\}$ lying in complementary open half-spaces, if and only if $\operatorname{cl}[K_1 - K_2]$ is pointed. This is equivalent to $[\operatorname{int} K_1^*] \cap [-\operatorname{int} K_2^*] \ne \emptyset$, and it holds if and only if K_1 and K_2 are pointed and $K_1 \cap K_2 = \{0\}$.*

Guide. Rely on 2.39, 3.8, 3.14 and 6.22. □

Finally, we record a curious consequence of the generic continuity result in Theorem 5.55.

6.49 Proposition (generic continuity of normal-cone mappings). *For any set $C \subset \mathbb{R}^n$, the set of points $\bar{x} \in \operatorname{bdry} C$ where the mapping N_C fails to be continuous relative to* bdry C *is meager in* bdry C*. In particular, the points where N_C is continuous relative to* bdry C *are dense in* bdry C*.*

Proof. We simply apply 5.55 to the mapping $S = N_C$ and the set $X = \operatorname{bdry} C$, which is closed. It's known that a meager subset of a closed set $X \subset \mathbb{R}^n$ has dense relative complement in X. □

Commentary

Tangent and normal cones have been introduced over the years in great variety, and the names for these objects have evolved in several directions. Rather than displaying the full spectrum of possibilities, our overriding goal in this chapter has been to identify and concentrate on a minimal family of such cones which is now known to be adequate for most applications, at least in a finite-dimensional context. We have kept notation and terminology as simple and coherent as we could, in the hope that this would help in making the subject attractive to a broader group of potential users.

It seemed desirable in this respect to tie into the classical pattern of tangent and normal subspaces to a smooth manifold by bestowing the plainest symbols $T_C(\bar{x})$ and $N_C(\bar{x})$ and the basic names 'tangent cone' and 'normal cone' on the objects now understood to go the farthest in conveying the basic theory. At the same time we thought it would be good to trace as clearly as possible in notation and terminology the role of Clarke regularity, a property which has come to be recognized as being among the most fundamental.

These motives have led us to a new outlook on variational geometry, which emerges in the scheme of Figure 6–17. To the extent that we have deviated in our development from the patterns that others have followed, the reason can almost always be found in our focus on this scheme and the need to present it most clearly and naturally.

One-sided notions of tangency closely related to the tangent cone $T_C(\bar{x})$ of 6.1 were first developed by Bouligand [1930], [1932b], [1935] and independently by Severi [1930a], [1930b], [1934], but from a somewhat different point of view. Bouligand and Severi considered sequences of points $x^\nu \underset{C}{\to} \bar{x}$ in C, with $x^\nu \neq \bar{x}$, such that the corresponding sequence of half-lines L^ν, starting at $\bar{x}$ and passing through x^ν, converges to some half-line L, likewise starting at $\bar{x}$. They viewed this as signaling directional convergence to $\bar{x}$ with the direction indicated by L and defined the union of all the half-lines L so obtainable, in Bouligand's terminology, as the *contingent* of C at $\bar{x}$ which Severi designated as the set of *semitangents* at $\bar{x}$. The contingent in this sense is the set $\bar{x} + T_C(\bar{x})$, unless $\bar{x}$ is an isolated point of C, in which case the contingent is the empty set. These differences with $T_C(\bar{x})$ have been glossed over in more recent literature, where $T_C(\bar{x})$ has often been called the 'contingent cone' to C at $\bar{x}$.

In Bouligand's and Severi's time, the algebraic notion of a 'cone' of vectors emanating from the origin wasn't current. It's clear from their writings that they really viewed his contingent as standing for a set of *directions*, i.e., as being the subset $\operatorname{dir} T_C(\bar{x})$ of $\operatorname{hzn} \mathbb{R}^n$ in our cosmic framework. To maintain this distinction and for the general reasons already mentioned, we prefer to speak of $T_C(\bar{x})$ as the *tangent cone*, which opens the way to referring to its elements as *tangent vectors*. This conforms also to the usage of Hestenes [1966] and other pioneers in the development of optimality conditions relative to an abstract constraint $x \in C$, as well as to the language of convex analysis, cf. Rockafellar [1970a].

Cones of a more special sort were invoked in the landmark paper of Kuhn and Tucker [1951] on Lagrange multipliers, which was later found to have been anticipated by the unpublished thesis of Karush [1939] (see Kuhn [1976]). There, the only tangent vectors w considered at $\bar{x}$ were ones representable as tangent to some smooth curve proceeding from $\bar{x}$ into C, at least for a short way. Such tangent vectors are in particular derivable as defined in 6.1, but derivability doesn't require the function $\xi(\tau)$ in the definition to be have any property of differentiability for $\tau > 0$. This restricted class of tangent vectors remains even now the one customarily presented in textbooks on optimization and made the basis for the development of optimality conditions by techniques that rely on the standard implicit function theorem of calculus. Such an approach has many limitations, however, especially in treating abstract constraints. These can be avoided by working in the broader framework provided here.

The subcone of $T_C(\bar{x})$ consisting of all derivable tangent vectors has been used to advantage especially by Frankowska in her work in control theory; see the book of Aubin and Frankowska [1990] for references and further developments. Rather than studying this cone from a general perspective, we have emphasized the typically occurring case of 'geometric derivability', where it actually coincides with the tangent cone $T_C(\bar{x})$ and affords a tangential approximation to C at $\bar{x}$. Such a mode of conical approximation was first investigated by Chernoff [1954] for the sake of an application in statistics. He expressed it by distance functions, but a description in terms of set convergence, like ours, follows at once from the role of distance functions in the theory of such convergence in Chapter 4 (cf. 4.7). For more on geometric derivability or its absence, see Jimenez and Novo [2006].

The limit formula we've used in defining the regular tangent cone $\widehat{T}_C(\bar{x})$ in 6.25 was devised by Rockafellar [1979b], but the importance of this cone was revealed earlier by Clarke [1973], [1975], who introduced it as the polar of his normal cone (discussed below), thus in effect taking the relation in 6.28(b) as its definition. The expression of $\widehat{T}_C(\bar{x})$ as the inner limit of the cones $T_C(x)$ at neighboring points $x \in C$ was effectively established by Aubin and Clarke [1977]; the idea was pursued by Cornet in unpublished work and by Penot [1981]. See Treiman [1983a] for a full proof that works to some extent for Banach spaces too.

The normal cone to a *convex* set C at a point $\bar{x}$ was already defined in essence by Minkowski [1911]. It was treated by Fenchel [1951] in modern cone fashion as consisting of the outward normals to the supporting half-spaces to C at $\bar{x}$. The concept was developed extensively in convex analysis as a geometric companion to subgradients of convex functions and for its usefulness in expressing conditions of optimality; cf. Rockafellar [1970a]. This case of normal cones $N_C(\bar{x})$ served as a prototype for all subsequent explorations of such concepts. The polar relationship between $N_C(\bar{x})$ and $T_C(\bar{x})$ when C is convex was taken as a role model as well, although rethinking had to be done later in that respect. The theory of polar cones in general began with Steinitz [1913-16]; earlier, Minkowski [1911] had developed for closed, bounded, convex neighborhoods of the origin the polarity correspondence that eventually furnished the basis for norms and the topological study of vector spaces (see 11.19).

The first robust generalization of normal cones beyond convexity was effected by Clarke [1973], [1975]. His normal cone to a nonconvex (but closed) set C at a point $\bar{x}$ was defined in stages, starting from a notion of subgradient for Lipschitz continuous functions and applying it to the distance function d_C, but he showed that in the end his cone could be realized by limits of proximal normals at neighboring points x of C (although the term 'proximal normal' itself didn't come until later). Thus, Clarke's normal cone had the expression

$$\operatorname{cl}\operatorname{con}\Big[\limsup_{x \xrightarrow[C]{} \bar{x}} \hat{\hat{N}}_C(x)\Big], \quad \hat{\hat{N}}_C(x) := [\text{cone of proximal normals to } C \text{ at } x]. \tag{6(20)}$$

This equals $\operatorname{cl}\operatorname{con} N_C(\bar{x})$ in the light of our present knowledge that the outer limit in 6(20) would be unchanged if $\hat{N}_C(x)$ were substituted for $\hat{\hat{N}}_C(x)$ (see 6.18(a)), which is a fact due to Kruger and Mordukhovich [1980]. Clarke's normal cone is identical therefore to the convexified cone $\overline{N}_C(\bar{x})$ in 6(19) and 6.38. Indeed, the achievement of the polarity relation in 6.38(a) was what compelled him to take the closed convex hull. The operation seemed harmless at the time, because many applications concerned situations covered by Clarke regularity, where the outer limit itself, $N_C(\bar{x})$, was sure to be convex anyway. Where convexity wasn't assured, as in efforts at generalizing Euler-Lagrange and Hamiltonian equations as necessary conditions for optimality on the calculus of variations, cf. Clarke [1973], [1983], [1989], it appeared necessary to add it for technical reasons connected with weak convergence in Lebesgue spaces.

The limit cone in 6(20), unconvexified, was studied for years as the cone of 'limiting proximal normal vectors' and a degree of calculus for it was derived, cf. Rockafellar [1985a] and the works of Clarke already cited. It was viewed however as a preliminary object, not a centerpiece of the theory as here, in its designation as $N_C(\bar{x})$ with a broader description through limits of regular normal vectors. Increasingly, though, the price paid for moving to the convexified cone was becoming apparent, especially in the parametric analysis of optimal solutions and related attempts at setting up a second-order theory of generalized differentiation.

A key geometric consideration for that purpose turned out to be the investigation of graphs of Lipschitz continuous mappings as 'nonsmooth manifolds'. Rockafellar [1985b] showed that Clarke's normal cone at a point of such a manifold of dimension d in $\mathbb{R}^n$ inevitably had to be a linear subspace, moreover one with dimension greater than $n-d$ unless the manifold was 'strictly smooth' at the point in question. Correspondingly, Clarke's tangent cone, defined through polarity, had to be a linear subspace, and its dimension had to be less than d except in the strictly smooth case. For such sets C, and others representable by them, like the graphs of basic subgradient mappings (see Chapters 8 and 12), it was hopeless then for variational geometry to make significant progress unless convexification of normal cones were relinquished.

Meanwhile, the possibility emerged of using the cone of 'limiting proximal normals' directly, without convexification, in the statement and derivation of optimality conditions. This approach, started by Mordukhovich [1976], required no appeal to convex analysis at all when just necessary conditions were in question. It contrasted not only with techniques based on the ideas already described, but also with those in the school of Dubovitzkii and Miliutin [1965], which relied on selecting convex subcones of certain tangent cones and then applying a separation theorem. Instead, auxiliary optimization problems were introduced in terms of penalties relative to a sort of approximate separation, and limits were taken. The idea was explored in depth by Mordukhovich [1980], [1984], [1988], Kruger [1985] and Ioffe [1981a], [1981b], [1981c], [1984b]. They were able to get around previously perceived obstacles and establish complementary rules of calculus which opened the door to new configurations of variational analysis in general.

The regular normal cone $\widehat{N}_C(\bar{x})$ as defined in 6.3 has not until now been placed in such a front-line position in finite-dimensional variational geometry, although it has been recognized as a serviceable tool. Its advantage over the proximal normal cone, which likewise is convex and serves equally well in the limit definition of $N_C(\bar{x})$, is that $\widehat{N}_C(\bar{x})$ is always closed. Furthermore, $\widehat{N}_C(\bar{x})$ is the polar of $T_C(\bar{x})$, whereas no useful polarity relation comes along with the proximal normal cone, which needn't even have $\widehat{N}_C(\bar{x})$ as its closure.

As the polar cone $T_C(\bar{x})^*$, $\widehat{N}_C(\bar{x})$ has played a role for a long time in optimization theory; cf. Gould and Tolle [1971], Bazaraa, Gould, and Nashed [1974], Hestenes [1975], Penot [1978]. The first paper in this list contains (within its proofs) a characterization of $\widehat{N}_C(\bar{x})$ as consisting of all vectors v realizable as $\nabla h(\bar{x})$ for a function h that is differentiable at $\bar{x}$ and has a local maximum there relative to C. Theorem 6.11 goes well beyond that characterization in constructing h to be a $\mathcal{C}^1$ function on $\mathbb{R}^n$ that attains its global maximum over C uniquely at $\bar{x}$. A result close to this was stated by Rockafellar [1993a], but the proof was sketchy in some details, which have been filled in here.

The important property of regularity of C at $\bar{x}$ developed by Clarke [1973], [1975] was formulated by him to mean, when C is closed, that the cone $T_C(\bar{x})$ in 6.1 lies in the polar of his normal cone, this being the same as the polar of the 'cone of limiting proximal normals'. Since the 'cone of limiting proximal normals' is the same as $N_C(\bar{x})$, we can write Clarke's defining relation as the inclusion $T_C(\bar{x}) \subset N_C(\bar{x})^*$. It's elementary that this is equivalent to $N_C(\bar{x}) \subset T_C(\bar{x})^*$ and therefore, because $T_C(\bar{x})^* = \widehat{N}_C(\bar{x})$, corresponds to the relation $N_C(\bar{x}) = \widehat{N}_C(\bar{x})$ taken to be the definition of Clarke regularity in 6.4 (with the extension that C need only be closed locally). This way of looking at regularity, adopted first by Mordukhovich, allows us to portray the regularity of C at $\bar{x}$ as referring to the case where every normal vector is regular. Of course, it was precisely for this purpose that we were led to calling $\widehat{N}_C(\bar{x})$ the *regular* normal cone and its elements *regular* normal vectors. At the same time, because $N_C(\bar{x})^* = \widehat{T}_C(\bar{x})$, the Clarke regularity of C at $\bar{x}$ corresponds to $T_C(\bar{x}) = \widehat{T}_C(\bar{x})$ and the case where every tangent vector is regular, as long as we speak of $\widehat{T}_C(\bar{x})$ as the regular tangent cone and its elements as regular tangent vectors as in 6.25.

Besides underscoring the dual nature of regularity, this pattern is appealing in the way the two 'regular' cones $\widehat{T}_C(\bar{x})$ and $\widehat{N}_C(\bar{x})$ come out as convex subcones of the general cones $T_C(\bar{x})$ and $N_C(\bar{x})$. Also, a major advantage for developments later in the book is that the 'hat' notation and terminology of regularity can be propagated to subderivatives and subgradients, and on to graphical derivatives and coderivatives, without any conflicts. None of the other ways we tried to capture the scheme in Figure 6–17 worked out satisfactorily. For instance, 'bars' instead of 'hats' might suggest closure or taking limits, which would clash with the desired pattern. Subscripts or superscripts would run into trouble when applied to subgradients, where a superscript ∞ is frequently needed, and so forth.

We say all this with full knowledge that departures from names and symbols which have already achieved a measure of acceptance in the literature are always controversial, even when made with the best of intentions. Again we point to the virtual necessity of this move in bringing the scheme in Figure 6–17 to the foreground of the subject.

Of course, this scheme is finite-dimensional. In infinite-dimensional spaces, the elements of $\widehat{N}_C(\bar{x})$, called 'Fréchet normals' because of the parallel between their defining condition and Fréchet differentiability, are typically more special than those of $T_C(\bar{x})^*$, which are often called 'Dini normals'. Also, the cone $\widehat{T}_C(\bar{x})$ need not coincide with the inner limit of the cones $T_C(x)$ as $x \xrightarrow[C]{} \bar{x}$.

The usefulness of normal cones in characterizing solutions to problems of optimization was the motivation behind much of the work we have already cited in the theory of tangent and normal cones, but the explicit appearance of such cones in the very statement of optimality conditions is another matter. That route was first taken in convex analysis; cf. Rockafellar [1970a]. In parallel, researchers studying problems in partial differential equations with unilateral constraints came up with the idea of a 'variational inequality', which eventually generated a whole school of mathematics of its own. It was Stampacchia [1964] who made the initial start, followed up by Lions and Stampacchia [1965], [1967]. A related subject, growing out of optimality conditions in linear and quadratic programming, was that of 'complementarity'. A combined overview of both subjects is provided by the book of Cottle, Giannessi and Lions [1980]. But despite the fact, known at least since the late 1960s, that variational inequalities and complementarity conditions are special convex cases of the fundamental condition in 6.13 relating a mapping to the normal cones to a set C, with strong ties to the optimality rule in 6.12, little has been made of this perspective, although it supplies numerous additional tools of analysis and calculation and suggests how to proceed in relaxing the convexity of C.

The book of Cottle, Pang and Stone [1992] covers the extensive theory of linear complementarity problems, while the book of Luo, Pang and Ralph [1996] is a source for recent developments on variational inequalities as constraints with other optimization problems.

The normal cone formula in 6.14 can be regarded as a special case of a chain rule for subdifferentiation which will be developed and discussed later (see 10.6 and 10.49, as well as the Commentary of Chapter 10). It offers a method of generating Lagrange multiplier rules which, like the one in 6.15, are more versatile and powerful than the famous rule of Kuhn and Tucker [1951] (and Karush [1939]). The derivation through quadratic penalties, from Rockafellar [1993a], resembles that of McShane [1973] but with key differences in scope and detail. Related 'chain rule' formulas for normal cones as well as for tangent cones, as in 6.31, were developed by Rockafellar [1979b], [1985a], but in terms of Clarke's convexified normal cone.

The constraint qualification in 6.14 in normal cone form has partial precedent in that earlier work (cf. also Rockafellar [1988]), as does the tangent cone form in 6.39(b). In the case of the nonlinear programming problem of 6.15 specialized to $X = \mathbb{R}^n$, the normal and tangent versions are equivalent and reduce respectively to the conditions posed by John [1948] and by Mangasarian and Fromovitz [1967] (theirs being the same as the condition used by Karush [1939] if only inequality constraints are present); cf. 6.40. For other tangent cone constraint qualifications like 6.39(a)(b), see Penot [1982] and Robinson [1983].

In the tradition of convex programming with inequality constraints only (the special case of $X = \mathbb{R}^n$, $D = \mathbb{R}^m_-$, and convex functions f_i as the components of F), the constraint qualification of Slater [1950] is equivalent to all these tangent and normal cone conditions as well. It merely requires the existence of a point $\hat{x}$ such that $f_i(\hat{x}) < 0$ for all i. Extensions of Slater's condition to cover equality constraints and a convex set $X \neq \mathbb{R}^n$ can be found in Rockafellar [1970a] (§28).

Recession vectors in the global sense were originally introduced by Rockafellar [1970a] for convex sets C. In that context, with C closed, there are now the same as the horizon vectors in C^∞. For nonconvex sets C, the 'recession' concept splits away from the 'horizon' concept and, as with many extensions beyond convexity, needs to take on a local character. We have tried to provide the appropriate generalization in the definition in 6.33 of the recession cone $R_C(\bar{x})$ to C at $\bar{x}$. The facts about such cones in 6.35 are new. The identification of the interior of $R_C(\bar{x})$ with that of $\widehat{T}_C(\bar{x})$ in 6.36 was discovered earlier by Rockafellar [1979b], although the 'recession' terminology wasn't used there. The vectors in this interior, characterized by the property in 6.35(a), were called 'hypertangents' in Rockafellar [1981a].

A notation related to local recession was central to the early research of Dubovitzkii and Miliutin [1965] on necessary conditions in constrained optimization. At a point $\bar{x} \in C$ they worked with vectors $\bar{w}$ for which there exist $\varepsilon > 0$ and $\delta > 0$ such that $\bar{x} + \tau w \in C$ for all $\tau \in [0, \varepsilon]$ and $w \in \mathbb{B}(\bar{w}, \delta)$. These are 'interior tangents' of a sort, but not local recession vectors in our sense because they work only at $\bar{x}$ and not also at neighboring points $x \in C$.

The calculus rule in 6.42 for normal cones to an intersection of finitely many closed sets C_i is one of the crucial contributions of Mordukhovich [1984], [1988], which helped to propel variational geometry away from convexification. The statement for convex sets was known much earlier, cf. Rockafellar [1970a], and there were nonconvex versions, less satisfactory than 6.42, in terms of Clarke's normal cones $\overline{N}_{C_i}(\bar{x})$, cf. Rockafellar [1979b], [1985a], as well other versions that did apply to the normal cones $N_{C_i}(\bar{x})$ but under a more stringent constraint qualification; cf. Ioffe [1981a], [1981c], [1984b]. (It's interesting that the pattern of proof adopted early on by Ioffe essentially passed by way of the constraint qualification now known to be the best, but, as a sign of the times, he didn't make this explicit and aimed instead at formulating his results in parallel with those of others.)

The rule for image sets in 6.43 and its corollary in 6.44 for sums of sets were established in Rockafellar [1985a].

The graphical limit result in 6.18(b) is new. The characterization of boundary points in 6.19 comes from Rockafellar [1979b]. The envelope description of convex sets in 6.20 is classical and goes back to Minkowski [1911]; cf. Rockafellar [1970a] for more on this and the properties of polyhedral sets in 6.45 and 6.46. The separation properties of cones in 6.48 are widely known as well.

The facts in 6.27 about normals to tangent cones appear here for the first time, as does the genericity result in 6.49. The latter illustrates the desirability of the extension made in 5.55 of earlier work on the generic semicontinuity of set-valued mappings to the case where such mappings may have unbounded values.

7. Epigraphical Limits

Familiar notions of convergence for real-valued functions on $\mathbb{R}^n$ require a bit of rethinking before they can be applied to possibly unbounded functions that might take on ∞ and $-\infty$ as values. Even then, they may fall short of meeting the basic needs in variational analysis.

A serious challenge comes up, for example, in the approximation of problems of constrained minimization. We've seen that a single such problem can be represented by a single function $f : \mathbb{R}^n \to \overline{\mathbb{R}}$. A sequence of approximating problems may be envisioned then as a sequence of functions $f^\nu : \mathbb{R}^n \to \overline{\mathbb{R}}$ that converges to f, but what sense of convergence is most appropriate? Clearly, it ought to be one for which the values $\inf f^\nu$ and sets $\operatorname{argmin} f^\nu$ behave well in relation to $\inf f$ and $\operatorname{argmin} f$ as $\nu \to \infty$, but this is a consideration that didn't influence convergence definitions of the past. Because of the way ∞ is employed in variational analysis, the convergence of f^ν to f has to bring with it a form of constraint approximation, in geometry or through penalties. This suggests that some appeal to the theory of set convergence may be helpful in finding the right approximation framework.

Another challenge is encountered in trying to break free of the limitations of 'differentiation' as usually conceived, which is only in terms of *linearizing* a given function locally. For a function $f : \mathbb{R}^n \to \overline{\mathbb{R}}$, linearization at $\bar{x}$ could never adequately reflect the local properties of f when $\bar{x}$ is a boundary point of $\operatorname{dom} f$, even if sense could be made out of it, but it's also too narrow an idea when $\bar{x}$ lies in the interior of $\operatorname{dom} f$ because of the prevalence of 'nonsmoothness' in many situations of importance in variational analysis. Convergence questions are central in differentiation because derivatives are usually defined through convergence of difference quotient functions. When the difference quotient functions may be discontinuous and extended-real-valued, new issues arise which require reconsideration of traditional approaches.

Both of these challenges will be taken up here, although the full treatment of generalized differentiation won't come until Chapter 8, where the variational geometry of epigraphs will be studied in detail. Our main tool now will be the application of set convergence theory to epigraphs. This will provide the correct notion of function convergence for our purposes. To set the stage so that comparisons with other notions of convergence can be made, we first discuss the obvious extension of pointwise convergence of functions to the case where the functions may have infinite values.

A. Pointwise Convergence

A sequence of functions $f^\nu : I\!R^n \to \overline{I\!R}$ is said to converge *pointwise* to a function $f : I\!R^n \to \overline{I\!R}$ at a point x if $f^\nu(x) \to f(x)$, and to converge pointwise on a set X if this is true at every $x \in X$. In the case of $X = I\!R^n$, f is said to be the *pointwise limit* of the sequence, which is expressed notationally by

$$f = \text{p-lim}_\nu f^\nu, \quad \text{or} \quad f^\nu \xrightarrow{\text{p}} f.$$

Whether or not a sequence of functions converges pointwise, we can associate with it a *lower pointwise limit function* p-lim $\inf_\nu f^\nu$ and an *upper pointwise limit function* p-lim $\sup_\nu f^\nu$, these being functions from $I\!R^n$ to $\overline{I\!R}$ defined by

$$\begin{aligned}(\text{p-lim inf}_\nu f^\nu)(x) &:= \liminf_\nu f^\nu(x),\\ (\text{p-lim sup}_\nu f^\nu)(x) &:= \limsup_\nu f^\nu(x).\end{aligned}$$

Obviously then, $f^\nu \xrightarrow{\text{p}} f$ if and only if p-lim $\inf_\nu f^\nu =$ p-lim $\sup_\nu f^\nu = f$.

Pointwise convergence, while useful in many ways, is inadequate for such purposes as setting up approximations of problems of optimization. For one thing, it fails to preserve lower semicontinuity of functions. The continuous functions $f^\nu(x) = \min\{1, |x|^{2\nu}\}$, for instance, converge pointwise to the function $f = \min\{1, \delta_{\text{int}\, B}\}$, which isn't lsc. More importantly, pointwise convergence relates poorly to 'max' and 'min'. *From* $f^\nu \xrightarrow{\text{p}} f$ *it doesn't follow that* $\inf f^\nu \to \inf f$, *or that sequences of points* x^ν *selected from the sets* $\operatorname{argmin} f^\nu$ *have all their cluster points in* $\operatorname{argmin} f$, *i.e., that* $\limsup_\nu \operatorname{argmin} f^\nu \subset \operatorname{argmin} f$.

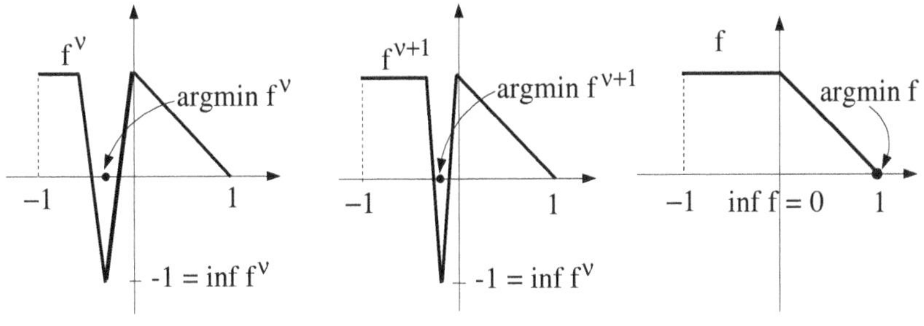

Fig. 7–1. Pointwise convergent functions with optimal values not converging.

An example of this deficiency of pointwise convergence is displayed in Figure 7–1. For each $\nu \in I\!N$ the function f^ν on $I\!R^1$ has $\operatorname{dom} f^\nu = [-1, 1]$. On this interval, $f^\nu(x)$ is the lowest of the quantities $1 - x$, 1, and $2\nu|x + 1/\nu| - 1$. Obviously f^ν is lsc and level-bounded, and it attains its minimum uniquely at $\bar{x}^\nu := -1/\nu$, which yields the value -1. We have pointwise convergence of f^ν to f, where $f(x) = \min\{1 - x, 1\}$ for $x \in [-1, 1]$ but $f(x) = \infty$ otherwise. The limit function f is again lsc and level-bounded, but its infimum is 0, not -1, and this is attained uniquely at $\bar{x} = 1$, not at the limit point of the sequence $\{\bar{x}^\nu\}$. Thus, even though $\text{p-lim}_\nu f^\nu = f$ and both $\lim_\nu(\inf f^\nu)$ and $\lim_\nu(\operatorname{argmin} f^\nu)$ exist, we don't have $\lim_\nu(\inf f^\nu) = \inf f$ or $\lim_\nu(\operatorname{argmin} f^\nu) \subset \operatorname{argmin} f$.

The fault in this example appears to lie in the fact that the functions f^ν don't converge *uniformly* to f on $[-1, 1]$. But a satisfactory remedy can't be sought in turning away from such a situation and restricting attention to sequences where uniform convergence is present. That would eliminate the wide territory of applications where ∞ is a handy device for expressing constraints, themselves possibly undergoing a kind of convergence which may be reflected in the functions f^ν and f not having the same effective domain.

The geometry in this example offers clues about how to proceed in the face of the difficulty. The source of the trouble can really be seen in what happens to the epigraphs. The sets $\operatorname{epi} f^\nu$ in Figure 7–1 don't converge to $\operatorname{epi} f$, but rather to $\operatorname{epi} \bar{f}$, where $\bar{f}$ is the lsc, level-bounded function with $\operatorname{dom} \bar{f} = [-1, 1]$ such that $\bar{f}(x) = 1$ on $[-1, 0)$, $\bar{f}(x) = 1 - x$ on $(0, 1]$ but $\bar{f}(0) = -1$. Moreover $\inf f^\nu \to \inf \bar{f}$ and $\operatorname{argmin} f^\nu \to \operatorname{argmin} \bar{f}$. From this perspective it's apparent that the functions f^ν should be thought of as approaching $\bar{f}$, not f, as far as the behavior of $\inf f^\nu$ and $\operatorname{argmin} f^\nu$ is concerned.

B. Epi-Convergence

The goal now is to build up the theory of such 'epigraphical' function convergence as an alternative and adjunct to pointwise convergence. As with pointwise convergence, it's convenient to follow a pattern of first introducing lower and upper limits of a sort. We approach this geometrically and then proceed to characterize the limits through special formulas which reveal, among other things, how the notions can be treated more flexibly in terms of f^ν converging epigraphically to f 'at an individual point'.

We start from a simple observation. For a sequence of sets $E^\nu \subset \mathbb{R}^{n+1}$ that are *epigraphs*, both the outer limit set $\limsup_\nu E^\nu$ and the inner limit set $\liminf_\nu E^\nu$ are again *epigraphs*. Indeed, the definitions of outer and inner limits (in 4.1) make it apparent that if either set contains (x, α), then it also contains (x, α') for all $\alpha' \in [\alpha, \infty)$. On the other hand, because both limit sets are closed (cf. 4.4), they intersect the vertical line $\{x\} \times \mathbb{R}$ in a closed interval. The criteria for being an epigraph are thus fulfilled.

7.1 Definition (lower and upper epi-limits). *For any sequence $\{f^\nu\}_{\nu \in \mathbb{N}}$ of functions on $\mathbb{R}^n$, the lower epi-limit* $\text{e-}\liminf_\nu f^\nu$ *is the function having as its epigraph the outer limit of the sequence of sets* $\operatorname{epi} f^\nu$*:*

$$\operatorname{epi}(\,\text{e-}\liminf_\nu f^\nu) := \limsup_\nu (\operatorname{epi} f^\nu).$$

The upper epi-limit $\text{e-}\limsup_\nu f^\nu$ *is the function having as its epigraph the inner limit of the sets* $\operatorname{epi} f^\nu$*:*

$$\operatorname{epi}(\,\text{e-}\limsup_\nu f^\nu) := \liminf_\nu (\operatorname{epi} f^\nu).$$

Thus $\text{e-lim inf}_\nu f^\nu \leq \text{e-lim sup}_\nu f^\nu$ *in general. When these two functions coincide, the epi-limit function* $\text{e-lim}_\nu f^\nu$ *is said to exist:* $\text{e-lim}_\nu f^\nu := \text{e-lim inf}_\nu f^\nu = \text{e-lim sup}_\nu f^\nu$. *In this event the functions* f^ν *are said to epi-converge to* f, *a condition symbolized by* $f^\nu \xrightarrow{e} f$. *Thus,*

$$f^\nu \xrightarrow{e} f \quad \Longleftrightarrow \quad \operatorname{epi} f^\nu \to \operatorname{epi} f.$$

This definition opens the door to applying set convergence directly to epigraphs, but because of the effort we've already put into applying set convergence to the graphical convergence of set-valued mappings in Chapter 5, it will often be expedient to use graphical convergence as an intermediary. The key to this is found in Example 5.5, which associates with a function $f : \mathbb{R}^n \to \overline{\mathbb{R}}$ the epigraphical profile mapping

$$E_f : x \mapsto \{\alpha \in \mathbb{R} \mid \alpha \geq f(x)\}$$

having $\operatorname{gph} E_f = \operatorname{epi} f$. Epigraphical limits of sequences of functions f^ν correspond to graphical limits of sequences of such mappings:

$$f^\nu \xrightarrow{e} f \quad \Longleftrightarrow \quad E_{f^\nu} \xrightarrow{g} E_f, \tag{7(1)}$$

$$\begin{aligned} \operatorname{epi}(\text{e-lim inf}_\nu f^\nu) &= \operatorname{gph}(\text{g-lim sup}_\nu E_{f^\nu}), \\ \operatorname{epi}(\text{e-lim sup}_\nu f^\nu) &= \operatorname{gph}(\text{g-lim inf}_\nu E_{f^\nu}). \end{aligned} \tag{7(2)}$$

We'll exploit this relationship repeatedly.

The geometric definition of epi-limits must be supplemented by other expressions in terms of limits of function values so as to facilitate calculations.

7.2 Proposition (characterization of epi-limits). *Let* $\{f^\nu\}_{\nu\in\mathbb{N}}$ *be any sequence of functions on* $\mathbb{R}^n$, *and let* x *be any point of* $\mathbb{R}^n$. *Then*

$$\begin{aligned} (\text{e-lim inf}_\nu f^\nu)(x) &= \min\{\alpha \in \overline{\mathbb{R}} \mid \exists\, x^\nu \to x \ \textit{ with } \ \liminf_\nu f^\nu(x^\nu) = \alpha\}, \\ (\text{e-lim sup}_\nu f^\nu)(x) &= \min\{\alpha \in \overline{\mathbb{R}} \mid \exists\, x^\nu \to x \ \textit{ with } \ \limsup_\nu f^\nu(x^\nu) = \alpha\}. \end{aligned}$$

Thus, $f^\nu \xrightarrow{e} f$ *if and only if at each point* x *one has*

$$\begin{cases} \liminf_\nu f^\nu(x^\nu) \geq f(x) & \textit{for every sequence } x^\nu \to x, \\ \limsup_\nu f^\nu(x^\nu) \leq f(x) & \textit{for some sequence } x^\nu \to x. \end{cases} \tag{7(3)}$$

Proof. From Definition 7.1, a value $\alpha \in \mathbb{R}$ satisfies $\alpha \geq (\text{e-lim inf}_\nu f^\nu)(x)$ if and only if for some choice of index set $N \in \mathcal{N}_\infty^\#$ there exist sequences $x^\nu \underset{N}{\to} x$ and $\alpha^\nu \underset{N}{\to} \alpha$ with $\alpha^\nu \in \mathbb{R}$, $\alpha^\nu \geq f^\nu(x^\nu)$. The same then holds with $\overline{\mathbb{R}}$ in place of $\mathbb{R}$, and the first formula is thereby made obvious. The argument for the second formula is identical, except for having $N \in \mathcal{N}_\infty$ instead of $N \in \mathcal{N}_\infty^\#$. ☐

Obviously, 7(3) implies that actually $f^\nu(x^\nu) \to f(x)$ for at least one sequence $x^\nu \to x$. Cases where this is true with $x^\nu \equiv x$ will be taken up in 7.10. A convexity-dependent criterion for ascertaining when $x^\nu \to x$ implies $f^\nu(x^\nu) \to f(x)$ will come up in 12.36.

The pair of inequalities in 7(3) is one of the most convenient means of verifying epi-convergence. To establish the first of the inequalities at a point x it would suffice to show that whenever an index set $N \in \mathcal{N}_\infty^\#$ is such that $x^\nu \underset{N}{\to} x$ and at the same time $f^\nu(x^\nu) \underset{N}{\to} \alpha$, then $f(x) \leq \alpha$. Once that is accomplished, the verification of the second inequality is equivalent to displaying a sequence $x^\nu \underset{N}{\to} x$ such that actually $f^\nu(x^\nu) \to f(x)$.

7.3 Exercise (alternative epi-limit formulas). *For any sequence $\{f^\nu\}_{\nu \in \mathbb{N}}$ of functions on $\mathbb{R}^n$ and any point x in $\mathbb{R}^n$, one has*

$$\begin{aligned}(\text{e-lim}\inf_\nu f^\nu)(x) &= \sup_{V \in \mathcal{N}(x)} \sup_{N \in \mathcal{N}_\infty} \inf_{\nu \in N} \inf_{x' \in V} f^\nu(x') \\ &= \lim_{\delta \searrow 0} \left[\liminf_{\nu \to \infty} \left[\inf_{x' \in \mathbb{B}(x,\delta)} f^\nu(x') \right]\right],\end{aligned} \qquad 7(4)$$

$$\begin{aligned}(\text{e-lim}\sup_\nu f^\nu)(x) &= \sup_{V \in \mathcal{N}(x)} \inf_{N \in \mathcal{N}_\infty} \sup_{\nu \in N} \inf_{x' \in V} f^\nu(x') \\ &= \lim_{\delta \searrow 0} \left[\limsup_{\nu \to \infty} \left[\inf_{x' \in \mathbb{B}(x,\delta)} f^\nu(x') \right]\right].\end{aligned} \qquad 7(5)$$

In the light of 7(4), the lower epi-limit of $\{f^\nu\}_{\nu \in \mathbb{N}}$ can be viewed as coming from an ordinary 'lim inf' with respect to convergence in two arguments:

$$(\text{e-lim}\inf_\nu f^\nu)(x) = \liminf_{\substack{x' \to x \\ \nu \to \infty}} f^\nu(x'). \qquad 7(6)$$

The upper epi-limit isn't similarly reducible, but it can be expressed by a formula parallel to one for the lower epi-limit in 7(4), namely

$$(\text{e-lim}\sup_\nu f^\nu)(x) = \sup_{V \in \mathcal{N}(x)} \sup_{N \in \mathcal{N}_\infty^\#} \inf_{\nu \in N} \inf_{x' \in V} f^\nu(x'). \qquad 7(7)$$

This formula is based on the following duality relation between $\mathcal{N}_\infty$ and $\mathcal{N}_\infty^\#$ in representing lower and upper limits of sequences of numbers as defined in 1(4):

$$\begin{aligned}\inf_{N \in \mathcal{N}_\infty^\#} \sup_{\nu \in N} \alpha^\nu &= \sup_{N \in \mathcal{N}_\infty} \inf_{\nu \in N} \alpha^\nu =: \liminf_{\nu \to \infty} \alpha^\nu, \\ \sup_{N \in \mathcal{N}_\infty^\#} \inf_{\nu \in N} \alpha^\nu &= \inf_{N \in \mathcal{N}_\infty} \sup_{\nu \in N} \alpha^\nu =: \limsup_{\nu \to \infty} \alpha^\nu.\end{aligned} \qquad 7(8)$$

Just as epi-convergence corresponds to convergence of epigraphs, *hypo-convergence* corresponds to the convergence of hypographs. There's little we need to say about hypo-convergence that isn't obvious in this respect. The notation for the upper and lower hypo-limits of $\{f^\nu\}_{\nu \in \mathbb{N}}$ is $\text{h-lim}\sup_\nu f^\nu$ and $\text{h-lim}\inf_\nu f^\nu$. When these equal the same function f, we write $f = \text{h-lim}_\nu f^\nu$ or $f^\nu \xrightarrow{\text{h}} f$. In parallel with the criterion in 7.2, we have $f^\nu \xrightarrow{\text{h}} f$ if and only if

$$\begin{cases} \liminf_\nu f^\nu(x^\nu) \geq f(x) & \text{for some sequence } x^\nu \to x, \\ \limsup_\nu f^\nu(x^\nu) \leq f(x) & \text{for every sequence } x^\nu \to x. \end{cases} \qquad 7(9)$$

The basic fact is that f^ν hypo-converges to f if and only if $-f^\nu$ epi-converges to $-f$. Where epi-convergence leads to lower semicontinuity, hypo-convergence leads to upper semicontinuity. Where epi-convergence corresponds to graphical convergence of epigraphical profile mappings, hypo-convergence corresponds to graphical convergence of hypographical profile mappings

$$H_f : x \mapsto \{\alpha \in I\!R \mid \alpha \le f(x)\}.$$

One has

$$f^\nu \xrightarrow{\text{h}} f \quad \Longleftrightarrow \quad H_{f^\nu} \xrightarrow{\text{g}} H_f.$$

In practice it's useful to have an expanded notation which can be brought in to highlight aspects of epi-convergence and identify the epigraphical variable that is involved. To this end, we give ourselves the option of writing

$$\begin{aligned} &\liminf_{\nu\to\infty}\text{-epi}_{x'\to x}\ f^\nu(x') \quad \text{for} \quad (\text{e-}\liminf_\nu f^\nu)(x),\\ &\limsup_{\nu\to\infty}\text{-epi}_{x'\to x}\ f^\nu(x') \quad \text{for} \quad (\text{e-}\limsup_\nu f^\nu)(x). \end{aligned} \tag{7(10)}$$

Such notation will eventually help for instance in dealing with convergence of difference quotients, cf. 7.23.

Definition 7.1 speaks only of epi-limit *functions*, but the notation on the left in 7(10) can be taken as referring to what we'll call the lower and upper epi-limit *values* of the sequence $\{f^\nu\}_{\nu\in I\!N}$ at the point x. These values are obtainable equivalently from any of the corresponding expressions on the right sides of the formulas in 7.2 or 7.3, or in 7(6) and 7(7). Accordingly, we'll say that the *epi-limit of* $\{f^\nu\}_{\nu\in I\!N}$ *exists at* x if the upper and lower epi-limit values at x coincide, writing

$$\lim_{\nu\to\infty}\text{-epi}_{x'\to x}\ f^\nu(x') := \liminf_{\nu\to\infty}\text{-epi}_{x'\to x}\ f^\nu(x') = \limsup_{\nu\to\infty}\text{-epi}_{x'\to x}\ f^\nu(x'). \tag{7(11)}$$

In other words, the epi-limit exists at x if and only if there is a value $\alpha \in \overline{I\!R}$ such that

$$\begin{cases} \liminf_\nu f^\nu(x^\nu) \ge \alpha & \text{for every sequence } x^\nu \to x,\\ \limsup_\nu f^\nu(x^\nu) \le \alpha & \text{for some sequence } x^\nu \to x. \end{cases}$$

This approach gives us room to maneuver in exploring convergence properties locally. We can speak of the functions f^ν *epi-converging at the point* x regardless of whether they may do so at other points. We aren't required to worry about the existence of the limit function e-$\lim_\nu f^\nu$, which might not be defined on all of $I\!R^n$ in the full sense of Definition 7.1. The condition for epi-convergence at x comes down to the two properties in 7(3) holding at x.

Through the geometry in Definition 7.1, the theory of set convergence can be applied to epigraphs to obtain many useful properties of epi-convergence.

7.4 Proposition (properties of epi-limits). *The following properties hold for any sequence* $\{f^\nu\}_{\nu\in I\!N}$ *of functions on* $I\!R^n$.

(a) *The functions* e-lim inf$_\nu f^\nu$ *and* e-lim sup$_\nu f^\nu$ *are lower semicontinuous, and so too is* e-lim$_\nu f^\nu$ *when it exists.*

(b) *The functions* e-lim inf$_\nu f^\nu$ *and* e-lim sup$_\nu f^\nu$ *depend only on the sequence* $\{\text{cl}\, f^\nu\}_{\nu\in I\!N}$; *thus, if* $\text{cl}\, g^\nu = \text{cl}\, f^\nu$ *for all* ν, *one has both* e-lim inf$_\nu g^\nu =$ e-lim inf$_\nu f^\nu$ *and* e-lim sup$_\nu g^\nu =$ e-lim sup$_\nu f^\nu$.

(c) *If the sequence* $\{f^\nu\}_{\nu\in I\!N}$ *is nonincreasing* ($f^\nu \geq f^{\nu+1}$), *then* e-lim$_\nu f^\nu$ *exists and equals* $\text{cl}[\inf_\nu f^\nu]$;

(d) *If the sequence* $\{f^\nu\}_{\nu\in I\!N}$ *is nondecreasing* ($f^\nu \leq f^{\nu+1}$), *then* e-lim$_\nu f^\nu$ *exists and equals* $\sup_\nu[\text{cl}\, f^\nu]$ *(rather than* $\text{cl}[\sup_\nu f^\nu]$).

(e) *If* f^ν *is positively homogeneous for all* ν, *then the functions* e-lim inf$_\nu f^\nu$ *and* e-lim sup$_\nu f^\nu$ *are positively homogeneous as well.*

(f) *For subsets* C^ν *and* C *of* $I\!R^n$, *one has* $C = \liminf_\nu C^\nu$ *if and only if* $\delta_C =$ e-lim sup$_\nu \delta_{C^\nu}$, *while* $C = \limsup_\nu C^\nu$ *if and only if* $\delta_C =$ e-lim inf$_\nu \delta_{C^\nu}$.

(g) *If* $f_1^\nu \leq f^\nu \leq f_2^\nu$ *with* $f_1^\nu \xrightarrow{e} f$ *and* $f_2^\nu \xrightarrow{e} f$, *then* $f^\nu \xrightarrow{e} f$.

(h) *If* $f^\nu \xrightarrow{e} f$, *or just* $f =$ e-lim inf$_\nu f^\nu$, *then* $\text{dom}\, f \subset \limsup_\nu[\text{dom}\, f^\nu]$.

Proof. We merely need to use the connections between function properties and epigraph properties, in particular 1.6 (lower semicontinuity) together with the facts about set limits in 4.4 for (a) and (b), 4.3 for (c) and (d), and 4.14 for (e). For (d) we rely also on the result in 1.26 on the lower semicontinuity of the supremum of a family of lsc functions. The relationships in (f) are obvious. The implication in (g) applies to epigraphs the sandwich rule in 4.3(c). The inclusion in (h) comes from the first inclusion in 4.26 as invoked for the mapping $L : (x, \alpha) \mapsto x$, which carries epigraphs to effective domains. □

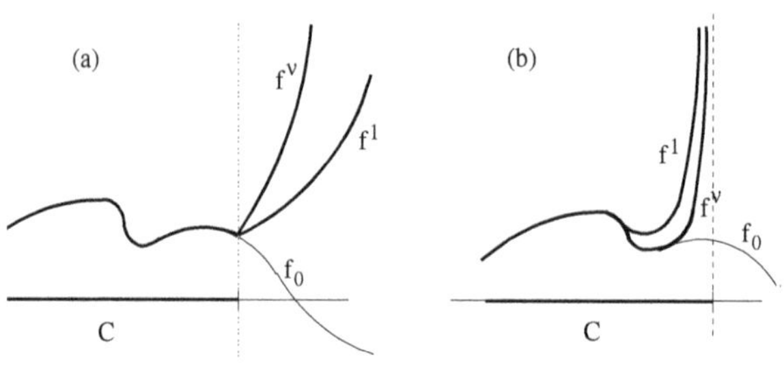

Fig. 7–2. Examples of monotone sequences: (a) penalty and (b) barrier functions.

As an illustration of 7.4(d), the Moreau envelopes $e_\lambda f$ associated with an lsc, prox-bounded function f as in 1.22 have $e_\lambda f \xrightarrow{e} f$ as $\lambda \searrow 0$. This is apparent from the monotonicity of $e_\lambda f(x)$ with respect to λ in 1.25. The Lasry-Lions double envelopes in 1.46 likewise then have $e_{\lambda,\mu} f \xrightarrow{e} f$ as $\lambda \searrow 0$ and $\mu \searrow 0$ with $0 < \mu < \lambda$, because of the sandwiching $e_\lambda f \leq e_{\lambda,\mu} f \leq e_{\lambda-\mu} f$ that was established in that example. In this case, 7.4(g) is applied.

The inclusion in 7.4(h) can be strict even when $f^\nu \xrightarrow{e} f$, as seen in the case where $f^\nu(x) = \nu|x|$ and $f(x) = \delta_{\{0\}}$.

The use of penalty or barrier functions as substitutes for constraints, as in Example 1.3, ordinarily leads to monotone sequences of functions which pointwise converge to the essential objective function f of the given optimization problem and therefore epi-converge to f by 7.4(c) and 7.4(d). For instance, let

$$f(x) = \begin{cases} f_0(x) & \text{if } f_i(x) \leq 0 \text{ for } i = 1, \dots m, \\ \infty & \text{otherwise,} \end{cases}$$

where all the functions $f_i : \mathbb{R}^n \to \mathbb{R}$ are continuous. In the case of penalty functions, say with

$$f^\nu(x) = f_0(x) + \sum_{i=1}^m \theta^\nu\big(f_i(x)\big) \text{ for } \theta^\nu(t) = \frac{\nu}{2}\max\{0, t\}^2,$$

the functions f^ν increase pointwise and epi-converge to f, as illustrated in Figure 7–2(a). In the case of barrier functions instead, say with

$$f^\nu(x) = f_0(x) + \sum_{i=1}^m \theta^\nu(f_i(x)) \text{ for } \theta^\nu(t) = \begin{cases} (-\nu t)^{-1} & \text{if } t < 0, \\ \infty & \text{if } t \geq 0, \end{cases}$$

the functions f^ν decrease pointwise and epi-converge to f; see Figure 7–2(b).

Because epigraphical limits are always lsc and are unaffected by passing to the lsc-regularizations of the functions in question, as asserted in 7.4(a)(b), the natural domain for the study of such limits is really the space of all *lsc* functions on $\mathbb{R}^n$. Indeed, if f isn't lsc, awkward situations can arise; for instance, the constant sequence $f^\nu \equiv f$ fails to be such that $f^\nu \xrightarrow{e} f$. Nonetheless, in stating results about epigraphical limits it isn't convenient to insist always on restriction of the wording to the case of lsc functions, because such restriction isn't obligatory and may appear in some circumstances as imposing the burden of checking an additional assumption.

Special consideration needs to be given to the concept of a sequence of functions *escaping epigraphically to the horizon* in the sense that the sets epi f^ν escape to the horizon of $\mathbb{R}^{n+1}$.

7.5 Exercise (epigraphical escape to the horizon). *The following conditions are equivalent to the sequence $\{f^\nu\}_{\nu\in\mathbb{N}}$ escaping epigraphically to the horizon:*

(a) $f^\nu \xrightarrow{e} \infty$;

(b) e-lim $\inf_\nu f^\nu \equiv \infty$;

(c) *for every $\rho > 0$ there is an index set $N \in \mathcal{N}_\infty$ such that $f^\nu(x) \geq \rho$ for all $x \in \rho\mathbb{B}$ when $\nu \in N$.*

We use this idea of epigraphical escape to the horizon in stating the fundamental compactness theorem for epi-convergence.

7.6 Theorem (extraction of epi-convergent subsequences). *Every sequence of functions f^ν on $\mathbb{R}^n$ that does not escape epigraphically to the horizon has a subsequence epi-converging to a function $f \not\equiv \infty$ (possibly taking on $-\infty$).*

Proof. Applying the compactness property of set convergence (in 4.18) to the space $I\!R^{n+1}$, we need only recall that the limit of a sequence of epigraphs is another epigraph, as explained before Definition 7.1. □

Epi-convergence of functions can be characterized in terms of a kind of set convergence of their level sets.

7.7 Proposition (level sets of epi-convergent functions). *For functions f^ν and f on $I\!R^n$, one has:*

(a) $f \leq \text{e-lim inf}_\nu f^\nu$ *if and only if* $\limsup_\nu(\text{lev}_{\leq\alpha^\nu} f^\nu) \subset \text{lev}_{\leq\alpha} f$ *for all sequences* $\alpha^\nu \to \alpha$;

(b) $f \geq \text{e-lim sup}_\nu f^\nu$ *if and only if* $\liminf_\nu(\text{lev}_{\leq\alpha^\nu} f^\nu) \supset \text{lev}_{\leq\alpha} f$ *for some sequence* $\alpha^\nu \to \alpha$*, in which case such a sequence can be chosen with* $\alpha^\nu \searrow \alpha$.

(c) $f = \text{e-lim}_\nu f^\nu$ *if and only if both conditions hold.*

Proof. Epi-convergence of f^ν to f has already been seen to be equivalent to graphical convergence of the profile mappings E_{f^ν} to E_f. But that is the same as graphical convergence of the inverses $E_{f^\nu}^{-1}$ to E_f^{-1}. We have $E_{f^\nu}^{-1}(\alpha^\nu) = \text{lev}_{\leq\alpha^\nu} f^\nu$ and $E_f^{-1}(\alpha) = \text{lev}_{\leq\alpha} f$, so the claims here are immediate from the general graphical limit formulas in 5.33, except for the assertion that for any $\alpha \in I\!R$ one can produce $\alpha^\nu \searrow \alpha$ such that $\liminf_\nu(\text{lev}_{\leq\alpha^\nu} f^\nu) \supset \text{lev}_{\leq\alpha} f$.

We do know that for every $x \in \text{lev}_{\leq\alpha} f$ there's a sequence $\hat{\alpha}^\nu \to \alpha$ with points $\hat{x}^\nu \in \text{lev}_{\leq\hat{\alpha}^\nu} f$ such that $\hat{x}^\nu \to x$. Although this sequence of values $\hat{\alpha}^\nu$ might not meet the prescription, because it depends on the particular x, we're able at least to conclude from it the following: for any $\varepsilon > 0$ the index set

$$N_{x,\varepsilon} := \left\{ \nu \in I\!N \,\middle|\, I\!B(x,\varepsilon) \cap \text{lev}_{\leq(\alpha+\varepsilon)} f^\nu \neq \emptyset \right\}$$

belongs to $\mathcal{N}_\infty$. For any $\varepsilon > 0$ the compact set $(\text{lev}_{\leq\alpha} f) \cap \varepsilon^{-1} I\!B$, if not empty, can be covered by balls $I\!B(x,\varepsilon)$ corresponding to a finite subset X of $\text{lev}_{\leq\alpha} f$, and then for ν in $\bigcap_{x\in X} N_{x,\varepsilon}$, which is an index set still belonging to $\mathcal{N}_\infty$, we have that the ball $I\!B(x, 2\varepsilon)$ meets $\text{lev}_{\leq(\alpha+\varepsilon)} f^\nu$ for all $x \in (\text{lev}_{\leq\alpha} f) \cap \varepsilon^{-1} I\!B$. Therefore, for every $\varepsilon > 0$ the index set

$$N_\varepsilon := \left\{ \nu \in I\!N \,\middle|\, \big(\text{lev}_{\leq\alpha} f\big) \cap \varepsilon^{-1} I\!B \subset \big(\text{lev}_{\leq(\alpha+\varepsilon)} f^\nu\big) + 2\varepsilon I\!B \right\}$$

belongs to $\mathcal{N}_\infty$. Clearly, N_ε gets smaller, if anything, as ε decreases.

If the intersection $\bigcap_{\varepsilon>0} N_\varepsilon$ happens actually to be nonempty, it's an index set $N \in \mathcal{N}_\infty$, and we have the simple case where $\text{lev}_{\leq\alpha} f \subset \text{cl}(\text{lev}_{\leq\beta} f^\nu)$ for all $\nu \in N$ when $\beta > \alpha$. Then the desired inclusion holds trivially for any choice of $\alpha^\nu \searrow \alpha$. Otherwise, the intersection $\bigcap_{\varepsilon>0} N_\varepsilon$ is empty. In this case let $N := \bigcup_{\varepsilon>0} N_\varepsilon$, and for each $\nu \in N$ let ε^ν be the smallest value of $\varepsilon > 0$ such that $\nu \in N_\varepsilon$. Then $\varepsilon^\nu \searrow 0$. Taking $\alpha^\nu = \alpha + \varepsilon^\nu$, we have $\alpha^\nu \searrow \alpha$ and

$$\big(\text{lev}_{\leq\alpha} f\big) \cap (\varepsilon^\nu)^{-1} I\!B \subset \big(\text{lev}_{\leq\alpha^\nu} f^\nu\big) + 2\varepsilon^\nu I\!B \quad \text{for all } \nu \in N.$$

This gives us the 'lim inf' inclusion by the criterion in 4.10(a). □

The second inclusion in 7.7, in combination with the first, tells us that for each $\alpha \in I\!R$ we have

$$\operatorname{lev}_{\leq \alpha^\nu} f^\nu \to \operatorname{lev}_{\leq \alpha} f \text{ for at least one sequence } \alpha^\nu \to \alpha.$$

The fact that this can't be claimed for every sequence $\alpha^\nu \to \alpha$, even under the assumption that $\alpha^\nu > \alpha$, is illustrated in Figure 7–3, where $\alpha = 1$ and the functions f^ν on $I\!R^1$ are given by $f^\nu(x) := 1 + \min\{x, \nu^{-1}x\}$, while $f(x) := 1 + \min\{x, 0\}$. Obviously $f^\nu \xrightarrow{e} f$, but we don't have $\operatorname{lev}_{\leq 1} f^\nu \to \operatorname{lev}_{\leq 1} f$, the latter set being all of $(-\infty, \infty)$. For sequences $\alpha^\nu \searrow 1$ that converge too rapidly, for instance with $\alpha^\nu = 1 + 1/\nu^2$, we merely have $\operatorname{lev}_{\leq \alpha^\nu} f^\nu \to (-\infty, 0]$. For slower sequences like $\alpha^\nu = 1 + 1/\sqrt{\nu}$, we do get $\operatorname{lev}_{\leq \alpha^\nu} f^\nu \to \operatorname{lev}_{\leq 1} f$.

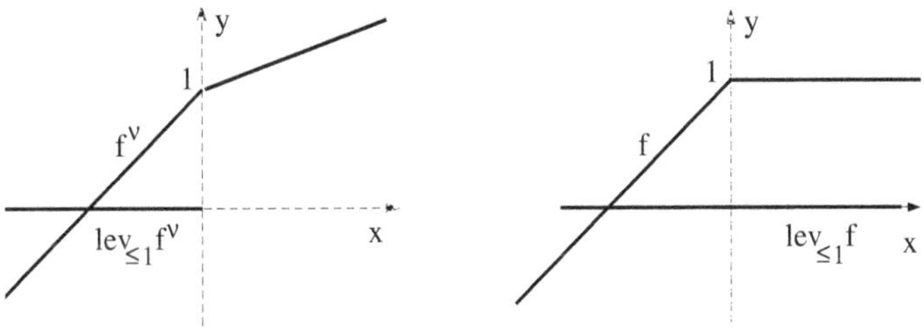

Fig. 7–3. Counterexample in the convergence of level sets.

7.8 Exercise (convergence-preserving operations). *Let f^ν and f be functions on $I\!R^n$ such that $f^\nu \xrightarrow{e} f$.*

(a) *If $g : I\!R^n \to I\!R$ is continuous and finite, then $f^\nu + g \xrightarrow{e} f + g$.*

(b) *If $F : I\!R^n \to I\!R^n$ is continuous and bijective, then $f^\nu \circ F \xrightarrow{e} f \circ F$.*

(c) *If $\theta : I\!R \to I\!R$ is continuous and increasing, and θ is extended to $\overline{I\!R}$ by setting $\theta(\infty) = \sup \theta$ and $\theta(-\infty) = \inf \theta$, then $\theta \circ f^\nu \xrightarrow{e} \theta \circ f$.*

(d) *If $g^\nu(x) = \lambda^\nu f^\nu(x + a^\nu) + \alpha^\nu$ and $g(x) = \lambda f(x + a) + \alpha$ with $\lambda^\nu \to \lambda > 0$, $a^\nu \to a$ and $\alpha^\nu \to \alpha$, then $g^\nu \xrightarrow{e} g$.*

Guide. Rely on the characterization of epi-convergence in 7.2. In (b), recall that the assumptions on F imply F^{-1} is continuous too. □

In general, epi-convergence neither implies nor is implied by pointwise convergence. The example in Figure 7–4 demonstrates how a sequence of functions f^ν can have both an epi-limit and a pointwise limit, but different. The concepts are therefore independent. Much can be said nonetheless about their relationship. A basic pattern of inequalities among inner and outer limits is available through comparison between the formulas in 7.2 and 7.3:

$$\operatorname{e-lim\,inf}_\nu f^\nu \quad \leq \quad \left\{ \begin{array}{c} \operatorname{p-lim\,inf}_\nu f^\nu \\ \operatorname{e-lim\,sup}_\nu f^\nu \end{array} \right\} \quad \leq \quad \operatorname{p-lim\,sup}_\nu f^\nu. \qquad 7(12)$$

In particular, $\operatorname{e-lim}_\nu f^\nu \leq \operatorname{p-lim}_\nu f^\nu$ when both limits exist. The circumstances in which $\operatorname{e-lim}_\nu f^\nu = \operatorname{p-lim}_\nu f^\nu$ can be ascertained with the help of a concept of equi-semicontinuity which we develop for extended-real-valued functions through their association with profile mappings. Here we utilize the definition in 5.38 of what it means for a set-valued mapping to be equi-osc.

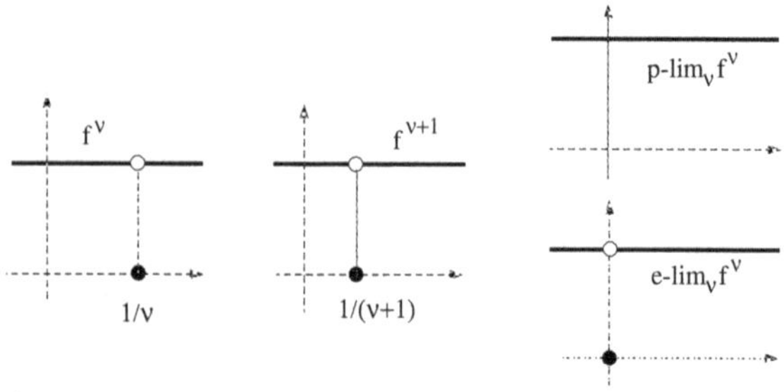

Fig. 7–4. A sequence having pointwise and epigraphical limits that differ.

7.9 Exercise (equi-semicontinuity properties of sequences). *Consider a sequence of functions $f^\nu : I\!R^n \to \overline{I\!R}$, a set $X \subset I\!R^n$, and a point $\bar{x} \in X$.*

(a) *The sequence of epigraphical profile mappings E_{f^ν} is equi-osc at $\bar{x}$ relative to X if and only if for every $\rho > 0$ and $\varepsilon > 0$ there is a $\delta > 0$ with*

$$\inf_{x \in \boldsymbol{B}(\bar{x},\delta)\cap X} f^\nu(x) \geq \min\{f^\nu(\bar{x}) - \varepsilon, \rho\} \ \text{ for all } \nu \in I\!N.$$

It is asymptotically equi-osc if this holds for all ν in an index set $N \in \mathcal{N}_\infty$, which like δ can depend on ρ and ε, or when $\limsup_\nu f^\nu(\bar{x}) = -\infty$

(b) *The sequence of hypographical profile mappings H_{f^ν} is equi-osc at $\bar{x}$ relative to X if and only if for every $\rho > 0$ and $\varepsilon > 0$ there is a $\delta > 0$ with*

$$\sup_{x \in \boldsymbol{B}(\bar{x},\delta)\cap X} f^\nu(x) \leq \max\{f^\nu(\bar{x}) + \varepsilon, -\rho\} \ \text{ for all } \nu \in I\!N.$$

It is asymptotically equi-osc if holds for all ν in an index set $N \in \mathcal{N}_\infty$, which like δ can depend on ρ and ε, or when $\liminf_\nu f^\nu(\bar{x}) = \infty$.

We define a sequence $\{f^\nu\}_{\nu \in I\!N}$ to be *equi-lsc* at $\bar{x}$ relative to X when the equivalent properties in 7.9(a) hold, and to be *equi-usc* when the ones in 7.9(b) hold. We define it to be *equicontinuous* when it is both equi-lsc and equi-usc. The sequence is equi-lsc, equi-usc, or equi-continuous *relative to* X if these properties hold at every $\bar{x}$ in X. In all these instances, the addition of the qualification *asymptotically* refers to the asymptotic properties in 7.9, i.e., the property holding for all ν in an index set $N \in \mathcal{N}_\infty$ rather than for all $\nu \in I\!N$.

As with set-valued mappings, the equicontinuity of a sequence of functions $f^\nu : I\!R^n \to \overline{I\!R}$ entails its asymptotic equicontinuity. But equicontinuity is a much stronger property, implying in particular that all the functions are continuous at the point in question. A sequence $\{f^\nu\}_{\nu \in I\!N}$ can be asymptotically equicontinuous at $\bar{x}$ in situations where every f^ν happens to be discontinuous at $\bar{x}$, as exemplified at $\bar{x} = 0$ by the sequence on $I\!R$ defined by $f^\nu(x) = \varepsilon^\nu$ for

$x > 0$ but $f^\nu(x) = 0$ for $x \leq 0$, with $\varepsilon^\nu \searrow 0$. Similar observations can be made about the asymptotic lsc and usc properties.

The various properties take a simpler form when the sequence $\{f^\nu\}_{\nu\in \mathbb{N}}$ is *locally bounded* at $\bar{x}$, in the sense that for some $\rho \in \mathbb{R}_+$ and neighborhood V of $\bar{x}$ one has $|f^\nu(x)| \leq \rho$ for all ν when $x \in V$. In that case it is equi-lsc at $\bar{x}$ relative to X if for every $\varepsilon > 0$ there exists $\delta > 0$ with

$$f^\nu(x) \geq f^\nu(\bar{x}) - \varepsilon \;\text{ for all }\; \nu \in \mathbb{N} \;\text{ when }\; x \in X,\; |x - \bar{x}| \leq \delta,$$

and it is equi-usc at $\bar{x}$ relative to X if for every $\varepsilon > 0$ there exists $\delta > 0$ with

$$f^\nu(x) \leq f^\nu(\bar{x}) + \varepsilon \;\text{ for all }\; \nu \in \mathbb{N} \;\text{ when }\; x \in X,\; |x - \bar{x}| \leq \delta.$$

Also in that case, equicontinuity at $\bar{x}$ relative to X reduces to the classical property that for every $\varepsilon > 0$ there exists $\delta > 0$ with

$$\big|f^\nu(x) - f^\nu(\bar{x})\big| \leq \varepsilon \;\text{ for all }\; \nu \in \mathbb{N} \;\text{ when }\; x \in X,\; |x - \bar{x}| \leq \delta.$$

The asymptotic properties simplify in such a manner as well, moreover with the sequence only having to be *eventually* locally bounded at $\bar{x}$, i.e., such that for some $\rho \in \mathbb{R}_+$, neighborhood V of $\bar{x}$ and index set $N \in \mathcal{N}_\infty$ one has $|f^\nu(x)| \leq \rho$ for all $\nu \in N$ when $x \in V$. But it is important for many purposes to admit sequences of extended-real-valued functions that don't necessarily exhibit such local boundedness, and for that case the more general formulation of the various equicontinuity properties is essential.

7.10 Theorem (epi-convergence versus pointwise convergence). *Consider any sequence of lsc functions $f^\nu : \mathbb{R}^n \to \overline{\mathbb{R}}$ and a point $\bar{x} \in \mathbb{R}^n$. If the sequence is asymptotically equi-lsc at $\bar{x}$, then*

$$\begin{aligned}(\text{e-lim inf}_\nu f^\nu)(\bar{x}) &= (\text{p-lim inf}_\nu f^\nu)(\bar{x}) = \liminf_{\nu\to\infty} f^\nu(\bar{x}),\\ (\text{e-lim sup}_\nu f^\nu)(\bar{x}) &= (\text{p-lim sup}_\nu f^\nu)(\bar{x}) = \limsup_{\nu\to\infty} f^\nu(\bar{x}).\end{aligned}$$

Thus, if it is asymptotically equi-lsc everywhere, $f^\nu \xrightarrow{e} f$ if and only if $f^\nu \xrightarrow{p} f$.

More generally, relative to an arbitrary set $X \subset \mathbb{R}^n$ containing $\bar{x}$, any two of the following conditions implies the third:

(a) *the sequence is asymptotically equi-lsc at $\bar{x}$ relative to X;*

(b) *$f^\nu \xrightarrow{e} f$ at $\bar{x}$ relative to X;*

(c) *$f^\nu \xrightarrow{p} f$ at $\bar{x}$ relative to X.*

Proof. This is immediate from the corresponding result for set-valued mappings in Theorem 5.40 by virtue of the equivalences in 7.9(a). ☐

For example, a sequence $\{\delta_{C^\nu}\}_{\nu\in N}$ of indicator functions of closed sets C^ν converging to C is equi-lsc if and only if for every $x \in C$ there exists $N \in \mathcal{N}_\infty$ such that $x \in C^\nu$ for all $\nu \in N$.

C. Continuous and Uniform Convergence

Epi-convergence has an enlightening connection with continuous convergence. A sequence of functions f^ν *converges continuously to* f if $f^\nu(x^\nu) \to f(x)$ whenever $x^\nu \to x$. We speak of f^ν converging continuously to f *at an individual point* $\bar{x}$ if this holds for sequences $x^\nu \to \bar{x}$, although not necessarily for $x^\nu \to x \neq \bar{x}$. Similarly, continuous convergence of f^ν to f *relative to a set* X refers to having $f^\nu(x^\nu) \to f(x)$ whenever $x^\nu \underset{X}{\to} x$. Continuous convergence of functions is consistent under this definition with continuous convergence of mappings as defined in 5.41 when the functions are regarded as single-valued mappings into the space $\overline{I\!R}$, but also when the functions are viewed through their associated epigraphical or hypographical profile mappings: f^ν converges continuously to f if and only if E_{f^ν} converges continuously to E_f, or for that matter if and only if H_{f^ν} converges continuously to H_f.

7.11 Theorem (epi-convergence versus continuous convergence). *For a sequence of functions $f^\nu : I\!R^n \to \overline{I\!R}$, the following properties are equivalent at any point $\bar{x} \in X \subset I\!R^n$:*

(a) *f^ν converges continuously to f at $\bar{x}$ relative to X;*

(b) *f^ν both epi-converges and hypo-converges to f at $\bar{x}$ relative to X;*

(c) *f^ν epi-converges to f at $\bar{x}$ relative to X, and the sequence is asymptotically equicontinuous at $\bar{x}$ relative to X;*

(d) *f^ν hypo-converges to f at $\bar{x}$ relative to X, and the sequence is asymptotically equicontinuous at $\bar{x}$ relative to X.*

When f is finite around $\bar{x}$ relative to X, these properties are equivalent to each other and also to the existence of $V \in \mathcal{N}(\bar{x})$ such that the sequence of functions f^ν is eventually locally bounded on $V \cap X$ and converges graphically to f relative to $V \cap X$.

Proof. This is obvious from Theorem 5.44 as applied to the mappings E_{f^ν} and H_{f^ν} in the light of the equivalences just explained. The assertion about graphical convergence is based on 5.45. □

In the standard analysis of real-valued functions, continuous convergence is intimately related to uniform convergence. The question of what uniform convergence ought to signify in the nonstandard context of functions that might be unbounded and even take on $\pm\infty$ requires some reflection, however. For *bounded* (and therefore finite-valued) functions f^ν and f, uniform convergence of f^ν to f on a set X requires the existence for each $\varepsilon > 0$ of an index set $N \in \mathcal{N}_\infty$ such that

$$\left|f^\nu(x) - f(x)\right| \leq \varepsilon \text{ for all } x \in X \text{ when } \nu \in N. \qquad 7(13)$$

We'll refer to this property as uniform convergence *in the bounded sense*. The kind of uniformity it invokes isn't appropriate for functions that aren't bounded, not to mention troubles of interpretation for infinite values of $f(x)$ and $f^\nu(x)$. The notion needs to be extended.

7.12 Definition (uniform convergence, extended). *For any function $f : I\!R^n \to \overline{I\!R}$ and any $\rho \in (0, \infty)$, the ρ-truncation of f is the function $f_{\wedge\rho}$ defined by*

$$f_{\wedge\rho}(x) = \begin{cases} -\rho & \text{if } f(x) \in [-\infty, -\rho), \\ f(x) & \text{if } f(x) \in [-\rho, \rho], \\ \rho & \text{if } f(x) \in (\rho, \infty]. \end{cases}$$

A sequence of functions f^ν will be said to converge uniformly to f on a set $X \subset I\!R^n$ if, for every $\rho > 0$, their truncations $f^\nu_{\wedge\rho}$ converge uniformly to $f_{\wedge\rho}$ on X in the bounded sense.

This extended notion of uniform convergence is only novel relative to the traditional one in cases where f itself is unbounded. As long as f is bounded on X, it's obvious (from considering any ρ with $|f(x)| < \rho$ on X) that f^ν must eventually be bounded on X as well, so that one simply reverts to the classical property 7(13). Another way of looking at the extended notion is that f^ν converges uniformly to f in the general sense if and only if the functions $g^\nu = \arctan f^\nu$ converge uniformly to $g = \arctan f$ in the bounded sense (with $\arctan(\pm\infty) = \pm\pi/2$). This reflects the fact that the topology of $\overline{I\!R}$ can be identified with that of $[-\pi/2, \pi/2]$ under the arctan mapping.

The following proposition provides other perspectives on uniform convergence of functions. It appeals to uniform convergence of set-valued mappings as defined in 5.41, in particular of profile mappings as in 5.5.

7.13 Proposition (characterizations of uniform convergence). *For extended-real-valued functions f^ν and f on $X \subset I\!R^n$, the following are equivalent:*

(a) *the functions f^ν converge uniformly to f;*

(b) *the profile mappings E_{f^ν} converge uniformly to E_f on X;*

(c) *the profile mappings H_{f^ν} converge uniformly to H_f on X;*

(d) *for each $\rho > 0$ and $\varepsilon \in (0, \rho)$ there is an index set $N \in \mathcal{N}_\infty$ such that, for all $\nu \in N$, one has*

$$\begin{aligned} f^\nu(x) - \varepsilon &\le \begin{cases} f(x) & \text{for } x \in X \text{ with } -\rho \le f(x) \le \rho - \varepsilon, \\ -\rho & \text{for } x \in X \text{ with } -\rho > f(x), \end{cases} \\ f^\nu(x) + \varepsilon &\ge \begin{cases} f(x) & \text{for } x \in X \text{ with } -\rho + \varepsilon \le f(x) \le \rho, \\ \rho & \text{for } x \in X \text{ with } f(x) > \rho. \end{cases} \end{aligned} \tag{7(14)}$$

Proof. The equivalence of (a) with (d) is seen at once from the definition of the truncations. Condition (b) translates directly (through Definition 5.41 as applied in the special case of E_{f^ν} and E_f) into something very close to (d), where 7(14) is replaced by

$$\begin{aligned} f^\nu(x) - \varepsilon &\le \begin{cases} f(x) & \text{for } x \in X \text{ with } -\rho \le f(x) \le \rho, \\ -\rho & \text{for } x \in X \text{ with } -\rho > f(x), \end{cases} \\ f^\nu(x) + \varepsilon &\ge \begin{cases} f(x) & \text{for } x \in X \text{ with } -\rho + \varepsilon \le f(x) \le \rho + \varepsilon, \\ \rho & \text{for } x \in X \text{ with } f(x) > \rho + \varepsilon. \end{cases} \end{aligned}$$

The type of convergence described by this variant is the same as that in (d); thus, (b) is equivalent to (d). Likewise (c) is equivalent to (d), because (d) is unaffected when the functions in question are multiplied by -1. □

7.14 Theorem (continuous versus uniform convergence). *A sequence of functions $f^\nu : \mathbb{R}^n \to \overline{\mathbb{R}}$ converges continuously to f relative to a set X if and only if f is continuous relative to X and f^ν converges uniformly to f on all compact subsets of X.*

In general, whenever f^ν converges continuously to f relative to a set X at $\bar{x} \in X$ and at the same time converges pointwise to f on a neighborhood of $\bar{x}$ relative to X, f must be continuous at $\bar{x}$ relative to X.

Proof. We apply the corresponding result in 5.43 to the set-valued mappings E_{f^ν} and E_f, utilizing the equivalences that have been pointed out. □

7.15 Proposition (epi-convergence from uniform convergence). *Consider $f^\nu, f : \mathbb{R}^n \to \overline{\mathbb{R}}$ and a set $X \subset \mathbb{R}^n$.*

(a) *If the functions f^ν are lsc relative to X and converge uniformly to f on X, then f is lsc relative to X and $f^\nu \xrightarrow{e} f$ relative to X.*

(b) *If the functions f^ν are finite on X, continuous relative to X and converge uniformly to f on X, then f is continuous relative to X and f^ν both epi-converges and hypo-converges to f relative to X and thus also converges graphically to f relative to X.*

Proof. In view of 7(2) and 7.13(a), one can translate (a) so that it reads: If the profile mappings E_{f^ν} are osc relative to X and converge uniformly to E_f relative to X, then the profile mapping E_f is osc relative to X and $E_{f^\nu} \xrightarrow{g} E_f$ relative to X. This is a special case of Theorem 5.46(a). Part (b) is obtained then by applying part (a) to f and $-f$ in the light of 7.11. □

7.16 Exercise (Arzelà-Ascoli theorem, extended). *Consider a sequence of functions $f^\nu : \mathbb{R}^n \to \overline{\mathbb{R}}$ and a set $X \subset \mathbb{R}^n$.*

(a) *If the sequence is asymptotically equi-lsc relative to X, it has a subsequence converging both pointwise and epigraphically relative to X to some lsc function f.*

(b) *If the sequence is asymptotically equi-usc relative to X, it has a subsequence converging both pointwise and hypographically relative to X to some usc function f.*

(c) *If the sequence is asymptotically equicontinuous relative to X, it has a subsequence converging uniformly on all compact subsets of X to a function f that is continuous relative to X.*

Guide. Obtain (a) from 7.6 and 7.10. Get (b) by symmetry. Then combine (a) and (b) to obtain (c) through 7.11 and 7.14. □

7.17 Theorem (epi-limits of convex functions). *For any sequence $\{f^\nu\}_{\nu\in\mathbb{N}}$ of convex functions on $\mathbb{R}^n$, the function $\operatorname{e-lim\,sup}_\nu f^\nu$ is convex, and so too is the function $\operatorname{e-lim}_\nu f^\nu$ when it exists.*

Moreover, under the assumption that f is a convex, lsc function on $\mathbb{R}^n$ such that $\operatorname{dom} f$ *has nonempty interior, the following are equivalent:*

(a) $f = \text{e-lim}_\nu f^\nu$;

(b) *there is a dense subset D of $\mathbb{R}^n$ such that $f^\nu(x) \to f(x)$ for all x in D;*

(c) *f^ν converges uniformly to f on every compact set C that does not contain a boundary point of* $\operatorname{dom} f$.

Proof. The initial claims follow from the corresponding facts about convergence of convex sets (in 4.15) through the epigraphical connections in 2.4 and Definition 7.1.

For the claimed equivalence under the assumptions on the candidate limit function f, we observe first that (c)⇒(b), because the set of all points of $\mathbb{R}^n$ that aren't boundary points of $\operatorname{dom} f$ is dense in $\mathbb{R}^n$. This follows from the line segment principle in 2.33 as applied to the convex set $\operatorname{dom} f$.

We demonstrate next that (a)⇒(c). Any compact set not containing a boundary point of $\operatorname{dom} f$ is the union of a compact set lying within the interior of $\operatorname{dom} f$ and a compact set not meeting the closure of $\operatorname{dom} f$. It suffices to consider these two cases separately. Suppose first that C lies outside the closure of $\operatorname{dom} f$. Then $f \equiv \infty$ on C. For any $\rho > 0$, the compact set $B = C \times \{\rho\}$ in $\mathbb{R}^{n+1}$ does not meet $\operatorname{epi} f$. Invoking the hit-and-miss criterion in 4.5(b) in the epigraphical framework of Definition 7.1, we see that for all ν in some index set $N \in \mathcal{N}_\infty$ the set B does not meet $\operatorname{epi} f^\nu$, so that $f^\nu(x) \geq \rho$ for every $x \in C$. This gives the uniform convergence to f on C.

Suppose next that C is a subset of $\operatorname{int}(\operatorname{dom} f)$ but that f is improper. Then the argument is quite similar. We have $f \equiv -\infty$ on C, and for every $\rho > 0$ the set $B = C \times \{-\rho\}$ lies in the interior of $\operatorname{epi} f$. The internal uniformity property of convergence of convex sets (in 4.15) tells us that for all ν in some index set $N \in \mathcal{N}_\infty$, the set B lies in the interior of $\operatorname{epi} f^\nu$, implying that $f^\nu(x) \leq -\rho$ for all $x \in C$. Again, we see that $f^\nu \to f$ uniformly on C.

We address now the case where C is a subset of $\operatorname{int}(\operatorname{dom} f)$ and f is proper. Then f is finite and continuous on C (see 2.35). For arbitrary $\varepsilon > 0$ consider the compact sets

$$B_+ = \{(x, \alpha) \in C \times \mathbb{R} \mid \alpha = f(x) + \varepsilon\},$$
$$B_- = \{(x, \alpha) \in C \times \mathbb{R} \mid \alpha = f(x) - \varepsilon\}.$$

The first lies in $\operatorname{int}(\operatorname{epi} f)$, while the second lies outside of $\operatorname{cl}(\operatorname{epi} f)$. According to the hit-and-miss criterion in 4.5(b), there is an index set $N_- \in \mathcal{N}_\infty$ such that for $\nu \in N_-$ the set B_- does not meet $\operatorname{epi} f^\nu$. At the same time, by the internal uniformity property for convergence of convex sets in 4.15 there is an index set $N_+ \in \mathcal{N}_\infty$ such that for $\nu \in N_+$ the set B_+ lies within $\operatorname{epi} f^\nu$. Then, for all $x \in C$ and ν in the set $N := N_- \cap N_+ \in \mathcal{N}_\infty$, we have both $f^\nu(x) \geq f(x) - \varepsilon$ and $f^\nu(x) \leq f(x) + \varepsilon$. Thus, uniform convergence on C is established again.

Our task now is to show that (b)⇒(a). For this, the compactness principle in 7.6 will be handy. To conclude that $f^\nu \xrightarrow{e} f$, it suffices to verify that if f_0 is any cluster point of the sequence $\{f^\nu\}_{\nu\in\mathbb{N}}$ in the sense of epi-convergence, then $f_0 = f$. For such a cluster point f_0, which necessarily is a convex function, we have $f_0 = \text{e-lim}_{\nu\in N} f^\nu$ for an index set $N \in \mathcal{N}_\infty^\#$. From the implication (a)⇒(c) already proved, the subsequence $\{f^\nu\}_{\nu\in N}$ must converge uniformly to f_0 on all compact sets not meeting the boundary of $\operatorname{dom} f_0$. But by assumption, this subsequence also converges pointwise to f on a dense set D. It follows that f_0 and f must agree on a dense subset of $\mathbb{R}^n$. Since both functions are convex and lsc, this necessitates $f_0 = f$ (cf. 2.35). □

7.18 Corollary (uniform convergence of finite convex functions). *Let f^ν and f be finite, convex functions on an open, convex set $O \subset \mathbb{R}^n$, and suppose that $f^\nu(x) \to f(x)$ for all x in some dense subset of O. Then f^ν converges uniformly to f on every compact set $B \subset O$.*

Proof. We can suppose that O is nonempty and that f^ν and f have the value ∞ outside of O. Then $\operatorname{cl} f$ is a proper, lsc, convex function on $\mathbb{R}^n$ which agrees with f on O and has $\operatorname{int}(\operatorname{dom} \operatorname{cl} f) = O$ (see 2.35). It follows from Theorem 7.17 that $f^\nu \to \operatorname{cl} f$, the convergence being uniform with respect to every compact set that does not meet the boundary of O. □

The approximation of a function f by 'averaging' provides another example of how the various convergence relationships can help to pin down the properties of an approximating sequence of functions f^ν. Here we recall that a function $f : \mathbb{R}^n \to \mathbb{R}$ is *locally summable* if it is measurable and each point $\bar{x} \in \mathbb{R}^n$ has a neighborhood V such that $\int_V |f(x)|dx < \infty$, where dx stands for n-dimensional Lebesgue measure. (Then actually $\int_B |f(x)|dx < \infty$ for any bounded set B.)

7.19 Example (functions mollified by averaging). *Let $f : \mathbb{R}^n \to \mathbb{R}$ be locally summable, and let $C \subset \mathbb{R}^n$ consist of the points, if any, where f is continuous. Consider any* bounded mollifier sequence, *i.e., a sequence of bounded, measurable functions $\psi^\nu \geq 0$ with $\int_{\mathbb{R}^n} \psi^\nu(z)dz = 1$ such that the sets $B^\nu = \{z \mid \psi^\nu(z) > 0\}$ form a bounded sequence converging to $\{0\}$. The corresponding sequence of* averaged functions *$f^\nu : \mathbb{R}^n \to \mathbb{R}$, defined by*

$$f^\nu(x) := \int_{\mathbb{R}^n} f(x-z)\psi^\nu(z)dz = \int_{B^\nu} f(x-z)\psi^\nu(z)dz,$$

then has the following properties.

(a) *f^ν converges continuously to f at every point of C and thus converges uniformly to f on every compact subset of C.*

(b) *f^ν epi-converges to f on $\mathbb{R}^n$ if f is lsc and is determined in this respect by its values on C, or in other words, if*

$$f(x) = \liminf_{x' \underset{C}{\to} x} f(x') \quad \text{for all } x.$$

Detail. Consider any point $\bar{x}$ where f is lsc and any sequence $x^\nu \to \bar{x}$. For any $\varepsilon > 0$ there exists $\delta > 0$ such that $f(\bar{x}) - f(x) \leq \varepsilon$ when $|x - \bar{x}| \leq 2\delta$. Next there exists ν_0 such that $|x^\nu - \bar{x}| \leq \delta$ and $B^\nu \subset \delta\mathbb{B}$ when $\nu \geq \nu_0$; for the latter, see 4.10(b). Since $f^\nu(x^\nu) = \int_{B^\nu} f(x^\nu - z)\psi^\nu(z)dz$ and $f(\bar{x}) = \int_{B^\nu} f(\bar{x})\psi^\nu(z)dz$, we then have for $\nu \geq \nu_0$ that

$$f^\nu(x^\nu) - f(\bar{x}) \geq \int_{B^\nu} [f(x^\nu - z) - f(\bar{x})]\psi^\nu(z)dz \geq -\varepsilon \int_{B^\nu} \psi^\nu(z)dz = -\varepsilon.$$

This demonstrates that $\liminf_\nu f^\nu(x^\nu) \geq f(\bar{x})$. When $\bar{x} \in C$ we can also argue the opposite inequality to get $\lim_\nu f^\nu(x^\nu) = f(\bar{x})$. This takes care of (a), except for an appeal to 7.14 for the relationship between continuous and uniform convergence. It also takes care of the half of (b) concerned with the lower epi-limit, since f is everywhere lsc in (b).

The other half of (b) requires, in effect, that $\operatorname{gph} f \subset \liminf_\nu \operatorname{gph} f^\nu$. By assumption, $\operatorname{gph} f \subset \operatorname{cl}\{(x, f(x)) \mid x \in C\}$. But $f^\nu \xrightarrow{p} f$ on C by (a), so $\{(x, f(x)) \mid x \in C\} \subset \liminf_\nu \{(x, f^\nu(x)) \mid x \in C\}$, the latter being a closed set. Therefore, $\operatorname{cl}\{(x, f(x)) \mid x \in C\} \subset \liminf_\nu \operatorname{gph} f^\nu$. □

A common way of generating a bounded mollifier sequence that meets the prescription of Example 7.19 is to start from a compact neighborhood B of the origin and any continuous function $\psi : B \to [0, \infty]$ with $\int_B \psi(x)dx = 1$, and to define $\psi^\nu(x)$ to be $\nu^n \psi(\nu x)$ when $x \in \nu^{-1}B$, but 0 when $x \notin \nu^{-1}B$.

D. Generalized Differentiability

Limit notions are essential in any attempt at defining derivatives of functions more general than those treated classically, and the theory laid out so far in this chapter has important applications in that way. For the study of $f : \mathbb{R}^n \to \overline{\mathbb{R}}$ at a point $\bar{x}$ where it's finite, the key is what happens to the *difference quotient functions* $\Delta_\tau f(\bar{x}) : \mathbb{R}^n \to \overline{\mathbb{R}}$ defined by

$$\Delta_\tau f(\bar{x})(w) := \frac{f(\bar{x} + \tau w) - f(\bar{x})}{\tau} \quad \text{for } \tau \neq 0 \qquad 7(15)$$

as τ tends to 0. Different ideas about derivatives arise from different choices of a mode of convergence. For pointwise convergence and uniform convergence there's a long tradition, but we're now in the position of being able also to explore the potential of continuous convergence, graphical convergence and epi-convergence, moreover in a context of extended-real-valuedness. Further, we know how to work with these notions point by point, so that derivatives of various kinds can be developed in terms of limits of the functions $\Delta_\tau f(\bar{x})$ at a particular w without insisting on a corresponding global limit for all w.

For a beginning, let's look at how the standard concept of differentiability can be embedded in *semi*differentiability, which deals robustly with *one-sided* limits of difference quotients. To say that f is *differentiable* at $\bar{x}$ is to assert the existence of a vector $v \in \mathbb{R}^n$ such that

$$f(x) = f(\bar{x}) + \langle v, x - \bar{x}\rangle + o(|x - \bar{x}|), \qquad 7(16)$$

where the '$o(t)$' notation, as always, indicates a term with the property that

$$\frac{o(t)}{t} \to 0 \ \text{ as } \ t \to 0,\, t \neq 0.$$

There can be at most one vector v for which this holds. When it exists, it's called the *gradient* of f at $\bar{x}$ and denoted by $\nabla f(\bar{x})$. In this case one has

$$\lim_{\tau \to 0} \frac{f(\bar{x}+\tau w) - f(\bar{x})}{\tau} = \langle \nabla f(\bar{x}), w \rangle, \tag{7(17)}$$

the limit on the left being the *directional derivative* of f at $\bar{x}$ for w.

Differentiability isn't merely equivalent, though, to the existence of all such directional derivative limits and their linear dependence on w, which would be tantamount to the *pointwise convergence* of the functions $\Delta_\tau f(\bar{x})$ to a linear function h as $\tau \to 0$. It's a stronger property requiring a degree of uniformity in the limits. This is evident from writing 7(16) as

$$\Delta_\tau f(\bar{x})(w) = \langle v, w \rangle + \frac{o\big(|\tau w|\big)}{\tau} \ \text{ for } \tau \neq 0, \tag{7(18)}$$

which describes a type of approximation of the difference quotients on the left, as functions of w, by the linear function $h(w) = \langle v, w \rangle$. The functions $\Delta_\tau f(\bar{x})$ must in fact converge uniformly to h on $\rho I\!B$ for any ρ, arbitrarily large.

An easy step toward greater generality in directional derivatives would be to replace the linear expression on the right side of 7(17) by some other expression $h(w)$ while at the same time relaxing the limit on the left side from the two-sided form $\tau \to 0$ to the one-sided form $\tau \searrow 0$. This by itself would have the shortcoming, however, of only furnishing an extension of 7(17) without capturing the uniformity of approximation embodied in 7(18). For that, one can aim directly instead at a one-sided extension of 7(18), yet there would still be the problem that in many applications the difference quotients behave well in the limit for some w's but not others. What's needed is an approach to directional differentiation that can capture the essence of the desired uniformity one direction at a time.

7.20 Definition (semiderivatives). *Consider $f : I\!R^n \to \overline{I\!R}$ and a point $\bar{x}$ with $f(\bar{x})$ finite. If the (possibly infinite) limit*

$$\lim_{\substack{\tau \searrow 0 \\ w' \to w}} \frac{f(\bar{x}+\tau w') - f(\bar{x})}{\tau} \tag{7(19)}$$

exists, it is the semiderivative of f at $\bar{x}$ for w, and f is semidifferentiable at $\bar{x}$ for w. If this holds for every w, f is semidifferentiable at $\bar{x}$.

The crucial distinction between semiderivatives and ordinary directional derivatives, of course, is that the behavior of the difference quotient is tested not only along the line $\{\bar{x} + \tau w \mid \tau \in I\!R\}$ or, for the sake of one-sidedness the half-line $\{\bar{x} + \tau w \mid \tau \in I\!R_+\}$, when $w \neq 0$, but on all 'curves' emanating from $\bar{x}$

in the direction of w. Indeed, to say that the semiderivative limit 7(19) exists and equals β is to say that whenever x^ν converges to $\bar{x}$ from the direction of w, in the sense that $[x^\nu - \bar{x}]/\tau^\nu \to w$ for some choice of $\tau^\nu \searrow 0$, one has $\big[f(x^\nu) - f(\bar{x})\big]/\tau^\nu \to \beta$.

The ability of semiderivatives to work along curves instead of just straight line segments provides a robustness to this concept which is critical for handling the situations in variational analysis where movement has to be confined to a set having curvilinear boundary, or a function has derivative discontinuities along such a boundary. It's obvious that if the semiderivative of f at $\bar{x}$ exists for w and equals β, the *one-sided directional derivative*

$$f'(x;w) := \lim_{\tau \searrow 0} \frac{f(\bar{x}+\tau w) - f(\bar{x})}{\tau} \tag{7(20)}$$

also exists and equals β. But the converse can't be counted on unless something more is known about f, ensuring that $\big[f(\bar{x}+\tau w') - f(\bar{x}+\tau w)\big]/\tau \to 0$ as $\tau \searrow 0$ while $w' \to w$. Actually, in many of the circumstances where the existence of simple one-sided directional derivatives can be established, the more powerful property of semidifferentiability does turn out to be available automatically. The identification of such circumstances is one of our aims here.

We won't introduce a special symbol here for semiderivatives. Instead, we'll appeal to the more general notation

$$df(\bar{x})(w) := \liminf_{\substack{\tau \searrow 0 \\ w' \to w}} \frac{f(\bar{x}+\tau w') - f(\bar{x})}{\tau}, \tag{7(21)}$$

which will be important from Chapter 8 on (see 8.1 and the surrounding discussion), but can already serve here, a bit ahead of schedule. Clearly the semiderivative of f at $\bar{x}$ for w, when it exists, equals $df(\bar{x})(w)$ in particular. Semidifferentiability of f at $\bar{x}$ for w refers to the case where the value in 7(21), which exists always, can be expressed by 'lim' instead of just 'liminf'. This value is then given also by the simpler directional derivative limit in 7(20).

7.21 Theorem (characterizations of semidifferentiability). *For any function $f : \mathbb{R}^n \to \overline{\mathbb{R}}$, finite at $\bar{x}$, the following properties are equivalent and entail f being continuous at $\bar{x}$:*

(a) *f is semidifferentiable at $\bar{x}$;*

(b) *as $\tau \searrow 0$, the functions $\Delta_\tau f(\bar{x})$ converge continuously to a function h;*

(c) *as $\tau \searrow 0$, the functions $\Delta_\tau f(\bar{x})$ converge uniformly on all bounded subsets of $\mathbb{R}^n$ to a continuous function h;*

(d) *for each $\rho > 0$ there exists $\mu > 0$ such that the functions $\Delta_\tau f(\bar{x})$ with $\tau \in (0,\mu)$ are equi-bounded on $\rho\mathbb{B}$ and, as $\tau \searrow 0$, converge graphically to a function h as mappings from $\rho\mathbb{B}$ to $\mathbb{R}$;*

(e) *there is a finite, continuous, positively homogeneous function h with*

$$f(x) \;=\; f(\bar{x}) + h(x-\bar{x}) + o\big(|x-\bar{x}|\big).$$

Then the semiderivative of f at $\bar{x}$ for w is finite and depends continuously and positively homogeneously on w; it is given by $h(w) = df(\bar{x})(w) = f'(\bar{x}; w)$.

Proof. We have (a) $\Leftrightarrow$ (b) by Definition 7.20 and the meaning of continuous convergence (explained before 7.11). Then $h(w) = df(\bar{x})(w)$ as already observed. On the other hand, we have (b) $\Leftrightarrow$ (c) by Theorem 7.14. The common function h in these cases must by (c) be continuous and by (a) be positively homogeneous with $h(0) = 0$ (as implied by 7(19), the semiderivative limit); then h has to be finite everywhere besides. Likewise in (d), h must be finite; the equivalence of (d) with (b) is seen from the scalar-valued case of 5.45. Putting $\bar{x} + \tau w'$ in place of x, we can transform the expansion in (e) into

$$\Delta_\tau f(\bar{x})(w') = h(w') + \frac{o(|\tau w'|)}{\tau} \quad \text{for } \tau \neq 0. \tag{7(22)}$$

This refers to the existence for each $\varepsilon > 0$ of $\delta > 0$ such that

$$\big|\Delta_\tau f(\bar{x})(w') - h(w')\big| \leq \varepsilon |w'| \text{ when } 0 < \tau |w'| \leq \delta.$$

When this property is present, the uniform convergence in (c) is at hand. Thus, (e) implies (c). Conversely, under (c) there exists for any $\varepsilon > 0$ and $\rho > 0$ a $\delta > 0$ with

$$\big|\Delta_\tau f(\bar{x})(w) - h(w)\big| \leq \varepsilon \text{ when } 0 < \tau \leq \delta \text{ and } |w| \leq \rho.$$

Writing a general $w \in \rho I\!B$ as θz with $|z| = \rho$ and $\theta = |w|/\rho \in [0, 1]$, and noting that $\big|\Delta_\tau f(\bar{x})(w) - h(w)\big|/\theta = \big|\Delta_{\theta\tau} f(\bar{x})(z) - h(z)\big|$, we can apply this inequality to the latter to get

$$\big|\Delta_\tau f(\bar{x})(w) - h(w)\big| \leq \varepsilon\big(|w|/\rho\big) \text{ when } 0 < \tau\big(|w|/\rho\big) \leq \delta.$$

The fact that for any $\varepsilon > 0$ and $\rho > 0$ there exists $\delta > 0$ with this property tells us that 7(22) holds. We conclude that (c) implies (e), and therefore that (e) can be added to the list of equivalence conditions. Finally, we note that (e) implies through the continuity of h at 0 the continuity of f at $\bar{x}$. □

7.22 Corollary (differentiability versus semidifferentiability). *A function $f : I\!R^n \to \overline{I\!R}$ is differentiable at $\bar{x}$, a point with $f(\bar{x})$ finite, if and only if f is semidifferentiable at $\bar{x}$ and the semiderivative for w depends linearly on w.*

Proof. This follows from comparing 7(16), the definition of differentiability, with 7.21(e) and invoking the equivalence of this with 7.21(a). □

When f is semidifferentiable at $\bar{x}$, its continuity there, as asserted in 7.21, forces it to be finite on a neighborhood of $\bar{x}$. Semidifferentiability, unless restricted only to special directions, therefore can't assist in the study of situations where, for instance, $\bar{x}$ is a boundary point of $\operatorname{dom} f$. This limitation in the theory of generalized differentiation can satisfactorily be avoided by an approach through epi-convergence instead of continuous convergence and uniform convergence.

7.23 Definition (epi-derivatives). *Consider $f : \mathbb{R}^n \to \overline{\mathbb{R}}$ and a point $\bar{x}$ with $f(\bar{x})$ finite. If the (possibly infinite) limit*

$$\operatorname*{lim-epi}_{\tau \searrow 0\ w' \to w} \frac{f(\bar{x} + \tau w') - f(\bar{x})}{\tau} \tag{7(23)}$$

exists, or in other words if the epi-limit of the functions $\Delta_\tau f(\bar{x})$ as $\tau \searrow 0$ exists at w, it is the epi-derivative of f at $\bar{x}$ for w, and f is epi-differentiable at $\bar{x}$ for w. If this holds for every w, so that the functions $\Delta_\tau f(\bar{x})$ epi-converge on $\mathbb{R}^n$ to some function h as $\tau \searrow 0$, f is epi-differentiable at $\bar{x}$; it is properly epi-differentiable if the epi-limit function is proper.

The notation in 7(23) has the meaning explained around 7(11) and refers to a value β with the property that for all sequences $\tau^\nu \searrow 0$ one has

$$\begin{cases} \liminf_\nu \dfrac{f(\bar{x} + \tau^\nu w^\nu) - f(\bar{x})}{\tau^\nu} \geq \beta \text{ for every sequence } w^\nu \to w, \\ \limsup_\nu \dfrac{f(\bar{x} + \tau^\nu w^\nu) - f(\bar{x})}{\tau^\nu} \leq \beta \text{ for some sequence } w^\nu \to \bar{w}. \end{cases}$$

Then β must coincide with the value $df(\bar{x})(w)$ in 7(21). Thus, as with semiderivatives, we aren't obliged to introduce a special notation for epi-derivatives but can make do with $df(\bar{x})(w)$ and the observation that the epi-differentiability of f at $\bar{x}$ for w corresponds to requiring the limit that defines this expression to hold with additional properties.

If f is semidifferentiable at $\bar{x}$ for w, it's also epi-differentiable, and from what we've just seen, the epi-derivative equals the semiderivative. But epi-differentiability is a broader property, not equivalent to semidifferentiability in general. As a matter of fact, epi-differentiability is the 'upper' part of semidifferentiability, in the sense that epi-convergence is the 'upper' part of continuous convergence, cf. 7.11. It's clear through 7.21 that f is semidifferentiable at $\bar{x}$ if and only if both f and $-f$ are properly epi-differentiable at $\bar{x}$. Epi-differentiability of $-f$ corresponds of course to *hypo-differentiability* of f, with an obvious definition.

It should be noted that although the epi-derivative of f at $\bar{x}$ for w, when it exists, has the value $df(\bar{x})(w)$ in particular, it need not have the same value as the one-sided derivative $f'(\bar{x}; w)$. A simple example is provided by the indicator $f = \delta_C$ of $C = \{(x_1, x_2) \in \mathbb{R}^2 \mid x_2 = x_1^2\}$. For $\bar{x} = (0, 0)$ and $w = (w_1, w_2)$ one has

$$df(\bar{x})(w) = \begin{cases} 0 & \text{if } w_2 = 0, \\ \infty & \text{if } w_2 \neq 0, \end{cases} \qquad f'(\bar{x}; w) = \begin{cases} 0 & \text{if } (w_1, w_2) = (0, 0), \\ \infty & \text{if } (w_1, w_2) \neq (0, 0). \end{cases}$$

This shows very clearly too the deficiencies of taking one-sided directional derivatives only along rays and it provides further motivation for looking at the epi-convergence approach.

Epi-differentiability can be interpreted geometrically through the fact that

$$\operatorname{epi} \Delta_\tau f(\bar{x}) = \frac{\operatorname{epi} f - (\bar{x}, f(\bar{x}))}{\tau} \quad \text{for any } \tau > 0. \tag{7(24)}$$

Epi-convergence of difference quotients relates through this to the geometric concept of derivability in 6.1, but as applied to the set $\operatorname{epi} f$.

7.24 Proposition (geometry of epi-differentiability). *A function $f : \mathbb{R}^n \to \overline{\mathbb{R}}$ is epi-differentiable at $\bar{x}$ if and only if $\operatorname{epi} f$ is geometrically derivable at $(\bar{x}, f(\bar{x}))$. The epi-derivative function, when it exists, has $T_{\operatorname{epi} f}(\bar{x}, f(\bar{x}))$ as its epigraph.*

Proof. Functions epi-converge on $\mathbb{R}^n$ if and only if their epigraphs converge as subsets of $\mathbb{R}^{n+1}$. Therefore, in view of 7(24), the functions $\Delta_\tau f(\bar{x})$ epi-converge to h as $\tau \searrow 0$ if and only if the subsets $[\operatorname{epi} f - (\bar{x}, f(\bar{x}))]/\tau$ converge to $\operatorname{epi} h$ as $\tau \searrow 0$. But if these sets converge at all, the limit has to be $T_{\operatorname{epi} f}(\bar{x}, f(\bar{x}))$, and this means by 6.2 that $\operatorname{epi} f$ is geometrically derivable at $(\bar{x}, f(\bar{x}))$. □

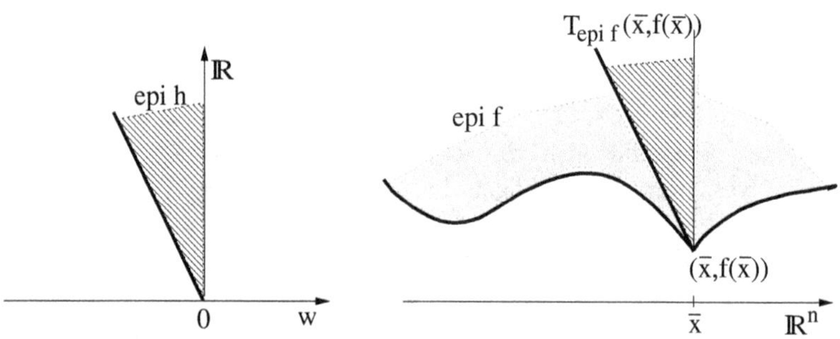

Fig. 7–5. Epi-differentiability as an epigraphical tangent cone condition.

The geometric concept of derivability has been seen in 6.30 to follow from that of Clarke regularity. A related property of regularity has a major role also in the study of functions.

7.25 Definition (subdifferential regularity). *A function $f : \mathbb{R}^n \to \overline{\mathbb{R}}$ is called subdifferentially regular at $\bar{x}$ if $f(\bar{x})$ is finite and $\operatorname{epi} f$ is Clarke regular at $(\bar{x}, f(\bar{x}))$ as a subset of $\mathbb{R}^n \times \mathbb{R}$.*

For short, we'll often just speak of this property as the 'regularity' of f at $\bar{x}$. More fully, if needed, it's subdifferential regularity in the *epigraphical* sense. A corresponding property in the *hypographical sense* is obtained in replacing $\operatorname{epi} f$ by $\operatorname{hypo} f$. The 'subdifferential' designation refers ultimately to the theory of 'subderivatives' and 'subgradients' in Chapter 8, where this notion will be especially significant.

7.26 Theorem (derivative consequences of regularity). *Let $f : \mathbb{R}^n \to \overline{\mathbb{R}}$ be subdifferentially regular at $\bar{x}$. Then f is epi-differentiable at $\bar{x}$, and its epi-derivative for w, coinciding with $df(\bar{x})(w)$, depends sublinearly as well as lower semicontinuously on w.*

Such a function f is semidifferentiable at $\bar{x}$ if and only if $df(\bar{x})(w) < \infty$ for all w. More generally, it is semidifferentiable at $\bar{x}$ for w if and only if w belongs to the interior of the cone $D = \{w \mid df(\bar{x})(w) < \infty\}$.

Proof. The first assertion is based on the geometric facts in 6.30 as translated here to epigraphs through 7.24. Because the function h giving the epi-derivatives has $T_{\text{epi}\, f}(\bar{x}, f(\bar{x}))$ as its epigraph, it's sublinear when that cone is convex (cf. 3.19).

We've already observed that $h(w)$ must be $df(\bar{x})(w)$, so that the set D in the theorem is $\operatorname{dom} h$. The sublinearity of h implies by 2.35 that h is continuous on $\operatorname{int}(\operatorname{dom} h)$. Hence, when $\bar{w} \in \operatorname{int}(\operatorname{dom} h)$ there exists for any $\beta \in (h(\bar{w}), \infty)$ a $\delta > 0$ such that the compact set $B = I\!B(\bar{w}, \delta) \times \{\beta\}$ lies in $\operatorname{int}(\operatorname{epi} h) = \operatorname{int} T_{\text{epi}\, f}(\bar{x}, f(\bar{x}))$. Then by 6.37 and the regularity of $\operatorname{epi} f$ at $(\bar{x}, f(\bar{x}))$ there exists $\varepsilon > 0$ such that $[\operatorname{epi} f - (\bar{x}, f(\bar{x}))]/\tau \supset B$ when $0 < \tau < \varepsilon$. This means by 7(24) that $\Delta_\tau f(\bar{x})(w) \leq \beta$ for all $w \in I\!B(\bar{w}, \delta)$ when $0 < \tau < \varepsilon$. From the arbitrary choice of $\beta > h(\bar{w})$ it follows that

$$\limsup_{\substack{\tau \searrow 0 \\ w \to \bar{w}}} \Delta_\tau f(\bar{x})(w) \leq h(\bar{w}).$$

The opposite inequality for lim inf holds by virtue of epi-differentiability, so the semiderivative limit in Definition 7.20 exists and equals $h(\bar{w})$.

When $\operatorname{dom} h$ is all of $I\!R^n$, this is true for every vector $\bar{w} \in I\!R^n$, so f is semidifferentiable at $\bar{x}$. □

7.27 Example (regularity of convex functions). *A proper convex function $f : I\!R^n \to \overline{I\!R}$ is subdifferentially regular at any point $\bar{x} \in \operatorname{dom} f$ where it is locally lsc, and it is epi-differentiable at every such point as well. These properties hold in particular when $\bar{x} \in \operatorname{int}(\operatorname{dom} f)$, and then f is semidifferentiable at $\bar{x}$.*

Detail. When f is locally lsc at $\bar{x}$, $\operatorname{epi} f$ is locally closed at $(\bar{x}, f(\bar{x}))$ (see 1.34). The convexity of $\operatorname{epi} f$ when f is convex (cf. 2.4) then yields the regularity of $\operatorname{epi} f$ at $(\bar{x}, f(\bar{x}))$ by 6.9, which is the regularity of f at $\bar{x}$ by Definition 7.25. The results in 7.26 now apply along with the fact that f is sure to be lsc (actually continuous) at points $\bar{x} \in \operatorname{int}(\operatorname{dom} f)$ (cf. 2.35). □

7.28 Example (regularity of max functions). *If $f(x) = \delta_C(x) + \max_{i \in I} f_i(x)$ for a set $C \subset I\!R^n$ and a finite collection $\{f_i\}_{i \in I}$ of smooth functions $f_i : I\!R^n \to \overline{I\!R}$, then f is subdifferentially regular and epi-differentiable at every point $\bar{x} \in C$ where C is regular, and f is semidifferentiable at every point $\bar{x} \in \operatorname{int} C$.*

Detail. In this case, $\operatorname{epi} f$ consists of the points $(x, \alpha) \in X := C \times I\!R$ satisfying the constraints $g_i(x, \alpha) \leq 0$ for all $i \in I$, where $g_i(x, \alpha) = f_i(x) - \alpha$. It follows from 6.14 that $\operatorname{epi} f$ is regular at any of its points $(\bar{x}, \bar{\alpha})$ such that C is regular at $\bar{x}$. The remaining assertions then follow from 7.26. □

A formula for the epi-derivative function in 7.28 will later be presented in the wider framework of subderivatives and subgradients; see 8.31. Many examples of subdifferentially regular functions can be generated from the calculus in

Chapter 6 as applied to epigraphs, and that topic will be taken up in Chapter 10. Further elaboration of generalized differentiability properties of functions f in terms of tangent cone properties of $\operatorname{epi} f$ will proceed with the theory of 'subderivatives' in Chapter 8.

E. Convergence in Minimization

Another prime source of interest in epigraphical limits, besides their use in local approximations corresponding to epi-differentiation, is their importance in understanding what happens in a problem of minimization when the data elements in the problem are varied or approximated in a global sense. A given problem is represented by a single function f, as explained in Chapter 1, and a sequence of functions f^ν that approaches f in some way can therefore be viewed as representing a sequence of nearby problems that get closer and closer. To what extent can it be determined that the optimal values $\inf f^\nu$ approach $\inf f$, and what can be said about the optimal solution sets $\operatorname{argmin} f^\nu$? The result we state next, while not yet providing full answers to these questions, reveals why epi-convergence is deeply involved.

7.29 Proposition (characterization of epi-convergence via minimization). *For functions f^ν and f on $\mathbb{R}^n$ with f lsc, one has*

(a) *$\operatorname{e-lim\,inf}_\nu f^\nu \geq f$ if and only if $\liminf_\nu(\inf_B f^\nu) \geq \inf_B f$ for every compact set $B \subset \mathbb{R}^n$;*

(b) *$\operatorname{e-lim\,sup}_\nu f^\nu \leq f$ if and only if $\limsup_\nu(\inf_O f^\nu) \leq \inf_O f$ for every open set $O \subset \mathbb{R}^n$.*

Thus, $\operatorname{e-lim}_\nu f^\nu = f$ if and only if both conditions hold.

Proof. Relying on the geometry in 7.1 to translate the assertions into a geometric context, we apply the hit-and-miss criteria in 4.5 to the epigraphs of the functions f^ν and f. We utilize also the fact that instead of taking the usual balls $\mathbb{B}\big((x,\alpha),\delta\big)$ in $\mathbb{R}^{n+1}$ when calling on the conditions in 4.5(a′) and (b′), we can just as well take the cylinders $\mathbb{B}^+\big((x,\alpha),\delta\big) := \mathbb{B}(x,\delta) \times [\alpha-\delta, \alpha+\delta]$, since $\mathbb{B}\big((x,\alpha),\delta\big) \subset \mathbb{B}^+\big((x,\alpha),\delta\big) \subset \sqrt{2}\mathbb{B}\big((x,\alpha),\delta\big)$.

Necessity in (a). The assumption is that $\limsup_\nu(\operatorname{epi} f^\nu) \subset \operatorname{epi} f$. Consider any compact set $B \subset \mathbb{R}^n$ and value $\alpha \in \mathbb{R}$ such that $\inf_B f > \alpha$. The compact set (B,α) in $\mathbb{R}^{n+1}$ doesn't meet $\operatorname{epi} f$, so by the criterion in 4.5(b) there is an index set $N \in \mathcal{N}_\infty$ such that, for all $\nu \in N$, (B,α) doesn't meet $\operatorname{epi} f^\nu$ either. That implies $\inf_B f^\nu \geq \alpha$. We conclude that $\liminf_\nu(\inf_B f^\nu)$ can't be less than $\inf_B f$.

Sufficiency in (a). Consider any cylinder $\mathbb{B}^+\big((x,\alpha),\delta\big)$ that doesn't meet $\operatorname{epi} f$. Because f is lsc, so that $\operatorname{epi} f$ is closed, we have $\inf_{\mathbb{B}(x,\delta)} f > \alpha+\delta$. Since the ball $\mathbb{B}(x,\delta)$ is a compact set in $\mathbb{R}^n$, we have by assumption that $\liminf_\nu\big(\inf_{\mathbb{B}(x,\delta)} f^\nu\big) > \alpha+\delta$. Therefore, an index set $N \in \mathcal{N}_\infty$ exists such that $\inf_{\mathbb{B}(x,\delta)} f^\nu > \alpha+\delta$ for all $\nu \in N$. Then the cylinder $\mathbb{B}^+\big((x,\alpha),\delta\big)$

doesn't meet $\operatorname{epi} f^\nu$ for any $\nu \in N$. This proves by the criterion in 4.5(b′) that $\limsup_\nu (\operatorname{epi} f^\nu) \subset \operatorname{epi} f$.

Necessity in (b). The assumption is that $\liminf_\nu (\operatorname{epi} f^\nu) \supset \operatorname{epi} f$. Consider any open set $O \subset \mathbb{R}^n$ and value $\alpha \in \mathbb{R}$ such that $\inf_O f < \alpha$. The set $O \times (-\infty, \alpha)$ is open in $\mathbb{R}^{n+1}$ and meets $\operatorname{epi} f$. Hence from the condition in 4.5(a) we know there is an index set $N \in \mathcal{N}_\infty$ such that, for all $\nu \in N$, this set meets $\operatorname{epi} f^\nu$, with the consequence that $\inf_O f^\nu < \alpha$. The upper limit of $\inf_O f^\nu$ as $\nu \to \infty$ therefore can't exceed $\inf_O f^\nu$.

Sufficiency in (b). Consider any open cylinder of the form $\operatorname{int} \mathbb{B}^+\big((x,\alpha),\delta\big)$ that meets $\operatorname{epi} f$; this condition means that $\inf_{\operatorname{int} \mathbb{B}(x,\delta)} f < \alpha+\delta$. Our hypothesis yields the existence of an index set $N \in \mathcal{N}_\infty$ such that $\inf_{\operatorname{int} \mathbb{B}(x,\delta)} f^\nu < \alpha+\delta$ for all $\nu \in N$. Then the set $\operatorname{int} \mathbb{B}^+\big((x,\alpha),\delta\big)$ meets $\operatorname{epi} f^\nu$ for all $\nu \in N$. The criterion in 4.5(a′) tells us on this basis that $\liminf_\nu (\operatorname{epi} f^\nu) \supset \operatorname{epi} f$. □

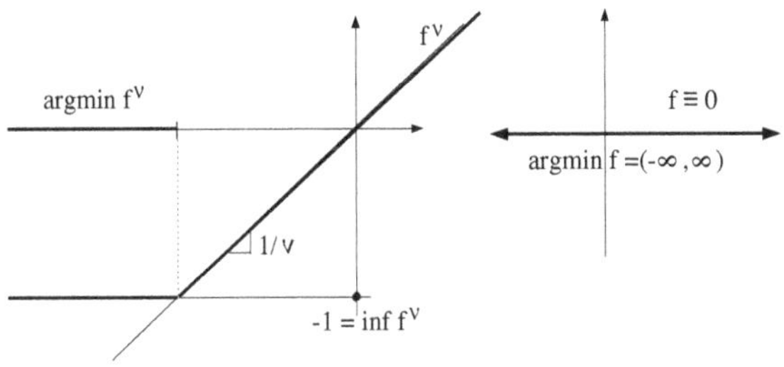

Fig. 7–6. Epi-convergent convex functions with optimal values not converging.

In general, epi-convergence has to be supplemented by some further condition, if one wishes to guarantee that $\inf f^\nu \to \inf f$. A simple example in Figure 7–6, involving only convex functions, illustrates how one may have $f^\nu \xrightarrow{e} f$ but $\inf f^\nu \not\to \inf f$. The functions $f^\nu(x) := \max\{\nu^{-1}x, -1\}$ on $\mathbb{R}^1$ have the property that $f^\nu \xrightarrow{e} f \equiv 0$, yet $\inf f^\nu = -1$ for all ν. The trouble clearly resides in the unbounded level sets, so it won't be surprising that some form of 'level-boundedness' will eventually be brought into the analysis of this kind of situation.

As far convergence of optimal solutions is concerned, it would be unrealistic to aim at finding conditions that guarantee that $\lim_\nu(\operatorname{argmin} f^\nu) = \operatorname{argmin} f$, except in cases where $\operatorname{argmin} f$ consists of a single point. Rather, we must be content with the inclusion

$$\operatorname{limsup}_\nu(\operatorname{argmin} f^\nu) \subset \operatorname{argmin} f,$$

which says that every cluster point x of a sequence $\{x^\nu\}_{\nu\in N}$ with x^ν optimal for f^ν will be optimal for f. An illustration in $\mathbb{R}$ is furnished in Figure 7–7. Although the functions f^ν epi-converge to f and even converge pointwise to f uniformly on all of $\mathbb{R}$, it isn't true that *every* point in $\operatorname{argmin} f$ is the limit of a sequence of points belonging to the sets $\operatorname{argmin} f^\nu$. Observe, though, that

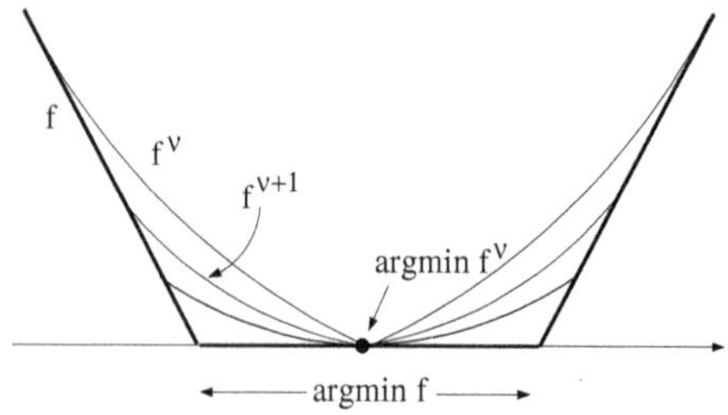

Fig. 7–7. Outer limit property in the convergence of optimal solutions.

every point $x \in \operatorname{argmin} f$ can be expressed as a limit of points x^ν that are 'approximately optimal' for the functions f^ν.

To state the basic facts about the convergence of optimal values and optimal solutions in a manner consonant with these observations, we recall from Chapter 1 the notion of ε-optimality: the set of points that minimize a function f to within ε make up the set

$$\varepsilon\text{-}\operatorname{argmin} f := \big\{ x \,\big|\, f(x) \le \inf f + \varepsilon \big\}.$$

We begin with a statement that only calls on 'epigraphical nesting', a requirement substantially weaker than epi-convergence because only the upper epi-limit is involved. Convergence proofs for algorithmic procedures can often be obtained as a consequence of epigraphical nesting alone.

7.30 Proposition (epigraphical nesting). *If* e-$\limsup_\nu f^\nu \le f$, *then*

$$\operatorname{limsup}_\nu (\inf f^\nu) \;\le\; \inf f.$$

Furthermore, the inclusion

$$\operatorname{limsup}_\nu (\varepsilon^\nu\text{-}\operatorname{argmin} f^\nu) \;\subset\; \operatorname{argmin} f$$

holds for any sequence $\varepsilon^\nu \searrow 0$ *such that whenever* $N \in \mathcal{N}_\infty^\#$ *and* $x^\nu \underset{N}{\to} x$ *with* $x^\nu \in \varepsilon^\nu$-argmin f^ν, *then* $f^\nu(x^\nu) \underset{N}{\to} f(x)$.

Proof. The first assertion merely specializes 7.29(b) to $O = I\!R^n$. For the second assertion, suppose the sequence $\varepsilon^\nu \searrow 0$ has the property mentioned; we have to verify that $\operatorname{argmin} f$ contains any x expressible as $\lim_{\nu \in N} x^\nu$ for an index set $N \in \mathcal{N}_\infty^\#$ and points $x^\nu \in \varepsilon^\nu$-argmin f^ν. Such points x^ν satisfy $f^\nu(x^\nu) - \varepsilon^\nu \le \inf f^\nu$, where by assumption also $\lim_{\nu \in N} f^\nu(x^\nu) = f(x)$. Then $f(x) \le \limsup_{\nu \in N} (\inf f^\nu)$, but we already know that the latter can't exceed $\inf f$. Hence $f(x) \le \inf f$, so $x \in \operatorname{argmin} f$. □

7.31 Theorem (inf and argmin). *Suppose* $f^\nu \xrightarrow{e} f$ *with* $-\infty < \inf f < \infty$.

(a) $\inf f^\nu \to \inf f$ *if and only if there exists for every* $\varepsilon > 0$ *a compact set* $B \subset I\!R^n$ *along with an index set* $N \in \mathcal{N}_\infty$ *such that*

$$\inf_B f^\nu \;\le\; \inf f^\nu + \varepsilon \;\text{ for all }\; \nu \in N.$$

(b) $\limsup_\nu(\varepsilon\text{-argmin}\, f^\nu) \subset \varepsilon\text{-argmin}\, f$ *for every* $\varepsilon \geq 0$ *and consequently*

$$\text{limsup}_\nu(\varepsilon^\nu\text{-argmin}\, f^\nu) \subset \text{argmin}\, f \quad \textit{whenever } \varepsilon^\nu \searrow 0.$$

(c) *Under the assumption that* $\inf f^\nu \to \inf f$, *there exists a sequence* $\varepsilon^\nu \searrow 0$ *such that* $\varepsilon^\nu\text{-argmin}\, f^\nu \to \text{argmin}\, f$. *Conversely, if such a sequence exists, and if* $\text{argmin}\, f \neq \emptyset$, *then* $\inf f^\nu \to \inf f$.

Proof. Sufficiency in (a). We already know that $\limsup_\nu \big(\inf f^\nu\big) \leq \inf f$ by 7.30, while on the other hand, the inequality in (a) when combined with the property in 7.29(a) yields

$$\liminf_\nu \big(\inf f^\nu\big) + \varepsilon \geq \liminf_\nu \big(\inf_B f^\nu\big) \geq \inf_B f \geq \inf f.$$

This being true for arbitrary ε, we conclude that $\lim_\nu \big(\inf f^\nu\big) = \inf f$.

Necessity in (a). Since $\inf f^\nu \to \inf f$ by assumption, it is enough to demonstrate for arbitrary $\delta > 0$ the existence of a compact set B such that $\limsup_\nu \big(\inf_B f^\nu\big) \leq \inf f + \delta$. Choose any point x such that $f(x) \leq \inf f + \delta$. Because $f^\nu \xrightarrow{e} f$, there exists by 7.2 a sequence $x^\nu \to x$ such that $\limsup_\nu f^\nu(x^\nu) \leq f(x)$. Let B be any compact set large enough to contain all the points x^ν. Then $\inf_B f^\nu \leq f^\nu(x^\nu)$ for all ν, so B has the desired property.

Proof of (b). Fix $\varepsilon \geq 0$. Let $x^\nu \in \varepsilon\text{-argmin}\, f^\nu$. Any cluster point $\bar{x}$ of the sequence $\{x^\nu\}_{\nu \in \mathbb{N}}$ has

$$\begin{aligned} f(\bar{x}) &\leq \liminf_\nu f^\nu(x^\nu) \leq \limsup_\nu f^\nu(x^\nu) \\ &\leq \limsup_\nu(\inf f^\nu + \varepsilon) \leq \inf f + \varepsilon, \end{aligned}$$

where the first inequality follows from epi-convergence, and the last via 7.30. Hence $\bar{x} \in \varepsilon\text{-argmin}\, f$. From this we get $\limsup_\nu(\varepsilon\text{-argmin}\, f^\nu) \subset \varepsilon\text{-argmin}\, f$. The inclusion $\limsup_\nu \varepsilon^\nu\text{-argmin}\, f^\nu \subset \text{argmin}\, f$ follows from the preceding one because the collection $\{\varepsilon\text{-argmin}\, f^\nu\}_{\varepsilon \geq 0}$ is nested with respect to inclusion.

Proof of (c). Suppose $\bar{\alpha}^\nu := \inf f^\nu \to \inf f =: \bar{\alpha}$. By assumption we have $\bar{\alpha} \in \mathbb{R}$; hence for ν large enough, $\bar{\alpha}^\nu$ is finite. From the convergence of level sets (in 7.7), we deduce the existence of $\alpha^\nu \searrow \bar{\alpha}$ such that $\text{lev}_{\leq \alpha^\nu} f^\nu \to \text{lev}_{\leq \bar{\alpha}} f = \text{argmin}\, f$. Simply set $\varepsilon^\nu := \alpha^\nu - \bar{\alpha}^\nu$.

For the converse, suppose there's a sequence $\varepsilon^\nu \searrow 0$ with $\varepsilon^\nu\text{-argmin}\, f^\nu \to \text{argmin}\, f \neq \emptyset$. For any $x \in \text{argmin}\, f$ one can select $x^\nu \in \varepsilon^\nu\text{-argmin}\, f^\nu$ with $x^\nu \to x$. Then because $f^\nu \xrightarrow{e} f$, we obtain

$$\begin{aligned} \inf f = f(x) &\leq \liminf_\nu f^\nu(x^\nu) \leq \liminf_\nu(\inf f^\nu + \varepsilon^\nu) \\ &\leq \liminf_\nu(\inf f^\nu) \leq \text{limsup}_\nu(\inf f^\nu) \leq \inf f, \end{aligned}$$

where the last inequality comes from 7.30. □

The need for assuming $\text{argmin}\, f \neq \emptyset$ in the converse part of 7.31(c) is illustrated in Figure 7–8 by the sequence of functions f^ν on $\mathbb{R}^1$ defined by $f^\nu(x) = 1 + e^{-x}$ when $x \neq \nu$, $f^\nu(x) = 0$ for $x = \nu$. These epi-converge to $f(x) := 1 + e^{-x}$, yet $\text{argmin}\, f^\nu \to \text{argmin}\, f = \emptyset$ while $0 = \inf f^\nu \not\to \inf f = 1$.

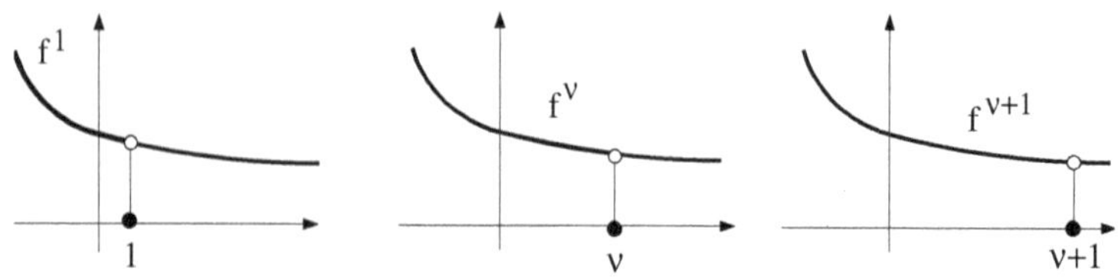

Fig. 7–8. For any compact set B, $\inf f^\nu < \inf_B f^\nu$ for ν large enough.

To provide a practical criterion for verifying the condition in 7.31(a) and obtaining additional properties of the kind usually demanded in optimization, we appeal to a simple generalization of the level boundedness concept in 1.8. We say that a sequence $\{f^\nu\}_{\nu\in\mathbb{N}}$ of functions is *eventually level-bounded* if for each $\alpha \in \mathbb{R}$ the sequence of sets $\text{lev}_{\leq\alpha} f^\nu$ is eventually bounded.

7.32 Exercise (eventual level boundedness).

(a) *The sequence $\{f^\nu\}_{\nu\in\mathbb{N}}$ is eventually level-bounded if there is a level-bounded function g such that eventually $f^\nu \geq g$, or if the sequence of sets $\operatorname{dom} f^\nu$ is eventually bounded.*

(b) *If the sequence $\{f^\nu\}_{\nu\in\mathbb{N}}$ is eventually level-bounded and $f^\nu \xrightarrow{e} f$, then f is level-bounded, in fact inf-compact.*

(c) *If $f^\nu \xrightarrow{e} f$ with f level-bounded, $f \not\equiv \infty$, and all the sets $\text{lev}_{\leq\alpha} f^\nu$ are connected (as for instance when the functions f^ν are convex), then the sequence $\{f^\nu\}_{\nu\in\mathbb{N}}$ is eventually level-bounded.*

Guide. Get (a) from the definitions and (b) as a corollary of 7.7. In deriving (c) make use also of 4.12. □

Note that the second case in 7.32(a) merely specializes the first case to a function g having $-\infty$ on a certain bounded set B but ∞ elsewhere. More generally in connection with the first case in 7.32(a), it may be recalled that for g to be level-bounded (as defined in 1.8), a sufficient condition is that g be level-coercive (cf. 3.27). This is true if $g^\infty(w) > 0$ for all $w \neq 0$ (cf. 3.26), that being actually equivalent to level boundedness when g is convex (cf. 3.27).

7.33 Theorem (convergence in minimization). *Suppose the sequence $\{f^\nu\}_{\nu\in\mathbb{N}}$ is eventually level-bounded, and $f^\nu \xrightarrow{e} f$ with f^ν and f lsc and proper. Then*

$$\inf f^\nu \to \inf f \quad \textit{(finite)},$$

while for ν in some index set $N \in \mathcal{N}_\infty$ the sets $\operatorname{argmin} f^\nu$ are nonempty and form a bounded sequence with

$$\limsup_\nu (\operatorname{argmin} f^\nu) \subset \operatorname{argmin} f.$$

Indeed, for any choice of $\varepsilon^\nu \searrow 0$ and $x^\nu \in \varepsilon^\nu\text{-}\operatorname{argmin} f^\nu$, the sequence $\{x^\nu\}_{\nu\in\mathbb{N}}$ is bounded and such that all its cluster points belong to $\operatorname{argmin} f$. If $\operatorname{argmin} f$ consists of a unique point $\bar{x}$, one must actually have $x^\nu \to \bar{x}$.

Proof. From 7.29(b) with $O = I\!R^n$, we have $\limsup_\nu(\inf f^\nu) \leq \inf f$. Hence there's a value $\hat{\alpha} \in I\!R$ such that $\inf f^\nu < \hat{\alpha}$ for all ν. Then by the definition of eventual level boundedness there's an index set $N \in \mathcal{N}_\infty$ along with a compact set B such that $\operatorname{lev}_{\leq\hat{\alpha}} f^\nu \subset B$ for all $\nu \in N$. We have $\inf f^\nu = \inf_B f^\nu$ for all $\nu \in N$, and this ensures through 7.30 that $\inf f^\nu \to \inf f$.

We also have $\operatorname{argmin} f^\nu = \operatorname{argmin}_B f^\nu = \operatorname{argmin}(f + \delta_B)$ for all $\nu \in N$, these sets being nonempty by 1.9. Likewise, $\operatorname{argmin} f \neq \emptyset$ because f is inf-compact by 7.32(b). For $\varepsilon^\nu \searrow 0$, we eventually have $\inf f^\nu + \varepsilon^\nu < \hat{\alpha}$; the sets ε^ν-$\operatorname{argmin} f^\nu$ all lie then in B. The conclusions follow then from 7.31(b). □

This convergence theorem can be useful not only directly, but also in technical constructions, as we now illustrate.

7.34 Example (bounds on converging convex functions). *If a sequence of lsc, proper, convex functions $f^\nu : I\!R^n \to \overline{I\!R}$ epi-converges to a proper function f, there exists $\rho \in (0, \infty)$ such that*

$$f \geq -\rho(|\cdot|+1) \quad \text{and} \quad f^\nu \geq -\rho(|\cdot|+1) \quad \text{for all } \nu \in I\!N.$$

Detail. We know that f inherits convexity from the f^ν's (cf. 7.17) and hence that all these functions are not counter-coercive: by 3.26 and 3.27, there are constants $\bar{\rho}$, $\rho^\nu \in (0, \infty)$ with $f \geq -\bar{\rho}(|\cdot|+1)$ and $f^\nu \geq -\rho^\nu(|\cdot|+1)$ for all $\nu \in I\!N$. The issue is whether a single ρ will work uniformly. Raising $\bar{\rho}$ if necessary, we can arrange that the function $g := f + \bar{\rho}(|\cdot|+1)$ is level-bounded. Let $\alpha = \inf g$ (this being finite by 1.9). The convex functions $g^\nu := f^\nu + \bar{\rho}(|\cdot|+1)$ epi-converge to g by 7.8(a), so that $\inf g^\nu \to \inf g$. Then for some $N \in \mathcal{N}_\infty$ we have $\inf g^\nu \geq \alpha - 1$, hence $f^\nu \geq -\bar{\rho}(|\cdot|+1) + \alpha - 1$. Taking ρ larger than $\bar{\rho} - \alpha + 1$, $\bar{\rho}$ and the constants ρ^ν for $\nu \in I\!N \setminus N$, we get the result. □

Theorem 7.33 can be applied to the Moreau envelopes in 1.22 in order to obtain yet another characterization of epi-convergence. For this purpose an extension of the notion of prox-boundedness in 1.23 is needed.

7.35 Definition (eventually prox-bounded sequences). *A sequence $\{f^\nu\}_{\nu \in I\!N}$ of functions on $I\!R^n$ is eventually prox-bounded if there exists $\lambda > 0$ such that $\liminf_\nu e_\lambda f^\nu(x) > -\infty$ for some x. The supremum of all such λ is then the threshold of eventual prox-boundedness of the sequence.*

7.36 Exercise (characterization of eventual prox-boundedness). *A sequence $\{f^\nu\}_{\nu \in I\!N}$ is eventually prox-bounded if and only if there is a quadratic function q such that $f^\nu \geq q$ for all ν in some index set $N \in \mathcal{N}_\infty$.*

Guide. Appeal to 1.24 and the meaning of an inequality $e_\lambda f^\nu(b) \geq \beta$. □

7.37 Theorem (epi-convergence through Moreau envelopes). *For proper, lsc functions f^ν and f on $I\!R^n$, the following are equivalent:*

(a) *the sequence $\{f^\nu\}_{\nu \in I\!N}$ is eventually prox-bounded, and $f^\nu \xrightarrow{e} f$;*

(b) *f is prox-bounded, and $e_\lambda f^\nu \xrightarrow{p} e_\lambda f$ for all λ in an interval $(0, \varepsilon)$, $\varepsilon > 0$.*

Then the pointwise convergence of $e_\lambda f^\nu$ *to* $e_\lambda f$ *for* $\lambda > 0$ *sufficiently small is uniform on all bounded subsets of* $\mathbb{R}^n$*, hence yields continuous convergence and epi-convergence too, and indeed* $e_{\lambda^\nu} f^\nu$ *converges in all these ways to* $e_\lambda f$ *whenever* $\lambda^\nu \to \lambda \in (0, \bar\lambda)$*, where* $\bar\lambda$ *is the threshold of eventual prox-boundedness.*

If the functions f^ν *and* f *are convex, then the threshold of eventual prox-boundedness is* $\bar\lambda = \infty$ *and condition (b) can be replaced by*

(b′) $e_\lambda f^\nu \xrightarrow{p} e_\lambda f$ *for some* $\lambda > 0$.

Proof. Under (a), we know that for arbitrary $\varepsilon \in (0, \bar\lambda)$ there exist $b \in \mathbb{R}^n$, $\beta \in \mathbb{R}$ and $N \in \mathcal{N}_\infty$ such that $e_\varepsilon f^\nu(b) \geq \beta$ for all $\nu \in N$. This means for such ν that $f^\nu(w) \geq \beta - (1/2\varepsilon)|w - b|^2$ for all $w \in \mathbb{R}^n$. Now consider any $x \in \mathbb{R}^n$ and any $\lambda \in (0, \varepsilon)$, as well as any sequences $x^\nu \to x$ and $\lambda^\nu \to \lambda$ with $\lambda^\nu > 0$. Let $j(w) = (1/2\lambda)|w - x|^2$ and $j^\nu(w) = (1/2\lambda^\nu)|w - x^\nu|^2$. The functions $f^\nu + j^\nu$ epi-converge to $f + j$ by the criterion in 7.2, inasmuch as $j^\nu(w^\nu) \to j(w)$ when $w^\nu \to w$. There's an index set $N' \in \mathcal{N}_\infty$ with $N' \subset N$ along with $\delta \in (\lambda, \varepsilon)$ such that $\lambda^\nu \in (0, \delta)$ when $\nu \in N'$. Then $(f^\nu + j^\nu)(w)$ is bounded below by

$$\begin{aligned}\beta - \frac{1}{2\varepsilon}|w - b|^2 &+ \frac{1}{2\delta}|(w - b) - (x^\nu - b)|^2 \\ &\geq \beta + \left[\frac{1}{2\delta} - \frac{1}{2\varepsilon}\right]|w - b|^2 - \frac{1}{\delta}\langle w - b,\, x^\nu - b\rangle + \frac{1}{2\delta}|x^\nu - b|^2 \\ &\geq \beta + \frac{[\varepsilon - \delta]}{2\varepsilon\delta}|w - b|^2 - \frac{1}{\delta}|w - b||x^\nu - b|.\end{aligned}$$

We see that for any upper bound ρ on $|x^\nu - b|$ for $\nu \in N'$ we have $f^\nu + j^\nu \geq h$ for $h(w) := \beta + ([\varepsilon - \delta]/2\varepsilon\delta)|w - b|^2 - (\rho/\delta)|w - b|$. The function h is level-bounded because $\varepsilon > \delta > 0$. Hence by 7.33 the values $\inf\{f^\nu + j^\nu\} =: e_{\lambda^\nu} f^\nu(x^\nu)$ converge to $\inf\{f + j\} =: e_\lambda f(x)$. From the special case of constant sequences $x^\nu \equiv x$ and $\lambda^\nu \equiv \lambda$ we deduce that (b) is true. Also, we see that $e_\lambda f(x)$ is finite, so that f must itself be prox-bounded (cf. Definition 1.23).

In demonstrating that $e_{\lambda^\nu} f^\nu(x^\nu) \to e_\lambda f(x)$ for general sequences $x^\nu \to x$ and $\lambda^\nu \to \lambda$ in $(0, \varepsilon)$ we've actually shown that the functions $e_{\lambda^\nu} f^\nu$ converge continuously to $e_\lambda f$. The convergence is then uniform on all bounded sets, cf. 7.14 and 1.25, and it also entails epi-convergence, cf. 7.11. Since ε was selected arbitrarily from $(0, \bar\lambda)$, this holds whenever $\lambda^\nu \to \lambda \in (0, \bar\lambda)$.

Conversely now, suppose (b) holds. The prox-boundedness of f ensures that $e_\lambda f$ is finite for $\lambda > 0$ sufficiently small (cf. 1.25). When this property is combined with pointwise convergence of $e_\lambda f^\nu$ to $e_\lambda f$, we see that the definition of eventual prox-boundedness is satisfied for the sequence $\{f^\nu\}_{\nu \in \mathbb{N}}$. This sequence can't escape to the horizon, so all its cluster points g with respect to epi-convergence are proper as well as lsc.

Let g be any such cluster point, say $g = \text{e-lim}_{\nu \in N} f^\nu$ for an index set $N \in \mathcal{N}_\infty^\#$. Applying (a) to the subsequence $\{f^\nu\}_{\nu \in N}$, which itself is eventually prox-bounded, we see that the subsequence $\{e_\lambda f^\nu\}_{\nu \in N}$ converges pointwise to $e_\lambda g$ for $\lambda > 0$ sufficiently small. But we already know it converges to $e_\lambda f$. Hence $e_\lambda g = e_\lambda f$ for all $\lambda > 0$ sufficiently small. Then $g = f$ by 1.25. From the fact that f is the unique cluster point of $\{f^\nu\}_{\nu\in\mathbb{N}}$ with respect to epi-convergence, we conclude through Theorem 7.6 that $f^\nu \xrightarrow{e} f$.

When the functions f^ν and f are convex, implying that every cluster point g of $\{f^\nu\}_{\nu\in\mathbb{N}}$ with respect to epi-convergence is convex as well (cf. 7.17), the argument just given can be strengthened: to conclude that $g = f$ we need only know that $e_\lambda g = e_\lambda f$ for some $\lambda > 0$ (cf. 3.36). Also in this case, the sequence of functions $f^\nu + j^\nu$ is eventually level-bounded for every $\lambda > 0$ (see 7.32(c)), so we obtain the convergence of $e_\lambda f^\nu$ to $e_\lambda f$ for every $\lambda > 0$. □

7.38 Exercise (graphical convergence of proximal mappings). *Suppose $f^\nu \xrightarrow{e} f$ for proper, lsc functions f^ν and f on $\mathbb{R}^n$ such that the sequence $\{f^\nu\}_{\nu\in\mathbb{N}}$ is eventually prox-bounded with threshold $\bar\lambda$. Then*

$$\text{g-lim sup}_\nu\, P_{\lambda^\nu} f^\nu \subset P_\lambda f \quad \textit{whenever} \quad \lambda^\nu \to \lambda \in (0, \bar\lambda),$$

and in this case the sequence of mappings $P_{\lambda^\nu} f^\nu$ is eventually locally bounded, in the sense that each point x has a neighborhood V for which the sequence of sets $P_{\lambda^\nu} f^\nu(V)$ is eventually bounded.

Furthermore, $\text{g-lim}_\nu\, P_{\lambda^\nu} f^\nu = P_\lambda f$ when f is convex, or more generally whenever $P_\lambda f$ is single-valued.

Guide. Adapt the initial argument in the proof of Theorem 7.37, paying heed to the implications of 7.33 for the sets $\operatorname{argmin}\{f^\nu + j^\nu\} =: P_{\lambda^\nu} f^\nu(x^\nu)$ and $\operatorname{argmin}\{f + j\} =: P_\lambda f(x)$ rather than just the values $\inf\{f^\nu + j^\nu\}$ and $\inf\{f + j\}$. In the case where f is convex, apply 2.26(a). □

Here's an example showing in the setting of 7.38 that $f^\nu \xrightarrow{e} f$ doesn't always ensure $P_\lambda f^\nu \xrightarrow{g} P_\lambda f$, regardless of the assumptions of eventual prox-boundedness. On $\mathbb{R}$, let $f(x) = \max\{0, 1 - x^2\}$ and $f^\nu = (1 + \varepsilon^\nu) f$ for a sequence $\varepsilon^\nu \searrow 0$. The functions f^ν converge continuously to f and in particular epi-converge, and their sequence is prox-bounded with threshold $\bar\lambda = \infty$. It's easy to verify that for $\lambda = 1/2$ one has

$$P_\lambda f(x) = \begin{cases} x & \text{for } -\infty < x \le -1, \\ -1 & \text{for } -1 \le x < 0, \\ [-1, 1] & \text{for } x = 0, \\ 1 & \text{for } 0 < x \le 1, \\ x & \text{for } 1 \le x < \infty, \end{cases}$$

and $P_\lambda f^\nu(x) = P_\lambda f(x)$ for all $x \ne 0$, but $P_\lambda f^\nu(0) = \{-1, 1\} \ne P_\lambda f(0)$. Thus, the sequence of mappings $P_\lambda f^\nu$ is constant, yet corresponds only to part of $P_\lambda f$ and therefore doesn't converge graphically to $P_\lambda f$.

F. Epi-Continuity of Function-Valued Mappings

Parametric minimization in x of an expression depending on a parameter element u was viewed in Chapter 1 in the framework of a bivariate expression

$f(x,u)$, possibly with infinite values. Another valuable perspective is afforded by parallels with the theory of set-valued mappings, however, and this too supports many applications of variational analysis.

Along with the concept of a set-valued mapping from a space U to the space sets(X) consisting of all subsets of X, it's useful to consider that of a *function-valued mapping* from U to the space

$$\text{fcns}(X) := \text{ collection of all extended-real-valued functions on } X.$$

Such a mapping, also called a *bifunction*, assigns to each $u \in U$ a function defined on X that has values in $\overline{\mathbb{R}}$. There is obviously a one-to-one correspondence between such mappings $U \to \text{fcns}(X)$ and bivariate functions $f : X \times U \to \overline{\mathbb{R}}$: the mapping has the form $u \mapsto f(\cdot,u)$, where $f(\cdot,u)$ denotes the function that assigns to x the value $f(x,u)$. This correspondence, whereby many questions concerning parametric dependence can be posed simply in terms of a single function on a product space, is very convenient, as already seen, but the mapping viewpoint is interesting as well, because it puts the spotlight on the 'dynamic' qualities of the dependence.

The analogy between function-valued mappings and set-valued mappings provides a number of insights from the start. Without significant loss of generality for our purposes here, we can reduce discussion to the case where $X = \mathbb{R}^n$ and $U = \mathbb{R}^m$.

Any set-valued mapping $S : \mathbb{R}^m \rightrightarrows \mathbb{R}^n$ can be identified with a special function-valued mapping from $\mathbb{R}^m$ to $\mathbb{R}^n$, namely the one that assigns to each $u \in \mathbb{R}^m$ the indicator of the set $S(u)$. In bivariate terms, this *indicator mapping* for S corresponds to $u \mapsto f(\cdot,u)$ with $f : \mathbb{R}^n \times \mathbb{R}^m \to \overline{\mathbb{R}}$ taken as the indicator function $\delta_{\text{gph}\, S}$. Thus, the bivariate approach to function-valued mappings parallels the graphical approach to set-valued mappings, which certainly can be very attractive but isn't the only good path of analysis.

On the other hand, each function-valued mapping $u \mapsto f(\cdot,u)$ from $\mathbb{R}^m$ to fcns$(\mathbb{R}^n)$ can be identified with a certain set-valued mapping from $\mathbb{R}^m$ to sets$(\mathbb{R}^n \times \mathbb{R})$, namely its *epigraphical mapping* $u \mapsto \text{epi}\, f(\cdot,u)$. This opens up ways of treating parametric function dependence that are distinctly different from the classical ones and are tied to the notions of epi-convergence.

7.39 Definition (epi-continuity properties). *For $f : \mathbb{R}^n \times \mathbb{R}^m \to \overline{\mathbb{R}}$, the function-valued mapping $u \mapsto f(\cdot,u)$ is epi-continuous at $\bar{u}$ if*

$$f(\cdot,u) \xrightarrow{\text{e}} f(\cdot,\bar{u}) \text{ as } u \to \bar{u},$$

which means that the set-valued mapping $u \mapsto \text{epi}\, f(\cdot,u)$ is continuous at $\bar{u}$. More generally, it is epi-usc if the mapping $u \mapsto \text{epi}\, f(\cdot,u)$ is isc, whereas it is epi-lsc if this mapping is osc.

7.40 Exercise (epi-continuity versus bivariate continuity). *For $f : \mathbb{R}^n \times \mathbb{R}^m \to \overline{\mathbb{R}}$, the function-valued mapping $u \mapsto f(\cdot,u)$ is epi-usc at $\bar{u}$ if and only if, for every sequence $u^\nu \to \bar{u}$ and point $\bar{x} \in \mathbb{R}^n$,*

$$\limsup_{\nu\to\infty} f(x^\nu, u^\nu) \leq f(\bar{x}, \bar{u}) \text{ for some sequence } x^\nu \to \bar{x}.$$

On the other hand, it is epi-lsc at $\bar{u}$ *if and only if, for every sequence* $u^\nu \to \bar{u}$ *and point* $\bar{x} \in \mathbb{R}^n$,

$$\liminf_{\nu\to\infty} f(x^\nu, u^\nu) \geq f(\bar{x}, \bar{u}) \text{ for every sequence } x^\nu \to \bar{x}.$$

Thus, it is epi-lsc (everywhere) if and only if f *is lsc on* $\mathbb{R}^n \times \mathbb{R}^m$. *It is epi-usc if* f *is usc; but it can also be epi-usc without* f *necessarily being usc on* $\mathbb{R}^n \times \mathbb{R}^m$. *Likewise, the mapping* $u \mapsto f(\cdot, u)$ *is epi-continuous in particular if* f *is continuous, but this condition is only sufficient, not necessary.*

As a refinement of the terminology in Definition 7.39, it is sometimes useful to speak of a function-valued mapping $u \mapsto f(\cdot, u)$ as being epi-continuous (or epi-usc, or epi-lsc) at $\bar{u}$ *relative to a set* U containing $\bar{u}$. This is taken to mean that the mapping $u \mapsto \operatorname{epi} f(\cdot, u)$ is continuous (or isc, or osc) at $\bar{u}$ relative to U, in the sense that only sequences $u^\nu \to \bar{u}$ with $u^\nu \in U$ are considered in the version of 5(1) for this situation. The characterizations in 7.40 carry over to such restricted continuity similarly.

We now study the application of epi-continuity properties of $u \mapsto f(\cdot, u)$ to parametric minimization. Substantial results were achieved in Theorem 1.17 for the optimal-value function $p(u) = \inf_x f(x, u)$, and we are now prepared for a more thorough analysis of optimal-set mapping $P(u) = \operatorname{argmin}_x f(x, u)$. For this purpose, and also in the theory of subgradients in Chapter 8, it will be valuable to have the notion of sequences of points that not only converge but do so in a manner that pays attention to making the values of some particular function converge, even though the function might not be continuous.

With respect to any function $p : \mathbb{R}^m \to \overline{\mathbb{R}}$ (the focus in a moment will be on an optimal-value function p), we'll say that a sequence of points u^ν in $\mathbb{R}^m$ *converges in the p-attentive sense* to $\bar{u}$, written $u^\nu \underset{p}{\to} \bar{u}$, when not only $u^\nu \to \bar{u}$ but $p(u^\nu) \to p(u)$:

$$u \underset{p}{\to} \bar{u} \iff u \to \bar{u} \text{ with } p(u) \to p(\bar{u}).$$

Thus, p-attentive convergence is convergence relative to the topology on $\mathbb{R}^m$ that's induced by the topology on $\mathbb{R}^m \times \overline{\mathbb{R}}$ in associating with each point u the pair $\big(u, p(u)\big)$. Of course, it's the same as ordinary convergence of u to $\bar{u}$ wherever p is continuous.

The continuity concepts introduced for set-valued mappings extend to p-attentive convergence in obvious ways. In particular, a mapping $T : \mathbb{R}^m \rightrightarrows \mathbb{R}^n$ is *osc with respect to p-attentive convergence* when the conditions $u^\nu \underset{p}{\to} \bar{u}$ and $x^\nu \to \bar{x}$ with $x^\nu \in T(u^\nu)$ imply $\bar{x} \in T(\bar{u})$.

7.41 Theorem (optimal-set mappings). *For* $f : \mathbb{R}^n \times \mathbb{R}^m \to \overline{\mathbb{R}}$ *proper, lsc, and such that* $f(x, u)$ *is level-bounded in* x *locally uniformly in* u, *let*

$$p(u) := \inf_x f(x, u), \qquad P(u) := \operatorname{argmin}_x f(x, u).$$

(a) *The mapping* P, *which is compact-valued with* $\operatorname{dom} P = \operatorname{dom} p$, *is osc with respect to p-attentive convergence* $\underset{p}{\to}$. *Furthermore, it is locally bounded relative to any set in* $\mathbb{R}^m$ *on which* p *is bounded from above.*

(b) *For* P *to be locally bounded and osc (not just in the p-attentive sense) relative to a set* $U \subset \mathbb{R}^m$ *at a point* $\bar{u} \in U$ *with* $P(\bar{u}) \neq \emptyset$, *it suffices to have* p *be continuous relative to* U *at* $\bar{u}$. *This is true not only under the condition in Theorem 1.17(c) but also when the function-valued mapping* $u \mapsto f(\cdot, u)$ *is epi-continuous at* $\bar{u}$ *relative to* U. *In particular it holds for* $U = \operatorname{int}(\operatorname{dom} P) = \operatorname{int}(\operatorname{dom} p)$ *when* f *is convex.*

Proof. For $u^\nu \underset{p}{\to} u$ and $x^\nu \to x$ with $x^\nu \in P(u^\nu) = \operatorname{argmin}_x f(\cdot, u^\nu)$, one has

$$f(x,u) \leq \operatorname{liminf}_\nu f(x^\nu, u^\nu) \leq \operatorname{limsup}_\nu \big[\inf f(\cdot, u^\nu)\big] = \inf f(\cdot, u),$$

where the first inequality comes from the lower semicontinuity of f, and the last equality from the p-attentive convergence, i.e., $x \in P(u)$. Thus, P is osc with respect to p-attentive convergence.

If $p(u) \leq \alpha$, then $P(u) \subset \operatorname{lev}_{\leq \alpha} f(\cdot, u)$. From this it follows that if $p(u) \leq \alpha$ for all u in a set U, then P is locally bounded on U, since for all $\alpha \in \mathbb{R}$, the mappings $u \mapsto \operatorname{lev}_{\leq \alpha} f(\cdot, u)$ are locally bounded by assumption. This establishes (a) and also the sufficient condition in the first assertion in (b), inasmuch as p-attentive convergence $u^\nu \underset{p}{\to} \bar{u}$ is equivalent to ordinary convergence $u^\nu \to \bar{u}$ when p is continuous at $\bar{u}$ relative to a set containing the sequence in question.

The epi-continuity condition in (b) is sufficient by Theorem 7.33. It holds when f is convex, because the set-valued mapping $u \mapsto \operatorname{epi} f(\cdot, u)$ is graph-convex, hence isc on the interior of its effective domain by Proposition 5.9(b). A more direct argument is that the convexity of f implies that of p by 2.22(a), and then p is continuous on $\operatorname{int}(\operatorname{dom} p)$ by Theorem 2.35. □

The sufficient condition in Theorem 1.17(c) for p to be continuous at $\bar{u}$ relative to U is independent of the epi-continuity condition in Theorem 7.41(b); in general, neither subsumes the other.

7.42 Corollary (fixed constraints). *Let* $P(u) = \operatorname{argmin}_{x \in X} f_0(x,u)$ *for* $u \in U$, *where the sets* $X \subset \mathbb{R}^n$ *and* $U \subset \mathbb{R}^m$ *are closed and the function* $f_0 : X \times U \to \mathbb{R}$ *is continuous. If for each bounded set* $B \subset U$ *and value* $\alpha \in \mathbb{R}$ *the set* $\{(x,u) \in X \times B \mid f_0(x,u) \leq \alpha\}$ *is bounded, then the mapping* $P : U \rightrightarrows X$ *is osc and locally bounded relative to* U.

Proof. This corresponds in the theorem to taking $f(x,u)$ to be $f_0(x,u)$ when $(x,u) \in X \times U$ but ∞ otherwise. The function p is continuous relative to U by the criterion in Theorem 1.17(c). □

7.43 Corollary (solution mappings in convex minimization). *Suppose* $P(u) = \operatorname{argmin}_x f(x,u)$ *with* $f : \mathbb{R}^n \times \mathbb{R}^m \to \overline{\mathbb{R}}$ *proper, lsc, convex, and such that* $f^\infty(x,0) > 0$ *for all* $x \neq 0$. *If* $f(x,u)$ *is strictly convex in* x, *then* P *is single-valued on* $\operatorname{dom} P$ *and continuous on* $\operatorname{int}(\operatorname{dom} P)$.

Proof. Because f is convex, the level-boundedness condition in the theorem is equivalent to the condition here on f^∞, cf. 3.27. The continuity of P, when single-valued, comes from 5.20. □

7.44 Example (projections and proximal mappings).

(a) *For any nonempty, closed set C in $\mathbb{R}^n$ the projection mapping P_C is osc and locally bounded.*

(b) *For any proper, lsc function f on $\mathbb{R}^n$ that is prox-bounded with threshold λ_f, and any $\lambda \in (0, \lambda_f)$, the associated proximal mapping $P_\lambda f$ is osc and locally bounded.*

Detail. These statements merely place some of the facts already developed in 1.20 and 1.25 into the context of Theorem 7.41 and the terminology of set-valued mappings in Chapter 5. □

7.45 Exercise (local minimization). *Let $f : \mathbb{R}^n \to \overline{\mathbb{R}}$ be proper and lsc, and for $x \in \mathbb{R}^n$ and $\varepsilon \geq 0$ define*

$$p(x, \varepsilon) = \inf_{y \in \boldsymbol{B}(x,\varepsilon)} f(y), \qquad P(x, \varepsilon) = \operatorname*{argmin}_{y \in \boldsymbol{B}(x,\varepsilon)} f(y).$$

Then p is lsc and proper, while P is closed-valued and locally bounded, and the common effective domain of p and P is

$$D := \left\{ (x, \varepsilon) \in \mathbb{R}^n \times \mathbb{R} \,\middle|\, \varepsilon \geq 0,\ (\operatorname{dom} f) \cap \mathbb{B}(x, \varepsilon) \neq \emptyset \right\}.$$

If f is convex, p is continuous on $\operatorname{int} D$ and P is osc on $\operatorname{int} D$; moreover

$$\operatorname{int} D = \left\{ (x, \varepsilon) \in \mathbb{R}^n \times \mathbb{R} \,\middle|\, \varepsilon > 0,\ (\operatorname{dom} f) \cap \operatorname{int} \mathbb{B}(x, \varepsilon) \neq \emptyset \right\}.$$

If f is continuous everywhere, then (regardless of whether f is convex or not) p is continuous relative to D and P is osc relative to D; if in addition f is strictly convex, P is actually single-valued and continuous relative to D.

Guide. Apply Theorem 7.41 to the function g on $\mathbb{R}^n \times \mathbb{R}^n \times \mathbb{R}$ defined by $g(y, x, \varepsilon) = f(y) + \delta\big(y \,|\, \mathbb{B}(x, \varepsilon)\big)$ when $\varepsilon \geq 0$, but $g(y, x, \varepsilon) = \infty$ otherwise. When f is convex, obtain the convexity of p from 2.22(a). Note that the set claimed to be $\operatorname{int} D$ is open and its closure includes D, and then invoke the properties of interiors of convex sets in 2.33. In the case where f is continuous everywhere, verify that the epi-continuity condition in Theorem 7.41(b) is fulfilled. Obtain the consequences of strict convexity from 2.6 and 5.20. □

The fact that lower semicontinuity of p is an inadequate basis, in itself, for concluding the outer semicontinuity of P in Theorem 7.41(b) is demonstrated by the following example, which is also an eye-opener in other respects. For $u \in \mathbb{R}^1$ let $f(\cdot, u) = f_0 + \delta_{T(u)}$ on $\mathbb{R}^2$ with $f_0(x_1, x_2) := -x_1$, where in terms of $\varphi(t) := 2t/(1 + t^2)$ one has

$$T(u) := \left\{ (x_1, x_2) \,\middle|\, x_1 \leq 3,\ \varphi(x_1) + x_2 \leq u,\ \varphi(x_1) - x_2 \leq u \right\}.$$

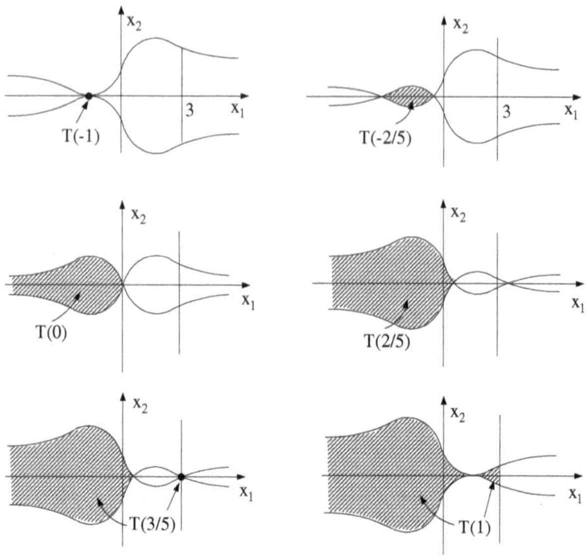

Fig. 7–9. A parameterized feasible set.

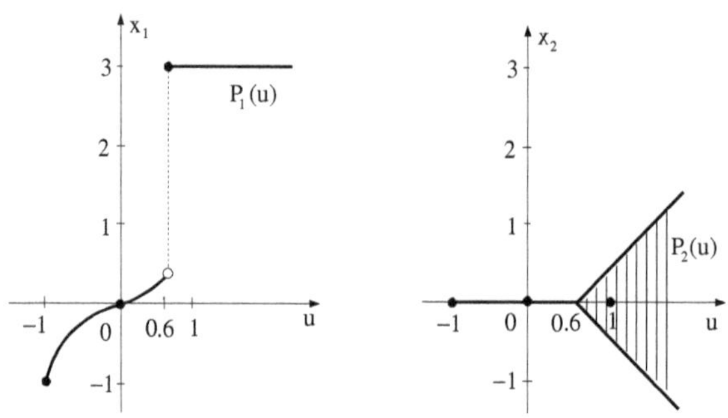

Fig. 7–10. Behavior of an optimal-set mapping $P = P_1 \times P_2$.

The nature of $T(u)$ is indicated in Figure 7–9 for representative choices of u.

Figure 7–10 shows the graph of the associated optimal-set mapping $P : \mathbb{R}^2 \rightrightarrows \mathbb{R}^2$; it happens to have the form $P(u) = P_1(u) \times P_2(u)$ for mappings P_1 and P_2 into $\mathbb{R}$, and the graphs of P_1 and P_2 suffice for a full description. The function f is proper and lsc, and $f(x,u)$ is level-bounded in x locally uniformly in u. For $u < -1$ there are no feasible solutions to the problem of minimizing $f(x,u)$ in x and therefore no optimal solutions; $P(u) = \emptyset$. For $-1 \leq u < .6$ the problem has the unique optimal solution $(x_1, x_2) = (\theta(u), 0)$, where $\theta(u) = u/(1+\sqrt{1-u^2})$; the optimal value is then $p(u) = -\theta(u)$. For $u \geq .6$, however, the optimal solution set is the nonempty line segment consisting of all the points (x_1, x_2) having $x_1 = 3$ and $|x_2| \leq u - .6$; then $p(u) = -3$. Although the function p is discontinuous at $u = .6$, it's lsc. Nonetheless, the set-valued mapping P fails to be osc at $u = .6$ relative to the set $\operatorname{dom} p = [-1, \infty)$. It may be observed that the function-valued mapping $u \mapsto f(u, \cdot)$ fails to be epi-continuous at $u = .6$, as must be the case because otherwise p would be continuous there by 7.41(b).

G.* Continuity of Operations

In practice, in order to apply basic criteria as in Theorem 7.41 to optimization problems represented in terms of minimizing a single extended-real-valued function on $I\!R^n$, it's crucial to have available some results on how epi-convergence carries through the constructions used in setting up such a representation. Some elementary operations have already been treated in 7.8, but we now investigate the matter more closely.

Epi-convergence isn't always preserved under addition of functions, but applications where the issue comes up tend to fall into one of the categories covered by the next couple of results.

7.46 Theorem (epi-convergence under addition). *For sequences of functions f_1^ν and f_2^ν on $I\!R^n$ one has*

$$\text{e-lim inf}_\nu f_1^\nu + \text{e-lim inf}_\nu f_2^\nu \leq \text{e-lim inf}_\nu (f_1^\nu + f_2^\nu).$$

When $f_1^\nu \xrightarrow{e} f_1$ and $f_2^\nu \xrightarrow{e} f_2$ with f_1 and f_2 proper, either one of the following conditions is sufficient to ensure that $f_1^\nu + f_2^\nu \xrightarrow{e} f_1 + f_2$:

(a) *$f_1^\nu \xrightarrow{p} f_1$ and $f_2^\nu \xrightarrow{p} f_2$;*

(b) *one of the two sequences converges continuously.*

Proof. For any x there exists by 7(3) a sequence $x^\nu \to x$ with $(f_1^\nu + f_2^\nu)(x^\nu) \to [\text{e-lim inf}_\nu (f_1^\nu + f_2^\nu)](x)$. For this sequence we have

$$\begin{aligned} \lim_\nu (f_1^\nu + f_2^\nu)(x^\nu) &\geq \liminf_\nu f_1^\nu(x^\nu) + \liminf_\nu f_2^\nu(x^\nu) \\ &\geq \text{e-lim inf}_\nu f_1^\nu(x) + \text{e-lim inf}_\nu f_2^\nu(x). \end{aligned}$$

This proves the first claim in the theorem and shows that in order to obtain epi-convergence of $f_1^\nu + f_2^\nu$ in a case where $f_1^\nu \xrightarrow{e} f_1$ and $f_2^\nu \xrightarrow{e} f_2$ it suffices to establish for each x the existence of a sequence $x^\nu \to x$ with the property that $\limsup_\nu (f_1^\nu + f_2^\nu)(x^\nu) \leq (\text{e-lim sup}_\nu f_1^\nu + \text{e-lim sup}_\nu f_2^\nu)(x)$.

Under the extra assumption in (a) of pointwise convergence, such a sequence is obtained by setting $x^\nu \equiv x$. Under the extra assumption in (b) instead, say with f_1^ν converging continuously to f_1, one can choose x^ν such that $\limsup_\nu f_2^\nu(x^\nu) \leq \text{e-lim sup}_\nu f_2^\nu(x)$. At least one such sequence exists by Proposition 7.2. □

Note that criterion (b) of Theorem 7.46 covers the case of $f^\nu + g \xrightarrow{e} f + g$ encountered earlier in 7.8(a). As an illustration of the requirements in Theorem 7.46 more generally, the functions $f_1^\nu(x) = \min\{0, |\nu x - 1| - 1\}$ and $f_2^\nu(x) =$

$f_1^\nu(-x)$ both epi-converge to the function f defined by $f(x) = 0$ when $x \neq 0$ and $f(0) = -1$. Their sum epi-converges to f, not to $f + f$; conditions (a) and (b) of the theorem fail to be satisfied.

A sequence of convex functions that epi-converges to f converges continuously to f except possibly on the boundary of $\operatorname{dom} f$; cf. 7.17(c). It won't be surprising therefore that for convex functions $f_1^\nu \xrightarrow{e} f_1$ and $f_2^\nu \xrightarrow{e} f_2$, epi-convergence is preserved under addition except possibly on the boundary of $\operatorname{dom}(f_1 + f_2)$. We begin with a broader statement for functions of the type: $g \circ L + h$ where L is a linear mapping.

7.47 Exercise (epi-convergence of sums of convex functions). *Suppose $f^\nu(x) = g^\nu\big(L^\nu(x)\big) + h^\nu(x)$ and $f(x) = g\big(L(x)\big) + h(x)$, where g^ν and g are convex functions on $I\!R^m$, h^ν and h are convex functions on $I\!R^n$, and L^ν and L are linear mappings from $I\!R^n$ to $I\!R^m$ such that*

$$\operatorname{e-lim\,sup}_\nu g^\nu \leq g, \qquad \operatorname{e-lim\,sup}_\nu h^\nu \leq h, \qquad L^\nu \to L.$$

If the convex sets $\operatorname{dom} g$ and $L(\operatorname{dom} h)$ cannot be separated, $\operatorname{e-lim\,sup}_\nu f^\nu \leq f$; if also $g^\nu \xrightarrow{e} g$ and $h^\nu \xrightarrow{e} h$, then $f^\nu \xrightarrow{e} f$. As special cases one has:

(a) $g^\nu \circ L^\nu \xrightarrow{e} g \circ L$ when $g^\nu \xrightarrow{e} g$, provided that $0 \in \operatorname{int}(\operatorname{dom} g - \operatorname{rge} L)$, or equivalently, that $\operatorname{dom} g$ and $\operatorname{rge} L$ cannot be separated.

(b) $g^\nu + h^\nu \xrightarrow{e} g + h$ when $m = n$, $g^\nu \xrightarrow{e} g$ and $h^\nu \xrightarrow{e} h$, provided that $0 \in \operatorname{int}(\operatorname{dom} g - \operatorname{dom} h)$, or equivalently, that $\operatorname{dom} g$ and $\operatorname{dom} h$ cannot be separated.

Guide. This is essentially an adaptation of Theorem 4.32 to epigraphs. Let

$$\begin{aligned} M^\nu(x,\alpha) &:= (x, L^\nu(x), \alpha) \text{ for } (x,\alpha) \in I\!R^n \times I\!R, \\ E^\nu &:= \big\{(x,u,\alpha) \,\big|\, \alpha \geq h^\nu(x) + g^\nu(u)\big\}, \\ C^\nu &:= \big\{(x,\alpha) \,\big|\, M^\nu(x,\alpha) \in E^\nu\big\} = \operatorname{epi} f^\nu, \end{aligned}$$

and define E, M and C similarly. To apply 4.32, one needs to show that $\liminf_\nu E^\nu \supset E$ and that the convex sets E and $\operatorname{rge} M$ can't be separated, the condition for that being $(0,0,0) \in \operatorname{int}\big[\operatorname{gph} L - \big\{(x,u,\alpha) \,\big|\, \alpha \geq h(x) + g(u)\big\}\big]$, cf. 2.39. Identify this with having $(0,0) \in \operatorname{int}\big[\operatorname{gph} L - (\operatorname{dom} h \times \operatorname{dom} g)\big]$ and reduce the latter to saying that the sets $\operatorname{gph} L$ and $\operatorname{dom} h \times \operatorname{dom} g$ can't be separated. Finally, verify that this holds if and only if $\operatorname{dom} g$ and $L(\operatorname{dom} h)$ can't be separated. □

From the definition of epi-limits (in 7.1) and the corresponding properties for set convergence, in particular the formulas in 4.31 and 4(7), one has for any sequence of functions f^ν and g^ν that

$$f^\nu \leq g^\nu \quad \Longrightarrow \quad \begin{cases} \operatorname{e-lim\,inf}_\nu f^\nu \leq \operatorname{e-lim\,inf}_\nu g^\nu, \\ \operatorname{e-lim\,sup}_\nu f^\nu \leq \operatorname{e-lim\,sup}_\nu g^\nu, \end{cases}$$

$$\begin{aligned} \text{e-lim inf}_\nu \min\{f_1^\nu, f_2^\nu\} &= \min\{\text{e-lim inf}_\nu f_1^\nu, \text{ e-lim inf}_\nu f_2^\nu\}, \\ \text{e-lim sup}_\nu \min\{f_1^\nu, f_2^\nu\} &= \min\{\text{e-lim sup}_\nu f_1^\nu, \text{ e-lim sup}_\nu f_2^\nu\}, \\ \text{e-lim inf}_\nu \max\{f_1^\nu, f_2^\nu\} &\geq \max\{\text{e-lim inf}_\nu f_1^\nu, \text{ e-lim inf}_\nu f_2^\nu\}. \end{aligned}$$

No inequality holds for $\text{e-lim sup}_\nu \max\{f_1, f_2\}$ in general, but for the case of convex functions the following is available.

7.48 Proposition (epi-convergence of max functions). *Let f_1^ν, f_2^ν, f_1, and f_2 be convex with $f_1 \geq \text{e-lim sup}_\nu f_1^\nu$ and $f_2 \geq \text{e-lim sup}_\nu f_2^\nu$. Suppose $\operatorname{dom} f_1$ and $\operatorname{dom} f_2$ cannot be separated (or equivalently, $0 \in \operatorname{int}(\operatorname{dom} f_1 - \operatorname{dom} f_2)$). Then*

$$\text{e-lim sup}_\nu \max\{f_1^\nu, f_2^\nu\} \leq \max\{f_1, f_2\},$$

so if actually $f_1^\nu \xrightarrow{e} f_1$ and $f_2^\nu \xrightarrow{e} f_2$, one has $\max\{f_1^\nu, f_2^\nu\} \xrightarrow{e} \max\{f_1, f_2\}$.

Proof. This follows directly from the definition of the upper epi-limit (in 7.1), Proposition 4.32, and the fact that the epigraphs of f_1 and f_2 can be separated if and only if $\operatorname{dom} f_1$ and $\operatorname{dom} f_2$ can be separated. □

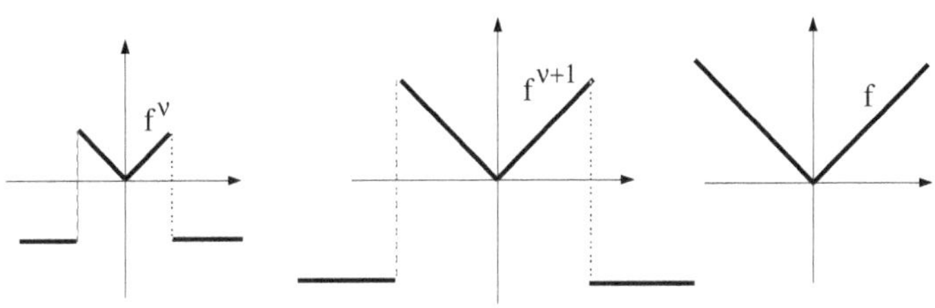

Fig. 7–11. A level-bounded epi-limit of functions that aren't level-bounded.

To what extent can it be said that the 'inf' functional $f \mapsto \inf f$ is continuous on the space $\operatorname{fcns}(\mathbb{R}^n)$, supplied with the topology of epi-convergence? Theorem 7.33 may be interpreted through 7.32(a) as providing this property relative to subsets of the form $\{f \text{ lsc} \mid f \geq g\}$ with g level-bounded. It also says that the set-valued mapping $f \mapsto \operatorname{argmin} f$ from $\operatorname{fcns}(\mathbb{R}^n)$ to $\mathbb{R}^n$ is osc relative to such a subset.

Subsets of the form $\{f \text{ lsc} \mid f \geq g\}$ always have empty interior with respect to epi-convergence: given any lsc, level-bounded function f, we can define a sequence $f^\nu \xrightarrow{e} f$ consisting of lsc functions that aren't level-bounded. This is indicated in Figure 7–11, where $f^\nu(x) = f(x)$ when $|x| < \nu$ but $f^\nu(x^\nu) = -\nu$ when $|x| \geq \nu$. It isn't possible, therefore, to assert on the basis of 7.33 the epi-continuity of the 'inf' functional at any f with $\inf f > -\infty$ when unrestricted neighborhoods of f are admitted.

The question arises as to whether there is a type of convergence stronger than epi-convergence with respect to which the 'inf' functional does exhibit continuity at various elements f. To achieve a satisfying answer to this question, as well as help elucidate issues related to the epi-continuity of various operations, we draw on the notions of cosmic convergence. A better understanding of conditions ensuring the level-boundedness needed in 7.33 will be a useful by-product.

H*. Total Epi-Convergence

Epi-convergence of a sequence of functions f^ν on $\mathbb{R}^n$ corresponds by definition to ordinary set convergence of associated sequence of epigraphs in $\mathbb{R}^{n+1}$. In parallel one can consider *cosmic* epi-convergence of f^ν to f, which is defined to mean that $\operatorname{epi} f^\nu \xrightarrow{c} \operatorname{csm} \operatorname{epi} f$ in the sense of the cosmic set convergence introduced in Chapter 4. Our main tool in working with this notion will be horizon limits of functions. It's evident from the definition of the inner and outer horizon limits of a sequence of sets that if the sets are epigraphs the limits are epigraphs as well.

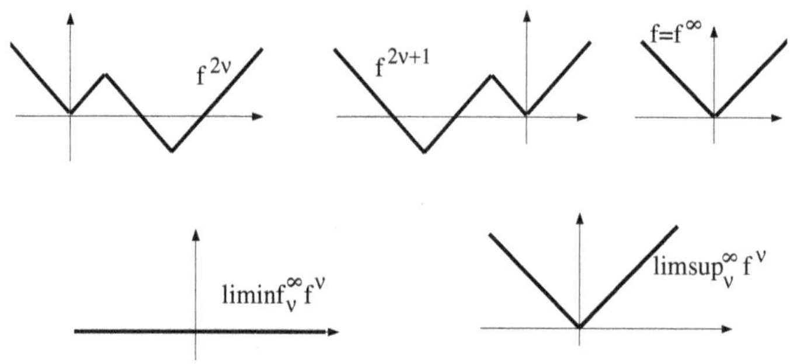

Fig. 7–12. Illustration of horizon limits.

7.49 Definition (horizon limits of functions). *For any sequence* $\{f^\nu neq\infty\}_{\nu\in\mathbb{N}}$, *the lower horizon epi-limit, denoted by* $\text{e-}\liminf_\nu^\infty f^\nu$, *is the function on* $\mathbb{R}^n$ *having as epigraph the cone* $\limsup_\nu^\infty (\operatorname{epi} f^\nu)$:

$$\operatorname{epi}(\text{e-}\liminf_\nu^\infty f^\nu) := \limsup_\nu^\infty (\operatorname{epi} f^\nu),$$

unless $f^\nu \equiv \inf$ *for* ν *sufficently large, then* $\text{e-}\liminf_\nu^\infty f^\nu = \delta_0$ *rather than just* $\{0\}$ *as would follow from the preceding formula. Similarly, the upper horizon epi-limit, denoted by* $\text{e-}\limsup_\nu^\infty f^\nu$, *is the function on* $\mathbb{R}^n$ *having as epigraph the cone* $\liminf_\nu^\infty (\operatorname{epi} f^\nu)$:

$$\operatorname{epi}(\text{e-}\limsup_\nu^\infty f^\nu) := \liminf_\nu^\infty (\operatorname{epi} f^\nu).$$

When these two functions coincide, the horizon limit function $\text{e-}\lim_\nu^\infty f^\nu$ *is said to exist:* $\text{e-}\lim_\nu^\infty f^\nu := \text{e-}\liminf_\nu^\infty f^\nu = \text{e-}\limsup_\nu^\infty f^\nu$.

The geometry in this definition can be translated into specific formulas resembling those in 7.2 and 7.3 in the case of plain epi-limits and proved in much the same way.

7.50 Exercise (horizon limit formulas). *For any sequence* $\{f^\nu\}_{\nu\in\mathbb{N}}$, *one has*

$$(\text{e-lim\,inf}_\nu^\infty f^\nu)(x) = \min\{\alpha \in \overline{\mathbb{R}} \mid \exists \lambda^\nu x^\nu \to x,\, \lambda^\nu \searrow 0,\, \liminf_\nu \lambda^\nu f^\nu(x^\nu) = \alpha\},$$
$$(\text{e-lim\,sup}_\nu^\infty f^\nu)(x) = \min\{\alpha \in \overline{\mathbb{R}} \mid \exists \lambda^\nu x^\nu \to x,\, \lambda^\nu \searrow 0,\, \limsup_\nu \lambda^\nu f^\nu(x^\nu) = \alpha\}.$$

Furthermore,

$$\begin{aligned}(\text{e-lim\,inf}_\nu^\infty f^\nu)(x) &= \sup_{\substack{V\in\mathcal{N}(x)\\ \delta>0}}\ \sup_{N\in\mathcal{N}_\infty}\ \inf_{\nu\in N}\ \inf_{\substack{x'\in V\\ 0<\lambda<\delta}} \lambda f^\nu(\lambda^{-1}x')\\ &= \lim_{\delta\searrow 0}\left[\liminf_{\nu\to\infty}\left[\inf_{\substack{x'\in\mathbb{B}(x,\delta)\\ 0<\lambda<\delta}} \lambda f^\nu(\lambda^{-1}x')\right]\right],\\ (\text{e-lim\,sup}_\nu^\infty f^\nu)(x) &= \sup_{\substack{V\in\mathcal{N}(x)\\ \delta>0}}\ \inf_{N\in\mathcal{N}_\infty}\ \sup_{\nu\in N}\ \inf_{\substack{x'\in V\\ 0<\lambda<\delta}} \lambda f^\nu(\lambda^{-1}x')\\ &= \lim_{\delta\searrow 0}\left[\limsup_{\nu\to\infty}\left[\inf_{\substack{x'\in\mathbb{B}(x,\delta)\\ 0<\lambda<\delta}} \lambda f^\nu(\lambda^{-1}x')\right]\right].\end{aligned}$$

Guide. This is parallel to 7.2 and 7.3. □

Rather than explore the full ramifications of cosmic epi-convergence using horizon epi-limits, we concentrate here on the counterpart to the version of cosmic set convergence that was defined in 4.23 as total set convergence.

7.51 Definition (total epi-convergence). *A sequence of functions* $f^\nu : \mathbb{R}^n \to \overline{\mathbb{R}}$ *totally epi-converges to a function* $f : \mathbb{R}^n \to \overline{\mathbb{R}}$, *written* $f^\nu \xrightarrow{t} f$, *if the epigraphs totally converge:* $\operatorname{epi} f^\nu \xrightarrow{t} \operatorname{epi} f$.

7.52 Proposition (properties of horizon limits). *For a sequence* $\{f^\nu\}_{\nu\in\mathbb{N}}$, *the functions* $\text{e-lim\,inf}_\nu^\infty f^\nu$ *and* $\text{e-lim\,sup}_\nu^\infty f^\nu$, *as well as* $\text{e-lim}_\nu^\infty f^\nu$ *when it exists, are lsc and positively homogeneous functions on* $\mathbb{R}^n$. *One has*

$$\text{e-lim\,inf}_\nu (f^\nu)^\infty \ \geq\ \text{e-lim\,inf}_\nu^\infty f^\nu, \qquad \text{e-lim\,sup}_\nu (f^\nu)^\infty \ \geq\ \text{e-lim\,sup}_\nu^\infty f^\nu,$$

where the function $\text{e-lim\,sup}_\nu^\infty f^\nu$ *is sublinear when the functions* f^ν *are convex. Furthermore, total epi-convergence is characterized by*

$$\begin{aligned} f^\nu \xrightarrow{t} f \quad &\Longleftrightarrow\quad f^\nu \xrightarrow{e} f,\quad \text{e-lim}_\nu^\infty f^\nu = f^\infty\\ &\Longleftrightarrow\quad f^\nu \xrightarrow{e} f,\quad \text{e-lim\,inf}_\nu^\infty f^\nu \geq f^\infty.\end{aligned}$$

Proof. This just translates the horizon limit properties in 4.24 to the epigraphical context of 7.49. □

A one-dimensional example displayed in Figure 7–12 furnishes some insight into horizon limits. For ν even, we take $f^\nu(x) = \min\{\,|x|, |x-\nu|-1\,\}$, while for ν odd, we take $f^\nu(x) = \min\{\,|x|, |x+\nu|-1\,\}$. This sequence of functions epi-converges to the absolute value function, $f(x) = |x|$. The lower horizon limit is given by $\big(\liminf_\nu^\infty f^\nu\big)(x) \equiv 0$, as is obvious from the fact that the outer horizon limit of the epigraphs is the entire upper half-plane. The upper horizon limit, however, is given by $\big(\limsup_\nu^\infty f^\nu\big)(x) = |x|$. The oscillation between even and odd ν destroys the possibility of anything greater. The horizon limit itself doesn't exist in this case.

Observe, by the way, that this is another example of misbehavior with respect to minimization. The individual functions f^ν are level-bounded, as is the epi-limit f, but $\inf f^\nu \equiv -1$ compared to $\inf f = 0$, and $\operatorname{argmin} f^\nu$ is either $\{\nu\}$ (for ν odd) or $\{-\nu\}$ (for ν even), compared to $\operatorname{argmin} f = \{0\}$. Of course, the *sequence* isn't eventually level-bounded, or it would come under the sway of Theorem 7.33, which forbids this from happening.

Interest in the possible merits of total epi-convergence is fueled by the fact that in many of the situations frequently encountered, this type of convergence is sure to be present without the need for any assumptions beyond those guaranteeing ordinary epi-convergence.

7.53 Theorem (automatic cases of total epi-convergence). *In the cases that follow, ordinary epi-convergence $f^\nu \xrightarrow{e} f$ for functions $f^\nu \not\equiv \infty$ and $f \not\equiv \infty$ always entails the stronger property of total epi-convergence $f^\nu \xrightarrow{t} f$:*

(a) *the functions f^ν are convex;*

(b) *the functions f^ν are positively homogeneous;*

(c) *the sequence is nonincreasing ($f^{\nu+1} \leq f^\nu$).*

(d) *the sequence is equi-coercive, i.e., there is a coercive function g such that for all ν one has $f^\nu \geq g$.*

Proof. Except for (d), these criteria arise from applying the corresponding result for sets (in 4.25) to epigraphs. In the case of (d), we have $\text{e-lim}\inf_\nu^\infty f^\nu \geq g^\infty$, where g^∞ is the indicator $\delta_{\{0\}}$ (cf. 3.25). On the other hand, $\text{e-lim}\sup_\nu^\infty f^\nu \leq \text{e-lim}\inf_\nu (f^\nu)^\infty$ by 4.25(d), where $(f^\nu)^\infty \leq \delta_{\{0\}}$ for all ν. Therefore $\text{e-lim}_\nu^\infty f^\nu = \delta_{\{0\}}$. If $f^\nu \xrightarrow{e} f$, we necessarily have $f \geq \operatorname{cl} g$ and therefore also f coercive, $f^\infty = \delta_{\{0\}}$. If follows then that $f^\nu \xrightarrow{t} f$. □

The content of Theorem 7.53 is especially significant for the approximation of *convex* types of optimization problems. For such problems, approximation in the sense of epi-convergence *automatically* entails the stronger property of approximation in the sense of total epi-convergence.

7.54 Theorem (horizon criterion for eventual level-boundedness). *A sequence of proper functions $f^\nu : \mathbb{R}^n \to \overline{\mathbb{R}}$ is eventually level-bounded if*

$$\big(\text{e-liminf}_\nu^\infty f^\nu\big)(x) > 0 \quad \text{for all } x \neq 0.$$

Thus, it is eventually level-bounded if $f^\nu \xrightarrow{t} f$ with f proper and level-coercive.

Proof. Let $h = \text{e-liminf}_\nu^\infty f^\nu$, and suppose that $h(x) > 0$ for all $x \neq 0$. Consider for each $N \in \mathcal{N}_\infty$ the function $g_N(x) := \inf_{\nu \in N} f^\nu(x)$, which has the property that $f^\nu \geq g_N$ for all $\nu \in N$. We'll show that one of these functions g_N is level-bounded, from which it will obviously follow that the sequence $\{f^\nu\}_{\nu \in N}$ is level-bounded.

An unbounded level set of g_N would have a direction point $\operatorname{dir} x$ in its cosmic closure, and the horizon of $\operatorname{epi} g_N$ would then contain a point $\operatorname{dir}(x,0)$. The horizon of $\operatorname{epi} g_N$ is the horizon of the set $E_N := \bigcup_{\nu\in N} \operatorname{epi} f^\nu$ and is a compact subset of $\operatorname{hzn} I\!R^{n+1}$, as is the set of all direction points of the form $\operatorname{dir}(x,0)$ in $\operatorname{hzn} I\!R^{n+1}$. Therefore, if none of the functions g_N were level-bounded there would exist $x \neq 0$ such that $\operatorname{dir}(x,0) \in \bigcap_{N\in\mathcal{N}_\infty} \operatorname{hzn} E_N$. This set is the horizon part of the cosmic limit of $\operatorname{epi} f^\nu$ as $\nu \to \infty$ and is represented by the cone $\operatorname{epi} h$. Thus, there would exist $x \neq 0$ such that $h(x) \leq 0$, which has been excluded.

When $f^\nu \xrightarrow{t} f$, the function h coincides with f^∞. As long as f is proper, the condition that $h(x) > 0$ when $x \neq 0$ reduces in that case to the level coercivity of f; cf. 3.26(a). □

The case of Theorem 7.54 where the functions are convex corresponds to the criterion for eventual level-boundedness already established in 7.32(c).

7.55 Corollary (continuity of the minimization operation). *On the function space* lsc-fcns$(I\!R^n)$ *equipped with the topology of total epi-convergence, the functional* $f \mapsto \inf f$ *is continuous at any element* f *that is proper and level-coercive, and the mapping* $f \mapsto \operatorname{argmin} f$ *is osc at any such element* f.

Indeed, the lsc, proper, level-coercive functions on $I\!R^n$ *form an open subset of* lsc-fcns$(I\!R^n)$ *with respect to total epi-convergence.*

Proof. This combines Theorem 7.54 with Theorem 7.33. □

Total epi-convergence also enters the picture in handling epi-sums, where it's accompanied by a counterpart for functions of the condition used in 4.29 in obtaining the convergence of sums of sets.

7.56 Proposition (convergence of epi-sums). *For functions* $f_i^\nu : I\!R^n \to \overline{I\!R}$, *one has* $\operatorname{e-lim\,sup}_\nu (f_1^\nu \# \cdots \# f_m^\nu) \leq \operatorname{e-lim\,sup}_\nu f_1^\nu \# \cdots \# \operatorname{e-lim\,sup}_\nu f_m^\nu$. *Also,*

(a) $f_1^\nu \# f_2^\nu \xrightarrow{e} f_1 \# f_2$ *when* $f_1^\nu \xrightarrow{t} f_1$ *and* $f_2^\nu \xrightarrow{t} f_2$, *as long as*

$$f_1^\infty(x) + f_2^\infty(-x) > 0 \ \text{ for all } x \neq 0;$$

(b) $f_1^\nu \# \cdots \# f_m^\nu \xrightarrow{t} f_1 \# \cdots \# f_m$ *when* $f_i^\nu \xrightarrow{t} f_i$, *the condition holds that*

$$\sum\nolimits_{i=1}^m f_i^\infty(x_i) \leq 0 \implies x_i = 0 \ \text{ for all } i,$$

and, in addition, $(\operatorname{epi} f_1 \times \cdots \times \operatorname{epi} f_m)^\infty = \operatorname{epi} f_1^\infty \times \cdots \times \operatorname{epi} f_m^\infty$. *The latter is fulfilled in particular when the functions* f_i *are proper and convex.*

Proof. We apply 4.29 to epigraphs after observing that the horizon condition in (b) is satisfied if and only if there's no choice of $(x_i, \alpha_i) \in \operatorname{epi} f_i^\infty$ with $(x_1,\alpha_1) + \cdots + (x_m,\alpha_m) = (0,0)$ other than $(x_i,\alpha_i) = (0,0)$ for all i. The assertion in (b) for the case of convex functions comes out of 3.11. □

7.57 Exercise (epi-convergence under inf-projection). *Consider lsc, proper, functions* $f, f^\nu : I\!R^n \times I\!R^m \to \overline{I\!R}$ *and let*

$$p(u) = \inf_x f(x,u), \qquad p^\nu(u) = \inf_x f^\nu(x,u).$$

Suppose that $f^\infty(x,0) > 0$ *for all* $x \neq 0$. *Then*

$$f^\nu \xrightarrow{\text{t}} f \implies p^\nu \xrightarrow{\text{t}} p.$$

Thus in particular, one has $p^\nu \xrightarrow{\text{e}} p$ *whenever* $f^\nu \xrightarrow{\text{e}} f$ *and the functions* f^ν *are convex or positively homogeneous, or the sequence* $\{f^\nu\}_{\nu\in\mathbb{N}}$ *is nondecreasing.*

Guide. Apply 4.27 to epigraphs in the context of 1.18. Invoke 7.53. (For a comparison, see 3.31.) □

I.* Epi-Distances

To quantify the rate at which a sequence of functions f^ν epi-converges to a function f, one needs to introduce suitable metrics or distance-like functions that are able to act in the role of metrics. The foundation has been laid in Chapter 4, where set convergence was quantified by means of the metric $d\!l$ along with the pseudo-metrics $d\!l_\rho$ and estimating expressions $\hat{d\!l}_\rho$; cf. 4(11) and 4(12). That pattern will be extended now to functions by way of their epigraphs.

Because epi-convergence doesn't distinguish between a function f and its lsc regularization $\operatorname{cl} f$, cf. 7.4(b), a full metric space interpretation of epi-convergence isn't possible without restricting to lsc functions. In the notation

$$\begin{aligned} \text{fcns}(\mathbb{R}^n) &:= \text{ the space of all } f : \mathbb{R}^n \to \overline{\mathbb{R}}, \\ \text{lsc-fcns}(\mathbb{R}^n) &:= \text{ the space of all lsc } f : \mathbb{R}^n \to \overline{\mathbb{R}}, \\ \text{lsc-fcns}_{\not\equiv\infty}(\mathbb{R}^n) &:= \text{ the space of all lsc } f : \mathbb{R}^n \to \overline{\mathbb{R}},\ f \not\equiv \infty, \end{aligned} \tag{7(25)}$$

our concern will be with lsc-fcns($\mathbb{R}^n$), or indeed lsc-fcns$_{\not\equiv\infty}(\mathbb{R}^n)$, rather than fcns($\mathbb{R}^n$). The exclusion of the constant function $f \equiv \infty$ is in line with our focus on cl-sets$_{\neq\emptyset}(\mathbb{R}^n)$ instead of cl-sets($\mathbb{R}^n$) in the metric theory of Chapter 4 and corresponds to the separate emphasis given in 7.5 and 7.6 to sequences of functions that epi-converge to the horizon.

The basic idea is to measure distances between functions on $\mathbb{R}^n$ in terms of distances between their epigraphs as subsets of $\mathbb{R}^{n+1}$. For any functions f and g on $\mathbb{R}^n$ that aren't identically ∞, we define

$$d\!l(f,g) := d\!l(\operatorname{epi} f, \operatorname{epi} g) \tag{7(26)}$$

and for $\rho \geq 0$ also

$$d\!l_\rho(f,g) := d\!l_\rho(\operatorname{epi} f, \operatorname{epi} g), \qquad \hat{d\!l}_\rho(f,g) := \hat{d\!l}_\rho(\operatorname{epi} f, \operatorname{epi} g). \tag{7(27)}$$

We call $d\!l(f,g)$ the *epi-distance* between f and g, and $d\!l_\rho(f,g)$ the *ρ-epi-distance*; the value $\hat{d\!l}_\rho(f,g)$ assists in estimating $d\!l_\rho(f,g)$.

7.58 Theorem (quantification of epi-convergence). *For each* $\rho \geq 0$, $d\!l_\rho$ *is a pseudo-metric on the space* lsc-fcns$_{\not\equiv\infty}(\mathbb{R}^n)$, *but* $\hat{d\!l}_\rho$ *is not. Both families* $\{d\!l_\rho\}_{\rho\geq 0}$ *and* $\{\hat{d\!l}_\rho\}_{\rho\geq 0}$ *characterize epi-convergence: for any* $\bar\rho \in \mathbb{R}_+$, *one has*

$$\begin{aligned} f^\nu \xrightarrow{e} f \quad &\Longleftrightarrow \quad d\!l_\rho(f^\nu, f) \to 0 \;\text{ for all }\; \rho \geq \bar\rho \\ &\Longleftrightarrow \quad \hat{d\!l}_\rho(f^\nu, f) \to 0 \;\text{ for all }\; \rho \geq \bar\rho. \end{aligned}$$

Further, $d\!l$ *is a metric on* lsc-fcns$_{\not\equiv\infty}(\mathbb{R}^n)$ *that characterizes epi-convergence:*

$$f^\nu \xrightarrow{e} f \quad \Longleftrightarrow \quad d\!l(f^\nu, f) \to 0.$$

Indeed, $\big(\text{lsc-fcns}_{\not\equiv\infty}(\mathbb{R}^n), d\!l\big)$ *is a complete metric space in which a sequence* $\{f^\nu\}_{\nu\in\mathbb{N}}$ *escapes to the horizon if and only if, for some* $f_0 \in$ lsc-fcns$_{\not\equiv\infty}(\mathbb{R}^n)$ *(and then for every* f_0 *in this space), one has* $d\!l(f^\nu, f_0) \to \infty$.

Moreover, this metric space has the property that, for every one of its elements f_0 *and every* $r > 0$, *the ball* $\big\{f \,\big|\, d\!l(f, f_0) \leq r\big\}$ *is compact.*

Proof. Epi-convergence is by definition the set convergence of the epigraphs, and since the epi-distance between functions in lsc-fcns$_{\not\equiv\infty}(\mathbb{R}^n)$ is just the set distances between these epigraphs, the assertions here are just translations of those of 4.36, 4.42 and 4.43. □

Total epi-convergence of functions can be characterized similarly by applying to epigraphs, as subsets of $\mathbb{R}^{n+1}$, the cosmic set metric in 4(15) for subsets of $\mathbb{R}^n$ (as subsets also of csm $\mathbb{R}^n$). We define the *cosmic epi-distance* between any two functions $f, g : \mathbb{R}^n \to \overline{\mathbb{R}}$ by

$$\begin{aligned} d\!l_{\text{csm}}(f, g) :&= d\!l_{\text{csm}}(\text{epi}\, f, \text{epi}\, g) \\ &= d\!l\big(\,\text{pos}(\text{epi}\, f, -1), \text{pos}(\text{epi}\, g, -1)\,\big), \end{aligned} \qquad 7(28)$$

where the second equation reflects the fact, extending 4(16), that the cosmic distance between epi f and epi g is the same as the set distance between the cones in $\mathbb{R}^{n+2}$ that correspond to these sets in the ray space model for csm $\mathbb{R}^{n+1}$.

7.59 Theorem (quantification of total epi-convergence). *On* lsc-fcns$_{\not\equiv\infty}(\mathbb{R}^n)$, *cosmic epi-distances* $d\!l_{\text{csm}}(f, g)$ *provide a metric that characterizes total epi-convergence:*

$$f^\nu \xrightarrow{t} f \quad \Longleftrightarrow \quad d\!l_{\text{csm}}(f^\nu, f) \to 0.$$

Proof. This comes immediately from 4.47 through definition 7(28). □

In working quantitatively with epi-convergence, the following facts can be brought to bear.

7.60 Exercise (epi-distance estimates). *For* $f_1, f_2 : \mathbb{R}^n \to \overline{\mathbb{R}}$ *not identically* ∞, *the following properties hold for* $d\!l(f_1, f_2)$, $d\!l_\rho(f_1, f_2)$ *and* $\hat{d\!l}_\rho(f_1, f_2)$ *in terms of* $d_1 = d_{\text{epi}\, f_1}(0, 0)$ *and* $d_2 = d_{\text{epi}\, f_2}(0, 0)$:

(a) $d\!l_\rho(f_1, f_2)$ *and* $\hat{d\!l}_\rho(f_1, f_2)$ *are nondecreasing functions of* ρ;

(b) $d\!l_\rho(f_1,f_2)$ *depends continuously on* ρ;
(c) $\hat{d\!l}_\rho(f_1,f_2) \le d\!l_\rho(f_1,f_2) \le \hat{d\!l}_{\rho'}(f_1,f_2)$ *when* $\rho' \ge 2\rho + \max\{d_1,d_2\}$;
(d) $d\!l_\rho(f_1,f_2) \le \max\{d_1,d_2\} + \rho$;
(e) $\big|d\!l_\rho(f_1,f_2) - d\!l_{\rho_0}(f_1,f_2)\big| \le 2|\rho-\rho_0|$ *for any* $\rho_0 \ge 0$;
(f) $d\!l(f_1,f_2) \ge (1-e^{-\rho})|d_1-d_2| + e^{-\rho} d\!l_\rho(f_1,f_2)$;
(g) $d\!l(f_1,f_2) \le (1-e^{-\rho}) d\!l_\rho(f_1,f_2) + e^{-\rho}\big(\max\{d_1,d_2\}+\rho+1\big)$;
(h) $|d_1-d_2| \le d\!l(f_1,f_2) \le \max\{d_1,d_2\}+1$.

When f_1 *and* f_2 *are convex,* 2ρ *can be replaced by* ρ *in (c). If in addition* $f_1(0)\le 0$ *and* $f_2(0) \le 0$, *then* ρ' *can be taken to be* ρ *in (c), so that*

$$\hat{d\!l}_\rho(f_1,f_2) = d\!l_\rho(f_1,f_2) \ \text{ for all } \ \rho \ge 0.$$

The latter holds even without convexity when f_1 *and* f_2 *are positively homogeneous, and then*

$$\hat{d\!l}_\rho(f_1,f_2) = \rho \hat{d\!l}_1(f_1,f_2), \qquad d\!l_\rho(f_1,f_2) = \rho d\!l_1(f_1,f_2).$$

Guide. Apply 4.37 and 4.41 to the epigraphs of f_1 and f_2. □

The use of these inequalities is facilitated by the following bounds in terms of an auxiliary value $\hat{d\!l}{}^+_\rho(f_1,f_2)$ that is sometimes easier to develop than $\hat{d\!l}_\rho(f_1,f_2)$ out of the properties of f_1 and f_2.

7.61 Proposition (auxiliary epi-distance estimate). *For lsc* $f_1, f_2 : I\!R^n \to \overline{I\!R}$ *not identically* ∞, *one has*

$$\hat{d\!l}{}^+_{\rho/\sqrt{2}}(f_1,f_2) \le \hat{d\!l}_\rho(f_1,f_2) \le \sqrt{2}\, \hat{d\!l}{}^+_\rho(f_1,f_2)$$

with $\hat{d\!l}{}^+_\rho(f_1,f_2)$ *defined as the infimum of all* $\eta \ge 0$ *such that*

$$\left.\begin{aligned} \min_{B(x,\eta)} f_2 &\le \max\{f_1(x),-\rho\}+\eta \\ \min_{B(x,\eta)} f_1 &\le \max\{f_2(x),-\rho\}+\eta \end{aligned}\right\} \ \text{for all } x \in \rho I\!B. \tag{7(29)}$$

Proof. Distinguish the Euclidean balls in $I\!R^n$ and $I\!R^{n+1}$ by $I\!B$ and $I\!B'$, observing that $I\!B' \subset I\!B \times [-1,1] \subset \sqrt{2}\, I\!B'$. By definition, $\hat{d\!l}_\rho(f_1,f_2)$ is the infimum of all $\eta \ge 0$ such that

$$\begin{aligned} (\operatorname{epi} f_1) \cap \rho I\!B' &\subset (\operatorname{epi} f_2) + \eta I\!B', \\ (\operatorname{epi} f_2) \cap \rho I\!B' &\subset (\operatorname{epi} f_1) + \eta I\!B'. \end{aligned}$$

Similarly let $\theta_\rho(f_1,f_2)$ denote the infimum of all $\eta \ge 0$ such that

$$\begin{aligned} (\operatorname{epi} f_1) \cap \rho(I\!B\times[-1,1]) &\subset (\operatorname{epi} f_2) + \eta(I\!B\times[-1,1]), \\ (\operatorname{epi} f_2) \cap \rho(I\!B\times[-1,1]) &\subset (\operatorname{epi} f_1) + \eta(I\!B\times[-1,1]). \end{aligned} \tag{7(30)}$$

The relations between $I\!B'$ and $I\!B\times[-1,1]$ yield

$$\theta_{\rho/\sqrt{2}}(f_1, f_2) \le \hat{d\!l}_\rho(f_1, f_2) \le \sqrt{2}\,\theta_\rho(f_1, f_2),$$

so we need only verify that $\theta_\rho(f_1, f_2) = \hat{d\!l}_\rho^+(f_1, f_2)$. Because of the epigraphs on both sides of the inclusions in 7(30), these inclusions aren't affected if the intervals $[-1, 1]$ are replaced in every case by $[-1, \infty)$. Hence the inclusions can be written equivalently as

$$\begin{aligned} \text{epi}\left(\max\{f_1, -\rho\} + \delta_{\rho\boldsymbol{B}}\right) &\subset (\text{epi}\, f_2) + \text{epi}(\delta_{\eta\boldsymbol{B}} - \eta), \\ \text{epi}\left(\max\{f_2, -\rho\} + \delta_{\rho\boldsymbol{B}}\right) &\subset (\text{epi}\, f_1) + \text{epi}(\delta_{\eta\boldsymbol{B}} - \eta). \end{aligned}$$

To identify this with 7(29), it remains only to observe through 1.28 that $(\text{epi}\, f_i) + \text{epi}(\delta_{\eta\boldsymbol{B}} - \eta)$ is $\text{epi}\, f_{i,\eta}$ for the function $f_{i,\eta} := f_i \,\#\, (\delta_{\eta\boldsymbol{B}} - \eta)$, which has $f_{i,\eta}(x) = \min_{\boldsymbol{B}(x,\eta)} f_i - \eta$. □

7.62 Example (functions satisfying a Lipschitz condition). *For $i = 1, 2$ suppose $f_i = f_{0i} + \delta_{C_i}$ for a nonempty, closed set $C_i \subset I\!R^n$ and a function $f_{0i} : I\!R^n \to I\!R$ having for each $\rho \in I\!R_+$ a constant $\kappa_i(\rho) \in I\!R_+$ such that*

$$|f_{0i}(x') - f_{0i}(x)| \le \kappa_i(\rho)|x' - x| \;\text{ when }\; x, x' \in \rho I\!B.$$

Then one has the estimate that

$$\hat{d\!l}_\rho^+(f_1, f_2) \le \max_{\rho\boldsymbol{B}} \left|f_{01} - f_{02}\right| + \left[1 + \max\{\kappa_1(\rho'), \kappa_2(\rho')\}\right] \hat{d\!l}_\rho(C_1, C_2)$$
$$\text{whenever } 0 < \rho < \rho' < \infty \text{ and } \hat{d\!l}_\rho(C_1, C_2) < \rho' - \rho.$$

Detail. We first investigate what it means to have $\min_{\boldsymbol{B}(x,\eta)} f_2 \le f_1(x) + \eta$. This inequality is trivial when $x \notin C_1$, since the right side is then ∞. Suppose $\eta \ge \eta'$ with $\hat{d\!l}_\rho(C_1, C_2) < \eta' \le \rho' - \rho$. Any $x \in C_1 \cap \rho I\!B$ belongs then to $C_2 + \eta' I\!B$. For such x we have $f_1(x) = f_{01}(x)$ but also

$$\min_{x' \in \boldsymbol{B}(x,\eta)} f_{02}(x') \le \min_{x' \in \boldsymbol{B}(x,\eta')} f_{02}(x') \le f_{02}(x) + \kappa_2(\rho')\eta',$$

and consequently $\min_{\boldsymbol{B}(x,\eta)} f_2 \le f_1(x) + \eta$ when η is also at least as large as $-f_{01}(x) + f_{02}(x) + \kappa_2(\rho')\eta'$. Obviously $f_1(x) \le \max\{f_1(x), -\rho\}$ always, and therefore the inequality $\min_{\boldsymbol{B}(x,\eta)} f_2 \le \max\{f_1(x), -\rho\} + \eta$ holds for all $x \in \rho I\!B$ when $\eta \ge \max_{\rho\boldsymbol{B}}\{f_{02} - f_{01}\} + \kappa_2(\rho')\eta' + \eta'$.

Similarly, the inequality $\min_{\boldsymbol{B}(x,\eta)} f_1 \le \max\{f_2(x), -\rho\} + \eta$ holds for all $x \in \rho I\!B$ if $\eta \ge \max_{\rho\boldsymbol{B}}\{f_{01} - f_{02}\} + \kappa_1(\rho')\eta' + \eta'$. Thus, $\hat{d\!l}_\rho^+(f_1, f_2) \le \eta$ when $\eta \ge \max_{\rho\boldsymbol{B}} |f_{01} - f_{02}| + \max\{\kappa_1(\rho'), \kappa_2(\rho')\}\eta' + \eta'$ with $\hat{d\!l}_\rho(C_1, C_2) < \eta' \le \rho' - \rho$. The conclusion is successfully reached now by letting both η and η' tend to their lower limits. □

As the special case of Example 7.62 in which $f_{01} \equiv f_{02} \equiv 0$, one sees that $\hat{d\!l}_\rho^+(\delta_{C_1}, \delta_{C_2}) \le \hat{d\!l}_\rho(C_1, C_2)$ for any $\rho > 0$. This can also be deduced directly from the formula for $\hat{d\!l}_\rho^+(\delta_{C_1}, \delta_{C_2})$, and indeed it holds always with equality.

J*. Solution Estimates

Consider two problems of optimization in $\mathbb{R}^n$, represented abstractly by two functions f and g; the second problem can be thought of as an approximation or perturbation of the first. The corresponding optimal values are $\inf f$ and $\inf g$ while the optimal solution sets are $\operatorname{argmin} f$ and $\operatorname{argmin} g$. Is it possible to give quantitative estimates of the closeness of $\inf g$ and $\operatorname{argmin} g$ to $\inf f$ and $\operatorname{argmin} f$ in terms of an epi-distance estimate of the closeness of g to f?

In principle that might be accomplished on the basis of 7.55 and 7.59 through estimates of $d\!l_{\rm csm}(f,g)$, but we'll focus here on the more accessible measure of closeness furnished by $\hat{d\!l}{}^+_\rho(f,g)$, as defined in 7.61. This requires us to work with a ball $\rho\mathbb{B}$ containing $\operatorname{argmin} f$ and to consider the minimization of g over $\rho\mathbb{B}$ instead of over all of $\mathbb{R}^n$.

To motivate the developments, we begin by formulating, in terms of the notion of a 'conditioning function' ψ for f, a result which will emerge as a consequence of the main theorem. This concerns the case where f has a unique minimizing point, considered to be known. The point can be identified then with the origin for the sake of simplifying the perturbation estimates.

7.63 Example (conditioning functions in solution stability). *Let $f:\mathbb{R}^n\to\overline{\mathbb{R}}$ be lsc and level-bounded with $f(0)=0=\min f$, and suppose f has a conditioning function ψ at this minimum in the sense that, for some $\varepsilon>0$, ψ is a continuous, increasing function on $[0,\varepsilon]$ with $\psi(0)=0$, such that*

$$f(x)\geq\psi(|x|)\ \text{ when }\ |x|<\varepsilon.$$

Define $\Psi(\eta):=\eta+\psi^{-1}(2\eta)$, so that Ψ likewise is a continuous, increasing function on $[0,\psi(\varepsilon)/2]$ with $\Psi(0)=0$. Fix any $\rho\in[\varepsilon,\infty)$. Then for functions $g:\mathbb{R}^n\to\overline{\mathbb{R}}$ that are lsc and proper,

$$\hat{d\!l}{}^+_\rho(f,g)<\min\{\rho,\psi(\varepsilon)/2\}\implies\begin{cases}|\min_{\rho B} g|\leq\hat{d\!l}{}^+_\rho(f,g),\\ |x|\leq\Psi\big(\hat{d\!l}{}^+_\rho(f,g)\big),\ \forall x\in\operatorname{argmin}_{\rho B} g.\end{cases}$$

Detail. This specializes Theorem 7.64 (below) to the case where $\min f=0$ and $\operatorname{argmin} f=\{0\}$. In the notation of that theorem, one takes $\bar\rho=0$; one has $\psi\leq\psi_f$ on $[0,\varepsilon)$, so $\psi_f^{-1}(2\eta)\leq\psi^{-1}(2\eta)$ when $2\eta<\psi(\varepsilon)$. Then $\Psi_f(\eta)\leq\Psi(\eta)$ as long as $\eta\leq\psi(\varepsilon)/2$. □

Figure 7–13 illustrates the notion of a conditioning function ψ for f in two common situations. When $\psi(\tau)=\gamma\tau$ for some $\gamma>0$, f must have a 'kink' at its minimum as in Figure 7–13(a), whereas if $\psi(\tau)=\gamma\tau^2$, f can be smooth there, as in Figure 7–13(b). These two cases will be considered further in 7.65. In general, of course, the higher the conditioning function ψ the lower (and therefore better) the associated function Ψ giving the perturbation estimate.

The main theorem on this subject, which follows, removes the special assumptions about $\min f$ and $\operatorname{argmin} f$ and works with an intrinsic expression

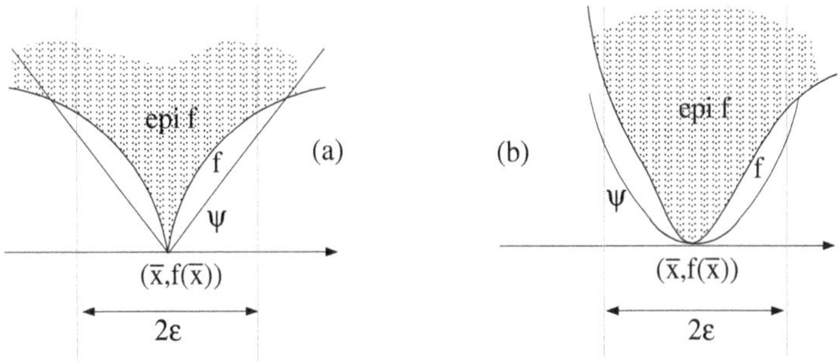

Fig. 7–13. Conditioning functions ψ.

of the growth of f away from $\operatorname{argmin} f$. A foundation is thereby provided for using the 'conditioning' concept more broadly.

7.64 Theorem (approximations of optimality). *Let $f : \mathbb{R}^n \to \overline{\mathbb{R}}$ be lsc, proper and level-bounded, and let $\bar\rho \geq 0$ be large enough that $\operatorname{argmin} f \subset \bar\rho\mathbb{B}$ and $\min f \geq -\bar\rho$. Fix $\rho \in (\bar\rho, \infty)$. Then for any lsc, proper function $g : \mathbb{R}^n \to \overline{\mathbb{R}}$ close enough to f that $\hat{d\!l}^+_\rho(f,g) < \rho - \bar\rho$, one has*

$$\left| \min\nolimits_{\rho\mathbf{B}} g - \min f \right| \;\leq\; \hat{d\!l}^+_\rho(f,g).$$

Furthermore, in terms of $\psi_f(\tau) := \min\{ f(x) - \min f \mid d(x, \operatorname{argmin} f) \geq \tau \}$ and $\Psi_f(\eta) := \eta + \psi_f^{-1}(2\eta)$, one has

$$\operatorname{argmin}_{\rho\mathbf{B}} g \;\subset\; \operatorname{argmin} f + \Psi_f\big(\hat{d\!l}^+_\rho(f,g)\big)\mathbb{B}.$$

(Here the function $\psi_f : [0,\infty) \to [0,\infty]$ is lsc and nondecreasing with $\psi_f(0) = 0$, but $\psi_f(\tau) > 0$ when $\tau > 0$; if ψ_f is not continuous, $\psi_f^{-1}(2\eta)$ is taken as denoting the $\tau \geq 0$ for which $\operatorname{lev}_{\leq 2\eta} \psi_f = [0,\tau]$. The function Ψ_f is lsc and increasing on $[0,\infty)$ with $\Psi_f(0) = 0$, and in addition, $\Psi_f(\eta) \searrow 0$ as $\eta \searrow 0$.)

In restricting f and g to be convex and imposing the tighter requirement that $\hat{d\!l}^+_\rho(f,g) < \min\{ \frac{1}{2}[\rho - \bar\rho], \frac{1}{2}\psi_f(\frac{1}{2}[\rho - \bar\rho]) \}$, one can replace $\min_{\rho\mathbf{B}} g$ and $\operatorname{argmin}_{\rho\mathbf{B}} g$ in these estimates by $\min g$ and $\operatorname{argmin} g$.

Proof. Let $\bar\eta = \hat{d\!l}^+_\rho(f,g)$ and consider $\eta \in (\bar\eta,\, \rho - \bar\rho)$. By definition we have

$$\left.\begin{aligned} \min_{\mathbf{B}(x,\eta)} f &\leq \max\{g(x), -\rho\} + \eta \\ \min_{\mathbf{B}(x,\eta)} g &\leq \max\{f(x), -\rho\} + \eta \end{aligned}\right\} \text{ for all } x \in \rho\mathbb{B}. \tag{7(31)}$$

Applying the second of these inequalities to any $\bar x \in \operatorname{argmin} f$ (so $\bar x \in \bar\rho\mathbb{B}$) and using $f(\bar x) = \min f \geq -\bar\rho > -\rho$, we see that $\min_{\mathbf{B}(\bar x,\eta)} g \leq \min f + \eta$. There exists then some $\tilde x \in \mathbb{B}(\bar x, \eta)$ such that $g(\tilde x) \leq \min f + \eta$. But $\mathbb{B}(\bar x,\eta) \subset \operatorname{argmin} f + \eta\mathbb{B} \subset \rho\mathbb{B}$ because $\operatorname{argmin} f \subset \bar\rho\mathbb{B}$ and $\eta < \rho - \bar\rho$. Therefore $g(\tilde x) \geq \min_{\rho\mathbf{B}} g$ and we get

$$\min_{\rho B} g - \min f \leq \eta. \tag{7(32)}$$

On the other hand, in the first of the inequalities in 7(31) we have $\min_{B(x,\eta)} f \geq \min f \geq -\bar{\rho}$. Hence $\min_{B(x,\eta)} f - \eta > -\bar{\rho} - (\rho - \bar{\rho}) = -\rho$ on the left side, so that necessarily $\max\{g(x), -\rho\} = g(x)$ on the right side. This implies that $\min f \leq g(x) + \eta$ for all $x \in \rho I\!B$ and confirms that $\min_{\rho B} g - \min f \geq -\eta$. In view of 7(32) and the arbitrary choice of $\eta \in (\bar{\eta}, \rho - \bar{\rho})$, we obtain $|\min_{\rho B} g - \min f| \leq \bar{\eta}$, as claimed.

Let $A :=$ argmin f; this set is compact as well as nonempty (cf. 1.9). From the definition of the function ψ_f it's obvious that ψ_f is nondecreasing with

$$f(x) - \min f \geq \psi_f\big(d_A(x)\big) \text{ for all } x. \tag{7(33)}$$

The fact that ψ_f is lsc can be deduced from Theorem 1.17: we have

$$\psi_f(\tau) = \inf_x h(x,\tau) \quad \text{for} \quad h(x,\tau) := \begin{cases} f(x) - \min f & \text{if } d_A(x) \geq \tau \geq 0, \\ \infty & \text{otherwise,} \end{cases}$$

the function h being proper and lsc on $I\!R^n \times I\!R$ with $h(x,\tau)$ level-bounded in x locally uniformly in τ (inasmuch as f is level-bounded). The corresponding properties of the function Ψ_f are then evident as well.

Recalling that necessarily $\max\{g(x), -\rho\} = g(x)$ in the first of the inequalities in 7(31), we combine this inequality with the one in 7(33) to get

$$g(x) - \min f + \eta \geq \min_{x' \in B(x,\eta)} \psi_f\big(d_A(x')\big) \text{ for all } x \in \rho I\!B.$$

When $x \in \operatorname{argmin}_{\rho B} g$, so that $g(x) = \min_{\rho B} g$, the expression on the left is bounded above by 2η; cf. 7(32) again. Then

$$2\eta \geq \min_{x' \in B(x,\eta)} \psi_f\big(d_A(x')\big) \geq \psi_f\Big(\min_{x' \in B(x,\eta)} d_A(x')\Big) = \psi_f\big(d_{A+\eta B}(x)\big),$$

so that $d_{A+\eta B}(x) \leq \psi_f^{-1}(2\eta)$. It follows that $d_A(x) \leq \eta + \psi_f^{-1}(2\eta) = \Psi_f(\eta)$. Because this holds for any $\eta \in (\bar{\eta}, \rho - \bar{\rho})$, we must have $d_A(x) \leq \Psi_f(\bar{\eta})$ as well. Thus, $\operatorname{argmin}_{\rho B} g \subset A + \Psi_f(\bar{\eta}) I\!B$.

In the convex case, any local minimum is a global minimum, and we have $\min_{\rho B} g = \min g$ and $\operatorname{argmin}_{\rho B} g = \operatorname{argmin} g$ whenever $\operatorname{argmin}_{\rho B} g \subset \operatorname{int} \rho I\!B$. From the relation between $\operatorname{argmin}_{\rho B} g$ and argmin f, which lies in $\bar{\rho} I\!B$, we see that this holds when, along with $\bar{\eta} < \rho - \bar{\rho}$, we also have $\Psi_f(\bar{\eta}) < \rho - \bar{\rho}$, or in other words, $\psi_f^{-1}(2\bar{\eta}) < \rho - \bar{\rho} - \bar{\eta}$. Under the assumption that actually $\bar{\eta} < \frac{1}{2}(\rho - \bar{\rho})$, we get this when $2\bar{\eta} < \psi_f(\frac{1}{2}[\rho - \bar{\rho}])$. □

7.65 Corollary (linear and quadratic growth at solution set). *Consider f, $\bar{\rho}$ and ρ as in Theorem 7.64. Let $\varepsilon > 0$ and $\gamma > 0$.*

(a) *If $f(x) - \min f \geq \gamma d(x, \operatorname{argmin} f)$ when $d(x, \operatorname{argmin} f) < \varepsilon$, then for every lsc, proper function g satisfying $\hat{d\!l}_\rho^+(f,g) < \min\{\rho - \bar{\rho}, \gamma\varepsilon/2\}$ one has*

$$d(x, \operatorname{argmin} f) \leq (1 + 2/\gamma)\hat{d\!l}^+_\rho(f,g) \;\; \textit{for all} \;\; x \in \operatorname{argmin}_{\rho B} g.$$

(b) *If* $f(x) - \min f \geq \gamma\, d(x, \operatorname{argmin} f)^2$ *when* $d(x, \operatorname{argmin} f) < \varepsilon$, *then for every lsc, proper function* g *satisfying* $\hat{d\!l}^+_\rho(f,g) < \min\{\rho - \bar\rho, \gamma\varepsilon^2/2\}$ *one has*

$$d(x, \operatorname{argmin} f) \leq \hat{d\!l}^+_\rho(f,g) + [2\hat{d\!l}^+_\rho(f,g)/\gamma]^{1/2} \;\; \textit{for all} \;\; x \in \operatorname{argmin}_{\rho B} g,$$

where $\hat{d\!l}^+_\rho(f,g) + [2\hat{d\!l}^+_\rho(f,g)/\gamma]^{1/2} \leq 2[\hat{d\!l}^+_\rho(f,g)/\gamma]^{1/2}$ *if also* $\hat{d\!l}^+_\rho(f,g) \leq 1/4\gamma$.

Proof. Both cases involve a continuous, increasing function ψ on $[0,\varepsilon]$ with $\psi(0) = 0$ such that $\psi_f \geq \psi$ on $[0,\varepsilon)$. Then, as in the explanation given for Example 7.63, we have $\psi_f^{-1}(2\eta) \leq \psi^{-1}(2\eta)$ when $2\eta < \psi(\varepsilon)$, which in the context of Theorem 7.64 yields $\Psi_f(\eta) \leq \eta + \psi^{-1}(2\eta)$ as long as $\eta \leq \psi(\varepsilon)/2$.

Case (a) has $\psi(\varepsilon) = \gamma\varepsilon$ and $\psi^{-1}(2\eta) = 2\eta/\gamma$, whereas case (b) has $\psi(\varepsilon) = \gamma\varepsilon^2$ and $\psi^{-1}(2\eta) = \sqrt{2\eta/\gamma}$. For the latter we note further that $\eta + \sqrt{2\eta/\gamma} = (\sqrt{\eta\gamma} + \sqrt{2})\sqrt{\eta/\gamma}$, and that $\sqrt{\eta\gamma} + \sqrt{2} \leq 2$ when $\eta\gamma \leq (2 - \sqrt{2})^2$ and hence in particular when $\eta\gamma \leq (.5)^2 = 1/4$. □

7.66 Example (proximal estimates). *Let* $f : I\!R^n \to \overline{I\!R}$ *be lsc, proper, and convex. Fix* $\lambda > 0$ *and any point* $a \in I\!R^n$, *and suppose* $\bar\rho$ *is large enough that*

$$e_\lambda f(a) \geq -\bar\rho, \qquad \big|P_\lambda f(a)\big| \leq \bar\rho.$$

Let $\rho > \bar\rho + 8\lambda$ *and* $\Psi(\eta) := \eta + 2\sqrt{\lambda\eta}$. *Then for functions* $g : I\!R^n \to \overline{I\!R}$ *that are lsc, proper and convex, one has the estimates*

$$\begin{aligned} \big|e_\lambda g(a) - e_\lambda f(a)\big| &\leq \hat{d\!l}^+_\rho(\bar f, \bar g), \\ \big|P_\lambda g(a) - P_\lambda f(a)\big| &\leq \Psi\big(\hat{d\!l}^+_\rho(\bar f, \bar g)\big), \end{aligned}$$

in terms of $\bar f(x) := f(x) + (1/2\lambda)|x-a|^2$ *and* $\bar g(x) := g(x) + (1/2\lambda)|x-a|^2$, *as long as these functions are near enough to each other that* $\hat{d\!l}^+_\rho(\bar f, \bar g) \leq 4\lambda$.

Detail. Theorem 7.64 will be applied to $\bar f$ and $\bar g$, since these functions have $\min \bar f = e_\lambda f(a)$, $\operatorname{argmin} \bar f = P_\lambda f(a)$, $\min \bar g = e_\lambda g(a)$ and $\operatorname{argmin} \bar g = P_\lambda g(a)$. For this purpose we have to verify that $\bar f$ satisfies a conditioning inequality of quadratic type. Noting from the convexity of f that $P_\lambda f(a)$ is a single point (cf. 2.26), and denoting this point by $\bar x$ for simplicity, we demonstrate that

$$\bar f(x) - \bar f(\bar x) \geq \frac{1}{2\lambda}|x - \bar x|^2 \;\; \text{for all} \;\; x \in I\!R^n. \qquad 7(34)$$

It suffices to consider x with $\bar f(x)$ and $f(x)$ finite; we have

$$0 \leq \bar f(x) - \bar f(\bar x) = f(x) - f(\bar x) - \frac{1}{\lambda}\langle a - \bar x,\, x - \bar x\rangle + \frac{1}{2\lambda}|x - \bar x|^2. \qquad 7(35)$$

Letting $\varphi(\tau) = f(\bar x + \tau[x - \bar x])$, so that φ is finite and convex on $[0,1]$, we get for all $\tau \in (0,1]$ that $\varphi(1) - \varphi(0) \geq [\varphi(\tau) - \varphi(0)]/\tau$ (by Lemma 2.12) and consequently through the nonnegativity in 7(35) that

$$f(x) - f(\bar{x}) \geq \frac{f(\bar{x} + \tau[x - \bar{x}]) - f(\bar{x})}{\tau} \geq \frac{1}{\lambda}\langle a - \bar{x},\, x - \bar{x}\rangle - \frac{\tau}{2\lambda}|x - \bar{x}|^2.$$

Because the outer inequality holds for all $\tau \in (0,1]$, it also holds for $\tau = 0$. Hence the expression $f(x) - f(\bar{x}) - (1/\lambda)\langle a - \bar{x},\, x - \bar{x}\rangle$ in 7(35) is sure to be nonnegative. In replacing it by 0, we obtain the inequality in 7(34).

On the basis of the conditioning inequality in 7(34), we can apply 7.64 (in the convex case at the end of this theorem) to $\bar{f}$ and $\bar{g}$ with the knowledge that $\psi_{\bar{f}}(\tau) \geq \tau^2/2\lambda$ and thus $\psi_{\bar{f}}^{-1}(2\eta) \leq 2\sqrt{\lambda\eta}$. This tells us that

$$\hat{d\kern-0.3em l}{}_\rho^+(\bar{f},\bar{g}) < \min\left\{\frac{\rho - \bar{\rho}}{2}, \frac{(\rho - \bar{\rho})^2}{16\lambda}\right\}$$
$$\implies \begin{cases} \big|e_\lambda g(a) - e_\lambda f(a)\big| \leq \hat{d\kern-0.3em l}{}_\rho^+(\bar{f},\bar{g}), \\ \big|P_\lambda g(a) - P_\lambda f(a)\big| \leq \Psi\big(\hat{d\kern-0.3em l}{}_\rho^+(\bar{f},\bar{g})\big). \end{cases}$$

Our choice of ρ has ensured that $(\rho - \bar{\rho})/2 \leq (\rho - \bar{\rho})^2/16\lambda$, so it's enough to require that $\hat{d\kern-0.3em l}{}_\rho^+(\bar{f},\bar{g}) < (\rho - \bar{\rho})/2$. Since $\rho - \bar{\rho} > 8\lambda$, this holds in particular when $\hat{d\kern-0.3em l}{}_\rho^+(\bar{f},\bar{g}) \leq 4\lambda$. □

This kind of application could be carried further by means of estimates of $\hat{d\kern-0.3em l}{}_\rho^+(\bar{f},\bar{g})$ in terms of the distance $\hat{d\kern-0.3em l}{}_{\rho'}^+(f,g)$ for some $\rho' \geq \rho$. The following exercise works out a special case.

7.67 Exercise (projection estimates). *Let $C \subset I\!R^n$ be nonempty, closed and convex. Fix any point $a \in I\!R^n$ and let $\rho > |P_C(a)| + 4$ and $\theta = 1 + |a| + \rho$. Then for nonempty, closed, convex sets D near enough to C that $\hat{d\kern-0.3em l}_\rho(C,D) < 1/4\theta$, one has*

$$\big|P_D(a) - P_C(a)\big| \leq 3\sqrt{\theta\, \hat{d\kern-0.3em l}_\rho(C,D)}.$$

Guide. Apply 7.66 to $f = \delta_C$ and $g = \delta_D$ with $\lambda = \frac{1}{2}$ and $\bar{\rho} = |P_C(a)|$. Then $\bar{f} = \delta_C + q$ and $\bar{g} = \delta_D + q$ for $q(x) = |x - a|^2$, and 7.62 can be used to estimate $\hat{d\kern-0.3em l}{}_\rho^+(\bar{f},\bar{g})$ in terms of $\hat{d\kern-0.3em l}{}_\rho^+(C,D)$. Specifically, verify for general $\rho > 0$ that

$$|q(x') - q(x)| \leq 2(|a| + \rho)|x' - x| \;\text{ when }\; x, x' \in \rho I\!B,$$

and then invoke 7.62 with $\rho' = \rho + \frac{1}{2}$ in order to establish that

$$\hat{d\kern-0.3em l}{}_\rho^+(\bar{f},\bar{g}) \leq 2\theta \hat{d\kern-0.3em l}{}_\rho^+(C,D) \;\text{ when }\; \hat{d\kern-0.3em l}{}_\rho^+(C,D) < \tfrac{1}{2}.$$

Work this into the result in 7.66, where $\Psi(2\theta\eta) = 2(\theta\eta + \sqrt{\theta\eta})$. Argue that $2(\tau + \sqrt{\tau}) \leq 3\sqrt{\tau}$ when $\tau \leq 1/4$. □

Alongside of optimal solutions to the problem of minimizing a function f one can investigate perturbations of ε-optimal solutions; here we refer to replacing $\operatorname{argmin} f$ by the level set

$$\varepsilon\text{-}\operatorname{argmin} f := \big\{x \,\big|\, f(x) \leq \min f + \varepsilon\big\}, \quad \varepsilon > 0.$$

For convex functions, a strong result can be obtained. To prepare the way for it, we first derive an estimate, of interest in its own right, for the distance between two different level sets of a single convex function.

7.68 Proposition (distances between convex level sets). *Let* $f : \mathbb{R}^n \to \overline{\mathbb{R}}$ *be proper, lsc and convex with* $\operatorname{argmin} f \neq \emptyset$. *Then for* $\alpha_2 > \alpha_1 > \alpha_0 := \min f$ *and any choice of* $\rho \geq \rho_0 := d(0, \operatorname{argmin} f)$, *one has*

$$\hat{d\!l}_\rho\big(\operatorname{lev}_{\leq\alpha_2} f, \operatorname{lev}_{\leq\alpha_1} f\big) \;\leq\; \left[\frac{\alpha_2-\alpha_1}{\alpha_2-\alpha_0}\right](\rho+\rho_0).$$

Proof. Let μ denote the bound on the right side of this inequality. Since $\operatorname{lev}_{\leq\alpha_1} f \subset \operatorname{lev}_{\leq\alpha_2} f$, we need only show for arbitrary $x_2 \in [\operatorname{lev}_{\leq\alpha_2} f] \cap \rho\mathbb{B}$ that there exists $x_1 \in \operatorname{lev}_{\leq\alpha_1} f$ with $|x_1 - x_2| \leq \mu$.

Consider $x_0 \in \operatorname{argmin} f$ such that $|x_0| = \rho_0$; one has $f(x_0) = \alpha_0$. In $\mathbb{R}^{n+1}$ the points (x_2, α_2) and (x_0, α_0) belong to $\operatorname{epi} f$, so by convexity the line segment joining them lies in $\operatorname{epi} f$ as well. This line segment intersects the hyperplane $H = \mathbb{R}^n \times \{\alpha_1\}$ at a point (x_1, α_1), namely the one given by $x_1 = (1-\tau)x_2 + \tau x_0$ with $\tau = (\alpha_2 - \alpha_1)/(\alpha_2 - \alpha_0)$. Then $x_1 \in \operatorname{lev}_{\leq\alpha_1} f$ and $|x_1 - x_2| = \tau|x_0 - x_2| \leq \tau(|x_2| + |x_0|) \leq \tau(\rho + \rho_0) = \mu$, as required. □

7.69 Theorem (estimates for approximately optimal solutions). *Let* f *and* g *be proper, lsc, convex functions on* $\mathbb{R}^n$ *with* $\operatorname{argmin} f$ *and* $\operatorname{argmin} g$ *nonempty. Let* ρ_0 *be large enough that* $\rho_0\mathbb{B}$ *meets these two sets, and also* $\min f \geq -\rho_0$ *and* $\min g \geq -\rho_0$. *Then, with* $\rho > \rho_0$, $\varepsilon > 0$ *and* $\bar\eta = \hat{d\!l}{}^+_\rho(f,g)$,

$$\hat{d\!l}_\rho\big(\varepsilon\text{-}\operatorname{argmin} f, \varepsilon\text{-}\operatorname{argmin} g\big) \;\leq\; \bar\eta\Big(1 + \frac{2\rho}{\bar\eta + \varepsilon/2}\Big) \;\leq\; \big(1 + 4\rho\varepsilon^{-1}\big)\hat{d\!l}{}^+_\rho(f,g).$$

Proof. First, let's establish that $|\min g - \min f| \leq \bar\eta$. Let $\bar x \in \operatorname{argmin} f \cap \rho\mathbb{B}$. Since $f(\bar x) = \min f \geq -\rho_0 > -\rho$, from the definition of $\hat{d\!l}{}^+_\rho(f,g)$ in 7.61, one has that for any $\eta > \bar\eta$

$$\min\nolimits_{x\in \mathbb{B}(\bar x,\eta)} g(x) \;\leq\; f(\bar x) + \eta.$$

Hence $\min g \leq \min f + \eta$. Reversing the roles of f and g, we deduce similarly that $\min f \leq \min g + \eta$ from which follows: $|\min g - \min f| \leq \bar\eta$.

Again, by the definition of $\hat{d\!l}{}^+_\rho(f,g)$, for any $x \in \varepsilon\text{-}\operatorname{argmin} f \cap \rho\mathbb{B}$ and $\eta > \bar\eta$, one has

$$\min\nolimits_{\mathbb{B}(x,\eta)} g \leq f(x) + \eta,$$

since $f(x) \geq \min f \geq -\rho_0 > -\rho$. Therefore, there exists x' with $|x' - x| \leq \eta$ and $g(x') \leq f(x) + \eta$. Since $x \in \varepsilon\text{-}\operatorname{argmin} f$, $f(x) \leq \min f + \varepsilon$. Moreover $\min f \leq \min g + \bar\eta$, as shown earlier, hence $g(x') \leq \min g + \varepsilon + 2\eta$, i.e., $x' \in (\varepsilon + 2\eta)\text{-}\operatorname{argmin} g$ and consequently,

$$\varepsilon\text{-}\operatorname{argmin} f \cap \rho\mathbb{B} \;\subset\; \big[(\varepsilon + 2\eta)\text{-}\operatorname{argmin} g + \eta\mathbb{B}\big] \cap \rho\mathbb{B}.$$

Applying Proposition 7.68 now to g with $\alpha_0 = \min g$, $\alpha_1 = \min g + \varepsilon$ and

$\alpha_2 = \min g + \varepsilon + 2\eta$, we obtain the estimates

$$\begin{aligned}&\big[(\varepsilon + 2\eta)\text{-}\operatorname{argmin} g\big] \cap \rho\mathbb{B} \\ &\qquad \subset \varepsilon\text{-}\operatorname{argmin} g + \left[\frac{(\min g + \varepsilon + 2\eta) - (\min g + \varepsilon)}{(\min g + \varepsilon + 2\eta) - \min g}\right] [\rho_0 + \rho]\mathbb{B} \\ &\qquad \subset \varepsilon\text{-}\operatorname{argmin} g + \eta\Big(\frac{2\rho}{\eta + \varepsilon/2}\Big)\mathbb{B}.\end{aligned}$$

From this inclusion and the earlier one, we deduce

$$\varepsilon\text{-}\operatorname{argmin} f \cap \rho\mathbb{B} \subset \varepsilon\text{-}\operatorname{argmin} g + \eta\Big(1 + \frac{2\rho}{\eta + \varepsilon/2}\Big)\mathbb{B}.$$

The same argument works when the roles of f and g are interchanged, and so $\hat{d\!l}_\rho(\varepsilon\text{-}\operatorname{argmin} f, \varepsilon\text{-}\operatorname{argmin} g) \leq \eta\big(1 + 2\rho(\eta + \varepsilon/2)^{-1}\big)$. The fact that this holds for every $\eta > \bar{\eta}$ yields the asserted inequalities. □

In these estimates as well as the previous ones, it should be borne in mind that because the epi-distance expressions are centered at the origin it may be advantageous in some situations to apply the results to certain translates of the given functions rather than to these functions directly.

Finally, observe that the preceding theorem actually implies the Lipschitz continuity for the ε–argmin-mappings of convex functions, whereas the results about the argmin-mappings of arbitrary functions only delivered Hölder-type continuity, and then under the stated conditioning requirements.

Commentary

The notion of epi-convergence, albeit under the name of *infimal convergence*, was introduced by Wijsman [1964], [1966], to bring in focus a fundamental relationship between the convergence of convex functions and that of their conjugates (as described later in Theorem 11.34). He was motivated by questions of efficiency and optimality in the theory of a certain sequential statistical test.

The importance of the connection with conjugate convex functions was so compelling that for many years after Wijsman's initial contribution epi-convergence was regarded primarily as a topic in convex analysis, affording tools suitable for dealing with approximations to *convex* optimization problems. This is reflected in the work of Mosco [1969] on variational inequalities, of Joly [1973] on topological structures compatible with epi-convergence, of Salinetti and Wets [1977] on equi-semicontinuous families of convex functions, of Attouch [1977] on the relationship between the epi-convergence of convex functions and the graphical convergence of their subgradient mappings, and of McLinden and Bergstrom [1981] on the preservation of epi-convergence under various operations performed on convex functions. The literature in this vein was surveyed to some extent in Wets [1980], where the term epi-convergence was probably used for the first time.

In the late 70's the umbilical cord with convexity was cut, and epi-convergence came to be seen more and more as the natural convergence notion for handling approximations of nonconvex problems of minimization as well, whether on the global or local level. The impetus for the shift arose from difficulties that had to be overcome in the infinite-dimensional calculus of variations, specifically in making use of the so-called 'indirect method', in which the solution to a given problem is derived by considering a sequence of approximate Euler equations and studying the limits of the solutions to those equations. De Giorgi and his students found that the classical version of the indirect method wasn't justifiable in application to some of the newer problems they were interested in. This led them to rediscover epi-convergence from a different angle; cf. De Giorgi and Franzoni [1975], [1979], Buttazzo [1977], De Giorgi and Dal Maso [1983]. They called it *Γ-convergence* in order to distinguish it from G-convergence, the name used at that time by Italian mathematicians for graphical convergence, cf. De Giorgi [1977]. The book of Dal Maso [1993] provides a fine overview of these developments.

In the definitions adopted for 'Γ-limits', convexity no longer played a central role. Soon that was the pattern too in other work on epi-convergence, such as that of Attouch and Wets [1981], [1983], on approximating nonlinear programming problems, of Dolecki, Salinetti and Wets [1983] on equi-lower semicontinuity, and in the quantitative theory developed by Attouch and Wets [1991], [1993a], [1993b] (see also Azé and Penot [1990] and Attouch, Lucchetti and Wets [1991]).

The 1983 conference in Catania (Italy) on "Multifunctions and Integrands: Stochastic Analysis, Approximation and Optimization" (cf. Salinetti [1984]), and especially the book by Attouch [1984], which was the first to provide a comprehensive and systematic exposition of the theory and some of its applications, were instrumental in popularizing epi-convergence and convincing many people of its significance. Meanwhile, still other researchers, such as Artstein and Hart [1981] and Kushner [1984], stumbled independently on the notion in their quest for good forms of approximation in extremal problems. This spotlighted all the more the inevitability of epi-convergence occupying a pivotal position, even though Kushner's analysis, for instance, dealt only with special situations where epi-convergence and pointwise convergence are available simultaneously to act in tandem. By now, a complete bibliography on epi-convergence would have thousands of entries, reflecting the spread of techniques based on such theory into numerous areas of mathematics, pure and applied.

The relation between the set convergence of epigraphs and the sequential characterization in Proposition 7.2 is already implicit in the results of Wijsman, at least in the convex case. Klee [1964] rendered it explicit in his review of Wijsman's paper. The alternative formulas in Exercise 7.3 and the properties in Proposition 7.4 come from De Giorgi and Franzoni [1979], Attouch and Wets [1983], and Rockafellar and Wets [1984]. The fact that epi-convergence techniques can be used to obtain general convergence results for penalty and barrier methods was first pointed out by Attouch and Wets [1981].

The sequential compactness of the space of lsc functions equipped with the epi-convergence topology (Theorem 7.6) is inherited at once from the parallel property for the space of closed sets; arguments of different type were provided by De Giorgi and Franzoni [1979], and by Dolecki, Salinetti and Wets [1983]. Theorem 7.7 on the convergence of level sets is taken from Beer, Rockafellar and Wets [1992], with the first formulas for the level sets of epi-limits appearing in Wets [1981]. The preservation of epi-convergence under continuous perturbation, as in 7.8(a), was first noted by Attouch and Sbordone [1980], but there seems to be no explicit record previously of the other properties listed in 7.8.

The strategy of using profile mappings to analyze the convergence of sequences of functions was systematically exploited by Aubin and Frankowska [1990]. Here it leads to a multitude of results, most of them readily extracted from the corresponding ones for convergent sequences of mappings when these are taken to be profile mappings. Theorem 7.10 owes its origin to Dolecki, Salinetti and Wets [1983], while Theorem 7.11 is taken from a paper of Kall [1986] which explores the relationship between epi-convergence and other, more classical, modes of function convergence. Proposition 7.13, Theorem 7.14 and Proposition 7.15, are new along with the accompanying Definition 7.12 of uniform convergence for extended real-valued functions; the result in 7.15(a) is based on an observation of L. Korf (unpublished). The generalization of the Arzelà-Ascoli Theorem that's stated in 7.16 can be found in Dolecki, Salinetti and Wets [1983].

Theorem 7.17 and Corollary 7.18 are built upon long-known facts about the uniformity of pointwise convergence of convex functions (see Rockafellar [1970a]) and results of Robert [1974] and Salinetti and Wets [1977] relating epi-convergence to uniform convergence.

The approximation of functions, in particular discontinuous functions, by averaging through integral convolution with mollifier functions as in 7.19 goes back to Steklov [1907] and lies at the foundations of the theory of 'distributions' as developed by Sobolev [1963] and Schwartz [1966]. This example comes from Ermoliev, Norkin and Wets [1995].

Epi-convergence notions as alternatives to pointwise convergence in the definition of generalized directional derivatives were first embraced in a paper of Rockafellar [1980]. That work concentrated however on taking an *upper* epi-limit of difference quotient functions rather than the full epi-limit, a tactic which leads instead to what will be studied as 'regular subderivatives' in Chapter 8. True epi-derivatives as defined in 7.23 didn't receive explicit treatment until their emergence in Rockafellar [1988] as a marker along the road to second-order theory of the kind that will be discussed in Chapter 13. Geometric connections with tangent cones such as in 7.24 were clear from the start, and indeed go back to the origins of convex analysis, but somewhat surprisingly there was no attention paid until much more recently to the particular case in which the usual 'lim sup' in the tangent cone limit coincides with the 'lim inf', thereby producing the geometric derivability of the epigraph that characterizes epi-differentiability (Proposition 7.24).

Meanwhile, Aubin [1981], [1984], took up the banner of generalized directional derivatives defined by the tangent cone $T_{\text{epi}\, f}(\bar{x}, f(\bar{x}))$ itself, without regard to the geometric derivability property of 6.1 as applied to epi f at $(\bar{x}, f(\bar{x}))$. These are the 'subderivatives' $df(\bar{x})(w)$ of 7(21), whose theory will come on line in Chapter 8. (Aubin also pursued this idea with respect to the graphs of vector-valued functions and set-valued mappings, which likewise will be studied in Chapter 8.) He spoke of *contingent derivatives* because of the tangent cone's connection with Bouligand's 'contingent' cone (cf. the notes to Chapter 6).

Robinson [1987a] addressed the special case where the tangent cone at a point $\bar{x}$ to the graph (not epigraph!) of a function, either real-valued or vector-valued, was itself the graph of a single-valued function—which he termed then the *Bouligand derivative* function at $\bar{x}$. He called this property *B-differentiability*. It's equivalent to semidifferentiability as defined in 7.20 for real-valued functions f; thus Robinson's B-derivatives, when they exist, are identical to the semiderivatives in 7.20. But semiderivatives make sense with respect to individual vectors w and don't require semidifferentiability/B-differentiability in order to be useful, nor need they always be finite in value.

Robinson's primary concern was with Lipschitz continuous functions (cf. Chapter 9), for which such distinctions don't matter and it's superfluous even to bother with directional derivatives other than the one-sided straight-line ones of 7(20). For such functions he showed that his B-differentiability (semidifferentiability) was equivalent to the approximation property in 7.21(e), which resembles classical differentiability except for replacing the usual linear function by a continuous, positively homogeneous function. (In contrast, Theorem 7.21 establishes this equivalence without any assumption of Lipschitz continuity.) In subsequent work of Robinson and his followers, likewise concentrating on Lipschitz continuous functions, it was this approximation property that was used as the definition of B-differentiability rather than the geometric property corresponding to semidifferentiability.

Actually, though, the definition of generalized directional derivatives through the approximation property in 7.21(e), i.e., through a first-order expansion with a possibly nonlinear term, has a much older history than this, which can be read in the survey articles of Averbukh and Smolyanov [1967], [1968] (see also Shapiro [1990]). In that setting, such generalized differentiability is 'bounded differentiability'. The equivalence between 7.21(a) and 7.21(c) thus links semidifferentiability to bounded differentiability. But that, of course, is a result for $I\!R^n$ which doesn't carry over to infinite-dimensional spaces.

The term 'semidifferentiability' goes back to Penot [1978]. He formulated this property in the abstract setting of single-valued mappings from a linear topological space into another such space furnished with an ordering to which elements '∞' and '$-\infty$' are adjoined in imitation of $\overline{I\!R}$, so that the semiderivative limit could be defined by requiring a certain limsup to coincide with a liminf. He didn't look at semiderivatives vector by vector, as here, or obtain a result like Theorem 7.21. Use of the term beyond a liminf/limsup context came with Rockafellar [1989b], who developed the concept for general set-valued mappings (cf. 8.43 and the paragraph that precedes it).

The regularity notion for functions in 7.25 was introduced by Clarke [1975]. In the scheme he followed, however, it was first defined in a different manner just for Lipschitz continuous functions and only later, through the intervention of distance functions, which themselves are Lipschitz continuous, shown to be equivalent to this epigraphical property and therefore extendible in terms of that. (The Lipschitzian version will come up in 9.16.) In that framework he established the subdifferential regularity of lsc convex functions in 7.27. The epi-differentiability of such functions follows at once from this, but no explicit notice of it was taken before Rockafellar [1990b]. The semiderivative properties in 7.27, although verging on various familiar facts, haven't previously been crystallized in this form.

The regularity and semidifferentiability result in 7.28 for a max function plus an indicator specializes facts proved by Rockafellar [1988]. For the case of a finite max function alone, the question of one-sided directional derivatives—without the varying direction that distinguishes semiderivatives—was treated by many authors, with Danskin [1967] and Demyanov and Malozemov [1971] the earliest.

Upper and lower epi-limits in the context of generalized directional derivatives will be an ever-present theme in Chapter 8.

The characterization of epi-convergence through the convergence of the infima on open and compact sets (Proposition 7.29) was recorded in Rockafellar and Wets [1984], and was used by Vervaat [1988] to define a topology on the space of lsc functions that is compatible with epi-convergence; cf. also Shunmugaraj and Pai [1991]. Proposition 7.30 and the concept of epigraphical nesting are due to Higle and Sen [1995], who came up with them in devising a convergence theory for algorithmic procedures; cf. Higle and Sen [1992]. Likewise, Lemaire [1986], Alart and Lemaire [1991], and Polak [1993], [1997], have relied on the guidelines provided by epi-convergence in designing algorithmic schemes for certain classes of optimization problems.

The first assertion in Theorem 7.31 is due to Salinetti (unpublished, but reported in Rockafellar and Wets [1984]), while other assertions stem from Attouch and Wets [1981]. Robinson [1987b] detailed an application of this theorem, more precisely of 7.31(a), to the case when the set of minimizers of a function f on an open set O is a 'complete local minimizing set', meaning that $\operatorname{argmin}_O f = \operatorname{argmin}_{\operatorname{cl} O} f$. Level-boundedness was already exploited in Wets [1981] to characterize, as in 7.32, certain properties of epi-limits. Theorem 7.33 just harvests the additional implications of level-boundedness in this setting of the convergence of optimal values and optimal solutions.

The use of Moreau envelopes in studying epi-convergent sequences of functions was pioneered by Attouch [1977] in the convex case and extended to the nonconvex case in Attouch and Wets [1983a] and more fully in Poliquin [1992]. Theorem 7.37 and Exercise 7.38, by utilizing the concept of prox-boundedness (Definition 7.36), slightly refine these earlier results.

One doesn't have to restrict oneself to Moreau envelopes to obtain convergence results along the lines of those in Theorem 7.37. Instead of $e_\lambda f = f \# (2\lambda)^{-1}|\cdot|^2$, one can work with the Pasch-Hausdorff envelopes $f \# \lambda^{-1}|\cdot|$ as defined in 9.11 (cf. Borwein and Vanderwerff [1993]), or more generally with any p-kernels generating the envelopes $f \# (p\lambda)^{-1}|\cdot|^p$ for $p \in [1, \infty)$ as in Attouch and Wets [1989]. For theoretical purposes one can even work with still more general kernels (cf. Wets [1980], Fougères and Truffert [1988]) for the sake of constructing envelopes that will have at least some of the properties exhibited in Theorem 7.37. A particularly interesting variant was proposed by Teboulle [1992]; it relies on the Csiszar φ-divergence (or generalized relative entropy).

The translation of epi-convergence to the framework where, instead of a *sequence* of functions one has a *family* parameterized by an element of $\mathbb{R}^m$, say, leads naturally, as in Rockafellar and Wets [1984], to the notions of epi-continuity and epi-semicontinuity. The parametric optimization results in Theorem 7.41 and its Corollaries 7.42 and 7.43 are new, under the specific conditions utilized. These results, their generality not withstanding, don't cover all the possible twists that have been explored to date in the abundant literature on parametric optimization, but they do capture the basic principles. A comprehensive survey of related results, focusing on the very active German school, can be found in Bank, Guddat, Klatte, Kummer and Tammer [1982]. For more recent developments see Fiacco and Ishizuka [1990].

The facts presented about the preservation of epi-convergence under various operations performed on functions are new, except for those in Exercise 7.47, which essentially come from McLinden and Bergstrom [1981], and for Proposition 7.56 about what happens to epi-convergence under epi-addition, which weakens the conditions initially given by Attouch and Wets [1989]. The definitions and results concerning horizon epi-limits and total epi-convergence are new as well, as is the epi-projection result in 7.57.

The distance measures $\{\hat{d\!l}_\rho^+\}_{\rho \geq 0}$ on $\text{lsc-fcns}_{\not\equiv\infty}(\mathbb{R}^n)$ were introduced by Attouch and Wets [1991], [1993a], [1993b], specifically to study the stability of optimal values and optimal solutions. The expression used in these papers is different from the one adopted here, which is tuned more directly to functions than their epigraphs, but the two versions are equivalent as seen from the proof of 7.61. Corresponding pseudo-metrics $\{d\!l_\rho^+\}_{\rho>0}$, like $d\!l_\rho$ but based on the norm for $I\!R^n \times I\!R$ that is generated by the cylinder $I\!B \times [-1,1]$, were developed earlier by Attouch and Wets [1986] as a limit case of another class of pseudo-metrics, namely

$$d\!l_{\lambda,\rho}(f,g) := \sup_{|x|\leq\rho} \left|e_\lambda f(x) - e_\lambda g(x)\right| \text{ for } \lambda > 0,\ \rho \geq 0.$$

The relationship between $\hat{d\!l}_\rho^+$ and $d\!l_\rho^+$ was clarified in Attouch and Wets [1991] and Attouch, Lucchetti and Wets [1991]. The $\hat{d\!l}_\rho^+$ distance estimate in 7.62 for sums of Lipschitz continuous functions and indicators hasn't been worked out until now. The characterization of total epi-convergence in Theorem 7.59 is new as well.

The stability results in 7.63–7.65 and 7.67–7.69 have ancestry in the work of Attouch and Wets [1993a] but show some improvements and original features, especially in the case of the fundamental theorem in 7.64. Most notable among these is the extension of solution estimates and 'conditioning' to problems where the 'argmin' set might not be a singleton. For proximal mappings, the estimate in 7.66 is the first to be recorded. In his analysis of the inequalities derived in Theorem 7.69, Georg Pflug concluded that although the rate coefficient $(1+4\rho\varepsilon^{-1})$ might not be sharp, it might be so in a 'practical' sense, i.e., one could find examples where the convergence of the $\varepsilon-$argmin is arbitrarily slow and approaches the given rate.

8. Subderivatives and Subgradients

Maximization and minimization are often useful in constructing new functions and mappings from given ones, but, in contrast to addition and composition, they commonly fail to preserve smoothness. These operations, and others of prime interest in variational analysis, fit poorly in the traditional environment of differential calculus. The conceptual platform for 'differentiation' needs to be enlarged in order to cope with such circumstances.

Notions of semidifferentiability and epi-differentiability have already been developed in Chapter 7 as a start to this project. The task is carried forward now in a thorough application of the variational geometry of Chapter 6 to epigraphs. 'Subderivatives' and 'subgradients' are introduced as counterparts to tangent and normal vectors and shown to enjoy various useful relationships. Alongside of general subderivatives and subgradients, there are 'regular' ones of more special character. These are intimately tied to the regular tangent and normal vectors of Chapter 6 and show aspects of convexity. The geometric paradigm of Figure 6–17 finds its reflection in Figure 8–9, which schematizes the framework in which all these entities hang together.

Subdifferential regularity of f at $\bar{x}$, as defined in 7.25 through Clarke regularity of epi f at $\big(\bar{x}, f(\bar{x})\big)$, comes to be identified with the property that the subderivatives of f at $\bar{x}$ are regular subderivatives and, in parallel, with the property that all of the subgradients of f at $\bar{x}$ are regular subgradients, in a cosmic sense based on the ideas of Chapter 3. This gives direct significance to the term 'regular', beyond its echo from variational geometry. In the presence of regularity, the subgradients and subderivatives of a function f are completely dual to each other. The classical duality between gradient vectors and the linear functions giving directional derivatives is vastly expanded through the one-to-one correspondence between closed convex sets and the sublinear functions that describe their supporting half-spaces.

For functions f that aren't subdifferentially regular, subderivatives and subgradients can have distinct and independent roles, and some of the duality must be relinquished. Even then, however, the two kinds of objects tell much about each other, and their theory has powerful consequences in areas such as the study of optimality conditions.

After exploring what variational geometry says for epigraphs of functions $f : I\!R^n \to \overline{I\!R}$, we apply it to graphs of mappings $S : I\!R^n \rightrightarrows I\!R^m$, substituting graphical convergence for epi-convergence. This leads to graphical 'derivative' and 'coderivative' mappings, which likewise exhibit a far-reaching duality, as depicted in Figure 8–12. The fundamental themes of classical differential analysis thus achieve a robust extension to realms of objects remote from the mathematics of the past, but recognized now as essential in the analysis of numerous problems of practical importance.

The study of tangent and normal cones in Chapter 6 was accompanied by substantial rules of calculus, making it possible to see right away how the results could be brought down to specific cases and used for instance in strengthening the theory of Lagrange multipliers. Here, however, we are faced with material so rich that most such elaborations, although easy with the tools at hand, have to be postponed. They will be covered in Chapter 10, where they can benefit too from the subdifferential theory of Lipschitzian properties that will be made available in Chapter 9.

A. Subderivatives of Functions

'Semiderivatives' and 'epi-derivatives' of f at $\bar{x}$ for a vector $\bar{w}$ have been defined in 7.20 and 7.23 in terms of certain limits of the difference quotient functions $\Delta_\tau f(x) : \mathbb{R}^n \to \overline{\mathbb{R}}$, where

$$\Delta_\tau f(x)(w) := \frac{f(x+\tau w) - f(x)}{\tau} \quad \text{for } \tau > 0.$$

For general purposes, even in the exploration of those concepts, it's useful to work with underlying 'subderivatives' whose existence is never in doubt.

8.1 Definition (subderivatives). *For a function $f : \mathbb{R}^n \to \overline{\mathbb{R}}$ and a point $\bar{x}$ with $f(\bar{x})$ finite, the subderivative function $df(\bar{x}) : \mathbb{R}^n \to \overline{\mathbb{R}}$ is defined by*

$$df(\bar{x})(\bar{w}) := \liminf_{\substack{\tau \searrow 0 \\ w \to \bar{w}}} \frac{f(\bar{x}+\tau w) - f(\bar{x})}{\tau}.$$

Thus, $df(\bar{x})$ is the lower epi-limit of the family of functions $\Delta_\tau f(\bar{x})$:

$$df(\bar{x}) := \operatorname*{e-lim\,inf}_{\tau \searrow 0} \Delta_\tau f(\bar{x}).$$

The value $df(\bar{x})(\bar{w})$ is more specifically the *lower* subderivative of f at $\bar{x}$ for $\bar{w}$. Corresponding *upper* subderivatives are defined with 'lim sup' in place of 'lim inf'. (We take the prefix 'sub' as indicating a logical order, rather than a spatial order.) Because the situations in which both upper and lower subderivatives have to be dealt with simultaneously are relatively infrequent, we find it better here not to weigh down the notation for subderivatives with compulsory distinctions, such as writing $d^- f(\bar{x})$ for $df(\bar{x})$ and $d^+ f(\bar{x})$ for its upper counterpart, although this could well be desirable in some other contexts. For simplicity in this exposition, we take 'lower' for granted and rely on the fact, when actually needed, that the upper subderivative function can be denoted by $-d(-f)(\bar{x})$: it's evident that

$$-d(-f)(\bar{x})(\bar{w}) = \limsup_{\substack{\tau \searrow 0 \\ w \to \bar{w}}} \frac{f(\bar{x}+\tau w) - f(\bar{x})}{\tau}.$$

The notation $df(\bar{x})(w)$ has already been employed ahead of its time in descriptions of semidifferentiability in 7.21 and epi-differentiability in 7.26, but the subderivative function $df(\bar{x})$ will now be studied for its own sake. As in Chapter 7, a bridge into the geometry of epi f is provided by the fact that

$$\text{epi } \Delta_\tau f(\bar{x}) = \frac{\text{epi} f - \big(\bar{x}, f(\bar{x})\big)}{\tau} \quad \text{for any } \tau > 0. \tag{8(1)}$$

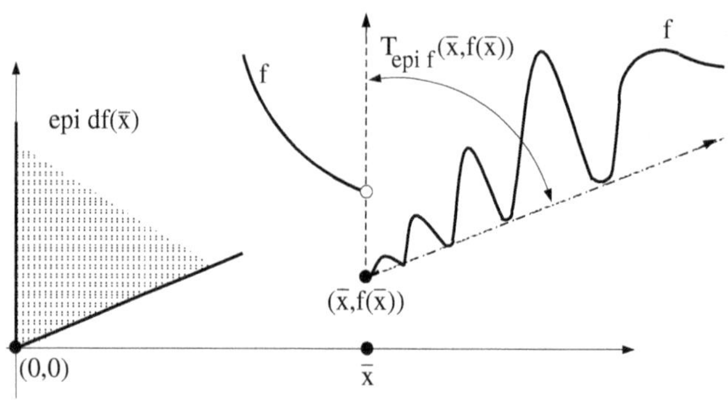

Fig. 8–1. Epigraphical interpretation of subderivatives.

8.2 Theorem (subderivatives versus tangents).

(a) *For $f : \mathbb{R}^n \to \overline{\mathbb{R}}$ and any point $\bar{x}$ with $f(\bar{x})$ finite, one has*

$$\text{epi}\, df(\bar{x}) = T_{\text{epi} f}\big(\bar{x}, f(\bar{x})\big).$$

(b) *For the indicator δ_C of a set $C \subset \mathbb{R}^n$ and any point $\bar{x} \in C$, one has*

$$d\delta_C(\bar{x}) = \delta_K \ \text{ for } K = T_C(\bar{x}).$$

Proof. The first relation is immediate from 8(1), Definition 8.1, and the formula for $T_{\text{epi} f}\big(\bar{x}, f(\bar{x})\big)$ as derived from Definition 6.1. The second is obvious from the definitions as well. □

B. Subgradients of Functions

Just as subderivatives express the tangent cone geometry of epigraphs, subgradients will express the normal cone geometry. In developing this idea and using it as a vehicle for translating the variational geometry of Chapter 6 into analysis, we'll sometimes make use of the notion of *local lower semicontinuity* of a function $f : \mathbb{R}^n \to \overline{\mathbb{R}}$ as defined in 1.33, since that corresponds to local closedness of the set epi f (cf. 1.34). We usually won't wish to assume f is continuous, so we'll often appeal to f-*attentive* convergence in the notation

$$x^\nu \underset{f}{\to} \bar{x} \iff x^\nu \to \bar{x} \text{ with } f(x^\nu) \to f(\bar{x}). \qquad 8(2)$$

8.3 Definition (subgradients). *Consider a function $f : I\!R^n \to \overline{I\!R}$ and a point $\bar{x}$ with $f(\bar{x})$ finite. For a vector $v \in I\!R^n$, one says that*

(a) v is a regular subgradient of f at $\bar{x}$, written $v \in \widehat{\partial} f(\bar{x})$, if

$$f(x) \geq f(\bar{x}) + \langle v, x - \bar{x} \rangle + o(|x - \bar{x}|); \qquad 8(3)$$

(b) v is a (general) subgradient of f at $\bar{x}$, written $v \in \partial f(\bar{x})$, if there are sequences $x^\nu \underset{f}{\to} \bar{x}$ and $v^\nu \in \widehat{\partial} f(x^\nu)$ with $v^\nu \to v$;

(c) v is a horizon subgradient of f at $\bar{x}$, written $v \in \partial^\infty f(\bar{x})$, if the same holds as in (b), except that instead of $v^\nu \to v$ one has $\lambda^\nu v^\nu \to v$ for some sequence $\lambda^\nu \searrow 0$, or in other words, $v^\nu \to \operatorname{dir} v$ (or $v = 0$).

The regular subgradient inequality 8(3) in 'o' form is shorthand for the one-sided limit condition

$$\liminf_{\substack{x \to \bar{x} \\ x \neq \bar{x}}} \frac{f(x) - f(\bar{x}) - \langle v, x - \bar{x} \rangle}{|x - \bar{x}|} \geq 0. \qquad 8(4)$$

Regular subgradients have immediate appeal but are inadequate in themselves for a satisfactory theory, in particular for a robust calculus covering all the properties needed. This is why limits are introduced in Definition 8.3.

The difference between $v \in \partial f(\bar{x})$ and $v \in \partial^\infty f(\bar{x})$ corresponds to the two modes of convergence that a sequence of vectors v^ν can have in the cosmic setting of Chapter 3. In the notation of ordinary and horizon set limits in Chapter 4, one has

$$\partial f(\bar{x}) = \limsup_{x \underset{f}{\to} \bar{x}} \widehat{\partial} f(x), \qquad \partial^\infty f(\bar{x}) = \limsup_{x \underset{f}{\to} \bar{x}}{}^{\infty} \widehat{\partial} f(x), \qquad 8(5)$$

cf. 4.1, 4(6), 5(1). This double formula amounts to taking a cosmic outer limit in the space csm $I\!R^n$ (cf. 4.20 and 5.27):

$$\partial f(\bar{x}) \cup \operatorname{dir} \partial^\infty f(\bar{x}) = \text{c-}\limsup_{x \underset{f}{\to} \bar{x}} \widehat{\partial} f(x). \qquad 8(6)$$

While $\partial f(\bar{x})$ consists of all ordinary limits of sequences of regular subgradients v^ν at points x^ν that approach $\bar{x}$ in such a way that $f(x^\nu)$ approaches $f(\bar{x})$, $\operatorname{dir} \partial^\infty f(\bar{x})$ consists of all points of hzn $I\!R^n$ that arise as limits when $|v^\nu| \to \infty$.

8.4 Exercise (regular subgradients from subderivatives). *For $f : I\!R^n \to \overline{I\!R}$ and any point $\bar{x}$ with $f(\bar{x})$ finite, one has*

$$\widehat{\partial} f(\bar{x}) = \{ v \mid \langle v, w \rangle \leq df(\bar{x})(w) \text{ for all } w \}. \qquad 8(7)$$

Guide. Deduce this from version 8(4) of the regular subgradient inequality through a compactness argument based on the definition of $df(\bar{x})$. □

The regular subgradients of f at $\bar{x}$ can be characterized as the gradients at $\bar{x}$ of the smooth functions $h \leq f$ with $h(\bar{x}) = f(\bar{x})$, as we prove next. This is a valuable tool in subgradient theory, because it opens the way to developing properties of vectors $v \in \widehat{\partial} f(\bar{x})$ out of optimality conditions associated in particular cases with having $\bar{x} \in \operatorname{argmin}\{f - h\}$.

8.5 Proposition (variational description of regular subgradients). *A vector v belongs to $\widehat{\partial} f(\bar{x})$ if and only if, on some neighborhood of $\bar{x}$, there is a function $h \leq f$ with $h(\bar{x}) = f(\bar{x})$ such that h is differentiable at $\bar{x}$ with $\nabla h(\bar{x}) = v$. Moreover h can be taken to be smooth with $h(x) < f(x)$ for all $x \neq \bar{x}$ near $\bar{x}$.*

Proof. The proof of Theorem 6.11 showed that a term is of type $o(|x - \bar{x}|)$ if and only if its absolute value is bounded on some neighborhood of $\bar{x}$ by a function k with $\nabla k(\bar{x}) = 0$, and then k can be taken actually to be smooth and positive away from $\bar{x}$. Here we let $h(x) = \langle v, x - \bar{x} \rangle - k(x)$. □

8.6 Theorem (subgradient relationships). *For a function $f : I\!R^n \to \overline{I\!R}$ and a point $\bar{x}$ where f is finite, the subgradient sets $\partial f(\bar{x})$ and $\widehat{\partial} f(\bar{x})$ are closed, with $\widehat{\partial} f(\bar{x})$ convex and $\widehat{\partial} f(\bar{x}) \subset \partial f(\bar{x})$. Furthermore, $\partial^\infty f(\bar{x})$ and $\widehat{\partial} f(\bar{x})^\infty$ are closed cones, with $\widehat{\partial} f(\bar{x})^\infty$ convex and $\widehat{\partial} f(\bar{x})^\infty \subset \partial^\infty f(\bar{x})$.*

Proof. The properties of $\partial f(\bar{x})$ and $\partial^\infty f(\bar{x})$ are immediate from Definition 8.3 in the background of the cosmic space csm $I\!R^n$ in Chapter 3 and the associated notions of set convergence in Chapter 4. The convexity of $\widehat{\partial} f(\bar{x})$ is evident from the defining inequality 8(3), so the horizon cone $\widehat{\partial} f(\bar{x})^\infty$ is closed and convex (by 3.6). The closedness of $\widehat{\partial} f(\bar{x})$ is apparent from 8.4. □

8.7 Proposition (subgradient semicontinuity). *For a function $f : I\!R^n \to \overline{I\!R}$ and a point $\bar{x}$ where f is finite, the mappings ∂f and $\partial^\infty f$ are osc at $\bar{x}$ with respect to f-attentive convergence $x \underset{f}{\to} \bar{x}$, while the mapping $x \mapsto \partial f(x) \cup \operatorname{dir} \partial^\infty f(x)$ is cosmically osc at $\bar{x}$ with respect to such convergence.*

Proof. This is clear from 8(6) and the meaning of cosmic limits; cf. 5.27. □

The subgradients in Definition 8.3 are more specifically *lower* subgradients. The opposite inequalities define *upper* subgradients. Obviously a vector v is both a regular lower subgradient and a regular upper subgradient of f at $\bar{x}$ if and only if f is differentiable at $\bar{x}$ and $v = \nabla f(\bar{x})$.

Although lower and upper subgradients sometimes need to be considered simultaneously, this is uncommon in our inherently one-sided context. For this reason we take 'lower' for granted as in the case of subderivatives, and rather than introducing a more complicated symbolism, prefer here to note that the various sets of upper subgradients can be identified with $-\widehat{\partial}(-f)(\bar{x})$, $-\partial(-f)(\bar{x})$, and $-\partial^\infty(-f)(\bar{x})$. A convenient alternative in other contexts is to write $\partial^- f(\bar{x})$ for lower subgradients and $\partial^+ f(\bar{x})$ for upper subgradients.

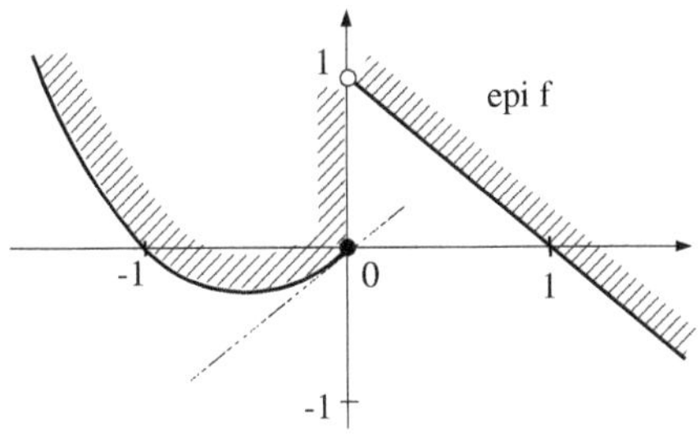

Fig. 8–2. The role of f-attentive convergence in generating subgradients.

The effect of insisting on $x^\nu \underset{f}{\to} \bar{x}$ instead of just $x^\nu \to \bar{x}$ when defining $\partial f(\bar{x})$ is illustrated in Figure 8–2 for the lsc function f on $I\!R^1$ given by

$$f(x) = x^2 + x \text{ when } x \leq 0, \qquad f(x) = 1 - x \text{ when } x > 0.$$

Here f is (subdifferentially) regular everywhere. We have $\partial f(0) = \widehat{\partial} f(0) = [1, \infty)$ and $\partial^\infty f(0) = [0, \infty)$. The requirement that $f(x^\nu) \to f(0)$ excludes sequences $x^\nu \searrow 0$, which with $v^\nu = -1$ would have limit $v = -1$. The role of f-attentive convergence is thus to ensure that the subgradients at $\bar{x}$ reflect no more than the local geometry of $\operatorname{epi} f$ around $(\bar{x}, f(\bar{x}))$. This example shows also how horizon subgradients may arise from what amounts to a kind of local constraint at $x = 0$ on the behavior of f.

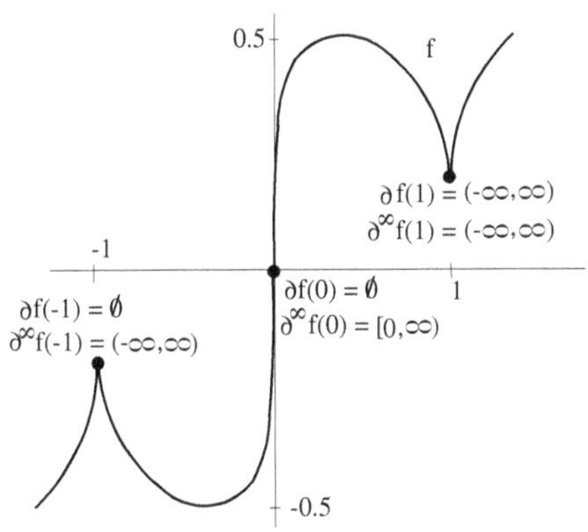

Fig. 8–3. Example of horizon subgradients.

A different sort of example with f is given in Figure 8–3 with

$$f(x) = 0.6\big(|x-1|^{1/2} - |x+1|^{1/2}\big) + x^{1/3} \text{ for } x \in I\!R.$$

In this case f is regular everywhere except -1. We have $\partial^\infty f(1) = \partial^\infty f(-1) = (-\infty, \infty)$ but $\partial^\infty f(0) = [0, \infty)$; elsewhere, $\partial^\infty f(x) = \{0\}$. Also, $\partial f(0) = \partial f(-1) = \emptyset$, whereas $\partial f(1) = (-\infty, \infty)$.

8.8 Exercise (subgradients versus gradients).

(a) *If f_0 is differentiable at $\bar{x}$, then $\widehat{\partial} f_0(\bar{x}) = \{\nabla f_0(\bar{x})\}$, so $\nabla f_0(\bar{x}) \in \partial f_0(\bar{x})$.*

(b) *If f_0 is smooth on a neighborhood of $\bar{x}$, then $\partial f_0(\bar{x}) = \{\nabla f_0(\bar{x})\}$ and $\partial^\infty f_0(\bar{x}) = \{0\}$, so the cosmic set $\partial f_0(\bar{x}) \cup \operatorname{dir} \partial^\infty f_0(\bar{x})$ is a singleton.*

(c) *If $f = g + f_0$ with g finite at $\bar{x}$ and f_0 smooth on a neighborhood of $\bar{x}$, then $\widehat{\partial} f(\bar{x}) = \widehat{\partial} g(\bar{x}) + \nabla f_0(\bar{x})$, $\partial f(\bar{x}) = \partial g(\bar{x}) + \nabla f_0(\bar{x})$, and $\partial^\infty f(\bar{x}) = \partial^\infty g(\bar{x})$.*

Guide. These facts follow closely from the definitions. Get the uniqueness in (a) by demonstrating that if v and v' satisfy $\langle v, x - \bar{x}\rangle \geq \langle v', x - \bar{x}\rangle + o(|x - \bar{x}|)$, then $v = v'$. Get the inclusions '$\supset$' in (c) directly. Then get the inclusions '$\subset$' by applying this rule to the representation $g = f + (-f_0)$. □

An example in which f is differentiable at $\bar{x}$, yet $\partial f(\bar{x})$ consists of more than just $\nabla f(\bar{x})$, is the following. Define $f : \mathbb{R}^1 \to \mathbb{R}$ by

$$f(x) = \begin{cases} x^2 \sin(1/x) & \text{when } x \neq 0, \\ 0 & \text{when } x = 0, \end{cases}$$

and take $\bar{x} = 0$. Then f is differentiable everywhere, but the derivative mapping (viewed here as the 'gradient' mapping) is discontinuous at $\bar{x}$, and f isn't regular there. Although $\widehat{\partial} f(0) = \{0\}$, we have $\partial f(0) = [-1, 1]$, and incidentally $\partial^\infty f(0) = \{0\}$. In substituting $x^2 \sin(1/x^2)$ for $x^2 \sin(1/x)$, one obtains an example with $\widehat{\partial} f(0) = \{0\}$, $\partial f(0) = (-\infty, \infty)$ and $\partial^\infty f(0) = (-\infty, \infty)$.

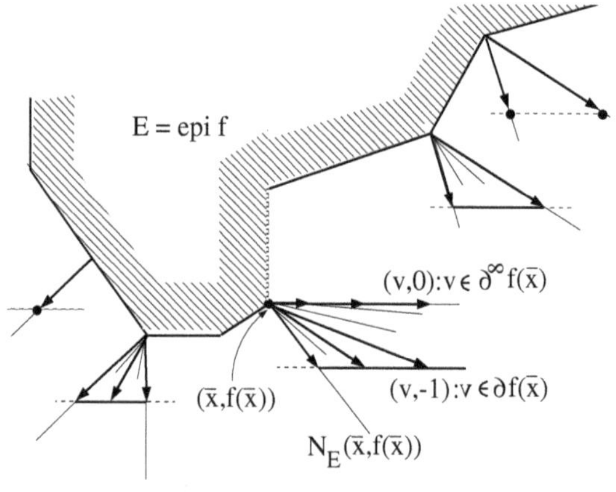

Fig. 8–4. Normals to epigraphs.

Subgradients have important connections to normal vectors through the variational geometry of epigraphs.

8.9 Theorem (subgradients from epigraphical normals). *For $f : \mathbb{R}^n \to \overline{\mathbb{R}}$ and any point $\bar{x}$ at which f is finite, one has*

$$\begin{aligned} \widehat{\partial} f(\bar{x}) &= \{v \mid (v,-1) \in \widehat{N}_{\operatorname{epi} f}(\bar{x}, f(\bar{x}))\}, \\ \partial f(\bar{x}) &= \{v \mid (v,-1) \in N_{\operatorname{epi} f}(\bar{x}, f(\bar{x}))\}, \\ \partial^\infty f(\bar{x}) &\subset \{v \mid (v,0) \in N_{\operatorname{epi} f}(\bar{x}, f(\bar{x}))\}. \end{aligned}$$

The last relationship holds with equality when f is locally lsc at $\bar{x}$, and then

$$N_{\operatorname{epi} f}(\bar{x}, f(\bar{x})) = \{\lambda(v,-1) \mid v \in \partial f(\bar{x}),\ \lambda > 0\} \cup \{(v,0) \mid v \in \partial^\infty f(\bar{x})\}.$$

On the other hand, whenever $\widehat{\partial} f(\bar{x}) \neq \emptyset$ one has

$$\widehat{N}_{\operatorname{epi} f}(\bar{x}, f(\bar{x})) = \{\lambda(v,-1) \mid v \in \widehat{\partial} f(\bar{x}),\ \lambda > 0\} \cup \{(v,0) \mid v \in \widehat{\partial} f(\bar{x})^\infty\}.$$

Proof. Let $E = \operatorname{epi} f$. The formula in 8.4 can be construed as saying that $v \in \widehat{\partial} f(\bar{x})$ if and only if $\langle (v,-1), (w,\beta) \rangle \leq 0$ for all $(w,\beta) \in \operatorname{epi} df(\bar{x})$. Through 8.2(a) and 6.5 this corresponds to $(v,-1) \in \widehat{N}_{\operatorname{epi} f}(\bar{x}, f(\bar{x}))$. The analogous description of $\partial f(\bar{x})$ is immediate then from Definition 8.3 and the definition of normals as limits of regular normals, and so too is the inclusion for $\partial^\infty f(\bar{x})$.

We see also through 8.2(a) and 6.5 that $(v,0) \in \widehat{N}_{\operatorname{epi} f}(\bar{x}, f(\bar{x}))$ if and only if $\langle (v,0), (w,\beta) \rangle \leq 0$ for all $(w,\beta) \in \operatorname{epi} df(\bar{x})$, or equivalently, $\langle v, w \rangle \leq 0$ for all $w \in \operatorname{dom} df(\bar{x})$. In view of 8(7), this is equivalent to having $v \in \widehat{\partial} f(\bar{x})^\infty$ as long as $\widehat{\partial} f(\bar{x}) \neq \emptyset$; cf. 3.24. In combination with the description achieved for $\widehat{\partial} f(\bar{x})$ this yields the formula for $\widehat{N}_{\operatorname{epi} f}(\bar{x}, f(\bar{x}))$.

The equation for $\partial f(\bar{x})$ and inclusion for $\partial^\infty f(\bar{x})$ yield at the same time the inclusion '$\supset$' in the formula asserted for $N_{\operatorname{epi} f}(\bar{x}, f(\bar{x}))$. To complete the proof of the theorem we henceforth assume that f is locally lsc at $\bar{x}$ and aim at verifying that whenever $(v,0) \in N_{\operatorname{epi} f}(\bar{x}, f(\bar{x}))$ we have $v \in \partial^\infty f(\bar{x})$. By definition, any normal $(v,0)$ can be generated as a limit of regular normals at nearby points, and by 6.18(a) we can even limit attention to proximal normals. Any proximal normal to E at a point (x,α) with $\alpha > f(x) > -\infty$ is obviously a proximal normal to E at $(x, f(x))$ as well. Thus any normal $(v,0)$ to E at $(\bar{x}, f(\bar{x}))$ can be generated as a limit of proximal normals at points $(x^\nu, f(x^\nu)) \to (\bar{x}, f(\bar{x}))$, i.e., with $x^\nu \xrightarrow[f]{} \bar{x}$. This can be reduced to two cases: in the first the proximal normals have the 'nonhorizontal' form $\lambda^\nu(v^\nu,-1)$ with $\lambda^\nu \searrow 0$, $\lambda^\nu v^\nu \to v$, whereas in the second they have the 'horizontal' form $(v^\nu, 0)$ with $v^\nu \to v$. In the first case we have $v^\nu \in \widehat{\partial} f(x^\nu)$ in particular, so that $v \in \partial^\infty f(\bar{x})$ by definition. Therefore, only the second case need concern us further. To handle it, we only have to demonstrate that the vectors v^ν in this case belong themselves to $\partial^\infty f(x^\nu)$, because the result will then follow from the semicontinuity in 8.7.

The question has come down to this. If $(\tilde{v}, 0)$ is a proximal normal to E at a point $(\tilde{x}, f(\tilde{x}))$, is $\tilde{v} \in \partial^\infty f(\tilde{x})$? In answering this we can harmlessly restrict to the case where f is lsc on all of $I\!R^n$ with $\operatorname{dom} f$ bounded (as can be arranged on the basis of the local lower semicontinuity of f at $\tilde{x}$ by adding to f the indicator of some neighborhood of $\tilde{x}$ intersected with some level set

$\operatorname{lev}_{\leq \alpha} f)$. To say that $(\tilde{v}, 0)$ is a proximal normal to E at $(\tilde{x}, f(\tilde{x}))$ is to say that for some $\varepsilon > 0$ there is a closed ball B around $(\tilde{x}, f(\tilde{x})) + \varepsilon(\tilde{v}, 0)$ that touches E only at $(\tilde{x}, f(\tilde{x}))$. Since we can replace $\tilde{v}$ by any of its positive scalar multiples without affecting the issue, and can even rescale the whole space as convenience dictates, we can simplify the investigation to the case where $|\tilde{v}| = 1$ and the ball B has radius 1, as depicted in Figure 8–5 with $f(\tilde{x}) = 0$. In terms of the function

$$g(x) = \theta(|x - [\tilde{x} + \tilde{v}]|) \ \text{ for } \ \theta(t) = \begin{cases} -\sqrt{1 - t^2} & \text{for } 0 \leq t \leq 1, \\ \infty & \text{for } t > 1, \end{cases}$$

whose graph is the 'lower surface' of $\mathbb{B}((\tilde{x} + \tilde{v}, 0), 1)$, we have $\operatorname{argmin}\{f + g\} = \{\tilde{x}\}$. Equivalently, $(\tilde{x}, \tilde{x})$ is the unique optimal solution to the problem

$$\text{minimize } \ f(x) + g(u) \ \text{ over } \ (x, u) \in \mathbb{R}^n \times \mathbb{R}^n \ \text{ subject to } \ x - u = 0.$$

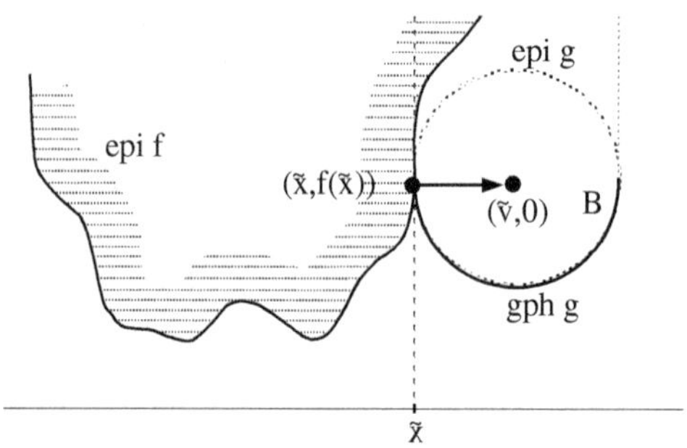

Fig. 8–5. Setting for the proximal normal argument.

For a sequence of positive penalty values $r^\nu \nearrow \infty$ consider the approximate problems

$$\text{minimize } \ f(x) + g(u) + r^\nu |x - u| \ \text{ over } \ (x, u) \in \mathbb{R}^n \times \mathbb{R}^n,$$

selecting for each an optimal solution (x^ν, u^ν), which must exist because of the boundedness of the effective domain of the function being minimized, this being $\operatorname{dom} f \times \operatorname{dom} g$. Here it's impossible for any ν that $x^\nu - u^\nu = 0$, for that would necessitate $(x^\nu, u^\nu) = (\tilde{x}, \tilde{x})$ and imply that

$$f(\tilde{x}) + g(u) + r^\nu |\tilde{x} - u| \geq f(\tilde{x}) + g(\tilde{x}) \ \text{ for all } \ u,$$

in contradiction to the nature of g. From 1.21 we know, however, that $(x^\nu, u^\nu) \to (\tilde{x}, \tilde{x})$ and also that $f(x^\nu) \to f(\tilde{x})$ and $g(u^\nu) \to g(\tilde{x}) = 0$. We have

$$\begin{aligned} f(x) + g(u^\nu) + r^\nu |x - u^\nu| \ &\geq \ f(x^\nu) + g(u^\nu) + r^\nu |x^\nu - u^\nu| \ \text{ for all } \ x, \\ f(x^\nu) + g(u) + r^\nu |x^\nu - u| \ &\geq \ f(x^\nu) + g(u^\nu) + r^\nu |x^\nu - u^\nu| \ \text{ for all } \ u, \end{aligned}$$

so that for the functions $h^\nu(x) := f(x^\nu) + r^\nu |x^\nu - u^\nu| - r^\nu |x - u^\nu|$ and $k^\nu(u) :=$

$g(x^\nu) + r^\nu|x^\nu - u^\nu| - r^\nu|x^\nu - u|$ we have

$$f(x) \geq h^\nu(x) \text{ for all } x, \qquad f(x^\nu) = h^\nu(x^\nu),$$
$$g(u) \geq k^\nu(u) \text{ for all } u, \qquad g(u^\nu) = k^\nu(u^\nu).$$

Let $v^\nu = \nabla h^\nu(x^\nu) = -r^\nu(x^\nu - u^\nu)/|x^\nu - u^\nu|$, observing that $-v^\nu = \nabla k^\nu(u^\nu)$. Then $v^\nu \in \widehat{\partial} f(x^\nu)$ by 8.5, but on the same basis also $-v^\nu \in \widehat{\partial} g(u^\nu)$, hence $(-v^\nu, -1) \in N_{\operatorname{epi} g}\big(u^\nu, g(u^\nu)\big)$ so that $(-v^\nu, -1) \in N_B\big(u^\nu, g(u^\nu)\big)$ and consequently $(-v^\nu, 1) \in N_B\big(u^\nu, -g(u^\nu)\big)$. Let $\lambda^\nu = 1/r^\nu$; then $\lambda^\nu \searrow 0$ and $(-\lambda^\nu v^\nu, \lambda^\nu) \in N_B\big(u^\nu, -g(u^\nu)\big)$ with $|\lambda^\nu v^\nu| = 1$. Since $\big(u^\nu, -g(u^\nu)\big) \to (\tilde{x}, 0)$, while the unique unit normal to B at $(\tilde{x}, 0)$ is $(-\tilde{v}, 0)$, it follows that $\lambda^\nu v^\nu \to \tilde{v}$. Since $v^\nu \in \widehat{\partial} f(x^\nu)$ and $\lambda^\nu \searrow 0$, we conclude that $\tilde{v} \in \partial^\infty f(\tilde{x})$. □

The description of the normal cone $N_{\operatorname{epi} f}\big(\bar{x}, f(\bar{x})\big)$ provided by Theorem 8.9 fits neatly into the ray space model for $\operatorname{csm} \mathbb{R}^n$ in Chapter 3. It's the ray bundle associated with the set $\partial f(\bar{x}) \cup \operatorname{dir} \partial^\infty f(\bar{x})$ in $\operatorname{csm} \mathbb{R}^n$.

8.10 Corollary (existence of subgradients). *Suppose $f : \mathbb{R}^n \to \overline{\mathbb{R}}$ is finite and locally lsc at $\bar{x}$. Then either $\partial f(\bar{x}) \neq \emptyset$ or $\partial^\infty f(\bar{x})$ contains a vector $v \neq 0$; thus, the set $\partial f(\bar{x}) \cup \operatorname{dir} \partial^\infty f(\bar{x}) \subset \operatorname{csm} \mathbb{R}^n$ has to be nonempty. Either way, there must be a sequence $x^\nu \underset{f}{\to} \bar{x}$ with $\widehat{\partial} f(x^\nu) \neq \emptyset$ (and hence $\partial f(x^\nu) \neq \emptyset$).*

Proof. The assumptions imply $\operatorname{epi} f$ is locally closed at its boundary point $\big(\bar{x}, f(\bar{x})\big)$, so the normal cone there can't be just the zero cone; cf. 6.19. The claims are evident then from 8.9 and the definitions of $\partial f(\bar{x})$ and $\partial^\infty f(\bar{x})$. □

8.11 Corollary (subgradient criterion for subdifferential regularity). *For a function $f : \mathbb{R}^n \to \overline{\mathbb{R}}$ and a point $\bar{x}$ with $f(\bar{x})$ finite and $\partial f(\bar{x}) \neq \emptyset$, one has f regular at $\bar{x}$ if and only f is locally lsc at $\bar{x}$ with*

$$\partial f(\bar{x}) = \widehat{\partial} f(\bar{x}), \qquad \partial^\infty f(\bar{x}) = \widehat{\partial} f(\bar{x})^\infty.$$

Proof. By 8.9, these conditions mean that $\operatorname{epi} f$ is locally closed at $\big(\bar{x}, f(\bar{x})\big)$ with $N_{\operatorname{epi} f}\big(\bar{x}, f(\bar{x})\big) = \widehat{N}_{\operatorname{epi} f}\big(\bar{x}, f(\bar{x})\big)$. Those properties are equivalent by 6.29 to the regularity of $\operatorname{epi} f$ at $\big(\bar{x}, f(\bar{x})\big)$, hence the regularity of f at $\bar{x}$. □

As the proof of Theorem 8.9 reveals, regardless of whether $\widehat{\partial} f(\bar{x}) \neq \emptyset$ the vectors $(v, 0)$ that are regular normals to $\operatorname{epi} f$ at $\big(\bar{x}, f(\bar{x})\big)$ are always those such that $\langle v, w \rangle \leq 0$ for all w with $df(\bar{x})(w) < \infty$. Therefore, in defining the set of *regular horizon subgradients* to be the polar cone

$$\widehat{\partial}^\infty f(\bar{x}) := \big[\operatorname{dom} df(\bar{x})\big]^*$$

and substituting this for $\widehat{\partial} f(\bar{x})^\infty$ we can obtain for $\widehat{N}_{\operatorname{epi} f}\big(\bar{x}, f(\bar{x})\big)$ an alternative formula to the one in 8.9 that is always valid, and likewise an alternative criterion for regularity in 8.11 that doesn't require subgradients to exist. For most purposes, however, this extension is hardly needed.

C. Convexity and Optimality

The geometrical insights from the epigraphical framework of Theorem 8.9 are especially illuminating in the case of convex functions.

8.12 Proposition (subgradients of convex functions). *For any proper, convex function $f : \mathbb{R}^n \to \overline{\mathbb{R}}$ and any point $\bar{x} \in \operatorname{dom} f$, one has*

$$\partial f(\bar{x}) = \Big\{ v \,\Big|\, f(x) \geq f(\bar{x}) + \langle v, x - \bar{x} \rangle \ \text{ for all } x \Big\} = \widehat{\partial} f(\bar{x}),$$
$$\partial^\infty f(\bar{x}) \subset \Big\{ v \,\Big|\, 0 \geq \langle v, x - \bar{x} \rangle \ \text{ for all } x \in \operatorname{dom} f \Big\} = N_{\operatorname{dom} f}(\bar{x}).$$

The horizon subgradient inclusion is an equation when f is locally lsc at $\bar{x}$ or when $\partial f(\bar{x}) \neq \emptyset$, and in the latter case one also has $\partial^\infty f(\bar{x}) = \widehat{\partial} f(\bar{x})^\infty$.

Proof. The convexity of f implies that of $\operatorname{epi} f$ (cf. 2.4), and the normal cone to $\operatorname{epi} f$ at $(\bar{x}, f(\bar{x}))$ can be determined therefore from 6.9:

$$N_{\operatorname{epi} f}(\bar{x}, f(\bar{x})) = \widehat{N}_{\operatorname{epi} f}(\bar{x}, f(\bar{x}))$$
$$= \Big\{ (v, \beta) \,\Big|\, \big\langle (v, \beta), (x, \alpha) - (\bar{x}, f(\bar{x})) \big\rangle \leq 0 \ \text{ for all } (x, \alpha) \in \operatorname{epi} f \Big\}.$$

We only have to invoke 8.9 and the ray space representation of $\widehat{\partial} f(\bar{x})^\infty$. □

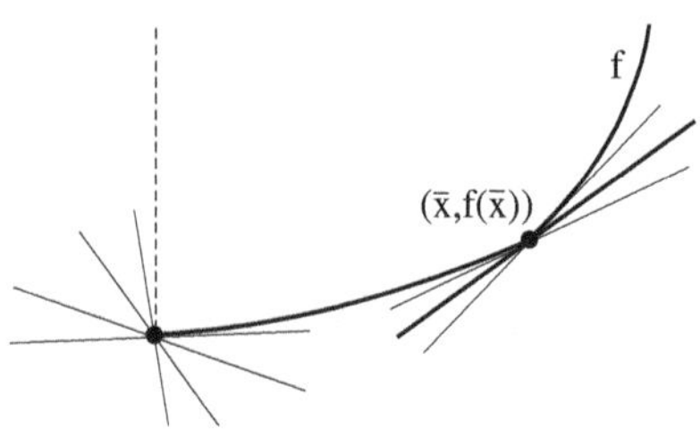

Fig. 8–6. Subgradient inequalities for convex functions.

The subgradient inequality for convex functions in 8.12 extends the gradient inequality for smooth convex functions in 2.14(b). It sets up a correspondence between subgradients and 'affine supports'. An affine function l is said to *support* a function $f : \mathbb{R}^n \to \overline{\mathbb{R}}$ at $\bar{x}$ if $l(x) \leq f(x)$ for all x, and $l(\bar{x}) = f(\bar{x})$. Proposition 8.12 says that *the subgradients of a proper convex function f at $\bar{x}$ are the gradients of its affine supports at $\bar{x}$.* On the other hand, nonzero horizon subgradients are nonzero normals to supporting half-spaces to $\operatorname{dom} f$ at $\bar{x}$; the converse is true as well if f is lsc or $\partial f(\bar{x}) \neq \emptyset$.

Subgradients of a convex function f correspond also, through their association in 8.9 with normals to the convex set $\operatorname{epi} f$, to supporting half-spaces to $\operatorname{epi} f$, and this has an important interpretation as well. In $\mathbb{R}^n \times \mathbb{R}$ the closed half-spaces fall in three categories: *upper, lower*, and *vertical.* The upper ones are the epigraphs of affine functions, and the lower ones the hypographs; in the first case a nonzero normal can be scaled to have the form $(a,-1)$, and in the second case $(a,1)$. The vertical ones have the form $H \times I\!R$ for a half-space H in $I\!R^n$; their normals take the form $(a,0)$. Clearly, the affine functions that support f at $\bar{x}$ correspond to the upper half-spaces that support $\operatorname{epi} f$ at $(\bar{x}, f(\bar{x}))$. In other words, the vectors of form $(v,-1)$ with $v \in \partial f(\bar{x})$ characterize such supporting half-spaces. The vectors of form $(v,0)$ with $v \in \partial^\infty f(\bar{x})$, $v \neq 0$, similarly give vertical supporting half-spaces to $\operatorname{epi} f$ at $(\bar{x}, f(\bar{x}))$ and characterize such half-spaces completely as long as f is lsc or $\partial f(\bar{x}) \neq \emptyset$. Of course, lower half-spaces can't support or even contain an epigraph.

8.13 Theorem (envelope representation of convex functions). *A proper, lsc, convex function $f : I\!R^n \to \overline{I\!R}$ is the pointwise supremum of its affine supports; it has such a support at every point of* $\operatorname{int}(\operatorname{dom} f)$ *when* $\operatorname{int}(\operatorname{dom} f) \neq \emptyset$. *Thus, a function on $I\!R^n$ is proper, lsc and convex, if and only if it is the pointwise supremum of a nonempty family of affine functions, but is not everywhere ∞.*

For a proper, lsc, sublinear function, supporting affine functions must be linear. On the other hand, the pointwise supremum of any nonempty collection of linear functions is a proper, lsc, sublinear function.

Proof. When f is proper, lsc and convex, $\operatorname{epi} f$ is nonempty, closed and convex and therefore by Theorem 6.20 it is the intersection of its supporting half-spaces, which exist at all its boundary points (by 6.19). The normals to such half-spaces are derived from the subgradients of f, as just seen. The normals to the vertical supporting half-spaces, if any, are limits of the normals to nonvertical supporting half-spaces at neighboring points of $\operatorname{epi} f$, because of the way horizon subgradients arise as limits of ordinary subgradients. The vertical half-spaces may therefore be omitted without affecting the intersection. Anyway, a supporting half-space at $(\bar{x}, f(\bar{x}))$ can't be vertical if $\bar{x} \in \operatorname{int}(\operatorname{dom} f)$. The nonvertical supporting half-spaces to $\operatorname{epi} f$ are the epigraphs of the affine functions that support f. An intersection of epigraphs corresponds to a pointwise supremum of the functions involved.

The case where f is sublinear corresponds to $\operatorname{epi} f$ being a closed, convex cone; cf. 3.19. Then the representations in 6.20 for cones can be used. □

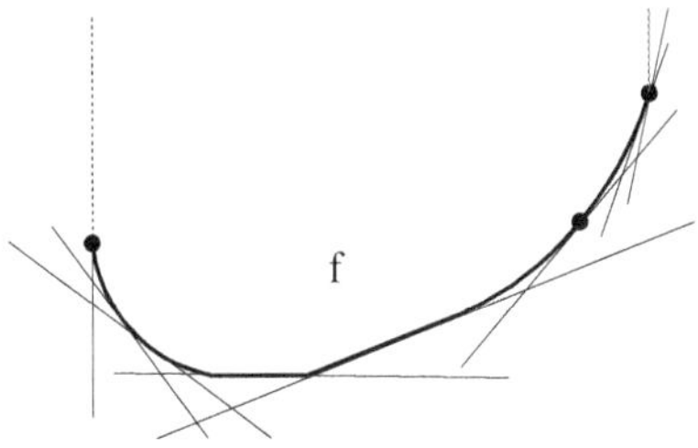

Fig. 8–7. A convex lsc function as the pointwise supremum of its affine supports.

The epigraphical connections with variational geometry in Theorem 8.9 are supplemented by connections through indicator functions.

8.14 Exercise (normal vectors as indicator subgradients). *The indicator δ_C of a set $C \subset I\!R^n$ is regular at $\bar{x}$ if and only if C is regular at $\bar{x}$. In general,*

$$\partial\delta_C(\bar{x}) = \partial^\infty\delta_C(\bar{x}) = N_C(\bar{x}), \qquad \widehat{\partial}\delta_C(\bar{x}) = \widehat{\partial}\delta_C(\bar{x})^\infty = \widehat{N}_C(\bar{x}).$$

Guide. Obtain these relations by applying the definitions to $f = \delta_C$. Convergence $x \underset{f}{\to} \bar{x}$ reduces in this case to $x \underset{C}{\to} \bar{x}$. □

The fact that the calculus of tangent and normal cones can be viewed through 8.14 as a special case of the calculus of subderivatives and subgradients is not only interesting for theory but crucial in applications such as the derivation of optimality conditions, since constraints in a problem of optimization are often expressed conveniently by indicator functions. While the power of this method won't be seen until Chapter 10, the following extension of the basic first-order conditions for optimality in Theorem 6.12 will give a taste of things to come. Where previously the conditions were limited to the minimization of a smooth function f_0 over a set C, now f_0 can be very general.

8.15 Theorem (optimality relative to a set). *Consider a problem of minimizing a proper, lsc function $f_0 : I\!R^n \to \overline{I\!R}$ over a closed set $C \subset I\!R^n$. Let $\bar{x}$ be a point of C at which the following constraint qualification is fulfilled: the set $\partial^\infty f_0(\bar{x})$ contains no vector $v \neq 0$ such that $-v \in N_C(\bar{x})$. Then for $\bar{x}$ to be locally optimal it is necessary that*

$$\partial f_0(\bar{x}) + N_C(\bar{x}) \ni 0,$$

which in the case of f_0 and C also being regular at $\bar{x}$ is equivalent to having

$$df_0(\bar{x})(w) \geq 0 \text{ for all } w \in T_C(\bar{x}).$$

When f_0 and C are convex (hence regular), either condition is sufficient for $\bar{x}$ to be globally optimal, even if the constraint qualification is not fulfilled.

Proof. Consider in $I\!R^{n+1}$ the closed sets $D = C \times I\!R$, $E_0 = \operatorname{epi} f_0$, and $E = D \cap E_0$, which are convex when f_0 and C are convex. To say that f_0 has a local minimum over C at $\bar{x}$ is to say that the function $l(x, \alpha) := \alpha$ has a local minimum over E at $\big(\bar{x}, f_0(\bar{x})\big)$; similarly for a global minimum. The optimality conditions in 6.12 are applicable in this setting, because l is smooth with $\nabla l\big(\bar{x}, f_0(\bar{x})\big) = (0, 1)$. To make use of 6.12, we have to know something about the variational geometry of E at $\big(\bar{x}, f_0(\bar{x})\big)$, but information is available through 6.42, which deals with set intersections. According to 6.42 as specialized here, the constraint qualification

$$\left.\begin{array}{r} (y, -\eta) \in N_{E_0}\big(\bar{x}, f_0(\bar{x})\big) \\ -(y, -\eta) \in N_D\big(\bar{x}, f_0(\bar{x})\big) \end{array}\right\} \implies (y, -\eta) = (0, 0)$$

guarantees the inclusions

$$\begin{aligned}\widehat{T}_E(\bar{x}, f_0(\bar{x})) &\supset \widehat{T}_D(\bar{x}, f_0(\bar{x})) \cap \widehat{T}_{E_0}(\bar{x}, f_0(\bar{x})),\\ N_E(\bar{x}, f_0(\bar{x})) &\subset N_D(\bar{x}, f_0(\bar{x})) + N_{E_0}(\bar{x}, f_0(\bar{x})).\end{aligned}$$

Also by 6.42, when D and E_0 are regular at $(\bar{x}, f_0(\bar{x}))$, the opposite inclusions hold (with $\widehat{T} = T$). Having $D = C \times \mathbb{R}$ implies $T_D(\bar{x}, f_0(\bar{x})) = T_C(\bar{x}) \times \mathbb{R}$ and $N_D(\bar{x}, f_0(\bar{x})) = N_C(\bar{x}) \times \{0\}$, cf. 6.41. Because f_0 is locally lsc, $T_{E_0}(\bar{x}, f_0(\bar{x}))$ and $N_{E_0}(\bar{x}, f_0(\bar{x}))$ are known from 8.2(a) and 8.9; $(y, 0) \in N_{E_0}(\bar{x}, f_0(\bar{x}))$ means $y \in \partial^\infty f_0(\bar{x})$. The constraint qualification of 6.42 thus translates to the one in the theorem, hence

$$\begin{aligned}T_D(\bar{x}, f_0(\bar{x})) \cap T_{E_0}(\bar{x}, f_0(\bar{x})) &= \big[T_C(\bar{x}) \times \mathbb{R}\big] \cap \big[\operatorname{epi} df_0(\bar{x})\big],\\ N_D(\bar{x}, f_0(\bar{x})) + N_{E_0}(\bar{x}, f_0(\bar{x})) &= \big\{(v', 0) \,\big|\, v' \in N_C(\bar{x})\big\} + N_{E_0}(\bar{x}, f_0(\bar{x}))\\ = \big\{(v', 0) \,\big|\, v' \in N_C(\bar{x})\big\}&\\ + \big\{\lambda(v, -1) \,\big|\, v \in \partial f_0(\bar{x}),\ \lambda > 0\big\} &\cup \big\{(v, 0) \,\big|\, v \in \partial^\infty f_0(\bar{x})\big\}.\end{aligned}$$

In light of this, the earlier conditions in 6.12 with respect to minimizing l on E immediately give the conditions claimed here. In the convex case the sufficiency can also be seen directly from the characterizations in 6.9 and 8.12. □

Theorem 8.15 can be applied to a set C with constraint structure by way of formulas such as those in 6.14 and 6.31. The earlier result in 6.12 for smooth f_0 follows from 8.15 and the regularity of smooth functions (cf. 7.28 and 8.20(a))—only as long as C is regular at $\bar{x}$. For general C, however, and f_0 *semidifferentiable* at $\bar{x}$, the necessity of the tangent condition in 8.15 can be seen at once from the definitions of $T_C(\bar{x})$ in 6.1 and semidifferentiability in 7.20.

D. Regular Subderivatives

To complete the picture of epigraphical geometry we still need to develop the subderivative counterpart to the theory of regular tangent cones.

8.16 Definition (regular subderivatives). *For $f : \mathbb{R}^n \to \overline{\mathbb{R}}$ and a point $\bar{x}$ with $f(\bar{x})$ finite, the regular subderivative function $\widehat{d}f(\bar{x}) : \mathbb{R}^n \to \overline{\mathbb{R}}$ is defined by*

$$\widehat{d}f(\bar{x}) := \operatorname*{e\text{-}lim\,sup}_{\substack{\tau \searrow 0\\ x \xrightarrow[f]{} \bar{x}}} \Delta_\tau f(x).$$

The regular subderivative $\widehat{d}f(\bar{x})(w)$ at $\bar{x}$ for a vector $\bar{w}$ thus has the formula

$$\begin{aligned}\widehat{d}f(\bar{x})(\bar{w}) &:= \operatorname*{lim\,sup\text{-}epi}_{\substack{x \xrightarrow[f]{} \bar{x} \ \ w \to \bar{w}\\ \tau \searrow 0}} \frac{f(x + \tau w) - f(x)}{\tau}\\ &= \lim_{\delta \searrow 0}\left(\limsup_{\substack{x \xrightarrow[f]{} \bar{x}\\ \tau \searrow 0}}\left[\inf_{w \in \mathbf{B}(\bar{w}, \delta)} \frac{f(x + \tau w) - f(x)}{\tau}\right]\right).\end{aligned} \qquad 8(8)$$

The special epi-limit notation in this definition comes from 7(10); see also 7(5). Although formula 8(8) may seem too formidable to be very useful, it's rooted deeply in the variational geometry of epigraphs. Other, more elementary expressions of regular subderivatives will usually be available, as we'll see in Theorem 8.22 and all the more in the theory of Lipschitz continuity in Chapter 9 (where regular subderivatives will be prominent in key facts like 9.13, 9.15 and 9.18). Here we'll especially be interested in regular subderivatives $\widehat{d}f(\bar{x})(w)$ to the extent that they reflect properties of $\partial f(\bar{x})$ and can be shown sometimes to coincide with the subderivatives $df(\bar{x})(w)$ and thereby furnish a characterization of subdifferential regularity of f.

8.17 Theorem (regular subderivatives versus regular tangents).

(a) *For $f : \mathbb{R}^n \to \overline{\mathbb{R}}$ and any point $\bar{x}$ at which f is lsc and finite, one has*

$$\operatorname{epi} \widehat{d}f(\bar{x}) = \widehat{T}_{\operatorname{epi} f}\big(\bar{x}, f(\bar{x})\big).$$

(b) *For the indicator δ_C of a set $C \subset \mathbb{R}^n$ and any point $\bar{x} \in C$, one has*

$$\widehat{d}\delta_C(\bar{x}) = \delta_K \text{ for } K = \widehat{T}_C(\bar{x}).$$

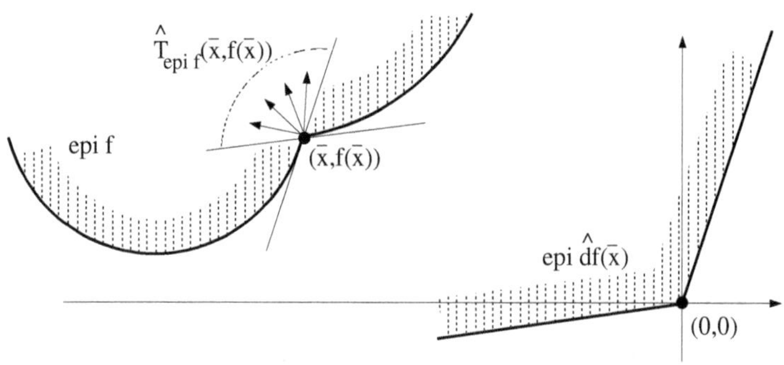

Fig. 8–8. Epigraphical interpretation of regular subderivatives.

Proof. The relation in (a) is almost immediate from 8(1), Definition 8.16, and the formula for $\widehat{T}_{\operatorname{epi} f}\big(\bar{x}, f(\bar{x})\big)$ as derived from Definition 6.25, but its proof must contend with the discrepancy that

$$\widehat{T}_{\operatorname{epi} f}\big(\bar{x}, f(\bar{x})\big) = \liminf_{\substack{\tau \searrow 0 \\ (x,\alpha)\to(\bar{x},f(\bar{x})) \\ f(x)\leq\alpha\in\mathbb{R}}} \frac{\operatorname{epi} f - (x,\alpha)}{\tau},$$

whereas $\operatorname{epi} \widehat{d}f(\bar{x})$ is through 8(1) the possibly larger inner limit set generated in the same manner but with $f(x) = \alpha$ instead of $f(x) \leq \alpha$. The assumption that f is lsc at $\bar{x}$, where its value is finite, guarantees whenever $(x^\nu, \alpha^\nu) \to \big(\bar{x}, f(\bar{x})\big)$ with $f(x^\nu) \leq \alpha^\nu$ (finite) one has $f(x^\nu) \to f(\bar{x})$, so that $x^\nu \underset{f}{\to} \bar{x}$. In particular $f(x^\nu)$ must be finite for all large enough ν, and then

$$\frac{\text{epi}\, f - \big(x^\nu, f(x^\nu)\big)}{\tau^\nu} \subset \frac{\text{epi}\, f - (x^\nu, \alpha^\nu)}{\tau^\nu}.$$

Sequences $(x^\nu, \alpha^\nu) \to \big(\bar{x}, f(\bar{x})\big)$ with $f(x^\nu) < \alpha^\nu$ can therefore be ignored when generating the inner limit; in restricting α^ν to equal $f(x^\nu)$, nothing is lost.

The indicator formula claimed in (b) is obvious from the definitions of regular subderivatives and regular tangents. □

8.18 Theorem (subderivative relationships). *For $f : \mathbb{R}^n \to \overline{\mathbb{R}}$ and any point $\bar{x}$ with $f(\bar{x})$ finite, the functions $df(\bar{x}) : \mathbb{R}^n \to \overline{\mathbb{R}}$ and $\widehat{d}f(\bar{x}) : \mathbb{R}^n \to \overline{\mathbb{R}}$ are lsc and positively homogeneous with $\widehat{d}f(\bar{x}) \geq df(\bar{x})$. When f is lsc at $\bar{x}$, $\widehat{d}f(\bar{x})$ is sublinear. As long as f is locally lsc at $\bar{x}$, one further has*

$$\widehat{d}f(\bar{x}) = \operatorname*{e-lim\,sup}_{x \xrightarrow[f]{} \bar{x}} df(x). \tag{8(9)}$$

Proof. The lower semicontinuity of $df(\bar{x})$ and $\widehat{d}f(\bar{x})$ is a direct consequence of these functions being defined by certain epi-limits. The positive homogeneity comes out of the difference quotients. These properties can also be deduced from the geometry in 8.2(a) and 8.17(a), but that requires in the case of $\widehat{d}f(\bar{x})$ the stronger assumption that f is lsc at $\bar{x}$. Then there is something more: because $\widehat{T}_{\text{epi}\, f}\big(\bar{x}, f(\bar{x})\big)$ is convex (by 6.26), $\widehat{d}f(\bar{x})$ is not just positively homogeneous but convex, hence sublinear. If f is locally lsc at $\bar{x}$, we can apply the limit relation in 6.26 to epi f to obtain the additional property that

$$\widehat{T}_{\text{epi}\, f}\big(\bar{x}, f(\bar{x})\big) = \liminf_{\substack{(x,\alpha) \to (\bar{x}, f(\bar{x})) \\ f(x) \leq \alpha \in \mathbb{R}}} T_{\text{epi}\, f}(x, \alpha).$$

Here the pairs $(x, \alpha) \in \text{epi}\, f$ with $\alpha > f(x)$ can be ignored in generating the inner limit, for the reasons laid down in the proof of Theorem 8.17. Hence

$$\text{epi}\, \widehat{d}f(\bar{x}) = \liminf_{x \xrightarrow[f]{} \bar{x}} \text{epi}\, df(x),$$

which is equivalent to the final formula in the theorem. □

8.19 Corollary (subderivative criterion for subdifferential regularity). *A function $f : \mathbb{R}^n \to \overline{\mathbb{R}}$ is regular at $\bar{x}$ if and only if f is both finite and locally lsc at $\bar{x}$ and has $df(\bar{x}) = \widehat{d}f(\bar{x})$, this equation being equivalent to*

$$\operatorname*{e-lim\,sup}_{x \xrightarrow[f]{} \bar{x}} df(x) = df(\bar{x}).$$

Proof. This applies the regularity criteria in 6.29(b) and 6.29(g) to epi f through the window provided by epi-convergence notation, as in 7(2). □

The expression of the regular subderivative function $\widehat{d}f(\bar{x})$ as an upper epi-limit of neighboring functions $df(x)$ in Theorem 8.18, whenever f is locally lsc at $\bar{x}$, has the pointwise meaning that

$$\widehat{d}f(\bar{x})(\bar{w}) := \limsup_{\substack{\tau \searrow 0 \\ x \xrightarrow[f]{} \bar{x}}} \text{epi}_{w \to \bar{w}}\; df(x)(w)$$

$$= \lim_{\delta \searrow 0} \left(\limsup_{\substack{\tau \searrow 0 \\ x \xrightarrow[f]{} \bar{x}}} \left[\inf_{w \in \mathbb{B}(\bar{w}, \delta)} df(x)(w) \right] \right).$$

When f is regular at $\bar{x}$, so that $\widehat{d}f(\bar{x})(\bar{w}) = df(\bar{x})(\bar{w})$, this formula expresses a semicontinuity property of subderivatives which can substitute in part for the continuity of directional derivatives of smooth functions. A stronger manifestation of this property in the presence of Lipschitz continuity will be seen in Theorem 9.16.

8.20 Exercise (subdifferential regularity from smoothness).

(a) *If f is smooth on an open set O, then f is regular on O with*

$$df(\bar{x})(w) = \widehat{d}f(\bar{x})(w) = \langle \nabla f(\bar{x}), w \rangle \;\; \textit{for all } w.$$

(b) *If $f = g + f_0$ with f_0 smooth, then f is regular at $\bar{x}$ if and only if g is regular at $\bar{x}$.*

Guide. Get both facts from 8.18 and the definitions. □

Convex functions f that are lsc and proper were seen in 7.27 to be regular and epi-differentiable on $\operatorname{dom} f$, even semidifferentiable on $\operatorname{int}(\operatorname{dom} f)$. Their subderivatives have the following properties as well.

8.21 Proposition (subderivatives of convex functions). *For a convex function $f : \mathbb{R}^n \to \overline{\mathbb{R}}$ and any point $\bar{x}$ where f is finite, the limit*

$$h(w) := \lim_{\tau \searrow 0} \frac{f(\bar{x} + \tau w) - f(\bar{x})}{\tau}$$

exists for all $w \in \mathbb{R}^n$ and defines a sublinear function $h : \mathbb{R}^n \to \overline{\mathbb{R}}$ such that $f(\bar{x} + \tau w) \geq f(\bar{x}) + \tau h(w)$ for all $\tau \geq 0$ and all w, and

$$\widehat{d}f(\bar{x})(w) = df(\bar{x})(w) = (\operatorname{cl} h)(w) \;\; \textit{for all } w.$$

Here actually $(\operatorname{cl} h)(w) = h(w)$ for all w in the set

$$\begin{aligned} W &= \{ w \mid \exists \tau \geq 0 \text{ with } \bar{x} + \tau w \in \operatorname{int}(\operatorname{dom} f) \} \\ &= \operatorname{int}(\operatorname{dom} \widehat{d}f(\bar{x})) = \operatorname{int}(\operatorname{dom} df(\bar{x})) = \operatorname{int}(\operatorname{dom} h). \end{aligned}$$

Proof. The limit defining $h(w)$ exists because the difference quotient is nondecreasing in $\tau > 0$ by virtue of convexity (cf. 2.12). This monotonicity also yields the inequality $f(\bar{x} + \tau w) \geq f(\bar{x}) + \tau h(w)$.

Let K be the convex cone in $\mathbb{R}^n \times \mathbb{R}$ consisting of the pairs (w, β) such that $(\bar{x}, f(\bar{x})) + \tau(w, \beta) \in \operatorname{epi} f$ for some $\tau > 0$. By 6.9, the cones $\widehat{T}_{\operatorname{epi} f}(\bar{x}, f(\bar{x}))$ and $T_{\operatorname{epi} f}(\bar{x}, f(\bar{x}))$ coincide with $\operatorname{cl} K$, and their interiors coincide with

$$K_0 := \{(w,\beta) \mid \exists \tau > 0 \text{ with } (\bar{x}, f(\bar{x})) + \tau(w,\beta) \in \operatorname{int}(\operatorname{epi} f)\}.$$

We have $h(w) = \inf\{\beta \mid (w,\beta) \in K\}$, so $\operatorname{cl}(\operatorname{epi} h) = \operatorname{cl} K$, $\operatorname{int}(\operatorname{epi} h) = K_0$. Since $\operatorname{cl}(\operatorname{epi} h) = \operatorname{epi}(\operatorname{cl} h)$ by definition, we conclude that $\widehat{d}f(\bar{x}) = df(\bar{x}) = \operatorname{cl} h$; here we use the geometry in 8.17(a) as extended slightly by the observation, made in the proof of that theorem, that even if f isn't lsc at $\bar{x}$ the inclusion $\operatorname{epi} \widehat{d}f(\bar{x}) \supset \widehat{T}_{\operatorname{epi} f}(\bar{x}, f(\bar{x}))$ is still valid. Since $\widehat{d}f(\bar{x}) \geq df(\bar{x})$ by 8.6, the inclusion in question can't be strict when $\widehat{T}_{\operatorname{epi} f}(\bar{x}, f(\bar{x})) = T_{\operatorname{epi} f}(\bar{x}, f(\bar{x}))$, as in this case.

Because h is sublinear, in particular convex, it must agree with $\operatorname{cl} h$ on $\operatorname{int}(\operatorname{dom} h)$ (cf. 2.35). Convexity ensures through 2.34, as applied to h, that $\operatorname{int}(\operatorname{dom} h)$ is the projection on $\mathbb{R}^n$ of $\operatorname{int}(\operatorname{epi} h)$, which we have seen to be K_0. But also by 2.35, as applied this time to f, the projection on $\mathbb{R}^n$ of K_0 is W. This establishes the interiority claims. □

Properties of semiderivatives established in 7.26 under the assumption of regularity, when subderivatives and regular subderivatives agree, carry over to a rule about regular subderivatives by themselves.

8.22 Theorem (interior formula for regular subderivatives). *Let $f : \mathbb{R}^n \to \overline{\mathbb{R}}$ be locally lsc at $\bar{x}$ with $f(\bar{x})$ finite. Let $O = \{\bar{w} \mid h(\bar{w}) < \infty\}$, where*

$$h(\bar{w}) := \limsup_{\substack{\tau \searrow 0 \\ x \to_f \bar{x} \\ w \to \bar{w}}} \frac{f(x+\tau w) - f(x)}{\tau}.$$

Then $\operatorname{int}(\operatorname{dom} \widehat{d}f(\bar{x})) = O$, and $\widehat{d}f(\bar{x})(\bar{w}) = h(\bar{w})$ for all $\bar{w} \in O$. As long as $O \neq \emptyset$, one has $\widehat{d}f(\bar{x})(\bar{w}) = \infty$ for $\bar{w} \notin \operatorname{cl} O$, while

$$\widehat{d}f(\bar{x})(\bar{w}) = \lim_{\tau \searrow 0} h\big((1-\tau)\bar{w} + \tau\hat{w}\big) \quad \textit{for any } \bar{w} \in \operatorname{bdry} O,\ \hat{w} \in O.$$

Proof. We have $h(\bar{w}) < \bar{\beta} \in \mathbb{R}$ if and only if there exists $\varepsilon > 0$ such that, for all $w \in \mathbb{B}(\bar{w}, \varepsilon)$ and $\beta > \bar{\beta} - \varepsilon$, one has $f(x+\tau w) \leq f(x) + \tau\beta$ when $\tau \in (0,\varepsilon)$, $x \in \mathbb{B}(\bar{x},\varepsilon)$, and $f(x) \leq f(\bar{x}) + \varepsilon$. This means that the pairs $(w,\beta) \in \mathbb{B}(\bar{w},\varepsilon) \times [\bar{\beta} - \varepsilon, \infty)$ have

$$\big(x, f(x)\big) + \tau(w,\beta) \in \operatorname{epi} f \ \text{ when } \tau \in (0,\varepsilon),\ x \in \mathbb{B}(\bar{x},\varepsilon),\ f(x) \leq f(\bar{x}) + \varepsilon.$$

Therefore, the condition $h(\bar{w}) < \bar{\beta} \in \mathbb{R}$ is equivalent to having $(\bar{w}, \bar{\beta})$ belong to the interior of the recession cone $R_{\operatorname{epi} f}(\bar{x}, f(\bar{x}))$; cf. Definition 6.33. But $\operatorname{epi} f$ is locally closed at $(\bar{x}, f(\bar{x}))$ through our assumption that f is locally lsc at $\bar{x}$ (cf. 1.34), so the cones $R_{\operatorname{epi} f}(\bar{x}, f(\bar{x}))$ and $\widehat{T}_{\operatorname{epi} f}(\bar{x}, f(\bar{x}))$ have the same interior by 6.36. Moreover $\widehat{T}_{\operatorname{epi} f}(\bar{x}, f(\bar{x})) = \operatorname{epi} \widehat{d}f(\bar{x})$ by 8.17(a) with $\widehat{d}f(\bar{x})$ a convex function by 8.18, so this interior consists of the pairs $(\bar{w}, \bar{\beta})$ such that $\bar{w} \in \operatorname{int}\big(\operatorname{dom} \widehat{d}f(\bar{x})\big)$ and $\widehat{d}f(\bar{x})(\bar{w}) < \bar{\beta} \in \mathbb{R}$. Thus, $h(\bar{w}) < \bar{\beta} \in \mathbb{R}$ holds if and only if $\bar{w} \in \operatorname{int}\big(\operatorname{dom} \widehat{d}f(\bar{x})\big)$ and $\widehat{d}f(\bar{x})(\bar{w}) < \bar{\beta} \in \mathbb{R}$. This yields everything except the boundary point formula claimed at the end of the theorem. That comes from the closure property of convex functions in Theorem 2.35. □

8.23 Exercise (regular subderivatives from subgradients). *For $f : \mathbb{R}^n \to \overline{\mathbb{R}}$ and any point $\bar{x}$ where f is finite and locally lsc, one has in terms of the polar cone $\partial^\infty f(\bar{x})^* = \{w \mid \langle v, w\rangle \le 0 \text{ for all } v \in \partial^\infty f(\bar{x})\}$ that*

$$\widehat{d}f(\bar{x})(w) = \begin{cases} \sup\{\langle v, w\rangle \mid v \in \partial f(\bar{x})\} & \text{when } w \in \partial^\infty f(\bar{x})^*, \\ \infty & \text{when } w \notin \partial^\infty f(\bar{x})^*. \end{cases}$$

Guide. Get this from having $\widehat{T}_{\operatorname{epi} f}(\bar{x}, f(\bar{x})) = N_{\operatorname{epi} f}(\bar{x}, f(\bar{x}))^*$ by 6.28 when epi f is closed at $(\bar{x}, f(\bar{x}))$; cf. 1.33, and the geometry in 8.9 and 8.17(a). □

The supremum in the regular subderivative formula in 8.23 need not always be finite when $w \in \partial^\infty f(\bar{x})^*$. For instance, when $\partial f(\bar{x}) = \emptyset$ the supremum is $-\infty$, so that $\widehat{d}f(\bar{x})(w) = \delta_K(w) - \infty$ for $K = \partial^\infty f(\bar{x})^*$. But the supremum can also be ∞, so that K is not necessarily the same as dom $\widehat{d}f(\bar{x})$, or for that matter even the closure of dom $\widehat{d}f(\bar{x})$. An example in two dimensions will shed light on this possibility. Consider at $\bar{x} = (0, 0)$ the function

$$f_0(x_1, x_2) := \begin{cases} x_2^2/2|x_1| & \text{if } x_1 \neq 0, \\ 0 & \text{if } x_1 = 0 \text{ and } x_2 = 0, \\ \infty & \text{for all other } x_1,\ x_2. \end{cases}$$

(The epigraph of f_0 is the union of two circular convex cones, the axes being the rays in $\mathbb{R}^3$ in the directions of $(1, 0, 1)$ and $(-1, 0, 1)$; each cone touches the x_1-axis and the x_3-axis.) This function is lsc and positively homogeneous, so that $df_0(0, 0)$ is f_0 itself. It is easily seen that $\widehat{\partial} f_0(0, 0) = \{0, 0\}$. Away from the x_2-axis, f_0 is differentiable with $\nabla f_0(x_1, x_2) = \pm(-x_2^2/2x_1^2, x_2/x_1)$, where the $+$ is taken when $x_1 > 0$ and the $-$ when $x_1 < 0$; at such places, $\widehat{\partial} f_0(x_1, x_2) = \{\nabla f_0(x_1, x_2)\}$. Thus, the mapping $\widehat{\partial} f_0$ is single-valued everywhere on its effective domain, which is the same as dom f_0 and consists of all points of $\mathbb{R}^2$ except the ones lying on the x_2-axis away from the origin. Furthermore, $\widehat{\partial} f_0$ is locally bounded at all such points except the origin, where it is definitely not locally bounded. From Definition 8.3 we get

$$\partial f_0(0, 0) = \{(\pm t^2/2, t) \mid -\infty < t < \infty\},$$
$$\partial^\infty f_0(0, 0) = \{(t, 0) \mid -\infty < t < \infty\}.$$

The formula in 8.23 then yields $\widehat{d}f_0(0, 0) = \delta_{\{(0,0)\}}$, so dom $\widehat{d}f_0(0, 0) = \{(0, 0)\}$. But the polar cone $\partial^\infty f_0(0, 0)^*$ is the x_2-axis, not $\{(0, 0)\}$.

This example has $\partial^\infty f_0(0, 0) = \partial f_0(0, 0)^\infty$, but the inclusion $\partial^\infty f(\bar{x}) \supset \partial f(\bar{x})^\infty$ can sometimes be strict when $\partial f(\bar{x}) \neq \emptyset$, which underscores the essential role of $\partial^\infty f(\bar{x})$ alongside of $\partial f(\bar{x})$ in the general case of the regular subderivative formula in 8.23. For an illustration of strict inclusion $\partial^\infty f(\bar{x}) \supset \partial f(\bar{x})^\infty$, again in $\mathbb{R}^2$ with $\bar{x} = (0, 0)$, consider

$$f_1(x_1,x_2) := \begin{cases} -\sqrt{x_1} - \sqrt{x_2} & \text{when } x_1 \geq 0,\ x_2 \geq 0, \\ 0 & \text{for all other } x_1,\ x_2. \end{cases}$$

It's easy to calculate that $df_1(0,0)$ has the value $-\infty$ on $\mathbb{R}^2_+$ but 0 everywhere else. Furthermore, $\partial f_1(0,0) = \{(0,0)\}$ while $\partial^\infty f_1(0,0) = \mathbb{R}^2_-$. Yet we have $\widehat{d}f_1(0,0)(w_1,w_2) = 0$ when $(w_1,w_2) \in \mathbb{R}^2_+$, and $\widehat{d}f_1(0,0)(w_1,w_2) = \infty$ otherwise. Knowledge of the set $\partial f_1(0,0)$ alone wouldn't pin this down.

E. Support Functions and Subdifferential Duality

The tightest connections between subderivatives and subgradients are seen when f is regular at $\bar{x}$. These hinge on a duality correspondence that generalizes the one for polar cones but goes between sets and certain functions. The *support function* of a set $D \subset \mathbb{R}^n$ is the function $\sigma_D : \mathbb{R}^n \to \overline{\mathbb{R}}$ of D defined by

$$\sigma_D(w) := \sup_{v \in D} \langle v, w \rangle =: \sup \langle D, w \rangle. \qquad 8(10)$$

This characterizes the closed half-spaces that include D:

$$D \subset \{ v \mid \langle v, w \rangle \leq \alpha \} \iff \sigma_D(w) \leq \alpha. \qquad 8(11)$$

If $D = \emptyset$, one has $\sigma_D \equiv -\infty$, but if $D \neq \emptyset$, one has $\sigma_D > -\infty$ and $\sigma_D(0) = 0$.

8.24 Theorem (sublinear functions as support functions). *For any set $D \subset \mathbb{R}^n$, the support function σ_D is sublinear and lsc (proper as well if $D \neq \emptyset$), and*

$$\mathrm{cl}(\mathrm{con}\, D) = \{ v \mid \langle v, w \rangle \leq \sigma_D(w) \text{ for all } w \}.$$

On the other hand, for any positively homogeneous function $h : \mathbb{R}^n \to \overline{\mathbb{R}}$, the set $D = \{ v \mid \langle v, w \rangle \leq h(w) \text{ for all } w \}$ is closed and convex, and

$$\sigma_D = \mathrm{cl}(\mathrm{con}\, h) \text{ as long as } D \neq \emptyset,$$

whereas if $D = \emptyset$ the function $\mathrm{cl}(\mathrm{con}\, h)$ is improper with no finite values.

Thus, there is a one-to-one correspondence between the nonempty, closed, convex subsets D of $\mathbb{R}^n$ and the proper, lsc, sublinear functions h on $\mathbb{R}^n$:

$$D \longleftrightarrow h \text{ with } h = \sigma_D \text{ and } D = \{ v \mid \langle v, \cdot \rangle \leq h \}.$$

Under this correspondence the closed, convex cones D^∞ and $\mathrm{cl}(\mathrm{dom}\, h)$ are polar to each other.

Proof. By definition, σ_D is the pointwise supremum of the collection of linear functions $\langle v, \cdot \rangle$ with $v \in D$. Except for the final polarity claim, all the assertions thus follow from the characterization in Theorem 8.13 of the functions described by such a pointwise supremum and the identification in Theorem 6.20 of the sets that are the intersections of families of closed half-spaces. An improper, lsc, convex function can't have any finite values (see 2.5).

For the polarity claim about corresponding D and h, observe from the expression of D in terms of the inequalities $\langle v, w\rangle \le h(w)$ for $w \in \operatorname{dom} h$ that, for any choice of $\bar{v} \in D$, a vector y has the property $\{\bar{v} + \tau y \mid \tau \ge 0\} \subset D$ if and only if $\langle y, w\rangle \le 0$ for all $w \in \operatorname{dom} h$. But the vectors y with this property comprise D^∞ by 3.6. Therefore, $D^\infty = (\operatorname{dom} h)^* = [\operatorname{cl}(\operatorname{dom} h)]^*$. Since $\operatorname{dom} h$ is convex when h is sublinear, it follows that $\operatorname{cl}(\operatorname{dom} h) = (D^\infty)^*$; cf. 6.21. □

8.25 Corollary (subgradients of sublinear functions). *Let h be a proper, lsc, sublinear function on $\mathbb{R}^n$ and let D be the unique closed, convex set in $\mathbb{R}^n$ such that $h = \sigma_D$. Then, for any $w \in \operatorname{dom} h$,*

$$\partial h(w) = \operatorname*{argmax}_{v \in D} \langle v, w\rangle = \{v \in D \mid w \in N_D(v)\}.$$

Proof. This follows from the description in 8.12 and 8.13 of the subgradients of a sublinear function as the gradients of its linear supports. □

The support function of a singleton set $\{a\}$ is the linear function $\langle a, \cdot\rangle$; in particular, $\sigma_{\{0\}} \equiv 0$. In this way the support function correspondence generalizes the traditional duality between elements of $\mathbb{R}^n$ and linear functions on $\mathbb{R}^n$. The support function of $\mathbb{R}^n$ is $\delta_{\{0\}}$, while the support function of $\emptyset$ is the constant function $-\infty$. On the other hand,

$$\sigma_D(w) = \max\{\langle a_1, w\rangle, \ldots, \langle a_m, w\rangle\} \;\text{ for }\; D = \operatorname{con}\{a_1, \ldots, a_m\}.$$

8.26 Example (vector-max and the canonical simplex). *The sublinear function $\operatorname{vecmax}(x) = \max\{x_1, \ldots, x_n\}$ for $x = (x_1, \ldots, x_n)$ is the support function of the canonical simplex $S = \{v \mid v_j \ge 0, \sum_{j=1}^n v_j = 1\}$ in $\mathbb{R}^n$. Denoting by $J(x)$ the set of indices j with $\operatorname{vecmax}(x) = x_j$, one has at any point $\bar{x}$ that*

$$v \in \partial[\operatorname{vecmax}](\bar{x}) \iff v_j \ge 0 \;\text{ for }\; j \in J(\bar{x}),\; v_j = 0 \;\text{ for }\; j \notin J(\bar{x}),\; \sum\nolimits_{j=1}^n v_j = 1.$$

On the other hand, $\partial^\infty[\operatorname{vecmax}](\bar{x}) = \{0\}$ for any $\bar{x}$.

Detail. This is the case of $D = \operatorname{con}\{e_1, \ldots, e_n\}$, where e_j is the vector with jth component 1 and all other components 0; the convex hull of these vectors consists of all their convex combinations, cf. 2.27. The subgradient formula specializes 8.25. No nonzero horizon subgradients can exist, because $\operatorname{dom}(\operatorname{vecmax}) = \mathbb{R}^n$; cf. 8.12. □

8.27 Exercise (subgradients of the Euclidean norm). *The norm $h(x) = |x|$ on $\mathbb{R}^n$ is the support function of the unit ball $\mathbb{B}$: $|x| = \sigma_{\mathbb{B}}(x) = \max_{v \in \mathbb{B}} \langle v, x\rangle$. It has $\partial^\infty h(\bar{x}) = \{0\}$ for all $\bar{x}$, and*

$$\partial h(\bar{x}) = \begin{cases} \{\bar{x}/|\bar{x}|\} & \text{if } \bar{x} \ne 0, \\ \mathbb{B} & \text{if } \bar{x} = 0. \end{cases}$$

In contrast, the function $k(x) = -|x|$, although positively homogeneous, is not *sublinear and is not the support function of any set, nor is it regular at 0. It has $\partial^\infty k(\bar{x}) = \{0\}$ for all $\bar{x}$, and*

$$\partial k(\bar{x}) = \begin{cases} \{-\bar{x}/|\bar{x}|\} & \text{if } \bar{x} \neq 0, \\ \text{bdry}\, \mathbb{B} & \text{if } \bar{x} = 0. \end{cases}$$

8.28 Example (support functions of cones). *For any cone $K \subset \mathbb{R}^n$, the support function σ_K is the indicator δ_{K^*} of the polar cone K^*. Thus, the polarity correspondence among closed, convex cones is imbedded within the correspondence between nonempty, closed, convex sets and proper, lsc, sublinear functions.*

8.29 Proposition (properties expressed by support functions).

(a) *A convex set C has $\operatorname{int} C \neq \emptyset$ if and only if there is no vector $v \neq 0$ such that $\sigma_C(v) = -\sigma_C(-v)$. In general,*

$$\operatorname{int} C = \{x \mid \langle x, v\rangle < \sigma_C(v) \;\text{ for all }\; v \neq 0\}.$$

(b) *The condition $0 \in \operatorname{int}(C_1 - C_2)$ holds for convex sets $C_1 \neq \emptyset$ and $C_2 \neq \emptyset$ if and only if there is no vector $v \neq 0$ such that $\sigma_{C_1}(v) \leq -\sigma_{C_2}(-v)$.*

(c) *A convex set $C \neq \emptyset$ is polyhedral if and only if σ_C is piecewise linear.*

(d) *A set C is nonempty and bounded if and only if σ_C is finite everywhere.*

Proof. In (a) there's no loss of generality in assuming C to be closed, because C and $\operatorname{cl} C$ have the same support function and also the same interior (by 2.33). The case of $C = \emptyset$ can be excluded as trivial; it has $\sigma_C \equiv -\infty$. For $C \neq \emptyset$, let $K = \operatorname{epi} \sigma_C$, this being a closed, convex cone in $\mathbb{R}^n \times \mathbb{R}$. The condition that no $v \neq 0$ has $\sigma_C(v) = -\sigma_C(-v)$ is equivalent to the condition that no $(v, \beta) \neq (0, 0)$ has both $(v, \beta) \in K$ and $-(v, \beta) \in K$. Thus, it means that K is pointed, which is equivalent by 6.22 to $\operatorname{int} K^* \neq \emptyset$. The polar cone K^* consists of pairs (x, γ) such that, for all $(v, \beta) \in K$, one has $\langle (x, \gamma), (v, \beta)\rangle \leq 0$, i.e., $\langle x, v\rangle + \gamma\beta \leq 0$. Clearly this property of (x, γ) requires $\gamma \leq 0$, and by inspecting the cases $\gamma = -1$ and $\gamma = 0$ separately we obtain

$$K^* = \{\lambda(x, -1) \mid x \in C,\, \lambda > 0\} \cup \{(x, 0) \mid x \in C^\infty\}.$$

The nonemptiness of $\operatorname{int} K^*$ comes down then to the nonemptiness of $\operatorname{int} C$. In particular, we have $(\bar{x}, -1) \in \operatorname{int} K^*$ if and only if $\bar{x} \in \operatorname{int} C$. But by 6.22, $(\bar{x}, -1) \in \operatorname{int} K$ if and only if $\langle (x, \gamma), (v, \beta)\rangle < 0$ for all $(v, \beta) \neq (0, 0)$ in K. This means that $\bar{x} \in \operatorname{int} C$ if and only if $\langle \bar{x}, v\rangle < \beta$ whenever $v \neq 0$ and $\beta \geq \sigma_C(v)$, which is the condition proposed.

In (b) we observe that a vector $v \neq 0$ satisfies $\sigma_{C_1}(v) \leq -\sigma_{C_2}(-v)$ if and only if there exists $\alpha \in \mathbb{R}$ such that $\sigma_{C_1}(v) \leq \alpha$ but $-\sigma_{C_2}(-v) \geq \alpha$. Here $\sigma_{C_1}(v)$ is the supremum of $\langle v, x\rangle$ over $x \in C_1$, whereas $-\sigma_{C_2}(-v)$ is the infimum of $\langle v, x\rangle$ over $x \in C_2$. The property thus corresponds to the existence of a separating hyperplane $\{x \mid \langle v, x\rangle = \alpha\}$. We know such a hyperplane exists if and only if $0 \notin \operatorname{int}(C_1 - C_2)$; cf. 2.39.

In (c), if C is nonempty and polyhedral it has a representation as a convex hull in the extended sense of 3.53:

$$C = \operatorname{con}\{a_1, \ldots, a_m\} + \operatorname{con}\big(\operatorname{pos}\{a_{m+1}, \ldots, a_r\}\big) \text{ with } m \geq 1.$$

Then σ_C is proper and given by $\sigma_C(v) = \max\{\langle a_i, v\rangle \,|\, i = 1, \ldots, m\}$ when v belongs to the polyhedral cone $K = \{v \,|\, \langle a_i, v\rangle \leq 0,\ i = m+1, \ldots, r\}$, whereas $\sigma_C(v) = \infty$ for $v \notin K$. In this case σ_C is piecewise linear by the characterization in 2.49. Conversely, if σ_C is proper and has such an expression, C must have the form described, because this description gives a certain polyhedral set having σ_C as its support function, as just seen, and the correspondence between closed, convex sets and their support functions is one-to-one.

Nonemptiness of a set C is equivalent to σ_C not taking on the value $-\infty$; of course, because σ_C is convex and lsc by 8.24, it can't take on $-\infty$ without being the constant $-\infty$ (cf. 2.5). When C is bounded, in addition to being nonempty, it's clear that the supremum of $\langle v, x\rangle$ over $x \in C$ is finite for every x; in other words, the function σ_C is finite everywhere. Conversely, when the latter holds, consider for the vectors $e^j = (0, \ldots, 1, 0 \ldots)$ (with the 1 in jth position) the finite values $\beta_j = \sigma_C(e^j)$ and $\alpha_j = -\sigma_C(-e^j)$. We have C nonempty with $\alpha_j \leq \langle e^j, x\rangle \leq \beta_j$ for $j = 1, \ldots, n$, so C lies in the box $[\alpha_1, \beta_1] \times \cdots \times [\alpha_n, \beta_n]$, which is bounded. □

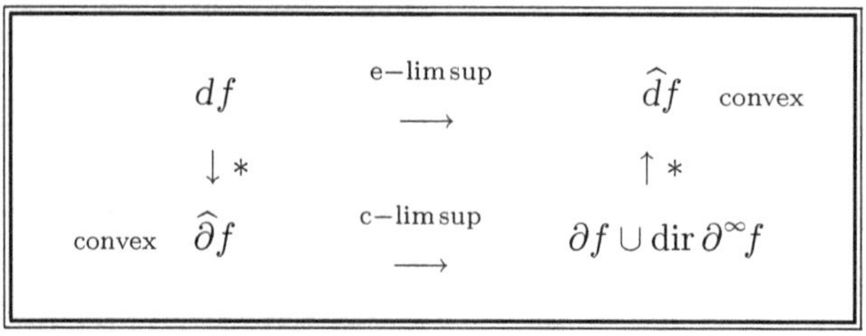

Fig. 8–9. Diagram of subderivative-subgradient relationships for lsc functions.

Just as polarity relates tangent cones with normal cones, the support function correspondence relates subderivatives with subgradients. The relations in 8.4 and 8.23 already fit in this framework, which is summarized in Figure 8–9. (Here the * refers schematically to support function duality, not the polarity operation.) The strongest connections are manifested in the presence of subdifferential regularity.

8.30 Theorem (subderivative-subgradient duality for regular functions). *If a function $f : \mathbb{R}^n \to \overline{\mathbb{R}}$ is subdifferentially regular at $\bar{x}$, one has*

$$\partial f(\bar{x}) \neq \emptyset \quad \Longleftrightarrow \quad df(\bar{x})(0) \neq -\infty,$$

and under these conditions, which in particular imply that f is properly epi-differentiable with $df(\bar{x})$ as the epi-derivative function, one has

$$\begin{aligned} df(\bar{x})(w) &= \sup\big\langle \partial f(\bar{x}), w\big\rangle = \sup\big\{\langle v, w\rangle \,\big|\, v \in \partial f(\bar{x})\big\}, \\ \partial f(\bar{x}) &= \big\{v \,\big|\, \langle v, w\rangle \le df(\bar{x})(w) \ \text{ for all } \ w\big\}, \end{aligned}$$

with $\partial f(\bar{x})$ closed and convex, and $df(\bar{x})$ lsc and sublinear. Also, regardless of whether $\partial f(\bar{x}) \neq \emptyset$ or $df(\bar{x})(0) \neq -\infty$, the cones $\partial^\infty f(\bar{x})$ and $\operatorname{dom} df(\bar{x})$ are in the polar relationship

$$\partial^\infty f(\bar{x}) = \big[\operatorname{dom} df(\bar{x})\big]^*, \qquad \partial^\infty f(\bar{x})^* = \operatorname{cl}\big[\operatorname{dom} df(\bar{x})\big].$$

Proof. When f is regular at $\bar{x}$, hence epi-differentiable there by 7.26, the cones $N_{\operatorname{epi} f}\big(\bar{x}, f(\bar{x})\big)$ and $T_{\operatorname{epi} f}\big(\bar{x}, f(\bar{x})\big)$ are polar to each other. Under this polarity, the normal cone contains only 'horizontal' vectors of form $(v, 0)$ if and only if the tangent cone includes the entire vertical line $\big\{(0, \beta) \,\big|\, \beta \in I\!R\big\}$. Through the epigraphical geometry in 8.2(a), 8.9 and 8.17(a), the first condition means $\partial f(\bar{x}) = \emptyset$, whereas the second means $df(\bar{x})(0) = -\infty$.

When $\partial f(\bar{x}) \neq \emptyset$ and $df(\bar{x})(0) = 0$, so that $df(\bar{x})$ must be proper (due to closedness and positive homogeneity), we in particular have $\partial^\infty f(\bar{x})^* = \big[\partial f(\bar{x})^\infty\big]^* = \operatorname{cl}(\operatorname{dom} h)$ for $h = \sigma_{\partial f(\bar{x})}$ by the polarity relation in 8.23, and there's no need in the subderivative formula of that result to make special provisions for $\partial^\infty f(\bar{x})^*$. □

The anomalies noted after 8.23 can't occur when f is regular, as assured by Theorem 8.30. In the regular case the subderivative function $df(\bar{x})$ is precisely the support function of the subgradient set $\partial f(\bar{x})$ under the correspondence between sublinear functions and convex sets in Theorem 8.24. This relationship provides a natural generalization of the formula $df(\bar{x})(w) = \big\langle \nabla f(\bar{x}), w\big\rangle$ that holds when f is smooth, since for any vector v the linear function $\langle v, \cdot\rangle$ is the support function of the singleton set $\{v\}$.

8.31 Exercise (elementary max functions). *Suppose $f = \max\{f_1, \ldots, f_m\}$ for a finite collection of smooth functions f_i on $I\!R^n$. Then with $I(\bar{x})$ denoting the set of indices i such that $f(\bar{x}) = f_i(\bar{x})$ one has*

$$\begin{aligned} \partial f(\bar{x}) &= \operatorname{con}\big\{\nabla f_i(\bar{x}) \,\big|\, i \in I(\bar{x})\big\}, \quad \partial^\infty f(\bar{x}) = \{0\}, \\ df(\bar{x})(w) &= \max_{i \in I(\bar{x})} \big\langle \nabla f_i(\bar{x}), w\big\rangle = \max_{i \in I(\bar{x})} df_i(\bar{x})(w). \end{aligned}$$

Guide. The set $E = \operatorname{epi} f$ is specified in $I\!R^{n+1}$ by the constraints $g_i(x, \alpha) \le 0$ for $i = 1, \ldots, m$, where $g_i(x, \alpha) = f_i(x) - \alpha$. Apply 6.14 together with the geometry in 8.2(a) and 8.9. □

The close tie between subderivatives and subgradients in 8.30 is remarkable because of the wide class of functions for which subdifferential regularity can be established. Exercise 8.31 furnishes only one illustration, which elaborates on 7.28. A full calculus of regularity will be built up in Chapter 10. In particular, a corresponding result for the pointwise maximum of an infinite collection of smooth functions will be obtained there, cf. 10.31.

F. Calmness

The concept of 'calmness' will further illustrate properties of subderivatives and their connection to subgradients. A function $f : \mathbb{R}^n \to \overline{\mathbb{R}}$ is *calm at $\bar{x}$ from below* with modulus $\kappa \in \mathbb{R}_+ = [0, \infty)$ if $f(\bar{x})$ is finite and on some neighborhood $V \in \mathcal{N}(\bar{x})$ one has

$$f(x) \geq f(\bar{x}) - \kappa|x - \bar{x}| \text{ when } x \in V. \tag{8(12)}$$

The property of f being *calm at $\bar{x}$ from above* is defined analogously; it corresponds to $-f$ being calm at $\bar{x}$ from below. Finally, f is *calm at $\bar{x}$* if it is calm both from above and below: for some $\kappa \in \mathbb{R}_+$ and $V \in \mathcal{N}(\bar{x})$ one has

$$\big|f(x) - f(\bar{x})\big| \leq \kappa|x - \bar{x}| \text{ when } x \in V. \tag{8(13)}$$

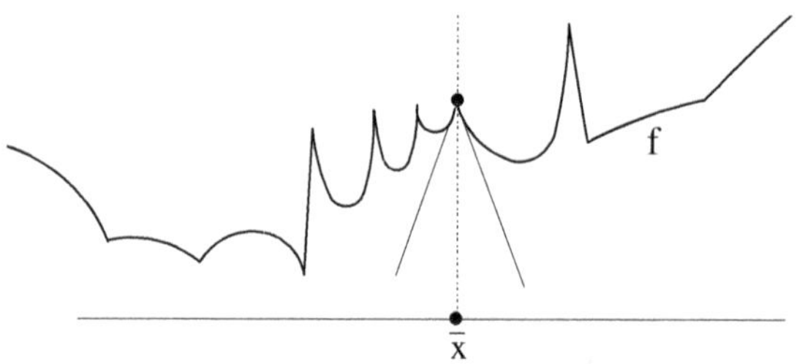

Fig. 8–10. Calmness from below.

8.32 Proposition (calmness from below). *A function $f : \mathbb{R}^n \to \overline{\mathbb{R}}$ is calm at $\bar{x}$ from below if and only if $df(\bar{x})(0) = 0$, or equivalently, $df(\bar{x})(w) > -\infty$ for all w. The infimum $\bar{\kappa} \in \mathbb{R}_+$ of all the constants $\kappa \in \mathbb{R}_+$ serving as a modulus of calmness for f at $\bar{x}$ is expressible by*

$$-\bar{\kappa} = \min_{w \in \mathbb{B}} df(\bar{x})(w).$$

As long as f is locally lsc at $\bar{x}$, one has $\partial f(\bar{x}) \neq \emptyset$ if and only if $\widehat{d}f(\bar{x})(0) = 0$, or equivalently $\widehat{d}f(\bar{x})(w) > -\infty$ for all $w \neq 0$. In particular these properties hold when f is calm at $\bar{x}$ in addition to being locally lsc at $\bar{x}$, and then there must actually exist $\bar{v} \in \partial f(\bar{x})$ such that $|\bar{v}| = \bar{\kappa}$.

When f is regular at $\bar{x}$, calmness at $\bar{x}$ from below is not only sufficient for having $\partial f(\bar{x}) \neq \emptyset$ but necessary. Then

$$\bar{\kappa} = d\big(0, \partial f(\bar{x})\big).$$

Proof. Let $\hat{\kappa} := -\inf_{|w|=1} df(\bar{x})(w)$. If f is calm at $\bar{x}$ from below, so that 8(12) holds for some κ, we clearly have $df(\bar{x})(w) \geq -\kappa|w|$ for all w. Then $\kappa \geq \hat{\kappa}$, so that necessarily $\bar{\kappa} \geq \max\{0, \hat{\kappa}\}$.

On the other hand, whenever $df(\bar{x})(w) > -\infty$ for all $w \neq 0$, the lower semicontinuity of $df(\bar{x})$ ensures that $\hat{\kappa}$ is finite. Consider in that case any finite

$\kappa > \max\{0, \hat{\kappa}\}$. There can't exist points $x^\nu \to \bar{x}$ with $f(x^\nu) < f(\bar{x}) - \kappa|x^\nu - \bar{x}|$, for if so we could write $x^\nu = \bar{x} + \tau^\nu w^\nu$ with $|w^\nu| = 1$ and $\tau^\nu \searrow 0$, getting $\big[f(\bar{x} + \tau^\nu w^\nu) - f(\bar{x})\big]/\tau^\nu < -\kappa$, and then a cluster point $\bar{w}$ of $\{w^\nu\}_{\nu \in \mathbb{N}}$ would yield $df(\bar{x})(\bar{w}) \leq -\kappa$, which is impossible by the choice of κ relative to $\hat{\kappa}$. For some $V \in \mathcal{N}(\bar{x})$, therefore, we have 8(12) and in particular $df(\bar{x})(0) \neq -\infty$, hence $df(\bar{x})(0) = 0$ (cf. 3.19). At the same time we obtain $\kappa \geq \bar{\kappa}$, hence $\max\{0, \hat{\kappa}\} \geq \bar{\kappa}$. It follows that $\bar{\kappa} = \max\{0, \hat{\kappa}\}$. Moreover, from the definition of $\hat{\kappa}$ and the fact that $df(\bar{x})$ is lsc and positively homogeneous with $df(\bar{x})(0) = 0$, we see that $\max\{0, \hat{\kappa}\} = -\min_{w \in \mathbb{B}} df(\bar{x})(w)$.

Our argument has further disclosed that $\bar{\kappa}$ is the lowest of the values $\kappa \geq 0$ such that $df(\bar{x}) \geq -\kappa|\cdot|$.

Assume from now on that f is locally lsc at $\bar{x}$. The formula for $\widehat{d}f(\bar{x})$ in terms of $\partial f(\bar{x})$ in 8.23 gives the equivalence between the nonemptiness of $\partial f(\bar{x})$ and the properties of $\widehat{d}f(\bar{x})$. Because $\widehat{d}f(\bar{x}) \geq df(\bar{x})$, calmness from below at $\bar{x}$ suffices for such properties to hold.

In the presence now of this calmness, we proceed toward verifying the existence of $\bar{v} \in \partial f(\bar{x})$ with $|\bar{v}| = \bar{\kappa}$. This is trivial when $\bar{\kappa} = 0$, since in that case $df(\bar{x})(w) \geq 0$ for all w and we have $0 \in \widehat{\partial} f(\bar{x})$ by 8.4, hence $0 \in \partial f(\bar{x})$ as well; we can take $\bar{v} = 0$. Supposing therefore that $\bar{\kappa} > 0$, let $\bar{w}$ be a point that minimizes $df(\bar{x})(w)$ subject to $w \in \mathbb{B}$; we have seen that such a point must exist and have $df(\bar{x})(\bar{w}) = -\bar{\kappa}$, $|\bar{w}| = 1$.

The ray through $(\bar{w}, -\bar{\kappa})$ lies in the cone $T := \operatorname{epi} df(\bar{x}) = T_{\operatorname{epi} f}\big(\bar{x}, f(\bar{x})\big)$ as well as in the convex cone $K := -\operatorname{epi}(\bar{\kappa}|\cdot|)$, whose interior however contains no points of T. A certain point $(0, -\mu)$ on the vertical axis in K has $(\bar{w}, -\bar{\kappa})$ as the point of T nearest to it; the value of μ can be determined from the condition that $(0, -\mu) - (\bar{w}, -\bar{\kappa})$ must be orthogonal to $(\bar{w}, -\bar{\kappa})$, and this yields $\mu = (1 + \bar{\kappa}^2)/\bar{\kappa}$. By 6.27(b) as applied to the set $C = \operatorname{epi} f$ at $\big(\bar{x}, f(\bar{x})\big)$, there's a vector $(v, \beta) \in N_{\operatorname{epi} f}\big(\bar{x}, f(\bar{x})\big)$ with $\big|(v, \beta)\big| = 1$ such that $d_T(0, -\mu) = \big\langle (0, -\mu), (v, \beta) \big\rangle = -\mu\beta$. We calculate that

$$d_T(0, -\mu)^2 = \big|(0, -\mu) - (\bar{w}, -\bar{\kappa})\big|^2 = |\bar{w}|^2 + |\mu - \bar{\kappa}|^2 = 1 + (1/\bar{\kappa})^2,$$

so that $d_T(0, -\mu) = \sqrt{1 + \bar{\kappa}^2}/\bar{\kappa}$ and consequently

$$\beta = -[\sqrt{1 + \bar{\kappa}^2}/\bar{\kappa}]/[(1 + \bar{\kappa}^2)/\bar{\kappa}] = -1/\sqrt{1 + \bar{\kappa}^2}.$$

On the other hand, from the fact that $1 = |(v, \beta)|^2 = |v|^2 + \beta^2$ we have $|v| = \bar{\kappa}/\sqrt{1 + \bar{\kappa}^2}$. Let $\bar{v} = -v/\beta$. Then $|\bar{v}| = \bar{\kappa}$ and $(\bar{v}, -1) \in N_{\operatorname{epi} f}\big(\bar{x}, f(\bar{x})\big)$, hence $\bar{v} \in \partial f(\bar{x})$ by the geometry in Theorem 8.9. This vector $\bar{v}$ thus meets the requirements.

When f is regular at $\bar{x}$, the conclusions so far can be combined, because $df(\bar{x}) = \widehat{d}f(\bar{x})$. Furthermore, $d\big(0, \partial f(\bar{x})\big) \leq |\bar{v}| = \bar{\kappa}$. To obtain the opposite inequality, observe that for any $v \in \partial f(\bar{x})$ we have $df(\bar{x}) \geq \langle v, \cdot \rangle$ by 8.30. Then $\min_{w \in \mathbb{B}} df(\bar{x})(w) \geq \min_{w \in \mathbb{B}} \langle v, w \rangle$, where the left side is $-\bar{\kappa}$ and the right side is $-|v|$. Therefore, $d\big(0, \partial f(\bar{x})\big)$ can't be less than $\bar{\kappa}$. □

G. Graphical Differentiation of Mappings

The application of variational geometry to the epigraph of a function $f : \mathbb{R}^n \to \overline{\mathbb{R}}$ has been a powerful principle in obtaining a theory of generalized differentiation that can be effective even when f is discontinuous or takes on $\pm\infty$. But how should one approach a vector-valued function $F : x \mapsto F(x) = \big(f_1(x), \ldots, f_m(x)\big)$? Although such a function doesn't have an epigraph, it does have a graph in $\mathbb{R}^n \times \mathbb{R}^m$, and this could be the geometric object to focus on, as in classical analysis—but with the more general notions of tangent and normal in Chapter 6.

If this geometric route to differentiation is to be taken, it can just as well be followed through the much larger territory of set-valued mappings $S : \mathbb{R}^n \rightrightarrows \mathbb{R}^m$, in which single-valued mappings merely constitute a special case. This is what we'll do next.

In dealing with multivaluedness there's an essential feature which may at first be disconcerting and needs to be appreciated from the outset. At any $\bar{x} \in \operatorname{dom} S$ where $S(\bar{x})$ isn't a singleton, there's a multiplicity of elements $\bar{u} \in S(\bar{x})$, hence a multiplicity of points $(\bar{x}, \bar{u})$ at which to examine the local geometry of $\operatorname{gph} S$. In consequence, whatever kind of generalized 'derivative mapping' one may wish to introduce graphically, there must be a separate such mapping at $\bar{x}$ for each choice of $\bar{u} \in S(\bar{x})$. On the other hand, even for single-valued S, for which this sort of multiplicity is avoided, the single 'derivative mapping' at $\bar{x}$ could be multivalued.

8.33 Definition (graphical derivatives and coderivatives). *Consider a mapping $S : \mathbb{R}^n \rightrightarrows \mathbb{R}^m$ and a point $\bar{x} \in \operatorname{dom} S$. The graphical derivative of S at $\bar{x}$ for any $\bar{u} \in S(\bar{x})$ is the mapping $DS(\bar{x}\,|\,\bar{u}) : \mathbb{R}^n \rightrightarrows \mathbb{R}^m$ defined by*

$$z \in DS(\bar{x}\,|\,\bar{u})(w) \iff (w, z) \in T_{\operatorname{gph} S}(\bar{x}, \bar{u}),$$

whereas the coderivative is the mapping $D^*S(\bar{x}\,|\,\bar{u}) : \mathbb{R}^m \rightrightarrows \mathbb{R}^n$ defined by

$$v \in D^*S(\bar{x}\,|\,\bar{u})(y) \iff (v, -y) \in N_{\operatorname{gph} S}(\bar{x}, \bar{u}).$$

Here the notation $DS(\bar{x}\,|\,\bar{u})$ and $D^*S(\bar{x}\,|\,\bar{u})$ is simplified to $DS(\bar{x})$ and $D^*S(\bar{x})$ when S is single-valued at $\bar{x}$, $S(\bar{x}) = \{\bar{u}\}$. Similarly, and with the same provision for simplified notation, the regular derivative $\widehat{D}S(\bar{x}\,|\,\bar{u}) : \mathbb{R}^n \rightrightarrows \mathbb{R}^m$ and the regular coderivative $\widehat{D}^*S(\bar{x}\,|\,\bar{u}) : \mathbb{R}^m \rightrightarrows \mathbb{R}^n$ are defined by

$$z \in \widehat{D}S(\bar{x}\,|\,\bar{u})(w) \iff (w, z) \in \widehat{T}_{\operatorname{gph} S}(\bar{x}, \bar{u}),$$
$$v \in \widehat{D}^*S(\bar{x}\,|\,\bar{u})(y) \iff (v, -y) \in \widehat{N}_{\operatorname{gph} S}(\bar{x}, \bar{u}).$$

An initial illustration of graphical differentiation is given in Figure 8–11, where it must be remembered that the tangent and normal cones are shown as translated from the origin to $(\bar{x}, \bar{u})$. While $DS(\bar{x}\,|\,\bar{u})$ has $T_{\operatorname{gph} S}(\bar{x}, \bar{u})$ as its graph, it's not true that $D^*S(\bar{x}\,|\,\bar{u})$ has $N_{\operatorname{gph} S}(\bar{x}, \bar{u})$ as its graph—for two reasons. First there is the minus sign in the formula for $D^*S(\bar{x}\,|\,\bar{u})$, but second, this mapping goes from $I\!R^m$ to $I\!R^n$ instead of from $I\!R^n$ to $I\!R^m$. These changes foster 'adjoint' relationships between graphical derivatives and coderivatives, as seen in the next example and more generally in 8.40.

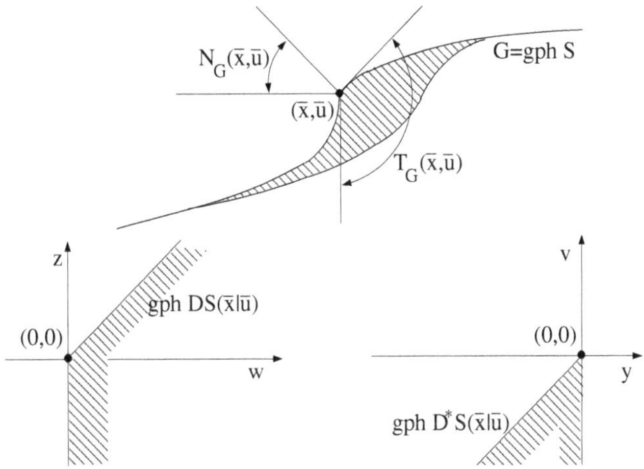

Fig. 8–11. Graphical differentiation.

Where do graphical derivatives and coderivatives fit in the classical picture of differentiation? Some understanding can be gained from the case of smooth, single-valued mappings.

8.34 Example (subdifferentiation of smooth mappings). *In the case of a smooth, single-valued mapping $F : I\!R^n \to I\!R^m$, one has*

$$DF(\bar{x})(w) = \widehat{D}F(\bar{x})(w) = \nabla F(\bar{x})w \;\; \textit{for all} \;\; w \in I\!R^n,$$
$$D^*F(\bar{x})(y) = \widehat{D}^*F(\bar{x})(y) = \nabla F(\bar{x})^* y \;\; \textit{for all} \;\; y \in I\!R^m.$$

Detail. These facts can be verified in more than one way, but here it is instructive to view them graphically. The equation $F(x) - u = 0$ represents $\operatorname{gph} F$ as a smooth manifold in the sense of 6.8. Setting $\bar{u} = F(\bar{x})$, we get

$$T_{\operatorname{gph} F}(\bar{x},\bar{u}) = \widehat{T}_{\operatorname{gph} F}(\bar{x},\bar{u}) = \big\{(w,z) \,\big|\, \nabla F(\bar{x})w - z = 0\big\},$$
$$N_{\operatorname{gph} F}(\bar{x},\bar{u}) = \widehat{N}_{\operatorname{gph} F}(\bar{x},\bar{u}) = \big\{(v,y) \,\big|\, v = -\nabla F(\bar{x})^* y\big\},$$

and the truth of the claims is then evident. □

At the opposite extreme, even subgradients and subderivatives can be seen as arising from graphical differentiation.

8.35 Example (subgradients and subderivatives via profile mappings). *For $f : I\!R^n \to \overline{I\!R}$ and the profile mapping $E_f : x \mapsto \big\{\alpha \in I\!R \,\big|\, f(x) \leq \alpha\big\}$, consider a point $\bar{x}$ where f is finite and locally lsc, and let $\bar{\alpha} = f(\bar{x}) \in E_f(\bar{x})$. Then*

$$
\begin{aligned}
DE_f(\bar{x}\,|\,\bar{\alpha})(w) &= \big\{\beta \in I\!R \,\big|\, df(\bar{x})(w) \le \beta\big\},\\
\widehat{D}E_f(\bar{x}\,|\,\bar{\alpha})(w) &= \big\{\beta \in I\!R \,\big|\, \widehat{d}f(\bar{x})(w) \le \beta\big\},\\
D^*E_f(\bar{x}\,|\,\bar{\alpha})(\lambda) &= \begin{cases} \lambda\partial f(\bar{x}) & \text{for } \lambda > 0,\\ \partial^\infty f(\bar{x}) & \text{for } \lambda = 0,\\ \emptyset & \text{for } \lambda < 0,\end{cases}\\
\widehat{D}^*E_f(\bar{x}\,|\,\bar{\alpha})(\lambda) &= \begin{cases} \lambda\widehat{\partial} f(\bar{x}) & \text{for } \lambda > 0,\\ \widehat{\partial} f(\bar{x})^\infty & \text{for } \lambda = 0 \text{ if } \widehat{\partial} f(\bar{x}) \ne \emptyset,\\ \{0\} & \text{for } \lambda = 0 \text{ if } \widehat{\partial} f(\bar{x}) = \emptyset,\\ \emptyset & \text{for } \lambda < 0.\end{cases}
\end{aligned}
$$

Detail. These formulas are immediate from Definition 8.33 in conjunction with Theorems 8.2(a), 8.9 and 8.17(a). □

A function $f : I\!R^n \to I\!R$ can be interpreted either as a special case of a function $f : I\!R^n \to \overline{I\!R}$, to be treated in an epigraphical manner with subgradients and subderivatives as up to now in this chapter, or as a special case of a mapping from $I\!R^n \to I\!R^1$ that happens to be single-valued, and thus treated graphically. The second approach, while seemingly more in harmony with traditional geometric motivations in analysis, encounters technical obstacles in the circumstance that the main results of variational geometry require closedness of the set to which they are applied.

For a single-valued mapping, closedness of the graph is tantamount to continuity when the mapping is locally bounded. Therefore, graphical differentiation isn't of much use for a real-valued function f unless f is continuous. It isn't good for such a wide range of applications as 'epigraphical differentiation'. Another drawback is that, even for a continuous function f on $I\!R^n$, the graphical derivative at $\bar{x}$ is a possibly multivalued mapping $Df(\bar{x}) : I\!R^n \rightrightarrows I\!R$ and in this respect suffers in comparison to $df(\bar{x})$, which is single-valued. But graphical differentiation in this special setting will nevertheless be important in the study of Lipschitz continuity in Chapter 9.

In the classical framework of 8.34 the derivative and coderivative mappings are linear, and each is the adjoint of the other—there is perfect duality between them. In general, however, derivatives and coderivatives won't be linear mappings, yet both will have essential roles to play.

Various expressions for the tangent and normal cones in Definition 8.33 yield other ways of looking at derivatives and coderivatives. For instance, 6(3) as applied to $T_{\operatorname{gph} S}(\bar{x}, \bar{u})$ produces the formula

$$
DS(\bar{x}\,|\,\bar{u})(\bar{w}) = \limsup_{\substack{\tau \searrow 0\\ w \to \bar{w}}} \frac{S(\bar{x} + \tau w) - \bar{u}}{\tau}. \tag{8(14)}
$$

The difference quotient is $\Delta_\tau S(\bar{x}\,|\,\bar{u})(w)$ for the mapping $\Delta_\tau S(\bar{x}\,|\,\bar{u}) : I\!R^n \rightrightarrows I\!R^m$ whose graph is $\big[\operatorname{gph} S - (\bar{x}, \bar{u})\big]/\tau$, and in terms of graphical limits we have

$$
DS(\bar{x}\,|\,\bar{u}) = \operatorname*{g\text{-}lim\,sup}_{\tau \searrow 0} \Delta_\tau S(\bar{x}\,|\,\bar{u}). \tag{8(15)}
$$

If $\operatorname{gph} S$ is locally closed at $(\bar{x}, \bar{u})$, we have through 6.26 that

$$\widehat{D}S(\bar{x} \,|\, \bar{u}) = \operatorname*{g\text{-lim\,inf}}_{(x,u) \xrightarrow[\operatorname{gph} S]{} (\bar{x},\bar{u})} DS(x \,|\, u). \tag{8(16)}$$

On the other hand, the definition of the cone $\widehat{N}_{\operatorname{gph} S}(\bar{x}, \bar{u})$ yields

$$\begin{aligned} &v \in \widehat{D}^*S(\bar{x} \,|\, \bar{u})(y) \iff \\ &\qquad \langle v, x - \bar{x} \rangle \leq \langle y, u - \bar{u} \rangle + o\big(|(x, u) - (\bar{x}, \bar{u})|\big) \text{ for } u \in S(x) \end{aligned} \tag{8(17)}$$

and the definition of $N_{\operatorname{gph} S}(\bar{x}, \bar{u})$ in combination with outer semicontinuity of normal cone mappings in 6.6 then gives

$$\begin{aligned} D^*S(\bar{x} \,|\, \bar{u}) &= \operatorname*{g\text{-lim\,sup}}_{(x,u) \xrightarrow[\operatorname{gph} S]{} (\bar{x},\bar{u})} \widehat{D}^*S(x \,|\, u) \\ &= \operatorname*{g\text{-lim\,sup}}_{(x,u) \xrightarrow[\operatorname{gph} S]{} (\bar{x},\bar{u})} D^*S(x \,|\, u). \end{aligned} \tag{8(18)}$$

Graphical differentiation behaves very simply in taking inverses of mappings:

$$\begin{aligned} z \in DS(\bar{x} \,|\, \bar{u})(w) &\iff w \in D(S^{-1})(\bar{u} \,|\, \bar{x})(z), \\ v \in D^*S(\bar{x} \,|\, \bar{u})(y) &\iff -y \in D^*(S^{-1})(\bar{u} \,|\, \bar{x})(-v). \end{aligned} \tag{8(19)}$$

When $F : \mathbb{R}^n \to \mathbb{R}^m$ is single-valued, formula 8(14) becomes

$$DF(\bar{x})(\bar{w}) = \limsup_{\substack{\tau \searrow 0 \\ w \to \bar{w}}} \frac{F(\bar{x} + \tau w) - F(\bar{x})}{\tau} \quad \text{(set of cluster points!)}, \tag{8(20)}$$

since 'lim sup' refers here to an outer limit in the sense of set convergence, but the sets happen to be singletons. Thus, $DF(\bar{x})$ is a *set-valued* mapping in principle, although it can be single-valued in particular situations. Single-valuedness at $\bar{w}$ occurs when there is a *single* cluster point, i.e., when the vector $[F(\bar{x} + \tau w) - F(\bar{x})]/\tau$ approaches a limit as $w \to \bar{w}$ and $\tau \searrow 0$. In general,

$$z \in DF(\bar{x})(\bar{w}) \iff \begin{cases} \exists\, w^\nu \to \bar{w},\ z^\nu \to z,\ \tau^\nu \searrow 0, \\ \text{with } F(\bar{x} + \tau^\nu w^\nu) = F(\bar{x}) + \tau^\nu z^\nu. \end{cases} \tag{8(21)}$$

When F is continuous, the graphical limits in 8(16) and 8(18) specialize to

$$\widehat{D}F(\bar{x}) = \operatorname*{g\text{-lim\,inf}}_{x \to \bar{x}} DF(x), \qquad D^*F(\bar{x}) = \operatorname*{g\text{-lim\,sup}}_{x \to \bar{x}} \widehat{D}^*F(x), \tag{8(22)}$$

where we have

$$\begin{aligned} &v \in \widehat{D}^*F(\bar{x})(y) \iff \\ &\qquad \big\langle v, x - \bar{x} \big\rangle \leq \big\langle y, F(x) - F(\bar{x}) \big\rangle + o\big(|(x, F(x)) - (\bar{x}, F(\bar{x}))|\big). \end{aligned} \tag{8(23)}$$

Such expressions can be analyzed further under assumptions of Lipschitz continuity on F, and this will be pursued in Chapter 9.

In order to understand better the nature of graphical derivatives and coderivatives, we'll need the concept of a *positively homogeneous* mapping H (set-valued in general), as distinguished by the properties

$$0 \in H(0) \text{ and } H(\lambda w) = \lambda H(w) \text{ for all } \lambda > 0 \text{ and } w. \tag{8(24)}$$

8.36 Proposition (positively homogeneous mappings). *For $H : \mathbb{R}^n \rightrightarrows \mathbb{R}^m$ to be positively homogeneous it is necessary and sufficient that* $\operatorname{gph} H$ *be a cone, and then both $H(0)$ and* $\operatorname{dom} H$ *must be cones as well (possibly with $H(0) = \{0\}$ or $\operatorname{dom} H = \mathbb{R}^n$). Such a mapping is graph-convex if and only if it also has*

$$H(w_1 + w_2) \supset H(w_1) + H(w_2) \text{ for all } w_1,\, w_2. \tag{8(25)}$$

If it is graph-convex with $H(0) = \{0\}$ and $\operatorname{dom} H = \mathbb{R}^n$, it must be a linear mapping: there must be an $(m \times n)$-matrix A such that $H(w) = Aw$ for all w.

Proof. The connection between positive homogeneity and the cone property for $\operatorname{gph} H$ is obvious, as are the general assertions about $\operatorname{dom} H$ and $H(0)$. The inclusion in the graph-convex case comes from the characterization of convex cones in 3.7. In particular then one has to have $H(0) \supset H(w) + H(-w)$ for all w. When $H(0) = \{0\}$ and $H(w) \neq \emptyset$ for every w, this implies that $H(w)$ is a singleton set for every w. Then it's not only true that $H(\lambda w) = \lambda H(w)$ for $\lambda > 0$ but also $H(-w) = -H(w)$ and $H(w_1 + w_2) = H(w_1) + H(w_2)$. This means that H is a linear mapping. □

A mapping H that is both positively homogeneous and graph-convex is said to be *sublinear*. Such a mapping is characterized by the combination of 8(24) and 8(25), or in geometric terms by its graph being a convex cone. An example from Chapter 5 would be the horizon mapping S^∞ for a graph-convex mapping S.

8.37 Proposition (basic derivative and coderivative properties). *For any mapping $S : \mathbb{R}^n \rightrightarrows \mathbb{R}^m$ and any points $\bar{x} \in \operatorname{dom} S$, $\bar{u} \in S(\bar{x})$, the mappings $DS(\bar{x}\,|\,\bar{u})$, $\widehat{D}S(\bar{x}\,|\,\bar{u})$, $D^*S(\bar{x}\,|\,\bar{u})$, and $\widehat{D}^*S(\bar{x}\,|\,\bar{u})$ are osc and positively homogeneous. Also, $\widehat{D}S(\bar{x}\,|\,\bar{u})$ and $\widehat{D}^*S(\bar{x}\,|\,\bar{u})$ are graph-convex, hence sublinear, with*

$$\widehat{D}S(\bar{x}\,|\,\bar{u})(w) \subset DS(\bar{x}\,|\,\bar{u})(w), \qquad \widehat{D}^*S(\bar{x}\,|\,\bar{u})(y) \subset D^*S(\bar{x}\,|\,\bar{u})(y).$$

If $\operatorname{gph} S$ *is locally closed at $(\bar{x}, \bar{u})$, in particular if S is osc, one has*

$$\begin{aligned} v \in \widehat{D}^*S(\bar{x}\,|\,\bar{u})(y) &\iff \langle v, w\rangle \leq \langle y, z\rangle \text{ when } z \in DS(\bar{x}\,|\,\bar{u})(w),\\ z \in \widehat{D}S(\bar{x}\,|\,\bar{u})(w) &\iff \langle v, w\rangle \leq \langle y, z\rangle \text{ when } v \in D^*S(\bar{x}\,|\,\bar{u})(y). \end{aligned} \tag{8(26)}$$

Proof. Here we simply restate the facts about normal and tangent cones in 6.2, 6.5 and 6.28, but in a graphical setting using 8.33. When S is osc its graph is locally closed at $(\bar{x}, \bar{u})$ for any $\bar{u} \in S(\bar{x})$. □

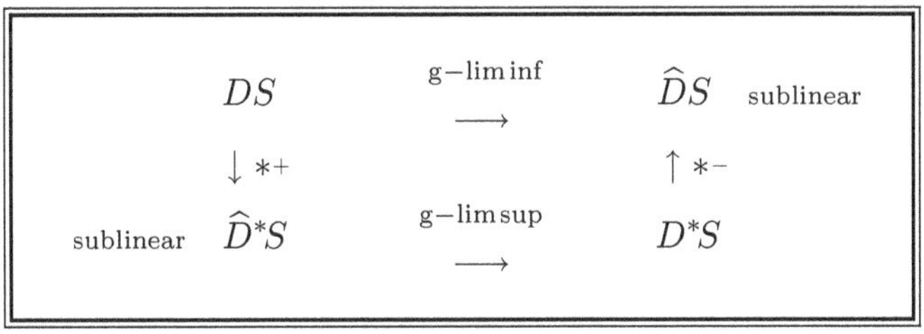

Fig. 8–12. Diagram of derivative-coderivative relationships for osc mappings.

The formulas in 8(26) can be understood as generalized 'adjoint' relations. For any positively homogeneous mapping $H : I\!R^n \rightrightarrows I\!R^m$ the *upper adjoint* $H^{*+} : I\!R^m \rightrightarrows I\!R^n$, always sublinear and osc, is defined by

$$H^{*+}(y) = \big\{ v \,\big|\, \langle v, w\rangle \leq \langle z, y\rangle \text{ when } z \in H(w) \big\}. \tag{8(27)}$$

The *lower adjoint* $H^{*-} : I\!R^m \rightrightarrows I\!R^n$, likewise always sublinear and osc, is defined instead by

$$H^{*-}(y) = \big\{ v \,\big|\, \langle v, w\rangle \geq \langle z, y\rangle \text{ when } z \in H(w) \big\}. \tag{8(28)}$$

The graphs of these adjoint mappings differ from the polar of the cone $\operatorname{gph} H$ only by sign changes in certain components and would be identical if this polar cone were a subspace, meaning the same for $\operatorname{gph} H$ itself, in which case H is called a *generalized linear* mapping. But anyway, even without this special property it's evident, when H is sublinear, that $(H^{*+})^{*-} = (H^{*-})^{*+} = \operatorname{cl} H$.

In the upper/lower adjoint notation, the formulas in 8(26) assert that

$$\widehat{D}^*S(\bar{x}\,|\,\bar{u}) = DS(\bar{x}\,|\,\bar{u})^{*+}, \qquad \widehat{D}S(\bar{x}\,|\,\bar{u}) = D^*S(\bar{x}\,|\,\bar{u})^{*-}.$$

These relationships and the others are schematized in Figure 8–12.

H.* Proto-Differentiability and Graphical Regularity

For the sake of a more complete duality, the relations in Figure 8–12 can be strengthened by invoking an appropriate version of Clarke regularity.

8.38 Definition (graphical regularity of mappings). *A mapping $S : I\!R^n \rightrightarrows I\!R^m$ is graphically regular at $\bar{x}$ for $\bar{u}$ if $\operatorname{gph} S$ is Clarke regular at $(\bar{x}, \bar{u})$.*

A manifestation of this property is seen in Figure 8–11. Obviously, a mapping S is graphically regular at $\bar{x}$ for $\bar{u} \in S(\bar{x})$ if and only if S^{-1} is graphically regular at $\bar{u}$ for the element $\bar{x} \in S^{-1}(\bar{u})$. Smooth mappings as in 8.34 are everywhere regular this way, in particular. Indeed, any mapping S such that $\operatorname{gph} S$ can be specified by a system of smooth constraints meeting the basic constraint qualification in 6.14 is graphically regular (cf. Example 8.42 below). Also in this category are mappings of the following kind.

8.39 Example (graphical regularity from graph-convexity). *If $S : \mathbb{R}^n \rightrightarrows \mathbb{R}^m$ is graph-convex and osc, then S is graphically regular at every $\bar{x} \in \operatorname{dom} S$ for every $\bar{u} \in S(\bar{x})$. In particular, linear mappings are graphically regular.*

Detail. This applies 6.9. □

8.40 Theorem (characterizations of graphically regular mappings). *Let $\bar{u} \in S(\bar{x})$ for a mapping $S : \mathbb{R}^n \to \mathbb{R}^m$ such that $\operatorname{gph} S$ is locally closed at $(\bar{x}, \bar{u})$. Then the following properties are equivalent and imply that the mappings $DS(\bar{x} \,|\, \bar{u})$ and $D^*S(\bar{x} \,|\, \bar{u})$ are graph-convex, hence sublinear:*

(a) *S is graphically regular at $\bar{x}$ for $\bar{u}$,*

(b) *$\langle v, w \rangle \leq \langle y, z \rangle$ when $v \in D^*S(\bar{x} \,|\, \bar{u})(y)$ and $z \in DS(\bar{x} \,|\, \bar{u})(w)$,*

(c) *$D^*S(\bar{x} \,|\, \bar{u}) = DS(\bar{x} \,|\, \bar{u})^{*+}$,*

(d) *$DS(\bar{x} \,|\, \bar{u}) = D^*S(\bar{x} \,|\, \bar{u})^{*-}$,*

(e) *$DS(\bar{x} \,|\, \bar{u}) = \widehat{D}S(\bar{x} \,|\, \bar{u})$,*

(f) *$D^*S(\bar{x} \,|\, \bar{u}) = \widehat{D}^*S(\bar{x} \,|\, \bar{u})$.*

Proof. These relations are immediate from 6.29 through 8.33 and 8.37. □

The tight derivative-coderivative relationship in 8.40(c)(d), where both mappings are osc and sublinear, generalizes the adjoint relationship in the classical framework of 8.34.

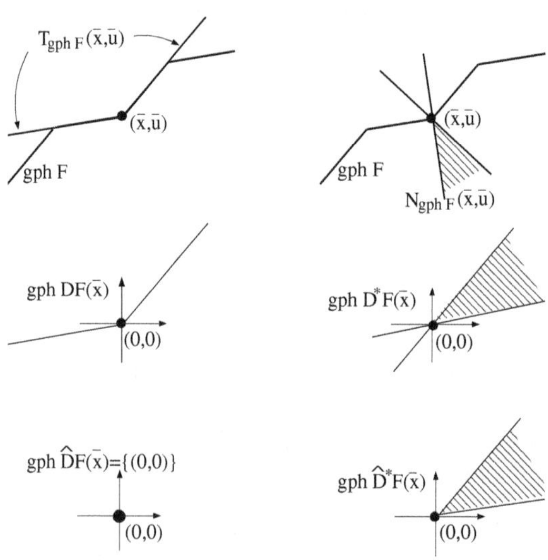

Fig. 8–13. An absence of graphical regularity.

This duality has many nice consequences, but it doesn't mark out the only territory in which derivatives and coderivatives are important. Despite the examples of graphically regular mappings that have been mentioned, many mappings of interest typically fail to have this property, for instance single-valued mappings that aren't smooth. This is illustrated in Figure 8–13 in the one-dimensional case of a piecewise linear mapping at one of the joints in its graph. Regularity is precluded because the derivative mapping obviously isn't graph-convex as would be guaranteed by Theorem 8.40 in that case. In particular the graphical derivative and coderivative mappings aren't sublinear mappings adjoint to each other.

A mapping $S : \mathbb{R}^n \rightrightarrows \mathbb{R}^m$ is said to be *proto-differentiable* at $\bar{x}$ for an element $\bar{u} \in S(\bar{x})$ if the outer graphical limit in 8(15) is a full limit:

$$\Delta_\tau S(\bar{x}\,|\,\bar{u}) \xrightarrow{g} DS(\bar{x}\,|\,\bar{u}) \quad \text{as} \quad \tau \searrow 0,$$

or in other words, if in addition to the 'lim sup' for $DS(\bar{x}\,|\,\bar{u})(\bar{w})$ in 8(14) there exist for each $\bar{z} \in DS(\bar{x}\,|\,\bar{u})(\bar{w})$ and choice of $\tau^\nu \searrow 0$ sequences

$$w^\nu \to \bar{w} \;\text{ and }\; z^\nu \to \bar{z} \;\text{ with }\; z^\nu \in \big[S(\bar{x} + \tau^\nu w^\nu) - \bar{u}\big]/\tau^\nu.$$

8.41 Proposition (proto-differentiability). *A mapping $S : \mathbb{R}^n \rightrightarrows \mathbb{R}^m$ is proto-differentiable at $\bar{x}$ for $\bar{u} \in S(\bar{x})$ if and only if $\operatorname{gph} S$ is geometrically derivable at $(\bar{x}, \bar{u})$. This is true in particular when S is graphically regular at $\bar{x}$ for $\bar{u}$.*

Whenever S is proto-differentiable at $\bar{x}$ for $\bar{u}$, its inverse S^{-1} is proto-differentiable at $\bar{u}$ for $\bar{x}$.

Proof. These facts are evident from 8.33, 6.2, and 6.30. The geometry of $\operatorname{gph} S^{-1}$ at $(\bar{u}, \bar{x})$ reflects that of $\operatorname{gph} S$ at $(\bar{x}, \bar{u})$. □

8.42 Exercise (proto-differentiability of feasible-set mappings). *Let $S(t) \subset \mathbb{R}^n$ depend on $t \in \mathbb{R}^d$ through a constraint representation*

$$S(t) = \big\{ x \in X \;\big|\; F(x,t) \in D \big\}$$

with $X \subset \mathbb{R}^n$ and $D \subset \mathbb{R}^m$ closed and $F : \mathbb{R}^n \times \mathbb{R}^d \to \mathbb{R}^m$ smooth. Suppose for given $\bar{t}$ and $\bar{x} \in S(\bar{t})$ that X is regular at $\bar{x}$, D is regular at $F(\bar{x}, \bar{t})$, and the following condition is satisfied:

$$\left.\begin{aligned} y \in N_D\big(F(\bar{x},\bar{t})\big) \\ \nabla_x F(\bar{x},\bar{t})^* y + N_X(\bar{x}) \ni 0 \\ \nabla_t F(\bar{x},\bar{t})^* y = 0 \end{aligned}\right\} \implies y = 0.$$

Then the mapping $S : t \mapsto S(t)$ is graphically regular at $\bar{t}$ for $\bar{x}$, hence also proto-differentiable at $\bar{t}$ for $\bar{x}$, with

$$\begin{aligned} DS(\bar{t}\,|\,\bar{x})(s) &= \Big\{ w \in T_X(\bar{x}) \;\Big|\; \nabla_x F(\bar{x},\bar{t})w + \nabla_t F(\bar{x},\bar{t})s \in T_D\big(F(\bar{x},\bar{t})\big) \Big\}, \\ D^*S(\bar{t}\,|\,\bar{x})(v) &= \Big\{ \nabla_t F(\bar{x},\bar{t})^* y \;\Big| \\ &\qquad y \in N_D\big(F(\bar{x},\bar{t})\big),\; v + \nabla_x F(\bar{x},\bar{t})^* y + N_X(\bar{x}) \ni 0 \Big\}. \end{aligned}$$

Guide. The tangents and normals to the set $\big\{(x,t) \in X \times \mathbb{R}^d \;\big|\; F(x,t) \in D\big\}$, a 'reflection' of $\operatorname{gph} S$, can be analyzed through 6.14 and 6.31. Use this to get the desired graphical derivatives and coderivatives, and invoke 8.41. □

The condition assumed in 8.42 is satisfied in particular under the constraint qualification for the set $S(\bar{t})$ in 6.14 and 6.31, and it reduces to that when $d = m$ and the matrix $\nabla_t F(\bar{x}, \bar{t})$ is nonsingular. The generally stronger condition given by the constraint qualification will be seen in 9.50 to yield the following property, more powerful than proto-differentiability.

For a set-valued mapping $S : \mathbb{R}^n \rightrightarrows \mathbb{R}^m$, a point $\bar{x} \in \operatorname{dom} S$ and an element $\bar{u} \in S(\bar{x})$, the limit

$$\lim_{\substack{\tau \searrow 0 \\ w \to \bar{w}}} \frac{S(\bar{x} + \tau w) - \bar{u}}{\tau}, \tag{8(29)}$$

if it exists, is the *semiderivative at $\bar{x}$ for $\bar{u}$ and $\bar{w}$*. If it exists for every vector $\bar{w} \in \mathbb{R}^n$, then S is *semidifferentiable at $\bar{x}$ for $\bar{u}$*.

8.43 Exercise (semidifferentiability of set-valued mappings). *For a mapping $S : \mathbb{R}^n \rightrightarrows \mathbb{R}^m$, any $\bar{x} \in \operatorname{dom} S$ and $\bar{u} \in S(\bar{x})$, the following four properties of a mapping $H : \mathbb{R}^n \rightrightarrows \mathbb{R}^m$ are equivalent and entail having $\bar{x} \in \operatorname{int}(\operatorname{dom} S)$ and H positively homogeneous and continuous with $\operatorname{dom} H = \mathbb{R}^n$:*

(a) *S is semidifferentiable at $\bar{x}$ for $\bar{u}$, the semiderivative for w being $H(w)$;*

(b) *as $\tau \searrow 0$, the mappings $\Delta_\tau S(\bar{x} \mid \bar{u})$ converge continuously to H;*

(c) *as $\tau \searrow 0$, the mappings $\Delta_\tau S(\bar{x} \mid \bar{u})$ converge uniformly to H on all bounded sets, and H is continuous;*

(d) *S is proto-differentiable at $\bar{x}$ for $\bar{u}$ with $DS(\bar{x} \mid \bar{u}) = H$, and there is a neighborhood $W \in \mathcal{N}(0)$ such that at each point $w \in W$ the mappings $\Delta_\tau S(\bar{x} \mid \bar{u})$ are asymptotically equicontinuous as $\tau \searrow 0$.*

These properties hold in particular under the following equivalent pair of conditions, which necessitate $S(\bar{x}) = \{\bar{u}\}$:

(e) *$DS(\bar{x} \mid \bar{u}) = H$ single-valued, and there exists $\kappa \in [0, \infty)$ such that, for all x in a neighborhood of $\bar{x}$, one has $\emptyset \neq S(x) \subset \bar{u} + \kappa|x - \bar{x}|\mathbb{B}$;*

(f) *H is single-valued, continuous and positively homogeneous, and for all x in a neighborhood of $\bar{x}$ one has*

$$\emptyset \neq S(x) \subset \bar{u} + H(x - \bar{x}) + o(|x - \bar{x}|)\mathbb{B}.$$

Guide. This parallels Theorem 7.21 in appealing to the convergence properties in 5.43 and 5.44. In dealing with (e) and (f), note that in both cases the assumptions imply the existence of $\varepsilon > 0$, a neighborhood $W \in \mathcal{N}(0)$ and a bounded set B, such that $\emptyset \neq \Delta_\tau DS(\bar{x} \mid \bar{u})(w) \subset B$ when $\tau \in (0, \varepsilon)$ and $w \in W$. Build an argument around the idea that if a sequence of mappings D^ν is nonempty-valued and uniformly locally bounded on an open set O, and $\operatorname{g-lim\,sup}_\nu D^\nu \subset D$ for a mapping D that is single-valued on O, then $D^\nu \xrightarrow{p} D$ uniformly on compact subsets of O. □

In the case of property 8.43(f) where H is a linear mapping, S is said to be *differentiable* at $\bar{x}$. One should appreciate that this property is definitely stronger than just having semidifferentiability with the semiderivatives behaving linearly. It wouldn't be appropriate to speak of 'differentiability' merely if the equivalent conditions 8.43(a)(b)(c)(d) hold and H happens to be single-valued and linear. For example, the mapping $S : I\!R \rightrightarrows I\!R$ defined by $S(x) = \{0\} \cup S_0(x)$, with $S_0(x)$ an arbitrarily chosen subset of $I\!R \setminus (-1, 1)$, is semidifferentiable at $\bar{x} = 0$ for $\bar{u} = 0$ and has $DS(\bar{x}\,|\,\bar{u}) \equiv 0$. But S fails to display the expansion property in 8.43(f) for $H \equiv 0$.

This example also shows that semidifferentiability of S at $\bar{x}$ for $\bar{u}$ doesn't necessarily imply continuity of S at $\bar{x}$. Semidifferentiability does imply

$$\bar{u} \in \liminf_{\substack{\tau \searrow 0 \\ w' \to w}} \Delta_\tau S(\bar{x}\,|\,\bar{u})(w'),$$

and for *single-valued* S this comes out as continuity at $\bar{x}$. Semidifferentiability of single-valued mappings will be pursued further in 9.25.

The 'growth condition' on S in 8.43(e) will later, in 9(30) with $\bar{u}$ replaced by $S(\bar{x})$ not necessarily a singleton, come under the heading of 'calmness' of set-valued mappings.

8.44 Exercise (subdifferentiation of derivatives).

(a) *Suppose $S : I\!R^n \rightrightarrows I\!R^m$ is osc at $\bar{x}$, and let $\bar{u} \in S(\bar{x})$. Then for the mapping $H = DS(\bar{x}\,|\,\bar{u})$ and any $\bar{w} \in \operatorname{dom} H$ and $\bar{z} \in H(\bar{w})$ one has*

$$D^*H(\bar{w}\,|\,\bar{z})(y) \subset D^*H(0\,|\,0)(y) \subset D^*S(\bar{x}\,|\,\bar{u})(y) \ \text{ for all } y.$$

(b) *Suppose $f : I\!R^n \to \overline{I\!R}$ is finite and locally lsc at $\bar{x}$. Then for the function $h = df(\bar{x})$ and any $\bar{w} \in \operatorname{dom} h$ one has*

$$\partial h(\bar{w}) \subset \partial h(0) \subset \partial f(\bar{x}), \qquad \partial^\infty h(\bar{w}) \subset \partial^\infty h(0) \subset \partial^\infty f(\bar{x}).$$

Guide. In (a) invoke 6.27(a) relative to the definition of H in 8.33. In (b) apply (a) through 8.35—or argue directly from 6.27(a) again in terms of the epigraphical geometry in 8.2(a), 8.9 and 8.17(a). □

I.* Proximal Subgradients

Proximal normals have been seen to be useful in some situations as a special case of regular normals, and there is a counterpart for subgradients.

8.45 Definition (proximal subgradients). *A vector v is called a proximal subgradient of a function $f : I\!R^n \to \overline{I\!R}$ at $\bar{x}$, a point where $f(\bar{x})$ is finite, if there exist $\rho > 0$ and $\delta > 0$ such that*

$$f(x) \geq f(\bar{x}) + \langle v, x - \bar{x}\rangle - \tfrac{1}{2}\rho|x - \bar{x}|^2 \ \text{ when } |x - \bar{x}| \leq \delta. \qquad 8(30)$$

The existence of a proximal subgradient at $\bar{x}$ corresponds to the existence of a 'local quadratic support' to f at $\bar{x}$. This relates to the proximal points and proximal hulls in Example 1.44, which concern a global version of the quadratic support property. It's easy to see that when f is prox-bounded the local inequality in 8(30) can be made to hold globally through an increase in ρ if necessary (cf. 8.46(f) below).

8.46 Proposition (characterizations of proximal subgradients).

(a) *A vector v is a proximal subgradient to f at $\bar{x}$ if and only if $(v,-1)$ is a proximal normal to* epi f *at* $(\bar{x}, f(\bar{x}))$.

(b) *A vector v is a proximal subgradient of δ_C at $\bar{x}$ if and only if v is a proximal normal to C at $\bar{x}$.*

(c) *A vector v is a proximal subgradient of f at $\bar{x}$ if and only if on some neighborhood of $\bar{x}$ there is a $\mathcal{C}^2$ function $h \leq f$ with $h(\bar{x}) = f(\bar{x})$, $\nabla h(\bar{x}) = v$.*

(d) *A vector v is a proximal normal to C at $\bar{x}$ if and only if on some neighborhood of $\bar{x}$ there is a $\mathcal{C}^2$ function h with $\nabla h(\bar{x}) = v$ such that h has a local maximum on C at $\bar{x}$.*

(e) *The proximal subgradients of f at $\bar{x}$ form a convex subset of $\widehat{\partial} f(\bar{x})$.*

(f) *As long as f is prox-bounded, the points at which f has a proximal subgradient are the points that are λ-proximal for some $\lambda > 0$, i.e., belong to $\bigcup_{\lambda>0} \operatorname{rge} P_\lambda f$. Indeed, if $x \in P_\lambda f(w)$ the vector $v = \lambda^{-1}[w - x]$ is a proximal subgradient at x and satisfies*

$$f(x') \geq f(x) + \langle v, x' - x \rangle - \tfrac{1}{2}\lambda^{-1}|x' - x|^2 \quad \text{for all } x' \in \mathbb{R}^n.$$

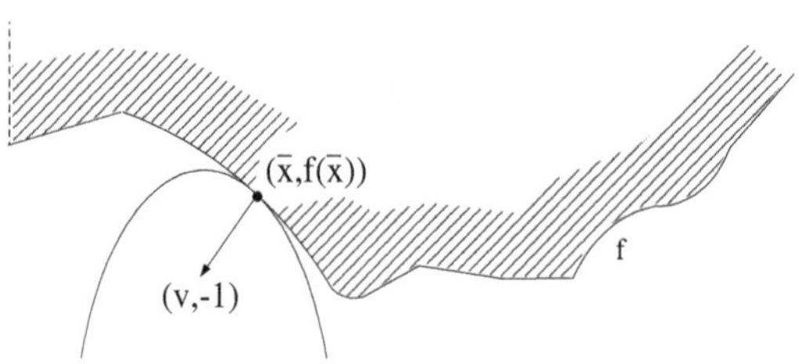

Fig. 8–14. Proximal subgradients generated from quadratic supports.

Proof. We start with (c), for which the necessity is clear from the defining condition 8(30). To prove the sufficiency, consider such h on $\mathbb{B}(\bar{x}, \delta)$ and let

$$\rho := -\min_{\substack{|x-\bar{x}|\leq\delta \\ |w|=1}} \langle w, \nabla^2 h(x) w \rangle.$$

For any w with $|w| = 1$ the function $\varphi(\tau) := h(\bar{x}+\tau w)$ has $\varphi(0) = f(\bar{x})$, $\varphi'(0) = \langle v, w \rangle$ and $\varphi''(\tau) = \langle w, \nabla^2 h(\bar{x} + \tau w) w \rangle \geq -\rho$, so $\varphi(\tau) \geq \varphi(0) + \varphi'(0)\tau - \frac{1}{2}\rho\tau^2$ for $0 \leq \tau \leq \delta$. Thus $h(x) \geq h(\bar{x}) + \langle v, x - \bar{x} \rangle - \frac{1}{2}\rho|x - \bar{x}|^2$ when $|x - \bar{x}| \leq \delta$. Since $f(x) \geq h(x)$ for $x \in \mathbb{B}(\bar{x}, \delta)$ and $f(\bar{x}) = h(\bar{x})$, we conclude that v is a proximal subgradient to f at $\bar{x}$.

We work next on (b) and (d) simultaneously. If v is a proximal subgradient of δ_C at $\bar{x}$, there exist $\rho > 0$ and $\delta > 0$ such that 8(30) holds for $f = \delta_C$. Then in terms of $\varepsilon := 1/\rho$ we have $2\varepsilon\langle v, x-\bar{x}\rangle \le |x-\bar{x}|^2$ when $x \in C$ and $|x-\bar{x}| \le \delta$. Since $\langle v, x-\bar{x}\rangle \le |x-\bar{x}||v|$, when $|x-\bar{x}| \ge \delta$, we get for $\tau := \min\{\varepsilon, \delta/2|v|\}$ that $2\tau\langle v, x-\bar{x}\rangle \le |x-\bar{x}|^2$ for all $x \in C$, so

$$0 \le |x-\bar{x}|^2 - 2\tau\langle v, x-\bar{x}\rangle = \big|x-(\bar{x}+\tau v)\big|^2 - \tau^2|v|^2 \quad \text{for all } x \in C. \qquad 8(31)$$

But this means $\bar{x} \in P_C(\bar{x}+\tau v)$; thus v is a proximal normal to C at $\bar{x}$. On the other hand, if the latter is true, so that 8(31) holds for a certain $\tau > 0$, the $\mathcal{C}^2$ function $h(x) := -(1/2\tau)\big|x-(\bar{x}+\tau v)\big|^2$ achieves its maximum over C at $\bar{x}$, and $\nabla h(\bar{x}) = v$. Then v satisfies the condition in (d). But in turn, if v satisfies that condition for any $\mathcal{C}^2$ function h, we have for $h_0(x) := h(x) - h(\bar{x})$ that $\delta_C(x) \ge h_0(x)$ for all x in a neighborhood of $\bar{x}$, while $\delta_C(\bar{x}) = h_0(\bar{x})$, and then from (c), as applied to δ_C and h_0, we conclude that v is a proximal subgradient of δ_C at $\bar{x}$. This establishes both (b) and (d).

Turning to the proof of (a), we observe that if v is a proximal subgradient of f at $\bar{x}$, so that 8(30) holds for some $\rho > 0$ and $\delta > 0$, the $\mathcal{C}^2$ function $h(x,\alpha) := f(\bar{x}) + \langle v, x-\bar{x}\rangle - \frac{1}{2}\rho|x-\bar{x}|^2 - \alpha$ achieves a local maximum over $\operatorname{epi} f$ at $\big(\bar{x}, f(\bar{x})\big)$, and $\nabla h\big(\bar{x}, f(\bar{x})\big) = (v,-1)$. Then $(v,-1)$ is a proximal normal to $\operatorname{epi} f$ at $\big(\bar{x}, f(\bar{x})\big)$ through the equivalence already established in (d), as applied to $\operatorname{epi} f$. Conversely, if $(v,-1)$ is a proximal normal to $\operatorname{epi} f$ at $\big(\bar{x}, f(\bar{x})\big)$ there exists $\tau > 0$ such that the ball of radius τ around $\big(\bar{x}, f(\bar{x})\big) + \tau(v,-1)$ touches $\operatorname{epi} f$ only at $\big(\bar{x}, f(\bar{x})\big)$ (cf. 6.16). The upper surface of this ball is the graph of a $\mathcal{C}^2$ function $h \le f$, defined on a neighborhood of $\bar{x}$, with $h(\bar{x}) = f(\bar{x})$ and $\nabla h(\bar{x}) = v$. If follows then from (c) that v is a proximal subgradient of f at $\bar{x}$.

The convexity property in (e) is elementary. As for the claims in (f), we can suppose $f \not\equiv \infty$. Prox-boundedness implies then that f is proper and the envelope function $e_\lambda f$ is finite for λ in an interval $(0, \lambda_f)$; cf. 1.24. Then for any $\bar{x}$ we have $f(x) \ge e_\lambda f(\bar{x}) - (1/2\lambda)|x-\bar{x}|^2$ for all x. If in addition v is a proximal subgradient at $\bar{x}$, so that 8(30) holds, we can get the global inequality

$$f(x) \ge f(\bar{x}) + \langle v, x-\bar{x}\rangle - \frac{1}{2\bar{\lambda}}|x-\bar{x}|^2 \quad \text{for all } x$$

by taking $\bar{\lambda} > 0$ small enough that $\bar{\lambda}^{-1} > \rho$ and $f(\bar{x}) + |v|\tau - (1/2\bar{\lambda}^2)\tau^2 \le e_\lambda f(\bar{x}) - (1/2\lambda)\tau^2$ for all $\tau \ge \delta$, which is always possible. Setting $w = \bar{x} + \bar{\lambda}v$, so $v = \bar{\lambda}^{-1}[w-\bar{x}]$, we can construe this global inequality as saying that $f(\bar{x}) + (1/2\bar{\lambda})|\bar{x}-w|^2 \le f(x) + (1/2\bar{\lambda})|x-w|^2$ for all x, or in other words, $\bar{x} \in P_{\bar{\lambda}} f(w)$.

Conversely, if $\bar{x} \in P_\lambda f(w)$, $\lambda > 0$, we have $f(\bar{x}) + (1/2\lambda)|\bar{x}-w|^2 \le f(x) + (1/2\lambda)|x-w|^2$ for all x, the value $f(\bar{x})$ being finite. In terms of $v = \lambda^{-1}[w-\bar{x}]$ this takes the form $f(x) \ge f(\bar{x}) + \langle v, x-\bar{x}\rangle - (1/2\lambda)|x-\bar{x}|^2$ and says that v is a proximal subgradient at $\bar{x}$. □

8.47 Corollary (approximation of subgradients). *Let f be lsc, proper, and finite at $\bar{x}$, and let $\bar{v} \in \partial f(\bar{x})$ and $\bar{v}' \in \partial^\infty f(\bar{x})$.*

(a) *For any $\delta > 0$ there exist $x \in \mathbb{B}(\bar{x}, \delta) \cap \operatorname{dom} f$ with $|f(x) - f(\bar{x})| \le \delta$ and a proximal subgradient $v \in \partial f(x)$ such that $v \in \mathbb{B}(\bar{v}, \delta)$. Likewise, there* *exist* $x' \in \mathbb{B}(\bar{x}, \delta) \cap \operatorname{dom} f$ *with* $|f(x') - f(\bar{x})| \leq \delta$ *and a proximal subgradient* $v' \in \partial f(x')$ *such that* $\lambda v' \in \mathbb{B}(\bar{v}, \delta)$ *for some* $\lambda \in (0, \delta)$.

(b) *For any sequence of lsc, proper, functions* f^ν *with* $\text{e-lim inf}_\nu f^\nu = f$, *there is a subsequence* $\{f^\nu\}_{\nu \in N}$ *(for* $N \in \mathcal{N}_\infty^\#$*) and points* $x^\nu \underset{N}{\to} \bar{x}$ *with* $f^\nu(x^\nu) \underset{N}{\to} f(\bar{x})$ *having proximal subgradients* $v^\nu \in \partial f^\nu(x^\nu)$ *such that* $v^\nu \underset{N}{\to} \bar{v}$.

Likewise, there exist $x'^\nu \underset{N}{\to} \bar{x}$ *with* $f^\nu(x'^\nu) \underset{N}{\to} f(\bar{x})$ *having proximal subgradients* $v'^\nu \in \partial f^\nu(x'^\nu)$ *such that* $\lambda^\nu v'^\nu \underset{N}{\to} \bar{v}'$ *for some choice of* $\lambda^\nu \searrow 0$.

In each case the index set N *can be taken in* $\mathcal{N}_\infty$ *if actually* $f^\nu \xrightarrow{e} f$.

Proof. Part (a) follows from 8.46(a) and the geometry in 8.9 by making use of the approximation in 6.18(a). Part (b) follows similarly from 6.18(b). □

Another way of generating subgradients $v \in \partial f(\bar{x})$ or $v \in \partial^\infty f(\bar{x})$ as limits, this time from sequences of nearby 'ε-regular subgradients', will be seen in 10.46. For convex functions a much stronger result than 8.47(b) will be available in 12.35: when such functions epi-converge, their subgradient mappings converge graphically.

8.48 Exercise (approximation of coderivatives). *Suppose* $\bar{u} \in S(\bar{x})$ *for a mapping* S *expressed as the graphical outer limit* $\text{g-lim sup}_\nu S^\nu$ *of a sequence of osc mappings* $S^\nu : \mathbb{R}^n \rightrightarrows \mathbb{R}^m$. *Let* $z \in D^*S(\bar{x} \mid \bar{u})(y)$. *Then there is a subsequence* $\{S^\nu\}_{\nu \in N}$ *(for* $N \in \mathcal{N}_\infty^\#$*) along with* $(x^\nu, u^\nu) \underset{N}{\to} \bar{x}$ *and* $(y^\nu, z^\nu) \underset{N}{\to} (y, z)$ *such that* $u^\nu \in S^\nu(x^\nu)$ *and* $z^\nu \in \widehat{D}^*S(x^\nu \mid u^\nu)(y^\nu)$, *hence in particular* $z^\nu \in D^*S(x^\nu \mid u^\nu)(y^\nu)$. *(The index set* N *can be taken in* $\mathcal{N}_\infty$ *if actually* $S^\nu \xrightarrow{g} S$.*)*

Guide. Derive this by applying 6.18(b) to the graphs of these mappings and using the fact that proximal normals are regular normals. □

J*. Other Results

The Clarke normal cone $\overline{N}_{\operatorname{epi} f}(\bar{x}, f(\bar{x})) = \operatorname{cl} \operatorname{con} N_{\operatorname{epi} f}(\bar{x}, f(\bar{x}))$ (cf. 6.38) leads to a concept of subgradients which captures a tighter duality for functions that aren't regular. The sets of *Clarke subgradients* and *Clarke horizon subgradients* of f at $\bar{x}$ are defined by

$$\begin{aligned} \bar{\partial} f(\bar{x}) &:= \left\{ v \,\middle|\, (v, -1) \in \overline{N}_{\operatorname{epi} f}(\bar{x}, f(\bar{x})) \right\}, \\ \bar{\partial}^\infty f(\bar{x}) &:= \left\{ v \,\middle|\, (v, 0) \in \overline{N}_{\operatorname{epi} f}(\bar{x}, f(\bar{x})) \right\}. \end{aligned} \qquad 8(32)$$

8.49 Theorem (Clarke subgradients). *Let* $f : \mathbb{R}^n \to \overline{\mathbb{R}}$ *be locally lsc and finite at* $\bar{x}$. *Then* $\bar{\partial} f(\bar{x})$ *is a closed convex set,* $\bar{\partial}^\infty f(\bar{x})$ *is a closed convex cone, and*

$$\begin{aligned} \bar{\partial} f(\bar{x}) &= \left\{ v \,\middle|\, \langle v, w \rangle \leq \widehat{d}f(\bar{x})(w) \text{ for all } w \in \mathbb{R}^n \right\}, \\ \bar{\partial}^\infty f(\bar{x}) &= \left\{ v \,\middle|\, \langle v, w \rangle \leq 0 \text{ for all } w \in \operatorname{dom} \widehat{d}f(\bar{x}) \right\}. \end{aligned}$$

Moreover $\bar{\partial}^\infty f(\bar{x}) = \bar{\partial} f(\bar{x})^\infty$ *as long as* $\bar{\partial} f(\bar{x}) \neq \emptyset$, *or equivalently* $\partial f(\bar{x}) \neq \emptyset$, *and in that case*

$$\widehat{d}f(\bar{x})(w) \;=\; \sup \big\langle \bar{\partial} f(\bar{x}), w \big\rangle \;\; \textit{for all } w \in I\!R^n.$$

When the cone $\bar{\partial}^\infty f(\bar{x})$ *is pointed, which is true if and only if the cone* $\partial^\infty f(\bar{x})$ *is pointed, one has the representations*

$$\bar{\partial} f(\bar{x}) \;=\; \operatorname{con} \partial f(\bar{x}) + \operatorname{con} \partial^\infty f(\bar{x}), \qquad \bar{\partial}^\infty f(\bar{x}) \;=\; \operatorname{con} \partial^\infty f(\bar{x}).$$

Proof. The initial assertions are immediate from the defining formulas in 8(32), because $\overline{N}_{\operatorname{epi} f}\big(\bar{x}, f(\bar{x})\big) = \widehat{T}_{\operatorname{epi} f}\big(\bar{x}, f(\bar{x})\big)^*$ by 6.38, while $\operatorname{epi} \widehat{d}f(\bar{x}) = \widehat{T}_{\operatorname{epi} f}\big(\bar{x}, f(\bar{x})\big)$ by 8.17(a). The function $\widehat{d}f(\bar{x})$ is lsc and sublinear (cf. 8.18), so the formula for $\bar{\partial} f(\bar{x})$ in terms of $\widehat{d}f(\bar{x})$ ensures through the theory of support functions in 8.24 that as long as $\bar{\partial} f(\bar{x})$ isn't empty, $\widehat{d}f(\bar{x})(w)$ is the indicated supremum. Since $\overline{N}_{\operatorname{epi} f}\big(\bar{x}, f(\bar{x})\big) = \operatorname{cl} \operatorname{con} N_{\operatorname{epi} f}\big(\bar{x}, f(\bar{x})\big)$ with

$$N_{\operatorname{epi} f}\big(\bar{x}, f(\bar{x})\big) = \big\{\lambda(v, -1) \,\big|\, v \in \partial f(\bar{x}),\ \lambda > 0\big\} \cup \big\{(v, 0) \,\big|\, v \in \partial^\infty f(\bar{x})\big\} \quad 8(33)$$

by 8.9, it's clear that $\bar{\partial} f(\bar{x}) \neq \emptyset$ if and only if $\partial f(\bar{x}) \neq \emptyset$. Also, $N_{\operatorname{epi} f}\big(\bar{x}, f(\bar{x})\big)$ is pointed if and only if $\partial^\infty f(\bar{x})$ is pointed. But then by 3.15, $\overline{N}_{\operatorname{epi} f}\big(\bar{x}, f(\bar{x})\big)$ is pointed too and simply equals $\operatorname{con} N_{\operatorname{epi} f}\big(\bar{x}, f(\bar{x})\big)$, so that $\bar{\partial}^\infty f(\bar{x})$ is pointed. Conversely, the pointedness of the cone $\bar{\partial}^\infty f(\bar{x})$ entails that of the cone $\partial^\infty f(\bar{x})$, because $\bar{\partial}^\infty f(\bar{x}) \supset \partial^\infty f(\bar{x})$.

The identification of $\overline{N}_{\operatorname{epi} f}\big(\bar{x}, f(\bar{x})\big)$ with $\operatorname{con} N_{\operatorname{epi} f}\big(\bar{x}, f(\bar{x})\big)$ yields at once through 8(33) the convex hull formulas for $\bar{\partial} f(\bar{x})$ and $\bar{\partial}^\infty f(\bar{x})$. □

In a variety of settings, including those in which f is subdifferentially regular, one actually has $\bar{\partial} f(\bar{x}) = \partial f(\bar{x})$ and $\bar{\partial}^\infty f(\bar{x}) = \partial^\infty f(\bar{x})$. When these equations don't hold, it is tempting to think that the replacement of $\partial f(\bar{x})$ and $\partial^\infty f(\bar{x})$ by $\bar{\partial} f(\bar{x})$ and $\bar{\partial}^\infty f(\bar{x})$ would be well rewarded by the close duality thereby achieved, as expressed in Theorem 8.49. There are some fundamental situations, however, where such replacement is definitely undesirable because it would lead to a serious loss of local information.

This is already apparent from the discussion of convexified normal cones in Chapter 6, since $\bar{\partial}\delta_C = \overline{N}_C = \operatorname{cl} \operatorname{con} N_C$ in contrast to $\partial\delta_C = N_C$, and there are many cases where convexification of a cone wipes out its structure. But the drawbacks are especially clear in the context of generalizing the subdifferentiation of functions $f : I\!R^n \to \overline{I\!R}$ to that of mappings $F : I\!R^n \to I\!R^m$.

Figure 8–13 shows a mapping $F : I\!R^1 \to I\!R^1$ and a point $(\bar{x}, \bar{u}) \in \operatorname{gph} F$ where $N_{\operatorname{gph} F}(\bar{x}, \bar{u})$ is a rather complicated cone comprised of two lines and one of the sectors between them. It embodies significant local information about F, but the convexified cone $\overline{N}_{\operatorname{gph} F}(\bar{x}, \bar{u})$ is all of $I\!R^2$ and says little. In this case the coderivative mapping $D^*F(\bar{x})$ is much richer than the convexified coderivative mapping $\overline{D}^*F(\bar{x})$ we would obtain through the general definition of $\overline{D}^*S(\bar{x} \,|\, \bar{u})$ for $S : I\!R^n \rightrightarrows I\!R^m$ by

$$v \in \overline{D}^*S(\bar{x}\,|\,\bar{u})(y) \iff (v,-y) \in \overline{N}_{\operatorname{gph} S}(\bar{x},\bar{u}).$$

Not only normal and tangent cones, but also recession cones to $\operatorname{epi} f$ have implications for the analysis of a function f.

8.50 Proposition (recession properties of epigraphs). *For a function $f : \mathbb{R}^n \to \overline{\mathbb{R}}$ and a point $\bar{x}$ where f is finite and locally lsc, the following properties of a pair $(w,\beta) \in \mathbb{R}^n \times \mathbb{R}$ are equivalent:*

(a) *(w,β) belongs to the local recession cone $R_{\operatorname{epi} f}\big(\bar{x}, f(\bar{x})\big)$;*

(b) *there exist $V \in \mathcal{N}(\bar{x})$ and $\delta > 0$ such that $df(x)(w) \le \beta$ when $x \in V$ and $f(x) \le f(\bar{x}) + \delta$;*

(c) *there exist $V \in \mathcal{N}(\bar{x})$ and $\delta > 0$ such that $\widehat{d}f(x)(w) \le \beta$ when $x \in V$ and $f(x) \le f(\bar{x}) + \delta$;*

(d) *there exist $V \in \mathcal{N}(\bar{x})$ and $\delta > 0$ such that $\langle v, w\rangle \le \beta$ for all $v \in \partial f(x)$ when $x \in V$ and $f(x) \le f(\bar{x}) + \delta$;*

(e) *there exist $V \in \mathcal{N}(\bar{x})$ and $\delta > 0$ such that $\langle v, w\rangle \le \beta$ for all $v \in \widehat{\partial} f(x)$ when $x \in V$ and $f(x) \le f(\bar{x}) + \delta$;*

(f) *for some $V \in \mathcal{N}(\bar{x})$, $\delta > 0$ and $\varepsilon > 0$ one has*

$$\frac{f(x+\tau w) - f(x)}{\tau} \le \beta \quad \text{when } \tau \in [0,\varepsilon],\ x \in V,\ f(x) \le f(\bar{x}) + \delta.$$

Proof. First we note the meaning of (a) on the basis of the definition of local recession vectors in 6.33: there is a neighborhood of $\big(\bar{x}, f(\bar{x})\big)$, which we may as well take in the form $V \times \big[f(\bar{x}) - \delta, f(\bar{x}) + \delta\big]$ with $V \in \mathcal{N}(\bar{x})$ and $\delta > 0$, along with $\varepsilon > 0$ such that for all points $(x,\alpha) \in \operatorname{epi} f$ lying in this neighborhood, and all $\tau \in [0,\varepsilon]$, one has $(x,\alpha) + \tau(w,\beta) \in \operatorname{epi} f$, or in other words,

$$\left.\begin{array}{r} \tau \in [0,\varepsilon],\ x \in V,\ \alpha \ge f(x) \\ f(\bar{x}) - \delta \le \alpha \le f(\bar{x}) + \delta \end{array}\right\} \implies f(x+\tau w) \le \alpha + \tau\beta.$$

This property is equivalent to (f) because f is locally lsc at $\bar{x}$: by choosing a smaller $V \in \mathcal{N}(\bar{x})$ relative to any δ, we can always ensure that whenever $x \in V$ and $f(x) \le \alpha \le f(\bar{x}) + \delta$, then also $\alpha \ge f(\bar{x}) - \delta$.

The equivalence of (a) with (b) and (c) is clear from the characterization of local recession vectors in 6.35(a)(b) and the identification of $T_{\operatorname{epi} f}\big(\bar{x}, f(\bar{x})\big)$ and $\widehat{T}_{\operatorname{epi} f}\big(\bar{x}, f(\bar{x})\big)$ with $\operatorname{epi} df(\bar{x})$ and $\operatorname{epi} \widehat{d}f(\bar{x})$ in 8.2(a) and 8.17(a).

Similarly, condition 6.35(c) tells us by way of the description of the normal cone $N_{\operatorname{epi} f}\big(\bar{x}, f(\bar{x})\big)$ in 8.9 that (a) is equivalent to having the existence of $V \in \mathcal{N}(\bar{x})$ and $\delta > 0$ such that

$$\left.\begin{array}{r} x \in V \\ f(x) \le f(\bar{x}) + \delta \end{array}\right\} \implies \left\{\begin{array}{l} \big\langle (v,-1),(w,\beta)\big\rangle \le 0 \ \text{ when } v \in \partial f(x) \\ \big\langle (v,0),(w,\beta)\big\rangle \le 0 \ \text{ when } v \in \partial^\infty f(x). \end{array}\right.$$

This condition obviously entails (d), hence in turn (e) because $\widehat{\partial} f(x) \subset \partial f(x)$, but it is at the same time a consequence of (e) through the limit definition of ∂f and $\partial^\infty f$ in terms of $\widehat{\partial} f$ in 8.3. Thus, (a) is equivalent to (d) and (e). □

8.51 Example (nonincreasing functions). *A function $f : \mathbb{R}^n \to \overline{\mathbb{R}}$ is said to be nonincreasing relative to K, a cone in $\mathbb{R}^n$, if $f(x+w) \leq f(x)$ for all $x \in \mathbb{R}^n$ and $w \in K$. For f lsc and proper, this holds if and only if*

$$\partial f(x) \subset K^* \quad \textit{for all } x \in \operatorname{dom} f \quad (K^* = \textit{polar cone}).$$

Detail. This comes from the equivalence of the recession properties in 8.50(d) and (f) in the case of $\beta = 0$. □

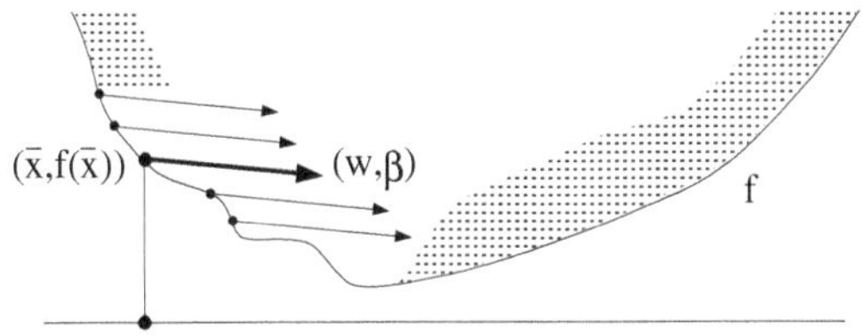

Fig. 8–15. Uniform descent indicated by an epigraphical recession vector.

Functions of a single real variable have special interest but also a number of peculiarities in the context of subgradients and subderivatives. In the case of $f : \mathbb{R}^1 \to \overline{\mathbb{R}}$ and a point $\bar{x}$ where f is finite, the elements of $\widehat{\partial} f(\bar{x})$, $\partial f(\bar{x})$ and $\partial^\infty f(\bar{x})$ are numbers rather than vectors, and they generalize slope values rather than gradients. By the convexity in 8.6, $\widehat{\partial} f(\bar{x})$ is a certain closed *interval* (perhaps empty) within the closed set $\partial f(\bar{x}) \subset \mathbb{R}$, while for $\partial^\infty f(\bar{x})$ there are only four possibilities: $[0,0]$, $[0,\infty)$, $(-\infty,0]$, or $(-\infty,\infty)$.

The subderivative functions $df(\bar{x})$ and $\widehat{d}f(\bar{x})$ are entirely determined then through positive homogeneity from their values at $\bar{w} = \pm 1$; limits $w \to \bar{w}$ don't have to be taken:

$$\begin{aligned}
\liminf_{\tau \searrow 0} \frac{f(\bar{x}+\tau) - f(\bar{x})}{\tau} &= df(\bar{x})(1), \\
\limsup_{\tau \searrow 0} \frac{f(\bar{x}+\tau) - f(\bar{x})}{\tau} &= -d(-f)(\bar{x})(1), \\
\liminf_{\tau \nearrow 0} \frac{f(\bar{x}+\tau) - f(\bar{x})}{\tau} &= -df(\bar{x})(-1), \\
\limsup_{\tau \nearrow 0} \frac{f(\bar{x}+\tau) - f(\bar{x})}{\tau} &= d(-f)(\bar{x})(-1),
\end{aligned} \qquad 8(34)$$

and through 8.18 one then has

$$\widehat{\partial} f(\bar{x})(1) = \limsup_{x \xrightarrow[f]{} \bar{x}} df(x)(1), \qquad \widehat{\partial} f(\bar{x})(-1) = \limsup_{x \xrightarrow[f]{} \bar{x}} df(x)(-1).$$

Differentiability of f at $\bar{x}$ corresponds to having all four of the values in 8(34) coincide and be finite; the common value is then $f'(\bar{x})$. More generally, the existence of the *right derivative* or the *left derivative* at $\bar{x}$, defined by

$$f'_+(\bar{x}) := \lim_{\tau \searrow 0} \frac{f(\bar{x}+\tau) - f(\bar{x})}{\tau}, \qquad f'_-(\bar{x}) := \lim_{\tau \nearrow 0} \frac{f(\bar{x}+\tau) - f(\bar{x})}{\tau},$$

where infinite values are allowed, corresponds to

$$\begin{aligned} f'_+(\bar{x}) &= df(\bar{x})(1) = -d(-f)(\bar{x})(1), \\ f'_-(\bar{x}) &= -df(\bar{x})(-1) = d(-f)(\bar{x})(-1). \end{aligned} \tag{8(35)}$$

When these one-sided derivatives exist, they fully determine $df(\bar{x})$ through

$$df(\bar{x})(w) = \begin{cases} f'_-(\bar{x})w & \text{for } w < 0, \\ f'_+(\bar{x})w & \text{for } w > 0, \\ 0 & \text{for } w = 0 \text{ if } f'_-(\bar{x}) < \infty \text{ and } f'_+(\bar{x}) > -\infty, \\ -\infty & \text{for } w = 0 \text{ if } f'_-(\bar{x}) = \infty \text{ or } f'_+(\bar{x}) = -\infty. \end{cases}$$

8.52 Example (convex functions of a single real variable). *For a proper, convex function $f : \mathbb{R}^1 \to \overline{\mathbb{R}}$ the right and left derivatives $f'_+(\bar{x})$ and $f'_-(\bar{x})$ exist at any $\bar{x} \in \operatorname{dom} f$ and satisfy $f'_-(\bar{x}) \le f'_+(\bar{x})$. When f is locally lsc at $\bar{x}$, one has*

$$\partial f(\bar{x}) = \widehat{\partial} f(\bar{x}) = \big\{ v \in \mathbb{R} \,\big|\, f'_-(\bar{x}) \le v \le f'_+(\bar{x}) \big\},$$

$$\partial^\infty f(\bar{x}) = \begin{cases} [0,0] & \text{if } f'_-(\bar{x}) > -\infty,\ f'_+(\bar{x}) < \infty, \\ [0,\infty) & \text{if } f'_-(\bar{x}) > -\infty,\ f'_+(\bar{x}) = \infty, \\ (-\infty,0] & \text{if } f'_-(\bar{x}) = -\infty,\ f'_+(\bar{x}) < \infty, \\ (-\infty,\infty) & \text{if } f'_-(\bar{x}) = -\infty,\ f'_+(\bar{x}) = \infty. \end{cases}$$

Detail. The existence of right and left derivatives is apparent from 8.21 (or even from 2.12), and the preceding facts can then be applied in combination with the regularity of f at points where it is locally lsc (cf. 7.27) and the duality between subgradients and subderivatives in such cases (cf. 8.30). □

A rich example, displaying nontrivial relationships among the subdifferentiation concepts we've been studying, is furnished by distance functions.

8.53 Example (subdifferentiation of distance functions). *For $f = d_C$ in the case of a closed set $C \neq \emptyset$ in $\mathbb{R}^n$, one has at any point $\bar{x} \in C$ that*

$$\partial f(\bar{x}) = N_C(\bar{x}) \cap \mathbb{B}, \qquad \widehat{\partial} f(\bar{x}) = \widehat{N}_C(\bar{x}) \cap \mathbb{B},$$
$$df(\bar{x})(w) = d\big(w, T_C(\bar{x})\big), \qquad \widehat{d} f(\bar{x})(w) = \max_{v \in N_C(\bar{x}) \cap \mathbb{B}} \langle v, w \rangle.$$

On the other hand, one has at any point $\bar{x} \notin C$ in terms of the projection $P_C(\bar{x}) = \big\{ \tilde{x} \in C \,\big|\, |\bar{x} - \tilde{x}| = d_C(\bar{x}) \big\}$ that

$$\partial f(\bar{x}) = \frac{\bar{x} - P_C(\bar{x})}{d_C(\bar{x})}, \qquad \widehat{\partial} f(\bar{x}) = \begin{cases} \left\{ \dfrac{\bar{x} - \tilde{x}}{d_C(\bar{x})} \right\} & \text{if } P_C(\bar{x}) = \{\tilde{x}\} \\ \emptyset & \text{otherwise,} \end{cases}$$

$$df(\bar{x})(w) = \min_{\tilde{x} \in P_C(\bar{x})} \frac{\langle \bar{x} - \tilde{x}, w \rangle}{d_C(\bar{x})}, \qquad \widehat{d}f(\bar{x})(w) = \max_{\tilde{x} \in P_C(\bar{x})} \frac{\langle \bar{x} - \tilde{x}, w \rangle}{d_C(\bar{x})}.$$

At all points $\bar{x} \in C$ and $\bar{x} \notin C$, one has both $\partial^\infty f(\bar{x}) = \{0\}$ and $\partial f(\bar{x})^\infty = \{0\}$.

Detail. The expressions for $df(\bar{x})$ will be verified first. If $\bar{x} \in C$, so $f(\bar{x}) = 0$, we have $[f(\bar{x} + \tau w) - f(\bar{x})]/\tau = d_C(\bar{x} + \tau w)/\tau = d\big(w, [C - \bar{x}]/\tau\big)$, where in addition $\big|d\big(w, [C - \bar{x}]/\tau\big) - d\big(\bar{w}, [C - \bar{x}]/\tau\big)\big| \leq |w - \bar{w}|$. From this we obtain by way of the connections between set limits and distance functions in 4.8 that

$$df(\bar{x})(\bar{w}) = \liminf_{\tau \searrow 0} d\Big(\bar{w}, [C - \bar{x}]/\tau\Big) = d\Big(w, \limsup_{\tau \searrow 0} [C - \bar{x}]/\tau\Big),$$

where the 'lim sup' is by definition $T_C(\bar{x})$. If instead $\bar{x} \notin C$, so $f(\bar{x}) > 0$, we have for any $\tilde{x}$ in $P_C(\bar{x})$ (this set being nonempty and compact; cf. 1.20) that

$$\begin{aligned} \liminf_{\substack{\tau \searrow 0 \\ w \to \bar{w}}} \frac{f(\bar{x} + \tau w) - f(\bar{x})}{\tau} &\leq \liminf_{\substack{\tau \searrow 0 \\ w \to \bar{w}}} \frac{|\bar{x} + \tau w - \tilde{x}| - |\bar{x} - \tilde{x}|}{\tau} \\ &= \langle \bar{x} - \tilde{x}, \bar{w} \rangle / |\bar{x} - \tilde{x}|. \end{aligned}$$

Here we utilize the fact that the norm function $h(z) := |z|$ is differentiable at points $z \neq 0$, namely with $\nabla h(z) = z/|z|$. This yields

$$df(\bar{x})(\bar{w}) \leq \min_{\tilde{x} \in P_C(\bar{x})} \langle \bar{x} - \tilde{x}, \bar{w} \rangle / d_C(\bar{x}).$$

To get the opposite inequality, we consider $w^\nu \to \bar{w}$ and $\tau^\nu \searrow 0$ with the property that $[f(\bar{x} + \tau^\nu w^\nu) - f(\bar{x})]/\tau^\nu \to df(\bar{x})(\bar{w})$. Selecting $x^\nu \in P_C(\bar{x} + \tau^\nu w^\nu)$, we obtain a bounded sequence with $f(\bar{x} + \tau^\nu w^\nu) = |\bar{x} + \tau^\nu w^\nu - x^\nu|$, and by passing to a subsequence if necessary we can suppose that x^ν converges to some $\tilde{x}$, which must be a point of $P_C(\bar{x})$ (cf. 1.20). In particular $x^\nu \in C$, so $|\bar{x} - x^\nu| \geq f(\bar{x}) = |\bar{x} - \tilde{x}|$ and we can estimate

$$\begin{aligned} [f(\bar{x} + \tau^\nu w^\nu) - f(\bar{x})]/\tau^\nu &\geq \big[\,|\bar{x} + \tau^\nu w^\nu - x^\nu| - |\bar{x} - x^\nu|\,\big]/\tau^\nu \\ &\geq \langle \bar{x} - x^\nu, w^\nu \rangle / |\bar{x} - x^\nu| \;\to\; \langle \bar{x} - \tilde{x}, \bar{w} \rangle / |\bar{x} - \tilde{x}|, \end{aligned}$$

where the convexity of $h(x) = |x|$ is used in writing $h(z') - h(z) \geq \langle \nabla h(z), z' - z \rangle$ (see 2.14 and 2.17). Thus, $df(\bar{x})(\bar{w}) \geq \langle \bar{x} - \tilde{x}, \bar{w} \rangle / |\bar{x} - \tilde{x}|$ and the general formula for $df(\bar{x})$ is therefore correct.

The regular subgradient set $\widehat{\partial} f(\bar{x})$ consists of the vectors v such that $\langle v, w \rangle \leq df(\bar{x})(w)$ for all w, cf. 8.4. In the case of $\bar{x} \in C$, where we have seen that $df(\bar{x})(w) = d\big(w, T_C(\bar{x})\big)$, this tells us that a vector v belongs to $\widehat{\partial} f(\bar{x})$ if and only if $\langle v, w\rangle \le |w - \tilde{w}|$ for all $\tilde{w} \in T_C(\bar{x})$ and $w \in I\!R^n$. Since for fixed v and $\bar{w}$ the inequality $\langle v, w\rangle \le |w - \bar{w}|$ holds for all w if and only if $|v| \le 1$ and $\langle v, \bar{w}\rangle \le 0$ (as seen from writing $\langle v, w\rangle = \langle v,\, w - \bar{w}\rangle + \langle v, \bar{w}\rangle$), we conclude in this case that $\widehat{\partial} f(\bar{x})$ consists of the vectors v belonging to both $I\!B$ and $\widehat{T}_C(\bar{x})^*$. The polar cone $T_C(\bar{x})^*$ is $\widehat{N}_C(\bar{x})$ by 6.28.

In the case of $\bar{x} \notin C$, the formula derived for $df(\bar{x})$ tells us instead that v belongs to $\widehat{\partial} f(\bar{x})$ if and only if

$$\langle v, w\rangle \;\le\; \langle \bar{x} - \tilde{x}, w\rangle / d_C(\bar{x}) \;\text{ for all }\; w \in I\!R^n \;\text{ and }\; \tilde{x} \in P_C(\bar{x}).$$

For fixed $\tilde{x}$, this inequality (for all w) holds if and only if $v = (\bar{x} - \tilde{x})/d_C(\bar{x})$. Therefore, if $P_C(\bar{x})$ consists of only one $\tilde{x}$, the corresponding v from this formula is the only element of $\widehat{\partial} f(\bar{x})$, whereas if $P_C(\bar{x})$ consists of more than one $\tilde{x}$, then $\widehat{\partial} f(\bar{x})$ has to be empty.

Armed with this knowledge of $\widehat{\partial} f$, we invoke the limits in 8(5) to generate ∂f and $\partial^\infty f$. Because f is continuous, we obtain

$$\partial f(\bar{x}) = \limsup_{x \to \bar{x}} \widehat{\partial} f(x), \qquad \partial^\infty f(\bar{x}) = \limsup_{x \to \bar{x}}{}^{\infty}\, \widehat{\partial} f(x).$$

We have determined that $\emptyset \ne \widehat{\partial} f(x) \subset I\!B$ for all $x \in I\!R^n$, so $\partial^\infty f(\bar{x}) = \{0\}$ and $\emptyset \ne \widehat{\partial} f(\bar{x}) \subset I\!B$ for all $\bar{x}$, hence also $\partial f(\bar{x})^\infty = \{0\}$. In particular we have

$$\widehat{\partial} f(x) \;\subset\; \big[x - P_C(x)\big]/d_C(x) \;\subset\; \bigcap\nolimits_{\tilde{x} \in P_C(x)} \widehat{N}_C(\tilde{x}) \;\text{ when }\; x \notin C,$$

because any vector of the form $v = (x - \tilde{x})/d_C(x)$ with $\tilde{x} \in P_C(x)$ is a proximal normal to C at $\tilde{x}$, hence a regular normal.

In the case of $\bar{x} \notin C$, the osc property of P_C (cf. 1.20, 7.44) implies through $\partial f(\bar{x}) = \limsup_{x \to \bar{x}} \widehat{\partial} f(x)$ that $\partial f(\bar{x}) \subset \big[\bar{x} - P_C(\bar{x})\big]/d_C(\bar{x})$. Equality actually has to hold, because for any $\tilde{x} \in P_C(\bar{x})$ and any $\tau \in (0,1)$ the point $x_\tau = (1 - \tau)\bar{x} + \tau\tilde{x}$ has $P_C(x_\tau) = \{\tilde{x}\}$ (see 6.16) so that $\widehat{\partial} f(x_\tau)$ consists uniquely of $(x_\tau - \tilde{x})/d_C(x_\tau)$, which approaches $(\bar{x} - \tilde{x})/d_C(\bar{x})$ as $\tau \searrow 0$. In the alternative case of $\bar{x} \in C$, we have for $x \to \bar{x}$ in the limit defining $\partial f(\bar{x})$ that $\widehat{\partial} f(x) = \widehat{N}_C(x) \cap I\!B$ when $x \in C$, but also $\widehat{\partial} f(x) \subset \widehat{N}_C(\tilde{x}) \cap I\!B$ for all $\tilde{x} \in P_C(x)$ when $x \notin C$. Because $P_C(x) \to \{\bar{x}\}$ as $x \to \bar{x} \in C$, we get

$$\partial f(\bar{x}) \;=\; \limsup_{x \,\underset{C}{\to}\, \bar{x}} \big[\widehat{N}_C(x) \cap I\!B\big] \;=\; N_C(\bar{x}) \cap I\!B$$

according to the definition of $N_C(\bar{x})$ in 6.3.

Only $\widehat{d}f(\bar{x})$ remains. We have $\widehat{d}f(\bar{x})(w) = \sup\big\{\langle v, w\rangle \,\big|\, v \in \partial f(\bar{x})\big\}$ from 8.23, since $\partial^\infty f(\bar{x}) = \{0\}$, and through this the expressions derived for $\partial f(\bar{x})$ when $\bar{x} \in C$ or $\bar{x} \notin C$ yield the ones claimed for $\widehat{d}f(\bar{x})(w)$. □

An observation about generic continuity concludes this chapter.

8.54 Exercise (generic continuity of subgradient mappings).

(a) *For any proper function* $f : \mathbb{R}^n \to \overline{\mathbb{R}}$, *the set of points where the mapping* $\partial f : \mathbb{R}^n \rightrightarrows \mathbb{R}^n$ *fails to be continuous relative to its domain in the topology of* f*-attentive convergence is meager in this topology. So too is the set of points where the mapping* $x \mapsto \partial f(x) \cup \operatorname{dir} \partial^\infty f(x)$ *into* $\operatorname{csm} \mathbb{R}^n$ *fails to be continuous.*

(b) *For any mapping* $S : \mathbb{R}^n \rightrightarrows \mathbb{R}^m$, *the set of points of* $\operatorname{gph} S$ *where the mapping-valued mapping* $(x, u) \mapsto D^*S(x \mid u)$ *fails to be continuous relative to its domain in the topology of graphical convergence is meager in this domain.*

Guide. Derive these properties from Theorem 5.55, recalling that the f-attentive topology on a subset of $\operatorname{dom} f$ can be identified with the ordinary topology on the corresponding subset of $\operatorname{gph} f$. □

Commentary

Subdifferential theory didn't grow out of variational geometry. Rather, the opposite was essentially what happened. Although tangent and normal cones can be traced back for many decades and attracted attention on their own with the advent of optimization in its modern form, the bulk of the effort that has been devoted to their study since the early 1960's came from the recognition that they provided the geometric testing ground for investigations of generalized differentiability.

Subgradients of convex functions marked the real beginning of the subject in the way it's now seen. This notion, in the sense of associating a *set* of vectors v with f at each point x through the affine support inequality in 8.12 and Figure 8–6, or

$$f(x + w) \geq f(x) + \langle v, w \rangle \text{ for all } w, \qquad 8(36)$$

and thus defining a *set-valued* mapping for which *calculus* rules could nonetheless be provided, first appeared in the dissertation of Rockafellar [1963]. The notation $\partial f(x)$ was introduced there as well. Affine supports had, of course, been considered before by others, but without any hint of building a set-valued calculus out of them.

Around the same time, Moreau [1963a] used the term 'subgradient' (French: sous-gradient), not for vectors v but for a *single-valued mapping* that selects for each x some element of $\partial f(x)$. After learning of Rockafellar's work, Moreau [1963b] proposed using 'subgradient' instead in the sense that thereafter became standard, and convex analysis started down its path of rapid development by those authors and soon others. The importance of multivaluedness in this context was also recognized early by Minty [1964], but he shied away from speaking of multivalued mappings and preferred instead the language of 'relations' as subsets of a product space. His work on maximal monotone relations in $\mathbb{R} \times \mathbb{R}$, cf. Minty [1960], was influential in Rockafellar's thinking. Such relations are now recognized as the graphs of the generally multivalued mappings ∂f associated with proper, lsc, convex functions f on $\mathbb{R}$ (as in 8.52).

It was well understood in those days that subgradients v correspond to epigraphical normal vectors $(v, -1)$, and that the subgradient set $\partial f(x)$ could be characterized with respect to the one-sided directional derivatives of f at x, which in turn are associated with tangent vectors to the convex set epi f. The support function duality between convex sets and sublinear functions (in 8.24), which afforded the means of expressing this relationship, goes back to Minkowski [1910], at least for bounded sets. The envelope representations in 8.13, on which this correspondence rests, have a long history as well but in their modern form (not just for convex functions that are finite on $I\!R^n$) they can be credited to Fenchel [1951]. See Hörmander [1954] for an early discussion of the corresponding facts beyond $I\!R^n$. The dualizations in Proposition 8.29 can be found in Rockafellar [1970a].

Subgradients in the convex case found widespread application in optimization and problems of control of ordinary and partial differential equations, as seen for instance in books of Rockafellar [1970a], Ekeland and Temam [1974], and Ioffe and Tikhomirov [1974]. Optimality conditions like the one in the convex case of Theorem 8.15 are just one example (to be augmented by others in Chapter 10). Applications not fitting within the domain of convexity had to wait for other ideas, however.

A major step forward came with the dissertation of Clarke [1973], who found a way of extending the theory to the astonishingly broad context of all lsc, proper functions f on $I\!R^n$. Clarke's approach was three-tiered. First he developed subgradients for Lipschitz continuous functions out of their almost everywhere differentiability, obtaining a duality with certain directional derivative expressions. Next he invoked that for distance functions d_C in order to get a new concept of normal cones to nonconvex sets C. Finally he applied that concept to epigraphs so as to obtain normals $(v, -1)$ whose v component could be regarded as a subgradient. This innovation, once it appeared in Clarke [1975], sparked years of efforts by many researchers in elaborating the ideas and applying them to a range of topics, most significantly optimal control; cf. Clarke [1983], [1989] for references. The notion of subdifferential regularity in 8.30, likewise from Clarke [1973], [1975], became a centerpiece.

Lipschitzian properties, which will be discussed in Chapter 9, so dominated the subject at this stage that relatively little was made of the remarkable extension that had been achieved to general functions $f : I\!R^n \to \overline{I\!R}$. For instance, although Clarke featured a duality between compact, convex sets posed as subgradient sets of Lipschitz continuous functions and finite sublinear functions expressed by certain limits of difference quotients, there was nothing analogous for the case of general f. Rockafellar [1979a], [1980], filled the gap by introducing a *subderivative* function which in the terminology adopted here is the one giving *regular* subderivatives (the notation used was $f^{\uparrow}(x; w)$, not $\hat{d}f(x)(w)$). He demonstrated that it gave Clarke's directional derivatives in the Lipschitzian case (cf. 9.16), yet furnished the full duality between subgradients and generalized directional derivatives that had so far been lacking.

Of course that duality differs from the more elaborate subderivative-subgradient pairing seen here in 8.4 and 8.23, apart from the case of subdifferential regularity in 8.30. The passage to those relationships, signaling a significant shift in viewpoint, was propelled by the growing realization that automatic convexification of subgradient sets (which came from Clarke's convexification of normal cones; cf. 6(19) and 8(32)) could often be a hindrance. This advance owes credit especially to Mordukhovich and Ioffe; see the Commentary of Chapter 6 and further comments below.

The importance and inevitability of making a comprehensive adjustment of the theory to the new paradigm has been our constant guide in putting this chapter together. It has molded not only our exposition but also our choices of terminology and notation, in particular in attaining full coordination with the format of variational geometry. The motivations explained in the Commentary to Chapter 6 carry over to this endeavor as well.

Much has already been said in the Commentary to Chapter 7 about one-sided directional derivatives, but here one-sidedness with respect to liminf and limsup too has been brought into the picture. The lower-limit expressions we've denoted by $df(x)(w)$ were considered by Penot [1978], who called them 'lower semiderivatives'. Aubin [1981] dubbed them 'contingent derivatives' (later 'contingent epi-derivatives' in Aubin and Frankowska [1990]), while Ioffe [1981b], [1984a], offered the name 'Dini derivatives'. One-sided derivatives defined with liminf and limsup were indeed studied long ago by Dini [1878], but only for functions on $I\!R^1$, where the complications of directional convergence don't surface. We prefer 'subderivative' for this concept because of its attractive parallel with 'subgradient' and the side benefit of harmonizing with the earlier terminology of Rockafellar [1980], as long as the subderivatives in that work are now designated 'regular', which is entirely appropriate.

Once nonconvex functions $f : I\!R^n \to \overline{I\!R}$ (not necessarily Lipschitz continuous) were addressed, the original definition of subgradients (through the affine support inequality enjoyed by convex functions) had to be replaced by something else. Clarke initially chose to appeal to normal vectors $(v, -1)$ to epigraphs. Rockafellar [1980] established that Clarke's subgradients at x could be described equally well as the vectors v satisfying

$$\widehat{d}f(x)(w) \geq \langle v, w \rangle \text{ for all } w \qquad 8(37)$$

(in present notation); this fact shows up here in 8.49. Although analogous to the subgradient inequality in the convex case, and able to serve directly as the definition of Clarke's subgradients in the general setting, this characterization was burdened by the complexity of the limits that are involved in expressing $\widehat{d}f(x)(w)$ unless f falls into some special category, or the direction vector w has an interior property as in Theorem 8.22, a result of Rockafellar [1979a]. For purposes such as the development of a strong calculus, alternative characterizations of subgradients were desirable.

That led to the introduction of 'proximal' subgradients in Rockafellar [1981b]. Such vectors v satisfy an inequality like 8(36), except that affine supports are replaced by quadratic supports; cf. 8(30). At the same time they correspond to $(v, -1)$ being a proximal normal to epi f, cf. 8.46, making them able to draw on the geometric constructions of Clarke. As in that geometry, limits then had to be taken. The definition reached was just like 8.3, but instead of 'regular subgradients' yielding 'general subgradients', proximal subgradients produced 'limiting proximal subgradients'. (Now, we know that these two kinds of limit vectors are actually the same; see 8.46 and other comments below.)

This wasn't the end of the process, because the goal at the time was to recover Clarke's convexified subgradient set, denoted here by $\bar{\partial}f(x)$. A serious complication from this angle was the fact that, in cases where f isn't Lipschitz continuous, $\bar{\partial}f(x)$ could well be more than just the convex hull of the set of 'limiting proximal subgradients'. Along with such subgradients, certain direction points had to be taken into account in the convexification.

The ideas were already available from convex analysis (see the Commentary to Chapter 3), but they hadn't previously been needed in subdifferential theory. What was required for their use was the introduction of a supplementary class of subgradients to serve in representing the crucial direction points. These 'singular limiting proximal subgradients' were generated just like the horizon subgradients in

8.3, but from sequences of proximal subgradients instead of regular subgradients. Not only did they yield in Rockafellar [1981b] the right convex hull formula for $\bar{\partial} f(x)$ (see 8.49), but also they soon supported new types of formulas by which subgradient sets could be expressed or estimated; cf. Rockafellar [1982a], [1985a] (the latter being where the notation $\partial^{\infty} f(x)$ first appeared, although for what we now write as $\bar{\partial}^{\infty} f(x)$). Even earlier in Rockafellar [1979a] there were subgradients representing direction points within the convexified framework of Clarke, and having a prominent role in the calculus of other subgradients, but they could be handled as elements of $\bar{\partial} f(x)^{\infty}$ or kept out of sight through the polarity between that convex cone and the closure of $\operatorname{dom} \widehat{d} f(x)$.

Meanwhile, other strategies were being tested. The idea of defining subgradients v in terms of 'limiting proximal normal vectors' to $\operatorname{epi} f$, and proceeding in that mode *without any move toward convexification*, began with Mordukhovich [1976]. He realized after a while that limiting proximal normal vectors could equally be generated from sequences of regular normal vectors; for a set C, they led to the same unconvexified normal cone $N_C(x)$ at x. His limit subgradients were thus the vectors v such that $(v, -1) \in N_{\operatorname{epi} f}(x, f(x))$. He also understood that regular subgradients v to f at x could be characterized by

$$f(x + w) \geq f(x) + \langle v, w \rangle + o\big(|w|\big), \tag{8(38)}$$

cf. Kruger and Mordukhovich [1980], as well as by

$$df(x)(w) \geq \langle v, w \rangle \text{ for all } w. \tag{8(39)}$$

Still more, he found that regular subgradients could be replaced in the limit process by the *ε-regular subgradients* $v \in \widehat{\partial}_{\varepsilon} f(x)$ for $\varepsilon > 0$, described by

$$df(x)(w) \geq \langle v, w \rangle - \varepsilon |w| \text{ for all } w, \tag{8(40)}$$

as long as ε was made to go to 0 as the target point was approached (see 10.46).

These elaborations, on the basis of which Mordukhovich's subgradients can be verified as being the same as the ones in $\partial f(x)$ in the present notation, were written up in 1980 in two working papers at the Belorussian State University in Minsk. The results were described in some published notes, e.g. Kruger and Mordukhovich [1980], Mordukhovich [1984], and they received full treatment in Mordukhovich [1988].

The first to contemplate an inequality like 8(39) as a potentially useful way of defining subgradients v of nonconvex functions f was Pshenichnyi [1971]. He didn't work with $df(\bar{x})$ but with the simpler directional derivative function $f'(\bar{x}; \cdot)$ of 7(20) (with limits taken only along half-lines from $\bar{x}$), and he asked that $f'(\bar{x}; w)$ be a finite, convex function of $w \in \mathbb{R}^n$, then calling f 'quasi-differentiable' at $\bar{x}$. Of course, in cases where f is semidifferentiable at $\bar{x}$, $f'(\bar{x}; \cdot)$ is a finite function agreeing with $df(\bar{x})$ (cf. 7.21), and it's sure to be convex for instance when f is also subdifferentially regular at $\bar{x}$. The inequality in 8(39) was used directly as a subgradient definition by Penot [1974], [1978], without any restrictions on f. Neither he nor Pshenichnyi followed up by taking limits of such subgradients with respect to sequences $x^{\nu} \underset{f}{\to} \bar{x}$ to construct other, more 'stable' sets of subgradients.

Vectors v satisfying instead the modified affine support inequality 8(38), which so aptly extends to nonconvex functions the subgradient inequality 8(36) associated with convex functions, were put to use as subgradients by Crandall and Lions [1981],

[1983], in work on 'viscosity solutions' to partial differential equations of Hamilton-Jacobi type; see also Crandall, Evans and Lions [1984] and other citations in these works. This approach was followed also in the Italian school of variational inequalities, e.g. as reported by Marino and Tosques [1990]. But these researchers too stopped short of the crucial step of taking limits of such vectors v to get other subgradients, without which only a relatively feeble calculus of subgradients can be achieved.

The fact in 8.5, that the modified affine support inequality 8(38) actually corresponds to v being the gradient of a 'smooth support' to f at x, is new here as a contribution to the basics of variational analysis, but a precedent in the theory of norms can be found in the book of Deville, Godefroy and Zizler [1993]. It's out of this fact and the immediate appeal of 8(38) in extending the convexity inequality 8(36) that we've adopted 8(38) [in the form 8(3)] as the hallmark of regular subgradients in Definition 8.3, instead of using the equivalent property in 8(39) or the geometric condition that $(v, -1)$ be a regular normal vector to $\operatorname{epi} f$ at $(x, f(x))$.

Substantial advances in the directions promoted by Mordukhovich were made by Ioffe [1981c], [1984a], [1984b], [1984c], [1986], [1989]. He used the term 'approximate subgradients' for the vectors $v \in \partial f(\bar{x})$, but this may have been a misnomer arising from language differences. Ioffe thought of this term as designating vectors obtained through a process of taking limits—like what had been called 'limiting subgradients' in English. While adjectives with the same root as 'approximate' in Russian and French might be construed in such a sense, the meaning of that word in English is different and has a numerical connotation, giving the impression that an 'approximate' subgradient is an 'estimate' calculated for some kind of 'true' subgradient, whose definition however had not in this case actually been formulated.

Ioffe's efforts went especially into the question of how to extend the theory of subgradients to spaces beyond $I\!R^n$, and this continues to be a topic of keen interest to many researchers. It's now clear that a major distinction has to be made between separable and nonseparable Banach spaces, and that separable Asplund spaces are particularly favorable territory, but even there the paradigm of relationships in Figure 8–9 (and on the geometric level in Figure 6–17) can't be maintained intact and requires some sacrifices. The current state of the infinite-dimensional theory can be gleaned from recent articles of Borwein and Ioffe [1996] and Mordukhovich and Shao [1996a]; for an exposition in the setting of Hilbert spaces see Loewen [1993]. Much of the difficulty in infinite-dimensional spaces has centered on the quest for an appropriate 'compactness' concept. Especially to be mentioned besides the items already cited are works of Borwein and Strojwas [1985], Loewen [1992], Jourani and Thibault [1995], Ginsburg and Ioffe [1996], Mordukhovich and Shao [1997], and Ioffe [1997].

In the infinite-dimensional setting the vectors v satisfying 8(38) can definitely be more special than the ones satisfying 8(39); the former are often called 'Fréchet subgradients', while the latter are called 'Hadamard subgradients'. (This terminology comes from parallels with differentiation; Fréchet and Hadamard had nothing to do with subgradients.) In spaces where neither of these kinds of vectors is adequately abundant, the ε-regular subgradients described by 8(40) can be substituted in limit processes. Besides their utilization in the work of Mordukhovich and Ioffe, ε-regular subgradients were developed independently in this context by Ekeland and Lebourg [1976], Treiman [1983b], [1986]. For still other ideas of 'subgradients' and their relatives, see Warga [1976], Michel and Penot [1984], [1992], and Treiman [1988].

For set-valued mappings $S : I\!R^n \rightrightarrows I\!R^m$, graphical derivatives $DS(x \,|\, u)$ with the cones $T_{\operatorname{gph} S}(x, u)$ as their graphs (cf. 8.33) were introduced by Aubin [1981], who made them the cornerstone for many developments; see Aubin and Ekeland [1984], Aubin and Frankowska [1990], and Aubin [1991]. This work, going back to reports distributed in 1979, included the mappings $\widehat{D}S(x \mid u)$ associated with the cones $\widehat{T}_{\mathrm{gph}\, S}(x \mid u)$, here called regular tangent cones, and also certain 'codifferential' mappings whose graphs corresponded to normal cones to gph S. Aubin mainly considered the latter for S graph-convex, in which case they are the coderivatives $D^*S(x \mid u)$; he noted that they can be obtained also by applying the upper adjoint operation to the mappings $DS(x \mid u)$, which are graph-convex like S, hence sublinear. The theory of adjoints of sublinear mappings H was invented by Rockafellar [1967c], [1970a]. Graphical derivatives and graphical regularity of set-valued mappings were explored in depth by Thibault [1983].

Independently, Mordukhovich [1980] introduced the unconvexified mappings $D^*S(x \mid u)$ for the purpose of deriving a maximum principle in the optimal control of differential inclusions. The systematic study of such mappings was begun by Ioffe [1984b], who was the first to employ the term 'coderivative' for them. The topic was pursued in Mordukhovich [1984], [1988], and more recently in Mordukhovich [1994d] and Mordukhovich and Shao [1997a].

Semidifferentiability of set-valued mappings was defined by Penot [1984] and proto-differentiability by Rockafellar [1989b], who also clarified the connections between the two concepts. An earlier contribution of Mignot [1976], predating even Aubin's introduction of graphical differentiation, was the definition of 'conical derivatives' of set-valued mappings, i.e., derivative mappings with conical graph, which were something like semiderivatives but with limits taken only along half-lines. Mignot didn't relate such derivatives to graphical geometry but employed them in the study of projections.

The results in 8.44 about the subdifferentiation of derivatives are new, and so too are the approximation results in 8.47(b) and 8.48; for facts connected with 8.47(b), see also Benoist [1992a], [1992b], [1994]. Convex functions, enjoy a more powerful property of subgradient convergence discovered by Attouch [1977]; see Theorem 12.35. For finite functions satisfying an assumption of 'uniform subdifferentiability' a stronger fact about convergence of subgradients than in 8.47(b), resembling the theorem of Attouch just mentioned, was proved by Zolezzi [1985]. Related results have been furnished by Poliquin [1992], Levy, Poliquin and Thibault [1995], and Penot [1995].

Recession properties of epigraphs were considered in Rockafellar [1979a], but not in the scope in which they appear in 8.50. An alternative approach to the characterization of nondecreasing functions in 8.51 has been developed by Clarke, Stern and Wolenski [1993]. The generic continuity fact in 8.54 hasn't previously been noted.

The property of calmness in 8.32 was developed by Clarke [1976a] in the context of optimal value functions, where it can be used to express a kind of constraint qualification; see also Clarke [1983].

The formulas in 8.53 for the subderivatives and subgradients of distance functions show that distance functions provide a vehicle capable of carrying all the basic notions of variational geometry. Indeed, distance functions were given such a role in Clarke's theory, as explained earlier, and they have remained prominent to this day in the research of others, notably Ioffe in his work at introducing appropriately robust normal cones and subgradients in infinite dimensional spaces.

9. Lipschitzian Properties

The notion of Lipschitz continuity is useful in many areas of analysis, but in variational analysis it takes on a fundamental role. To begin with, it singles out a class of functions which, although not necessarily differentiable, have a property akin to differentiability in furnishing estimates of the magnitudes, if not the directions, of change. For such functions, real-valued and vector-valued, subdifferentiation operates on an especially simple and powerful level. As a matter of fact, subdifferential theory even characterizes the presence of Lipschitz continuity and provides a calculus of the associated constants. It thereby supports a host of applications in which such constants serve to quantify the stability of a problem's solutions or the rate of convergence in a numerical method for determining a solution.

But the study of Lipschitzian properties doesn't stop there. It can be extended from single-valued mappings to general set-valued mappings as a means of obtaining quantitative results about continuity that go beyond the topological results obtained so far. In that context, Lipschitz continuity can be captured by coderivative conditions, which likewise pin down the associated constants. What's more, those conditions can be applied to basic objects of variational analysis such as profile mappings associated with functions, and this leads to important insights. For instance, the very concepts of normal vector and subgradient turn out to represent 'manifestations of singularity' in the Lipschitzian behavior of certain set-valued mappings.

A. Single-Valued Mappings

Single-valued mappings will absorb our attention at first, but groundwork will be laid at the same time for generalizations to set-valued mappings. In line with this program, we distinguish systematically from the start between 'Lipschitz continuity' as a property invoked with respect to an entire *set* in the domain space and 'strict continuity' as an allied concept that is tied instead to limit behavior at individual *points*. This distinction will be valuable later because, in handling set-valued mappings that aren't just locally bounded, we'll have to work not only with localizations in domain but also in range.

9.1 Definition (Lipschitz continuity and strict continuity). *Let F be a single-valued mapping defined on a set $D \subset \mathbb{R}^n$, with values in $\mathbb{R}^m$. Let $X \subset D$.*

(a) *F is Lipschitz continuous on X if there exists $\kappa \in I\!R_+ = [0,\infty)$ with*

$$|F(x') - F(x)| \;\leq\; \kappa|x' - x| \;\text{ for all }\; x, x' \in X.$$

Then κ is called a Lipschitz constant for F on X.

(b) *F is strictly continuous at $\bar{x}$ relative to X if $\bar{x} \in X$ and the value*

$$\operatorname{lip}_X F(\bar{x}) \;:=\; \limsup_{\substack{x,x' \xrightarrow[X]{} \bar{x} \\ x \neq x'}} \frac{|F(x') - F(x)|}{|x' - x|}$$

is finite. More simply, F is strictly continuous at $\bar{x}$ (without mention of X) if $\bar{x} \in \operatorname{int} D$ and the value

$$\operatorname{lip} F(\bar{x}) \;:=\; \limsup_{\substack{x,x' \to \bar{x} \\ x \neq x'}} \frac{|F(x') - F(x)|}{|x' - x|}$$

is finite. Here $\operatorname{lip} F(\bar{x})$ is the Lipschitz modulus of F at $\bar{x}$, whereas $\operatorname{lip}_X F(\bar{x})$ is this modulus relative to X.

(c) *F is strictly continuous relative to X if, for every point $\bar{x} \in X$, F is strictly continuous at $\bar{x}$ relative to X.*

Clearly, to say that F is strictly continuous at $\bar{x}$ relative to X is to assert the existence of a neighborhood $V \in \mathcal{N}(\bar{x})$ such that F is Lipschitz continuous on $X \cap V$. Therefore, strict continuity of F relative to X is identical to *local Lipschitz continuity* of F on X. When we get to set-valued mappings, however, the meanings of these terms will diverge, with local Lipschitz continuity being a comparatively narrow concept not implied by strict continuity, and with strict continuity at $\bar{x}$ no longer necessarily entailing even continuity beyond $\bar{x}$ itself. To aid in that transition, we prefer to speak already now of strict continuity rather than local Lipschitz continuity. That terminology also fits better with the 'pointwise' emphasis we'll be giving to the subject.

Whenever κ is a Lipschitz constant for F on X, the same holds for every $\kappa' \in (\kappa, \infty)$. The modulus $\operatorname{lip} F(\bar{x})$ captures a type of minimality, however; it's the lower limit of the Lipschitz constants for F that work on the balls $I\!B(\bar{x}, \delta)$ as $\delta \searrow 0$, and similarly for $\operatorname{lip}_X F(\bar{x})$. In developing properties of these expressions, we'll concentrate for simplicity on the function $\operatorname{lip} F$, since anyway $\operatorname{lip}_X F \leq \operatorname{lip} F$ where the latter is defined. Analogs for $\operatorname{lip}_X F$ are usually easy to obtain in parallel.

9.2 Theorem (Lipschitz continuity from bounded modulus). *For a single-valued mapping $F : D \to I\!R^m$, $D \subset I\!R^n$, the modulus function $\operatorname{lip} F : \operatorname{int} D \to \overline{I\!R}$ is usc, so that the set $O = \{x \in \operatorname{int} D \mid \operatorname{lip} F(x) < \infty\}$ is open.*

On any convex set $X \subset O$ where $\operatorname{lip} F$ is bounded from above, hence on any compact, convex set $X \subset O$, F is Lipschitz continuous with constant

$$\kappa = \sup_{x \in X} \{ \operatorname{lip} F(x) \}.$$

Proof. The upper semicontinuity of $\operatorname{lip} F$ on $\operatorname{int} D$ follows from the defining formula and implies that every set of the form $\{x \in \operatorname{int} D \mid \operatorname{lip} F(x) < \alpha\}$ is open. In particular, O is open. Then further, $\operatorname{lip} F$ is bounded above on every compact set $X \subset O$ (as may be seen for instance by applying Theorem 1.9 to the function $f : \mathbb{R}^n \to \overline{\mathbb{R}}$ that agrees with $-\operatorname{lip} F$ on X but has the value ∞ outside of X).

Suppose now that $\operatorname{lip} F$ is bounded above by κ on $X \subset \operatorname{int} D$, and consider any two points x and x' of X. Under the assumption that X is convex, the line segment $[x, x']$ lies in X. Let $\varepsilon > 0$. Around each point $y \in [x, x']$ there is a ball of some radius $\delta_y > 0$ on which F is Lipschitz continuous with constant $\kappa + \varepsilon$. The open balls $\operatorname{int} \mathbb{B}(y, \delta_y/2)$ cover $[x, x']$, and hence by the compactness principle a finite collection of them cover it. Thus, there exist points $y_k \in [x, x']$ and values $\delta_k > 0$ for $k = 1, \ldots, r$ such that $[x, x'] \subset \bigcup_{k=1}^r \mathbb{B}(y_k, \delta_k/2)$, while F is Lipschitz continuous on $\mathbb{B}(y_k, \delta_k)$ with constant $\kappa + \varepsilon$. Choose $\delta_0 > 0$ satisfying $\delta_0 < \delta_k/2$ for all k. Then for every $y \in [x, x']$ the ball $\mathbb{B}(y, \delta_0)$ will be included in one of the balls $\mathbb{B}(y_k, \delta_k)$, namely for an index k such that $y \in B(y_k, \delta_k/2)$. If we partition $[0, 1]$ into $0 = \tau_0 < \tau_1 < \cdots < \tau_m = 1$ in such a way that the corresponding points $x_i = x + \tau_i(x' - x)$ for $i = 0, 1, \ldots, m$ satisfy $|x_i - x_{i-1}| \le \delta_0$, we will have on each segment $\{x + \tau(x' - x) \mid \tau_{i-1} \le \tau \le \tau_i\}$ that F is Lipschitz continuous with constant $\kappa + \varepsilon$. In this case $|F(x_i) - F(x_{i-1})| \le (\kappa + \varepsilon)|x_i - x_{i-1}|$ and $\sum_{i=1}^m |x_i - x_{i-1}| = |x' - x|$. Therefore

$$\begin{aligned} \big|F(x') - F(x)\big| &\le \sum\nolimits_{i=1}^m \big|F(x_i) - F(x_{i-1})\big| \\ &\le \sum\nolimits_{i=1}^m (\kappa + \varepsilon)|x_i - x_{i-1}| \;=\; (\kappa + \varepsilon)|x' - x|, \end{aligned}$$

so F is Lipschitz continuous relative to X with constant $\kappa + \varepsilon$. Since $\varepsilon > 0$ was arbitrary, we conclude that F is Lipschitz continuous relative to X with constant κ. □

Lipschitz continuity is related to the concept of calmness introduced in Chapter 8 for functions $f : \mathbb{R}^n \to \overline{\mathbb{R}}$. A mapping $F : D \to \mathbb{R}^m$ is called *calm* at $\bar{x}$ relative to X if, in analogy with 8(13) in the real-valued case, there exist $\kappa \in \mathbb{R}_+$ and $V \in \mathcal{N}(\bar{x})$ such that

$$\big|F(x) - F(\bar{x})\big| \le \kappa|x - \bar{x}| \text{ for all } x \in X \cap V. \qquad 9(1)$$

This is like F being strictly continuous at $\bar{x}$, but it involves comparisons only between $\bar{x}$ and nearby points x, not between all possible pairs of points x and x' in some neighborhood of $\bar{x}$. In requiring the same constant κ to work for all such pairs, strict continuity is 'locally uniform calmness'. Similarly, Lipschitz continuity on a set X is 'uniform calmness' relative to X.

The possible absence of such uniformity is demonstrated in Figure 9–1, which depicts a mapping F that is calm at every point of $\mathbb{R}$, including the origin, but with constants that necessarily tend toward ∞ as the origin is approached (because the segments in the graph get steeper and steeper in the

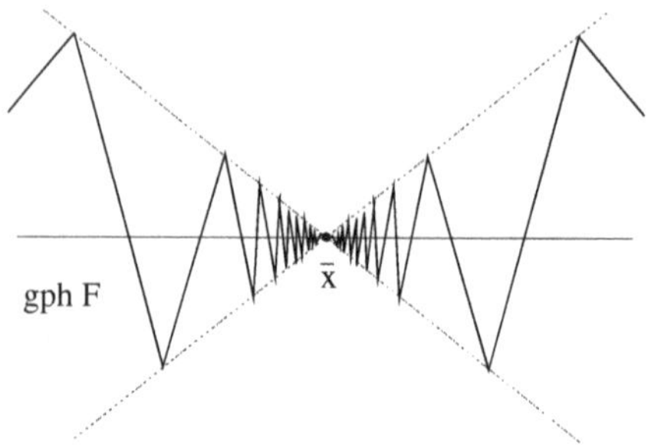

Fig. 9–1. A function that is calm everywhere but not strictly continuous everywhere.

vicinity of the origin). This mapping isn't strictly continuous at the origin.

In the case of linear mappings, Lipschitzian properties have a global character and can be expressed by a matrix norm.

9.3 Example (linear and affine mappings). *A mapping $F : \mathbb{R}^n \to \mathbb{R}^m$ is called affine if it differs from a linear mapping by at most a constant term, or in other words, if $F(x) = Ax + a$ for a matrix $A \in \mathbb{R}^{m\times n}$ and a vector $a \in \mathbb{R}^m$. Such a mapping F is Lipschitz continuous on $\mathbb{R}^n$ with constant $\kappa = |A|$, where $|A|$ is the norm of A, defined by*

$$|A| := \max_{w\neq 0} \frac{|Aw|}{|w|} = \max_{|w|=1} |Aw| = \max_{|y|=1,\,|w|=1} \langle y, Aw\rangle$$

and satisfying in terms of the components a_{ij} of A the inequalities

$$\max{}_{i,j=1}^{m,n} |a_{i,j}| \leq |A| \leq \big(\textstyle\sum_{i,j=1}^{m,n} a_{ij}^2\big)^{1/2} \leq \sqrt{mn}\max{}_{i,j=1}^{m,n} |a_{i,j}|.$$

Detail. Writing $x' = x + w$, we get $|F(x') - F(x)|/|x' - x| = |Aw|/|w|$. The maximum of the difference ratio is therefore $|A|$, and this serves then as a Lipschitz constant for F relative to $X = \mathbb{R}^n$. The alternate formula for $|A|$ in terms of maximization over both y and w is valid because $\langle y, w\rangle \leq |y||w|$ for all vectors y and w, with equality when $y = w$. The lower bound indicated for $|A|$ comes from the fact that special choices of y and w with $|y| = 1$ and $|w| = 1$ make $\langle y, Aw\rangle$ equal to either a_{ij} or $-a_{ij}$. The first upper bound is derived by writing $\langle y, Aw\rangle = \sum_{i,j=1}^{m,n} a_{ij}b_{ij}$ for $b_{ij} := y_i w_j$ and observing that this implies $\langle y, Aw\rangle \leq (\sum_{i,j=1}^{m,n} a_{ij}^2)^{1/2}(\sum_{i,j=1}^{m,n} b_{ij}^2)^{1/2}$ where $\sum_{i,j=1}^{m,n} b_{ij}^2 = (\sum_{i=1}^m y_i^2)(\sum_{j=1}^n w_i^2)$. The latter expression equals 1 when $|y| = 1$ and $|w| = 1$, and the upper bound then follows. The second upper bound is obtained by replacing each term a_{ij}^2 in the first by the upper estimate α^2, where $\alpha = \max{}_{i,j=1}^{m,n} |a_{ij}|$. □

A sequence of matrices $A^\nu \in \mathbb{R}^{m\times n}$ is said to *converge* to a matrix A if the components converge: $a_{ij}^\nu \to a_{ij}$ for all i and j. The bounds on $|A|$ in 9.3 show that this is true if and only if $\lim_\nu |A^\nu - A| = 0$. The standard matrix norm $|A|$ obeys the usual laws of norms: $|A| > 0$ for $A \neq 0$, $|A+B| \leq |A|+|B|$, and $|\lambda A| = |\lambda||A|$, and in addition it satisfies $|AB| \leq |A||B|$. The expression $(\sum_{i,j=1}^{m,n} a_{ij}^2)^{1/2}$, called the *Frobenius norm* of A, likewise has these properties. For affine mappings $L^\nu(x) = A^\nu x + a^\nu$ and $L(x) = Ax + a$, one has $L^\nu \xrightarrow{p} L$ (or equivalently $L^\nu \xrightarrow{g} L$) if and only if $|A^\nu - A| \to 0$ and $|a^\nu - a| \to 0$.

9.4 Example (nonexpansive and contractive mappings). *A mapping $F : D \to \mathbb{R}^n$, $D \subset \mathbb{R}^n$, is said to be nonexpansive on a set $X \subset D$ if*

$$\big|F(x') - F(x)\big| \leq |x' - x| \text{ for all } x, x' \in X,$$

and contractive on X if this inequality is strict as long as $x' \neq x$. The nonexpansive property is Lipschitz continuity with constant $\kappa = 1$, while a sufficient condition for the contractive property is Lipschitz continuity with $\kappa < 1$.

The property of being contractive is important in the study of iterative procedures for computing fixed points (cf. 5.3). The rate of convergence to a fixed point $\bar{x} = F(\bar{x})$ is governed by the value of $\operatorname{lip} F$ at $\bar{x}$.

9.5 Exercise (rate of convergence to a fixed point). *Let $F : D \to \mathbb{R}^n$, with $D \subset \mathbb{R}^n$, have a fixed point $\bar{x} \in \operatorname{int} D$ at which $\operatorname{lip} F(\bar{x}) < 1$.*

For all $\delta > 0$ sufficiently small, F is contractive on $\mathbb{B}(\bar{x}, \delta)$, moreover with $F\big(\mathbb{B}(\bar{x}, \delta)\big) \subset \mathbb{B}(\bar{x}, \delta)$. For any such δ and any choice of $x^0 \in \mathbb{B}(\bar{x}, \delta)$, the sequence $\{x^\nu\}_{\nu \in \mathbb{N}}$ generated by $x^\nu = F(x^{\nu-1})$ converges to $\bar{x}$ and, unless it eventually coincides with $\bar{x}$, does so at the rate

$$\limsup_{\nu \to \infty} \frac{|x^\nu - \bar{x}|}{|x^{\nu-1} - \bar{x}|} \leq \operatorname{lip} F(\bar{x}).$$

Guide. Apply the definition of $\operatorname{lip} F(\bar{x})$ in 9.1. □

While the language in 9.1 and 9.2 is that of vector-valued mappings $F : D \to \mathbb{R}^m$, an important special case to be kept in mind is that of a proper function $f : \mathbb{R}^n \to \overline{\mathbb{R}}$ and the set $D = \operatorname{dom} f$. Note that when F is expressed coordinatewise by $F(x) = \big(f_1(x), \ldots, f_m(x)\big)$, its Lipschitz continuity or strict continuity relative to a subset $X \subset D$ corresponds to the same property holding for each function $f_i : D \to \mathbb{R}$.

For a real-valued function f, Lipschitz continuity relative to X with constant κ amounts to the double inequality

$$f(x) - \kappa|x' - x| \leq f(x') \leq f(x) + \kappa|x' - x| \text{ for all } x, x' \in X.$$

This is illustrated in Figure 9–2 with $X \subset \mathbb{R}^1$ and $\kappa = 1$. Actually, if either half of the double inequality holds for all x and x' in X, then so does the other. Once more, this resembles calmness as defined ahead of 8.32, but it requires the same constant κ to work globally for all points in the set X.

9.6 Example (distance functions). *For any nonempty, closed set $C \subset \mathbb{R}^n$, the distance function d_C is Lipschitz continuous on $\mathbb{R}^n$ with constant $\kappa = 1$. It has $\operatorname{lip} d_C(x) = 1$ when $x \notin \operatorname{int} C$, but $\operatorname{lip} d_C(x) = 0$ when $x \in \operatorname{int} C$.*

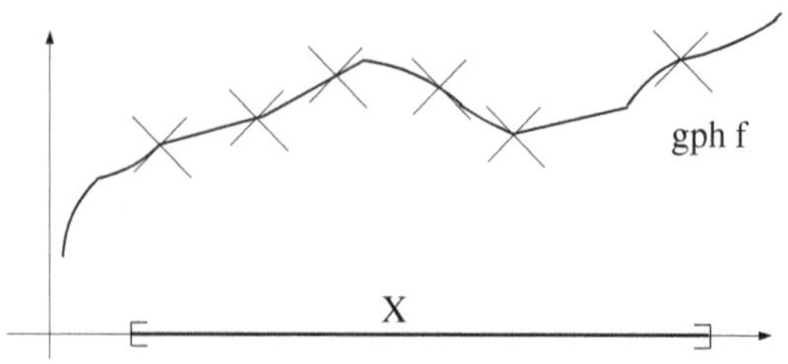

Fig. 9–2. A real-valued function f that is Lipschitz continuous on an interval X.

Detail. The Lipschitz continuity is ensured by 4(3). If $x \in \operatorname{int} C$, d_C is identically 0 on a neighborhood of x, so it's clear that $\operatorname{lip} d_C(x) = 0$. To verify that $\operatorname{lip} d_C(x) = 1$ when $x \notin \operatorname{int} C$, it suffices by the upper semicontinuity of $\operatorname{lip} d_C$ (in 9.2) to do so when $x \notin C$. For any point $\bar{x} \notin C$, there's a point $\tilde{x} \neq \bar{x}$ in the projection $P_C(\bar{x})$. On the line segment from $\tilde{x}$ to $\bar{x}$, the distance from C increases linearly: $d_C\big(\tilde{x} + \tau(\bar{x} - \tilde{x})\big) = \tau d_C(\bar{x})$ for $0 \leq \tau \leq 1$, where $d_C(\bar{x}) = |\bar{x} - \tilde{x}|$. It follows that for any two points $x = \tilde{x} + \tau(\bar{x} - \tilde{x})$ and $x' = \tilde{x} + \tau'(\bar{x} - \tilde{x})$ in this line segment we have $|d_C(x') - d_C(x)| = |\tau' - \tau||\bar{x} - \tilde{x}| = |x' - x|$. Since $\bar{x}$ can be approached by such pairs of points, $\operatorname{lip} d_C(\bar{x})$ can't be less than 1, but from earlier, it can't be greater than 1 either. □

It is convenient to define $\operatorname{lip} f$ also for *extended*-real-valued functions $f : I\!R^n \to \overline{I\!R}$ by interpreting $|f(x') - f(x)|$ as ∞ when either $f(x')$ or $f(x)$ is ∞ (or both) in the formula

$$\operatorname{lip} f(\bar{x}) \;:=\; \limsup_{\substack{x,x' \to \bar{x} \\ x \neq x'}} \frac{|f(x') - f(x)|}{|x' - x|}.$$

This convention fits the extended arithmetic in Chapter 1 and it makes $\operatorname{lip} f$ be usc on all of $I\!R^n$, with

$$\operatorname{lip} f(\bar{x}) := \infty \;\text{ for }\; \bar{x} \notin \operatorname{int}(\operatorname{dom} f). \qquad 9(2)$$

B. Estimates of the Lipschitz Modulus

The question of how to estimate the values of the modulus function $\operatorname{lip} F$ associated with a mapping F is crucial. In the case of differentiable mappings, which we take up first, bounds on partial derivatives can be utilized. While our aim in the large is to handle mappings that aren't necessarily differentiable, this case is a handy building block. Lipschitz constants for smooth mappings F can be derived through Theorem 9.2 from the following result, which relies on the matrix norm (as in 9.3) of the Jacobian of the mapping.

9.7 Theorem (strict continuity from smoothness). *Let $O \subset I\!R^n$ be open. If $F : O \to I\!R^m$ is of class $\mathcal{C}^1$, then F is strictly continuous on O with*

$$\operatorname{lip} F(\bar{x}) = \big|\nabla F(\bar{x})\big| \quad \textit{for any } \bar{x} \in O.$$

This holds in particular for real-valued $f : O \to I\!R$, in which case if f is actually of class $\mathcal{C}^2$, the mapping $\nabla f : O \to I\!R^n$ is strictly continuous too on O with

$$\operatorname{lip} \nabla f(\bar{x}) = \big|\nabla^2 f(\bar{x})\big| \quad \textit{for any } \bar{x} \in O.$$

Proof. The expression $|F(x') - F(x)|/|x' - x|$ can be written in the form $|F(x + \tau w) - F(x)|/\tau$ with $|w| = 1$ by taking $w = (x' - x)/|x' - x|$ and $\tau = |x' - x|$. Choose $\delta > 0$ such that $x + \tau w \in O$ when $|x - \bar{x}| \leq \delta$, $\tau \in [0, \delta]$, and $|w| = 1$. For any $y \in I\!R^m$ with $|y| = 1$ we have

$$\begin{aligned} \Big\langle y, \big[F(x + \tau w) - F(x)\big]/\tau \Big\rangle &= \Big[\big\langle y, F(x + \tau w)\big\rangle - \big\langle y, F(x)\big\rangle\Big]\Big/\tau \\ &= \big\langle y, \nabla F(x + \sigma w) w\big\rangle \end{aligned}$$

for some $\sigma \in (0, \tau)$ by the mean value theorem as applied to the smooth function $\varphi(\tau) = \big\langle y, F(x + \tau w)\big\rangle$. Consequently,

$$\Big\langle y, \big[F(x + \tau w) - F(x)\big]/\tau \Big\rangle \;\leq\; |y| \, \big|\nabla F(x + \sigma w)\big| \, |w| = \big|\nabla F(x + \sigma w)\big|,$$

where $\big|\nabla F(x + \sigma w)\big|$ is the norm of $\nabla F(x + \sigma w)$ as in 9.3. Maximizing over all y with $|y| = 1$ we get $\big|[F(x + \tau w) - F(x)]/\tau\big| \leq \big|\nabla F(x + \sigma w)\big|$, where the left side is $|F(x') - F(x)|/|x' - x|$ in the notation we started with. In passing next to the upper limit that defines $\operatorname{lip} F(\bar{x})$ in 9.1, we can't get more than the upper limit achievable by sequences $\big|\nabla F(x^\nu + \sigma^\nu w^\nu)\big|$ as $x^\nu \to \bar{x}$ and $\sigma^\nu \searrow 0$ with $|w^\nu| \equiv 1$. But the limit of all such sequences is $\big|\nabla F(\bar{x})\big|$ by the assumed smoothness of F, since the norm of a matrix A depends continuously on the components of A (cf. the final inequalities in 9.3). Thus, $\operatorname{lip} F(\bar{x}) \leq \big|\nabla F(\bar{x})\big|$.

To get equality, we take $\bar{w}$ with $|\bar{w}| = 1$ such that $|\nabla F(\bar{x})\bar{w}| = |\nabla F(\bar{x})|$, as is possible from the definition of $|\nabla F(\bar{x})|$. Then for $x_\tau = \bar{x} + \tau \bar{w}$ we have $|F(x_\tau) - F(\bar{x})|/|x_\tau - \bar{x}| \to |\nabla F(\bar{x})|$ as $\tau \searrow 0$.

The remaining assertions in the theorem are now immediate, first from the case of $F = f$ and then from the case of $F = \nabla f$. □

From this and Theorem 9.2, for instance, F is Lipschitz continuous relative to a convex set X with constant κ when F is smooth on an open set $O \supset X$ and $\big|\nabla F(x)\big| \leq \kappa$ for all $x \in X$.

Of course, a $\mathcal{C}^1$ function f can have ∇f strictly continuous without f having to be $\mathcal{C}^2$. Then f is called a $\mathcal{C}^{1+}$ function. In general for an open set $O \subset I\!R^n$, a mapping $F : O \to I\!R^m$ is said to be of *class* $\mathcal{C}^{k+}$ if it's of class $\mathcal{C}^k$ and the kth partial derivatives are not just continuous but strictly continuous on O. In accordance with this, one can think of continuous mappings as being of class $\mathcal{C}^0$ on O, while strictly continuous mappings are of class $\mathcal{C}^{0+}$.

Constants of Lipschitz continuity can often be obtained quite easily also for mappings that aren't smooth and therefore aren't covered by 9.7. The distance functions in 9.6 are an example; d_C is known to have 1 as such a constant even though it fails to be differentiable at boundary points of C. The following rules help in estimating the modulus for a mapping constructed from other mappings with known modulus, regardless of smoothness.

9.8 Exercise (calculus of the modulus). *For single-valued mappings F on open domains, the corresponding functions* $\operatorname{lip} F$ *obey the rules*

(a) $\operatorname{lip}(\lambda F)(x) = |\lambda| \operatorname{lip} F(x)$, *and in particular,* $\operatorname{lip}(-F)(x) = \operatorname{lip} F(x)$,

(b) $\operatorname{lip}(F_1 + F_2)(x) \leq \operatorname{lip} F_1(x) + \operatorname{lip} F_2(x)$,

(c) $\operatorname{lip}(F_2 \circ F_1)(x) \leq \operatorname{lip} F_2(F_1(x)) \cdot \operatorname{lip} F_1(x)$,

(d) $\operatorname{lip}(F_1, \ldots, F_r)(x) \leq \big|\big(\operatorname{lip} F_1(x), \ldots, \operatorname{lip} F_r(x)\big)\big|$, *where* $(F_1, \ldots, F_r)$ *is the mapping* $x \mapsto \big(F_1(x), \ldots, F_r(x)\big)$.

9.9 Exercise (scalarization of strict continuity). *Consider* $F : D \to \mathbb{R}^m$ *with* $D \subset \mathbb{R}^n$, *and write* $F(x) = \big(f_1(x), \ldots, f_m(x)\big)$. *For each vector* $y = (y_1, \ldots, y_m) \in \mathbb{R}^m$, *define the function* $yF : D \to \mathbb{R}$ *by*

$$(yF)(x) := \langle y, F(x) \rangle \;=\; y_1 f_1(x) + \cdots + y_m f_m(x).$$

Then F is strictly continuous at a point $\bar{x}$ if and only if all the functions yF are strictly continuous at $\bar{x}$. Moreover

$$\operatorname{lip} F(\bar{x}) = \sup_{y \in \mathbb{B}} \big\{ \operatorname{lip}(yF)(\bar{x}) \big\} = \sup_{|y|=1} \big\{ \operatorname{lip}(yF)(\bar{x}) \big\}.$$

Guide. The final formula is the key. The inequality $\geq$ follows from 9.8(c) in viewing yF as the composition of F with the linear function associated with y. Equality is obtained by arguing that, when $\operatorname{lip} F(\bar{x}) > 0$, there exist sequences $x^\nu \to \bar{x}$ and $x'^\nu \to \bar{x}$ with $\operatorname{lip} F(\bar{x}) = \lim_\nu \big|F(x'^\nu) - F(x^\nu)\big| / \big|x'^\nu - x^\nu\big|$ and for which the vectors $z^\nu := \big[F(x'^\nu) - F(x^\nu)\big] / \big|x'^\nu - x^\nu\big|$ converge to some $\bar{z} \neq 0$. With $\bar{y} := \bar{z}/|\bar{z}|$ one gets $\operatorname{lip}(\bar{y}F)(\bar{x}) = \operatorname{lip} F(\bar{x})$. □

For mappings into $\mathbb{R}$, the operations of pointwise minimization and maximization are available along with the ones in 9.8.

9.10 Proposition (strict continuity in pointwise max and min). *Consider any collection $\{f_i\}_{i \in I}$ of functions $f_i : O \to \mathbb{R}$, where O is open in $\mathbb{R}^n$. If the index set I is finite, one has*

$$\operatorname{lip}\big(\sup_{i \in I} f_i\big)(x) \leq \sup_{i \in I} \big(\operatorname{lip} f_i\big)(x),$$
$$\operatorname{lip}\big(\inf_{i \in I} f_i\big)(x) \leq \sup_{i \in I} \big(\operatorname{lip} f_i\big)(x),$$

so that the functions $\sup_{i \in I} f_i$ and $\inf_{i \in I} f_i$ are strictly continuous at any point where all the functions f_i are.

More generally, if I is possibly infinite but $\operatorname{lip} f_i(x) \leq k(x)$ for all $i \in I$ for a function $k : O \to \overline{\mathbb{R}}$ that is usc, then both $\operatorname{lip}\big(\sup_{i \in I} f_i\big)(x) \leq k(x)$ and $\operatorname{lip}\big(\inf_{i \in I} f_i\big)(x) \leq k(x)$. In this case, therefore, the functions $\sup_{i \in I} f_i$ and $\inf_{i \in I} f_i$ are strictly continuous at every point $\bar{x} \in O$ where $k(\bar{x}) < \infty$.

Proof. Each of the functions $\operatorname{lip} f_i$ is usc (cf. Theorem 9.2), and so too then is the function $\sup_i(\operatorname{lip} f_i)$ when I is finite (cf. 1.26). The assertions for I finite thus follow from the general case by taking $k(x) = \sup_i(\operatorname{lip} f_i)(x)$. The assertions about $\sup_i f_i$ are equivalent to those about $\inf_i f_i$, because $\operatorname{lip}(-f_i)(x) = \operatorname{lip} f_i(x)$.

Concentrating therefore on the general case of $\sup_i f_i$, we fix any $\bar{x} \in O$ and observe that if $\kappa > k(\bar{x})$ there exists by the upper semicontinuity of k a ball $\mathbb{B}(\bar{x}, \delta)$ on which $k < \kappa$. Then for all $x \in \mathbb{B}(\bar{x}, \delta)$ we have $\operatorname{lip} f_i(x) < \kappa$ by assumption, and because $\mathbb{B}(\bar{x}, \delta)$ is a convex set we may conclude from 9.2 that κ is a constant for f_i on $\mathbb{B}(\bar{x}, \delta)$. Thus,

$$\left.\begin{array}{l} f_i(x') \leq f_i(x) + \kappa|x' - x| \\ f_i(x) \leq f_i(x') + \kappa|x' - x| \end{array}\right\} \quad \text{for } x, x' \in \mathbb{B}(\bar{x}, \delta).$$

Taking the supremum over $i \in I$ on both sides of each inequality, we find that the same two inequalities hold for $f = \sup_i f_i$. Thus, κ is a constant for f on $\mathbb{B}(\bar{x}, \delta)$. Since for arbitrary $\kappa > k(\bar{x})$ this holds for some $\delta > 0$, we have by Definition 9.1 that $\operatorname{lip} f(\bar{x}) \leq k(\bar{x})$. □

9.11 Example (Pasch-Hausdorff envelopes). *Suppose $f : \mathbb{R}^n \to \overline{\mathbb{R}}$ is proper and lsc. For any $\kappa \in \mathbb{R}_+$ the function $f_\kappa := f \# \kappa|\cdot|$, with*

$$f_\kappa(x) := \inf_w \left\{ f(w) + \kappa|w - x| \right\},$$

is the Pasch-Hausdorff envelope of f for the value κ. Unless $f_\kappa \equiv -\infty$, f_κ is Lipschitz continuous on $\mathbb{R}^n$ with constant κ, and it is the greatest of all such functions majorized by f. (When $f_\kappa \equiv -\infty$, there is no function majorized by f that is Lipschitz continuous on $\mathbb{R}^n$ with constant κ.)

Furthermore, as long as there exists a $\bar{\kappa} \in \mathbb{R}_+$ with $f_{\bar{\kappa}} \not\equiv -\infty$, one has, when $\kappa \nearrow \infty$, that $f_\kappa(x) \nearrow f(x)$ for all x, and consequently also $f_\kappa \xrightarrow{e} f$.

Detail. From the definition we have for any x and x' that

$$\begin{aligned} f_\kappa(x') &:= \inf_w \{ f(w) + \kappa|w - x + x - x'| \} \\ &\leq \inf_w \{ f(w) + \kappa|w - x| + \kappa|x' - x| \} = f_\kappa(x) + \kappa|x' - x|. \end{aligned}$$

Therefore, either $f_\kappa \equiv -\infty$ or f_κ is finite and Lipschitz continuous with constant κ. But for any function $g \leq f$ with the latter properties, we have

$$g(x) = \inf_w \{ g(w) + \kappa|w - x| \} \leq \inf_w \{ f(w) + \kappa|w - x| \}$$

for all x, so that $g \leq f_\kappa$.

The convergence assertion at the end is based on observing that if $f_{\bar{\kappa}}(\bar{x}) \neq -\infty$ for some $\bar{\kappa}$ and $\bar{x}$, then $f(w) \geq f_{\bar{\kappa}}(\bar{x}) - \bar{\kappa}|w - \bar{x}|$, so that $f^\infty \geq -\bar{\kappa}|\cdot|$. Since $\kappa|\cdot| \xrightarrow{t} \delta_{\{0\}}$ as $\kappa \nearrow \infty$ (by 7.53), we then have $f \# \kappa|\cdot| \xrightarrow{e} f \# \delta_{\{0\}}$ by 7.56(a), and this means by definition that $f_\kappa \xrightarrow{e} f$. But $f_\kappa(x)$ is nondecreasing with respect to κ, so in this case we also have $f_\kappa \xrightarrow{p} f$ by 7.4(d). □

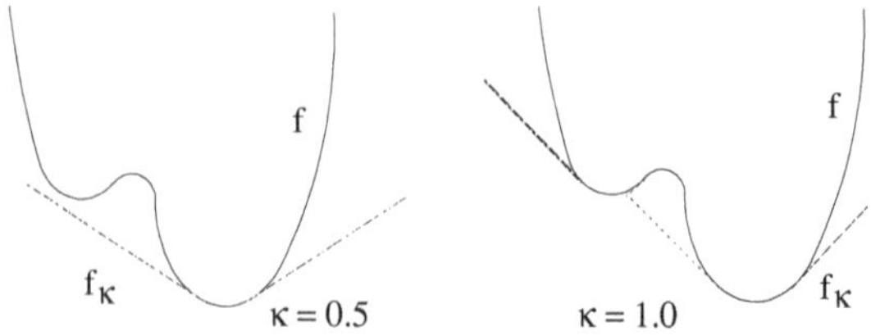

Fig. 9–3. Pasch-Hausdorff envelopes for different Lipschitz constants.

9.12 Exercise (extensions of Lipschitz continuity). *If $f : \mathbb{R}^n \to \overline{\mathbb{R}}$ is Lipschitz continuous on $C \neq \emptyset$ with constant κ, then the function*

$$\bar{f}(x) := \inf_{w \in C} \{ f(w) + \kappa |x - w| \}$$

agrees with f on C and is Lipschitz continuous on $\mathbb{R}^n$ with constant κ. Also,

$$\inf_{x \in C} f(x) = \inf_{x \in \mathbb{R}^n} \bar{f}(x), \qquad \operatorname*{argmin}_{x \in C} f(x) = C \cap \operatorname*{argmin}_{x \in \mathbb{R}^n} \bar{f}(x).$$

Guide. Consider the Pasch-Hausdorff envelope of $f + \delta_C$ for κ. □

Extensions that preserve Lipschitz constants will be treated in greater depth in Theorem 9.58.

C. Subdifferential Characterizations

The theorem proved next reveals a major strength of variational analysis through subgradients and subderivatives: the ability to characterize Lipschitzian properties even to the extent of determining the size of the modulus. Here again the virtues of the concept of horizon subgradients are evident.

Although we focus on an extended-real-valued function f defined on the whole space $\mathbb{R}^n$, this is just a convention having its roots in the way ∞ can be used in simplifying the expression of optimization problems, as explained in Chapter 1. The result, along with others subsequently derived from it, really just concerns the behavior of f on a neighborhood of a point $\bar{x}$. When f is only given locally, it can be applied by regarding f to be ∞ elsewhere.

9.13 Theorem (subdifferential characterization of strict continuity). *Suppose $f : \mathbb{R}^n \to \overline{\mathbb{R}}$ is locally lsc at $\bar{x}$ with $f(\bar{x})$ finite. Then the following conditions are equivalent:*

(a) *f is strictly continuous at $\bar{x}$,*

(b) $\partial^\infty f(\bar{x}) = \{0\}$,

(c) *the mapping $\widehat{\partial} f : x \mapsto \widehat{\partial} f(x)$ is locally bounded at $\bar{x}$,*

(d) *the mapping $\partial f : x \mapsto \partial f(x)$ is locally bounded at $\bar{x}$.*

(e) *the function $\widehat{d} f(\bar{x})$ is finite everywhere,*

(f) *f is finite around $\bar{x}$, and for each $w \in I\!R^n$ the function $x \mapsto df(x)(w)$ is bounded above on some neighborhood of $\bar{x}$.*

Moreover, when these conditions hold, $\partial f(\bar{x})$ is nonempty and compact and one has

$$\operatorname{lip} f(\bar{x}) \;=\; \max_{|w|=1} \widehat{d}f(\bar{x})(w) \;=\; \max_{v \in \partial f(\bar{x})} |v| \;=:\; \max|\partial f(\bar{x})|. \qquad 9(3)$$

Proof. Conditions (b) and (c) are equivalent by the definition of $\partial^\infty f(\bar{x})$ in 8.3. (It suffices to interpret local boundedness in (c) and (d) in the sense of f-attentive convergence, since the distinction drops away once the equivalence with (a) is established.) These conditions imply through 8.10 that $\partial f(\bar{x})$ is nonempty and through 8.6 that it is closed. Local boundedness of ∂f at $\bar{x}$ in (d) follows from that in (c) via the definition of ∂f, whereas the reverse implication comes out of 8.6. In view of the support function formula for regular subderivatives in 8.23, along with the finiteness criterion in 8.29(d), conditions (b), (c) and (d) are equivalent therefore to (e). This formula yields the second equation in 9(3).

Under (a) we have for any $\kappa > \operatorname{lip} f(\bar{x})$ that $\big[f(x+\tau w) - f(x)\big]/\tau \le \kappa|w|$, as long as τ is small enough and x is close enough to $\bar{x}$. Thus, (a) implies (f). But (f) in turn implies (e) through the formula in 8.18 for regular subderivatives as limits of ordinary subderivatives.

This argument reveals that $\widehat{d}f(\bar{x})(w) \le \kappa$ when $|w| = 1$ and $\kappa > \operatorname{lip} f(\bar{x})$. Therefore, $\max_{|w|=1} \widehat{d}f(\bar{x})(w) \le \operatorname{lip} f(\bar{x})$; here 'max' can appropriately be written in place of 'sup' because the function $\widehat{d}f(\bar{x})$, being convex (through sublinearity; cf. 8.6) is continuous when it is finite (cf. 2.36). The value $\operatorname{lip} f(\bar{x})$, however, is the limit of a certain sequence of difference quotients

$$\big[f(x^\nu + \tau^\nu w^\nu) - f(x^\nu)\big]/\tau^\nu \;\text{ with }\; x^\nu \to \bar{x},\; w^\nu \to \bar{w},\; \tau^\nu \searrow 0$$

having $|\bar{w}| = 1$, which is not less that $\widehat{d}f(\bar{x})(\bar{w})$, so in (a) we must have $\max_{|w|=1} \widehat{d}f(\bar{x})(w) = \operatorname{lip} f(\bar{x})$, hence 9(3). From 8.22, however, (a) is guaranteed by $0 \in \operatorname{int}\big(\operatorname{dom} \widehat{d}f(\bar{x})\big)$. That condition is identical to (e), because $\operatorname{dom} \widehat{d}f(\bar{x})$ is a cone. □

Although Theorem 9.13 asserts that $\partial f(\bar{x})$ is nonempty and compact when f is strictly continuous at $\bar{x}$, it's possible also for $\partial f(\bar{x})$ to have these properties without f necessarily being strictly continuous at $\bar{x}$. This is exemplified by the continuous function $f : I\!R \to I\!R$ with

$$f(x) = \begin{cases} -\sqrt{x} & \text{for } x \ge 0, \\ 0 & \text{for } x \le 0, \end{cases}$$

which has $\partial f(0) = \{0\}$ but $\partial^\infty f(0) = (-\infty, 0]$.

9.14 Example (Lipschitz continuity from convexity). *A finite convex function f on an open, convex set $O \subset I\!R^n$ is strictly continuous on O.*

In fact, for any proper, convex function $f : \mathbb{R}^n \to \overline{\mathbb{R}}$ and any set X on which f is bounded below by α, if there exists $\varepsilon > 0$ such that f is bounded above by β on $X + \varepsilon\mathbb{B}$, then f is Lipschitz continuous on X with constant $\kappa = (\beta - \alpha)/\varepsilon$ and therefore also has $\partial f(x) \subset \kappa\mathbb{B}$ for all $x \in X$.

Detail. To obtain the first assertion as a consequence of Theorem 9.13, extend f to be a proper, lsc, convex function on $\mathbb{R}^n$ with $O \subset \operatorname{int}(\operatorname{dom} f)$; cf. 2.36. At every point $\bar{x} \in O$ we have $\partial^\infty f(\bar{x}) = \{0\}$ by 8.12, so f is strictly continuous at such points $\bar{x}$. The second assertion can be confirmed on the basis of convexity alone, without the intervention of Theorem 9.13. For any two points x and x' in X, we can write $x' = (1-\tau)x + \tau x''$ with $x'' = x' + \varepsilon z$ for

$$z = \frac{1}{|x'-x|}(x'-x), \qquad \tau = \frac{|x'-x|}{\varepsilon + |x'-x|} \leq \frac{|x'-x|}{\varepsilon}.$$

Then $x'' \in X + \varepsilon\mathbb{B}$ and $f(x') \leq (1-\tau)f(x) + \tau f(x'')$, so that $f(x') - f(x) \leq \tau(\beta - \alpha) \leq \kappa|x'-x|$. From the formula at the end of Theorem 9.13, one obtains in this case the bound on $\partial f(x)$.

Note that the first assertion could be derived independently of 9.13 by invoking the second assertion in the case of X being a sufficiently small ball around any point $\bar{x} \in O$. Because f is to be continuous on O (by 2.36), it's bounded above and below on all compact subsets of O. □

9.15 Exercise (subderivatives under strict continuity). *Whenever f is strictly continuous at $\bar{x}$, one has*

$$\begin{aligned} df(\bar{x})(w) &= \liminf_{\tau \searrow 0} \frac{f(\bar{x}+\tau w) - f(\bar{x})}{\tau}, \\ \widehat{d}f(\bar{x})(w) &= \limsup_{\substack{\tau \searrow 0 \\ x \to \bar{x}}} \frac{f(x+\tau w) - f(x)}{\tau} = \limsup_{x \to \bar{x}} df(x)(w) \\ &= \max \langle \partial f(\bar{x}), w \rangle := \max\{\langle v, w\rangle \mid v \in \partial f(\bar{x})\}. \end{aligned}$$

Then the functions $df(\bar{x})$ and $\widehat{d}f(\bar{x})$ are globally Lipschitz continuous with constant $\kappa = \operatorname{lip} f(\bar{x})$.

Guide. Specialize Definitions 8.1 and 8.16, using a local constant $\kappa' > \operatorname{lip} f(\bar{x})$ to compare quotients: $\Delta_\tau f(x)(w') \leq \Delta_\tau f(x)(w) + \kappa'|w'-w|$ when x is close enough to $\bar{x}$ and $\tau > 0$ is sufficiently small. Invoke 8.18 and 8.23, utilizing criterion 9.13(b). □

9.16 Theorem (subdifferential regularity under strict continuity). *A function f that is finite on an open set $O \subset \mathbb{R}^n$ is both strictly continuous and regular on O if and only if for every $x \in O$ and $w \in \mathbb{R}^n$ the limit*

$$h(x,w) = \lim_{\tau \searrow 0} \frac{f(x+\tau w) - f(x)}{\tau}$$

exists, is finite, and depends upper semicontinuously on x for each fixed w.

Then $h(x,w)$ depends upper semicontinuously on x and w together, and f is semidifferentiable on O with

$$h(x,w) = \lim_{\substack{\tau \searrow 0 \\ w' \to w}} \frac{f(x+\tau w')-f(x)}{\tau} = \limsup_{\substack{\tau \searrow 0 \\ x' \to x \\ w' \to w}} \frac{f(x'+\tau w')-f(x')}{\tau}$$
$$= df(x)(w) = \widehat{d}f(x)(w) = \max\{\langle v,w\rangle \,|\, v\in\partial f(x)\}.$$

Also, $\partial f : O \rightrightarrows \mathbb{R}^n$ is nonempty-convex-valued, osc, and locally bounded.

Proof. First suppose the limit $h(x,w)$ always exists, and assume it's finite and is usc in x. Then for each w the function $x \mapsto h(x,w)$ is locally bounded from above on O. From the definition of $df(x)(w)$ we have $df(x)(w) \le h(x,w)$, so for each w the function $x \mapsto df(x)(w)$ likewise is locally bounded from above. This implies through criterion (f) of Theorem 9.13 that f is strictly continuous on O. Then by 9.15, $df(x)(w) = h(x,w)$, so $df(x)(w)$ is usc in x for each w. Also by 9.15, $df(x)(w) = \widehat{d}f(x)(w)$, hence f is regular by 8.19.

Restart now by assuming only that f is strictly continuous and regular on O; all the claims about $h(x,w)$ will turn out to follow from this. Continuity of f makes the mapping ∂f osc on O by 8.7, because f-attentive convergence can be replaced in that result by ordinary convergence. Strict continuity implies through Theorem 9.13 that ∂f is nonempty-compact-valued and locally bounded on O, and in addition that $df(x)(w)$ is locally bounded from above in $x \in O$ for each $w \in \mathbb{R}^n$. Then by regularity, $\partial f(x)$ is convex with $df(x)$ as its support function: $df(x)(w) = \max\{\langle v,w\rangle \,|\, v \in \partial f(x)\}$; cf. 8.30. Hence $df(x)(w)$ is convex in w and, because it is finite, also continuous in w; cf. 2.36. Regularity further ensures through 8.19 that $\widehat{d}f(x)(w) = df(x)(w)$.

The alternative limit expressions in 9.15 for $df(x)(w)$ and $\widehat{d}f(x)(w)$ then coincide and in particular yield the value $h(x,w)$, this being finite and usc in $x \in O$ for each w, and also given by the max expression above. The finiteness of $\widehat{d}f(x)(w)$ for all $w \in \mathbb{R}^n$ triggers through 8.18 the formula

$$\widehat{d}f(x)(w) = \limsup_{\substack{x' \underset{f}{\to} x \\ \tau \searrow 0 \\ w' \to w}} \frac{f(x'+\tau w')-f(x')}{\tau}.$$

This upper limit therefore gives $h(x,w)$ as well, with $x' \underset{f}{\to} x$ reducing to $x' \to x$ because f is continuous. It further implies that h is usc on $O \times \mathbb{R}^n$. □

The strong properties in 9.16 hold in particular for finite, convex functions, since such functions are strictly continuous (by 9.14) and regular (by 7.27). A much broader class of functions for which these properties always hold will be developed in Chapter 10 in terms of 'subsmoothness' (cf. 10.29).

We look now at a concept of 'strict differentiability' related to strict continuity and see how it too can be characterized in terms of subgradients.

9.17 Definition (strict differentiability). *A function $f : \mathbb{R}^n \to \overline{\mathbb{R}}$ is strictly* *differentiable at a point* $\bar{x}$ *if* $f(\bar{x})$ *is finite and there is a vector* v, *which will be the gradient* $\nabla f(\bar{x})$, *such that* $f(x') = f(x) + \langle v, x' - x\rangle + o(|x' - x|)$, *i.e.*,

$$\frac{f(x') - f(x) - \langle v, x' - x\rangle}{|x' - x|} \to 0 \quad \text{as} \quad x, x' \to \bar{x} \text{ with } x' \neq x.$$

Strict differentiability can be regarded as the pointwise localization of smoothness, as will be borne out by a corollary of the next theorem.

9.18 Theorem (subdifferential characterization of strict differentiability). *Let* $f : I\!R^n \to \overline{I\!R}$ *be finite at* $\bar{x}$. *Then the following conditions are equivalent:*

(a) f *is strictly differentiable at* $\bar{x}$;

(b) f *is strictly continuous at* $\bar{x}$ *and has at most one subgradient there;*

(c) f *is locally lsc at* $\bar{x}$, $\partial f(\bar{x})$ *is a singleton, and* $\partial^\infty f(\bar{x}) = \{0\}$;

(d) f *is continuous around* $\bar{x}$, *and* f *and* $-f$ *are regular at* $\bar{x}$, *with* $\widehat{\partial} f(\bar{x})$ *and* $\widehat{\partial}[-f](\bar{x})$ *nonempty;*

(e) f *is regular at* $\bar{x}$, *and on a neighborhood of* $\bar{x}$ *there exists* $h \geq f$ *with* $h(\bar{x}) = f(\bar{x})$, *such that* h *differentiable at* $\bar{x}$;

(f) f *is locally lsc at* $\bar{x}$, *and* $\widehat{d} f(\bar{x})$ *is a linear function;*

(g) f *is locally lsc at* $\bar{x}$, *and* $\widehat{d} f(\bar{x})(-w) = -\widehat{d} f(\bar{x})(w)$ *for all* w.

When these equivalent conditions hold, one has moreover that

$$\partial f(\bar{x}) = \{\nabla f(\bar{x})\}, \qquad df(\bar{x}) = \langle \nabla f(\bar{x}), \cdot\rangle = \widehat{d} f(\bar{x}), \qquad \operatorname{lip} f(\bar{x}) = \big|\nabla f(\bar{x})\big|.$$

Proof. By Definition 9.17, f is strictly differentiable at $\bar{x}$ with $\nabla f(\bar{x}) = v$ if and only if the function $g(x) := f(x) - \langle v, x\rangle$ satisfies

$$[g(x') - g(x)]/|x' - x| \to 0 \text{ as } x, x' \to \bar{x} \text{ with } x' \neq x.$$

This condition is identical to having $\operatorname{lip} g(\bar{x}) = 0$, which implies that g is strictly continuous at $\bar{x}$ and the same then for f. (In particular, all the conditions (a)–(g) do entail f being locally lsc at $\bar{x}$.) Applying Theorem 9.13 to g, we see that $\operatorname{lip} g(\bar{x}) = 0$ corresponds to g being not only strictly continuous at $\bar{x}$ but having $\partial g(\bar{x}) = \{0\}$. Obviously $\partial g(\bar{x}) = \partial f(\bar{x}) - v$, so we obtain in this way the equivalence of (a) with (b) and (c), along with the validity of the last line of the theorem when these properties are present.

To see how these properties imply (d), observe that since $\nabla f(\bar{x}) \in \widehat{\partial} f(\bar{x})$ by 8.8(a), the relations $\partial f(\bar{x}) = \{\nabla f(\bar{x})\}$ and $\partial^\infty f(\bar{x}) = \{0\}$ entail the regularity of f at $\bar{x}$; cf. 8.11. Since f is strictly differentiable if and only if $-f$ is, they imply the regularity of $-f$ at $\bar{x}$ also. Thus, (a) gives (d).

If (d) holds, the regularity of $-f$ at $\bar{x}$ yields a vector $\bar{v}$ with

$$f(x) \;\leq\; f(\bar{x}) + \langle \bar{v}, x - \bar{x}\rangle + o_1(|x - \bar{x}|).$$

But since f is regular at $\bar{x}$ as well, any vector $v \in \partial f(\bar{x})$ satisfies

$$f(x) \;\geq\; f(\bar{x}) + \langle v, x - \bar{x}\rangle + o_2(|x - \bar{x}|).$$

Then $\langle v - \bar{v}, x - \bar{x}\rangle \leq o_1(|x - \bar{x}|) + o_2(|x - \bar{x}|)$, which is only possible when $v = \bar{v}$. The set $\partial f(\bar{x}) = \widehat{\partial} f(\bar{x})$ must just be $\{\bar{v}\}$, so $\partial^\infty f(\bar{x})$ must be $\{0\}$.

The equivalence of (f) and (g) with (c) is furnished by the formula for $\widehat{\partial} f(\bar{x})$ in 8.23, which also shows that $\widehat{d}f(\bar{x})(w) = \langle \nabla f(\bar{x}), w\rangle$.

Next, (e) follows from (a) by taking $h = f$. But if (e) holds we have $df(\bar{x})(w) \leq dh(\bar{x})(w) = \langle \nabla h(\bar{x}), w\rangle < \infty$ with $df(\bar{x}) = \widehat{d}f(\bar{x})$ sublinear (by 8.18, 8.19). Then $\widehat{d}f(\bar{x})(w) + \widehat{d}f(\bar{x})(-w) \geq \widehat{d}f(\bar{x})(0)$ with $\widehat{d}f(\bar{x})(0) = 0$ (otherwise $\widehat{d}f(\bar{x}) \equiv -\infty$, so $(0, -1) \in \operatorname{int} R_{\operatorname{epi} f}(\bar{x}, f(\bar{x}))$; cf. 6.36), while

$$\widehat{d}f(\bar{x})(w) + \widehat{d}f(\bar{x})(-w) \;\leq\; \big\langle \nabla h(\bar{x}), w\big\rangle + \big\langle \nabla h(\bar{x}), -w\big\rangle \;=\; 0.$$

Thus, (e) implies (g). This completes the network of equivalences. The formula for $\operatorname{lip} f(\bar{x})$ follows from (b) and Theorem 9.13. □

The condition $\partial^\infty f(\bar{x}) = \{0\}$ isn't superfluous in 9.18(c). An example where $\partial f(\bar{x})$ is a singleton, but f isn't strictly differentiable at $\bar{x}$ and $\partial^\infty f(\bar{x})$ must therefore contain nonzero vectors, is furnished by the function f that was considered after the proof of Theorem 9.13.

9.19 Corollary (smoothness). *For a function $f : \mathbb{R}^n \to \overline{\mathbb{R}}$ and an open set O where f is finite, the following properties are equivalent:*

(a) *f is strictly differentiable on O;*

(b) *f is $\mathcal{C}^1$ on O;*

(c) *f is strictly continuous and regular on O with ∂f single-valued;*

(d) *f is lsc on O, ∂f is locally bounded, and ∂f is single-valued where it is nonempty-valued in O.*

Proof. The theorem immediately gives the equivalence of (a) with (c). The equivalence of (c) with (b) is apparent from 9.16, since f is differentiable at x if and only if it is semidifferentiable at x and $df(x)$ is a linear function; cf. 7.21; the support function of $\partial f(x)$ is linear if and only if $\partial f(x)$ is a singleton.

Obviously (c) implies (d). But if (d) holds, f is strictly continuous on O by criterion 9.13(d), and then $\partial f(x)$ is nonempty for all $x \in O$ by 9.13 also. We get (a) in this case through the equivalence of 9.18(b) with 9.18(a). □

9.20 Corollary (continuity of gradient mappings). *When f is regular on an open set $O \subset \mathbb{R}^n$, it is strictly differentiable at every point of O where it is differentiable. Relative to this set of points, the mapping ∇f is continuous.*

In particular, these properties hold when f is a finite convex function on an open convex set O.

Proof. The first assertion is evident from 9.18(e) with $h = f$. The second is then justified because ∂f reduces to ∇f on the set of points in question and yet is osc with respect to f-attentive convergence; cf. 8.7. Since f is continuous wherever it's differentiable, f-attentive convergence in this set is the same as ordinary convergence. Finite convex functions are regular by 7.27. □

9.21 Corollary (upper subgradient property). *If f is strictly continuous and regular on an open set O, as for instance when f is finite and convex, then at all points $\bar{x} \in O$ one has*

$$-\widehat{\partial}(-f)(\bar{x}) = \begin{cases} \{\nabla f(\bar{x})\} & \text{if } f \text{ is differentiable at } \bar{x}, \\ \emptyset & \text{if } f \text{ is not differentiable at } \bar{x}, \end{cases},$$

$$-\partial(-f)(\bar{x}) = \{v \mid \exists x^\nu \to \bar{x} \text{ with } \nabla f(x^\nu) \to v\}$$

and therefore in particular $-\partial(-f)(\bar{x}) \subset \partial f(\bar{x})$.

Proof. The assertion for $-\widehat{\partial}(-f)(\bar{x})$ follows from criterion (e) in the theorem along with the variation description of regular subgradients in 8.5. The assertion for $-\partial(-f)(\bar{x})$ then follows by definition. Finite convex functions are strictly continuous by 9.14 and regular by 7.27. □

9.22 Exercise (criterion for constancy). *Let $f : \mathbb{R}^n \to \overline{\mathbb{R}}$ be lsc and proper, and let $\bar{x} \in \operatorname{dom} f$. Then the following conditions are equivalent:*

(a) *f is constant on a neighborhood of $\bar{x}$,*

(b) *$\partial f(x) = \{0\}$ for all x in a neighborhood of $\bar{x}$ relative to $\operatorname{dom} f$,*

(c) *there exists $\varepsilon > 0$ such that whenever $|x - \bar{x}| \le \varepsilon$, $|f(x) - f(\bar{x})| \le \varepsilon$ and $\widehat{\partial} f(x) \ne \emptyset$, one has $\widehat{\partial} f(x) = \{0\}$.*

Guide. Through the basic facts about subgradients and the preceding results, these conditions can be identified with f being strictly differentiable at all x in some neighborhood of $\bar{x}$, with $\nabla f(x) = 0$. □

In 9.22(c), the regular subgradient set $\widehat{\partial} f(x)$ could be replaced by the set of proximal subgradients defined by 8(30), because every regular subgradient is a limit of proximal subgradients at nearby points; cf. 8.47(a).

D. Derivative Mappings and Their Norms

To extend some of these characterizations to the vector-valued case, i.e., from $f : \mathbb{R}^n \to \overline{\mathbb{R}}$ to $F : \mathbb{R}^n \to \mathbb{R}^m$, we must replace subderivatives and subgradients by the graphical derivative and coderivative mappings introduced in 8.33, which are positively homogeneous but possibly multivalued. As an aid in this development we define for any positively homogeneous mapping $H : \mathbb{R}^n \rightrightarrows \mathbb{R}^m$ the *outer norm* $|H|^+ \in [0, \infty]$ by

$$|H|^+ := \sup_{w \in \mathbb{B}} \sup_{z \in H(w)} |z|. \qquad 9(4)$$

This is the infimum over all constants $\kappa \ge 0$ such that $|z| \le \kappa |w|$ for all pairs $(w, z) \in \operatorname{gph} H$. Although we won't really make use of it here, the *inner norm* $|H|^- \in [0, \infty]$ is defined instead by

$$|H|^- := \sup_{w \in \mathbb{B}} \inf_{z \in H(w)} |z|. \qquad 9(5)$$

It gives the infimum of all constants $\kappa \geq 0$ such that $d\big(0, H(w)\big) \leq \kappa|w|$ for all w. Obviously $|H|^- \leq |H|^+$ always, and equality holds when H is single-valued.

When H is actually a linear mapping, its outer and inner norms agree with the norm of the corresponding matrix (as defined in 9.3):

$$|H|^+ = |H|^- = |A| \text{ when } H(w) = Aw.$$

In general, however, the positively homogeneous mappings $H : I\!R^n \rightrightarrows I\!R^m$ don't even form a vector space under addition and scalar multiplication, so neither $|H|^+$ nor $|H|^-$ is truly a 'norm' in the sense ordinarily accompanying that term. Nevertheless, elementary rules are valid like

$$\begin{aligned} &|\lambda H|^+ = |\lambda||H|^+, \quad |H_1 + H_2|^+ \leq |H_1|^+ + |H_2|^+, \quad |H_2 \circ H_1|^+ \leq |H_2|^+|H_1|^+, \\ &|\lambda H|^- = |\lambda||H|^-, \quad |H_1 + H_2|^- \leq |H_1|^- + |H_2|^-, \quad |H_2 \circ H_1|^- \leq |H_2|^-|H_1|^-. \end{aligned}$$

9.23 Proposition (local boundedness of positively homogeneous mappings). *For $H : I\!R^n \rightrightarrows I\!R^m$ positively homogeneous, the following are equivalent:*

(a) *H is osc and locally bounded;*

(b) *H is osc relative to some neighborhood of 0, and $H(0) = \{0\}$;*

(c) *H is osc relative to $I\!B$, and $|H|^+ < \infty$.*

Furthermore, whenever g-lim sup$_\nu$ $H^\nu \subset H$ *for positively homogeneous mappings $H^\nu : I\!R^n \rightrightarrows I\!R^m$, one has* $\limsup_\nu |H^\nu|^+ \leq |H|^+$*, so that if H is locally bounded the same must be true of H^ν for all ν in an index set $N \in \mathcal{N}_\infty$.*

Proof. Under (a) the set $H(I\!B)$ is compact (by 5.15 and 5.25(a)), so we have (c). On the other hand, (c) implies (b) because $H(0)$ is itself a cone (by 8.36) and thus can't be bounded unless it consists only of 0. Finally, if (b) holds we have H osc everywhere by positive homogeneity. If H weren't locally bounded, there would be a bounded sequence of points w^ν for which there exist $z^\nu \in H(w^\nu)$ with $0 < |z^\nu| \to \infty$. Let $\bar{z}^\nu = z^\nu/|z^\nu|$ and $\bar{w}^\nu = w^\nu/|z^\nu|$. Then $\bar{z}^\nu \in H(\bar{w}^\nu)$ with $|\bar{z}^\nu| = 1$ and $\bar{w}^\nu \to 0$. Since H is osc, any cluster point $\bar{z}$ of $\{\bar{z}^\nu\}_{\nu \in I\!N}$ would give us $\bar{z} \in H(0)$, $\bar{z} \neq 0$, which is forbidden in (b).

Suppose now that g-lim sup$_\nu$ $H^\nu \subset H$. Consider $\kappa < \limsup_\nu |H^\nu|^+$. There's an index set $N \in \mathcal{N}_\infty^\#$ along with $\varepsilon > 0$ such that $\kappa + \varepsilon < |H^\nu|^+$ for all $\nu \in N$, and accordingly there exist $(w^\nu, z^\nu) \in \operatorname{gph} H^\nu$ such that $|z^\nu| \geq (\kappa+\varepsilon)|w^\nu|$. Since $\operatorname{gph} H^\nu$ is a cone (cf. 8.36), we can assume $|w^\nu|^2 + |z^\nu|^2 = 1$. The sequence of pairs (w^ν, z^ν) then has a cluster point $(\bar{w}, \bar{z}) \neq (0,0)$, necessarily in $\operatorname{gph} H$, for which $|\bar{z}| \geq (\kappa+\varepsilon)|\bar{w}|$. Then $|H|^+ \geq \kappa + \varepsilon$. Having proceeded from any $\kappa < \limsup_\nu |H^\nu|^+$, we deduce that $|H|^+ \geq \limsup_\nu |H^\nu|^+$.

In particular then, if $|H|^+ < \infty$, this condition being the hallmark of local boundedness as determined earlier, we must have $|H^\nu|^+ < \infty$ for all ν sufficiently large. The osc mapping $\operatorname{cl} H^\nu$, which like H^ν is positively homogeneous (its graph is still a cone) obviously has $|\operatorname{cl} H^\nu|^+ = |H^\nu|^+$, so by the foregoing it too must be locally bounded for all ν sufficiently large. Local boundedness of $\operatorname{cl} H^\nu$ entails that of H^ν. □

Additional properties of $|H|^+$ and also $|H|^-$ will be developed in 11.29 and 11.30 for mappings H that are sublinear.

9.24 Proposition (derivatives and coderivatives of single-valued mappings). *For $F : D \to I\!R^m$ with $D \subset I\!R^n$, consider $\bar{x} \in \operatorname{int} D$.*

(a) *If F is calm at $\bar{x}$, as is true when F is strictly continuous at $\bar{x}$, its derivative mapping $DF(\bar{x})$ is nonempty-valued, osc and locally bounded, with*

$$\big|DF(\bar{x})\big|^+ = \limsup_{\substack{x \to \bar{x} \\ x \neq \bar{x}}} \frac{\big|F(x) - F(\bar{x})\big|}{|x - \bar{x}|} \leq \operatorname{lip} F(\bar{x}).$$

(b) *If F is strictly continuous at $\bar{x}$, its coderivative mapping $D^*F(\bar{x})$ is nonempty-valued, osc and locally bounded, with*

$$D^*F(\bar{x})(y) = \partial(yF)(\bar{x}) \ \text{ for all } y \ \text{ and } \ \big|D^*F(\bar{x})\big|^+ = \operatorname{lip} F(\bar{x}).$$

The regular coderivative $\widehat{D}^*F(\bar{x})$ likewise has $\widehat{D}^*F(\bar{x})(y) = \widehat{\partial}(yF)(\bar{x})$ for all y.

Proof. The limit expression for $\big|DF(\bar{x})\big|^+$ in (a) is clear from the definition of the outer norm in 9(4) and the characterization of $DF(\bar{x})(w)$ in 8(21), or equivalently 8(20); the boundedness of the difference quotients in the latter serves to establish the nonemptiness of $DF(\bar{x})(w)$ for every w. The other properties of $DF(\bar{x})$ then follow from 9.23 as applied to $H = DF(\bar{x})$.

For (b), let $V \in \mathcal{N}(\bar{x})$ be open and such that F is strictly continuous on $D \cap V$; cf. 9.1. For any $\tilde{x}$ we know from 8(23) that $v \in \widehat{D}^*F(\tilde{x})(y)$ if and only if

$$\langle v,\, x - \tilde{x} \rangle \leq \langle y,\, F(x) - F(\tilde{x}) \rangle + o\big(\big|(x, F(x)) - (\tilde{x}, F(\tilde{x}))\big|\big),$$

but when $\tilde{x} \in V$ the strict continuity of F induces the error term to be of the form $o\big(|x - \tilde{x}|\big)$, so that the condition can be written simply as

$$(yF)(x) \ \geq \ (yF)(\tilde{x}) + \langle v,\, x - \tilde{x} \rangle + o\big(|x - \tilde{x}|\big).$$

Thus, $\widehat{D}^*F(\tilde{x})(y) = \widehat{\partial}(yF)(\tilde{x})$ for all y. The graphical limit formula for $D^*F(\bar{x})$ in 8(22) then gives us $D^*F(\bar{x})(y) = \partial(yF)(\bar{x})$ for all y, utilizing the estimate that $\widehat{\partial}[y^\nu F](x^\nu) \subset \partial[yF + (y^\nu - y)F](x^\nu) \subset \partial[yF](x^\nu) + \partial[(y^\nu - y)F](x^\nu)$, where $\partial[(y^\nu - y)F](x^\nu) \to \{0\}$ when $x^\nu \to \bar{x}$ and $y^\nu \to y$. The formula for $\big|D^*F(\bar{x})\big|^+$ follows then, via 9(3), from the one in 9.9 and the definition. □

An example of a Lipschitz continuous mapping $F : I\!R \to I\!R$ which at 0 has a multivalued derivative mapping $DF(0) : I\!R \rightrightarrows I\!R$ is shown in Figure 9–4. For more on this kind of situation, see 9.49(b).

Some of the results about differentiability properties of real-valued functions can be extended to vector-valued functions through 9.24. Generalizing Definition 7.20 to the case of $F : D \to I\!R^m$, $D \subset I\!R^n$, we call

$$\lim_{\substack{w' \to w \\ \tau \searrow 0}} \frac{F(\bar{x} + \tau w') - F(\bar{x})}{\tau}$$

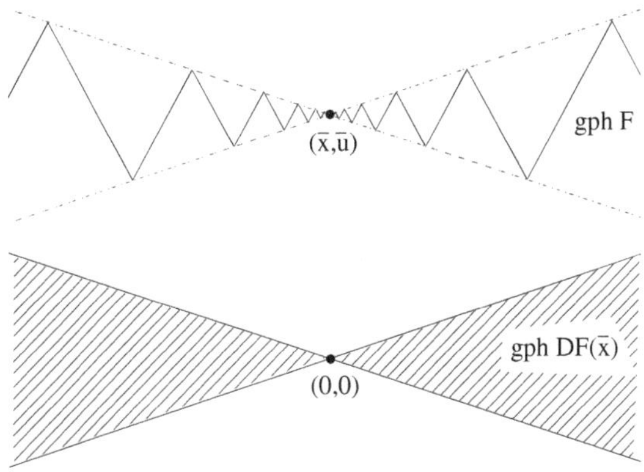

Fig. 9–4. A single-valued mapping with multivalued graphical derivative.

the *semiderivative* of F at $\bar{x}$ for w, if this limit vector exists. If it exists for every $w \in I\!R^n$, we say that F is *semidifferentiable at* $\bar{x}$. The same concept can also be reached by specializing to single-valued mappings the definition of semidifferentiability given for set-valued mappings ahead of 8.43.

In the notation $F(x) = \big(f_1(x), \ldots, f_m(x)\big)$, semidifferentiability of F at $\bar{x}$ corresponds to that property for each of the functions f_i. The characterizations of semidifferentiability in 7.21 carry over then to F. (For the sake of applying 7.21 to f_i, there's no harm in thinking of f_i as ∞ outside of D). In particular, F is seen to be semidifferentiable at $\bar{x}$ if and only if there's a continuous, positively homogeneous mapping $H : I\!R^n \to I\!R^m$ (single-valued) such that

$$F(x) = F(\bar{x}) + H(x - \bar{x}) + o\big(|x - \bar{x}|\big). \qquad 9(6)$$

This expansion, which corresponds to the one in 8.43(f) in the context of set-valued mappings, reduces to the *differentiability* of F at $\bar{x}$ when $H(w) = Aw$ for some $A \in I\!R^{m\times n}$; cf. 7.22. Then 9(6) can be written equivalently as

$$\frac{F(x') - F(\bar{x}) - A(x' - \bar{x})}{|x' - \bar{x}|} \to 0 \text{ as } x' \to \bar{x} \text{ with } x' \neq \bar{x}.$$

In emulation of 9.17 we go on to define F to be *strictly differentiable* at $\bar{x}$ if $F(x') = F(x) + A(x' - x) + o(|x' - x|)$, i.e.,

$$\frac{F(x') - F(x) - A(x' - x)}{|x' - x|} \to 0 \text{ as } x, x' \to \bar{x} \text{ with } x' \neq x. \qquad 9(7)$$

9.25 Exercise (differentiability properties of single-valued mappings). *Consider $F : D \to I\!R^m$, $D \subset I\!R^n$, and a point $\bar{x} \in \operatorname{int} D$.*

(a) *F is semidifferentiable at $\bar{x}$ if and only if F is calm at $\bar{x}$ and the mapping $DF(\bar{x})$ is single-valued. Then $F(x) = F(\bar{x}) + DF(\bar{x})(x - \bar{x}) + o\big(|x - \bar{x}|\big)$.*

(b) *F is differentiable at $\bar{x}$ if and only if F is calm at $\bar{x}$ and the mapping $DF(\bar{x})$ is not just single-valued but also linear. Then $DF(\bar{x})(w) = \nabla F(\bar{x})w$.*

(c) *F is strictly differentiable at $\bar{x}$ if and only if F is strictly continuous at $\bar{x}$ and the mapping $D^*F(\bar{x})$ is single-valued, in which case $D^*F(\bar{x})$ must be linear. Then $D^*F(\bar{x})(y) = \nabla F(\bar{x})^* y$.*

(d) *F is strictly differentiable at $\bar{x}$ if and only if F is both strictly continuous and graphically regular at $\bar{x}$.*

Guide. In (a) argue from the definition of semidifferentiability and the facts about it just mentioned. Then specialize to get (b). In (c) apply criterion 9.18(b) to $f = yF$, using the form for $D^*F(\bar{x})$ established in 9.24(b). Graphical regularity of F at $\bar{x}$ is equivalent to having $\widehat{D}^*F(\bar{x})(y) = D^*F(\bar{x})(y)$ for all y, cf. 8.40(f). Get the necessity of this in (d) through (c) and the form for $\widehat{D}^*F(\bar{x})$ in 9.24(b). Obtain the sufficiency by invoking 9.13 once more. □

The revelation in 9.25(d) that a strictly continuous mapping $F : I\!R^n \to I\!R^m$ can't be graphically regular without in fact being strictly differentiable underscores the special nature of the geometry of the graphs of such single-valued mappings. Regularity is absent at all 'nonsmooth' points and therefore can't begin to play the role that it does in the geometry of sets defined by smooth inequality constraints.

E. Lipschitzian Concepts for Set-Valued Mappings

Lipschitz continuity can be extended to set-valued mappings S in terms of Pompeiu-Hausdorff distance as defined in 4.13,

$$\begin{aligned} d\!l_\infty\big(S(x'), S(x)\big) &= \sup_{u \in I\!R^m} \Big| d\big(u, S(x')\big) - d\big(u, S(x)\big)\Big| \\ &= \sup_{u \in S(x') \cup S(x)} \Big| d\big(u, S(x')\big) - d\big(u, S(x)\big)\Big|. \end{aligned}$$

9.26 Definition (Lipschitz continuity of set-valued mappings). *A mapping $S : I\!R^n \rightrightarrows I\!R^m$ is Lipschitz continuous on X, a subset of $I\!R^n$, if it is nonempty-closed-valued on X and there exists $\kappa \in I\!R_+$, a Lipschitz constant, such that*

$$d\!l_\infty\big(S(x'), S(x)\big) \le \kappa |x' - x| \;\text{ for all } x, x' \in X,$$

or in equivalent geometric form,

$$S(x') \subset S(x) + \kappa |x' - x| I\!B \;\text{ for all } x, x' \in X.$$

Observe that in the notation $s_u(x) = d(u, S(x))$ the inequality in 9.26 refers to having $|s_u(x') - s_u(x)| \le \kappa |x' - x|$ for all $x, x' \in X$ and $u \in I\!R^m$. In other words, it requires all the functions s_u to be Lipschitz continuous on X with the same constant κ.

When S is actually single-valued, Lipschitz continuity on X in the sense of 9.26 reduces to Lipschitz continuity in the sense of 9.1. The new definition doesn't clash with the earlier one. But in relying on Pompeiu-Hausdorff distance, the set-valued version has serious limitations if the sets $S(x)$ are allowed to be unbounded, because it necessitates that these sets *have the same horizon cone* for all x; cf. 4.40. That's true in some situations of interest, such as when $S(x) = S_0(x) + K$ with S_0 locally bounded and K a fixed cone or other closed set, and also in the presence of polyhedral graph-convexity as will be seen in 9.35 and 9.47, but it leaves out a host of other potential applications.

Again the example of a ray rotating in $\mathbb{R}^2$ at a constant rate around the origin is instructive (see Figure 5–7 and the surrounding discussion). In interpreting the ray as a set $S(t)$ varying with time $t \in \mathbb{R}$, we get a mapping that doesn't even behave continuously with respect to Pompeiu-Hausdorff distance, much less in the manner laid out by Definition 9.26. Yet this mapping is continuous in the sense of set convergence as adopted in Chapter 5, and it's clearly a prime candidate for suggesting what kind of generalization ought to be made of Lipschitz continuity for the sake of wider relevance.

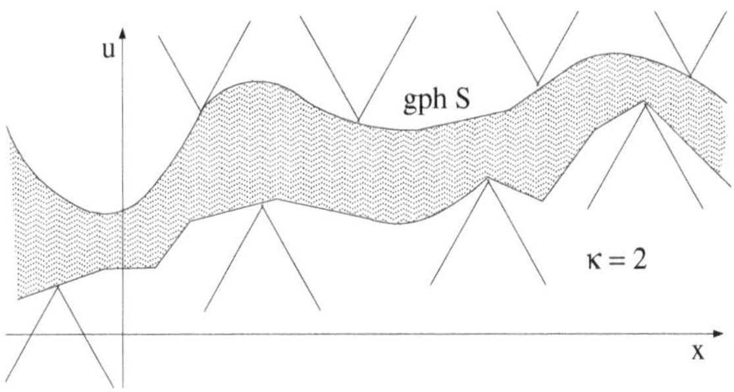

Fig. 9–5. A Lipschitz continuous set-valued mapping that is locally bounded.

To formulate such a property, appropriate in the analysis of a much larger class of mappings $S : \mathbb{R}^n \rightrightarrows \mathbb{R}^m$ than those covered by Definition 9.26, we can utilize the quantification of set convergence that was developed in Chapter 4 in terms of the distances $d\!l_\rho$ and $\hat{d\!l}_\rho$ in 4(11). We can apply these expressions to image sets $S(x)$ and $S(x')$ in $\mathbb{R}^m$, obtaining

$$\begin{aligned} &d\!l_\rho\big(S(x'), S(x)\big) = \max_{|u|\le\rho} \Big| d\big(u, S(x')\big) - d\big(u, S(x)\big)\Big|, \\ &\hat{d\!l}_\rho\big(S(x'), S(x)\big) = \\ &\qquad \inf\Big\{\eta \ge 0 \;\Big|\; S(x') \cap \rho\mathbb{B} \subset S(x) + \eta\mathbb{B},\;\; S(x) \cap \rho\mathbb{B} \subset S(x') + \eta\mathbb{B}\Big\}, \end{aligned} \qquad 9(8)$$

(where 'inf' can be replaced by 'min' when S is closed-valued).

Both $d\!l_\rho$ and $\hat{d\!l}_\rho$ are of interest, because, as explained in Chapter 4, $d\!l_\rho$ is a pseudo-metric, whereas $\hat{d\!l}_\rho$, although not a pseudo-metric, is typically easier to understand geometrically and to estimate. The double inequality

$$\hat{d\!l}_\rho\big(S(x'),S(x)\big) \le d\!l_\rho\big(S(x'),S(x)\big) \le \hat{d\!l}_{\rho'}\big(S(x'),S(x)\big) \\ \text{when } \ \rho' \ge 2\rho + \max\big\{d(0,S(x')),d(0,S(x))\big\}, \tag{9(9)}$$

which translates 4.37(a) to this context (with 2ρ replaceable by ρ when $S(x)$ and $S(x')$ are convex), ties the two expressions closely to each other. They relate to Pompeiu-Hausdorff distance $d\!l_\infty\big(S(x'),S(x)\big)$ through the fact, coming from 4.37(b) and 4.38, that

$$d\!l_\rho\big(S(x'),S(x)\big) \nearrow d\!l_\infty\big(S(x'),S(x)\big) \ \text{ as } \ \rho\nearrow\infty, \ \text{ with} \\ d\!l_\rho\big(S(x'),S(x)\big) = d\!l_\infty\big(S(x'),S(x)\big) \ \text{ when } \ S(x')\cup S(x)\subset \rho I\!\!B. \tag{9(10)}$$

It's clear from this that the definition of Lipschitz continuity in 9.26 could just as well have been stated without appeal to Pompeiu-Hausdorff distance, but instead in terms of the inequalities $d\!l_\rho(S(x'),S(x)) \le \kappa|x'-x|$ holding for all $x,x'\in X$, the crucial thing being that the same κ works for all ρ. The path to the concept we need for handling mappings S with $S(x)$ possibly unbounded is found in relinquishing such uniformity by allowing κ to depend on ρ.

9.27 Definition (sub-Lipschitz continuity). *A mapping $S: I\!\!R^n \rightrightarrows I\!\!R^m$ is sub-Lipschitz continuous on a set $X\subset I\!\!R^n$ if it is nonempty-closed-valued on X and for each $\rho\in I\!\!R_+$ there exists $\kappa\in I\!\!R_+$, a ρ-Lipschitz constant, such that*

$$d\!l_\rho\big(S(x'),S(x)\big) \ \le\ \kappa|x'-x| \ \text{ for all } \ x,x'\in X,$$

and hence also

$$S(x')\cap\rho I\!\!B \ \subset\ S(x)+\kappa|x'-x|I\!\!B \ \text{ for all } \ x,x'\in X.$$

In this case the inclusion is only implied by the inequality—for a given κ. But by virtue of 9(9), the inclusion can nonetheless replace the inequality in this definition as a whole; it's just that in passing from the inclusion to the inequality a shift to a larger κ is needed.

The fact that sub-Lipschitz continuity is more readily available than Lipschitz continuity in many practical situations involving set-valued mappings will be confirmed in 9.32. In particular, the example of the rotating ray will be seen to exhibit this property; this will be covered not only by 9.32, but because of convex-valuedness, also by a special truncation criterion in 9.33.

Together with Lipschitz and sub-Lipschitz continuity it will now be helpful, as in the analysis of single-valued mappings, to work with a concept of 'strict continuity' that focuses on individual points.

9.28 Definition (strict continuity of set-valued mappings). *Consider a mapping $S: I\!\!R^n \rightrightarrows I\!\!R^m$ and a set $X\subset I\!\!R^n$.*

(a) *S is strictly continuous at $\bar{x}$ relative to X if $\bar{x}\in X$ and S is nonempty-closed-valued on some neighborhood of $\bar{x}$ relative to X, and in addition, for each $\rho\in I\!\!R_+$ the value*

$$\operatorname{lip}_{X,\rho} S(\bar{x}) := \limsup_{\substack{x,x' \xrightarrow[X]{} \bar{x} \\ x \neq x'}} \frac{d\!\!l_\rho\big(S(x'), S(x)\big)}{|x' - x|}$$

is finite. More simply, S is strictly continuous at $\bar{x}$ (without mention of X) if $\bar{x} \in \operatorname{int}(\operatorname{dom} S)$, S is closed-valued around $\bar{x}$ and for each $\rho \in I\!R_+$ the value

$$\operatorname{lip}_\rho S(\bar{x}) := \limsup_{\substack{x,x' \to \bar{x} \\ x \neq x'}} \frac{d\!\!l_\rho\big(S(x'), S(x)\big)}{|x' - x|}$$

is finite. Here $\operatorname{lip}_\rho S(\bar{x})$ is the ρ-Lipschitz modulus for S at $\bar{x}$, and $\operatorname{lip}_{X,\rho} S(\bar{x})$ is this modulus relative to X. Similarly defined with $\rho = \infty$, $\operatorname{lip}_\infty S(\bar{x})$ is the Lipschitz modulus for S at $\bar{x}$, while $\operatorname{lip}_{X,\infty} S(\bar{x})$ is its relativization to X.

(b) *S is strictly continuous relative to X if, for every point $\bar{x} \in X$, S is strictly continuous at $\bar{x}$ relative to X*

9.29 Proposition (alternative descriptions of strict continuity). *Consider $S : I\!R^n \rightrightarrows I\!R^m$ and a set $X \subset I\!R^n$ on which S is nonempty-closed-valued. Then each of the following conditions at a point $\bar{x} \in X$ is equivalent to S being strictly continuous at $\bar{x}$ relative to X:*

(a) *for each $\rho \in I\!R_+$ there exist $\kappa \in I\!R_+$ and $V \in \mathcal{N}(\bar{x})$ such that*

$$d\!\!l_\rho\big(S(x'), S(x)\big) \leq \kappa |x' - x| \text{ for all } x, x' \in X \cap V; \qquad 9(11)$$

(b) *for each $\rho \in I\!R_+$ there exist $\kappa \in I\!R_+$ and $V \in \mathcal{N}(\bar{x})$ such that*

$$\hat{d\!\!l}_\rho\big(S(x'), S(x)\big) \leq \kappa |x' - x| \text{ for all } x, x' \in X \cap V, \qquad 9(12)$$

or equivalently, $S(x') \cap \rho I\!B \subset S(x) + \kappa |x' - x| I\!B$ for all $x, x' \in X \cap V$;

(c) *for each $\rho \in I\!R_+$ there exist $\kappa \in I\!R_+$ and $V \in \mathcal{N}(\bar{x})$ such that all the functions $x \mapsto d\big(u, S(x)\big)$ for $u \in \rho I\!B$ are Lipschitz continuous on $X \cap V$ with constant κ.*

Proof. The equivalence with (a) is immediate from the definition of $\operatorname{lip}_{X,\rho} S(\bar{x})$ in 9.28, while the equivalence with (c) comes from the $d\!\!l_\rho$ formula in 9(8). For the necessity of (b), note that 9(11) implies 9(12) by 9(9). For the sufficiency, assume that the inequality in (b) holds. Fixing any $\rho \in I\!R_+$, choose $\rho' > 2\rho + d\big(0, S(\bar{x})\big)$. By assumption there exist $\kappa \in I\!R_+$ and $V \in \mathcal{N}(\bar{x})$ such that 9(12) holds for ρ'. Looking at this property in its geometric form, we see in particular that $S(\bar{x}) \cap \rho' I\!B \subset S(x) + \kappa |x - \bar{x}| I\!B$ when $x \in X \cap V$ and consequently $S(x) \neq \emptyset$ for such x, in fact with $d\big(0, S(x)\big) \leq d\big(0, S(\bar{x})\big) + \kappa |x - \bar{x}|$. Take $\varepsilon > 0$ small enough that $d\big(0, S(\bar{x})\big) + \kappa\varepsilon < \rho' - 2\rho$ and $I\!B(\bar{x}, \varepsilon) \subset V$. Then $\rho' > 2\rho + d\big(0, S(x)\big)$ for all $x \in X \cap I\!B(\bar{x}, \varepsilon)$, so that, by 9(9) and the fact that 9(12) holds for ρ', we have 9(11) holding with $I\!B(\bar{x}, \varepsilon)$ in place of V. □

Strict continuity at a point entails continuity at that point, of course. This is clear from the characterization of set convergence in 4.36 in terms of the pseudo-metrics $d\!\!l_\rho$, or more simply from the distance function characterization of continuity in 5.11 when contrasted with the characterization of strict continuity in 9.29(c). In particular, if S is strictly continuous at $\bar{x}$, one must have $\bar{x} \in \operatorname{int}(\operatorname{dom} S)$. Continuity of S relative to X requires the functions $s_u : x \mapsto d\big(u, S(x)\big)$ for $u \in I\!R^m$ to be continuous relative to X. In comparison, strict continuity requires these functions s_u to be strictly continuous relative to X, moreover with certain uniformities in the modulus.

Just as the modulus $\operatorname{lip}_\rho S(\bar{x})$ defined in 9.28 corresponds to the limiting constant at $\bar{x}$ in the conditions 9(11), we can introduce a similar modulus with respect to the conditions in 9(12):

$$\operatorname{lip}^{\wedge}_{\rho} S(\bar{x}) := \limsup_{\substack{x,x' \to \bar{x} \\ x \neq x'}} \frac{\hat{d\!\!l}_\rho(S(x'), S(x))}{|x' - x|}. \qquad 9(13)$$

This is motivated by the fact that $\operatorname{lip}^{\wedge}_{\rho} S(\bar{x})$ is sometimes easier to estimate than $\operatorname{lip}_\rho S(\bar{x})$. Through 9(8) we deduce the useful relation that

$$\operatorname{lip}^{\wedge}_{\rho} S(\bar{x}) \leq \operatorname{lip}_\rho S(\bar{x}) \leq \operatorname{lip}^{\wedge}_{\rho'} S(\bar{x}) \text{ when } \rho' \geq 2\rho + d\big(0, S(\bar{x})\big). \qquad 9(14)$$

The corresponding modulus $\operatorname{lip}^{\wedge}_{X,\rho} S(\bar{x})$ relative to a set X is defined of course by restricting the convergence $x, x' \to \bar{x}$ to $x, x' \in X$.

Because the neighborhood V in 9.29(a) can shrink as ρ increases, strict continuity of S at a point $\bar{x}$ doesn't entail S necessarily being strictly continuous at all points nearby, much less that S is sub-Lipschitz continuous on a neighborhood of $\bar{x}$. This feature is crucial, however, to our being able later to characterize strict continuity in terms of coderivative mappings (in 9.38).

Of course, the problem with shrinking neighborhoods in the ρ-modulus limits of Definition 9.28 doesn't enter in considering the $\rho = \infty$ case by itself. From having $\operatorname{lip}_\infty S(\bar{x}) < \infty$, we do get that S is Lipschitz continuous in a neighborhood of $\bar{x}$. But local Lipschitz continuity is no longer identifiable with strict continuity, as it was for single-valued mappings. It's a narrower property with fewer applications. As part of the following theorem, however, local Lipschitz continuity does persist in being identifiable with strict continuity in the case of mappings that are *locally bounded.*

9.30 Theorem (Lipschitz versus sub-Lipschitz continuity). *If a mapping $S : I\!R^n \rightrightarrows I\!R^m$ is Lipschitz continuous on a set $X \subset I\!R^n$, then it is sub-Lipschitz continuous on X as well. The two properties are equivalent whenever S is bounded on X, i.e., the set $S(X)$ is bounded in $I\!R^m$.*

More generally, in the case where S is locally bounded on X, it is locally Lipschitz continuous on X if and only if it is locally sub-Lipschitz continuous on X. Furthermore, these properties are equivalent then to S being strictly continuous relative to X.

Indeed, if S is strictly continuous and locally bounded relative to X at a point $\bar{x}$, then S is Lipschitz continuous on a neighborhood of $\bar{x}$ relative to X.

Proof. The first part is obvious from 9(10). In the case of local boundedness, this is applied to a neighborhood of each point. When $\bar{x}$ has a neighborhood V_0 such that $S(X \cap V_0) \subset \rho I\!B$, the condition in 9.29(b) provides Lipschitz continuity of S on $X \cap V \cap V_0$. □

Without the boundedness assumption in Theorem 9.30, sub-Lipschitz continuity on X can differ from Lipschitz continuity, *even when specialized to single-valued mappings*. For an illustration of this phenomenon in the case of $X = I\!R$, consider the mapping $F : I\!R \to I\!R^2$ with $F(t) = (t, \sin t^2)$. As seen through 9.7, F is strictly continuous on $I\!R$ and therefore, by 9.2, Lipschitz continuous on any bounded interval of $I\!R$. But it's not Lipschitz continuous globally on $I\!R$: we have $F'(t) = (1, 2t\cos t^2)$, so that $\operatorname{lip} F(t) = (1 + 4t^2\cos^2 t^2)^{1/2}$, but this expression isn't bounded above as t ranges over $I\!R$. On the other hand, F is sub-Lipschitz continuous globally on $I\!R$. To see this, let $u = (u_1, u_2)$ and $s_u(t) = |u - F(t)| = [(u_1 - t)^2 + (u_2 - \sin t^2)^2]^{1/2}$, and observe on the basis of 9.7 that when $|u| \le \rho < |t|$ we have

$$\begin{aligned}\operatorname{lip} s_u(t) &= \frac{|\langle u - F(t), F'(t)\rangle|}{|u - F(t)|} \le \frac{|(u_1 - t) + 2t\cos t^2(u_2 - \sin t^2)|}{|u_1 - t|}\\ &\le 1 + \frac{2|\cos t^2||u_2 - \sin t^2|}{1 - \rho/|t|} \le 1 + 4(\rho + 1) \text{ when } |t| \ge 2\rho.\end{aligned}$$

This implies the existence for each $\rho \in I\!R_+$ of a constant $\kappa \in I\!R_+$ such that $\operatorname{lip} s_u(t) \le \kappa$ for all $t \in I\!R$ when $u \in \rho I\!B$, in which case s_u is Lipschitz continuous on $I\!R$ with this constant. Then $d\!l_\rho\big(F(t'), F(t)\big) \le \kappa|t' - t|$ for all $t, t' \in I\!R$.

9.31 Theorem (sub-Lipschitz continuity via strict continuity). *If $S : I\!R^n \rightrightarrows I\!R^m$ is strictly continuous on an open set $O \subset I\!R^n$, then for each $\rho \in I\!R_+$ the function $\operatorname{lip}_\rho S$ is finite and usc on O with $\operatorname{lip}_\rho S \le \operatorname{lip}_\infty S$.*

If $X \subset O$ is a convex set on which each of the functions $\operatorname{lip}_\rho S$ for $\rho \in I\!R_+$ is bounded from above, this being true in particular when X is also compact, then S is sub-Lipschitz continuous on X. Indeed, the value

$$\kappa_\rho = \sup_{x \in X}\{\operatorname{lip}_\rho S(x)\}$$

serves as a ρ-Lipschitz constant for S on X.

If these constants κ_ρ are all bounded from above by some κ, then S is Lipschitz continuous on X with constant κ. In particular, that holds when the function $\operatorname{lip}_\infty S$ is bounded from above on X, in which case one can take

$$\kappa = \sup_{x \in X}\{\operatorname{lip}_\infty S(x)\}.$$

Proof. The initial claims about the functions $\operatorname{lip}_\rho S$ are elementary consequences of Definition 9.28 and the first relation in 9(10). Note also that a finite osc function on a nonempty compact set attains its maximum and thus is bounded from above (cf. 1.9).

Through the $d\!\!l_\rho$ formula in 9(8), $\operatorname{lip}_\rho S(\bar{x})$ is the infimum of all κ serving on some neighborhood of $\bar{x}$ as a Lipschitz constant for all the functions $s_u : x \mapsto d\big(u, S(x)\big)$ with $u \in \rho I\!\!B$. These functions thus have $\operatorname{lip} s_u(\bar{x}) \le \operatorname{lip}_\rho S(\bar{x})$ at every point $\bar{x} \in X$, so that $\operatorname{lip} s_u \le \kappa_\rho$ on X. Applying 9.2, we obtain for every $u \in \rho I\!\!B$ that κ_ρ serves as a constant for s_u relative to all of X. Invoking 9(8) once more, we get $d\!\!l_\rho\big(S(x'), S(x)\big) \le \kappa_\rho |x' - x|$ for all $x, x' \in X$.

The remainder, about the case of Lipschitz continuity, is evident from 9(10) again and the fact that $\operatorname{lip}_\rho S(x) \le \operatorname{lip}_\infty S(x)$. □

The rotating ray exhibits sub-Lipschitz continuity, although not Lipschitz continuity, as already mentioned. This fits more broadly with the following example, in which a major class of sub-Lipschitz continuous mappings is identified. The example, based on the theorem just proved, also provides an illustration of how the constants involved in this property may be estimated in a particular situation.

9.32 Example (transformations acting on a set). *Let $S(x) = A(x)C + a(x)$ for a nonempty, closed set $C \subset I\!\!R^m$ and nonsingular matrices $A(x) \in I\!\!R^{m\times m}$ and vectors $a(x) \in I\!\!R^m$ depending on a parameter vector $x \in O$, where O is an open subset of $I\!\!R^n$. Suppose the functions $a : x \mapsto a(x)$ and $A : x \mapsto A(x)$ are strictly continuous with respect to $x \in O$. Then the mapping $S : O \rightrightarrows I\!\!R^m$ is strictly continuous at any point $\bar{x} \in O$, with*

$$\widehat{\operatorname{lip}}_\rho S(\bar{x}) \;\le\; |A(\bar{x})^{-1}|\big(\rho + |a(\bar{x})|\big) \operatorname{lip} A(\bar{x}) + \operatorname{lip} a(\bar{x}), \qquad 9(15)$$

so S is sub-Lipschitz continuous on any compact, convex set $X \subset O$. In fact, if C is bounded with $C \subset \lambda I\!\!B$, one has

$$\operatorname{lip}_\infty S(\bar{x}) \;\le\; \lambda \operatorname{lip} A(\bar{x}) + \operatorname{lip} a(\bar{x}),$$

and in that case S is Lipschitz continuous on such a set X. But if C is not bounded and the cone $A(x)C^\infty$ fails to be the same for all $x \in X$, then it is impossible for S to be Lipschitz continuous on X.

Detail. The nonsingularity of $A(x)$ ensures that S is closed-valued. Consider any x and x' in O and any $u' \in [A(x')C + a(x')] \cap \rho I\!\!B$; we have $|u'| \le \rho$ and there exists $z \in C$ with $u' = A(x')z + a(x')$. Let $u = A(x)z + a(x)$. Then $u' - u = [A(x') - A(x)]z + [a(x') - a(x)]$, so $|u' - u| \le |A(x') - A(x)||z| + |a(x') - a(x)|$. Also, $z = A(x')^{-1}[u' - a(x')]$, so that $|z| \le |A(x')^{-1}|\big(|u'| + |a(x')|\big)$. Thus,

$$\begin{aligned}&[A(x')C + a(x')] \cap \rho I\!\!B \subset [A(x)C + a(x)] + \lambda(x, x')I\!\!B\\ &\quad \text{for } \lambda(x, x') = |A(x') - A(x)||A(x')^{-1}|\big(\rho + |a(x')|\big) + |a(x) - a(x')|,\end{aligned}$$

hence $\hat{d\!\!l}\big(S(x'), S(x)\big) \le \bar{\lambda}(x, x')|x' - x|$ for $\bar{\lambda}(x, x') := \min\big\{\lambda(x, x'), \lambda(x', x)\big\}$. By dividing both sides of this inequality by $|x' - x|$ and taking the limit as $x, x' \to \bar{x}$ with $x \ne x'$, we obtain 9(15) through 9(13). Then S is strictly continuous at $\bar{x}$ by 9.29(a), hence sub-Lipschitz continuous by 9.31 relative to any compact, convex set $X \subset O$.

When C lies in $\lambda\mathbb{B}$, there exists $\bar{\rho}$ such that $S(x)$ and $S(x')$ lie in $\bar{\rho}\mathbb{B}$ when x and x' are near enough to $\bar{x}$. The estimate of the constant goes through with the inequality $|z| \leq |A(x')^{-1}|\big(|u'| + |a(x')|\big)$ simplified to $|z| \leq \lambda$. This yields $\operatorname{lip} S(\bar{x}) \leq \lambda \operatorname{lip} A(\bar{x}) + \operatorname{lip} a(\bar{x})$.

When C isn't bounded, yet S is Lipschitz continuous on X with constant κ, the inclusion $S(x') \subset S(x) + \kappa|x' - x|\mathbb{B}$ entails $S(x')^\infty \subset S(x)^\infty$ for all $x, x' \in X$ (cf. 3.12), hence actually $S(x')^\infty = S(x)^\infty$. But $S(x)^\infty = A(x)C^\infty$ by the inclusion in 3.10 and the invertibility of $A(x)$. It's essential then that $A(x')C^\infty = A(x)C^\infty$ for all $x, x' \in X$. □

Convexity, as always, brings simplifications and additional properties. The example of the rotating ray, because it concerns a set-valued mapping, gets its sub-Lipschitz continuity also by the following criterion.

9.33 Theorem (truncations of convex-valued mappings). *For a closed-valued mapping $S : \mathbb{R}^n \rightrightarrows \mathbb{R}^m$ and a set $X \subset \operatorname{dom} S$, let $\rho_0 = \sup_{x \in X} d\big(0, S(x)\big)$ and consider for each $\rho \in \mathbb{R}_+$ the truncation $S_\rho : \mathbb{R}^n \rightrightarrows \mathbb{R}^m$ given by*

$$S_\rho(x) := S(x) \cap \rho\mathbb{B}.$$

(a) *Suppose S is convex-valued and $\rho_0 < \infty$. Take any $\bar{\rho} \in (\rho_0, \infty)$. Then S is strictly continuous relative to X if and only if, for every $\rho \in (2\bar{\rho}, \infty)$, the bounded mapping S_ρ is locally Lipschitz continuous relative to X.*

(b) *Suppose S is graph-convex and X is compact with $X \subset \operatorname{int}(\operatorname{dom} S)$. Then $\rho_0 < \infty$, and there exists $\lambda \in \mathbb{R}_+$ such that, for $\rho \geq \rho_0 + 1$, one has*

$$S_\rho(x') \subset S_\rho(x) + \lambda\rho|x' - x|\mathbb{B} \text{ for all } x \in X \text{ and } x' \in \mathbb{R}^n, \tag{9(16)}$$

and therefore in particular $\hat{d\!l}_\rho\big(S(x'), S(x)\big) \leq \lambda\rho|x' - x|$ for all $x, x' \in X$. Moreover, one can take $\lambda = 2/\inf\big\{d(x, X) \,\big|\, d(0, S(x)) \geq \rho_0 + 1\big\}$.

Proof. The equivalence in (a) is immediate from the bounds in 4.39. The verification of (b) begins with the recollection from 5.9(b) that graph-convexity implies S is isc on $\operatorname{int}(\operatorname{dom} S)$ and therefore by 5.11(b) that $d\big(0, S(x)\big)$ is usc on $\operatorname{int}(\operatorname{dom} S)$. This guarantees that the supremum defining ρ_0 is finite and the value $\bar{\varepsilon} := \inf\big\{d(x, X) \,\big|\, d(0, S(x)) \geq \rho_0 + 1\big\}$ is positive. Take any $\varepsilon \in (0, \bar{\varepsilon})$; for $x \in X + \varepsilon\mathbb{B}$ we have $S_\rho(x) \neq \emptyset$ as long as $\rho \geq \rho_0 + 1$. Fix $\rho \in [\rho_0 + 1, \infty)$, $x \in X$ and $x' \neq x$. Let $e = (x' - x)/|x' - x|$ and $x'' = x - \varepsilon e$. Then $x'' \in \operatorname{dom} S$ and $x = (1 - \tau)x' + \tau x''$ for $\tau = |x' - x|/(\varepsilon + |x' - x|)$. The graph-convexity of S carries over to S_ρ and gives the inequality $S_\rho(x) \supset (1 - \tau)S_\rho(x') + \tau S_\rho(x'')$; cf. 5(4). Hence

$$S_\rho(x) + \tau S_\rho(x') \supset (1 - \tau)S_\rho(x') + \tau S_\rho(x') + \tau S_\rho(x'') = S_\rho(x') + \tau S_\rho(x''),$$

where we use the special distributive law in 2.23(c) for scalar multiplication of convex sets. But $S_\rho(x') \subset S_\rho(x'') + 2\rho\mathbb{B}$. Therefore

$$S_\rho(x) + \tau S_\rho(x'') + 2\tau\rho\mathbb{B} \supset S_\rho(x') + \tau S_\rho(x'').$$

Now we can invoke the cancellation law in 3.35 to drop $\tau S_\rho(x'')$ from both sides. This yields $S_\rho(x') \subset S_\rho(x) + 2\tau\rho\mathbb{B}$. Setting $\lambda = 2/\varepsilon$ we get $2\tau \le \lambda|x'-x|$ and the desired inclusion, except with ε in place of $\bar\varepsilon$. The result follows then for $\bar\varepsilon$ as well, due to the arbitrary choice of $\varepsilon \in (0,\bar\varepsilon)$. □

9.34 Corollary (strict continuity from graph-convexity). *If $S : \mathbb{R}^n \rightrightarrows \mathbb{R}^m$ is closed-valued and graph-convex, it is strictly continuous at any $\bar x \in \operatorname{int}(\operatorname{dom} S)$. Indeed, for $\rho > d\big(0, S(\bar x)\big)+1$ one has $\operatorname{lip}^{\wedge}_{\rho} S(\bar x) \le 2\rho/\bar\eta$, where $\bar\eta$ is the positive distance from $\bar x$ to $\big\{x \,\big|\, d(0,S(x)) \ge d(0,S(\bar x))+1\big\}$.*

Proof. Apply Theorem 9.33(b) to arbitrarily small neighborhoods X of $\bar x$. □

A still more special case, significant in treating linear constraint systems, is worth recording here as well.

9.35 Example (polyhedral graph-convex mappings). *If $S : \mathbb{R}^n \rightrightarrows \mathbb{R}^m$ is not just graph-convex but such that $\operatorname{gph} S$ is polyhedral in $\mathbb{R}^n \times \mathbb{R}^m$, then S is Lipschitz continuous on $\operatorname{dom} S$, even if its values $S(x)$ are unbounded sets.*

Detail. The polyhedral property means that $\operatorname{gph} S$ can be described by a finite system of linear inequalities. For some dimension d there exist $A \in \mathbb{R}^{d\times n}$, $B \in \mathbb{R}^{d\times m}$, and $c \in \mathbb{R}^d$ such that

$$u \in S(x) \quad\Longleftrightarrow\quad Ax + Bu + c \in \mathbb{R}^d_+.$$

Let K and R be the kernel and range, respectively, of the linear transformation $L : \mathbb{R}^m \to \mathbb{R}^d$ corresponding to the matrix B, and let $M : \mathbb{R}^d \to \mathbb{R}^m$ be the pseudo-inverse of L: in terms of the projection $P_R(w)$ of a point $w \in \mathbb{R}^d$ on the subspace $R \subset \mathbb{R}^d$, $M(w)$ is the point of the affine set $L^{-1}\big(P_R(w)\big)$ that is nearest to the origin of $\mathbb{R}^m$. The mapping M is linear, hence representable by a matrix D; we have $L^{-1}(w) = Dw + K$ when $w \in R$, whereas $L^{-1}(w) = \emptyset$ when $w \notin R$. In this notation,

$$S(x) = D\big[(C - Ax)\cap R\big] + K \;\text{ for all }\; x, \quad\text{where }\; C = \mathbb{R}^d_+ - c.$$

Since $(C - Ax') \subset (C - Ax) + |A||x'-x|\mathbb{B}$, we obtain for $x, x' \in \operatorname{dom} S$ that $S(x') \subset S(x) + \kappa|x'-x|\mathbb{B}$ for $\kappa = |A||D|$. Thus, S is Lipschitz continuous on the set $\operatorname{dom} S$. □

F. Aubin Property and Mordukhovich Criterion

A broader framework for Lipschitzian estimates will rise from the concepts we develop next. These allow the essence of strict continuity to be localized from points in the domain of a mapping to points in its graph.

9.36 Definition (Aubin property and graphical modulus). *A mapping $S : \mathbb{R}^n \rightrightarrows \mathbb{R}^m$ has the Aubin property relative to X at $\bar x$ for $\bar u$, where $\bar x \in X$*

and $\bar{u} \in S(\bar{x})$, if $\operatorname{gph} S$ *is locally closed at $(\bar{x}, \bar{u})$ and there are neighborhoods $V \in \mathcal{N}(\bar{x})$, $W \in \mathcal{N}(\bar{u})$, and a constant $\kappa \in I\!R_+$ such that*

$$S(x') \cap W \subset S(x) + \kappa|x' - x|I\!B \text{ for all } x, x' \in X \cap V. \tag{9(17)}$$

This condition with V in place of $X \cap V$ is simply the Aubin property at $\bar{x}$ for $\bar{u}$. The graphical modulus of S at $\bar{x}$ for $\bar{u}$ is then

$$\operatorname{lip} S(\bar{x} \mid \bar{u}) := \inf\{\kappa \mid \exists V \in \mathcal{N}(\bar{x}),\ W \in \mathcal{N}(\bar{u}), \text{ such that } S(x') \cap W \subset S(x) + \kappa|x' - x|I\!B \text{ for all } x, x' \in V\}.$$

The difference between strict continuity and the Aubin property is that an arbitrarily large ball $\rho I\!B$ is replaced by a sufficiently small neighborhood W of a particular vector $\bar{u} \in S(\bar{x})$, say $W = I\!B(\bar{u}, \rho)$ for ρ sufficiently small.

The Aubin property is stable in the sense that, if it holds at $\bar{x}$ for $\bar{u}$, then for all $(x, u) \in \operatorname{gph} S$ near enough to $(\bar{x}, \bar{u})$, it holds at x for u. Moreover the function $(x, u) \mapsto \operatorname{lip} S(x \mid u)$ on $\operatorname{gph} S$ is usc relative to such a neighborhood.

9.37 Exercise (distance characterization of the Aubin property). *Let $\bar{u} \in S(\bar{x})$ with $\bar{x} \in X \subset I\!R^n$, and suppose* $\operatorname{gph} S$ *is locally closed at $(\bar{x}, \bar{u})$. Then the following are equivalent:*

(a) *S has the Aubin property relative to X at $\bar{x}$ for $\bar{u}$;*

(b) *$(x, u) \mapsto d(u, S(x))$ is strictly continuous relative to $X \times I\!R^m$ at $(\bar{x}, \bar{u})$;*

(c) *there exist $\kappa \in I\!R_+$, $W \in \mathcal{N}(\bar{u})$ and $V \in \mathcal{N}(\bar{x})$ such that the functions $x \mapsto d(u, S(x))$ for $u \in W$ are Lipschitz continuous on $X \cap V$ with constant κ;*

(d) *there exist $\kappa \in I\!R_+$, $\rho \in I\!R_+$ and $V \in \mathcal{N}(\bar{x})$ such that*

$$d\!l_\rho\big(S(x') - \bar{u},\ S(x) - \bar{u}\big) \le \kappa|x' - x| \text{ for all } x, x' \in X \cap V;$$

(e) *there exist $\kappa \in I\!R_+$, $\rho \in I\!R_+$ and $V \in \mathcal{N}(\bar{x})$ such that*

$$\hat{d\!l}_\rho\big(S(x') - \bar{u},\ S(x) - \bar{u}\big) \le \kappa|x' - x| \text{ for all } x, x' \in X \cap V.$$

Moreover, in the case of $\bar{x} \in \operatorname{int} X$, the graphical modulus $\operatorname{lip} S(\bar{x} \mid \bar{u})$ is the infimum of all κ fitting the pattern in (c). Indeed,

$$\begin{aligned}
\operatorname{lip} S(\bar{x} \mid \bar{u}) &= \limsup_{(x,u) \to (\bar{x}, \bar{u})} \operatorname{lip} s_u(x) \quad \text{for } s_u(x) = d(u, S(x)) \\
&= \limsup_{\substack{x, x' \to \bar{x},\ x \neq x' \\ \rho \searrow 0}} \frac{d\!l_\rho\big(S(x') - \bar{u},\ S(x) - \bar{u}\big)}{|x' - x|} \\
&= \limsup_{\substack{x, x' \to \bar{x},\ x \neq x' \\ \rho \searrow 0}} \frac{\hat{d\!l}_\rho\big(S(x') - \bar{u},\ S(x) - \bar{u}\big)}{|x' - x|}.
\end{aligned}$$

Guide. In taking W to be of the form $I\!B(\bar{u}, \rho)$, one sees that (e) is merely a restatement of (a), while (d) is a restatement of (c). To establish the equivalence between (a) and (c), reduce to $\bar{u} = 0$. Work with the inequalities in 9(9) as ρ and ρ' tend to 0. In passing from (c) to (b), utilize the Lipschitz continuity of distance functions in 9.6. □

9.38 Theorem (graphical localization of strict continuity). *A closed-valued mapping $S : \mathbb{R}^n \rightrightarrows \mathbb{R}^m$ is strictly continuous at $\bar{x}$ relative to X if and only if it is osc relative to X at $\bar{x}$ and, for every $\bar{u} \in S(\bar{x})$, has the Aubin property relative to X at $\bar{x}$ for $\bar{u}$. Indeed, when $\bar{x} \in \operatorname{int} X$ and $\rho > d\big(0, S(\bar{x})\big)$, one has*

$$\operatorname{lip}_\rho S(\bar{x}) \geq \sup_{\substack{\bar{u} \in S(\bar{x}) \\ |\bar{u}| < \rho}} \big\{ \operatorname{lip} S(\bar{x} \,|\, \bar{u}) \big\}, \qquad \widehat{\operatorname{lip}}_\rho S(\bar{x}) \leq \max_{\substack{\bar{u} \in S(\bar{x}) \\ |\bar{u}| \leq \rho}} \big\{ \operatorname{lip} S(\bar{x} \,|\, \bar{u}) \big\}.$$

Proof. Suppose first that strict continuity holds. Since this entails continuity, we in particular have S osc relative to X at $\bar{x}$. Consider $\bar{u} \in S(\bar{x})$ and $\rho > |\bar{u}|$. Appealing to the equivalences in 9.29, let $V \in \mathcal{N}(\bar{x})$ be a corresponding neighborhood such that the condition in 9.29(b) is fulfilled for some $\kappa \in \mathbb{R}_+$; there's no loss of generality in taking V to be closed and such that S is closed-valued on V. Then the neighborhood $V \times \rho\mathbb{B}$ of $(\bar{x}, \bar{u})$ has closed intersection with $\operatorname{gph} S$. Thus, $\operatorname{gph} S$ is locally closed at $\bar{x}$. We have 9(17) for $W = \rho\mathbb{B}$, so the Aubin property holds relative to X at $\bar{x}$ for $\bar{u}$. When $\bar{x}$ is interior to X, we can conclude further that $\operatorname{lip} S(\bar{x} \,|\, \bar{u}) \leq \kappa$ and establish thereby that $\operatorname{lip} S(\bar{x} \,|\, \bar{u}) \leq \operatorname{lip}_\rho S(\bar{x})$. This yields the estimate claimed for $\operatorname{lip}_\rho S(\bar{x})$.

Suppose now instead that S is osc relative to X at $\bar{x}$ and has the Aubin property there for every $\bar{u} \in S(\bar{x})$. For each $\bar{u} \in S(\bar{x})$ we have the existence of $W_{\bar{u}} \in \mathcal{N}(\bar{u})$, $V_{\bar{u}} \in \mathcal{N}(\bar{x})$, and $\kappa_{\bar{u}} \in \mathbb{R}_+$ such that $V_{\bar{u}} \times W_{\bar{u}}$ has closed intersection with $\operatorname{gph} S$ and

$$S(x') \cap W_{\bar{u}} \subset S(x) + \kappa_{\bar{u}} |x' - x| \mathbb{B} \text{ for all } x, x' \in X \cap V_{\bar{u}}. \tag{9(18)}$$

Let $\rho > d\big(0, S(\bar{x})\big)$. The set $S(\bar{x}) \cap \rho\mathbb{B}$ is nonempty and compact, so we can cover it by finitely many of the open sets $\operatorname{int} W_{\bar{u}}$, corresponding say to points $\bar{u}_k \in S(\bar{x}) \cap \rho\mathbb{B}$ for $k = 1, \ldots, r$. Let $V = \bigcap_{k=1}^r V_{\bar{u}_k}$, $W = \bigcup_{k=1}^r W_{\bar{u}_k}$ and $\kappa = \max_{k=1}^r \kappa_{\bar{u}_k}$. Then $S(\bar{x}) \cap \rho\mathbb{B} \subset \operatorname{int} W$, and $V \times W$ has closed intersection with $\operatorname{gph} S$. Further, from 9(18) we have 9(17) for these elements.

There can't exist $x^\nu \to \bar{x}$ in $X \cap V$ with elements $u^\nu \in [S(x^\nu) \cap \rho\mathbb{B}] \setminus \operatorname{int} W$, for the sequence $\{u^\nu\}_{\nu \in \mathbb{N}}$ would have a cluster point $\bar{u} \in \rho\mathbb{B} \setminus \operatorname{int} W$, and then $\bar{u} \in [S(\bar{x}) \cap \rho\mathbb{B}] \setminus \operatorname{int} W$ because S is osc relative to X at $\bar{x}$, a contradiction. Hence, by shrinking V somewhat if necessary, we can arrange that $S(x) \cap \rho\mathbb{B} \subset W$ when $x \in V$. Then $[V \times W] \cap \operatorname{gph} S$ is closed. Also, the inclusion in 9(17) persists for the smaller V and with W replaced by $\rho\mathbb{B}$. Hence S is strictly continuous relative to X at $\bar{x}$, indeed with $\widehat{\operatorname{lip}}_\rho S(\bar{x}) \leq \kappa$.

When $\bar{x} \in \operatorname{int} X$, we can refine the argument by introducing $\varepsilon > 0$ and taking $\kappa_{\bar{u}} = \operatorname{lip} S(\bar{x} \,|\, \bar{u}) + \varepsilon$. Then the κ we get is $\max_{k=1}^r \operatorname{lip} S(\bar{x} \,|\, \bar{u}_k) + \varepsilon$, so that $\kappa \leq \hat{\kappa} + \varepsilon$ with $\hat{\kappa}$ the maximum of $\operatorname{lip} S(\bar{x} \,|\, \bar{u})$ over all $\bar{u} \in S(\bar{x}) \cap \rho\mathbb{B}$ (inasmuch as each $\bar{u}_k$ lies in this set). We obtain $\widehat{\operatorname{lip}}_\rho S(\bar{x}) \leq \hat{\kappa} + \varepsilon$, and through the arbitrary choice of ε we conclude that $\widehat{\operatorname{lip}}_\rho S(\bar{x}) \leq \hat{\kappa}$. □

The Aubin property has been depicted as a Lipschitzian property localized in the range space as well as the domain space, but the localization in range actually allows for a certain relaxation of the localization in domain that will play a crucial role in some later developments.

9.39 Lemma (extended formulation of the Aubin property). *For a mapping $S : \mathbb{R}^n \rightrightarrows \mathbb{R}^m$, a pair $(\bar{x}, \bar{u}) \in \operatorname{gph} S$ where $\operatorname{gph} S$ is locally closed, and a set $X \subset \mathbb{R}^n$ containing $\bar{x}$, the following conditions on $\kappa \in \mathbb{R}_+$ are equivalent:*

(a) *there exist $V \in \mathcal{N}(\bar{x})$ and $W \in \mathcal{N}(\bar{u})$ such that*

$$S(x') \cap W \;\subset\; S(x) + \kappa|x' - x|\mathbb{B} \;\text{ for all }\; x, x' \in X \cap V.$$

(b) *there exist $V \in \mathcal{N}(\bar{x})$ and $W \in \mathcal{N}(\bar{u})$ such that*

$$S(x') \cap W \;\subset\; S(x) + \kappa|x' - x|\mathbb{B} \;\text{ for all }\; x \in X \cap V,\;\; x' \in X.$$

Thus, the inclusion in (b) can be substituted for the inclusion in (a) in the definition of S having the Aubin property relative to X at $\bar{x}$ for $\bar{u}$. Moreover, in the case where X is suppressed (with $X \cap V$ simplified to V), the infimum of the constants κ with respect to which the extended inclusion in (b) holds, for various V and W, still gives the modulus $\operatorname{lip} S(\bar{x}\,|\,\bar{u})$.

Proof. Trivially (b) implies (a), so assume that (a) holds for neighborhoods $V = \mathbb{B}(\bar{x}, \delta)$ and $W = \mathbb{B}(\bar{u}, \varepsilon)$. We'll verify that (b) holds for $V' = \mathbb{B}(\bar{x}, \delta')$ and $W' = \mathbb{B}(\bar{u}, \varepsilon')$ when $0 < \delta' < \delta$, $0 < \varepsilon' < \varepsilon$, and $2\kappa\delta' + \varepsilon' \leq \kappa\delta$. Fix any $x \in X \cap \mathbb{B}(\bar{x}, \delta')$. Our assumption gives us

$$S(x') \cap \mathbb{B}(\bar{u}, \varepsilon') \;\subset\; S(x) + \kappa|x' - x|\mathbb{B} \;\text{ when }\; x' \in X \cap \mathbb{B}(\bar{x}, \delta),$$

and our goal is to demonstrate that this holds also when $x' \in X \setminus \mathbb{B}(\bar{x}, \delta)$. From applying (a) to $x' = \bar{x}$, we see that $\bar{u} \in S(x) + \kappa|x - \bar{x}|\mathbb{B}$ and consequently $\mathbb{B}(\bar{u}, \varepsilon') \subset S(x) + (\kappa\delta' + \varepsilon')\mathbb{B}$, where $\kappa\delta' + \varepsilon' \leq \kappa\delta - \kappa\delta'$. But $|x' - x| > \delta - \delta'$ when $x' \in X \setminus \mathbb{B}(\bar{x}, \delta)$, so for such points x' we have $\kappa\delta - \kappa\delta' \leq \kappa|x' - x|$, hence $S(x') \cap \mathbb{B}(\bar{u}, \varepsilon') \subset S(x) + \kappa|x' - x|\mathbb{B}$, as required. □

The next theorem provides not only a handy test for the Aubin property but also a means of estimating the graphical modulus $\operatorname{lip} S(\bar{x}\,|\,\bar{u})$ through knowledge of the coderivative mapping $D^*S(\bar{x}\,|\,\bar{u})$. It will be possible on the platform of the calculus of operations in Chapter 10 to gain such knowledge from the way that a given mapping S is put together from simpler mappings. Once estimates are at hand for the graphical modulus, estimates can also be obtained for other values like $\operatorname{lip}_\rho S(\bar{x})$, as seen in 9.38.

9.40 Theorem (Mordukhovich criterion). *Consider $S : \mathbb{R}^n \rightrightarrows \mathbb{R}^m$, $\bar{x} \in \operatorname{dom} S$ and $\bar{u} \in S(\bar{x})$. Suppose that $\operatorname{gph} S$ is locally closed at $(\bar{x}, \bar{u})$. Then S has the Aubin property at $\bar{x}$ for $\bar{u}$ if and only if*

$$D^*S(\bar{x}\,|\,\bar{u})(0) = \{0\},$$

or equivalently $|D^*S(\bar{x}\,|\,\bar{u})|^+ < \infty$, *and in that case* $\operatorname{lip} S(\bar{x}\,|\,\bar{u}) = |D^*S(\bar{x}\,|\,\bar{u})|^+$.

This condition holds if $\operatorname{dom} \widehat{D}S(\bar{x}\,|\,\bar{u}) = I\!R^n$, *and it is equivalent to that when* $\widehat{D}S(\bar{x}\,|\,\bar{u}) = DS(\bar{x}\,|\,\bar{u})$, *i.e., when* S *is graphically regular at* $\bar{x}$ *for* $\bar{u}$.

Proof. First we take care of the relationship with the regular graphical derivative mapping $\widehat{D}S(\bar{x}\,|\,\bar{u})$, whose graph is the regular tangent cone $\widehat{T}_{\operatorname{gph} S}(\bar{x},\bar{u})$. This cone is convex (cf. 6.26), as is its projection on $I\!R^n$, which is the cone $D := \operatorname{dom} \widehat{D}S(\bar{x}\,|\,\bar{u})$. We have $D = I\!R^n$ if and only if $D^* = \{0\}$ (through the polarity facts in 6.21 and the relative interior properties in 2.40). A vector v belongs to D^* if and only if $\langle v, w\rangle \leq 0$ for all $w \in D$, or equivalently, $\langle (v,0),(w,z)\rangle \leq 0$ for all $(w,z) \in \widehat{T}_{\operatorname{gph} S}(\bar{x},\bar{u})$, which is the same as $(v,0) \in \widehat{T}_{\operatorname{gph} S}(\bar{x},\bar{u})^*$. But $\widehat{T}_{\operatorname{gph} S}(\bar{x},\bar{u})^* = N_{\operatorname{gph} S}(\bar{x},\bar{u})^{**} = \operatorname{cl} \operatorname{con} N_{\operatorname{gph} S}(\bar{x},\bar{u})$ by 6.28(b) and 6.21. Thus, $\operatorname{dom} \widehat{D}S(\bar{x}\,|\,\bar{u}) = I\!R^n$ if and only if

$$(v,0) \in \operatorname{cl} \operatorname{con} N_{\operatorname{gph} S}(\bar{x},\bar{u}) \quad\Longrightarrow\quad v = 0.$$

In contrast, we have by definition that $D^*S(\bar{x}\,|\,\bar{u})(0) = \{0\}$ if and only if

$$(v,0) \in N_{\operatorname{gph} S}(\bar{x},\bar{u}) \quad\Longrightarrow\quad v = 0.$$

The claim in the theorem about $\widehat{D}S(\bar{x}\,|\,\bar{u})$ is justified therefore through the fact that $\operatorname{cl} \operatorname{con} N_{\operatorname{gph} S}(\bar{x},\bar{u}) \supset N_{\operatorname{gph} S}(\bar{x},\bar{u})$, with equality when $\operatorname{gph} S$ is Clarke regular at $(\bar{x},\bar{u})$ also characterized by having $\widehat{D}S(\bar{x}\,|\,\bar{u}) = DS(\bar{x}\,|\,\bar{u})$; cf. 8.40.

Moving on now to the proof of the main result, let's observe that the condition $D^*S(\bar{x}\,|\,\bar{u})(0) = \{0\}$, in relating only to the normal cone $N_{\operatorname{gph} S}(\bar{x},\bar{u})$, depends only on the nature of $\operatorname{gph} S$ in an arbitrarily small neighborhood of $(\bar{x},\bar{u})$. The same is true of the Aubin property, but an argument needs to be made. Without loss of generality, the inclusion 9(17) in the definition of this property can be focused on neighborhoods of type $V = I\!B(\bar{x},\delta)$ and $W = I\!B(\bar{u},\varepsilon)$, thus taking the form

$$S(x') \cap I\!B(\bar{u},\varepsilon) \;\subset\; S(x) + \kappa|x'-x|I\!B \;\text{ for all }\; x,x' \in I\!B(\bar{x},\delta). \qquad 9(19)$$

Then it never involves more on the right side than $\big[S(x) + 2\kappa\delta I\!B\big] \cap I\!B(\bar{u},\varepsilon)$ and thus stands or falls according to the behavior of $\operatorname{gph} S$ in the neighborhood $I\!B(\bar{x},\delta) \times I\!B(\bar{u},\varepsilon + 2\kappa\delta)$ of $(\bar{x},\bar{u})$.

There's no harm, therefore, in assuming from now on that $\operatorname{gph} S$ is closed in its entirety. Then in particular, $S(x)$ is a closed set for all x. We'll argue in this context that the coderivative condition is equivalent to the distance version of the Aubin property in 9.37(c), using this also to verify the equality between

$$\kappa_1 := \operatorname{lip} S(\bar{x}\,|\,\bar{u}), \qquad \kappa_2 := |D^*S(\bar{x}\,|\,\bar{u})|^+.$$

Necessity. Suppose the Aubin property holds, and view it and the value κ_1 in the manner of 9.37. The property $D^*S(\bar{x}\,|\,\bar{u}) = \{0\}$ to be confirmed is identical to having $\kappa_2 < \infty$, because κ_2 is the infimum of all $\kappa \in I\!R_+$ such that $|v| \leq \kappa|y|$ whenever $v \in D^*S(\bar{x}\,|\,\bar{u})(y)$, or equivalently from 8.33, $(v,-y) \in$

$N_{\mathrm{gph}\,S}(\bar{x},\bar{u})$. But $(v,-y) \in N_{\mathrm{gph}\,S}(\bar{x},\bar{u})$ if and only if there exist $(x^\nu,u^\nu) \to (\bar{x},\bar{u})$ in $\mathrm{gph}\,S$ and $(v^\nu,-y^\nu) \to (v,-y)$ with $(v^\nu,-y^\nu) \in \widehat{N}_{\mathrm{gph}\,S}(x^\nu,u^\nu)$. In place of such regular normals $(v^\nu,-y^\nu)$ we can restrict our attention to proximal normals by 6.18(a), inasmuch as $\mathrm{gph}\,S$ is a closed set. Therefore, κ_2 is also the infimum of all $\kappa \in \mathbb{R}_+$ such that there exist $\delta > 0$ and $\varepsilon > 0$ for which

$$\left.\begin{array}{c} |v| \leq \kappa|y| \;\text{ whenever }\; (v,-y) \;\text{ is a proximal normal vector to} \\ \mathrm{gph}\,S \;\text{ at a point }\; (\tilde{x},\tilde{u}) \;\text{ with }\; |\tilde{x}-\bar{x}| < \delta \;\text{ and }\; |\tilde{u}-\bar{u}| < \varepsilon. \end{array}\right\} \qquad 9(20)$$

We therefore consider any κ, δ, and ε satisfying 9(19) and aim at showing they satisfy 9(20). That will prove necessity along with the inequality $\kappa_1 \geq \kappa_2$.

Fix any $(\tilde{x},\tilde{u}) \in \mathrm{gph}\,S$ with $|\tilde{x}-\bar{x}| < \delta$ and $|\tilde{u}-\bar{u}| < \varepsilon$. To say that $(v,-y)$ is a proximal normal vector to $\mathrm{gph}\,S$ at $(\tilde{x},\tilde{u})$ is to assert the existence of $\tau > 0$ such that $(\tilde{x},\tilde{u})$ belongs to the projection $P_{\mathrm{gph}\,S}\big((\tilde{x},\tilde{u}) + \tau(v,-y)\big)$; cf. 6.16. Without affecting this projection property, we can reduce the size of τ so as to arrange that $\tilde{u} - \tau y \in \mathrm{int}\,\mathbb{B}(\bar{u},\varepsilon)$; we'll denote $\tilde{u} - \tau y$ by $\hat{u}$; then $|\hat{u}-\bar{u}| < \varepsilon$. Then $|(x,u) - (\tilde{x}+\tau v,\, \hat{u})| \geq |(\tau v, -\tau y)|$ for all $(x,u) \in \mathrm{gph}\,S$, i.e.,

$$|x - (\tilde{x}+\tau v)|^2 + |u-\hat{u}|^2 \;\geq\; \tau^2|v|^2 + \tau^2|y|^2 \;\text{ whenever }\; u \in S(x),$$

an inequality which turns into an equation when $u = \tilde{u}$ and $x = \tilde{x}$. We get

$$\begin{aligned} \big|x - (\tilde{x}+\tau v)\big|^2 + d\big(\hat{u}, S(x)\big)^2 \;&\geq\; \tau^2|v|^2 + \tau^2|y|^2 \;\text{ for all }\; x, \\ \big|\tilde{x} - (\tilde{x}+\tau v)\big|^2 + d\big(\hat{u}, S(\tilde{x})\big)^2 \;&=\; \tau^2|v|^2 + \tau^2|y|^2, \end{aligned}$$

and in this way we obtain

$$\begin{aligned} \tau^2|v|^2 - |x - (\tilde{x}+\tau v)|^2 \;&\leq\; d\big(\hat{u},S(x)\big)^2 - d\big(\hat{u},S(\tilde{x})\big)^2 \\ &=\; \big[d\big(\hat{u},S(x)\big) - d\big(\hat{u},S(\tilde{x})\big)\big]\big[d\big(\hat{u},S(x)\big) + d\big(\hat{u},S(\tilde{x})\big)\big]. \end{aligned} \qquad 9(21)$$

Due to having $|\tilde{x}-\bar{x}| < \delta$ and $|\hat{u}-\bar{u}| < \varepsilon$, we know from assuming 9(19) that

$$d\big(\hat{u},S(x)\big) - d\big(\hat{u},S(\tilde{x})\big) \;\leq\; \kappa|x-\tilde{x}| \;\text{ as long as }\; x \in \mathbb{B}(\bar{x},\delta).$$

On the other hand the Lipschitz continuity of $d\big(u,S(\tilde{x})\big)$ in u with constant 1 (in 9.6) gives $d\big(\hat{u},S(\tilde{x})\big) \leq d\big(\tilde{u},S(\tilde{x})\big) + \tau|y|$, where $d\big(\tilde{u},S(\tilde{x})\big) = 0$. Similarly we have $d\big(\hat{u},S(x)\big) \leq d\big(\tilde{u},S(x)\big) + \tau|y|$ with $d\big(\tilde{u},S(x)\big) \leq d\big(\tilde{u},S(\tilde{x})\big) + \kappa|x-\tilde{x}|$ as long as $x \in \mathbb{B}(\bar{x},\delta)$. Altogether then from 9(21), we get

$$2\tau\langle v,\, x-\tilde{x}\rangle - |x-\tilde{x}|^2 \;\leq\; \kappa|x-\tilde{x}|\big(\kappa|x-\tilde{x}| + 2\tau|y|\big) \;\text{ when }\; x \in \mathbb{B}(\bar{x},\delta).$$

Applying this to $x = \tilde{x} + \lambda v$ for $\lambda > 0$ small enough to ensure that this point lies in $\mathbb{B}(\bar{x},\delta)$ (which is possible because $|\tilde{x}-\bar{x}| < \delta$), we obtain

$$2\tau\lambda|v|^2 - \lambda^2|v|^2 \;\leq\; \kappa\lambda|v|\big(\kappa\lambda|v| + 2\tau|y|\big).$$

Unless $|v| = 0$, this implies $(2\tau-\lambda)|v| \leq \kappa\big(\kappa\lambda|v| + 2\tau|y|\big)$, and then in the limit as $\lambda \searrow 0$ that $2\tau|v| \leq 2\tau\kappa|y|$. Hence certainly $|v| \leq \kappa|y|$, which is the property in 9(20) that we needed.

Sufficiency. The description of κ_2 at the beginning of our proof of necessity can be given another twist. On the basis of 6.6 and the definition of the normal cone $N_{\mathrm{gph}\, S}(\bar{x}, \bar{u})$, we have $(v, -y)$ in this cone if and only if there exist $(x^\nu, u^\nu) \to (\bar{x}, \bar{u})$ in $\mathrm{gph}\, S$ and $(v^\nu, -y^\nu) \to (v, -y)$ with $(v^\nu, -y^\nu) \in N_{\mathrm{gph}\, S}(x^\nu, u^\nu)$. Therefore, κ_2 is also the infimum of all $\kappa \in I\!R_+$ such that there exist $\delta_0 > 0$ and $\varepsilon_0 > 0$ for which

$$|v| \leq \kappa|y| \;\text{ when }\; (v, -y) \in N_{\mathrm{gph}\, S}(\hat{x}, \hat{u}),\; \hat{x} \in I\!B(\bar{x}, \delta_0),\; \hat{u} \in I\!B(\bar{u}, \varepsilon_0). \qquad 9(22)$$

From the assumption that this holds, we'll demonstrate the existence of $\delta > 0$ and $\varepsilon > 0$ for which 9(19) holds. This will establish sufficiency and the complementary inequality $\kappa_2 \geq \kappa_1$.

Our context having been reduced to that of an osc mapping S, we know that the function $(x, u) \mapsto d(u, S(x))$ is lsc on $I\!R^n \times I\!R^m$ (cf. 5.11 and 9.6). Temporarily fixing any $\tilde{u}$, let's see what can be said about regular subgradients of the lsc function $d_{\tilde{u}}(x) := d(\tilde{u}, S(x))$, since those can be used to generate the other subgradients of $d_{\tilde{u}}$ and ascertain strict continuity of $d_{\tilde{u}}$ on the basis of Theorem 9.13.

Suppose $v \in \widehat{\partial} d_{\tilde{u}}(\hat{x})$, where $\hat{x}$ is a point with $d_{\tilde{u}}(\hat{x})$ finite, so that $S(\hat{x}) \neq \emptyset$. By the variational description of regular subgradients in 8.5, there exists on some open neighborhood O of $\hat{x}$ a smooth function $h \leq d_{\tilde{u}}$ with $h(\hat{x}) = d_{\tilde{u}}(\hat{x})$ and $\nabla h(\hat{x}) = v$. Then $\hat{x} \in \mathrm{argmin}_O (d_{\tilde{u}} - h)$. Let $\hat{u}$ be a point of $S(\hat{x})$ nearest to $\tilde{u}$; we have $|\hat{u} - \tilde{u}| = d_{\tilde{u}}(\hat{x})$. Then $(\hat{x}, \hat{u})$ gives the minimum of $-h(x) + |u - \tilde{u}|$ over all $(x, u) \in \mathrm{gph}\, S$ with $x \in O$.

We can rephrase this as follows. Taking V to be any closed neighborhood of $\bar{x}$ in O, define the lsc, proper function f_0 on $I\!R^n \times I\!R^m$ by $f_0(x, u) := -h(x) + |u - \tilde{u}|$ when $x \in V$ but $f_0(x, u) := \infty$ when $x \notin V$. Then f_0 achieves its minimum relative to the closed set $\mathrm{gph}\, S$ at $(\hat{x}, \hat{u})$. The optimality condition in Theorem 8.15 holds sway here: we have

$$\partial f_0(\hat{x}, \hat{u}) \subset \big\{(-v, y) \,\big|\, y \in I\!B\big\}, \qquad \partial^\infty f_0(\hat{x}, \hat{u}) = \big\{(0, 0)\big\}$$

(cf. 8.8, 8.27), where the expression for $\partial^\infty f_0(\hat{x}, \hat{u})$ confirms that the constraint qualification in 8.15 is fulfilled. The optimality condition takes the form

$$\partial f_0(\hat{x}, \hat{u}) + N_{\mathrm{gph}\, S}(\hat{x}, \hat{u}) \;\ni\; (0, 0).$$

We conclude from it, through the relation derived for $\partial f_0(\hat{x}, \hat{u})$, that our vector v must be such that there exists $y \in I\!B$ with $(v, -y) \in N_{\mathrm{gph}\, S}(\hat{x}, \hat{u})$.

Here is where we might be able to apply the property stemming from our assumption of the coderivative condition: if we knew at this stage that $\hat{x} \in I\!B(\bar{x}, \delta_0)$ and $\hat{u} \in I\!B(\bar{u}, \varepsilon_0)$, we would be sure from 9(22) that $|v| \leq \kappa|y| \leq \kappa$. We do, of course, have at least the estimate $|\hat{u} - \bar{u}| \leq |\hat{u} - \tilde{u}| + |\tilde{u} - \bar{u}| = d_{\tilde{u}}(\hat{x}) + |\tilde{u} - \bar{u}|$. Thus, to summarize what we have learned so far,

$$\begin{cases} \text{if } v \in \widehat{\partial} d_{\tilde{u}}(\hat{x}) \text{ with } |\hat{x} - \bar{x}| \le \delta_0 \\ \text{and } d_{\tilde{u}}(\hat{x}) + |\tilde{u} - \bar{u}| \le \varepsilon_0, \text{ then } |v| \le \kappa. \end{cases} \tag{9(23)}$$

We'll use this stepping stone now to reach our goal. We start with the case of $\tilde{u} = \bar{u}$, having $d_{\bar{u}}(\bar{x}) = 0$. By Definition 8.3, $\partial^\infty d_{\bar{u}}(\bar{x})$ is the cone giving the direction points in hzn $\mathbb{R}^n$ achievable as limits of unbounded sequences of subgradients $v^\nu \in \widehat{\partial} d_{\bar{u}}(x^\nu)$ associated with points $x^\nu \to \bar{x}$ such that $d_{\bar{u}}(x^\nu) \to d_{\bar{u}}(\bar{x})$. But in taking $\hat{x} = x^\nu$ and $v = v^\nu$ in 9(22), as well as $\tilde{u} = \bar{u}$, we see that for large enough ν in such a sequence the inequality $|v^\nu| \le \kappa$ has to hold. Therefore $\partial^\infty d_{\bar{u}}(\bar{x}) = \{0\}$. It follows from 9.13 that $d_{\bar{u}}$ is strictly continuous at $\bar{x}$, hence on some neighborhood of $\bar{x}$ and finite there, implying $S(x)$ is nonempty for x in such a neighborhood. At the same time the function $u \mapsto d\big(u, S(x)\big)$ is known from 9.6 to be Lipschitz continuous with constant 1 when $S(x) \neq \emptyset$. Hence the function $(x, u) \mapsto d\big(u, S(x)\big)$ is finite around $(\bar{x}, \bar{u})$, a point at which it vanishes and is calm.

The size of the constant doesn't yet matter; all we need for the moment is the existence of some $\lambda \in \mathbb{R}_+$ and $\mu \in \mathbb{R}_+$ such that

$$d\big(u, S(x)\big) \le \lambda\big(|x - \bar{x}| + |u - \bar{u}|\big) \text{ when } |x - \bar{x}| \le \mu \text{ and } |u - \bar{u}| \le \mu.$$

Furnished with this, we can choose $\delta > 0$ and $\varepsilon > 0$ small enough that $2\delta \le \min\{\mu, \delta_0\}$, $\varepsilon \le \min\{\mu, \varepsilon_0/2\}$, and $\lambda(2\delta + \varepsilon) \le \varepsilon_0/2$. Then in applying 9(23) to any $\tilde{u} \in \mathbb{B}(\bar{u}, \varepsilon)$ with $\hat{x} \in \mathbb{B}(\bar{x}, 2\delta)$ we obtain $\widehat{\partial} d_{\tilde{u}}(\hat{x}) \subset \kappa\mathbb{B}$. From that and the definition of $\partial d_{\tilde{u}}(x)$ we get $\partial d_{\tilde{u}}(x) \subset \kappa\mathbb{B}$ for all $x \in \operatorname{int} \mathbb{B}(\bar{x}, 2\delta)$. Then $\operatorname{lip} d_{\tilde{u}}(x) \le \kappa$ for all $x \in \operatorname{int} \mathbb{B}(\bar{x}, 2\delta)$ by Theorem 9.13, which implies through Theorem 9.2 that $d_{\tilde{u}}$ is Lipschitz continuous on $\operatorname{int} \mathbb{B}(\bar{x}, 2\delta)$ with constant κ. In particular, then, $d_{\tilde{u}}$ is Lipschitz continuous on $\mathbb{B}(\bar{x}, \delta)$ with constant κ. Thus, 9(19) holds, as demanded. □

The Mordukhovich criterion in Theorem 9.40 for the Aubin property to hold deserves special notice as a bridge to many other topics and results. It's equivalent to the local boundedness of the coderivative mapping $D^*S(\bar{x} \,|\, \bar{u})$ (see 9.23). From the calculus of coderivatives in Chapter 10, formulas will be available for checking whether indeed 0 is the only element of $D^*S(\bar{x} \,|\, \bar{u})(0)$, and if so, obtaining estimates of the outer norm $|D^*S(\bar{x} \,|\, \bar{u})|^+$, which in the case of graphical regularity, by the way, will be seen in 11.29 to agree with $|DS(\bar{x} \,|\, \bar{u})|^-$. This will provide a strong handle on Lipschitzian properties in general.

The Mordukhovich criterion opens a remarkable perspective on the very meaning of normal vectors and subgradients, teaching us that local aspects of Lipschitzian behavior of general set-valued mappings, as captured in the Aubin property, lie at the core of variational analysis.

9.41 Theorem (reinterpretation of normal vectors and subgradients).

(a) *For a set $C \subset \mathbb{R}^n$ and a point $\bar{x} \in C$ where C is locally closed, a vector $\bar{v}$ belongs to the normal cone $N_C(\bar{x})$ if and only if the mapping*

$$\alpha \mapsto \big\{x \in C \,\big|\, \langle \bar{v}, x - \bar{x}\rangle \ge \alpha\big\}$$

fails to have the Aubin property at $\alpha = 0$ *for* $x = \bar{x}$.

(b) *For a function* $f : \mathbb{R}^n \to \overline{\mathbb{R}}$ *and a point* $\bar{x}$ *where* f *is finite and locally lsc, a vector* $\bar{v}$ *belongs to the subgradient set* $\partial f(\bar{x})$ *if and only if the mapping*

$$\alpha \mapsto \{x \mid f(x) - f(\bar{x}) - \langle \bar{v}, x - \bar{x} \rangle \leq \alpha\}$$

fails to have the Aubin property at $\alpha = 0$ *for* $\bar{x}$.

Thus in particular, the condition $0 \in \partial f(\bar{x})$ *holds if and only if the level-set mapping* $\alpha \mapsto \operatorname{lev}_{\leq\alpha} f$ *fails to have the Aubin property at* $f(\bar{x})$ *for* $\bar{x}$.

Proof. Assertion (a) is merely the special case of (b) where $f = \delta_C$, inasmuch as $\partial\delta_C(\bar{x}) = N_C(\bar{x})$ (cf. 8.14), so it's enough to look at (b). Let $\bar{f}(x) = f(x) - \langle \bar{v}, x - \bar{x} \rangle$. The question is whether the condition $0 \in \partial\bar{f}(\bar{x})$ is equivalent to the mapping $S : \alpha \mapsto \operatorname{lev}_{\leq\alpha} \bar{f}$ not having the Aubin property at $\bar{\alpha} = \bar{f}(\bar{x})$ for $\bar{x}$. By the Mordukhovich criterion in 9.40, the Aubin property fails if and only if the set $D^*S(\bar{\alpha} \mid \bar{x})(0)$ contains more than just 0, which by definition of this coderivative mapping in 8.33 means that the cone $N_{\operatorname{gph} S}(\bar{\alpha}, \bar{x})$ in $\mathbb{R} \times \mathbb{R}^n$ contains a vector $(\eta, 0)$ with $\eta \neq 0$. But S is the inverse of the profile mapping $E_{\bar{f}} : \mathbb{R}^n \to \mathbb{R}$, which has $\operatorname{epi} \bar{f}$ as its graph (cf. 5.5), so this is the same as saying that the cone $N_{\operatorname{epi} \bar{f}}(\bar{x}, \bar{f}(\bar{x}))$ contains a vector $(0, \eta)$ with $\eta \neq 0$. That's equivalent by the variational geometry of $\operatorname{epi} \bar{f}$ in 8.9 to having $0 \in \partial\bar{f}(\bar{x})$. □

The normal vector characterization in 9.41(a) is illustrated in Figure 9–6. When C is locally closed at $\bar{x}$, we can tell whether a vector $\bar{v} \neq 0$ belongs to $N_C(\bar{x})$ by looking at the intersection of C with the family of parallel half-spaces $H_\alpha := \{x \mid \langle \bar{v}, x - \bar{x} \rangle \geq \alpha\}$. These half-spaces all have $-\bar{v}$ as (outward) normal, and H_0 is the one among them whose boundary hyperplane passes through $\bar{x}$. We inspect the behavior of $C \cap H_\alpha$ around $\bar{x}$ as α varies around 0. A lack of Lipschitzian behavior, in the sense of the Aubin property being absent, is the telltale sign that $\bar{v}$ is normal to C at $\bar{x}$. (It's important to keep in mind that the Aubin property requires nonemptiness of the intersections $C \cap H_\alpha$ for α near 0, whether above or below.)

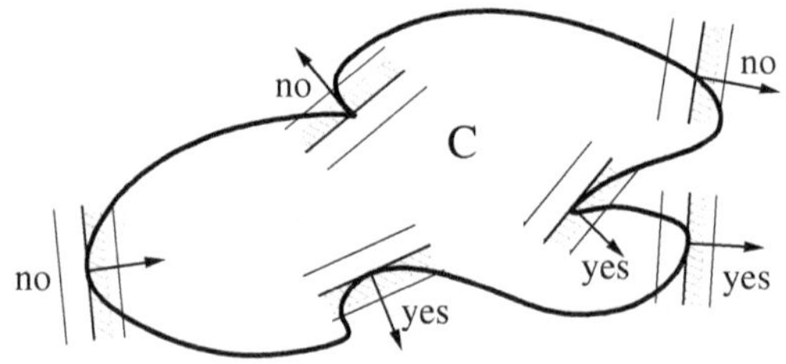

Fig. 9–6. Normality as reflecting a lack of Lipschitzian behavior of local sections.

The subgradient characterization in 9.41(b) corresponds to essentially the same picture, but for an epigraph in the setting of 8.9 and Figure 8–4.

Theorem 9.41 has far-ranging implications, beyond what might at once be apparent. The concepts of normal vector and subgradient that it interprets are central not only to the theory of generalized differentiation, as laid out in this book, but also even to classical differentiation by way of the subgradient characterization of strict differentiability in 9.18. The recognition that these concepts can be identified with the continuity properties of certain mappings brings with it the realization that continuity and differentiability, as basic themes of analysis, are intertwined more profoundly than has previously been revealed. It's interesting too that this insight requires an acceptance of set-valued mappings as basic mathematical objects; it wouldn't be available in the confines of a kind of analysis that looks only to single-valuedness.

Lipschitzian properties of set boundaries can also be studied by the methods of variational analysis. A set $C \subset \mathbb{R}^n$ is called *epi-Lipschitzian* at $\bar{x}$, one of its boundary points, if 'locally around $\bar{x}$ it can be viewed from some angle as the epigraph of a Lipschitz continuous function' as in Figure 9–7; this refers to the existence of a neighborhood $V \in \mathcal{N}(\bar{x})$, a vector e of unit length and, for the orthogonal hyperplane $H : \{x \mid \langle e, x - \bar{x}\rangle = 0\}$ through $\bar{x}$, a Lipschitz continuous function $g : H \cap V \to \mathbb{R}$ such that C agrees in some neighborhood of $\bar{x}$ with the set $\{x + \alpha e \mid x \in H,\ g(x) \leq \alpha < \infty\}$.

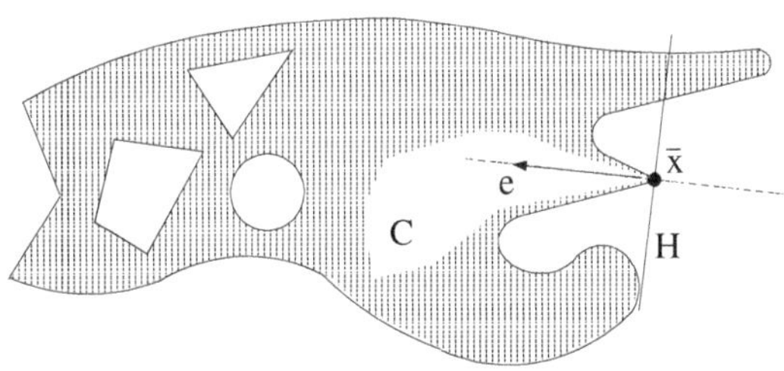

Fig. 9–7. A set that is epi-Lipschitzian at a certain boundary point.

A companion concept is that a function $f : \mathbb{R}^n \to \overline{\mathbb{R}}$ is *directionally Lipschitzian* at $\bar{x}$, a point where it is finite, if the set $\operatorname{epi} f$ is epi-Lipschitzian at $(\bar{x}, f(\bar{x}))$; cf. Figure 9–8. Note that this doesn't require f to be Lipschitz continuous around $\bar{x}$, or for that matter even continuous at $\bar{x}$.

9.42 Exercise (epi-Lipschitzian sets and directionally Lipschitzian functions).

(a) *A set $C \subset \mathbb{R}^n$ with boundary point $\bar{x}$ is epi-Lipschitzian at $\bar{x}$ if and only if C is locally closed at $\bar{x}$ and the normal cone $N_C(\bar{x})$ is pointed, the latter being equivalent to having $\operatorname{int} \widehat{T}_C(\bar{x}) \neq \emptyset$.*

(b) *A function $f : \mathbb{R}^n \to \overline{\mathbb{R}}$, finite at $\bar{x}$, is directionally Lipschitzian at $\bar{x}$ if and only if it is locally lsc at $\bar{x}$ and the cone $\partial^\infty f(\bar{x})$ is pointed, the latter being equivalent to having $\operatorname{int}(\operatorname{dom} \widehat{d}f(\bar{x})) \neq \emptyset$, or in other words the existence of a vector $w \neq 0$ such that*

$$\limsup_{\substack{x \underset{f}{\to} \bar{x} \\ x'-x \underset{\mathrm{dir}\, w}{\longrightarrow} 0}} \frac{f(x')-f(x)}{|x'-x|} < \infty.$$

(c) *A function $f : \mathbb{R}^n \to \overline{\mathbb{R}}$ that is finite and locally lsc at $\bar{x}$ is directionally Lipschitzian at $\bar{x}$ in particular whenever there is a convex cone $K \subset \mathbb{R}^n$ with* $\operatorname{int} K \neq \emptyset$ *such that f is nonincreasing relative to K.*

Guide. Important for (a) are the recession properties in 6.36 and, with respect to the desired function g, the epigraphical geometry in 8.9 along with 9.13. Get (b) by applying (a) to $C = \operatorname{epi} f$ and bringing in the recession properties in 8.50, or for that matter 8.22. Derive (c) as a corollary of (b); cf. 8.51. □

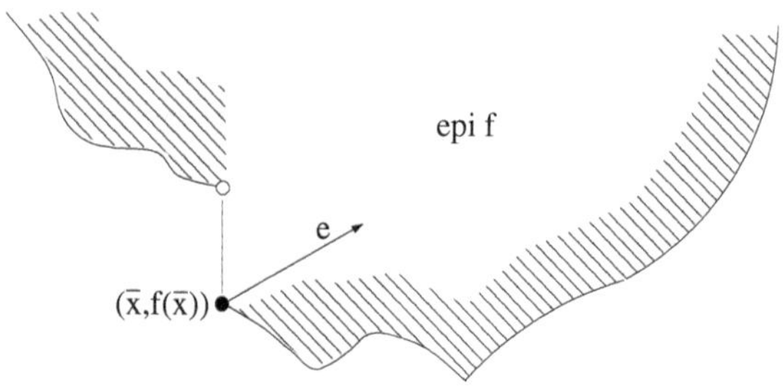

Fig. 9–8. A directionally Lipschitzian function that is not Lipschitz continuous.

G. Metric Regularity and Openness

Lipschitzian properties of S^{-1} come out as important properties of S itself. They furnish useful estimates of a different kind. In the following theorem, the property in (b) is the *metric regularity of S at $\bar{x}$ for $\bar{u}$ with constant κ*, whereas the property in (c) is *openness at $\bar{x}$ for $\bar{u}$ with linear rate κ*. Metric regularity is related to estimates of how far a given point x may be from solving a generalized equation $S(x) \ni \bar{u}$ for a given element $\bar{u}$, i.e., estimates of the distance of x from $S^{-1}(\bar{u})$. The estimates are expressed in terms of the 'residual' in the generalized equation, namely the distance of $\bar{u}$ from $S(x)$.

9.43 Theorem (metric regularity and openness). *For a mapping $S : \mathbb{R}^n \rightrightarrows \mathbb{R}^m$ and any $\bar{x}$ and $\bar{u} \in S(\bar{x})$ such that* $\operatorname{gph} S$ *is locally closed at $(\bar{x}, \bar{u})$, the following conditions are equivalent:*

(a) (inverse Aubin property): *S^{-1} has the Aubin property at $\bar{u}$ for $\bar{x}$;*

(b) (metric regularity): *$\exists V \in \mathcal{N}(\bar{x})$, $W \in \mathcal{N}(\bar{u})$, $\kappa \in \mathbb{R}_+$, such that*

$$d\big(x, S^{-1}(u)\big) \leq \kappa d\big(u, S(x)\big) \text{ when } x \in V,\ u \in W;$$

(c) (linear openness): *$\exists V \in \mathcal{N}(\bar{x})$, $W \in \mathcal{N}(\bar{u})$, $\kappa \in \mathbb{R}_+$, such that*

$$S(x+\kappa\varepsilon I\!B) \supset \big[S(x)+\varepsilon I\!B\big] \cap W \text{ for all } x \in V, \varepsilon > 0;$$

(d) (coderivative nonsingularity): $0 \in D^*S(\bar{x}\,|\,\bar{u})(y)$ *only for* $y = 0$,

where (d) holds in particular when $\operatorname{rge}\widehat{D}S(\bar{x}\,|\,\bar{u}) = I\!R^m$, *and is equivalent to that when* $\widehat{D}S(\bar{x}\,|\,\bar{u}) = DS(\bar{x}\,|\,\bar{u})$, *i.e., when* S *is graphically regular at* $\bar{x}$ *for* $\bar{u}$.

Furthermore, the infimum of all κ *fitting the circumstances in (b) agrees with the similar infimum in (c) and has the value*

$$\begin{aligned}\operatorname{lip} S^{-1}(\bar{u}\,|\,\bar{x}) &= \big|D^*S^{-1}(\bar{u}\,|\,\bar{x})\big|^+ = \big|D^*S(\bar{u}\,|\,\bar{x})^{-1}\big|^+ \\ &= \max\Big\{|y| \,\Big|\, D^*S(\bar{x}\,|\,\bar{u})(y) \cap I\!B \neq \emptyset\Big\} = 1\Big/\min_{|y|=1} d\big(0, D^*S(\bar{x}\,|\,\bar{u})(y)\big).\end{aligned}$$

Proof. The equivalence between (a) and (d) stems from the Mordukhovich criterion 9.40 as applied to S^{-1} instead of S; this also gives $\operatorname{lip}S^{-1}(\bar{u}\,|\,\bar{x}) = |D^*S^{-1}(\bar{u}\,|\,\bar{x})|^+$. By definition, the graph of $D^*S^{-1}(\bar{u}\,|\,\bar{x})$ consists of all pairs $(-v, y)$ such that $(y, v) \in N_{\operatorname{gph} S^{-1}}(\bar{u}, \bar{x})$, or equivalently, $(v, y) \in N_{\operatorname{gph} S}(\bar{x}, \bar{u})$, while that of $D^*S(\bar{u}\,|\,\bar{x})^{-1}$ consists of all the pairs $(v, -y)$ such that $(v, y) \in N_{\operatorname{gph} S}(\bar{x}, \bar{u})$. Thus, $y \in D^*S^{-1}(\bar{u}\,|\,\bar{x})(-v)$ if and only if $v \in D^*S(\bar{x}\,|\,\bar{u})^{-1}(-y)$. From this we obtain $|D^*S^{-1}(\bar{u}\,|\,\bar{x})|^+ = |D^*S(\bar{u}\,|\,\bar{x})^{-1}|^+$. But the latter value equals $\max\big\{|y| \,\big|\, y \in D^*S(\bar{u}\,|\,\bar{x})^{-1}(I\!B)\big\}$, cf. 9(4). (The supremum is attained because the mapping $H = D^*S(\bar{u}\,|\,\bar{x})^{-1}$ is osc and locally bounded (see 9.23), so that $H(I\!B)$ is compact; see 5.25(a).) This maximum agrees with the one in the theorem, since $y \in H(I\!B)$ if and only if $H^{-1}(y) \cap I\!B \neq \emptyset$. This establishes the alternative formulas claimed for $\operatorname{lip}S^{-1}(\bar{u}\,|\,\bar{x})$.

As the key to the rest of the proof of Theorem 9.43, we apply Lemma 9.39 to S^{-1} to interpret the property in (a) as referring to the existence of $V \in \mathcal{N}(\bar{x})$, $W \in \mathcal{N}(\bar{u})$ and $\kappa \in I\!R_+$ such that

$$S^{-1}(u') \cap V \subset S^{-1}(u) + \kappa|u'-u|I\!B \text{ for all } u \in W,\ u' \in I\!R^m. \tag{9(24)}$$

Clearly $x' \in S^{-1}(u') \cap V$ if and only if $u' \in S(x')$ with $x' \in V$. On the other hand, $x' \in S^{-1}(u) + \kappa|u'-u|I\!B$ if and only if there exists $x \in S^{-1}(u)$ with $|x'-x| \leq \kappa|u'-u|$. The latter is the same as having $d\big(x', S^{-1}(u)\big) \leq \kappa|u'-u|$ when x' has a nearest point in $S^{-1}(u)$. Thus, under that proviso, which is sure to hold when the neighborhoods V and W containing x' and u are sufficiently small, we can equally well think of 9(24) as asserting that

$$x' \in V,\ u \in W,\ u' \in S(x') \implies d\big(x', S^{-1}(u)\big) \leq \kappa|u'-u|.$$

We can identify this with (b) by minimizing with respect to $u' \in S(x')$. Next, observe that we can also recast 9(24) as the implication

$$x' \in V,\ u \in W,\ u' \in S(x'),\ \varepsilon \geq |u'-u| \implies u \in S(x'+\kappa\varepsilon I\!B).$$

This means that when $x' \in V$ and $\varepsilon > 0$ the set $S(x'+\kappa\varepsilon I\!B)$ contains all $u \in W$ belonging to $S(x') + \varepsilon I\!B$, which is the condition in (c). □

It will later be seen from 11.29 that in the case of graphical regularity the constant in Theorem 9.43 also equals $|DS(\bar{x}\,|\,\bar{u})^{-1}|^{-}$.

9.44 Example (metric regularity in constraint systems). *Let $S(x) = F(x) - D$ for smooth $F : \mathbb{R}^n \to \mathbb{R}^m$ and closed $D \subset \mathbb{R}^m$; then $S^{-1}(u)$ consists of all x satisfying the constraint system $F(x) - u \in D$, with u as parameter.*

Metric regularity of S for $\bar{u} = 0$ at a point $\bar{x} \in S^{-1}(\bar{u})$ means the existence of $\kappa \geq 0$ furnishing for some $\varepsilon > 0$ and $\delta > 0$ the estimate

$$d\big(x, S^{-1}(u)\big) \leq \kappa\, d\big(F(x) - u,\, D\big) \text{ when } |x - \bar{x}| \leq \varepsilon,\ |u| \leq \delta,$$

where $d(F(x) - u, D)$ measures the extent to which x might fail to satisfy the system. Taking $u = 0$ in particular, one gets

$$d(x, C) \leq \kappa d(F(x), D) \text{ for } C = F^{-1}(D) \text{ when } |x - \bar{x}| \leq \varepsilon.$$

A condition both necessary and sufficient for this metric regularity to hold is the constraint qualification

$$y \in N_D\big(F(\bar{x})\big),\ \nabla F(\bar{x})^* y = 0 \implies y = 0$$

(which is equivalent to having $T_D(F(\bar{x})) + \nabla F(\bar{x})\mathbb{R}^n = \mathbb{R}^m$, provided that D is Clarke regular at $\bar{x}$, as when D is convex). The infimum of the constants κ for which the metric regularity property holds is equal to

$$\max_{\substack{y \in N_D(F(\bar{x})) \\ |y|=1}} \frac{1}{|\nabla F(\bar{x})^* y|}.$$

Detail. The set $\operatorname{gph} S$ is specified by $F_0(x, u) \in D$ with $F_0(x, u) = F(x) - u$. The mapping F_0 is smooth from $\mathbb{R}^n \times \mathbb{R}^m$ to $\mathbb{R}^m$, and its Jacobian $\nabla F_0(\bar{x}, 0) = \big[\nabla F(\bar{x}), -I\big]$ has full rank m. Applying the rule in 6.7, we see that

$$N_{\operatorname{gph} S}(\bar{x}, 0) = \big\{(v, -y)\,\big|\, y \in N_D(F(\bar{x})),\ v = \nabla F(\bar{x})^* y\big\}.$$

Thus, $D^*S(\bar{x}\,|\,0)$ is the mapping $y \mapsto \nabla F(\bar{x})^* y$ as restricted to $y \in N_D(F(\bar{x}))$ (empty-valued elsewhere). Theorem 9.43 then justifies the claims. The tangent cone alternative to the constraint qualification comes from 6.39. □

When $D = \{0\}$ in this example, the constraint system $F(x) - u \in D$ reduces to the equation $F(x) = u$ in which x is unknown and the vector u is a parameter; the constraint qualification then is the nonsingularity of $\nabla F(\bar{x})$. When D is a box, e.g., $\mathbb{R}^m_+$, one is dealing with an inequality system instead.

9.45 Example (openness from graph-convexity). *Let $S : \mathbb{R}^n \rightrightarrows \mathbb{R}^m$ be graph-convex and osc, and let $\bar{u} \in S(\bar{x})$. In this case the properties 9.43(a)–(d) hold for $\bar{x}$ and $\bar{u}$ if and only if $\bar{u} \in \operatorname{int}(\operatorname{rge} S)$, or equivalently $\operatorname{rge} DS(\bar{x}\,|\,\bar{u}) = \mathbb{R}^m$. In particular, one then has the existence of $\kappa \in \mathbb{R}_+$ such that*

$$S(\bar{x} + \kappa\varepsilon\mathbb{B}) \supset \bar{u} + \varepsilon\mathbb{B} \text{ for all } \varepsilon > 0 \text{ sufficiently small.}$$

Detail. Since $\operatorname{gph} S$ is a closed, convex set under these assumptions, S is graphically regular at $\bar{x}$ for $\bar{u}$. We know from Theorem 9.43 that in this case the properties (a)–(d) are equivalent to $\operatorname{rge} DS(\bar{x}\,|\,\bar{u}) = I\!R^m$. It's elementary that the Aubin property of S^{-1} in 9.43(a) requires $\bar{u} \in \operatorname{int} \operatorname{dom} S^{-1} = \operatorname{int} \operatorname{rge} S$. On the other hand, the latter implies that property through 9.34 as applied to S^{-1}. The 'in particular' assertion just specializes 9.43(c) to $x = \bar{x}$. □

A more powerful, 'global' consequence of graph-convexity will be uncovered below in Theorem 9.48.

The equivalences developed in the proof of Theorem 9.43 also lead to 'global' results related to metric regularity and openness.

9.46 Proposition (inverse properties). *If $S : I\!R^n \rightrightarrows I\!R^m$ is osc and $U \subset \operatorname{rge} S$, the following conditions are equivalent (with $S(x) \cap U = S(x)$ if $U = \operatorname{rge} S$):*

(a) *S^{-1} is Lipschitz continuous on U with constant κ;*

(b) *$d\big(x, S^{-1}(u)\big) \leq \kappa d\big(u, S(x) \cap U\big)$ for all $u \in U$ and $x \in I\!R^n$;*

(c) *$S(x + \kappa r I\!B) \supset \big[[S(x) \cap U] + r I\!B\big] \cap U$ for all $x \in I\!R^n$, $r > 0$.*

More generally, one has the equivalence of the following conditions (which reduce to the preceding conditions when $S^{-1}(U)$ is bounded):

(a′) *S^{-1} is sub-Lipschitz continuous on U;*

(b′) *for every $\rho \in I\!R_+$ there exists $\kappa \in I\!R_+$ such that*

$$d\big(x, S^{-1}(u)\big) \leq \kappa d\big(u, S(x) \cap U\big) \;\; \text{when } u \in U \text{ and } |x| \leq \rho;$$

(c′) *for every $\rho \in I\!R_+$ there exists $\kappa \in I\!R_+$ such that*

$$S(x + \kappa r I\!B) \supset \big[[S(x) \cap U] + r I\!B\big] \cap U \;\; \text{when } r > 0 \text{ and } |x| \leq \rho.$$

Proof. This rests on recognizing first the equivalence of the following three conditions on V, U, and κ:

$$S^{-1}(u') \cap V \subset S^{-1}(u) + \kappa|u' - u| I\!B \;\text{ for all } u, u' \in U; \tag{9(25a)}$$

$$d\big(x, S^{-1}(u)\big) \leq \kappa d\big(u, S(x) \cap U\big) \;\text{ for all } x \in V,\ u \in U; \tag{9(25b)}$$

$$S(x + \kappa r I\!B) \supset \big[[S(x) \cap U] + r I\!B\big] \cap U \;\text{ for all } x \in V,\ r > 0. \tag{9(25c)}$$

The equivalence can be established in the same way that conditions 9.43(b) and 9.43(c) were identified with 9(24) in the proof of Theorem 9.43. The argument requires S^{-1} to be closed-valued, and that is certainly the case when S is osc (and therefore has closed graph), as here.

To pass to the equivalence of the present (a), (b), (c), take $V = I\!R^n$. For (a′), (b′), (c′), take $V = \rho I\!B$ instead. □

9.47 Example (global metric regularity from polyhedral graph-convexity). *Let $S : I\!R^n \rightrightarrows I\!R^m$ be such that $\operatorname{gph} S$ is not just convex but polyhedral. Then there exists $\kappa \in I\!R_+$ such that*

$$d\big(x, S^{-1}(u)\big) \leq \kappa d\big(u, S(x)\big) \;\; \text{for all } x \text{ when } u \in \operatorname{rge} S,$$

and then also $S(x+\kappa r\mathbb{B}) \supset [S(x)+r\mathbb{B}] \cap \operatorname{rge} S$ *for all* $r>0$ *and all* x.

Detail. This follows from Proposition 9.46 by way of the Lipschitz continuity of S^{-1} relative to $\operatorname{rge} S$ in 9.35. There's no need to require $x \in \operatorname{dom} S$, since points $x \notin \operatorname{dom} S$ have $S(x)=\emptyset$ and $d\big(u,S(x)\big)=\infty$. □

Note that Example 9.47 fits the constraint system framework of Example 9.44 when $S(x)=F(x)-D$ with F linear and D polyhedral.

For graph-convexity that isn't polyhedral, there's the following estimate. It's very close to what comes from putting Proposition 9.46 together with Theorem 9.33(b), but because the latter goes slightly beyond Lipschitz continuity in its formulation (with only one of the two points in the inclusion being restricted), a direct derivation from 9.33(b) is needed.

9.48 Theorem (metric regularity from graph-convexity; Robinson-Ursescu). *Let* $S:\mathbb{R}^n \rightrightarrows \mathbb{R}^m$ *be osc and graph-convex. Let* $\bar{u} \in \operatorname{int}(\operatorname{rge} S)$, $\bar{x} \in S^{-1}(\bar{u})$. *Then there exists* $\varepsilon>0$ *along with coefficients* $\alpha,\beta \in \mathbb{R}_+$ *such that*

$$d\big(x,S^{-1}(u)\big) \;\leq\; \big(\alpha|x-\bar{x}|+\beta\big)\, d\big(u,S(x)\big) \;\; \textit{when} \;\; |u-\bar{u}| \leq \varepsilon.$$

Proof. For simplicity and without loss of generality, we can assume that $\bar{u}=0$ and $\bar{x}=0$. Then S^{-1} is continuous at 0 (by its graph-convexity; cf. 9.34), and $0 \in S^{-1}(0)$. Take $\varepsilon>0$ small enough that $2\varepsilon\mathbb{B} \subset \operatorname{int}(\operatorname{rge} S)$ and every $u \in 2\varepsilon\mathbb{B}$ has $S^{-1}(u)\cap\mathbb{B} \neq \emptyset$. Let $T=S^{-1}$, and for each $\rho \geq 0$ define $T_\rho(u)=T(u)\cap\rho\mathbb{B}$. We have $d\big(0,T(u)\big) \leq 1$ for all $u \in 2\varepsilon\mathbb{B}$, and from 9.33(b) this gives us $\lambda \in \mathbb{R}_+$ such that $T_\rho(u') \subset T_\rho(u)+\lambda\rho|u'-u|\mathbb{B}$ when $u \in \varepsilon\mathbb{B}$, $\rho \geq 2$. This inclusion means that whenever $u' \in T_\rho^{-1}(x)$ there exists $x' \in T_\rho(u)$ with $|x'-x| \leq \lambda\rho|u'-u|$. Thus, $d\big(x,T_\rho(u)\big) \leq \lambda\rho d\big(u,T_\rho^{-1}(x)\big)$ when $u \in \varepsilon\mathbb{B}$, $\rho \geq 2$. We note next that $T_\rho^{-1}(x)=S(x)$ when $x \in \rho\mathbb{B}$, but otherwise $T_\rho^{-1}(x)=\emptyset$. Therefore, the property we have obtained is that

$$d\big(x,S^{-1}(u)\cap\rho\mathbb{B}\big) \;\leq\; \lambda\rho d\big(u,S(x)\big) \;\text{ when }\; u \in \varepsilon\mathbb{B},\; x \in \rho\mathbb{B},\; \rho \geq 2.$$

Here $d\big(x,S^{-1}(u)\cap\rho\mathbb{B}\big)=d\big(x,S^{-1}(u)\big)$ as long as $|x| \leq (\rho-1)/2$, inasmuch as $S^{-1}(u)\cap\mathbb{B} \neq \emptyset$. (Namely, this inequality guarantees that the distance of x to $S^{-1}(u)$ is no more than $1+(\rho-1)/2=(1+\rho)/2$, while the distance of x from the exterior of $\rho\mathbb{B}$ is at least $\rho-(\rho-1)/2=(1+\rho)/2$.) Hence

$$d\big(x,S^{-1}(u)\big) \;\leq\; \lambda\rho d\big(u,S(x)\big) \;\text{ when }\; |u| \leq \varepsilon,\; \rho \geq 2|x|+2.$$

The conclusion is reached then with $\alpha=2\lambda$ and $\beta=2\lambda$. □

H* Semiderivatives and Strict Graphical Derivatives

The Aubin property, along with Lipschitz or sub-Lipschitz continuity more generally, has interesting consequences for derivatives.

9.49 Exercise (Lipschitzian properties of derivative mappings).

(a) *If* $S : \mathbb{R}^n \rightrightarrows \mathbb{R}^m$ *has the Aubin property at* $\bar{x}$ *for* $\bar{u} \in S(\bar{x})$, *its derivative mapping* $DS(\bar{x}\,|\,\bar{u}) : \mathbb{R}^n \rightrightarrows \mathbb{R}^m$ *is nonempty-valued and Lipschitz continuous globally with constant* $\kappa = \operatorname{lip} S(\bar{x}\,|\,\bar{u})$*: one has*

$$DS(\bar{x}\,|\,\bar{u})(w') \subset DS(\bar{x}\,|\,\bar{u})(w) + \operatorname{lip} S(\bar{x}\,|\,\bar{u})\,|w' - w|\mathbb{B} \text{ for all } w, w'.$$

Furthermore $DS(\bar{x}\,|\,\bar{u})(w) = \limsup_{\tau \searrow 0} \Delta_\tau S(\bar{x}\,|\,\bar{u})(w)$ *for all* w, *with*

$$DS(\bar{x}\,|\,\bar{u})(w)^\infty = DS(\bar{x}\,|\,\bar{u})(0) = T_{S(\bar{x})}(\bar{u}).$$

(b) *If* $F : \mathbb{R}^n \to \mathbb{R}^m$ *is strictly continuous at* $\bar{x}$, *its derivative mapping* $DF(\bar{x}) : \mathbb{R}^n \rightrightarrows \mathbb{R}^m$ *is nonempty-valued, locally bounded and Lipschitz continuous globally with constant* $\kappa = \operatorname{lip} F(\bar{x})$*: one has*

$$DF(\bar{x})(w') \subset DF(\bar{x})(w) + \operatorname{lip} F(\bar{x})\,|w' - w|\mathbb{B} \text{ for all } w, w'.$$

Furthermore, $DF(\bar{x})(w) = \limsup_{\tau \searrow 0} \Delta_\tau F(\bar{x})(w)$ *for all* w.

Guide. In (a), consider $\kappa' > \operatorname{lip} S(\bar{x}\,|\,\bar{u})$ and verify the existence of neighborhoods $V \in \mathcal{N}(\bar{x})$ and $W \in \mathcal{N}(0)$ such that

$$\Delta_\tau S(\bar{x}\,|\,\bar{u})(w') \cap \tau^{-1} W \subset \Delta_\tau S(\bar{x}\,|\,\bar{u})(w) + \kappa' |w' - w|\mathbb{B}$$

when $\bar{x} + \tau w$ and $\bar{x} + \tau w$ belong to V. Get not only the Lipschitz continuity from this, but also the fact that $DS(\bar{x}\,|\,\bar{u})$ is the pointwise limsup of the mappings $\Delta_\tau S(\bar{x}\,|\,\bar{u})$ as $\tau \searrow 0$, rather than just the graphical limsup (having provision for $w' \to w$ as $\tau \searrow 0$). Applying this at $w = 0$, deduce the formula for $DS(\bar{x}\,|\,\bar{u})(0)$. Use the inclusion of Lipschitz continuity for $DS(\bar{x}\,|\,\bar{u})$ along with 3.12 to see that $DS(\bar{x}\,|\,\bar{u})(w')^\infty = DS(\bar{x}\,|\,\bar{u})(w)^\infty$ for all w, w', checking also that $DS(\bar{x}\,|\,\bar{u})(0)^\infty = DS(\bar{x}\,|\,\bar{u})(0)$. Finally, specialize (a) to obtain (b). □

The Aubin property also has significance for the semidifferentiability of mappings, which for general set-valued case was defined ahead of 8.43.

9.50 Proposition (semidifferentiability from proto-differentiability).

(a) *If* $S : \mathbb{R}^n \rightrightarrows \mathbb{R}^m$ *is proto-differentiable at* $\bar{x}$ *for* $\bar{u}$ *and has the Aubin property there, it is semidifferentiable at* $\bar{x}$ *for* $\bar{u}$. *In particular,* S *is semidifferentiable anywhere that it is strictly continuous and graphically regular.*

(b) *For single-valued* $F : D \to \mathbb{R}^m$, $D \subset \mathbb{R}^n$, *and any point* $\bar{x} \in \operatorname{int} D$ *where* F *is strictly continuous, semidifferentiability is equivalent to proto-differentiability. Indeed, calmness of* F *at* $\bar{x}$ *suffices for this equivalence when* $DF(\bar{x})$ *is single-valued.*

Proof. The second statement in (a) will immediately result from combining 8.41 with the first. The first statement could itself be derived from the semidifferentiability characterization in 8.43(d); the Aubin property can be seen to ensure that the mappings $\Delta_\tau S(\bar{x}\,|\,\bar{u})$ are asymptotically equicontinuous around the origin as $\tau \searrow 0$. An alternative argument can be based, however, on the characterization in 8.43(c), as follows.

Let $H = DS(\bar{x} \,|\, \bar{u})$, noting that $\operatorname{dom} H = I\!R^n$ by 9.49(a). The outer limit aspect of proto-differentiability already gives

$$\limsup_{\substack{w \to \bar{w} \\ \tau \searrow 0}} \tau^{-1}\big(S(\bar{x}+\tau w) - \bar{u}\big) \;=\; H(\bar{w}).$$

To verify semidifferentiability, we have to show that the same formula holds for the inner limit. Thus, we have to verify for arbitrary choice of $\bar{z} \in H(\bar{w})$, $\tau^\nu \searrow 0$ and $w^\nu \to \bar{w}$, the existence of $z^\nu \to \bar{z}$ with $\bar{u} + \tau^\nu z^\nu \in S(\bar{x} + \tau^\nu w^\nu)$.

The inner limit aspect of proto-differentiability merely furnishes for $\bar{z}$ and the sequence $\tau^\nu \searrow 0$ the existence of some $\bar{w}^\nu \to \bar{w}$ and $\bar{z}^\nu \to \bar{z}$ with $\bar{u} + \tau^\nu \bar{z}^\nu \in S(\bar{x} + \tau^\nu \bar{w}^\nu)$. But due to the Aubin property, there's a constant $\kappa \geq 0$ together with neighborhoods $V \in \mathcal{N}(\bar{x})$ and $W \in \mathcal{N}(\bar{u})$ such that

$$\begin{aligned} S(\bar{x}+\tau^\nu \bar{w}^\nu) \cap W \;\subset\; S(\bar{x}+\tau^\nu w^\nu) + \kappa\big|(\bar{x}+\tau^\nu \bar{w}^\nu) - (\bar{x}+\tau^\nu w^\nu)\big| I\!B \\ \text{when } \bar{x}+\tau^\nu \bar{w}^\nu \in V,\;\; \bar{x}+\tau^\nu w^\nu \in V. \end{aligned}$$

Because $\bar{u} + \tau^\nu \bar{z}^\nu$ belongs eventually to the set on the left, it must belong eventually to the set on the right. Thus, for large ν there must be an element of $S(\bar{x} + \tau^\nu w^\nu)$, which we can write in the form $\bar{u} + \tau^\nu z^\nu$, such that

$$\big|(\bar{u}+\tau^\nu \bar{z}^\nu) - (\bar{u}+\tau^\nu z^\nu)\big| \;\leq\; \kappa\big|(\bar{x}+\tau^\nu \bar{w}^\nu) - (\bar{x}+\tau^\nu w^\nu)\big|.$$

Then $|\bar{z}^\nu - z^\nu| \leq |\bar{w}^\nu - w^\nu| \to 0$, so we have $z^\nu \to \bar{z}$ as required.

Part (b) specializes (a) and recalls the equivalence of 8.43(e) and (f). □

Proposition 9.50 makes it possible to detect semidifferentiability through the same calculus that can be used to establish the Aubin property and graphical regularity. It provides a powerful tool in the study of 'rates of change' in mathematical models involving set-valued mappings.

9.51 Example (semidifferentiability of feasible-set mappings). *Consider a set $C(p) \subset I\!R^n$ depending on $p \in I\!R^d$ through a constraint representation*

$$C(p) \;=\; \big\{ x \in X \,\big|\, F(x,p) \in D \big\}$$

with $X \subset I\!R^n$ and $D \subset I\!R^m$ closed and $F : I\!R^n \times I\!R^d \to I\!R^m$ smooth. Suppose for a given $\bar{p}$ that the constraint qualification for this representation of $C(\bar{p})$ is satisfied at the point $\bar{x} \in C(\bar{p})$, namely

$$\left.\begin{aligned} y \in N_D\big(F(\bar{x},\bar{p})\big) \\ -\nabla_x F(\bar{x},\bar{p})^* y \in N_X(\bar{x}) \end{aligned}\right\} \;\Longrightarrow\; y = 0, \qquad 9(26)$$

moreover with X regular at $\bar{x}$ and D regular at $F(\bar{x},\bar{p})$. Then the mapping $C : p \mapsto C(p)$ has the Aubin property at $\bar{p}$ for $\bar{x}$, with

$$\operatorname{lip} C(\bar{p}\,|\,\bar{x}) \;=\; \max_{\substack{y \in N_D(F(\bar{x},\bar{p}))\\ |y|=1}} \frac{\left|\nabla_p F(\bar{x},\bar{p})^* y\right|}{d\big(-\nabla_x F(\bar{x},\bar{p})^* y,\, N_X(\bar{x})\big)},$$

and it is semidifferentiable at $\bar{p}$ for $\bar{x}$ with derivative mapping

$$DC(\bar{p}\,|\,\bar{x})(q) \;=\; \Big\{ w \in T_X(\bar{x}) \,\Big|\, \nabla_x F(\bar{x},\bar{p})w + \nabla_p F(\bar{x},\bar{p})q \,\in\, T_D\big(F(\bar{x},\bar{p})\big) \Big\}.$$

Detail. The regularity assumptions and constraint qualification 9(26) imply the condition in 8.42, which led there to the conclusion that C is proto-differentiable at $\bar{p}$ for $\bar{x}$ with $DC(\bar{p}\,|\,\bar{x})$ given by the formula restated above, and also provided a formula for $D^*C(\bar{p}\,|\,\bar{x})$. Through the latter, 9(26) is seen to correspond to $D^*C(\bar{p}\,|\,\bar{x})(0) = \{0\}$. We know then from the Mordukhovich criterion in 9.40 that C has the Aubin property at $\bar{p}$ for $\bar{x}$, and that the modulus has the value described. Finally, we note from 9.50 that in this case the proto-differentiability becomes semidifferentiability. □

9.52 Example (semidifferentiability from graph-convexity). *If $S : \mathbb{R}^n \rightrightarrows \mathbb{R}^m$ is osc and graph-convex, and if $\bar{x} \in \operatorname{int}(\operatorname{dom} S)$, then S is semidifferentiable at $\bar{x}$ for every $\bar{u} \in S(\bar{x})$.*

Detail. We have S strictly continuous at $\bar{x}$ by 9.34, so the Aubin property holds at $\bar{x}$ for every $\bar{u} \in S(\bar{x})$. Furthermore, $\operatorname{gph} S$ is regular because it is closed and convex, cf. 6.9. The semidifferentiability follows then from 9.50. □

The Aubin property provides a means of localizing strict continuity to a point $(\bar{x}, \bar{u})$ of the graph of a mapping S, but often there's interest in knowing whether, in addition, single-valuedness emerges when the graph is viewed in a sufficiently small neighborhood. The concept here is that S *has a single-valued localization at $\bar{x}$ for $\bar{u}$* if there are neighborhoods $V \in \mathcal{N}(\bar{x})$ and $W \in \mathcal{N}(\bar{u})$ such that the mapping $x \mapsto S(x) \cap W$, when restricted to V, is single-valued; specifically, this restricted mapping T is then the single-valued localization in question. To what extent can this property be characterized by 'differentiation', especially in getting T also to be Lipschitz continuous on V? For this we need to appeal to another type of graphical derivative.

9.53 Definition (strict graphical derivatives). *For a mapping $S : \mathbb{R}^n \rightrightarrows \mathbb{R}^m$, the strict derivative mapping $D_*S(\bar{x}|\bar{u}) : \mathbb{R}^n \rightrightarrows \mathbb{R}^m$ for S at $\bar{x}$ for $\bar{u}$, where $\bar{u} \in S(\bar{x})$, is defined by*

$$D_*S(\bar{x}\,|\,\bar{u})(w) := \Big\{ z \,\Big|\, \exists\, \tau^\nu \searrow 0,\ (x^\nu, u^\nu) \underset{\operatorname{gph} S}{\longrightarrow} (\bar{x}, \bar{u}),\ w^\nu \to w, \quad \text{with } z^\nu \in [S(x^\nu + \tau^\nu w^\nu) - u^\nu]/\tau^\nu,\ z^\nu \to z \Big\}, \tag{9(27)}$$

or in other words,

$$D_*S(\bar{x}\,|\,\bar{u}) := \mathop{\text{g-lim sup}}_{\substack{\tau \searrow 0\\ (x,u) \underset{\operatorname{gph} S}{\longrightarrow} (\bar{x},\bar{u})}} \Delta_\tau S(x\,|\,u).$$

The notation reduces to $D_*S(\bar{x})$ when $S(\bar{x}) = \{\bar{u}\}$. In the case of a continuous single-valued mapping F, *strict differentiability of F corresponds to $D_*F(\bar{x})$ being a linear mapping.* This is the source of the term 'strict' for the graphical derivatives in Definition 9.53 in general. Note that if F is strictly continuous, sequences $w^\nu \to w$ are superfluous in the definition of $D_*F(\bar{x})$, because $|\Delta_\tau F(x)(w') - \Delta_\tau F(x)(w)| \leq \kappa |w' - w|$ locally for some constant κ; one then has the simpler formula

$$D_*F(\bar{x})(w) = \big\{ z \,\big|\, \exists\, \tau^\nu \searrow 0,\ x^\nu \to \bar{x} \text{ with } \Delta_{\tau^\nu} F(x^\nu)(w) \to z \big\}. \tag{9(28)}$$

It's evident that $D_*S(\bar{x} \,|\, \bar{u})$ is always osc and positively homogeneous. Moreover

$$D_*S(\bar{x} \,|\, \bar{u}) \;\supset\; \operatorname*{g-lim\,sup}_{\substack{\tau \searrow 0 \\ (x,u) \xrightarrow[\text{gph}\, S]{} (\bar{x},\bar{u})}} DS(x \,|\, u),$$

but the latter limit mapping can be smaller. This is illustrated in Figure 9–9 by the case of $S : \mathbb{R} \rightrightarrows \mathbb{R}$ with $S(x) = \{x^2, -x^2\}$, which for $(\bar{x}, \bar{u}) = (0,0)$ has $D_*S(\bar{x} \,|\, \bar{u})(0) = \mathbb{R}$ even though all the mappings $DS(x \,|\, u)$ are single-valued and linear, and converge uniformly on bounded sets to $DS(0 \,|\, 0) = 0$ as $(x, u) \to (0,0)$ in $\text{gph}\, S$.

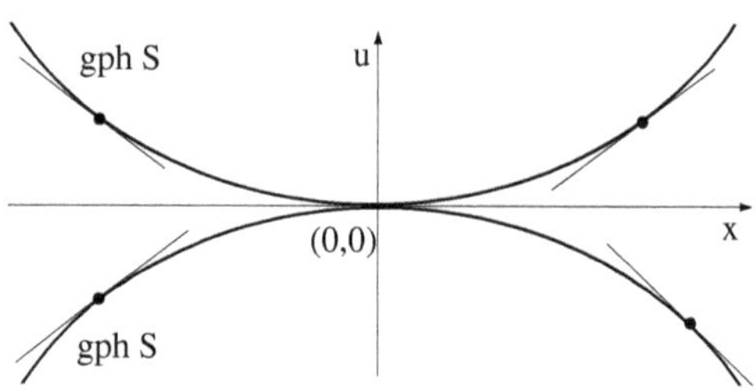

Fig. 9–9. A mapping with strict derivative larger than the limit of nearby derivatives.

In the theorem on Lipschitz continuous single-valued localizations that we're about to state, strict derivatives are the key. The part of the proof concerned with inverse mappings will make use of *Brouwer's theorem on invariance of domains.* This says that when two subsets of $\mathbb{R}^n$ are homeomorphic, and one of them is open, the other must be open as well. Indeed, a nonempty open subset of $\mathbb{R}^n$ can't be homeomorphic to such a subset of $\mathbb{R}^m$ unless $m = n$; for a reference, see the Commentary at the end of this chapter.

9.54 Theorem (single-valued localizations). *Let $\bar{u} \in S(\bar{x})$ for a mapping $S : \mathbb{R}^n \rightrightarrows \mathbb{R}^m$, and suppose that S is locally isc around $\bar{x}$ for $\bar{u}$, in the sense of there existing neighborhoods $V \in \mathcal{N}(\bar{x})$ and $W \in \mathcal{N}(\bar{u})$ such that, for any $x \in V$ and $\varepsilon > 0$, one can find $\delta > 0$ with*

$$\left.\begin{aligned} S(x) \cap W &\subset S(x') + \varepsilon \mathbb{B} \\ S(x') \cap W &\subset S(x) + \varepsilon \mathbb{B} \end{aligned}\right\} \quad \text{when } x' \in V \cap \mathbb{B}(x, \delta).$$

(a) *S has a Lipschitz continuous single-valued localization T at $\bar{x}$ for $\bar{u}$ if and only if $D_*S(\bar{x}\,|\,\bar{u})(0) = \{0\}$. Then $D^*S(\bar{x}\,|\,\bar{u})(0) = \{0\}$ as well, and*

$$\operatorname{lip} T(\bar{x}) = |D_*S(\bar{x}\,|\,\bar{u})|^+ = |D^*S(\bar{x}\,|\,\bar{u})|^+.$$

If S is proto-differentiable at $\bar{x}$ for $\bar{u}$, then also T is semidifferentiable at $\bar{x}$ with $DT(\bar{x}) = DS(\bar{x}\,|\,\bar{u})$.

(b) *For S convex-valued, S^{-1} has a Lipschitz continuous single-valued localization T at $\bar{u}$ for $\bar{x}$ if and only if $D_*S(\bar{x}\,|\,\bar{u})^{-1}(0) = \{0\}$. Then $m = n$, $D^*S(\bar{x}\,|\,\bar{u})^{-1}(0) = \{0\}$, and*

$$\operatorname{lip} T(\bar{u}) = |D_*S(\bar{x}\,|\,\bar{u})^{-1}|^+ = |D^*S(\bar{x}\,|\,\bar{u})^{-1}|^+.$$

If S is proto-differentiable at $\bar{x}$ for $\bar{u}$, then also T is semidifferentiable at $\bar{u}$ with $DT(\bar{u}) = DS(\bar{x}\,|\,\bar{u})^{-1}$.

Proof. As a preliminary we demonstrate, without invoking any continuity hypothesis, that the condition $D_*S(\bar{x}\,|\,\bar{u})(0) = \{0\}$ is equivalent to the existence of $V_0 \in \mathcal{N}(\bar{x})$ and $W_0 \in \mathcal{N}(\bar{u})$ such that the truncated mapping $S_0 : x \mapsto S(x) \cap W_0$ single-valued and Lipschitz continuous relative to $V_0 \cap \operatorname{dom} S_0$. From 9.23 we have $D_*S(\bar{x}\,|\,\bar{u})(0) = \{0\}$ if and only if $D_*S(\bar{x}\,|\,\bar{u})$ is locally bounded, $|D_*S(\bar{x}\,|\,\bar{u})|^+ < \infty$. That corresponds to the existence of $\kappa \in I\!R_+$ and $W_0 \in \mathcal{N}(\bar{u})$ such that $|z| \leq \kappa$ whenever $z = [u' - u]/\tau$ with $u' \in S(x + \tau w) \cap W_0$, $w \in I\!B$, (x, u) close enough to $(\bar{x}, \bar{u})$ in $\operatorname{gph} S$, and τ near to 0; thus, $|u' - u| \leq \kappa|x' - x|$ when $u \in S_0(x)$ and $u' \in S_0(x')$ with (x, u) and (x', u') close enough to $(\bar{x}, \bar{u})$. Then S_0 can't be multivalued near $\bar{x}$ (as seen by taking $x' = x$); for some $V_0 \in \mathcal{N}(\bar{x})$, S_0 is single-valued and Lipschitz continuous relative to $V_0 \cap \operatorname{dom} S_0$.

Sufficiency in (a). When $D_*S(\bar{x}\,|\,\bar{u})(0) = \{0\}$, we can take T to be the restriction, to a neighborhood of $\bar{x}$, of a truncation S_0 of the kind in our preliminary analysis. Our isc assumption makes S_0 be nonempty-valued there.

Necessity in (a). When S has a localization T as in (a), S in particular has the Aubin property at $\bar{x}$ for $\bar{u}$. Then $D^*S(\bar{x}\,|\,\bar{u})(0) = \{0\}$ and $\operatorname{lip} T(\bar{x}) = |D^*S(\bar{x}\,|\,\bar{u})|^+$ by Theorem 9.40. On the other hand, T must coincide around $\bar{x}$ with a truncation S_0 of the kind above, hence $D_*S(\bar{x}\,|\,\bar{u})(0) = \{0\}$. Moreover

$$\operatorname{lip} T(\bar{x}) \;=\; \limsup_{\substack{x \to \bar{x} \\ \tau \searrow 0}} \left(\sup_{w \in B} \left| \frac{T(x + \tau w) - T(x)}{\tau} \right| \right), \qquad 9(29)$$

where the difference quotients can be identified with ones for S, as in 9.53. The cluster points of such quotients as $x \to \bar{x}$ and $\tau \searrow 0$ are the elements of $D_*S(\bar{x}\,|\,\bar{u})(w)$. Thus by 9(29), $\operatorname{lip} T(\bar{x}) = \max\{|z| \,\big|\, z \in D_*S(\bar{x}\,|\,\bar{u})(I\!B)\} =: |D_*S(\bar{x}\,|\,\bar{u})|^+$. The claims about semidifferentiability are supported by 9.50(b).

Necessity in (b). The argument is the same as for necessity in (a), we apply it to S^{-1} instead of S and, in stating the result, utilize the fact that $DS^{-1}(\bar{u}\,|\,\bar{x}) = DS(\bar{x}|\bar{u})^{-1}$ and $D_*S^{-1}(\bar{u}\,|\,\bar{x}) = D_*S(\bar{x}|\bar{u})^{-1}$, whereas $y \in D^*S^{-1}(\bar{u}\,|\,\bar{x})(v)$ means $-y \in D^*S(\bar{x}\,|\,\bar{u})^{-1}(-v)$. The proto-differentiability of S at $\bar{x}$ for $\bar{u}$ is equivalent to that of S^{-1} at $\bar{u}$ for $\bar{x}$ through its identification with the geometric derivability of $\operatorname{gph} S$ at $(\bar{x}, \bar{u})$.

Sufficiency in (b). Likewise on the basis of our preliminary analysis but as applied to S^{-1} in place of S, we get from $D_*S(\bar{x}\,|\,\bar{u})^{-1}(0) = \{0\}$ the existence of neighborhoods $V_0 \in \mathcal{N}(\bar{x})$ and $W_0 \in \mathcal{N}(\bar{u})$ such that the truncation $T_0 : u \mapsto S^{-1}(u) \cap V_0$ is single-valued and Lipschitz continuous relative to $W_0 \cap \operatorname{dom} T_0$. We need to verify, however, that the latter set has $\bar{u}$ in its interior.

In combination with our isc assumption on S, the convex-valuedness of S yields the existence around $\bar{x}$ of a continuous selection $s(x) \in S(x)$ with $s(\bar{x}) = \bar{u}$. Indeed, consider neighborhoods V and W as in the isc property and choose $\lambda > 0$ with $\mathbb{B}(\bar{u}, 2\lambda) \subset W$. Take $V_1 \in \mathcal{N}(\bar{x})$ within V such that $S(x) \cap \operatorname{int} \mathbb{B}(\bar{u}, \lambda) \neq \emptyset$ when $x \in V_1$. Then we have for any $x \in V_1$ and $\varepsilon \in (0, \lambda)$ the existence of $\delta > 0$ such that $x' \in V_1 \cap \mathbb{B}(\bar{x}, \delta)$ implies

$$\emptyset \neq S(x) \cap \mathbb{B}(\bar{u}, 2\lambda - \varepsilon) \subset \big[S(x') + \varepsilon\mathbb{B}\big] \cap \mathbb{B}(\bar{u}, 2\lambda - \varepsilon) \subset S(x') \cap \mathbb{B}(\bar{u}, 2\lambda) + \varepsilon\mathbb{B}.$$

This tells us that $\lim_{\varepsilon \searrow 0} S(x) \cap \mathbb{B}(\bar{u}, 2\lambda - \varepsilon) \subset \liminf_{x' \to x} S(x') \cap \mathbb{B}(\bar{u}, 2\lambda)$, where the limit on the left includes $S(x) \cap \mathbb{B}(\bar{u}, 2\lambda)$ because $S(x)$ is convex and meets $\operatorname{int} \mathbb{B}(\bar{u}, 2\lambda)$. It follows that the truncation $S_0 : x \mapsto S(x) \cap \mathbb{B}(\bar{u}, 2\lambda)$ is isc on V_1 as well as convex-valued there. Let $S_1(x) = S_0(x)$ when $x \neq \bar{x}$ but $S_1(\bar{x}) = \{\bar{u}\}$. Then S_1 too is isc and convex-valued on V_1. By applying Theorem 5.58 to S_1, we get the desired selection s.

Near $(\bar{u}, \bar{x})$, the graph of s^{-1} for this selection s must lie in $\operatorname{gph} T_0$. Hence, on some open set O that contains $\bar{x}$ and is small enough that $s(O) \subset W_0$, s is one-to-one with continuous inverse and thus provides a homeomorphism of O with a set $O' \subset W_0 \cap \operatorname{dom} T_0$ containing $\bar{u}$. By Brouwer's theorem, quoted above, O' is open and $m = n$. □

9.55 Corollary (single-valued Lipschitzian invertibility). *Let $O \subset \mathbb{R}^n$ be open, and let $F : O \to \mathbb{R}^n$ be continuous. For F^{-1} to have a Lipschitz continuous single-valued localization at $\bar{u} = F(\bar{x})$ for a point $\bar{x} \in O$, it is necessary and sufficient that F satisfy the nonsingular strict derivative condition*

$$D_*F(\bar{x})(w) = 0 \quad \Longrightarrow \quad w = 0,$$

in which case F also satisfies the nonsingular coderivative condition

$$D^*F(\bar{x})(y) = 0 \quad \Longrightarrow \quad y = 0,$$

and one has $\operatorname{lip} F^{-1}(\bar{u}\,|\,\bar{x}) = |D_*F(\bar{x})^{-1}|^+ = |D^*F(\bar{x})^{-1}|^+$. Moreover, if F is semidifferentiable at $\bar{x}$, the localized inverse is semidifferentiable at $\bar{u}$ with single-valued Lipschitz continuous derivative mapping equal to $DF(\bar{x})^{-1}$.

When F is strictly differentiable at $\bar{x}$, the two conditions are equivalent and mean $\nabla F(\bar{x})$ is nonsingular. Then $\operatorname{lip} F^{-1}(\bar{u}\,|\,\bar{x}) = |\nabla F(\bar{x})^{-1}|$, and the localized inverse is strictly differentiable at $\bar{u}$ with $\nabla F(\bar{x})^{-1}$ as its Jacobian.

Proof. This just specializes 9.54(b) from set-valued S to single-valued F. The claims in the case of strict differentiability are based on the identification of that property with the linearity of $D_*F(\bar{x})$. The strict derivatives are given then by $D_*F(\bar{x})(w) = \nabla F(\bar{x})w$. But one also has $D^*F(\bar{x})(y) = \nabla F(\bar{x})^*y$ in that case by cf. 9.25(c). □

It's evident that the nonsingular strict derivative condition in 9.55, namely that $D_*F(\bar{x})(w) = 0$ only for $w = 0$, holds if and only if

$$\liminf_{\substack{x,x' \to \bar{x} \\ x' \neq x}} \frac{|F(x') - F(x)|}{|x' - x|} > 0.$$

The reciprocal of this limit quantity equals $\operatorname{lip} F^{-1}(\bar{u}\,|\,\bar{x})$ for $\bar{u} = F(\bar{x})$.

When F is strictly continuous in 9.55, the nonsingular coderivative condition can be expressed through 9.24(b) in the form that

$$0 \in \partial(yF)(\bar{x}) \quad \text{only for} \quad y = 0.$$

It might be wondered whether, at least in this case along with the strictly differentiable one, coderivative nonsingularity isn't merely implied by strict derivative nonsingularity, but is equivalent to it and thus sufficient on its own for *single-valued* Lipschitzian invertibility.

Of course, coderivative nonsingularity at $\bar{x}$ is already known to be equivalent to the Aubin property of F^{-1} at $F(\bar{x})$ for $\bar{x}$ (through the Mordukhovich criterion in 9.40), so the question revolves around whether localized single-valuedness of F^{-1} might be automatic in this situation. If true, that would be advantageous for applications of 9.55, because of the calculus of subgradients and coderivatives to be developed in Chapter 10; much less can be said about strict derivatives beyond the classical domain of smooth mappings. But the answer is negative in general, even for Lipschitz continuous F, as seen from the following counterexample.

Take $F : \mathbb{R}^2 \to \mathbb{R}^2$ to be the mapping that assigns to a point having *polar* coordinates (r, θ) the point having *polar* coordinates $(r, 2\theta)$. Let $(\bar{x}, \bar{u}) = (0, 0)$. It's easy to see that F^{-1} is Lipschitz continuous globally, and that $F^{-1}(0) = \{0\}$ but $F^{-1}(u)$ is a doubleton at all $u \neq 0$. The Lipschitz continuity tells us through 9.40 that $D^*F(0)^{-1} = \{0\}$. However, we don't have $D_*F(0)^{-1}(0) = \{0\}$, as indeed we can't without contradicting 9.54(b). Instead, actually $D_*F(0)^{-1}(0) = \mathbb{R}^2$.

This example doesn't preclude the possibility of identifying *special* classes of strictly continuous F or related mappings S for which the Aubin property of the inverse automatically entails localized single-valuedness of the inverse.

The theory of Lipschitzian properties of inverse mappings readily extends to cover implicit mappings that are defined by solving generalized equations.

9.56 Theorem (implicit mappings). *For an osc mapping $S : \mathbb{R}^n \times \mathbb{R}^d \rightrightarrows \mathbb{R}^m$, consider the generalized equation $S(x, p) \ni 0$, with parameter element p, and let $\bar{x} \in R(\bar{p})$, where the implicit solution mapping $R : \mathbb{R}^d \rightrightarrows \mathbb{R}^n$ is defined by*

$$R(p) = \{x \mid S(x, p) \ni 0\}.$$

(a) *For R to have the Aubin property at $\bar{p}$ for $\bar{x}$, it is sufficient that S satisfy the condition*

$$(0,r) \in D^*S(\bar{x},\bar{p}\,|\,0)(y) \implies y=0,\ r=0.$$

(b) *For R to have a Lipschitz continuous single-valued localization at $\bar{p}$ for $\bar{x}$, it is sufficient that S satisfy the condition in (a) along with*

$$0 \in D_*S(\bar{x},\bar{p}\,|\,0)(w,0) \implies w=0.$$

(c) *Under the condition in (a), if S is proto-differentiable at $(\bar{x},\bar{p})$ for 0, then R is semidifferentiable at $\bar{p}$ with*

$$DR(\bar{p}\,|\,\bar{x})(q) = \big\{ w \,\big|\, DS(\bar{x},\bar{p}\,|\,0)(w,q) \ni 0 \big\}.$$

When the condition in (b) is satisfied as well, this derivative mapping is single-valued and Lipschitz continuous.

(d) *If S is single-valued at $(\bar{x},\bar{p})$ and strictly differentiable there, in the sense of $D_*S(\bar{x},\bar{p})$ being a linear mapping with $n\times(n+p)$ matrix $\nabla S(\bar{x},\bar{p}) = [\nabla_x S(\bar{x},\bar{p}), \nabla_p S(\bar{x},\bar{p})]$, the conditions in (a) and (b) are equivalent and mean the nonsingularity of $\nabla_x S(\bar{x},\bar{p})$. Then R is strictly differentiable at $\bar{p}$ with*

$$\nabla R(\bar{p}) = \nabla_x S(\bar{x},\bar{p})^{-1} \nabla_p S(\bar{x},\bar{p}).$$

Proof. Let $T(u,p) = \big\{ x \,\big|\, S(x,p) \ni u \big\}$, so that $R(p) = T(0,p)$. The graph of T differs from that of S only by a permutation of arguments, so that $(y,r,v) \in N_{\operatorname{gph} T}(0,\bar{p},\bar{x})$ if and only if $(v,r,y) \in N_{\operatorname{gph} S}(\bar{x},\bar{p},0)$. Thus, $(0,r) \in D^*S(\bar{x},\bar{p}\,|\,0)(y)$ corresponds to $(-y,r) \in D^*T(0,\bar{p}\,|\,\bar{x})(0)$. The condition in (a) is equivalent therefore to the Mordukhovich criterion in 9.40 for T to have the Aubin property at $(0,\bar{p})$ for $\bar{x}$, in which case R has the Aubin property at $\bar{p}$ for $\bar{x}$.

Through graph considerations it's elementary that $w \in D_*T(0,\bar{p}\,|\,\bar{x})(z,q)$ if and only if $z \in D_*S(\bar{x},\bar{p}\,|\,0)(w,q)$. Also, though, $w \in D_*R(\bar{p}\,|\,\bar{x})(q)$ implies $w \in D_*T(0,\bar{p}\,|\,\bar{x})(0,q)$. The condition in (b) thereby corresponds to having $D_*R(\bar{p}\,|\,\bar{x})(0) = \{0\}$. When this is satisfied in tandem with the condition in (a), which makes R be locally continuous around $\bar{p}$ for $\bar{x}$ by virtue of the Aubin property, we are able to apply Theorem 9.54(a) to get the existence of a Lipschitz continuous single-valued localization.

The semidifferentiability result in (c) is obtained then from 9.54(a) and the fact, again seen through graphs, that T is proto-differentiable at $(0,\bar{p})$ for $\bar{x}$ if and only if S is proto-differentiable at $(\bar{x},\bar{p})$ for 0. The proto-differentiability of T is semidifferentiability when T has the Aubin property at $(0,\bar{p})$ for $\bar{x}$ (cf. 9.50), and such semidifferentiability of T yields, as a special case, the claimed semidifferentiability of R.

Under the assumption in (d), not only is $D_*S(\bar{x},\bar{p})$ the linear mapping associated with $\nabla S(\bar{x},\bar{p})$, but also, $D^*S(\bar{x},\bar{p})$ is the linear mapping associated with $\nabla S(\bar{x},\bar{p})^*$. The assertions then follow from the earlier ones. □

Observe that Theorem 9.56(d) covers the classical implicit function theorem as the special case where S is locally a smooth, single-valued mapping, just as 9.55 covers the classical inverse function theorem.

I.* Other Properties

Calmness, like Lipschitz continuity and strict continuity, can be generalized to set-valued mappings. A mapping $S : \mathbb{R}^n \rightrightarrows \mathbb{R}^m$ is *calm* at $\bar{x}$ if $S(\bar{x}) \neq \emptyset$ and there's a constant $\kappa \in \mathbb{R}_+$ along with a neighborhood $V \in \mathcal{N}(\bar{x})$ such that

$$S(x) \subset S(\bar{x}) + \kappa|x - \bar{x}|\mathbb{B} \text{ for all } x \in V. \qquad 9(30)$$

Note that although $\bar{x}$ must be an interior point of $\operatorname{dom} S$ when S is strictly continuous at $\bar{x}$, calmness at $\bar{x}$ has no such implication. A calmness variant of the Aubin property is the following: S is *calm at* $\bar{x}$ *for* $\bar{u}$, where $\bar{u} \in S(\bar{x})$, if there's a constant $\kappa \in \mathbb{R}_+$ along with neighborhoods $V \in \mathcal{N}(\bar{x})$ and $W \in \mathcal{N}(\bar{u})$ such that

$$S(x) \cap W \subset S(\bar{x}) + \kappa|x - x|\mathbb{B} \text{ for all } x \in V. \qquad 9(31)$$

9.57 Example (calmness of piecewise polyhedral mappings). *A mapping* $S : \mathbb{R}^n \rightrightarrows \mathbb{R}^m$ *is piecewise polyhedral if* $\operatorname{gph} S$ *is piecewise polyhedral, i.e., expressible as the union of finitely many polyhedral sets. Then S is calm at every point* $\bar{x} \in \operatorname{dom} S$. *Piecewise linear mappings, being the piecewise polyhedral mappings S that are single-valued on* $\operatorname{dom} S$, *have this property in particular.*

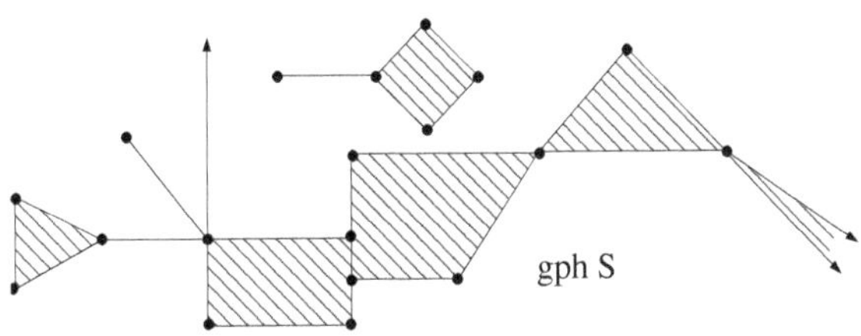

Fig. 9–10. A piecewise polyhedral mapping.

Detail. Writing $\operatorname{gph} S = \bigcup_{i=1}^r G_i$ for polyhedral sets G_i, we can introduce for each index i the mapping $S_i : \mathbb{R}^n \rightrightarrows \mathbb{R}^m$ having $\operatorname{gph} S_i = G_i$. By 9.35, each such mapping S_i is Lipschitz continuous on $X_i := \operatorname{dom} S_i$, say with constant κ_i. Taking $\kappa = \max_{i=1}^r \kappa_i$ we see that at any point $\bar{x} \in \operatorname{dom} S = \bigcup_{i=1}^r X_i$ that every mapping S_i is in particular calm at $\bar{x}$ with constant κ, and therefore that S has this property at $\bar{x}$ too.

Piecewise linear mappings are piecewise polyhedral in consequence of the characterization of their graphs in 2.48. □

The extension of a real-valued Lipschitz continuous function f from a subset $X \subset \mathbb{R}^n$ to all of $\mathbb{R}^n$ has been studied in 9.12. For a Lipschitz continuous vector-valued function $F : X \to \mathbb{R}^m$, written with $F(x) = \big(f_1(x), \ldots, f_m(x)\big)$, the device in 9.12 can be applied to each component function f_i so as likewise to obtain a Lipschitz continuous extension of F beyond X. But unfortunately, that approach won't ensure that the Lipschitz constant is preserved. It's quite important for some purposes, as in the theory of nonexpansive mappings, that an extension be obtained without increasing the constant.

9.58 Theorem (Lipschitz continuous extensions of single-valued mappings). *If a mapping $F : X \to \mathbb{R}^m$, $X \subset \mathbb{R}^n$, is Lipschitz continuous relative to X with constant κ, it has a unique continuous extension to $\operatorname{cl} X$, necessarily Lipschitz continuous with constant κ, and it can even be extended in this mode beyond $\operatorname{cl} X$, although nonuniquely.*

Indeed, there is a mapping $\overline{F} : \mathbb{R}^n \to \mathbb{R}^m$ that agrees with F on X and is Lipschitz continuous on all of $\mathbb{R}^n$ with the same constant κ.

Proof. The extension just to $\operatorname{cl} X$ is relatively elementary. Consider any $a \in X$ and $r > 0$, along with arbitrary points $\bar{x}_1, \bar{x}_2 \in \mathbb{B}(a, r) \cap \operatorname{cl} X$. We have $|F(x)| \leq |F(a)| + \kappa|x - a|$ when $x \in \mathbb{B}(a, r) \cap X$, so that for $i = 1, 2$ the set U_i consisting of the cluster points of $F(x)$ as $x \underset{X}{\to} \bar{x}_i$ is nonempty. Take any $\bar{u}_i \in U_i$; there exist sequences $x_i^\nu \underset{X}{\to} \bar{x}_i$ with $F(x_i^\nu) \to \bar{u}_i$. We have $|F(x_2^\nu) - F(x_1^\nu)| \leq \kappa|x_2^\nu - x_1^\nu|$, and hence in the limit that $|\bar{u}_2 - \bar{u}_1| \leq \kappa|\bar{x}_2 - \bar{x}_1|$. In particular this implies that $\bar{u}_2 = \bar{u}_1$ when $\bar{x}_2 = \bar{x}_1$ and therefore that for any $\bar{x} \in \operatorname{cl} X$ there is a unique limit to $F(x)$ as $x \underset{X}{\to} \bar{x}$. Thus, F has a unique continuous extension to $\operatorname{cl} X$. Denoting this extension still by F, we get $F(\bar{x}_i) = \bar{u}_i$ and consequently see that $|F(\bar{x}_2) - F(\bar{x}_1)| \leq \kappa|\bar{x}_2 - \bar{x}_1|$. Hence the extended F is Lipschitz continuous on $\operatorname{cl} X$ with constant κ.

For the extension beyond $\operatorname{cl} X$, consider now the collection of all pairs (X', F') where $X' \supset X$ in $\mathbb{R}^n$ and $F' : X' \to \mathbb{R}^m$ is Lipschitz continuous on X' with constant κ and agrees with F on X. Applying Zorn's Lemma relative to the inclusion relation for graphs, we obtain a subcollection which is totally ordered: for any of its elements (X_1, F_1) and (X_2, F_2), either $X_2 \supset X_1$ and F_2 agrees with F_1 on X_1, or the reverse. Take the union of the graphs in the subcollection to get a set $\overline{X}$ and a mapping $\overline{F} : \overline{X} \to \mathbb{R}^m$. Evidently $\overline{F}$ is Lipschitz continuous on $\overline{X}$ with constant κ. Our task is to prove that $\overline{X}$ is all of $\mathbb{R}^n$, and it can be accomplished by demonstrating that if this were not true, the mapping $\overline{F}$ could be extended beyond $\overline{X}$, contrary to the supposed maximality of the pair $(\overline{X}, \overline{F})$. Dividing $\overline{F}$ by κ if necessary, we can simplify considerations to the case of $\kappa = 1$.

Suppose $x \notin \overline{X}$. We aim at demonstrating the existence of $y \in \mathbb{R}^m$ such that $|y - F(x')| \leq |x - x'|$ for all $x' \in \overline{X}$, or in other words, such that the intersection of all the balls $\mathbb{B}\big(F(x'), |x - x'|\big)$ for $x' \in \overline{X}$ is nonempty. Since these balls are compact sets, it suffices by the finite intersection property of compactness to show that the intersection of every finite collection of them is nonempty. Thus, we need only prove that if for $i = 1, \ldots, k$ we have pairs $(x_i, y_i) \in \mathbb{R}^n \times \mathbb{R}^m$ satisfying

$$|y_i - y_j| \le |x_i - x_j| \;\text{ for } i,j \in \{1,\ldots,k\}, \tag{9(32)}$$

as would be true for any finite set of elements of gph F, then for any x there is a y such that $|y_i - y| \le |x_i - x|$ for $i = 1,\ldots,k$. The case of $x = 0$ is enough, because we can always reduce to it by shifting x_i to $x_i' = x_i - x$ in 9(32).

An elementary argument based on minimization will work. Consider $\varphi(y) := \sum_{i=1}^k \theta\big(|y - y_i|^2 - |x_i|^2\big)$ with $\theta(\tau) = |\tau|_+^2$, where $|\tau|_+ := \max\{0,\tau\}$. We have $\varphi \ge 0$, and $\varphi(y) = 0$ if and only if y has the desired property. Clearly φ is continuous, and it is also level-bounded because $\varphi(y) \le \beta$ implies $|y - y_i|^2 \le |x_i|^2 + \sqrt{\beta}$ for $i = 1,\ldots,k$. Hence $\operatorname{argmin} \varphi \ne \emptyset$ (by 1.9). Because θ is differentiable with $\theta'(\tau) = 2|\tau|_+$, φ is differentiable, and in fact its gradient can be calculated to be

$$\nabla\varphi(y) = 4\sum_{i=1}^k \mu_i(y)(y - y_i) \;\text{ for } \mu_i(y) := \Big|\, |y - y_i|^2 - |x_i|^2 \Big|_+. \tag{9(33)}$$

Fix any $\bar y \in \operatorname{argmin} \varphi$, so that $\nabla\varphi(\bar y) = 0$. If $\mu_i(\bar y) = 0$ for all i, then $\bar y$ meets our specifications and we are done. Otherwise, by dropping the pairs (x_i, y_i) with $\mu_i(\bar y) = 0$ (which have no role to play), we may as well suppose that $\mu_i(\bar y) > 0$ for all i, or equivalently, that

$$|x_i| < |y_i - \bar y| \;\text{ for } i = 1,\ldots,k. \tag{9(34)}$$

It will be demonstrated that this case is untenable.

Let $\lambda_i = \mu_i(\bar y)/\big[\mu_1(\bar y) + \cdots + \mu_k(\bar y)\big]$, so that $\lambda_i > 0$ and $\lambda_1 + \cdots + \lambda_k = 1$. From having $\nabla\varphi(\bar y) = 0$ in 9(33) we get $\bar y = \sum_{i=1}^k \lambda_i y_i$. Then from 9(34),

$$\begin{aligned}\sum_{i=1}^k \lambda_i |x_i|^2 &< \sum_{i=1}^k \lambda_i |y_i - \bar y|^2 = \sum_{i=1}^k \lambda_i\Big(|y_i|^2 - 2\langle y_i, \bar y\rangle + |\bar y|^2\Big)\\ &= \sum_{i=1}^k \lambda_i |y_i|^2 - 2\Big\langle \sum_{i=1}^k \lambda_i y_i, \bar y\Big\rangle + \Big(\sum_{i=1}^k \lambda_i\Big)|\bar y|^2 = \sum_{i=1}^k \lambda_i |y_i|^2 - |\bar y|^2.\end{aligned}$$

But 9(32) gives us $\sum_{i,j=1}^{k,k} \lambda_i\lambda_j |y_i - y_j|^2 \le \sum_{i,j=1}^{k,k} \lambda_i\lambda_j |x_i - x_j|^2$, which in terms of $|y_i - y_j|^2 = |y_i|^2 - 2\langle y_i, y_j\rangle + |y_j|^2$ on the left and the corresponding identity on the right reduces to $\sum_{i=1}^k \lambda_i |y_i|^2 - |\bar y|^2 \le \sum_{i=1}^k \lambda_i |x_i|^2 - |\bar x|^2$ with $\bar x := \sum_{i=1}^k \lambda_i x_i$. In combining this with the inequality just displayed, we obtain $\sum_{i=1}^k \lambda_i |x_i|^2 < \sum_{i=1}^k \lambda_i |x_i|^2 - |\bar x|^2$, which is impossible. □

9.59 Proposition (Lipschitzian properties in limits).

(a) *Suppose for $X \subset \mathbb{R}^n$ and $f^\nu : X \to \mathbb{R}$ that the mappings f^ν are Lipschitz continuous on X with constant κ and converge pointwise on a dense subset D of X. They then converge uniformly on bounded subsets of X to a function $f : X \to \mathbb{R}$ which is Lipschitz continuous on X with constant κ. Moreover, for any $\bar x \in \operatorname{int} X$,*

$$v \in \partial f(\bar{x}) \implies \exists\, x^\nu \to \bar{x},\ v^\nu \in \partial f^\nu(x^\nu) \text{ with } v^\nu \to v.$$

(b) *Suppose for $X \subset \mathbb{R}^n$ and $F^\nu : X \to \mathbb{R}^m$ that the mappings F^ν are Lipschitz continuous on X with constant κ and converge pointwise on a dense subset D of X. They then converge uniformly on bounded subsets of X to a mapping $F : X \to \mathbb{R}^n$ which is Lipschitz continuous relative to X with constant κ. Moreover, for any $\bar{x} \in \operatorname{int} X$ and $\bar{w} \in \mathbb{R}^n$,*

$$v \in D^*F(\bar{x})(\bar{w}) \implies \exists\, x^\nu \to \bar{x},\ v^\nu \in D^*F^\nu(x^\nu)(\bar{w}) \text{ with } v^\nu \to v.$$

(c) *Suppose $S = \text{g-lim sup}_\nu S^\nu$ for osc mappings $S^\nu : \mathbb{R}^n \rightrightarrows \mathbb{R}^m$, and let $\bar{u} \in S(\bar{x})$. Suppose there exist neighborhoods $V \in \mathcal{N}(\bar{x})$ and $W \in \mathcal{N}(\bar{u})$ along with $\kappa \in \mathbb{R}_+$ such that, whenever $(x,u) \in \operatorname{gph} S^\nu \cap [V \times W]$, the mapping S^ν has the Aubin property at x for u with constant κ. Then S has the Aubin property at $\bar{x}$ for $\bar{u}$ with constant κ.*

Proof. The claims in (a) can be identified with the special case of (b) where $m = 1$; the subgradient inclusion in (a) comes from the inclusion in (b) by taking $\bar{w} = 1$ in the context of the profile mappings in 8.35.

In (b), we can suppose without loss of generality, on the basis of the first part of Theorem 9.58, that X is closed. By assumption, $\lim_\nu F^\nu(x)$ exists for each $x \in D$; denote this limit by $F(x)$, obtaining a mapping $F : D \to \mathbb{R}^m$. For any points $x_1, x_2 \in D$ we have $|F^\nu(x_2) - F^\nu(x_1)| \le \kappa|x_2 - x_1|$ for all ν, hence in the limit $|F(x_2) - F(x_1)| \le \kappa|x_2 - x_1|$. Thus F is Lipschitz continuous on D with constant κ and, again by the first part of Theorem 9.58, has a unique extension to such a mapping on all of $X = \operatorname{cl} D$. With this extension denoted still by F, consider any $x \in X$ and any sequence $x^\nu \underset{D}{\to} x$. For any $\mu \in \mathbb{N}$,

$$\begin{aligned}\big|F^\nu(x^\nu) - F(x)\big| &\le \big|F^\nu(x^\nu) - F^\nu(x^\mu)\big| + \big|F^\nu(x^\mu) - F(x^\mu)\big| + \big|F(x^\mu) - F(x)\big| \\ &\le \kappa|x^\nu - x^\mu| + \big|F^\nu(x^\mu) - F(x^\mu)\big| + \kappa|x - x^\mu|.\end{aligned}$$

This gives us $\limsup_\nu |F^\nu(x^\nu) - F(x)| \le 2\kappa|x - x^\mu|$, and in letting $\mu \to \infty$ we then get $\limsup_\nu |F^\nu(x^\nu) - F(x)| = 0$. Therefore, $F^\nu(x^\nu) \to F(x)$. In particular this tells us that F^ν converges pointwise to F on all of X, so the argument can be repeated with $D = X$ to conclude that $F^\nu(x^\nu) \to F(x)$ whenever $x^\nu \to x$ in X. Such continuous convergence of F^ν to F automatically entails uniform convergence on all compact subsets of X, hence on all bounded subsets, since X is closed; cf. 5.43 as specialized to single-valued mappings.

The coderivative inclusion in (b) follows from the subgradient approximation rule in 8.47(b) as applied to the functions $f(x) = \langle \bar{w}, F(x)\rangle + \delta_B(x)$ and $f^\nu(x) = \langle \bar{w}, F^\nu(x)\rangle + \delta_B(x)$ for a closed neighborhood B of $\bar{x}$ in X; cf. 9.24(b). The locally uniform convergence of F^ν to F on X entails that $f^\nu \overset{e}{\to} f$.

In (c), we have by 9.40 that $|D^*S^\nu(x \,|\, u)|^+ \le \kappa$ when $u \in S^\nu(x)$ and (x,u) is close enough to $(\bar{x}, \bar{u})$. It follows then from the approximation rule in 8.48 that $|D^*S(x \,|\, u)|^+ \le \kappa$, and this yields the result, again by 9.40. $\square$

J.* Rademacher's Theorem and Consequences

Lipschitz continuity has been viewed as intermediate between continuity and differentiability, and this is borne out in the study of the extent to which a single-valued Lipschitz continuous (or strictly continuous) mapping must actually be differentiable. It turns out that this must be true at 'most' points. To render a precise description of the exceptional set where differentiability fails, we recall the following concept of measure-theoretic analysis.

A set $A \subset \mathbb{R}^n$ is called *negligible* if for every $\varepsilon > 0$ there is a family of boxes $\{B^\nu\}_{\nu \in \mathbb{N}}$ with n-dimensional volumes ε^ν such that

$$A \subset \bigcup_{\nu=1}^{\infty} B^\nu, \qquad \sum_{\nu=1}^{\infty} \varepsilon^\nu < \varepsilon.$$

For instance, any countable (or finite) set of points in $\mathbb{R}^n$ is negligible, or any line segment in $\mathbb{R}^n$ when $n \geq 2$ is negligible. If a set $A \subset \mathbb{R}^n$ is negligible, then so is every subset of A. In particular, the intersection of any collection of negligible sets is negligible. Likewise, the union of any countable collection of negligible sets is again negligible.

A useful criterion for negligibility is this: with respect to any vector $w \neq 0$, a measurable set $D \subset \mathbb{R}^n$ is negligible if and only if the set $\{\tau \mid x + \tau w \in D\} \subset \mathbb{R}$ is negligible (in the one-dimensional sense) for all but a negligible set of points $x \in \mathbb{R}^n$. Another standard fact to be noted concerns Lipschitz continuous (or more generally 'absolutely continuous') real-valued functions φ on intervals $I \subset \mathbb{R}$. Any such function is differentiable on I except at a negligible set of points, and it's the integral of its derivative function; in other words, one has $\varphi(t_1) = \varphi(t_0) + \int_{t_0}^{t_1} \varphi'(t)dt$ when $[t_0, t_1] \subset I$. The theorem stated next provides a generalization in part to higher dimensions. The proof is omitted because it requires too lengthy an excursion into measure theory beyond the needs of our subject; see the Commentary at the end of this chapter for a reference.

9.60 Theorem (almost everywhere differentiability; Rademacher). *Let $O \subset \mathbb{R}^n$ be open, and let $F : O \to \mathbb{R}^m$ be strictly continuous. Let D be the subset of O consisting of the points where F is differentiable. Then $O \setminus D$ is a negligible set. In particular D is dense in O, i.e., $\operatorname{cl} D \supset O$.*

This prevalence of differentiability has powerful consequences for the subdifferential theory of strictly continuous mappings.

9.61 Theorem (Clarke subgradients of strictly continuous functions). *Let f be strictly continuous on an open set $O \subset \mathbb{R}^n$, and let D be the subset of O where f is differentiable. At each point $\bar{x} \in O$ let*

$$\overline{\nabla} f(\bar{x}) := \limsup_{x \xrightarrow[D]{} \bar{x}} \{\nabla f(x)\} = \{v \mid \exists x^\nu \to \bar{x} \text{ with } x^\nu \in D,\ \nabla f(x^\nu) \to v\}.$$

Then $\overline{\nabla} f(\bar{x})$ is a nonempty, compact subset of $\partial f(\bar{x})$ such that

$$\operatorname{con} \overline{\nabla} f(\bar{x}) = \operatorname{con} \partial f(\bar{x}) = \bar{\partial} f(\bar{x}),$$
$$\widehat{d}f(\bar{x})(w) = \max \langle \overline{\nabla} f(\bar{x}), w \rangle = \max\big\{ \langle v, w \rangle \,\big|\, v \in \overline{\nabla} f(\bar{x}) \big\},$$

hence $\operatorname{con} \overline{\nabla} f(\bar{x}) = \partial f(\bar{x})$ *when* f *is regular at* $\bar{x}$. *These relations persist when* D *is replaced by any set* $D' \subset D$ *such that* $D \setminus D'$ *is negligible.*

Proof. At any $\bar{x} \in O$, the Clarke subgradient set $\bar{\partial} f(\bar{x})$ is $\operatorname{con} \partial f(\bar{x})$ by 8.49, since $\partial^\infty f(\bar{x}) = \{0\}$ by 9.13. Also, $\partial f(\bar{x})$ is nonempty and compact by 9.13.

Let D be the subset of O where f is differentiable. This set is dense in O by Theorem 9.60, so there do exist sequences $x^\nu \underset{D}{\to} \bar{x}$ as called for in the definition of $\overline{\nabla} f(\bar{x})$. At any $x \in D$ we have $\nabla f(x) \in \widehat{\partial} f(x) \subset \partial f(x)$ and in particular $|\nabla f(x)| \le \operatorname{lip} f(x)$. Hence any sequence of gradients $\nabla f(x^\nu)$ as $x^\nu \underset{D}{\to} \bar{x}$ is bounded (due to $\operatorname{lip} f$ being osc; cf. 9.2) and has all its cluster points in $\partial f(\bar{x})$ (by the definition of this subgradient set in 8.3, when the continuity of f is taken into account). This tells us that $\overline{\nabla} f(\bar{x}) \subset \partial f(\bar{x})$, with both sets nonempty and compact.

The convex sets $\operatorname{con} \overline{\nabla} f(\bar{x})$ and $\operatorname{con} \partial f(\bar{x})$ are closed (even compact by 2.30) and so they coincide if and only if they have the same support function (cf. 8.24). The support function of $\operatorname{con} \partial f(\bar{x})$ is $\widehat{d}f(\bar{x})$ by 9.15, while the support function of $\operatorname{con} \overline{\nabla} f(\bar{x})$ is

$$h(w) := \sup\big\{ \langle v, w \rangle \,\big|\, v \in \operatorname{con} \overline{\nabla} f(\bar{x}) \big\} = \sup\big\{ \langle v, w \rangle \,\big|\, v \in \overline{\nabla} f(\bar{x}) \big\}.$$

Because $\operatorname{con} \overline{\nabla} f(\bar{x}) \subset \operatorname{con} \partial f(\bar{x})$, we already know that $h(w) \le \widehat{d}f(\bar{x})(w)$. Our task is to demonstrate that $\widehat{d}f(\bar{x})(w) \le h(w)$. Moreover, it really suffices to consider vectors w of unit length, inasmuch as both sides are positively homogeneous.

Fix any subset $D' \subset D$ with $D \setminus D'$ negligible and choose any vector w with $|w| = 1$. The set $A = O \setminus D' = (O \setminus D) \cup (D \setminus D')$ is negligible, so its intersection with the line $\big\{ x + \tau w \,\big|\, \tau \in \mathbb{R} \big\}$ is negligible in the one-dimensional sense for all but a negligible set of x's (cf. the remark ahead of 9.60.). Thus, there is a dense subset $X \subset O$ such that for each $x \in X$ the set of $\tau \in \mathbb{R}$ with $x + \tau w \in A$ is negligible in $\mathbb{R}$. We know from 9.15, along with the continuity of f and the density of X, that

$$\widehat{d}f(\bar{x}) = \limsup_{\substack{x \underset{X}{\to} \bar{x} \\ \tau \searrow 0}} \frac{f(x + \tau w) - f(x)}{\tau}. \tag{9(35)}$$

When $x \in X$ and τ is small enough that the line segment from x to $x + \tau w$ lies in O, the function $\varphi(t) = f(x + tw)$ is Lipschitz continuous for $t \in [0, \tau]$ and such that $x + tw \in D'$ for all but a negligible set of $t \in [0, \tau]$. The Lipschitz continuity guarantees, as noted before 9.60, that $\varphi(\tau) = \varphi(0) + \int_0^\tau \varphi'(t) dt$, and in this integral a negligible set of t values can be disregarded. Thus, the integral is unaffected if we concentrate on t values such that $x + tw \in D'$, in which case $\varphi'(t) = \langle \nabla f(x + tw), w \rangle$. From this we see that

$$\frac{f(x+\tau w)-f(x)}{\tau}=\frac{\varphi(\tau)-\varphi(0)}{\tau}=\frac{1}{\tau}\int_0^\tau \langle \nabla f(x+tw), w\rangle\, dt$$
$$\leq \sup_{\substack{t\in[0,\tau]\\ x+tw\in D'}} \langle \nabla f(x+tw), w\rangle \leq \sup_{x'\in \boldsymbol{B}(x,\tau)\cap D'} \langle \nabla f(x'), w\rangle.$$

Because $|\nabla f(x')|$ remains bounded when $x' \underset{D'}{\to} \bar{x}$, as noted earlier in our argument, we obtain from 9(35) that

$$\widehat{d}f(\bar{x}) \leq \limsup_{\substack{x' \underset{D'}{\to} \bar{x}\\ \tau \searrow 0}} \langle \nabla f(x'), w\rangle \leq \max_{v\in \overline{\nabla}' f(\bar{x})} \langle v, w\rangle,$$

where $\overline{\nabla}' f(\bar{x})$ is the set defined in the same manner as $\overline{\nabla} f(\bar{x})$ but with D' substituted for D. Since $D' \subset D$ we have $\overline{\nabla}' f(\bar{x}) \subset \overline{\nabla} f(\bar{x})$, so this inequality establishes the result. The assertion about regularity refers to the convexity of $\partial f(\bar{x})$ in that case; cf. 8.11. □

The corresponding result for strictly continuous mappings $F: \mathbb{R}^n \to \mathbb{R}^m$ brings to light an 'adjoint' relationship between coderivatives and the strict derivatives in 9.53 and in particular 9(28).

9.62 Theorem (coderivative duality for strictly continuous mappings). *Let $F: O \to \mathbb{R}^m$ be strictly continuous, with $O \subset \mathbb{R}^n$ open, and let $D \subset O$ consist of the points where F is differentiable. For $\bar{x} \in O$, define*

$$\overline{\nabla} F(\bar{x}) := \{A \in \mathbb{R}^{m\times n} \mid \exists x^\nu \to \bar{x} \text{ with } x^\nu \in D,\ \nabla F(x^\nu) \to A\}.$$

Then $\overline{\nabla} F(\bar{x})$ is a nonempty, compact set of matrices, and for every $w \in \mathbb{R}^n$ and $y \in \mathbb{R}^m$ one has

$$\max \langle y, D_* F(\bar{x})(w)\rangle = \max \langle D^* F(\bar{x})(y), w\rangle = \max \langle y, \overline{\nabla} F(\bar{x}) w\rangle$$

for the expressions

$$\max \langle y, D_* F(\bar{x})(w)\rangle := \max\{\langle y, z\rangle \mid z \in D_* F(\bar{x})(w)\},$$
$$\max \langle D^* F(\bar{x})(y), w\rangle := \max\{\langle v, w\rangle \mid v \in D^* F(\bar{x})(y)\},$$
$$\max \langle y, \overline{\nabla} F(\bar{x}) w\rangle := \max\{\langle y, Aw\rangle \mid A \in \overline{\nabla} F(\bar{x})\},$$

and therefore also

$$\operatorname{con} D_* F(\bar{x})(w) = \operatorname{con}\{Aw \mid A \in \overline{\nabla} F(\bar{x})\} = \{Aw \mid A \in \operatorname{con} \overline{\nabla} F(\bar{x})\},$$
$$\operatorname{con} D^* F(\bar{x})(y) = \operatorname{con}\{A^* y \mid A \in \overline{\nabla} F(\bar{x})\} = \{A^* y \mid A \in \operatorname{con} \overline{\nabla} F(\bar{x})\}.$$

These relations persist when D is replaced in the formula for $\overline{\nabla} F(\bar{x})$ by any set $D' \subset D$ such that $D \setminus D'$ is negligible.

Proof. For any choice of $y \in \mathbb{R}^m$ and $x \in D$, the function yF is differentiable at x with $\nabla(yF)(x) = \nabla F(x)^* y$; moreover $\operatorname{dom} \nabla(yF) \setminus D$ is negligible. Hence by 9.61, $\overline{\nabla}(yF)(\bar{x}) = \{A^* y \mid A \in \overline{\nabla} F(\bar{x})\}$ for any $\bar{x} \in O$. We have $D^*F(\bar{x})(y) = \partial(yF)(\bar{x})$ by 9.24(b) and $\operatorname{con}\partial(yF)(\bar{x}) = \operatorname{con}\overline{\nabla}(yF)(\bar{x})$ by 9.61, so this yields the formula for $\operatorname{con} D^*F(\bar{x})(y)$ and the fact that $\max\langle D^*F(\bar{x})(y), w\rangle = \max\langle\overline{\nabla}(yF)(\bar{x}), w\rangle = \max\{\langle A^*y, w\rangle \mid A \in \overline{\nabla}F(\bar{x})\}$, where of course $\langle A^*y, w\rangle = \langle y, Aw\rangle$. At the same time, we know from 9.61 that $\max\langle\overline{\nabla}(yF)(\bar{x}), w\rangle = \widehat{d}(yF)(\bar{x})(w)$. But

$$\widehat{d}(yF)(\bar{x})(w) = \limsup_{\substack{x \to \bar{x} \\ \tau \searrow 0}} \Delta_\tau(yF)(x)(w) = \limsup_{\substack{x \to \bar{x} \\ \tau \searrow 0}} \langle y, \Delta_\tau F(x)(w)\rangle.$$

Because $\Delta_\tau F(x)(w)$ remains bounded as $x \to \bar{x}$ and $\tau \searrow 0$, due to the Lipschitz continuity of F, this expression coincides with the max of $\langle y, z\rangle$ over all vectors z achievable as cluster points of $\Delta_\tau F(x)(w)$ with respect to such convergence, i.e., all $z \in D_*F(\bar{x})(w)$; cf. 9(28). Hence $\max\langle\overline{\nabla}(yF)(\bar{x}), w\rangle = \max\langle y, D_*F(\bar{x})(w)\rangle$. Since the left side of this equation gives the max of $\langle y, Aw\rangle$ over all $A \in \overline{\nabla}F(\bar{x})$, i.e., the support function of the compact set $\overline{\nabla}F(\bar{x})w$, the convex hull of that set must agree with the convex hull of $D_*F(\bar{x})(w)$, see 8.24 (and, for the closedness of these convex hulls, 2.30). □

9.63 Corollary (coderivative criterion for Lipschitzian invertibility). *For a mapping $F : \mathbb{R}^n \to \mathbb{R}^n$ and a point $\bar{x}$ where F is strictly continuous, suppose that*

$$0 \in \operatorname{con}\partial(yF)(\bar{x}) \implies y = 0.$$

Then F^{-1} has a Lipschitz continuous single-valued localization at $F(\bar{x})$ for $\bar{x}$.

Proof. Because $\partial(yF)(\bar{x}) = D^*F(\bar{x})(y)$ (cf. 9.24(b)), this condition implies that $0 \in D^*F(\bar{x})(y)$ only for $y = 0$, but also through 9.62 that $0 \in D_*F(\bar{x})(w)$ only for $w = 0$. The result is then obtained as a specialization of 9.55. □

9.64 Exercise (strict differentiability and derivative continuity). *Let $O \subset \mathbb{R}^n$ be open, and let $F : O \to \mathbb{R}^m$ be strictly continuous. Let $D \subset O$ be the domain of the mapping $\nabla F : x \mapsto \nabla F(x)$, i.e., the set of points where F is differentiable. Then the following are equivalent:*

(a) *F is strictly differentiable at $\bar{x}$;*

(b) *F is differentiable at $\bar{x}$ and ∇F is continuous at $\bar{x}$ relative to D.*

Guide. Get this from 9.25(c) and 9.62. □

The following fact complements Rademacher's theorem (in 9.60) in being applicable to F^{-1} instead of F when the dimensions of the range space agree.

9.65 Theorem (singularities of Lipschitzian equations; Mignot). *Let $O \subset \mathbb{R}^n$ be open, and let $F : O \to \mathbb{R}^n$ be strictly continuous. Let $R \subset F(O)$ consist of the points $u \in \mathbb{R}^n$ such that, for every $x \in O$ with $F(x) = u$, F is differentiable at x and $\nabla F(x)$ is nonsingular. Then $F(O) \setminus R$ is negligible. In particular, R must be dense in the interior of $F(O)$ (when that is nonempty).*

Proof. Let D be the set of points $x \in O$ where F is differentiable, and let E be the set of $x \in D$ where $\nabla F(x)$ is nonsingular. Then $F(0) \setminus R = F(O \setminus E) =$

$F(O \setminus D) \cup F(D \setminus E)$. We'll demonstrate first that $F(O \setminus D)$ is negligible and second that $F(D \setminus E)$ is negligible.

The boxes in the definition of negligibility (ahead of 9.60) can just as well be taken to be cubes. Let's estimate what F does to a cube. Without loss of generality we can assume that F is globally Lipschitz continuous on O with constant κ. Any (closed) cube $C \subset O$ of side s, with n-dimensional volume $\mathrm{vol}(C) = s^n$, has diameter $s\sqrt{n}$. It lies in the ball around its center c whose radius is $\frac{1}{2}s\sqrt{n}$, and its image $F(C)$ thus lies in a ball in $I\!R^n$ whose center is $F(c)$ and whose radius is $\frac{1}{2}\kappa s\sqrt{n}$. This in turn lies in a cube $\widetilde{C}$ of side $\kappa s\sqrt{n}$, having $\mathrm{vol}(\widetilde{C}) = (\kappa s\sqrt{n})^n = (\kappa\sqrt{n})^n \mathrm{vol}(C)$.

Because $O \setminus D$ is negligible, there exists for any $\varepsilon > 0$ a covering of $O \setminus D$ by cubes C^ν with $\sum_{\nu=1}^{\infty} \mathrm{vol}(C^\nu) < \varepsilon$. In the manner explained, this yields a covering of $F(O \setminus D)$ by cubes $\widetilde{C}^\nu$ with $\sum_{\nu=1}^{\infty} \mathrm{vol}(\widetilde{C}^\nu) < \varepsilon(\kappa\sqrt{n})^n$. This upper bound can be made arbitrarily small, so $F(O \setminus D)$ is negligible.

Seeking now to establish the negligibility of $F(D \setminus E)$, it's enough to focus on the negligibility of $F\big(B \cap [D \setminus E]\big)$ for any box B. We can take B to be the unit cube, $[0,1]^n$. For each $p \in I\!N$, B can be evenly divided up into p^n subcubes of side 2^{-p}, hence volume 2^{-pn} and diameter $2^{-p}\sqrt{n}$. Denote the collection of these cubes by $\mathcal{C}_p$. Fixing $\varepsilon > 0$, construct a covering of $B \cap [D \setminus E]$ as follows.

At each $x \in D \setminus E$ the differentiability of F provides the existence of some $\delta(x) > 0$ such that

$$\begin{aligned} \big|F(x') - F(x) - \nabla F(x)(x'-x)\big| \;&\leq\; \varepsilon|x'-x| \\ &\text{for all } x' \in O \text{ with } |x'-x| \leq \delta(x). \end{aligned} \tag{9(36)}$$

Denote by $\mathcal{C}_1^*$ the collection of all the cubes $C \in \mathcal{C}_1$ such that C contains an $x \in D \setminus B$ with $\delta(x) > \frac{1}{2}\sqrt{n}$. Recursively, having gotten $\mathcal{C}_1^*, \ldots, \mathcal{C}_{p-1}^*$, let $\mathcal{C}_p^*$ consist of the cubes $C \in \mathcal{C}_p$ such that C contains a point $x \in D \setminus B$ with $\delta(x) > 2^{-p}\sqrt{n}$, but C isn't covered by the union of the cubes in $\mathcal{C}_1^*, \ldots, \mathcal{C}_{p-1}^*$. Let $\mathcal{C}^* = \bigcup_{p=1}^{\infty} \mathcal{C}_p^*$, noting that this is a countable collection that covers $B \cap [D \setminus E]$. The interiors of the cubes in this countable collection are mutually disjoint, so

$$\sum_{p=1}^{\infty} \sum_{C \in \mathcal{C}_p^*} \frac{1}{2^{pn}} \;=\; \sum_{C \in \mathcal{C}^*} \mathrm{vol}(C) \;\leq\; \mathrm{vol}(B) \;=\; 1. \tag{9(37)}$$

Consider any one of the cubes in $\mathcal{C}^*$, say $C \in \mathcal{C}_p^*$, and let x be a point of $C \cap [D \setminus E]$ with $\delta(x) > 2^{-p}\sqrt{n}$, as exists by our construction. Since $2^{-p}\sqrt{n}$ is the diameter of C, the inequality in 9(36) holds for all $x' \in C$. Let L be the 'linearization' of F at x, i.e., the mapping $x' \mapsto F(x) + \nabla F(x)(x'-x)$. Because $x \notin E$, the range of L isn't all of $I\!R^n$ and is contained in some hyperplane H with unit normal e. The inequality in 9(36) tells us that $|F(x') - L(x')| \leq \varepsilon 2^{-p}\sqrt{n}$, and this translates geometrically into $F(C) \subset L(C) + [-\lambda, \lambda]^n$ for $\lambda = \varepsilon 2^{-p}\sqrt{n}$. On the other hand L, like F, is Lipschitz continuous with constant κ, so we can argue that $L(C) \subset H \cap I\!B\big(F(x), \rho\big)$ for $\rho = \kappa 2^{-p}\sqrt{n}$. Relative to changing to an orthonormal basis that includes e, we see that $L(C)$ can be estimated as lying in an $(n-1)$-dimensional cube of side 2ρ. Thus, $F(C)$ lies in a (rotated) box $\widetilde{C}$ with $\operatorname{vol}(\widetilde{C}) = (2\rho)^{n-1}(2\lambda) = (2^{1-p}\kappa\sqrt{n})^{n-1}(2^{1-p}\varepsilon\sqrt{n})$ and consequently $\operatorname{vol}(\widetilde{C}) = \varepsilon\mu \operatorname{vol}(C)$ for $\mu := \kappa^{n-1}(\sqrt{n})^n 2^n$. Associating each cube $C \in \mathcal{C}^*$ with a box $\widetilde{C}$ in this manner, we obtain a collection $\widetilde{\mathcal{C}}^*$ of boxes that covers $F\big(B \cap [D \setminus E]\big)$ and, by 9(37), has total volume no more than $\varepsilon\mu$. The factor μ being independent of ε, which can be chosen arbitrarily small, we conclude that $F\big(B \cap [D \setminus E]\big)$ is negligible and the same for $F(D \setminus E)$. □

Surprisingly, much of the special geometry of the graphs of Lipschitz continuous mappings carries over to a vastly larger class of mappings. This comes from the fact that many mappings of keen interest, although neither single-valued nor Lipschitz continuous, have graphs that locally, 'from another angle', appear like graphs of Lipschitz continuous mappings.

9.66 Definition (graphically Lipschitzian mappings). *A mapping $S : \mathbb{R}^n \rightrightarrows \mathbb{R}^m$ is graphically Lipschitzian at $(\bar{x}, \bar{u})$, where $\bar{u} \in S(\bar{x})$, and of dimension d in this respect, if there is a change of coordinates around $(\bar{x}, \bar{u})$, $\mathcal{C}^1$ in both directions, under which* $\operatorname{gph} S$ *can be identified locally with the graph in $\mathbb{R}^d \times \mathbb{R}^{m+n-d}$ of a Lipschitz continuous mapping defined on a neighborhood of a point $\bar{w} \in \mathbb{R}^d$.*

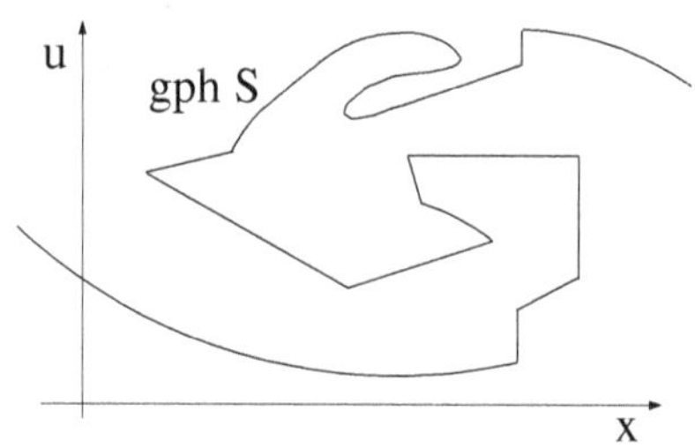

Fig. 9–11. A graphically Lipschitzian mapping S with neither S nor S^{-1} single-valued.

Any single-valued strictly continuous mapping is graphically Lipschitzian everywhere. (No change of coordinates is needed at all.) So too is the *inverse* of any single-valued strictly continuous mapping. (The change of coordinates is then just the reversal of the two arguments in the graph.) An example fitting neither of these simple cases is shown in Figure 9–11. Highly important later as graphically Lipschitzian mappings will be maximal monotone mappings (Chapter 12) and the subgradient mappings associated with 'prox-regular' functions (Chapter 13), as well as various solution mappings.

K*. Mollifiers and Extremals

Another approach to generalized differentiation is that of approximating a Lipschitz continuous function f by a sequence of smooth functions f^ν and looking at the vectors v that can be generated at a point $\bar{x}$ as limits of gradients

$\nabla f^\nu(x^\nu)$ at points $x^\nu \to \bar{x}$. In the mollifier context of 7.19, this idea leads to a characterization of the set $\operatorname{con} \partial f(\bar{x})$ that supplements the one in 9.61.

9.67 Theorem (subgradients via mollifiers). *Let the function $f : \mathbb{R}^n \to \mathbb{R}$ be strictly continuous. For any sequence of bounded and continuous mollifiers $\{\psi^\nu\}_{\nu \in \mathbb{N}}$ on $\mathbb{R}^n$ as described in 7.19, the corresponding averaged functions*

$$f^\nu(x) := \int_{\mathbb{R}^n} f(x-z)\psi^\nu(z)dz$$

are smooth and converge uniformly to f on compact sets; moreover, at every point $\bar{x}$ the set

$$G(\bar{x}) := \big\{ v \,\big|\, \exists N \in \mathcal{N}_\infty^\#,\ x^\nu \underset{N}{\to} \bar{x} \text{ with } \nabla f^\nu(x^\nu) \underset{N}{\to} v \big\}$$

is such that

$$\partial f(\bar{x}) \subset G(\bar{x}) \quad \text{and} \quad \operatorname{con} \partial f(\bar{x}) = \operatorname{con} G(\bar{x}),$$

hence $\partial f(\bar{x}) = G(\bar{x})$ when f is subdifferentially regular at $\bar{x}$.

In general, a vector v belongs to the set $\bar{\partial} f(\bar{x}) = \operatorname{con} \partial f(\bar{x})$ if and only if for some choice of a bounded mollifier sequence $\{\psi^\nu\}_{\nu \in \mathbb{N}}$ one has $\nabla f^\nu(\bar{x}) \to v$.

Proof. As specified in 7.19, the functions ψ^ν are nonnegative, measurable and bounded with $\int_{\mathbb{R}^n} \psi^\nu(z)dz = 1$, and the sets $B^\nu = \big\{ z \,\big|\, \psi^\nu(z) > 0 \big\}$ form a bounded sequence that converges to $\{0\}$ and thus, for any $\delta > 0$, eventually lies in $\delta\mathbb{B}$. Since f is strictly continuous, it's Lipschitz continuous relative to any bounded subset of $\mathbb{R}^n$ (cf. 9.2). By Rademacher's theorem in 9.60, the set D that is comprised of the points x where f is differentiable has negligible complement in $\mathbb{R}^n$. When O is an open set O where f is Lipschitz continuous with constant κ, the points $x \in D \cap O$ satisfy $|\nabla f(x)| \leq \kappa$.

Consider ν and any x. We'll verify that f^ν is differentiable at x with

$$\nabla f^\nu(x) = \int_{\mathbb{R}^n} \nabla f(x-z)\psi^\nu(z)dz = \int_{B^\nu} \nabla f(x-z)\psi^\nu(z)dz, \tag{9(38)}$$

where the integral makes sense because of the properties of ∇f just mentioned. Observe first that

$$\Delta_\tau f^\nu(x)(w) = \int_{B^\nu} \Delta_\tau f(x-z)(w)\psi^\nu(z)dz. \tag{9(39)}$$

When $x - z \in D$, we have $\Delta_\tau f(x-z)(w') \to \langle \nabla f(x-z), w \rangle$ as $w' \to w$ and $\tau \searrow 0$. Let λ be a Lipschitz constant for f on the (bounded) set $x - B^\nu + 2\mathbb{B}$. Then $|\Delta_\tau f(x-z)(w)| \leq \lambda$ when $z \in B^\nu$, $w \in \mathbb{B}$, $\tau \in (0,1)$. Since the function ψ^ν is bounded, say by β^ν, we have

$$\big|\Delta_\tau f(x-z)(w)\psi^\nu(z)\big| \leq \lambda\beta^\nu \text{ for } z \in B^\nu,\ w \in \mathbb{B},\ \tau \in (0,1),$$

and consequently, by the Lebesgue dominated convergence theorem,

$$\lim_{\substack{\tau \searrow 0 \\ w' \overrightarrow{B} w}} \int_{B^\nu} \Delta_\tau f(x-z)(w')\psi^\nu(z)dz = \int_{B^\nu} \langle \nabla f(x-z), w\rangle \psi^\nu(z)dz$$

$$= \langle v, w\rangle \text{ for } v = \int_{B^\nu} \nabla f(x-z)\psi^\nu(z)dz.$$

In combination with 9(39), this establishes the formula in 9(38) and in particular the existence of $\nabla f^\nu(x)$ for any $x \in I\!R^n$. The continuity of $\nabla f^\nu(x)$ with respect to x, giving the smoothness of f^ν, is evident from this formula, the continuity of the mollifier and elementary properties of the integral.

The uniform convergence on bounded sets of the functions f^ν to f comes from 7.19(a). As far as the gradient mappings ∇f^ν are concerned, we know that for any bounded set X and any $\delta > 0$ there exists κ such that $|\nabla f(x-z)| \leq \kappa$ when $x - z \in D$, $x \in X$ and $z \in \delta I\!B$, and therefore from 9(38) that

$$|\nabla f^\nu(x)| \leq \int_{B^\nu} |\nabla f(x-z)|\psi(z)dz \leq \kappa \int_{B^\nu} \psi(z)dz = \kappa \text{ for } x \in X,$$

as long as ν is large enough that $B^\nu \subset \delta I\!B$. This shows that for any sequence of points $x^\nu \in X$ converging to some $\bar{x}$, the sequence of gradients $\nabla f^\nu(x^\nu)$ is bounded, and all of its cluster points v have $|v| \leq \kappa$.

We now take up the claim that $\partial f(\bar{x}) \subset G(\bar{x})$. For any $\bar{x}$, the set $G(\bar{x})$ is nonempty and bounded. Indeed, the mapping $G : I\!R^n \rightrightarrows I\!R^n$ is locally bounded, and the limit process in its definition ensures that it's osc. Therefore, to know that $\partial f \subset G$, we only need to check that $\widehat{\partial} f \subset G$, or on the basis of 8.47(a), that whenever $\bar{v}$ is a proximal subgradient of f at $\bar{x}$, one has $\bar{v} \in G(\bar{x})$. In verifying the latter, it's harmless to assume that f is bounded below on $I\!R^n$, inasmuch as our analysis is local anyway. In that case there's a $\bar{\rho} > 0$ with

$$f(x) \geq f(\bar{x}) + \langle \bar{v}, x - \bar{x}\rangle - \tfrac{1}{2}\bar{\rho}|x - \bar{x}|^2 \text{ for all } x.$$

For any $\rho \in (\bar{\rho}, \infty)$, let $g(x) = f(x) - f(\bar{x}) - \langle \bar{v}, x - \bar{x}\rangle + \frac{1}{2}\rho|x - \bar{x}|^2$. Then $g(x) \geq \frac{1}{2}(\rho - \bar{\rho})|x - \bar{x}|^2$ and $g(\bar{x}) = 0$. Hence $\operatorname{argmin} g = \{\bar{x}\}$. As with f above, the averaged functions g^ν obtained from the mollifiers ψ^ν are smooth and converge to g uniformly on compact sets; also, $g^\nu \xrightarrow{e} g$. We have

$$\begin{aligned} g^\nu(x) &\geq \int_{B^\nu} \tfrac{1}{2}(\rho - \bar{\rho})|x - z - \bar{x}|^2 \psi^\nu(z)dz \\ &= \tfrac{1}{2}(\rho - \bar{\rho}) \int_{B^\nu} \big(|x - \bar{x}|^2 - 2\langle x - \bar{x}, z\rangle + |z|^2\big)\psi^\nu(z)dz \\ &\geq \tfrac{1}{2}(\rho - \bar{\rho})\big(|x - \bar{x}|^2 - 2\langle x - \bar{x}, z^\nu\rangle\big) \text{ for } z^\nu := \int_{B^\nu} z\psi^\nu(z)dz, \end{aligned}$$

where $z^\nu \to 0$, so that $\langle x - \bar{x}, z^\nu\rangle \leq |x - \bar{x}|$ once ν is sufficiently large. For such ν we have $g^\nu \geq g_0$ for the function

$$g_0(x) = \tfrac{1}{2}(\rho - \bar{\rho})\big[(|x - \bar{x}|^2 - 2|x - \bar{x}|\big] = \tfrac{1}{2}(\rho - \bar{\rho})\big[(|x - \bar{x}|^2 - 1)^2 - 1\big],$$

which is level-bounded. Thus, the sequence $\{g^\nu\}_{\nu\in\mathbb{N}}$ is eventually level-bounded by 7.32(a). Then by Theorem 7.33, since $g^\nu \xrightarrow{e} g$, the sets $\operatorname{argmin} g^\nu$ are nonempty for all ν in some index set $N \in \mathcal{N}_\infty \subset \mathcal{N}_\infty^\#$; furthermore, for an arbitrary choice of points $x^\nu \in \operatorname{argmin} g^\nu$ for $\nu \in N$ we have $x^\nu \underset{N}{\to} \bar{x}$.

Along such a sequence $\{x^\nu\}_{\nu\in N}$, we have $\nabla g^\nu(x^\nu) = 0$. But from the definition of g^ν and the gradient formula in 9(38) we have $\nabla g^\nu(x^\nu) = \nabla f^\nu(x^\nu) - \nabla h^\nu(x^\nu)$ for the sequence of mollified functions h^ν obtained from $h(x) := f(\bar{x}) + \langle \bar{v}, x - \bar{x}\rangle - \frac{1}{2}\rho|x-\bar{x}|^2$. Thus, again via 9(38),

$$\begin{aligned}\nabla f^\nu(x^\nu) = \nabla h^\nu(x^\nu) &= \int_{B^\nu} \nabla h(x^\nu - z)\psi^\nu(z)dz \\ &= \int_{B^\nu} \big[\bar{v} - \rho(x^\nu - z + \bar{x})\big]\psi^\nu(z)dz = \bar{v} - \rho(x^\nu - z^\nu - \bar{x}).\end{aligned}$$

Since $x^\nu \underset{N}{\to} \bar{x}$ and $z^\nu \to 0$, we conclude that $\nabla f^\nu(x^\nu) \underset{N}{\to} \bar{v}$, as required.

Next we must demonstrate that $\operatorname{con}\partial f(\bar{x}) = \operatorname{con} G(\bar{x})$. Both $\partial f(\bar{x})$ and $G(\bar{x})$ are compact, and the same for their convex hulls (by Corollary 2.30), so the convex hulls coincide when $\partial f(\bar{x})$ and $G(\bar{x})$ have the same support function (cf. Theorem 8.24). The support function of $\partial f(\bar{x})$ is $\widehat{d}f(\bar{x})$ (cf. 9.15), so our task is to show that

$$\sup\big\{\langle v, w\rangle \,\big|\, v \in G(\bar{x})\big\} = \widehat{d}f(\bar{x})(w) \text{ for all } w. \tag{9(40)}$$

Consider any $\bar{v} \in G(\bar{x})$ along with $N \in \mathcal{N}_\infty^\#$ and $x^\nu \underset{N}{\to} \bar{x}$ such that $\nabla f^\nu(x^\nu) \underset{N}{\to} \bar{v}$. We have

$$\big\langle \bar{v}, w\big\rangle = \lim_{\nu\in N} \big\langle \nabla f^\nu(x^\nu), w\big\rangle = \lim_{\nu\in N} \int_{B^\nu} \big\langle \nabla f(x^\nu - z), w\big\rangle \psi^\nu(z)dz$$

with $\big\langle \nabla f(x^\nu - z), w\big\rangle \le \widehat{d}f(x^\nu - z)(w)$ because the gradient $\nabla f(x)$ belongs to $\partial f(x)$ when it exists, and $\widehat{d}f(x)$ is the support function of $\partial f(x)$. We know from 9.16 that $\widehat{d}f(x)(w)$ is usc with respect to x, so for any w and $\varepsilon > 0$ there exists $\delta > 0$ such that

$$\widehat{d}f(x-z)(w) \le \widehat{d}f(\bar{x})(w) + \varepsilon \text{ for } z \in \delta\mathbb{B} \text{ when } |x - \bar{x}| \le \delta.$$

The latter holds for $\nu \in N$ by $x = x^\nu - z$ when $z \in B^\nu$, and then

$$\int_{B^\nu} \big\langle \nabla f(x^\nu - z), w\big\rangle \psi^\nu(z)dz \le \int_{B^\nu} \big[\widehat{d}f(\bar{x})(w) + \varepsilon\big]\psi^\nu(z)dz = \widehat{d}f(\bar{x})(w) + \varepsilon.$$

Thus, $\langle \bar{v}, w\rangle \le \widehat{d}f(\bar{x})(w) + \varepsilon$ for all $w \in \mathbb{R}^n$ and $\varepsilon > 0$. Since $\bar{v}$ was an arbitrary element of $G(\bar{x})$, this tells us that '$\le$' holds in 9(40).

To verify also that '$\ge$' holds in 9(40), fix $\bar{w}$ and again pick any $\varepsilon > 0$. According to the formula for $\widehat{d}f(\bar{x})(\bar{w})$ in 9.15, there exist $\tau_\varepsilon \in (0, \varepsilon)$ and $x_\varepsilon \in \mathbb{B}(\bar{x}, \varepsilon)$ such that $\big[f(x_\varepsilon + \tau_\varepsilon \bar{w}) - f(x_\varepsilon)\big]/\tau_\varepsilon > \widehat{d}f(\bar{x})(\bar{w}) - \varepsilon$. From the convergence of f^ν to f, there then exists $N_\varepsilon \in \mathcal{N}_\infty$ such that

$$\big[f^\nu(x_\varepsilon+\tau_\varepsilon\bar w)-f^\nu(x_\varepsilon)\big]/\tau_\varepsilon>\widehat{d}f(\bar x)(\bar w)-\varepsilon \text{ when } \nu\in N_\varepsilon.$$

The mean value theorem, which is applicable to f^ν because of its smoothness, allows us to write $\big[f^\nu(x_\varepsilon+\tau_\varepsilon\bar w)-f^\nu(x_\varepsilon)\big]/\tau_\varepsilon$ as $\langle\nabla f^\nu(x_\varepsilon^\nu),\bar w\rangle$ for some x_ε^ν on the line segment between x_ε and $x_\varepsilon+\tau_\varepsilon\bar w$. It follows that, through consideration of smaller and smaller values of ε, one can come up with and index set $N_0\in\mathcal{N}_\infty^\#$ and points $x^\nu \underset{N_0}{\to} \bar x$ such that $\liminf_{\nu\in N_0}\langle\nabla f^\nu(x^\nu),\bar w\rangle\geq\widehat{d}f(\bar x)(\bar w)$. Any cluster point of $\{\nabla f^\nu(x^\nu)\}_{\nu\in N_0}$ is then a vector $v\in G(\bar x)$ with $\langle v,\bar w\rangle\geq\widehat{d}f(\bar x)(\bar w)$, as needed.

The claim that $\partial f(\bar x)=G(\bar x)$ when f is subdifferentially regular at $\bar x$ merely reflects the fact that in this case $\partial f(\bar x)$ is convex (cf. 8.11), so the inclusions $\partial f(\bar x)\subset G(\bar x)$ and $\operatorname{con}G(\bar x)\subset\operatorname{con}\partial f(\bar x)$ collapse to an equation.

For the last assertion of the theorem, observe that any $v\in G(\bar x)$, generated as $\lim_{\nu\in N}\nabla f^\nu(x^\nu)$ in terms of an index set $N\in\mathcal{N}_\infty^\#$ and points $x^\nu \underset{N}{\to} \bar x$, can also be generated as $\lim_{\nu\in N}\nabla\tilde f^\nu(\bar x)$ for the averaged functions $\tilde f^\nu(x)=\int_{\mathbb{R}^n}f(x-\tilde z)\tilde\psi^\nu(\tilde z)d\tilde z$ corresponding to a certain mollifier sequence $\{\tilde\psi^\nu\}_{\nu\in\mathbb{N}}$. This comes from writing

$$\nabla f^\nu(x^\nu)=\int_{B^\nu}\nabla f(x^\nu-z)\psi^\nu(z)dz=\int_{\tilde B^\nu}\nabla f(\bar x-\tilde z)\tilde\psi^\nu(\tilde z)\,d\tilde z$$

with $\tilde\psi^\nu(\tilde z)=\psi^\nu(\tilde z-\bar x+x^\nu)$ and $\tilde B^\nu=\big\{\tilde z'\,\big|\,\tilde\psi^\nu(\tilde z)>0\big\}=B^\nu-x^\nu+\bar x$. By discarding the mollifiers $\tilde\psi^\nu$ for $\nu\notin N$ and re-indexing, we can arrange to have v be the full limit $\lim_\nu\nabla\tilde f^\nu(\bar x)$.

This informs us that the set $G(\bar x)$, arising at $\bar x$ from one given mollifier sequence $\{\psi^\nu\}_{\nu\in\mathbb{N}}$, is included within the set $\overline{G}(\bar x)$ consisting of the vectors v that arise at $\bar x$ through consideration of all possible mollifier sequences, but with restriction to full limits (not just cluster points) of gradients evaluated only at $\bar x$ itself. Then $\operatorname{con}G(\bar x)\subset\operatorname{con}\overline{G}(\bar x)$. But every element of $\overline{G}(\bar x)$ comes from some mollifier sequence and hence, by the preceding analysis, belongs to $\operatorname{con}\partial f(\bar x)$. From the already established fact that $\operatorname{con}G(\bar x)=\operatorname{con}\partial f(\bar x)$, we deduce that $\operatorname{con}\overline{G}(\bar x)=\operatorname{con}\partial f(\bar x)$.

We want to confirm that actually $\overline{G}(\bar x)=\operatorname{con}\partial f(\bar x)$. This can be accomplished now by ascertaining that $\overline{G}(\bar x)$ is convex. Consider any v_0 and v_1 in $\overline{G}(\bar x)$ and $\tau\in(0,1)$. We must check that the vector $v_\tau=(1-\tau)v_0+\tau v_1$ belongs to $G(\bar x)$ as well. By the definition of $\overline{G}(\bar x)$, there are mollifier sequences $\{\psi_0^\nu\}_{\nu\in\mathbb{N}}$ and $\{\psi_1^\nu\}_{\nu\in\mathbb{N}}$ such that the associated sequences of averaged functions f_0^ν and f_1^ν have $v_0=\lim_\nu\nabla f_0^\nu(\bar x)$ and $v_1=\lim_\nu\nabla f_1^\nu(\bar x)$. Let B_0^ν and B_1^ν give the sets of points where ψ_0^ν and ψ_1^ν are positive, and let $v_\tau^\nu=(1-\tau)\nabla f_0^\nu(\bar x)+\tau\nabla f_1^\nu(\bar x)$, so that $v_\tau^\nu\to v_\tau$. We have from 9(38) that

$$\begin{aligned}v_\tau^\nu&=(1-\tau)\int_{B_0^\nu}\nabla f(\bar x-z)\psi_0^\nu(z)dz+\tau\int_{B_1^\nu}\nabla f(\bar x-z)\psi_1^\nu(z)dz\\&=\int_{B_0^\nu\cup B_1^\nu}\nabla f(\bar x-z)\psi_\tau^\nu(z)dz\ \text{ for }\ \psi_\tau^\nu=(1-\tau)\psi_0^\nu+\tau\psi_1^\nu.\end{aligned}$$

The sets $B_\tau^\nu = \{z \,|\, \psi_\tau^\nu(z) > 0\} = B_0^\nu \cup B_1^\nu$ converge to $\{0\}$, and the functions $\psi_\tau^\nu \geq 0$ are still measurable and bounded with $\int_{\mathbb{R}}^n \psi_\tau^\nu(z)dz = 1$. We thus have a mollifier sequence $\{\psi_\tau^\nu\}_{\nu\in\mathbb{N}}$ yielding functions $f_\tau^\nu(x) = \int_{\mathbb{R}^n} f(x-z)\psi_\tau^\nu(z)dz$ with $v_\tau^\nu = \nabla f_\tau^\nu(\bar{x})$. Then $v_\tau = \lim_\nu \nabla f_\tau^\nu(\bar{x})$, so $v \in \overline{G}(\bar{x})$. □

Distance functions are sometimes used in reducing a constrained minimization problem to an unconstrained one, although at the price of nonsmoothness. Here too, Lipschitzian properties are essential.

9.68 Proposition (constraint relaxation). *Suppose $f : \mathbb{R}^n \to \overline{\mathbb{R}}$ has $\operatorname{lip} f \leq \kappa$ on C, where the set $C \subset \mathbb{R}^n$ is closed and nonempty. Then for any $\lambda > \kappa$ there is an open set $O \supset C$ such that $\operatorname{lip} f < \lambda$ on O and*

$$\inf\nolimits_O (f + \lambda d_C) = \inf\nolimits_C f, \qquad \operatorname{argmin}_O (f + \lambda d_C) = \operatorname{argmin}_C f. \qquad 9(41)$$

Proof. Fix $\lambda > 0$ and any $\varepsilon > 0$ such that $\kappa + \varepsilon < \lambda$. For each $x \in C$ there is an open neighborhood $V \in \mathcal{N}(x)$ such that $\operatorname{lip} f \leq \kappa + \varepsilon$ on V. The union of all such neighborhoods as x ranges over C is an open set $O_1 \supset C$ on which $\operatorname{lip} f \leq \kappa + \varepsilon$. Let D be the complement of O_1 in $\mathbb{R}^n$. Then D is closed and $d_D > 0$ on C. In terms of the projection mapping P_C for C (cf. 1.20), define

$$p(x) = \inf_{w \in P_C(x)} d_D(w).$$

Obviously $0 < p(x) < \infty$ for all $x \in \mathbb{R}^n$, because d_D is continuous while P_C is osc and locally bounded with $\operatorname{dom} P_C = \mathbb{R}^n$ (cf. 7.44) these properties implying in particular that $P_C(x)$ is a nonempty, compact set for every x. By interpreting p as arising from parametric optimization by minimizing $g(x,w) = d_D(w) + \delta_{P_C(x)}(w)$ in w, we can apply 1.17 to ascertain that p is lsc. Let

$$O = \{x \in \mathbb{R}^n \,|\, d_C(x) < p(x)\}.$$

Since d_C is continuous and $p > 0$, we have $p - d_C$ lsc and O open with $O \supset C$.

For any $x \in O$ and any $w \in P_C(x)$ the value $\rho = |x - w|$ satisfies $\rho = d_C(x) < p(x) \leq d_D(w)$, so that x and the ball $\mathbb{B}(w,\rho)$ lie in O_1. Then $\operatorname{lip} f \leq \kappa + \varepsilon$ on $\mathbb{B}(w,\rho)$, and in particular $\operatorname{lip} f \leq \lambda$ at x by our arrangement that $\kappa + \varepsilon < \lambda$ at the beginning of the proof. Because $\mathbb{B}(w,\rho)$ is a convex set, we know from 9.2 that f is Lipschitz continuous on $\mathbb{B}(w,\rho)$ with constant $\kappa + \varepsilon$. Hence $f(w) - f(x) \leq (\kappa+\varepsilon)|x - w| = (\kappa + \varepsilon)d_C(x)$ and

$$(f + \lambda d_C)(x) \;\geq\; f(w) + \big[\lambda - (\kappa+\varepsilon)\big] d_C(x) \;\text{ for all } x \in O,$$

where $\lambda - (\kappa + \varepsilon) > 0$. From this we know that for any $x \in O \setminus C$ there is a point $w \in C$ with $f(w) < (f + \lambda d_C)(x)$, while on C itself $f + \lambda d_C$ agrees with f. The assertions about 'inf' and 'argmin' follow immediately. □

The relations in 9(41) tell us that the constrained problem of minimizing f over C can be reduced, in a localized sense specified by the open set $O \supset C$, to that of minimizing $g = f + \lambda d_C$ over $\mathbb{R}^n$: the two problems have the same optimal value and the same optimal solutions. Under these circumstances the term λd_C is called an *exact penalty function* for the given problem. Note that the modified objective function $g = f + \lambda d_C$ in 9.68 is such that $\operatorname{lip} g \leq \kappa + \lambda$ on O. This follows from 9.6 and the calculus in 9.8.

The result in 9.68 can be applied to locally optimal solutions $\bar{x}$ relative to C by substituting a subset $C \cap \mathbb{B}(\bar{x}, \varepsilon)$ for C.

Among the many interesting consequences of the theory of openness of mappings is the following collection of 'extremal principles', so called because they identify consequences of certain extreme, or boundary aspects of $\bar{x}$.

9.69 Proposition (extremal principles).

(a) *Suppose $\bar{x} \in X$ but $F(\bar{x}) \notin \operatorname{int} F(X)$, where $X \subset \mathbb{R}^n$ is a closed set and $F : \mathbb{R}^n \to \mathbb{R}^m$ is strictly continuous. Then there must be a vector $y \neq 0$ such that $N_X(\bar{x}) + \partial(yF)(\bar{x}) \ni 0$.*

(b) *Let $C = C_1 + \ldots + C_r$ for closed sets $C_i \subset \mathbb{R}^n$. Consider a point $\bar{x} \in C$ along with any points $\bar{x}_i \in C_i$ such that $\bar{x} = \bar{x}_1 + \cdots + \bar{x}_r$. If $\bar{x} \notin \operatorname{int} C$, there must be a vector $v \neq 0$ such that $v \in N_{C_i}(\bar{x}_i)$ for all $i = 1, \ldots, r$.*

(c) *Suppose $\bar{x} \in C \cap D$ for closed sets $C, D \subset \mathbb{R}^n$ that are nonoverlapping at $\bar{x}$ in the sense of having $0 \notin \operatorname{int} \big[C \cap \mathbb{B}(\bar{x}, \delta) - D \cap \mathbb{B}(\bar{x}, \delta)\big]$ for some $\delta > 0$. Then there must exist $v \neq 0$ such that*

$$v \in N_C(\bar{x}), \qquad -v \in N_D(\bar{x}),$$

in which case in particular the hyperplane $H = \big\{w \,\big|\, \langle v, w\rangle = 0\big\} = v^{\perp}$ separates the regular tangent cones $\widehat{T}_C(\bar{x})$ and $\widehat{T}_D(\bar{x})$ from each other.

Proof. To verify (a), let $\bar{u} = F(\bar{x})$ and look at the mapping S defined by $S(x) = \{F(x)\}$ when $x \in X$ but $S(x) = \emptyset$ when $x \notin X$. Obviously $\bar{u} \in S(\bar{x})$, but $\bar{u} \notin \operatorname{int}(\operatorname{rge} S)$ because $\operatorname{rge} S = F(X)$. In particular, S can't be open at $\bar{u}$ for $\bar{x}$. Our strategy will be to explore what this implies through Theorem 9.43, which is applicable because S is osc.

According to that theorem, the failure of S to be open at $\bar{u}$ for $\bar{x}$ is associated with the existence of a vector $y \neq 0$ such that $0 \in D^*S(\bar{x} \,|\, \bar{u})(y)$, a condition meaning by definition that $(0, -y) \in N_{\operatorname{gph} S}(\bar{x}, \bar{u})$. We have $\operatorname{gph} S = (X \times \mathbb{R}^m) \cap \operatorname{gph} F$, hence by the normal cone rules for intersections in 6.42 and products in 6.41 also $N_{\operatorname{gph} S}(\bar{x}, \bar{u}) \subset \big[N_X(\bar{x}) \times \{0\}\big] + N_{\operatorname{gph} F}(\bar{x}, \bar{u})$. Thus, there must exist $v \in N_X(\bar{x})$ with $(-v, -y) \in N_{\operatorname{gph} F}(\bar{x}, \bar{u})$, or in other words $-v \in D^*F(\bar{x})(y)$, which by 9.24(b) is the same as $-v \in \partial(yF)(\bar{x})$. Then $N_X(\bar{x}) + \partial(yF)(\bar{x}) \ni v - v = 0$. This is what was claimed.

Part (b) is obtained from part (a) by specialization to $X = C_1 \times \cdots \times C_r$ and $F : (x_1, \ldots, x_r) \mapsto x_1 + \cdots + x_r$. Part (c) falls out of (b) by choosing $C_1 = C \cap \mathbb{B}(\bar{x}, \delta)$ and $C_2 = -D \cap \mathbb{B}(\bar{x}, \mathbb{B})$, with the $\bar{x}$ in (b) identified not with the $\bar{x}$ in (c), but with 0. Because $\widehat{T}_C(\bar{x})$ is the polar cone $N_C(\bar{x})^*$ (see 6.28(b)), we have $\widehat{T}_C(\bar{x}) \subset \big\{w \,\big|\, \langle v, w\rangle \leq 0\big\}$ when $v \in N_C(\bar{x})$. On the other hand, $\widehat{T}_D(\bar{x}) \subset \big\{w \,\big|\, \langle v, w\rangle \geq 0\big\}$ when $-v \in N_D(\bar{x})$. Therefore, in this situation we get separation of the cones $\widehat{T}_C(\bar{x})$ and $\widehat{T}_D(\bar{x})$ from each other. □

The result in 9.69(c) is a sort of *nonconvex separation principle.* Most appealing is the case where C and D are regular at $\bar{x}$. Then the regular tangent cones reduce to $T_C(\bar{x})$ and $T_D(\bar{x})$, while v and $-v$ are regular normals to C and D at $\bar{x}$. The hyperplane H has $T_C(\bar{x})$ on one side and $T_D(\bar{x})$ on the other, a property expressible also by the inequality characterizing regular normals to C and D, and which can be construed as 'infinitesimal separation'. It reduces to true separation of C and D when they're convex.

Commentary

Lipschitz continuous functions, stemming from Lipschitz [1877], have long been important in real analysis, especially in measure-theoretic developments and the existence theory for solutions to differential equations. In variational analysis they came to the fore with Clarke's definition of subgradients by way of Rademacher's theorem (in 9.60), a modern proof of which can be found in the book of Evans and Gariepy [1992]. Such functions have been central to the subject ever since. They have offered prime territory for explorations of nonsmoothness and have given rise to a kind of 'Lipschitzian analysis' comparable to convex analysis.

But Lipschitzian properties haven't only been important as favorable *assumptions*; they have also been sought as favorable *conclusions* in quantifying the variations of optimal values, optimal solutions, and the behavior of solution elements more generally. In this, their role can be traced from the classical literature of perturbations to modern nonsmooth extensions of the implicit function theorem, and on to quantitative studies of the continuity of general set-valued mappings, which tie in with 'metric regularity'.

One of the principle achievements of researchers in variational analysis in recent years has been the recognition of the ways that these different lines of development can be brought together and supported by an effective *calculus*. This chapter and the next present this achievement along with new insights such as the continuity interpretation of normal vectors and subgradients in 9.41.

Clarke's approach to generalizing subgradients beyond convex analysis was to start with Lipschitz continuous functions f, taking the set of subgradients at $\bar{x}$ to be what we've denoted in 9.61 by $\operatorname{con} \overline{\nabla} f(\bar{x})$. As mentioned in the Commentary for Chapter 8, he went on then to use the subgradients of the distance function d_C (which is Lipschitz continuous by 9.6) to get normal vectors to a closed set C at a point $\bar{x}$. Specifically, he took the normal cone to be $\operatorname{cl} \operatorname{pos}[\operatorname{con} \overline{\nabla} d_C(\bar{x})]$, and he relied then on epigraphical normal vectors of such type in defining subgradients of lsc functions f in general. In the notation used here, the normal cone obtained by this route is $\overline{N}_C(\bar{x})$ (cf. 6(19), 6.38 and 8.53) while the subgradient set is $\bar{\partial} f(\bar{x})$ (cf. 8(32), 8.49 and 9.61).

Although a different approach to subgradients and normal vectors has been followed in this book, Clarke's pattern still has echoes in the infinite-dimensional theory (with some modifications, such as dropping the convexification and placing special attention on different modes of convergence and closure); see Borwein and Ioffe [1996], for example. But anyway the importance of Lipschitzian properties hasn't been diminished even for finite-dimensional spaces. The results in 9.41 on the reinterpretation of normal vectors and subgradients in terms of such properties, which are stated here for the first time, reinforce the significance on the deepest level.

There's much to say about this, but before going further we have to comment on the terminology that we've adopted. Traditionally, Lipschitzian continuity of a mapping F has been depicted in relation to a set X and the specification of a particular constant κ acting on X. While maintaining this on the side, we've worked at freeing the concept into one capable also of being applied *pointwise*, hence our introduction of the term 'strict continuity', which can readily be handled in this manner. Although strict continuity of F relative to a set X is the same as local Lipschitz continuity of F relative to X, as long as a *single-valued* mapping is in question, local Lipschitz continuity tends to blur this focus and falls short of the challenges when set-valued generalizations are sought. (The choice of 'strict' harmonizes with strict differentiability and other usages where a condition at a single point is strengthened to pairs of nearby points.)

The philosophy is evident in Theorem 9.2 as well as elsewhere in the chapter, and the advantages gained are both stylistic and technical. It's no longer necessary, whenever the Lipschitz continuity comes up locally, to specify an explicit background set X when the particular choice of X (other than as a neighborhood, say) may be irrelevant. More significantly, a path is laid to working with the modulus *function* $\operatorname{lip} F$ and its characteristics. The theme of calculating and estimating $\operatorname{lip} F(\bar{x})$ from the available structure of F runs through many of the results we obtain.

A key result on these lines is Theorem 9.13, which characterizes the property of f being strictly continuous *at a point* $\bar{x}$ as corresponding to the situation where the 'cosmic' subgradient set $\partial f(\bar{x}) \cup \operatorname{dir} \partial^\infty f(\bar{x})$ actually contains no direction points (this being describable in several different ways), and expresses the modulus $\operatorname{lip} f(\bar{x})$ as the 'outer norm' of this set. The fact that functions satisfying Lipschitz conditions have bounded sets of subgradients from which the values of local constants can be derived was apparent from the start in Clarke's scheme. The implication from the boundedness of the convexified subgradient set $\bar{\partial} f(\bar{x})$ to f being Lipschitz continuous on a neighborhood of $\bar{x}$ was established, however, by Rockafellar [1979b]. This translates to the conditions in 9.13 through the relations in 8.49, which originated in Rockafellar [1981b].

The Lipschitz properties of finite convex functions in 9.14 were well appreciated in convex analysis along with the fact that such functions f are differentiable precisely at the points $\bar{x}$ where $\partial f(\bar{x})$ is a singleton; cf. Rockafellar [1970a]. Recognition that this singleton property corresponds more generally as in 9.18, regardless of convexity, to *strict* differentiability came in Rockafellar [1979a]. Strict differentiability is a concept attributable to Peano in the late 19th century and discussed for example in Bourbaki [1967]. Little was made of it until the idea was reinvented by Leach [1961], who called it 'strong differentiability'.

The first of the 'lim sup' formulas in 9.15 for the regular derivatives $\widehat{d} f(\bar{x})(w)$ was originally Clarke's definition of a new kind of directional derivative $f^\circ(\bar{x}; w)$, yielding insights about Lipschitz continuous functions. He showed that this gave the support function of $\bar{\partial} f(\bar{x})$ (which is the same, in this setting, as the support function of either $\partial f(\bar{x})$ or $\overline{\nabla} f(\bar{x})$); cf. Clarke [1975]. Rockafellar [1980] extended Clarke's directional derivatives $f^\circ(\bar{x}; w)$ beyond Lipschitz continuous functions by means of the formula in Definition 8.16, using the notation $f^\uparrow(\bar{x}, w)$ instead of $\widehat{d} f(\bar{x})(w)$.

Theorem 9.16, characterizing the combination of Lipschitz continuity with subdifferential regularity, was proved in Rockafellar [1982b]. This is also the source for the identification of differentiability with strict differentiability of such functions through 9.18(e) and for the equivalences of continuous differentiability with 9.19(c) and 9.19(d); likewise for the continuity fact in 9.20. The constancy criterion in 9.22, in the proximal subgradient form mentioned afterward, was brought out by Clarke and Redheffer [1993]. For more on this and also characterizations of Lipschitz continuity in terms of directional derivatives, see Clarke, Stern and Wolenski [1993].

The outer and inner 'norms' defined for positively homogeneous mappings in 9(4)–9(5) go back to the days of convex analysis; see Rockafellar [1967a], [1976b]. They were used in estimates already by Robinson [1972].

The 'scalarization formula' in 9.24(b) for expressing the coderivatives of single-valued strictly continuous mappings F appeared in Ioffe [1984a], [1984b]; cf. also Mordukhovich [1984]. In geometric formulation, this fact was stated in the dissertation of Kruger [1981]; see Kruger [1985]. The contrast in 9.25 between single-valuedness of $DF(\bar{x})$ and that of $D^*F(\bar{x})$ in characterizing semidifferentiability versus strict differentiability at $\bar{x}$ hasn't previously been recorded.

In developing Lipschitzian properties of set-valued mappings, we were motivated by the desirability of a 'pointwise' calculus of the associated constants, like the one in the single-valued case, and by the shape of the quantitative theory of set convergence and set-valued mappings in Chapters 4 and 5. In that theory, Hausdorff distance is a concept effectively limited to a context of bounded sets and locally bounded mappings, with minor exceptions. While Lipschitz continuity of locally bounded mappings is important in areas like the study of differential inclusions, cf. Aubin and Cellina [1984], many other applications need to contend with mappings having unbounded images. This made it imperative for us to adopt the tactic, fresh to this topic of mappings, of working with $d\!l_\rho$ and $\hat{d\!l}_\rho$ in tandem. It's in this spirit that 9.26–9.31 should be received.

The estimates in Example 9.32 are new and serve to point out the differences between 'Lipschitz' and 'sub-Lipschitz' continuity in practice when dealing with mappings that aren't locally bounded. Part (a) of Theorem 9.33 is new too; part (b), however, is closely related to the Robinson-Ursescu theorem in 9.48, discussed below. The statement in 9.35 for mappings with $\operatorname{gph} S$ polyhedral can be attributed to Walkup and Wets [1969a].

The concept of sub-Lipschitz continuity as a way out of the limitations of Hausdorff distance first surfaced in Rockafellar [1985c], where it was characterized relative to Lipschitz continuity, distance functions and the Aubin property (much as in 9.29, 9.31, 9.37 and 9.38). The latter property was designated 'pseudo-Lipschitz' by Aubin [1984]. Noting that *pseudo* has the connotation of 'false', we prefer to name this after Aubin himself as the inventor.

The characterizations of epi-Lipschitzian sets in 9.42(a) and directionally Lipschitzian functions in 9.42(b) come from Rockafellar [1979a] and [1979b], respectively.

The equivalence in Theorem 9.43 between the metric regularity property in (b) and the openness property in (c) was known to Dmitruk, Miliutin and Osmolovskii [1980] and Ioffe [1981]. Connections between these and (a), the Aubin property for S^{-1}, came out in Borwein and Zhuang [1988] and a similar result of Penot [1989]. For subsequent work on the relations between such properties, see Cominetti [1990], Dontchev and Hager [1994], Mordukhovich [1992], [1993], [1994c], Mordukhovich and Shao [1995], [1997b], and Ioffe [1997]. Here we have relied on the extended formulation of the Aubin property in Lemma 9.39 to simplify the statements; this improvement is owed to Henrion [1997].

The Mordukhovich criterion in Theorem 9.40 is the key to a Lipschitzian *calculus* for set-valued mappings and indeed to understanding the profound role of Lipschitzian properties in variational analysis quite generally. Its original context was not that of the Aubin property, but the property of 'openness at a linear rate' in 9.43(c). In such terms, the coderivative criterion was announced in Mordukhovich [1984] along with an estimate for the corresponding constant; the proofs were furnished in Mordukhovich [1988], where the estimate was shown to be exact, at least for S local bounded. Ioffe [1981b] had earlier obtained the same estimate (but not the criterion itself) through his theory of 'fans', certain predecessors to graphical coderivatives; in Ioffe [1987] he connected it with coderivative norms. Other early research related to this matter was undertaken by Dmitruk, Miliutin and Osmolovskii [1980] and Warga [1981]; see also Kruger [1988].

The equivalences of Borwein and Zhuang [1988] paved the way for translating Mordukhovich's criterion for the openness property to one for the Aubin property in the form that we have adopted in Theorem 9.40. Mordukhovich [1992] later provided a direct derivation of the criterion in this form and the associated estimates for the modulus. Upper bounds for the modulus in the Aubin property had previously been derived by Rockafellar [1985c], but in essence relative to the sublinearized coderivative mapping $\overline{D}^*S(\bar{x}\,|\,\bar{u})$, arising from the Clarke normal cone $\overline{N}_{\mathrm{gph}\,S}(\bar{x},\bar{u})$, instead of $D^*S(\bar{x}\,|\,\bar{u})$. Such bounds can be far from exact, unless S is graphically regular.

For single-valued mappings the concepts of metric regularity and openness at a linear rate really go back Ljusternik [1934] and in more robust form to Graves [1950]; see Dontchev [1996] for discussion and history. The early context could be viewed as one occupied with equality constraints, especially in infinite-dimensional spaces, since the finite-dimensional case falls into the framework of the classical implicit-function theorem.

Once inequality constraints attracted interest in the mathematical programming community, a different line of development started up with the result of Hoffman [1952] about estimating errors in systems of linear inequalities. Hoffman's result, which can be identified with 9.47 as applied to $S(x) = Ax - b + I\!R^m_+$ for $A \in I\!R^{n\times n}$ and $b \in I\!R^m$, continues to be the object of much research in getting estimates of the modulus κ; cf. Luo and Tseng [1994], Klatte and Thiere [1995], and Burke and Tseng [1996] for recent work and additional references. The 'piecewise polyhedral' formulation in 9.47 is new, but it doesn't provide much of a handle on κ.

Robinson was a key player in that line of development from the beginning. After applying and extending Hoffman's theorem in several ways, cf. Robinson [1973a], [1973b], [1975], he moved on to nonlinear generalizations, first in the context of convexity. Theorem 9.48, one of the landmark results on the metric regularity of set-valued mappings, was independently put together by Ursescu [1975] and Robinson [1972], [1976a], for mappings with closed convex graph from one Banach space to another. In Robinson [1976b], error bounds for nonconvex inequality systems were obtained. For an infinite-dimensional extension of Hoffman's theorem with a tie-in to Clarke subgradients, see Ioffe [1979a].

The conditions of metric regularity and openness of mappings that have been treated here are, in a sense, 'first-order' properties. Higher-order versions have been developed by Frankowska [1986], [1987a], [1987b], [1989], [1990], even for mappings between metric spaces.

The Lipschitzian properties of derivative mappings in 9.49 are new. The result in 9.50 on the semidifferentiability consequences of proto-differentiability could be put together from formulas of Thibault [1983]. It was stated explicitly in this form by Rockafellar [1989b], where much of the example in 9.51 of feasible-set mappings was developed as well.

The strict derivative mapping $D_*S(\bar{x}\,|\,\bar{u})$ in Definition 9.53 is the 'paratingent' derivative of Aubin and Frankowska [1990] (corresponding graphically to their 'paratingent cone' in variational geometry); our term brings out the evident similarity to the kind of limit taken in the definition of strict differentiability.

Strict derivatives of single-valued Lipschitz continuous mappings F were investigated by Kummer [1991]. He noted that the existence of a single-valued Lipschitz continuous inverse in the circumstances of 9.55 is an immediate consequence of the definition of such derivatives in combination with Brouwer's theorem (which itself can be found in Spanier [1966], for instance), and he developed some calculus for taking advantage of that fact. The special case of 9.55 in which F is assumed to be strictly differentiable at $\bar{x}$ gives an earlier result of Leach [1961], [1963], generalizing the inverse function theorem. An extension in that mode to an implicit function theorem was achieved by Nijenhuis [1974] in a statement resembling 9.56(d), but as applied to a single-valued mapping in place of the generally set-valued S.

To the extent that a set-valued mapping S is involved from the start, the localization results in Theorem 9.54 are new, and the same is true of the implicit mapping results in 9.56, except for part (a) of that theorem in giving a sufficient condition for the Aubin property to hold. Precedents for the latter exist in papers of Mordukhovich [1994a], [1994b], [1994c], which also go into detail about many special cases. Earlier work of Rockafellar [1985c] and Aubin and Frankowska [1987], [1990] in providing sufficient conditions relied on stronger assumptions in terms of the convexified coderivative mapping obtained when the normal cone to $\operatorname{gph} S$ in Definition 8.53 is replaced by its closed, convex hull, the Clarke normal cone.

A common case where the Aubin property of the implicit mapping in 9.56(a) yields a Lipschitz continuous single-valued localization without any need for the strict derivative assumption in 9.56(b) has been uncovered by Dontchev and Rockafellar [1996]. This case concerns variational inequalities over polyhedral sets. The same paper indicates a significant class of nonsmooth single-valued mappings (the 'normal maps' associated with such variational inequalities) for which the Aubin property suffices for single-valued Lipschitzian invertibility, and the strict derivative test of 9.55 can be bypassed. The book of Dontchev and Rockafellar [2009] presents much more fully the modern generalizations of the implicit function theorem that are now available through variational analysis.

Because strict derivatives of strictly continuous mappings F can be hard to calculate, the coderivative criterion in 9.63, although generally only sufficient for single-valued Lipschitzian invertibility, may sometimes be more practical than the strict derivative criterion in 9.55. This version of the inverse function theorem was discovered by Clarke [1976a]. The corresponding version of the implicit function theorem was laid out by Hiriart-Urruty [1979]. Levy [1996] provided estimates for the graphical derivatives of 'implicit' mappings that might not be single-valued.

The matrix set $\operatorname{con}\overline{\nabla}F(\bar{x})$ in 9.62 and 9.63 is the *generalized Jacobian* of F at $\bar{x}$ introduced by Clarke; see Clarke [1983] for more on this object and its behavior. The results in 9.62 about $D_*F(\bar{x})$ and its 'adjoint' relationship to $D^*F(\bar{x})$ appear here for the first time in terms of $D_*F(\bar{x})$ but are closely related to the results of Ioffe [1981a] on fans. Borwein, Borwein and Wang [1996] have demonstrated the existence of Lipschitz continuous mappings F such that, for almost every x, $D^*F(x)$ fails to be convex-valued. They have also shown that it's possible to have $D^*F(x) = D^*G(x)$ for all x even though the Lipschitz continuous mappings F and G differ by more than just a constant. For mappings from $\mathbb{R}^n$ to $\mathbb{R}$, this was demonstrated earlier by Rockafellar [1982b] for the sublinearized coderivative $\overline{D}^* F(x)$.

The calmness property in 9(31) was called 'upper Lipschitzian' by Robinson [1979], who established the fact about piecewise polyhedral mappings in 9.57; cf. Robinson [1981]. We prefer the terminology of calmness because it's consonant with other uses of that word and avoids the discrepancy with true Lipschitzian concepts, which always concern uniformities with respect to variable *pairs* of points (without one of them being fixed, as in this case). Other Lipschitzian properties of piecewise polyhedral mappings, in particular the inverses of piecewise linear mappings, have been studied by Gowda and Sznajder [1996].

The envelopes in 9.11 entered the literature through Hausdorff [1919], who credited them to Pasch; see also Hiriart-Urruty [1980a]. The extension theorem for Lipschitz continuous mappings in 9.58 can be attributed to Mickle [1949], but our proof, based on minimization, is new. The singularity result in 9.65, a Lipschitzian version of a famous theorem of Sard, was obtained by Mignot [1976]. For related results about generic Lipschitzian invertibility, see Jofré [1989]. For infinite-dimensional generalizations of Rademacher's theorem in 9.60, see Preiss [1990].

Graphically Lipschitzian mappings as in 9.66 were studied by Rockafellar [1985b]. The constraint relaxation principle in 9.68 was pioneered by Clarke and has often been helpful in the derivation of optimality conditions; cf. Clarke [1976b], [1983].

The rule in 9.67 for generating subgradients by 'mollification' fits with the broader idea of defining classes of subgradients of a Lipschitz continuous function f, at a point $\bar{x}$ where f may fail to be differentiable, as limits of gradients $\nabla f^\nu(x^\nu)$ obtained from sequences of smooth functions f^ν that approximate f in one way or another. The use of that process was initiated by Halkin [1976a], [1976b], and Warga [1975], [1976], [1981], and continued by Frankowska [1984]. The version in 9.67, where the approximating functions f^ν are produced by 'averaging', was developed by Ermoliev, Norkin and Wets [1985], who also investigated what might happen when f is discontinuous as in the mollification framework of 7.19. Here we have sharpened their results for Lipschitz continuous f by dropping the smoothness of the mollifiers (through an appeal to the generic differentiability of f in Rademacher's theorem) and by bringing to light the fact that *every* subgradient $v \in \partial f(\bar{x})$ can be generated as a cluster point of the gradients $\nabla f^\nu(x^\nu)$ associated with some sequence $x^\nu \to \bar{x}$.

The 'nonconvex separation' result in 9.69(c), which can be found in Borwein and Jofré [1996] except for the assertion about regular tangent cones, traces back to the idea that led Mordukhovich to some of his earliest successes in bypassing convex analysis. For more on this kind of extremal principal and its history see Mordukhovich [1996], Mordukhovich and Shao [1996a]. The results in 9.69(a)(b) were obtained by Jofré and Rivera [1995a], although in slightly different form. Applications to economic models were provided in Jofré and Rivera [1995b].

A final comment goes back to the discussion around Fig. 5–7 and why we work with "osc" instead of "usc" as a property of mappings. Let $S(t) = C - ta$ with $C \subset \mathbb{R}^2$ being $\{(x_1, x_2) \in O \mid x_1 x_2 \geq 0\}$ for $O = \operatorname{int} \mathbb{R}^2_+$, and $a = (1,1)$. Then S is Lipschitz continuous but not usc (although it is osc)! For instance, $S(0) \subset O$, yet $S(t) \not\subset O$ for all $t > 0$.

10. Subdifferential Calculus

Numerous facts about functions $f : \mathbb{R}^n \to \overline{\mathbb{R}}$ and mappings $F : \mathbb{R}^n \to \mathbb{R}^m$ and $S : \mathbb{R}^n \rightrightarrows \mathbb{R}^m$ have been developed in Chapters 7, 8, and 9 by way of the variational geometry in Chapter 6 and characterized through subdifferentiation. In order to take advantage of this body of results, bringing the theory down from an abstract level to workhorse use in practice, one needs to have effective machinery for determining subderivatives, subgradients, and graphical derivatives and coderivatives in individual situations. Just as in classical analysis, contemplation of ε's and δ's only goes so far. In the end, the vitality of the subject rests on tools like the chain rule.

In variational analysis, though, calculus serves additional purposes. While classically the calculation of derivatives can't proceed without first assuming that the functions to be differentiated are differentiable, the subdifferentiation concepts of variational analysis require no such preconditions. Their rules of calculation operate in inequality or inclusion form with little more needed than closedness or semicontinuity, and they give a means of *establishing* whether a differentiability property or Lipschitzian property is present or not.

One of the motivating ideas is that of deriving conditions of optimality in problems of constrained minimization and understanding their behavior in relation to stability and perturbation. A reexamination of what that involves, and what we can now bring to bear on it, will help to light the way.

A. Optimality and Normals to Level Sets

Powerful results in the study of optimal solutions have already been obtained in the first-order conditions of Theorem 6.12, leading to a Lagrange multiplier rule in 6.15, and through their extension in Theorem 8.15 to the minimization of nonsmooth, possibly even discontinuous functions. These results will be carried further in this chapter and combined with others, but it's instructive to begin by returning to first principles.

We've taken the view that any problem of minimization in n variables can be represented as the minimization, over all of $\mathbb{R}^n$, of a certain function $f : \mathbb{R}^n \to \overline{\mathbb{R}}$, which without loss of significant generality could be taken to be proper and lower semicontinuous. If f were a differentiable function, we could immediately appeal to the classical *rule of Fermat*, according to which a function's derivatives must vanish at a local minimum. This rule has the following extension to our abstract framework.

10.1 Theorem (Fermat's rule generalized). *If a proper function $f : \mathbb{R}^n \to \overline{\mathbb{R}}$ has a local minimum at $\bar{x}$, then*

$$df(\bar{x}) \geq 0, \qquad \partial f(\bar{x}) \ni 0,$$

where the first condition corresponds to $\widehat{\partial} f(\bar{x}) \ni 0$ and thus implies the second.

If f is subdifferentially regular or in particular convex, these conditions are equivalent. In the convex case they are not just necessary for a local minimum but sufficient for a global minimum, i.e., for having $\bar{x} \in \operatorname{argmin} f$.

If $f = f_0 + g$ with f_0 smooth, the condition $df(\bar{x}) \geq 0$ takes the form that $-\langle \nabla f_0(\bar{x}), \cdot \rangle \leq dg(\bar{x})$, whereas $\partial f(\bar{x}) \ni 0$ comes out as $-\nabla f_0(\bar{x}) \in \partial g(\bar{x})$.

Proof. The subgradient condition is the case of Theorem 8.15 with $C = \mathbb{R}^n$, but it warrants a direct proof here. If $f(x) \geq f(\bar{x})$ around $\bar{x}$ it's evident that $df(\bar{x})(w) \geq 0$ for all w by Definition 8.1. In fact, $f(x) \geq f(\bar{x}) + o(|x - \bar{x}|)$, so by Definition 8.3 we have $0 \in \widehat{\partial} f(\bar{x})$, where always $\widehat{\partial} f(\bar{x}) \subset \partial f(\bar{x})$. The equivalence of $df(\bar{x}) \geq 0$ with $0 \in \widehat{\partial} f(\bar{x})$ is known from 8.4, while the equality between $\partial f(\bar{x})$ and $\widehat{\partial} f(\bar{x})$ under regularity is known from 8.11. As seen in 8.12, this equality always holds for convex functions, whose subgradient characterization shows that $0 \in \partial f(\bar{x})$ implies that $f(x) \geq f(\bar{x})$ for all x.

The rest of the theorem relies on the formula $\partial f(\bar{x}) = \nabla f_0(\bar{x}) + \partial g(\bar{x})$ in 8.8(c) and the defining formula for $df(\bar{x})(w)$ in 8.1. □

When f is differentiable, this rule gives the classical condition $\nabla f(\bar{x}) = 0$. The case of $f = f_0 + g$ at the end of the theorem specializes on choosing $g = \delta_C$ to the conditions in 6.12 for the minimization of f_0 over C, because $dg(\bar{x}) = \delta\big(\cdot\,|T_C(\bar{x})\big)$ and $\partial g(\bar{x}) = N_C(\bar{x})$; cf. 8.2, 8.14. Another illustration of the last part of Theorem 10.1 is seen in connection with the proximal mappings

$$P_\lambda f(x) := \operatorname{argmin}_w \Big\{ f(w) + \frac{1}{2\lambda}|w - x|^2 \Big\}.$$

10.2 Example (subgradients in proximal mappings). *For any proper function $f : \mathbb{R}^n \to \overline{\mathbb{R}}$ and any $\lambda > 0$, one has*

$$P_\lambda f(x) \subset (I + \lambda \partial f)^{-1}(x) \quad \textit{for all } x.$$

When f is convex, this inclusion holds as an equation.

Detail. Fixing x and λ, let $f_0(w) = (1/2\lambda)|w - x|^2$. The function f_0 is convex and smooth with $\nabla f_0(w) = (w - x)/\lambda$. In the problem that defines $P_\lambda f(x)$, we minimize $f_0 + f$ over $\mathbb{R}^n$, and this is covered by the last part of Theorem 10.1 (with f acting as g). The condition $-\nabla f_0(w) \in \partial f(w)$ is necessary for w to belong to $P_\lambda f(x)$, and when f is convex it is sufficient. Because $\nabla f_0(w) = (w - x)/\lambda$, this condition can be written as $x \in (I + \lambda \partial f)(w)$, or equivalently, $w \in (I + \lambda \partial f)^{-1}(x)$. □

It's not yet evident that the optimality conditions in 8.15 can likewise be recovered from the fundamentals in Theorem 10.1, although the minimization of f_0 over C clearly corresponds to the minimization of $f = f_0 + \delta_C$ over $I\!R^n$ regardless of any smoothness assumptions on f_0. But that will emerge once we are able to handle the subdifferentiation of a sum like $f_0 + \delta_C$ with greater generality.

Anyway, the extended statement of Fermat's rule brings to the forefront the desirability of being able to evaluate $df(\bar{x})$ and $\partial f(\bar{x})$ in very broad circumstances, taking into account the particular manner in which f may be constructed out of other functions, or even various sets or set-valued mappings. This is the goal we are headed toward here.

Often the formulas we'll arrive at won't be expressed as equations but inclusions. They can be high in their content of information nonetheless, because many applications merely require verification of an inclusion or inequality in one direction. For instance, constraint qualifications and tests of Lipschitz continuity typically require verifying that some set contains no more than the zero vector, and for that purpose it's obviously enough to show that some other set, known at least to include the one in question, doesn't contain more than the zero vector.

A feature soon to be apparent is that subderivatives and subgradients won't be treated symmetrically on an equal theoretical footing. Despite the temptation to think of subgradients as 'dual' objects, their role will more and more be 'primary'. The seeds of this development can be seen already in 8.15. Although the relation $df(\bar{x}) \geq 0$ in Theorem 10.1 can in principle be sharper as a necessary condition for a minimum than $\partial f(\bar{x}) \ni 0$ when f isn't regular at $\bar{x}$, it tends to be less robust because of instabilities. This has to be expected from its equivalence with $\widehat{\partial} f(\bar{x}) \ni 0$ and the fact that the mapping $\widehat{\partial} f$ lacks the outer semicontinuity properties built into ∂f, as recorded in 8.7.

Indeed, the condition $\partial f(\bar{x}) \ni 0$ has a geometric interpretation in the behavior of f locally around $\bar{x}$ rather than just at $\bar{x}$ itself. As found in 9.41(b), it means exactly that the level-set mapping $\alpha \mapsto \operatorname{lev}_{\leq \alpha} f$ fails to have the Aubin property at $\bar{\alpha} = f(\bar{x})$ for the point $\bar{x}$. In other words, the condition $\partial f(\bar{x}) \ni 0$ signifies that $\bar{x}$ is a sort of generalized 'stationary point' of f in the sense that a specific mode of non-Lipschitzian behavior is exhibited at $\bar{x}$ by the nest of level sets of f.

Out of such considerations, the calculus of subgradients, along with that of normal vectors and coderivative mappings, ultimately dominates over the calculus of subderivatives, tangent vectors and graphical derivative mappings.

Two examples furnish additional illustration of this trend. First, recall that in Chapter 6 there were two versions of the constraint qualification needed for the analysis of a set defined by a system of constraints, one in terms of normal cones (in 6.14) and the other in terms of tangent cones (in 6.39). It was seen that while these two conditions were equivalent in the presence of Clarke regularity, the normal cone version was better in general (cf. 6.39). Second, recall that in Chapter 9 a key result was the derivation of the Mordukhovich criterion for the Aubin property (in 9.38), which was shown to have deep consequences in many directions, in particular for strict continuity and metric regularity of mappings. That criterion, involving a coderivative mapping, has no convenient counterpart in terms of derivative mappings.

Key relations between the variational geometry of sets and the subdifferentiation of functions have already been seen in terms of epigraphs in 8.2 and 8.9 and indicators in 8.2 and 8.14. These allow us to pass readily between the two contexts in ways that can be very useful. On the one hand, we are able to use formulas for tangent and normal cones in Chapter 6 to get formulas now for subderivatives and subgradients. But on the other hand, any formulas we get for subderivatives or subgradients can be specialized back to tangent vectors and normal vectors by taking the functions to be indicators.

Still another connection between sets and functions, also in relation to the understanding of Fermat's rule, is the following.

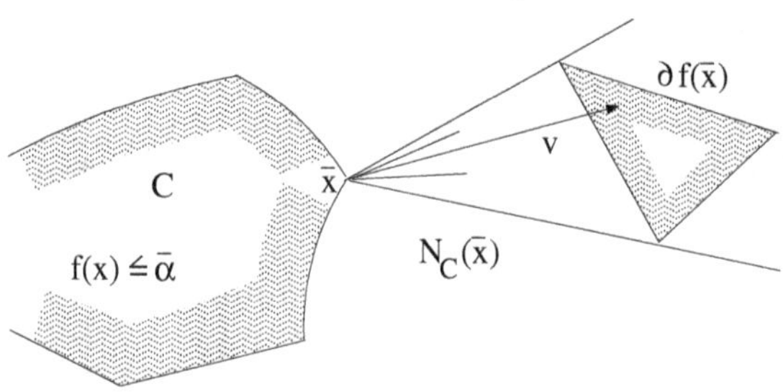

Fig. 10–1. Normal cone to a level set.

10.3 Proposition (normals to level sets). *Suppose $C = \{x \mid f(x) \leq \bar{\alpha}\}$ for a proper, lsc function $f : \mathbb{R}^n \to \overline{\mathbb{R}}$, and let $\bar{x}$ be a point with $f(\bar{x}) = \bar{\alpha}$. Then*

$$T_C(\bar{x}) \subset \{w \mid df(\bar{x})(w) \leq 0\}, \qquad \widehat{N}_C(\bar{x}) \supset \operatorname{pos} \widehat{\partial} f(\bar{x}).$$

If $\partial f(\bar{x}) \not\ni 0$, then also

$$\widehat{T}_C(\bar{x}) \supset \{w \mid \widehat{d}f(\bar{x})(w) \leq 0\}, \qquad N_C(\bar{x}) \subset \operatorname{pos} \partial f(\bar{x}) \cup \partial^\infty f(\bar{x}).$$

If f is regular at $\bar{x}$ with $\partial f(\bar{x}) \not\ni 0$, then C is regular at $\bar{x}$ and

$$T_C(\bar{x}) = \{w \mid df(\bar{x})(w) \leq 0\}, \qquad N_C(\bar{x}) = \operatorname{pos} \partial f(\bar{x}) \cup \partial^\infty f(\bar{x}).$$

Proof. We can think of C as defined by the constraint $F(x) \in D$ for $D = \operatorname{epi} f \subset \mathbb{R}^{n+1}$ and $F : x \mapsto (x, \bar{\alpha}) = \big(x, f(\bar{x})\big)$. Then all we have to do is apply Theorems 6.14 and 6.31 with $X = \mathbb{R}^n$, calling forth the description in Theorems 8.2 and 8.9 of the tangent and normal cones to D at $F(\bar{x}) = \big(\bar{x}, f(\bar{x})\big)$. Since $\nabla F(\bar{x})w = (w, 0)$ and $\nabla F(\bar{x})^*(z, \zeta) = z$, the inclusions in 8.2 and 8.9 yield the inclusions here. The constraint qualification in 6.14 and 6.31 is the condition that $N_{\operatorname{epi} f}\big(\bar{x}, f(\bar{x})\big)$ shouldn't contain any (z, ζ) with $z = 0$ but

$\zeta \neq 0$, and according to the normal cone formula in 8.9 this corresponds to having $0 \notin \partial f(\bar{x})$. Regularity of f at $\bar{x}$ is regularity of D at $F(\bar{x})$ by definition (see 7.25). □

When f is smooth, of course, f is regular (cf. 7.28) and the condition $\partial f(\bar{x}) \not\ni 0$ means $\nabla f(\bar{x}) \neq 0$ (cf. 8.8). The formulas in 10.3 reduce then to

$$T_C(\bar{x}) = \{w \mid \langle \nabla f(\bar{x}), w \rangle \leq 0\}, \qquad N_C(\bar{x}) = \{\lambda \nabla f(\bar{x}) \mid \lambda \geq 0\}.$$

This could already have been gleaned from the formulas in Theorems 6.14 and 6.31 in the special case of a single inequality constraint. On the other hand, we'll later see in 10.50 how such formulas can be extended to a wider class of constraint systems in which the constraint functions don't have to be smooth.

10.4 Example (level sets with epi-Lipschitzian boundary). *For a strictly continuous function $f : \mathbb{R}^n \to \mathbb{R}$, let $C = \{x \mid f(x) \leq \bar{\alpha}\}$ and consider a point $\bar{x}$ of C with $f(\bar{x}) = \bar{\alpha}$. In order that C have epi-Lipschitzian boundary at $\bar{x}$ in the sense of 9.42, the condition*

$$0 \notin \operatorname{con} \partial f(\bar{x})$$

is always sufficient. Under this condition there is a neighborhood $V \in \mathcal{N}(\bar{x})$ such that

$$V \cap \operatorname{bdry} C \;=\; V \cap \{x \mid f(x) = \bar{\alpha}\}, \qquad V \cap \operatorname{int} C \;=\; V \cap \{x \mid f(x) < \bar{\alpha}\}.$$

In fact there is a vector $\bar{w} \neq 0$ along with $V \in \mathcal{N}(\bar{x})$ and $\delta > 0$ such that

$$\left.\begin{aligned} f(x + \tau\bar{w}) &\leq f(x) - \lambda\tau \\ f(x - \tau\bar{w}) &\geq f(x) + \lambda\tau \end{aligned}\right\} \text{ for all } \tau \in (0, \delta) \text{ when } x \in V. \qquad 10(1)$$

Detail. Strict continuity at $\bar{x}$ corresponds in 9.13 to having $\partial^\infty f(\bar{x}) = \{0\}$, in which case $\partial f(\bar{x})$ is nonempty and compact. Then, already just from $0 \notin \partial f(\bar{x})$, we have $N_C(\bar{x}) \subset \operatorname{pos} \partial f(\bar{x})$ by 10.3, with equality when f is regular at $\bar{x}$. According to 9.42, C has the epi-Lipschitzian boundary property if and only if $N_C(\bar{x})$ is pointed, and that's true then as long as $\operatorname{pos} \partial f(\bar{x})$ is pointed. But if $\operatorname{pos} \partial f(\bar{x})$ weren't pointed, we would have $0 \in \operatorname{con} \partial f(\bar{x})$; see 3.15, 2.27.

The pointedness of a cone implies the nonemptiness of the interior of its polar (see 6.22), so the condition $0 \notin \operatorname{con} \partial f(\bar{x})$ also gives us a vector $\bar{w}$ such that $\langle v, w \rangle < 0$ for all $v \in \partial f(\bar{x})$. Then $\widehat{d}f(\bar{x})(\bar{w}) < 0$ by 8.23 and the compactness of $\partial f(\bar{x})$, and we get 10(1) from 8.22 (in the light of $\operatorname{dom} \widehat{d}f(\bar{x})$ being all of $\mathbb{R}^n$; cf. 9.13). It's evident from this local property of descent in the direction of $\bar{w}$ that all points $\tilde{x} \in V$ with $f(\tilde{x}) = \bar{\alpha}$ belong to the closures of $\{x \mid f(x) < \bar{\alpha}\}$ and $\{x \mid f(x) > \bar{\alpha}\}$ and therefore to $\operatorname{bdry} C$. Hence the local descriptions of $\operatorname{bdry} C$ and $\operatorname{int} C$ are correct. □

An example showing the difference between the conditions $\partial f(\bar{x}) \not\ni 0$ and $\operatorname{con} \partial f(\bar{x}) \not\ni 0$ in their effect on the boundary of $C = \{x \mid f(x) \leq f(\bar{x})\}$ is that of the Lipschitz continuous function $f(x_1,x_2) = |x_1| - |x_2|$ on $\mathbb{R}^2$ at $\bar{x} = (0,0)$. The set $\partial f(\bar{x})$ is the union of the segments $[-1,1] \times \{1\}$ and $[-1,1] \times \{-1\}$, so $\partial f(\bar{x}) \not\ni 0 = (0,0)$. In this case, as predicted from our analysis of Fermat's rule, the mapping $\alpha \mapsto \operatorname{lev}_{\leq \alpha} f$ has the Aubin property at $\bar{\alpha} = f(\bar{x}) = 0$ for $\bar{x}$. But $\operatorname{con} \partial f(\bar{x})$ is the square $[-1,1] \times [-1,1]$, so $\operatorname{con} \partial f(\bar{x}) \ni 0 = (0,0)$. The set $C = \operatorname{lev}_{\leq 0} f$ is the union of two quadrants, and it doesn't have epi-Lipschitzian boundary property that's the concern of 10.4.

B. Basic Chain Rule

Moving on now to the systematic development of the extended calculus that is needed, we start with rules that can readily be derived as extensions of the ones for normals and tangents in Chapter 6.

10.5 Proposition (separable functions). *Let $f(x) = f_1(x_1) + \cdots + f_m(x_m)$ for lsc functions $f_i : \mathbb{R}^{n_i} \to \overline{\mathbb{R}}$, where $x \in \mathbb{R}^n$ is expressed as $(x_1,\ldots,x_m)$ with $x_i \in \mathbb{R}^{n_i}$. Then at any $\bar{x} = (\bar{x}_1,\ldots,\bar{x}_m)$ with $f(\bar{x})$ finite and $df_i(\bar{x}_i)(0) = 0$, one has*

$$\begin{aligned} \widehat{\partial} f(\bar{x}) &= \widehat{\partial} f_1(\bar{x}_1) \times \cdots \times \widehat{\partial} f_m(\bar{x}_m), \\ \partial f(\bar{x}) &= \partial f_1(\bar{x}_1) \times \cdots \times \partial f_m(\bar{x}_m), \\ \partial^\infty f(\bar{x}) &\subset \partial^\infty f_1(\bar{x}_1) \times \cdots \times \partial^\infty f_m(\bar{x}_m), \end{aligned}$$

while on the other hand

$$\begin{aligned} df(\bar{x}) &\geq df_1(\bar{x}_1) + \cdots + df_m(\bar{x}_m), \\ \widehat{d}f(\bar{x}) &\leq \widehat{d}f_1(\bar{x}_1) + \cdots + \widehat{d}f_m(\bar{x}_m). \end{aligned}$$

Moreover, f is regular at $\bar{x}$ when f_i is regular at $\bar{x}_i$ for each i. Then the inclusions and inequalities become equations.

Proof. The formula for $\widehat{\partial} f(\bar{x})$ is justified by the variational description of regular subgradients in 8.5. The formula for $\partial f(\bar{x})$ then follows from the definition of general subgradients as limits of regular subgradients; obviously $x^\nu \underset{f}{\to} \bar{x}$ if and only if $x_i^\nu \underset{f_i}{\to} \bar{x}_i$ for each i. Next consider any $v \neq 0$ in $\partial^\infty f(\bar{x})$. By definition we have $\lambda^\nu v^\nu \to v$ for some choice of $\lambda^\nu \searrow 0$, $v^\nu \in \widehat{\partial} f(x^\nu)$, $x^\nu \underset{f}{\to} \bar{x}$. The formula for $\widehat{\partial} f(\bar{x})$ translates this into having $\lambda^\nu v_i^\nu \to v_i$ with $v_i^\nu \in \widehat{\partial} f_i(\bar{x}_i)$, $x_i^\nu \underset{f_i}{\to} \bar{x}_i$. Then $v_i \in \partial^\infty f_i(\bar{x}_i)$ for each i. The inclusion for $\partial^\infty f(\bar{x})$ is therefore correct. The inequality for $df(\bar{x})$ comes from the fact that

$$\begin{aligned} \liminf_{\substack{\tau \searrow 0 \\ w \to \bar{w}}} \frac{f(\bar{x} + \tau w) - f(\bar{x})}{\tau} &= \liminf_{\substack{w_i \to \bar{w}_i \\ \tau \searrow 0}} \sum_{i=1}^m \frac{f_i(\bar{x}_i + \tau w_i) - f(\bar{x}_i)}{\tau} \\ &\geq \sum_{i=1}^m \liminf_{\substack{w_i \to \bar{w}_i \\ \tau_i \searrow 0}} \frac{f_i(\bar{x}_i + \tau w_i) - f(\bar{x}_i)}{\tau_i}. \end{aligned}$$

The inequality for $\widehat{d}f(\bar{x})$, on the other hand, is a consequence of the equation for $\partial f(\bar{x})$ and inclusion for $\partial^\infty f(\bar{x})$ through the expression for $\widehat{d}f(\bar{x})$ in 8.23.

Now suppose each f_i is regular at $\bar{x}_i$. We have $df_i(\bar{x}_i) = \widehat{d}f_i(\bar{x}_i)$ and $df_i(0)(\bar{x}_i) = 0$ (cf. 8.19). Since in general $df(\bar{x}) \leq \widehat{d}f(\bar{x})$, the inequalities we have obtained for $df(\bar{x})$ and $\widehat{d}f(\bar{x})$ must be equations, and these two functions must coincide and have the value 0 at the origin. In particular, f itself must be regular at $\bar{x}$. We further have $\partial f_i(\bar{x}_i) = \widehat{\partial} f_i(\bar{x}_i) \neq \emptyset$ and $\partial^\infty f_i(\bar{x}_i) = \widehat{\partial} f_i(\bar{x}_i)^\infty$ for each i (by 8.11), hence $\partial f(\bar{x}) = \widehat{\partial} f(\bar{x}) \neq \emptyset$ and also $\partial^\infty f(\bar{x}) \subset \widehat{\partial} f_1(\bar{x}_1)^\infty \times \cdots \times \widehat{\partial} f_m(\bar{x}_m)^\infty$. But this product set is $[\widehat{\partial} f_1(\bar{x}_1) \times \cdots \times \widehat{\partial} f_m(\bar{x}_m)]^\infty$ by 3.11, because the sets $\widehat{\partial} f_i(\bar{x}_i)$ are convex. Therefore it equals $\widehat{\partial} f(\bar{x})^\infty$. Thus, $\partial^\infty f(\bar{x}) \subset \widehat{\partial} f(\bar{x})^\infty$. The opposite inclusion holds always by 8.6, so $\partial^\infty f(\bar{x}) = \partial^\infty f_1(\bar{x}_1) \times \cdots \times \partial^\infty f_m(\bar{x}_m)$. □

When applied to indicators $f_i = \delta_{C_i}$, the rules for separable functions in 10.5 reduce to the ones for product sets in 6.41. This illustrates once more the geometric underpinnings of the subject.

A master key to calculus is the following chain rule. In this we continue to make use of a smooth mapping $F : \mathbb{R}^n \to \mathbb{R}^m$ and its Jacobian

$$\nabla F(x) := \left[\frac{\partial f_i}{\partial x_j}(x)\right]_{i,j=1}^{m,n} \in \mathbb{R}^{m \times n},$$

but an extension will later be achieved in 10.49 for mappings F that aren't necessarily even differentiable but just strictly continuous.

10.6 Theorem (basic chain rule). *Suppose $f(x) = g\big(F(x)\big)$ for a proper, lsc function $g : \mathbb{R}^m \to \overline{\mathbb{R}}$ and a smooth mapping $F : \mathbb{R}^n \to \mathbb{R}^m$. Then at any point $\bar{x} \in \operatorname{dom} f = F^{-1}(\operatorname{dom} g)$ one has*

$$\begin{aligned} \widehat{\partial} f(\bar{x}) &\supset \nabla F(\bar{x})^* \widehat{\partial} g\big(F(\bar{x})\big), \\ df(\bar{x})(w) &\geq dg\big(F(\bar{x})\big)\big(\nabla F(\bar{x})w\big). \end{aligned}$$

If the only vector $y \in \partial^\infty g(F(\bar{x}))$ with $\nabla F(\bar{x})^* y = 0$ is $y = 0$ (this being true for convex g when $\operatorname{dom} g$ cannot be separated from the range of the linearized mapping $w \mapsto F(\bar{x}) + \nabla F(\bar{x})w$) one also has

$$\begin{aligned} \partial f(\bar{x}) \subset \nabla F(\bar{x})^* \partial g\big(F(\bar{x})\big), \qquad \partial^\infty f(\bar{x}) \subset \nabla F(\bar{x})^* \partial^\infty g\big(F(\bar{x})\big), \\ \widehat{d}f(\bar{x})(w) \leq \widehat{d}g\big(F(\bar{x})\big)\big(\nabla F(\bar{x})w\big). \end{aligned}$$

If in addition g is regular at $F(\bar{x})$, then f is regular at $\bar{x}$ and

$$\begin{aligned} \partial f(\bar{x}) = \nabla F(\bar{x})^* \partial g\big(F(\bar{x})\big), \qquad \partial^\infty f(\bar{x}) = \nabla F(\bar{x})^* \partial^\infty g\big(F(\bar{x})\big), \\ df(\bar{x})(w) = dg\big(F(\bar{x})\big)\big(\nabla F(\bar{x})w\big). \end{aligned}$$

Proof. We have $\operatorname{epi} f = \overline{F}^{-1}(\operatorname{epi} g)$ for the smooth mapping $\overline{F} : \mathbb{R}^n \times \mathbb{R} \to \mathbb{R}^m \times \mathbb{R}$ defined by $\overline{F}(x, \alpha) = \big(F(x), \alpha\big)$. The normal and tangent cones to $\operatorname{epi} f$

at $(\bar{x}, f(\bar{x}))$ can therefore be computed from those to epi f at $\big(F(\bar{x}), g(F(\bar{x}))\big)$ by the rules in 6.14 and 6.31 in terms of the Jacobian

$$\nabla \overline{F}(\bar{x}, f(\bar{x})) = \begin{bmatrix} \nabla F(\bar{x}) & 0 \\ 0 & 1 \end{bmatrix}.$$

The formulas epigraphical tangent and normal cones in 8.2 and 8.9 translate these cone results into the statements here.

The special comment about the convex case is founded on the fact that when g is convex the vectors $y \neq 0$ in $\partial^\infty g\big(F(\bar{x})\big)$, if any, are those that are normal to supporting hyperplanes to the convex set dom g at $F(\bar{x})$; cf. 6.9. Such a vector satisfies $\nabla F(\bar{x})^* y = 0$ if and only if it is normal to the affine set $M = \big\{F(\bar{x}) + \nabla F(\bar{x})w \,\big|\, w \in \mathbb{R}^n\big\}$ (which passes through $\bar{x}$)), and this corresponds to having the hyperplane $H = \big\{w \,\big|\, \langle y,\, u - F(\bar{x})\rangle\big\}$ furnish separation between dom g and M. □

When g is taken in 10.6 to be the indicator δ_D of a closed set $D \subset \mathbb{R}^m$, we have $f = \delta_C$ for the set $C = F^{-1}(D)$. The chain rule in this case yields the variational geometry results in 6.14 and 6.31 that correspond to choosing $X = \mathbb{R}^n$ in those theorems.

An alternative chain rule, which is sharper in some respects but more limited in its applicability, grows out of 6.7 instead of 6.14 and 6.31. This has the following statement.

10.7 Exercise (change of coordinates). *Suppose $f(x) = g\big(F(x)\big)$ for a proper, lsc function $g : \mathbb{R}^m \to \overline{\mathbb{R}}$ and a smooth mapping $F : \mathbb{R}^n \to \mathbb{R}^m$, and let $\bar{x}$ be a point where f is finite and the Jacobian $\nabla F(\bar{x})$ has rank m. Then*

$$\begin{aligned}
\widehat{\partial} f(\bar{x}) &= \nabla F(\bar{x})^* \widehat{\partial} g\big(F(\bar{x})\big), \\
\partial f(\bar{x}) &= \nabla F(\bar{x})^* \partial g\big(F(\bar{x})\big), \\
\partial^\infty f(\bar{x}) &= \nabla F(\bar{x})^* \partial^\infty g\big(F(\bar{x})\big), \\
df(\bar{x})(w) &= dg\big(F(\bar{x})\big)\big(\nabla F(\bar{x})w\big), \\
\widehat{d} f(\bar{x})(w) &= \widehat{d} g\big(F(\bar{x})\big)\big(\nabla F(\bar{x})w\big).
\end{aligned}$$

In particular, f is regular at $\bar{x}$ if and only if g is regular at $F(\bar{x})$.

Guide. This is patterned on 6.7 and 6.32 and can be derived similarly, or by adopting the pattern in the proof of 10.6 but appealing in the epigraphical framework to 6.7 and 6.32 instead of 6.14 and 6.31. □

Before proceeding with the many consequences of Theorem 10.6, we look at a typical application to necessary conditions for a minimum.

10.8 Example (composite Lagrange multiplier rule). *Consider the problem:*

$$\textit{minimize } f_0(x) + \theta\big(f_1(x), \ldots, f_m(x)\big) \textit{ over all } x \in X,$$

where X is a closed subset of $\mathbb{R}^n$, the functions f_i are $\mathcal{C}^1$ on $\mathbb{R}^n$, and θ is lsc, proper, and convex on $\mathbb{R}^m$. This corresponds to minimizing

$$f(x) := \delta_X(x) + f_0(x) + \theta\big(F(x)\big)$$

over all of $\mathbb{R}^n$, where $F = (f_1, \dots, f_m)$. Let $D = \operatorname{dom}\theta$ (convex), so that the set of feasible solutions to the problem is $C := \big\{x \in X \,\big|\, F(x) \in D\big\}$. Then at any point $\bar{x} \in C$ satisfying the constraint qualification

$$-\nabla F(x)^* y \in N_X(x), \;\; y \in N_D\big(F(x)\big) \quad\Longrightarrow\quad y = 0,$$

one has the inclusions

$$\begin{aligned} \partial f(\bar{x}) &\subset \big\{\nabla f_0(\bar{x}) + \nabla F(\bar{x})^* y \,\big|\, y \in \partial\theta\big(F(\bar{x})\big)\big\} + N_X(\bar{x}), \\ \partial^\infty f(\bar{x}) &\subset \big\{\nabla F(\bar{x})^* y \,\big|\, y \in N_D\big(F(\bar{x})\big)\big\} + N_X(\bar{x}), \end{aligned}$$

where equality holds when X is regular at $\bar{x}$ (as when X is convex, or in particular when $X = \mathbb{R}^n$).

Thus, a necessary condition for f to attain a local minimum at a point $\bar{x}$ satisfying the constraint qualification is the existence of a vector $\bar{y}$ such that

$$-\big[\nabla f_0(\bar{x}) + \nabla F(\bar{x})^* \bar{y}\big] \in N_X(\bar{x}) \;\; \text{with} \;\; \bar{y} \in \partial\theta\big(F(\bar{x})\big).$$

When X and f are convex, the existence of such a vector $\bar{y}$ is sufficient for f to attain a global minimum at $\bar{x}$, even in the absence of the constraint qualification being satisfied at $\bar{x}$.

Detail. Define $G(x) = \big(x, f_0(x), f_1(x), \dots, f_m(x)\big)$ and $g(x, u_0, u_1, \dots, u_m) = \delta_X(x) + u_0 + \theta(u_1, \dots, u_m)$. Then $f(x) = g\big(G(x)\big)$ with G smooth and g lsc and proper. Subgradients of g can be determined from 10.5 and the fact (from 8.12) that $\partial^\infty\theta(u) = N_D(u)$. The chain rule in Theorem 10.6 then yields the claimed inclusions and equations for $\partial f(\bar{x})$ and $\partial^\infty f(\bar{x})$; the constraint qualification here corresponds to the nonexistence of $y \in \partial^\infty g\big(G(\bar{x})\big)$ such that $\nabla G(\bar{x})^* y = 0$, except for $y = 0$. The necessary condition is immediate from these formulas and the general version of Fermat's rule in 10.1. Even without the constraint qualification we have

$$\widehat{\partial} f(\bar{x}) \supset \big\{\nabla f_0(\bar{x}) + \nabla F(\bar{x})^* y \,\big|\, y \in N_D\big(F(\bar{x})\big)\big\} + N_X(\bar{x}),$$

when X is regular at $\bar{x}$, in which case the multiplier rule says that $0 \in \widehat{\partial} f(\bar{x})$. If f is convex, this is equivalent to having $\bar{x} \in \operatorname{argmin} f$; cf. 10.1. □

The optimality condition in this example generalizes the multiplier rule in 6.15, which corresponds to $\theta = \delta_D$ for a box $D = D_1 \times \cdots \times D_m$. Although the framework now is much broader, and more than just constraints are involved in the choice of θ, it's still appropriate to speak of $\bar{y} = (\bar{y}_1, \dots, \bar{y}_m)$ as a *Lagrange multiplier vector*, recalling that

$$\nabla F(\bar{x})^* \bar{y} \;=\; \bar{y}_1 \nabla f_1(\bar{x}) + \cdots + \bar{y}_m \nabla f_m(\bar{x}).$$

The theory of generalized Lagrangian functions will later provide reinforcement for such terminology (cf. 11.45, 11.48). A generalization to functions $f_0, f_1, \ldots, f_m$ that aren't necessarily smooth, but only strictly continuous, will be furnished in 10.52.

The possibility of handling penalty terms and other composite cost expressions alongside of exact constraints reveals the power of Example 10.8 in more detail. The problem of minimizing

$$f_0(x) + \theta_1\big(f_1(x)\big) + \cdots + \theta_m\big(f_m(x)\big)$$

over $x \in X$ is the same as that of minimizing $f_0(x)$ over $x \in X$ subject to $f_i(x) \in D_i$ for $i = 1, \ldots, m$ when each θ_i is the indicator of a closed interval D_i, but more generally one could take at least some of the θ_i's merely to be proper, lsc, convex functions on $I\!R$ with effective domains D_i. Then in 10.8 one would have $\theta(u_1, \ldots, u_m) = \theta_1(u_1) + \cdots + \theta_m(u_m)$ with effective domain $D = D_1 \times \cdots \times D_m$. The condition $\bar{y} \in \partial\theta\big(F(\bar{x})\big)$ would come down to $\bar{y}_i \in \partial\theta_i\big(f_i(\bar{x})\big)$ for $i = 1, \ldots, m$. Each multiplier $\bar{y}_i$ is thereby restricted to lie in a certain closed interval, which depends on θ_i and the value $f_i(\bar{x})$. This is analogous to the sign restrictions on Lagrange multipliers seen in the constraint setting of 6.15.

In the framework of variational analysis, the chain rule for $f = g \circ F$ is the platform for many other rules of calculus. We have already noted that the rules for normals and tangents to inverse images are a special case; cf. 6.7 and 6.32 and more significantly Theorems 6.14 and 6.31. So too are the rules for set intersections in 6.42; they correspond in the following to the special case where $f_i = \delta_{C_i}$.

10.9 Corollary (addition of functions). *Suppose $f = f_1 + \cdots + f_m$ for proper, lsc functions $f_i : I\!R^n \to \overline{I\!R}$, and let $\bar{x} \in \operatorname{dom} f$. Then*

$$\begin{aligned} \widehat{\partial} f(\bar{x}) &\supset \widehat{\partial} f_1(\bar{x}) + \cdots + \widehat{\partial} f_m(\bar{x}), \\ df(\bar{x}) &\geq df_1(\bar{x}) + \cdots + df_m(\bar{x}). \end{aligned}$$

Under the condition that the only combination of vectors $v_i \in \partial^\infty f_i(\bar{x})$ with $v_1 + \cdots + v_m = 0$ is $v_1 = v_2 = \cdots = v_m = 0$ (this being true in the case of convex functions f_1, f_2 when $\operatorname{dom} f_1$ and $\operatorname{dom} f_2$ cannot be separated), one also has that

$$\begin{aligned} \partial f(\bar{x}) &\subset \partial f_1(\bar{x}) + \cdots + \partial f_m(\bar{x}), \\ \partial^\infty f(\bar{x}) &\subset \partial^\infty f_1(\bar{x}) + \cdots + \partial^\infty f_m(\bar{x}), \\ \widehat{d} f(\bar{x}) &\leq \widehat{d} f_1(\bar{x}) + \cdots + \widehat{d} f_m(\bar{x}). \end{aligned}$$

If also each f_i is regular at $\bar{x}$, then f is regular at $\bar{x}$ and

$$\begin{aligned} \partial f(\bar{x}) &= \partial f_1(\bar{x}) + \cdots + \partial f_m(\bar{x}), \\ \partial^\infty f(\bar{x}) &= \partial^\infty f_1(\bar{x}) + \cdots + \partial^\infty f_m(\bar{x}), \\ df(\bar{x}) &= df_1(\bar{x}) + \cdots + df_m(\bar{x}). \end{aligned}$$

Proof. Let $F : I\!R^n \to (I\!R^n)^m$ be the mapping that takes x to $(x, \ldots, x)$, and define the function $g : (I\!R^n)^m \to \overline{I\!R}$ by $g(x_1, \ldots, x_m) = f_1(x_1) + \cdots + f_m(x_m)$. Then $f(x) = g\big(F(x)\big)$. The chain rule in Theorem 10.6 and the facts about separable functions in 10.5, as applied to g, yield all. In the convex case, the separation condition mentioned in Theorem 10.6 turns into the one for the effective domains of f_1 and f_2 here. □

On the basis of this calculus result for sums, the optimality conditions in Theorem 8.15, with respect to the minimization of a possibly nonsmooth function f_0 over a set C, find their place as a special case of the optimality condition in the extended version of Fermat's rule in 10.1. They correspond to taking $f = f_0 + \delta_C$ in 10.1 and applying the calculus in 10.9.

10.10 Exercise (subgradients of Lipschitzian sums). *If $f = f_1 + f_2$ with f_1 strictly continuous at $\bar{x}$ while f_2 is lsc and proper with $f_2(\bar{x})$ finite, then*

$$\partial f(\bar{x}) \subset \partial f_1(\bar{x}) + \partial f_2(\bar{x}), \qquad \partial^\infty f(\bar{x}) \subset \partial^\infty f_2(\bar{x}).$$

If f_1 is strictly differentiable at $\bar{x}$, these inclusions hold as equations.

Guide. Adding to f_1 the indicator of some neighborhood of $\bar{x}$ so as to make f_1 lsc and proper, apply 10.9 with 9.13(b) to get the inclusions. The version with equality doesn't follow from 10.9, because f_2 isn't assumed regular. For that instead apply the inclusion result to $f_2 = f + (-f_1)$. □

Also fitting into the chain rule in 10.6 are the formulas given earlier in 8.31 for subderivatives and subgradients of elementary max functions. This is clear from the fact that

$$f = \max\{f_1, \ldots, f_m\} \iff f = \text{vecmax} \circ F \;\text{ for }\; F = (f_1, \ldots, f_m).$$

(For subgradients of $g = \text{vecmax}$, see 8.26; subderivatives of this convex function are readily determined as well.) More general rules applicable to the pointwise maximum of a possibly *infinite* collection of smooth functions will be developed below under the topic of 'subsmoothness'. (See Definition 10.29 and Theorem 10.31).

Still another chain rule application handles 'partial subdifferentiation'.

10.11 Corollary (partial subgradients and subderivatives). *For a proper, lsc function $f : I\!R^n \times I\!R^m \to \overline{I\!R}$ and a point $(\bar{x}, \bar{u}) \in \text{dom}\, f$, let $\partial_x f(\bar{x}, \bar{u})$ denote the subgradients of $f(\cdot, \bar{u})$ at $\bar{x}$, and similarly $\widehat{\partial}_x f(\bar{x}, \bar{u})$ and $\partial_x^\infty f(\bar{x}, \bar{u})$. Likewise, let $d_x f(\bar{x}, \bar{u})$ and $\widehat{d}_x f(\bar{x}, \bar{u})$ denote the subderivative functions associated with $f(\cdot, \bar{u})$ at $\bar{x}$. One always has*

$$\begin{aligned} \widehat{\partial}_x f(\bar{x}, \bar{u}) \;&\supset\; \big\{ v \,\big|\, \exists y \text{ with } (v, y) \in \widehat{\partial} f(\bar{x}, \bar{u}) \big\}, \\ d_x f(\bar{x}, \bar{u})(w) \;&\geq\; df(\bar{x}, \bar{u})(w, 0) \text{ for all } w. \end{aligned}$$

Under the condition that $(0, y) \in \partial^\infty f(\bar{x}, \bar{u})$ implies $y = 0$ (this being true for convex f when $\text{dom}\, f$ cannot be separated from $(I\!R^n, \bar{u})$), one also has

$$\begin{aligned}\partial_x f(\bar{x},\bar{u}) &\subset \{v \mid \exists y \text{ with } (v,y)\in\partial f(\bar{x},\bar{u})\},\\ \partial_x^\infty f(\bar{x},\bar{u}) &\subset \{v \mid \exists y \text{ with } (v,y)\in\partial^\infty f(\bar{x},\bar{u})\},\\ \hat{d}_x f(\bar{x},\bar{u})(w) &\le \hat{d}f(\bar{x},\bar{u})(w,0) \text{ for all } w.\end{aligned}$$

If also f is regular at $(\bar{x},\bar{u})$, then $f(\cdot,\bar{u})$ is regular at $\bar{x}$ and

$$\begin{aligned}\partial_x f(\bar{x},\bar{u}) &= \{v \mid \exists y \text{ with } (v,y)\in\partial f(\bar{x},\bar{u})\},\\ \partial_x^\infty f(\bar{x},\bar{u}) &= \{v \mid \exists y \text{ with } (v,y)\in\partial^\infty f(\bar{x},\bar{u})\},\\ d_x f(\bar{x},\bar{u})(w) &= df(\bar{x},\bar{u})(w,0) \text{ for all } w.\end{aligned}$$

Proof. Write $f(\cdot,\bar{u}) = f\circ F$ with $F(x) = (x,\bar{u})$ and apply Theorem 10.6. □

C. Parametric Optimality

An interesting and useful consequence of the formulas for partial subdifferentiation is a form of Fermat's rule that addresses the role parameters may, and typically do, have in a problem of optimization.

10.12 Example (parametric version of Fermat's rule). *For a proper, lsc function $f : \mathbb{R}^n \times \mathbb{R}^m \to \overline{\mathbb{R}}$ and vector $\bar{u} \in \mathbb{R}^m$, consider the problem of minimizing $f(x,\bar{u})$ over $x \in \mathbb{R}^n$. Suppose $\bar{x}$ is locally optimal and*

$$\nexists\, y \neq 0 \text{ with } (0,y)\in\partial^\infty f(\bar{x},\bar{u}), \tag{10(2)}$$

which in the case of convex f means that $\operatorname{dom} f$ *does not have a supporting half-space at $(\bar{x},\bar{u})$ with normal vector of the form $(0,y)\neq(0,0)$. Then*

$$\exists\, \bar{y} \text{ with } (0,\bar{y})\in\partial f(\bar{x},\bar{u}). \tag{10(3)}$$

If f is regular at $(\bar{x},\bar{u})$ and $f(x,\bar{u})$ is convex in x, this condition is sufficient for $\bar{x}$ to be globally optimal.

Detail. This combines the partial subgradient rule in 10.11 with the version of Fermat's rule already in 10.1. The convex case reflects 8.12. □

The vectors $\bar{y}$ in this parametric version of Fermat's rule can be regarded as generalized *Lagrange multiplier* elements for the problem of minimizing $f(x,\bar{u})$ in x. Condition 10(2) is the corresponding *abstract constraint qualification*. (A weaker condition giving the same conclusion, the 'calmness' constraint qualification, will be developed in 10.47.)

The connection of Fermat's rule with Lagrange multipliers and constraints can be seen from more than one angle. The simplest interpretation stems from the observation that the problem of minimizing $f(x,\bar{u})$ in x for a given $\bar{u} = (\bar{u}_1,\ldots,\bar{u}_m)$ is the same as the problem of minimizing $f(x,u_1,\ldots,u_m)$ with respect to all $(x,u_1,\ldots,u_m)\in\mathbb{R}^n\times\mathbb{R}^m$ subject to

$$f_i(x, u_1, \ldots, u_m) = 0 \text{ for } i = 1, \ldots, m,$$
$$\text{where } f_i(x, u_1, \ldots, u_m) := \bar{u}_i - u_i.$$

These equality constraints define a set $C \subset I\!R^n \times I\!R^m$, and in applying 8.15 to the minimization of f over C, and viewing the associated normal cone condition for optimality in the context of 6.14, we find that the components $\bar{y}_i$ of $\bar{y}$ are Lagrange multipliers for these constraints quite specifically.

This simple interpretation isn't the only one, however, and in focusing too closely on it one could miss something of deep conceptual importance. The formulation of Fermat's rule in 10.12 is ideally suited to the framework of parametric optimization very generally, such as we have been building up around Theorems 1.17, 7.41, and elsewhere. Explicit constraints and their parameterization are only one theme in the opus of perturbation and stability with respect to max and min.

The message in 10.12 is that *in passing from a model of minimization in terms of a single problem, by itself, to one in which a problem is regarded as belonging to a family with a particular parameterization, the description of optimality is opened up to the incorporation of a dual element in the space of parameters*, as expressed abstractly in condition 10(3). The significance of such dual elements $\bar{y}$ is revealed through their relation to subgradients of the optimal value function $p(u)$ in the chosen parameterization.

10.13 Theorem (subdifferentiation in parametric minimization). *Consider*

$$p(u) := \inf_x f(x,u), \qquad P(u) := \operatorname{argmin}_x f(x,u),$$

for a proper, lsc function $f : I\!R^n \times I\!R^m \to \overline{I\!R}$, and suppose $f(x,u)$ is level-bounded in x locally uniformly in u. Then at any $\bar{u} \in \operatorname{dom} p$ the sets

$$\begin{aligned}
\widehat{Y}(\bar{u}) &:= \bigcap\nolimits_{\bar{x}\in P(\bar{u})} \widehat{M}(\bar{x},\bar{u}) \quad \text{for } \widehat{M}(\bar{x},\bar{u}) := \big\{ y \,\big|\, (0,y) \in \widehat{\partial} f(\bar{x},\bar{u}) \big\},\\
Y(\bar{u}) &:= \bigcup\nolimits_{\bar{x}\in P(\bar{u})} M(\bar{x},\bar{u}) \quad \text{for } M(\bar{x},\bar{u}) := \big\{ y \,\big|\, (0,y) \in \partial f(\bar{x},\bar{u}) \big\},\\
Y_\infty(\bar{u}) &:= \bigcup\nolimits_{\bar{x}\in P(\bar{u})} M_\infty(\bar{x},\bar{u}) \quad \text{for } M_\infty(\bar{x},\bar{u}) := \big\{ y \,\big|\, (0,y) \in \partial^\infty f(\bar{x},\bar{u}) \big\},
\end{aligned}$$

are closed with $\widehat{Y}(\bar{u})$ convex, $\widehat{Y}(\bar{u}) \subset Y(\bar{u})$, $Y(\bar{u})^\infty \subset Y_\infty(\bar{u})$, and one has

$$\widehat{\partial} p(\bar{u}) \subset \widehat{Y}(\bar{u}), \qquad \partial p(\bar{u}) \subset Y(\bar{u}), \qquad \partial^\infty p(\bar{u}) \subset Y_\infty(\bar{u}),$$

$$\begin{aligned}
\widehat{d}p(\bar{u})(z) &\le \sup_{y\in Y(\bar{u})} \langle y,z\rangle + \delta_{Y_\infty(\bar{u})^*}(z)\\
&\le \sup_{\bar{x}\in P(\bar{u})} \Big\{ \inf_w \widehat{d}f(\bar{x},\bar{u})(w,z) \Big\},\\
dp(\bar{u})(z) &\le \inf_{\bar{x}\in P(\bar{u})} \Big\{ \inf_w df(\bar{x},\bar{u})(w,z) \Big\}.
\end{aligned}$$

When f is convex, so p is convex as well, one has for any $\bar{x} \in P(\bar{u})$ that

$$\partial p(\bar{u}) = Y(\bar{u}) = M(\bar{x},\bar{u}), \qquad \partial^{\infty} p(\bar{u}) = Y_{\infty}(\bar{u}) = M_{\infty}(\bar{x},\bar{u}),$$

$$dp(\bar{u})(z) = \operatorname{cl}_z \inf_w df(\bar{x},\bar{u})(w,z),$$

and as long as $\partial p(\bar{u}) \neq \emptyset$, or equivalently $dp(\bar{u})(0) \neq -\infty$, also

$$dp(\bar{u})(z) = \sup_{y \in M(\bar{x},\bar{u})} \langle y, z \rangle.$$

Proof. The assumptions on f ensure that the image of $C := \operatorname{epi} f$ under the projection mapping $F : (x,u,\alpha) \mapsto (u,\alpha)$ is $D := \operatorname{epi} p$, cf. 1.18; indeed, the set D is closed, and F^{-1} is locally bounded on D (as seen through 1.17). Theorem 6.43 can then be applied. The facts asserted here are simply the consequences of that theorem in the background of the variational geometry of epigraphs in 8.2 and 8.9. In particular, the cone asserted to include the normal cone in 6.43 is the one representing the set $Y(\bar{u}) \cup \operatorname{dir} Y_{\infty}(\bar{u})$ in the ray space model for $\operatorname{csm} \mathbb{R}^m$, so its closedness corresponds to having $Y(\bar{u})$ and $Y_{\infty}(\bar{u})$ closed with $Y(\bar{u})^{\infty} \subset Y_{\infty}(\bar{u})$; cf. 3.4. The subgradient inclusions yield the first inequality in the estimate for $\widehat{d}p(\bar{u})(z)$ by the rule (valid for any lsc function; cf. 8.23) that

$$\widehat{d}p(\bar{u})(z) = \sup_{y \in \partial p(\bar{u})} \langle y, z \rangle + \delta_{\partial^{\infty} p(\bar{u})^*}(z),$$

and the second inequality in this estimate comes then from the same rule as applied to $\widehat{d}f(\bar{x},\bar{u})(w,z)$, which gives

$$\widehat{d}f(\bar{x},\bar{u})(w,z) \geq \sup_{y \in M(\bar{x},\bar{u})} \langle (0,y),(w,z) \rangle + \delta_{M_{\infty}(\bar{x},\bar{u})^*}(z) \quad \text{for all } w.$$

In the convex case, 8.21 and 8.30 provide the additional simplifications beyond those coming from 6.43. The convexity of p follows from that of f by 2.22. □

A result that supplements Theorem 10.13 in some respects, giving an equation formula for $dp(\bar{u})(z)$ in certain cases where p may not be convex, will be provided in 10.58.

The result about image sets in 6.43 can be identified with the case of Theorem 10.13 where f is the indicator of $\{(x, F(x)) \mid x \in C\}$ for a set $C \subset \mathbb{R}^n$ and a mapping $F : \mathbb{R}^n \to \mathbb{R}^m$. Then p is the indicator of $D = F(C)$.

In particular, Theorem 10.13 says that *every subgradient vector $\bar{y} \in \partial p(\bar{u})$ must be a vector associated with an optimal solution $\bar{x} \in P(\bar{u})$* in the manner of the parametric form of Fermat's rule in 10.11—and when there's enough convexity the converse holds as well. Furthermore, the vectors y of concern in the abstract constraint qualification associated with this rule are closely tied to the horizon subgradients of p. The facts we state next turn the spotlight on the case that comes out most strongly in relating properties of p to these ideas.

10.14 Corollary (parametric Lipschitz continuity and differentiability). *Let $p(u) = \inf_x f(x,u)$ and $P(u) = \operatorname{argmin}_x f(x,u)$ as in Theorem 10.13 for a proper, lsc function $f : \mathbb{R}^n \times \mathbb{R}^m \to \overline{\mathbb{R}}$ such that $f(x,u)$ is level-bounded in x*

locally uniformly in u, and let $\bar{u}$ be a point of $\operatorname{dom} p$. *Let $Y(\bar{u})$ and $Y_\infty(\bar{u})$ be the associated sets of multiplier vectors as defined in Theorem 10.13.*

(a) *If $Y_\infty(\bar{u}) = \{0\}$, then p is strictly continuous at $\bar{u}$ with*

$$\operatorname{lip} p(\bar{u}) \leq \max_{y \in Y(\bar{u})} |y| < \infty.$$

(b) *If $Y(\bar{u}) = \{\bar{y}\}$ too, then p is strictly differentiable at $\bar{u}$ with $\nabla p(\bar{u}) = \bar{y}$.*

When f is convex, with corresponding simplification of $Y(\bar{u})$ and $Y_\infty(\bar{u})$ in 10.13, the condition $Y_\infty(\bar{u}) = \{0\}$ is also necessary for p to be strictly continuous at $\bar{u}$. Then too, the inequality for $\operatorname{lip} p(\bar{u})$ *is an equation.*

Proof. This just combines 9.13 and 9.18 with the facts in Theorem 10.13. The assumptions imply through 1.17 that p is lsc and proper. □

These insights help in particular in understanding the significance of the Lagrange multipliers appearing in the multiplier rule in 10.8, and as a special case, the earlier one in 6.15.

10.15 Example (subdifferential interpretation of Lagrange multipliers). *With $u = (u_1, \ldots, u_m) \in \mathbb{R}^m$ as parameter vector, let $p(u)$ denote the optimal value and $P(u)$ the optimal solution set in the problem*

$$\text{minimize } f_0(x) + \theta\big(f_1(x) + u_1, \ldots, f_m(x) + u_m\big) \text{ over all } x \in X,$$

where $X \subset \mathbb{R}^n$ is nonempty and closed, each $f_i : \mathbb{R}^n \to \mathbb{R}$ is smooth, and $\theta : \mathbb{R}^m \to \overline{\mathbb{R}}$ is proper, lsc and convex. This problem corresponds to minimizing

$$f(x,u) := \delta_X(x) + f_0(x) + \theta\big(F(x) + u\big)$$

over $x \in \mathbb{R}^n$, where $F(x) = \big(f_1(x), \ldots, f_m(x)\big)$. Assume that $f(x,u)$ is level-bounded in x locally uniformly in u, i.e., that for each $\alpha \in \mathbb{R}$ and $r \in \mathbb{R}_+$ the set $\{(x,u) \in X \times r\mathbb{B} \mid f_0(x) + \theta\big(F(x)+u\big) \leq \alpha\}$ is bounded. Let $D := \operatorname{dom}\theta$, and associate with $\bar{x} \in P(\bar{u})$ the multiplier sets

$$M(\bar{x},\bar{u}) := \Big\{ y \in \partial\theta\big(F(\bar{x}) + \bar{u}\big) \;\Big|\; -\nabla f_0(\bar{x}) - \nabla F(\bar{x})^* y \in N_X(\bar{x}) \Big\},$$
$$M_\infty(\bar{x},\bar{u}) := \Big\{ y \in N_D\big(F(\bar{x}) + \bar{u}\big) \;\Big|\; -\nabla F(\bar{x})^* y \in N_X(\bar{x}) \Big\}.$$

Consider a vector $\bar{u} \in \operatorname{dom} p$. If at every $\bar{x} \in P(\bar{u})$ the set X is regular and $M_\infty(\bar{x},\bar{u}) = \{0\}$ (constraint qualification), then p is strictly continuous at $\bar{u}$ with

$$\partial p(\bar{u}) \subset \bigcup_{\bar{x} \in P(\bar{u})} M(\bar{x},\bar{u}), \qquad \operatorname{lip} p(\bar{u}) \leq \max_{\substack{y \in M(\bar{x},\bar{u}) \\ \bar{x} \in P(\bar{u})}} |y| < \infty.$$

If besides there happens to be only one vector $\bar{y} \in \bigcup_{\bar{x} \in P(\bar{u})} M(\bar{x},\bar{u})$, then p must be strictly differentiable at $\bar{u}$ with $\nabla p(\bar{u}) = \bar{y}$.

Detail. Let $g(x, w_0, w_1, \ldots, w_m) := \delta_X(x) + w_0 + \theta(w_1, \ldots, w_m)$ and $G(x, u) := \big(x, f_0(x), f_1(x)+u_1, \ldots, f_m(x)+u_m\big)$, so that $f(x, u) = g\big(G(x, u)\big)$. This yields

$$\big\{y \,\big|\, (0, y) \in \partial f(\bar{x}, \bar{u})\big\} = M(\bar{x}, \bar{u}), \quad \big\{y \,\big|\, (0, y) \in \partial^\infty f(\bar{x}, \bar{u})\big\} = M_\infty(\bar{x}, \bar{u}),$$

through the chain rule in 10.6. The results then follow from 10.14. □

A remarkable feature of this application is that despite the potential nonsmoothness of the optimal value function p, which is inherent in constrained minimization regardless of any smoothness assumed for the functions f_i, it's possible to identify circumstances in which p has Lipschitzian properties and can even be strictly differentiable. In the latter case the components $\bar{y}_i$ of the Lagrange multiplier vector $\bar{y}$ are interpreted as giving the rates of change of the optimal value with respect to perturbations in the parameters. Each multiplier $\bar{y}_i$ signals the marginal effect of raising or lowering the value $f_i(\bar{x})$.

When the multiplier vector isn't unique in 10.15, or 10.14, its components can't be identified with partial derivatives of p, but the connection with rates of change of p—directional derivatives or subderivatives—is nevertheless very close. This is borne out by the formulas that apply to $dp(\bar{u})$ and $\widehat{d}p(\bar{u})$ in Theorem 10.13. For instance in the convex case with $\bar{u} \in \operatorname{int} \operatorname{dom} p$, we get

$$\lim_{\substack{\tau \searrow 0 \\ z \to \bar{z}}} \frac{p(\bar{u} + \tau z) - p(\bar{u})}{\tau} = \max_{y \in M(\bar{x}, \bar{u})} \big\langle y, \bar{z} \big\rangle \quad \text{for any } \bar{x} \in P(\bar{u}),$$

because convex functions are semidifferentiable at points in the interior of their effective domain; cf. 9.14 and 9.15.

Lipschitzian properties not only describe the behavior of the optimal value function p when the constraint qualification in the version of Fermat's rule in 10.12 is fulfilled, but even explain what the constraint qualification means.

10.16 Proposition (Lipschitzian meaning of constraint qualification)**.** *For a proper, lsc function $f : \mathbb{R}^n \times \mathbb{R}^m \to \overline{\mathbb{R}}$ and a point $(\bar{x}, \bar{u}) \in \operatorname{dom} f$, the condition*

$$(0, y) \in \partial^\infty f(\bar{x}, \bar{u}) \implies y = 0 \qquad 10(4)$$

is equivalent to the set-valued mapping $E : u \mapsto \operatorname{epi} f(\cdot, u)$ having the Aubin property at $\bar{u}$ for $\big(\bar{x}, f(\bar{x}, \bar{u})\big)$.

Proof. The Mordukhovich criterion in 9.40 for the Aubin property of E at $\bar{u}$ for $\big(\bar{x}, f(\bar{x}, \bar{u})\big)$ takes the form that the coderivative mapping $D^*E\big(\bar{u} \,|\, \bar{x}, f(\bar{x}, \bar{u})\big)$ should have image $\{0\}$ at 0. The graph of this coderivative mapping consists of the elements $(-v, -\beta, y)$ such that (v, β, y) belongs to the normal cone to $\operatorname{gph} E$ at $\big(\bar{u}, \bar{x}, f(\bar{x}, \bar{u})\big)$, but that's the same as (y, v, β) belonging to the normal cone to $\operatorname{epi} f$ at $\big(\bar{x}, \bar{u}, f(\bar{x}, \bar{u})\big)$. The Mordukhovich criterion works out therefore to requiring that the only element of form $(0, y, 0)$ in the epigraphical normal cone be $(0, 0, 0)$. But from Theorem 8.9, the elements $(v, y, 0)$ in this cone correspond to horizon subgradients $(v, y) \in \partial^\infty f(\bar{x}, \bar{u})$. □

10.17 Exercise (geometry of sets depending on parameters). *For a mapping $S : \mathbb{R}^n \rightrightarrows \mathbb{R}^m$ consider $S^{-1}(u) = \{x \mid S(x) \ni u\}$ as a variable subset of $\mathbb{R}^n$ depending on $u \in \mathbb{R}^m$ as parameter element. Suppose S is osc and let $\bar{u} \in \operatorname{rge} S$ and $C := S^{-1}(\bar{u})$. Then for $\bar{x} \in C$ one has*

$$T_C(\bar{x}) \subset \{w \mid DS(\bar{x}\,|\,\bar{u})(w) \ni 0\}, \qquad \widehat{N}_C(\bar{x}) \supset \operatorname{rge} \widehat{D}^*S(\bar{x}\,|\,\bar{u}),$$

while if the constraint qualification

$$D^*S(\bar{x}\,|\,\bar{u})(y) \ni 0 \implies y = 0 \tag{10(5)}$$

is fulfilled, one also has

$$\widehat{T}_C(\bar{x}) \supset \{w \mid \widehat{D}S(\bar{x}\,|\,\bar{u})(w) \ni 0\}, \qquad N_C(\bar{x}) \subset \operatorname{rge} D^*S(\bar{x}\,|\,\bar{u}).$$

If in addition S is graphically regular at $(\bar{x}, \bar{u})$, then

$$T_C(\bar{x}) = \{w \mid DS(\bar{x}\,|\,\bar{u})(w) \ni 0\}, \qquad N_C(\bar{x}) = \operatorname{rge} D^*S(\bar{x}\,|\,\bar{u}).$$

Guide. Apply 10.11 to the indicator of the graph of S. □

This result fits with the interpretation in 10.16, because the constraint qualification 10(5) means that S^{-1} has the Aubin property at $\bar{u}$ for $\bar{x}$, or equivalently that S is metrically regular for $\bar{u}$ at $\bar{x}$; see 9.43.

Another result related to parametric minimization is a calculus rule for functions generated by epi-addition.

10.18 Exercise (epi-addition). *Suppose $f = f_1 \,\#\, \cdots \,\#\, f_m$ for proper, lsc functions $f_i : \mathbb{R}^n \to \overline{\mathbb{R}}$ such that for each $\rho > 0$ the set*

$$\Big\{(x_1, \ldots, x_m) \in (\mathbb{R}^n)^m \;\Big|\; |x_1 + \cdots + x_m| \le \rho,\; f_1(x_1) + \cdots + f_m(x_m) \le \rho\Big\}$$

is bounded. Let $P(x) = \{(x_1, \ldots, x_m) \mid f_1(x_1) + \cdots + f_m(x_m) = f(x)\}$. Then at any $\bar{x} \in \operatorname{dom} f$ the sets

$$\begin{aligned}
\widehat{V}(\bar{x}) &= \bigcap_{(\bar{x}_1,\ldots,\bar{x}_m)\in P(\bar{x})} \; \bigcup_{i=1,\ldots,m} \widehat{\partial} f_i(\bar{x}_i),\\
V(\bar{x}) &= \bigcup_{(\bar{x}_1,\ldots,\bar{x}_m)\in P(\bar{x})} \; \bigcup_{i=1,\ldots,m} \partial f_i(\bar{x}_i),\\
V_\infty(\bar{x}) &= \bigcup_{(\bar{x}_1,\ldots,\bar{x}_m)\in P(\bar{x})} \; \bigcup_{i=1,\ldots,m} \partial^\infty f_i(\bar{x}_i),
\end{aligned}$$

are closed with $\widehat{V}(\bar{x})$ convex. Moreover, $\widehat{V}(\bar{x}) \subset V(\bar{x})$, and $V(\bar{x})^\infty \subset V_\infty(\bar{x})$, and one has

$$\widehat{\partial} f(\bar{x}) \subset \widehat{V}(\bar{x}), \qquad \partial f(\bar{x}) \subset V(\bar{x}), \qquad \partial^\infty f(\bar{x}) \subset V_\infty(\bar{x}),$$

and when each function f_i is regular at $\bar{x}_i$, also

$$df(\bar{x})(w) \le \inf_{\substack{(\bar{x}_1,\ldots,\bar{x}_m)\in P(\bar{x})\\ w_1+\cdots+w_m=w}} \sum\nolimits_{i=1}^m df(\bar{x}_i)(w_i),$$

$$\hat{d}f(\bar{x})(w) \le \sup_{(\bar{x}_1,\ldots,\bar{x}_m)\in P(\bar{x})} \Big\{ \inf_{w_1+\cdots+w_m=w} \sum\nolimits_{i=1}^m \hat{d}f(\bar{x}_i)(w_i) \Big\}.$$

When each f_i is convex, implying f is convex, the unions are superfluous in the definition of $V(\bar{x})$ and $V_\infty(\bar{x})$, and one has for any $(\bar{x}_1,\ldots,\bar{x}_m) \in P(\bar{x})$ that

$$\partial f(\bar{x}) = V(\bar{x}) = \bigcap_{i=1,\ldots,m} \partial f_i(\bar{x}_i),$$

$$\partial^\infty f(\bar{x}) = V_\infty(\bar{x}) = \bigcap_{i=1,\ldots,m} \partial^\infty f_i(\bar{x}_i),$$

$$df(\bar{x})(w) = \mathrm{cl}_w \Big\{ \inf_{w_1+\cdots+w_m=w} \sum\nolimits_{i=1}^m df(\bar{x}_i)(w_i) \Big\}.$$

Guide. For the function $\varphi(x, x_1, \ldots, x_m) := \delta_{\{0\}}\big(x - \sum_{i=1}^m x_i\big) + \sum_{i=1}^m f_i(x_i)$ on $(\mathbb{R}^n)^{m+1}$ one has in (b) that

$$f(x) = \inf_{(x_1,\ldots,x_m)} \varphi(x, x_1, \ldots, x_m), \qquad P(x) = \operatorname*{argmin}_{(x_1,\ldots,x_m)} \varphi(x, x_1, \ldots, x_m).$$

Apply Theorem 10.13, getting the subgradients of φ from the representation $\varphi = g \circ L$ with $g(u, x_1, \ldots, x_m) = \delta_{\{0\}}(u) + \sum_{i=1}^m f_i(x_i)$ and $L : (x, x_1, \ldots, x_m) \mapsto \big(x - \sum_{i=1}^m x_i, x_1, \ldots, x_m\big)$. The linear transformation L is nonsingular and merely provides a change of coordinates, so 10.7 can be used with 10.5. □

D. Rescaling

Continuing now with other rules of calculus, we make note of an obvious one which hasn't been mentioned until now:

$$\left.\begin{aligned} \hat{\partial}(\lambda f)(\bar{x}) &= \lambda \hat{\partial} f(\bar{x}) \\ \partial(\lambda f)(\bar{x}) &= \lambda \partial f(\bar{x}) \\ \partial^\infty(\lambda f)(\bar{x}) &= \partial^\infty f(\bar{x}) \end{aligned}\right\} \text{ when } \lambda > 0. \tag{10(6)}$$

Similarly, of course, $d(\lambda f)(\bar{x}) = \lambda df(\bar{x})$ and $\hat{d}(\lambda f)(\bar{x}) = \lambda \hat{d}f(\bar{x})$ when $\lambda > 0$. (When $\lambda = 0$ one has to be careful about $0 \cdot \pm\infty$.) The formulas can be viewed as an instance of a chain rule in which f is composed with the increasing function $t \to \lambda t$. The following is a general statement of such a rule in which the composition may be nonlinear. It has two forms, according to two different approaches that can be taken.

10.19 Proposition (nonlinear rescaling). *Suppose $h(x) = \theta\big(f(x)\big)$ for a proper, lsc function $f : \mathbb{R}^n \to \overline{\mathbb{R}}$ and a proper, lsc function $\theta : \mathbb{R} \to \overline{\mathbb{R}}$, with $\theta(\infty)$ interpreted as ∞. Let $\bar{x}$ be any point of $\operatorname{dom} h$.*

(a) *Assuming that* f *is smooth around* $\bar{x}$, *suppose that either* $\nabla f(\bar{x}) \neq 0$ *or* $\partial^\infty \theta\big(f(\bar{x})\big) = \{0\}$. *Then*

$$\begin{aligned} \partial h(\bar{x}) \;&\subset\; \big\{ \lambda \nabla f(\bar{x}) \,\big|\, \lambda \in \partial \theta\big(f(\bar{x})\big) \big\}, \\ \partial^\infty h(\bar{x}) \;&\subset\; \big\{ \lambda \nabla f(\bar{x}) \,\big|\, \lambda \in \partial^\infty \theta\big(f(\bar{x})\big) \big\}. \end{aligned}$$

Equality holds in this case if θ *is regular at* $f(\bar{x})$ *(as when* θ *is convex), and then* h *is regular at* $\bar{x}$.

(b) *Assuming that* θ *is nondecreasing with* $\sup \theta = \infty$, *and that* $\theta(\alpha) > \theta\big(f(\bar{x})\big)$ *for all* $\alpha > f(\bar{x})$, *suppose either* $\partial f(\bar{x}) \not\ni 0$ *or* $\partial^\infty \theta\big(f(\bar{x})\big) = \{0\}$. *Then*

$$\begin{aligned} \partial h(\bar{x}) \;&\subset\; \big\{ \lambda v \,\big|\, \lambda \in \partial \theta\big(f(\bar{x})\big),\ v \in \partial f(\bar{x}) \big\} \cup \partial^\infty f(\bar{x}), \\ \partial^\infty h(\bar{x}) \;&\subset\; \big\{ \lambda v \,\big|\, \lambda \in \partial^\infty \theta\big(f(\bar{x})\big),\ v \in \partial f(\bar{x}) \big\} \cup \partial^\infty f(\bar{x}). \end{aligned}$$

Equality holds in this case when f *and* θ *are convex (and then* h *is convex).*

Proof. In case (a) we simply apply the chain rule in Theorem 10.6. Although that rule is stated for a smooth mapping $F : I\!R^n \to I\!R^m$, only a neighborhood of $\bar{x}$ is really involved. In the present situation we take $m = 1$ and identify F with the function f, this being smooth on a neighborhood of $\bar{x}$, and identify g in that theorem with θ. The constraint qualification in 10.6, that only $y \in \partial^\infty g\big(F(\bar{x})\big)$ with $\nabla F(\bar{x})^* y = 0$ is $y = 0$, emerges as the requirement that the only $\lambda \in \partial^\infty \theta\big(f(\bar{x})\big)$ with $\lambda \nabla f(\bar{x}) = 0$ is $\lambda = 0$. Obviously that holds under our assumption in (a). The claimed inclusions then follow from 10.6 along with the fact that equality holds when θ is regular at $f(\bar{x})$.

In case (b) we argue entirely differently. Let $E = \operatorname{epi} f$ and $g(x, \alpha) = \delta_E(x, \alpha) + \theta(\alpha)$. Then $h(x) = \inf_\alpha g(x, \alpha)$, where for each $x \in \operatorname{dom} h$ the infimum is attained at $\alpha = f(x)$, in fact in the case of $\bar{x}$, *uniquely* at $\bar{\alpha} := f(\bar{x})$. We'll apply Theorem 10.13 to this instance of parametric minimization, but first we must make sure the hypothesis of that theorem is satisfied.

It's clear that g is proper and lsc. It also has to be level-bounded in α locally uniformly in x. This property holds if and only if, for each $\beta \in I\!R$ and bounded set $X \subset I\!R^n$, the set

$$B = \big\{ (x, \alpha) \in E \,\big|\, x \in X,\ \theta(\alpha) \leq \beta \big\}$$

is bounded. Our assumption that f is proper and lsc gives a finite lower bound α_1 to f on X (see 1.10), while our assumption that $\sup \theta = \infty$ gives a finite upper bound α_2 to the interval $\big\{ \alpha \,\big|\, \theta(\alpha) \leq \beta \big\}$. Then B lies in the bounded set $X \times [\alpha_1, \alpha_2]$. This establishes that Theorem 10.13 is applicable to the situation at hand. We deduce first by that route that

$$\partial h(\bar{x}) \subset \big\{ v \,\big|\, (v, 0) \in \partial g(\bar{x}, \bar{\alpha}) \big\}, \qquad \partial^\infty h(\bar{x}) \subset \big\{ v \,\big|\, (v, 0) \in \partial^\infty g(\bar{x}, \bar{\alpha}) \big\},$$

these inclusions being equations when g is convex, e.g. f and θ are convex.

The subgradients of g can be analyzed next from 10.9 and the fact that the subgradients of δ_E are normals to E which have a subgradient interpretation through Theorem 8.9. The constraint qualification in 10.9 comes out in this case as the requirement that there should be no $\lambda \neq 0$ in $\partial^\infty\theta(\bar{\alpha})$ such that $(0, -\lambda) \in N_E(\bar{x}, \bar{\alpha})$; the latter is possible with $\lambda \neq 0$ if and only if $(0, -1) \in N_E(\bar{x}, \bar{\alpha})$, which corresponds by 8.9 to having $0 \in \partial f(\bar{x})$. Our assumptions in (b) exclude this. We obtain then from the calculus rule in 10.9 that

$$\begin{aligned}\partial g(\bar{x}, \bar{\alpha}) &\subset N_E(\bar{x}, \bar{\alpha}) + \{0\} \times \partial\theta(\alpha),\\ \partial^\infty g(\bar{x}, \bar{\alpha}) &\subset N_E(\bar{x}, \bar{\alpha}) + \{0\} \times \partial^\infty\theta(\alpha),\end{aligned}$$

where equality holds in the convex case. All that's left to do is to take for expression for $N_E(\bar{x}, \bar{\alpha})$ in 8.9 and put these formulas together. □

E. Piecewise Linear-Quadratic Functions

An important class of functions enjoying especially simple rules of calculus is the following.

10.20 Definition (piecewise linear-quadratic functions). *A function $f : \mathbb{R}^n \to \overline{\mathbb{R}}$ is called piecewise linear-quadratic if* $\operatorname{dom} f$ *can be represented as the union of finitely many polyhedral sets, relative to each of which $f(x)$ is given by an expression of the form $\frac{1}{2}\langle x, Ax\rangle + \langle a, x\rangle + \alpha$ for some scalar $\alpha \in \mathbb{R}$, vector $a \in \mathbb{R}^n$, and symmetric matrix $A \in \mathbb{R}^{n\times n}$. (As a special case, $f(x)$ can be given by a such expression on a single polyhedral set, this set being* $\operatorname{dom} f$.*)*

The *piecewise linear* functions defined previously in 2.47 have expressions just of the form $\langle a, x\rangle + \alpha$. The term piecewise linear-quadratic is preferred here to piecewise quadratic because it helps to head off any false impression that quadratic terms have to be present. The formula on some 'pieces' may well just be affine.

Note that a function given by different linear-quadratic formulas on different pieces of $\operatorname{dom} f$ doesn't fit the definition of piecewise linear-quadratic unless the pieces can be arranged as a polyhedral union. For instance, the function $f(x_1, x_2) = |x_1^2 + x_2^2 - 1|$ on $\mathbb{R}^2$ isn't piecewise linear-quadratic, although its values are given by $1 - x_1^2 - x_2^2$ on $C_1 = \{(x_1, x_2) \,|\, x_1^2 + x_2^2 \leq 1\}$ and by $x_1^2 + x_2^2 - 1$ on $C_2 = \{(x_1, x_2) \,|\, x_1^2 + x_2^2 \geq 1\}$. This may seem unnecessarily restrictive, but it's crucial to many of the applications made of the concept.

10.21 Proposition (properties of piecewise linear-quadratic functions). *If a function $f : \mathbb{R}^n \to \overline{\mathbb{R}}$ is piecewise linear-quadratic, then* $\operatorname{dom} f$ *is closed and f is continuous relative to* $\operatorname{dom} f$*, hence lsc on $\mathbb{R}^n$. At any point $\bar{x} \in \operatorname{dom} f$ the subderivative function $df(\bar{x})$ is piecewise linear with $\operatorname{dom} df(\bar{x}) = T_{\operatorname{dom} f}(\bar{x})$, and it is given simply by*

$$df(\bar{x})(\bar{w}) = \lim_{\tau \searrow 0} \frac{f(\bar{x} + \tau\bar{w}) - f(\bar{x})}{\tau}$$

When f is convex, $\operatorname{dom} f$ *is polyhedral. The subgradient sets* $\partial f(\bar{x})$ *and* $\partial^\infty f(\bar{x})$ *at any* $\bar{x} \in \operatorname{dom} f$ *are then polyhedral as well, and* $\partial f(\bar{x}) \neq \emptyset$.

Proof. Consider a representation of $\operatorname{dom} f$ as the union of a finite collection of polyhedral sets C_k, $k \in K$, on each of which there is an expression $\frac{1}{2}\langle x, A_k x\rangle + \langle a_k, x\rangle + \alpha_k$ for $f(x)$. The sets C_k are closed, and f is continuous relative to them, so because only finitely many are involved, f is continuous relative to their union $\operatorname{dom} f$, which itself then is closed.

Consider a point $\bar{x} \in \operatorname{dom} f$, and let $K(\bar{x})$ denote the set of indices $k \in K$ such that $\bar{x} \in C_k$. For any sequence of points $\bar{x} + \tau^\nu w^\nu$ in $\operatorname{dom} f$ with $\tau^\nu \searrow 0$ and $w^\nu \to \bar{w}$ there must exist $\bar{k} \in K(\bar{x})$ and $N \in \mathcal{N}_\infty^\#$ such that $\bar{x} + \tau^\nu w^\nu \in C_{\bar{k}}$ for all $\nu \in N$. Then $\bar{w}$ belongs to the tangent cone $T_{C_{\bar{k}}}(\bar{x})$, which is polyhedral (see 6.46), and

$$\lim_{\nu \in N} \frac{f(\bar{x} + \tau^\nu w^\nu) - f(\bar{x})}{\tau^\nu} = \big\langle A_{\bar{k}}\bar{x} + a_{\bar{k}}, \bar{w}\big\rangle.$$

On the other hand, if $\bar{w} \in T_{C_{\bar{k}}}(\bar{x})$ we have $\bar{x} + \tau\bar{w} \in C_{\bar{k}}$ for all $\tau > 0$ sufficiently small because of the special approximation property of tangent cones to polyhedral sets in 6.47, and then

$$\lim_{\tau \searrow 0} \frac{f(\bar{x} + \tau\bar{w}) - f(\bar{x})}{\tau} = \big\langle A_{\bar{k}}\bar{x} + a_{\bar{k}}, \bar{w}\big\rangle.$$

It follows that $\operatorname{dom} df(\bar{x})$ is the union of the cones $T_{C_k}(\bar{x})$ for $k \in K(\bar{x})$, and that on each of these polyhedral cones one has $df(\bar{x})(w) = \langle A_k\bar{x} + a_k, w\rangle$. Thus, $df(\bar{x})$ is piecewise linear and $\operatorname{dom} df(\bar{x})$ is all of $T_{\operatorname{dom} f}(\bar{x})$. It's evident then as well that the formula for $df(\bar{x})(w)$ is valid when $w \notin \operatorname{dom} df(\bar{x})$.

When f is convex, $\operatorname{dom} f$ is a convex set; the representation of $\operatorname{dom} f$ as a union of finitely many polyhedral sets implies then that $\operatorname{dom} f$ itself is polyhedral; cf. 2.50. For any $\bar{x} \in \operatorname{dom} f$, the proper, piecewise linear function $df(\bar{x})$ is the support function of $\partial f(\bar{x})$ (by 8.30, 7.27), hence convex. It follows that $\partial f(\bar{x}) \neq \emptyset$. The cone $\operatorname{dom} df(\bar{x}) = T_{\operatorname{dom} f}(\bar{x})$ is polar to the cone $N_{\operatorname{dom} f}(\bar{x})$, which coincides with $\partial^\infty f(\bar{x})$ by 8.12 and is polyhedral by 6.46. The analysis above shows that $df(\bar{x})$ is the support function of the polyhedral set generated by this cone and finitely many points of the form $A_k\bar{x} + a_k$. Because the correspondence between nonempty, closed, convex sets and their support functions is one to one (cf. 8.24), this polyhedral set must be $\partial f(\bar{x})$. □

The main consequence of the piecewise linear-quadratic property is to remove the need for any constraint qualification when determining subderivatives and subgradients in certain situations.

10.22 Exercise (piecewise linear-quadratic calculus).

(a) *If* $f = f_1 + \cdot + f_m$ *for piecewise linear-quadratic functions* $f_i : \mathbb{R}^n \to \overline{\mathbb{R}}$, *then f is piecewise linear-quadratic, and at any point* $\bar{x} \in \operatorname{dom} f$ *one has* $df(\bar{x}) = df_1(\bar{x}) + \cdots + df_m(\bar{x})$. *When each f_i is convex, so that f is convex, one has in addition that* $\partial f(\bar{x}) = \partial f_1(\bar{x}) + \cdots + \partial f_m(\bar{x})$.

(b) *If* $f(x) = g(Ax + a)$ *for* $g : \mathbb{R}^m \to \overline{\mathbb{R}}$ *piecewise linear-quadratic and some* $A \in \mathbb{R}^{m \times n}$, *then* f *is piecewise linear-quadratic, and at any point* $\bar{x} \in \operatorname{dom} f$ *one has* $df(\bar{x})(w) = dg(A\bar{x} + a)(Aw)$. *When* g *is convex, so that* f *is convex, one has in addition that* $\partial f(\bar{x}) = A^* \partial g(A\bar{x} + a)$.

(c) *If* $\varphi(x) = f(x, \bar{u})$ *with* $f : \mathbb{R}^n \times \mathbb{R}^m \to \overline{\mathbb{R}}$ *piecewise linear-quadratic and* $\bar{u} \in \mathbb{R}^m$, *then* φ *is piecewise linear-quadratic and at any* $\bar{x} \in \operatorname{dom} \varphi$ *one has* $d\varphi(\bar{x})(w) = df(\bar{x}, \bar{u})(w, 0)$. *When* f *is convex, so that* φ *is convex, one has in addition that* $\partial \varphi(\bar{x}) = \{v \mid \exists y,\ (v, y) \in \partial f(\bar{x}, \bar{u})\}$.

Guide. In both (a) and (b), verify the piecewise linear-quadratic property from the definition in 10.20. (For the behavior of polyhedral sets under operations, see also 3.52(a).) Then use the formula in 10.21 to get the formula for $df(\bar{x})$. In the convex case, show that the set on the right side of the formula for $\partial f(\bar{x})$ is a nonempty, polyhedral set C (cf. 10.21, 3.52(a)) which, by direct verification, has for its support function σ_C the formula identified as giving $df(\bar{x})$. Argue then from the regularity of f (in 7.27, 8.30) and the one-to-oneness of the support function correspondence (in 8.24) that necessarily $C = \partial f(\bar{x})$. Identify (c) with a special case of (b) in which $g(x, u) = f(x, \bar{u} + u)$ and A is the matrix of the linear mapping $x \mapsto (x, 0)$. □

Other operations on piecewise linear-quadratic functions will be treated in 11.32 and 11.33. Note that the pointwise max of a finite collection of piecewise linear-quadratic functions isn't necessarily piecewise linear-quadratic. An example has already been provided in the function $f(x_1, x_2) = |x_1^2 + x_2^2 - 1|$ on $\mathbb{R}^2$, which is the maximum of $f_1(x_1, x_2) = x_1^2 + x_2^2 - 1$ and $f_2(x_1, x_2) = 1 - x_1^2 - x_2^2$, but such that the 'pieces' on which one or the other of these functions is dominant aren't polyhedral. This is the reason why the operation of pointwise maximization doesn't appear in 10.22.

F. Amenable Sets and Functions

Chain rules support the very definition of a major class of sets and functions, which also draws on piecewise linear-quadratic properties.

10.23 Definition (amenable sets and functions).

(a) *A function* $f : \mathbb{R}^n \to \overline{\mathbb{R}}$ *is amenable at* $\bar{x}$ *if* f *is finite at* $\bar{x}$ *and there is an open neighborhood* V *of* $\bar{x}$ *on which* f *can be represented in the form* $f = g \circ F$ *for a* $\mathcal{C}^1$ *mapping* F *from* V *into a space* $\mathbb{R}^m$ *and a proper, lsc, convex function* $g : \mathbb{R}^m \to \overline{\mathbb{R}}$ *such that, in terms of* $D = \operatorname{cl}(\operatorname{dom} g)$,

$$\text{the only vector } y \in N_D\big(F(\bar{x})\big) \text{ with } \nabla F(\bar{x})^* y = 0 \text{ is } y = 0.$$

It is strongly amenable if such a representation exists with F *not just* $\mathcal{C}^1$ *but* $\mathcal{C}^2$, *and it is fully amenable if, in addition to this,* g *can be taken to be piecewise linear-quadratic.*

(b) *A set $C \subset \mathbb{R}^n$ is amenable at one of its points $\bar{x}$ if there is an open neighborhood V of $\bar{x}$ along with a $\mathcal{C}^1$ mapping F from V into a space $\mathbb{R}^m$ and along with a closed, convex set $D \subset \mathbb{R}^m$ such that*

$$C \cap V = \{x \in V \mid F(x) \in D\}$$

and the constraint qualification in (a) is satisfied. It is strongly amenable if this can be arranged with F of class $\mathcal{C}^2$, and fully amenable if, in addition, D can be taken to be polyhedral.

The constraint qualification in this definition matches the one utilized in the variational geometry of normal and tangent cones in 6.14 and 6.31; cf. 6.39 for an alternate form. Obviously, C is amenable as defined in 10.23(b) if and only if its indicator function δ_C is amenable as defined in 10.23(a). On the other hand, if f is amenable the set $C = \text{cl}(\text{dom}\, f)$ is amenable. In both cases the same goes for strong amenability and full amenability. These stricter properties will be especially of interest later in the development of the theory of second-order subdifferentiation in Chapter 13.

Amenability is a property that bridges between smoothness and convexity while covering at the same time a great many of the functions that are of interest as the essential objective in a problem of minimization. It assists in analyzing terms of the type $\theta\big(f_1(x), \ldots, f_m(x)\big)$ that enter optimization formats like the one in 10.15, but it also comes up in situations where 'composition' doesn't appear on the surface. In the list given next, we refer to a function f simply as *amenable* if it is amenable at every point of $\text{dom}\, f$, and so forth.

10.24 Example (prevalence of amenability).

(a) *Every $\mathcal{C}^1$ function is amenable; every $\mathcal{C}^2$ function is fully amenable.*

(b) *Every proper, lsc, convex function is strongly amenable; every piecewise linear-quadratic convex function is fully amenable.*

(c) *Every closed, convex set is strongly amenable; every polyhedral set is fully amenable.*

(d) *A set $C = \{x \in X \mid F(x) \in D\}$ for closed, convex sets X, D, and a $\mathcal{C}^1$ mapping F is amenable at any of its points $\bar{x}$ where the constraint qualification holds that the only $y \in N_D\big(F(\bar{x})\big)$ with $-\nabla F(\bar{x})^* y \in N_X(\bar{x})$ is $y = 0$. It is strongly amenable at such a point when F is $\mathcal{C}^2$, and fully amenable if X and D are polyhedral besides.*

(e) *A function $f = \max\{f_1, \ldots, f_m\}$ is amenable if each f_i is $\mathcal{C}^1$, and it is fully amenable if each f_i is $\mathcal{C}^2$.*

(f) *A function $f = f_0 + \delta_C$ with f_0 finite is amenable at a point $\bar{x} \in C$ if f_0 and C are amenable at $\bar{x}$. The same goes for strong or full amenability.*

(g) *A function $f = \varphi + f_0$ with φ proper, lsc and convex and $f_0 \in \mathcal{C}^1$ is amenable. It is strongly amenable when $f_0 \in \mathcal{C}^2$, and fully amenable if, in addition, φ is piecewise linear-quadratic.*

Detail. In (a) we can regard f as the composition of the mapping $f : \mathbb{R}^n \to \mathbb{R}^1$ with $g(u) = u$. In (b) we can take F to be the identity mapping and let $g = f$.

Likewise F is the identity in (c). In (e) the mapping $F(x) = \{f_1(x, \ldots, f_m(x)\}$ is composed with $g(u_1, \ldots, u_m) = \max\{u_1, \ldots, u_m\}$, which is not only convex but piecewise linear. In all these cases the constraint qualification is satisfied trivially. The constraint qualification cited in (d) translates to the one in the definition of amenability when C is viewed as $F_0^{-1}(D_0)$ for $F_0 : x \mapsto \big(x, F(x)\big)$ and $D_0 = X \times D$, a representation which supports all the claims.

Closer scrutiny is required in (f), where it is merely assumed that on an open neighborhood V of $\bar{x}$ there are amenable representations $f_0 = g_0 \circ F_0$ and $C \cap V = \big\{x \in V \,\big|\, F_1(x) \in D_1\big\}$, where F_i maps $I\!R^n$ into $I\!R^{m_i}$. Observe first that the finiteness of f_0 along with its amenability at $\bar{x}$ implies that $F_0(\bar{x})$ belongs to the interior of $D_0 = \operatorname{cl}(\operatorname{dom} g_0)$, for otherwise there would exist a vector $y_0 \neq 0$ in $N_{D_0}\big(F(\bar{x})\big)$, and $\nabla F_0(\bar{x})^* y_0$ would then be a nonzero vector in $N_{\operatorname{dom} f_0}(\bar{x})$ by 6.14, contrary to $\bar{x}$ being an interior point of $\operatorname{dom} f_0$.

Let $F(x) = \big(F_0(x), F_1(x)\big)$ and $g(u) = g_0(u_0) + \delta_{D_1}(u_1)$ for $u = (u_0, u_1) \in I\!R^m = I\!R^{m_0} \times I\!R^{m_1}$. The set $D = \operatorname{cl}(\operatorname{dom} g)$ is $D_0 \times D_1$ and has $N_D\big(F(\bar{x})\big) = N_{D_0}\big(F_0(\bar{x})\big) \times N_{D_1}\big(F_1(\bar{x})\big)$ by 6.41. To say that a vector $y \in N_D\big(F(\bar{x})\big)$ satisfies $\nabla F(\bar{x})^* y = 0$ is to say that $y = (y_0, y_1)$ with $y_i \in N_{D_i}\big(F_i(\bar{x})\big)$ satisfying $\nabla F_0(\bar{x})^* y_0 + \nabla F_1(\bar{x})^* y_1 = 0$. But $N_{D_0}\big(F_0(\bar{x})\big) = \{0\}$ because $F_0(\bar{x}) \in \operatorname{int} D_0$, so $y_0 = 0$, and then also $y_1 = 0$ by the constraint qualification in the representation of C. This shows that $y = 0$ and confirms the amenability of f at $\bar{x}$.

For strong amenability in (f) we note that F is $\mathcal{C}^2$ when both F_0 and F_1 are $\mathcal{C}^2$. For full amenability, the observation is that g is piecewise linear-quadratic when g_0 has this property and D_1 is polyhedral.

The conclusions in (g) are obtained by writing $f = g \circ F$ for the mapping $F : I\!R^n \to I\!R^n \times I\!R$ with $F(x) = \big(x, f_0(x)\big)$ and the function $g : I\!R^n \times I\!R \to \overline{I\!R}$ with $g(u, u_0) = \varphi(u) + u_0$. ☐

A major class of convex, piecewise linear-quadratic functions θ that is often utilized in expressions like $\theta\big(f_1(x), \ldots, f_m(x)\big)$ in 10.15, rendering them fully amenable, will be described in Example 11.18.

10.25 Exercise (consequences of amenability).

(a) *If f is amenable at $\bar{x}$, it is regular at $\bar{x}$ and has $\partial^\infty f(\bar{x}) = N_{\operatorname{dom} f}(\bar{x})$. If f is fully amenable at $\bar{x}$, $\operatorname{dom} f$ is locally closed at $\bar{x}$ and $\partial f(\bar{x}) \neq \emptyset$. Then too, $\partial f(\bar{x})$ and $\partial^\infty f(\bar{x})$ are polyhedral, and $df(\bar{x})$ is piecewise linear.*

(b) *If f is amenable at $\bar{x}$, there is an open neighborhood V of $\bar{x}$ such that f is amenable at every point $x \in V \cap \operatorname{dom} f$, and the set $\operatorname{gph} \partial f$ is closed relative to $V \times I\!R^n$. Likewise, strong amenability and full amenability of f, if present at $\bar{x}$, hold also for all $x \in \operatorname{dom} f$ in a neighborhood of $\bar{x}$.*

(c) *If C is amenable at $\bar{x}$, it is regular at $\bar{x}$. If C is fully amenable at $\bar{x}$, it is locally closed at $\bar{x}$ and the cones $N_C(\bar{x})$ and $T_C(\bar{x})$ are polyhedral.*

(d) *If C is amenable at $\bar{x}$, there is an open neighborhood V of $\bar{x}$ such that C is amenable at every point $x \in V \cap C$, and the set $\operatorname{gph} N_C$ is closed relative to $V \times I\!R^n$. Likewise, strong amenability and full amenability of C, if present at $\bar{x}$, hold also for all $x \in C$ in a neighborhood of $\bar{x}$.*

Guide. Obtain the first part of (a) from the chain rule in 10.6 along with some facts in 6.9. For the rest of (a) utilize the formulas provided by the chain rule in conjunction with 10.21 and 10.22; see also 3.55. For the first claim in (b) argue that if the constraint qualification holds at $\bar{x}$ it has to hold at neighboring points as well. Get the closed graph property from the corresponding fact convex functions as applied to the outer function g in a representation $f = g \circ F$, which comes from the subgradient inequality enjoyed by such functions. Finally, obtain (c) and (d) by specializing (a) and (b). □

10.26 Exercise (calculus of amenability).

(a) *Suppose $f = f_1 + \cdots + f_m$ with each f_i amenable at $\bar{x}$ and such that the only choice of $v_i \in N_{\text{dom}\, f_i}(\bar{x})$ with $v_1 + \cdots + v_m = 0$ is $v_1 = \cdots = v_m = 0$. Then f is amenable at $\bar{x}$. The same holds for strong or full amenability. In all cases, one has for all $x \in \text{dom}\, f$ near $\bar{x}$ that*

$$\partial f(x) = \partial f_1(x) + \cdots + \partial f_m(x), \qquad df(x) = df_1(x) + \cdots + df_m(x).$$

(b) *Suppose $f = g \circ F$ with F a $\mathcal{C}^1$ mapping and g a function that is amenable at $F(\bar{x})$ and such that the only choice of $y \in N_{\text{dom}\, g}\big(F(\bar{x})\big)$ with $\nabla F(\bar{x})^* y = 0$ is $y = 0$. Then f is amenable at $\bar{x}$. Likewise, f is strongly amenable at $\bar{x}$ if F is $\mathcal{C}^2$ rather than just $\mathcal{C}^1$ and g is strongly amenable at $F(\bar{x})$, and it is fully amenable at $\bar{x}$ if F is $\mathcal{C}^2$ and g is fully amenable at $F(\bar{x})$. In all cases, one has for all $x \in \text{dom}\, f$ near $\bar{x}$ that*

$$\partial f(x) = \nabla F(x)^* \partial g\big(F(x)\big), \qquad df(x)(w) = dg\big(F(x)\big)\big(\nabla F(x) w\big).$$

(c) *Suppose $C = C_1 \cap \cdots \cap C_m$ with each C_i amenable at $\bar{x}$ and such that the only choice of $v_i \in N_{C_i}(\bar{x})$ with $v_1 + \cdots + v_m = 0$ is $v_1 = \cdots = v_m = 0$. Then C is amenable at $\bar{x}$. The same holds for strong or full amenability. In all cases, one has for all $x \in C$ near $\bar{x}$ that*

$$N_C(x) = N_{C_1}(x) + \cdots + N_{C_m}(x), \qquad T_C(x) = T_{C_1}(x) \cap \cdots \cap T_{C_m}(x).$$

(d) *Suppose $C = F^{-1}(D)$ with F a $\mathcal{C}^1$ mapping and D a set that is amenable at $F(\bar{x})$ and such that the only choice of $y \in N_D\big(F(\bar{x})\big)$ with $\nabla F(\bar{x})^* y = 0$ is $y = 0$. Then C is amenable at $\bar{x}$. Likewise, C is strongly amenable at $\bar{x}$ if F is $\mathcal{C}^2$ rather than just $\mathcal{C}^1$ and D is strongly amenable at $F(\bar{x})$, and it is fully amenable at $\bar{x}$ if F is $\mathcal{C}^2$ and D is fully amenable at $F(\bar{x})$. In all cases, one has for all $x \in C$ near $\bar{x}$ that*

$$N_C(x) = \nabla F(x)^* N_D\big(F(x)\big), \qquad T_C(x) = \big\{ w \,\big|\, \nabla F(x) w \in T_D\big(F(x)\big) \big\}.$$

Guide. The key is (b); one can derive (a) from (b) in the manner that the addition rule in 10.9 was derived from the chain rule in 10.6, and then (c) and (d) can be deduced through specialization to indicator functions. To get (b), consider a local representation $g = g_0 \circ F_0$ around $F(\bar{x})$ as provided by the amenability of g, and observes that this gives a local representation $f = g_0 \circ G$

of f with $G = F_0 \circ F$. Show that this representation meets the requirements for f to be amenable at $\bar{x}$. Facts in 10.25 can be used, if needed. □

G. Semiderivatives and Subsmoothness

For subderivatives in general, the calculus rules given so far in this chapter can be supplemented by special ones in the case of semidifferentiability.

10.27 Exercise (calculus of semidifferentiability).

(a) *If $f = \lambda_1 f_1 + \ldots + \lambda_m f_m$ for functions f_i that are semidifferentiable at $\bar{x}$, and any coefficients $\lambda_i \in I\!R$, then f is semidifferentiable at $\bar{x}$ with $df(\bar{x})(w) = \lambda_1 df_1(\bar{x})(w) + \cdots + \lambda_m df_m(\bar{x})(w)$.*

(b) *If $f = g \circ F$ for F smooth and g semidifferentiable at $F(\bar{x})$, then f is semidifferentiable at $\bar{x}$ with $df(\bar{x})(w) = dg\big(F(\bar{x})\big)\big(\nabla F(\bar{x})w\big)$. The same holds if F is merely semidifferentiable at $\bar{x}$, as long as $\nabla F(\bar{x})w$ replaced by $DF(\bar{x})(w)$.*

(c) *If $f = \max\{f_1, \ldots, f_m\}$ with f_i semidifferentiable at $\bar{x}$, then f too is semidifferentiable at $\bar{x}$ with $df(\bar{x})(w) = \max_{i \in I(\bar{x})} df_i(\bar{x})(w)$ for $I(\bar{x}) = \{i \mid f_i(\bar{x}) = f(\bar{x})\}$. The same holds with min in place of max.*

Guide. Work from 7.20 and 7.21. (Incidentally, the chain rule in (b) wouldn't hold if semidifferentiability didn't entail taking the limit as $w' \to w$ as well as $t \searrow 0$ in Definition 7.20.) It's possible to derive (c) from (b) by taking $g =$ vecmax, which is semidifferentiable by 7.27. □

The results in 10.27 can readily be brought down also to the level of semidifferentiability at $\bar{x}$ *for a vector w*. The given statements correspond to that property holding for every $w \in I\!R^n$; cf. Definition 7.20.

As an illustration of 10.27, any functions f generated from finite convex or concave functions by addition, scalar multiplication, composition with a smooth mapping, or (finitary) pointwise max or min, are semidifferentiable because finite, convex functions are themselves semidifferentiable (see 7.27). The following is a specific case.

10.28 Example (semidifferentiability of eigenvalue functions). *For an $m \times m$ symmetric matrix $A(x)$ whose components are $\mathcal{C}^1$ functions of x as a parameter vector in an open set $O \subset I\!R^n$, let $\lambda_1(x) \geq \lambda_2(x) \geq \cdots \geq \lambda_m(x)$ be the eigenvalues of $A(x)$ (with repetition according to multiplicity). Then λ_k is strictly continuous and semidifferentiable on O.*

Detail. For the finite, convex functions Λ_k defined in 2.54 on the space $I\!R^{m \times m}_{\text{sym}}$ of symmetric matrices of order m, we have

$$\lambda_1(x) = \Lambda_1\big(A(x)\big), \qquad \lambda_k(x) = \Lambda_k\big(A(x)\big) - \Lambda_{k-1}\big(A(x)\big) \text{ for } k = 2, \ldots, m.$$

Since the mapping $x \mapsto A(x)$ from $I\!R^n$ into $I\!R^{m \times m}_{\text{sym}}$ is smooth, while finite, convex functions are semidifferentiable (by 7.27), we get the semidifferentiability of these eigenvalue functions from rules 10.27(a)(b). The strict continuity comes similarly from 9.8 and 9.14. □

The pointwise max or min of finitely many smooth functions, while generally not differentiable, is semidifferentiable according to 10.27(c). Many functions expressed by pointwise max or min of *infinite* collections of smooth functions are semidifferentiable as well. To develop this fact with other properties of such functions and their calculus, we introduce classes of 'subsmoothness' between strict continuity and strict differentiability.

10.29 Definition (subsmooth functions). *A function $f : O \to \mathbb{R}$, where O is an open set in $\mathbb{R}^n$, is said to be lower-$\mathcal{C}^1$ on O, if on some neighborhood V of each $\bar{x} \in O$ there is a representation*

$$f(x) = \max_{t \in T} f_t(x) \qquad 10(7)$$

in which the functions f_t are of class $\mathcal{C}^1$ on V and the index set T is a compact space such that $f_t(x)$ and $\nabla f_t(x)$ depend continuously not just on $x \in V$ but jointly on $(t, x) \in T \times V$.

More generally, f is lower-$\mathcal{C}^k$ on O if such a local representation can be arranged in which the functions f_t are of class $\mathcal{C}^k$, with $f_t(x)$ and all its partial derivatives through order k depending continuously not just on x but on (t, x).

In particular, a function $f(x) = \max\{f_1(x), \ldots, f_m(x)\}$ is lower-$\mathcal{C}^k$ when each f_i is of class $\mathcal{C}^k$. (This is the case where T is the index set $\{1, \ldots, m\}$ in the discrete topology.)

Similarly, upper-$\mathcal{C}^k$ functions are defined in terms of a minimum in place of a maximum; thus, f is upper-$\mathcal{C}^k$ when $-f$ is lower-$\mathcal{C}^k$.

A prime example of a compact index space T other than $\{1, \ldots, m\}$ would be a closed, bounded subset of $\mathbb{R}^m$. Thus if

$$f(x) = \max_{t \in T} g(t, x) \ \text{ for all } x \in O \qquad 10(8)$$

where T is compact in $\mathbb{R}^m$, one has that f is lower-$\mathcal{C}^k$ if g and its partial derivatives in x up through order k depend continuously on (t, x).

Often the functions of interest in connection with subsmoothness have a special such 'max' representation around which all considerations revolve, but in requiring just the *existence* of some such representation *locally*, Definition 10.29 sets the stage for identifying and characterizing properties that this class of functions has in general.

Any $\mathcal{C}^k$ function is in particular lower-$\mathcal{C}^k$, since it can be regarded trivially as given by a representation of the form in Definition 10.29 in which the index set consists of a single element.

10.30 Proposition (smoothness versus subsmoothness). *Consider an open set $O \subset \mathbb{R}^n$ and a function $f : O \to \mathbb{R}$. For f to be of class $\mathcal{C}^1$ on O, it is necessary and sufficient that f be both lower-$\mathcal{C}^1$ and upper-$\mathcal{C}^1$ on O.*

Proof. The necessity is clear from the elementary fact stated just before the proposition. For the sufficiency, consider in a neighborhood of $\bar{x} \in O$ both a lower representation (with max) and an upper representation (with min). In particular these furnish on some neighborhood $V_1 \in \mathcal{N}(\bar{x})$ a smooth function $g_1 \leq f$ satisfying $g_1(\bar{x}) = f(\bar{x})$ as well as on some neighborhood $V_2 \in \mathcal{N}(\bar{x})$ a smooth function $g_2 \geq f$ satisfying $g_2(\bar{x}) = f(\bar{x})$. From the local expansions $g_i(x) = g_i(\bar{x}) + \langle \nabla g_i(\bar{x}), x - \bar{x}\rangle + o_i(x - \bar{x})$ with $o_i(x-\bar{x})/|x - \bar{x}| \to 0$ as $x \to \bar{x}$, we obtain in some neighborhood $V \in \mathcal{N}(\bar{x})$ that

$$\langle \nabla g_1(\bar{x}), x - \bar{x}\rangle + o_1(x - \bar{x}) \leq f(x) - f(\bar{x}) \leq \langle \nabla g_2(\bar{x}), x - \bar{x}\rangle + o_2(x - \bar{x}).$$

This implies that $\langle \nabla g_1(\bar{x}) - \nabla g_2(\bar{x}), x - \bar{x}\rangle \leq o_2(x - \bar{x}) - o_1(x - \bar{x})$ and therefore that $\nabla g_1(\bar{x}) = \nabla g_2(\bar{x})$. Denoting the common gradient vector by v, we get

$$o_1(x - \bar{x}) \leq f(x) - f(\bar{x}) - \langle v, x - \bar{x}\rangle \leq o_2(x - \bar{x}).$$

Thus, f is differentiable at $\bar{x}$ with $\nabla f(\bar{x}) = v$. □

Proposition 10.30 will be supplemented in 13.34 by the characterization that f belongs to $\mathcal{C}^{1+}$, the class of smooth functions with strictly continuous gradient, if and only if f is both lower-$\mathcal{C}^2$ and upper-$\mathcal{C}^2$.

10.31 Theorem (consequences of subsmoothness). *Any lower-$\mathcal{C}^1$ function f on an open set $O \subset \mathbb{R}^n$ is both strictly continuous and regular on O and thus is semidifferentiable and exhibits the properties in 9.16. Further, f is strictly differentiable where it is differentiable; ∇f is continuous relative to the set D consisting of such points of differentiability, and $O \setminus D$ is negligible.*

Moreover, in terms of $\overline{\nabla} f(\bar{x}) := \{v \mid \exists x^\nu \underset{D}{\to} \bar{x} \text{ with } \nabla f(x^\nu) \to v\}$ and the expressions

$$f(x) = \max_{t \in T} f_t(x), \qquad T(x) = \operatorname*{argmax}_{t \in T} f_t(x),$$

that correspond to a local representation of f around $\bar{x}$ under the criterion for subsmoothness in Definition 10.29, one has

$$\begin{aligned}
\operatorname{lip} f(\bar{x}) &= \max_{v \in \overline{\nabla} f(\bar{x})} |v| = \max_{t \in T(\bar{x})} \big|\nabla f_t(\bar{x})\big|, \\
df(\bar{x})(w) &= \max_{v \in \overline{\nabla} f(\bar{x})} \big\langle v, w\big\rangle = \max_{t \in T(\bar{x})} \big\langle \nabla f_t(\bar{x}), w\big\rangle, \\
\partial f(\bar{x}) &= \operatorname{con} \overline{\nabla} f(\bar{x}) = \operatorname{con}\big\{\nabla f_t(\bar{x}) \,\big|\, t \in T(\bar{x})\big\}, \\
-\partial(-f)(\bar{x}) &= \overline{\nabla} f(\bar{x}) \subset \big\{\nabla f_t(\bar{x}) \,\big|\, t \in T(\bar{x})\big\}.
\end{aligned}$$

Proof. Let V be a compact, convex neighborhood of $\bar{x}$ on which the specified local representation is valid. For each $t \in T$ the function f_t has $\operatorname{lip} f_t(x) = \big|\nabla f_t(x)\big|$ by 9.7, hence $\operatorname{lip} f(x) \leq \sup_{t \in T} \big|\nabla f_t(x)\big|$ by 9.10. The supremum in this estimate is finite, because $\nabla f_t(x)$ depends continuously on t in the compact space T (and then $\big|\nabla f_t(x)\big|$ likewise). Hence f is Lipschitz continuous on V.

Since $t \in T(x)$ if and only if $f_t(x) - f(x) = 0$, the continuity of $f(x)$ in x

and the continuity of $f_t(x)$ in (t,x) imply that the set

$$G := \{(x,t) \in V \times T \,|\, t \in T(x)\}$$

is closed in the compact space $V \times T$. In particular, $T(x)$ is compact—and nonempty—for each $x \in V$.

Let $k(x,w) := \max_{t \in T(x)} \langle \nabla f_t(x), w \rangle$. For $\alpha \in I\!R$ and any $w \in I\!R^n$, the set $\{x \in V \,|\, k(x,w) \geq \alpha\}$ is compact because it is the image of the compact set $\{(x,t) \in G \,|\, \langle \nabla f_t(x), w \rangle \geq \alpha\}$ under the projection $(x,t) \mapsto x$. Hence $k(x,w)$ is usc in x. By showing next that $df(x)(w) = k(x,w)$ for $x \in \operatorname{int} V$, we'll not only get the subderivative formula claimed in terms of the gradients ∇f_t but also establish through 9.15 that $\widehat{d}f(\bar{x})(w) = df(\bar{x})(w)$ and therefore that f is regular at $\bar{x}$ (cf. 8.19). This regularity, in combination with f being strictly continuous at $\bar{x}$, will immediately yield also, through Corollary 9.20 and Theorem 9.61, the properties claimed for ∇f and $\overline{\nabla} f$.

With $x \in \operatorname{int} V$ and $w \in I\!R^n$ fixed for now, observe that for any $t \in T(x)$ we have $f_t(x) = f(x)$, but on the other hand $f_t(x + \tau w) \leq f(x + \tau w)$ as long as $\tau > 0$ is small enough that $x + \tau w \in V$. It follows that

$$\liminf_{\tau \to 0} \frac{f(x+\tau w) - f(x)}{\tau} \;\geq\; \liminf_{\tau \to 0} \frac{f_t(x+\tau w) - f_t(x)}{\tau} \;=\; \langle \nabla f_t(x), w \rangle,$$

where the first expression is $df(x)(w)$ by 9.16 and the strict continuity of f. This being true for any $t \in T(x)$, we conclude that $df(x)(w) \geq k(x,w)$. For the opposite inequality, we note that when $x + \tau w \in V$ and $t \in T(x + \tau w)$ we have $f(x + \tau w) = f_t(x + \tau w)$ and $f(x) \geq f_t(x)$, so

$$\begin{aligned}
\frac{f(x+\tau w) - f(x)}{\tau} \;&\leq\; \frac{f_t(x+\tau w) - f_t(x)}{\tau} \\
&\leq\; \max_{\mu \in (0,1)} \langle \nabla f_t(x + \mu\tau w), w \rangle \;\leq\; \max_{\mu \in (0,1)} k(x + \mu\tau w, w).
\end{aligned}$$

This yields the estimate

$$\begin{aligned}
df(x)(w) \;&\leq\; \limsup_{\tau \searrow 0} \frac{f(x+\tau w) - f(x)}{\tau} \\
&\leq\; \limsup_{\tau \searrow 0} \Big[\max_{\mu \in (0,1)} k(x + \mu\tau w, w) \Big] \;\leq\; k(x,w),
\end{aligned}$$

which comes out of the standard mean value theorem and the upper semicontinuity of the function $k(\cdot, w)$. Thus it's true, as claimed, that $df(x)(w) = k(x,w)$ and f is regular. In particular, we have from formula 9(3) in 9.13 that

$$\begin{aligned}
\operatorname{lip} f(\bar{x}) \;&=\; \max_{|w|=1} \Big[\max_{t \in T(\bar{x})} \langle \nabla f_t(\bar{x}), w \rangle \Big] \\
&=\; \max_{t \in T(\bar{x})} \Big[\max_{|w|=1} \langle \nabla f_t(\bar{x}), w \rangle \Big] \;=\; \max_{t \in T(\bar{x})} \big| \nabla f_t(\bar{x}) \big|.
\end{aligned}$$

Next, let $C = \{\nabla f_t(\bar{x}) \mid t \in T(\bar{x})\}$. This is a nonempty, compact subset of $\mathbb{R}^n$, because it is the image of $T(\bar{x})$ under the continuous mapping $t \mapsto \nabla f_t(\bar{x})$. We have seen that $df(\bar{x})$ is the support function of C, but $df(\bar{x})$ is known from Theorem 9.16 to be the support function of the compact, convex set $\partial f(\bar{x})$. Then $\partial f(\bar{x})$ must be $\mathrm{cl}(\mathrm{con}\, C)$; cf. 8.24. The closure operation can be dropped, because the convex hull of a compact set is compact by 2.30. This verifies the formula for $\partial f(\bar{x})$ as the convex hull of gradients $\nabla f_t(\bar{x})$.

The formula for $\partial(-f)(\bar{x})$ follows now by observing that $d(-f)(\bar{x}) = -df(\bar{x})$ by semidifferentiability and applying the formula already derived for $df(\bar{x})$. One has $\partial(-f)(\bar{x}) = \{-\nabla f(\bar{x})\}$ if f is differentiable at $\bar{x}$, but $\partial(-f)(\bar{x}) = \emptyset$ otherwise. □

10.32 Example (subsmoothness of Moreau envelopes). *Let $f : \mathbb{R}^n \to \overline{\mathbb{R}}$ be lsc, proper, and prox-bounded with threshold λ_f. Then for $\lambda \in (0, \lambda_f)$ the function $-e_\lambda f$ is lower-$\mathcal{C}^2$, hence semidifferentiable, strictly continuous and regular, and*

$$\begin{aligned}
\mathrm{lip}[-e_\lambda f](x) &= \lambda^{-1} \max\{|y - x| \mid y \in P_\lambda f(x)\},\\
d[-e_\lambda f](x)(w) &= \lambda^{-1} \max\{\langle y - x, w\rangle \mid y \in P_\lambda f(x)\},\\
\partial[-e_\lambda f](x) &= \lambda^{-1}\big[\mathrm{con}\, P_\lambda f(x) - x\big],\\
\partial[e_\lambda f](x) &\subset \lambda^{-1}\big[x - P_\lambda f(x)\big].
\end{aligned}$$

Proof. Let $h = -e_\lambda f$. We know from 1.25 that $h(x) = \max_y g(y, x)$ for the function $g(y, x) := -f(y) - (1/2\lambda)|y - x|^2$. The associated argmax set is $P_\lambda f(x)$. The mapping $P_\lambda f$ is nonempty-valued, osc and locally bounded by 1.25, so for any $\bar{x}$ there is an open neighborhood $O \in \mathcal{N}(\bar{x})$ along with a compact set Y such that, for all $x \in O$, one has $h(x) = \max_{y \in Y} g(y, x)$ and $P_\lambda f(x) \subset Y$. Clearly $g(y, x)$ is $\mathcal{C}^\infty$ in x with all derivatives depending continuously on (y, x), so it follows that h is lower-$\mathcal{C}^\infty$, in particular lower-$\mathcal{C}^2$. The three equations then follow from the ones in Theorem 10.31, since $\nabla_x g(y, x) = \lambda^{-1}(y - x)$, and the inclusion comes similarly from having $\partial[e_\lambda f](\bar{x}) = -\partial(-h)(\bar{x})$. □

It will be demonstrated in 12.62 that, in contrast to the Moreau envelopes $e_\lambda f$, the double envelopes $e_{\lambda,\mu} f$ in 1.46 are $\mathcal{C}^{1+}$ even when f isn't convex.

10.33 Theorem (subsmoothness and convexity). *Any finite, convex function is lower-$\mathcal{C}^2$. In fact a function f is lower-$\mathcal{C}^2$ on an open set $O \subset \mathbb{R}^n$ if and only if, relative to some neighborhood of each point of O, there is an expression $f = g - h$ in which g is a finite, convex function and h is $\mathcal{C}^2$; indeed, h can be taken to be a convex, quadratic function, even of form $\frac{1}{2}\rho|\cdot|^2$.*

In this way f can be given local representations $f(x) = \max_{t \in T} f_t(x)$ in which the compact index space T and functions f_t are chosen in such a manner that $f_t(x)$ is a quadratic (although not necessarily convex) function of x having coefficients that depend continuously on $t \in T$. When f is convex, such representations can be set up with $f_t(x)$ affine in x.

Proof. Consider any lower-$\mathcal{C}^2$ function f on O and a local representation of f on a neighborhood V of $\bar{x} \in O$ in accordance with Definition 10.29. Choose

$\delta > 0$ with $\mathbb{B}(\bar{x}, \delta) \subset V$, and take $\rho > 0$ to be an upper bound to $|\nabla^2 f_t(x)|$ on $T \times \mathbb{B}(\bar{x}, \delta)$. It will be shown first that

$$f_t(x) \geq f_t(w) + \langle \nabla f_t(w), x - w \rangle - \tfrac{1}{2}\rho|x - w|^2 \text{ for all } x, w \in \mathbb{B}(\bar{x}, \delta). \quad 10(9)$$

Specifically, for any points x and w in $\mathbb{B}(\bar{x}, \delta)$, we have $w + \tau(x - w) \in \mathbb{B}(\bar{x}, \delta)$ for $\tau \in [0, 1]$ by the convexity of the ball. The function $\varphi(\tau) := f_t\big(w + \tau(x - w)\big)$ has $|\varphi''(\tau)| = \big|\big\langle (x - w), \nabla^2 f_t(w + \tau(x - w))(x - w) \big\rangle\big| \leq \rho|x - w|^2$, so that in integrating φ back from its second derivative we get $\varphi(\tau) \geq \varphi(0) + \varphi(\tau) - \frac{1}{2}\rho\tau|x - w|^2$ for $\tau \in [0, 1]$. This inequality gives 10(9). Since 10(9) holds with equality when $w = x$, we obtain in terms of $B := \mathbb{B}(\bar{x}, \delta)$ that

$$f(x) = \max_{(t,w) \in T \times B} \Big\{ f_t(w) + \big\langle \nabla f_t(w), x - w \big\rangle - \tfrac{1}{2}\rho|x - w|^2 \Big\} \text{ for all } x \in B.$$

It follows then that $f = g - h$ on B with $h(x) = \frac{1}{2}\rho|x|^2$ and

$$g(x) = \max_{(t,w) \in T \times B} \Big\{ f_t(w) + \big\langle \nabla f_t(w), x - w \big\rangle - \tfrac{1}{2}\rho|w|^2 + \rho\big\langle x, w \big\rangle \Big\}.$$

As the pointwise supremum of a collection of affine functions of x, g is convex. Thus, f is locally of the form $g - h$ with g a finite, convex function and h a quadratic convex function, hence a $\mathcal{C}^\infty$ function.

To complete the proof of the theorem, it suffices now to consider any finite, convex function g on an open, convex set $C \subset \mathbb{R}^n$ and show that on any compact set $B \subset C$ there's a representation $g(x) = \max_{s \in S} \big\{ \langle a(s), x \rangle + \alpha(s) \big\}$ in which S is a compact space and the vector $a(s) \in \mathbb{R}^n$ and scalar $\alpha(s) \in \mathbb{R}$ depend continuously on $s \in S$. By invoking the subgradient inequality

$$g(x) \geq g(w) + \langle v, x - w \rangle \text{ when } v \in \partial g(w)$$

(see 8.12) we get a representation of the desired kind in which

$$S = \big\{ s = (w, v) \in B \times \mathbb{R}^n \,\big|\, v \in \partial g(w) \big\},$$
$$a(s) = v, \qquad \alpha(s) = g(w) - \langle v, w \rangle.$$

Here $a(s)$ depends continuously on $s = (w, v)$, and so too does $\alpha(s)$, because g is continuous, even strictly continuous and regular by 9.14 (and 7.27). These properties of g imply the local boundedness and outer semicontinuity of ∂g on C through 9.16, and the compactness of S then follows. □

10.34 Corollary (higher-order subsmoothness). *If f is lower-$\mathcal{C}^2$ on an open set O, then f is also lower-$\mathcal{C}^\infty$ on O. Thus, for all $k > 2$, the class of lower-$\mathcal{C}^k$ functions is indistinguishable from the class of lower-$\mathcal{C}^2$ functions.*

Proof. The special representation of a lower-$\mathcal{C}^2$ function furnished by Theorem 10.33 in terms of quadratic functions is a representation that fits the definition of lower-$\mathcal{C}^k$ for all k. □

Although the lower-$\mathcal{C}^k$ classes coincide for $k \geq 2$, there do exist lower-$\mathcal{C}^1$ functions that are not lower-$\mathcal{C}^2$. An example is $f(x) = -|x|^{3/2}$ on $\mathbb{R}^1$. This cannot be lower-$\mathcal{C}^2$, for that would imply the existence of a $\mathcal{C}^2$ function g on a neighborhood of 0 such that $g(x) \leq f(x)$ and $g(0) = f(0)$. Then $g'(0) = f'(0) = 0$ too, so that as $|x| \to 0$ one would have $g(x)/|x|^2 \to g''(0)/2$ in contradiction to $f(x)/|x|^2 \to -\infty$.

Note that the index set and representation used to achieve the stronger properties asserted in Theorem 10.33 may be different from the ones initially given in verifying that f is lower-$\mathcal{C}^2$. For instance, if f were given as the pointwise maximum of a *finite* family of $\mathcal{C}^2$ functions it would typically be necessary to pass to an auxiliary representation involving an *infinite* family.

10.35 Exercise (calculus of subsmoothness).

(a) *If $f = \sum_{i=1}^m \lambda_i f_i$ on an open set $O \subset \mathbb{R}^n$ where the functions $f_i : O \to \mathbb{R}$ are lower-$\mathcal{C}^1$ and $\lambda_i \geq 0$, then f is lower-$\mathcal{C}^1$ on O. Likewise, f is lower-$\mathcal{C}^2$ if every f_i is lower-$\mathcal{C}^2$.*

(b) *If $f = g \circ F$, where g is lower-$\mathcal{C}^1$ on an open set $O \subset \mathbb{R}^m$ and the mapping $F : \mathbb{R}^n \to \mathbb{R}^m$ is $\mathcal{C}^1$, then f is lower-$\mathcal{C}^1$ on $F^{-1}(O)$. Likewise, f is lower-$\mathcal{C}^2$ on $F^{-1}(O)$ when g is lower-$\mathcal{C}^2$ on O and F is $\mathcal{C}^2$.*

10.36 Exercise (subsmoothness and amenability). *A function f is lower-$\mathcal{C}^2$ around $\bar{x}$ if and only if f is strongly amenable at $\bar{x}$ and $\bar{x} \in \operatorname{int}(\operatorname{dom} f)$.*

Guide. For the 'if' direction argue that, in the amenability representation $f = g \circ F$ with its constraint qualification, the finiteness of f around $\bar{x}$ forces $F(\bar{x})$ to belong to the interior of the convex set $\operatorname{dom} g$. Then apply 10.35(b). For the 'only if' direction utilize Theorem 10.33 and Example 10.24(g). □

H.* Coderivative Calculus

So far we have focused on subderivatives and subgradients of functions on $\mathbb{R}^n$, but the calculus of coderivatives of mappings from $\mathbb{R}^n$ to $\mathbb{R}^m$ is next. Because many of the rules come out as inclusions, their statement can be made simpler by adopting the notation that

$$H \subset H' \text{ for mappings } H \text{ and } H' \text{ with } \operatorname{gph} H \subset \operatorname{gph} H'.$$

Likewise it's convenient to write $H = \bigcup_{i \in I} H_i$ for a mapping H defined by $\operatorname{gph} H = \bigcup_{i \in I} \operatorname{gph} H_i$, and so forth.

10.37 Theorem (coderivative chain rule). *Suppose $S = S_2 \circ S_1$ for osc mappings $S_1 : \mathbb{R}^n \rightrightarrows \mathbb{R}^p$ and $S_2 : \mathbb{R}^p \rightrightarrows \mathbb{R}^m$. Let $\bar{x} \in \operatorname{dom} S$, $\bar{u} \in S(\bar{x})$, and assume:*

(a) *the mapping $(x, u) \mapsto S_1(x) \cap S_2^{-1}(u)$ is locally bounded at $(\bar{x}, \bar{u})$ (this being true in particular if either S_1 is locally bounded at $\bar{x}$ or S_2^{-1} is locally bounded at $\bar{u}$),*

(b) $D^*S_2(\bar{w}\,|\,\bar{u})(0) \cap D^*S_1(\bar{x}\,|\,\bar{w})^{-1}(0) = \{0\}$ *for every* $\bar{w} \in S_1(\bar{x}) \cap S_2^{-1}(\bar{u})$ *(this being true in particular if either* S_2 *is strictly continuous at* $\bar{x}$ *or* S_1^{-1} *is strictly continuous at* $\bar{u}$*).*

Then $\operatorname{gph} S$ *is locally closed at* $(\bar{x}, \bar{u})$*, and*

$$D^*S(\bar{x}\,|\,\bar{u}) \subset \bigcup_{\bar{w}\in S_1(\bar{x})\cap S_2^{-1}(\bar{u})} D^*S_1(\bar{x}\,|\,\bar{w})\circ D^*S_2(\bar{w}\,|\,\bar{u}).$$

When S_1 *and* S_2 *are graph-convex,* S *is graph-convex as well and this inclusion becomes an equation in which the union is superfluous: one has*

$$D^*S(\bar{x}\,|\,\bar{u}) \;=\; D^*S_1(\bar{x}\,|\,\bar{w})\circ D^*S_2(\bar{w}\,|\,\bar{u}) \;\text{ for any }\; \bar{w} \in S(\bar{x}) \cap S_2^{-1}(\bar{u}).$$

Proof. Let $C = \big\{(x,w,u) \,\big|\, (x,w) \in \operatorname{gph} S_1,\ (w,u) \in \operatorname{gph} S_2\big\}$. We have $C = F^{-1}(D)$ for $D = (\operatorname{gph} S_1) \times (\operatorname{gph} S_2)$ and $F : (x,w,u) \mapsto (x,w,w,u)$, whereas $\operatorname{gph} S = G(C)$ for $G : (x,w,u) \mapsto (x,u)$. To gain insights from the representation $\operatorname{gph} S = G(C)$, we can apply 6.43, but we don't want to do this to the entire set $G(C)$, just locally around $(\bar{x}, \bar{u})$. Condition (a) ensures that this is justified. We obtain the local closedness of $\operatorname{gph} S$ at $(\bar{x}, \bar{u})$, and

$$N_{\operatorname{gph} S}(\bar{x}, \bar{u}) \subset \bigcup_{\bar{w}\in S_1(\bar{x})\cap S_2^{-1}(\bar{u})} \Big\{(v,-y) \,\Big|\, (v,0,-y) \in N_C(\bar{x},\bar{w},\bar{u})\Big\}.$$

On the other hand, from the representation $C = F^{-1}(D)$ we get by 6.14 that $N_C(\bar{x},\bar{w},\bar{u})$ is included in the union of

$$\Big\{(v,-z+z',-y) \,\Big|\, (v,-z) \in N_{\operatorname{gph} S_1}(\bar{x},\bar{w}),\ (z',-y) \in N_{\operatorname{gph} S_2}(\bar{w},\bar{u})\Big\}$$

over all $\bar{w} \in S_1(\bar{x}) \cap S_2^{-1}(\bar{u})$, provided that the relations $(0,-z) \in \operatorname{gph} S_1(\bar{x},\bar{w})$ and $(z',0) \in \operatorname{gph} S_2(\bar{w},\bar{u})$ with $\bar{w} \in S_1(\bar{x}) \cap S_2^{-1}(\bar{u})$ occur only for $z = z' = 0$. In view of the definition of coderivative mappings in terms of normal cones, this is the condition assumed in (b). Through the two inclusions we have developed it yields the desired relation for $D^*S(\bar{x}\,|\,\bar{u})$.

When S_1 and S_2 are graph-convex, the sets C and D in this argument are convex, hence so is $\operatorname{gph} S$. Theorems 6.43 and 6.14 furnish equations rather than just inclusions in this case, the union in the one for $N_{\operatorname{gph} S}(\bar{x},\bar{u})$ being superfluous. The result then has this character too. □

10.38 Corollary (Lipschitzian properties under composition). *Let* $S = S_2\circ S_1$ *for osc mappings* $S_1 : \mathbb{R}^n \rightrightarrows \mathbb{R}^p$ *and* $S_2 : \mathbb{R}^p \rightrightarrows \mathbb{R}^m$*. Let* $\bar{x} \in \operatorname{dom} S$ *and* $\bar{u} \in S(\bar{x})$ *be such that the boundedness condition in 10.37(a) holds, and suppose for all* $\bar{w} \in S_1(\bar{x}) \cap S_2^{-1}(\bar{u})$ *that* S_1 *has the Aubin property at* $\bar{x}$ *for* $\bar{w}$*, while* S_2 *has the Aubin property at* $\bar{w}$ *for* $\bar{u}$*. Then* S *has the Aubin property at* $\bar{x}$ *for* $\bar{u}$*, and*

$$\operatorname{lip} S(\bar{x}\,|\,\bar{u}) \;\le\; \max_{\bar{w}\in S_1(\bar{x})\cap S_2^{-1}(\bar{u})} \operatorname{lip} S_1(\bar{x}\,|\,\bar{w})\cdot \operatorname{lip} S_2(\bar{w}\,|\,\bar{u}),$$

which in the case of graph-convex S_1 *and* S_2 *can be sharpened to*

$$\operatorname{lip} S(\bar{x} \,|\, \bar{u}) \le \min_{\bar{w} \in S_1(\bar{x}) \cap S_2^{-1}(\bar{u})} \operatorname{lip} S_1(\bar{x} \,|\, \bar{w}) \cdot \operatorname{lip} S_2(\bar{w} \,|\, \bar{u}).$$

In particular, if S_1 *is strictly continuous at* $\bar{x}$ *while* S_2 *is strictly continuous at every point of* $S_1(\bar{x})$*, and if the boundedness condition in 10.37(a) holds at* $\bar{x}$ *for every* $\bar{u} \in S(\bar{x})$*, then* S *is strictly continuous at* $\bar{x}$*.*

If S_1 *is strictly continuous and locally bounded at* $\bar{x}$*, while* S_2 *has these properties at every point of* $S_1(\bar{x})$*, then* S *is strictly continuous and locally bounded at* $\bar{x}$ *with*

$$\operatorname{lip}_\infty S(\bar{x}) \le \operatorname{lip}_\infty S_1(\bar{x}) \cdot \max\big\{ \operatorname{lip}_\infty S_2(\bar{w}) \,\big|\, \bar{w} \in S_1(\bar{x}) \big\}.$$

Proof. The assertions follow from Theorem 10.37 via 9.38 and 9.40; S is osc at $\bar{x}$ when $\operatorname{gph} S$ is locally closed at $(\bar{x}, \bar{u})$ for all $\bar{u} \in S(\bar{x})$. Also used is the fact that $|H_2 \circ H_1|^+ \le |H_2|_+ |H_1|_+$ for positive homogeneous mappings H_i. □

Neither Theorem 10.37 nor Corollary 10.38 mentions the graphical derivative mapping $DS(\bar{x} \,|\, \bar{u})$, and that's for a good reason. The coderivative result works because, in the two representations utilized in the proof of 10.37, the normal cone inclusions obtainable from 6.14 and 6.43 go in the same direction. The tangent cone inclusions in these results go in opposite directions, however, and can't be put together helpfully.

10.39 Exercise (outer composition with a single-valued mapping). *Let* $S = F \circ S_0$ *for osc* $S_0 : \mathbb{R}^n \rightrightarrows \mathbb{R}^p$ *and single-valued* $F : \mathbb{R}^p \to \mathbb{R}^m$*. Let* $\bar{u} \in S(\bar{x})$ *and suppose* F *is strictly continuous at every* $\bar{w} \in S_0(\bar{x})$*. Suppose also that the mapping* $(x, u) \mapsto S_0(x) \cap F^{-1}(u)$ *is locally bounded at* $(\bar{x}, \bar{u})$*. Then*

$$D^*S(\bar{x} \,|\, \bar{u}) \subset \bigcup_{\bar{w} \in S_0(\bar{x}) \cap F^{-1}(\bar{u})} D^*S_0(\bar{x} \,|\, \bar{w}) D^*F(\bar{w}).$$

If S_0 *is strictly continuous at* $\bar{x}$*, then* S *is strictly continuous at* $\bar{x}$ *as well, and*

$$\operatorname{lip} S(\bar{x} \,|\, \bar{u}) \le \max_{\bar{w} \in S_0(\bar{x}) \cap F^{-1}(\bar{u})} \operatorname{lip} S_0(\bar{x} \,|\, \bar{w}) \cdot \operatorname{lip} F(\bar{w}).$$

If S_0 *is also locally bounded at* $\bar{x}$*, then* S *too is locally bounded at* $\bar{x}$*, with*

$$\operatorname{lip}_\infty S(\bar{x}) \le \max_{\bar{w} \in S_0(\bar{x})} \operatorname{lip} S_0(\bar{x} \,|\, \bar{w}) \cdot \operatorname{lip} F(\bar{w}) \le \operatorname{lip}_\infty S_0(\bar{x}) \cdot \max_{\bar{w} \in S_0(\bar{x})} \operatorname{lip} F(\bar{w}).$$

Guide. Derive these facts from 10.37 and 10.38. □

In the special case of 10.37 and 10.38 when the inner mapping is single-valued, in contrast to the outer mapping as in 10.39, additional conclusions can be drawn.

10.40 Theorem (inner composition with a single-valued mapping). *Let* $S = S_0 \circ F$ *for an osc mapping* $S_0 : \mathbb{R}^n \rightrightarrows \mathbb{R}^p$ *and a single-valued mapping* F :

$I\!R^p \to I\!R^m$ that is strictly continuous at $\bar{x}$. Let $\bar{u} \in S(\bar{x})$. Then

$$\widehat{D}^*S(\bar{x}\,|\,\bar{u}) \;\supset\; \widehat{D}^*F(\bar{x})\circ\widehat{D}^*S_0(F(\bar{x})\,|\,\bar{u}).$$

Under the condition that there exists no nonzero vector $z \in D^*S_0(F(\bar{x})\,|\,\bar{u})(0)$ with $0 \in \partial(zF)(\bar{x})$, it is also true that

$$D^*S(\bar{x}\,|\,\bar{u}) \;\subset\; D^*F(\bar{x})\circ D^*S_0(F(\bar{x})\,|\,\bar{u}).$$

This holds in particular when S_0 has the Aubin property at $F(\bar{x})$ for $\bar{u}$, and then S has the Aubin property at $\bar{x}$ for $\bar{u}$ with

$$\operatorname{lip} S(\bar{x}\,|\,\bar{u}) \;\leq\; \operatorname{lip} F(\bar{x})\cdot \operatorname{lip} S_0(F(\bar{x})\,|\,\bar{u}).$$

Furthermore in that case, if F is semidifferentiable at $\bar{x}$, one has

$$DS(\bar{x}\,|\,\bar{u}) \;=\; DS_0(F(\bar{x})\,|\,\bar{u})\circ DF(\bar{x}).$$

If in addition S_0 is semidifferentiable at $F(\bar{x})$ for $\bar{u}$, so too is S at $\bar{x}$ for $\bar{u}$.

On the other hand, still under the condition that there exists no nonzero vector $z \in D^*S_0(F(\bar{x})\,|\,\bar{u})(0)$ with $0 \in \partial(zF)(\bar{x})$, suppose that S_0 is graphically regular at $F(\bar{x})$ for $\bar{u}$, while the function zF is regular at $\bar{x}$ for all $z \in \operatorname{rge} D^*S_0\big(F(\bar{x})\,|\,\bar{u}\big)$ (as holds when F is strictly differentiable at $\bar{x}$). Then S likewise is graphically regular at $\bar{x}$ for $\bar{u}$, and

$$D^*S(\bar{x}\,|\,\bar{u}) \;=\; D^*F(\bar{x})\circ D^*S_0(F(\bar{x})\,|\,\bar{u}).$$

Proof. For the first inclusion, let $v \in \widehat{D}^*F(\bar{x})(z)$ for $z \in \widehat{D}^*S_0(F(\bar{x})\,|\,\bar{u})(y)$. The latter means by definition that $(z, -y)$ is a regular normal vector to $\operatorname{gph} S_0$ at $(F(\bar{x}), \bar{u})$, hence for pairs $(F(x), u) \in \operatorname{gph} S_0$ that

$$\big\langle z,\, F(x) - F(\bar{x})\big\rangle - \big\langle y,\, u - \bar{u}\big\rangle \;\leq\; o\big(\big|(F(x), u) - (F(\bar{x}), \bar{u})\big|\big).$$

On the other hand we know from 9.24(b) that $v \in \widehat{\partial}(zF)(\bar{x})$ and consequently that $\big\langle z,\, F(x) - F(\bar{x})\big\rangle \geq \big\langle v,\, x - \bar{x}\big\rangle + o\big(|x - \bar{x}|\big)$. Also, for x in some neighborhood of $\bar{x}$ we have $|F(x) - F(\bar{x})| \leq \kappa|x - \bar{x}|$ for a certain constant κ. The combination of the preceding inequalities tells us that for $(x, u) \in \operatorname{gph} S$ one has

$$\big\langle v,\, x - \bar{x}\big\rangle - \big\langle y,\, u - \bar{u}\big\rangle \;\leq\; o\big(\big|(x, u) - (\bar{x}, \bar{u})\big|\big).$$

Thus, we have arrived at the inequality that means $v \in \widehat{D}^*S(\bar{x}\,|\,\bar{u})(y)$.

The second inclusion is immediate from Theorem 10.37 as specialized to $S_1 = F$ and $S_2 = S_0$ with the observation that $S_1(\bar{x}) \cap S_2^{-1}(\bar{u}) = \big\{F(\bar{x})\big\}$. The Aubin property corresponds by 9.40 to $D^*S_0(F(\bar{x})\,|\,\bar{u})(0) = \{0\}$, which ensures that no nonzero vector z of the forbidden kind can exist. Then through the inclusion for $D^*S(\bar{x}\,|\,\bar{u})$ we get $D^*S(\bar{x}\,|\,\bar{u})(0) = \{0\}$ and may conclude from 9.40 that S has the Aubin property at $\bar{x}$ for $\bar{u}$. Further, for a constant κ_0 associated with the Aubin property of S_0 at $F(\bar{x})$ for $\bar{u}$ we have

$$\Delta_\tau S_0\big(F(\bar{x})\,|\,\bar{u}\big)(p') \cap \tau^{-1}\mathbb{B} \subset \Delta_\tau S_0\big(F(\bar{x})\,|\,\bar{u}\big)(p) + \kappa_0|p' - p|\mathbb{B}$$

when τ is sufficiently small. When F is semidifferentiable at $\bar{x}$, we can join this to the fact that

$$\begin{aligned}\Delta_\tau S(\bar{x}\,|\,\bar{u})(w) &= \frac{1}{\tau}\big[S(\bar{x}+\tau w) - \bar{u}\big] = \frac{1}{\tau}\big[S_0\big(F(\bar{x}+\tau w)\big) - \bar{u}\big] \\ &= \frac{1}{\tau}\left[S_0\left(F(\bar{x}) + \tau\frac{F(\bar{x}+\tau w) - F(\bar{x})}{\tau}\right) - \bar{u}\right] \\ &= \Delta_\tau S_0\big(F(\bar{x})\,|\,\bar{u}\big)\big(\Delta_\tau F(\bar{x})(w)\big)\end{aligned}$$

to derive as $\tau \searrow 0$ the formula for $DS(\bar{x}|\bar{u})(w)$ as well as, when S_0 too is semidifferentiable, the semidifferentiability of S.

For the last part about regularity, it suffices on the basis of the inclusions already established to show that

$$D^*F(\bar{x})\circ D^*S_0(F(\bar{x})\,|\,\bar{u}) = \widehat{D}^*F(\bar{x})\circ\widehat{D}^*S_0(F(\bar{x})\,|\,\bar{u}).$$

The graphical regularity of S_0 gives us $D^*S_0(F(\bar{x})\,|\,\bar{u}) = \widehat{D}^*S_0(F(\bar{x})\,|\,\bar{u})$, so we need only verify that $D^*F(\bar{x})(z) = \widehat{D}^*F(\bar{x})(z)$ for all $z \in \operatorname{rge} D^*S_0(F(\bar{x})\,|\,\bar{u})$. But this is true under our hypothesis because $D^*F(\bar{x})(z) = \partial(zF)(\bar{x})$ and $\widehat{D}^*F(\bar{x})(z) = \widehat{\partial}(zF)(\bar{x})$ by 9.24(b), whereas the equation $\partial(zF)(\bar{x}) = \widehat{\partial}(zF)(\bar{x})$ is one of the criteria for the subdifferential regularity of zF at $\bar{x}$; cf. 8.19. □

10.41 Theorem (coderivatives of sums of mappings). *Suppose $S = S_1 + \cdots + S_p$ for osc mappings $S_i : \mathbb{R}^n \rightrightarrows \mathbb{R}^m$, and let $\bar{x} \in \operatorname{dom} S$, $\bar{u} \in S(\bar{u})$. Assume both of the following conditions are satisfied.*

(a) (boundedness condition): *there exist $V \in \mathcal{N}(\bar{x})$ and $W \in \mathcal{N}(\bar{u})$ along with $\rho \in \mathbb{R}_+$ such that whenever $x \in V$, $u_i \in S_i(x)$ and $u_1 + \cdots + u_p \in W$, one has $|u_i| < \rho$ for $i = 1, \ldots, p$. (This is true in particular, regardless of the choice of $\bar{u}$, if all but at most one of the mappings S_i are locally bounded at $\bar{x}$, or more generally if there are no vectors $z_i \in \limsup^\infty_{x \to \bar{x}} S_i(x)$ with $z_1 + \cdots + z_p = 0$ except for $z_1 = \cdots = z_p = 0$.)*

(b) (constraint qualification):

$$\left.\begin{aligned} u_i \in S_i(\bar{x}), \quad u_1 + \cdots + u_p = \bar{u} \\ v_i \in D^*S_i(\bar{x}\,|\,u_i)(0), \quad v_1 + \ldots + v_p = 0 \end{aligned}\right\} \implies v_i = 0 \ \text{ for } i = 1, \ldots, p.$$

(This is true in particular, regardless of the choice of $\bar{u}$, if all but at most one of the mappings S_i are strictly continuous at $\bar{x}$).

Then $\operatorname{gph} S$ is locally closed at $(\bar{x}, \bar{u})$, and one has

$$D^*S(\bar{x}\,|\,\bar{u}) \subset \bigcup_{\substack{u_1+\cdots+u_p=\bar{u} \\ u_i \in S_i(\bar{x})}} D^*S_1(\bar{x}\,|\,u_1) + \cdots + D^*S_p(\bar{x}\,|\,u_p).$$

When every S_i is graph-convex, S is graph-convex as well and this inclusion *becomes an equation in which the union is superfluous: one has*

$$D^*S(\bar{x}\,|\,\bar{u}) = D^*S_1(\bar{x}\,|\,\bar{u}_1) + \cdots + D^*S_p(\bar{x}\,|\,\bar{u}_p) \quad \text{for any } \bar{u}_i \in S_i(\bar{x}) \text{ with } \bar{u}_1 + \cdots + \bar{u}_p = \bar{u}.$$

Proof. Clearly S is $F_2 \circ S_0 \circ F_1$ for $S_0 : (x_1, \ldots, x_p) \mapsto S_1(x_1) \times \cdots \times S_p(x_p)$, $F_1 : x \mapsto (x, \ldots, x)$ (p copies) and $F_2 : (u_1, \ldots, u_p) \mapsto u_1 + \cdots + u_p$. The result is obtained by first applying the special chain rule in 10.40 to $T = S_0 \circ F_1$ and then the one in 10.39 to $S = F_2 \circ T$. □

10.42 Corollary (Lipschitzian properties under addition). *Let $S = S_1 + \cdots + S_p$ for osc mappings $S_i : \mathbb{R}^n \rightrightarrows \mathbb{R}^m$, all of which are strictly continuous at $\bar{x}$. If the boundedness condition in 10.41(a) is fulfilled for $\bar{u} \in S(\bar{x})$, then S has the Aubin property at $\bar{x}$ for $\bar{u}$, and*

$$\operatorname{lip} S(\bar{x}\,|\,\bar{u}) \le \max_{\substack{u_1 + \cdots + u_p = \bar{u} \\ u_i \in S_i(\bar{x})}} \Big\{ \operatorname{lip} S_1(\bar{x}\,|\,u_1) + \cdots + \operatorname{lip} S_p(\bar{x}\,|\,u_p) \Big\},$$

which can be sharpened for graph-convex mappings S_i to

$$\operatorname{lip} S(\bar{x}\,|\,\bar{u}) \le \min_{\substack{u_1 + \cdots + u_p = \bar{u} \\ u_i \in S_i(\bar{x})}} \Big\{ \operatorname{lip} S_1(\bar{x}\,|\,u_1) + \cdots + \operatorname{lip} S_p(\bar{x}\,|\,u_p) \Big\},$$

When the boundedness condition in 10.41(a) holds for all $\bar{u} \in S(\bar{x})$, S is strictly continuous at $\bar{x}$. If the mappings S_i are also locally bounded at $\bar{x}$, then S too is locally bounded at $\bar{x}$, and

$$\operatorname{lip}_\infty S(\bar{x}) \le \operatorname{lip}_\infty S_1(\bar{x}) + \cdots + \operatorname{lip}_\infty S_p(\bar{x}).$$

Proof. Again we appeal to 9.38 and the criterion in 9.40. The estimate for the graphical modulus comes via the fact that $|H_1 + \cdots + H_p|^+ \le |H_1|^+ + \cdots + |H_p|^+$ for positively homogeneous mappings H_i. The assertion in the locally bounded case is immediate from the rule that $(\lambda_1 + \cdots + \lambda_p)\mathbb{B} = \lambda_1 \mathbb{B} + \cdots + \lambda_p \mathbb{B}$ when $\lambda_i \ge 0$; cf. 2.23. □

10.43 Exercise (addition of a single-valued mapping). *Suppose $S = S_0 + F$ for $S_0 : \mathbb{R}^n \rightrightarrows \mathbb{R}^m$ and a single-valued mapping $F : \mathbb{R}^n \to \mathbb{R}^m$. Let $\bar{x} \in \operatorname{dom} S$, $\bar{u} \in S(\bar{x})$, and $\bar{u}_0 = \bar{u} - F(\bar{x})$.*

(a) *If F is semidifferentiable at $\bar{x}$, one has*

$$DS(\bar{x}\,|\,\bar{u})(w) = DS_0(\bar{x}\,|\,\bar{u}_0)(w) + DF(\bar{x})(w) \quad \text{for all } w.$$

(b) *If F is strictly differentiable at $\bar{x}$, one has*

$$\begin{aligned} DS(\bar{x}\,|\,\bar{u})(w) &= DS_0(\bar{x}\,|\,\bar{u}_0)(w) + \nabla F(\bar{x})w \quad \text{for all } w, \\ D_*S(\bar{x}\,|\,\bar{u})(w) &= D_*S_0(\bar{x}\,|\,\bar{u}_0)(w) + \nabla F(\bar{x})w \quad \text{for all } w, \\ D^*S(\bar{x}\,|\,\bar{u})(y) &= D^*S_0(\bar{x}\,|\,\bar{u}_0)(y) + \nabla F(\bar{x})^*y \quad \text{for all } y. \end{aligned}$$

Proof. In (a), apply the limit expression for $DS(\bar{x}\,|\,\bar{u})(w)$ directly. This result specializes to the first formula in (b). Obtain the second formula in (b) likewise from the limit expression for $D_*S(\bar{x}\,|\,\bar{u})(w)$ and the definition of strict differentiability of F. For the third formula in (b), get the inclusion $\subset$ from Theorem 10.41. Then apply this result to the representation $S_0 = S + [-F]$ to get the opposite inclusion. □

I* Extensions

The special rule for subgradients of sums of functions in 10.10 has an interesting consequence for subgradient theory itself, when applied to Ekeland's variational principle in 1.43.

10.44 Proposition (variational principle for subgradients). *Let $f : \mathbb{R}^n \to \overline{\mathbb{R}}$ be lsc with $\inf f$ finite, and let $\bar{x}$ satisfy $f(\bar{x}) \leq \inf f + \varepsilon$, where $\varepsilon > 0$. Then for any $\delta > 0$ there exists $\tilde{x}$ with $|\tilde{x} - \bar{x}| \leq \varepsilon/\delta$ and $f(\tilde{x}) \leq f(\bar{x})$ for which there is a subgradient $\tilde{v} \in \partial f(\tilde{x})$ with $|\tilde{v}| \leq \delta$.*

Proof. The variational principle in 1.43 yields a point $\tilde{x} \in \mathbb{B}(\bar{x}, \varepsilon/\delta)$ such that $f(\tilde{x}) \leq f(\bar{x})$ and $\tilde{x} \in \operatorname{argmin}_x \{f(x) + \delta|x - \tilde{x}|\}$. Then for the Lipschitz continuous function $g(x) := \delta|x - \tilde{x}|$ we have $0 \in \partial(f+g)(\tilde{x})$ by the generalized version of Fermat's rule in 10.1, hence $0 \in \partial f(\tilde{x}) + \partial g(\tilde{x})$ by 10.10. But $\partial g(\tilde{x}) = \delta\mathbb{B}$ by 8.27. Hence there exists $\tilde{v} \in \partial f(\tilde{x})$ such that $-\tilde{v} \in \delta\mathbb{B}$, i.e., $|\tilde{v}| \leq \delta$. □

10.45 Exercise (variational principle for regular subgradients). *Let $f : \mathbb{R}^n \to \overline{\mathbb{R}}$ be lsc with $\inf f$ finite, and let $\bar{x}$ satisfy $f(\bar{x}) < \inf f + \varepsilon$, where $\varepsilon > 0$. Then for any $\delta > 0$ there exists $\tilde{x}$ with $|\tilde{x} - \bar{x}| < \varepsilon/\delta$ and $f(\tilde{x}) < \inf f + \varepsilon$ for which there is a regular subgradient $\tilde{v} \in \widehat{\partial} f(\tilde{x})$ with $|\tilde{v}| < \delta$.*

Guide. Obtain this by combining the fact in 10.44 with the approximation of general subgradients by regular subgradients in Definition 8.3. □

These ideas relate to the following enlargement of the category of limits that was used in Definition 8.3 to produce the sets $\partial f(\bar{x})$ and $\partial^\infty f(\bar{x})$.

10.46 Proposition (ε-regular subgradients). *At points where $f : \mathbb{R}^n \rightrightarrows \overline{\mathbb{R}}$ is finite and for arbitrary $\varepsilon > 0$, consider the ε-regular subgradient set*

$$\begin{aligned}\widehat{\partial}_\varepsilon f(\bar{x}) : &= \big\{ v \,\big|\, f(x) \geq f(\bar{x}) + \langle v, x - \bar{x}\rangle - \varepsilon|x - \bar{x}| + o(|x - \bar{x}|)\big\} \\ &= \big\{ v \,\big|\, df(\bar{x})(w) \geq \langle v, w\rangle - \varepsilon|w| \ \text{ for all } w\big\}.\end{aligned}$$

If f is locally lsc at $\bar{x}$, the conditions $x^\nu \underset{f}{\to} \bar{x}$, $\varepsilon^\nu \searrow 0$, $v^\nu \in \widehat{\partial}_{\varepsilon^\nu} f(x^\nu)$, $v^\nu \to v$, imply that $v \in \partial f(\bar{x})$; or instead $v \in \partial^\infty f(\bar{x})$ if $\lambda^\nu v^\nu \to v$ with $\lambda^\nu \searrow 0$. Thus,

$$\limsup_{\substack{x \underset{f}{\to} \bar{x} \\ \varepsilon \searrow 0}} \widehat{\partial}_\varepsilon f(x) \;=\; \partial f(\bar{x}) \cup \operatorname{dir} \partial^\infty f(\bar{x}).$$

Proof. The equivalence between the two expressions given for $\widehat{\partial}_\varepsilon f(\bar{x})$ is elementary; it follows for instance from the formula in 8(7) as applied to $g(x) := f(x) + \varepsilon|x - \bar{x}|$. Also elementary is the fact that every vector v in $\partial f(\bar{x})$ or $\partial^\infty f(\bar{x})$ can somehow or other be generated as limit of the sort described, since in particular $\widehat{\partial} f(x) \subset \widehat{\partial}_\varepsilon f(x)$ for all $\varepsilon > 0$, and limits employing vectors $v^\nu \in \widehat{\partial} f(x^\nu)$ are available by Definition 8.3.

It remains only to verify that, when f is locally lsc at $\bar{x}$, no other kinds of vectors v can show up as limits. Let $g^\nu(x) = f(x) + \varepsilon|x - x^\nu|$. We have $v^\nu \in \widehat{\partial} g^\nu(x^\nu)$, hence $v^\nu \in \partial g^\nu(x^\nu)$. The calculus rule in 10.10, in combination with the formula in 8.27 for the subgradients of the Euclidean norm, then yields $v^\nu \in \partial f(x^\nu) + \varepsilon^\nu I\!B$. This furnishes the existence of $\hat{v}^\nu \in \partial f(x^\nu)$ with $|\hat{v}^\nu - v^\nu| \le \varepsilon^\nu$. Then also $\hat{v}^\nu \to v$, so $v \in \partial f(\bar{x})$ by definition. The argument concerning $\partial^\infty f(\bar{x})$ is parallel. □

Horizon subgradients have been utilized most importantly in expressing the various 'constraint qualifications' that underlie the many subdifferentiation formulas. An advantage is that horizon subgradients can themselves be calculated through such formulas, which helps materially in achieving a robust calculus that can be applied not just once but in a series of steps. Furthermore, the conditions that come up are closely related to Lipschitzian properties and thus have a side benefit, as viewed in 10.14 and 10.16.

In some situations, however, the conditions that are sufficient in terms of horizon subgradients in order to reach a desired conclusion may be too strong. It's good to know that a refinement in terms of 'calmness' may then work instead. The parametric version of Fermat's rule in 10.12 provides a focus for the idea.

10.47 Proposition (calmness as a constraint qualification). *For a proper, lsc function $f : I\!R^n \times I\!R^m \to \overline{I\!R}$ and a vector $\bar{u} \in I\!R^m$, consider the problem of minimizing $f(x, \bar{u})$ over $x \in I\!R^n$. Suppose $\bar{x}$ is locally optimal and satisfies the constraint qualification that there exists $\kappa \ge 0$ with*

$$\liminf_{\substack{u \to \bar{u},\, u \neq \bar{u} \\ x \to \bar{x}}} \frac{f(x,u) - f(\bar{x},\bar{u})}{|u - \bar{u}|} \ge -\kappa > -\infty, \tag{10(10)}$$

as is true in particular if $\bar{x}$ is globally optimal and the function $p(u) := \inf_x f(x,u)$ is calm from below at $\bar{u}$ with constant κ. Then

$$\exists\, \bar{y} \ \text{ with } \ (0, \bar{y}) \in \partial f(\bar{x}, \bar{u}), \ \ |\bar{y}| \le \kappa.$$

Proof. First we note that $\bar{x}$ being globally optimal corresponds to having $f(\bar{x}, \bar{u}) = p(\bar{u})$, in which case

$$\frac{f(x,u) - f(\bar{x},\bar{u})}{|u - \bar{u}|} \ge \frac{p(u) - p(\bar{u})}{|u - \bar{u}|}.$$

Calmness of p from below at $\bar{u}$ with constant κ means by definition that the quotient on right side of this inequality is bounded from below by $-\kappa$ for u in some neighborhood of $\bar{u}$, and then 10(10) must obviously hold.

Turning to the general case where only 10(10) is assumed, we observe that this condition refers to the existence for any $\varepsilon > 0$ of a $\delta > 0$ such that

$$\frac{f(x,u) - f(\bar{x},\bar{u})}{|u - \bar{u}|} \geq -(\kappa + \varepsilon) \quad \text{when} \quad |x - \bar{x}| \leq \delta,\ |u - \bar{u}| \leq \delta.$$

Taking $V = \mathbb{B}(\bar{x},\delta)$ and $W = \mathbb{B}(\bar{u},\delta)$, we get $(\bar{x},\bar{u}) \in \operatorname{argmin}_{x,u} g_\varepsilon(x,u)$ for

$$g_\varepsilon(x,u) := f(x,u) + (\kappa + \varepsilon)|u - \bar{u}| + \delta_{V\times W}(x,u).$$

The basic version of Fermat's rule in 10.1 applies to this minimization of g_ε: we must have $(0,0) \in \partial g_\varepsilon(\bar{x},\bar{u})$. But by the rule in 10.10,

$$\partial g_\varepsilon(\bar{x},\bar{u}) \subset \partial f(\bar{x},\bar{u}) + \{(0,y) \mid |y| \leq \kappa + \varepsilon\}.$$

Hence there exists y_ε such that $(0,y_\varepsilon) \in \partial f(\bar{x},\bar{u})$ and $|y_\varepsilon| \leq \kappa + \varepsilon$.

Considering now a sequence $\varepsilon^\nu \searrow 0$, we get corresponding vectors y_{ε^ν} which form a bounded sequence having a cluster point $\bar{y}$, necessarily with $|\bar{y}| \leq \kappa$. The fact that $(0,y_{\varepsilon^\nu}) \in \partial f(\bar{x},\bar{u})$ ensures through the closedness of subgradient sets (cf. 8.6) that likewise $(0,\bar{y}) \in \partial f(\bar{x},\bar{u})$, as required. □

The optimization problem in Proposition 10.47 is said to be *calm at* $\bar{x}$ (relative to its parameterization in u at $\bar{u}$) when 10(10) holds for some $\kappa > 0$. This is a decidedly weaker constraint qualification for Fermat's rule than the one in Example 10.12. That constraint qualification excluded the existence of $y \neq 0$ with $(0,y) \in \partial^\infty f(\bar{x},\bar{u})$. Indeed, when $\bar{x}$ is globally optimal (as can always be arranged by restriction of the problem to some neighborhood of $\bar{x}$), this earlier condition guarantees that p is Lipschitz continuous on a neighborhood of $\bar{u}$ (see 10.14), hence certainly calm at $\bar{u}$ from below. That implies 10(10), as we've just seen.

The calmness constraint qualification pushes the existence of a multiplier vector $\bar{y}$ in this basic setting to a natural frontier. If it weren't satisfied at $\bar{x}$, there would be sequences $x^\nu \to \bar{x}$, $u^\nu \to \bar{u}$ and $\kappa^\nu \to \infty$ with

$$f(x^\nu,u^\nu) \to f(\bar{x},\bar{u}) \ \text{ but } \ f(x^\nu,u^\nu) \leq f(\bar{x},\bar{u}) - \kappa^\nu|u^\nu - \bar{u}|.$$

This means that tiny perturbations of the parameter vector $\bar{u}$ make it possible for tiny perturbations of $\bar{x}$ to decrease the objective value at an arbitrarily high rate. Heuristically, a multiplier vector $\bar{y}$ should reflect something about the rate of change of the optimal objective value with respect to perturbations of $\bar{u}$ (cf. Theorem 10.13 and the remarks after 10.15), but here the rate is infinite in a local sense, and an instability is thus revealed which militates against the existence of such $\bar{y}$.

An illustration is provided by the lsc, convex function $f : \mathbb{R} \times \mathbb{R} \to \overline{\mathbb{R}}$ with formula

$$f(x,u) := \begin{cases} \frac{1}{2}x^2 - 2\sqrt{xu} & \text{if } x \geq 0,\ u \geq 0, \\ \infty & \text{otherwise,} \end{cases}$$

at $(\bar{x},\bar{u}) = (0,0)$. The problem of minimizing $f(x,\bar{u}) = f(x,0) = \frac{1}{2}x^2 + \delta_{\mathbb{R}_+}(x)$ has the unique globally optimal solution $\bar{x} = 0$. But for $u > 0$ the problem of minimizing $f(x,u)$ in x has optimal solution $x_u = u^{1/3}$. We have

$$\frac{f(x_u,u) - f(0,0)}{u-0} = \frac{\frac{1}{2}u^{2/3} - 2(u^{1/3}u)^{1/2}}{u} = -\frac{3}{2u^{1/3}} \to -\infty \text{ as } u \searrow 0.$$

Therefore, the problem in question isn't calm at $\bar{x}$ (relative to its parameterization in u at $\bar{u}$). In particular, the function $p(u) = \inf_x f(x,u)$ isn't calm from below at $\bar{u}$.

The chain rule in Theorem 10.6 leads to a generalized form of the mean-value theorem, which holds for strictly continuous functions.

10.48 Theorem (mean-value, extended). *Suppose f is strictly continuous on an open, convex set O, and let x_0 and x_1 be points of O. Then for some $\tau \in (0,1)$ and the corresponding point $x_\tau = (1-\tau)x_0 + \tau x_1$ there is a vector v satisfying*

$$f(x_1) - f(x_0) = \langle v, x_1 - x_0 \rangle \text{ with } v \in \partial f(x_\tau) \text{ or } -v \in \partial(-f)(x_\tau).$$

If f is regular on O (as when f is lower-$\mathcal{C}^1$, or in particular when f is convex), this is sure to be true with $v \in \partial f(x_\tau)$.

Proof. Let $\varphi(t) = f\big(F(t)\big) - (1-t)f(x_0) - tf(x_1)$, where $F(t) := (1-t)x_0 + tx_1$; φ is Lipschitz continuous on the open interval of t values for which $F(t) \in O$, which includes $[0,1]$; it has $\varphi(0) = \varphi(1) = 0$. Because F is smooth and $\partial^\infty f(x) = \{0\}$ for all $x \in O$ (by 9.13), the chain rule in 10.6 applies:

$$\begin{aligned} \partial\varphi(t) &\subset \big\{ \langle v, x_1 - x_0 \rangle \,\big|\, v \in \partial f(F(t)) \big\} + f(x_0) - f(x_1), \\ \partial(-\varphi)(t) &\subset \big\{ \langle v, x_1 - x_0 \rangle \,\big|\, v \in \partial(-f)(F(t)) \big\} + f(x_0) - f(x_1). \end{aligned}$$

By the continuity of φ alone, together with the fact that $\varphi(0) = \varphi(1) = 0$, there must either be a point $\tau \in (0,1)$ where φ attains its minimum relative to $[0,1]$, or one where φ attains its maximum relative to $[0,1]$. In the first case one has $0 \in \partial\varphi(\tau)$, whereas in the second case the conclusion is that $0 \in \partial(-\varphi)(\tau)$. This argument gives the general result. □

With the technical results now available, an extended version of the chain rule of Theorem 10.6 for $f = g \circ F$ can be proved in which the smoothness of the mapping F is replaced by strict continuity. This opens the door to applications in which $F(x) = \big(f_1(x), \ldots, f_m(x)\big)$ with component functions f_i that are convex, concave, lower- or upper-$\mathcal{C}^1$, and so forth. The extended chain rule makes use also of the concept of strict differentiability of a mapping F at $\bar{x}$ as defined in 9(10) and characterized in 9.25. In a coordinate representation $F(x) = \big(f_1(x), \ldots, f_m(x)\big)$, F is strictly differentiable at $\bar{x}$ if and only if each component function f_i is strictly differentiable at $\bar{x}$.

10.49 Theorem (extended chain rule for subgradients). *Let $f(x) = g\big(F(x)\big)$ for a proper, lsc function $g : \mathbb{R}^m \to \overline{\mathbb{R}}$ and a strictly continuous vector-valued function $F : \mathbb{R}^n \to \mathbb{R}^m$. Let $\bar{x}$ be a point where f is finite. Then*

$$\widehat{\partial} f(\bar{x}) \supset \widehat{D}^* F(\bar{x})\big[\widehat{\partial} g(F(\bar{x}))\big] = \bigcup\Big\{\widehat{\partial}(yF)(\bar{x}) \,\Big|\, y \in \widehat{\partial} g\big(F(\bar{x})\big)\Big\}.$$

If the only vector $y \in \partial^\infty g(F(\bar{x}))$ with $0 \in \partial(yF)(\bar{x})$ is $y = 0$, one also has

$$\partial f(\bar{x}) \subset D^* F(\bar{x})\big[\partial g\big(F(\bar{x})\big)\big] = \bigcup\Big\{\partial(yF)(\bar{x}) \,\Big|\, y \in \partial g\big(F(\bar{x})\big)\Big\},$$

$$\partial^\infty f(\bar{x}) \subset D^* F(\bar{x})\big[\partial^\infty g\big(F(\bar{x})\big)\big] = \bigcup\Big\{\partial(yF)(\bar{x}) \,\Big|\, y \in \partial^\infty g\big(F(\bar{x})\big)\Big\},$$

and the mapping $x \mapsto D^* F(x)\big[\partial g\big(F(x)\big)\big] \cup D^* F(x)\big[\partial^\infty g\big(F(x)\big)\big]$ is cosmically osc at $\bar{x}$ with respect to f-attentive convergence.

If in addition g is regular at $F(\bar{x})$ and yF is regular at $\bar{x}$ for each $y \in \partial g(\bar{x})$ (as is true if F is strictly differentiable at $\bar{x}$), then f is regular at $\bar{x}$ and

$$\partial f(\bar{x}) = D^* F(\bar{x})\big[\partial g\big(F(\bar{x})\big)\big], \qquad \partial^\infty f(\bar{x}) = D^* F(\bar{x})\big[\partial^\infty g\big(F(\bar{x})\big)\big].$$

Proof. To simplify notation in what follows, let $\widehat{C}(\bar{x}) = \widehat{D}^* F(\bar{x})\big[\partial g\big(F(\bar{x})\big)\big]$, $C(\bar{x}) = D^* F(\bar{x})\big[\partial g\big(F(\bar{x})\big)\big]$, and $K(\bar{x}) = D^* F(\bar{x})\big[\partial^\infty g\big(F(\bar{x})\big)\big]$. The alternative formulas for these sets in terms of yF come from 9.24(b). Suppose first that $v \in \widehat{C}(\bar{x})$; we have $v \in \widehat{\partial}(yF)(\bar{x})$ for some $y \in \widehat{\partial} g\big(F(\bar{x})\big)$. Then

$$g\big(F(x)\big) \geq g\big(F(\bar{x})\big) + \big\langle y,\, F(x) - F(\bar{x})\big\rangle + o\big(|F(x) - F(\bar{x})|\big),$$

$$\big\langle y, F(x)\big\rangle \geq \big\langle y, F(\bar{x})\big\rangle + \big\langle v,\, x - \bar{x}\big\rangle + o\big(|x - \bar{x}|\big),$$

where $|F(x) - F(\bar{x})| \leq \kappa|x - \bar{x}|$ for a certain constant $\kappa \geq 0$. These conditions together yield $f(x) \geq f(\bar{x}) + \langle v,\, x - \bar{x}\rangle + o\big(|x - \bar{x}|\big)$. Thus $v \in \widehat{\partial} f(\bar{x})$, so that $\widehat{C}(\bar{x}) \subset \widehat{\partial} f(\bar{x})$ as claimed.

Next we apply the coderivative chain rule in 10.40 to the representation $E_f = E_g \circ F$, where E_f and E_g are the profile mappings associated with f and g (cf. 5.5). The characterization in 8.35 of the subgradients of f and g in terms of these profile mappings yields the inclusions $\partial f(\bar{x}) \subset C(\bar{x})$ and $\partial^\infty f(\bar{x}) \subset K(\bar{x})$. The assertions about the regular case follow from the corresponding ones in 10.40 as well.

The remaining task is to show that the mapping $x \mapsto C(x) \cup \operatorname{dir} K(x)$ is cosmically osc at $\bar{x}$ with respect to f-attentive convergence. We know that the mapping $u \mapsto \partial g(u) \cup \operatorname{dir} \partial^\infty g(u)$ is cosmically osc at $F(\bar{x})$ with respect to g-attentive convergence (see 8.7). Suppose $v^\nu \in C(x^\nu)$ and $x^\nu \underset{f}{\to} \bar{x}$. There exist vectors $y^\nu \in \partial g\big(F(x^\nu)\big)$ such that $v^\nu \in \partial(y^\nu F)(x^\nu)$; here $F(x^\nu) \underset{g}{\to} F(\bar{x})$. If $\{y^\nu\}$ were unbounded, a contradiction could be produced for the assumption that the only $y \in \partial^\infty g(F(\bar{x})$ with $0 \in (yF)(\bar{x})$ is $y = 0$. We can suppose therefore that y^ν converges to some y. If v^ν converges to a vector v, we have $y \in \partial g\big(F(\bar{x})\big)$, while on the other hand $v \in \partial(yF)(\bar{x})$ by the same argument used earlier, based on the estimate $\partial(y^\nu F)(x^\nu) \subset \partial(yF)(x^\nu) + \partial\big([y^\nu - y]F\big)(x^\nu)$. Then $v \in C(\bar{x})$. Likewise, if $v^\nu \to \operatorname{dir} v$ for a vector $v \neq 0$ we get by a modification of this argument that $v \in K(\bar{x})$. The proof that the mapping $x \mapsto K(x)$ is osc at $\bar{x}$ with respect to f-attentive convergence is entirely parallel. □

As far as subgradients are concerned, the extended chain rule in Theorem 10.49 fully encompasses the one in Theorem 10.6, which can be seen as the case where F is strictly differentiable everywhere; cf. 9.18.

10.50 Corollary (normals for Lipschitzian constraints). *Let $C = F^{-1}(D)$ for a closed set $D \subset I\!R^m$ and a strictly continuous mapping $F : I\!R^n \to I\!R^m$. At any $\bar{x} \in C$ one has*

$$\widehat{N}_C(\bar{x}) \supset \bigcup \Big\{ \widehat{\partial}(yF)(\bar{x}) \,\Big|\, y \in \widehat{N}_D\big(F(\bar{x})\big) \Big\}.$$

If the only vector $y \in N_D\big(F(\bar{x})\big)$ with $0 \in \partial(yF)(\bar{x})$ is $y = 0$, one also has

$$N_C(\bar{x}) \subset \bigcup \Big\{ \partial(yF)(\bar{x}) \,\Big|\, y \in N_D\big(F(\bar{x})\big) \Big\} \;=:\; K(\bar{x}),$$

and the mapping $x \mapsto K(x)$ is osc. If in addition D is regular at $F(\bar{x})$ and yF is regular at $\bar{x}$ for every $y \in N_D\big(F(\bar{x})\big)$ (as is true in particular when F is strictly differentiable at $\bar{x}$), then C is regular at $\bar{x}$ and

$$N_C(\bar{x}) \;=\; \bigcup \Big\{ \partial(yF)(\bar{x}) \,\Big|\, y \in N_D\big(F(\bar{x})\big) \Big\}.$$

Proof. Apply Theorem 10.49 with $g = \delta_D$. □

This description of normal vectors can be used for example in connection with the normal cone condition for optimality in 6.12 and 8.15, where a function f_0 is minimized over a set C. To say that $C = F^{-1}(D)$ is to say that C consists of all vectors x satisfying a constraint system of the form $F(x) \in D$.

10.51 Corollary (pointwise max of strictly continuous functions). *Suppose that $f(x) = \max\{f_1(x), \ldots, f_m(x)\}$ with each f_i strictly continuous on $I\!R^n$, f itself then being strictly continuous. In denoting by $I(\bar{x})$ the set of indices i for which the max is attained for $\bar{x}$, one has*

$$\partial f(\bar{x}) \subset \bigcup \Big\{ \sum\nolimits_{i \in I(\bar{x})} \lambda_i v_i \,\Big|\, v_i \in \partial f_i(\bar{x}),\ \lambda_i \geq 0,\ \sum\nolimits_{i \in I(\bar{x})} \lambda_i = 1 \Big\}.$$

The inclusion holds as an equation if f_i is strictly differentiable at $\bar{x}$ for each $i \in I(\bar{x})$, and then f is regular at $\bar{x}$.

Proof. Apply 10.49 with $F(x) = \big(f_1(x), \ldots, f_m(x)\big)$ and $g = \operatorname{vecmax}$, being mindful of the subgradient formula for g in 8.26. Because g is convex and finite, it is strictly continuous and regular; cf. 9.14. To get regularity of f it's only necessary to assume strict differentiability of the f_i's with $i \in I(\bar{x})$, because the other f_i's are out of action in a neighborhood of $\bar{x}$. □

The chain rule in Theorem 10.49 leads also to a generalization of the multiplier rule in 10.8 from smooth to strictly continuous mappings F.

10.52 Exercise (nonsmooth Lagrange multiplier rule). *For a nonempty, closed set $X \subset I\!R^n$ and strictly continuous functions $f_0 : I\!R^n \to I\!R$ and $F : I\!R^n \to I\!R^m$ with $F = (f_1, \ldots, f_m)$, consider the problem*

$$\text{minimize } f_0(x) + \theta\big(F(x)\big) \text{ over } x \in X,$$

where $\theta : I\!R^m \to \overline{I\!R}$ is proper, lsc and convex with effective domain D. (As a special case, one could have $\theta = \delta_D$.) Suppose $\bar{x}$ is a locally optimal solution at which the following constraint qualification is satisfied:

$$0 \in \partial(yF)(\bar{x}) + N_X(\bar{x}), \quad y \in N_D\big(F(\bar{x})\big) \quad \Longrightarrow \quad y = 0.$$

Then there exists a vector $\bar{y}$ such that

$$0 \in \partial(f_0 + \bar{y}F)(\bar{x}) + N_X(\bar{x}), \qquad \bar{y} \in \partial\theta\big(F(\bar{x})\big).$$

Moreover, the set of such vectors $\bar{y}$ is compact.

Guide. Represent this as the problem of minimizing $f = g \circ G$ over $I\!R^n$ where

$$G(x) = (f_0(x), f_1(x), \ldots, f_m(x), x),$$
$$g(w_0, w_1, \ldots, w_m, x) = w_0 + \theta(w_1, \ldots, w_m) + \delta_X(x).$$

Verify that G is strictly continuous and that $\partial^\infty g(\bar{w}_0, \bar{w}_1, \ldots, \bar{w}_m, \bar{x})$ and $\partial g(\bar{w}_0, \bar{w}_1, \ldots, \bar{w}_m, \bar{x})$ are respectively the sets

$$\Big\{(0, y_1, \ldots, y_m, v) \,\Big|\, (y_1, \ldots, y_m) \in N_D(\bar{w}_1, \ldots, \bar{w}_m),\; v \in N_X(\bar{x})\Big\},$$
$$\Big\{(1, y_1, \ldots, y_m, v) \,\Big|\, (y_1, \ldots, y_m) \in \partial\theta(\bar{w}_1, \ldots, \bar{w}_m),\; v \in N_X(\bar{x})\Big\}.$$

This uses the fact that $\partial^\infty\theta = N_D$ by convexity; cf. 8.12. Invoke 10.49 and then apply the basic necessary condition $0 \in \partial f(\bar{x})$ in 10.1. The cosmic osc property in 10.49 yields the compactness assertion at the end. □

The Lagrange multiplier interpretation in 10.15 could be extended to the setting of 10.52. This would involve replacing $F(x)$ by $F(x) + u$.

10.53 Exercise (composition with a strictly continuous mapping). *Suppose $S = S_0 \circ F$ for $F : I\!R^n \to I\!R^p$ strictly continuous and $S_0 : I\!R^p \rightrightarrows I\!R^m$ osc, and let $\bar{u} \in S(\bar{x})$. If F is strictly differentiable at $\bar{x}$ and S_0 is graphically regular at $F(\bar{x})$ for $\bar{u}$, then S is graphically regular at $\bar{x}$ for $\bar{u}$, and*

$$D^*S(\bar{x} \,|\, \bar{u}) \;=\; D^*F(\bar{x}) \circ D^*S_0(F(\bar{x}) \,|\, \bar{u}).$$

Guide. View $\operatorname{gph} S$ as $G^{-1}(\operatorname{gph} S_0)$ under $G : (x, u) \mapsto \big(F(x), u\big)$, and then apply 10.50. □

A fact of interest in connection with Theorem 10.33 and the ways that a subsmooth function can be represented is the following.

10.54 Proposition (representations of subsmoothness). *If f is lower-$\mathcal{C}^1$ on an open set $O \subset \mathbb{R}^n$, then there is not only a local representation of the kind in Definition 10.29 around each point $x \in O$, but, for any compact set $B \subset O$, a common representation valid at all points x in some open set O' satisfying $B \subset O' \subset O$. The same holds for the lower-$\mathcal{C}^2$ case.*

The proof of this result will make use of a lower bounding property.

10.55 Lemma (smooth bounds on semicontinuous functions). *Let $O \subset \mathbb{R}^n$ be nonempty and open, and let $f : O \to \overline{\mathbb{R}}$ be lsc with $f(x) > -\infty$ for all $x \in O$. Then there is a function $h : O \to \mathbb{R}$ of class $\mathcal{C}^\infty$ such that $f \geq h$ on O.*

Proof. First we construct a sequence of compact sets B^ν such that $B^\nu \subset \operatorname{int} B^{\nu+1}$ and $\bigcup_{\nu \in \mathbb{N}} B^\nu = O$. This can be done by fixing any point $\bar{x}$ in O and taking $B^\nu = \mathbb{B}(\bar{x}, \nu) \cap \{x \mid d_C(x) \geq 1/\nu\}$, where C is the complement of O; if $O = \mathbb{R}^n$, the simpler formula $B^\nu = \mathbb{B}(\bar{x}, \nu)$ suffices. Next we define $\beta^\nu = \inf_{B^\nu} f$. From the compactness of B^ν we have $\beta^\nu > -\infty$, because f is lsc and does not take the value $-\infty$ (one can apply 1.10 to $f + \delta_{B^\nu}$). Since the assertion of the lemma is trivially true when $f \equiv \infty$, we can suppose that the point $\bar{x}$ chosen in the construction of the sets B^ν satisfies $f(\bar{x}) < \infty$, so that $\beta^\nu < \infty$ as well. Clearly $\beta^{\nu+1} \leq \beta^\nu$ because $B^{\nu+1} \supset B^\nu$.

It will suffice now to construct a sequence of $\mathcal{C}^\infty$ functions $\varphi^\nu : \mathbb{R}^n \to [0,1]$ such that $\varphi^\nu(x) \equiv 0$ on B^ν but $\varphi^\nu(x) \equiv 1$ outside of $\operatorname{int} B^{\nu+1}$, since for such a sequence we can obtain the desired lower bounding function from the formula $h := \beta_1 - \sum_{\nu=1}^{\infty} (\beta^\nu - \beta^{\nu+1}) \varphi^\nu$. (Although an infinite series is involved, all but finitely many terms vanish in a neighborhood of any given point; cf. Figure 10–2.) To simplify the notation, we can concentrate henceforth on the case of nonempty, compact set B and a nonempty, closed set D not meeting B, and demonstrate the existence of a $\mathcal{C}^\infty$ function $\varphi : \mathbb{R}^n \to [0,1]$ such that $\varphi(x) \equiv 0$ on B but $\varphi(x) \equiv 1$ on D.

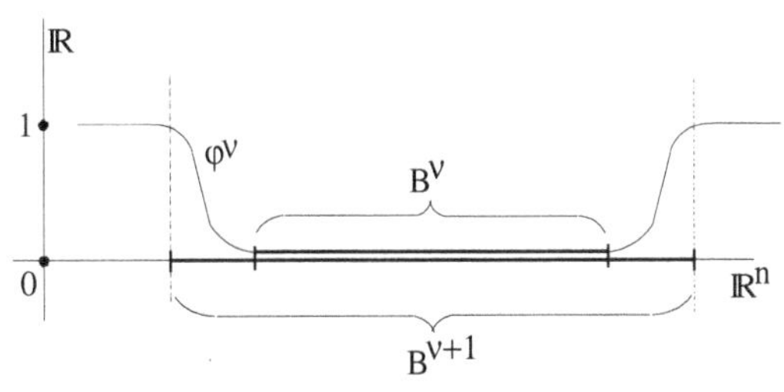

Fig. 10–2. Construction of an infinitely smooth lower bounding function.

The distance function d_D is continuous (cf. 1.20) and by assumption is positive at every point of the compact set B, so the value $\delta := \min_B d_D$ is positive. The collection of open balls $\{\mathbb{B}(x, \delta/2)\}_{x \in B}$ covers B, so by compactness a finite subcollection covers B: there exist points $x_i \in B$ for $i = 1, \ldots, m$ such that $B \subset \bigcup_{i=1}^m I\!B(x_i, \delta/2)$. Let $\theta : I\!R \to [0,1]$ be a $\mathcal{C}^\infty$ function such that $\theta(t) \equiv 0$ for $t \le 0$ but $\theta(t) \equiv 1$ for $t \ge 1$. (For instance, one can take

$$\theta(t) = c^{-1} \int_{-\infty}^{t} \big[\theta_0(s) + \theta_0(1-s)\big] ds \;\text{ with }\; c = \int_{-\infty}^{\infty} \big[\theta_0(s) + \theta_0(1-s)\big] ds,$$

where $\theta_0(s) = 0$ for $s \le 0$ and $\theta(s) = \exp(-1/s)$ for $s > 0$.) Then by defining $\psi(t) = \theta\big(2t/\delta - 1\big)$ we get a $\mathcal{C}^\infty$ function from $I\!R$ to $[0,1]$ with $\psi(t) \equiv 0$ for $t \le \delta/2$ but $\psi(t) \equiv 1$ for $t \ge \delta$. The desired function φ is obtained finally from the formula $\varphi(x) := \prod_{i=1}^m \psi\big(\,|x - x_i|\,\big)$. □

Proof of 10.54. For each point $w \in B$ we have by assumption a representation

$$f(x) = \max_{t \in T_w} f_{w,t}(x) \;\text{ for all }\; x \in \operatorname{int} I\!B(w, \delta_w), \tag{10(11)}$$

where $f_{w,t}(x)$ and $\nabla f_{w,t}(x)$ depend continuously on (t,x), and the open ball $\operatorname{int} I\!B(w, \delta_w)$ is included in O. Fixing w, we shall argue first that the functions $f_{w,t}$ can be modified outside of the smaller ball $I\!B(w, \delta_w/3)$ and extended to O in such a way that the continuity properties are maintained, the representation 10(11) persists on $I\!B(w, \delta_w/3)$, but the maximum gives a value at most $f(x)$ at all points $x \in O \setminus I\!B(w, \delta_w/3)$.

Since f is lower-$\mathcal{C}^1$, it's in particular continuous (see 10.31), and by the preceding lemma we are able to find a $\mathcal{C}^\infty$ function h such that $h \le f$ on O. Let β be the minimum value of the continuous function $(t,x) \mapsto f_{w,t}(x)$ on $T_w \times I\!B(w, 2\delta_w/3)$, and α be the maximum of h on $I\!B(w, 2\delta_w/3)$. The $\mathcal{C}^\infty$ function $h_0 := h - \max\{\alpha - \beta, 0\}$ then satisfies not only $h_0 \le f$ on O but $h_0 \le f_{w,t}$ on $I\!B(w, 2\delta_w/3)$ for all $t \in T_w$. Let $\psi_w : I\!R \to [0,1]$ be a $\mathcal{C}^\infty$ function such that $\psi_w(t) \equiv 0$ when $t \le \delta_w/3$ but $\psi_w(t) \equiv 1$ when $t \ge 2\delta_w/3$ (for way to the construct such a function, see the end of the proof of the preceding lemma). The functions

$$\bar f_{w,t}(x) := \begin{cases} \psi_w\big(f_{w,t}(x) - h_0(x)\big) + h_0(x) & \text{for } x \in I\!B(w, 2\delta_w/3), \\ h_0(x) & \text{for } x \notin I\!B(w, 2\delta_w/3), \end{cases}$$

then have the property that $\bar f_{w,t}(x)$ and $\nabla \bar f_{w,t}(x)$ depend continuously on $(t,x) \in T_w \times O$ and give

$$\max_{t \in T_w} \bar f_{w,t}(x) \begin{cases} = f(x) & \text{for all } x \in I\!B(w, \delta_w/3), \\ \ge f(x) & \text{for all } x \in O \setminus I\!B(w, \delta_w/3). \end{cases} \tag{10(12)}$$

Moving into the final stage, we now use the compactness of B to extract from the family of open balls $\big\{ \operatorname{int} I\!B(w, \delta_w/3) \big\}_{w \in B}$ covering B a finite subcollection that likewise covers B, say for the points $w_1, \ldots, w_m$. The union of this finite subcollection of open balls is taken to be the open set O'; it is included in O because the balls $\operatorname{int} I\!B(w, \delta_w)$ are all in O. For each of the points w_i we construct a modified family of functions $\bar f_{w_i,t}$ as in 10(12). Defining the compact index space T to be the union of (disjoint copies of) the compact sets T_{w_i}, we let $\bar{f}_t(x) = \bar{f}_{w_i,t}(x)$ for all $x \in O'$ when $t \in T_{w_i}$. Then $\bar{f}_t(x)$ and $\nabla \bar{f}_t(x)$ depend continuously on $(t,x) \in T \times O'$, and they give the representation $f(x) = \max_{t\in T} \bar{f}_t(x)$ for all $x \in O'$. This is what we needed.

The same argument carries over to the case where f is lower-$\mathcal{C}^2$. One just has to observe that the individual families in 10(12) maintain the property of the second partial derivatives depending continuously on (t,x). □

Epi-addition is an important source of subsmooth functions. The next theorem combines the main result along those lines with another criterion for Lipschitz continuity.

10.56 Theorem (properties under epi-addition). *Let $f = g \# h$ for a proper, lsc function $g : \mathbb{R}^n \to \overline{\mathbb{R}}$ and a function $h : \mathbb{R}^n \to \mathbb{R}$. Suppose that for each $\alpha \in \mathbb{R}$ the mapping $x \mapsto \{w \mid g(w) + h(x-w) \le \alpha\}$ is locally bounded. If h is strictly continuous, then f is strictly continuous with*

$$\operatorname{lip} f(x) \le \max_{w \in P(x)} \operatorname{lip} h(x-w) \quad \textit{for} \quad P(x) := \operatorname*{argmin}_{w \in \mathbb{R}^n} \{g(w) + h(x-w)\}.$$

If h is upper-$\mathcal{C}^1$, then f is upper-$\mathcal{C}^1$, while if h is upper-$\mathcal{C}^2$ then f is upper-$\mathcal{C}^2$.

Proof. We have $f(x) = \inf_w k(w,x)$ and $P(x) = \operatorname{argmin}_w k(w,x)$ for the function $k(w,x) = g(w) + h(x-w)$. The assumptions ensure that k is lsc and proper, and that $k(w,x)$ is level-bounded in w locally uniformly in x. Also, $k(w,x)$ is continuous in x for fixed w. From basic results on parametric minimization (cf. 1.17 and 10.13) we conclude that f is finite and continuous on $\mathbb{R}^n$, while P is nonempty-valued as well as osc and locally bounded, with

$$f(x) = \min_{w \in \mathbb{R}^n} \{g(w) + h(x-w)\} = g(\bar{w}) + h(x - \bar{w}) \text{ when } \bar{w} \in P(x). \quad 10(13)$$

Fix any $\bar{x} \in \mathbb{R}^n$ and any compact neighborhood $V \in \mathcal{N}(\bar{x})$. Let

$$T = \{(z,w) \mid z \in V,\ w \in P(z)\},$$
$$f_t(x) = f(z) - h(z-w) + h(x-w) \text{ for } t = (z,w) \in T.$$

The set T is compact because of the properties cited for P: it is closed because V is closed and P is osc, and it is bounded because V is bounded and P is locally bounded. For any point x and any $t = (z,w) \in T$ we have by 10(13) that $h(x-w) \ge f(x) - g(w)$ whereas $f(z) = g(w) + h(z-w)$, and from this it follows that $f_t(x) \le f(x)$. Furthermore, we get $f_t(x) = f(x)$ by taking $z = x$, $w \in P(x)$. We thus have a representation $f(x) = \min_{t \in T} f_t(x)$ in which $\operatorname{lip} f_t(x) = \operatorname{lip} h(x-w)$ for each $t = (z,w) \in T$ and $x \in \mathbb{R}^n$.

Under the assumption that h is strictly continuous we obtain from the last part of Proposition 9.10 on the behavior of strictly continuity under the pointwise min operation that

$$\operatorname{lip} f(\bar{x}) \le k_V(\bar{x}) := \max_{t \in T} \operatorname{lip} h(\bar{x} - w) = \max_{w \in P(V)} \operatorname{lip} h(\bar{x} - w).$$

The maximum in this expression is finite and attained because the function $\operatorname{lip} h$ is usc (see 9.2). This establishes in particular that f is strictly continuous at x. Because the estimate for $\operatorname{lip} f(\bar{x})$ is valid for any compact neighborhood V of $\bar{x}$, and the mapping is P is osc and locally bounded, it follows that the estimate actually holds with V replaced by $\{\bar{x}\}$. The estimate given for $\operatorname{lip} f$ in the theorem is therefore correct.

We consider now the case where h is upper-$\mathcal{C}^1$. Relative to a compact neighborhood V of a point $\bar{x}$, we again take the representation $f(x) = \min_{t\in T} f_t(x)$ constructed above, writing it as

$$f(x) = \min_{t\in T}\big\{f(z) - h(z-w) + h(x-w)\big\} \qquad 10(14)$$

and noting that as the index $t = (z,w)$ ranges over T, w ranges over the set $P(V)$, which we know to be compact because V is compact while P is osc and locally bounded (see 5.15, 5.25(a)). Let $B = V - P(V)$; this is the set over which the argument $x-w$ ranges in 10(14), and as the difference of two compact sets it too is compact (cf. 3.12). The upper-$\mathcal{C}^1$ property of h implies not only the existence of local representations of h as required by Definition 10.29, but according to 10.54, such a representation on some open set $O \supset B$: one has $h(w) = \min_{s\in S} l_s(w)$ for all $w \in O$, where S is a compact space and both $l_s(w)$ and $\nabla l_s(w)$ depend continuously on $(s,w) \in S \times O$. With this substituted into 10(14) we arrive at the expression

$$f(x) = \min_{s\in S,\, t\in T}\big\{f(z) - h(z-w) + l_s(x-w)\big\} \ \text{ for all } \ x \in V,$$

where as always t denotes the pair (z,w). By defining a new index set $R = S \times T$, again compact, and setting $g_r(x) = f(z) - h(z-w) + l_s(x-w)$ for $r = (s,t) = (t,z,w)$, we get a representation $f(x) = \min_{r\in R} g_r(x)$ for all $x \in V$, where $g_r(x)$ and $\nabla g_r(x)$ depend continuously on (r,x). This demonstrates that f is upper-$\mathcal{C}^1$.

For the upper-$\mathcal{C}^2$ case the argument is the same, except for the observation that at each step one also has the continuity of the second derivatives of the functions in the representations. □

10.57 Example (subsmoothness of distance functions). *For a nonempty, closed set $C \subset \mathbb{R}^n$, the function d_C^2 is upper-$\mathcal{C}^2$. On the complement of C, the function d_C itself is upper-$\mathcal{C}^2$.*

Detail. For d_C^2 this is the case of 10.56 where $f = \delta_C$. The assertion for d_C follows by taking the square root. □

The example of Moreau envelopes already treated in 10.32 fits as a special case of Theorem 10.56 too.

10.58 Theorem (subsmoothness in parametric optimization). *Let*

$$p(u) := \inf_x f(x,u), \qquad P(u) := \operatorname{argmin}_x f(x,u),$$

for a proper, lsc function $f : \mathbb{R}^n \times \mathbb{R}^m \to \overline{\mathbb{R}}$ *such that* $f(x,u)$ *is level-bounded in* x *locally uniformly in* u. *Let* $\bar{u} \in \operatorname{dom} p$ *and suppose there are open sets* $W \in \mathcal{N}(\bar{u})$ *and* $V \supset P(\bar{u})$ *such that* $\nabla_u f(x,u)$ *exists for* $(x,u) \in V \times W$ *and, along with* $f(x,u)$, *depends continuously on* (x,u). *Then* p *is upper-*$\mathcal{C}^1$ *on* W *and is strictly differentiable on a subset* $D \subset W$ *with* $W \setminus D$ *negligible.*

Here $u \in D$ *if and only if the set* $Y(u) := \{\nabla_u f(x,u) \mid x \in P(u)\}$ *is a singleton. In terms of* $\overline{\nabla} p(u) := \{y \mid \exists\, u^\nu \underset{D}{\to} u \text{ with } \nabla p(u^\nu) \to y\}$, *one has*

$$\begin{aligned} dp(\bar{u})(z) &= \min_{y \in \overline{\nabla} p(\bar{u})} \langle y, z\rangle = \min_{y \in Y(\bar{u})} \langle y, z\rangle, \\ \partial p(\bar{u}) &= \overline{\nabla} p(\bar{u}) \subset Y(\bar{u}), \end{aligned}$$

with $\operatorname{con} \overline{\nabla} p(\bar{u}) = \operatorname{con} Y(\bar{u})$. *In particular,* $\partial p(\bar{u}) = Y(\bar{u})$ *when* $\bar{u} \in D$.

Proof. The assumptions guarantee through Theorems 1.17 and 5.22 that the mapping P is osc and locally bounded on W. There exists then by 5.19 a compact neighborhood B of $\bar{u}$ within W such that $P(u) \subset V$ for all $u \in B$. Let $X = P(B)$. The set $X \subset V$ is compact by 5.15 and 5.25(a). For each $x \in X$ the function $f_x := f(x,\cdot)$ is $\mathcal{C}^1$ on $O := \operatorname{int} B$, with $f_x(u)$ and $\nabla f_x(u)$ depending continuously on the pair $(x,u) \in X \times O$. The representation $p(u) = \min_{x \in X} f_x(u)$ for $u \in O$ thus fits the pattern in Definition 10.29 for p to be upper-$\mathcal{C}^1$ on O. The claims are justified then through application of Theorem 10.31 to $-p$. □

On a final note, we record a useful fact about convex functions that follows from the rules for calculating subgradients.

10.59 Proposition (minimization estimate for a convex function). *For a proper, lsc, convex function* $f : \mathbb{R}^n \to \overline{\mathbb{R}}$ *with* $\operatorname{argmin} f \neq \emptyset$ *and any* $x \in \mathbb{R}^n$, *one has*

$$f(x) - \inf f \;\le\; d\big(x, \operatorname{argmin} f\big)\, d\big(0, \partial f(x)\big).$$

Proof. We have $\inf f$ finite by assumption. Fix any $\bar{x} \in \mathbb{R}^n$ and let $\rho = d\big(0, \partial f(\bar{x})\big)$. The aim is to show that $f(\bar{x}) \le \inf f + \rho d\big(x, \operatorname{argmin} f\big)$. This is trivially true when $\rho = \infty$, so suppose $\rho < \infty$, i.e., $\partial f(\bar{x}) \neq \emptyset$. Since $\partial f(\bar{x})$ is closed, there exists $\bar{v} \in \partial f(\bar{x})$ with $|\bar{v}| = \rho$. Then $\bar{v} \in \partial f(\bar{x}) \cap \rho \mathbb{B}$, so $0 \in \partial f(\bar{x}) + \rho\mathbb{B}$. Let $g(x) = f(x) + \rho|x - \bar{x}|$. We have $\partial g(\bar{x}) = \partial f(\bar{x}) + \rho\mathbb{B}$ by 10.9 and 8.27, hence $0 \in \partial g(\bar{x})$, so that $\bar{x} \in \operatorname{argmin} g$ by 10.1. Thus, $g(\bar{x}) \le g(x)$ for all x, or in other words, $f(\bar{x}) \le f(x) + \rho|x - \bar{x}|$ for all $x \in \mathbb{R}^n$. In particular this says that $f(\bar{x}) \le \inf f + \rho|x - \bar{x}|$ for all $x \in \operatorname{argmin} f$. Therefore, $f(\bar{x}) - \inf f \le \rho d(\bar{x}, \operatorname{argmin} f)$, as required. □

Commentary

Fermat did more than formulate his famous 'theorem' about numbers. He was involved in the very early exploration of concepts of calculus, in particular in the recognition that the minimum or maximum values of a function can be detected by looking for the points of the function's graph that exhibit a horizontal tangent line. It's this principle, as translated to epigraphs and their normal vectors, that has propelled many of the developments in variational analysis and distinguished them from the efforts of other schools of generalized differentiation. The theory of distributions, for instance, features 'almost everywhere' statements which suppress the exceptional points where a maximum or minimum might be attained, instead of seeking them out.

The idea of looking for exceptions in the behavior of a function or mapping has also been the motive for investigating critical points and singularities, earlier for instance in the theory of Morse [1976], still in the pattern of smoothness without constraints, but nowadays in variational analysis also in terms of the generalized 'stationary points.' Such points arise from the extended version of Fermat's principle in 10.1 and its many elaborations into variational inequalities, multiplier rules, and the like.

In all cases, the need for a *calculus* is clear. This is also the prerequisite to successful application of results such as those treating Lipschitzian properties and metric regularity. The notion that such a calculus is feasible and valuable, despite the unprecedented hurdle of set-valuedness, goes back to the early days of convex analysis in Rockafellar [1963].

Level sets are intimate partners to constraint representation. Formulas for normal and tangent cones to level sets of convex functions were derived in Rockafellar [1970a] and in convexified Clarke mode in Rockafellar [1979a], [1985a], with the 1979 reference providing the characterization of epi-Lipschitzian sets in 10.4. The formulas of 10.3 in their full scope are new, however. The condition on f at the end of 10.4 was taken in Rockafellar [1981a] as describing a type of 'nonstationarity' especially relevant to algorithms for minimizing strictly continuous functions.

The results for separable functions in 10.5 are elementary and straightforward, but the observation that only an inclusion is available for $\partial^{\infty} f(\bar{x})$ hasn't been made before. The analogous relation in Clarke mode does hold as an equation; cf. Rockafellar [1985a].

Chain rules for $f = g \circ F$ as in 10.6, 10.19 and 10.49 mark an major boundary between the territory covered by convex analysis and the convexified subgradient sets of Clarke in comparison to the new territory that could satisfyingly be covered once convexification was relinquished. Convex analysis itself could only handle the composition of convex functions g with linear (or affine) mappings F, and this was well addressed for instance in Rockafellar [1970a], [1974a]. Beyond that, a chain rule essentially focusing on the inclusion '$\supset$' for regular subgradients was furnished by Penot [1978]. Subgradient inclusions of the harder type '$\subset$' in the composition of nonconvex functions g with nonlinear mappings F were obtained in the Clarke mode by Aubin and Clarke [1979] for g and F Lipschitz continuous, and by Rockafellar [1979a], [1985a], more broadly. But such results had inherent limitations.

Ioffe [1984b] produced the first unconvexified version, although it was formulated with assumptions resembling the ones in the convexified case. Mordukhovich [1984], [1988] came up with a corresponding result on the level of the chain rule in 10.49 in its assumptions, and this even allowed for F to be multivalued in a sense. Further advances were made by Ioffe [1989] and by Jourani and Thibault [1993]. The statements in 10.37, 10.39 and 10.40 on graphical coderivatives in the composition of multivalued mappings correspond largely to Mordukhovich [1994d]. For extensions see Mordukhovich and Shao [1996b], [1997a], and Jourani and Thibault [1997].

The tactic of relying on a chain rule to deduce rules for subgradients of a sum, partial subgradients, and related normal cone formulas, was developed by Rockafellar [1979a], [1985a], although within the context of convexified subgradients that prevailed then. The opposite tactic, of first proving a rule for addition and using that to get a chain rule, can be effective as well and was the standard pattern in convex analysis. In a contination of that pattern, see Ioffe [1984b], [1989], for example, and more recently Ioffe and Penot [1996], for a family of more subtle results in which the 'constraint qualification' in the hypothesis is replaced by a distance condition. For supplementary rules of calculus that put more emphasis on various tangent cones and subderivatives, see Ward and Borwein [1987] and Ward [1991]. Other results of subgradient calculus, special to convex functions, are in the survey of Hiriart-Urruty, Moussaoui, Seeger and Volle [1995].

The parametric version of Fermat's rule in 10.12 can in essence already be found in Theorem 5.1 of Rockafellar [1985a]. There it's expressed in terms of convexified subgradient sets, because that was the language then, but the proof actually covers the sharper, unconvexified formulation now given to this result. Likewise coming from that paper is the connection in 10.13 between the multiplier vectors in this rule and subgradients of the marginal function $p(u) = \inf_x f(x,u)$. The case where $f(x,u)$ is convex in (x,u) has a longer history, going back to Rockafellar [1970a], [1974a]. Beyond the convex case, but in the context of nonlinear programming, the analysis of multiplier vectors as subgradients first surfaced in Rockafellar [1982a]. The subdifferential interpretation of multiplier vectors in 10.14 and 10.15 comes from these works as well.

The Lipschitzian meaning given in 10.16 for the constraint qualification in the general statement of Fermat's rule in 10.12 is freshly provided here. But the close tie between constraint qualifications and the Lipschitzian behavior of certain set-valued mappings has well been elucidated in other, related contexts by Mordukhovich [1994c].

Constraint qualifications in terms of 'calmness' were introduced by Clarke [1976a], but the limit version in 10.47 was devised by Rockafellar [1985a]. The formula in 10.18 for the subgradients of an epi-sum of functions likewise comes from the latter paper.

Piecewise linear-quadratic functions and fully amenable functions were introduced by Rockafellar [1988] for the sake of developments in the theory of second-order subdifferentiation, and they will be important in that role in Chapter 13. The terminology of 'amenability' actually came later; it was adopted in works of Poliquin and Rockafellar [1992], [1993]. Most of the facts in 10.21, 10.24, 10.25 and 10.26 were exposed in those papers.

The semidifferentiability results in 10.27 and the eigenvalue application of them in 10.28 appear not to have been made explicit before. Most discussions of such matters have concerned one-sided directional derivatives taken along half-lines, rather than with varying direction, as here. For such derivatives, however, a chain rule like the one in 10.27(b) isn't available. This is one of the major incentives behind semidifferentiability, along with the fact that in practice this stronger property is often present anyway. Subgradients of eigenvalue functions have been determined by Lewis [1996]. For results on the variational analysis of eigenvalues of matrices that aren't symmetric, see Burke and Overton [1994].

The subsmooth functions of 10.29 originated in Rockafellar [1982b], and the results about them in 10.33, 10.34 and 10.54 were established there as well along with the generic strict differentiability in 10.31. (The 'subsmooth' name was introduced in Rockafellar [1983].) To some degree, though, this concept formalizes the operation of pointwise maximization of smooth functions, and in this way it relates to the calculus of subderivatives and subgradients under that operation. Many of the facts in 10.31 can be traced in that sense to Clarke [1975]; this is also the source of 10.51, where the functions in the maximization aren't smooth. Even earlier, the existence of one-sided directional derivatives for functions defined by pointwise maximization of a compactly parameterized family of smooth functions was proved by Danskin [1967]; see also Pshenichnyi [1971]. (Those researchers overlooked the stronger property of semidifferentiability.) Functions expressible locally as a convex function minus a quadratic function were briefly considered by Janin [1973], but he didn't identify that property with a lower-$\mathcal{C}^2$ representation, as in 10.33. The characterization of the upper subgradient set $-\partial(-f)(\bar{x})$ in 10.31 for a lower-$\mathcal{C}^1$ function f is new.

The 'upper' subsmoothness of Moreau envelopes in 10.32 hasn't previously been noted or exploited as such, although some of its consequences have been known.

For convex functions, the subgradient variational principle in 10.44 predates Ekeland's principle 1.43, from which we've derived it here. It was proved in that case by Brøndsted and Rockafellar [1965]. An extension to nonconvex functions, with convexified subgradient sets, was given by Rockafellar [1985a], but the present, unconvexified version hasn't been stated before, nor has the variant in 10.45.

The ε-regular subgradient facts in 10.46 relate to the way this notion was utilized in early work of Kruger and Mordukhovich [1980]. The extended mean-value theorem in 10.48 goes back essentially to those authors as well, cf. Mordukhovich [1988], but for convexified subgradient sets it has a major precedent in Lebourg [1975]. That version was long influential in sustaining convexification as the seemingly right mode for subdifferential calculus. Other mean-value theorems that don't require strict continuity have since been evolving; see Zagrodny [1988], [1992], Gajek and Zagrodny [1995], and Loewen [1995].

11. Dualization

In the realm of convexity, almost every mathematical object can be paired with another, said to be *dual* to it. The pairing between convex cones and their polars has already been fundamental in the variational geometry of Chapter 6 in relating tangent vectors to normal vectors. The pairing between convex sets and sublinear functions in Chapter 8 has served as the vehicle for expressing connections between subgradients and subderivatives. Both correspondences are rooted in a deeper principle of duality for 'conjugate' pairs of convex functions, which will emerge fully here.

On the basis of this duality, close connections between otherwise disparate properties are revealed. It will be seen for instance that the level boundedness of one function in a conjugate pair corresponds to the finiteness of the other function around the origin. A catalog of such surprising linkages can be put together, and lists of dual operations and constructions to go with them.

In this way the analysis of a given situation can often be translated into an equivalent yet very different context. This can be a major source of insights as well as a means of unifying seemingly divergent parts of theory. The consequences go far beyond situations ruled by pure convexity, because many problems, although nonconvex, have crucial aspects of convexity in their structure, and the dualization of these can already be very fruitful. Among other things, we'll be able to apply such ideas to the general expression of optimality conditions in terms of a Lagrangian function, and even to the dualization of optimization problems themselves.

A. Legendre-Fenchel Transform

The general framework for duality is built around a 'transform' that gives an operational form to the envelope representations of convex functions. For any function $f : I\!R^n \to \overline{I\!R}$, the function $f^* : I\!R^n \to \overline{I\!R}$ defined by

$$f^*(v) := \sup_x \{\langle v, x\rangle - f(x)\} \qquad 11(1)$$

is *conjugate* to f, while the function $f^{**} = (f^*)^*$ defined by

$$f^{**}(x) := \sup_v \{\langle v, x\rangle - f^*(v)\} \qquad 11(2)$$

is *biconjugate* to f. The mapping $f \mapsto f^*$ from $\mathrm{fcns}(\mathbb{R}^n)$ into $\mathrm{fcns}(\mathbb{R}^n)$ is the *Legendre-Fenchel transform*.

The significance of the conjugate function f^* can easily be understood in terms of epigraph relationships. Formula 11(1) says that

$$(v,\beta) \in \operatorname{epi} f^* \iff \beta \geq \langle v,x\rangle - \alpha \text{ for all } (x,\alpha) \in \operatorname{epi} f.$$

If we write the inequality as $\alpha \geq \langle v,x\rangle - \beta$ and think of the affine functions on $\mathbb{R}^n$ as parameterized by pairs $(v,\beta) \in \mathbb{R}^n \times \mathbb{R}$, we can express this as

$$(v,\beta) \in \operatorname{epi} f^* \iff l_{v,\beta} \leq f, \text{ where } l_{v,\beta}(x) := \langle v,x\rangle - \beta.$$

Since the specification of a function on $\mathbb{R}^n$ is tantamount to the specification of its epigraph, this means that f^* *describes the family of all affine functions majorized by* f. Simultaneously, though, our calculation reveals that

$$\beta \geq f^*(v) \iff \beta \geq l_{x,\alpha}(v) \text{ for all } (x,\alpha) \in \operatorname{epi} f,$$

In other words, f^* *is the pointwise supremum of the family of all affine functions* $l_{x,\alpha}$ *for* $(x,\alpha) \in \operatorname{epi} f$. By the same token then, formula 11(2) means that f^{**} *is the pointwise supremum of all the affine functions majorized by* f.

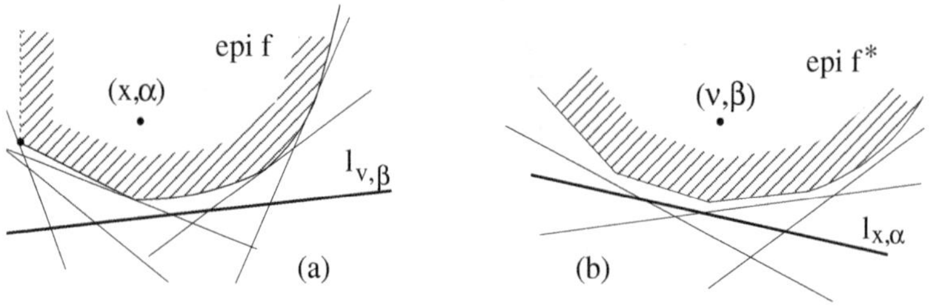

Fig. 11–1. (a) Affine functions majorized by f. (b) Affine functions majorized by f^*.

Recalling the facts about envelope representations in Theorem 8.13 and making use of the notion of the *convex hull* $\operatorname{con} f$ of an arbitrary function $f : \mathbb{R}^n \to \overline{\mathbb{R}}$ (see 2.31), we can summarize these relationships as follows.

11.1 Theorem (Legendre-Fenchel transform). *For any function* $f : \mathbb{R}^n \to \overline{\mathbb{R}}$ *with* $\operatorname{con} f$ *proper, both* f^* *and* f^{**} *are proper, lsc and convex, and*

$$f^{**} = \operatorname{cl} \operatorname{con} f.$$

Thus $f^{**} \leq f$, *and when* f *is itself proper, lsc and convex, one has* $f^{**} = f$. *Anyway, regardless of such assumptions, one always has*

$$f^* = (\operatorname{con} f)^* = (\operatorname{cl} f)^* = (\operatorname{cl} \operatorname{con} f)^*.$$

Proof. In the light of the preceding explanation of the meaning of the Legendre-Fenchel transform, this is immediate from Theorem 8.13; see 2.32 for the properness of $\operatorname{cl} f$ when f is convex and proper. □

11.2 Exercise (improper conjugates). *For a function $f : \mathbb{R}^n \to \overline{\mathbb{R}}$ with* con f *improper in the sense of taking on $-\infty$, one has $f^* \equiv \infty$ and $f^{**} \equiv -\infty$, while* cl con f *has the value $-\infty$ on the set* cl dom(con f) *but the value ∞ outside this set. For the improper function $f \equiv \infty$, one has $f^* \equiv -\infty$ and $f^{**} \equiv \infty$.*

Guide. Make use of 2.5. □

The Legendre-Fenchel transform obviously reverses ordering among the functions to which it is applied:

$$f_1 \le f_2 \implies f_1^* \ge f_2^*.$$

The fact that $f = f^{**}$ when f is proper, lsc and convex means that the Legendre-Fenchel transform sets up a one-to-one correspondence in the class of all such functions: if g is conjugate to f, then f is conjugate to g:

$$f \underset{*}{\longleftrightarrow} g \quad \text{when} \quad \begin{cases} g(v) = \sup_x \{ \langle v, x \rangle - f(x) \}, \\ f(x) = \sup_v \{ \langle v, x \rangle - g(v) \}. \end{cases}$$

This is called the *conjugacy* correspondence. Every property of one function in a conjugate pair must mirror some property of the other function. Every construction or operation must have its conjugate counterpart. This far-reaching principle of duality allows everything to be viewed from two different angles, often with remarkable consequences.

An initial illustration of the duality of operations is seen in the following relations, which immediately fall out of the definition of conjugacy. In each case the expression on the left gives a function of x while the one on the right gives the corresponding function of v under the assumption that $f \underset{*}{\longleftrightarrow} g$:

$$\begin{aligned} f(x) - \langle a, x \rangle &\underset{*}{\longleftrightarrow} g(v + a), \\ f(x + b) &\underset{*}{\longleftrightarrow} g(v) - \langle v, b \rangle, \\ f(x) + c &\underset{*}{\longleftrightarrow} g(v) - c, \\ \lambda f(x) &\underset{*}{\longleftrightarrow} \lambda g(\lambda^{-1} v) \quad (\text{for } \lambda > 0), \\ \lambda f(\lambda^{-1} x) &\underset{*}{\longleftrightarrow} \lambda g(v) \quad (\text{for } \lambda > 0). \end{aligned} \qquad 11(3)$$

Interestingly, the last two relations pair *multiplication* with *epi-multiplication*:

$$(\lambda f)^* = \lambda \star f^*, \qquad (\lambda \star f)^* = \lambda f^*,$$

for positive scalars λ. (An extension to $\lambda = 0$ will come out in Theorem 11.5.) Later we'll see a similar duality between *addition* and *epi-addition* of functions (in Theorem 11.23(a)).

One of the most important dualization rules operates on subgradients. It stems from the fact that the subgradients of a convex function correspond to its affine supports (as described after 8.12). To say that the affine function $l_{\bar{v},\bar{\beta}}$ supports f at $\bar{x}$, with $\bar{\alpha} = f(\bar{x})$, is to say that the affine function $l_{\bar{x},\bar{\alpha}}$ supports

f^* at $\bar{v}$, with $\bar{\beta} = f^*(\bar{v})$; cf. Figure 11–1 again. This gives us a relationship between subgradients of f and those of f^*.

11.3 Proposition (inversion rule for subgradient relations). *For any proper, lsc, convex function f, one has $\partial f^* = (\partial f)^{-1}$ and $\partial f = (\partial f^*)^{-1}$. Indeed,*

$$\bar{v} \in \partial f(\bar{x}) \iff \bar{x} \in \partial f^*(\bar{v}) \iff f(\bar{x}) + f^*(\bar{v}) = \langle \bar{v}, \bar{x} \rangle,$$

whereas $f(x) + f^*(v) \geq \langle v, x \rangle$ for all x, v. Hence $\operatorname{gph} \partial f$ is closed and

$$\partial f(\bar{x}) = \operatorname{argmax}_v \big\{ \langle v, \bar{x} \rangle - f^*(v) \big\}, \qquad \partial f^*(\bar{v}) = \operatorname{argmax}_x \big\{ \langle \bar{v}, x \rangle - f(x) \big\}.$$

Proof. The argument just given would suffice, but here's another view of why the relations hold. From the first formula in 11(3) we know that for any $\bar{v}$ the points $\bar{x}$ furnishing the minimum of the convex function $f_{\bar{v}}(x) := f(x) - \langle \bar{v}, x \rangle$, if any, are the ones such that $f(\bar{x}) - \langle \bar{v}, x \rangle = -f^*(\bar{v})$, finite. But by the version of Fermat's principle in 10.1 they are also the ones such that $0 \in \partial f_{\bar{v}}(\bar{x})$, this subgradient set being the same as $\partial f(\bar{x}) - \bar{v}$ (cf. 8.8(c)). Thus, $f(\bar{x}) + f^*(\bar{v}) = \langle \bar{v}, \bar{x} \rangle$ if and only if $\bar{v} \in \partial f(\bar{x})$. The rest follows now by symmetry. □

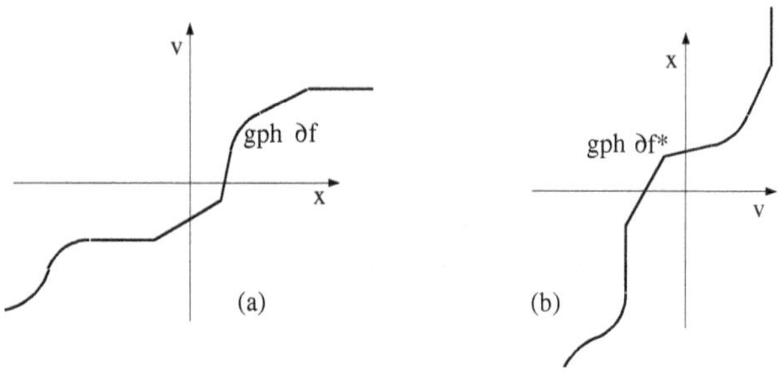

Fig. 11–2. Subgradient inversion for conjugate functions.

Subgradient relations, with normal cone relations as a special case, are widespread in the statement of optimality conditions. The inversion rule in 11.3 (illustrated in Figure 11–2) is therefore a key to writing such conditions in alternative ways and gaining other interpretations of them. That pattern will be prominent in our work with generalized Lagrangian functions and dual problems of optimization later in this chapter.

B. Special Cases of Conjugacy

Before proceeding with other features of the Legendre-Fenchel transform, let's observe that Theorem 11.1 covers, as special instances of the conjugacy correspondence, the fundamentals of cone polarity in 6.21 and support function theory in 8.24. This confirms that those earlier modes of dualization fit squarely in the picture now being unveiled.

11.4 Example (support functions and cone polarity).

(a) *For any set $C \subset \mathbb{R}^n$, the conjugate of the indicator function δ_C is the support function σ_C. On the other hand, for any positively homogeneous function h on $\mathbb{R}^n$ the conjugate h^* is the indicator δ_C of the set $C = \{x \mid \langle v, x\rangle \leq h(v) \text{ for all } v\}$. In this sense, the correspondence between closed, convex sets and their support functions is imbedded within conjugacy:*

$$\delta_C \underset{*}{\longleftrightarrow} \sigma_C \quad \text{for } C \text{ a closed, convex, set.}$$

Under this correspondence one has

$$\bar{v} \in N_C(\bar{x}) \iff \bar{x} \in \partial\sigma_C(\bar{v}) \iff \bar{x} \in C,\ \langle \bar{v}, \bar{x}\rangle = \sigma_C(\bar{v}). \qquad 11(4)$$

(b) *For a cone $K \subset \mathbb{R}^n$, the conjugate of the indicator function δ_K is the indicator function δ_{K^*}. In this sense, the polarity correspondence for closed, convex cones is imbedded within conjugacy:*

$$\delta_K \underset{*}{\longleftrightarrow} \delta_{K^*} \quad \text{for } K \text{ a closed, convex cone.}$$

Under this correspondence one has

$$\bar{v} \in N_K(\bar{x}) \iff \bar{x} \in N_{K^*}(\bar{v}) \iff \bar{x} \in K,\ \bar{v} \in K^*,\ \bar{x} \perp \bar{v}. \qquad 11(5)$$

Detail. The formulas for the conjugate functions immediately reduce in these ways, and then 11.3 can be applied. □

In particular, the orthogonal subspace correspondence $M \leftrightarrow M^\perp$ is imbedded within conjugacy through $\delta_M^* = \delta_{M^\perp}$.

The support function correspondence has a bearing on the Legendre-Fenchel transform from a different angle too, namely in characterizing the effective domains of functions conjugate to each other.

11.5 Theorem (horizon functions as support functions). *Let $f : \mathbb{R}^n \to \overline{\mathbb{R}}$ be proper, lsc and convex. The horizon function f^∞ is then the support function of $\operatorname{dom} f^*$, whereas $f^{*\infty}$ is the support function of $\operatorname{dom} f$.*

Proof. We have $(f^*)^* = f$ (by 11.1), and because of this symmetry it suffices to prove that the function $f^{*\infty} = (f^*)^\infty$ is the support function of $D := \operatorname{dom} f$. Fix any $v_0 \in \operatorname{dom} f^*$. We have for arbitrary $v \in \mathbb{R}^n$ and $\tau > 0$ that

$$\begin{aligned} f^*(v_0 + \tau v) &= \sup_{x \in D}\big\{\langle v_0 + \tau v, x\rangle - f(x)\big\} \\ &\leq \sup_{x \in D}\big\{\langle v_0, x\rangle - f(x)\big\} + \tau \sup_{x \in D}\langle v, x\rangle \;=\; f^*(v_0) + \tau\sigma_D(v), \end{aligned}$$

hence $\big[f^*(v_0 + \tau v) - f^*(v_0)\big]/\tau \leq \sigma_D(v)$ for all $v \in \mathbb{R}^n$, $\tau > 0$. Through 3(4) this guarantees that $f^{*\infty} \leq \sigma_D$. On the other hand, for $v \in \mathbb{R}^n$ and $\beta \in \mathbb{R}$ with $f^{*\infty}(v) \leq \beta$ one has $f^*(v_0 + \tau v) \leq f^*(v_0) + \tau\beta$ for all $\tau > 0$, hence for any $x \in \mathbb{R}^n$ that

$$\begin{aligned} f(x) &\geq \langle v_0 + \tau v, x\rangle - f^*(v_0 + \tau v) \\ &\geq \langle v_0, x\rangle - f^*(v_0) + \tau\big(\langle v, x\rangle - \beta\big) \ \text{ for all } \ \tau > 0. \end{aligned}$$

This implies that $\langle v, x\rangle \leq \beta$ for all x with $f(x) < \infty$, so $D \subset \{x \mid \langle v, x\rangle \leq \beta\}$. Thus $\sigma_D(v) \leq \beta$, and we conclude that also $\sigma_D \leq f^{*\infty}$, so $\sigma_D = f^{*\infty}$. □

Another way that support functions come up in the dualization of properties of f and f^* is seen in connection with level sets. For simplicity in the following, we look only at 0-level sets, since $\operatorname{lev}_{\leq\alpha} f$ can be studied as $\operatorname{lev}_{\leq 0} f_\alpha$ for $f_\alpha = f - \alpha$, with $f_\alpha^* = f^* + \alpha$ by 11(3).

11.6 Exercise (support functions of level sets). *If* $C = \{x \mid f(x) \leq 0\}$ *for a finite, convex function* f *such that* $\inf f < 0$, *then*

$$\sigma_C(v) \;=\; \inf_{\lambda > 0} \lambda f^*(\lambda^{-1} v) \ \text{ for all } \ v \neq 0.$$

Guide. Let h denote the function of v on the right side of the equation; take $h(0) = 0$. Show in terms of the 'pos' operation defined ahead of 3.48 that h is a positively homogeneous, convex function for which the points x satisfying $\langle v, x\rangle \leq h(v)$ for all v are the ones such that $f^{**}(x) \leq 0$. Argue that $f^{**} = f$ and hence via support function theory that $\sigma_C = \operatorname{cl} h$. Verify through 11.5 that $f^{*\infty} = \delta_{\{0\}}$ and in this way deduce from 3.48(b) that $\operatorname{cl} h = h$. □

Note that the roles of f and f^* in 11.6 could be reversed: the support functions for the level sets of f^*, when that function is finite, can be derived from f (as long as the convex function f is proper and lsc, so that $(f^*)^* = f$.)

While the polarity of cones is a special case of conjugacy of functions, the opposite is true as well, in a certain sense. This is not only interesting but valuable for certain theoretical purposes.

11.7 Exercise (conjugacy as cone polarity). *For proper functions* f *and* g *on* $\mathbb{R}^n$, *consider in* $\mathbb{R}^{n+2}$ *the cones*

$$\begin{aligned} K_f &= \Big\{(x, \alpha, -\lambda) \,\Big|\, \lambda > 0,\ (x, \alpha) \in \lambda \operatorname{epi} f; \ \text{ or } \ \lambda = 0,\ (x, \alpha) \in \operatorname{epi} f^\infty\Big\}, \\ K_g &= \Big\{(v, -\mu, \beta) \,\Big|\, \mu > 0,\ (v, \beta) \in \mu \operatorname{epi} g; \ \text{ or } \ \mu = 0,\ (v, \beta) \in \operatorname{epi} g^\infty\Big\}. \end{aligned}$$

Then f *and* g *are conjugate to each other if and only if* K_f *and* K_g *are polar to each other.*

Guide. Verify that K_f is convex and closed if and only if f is convex and lsc; indeed, K_f is then the cone representing $\operatorname{epi} f$ in the ray space model for $\operatorname{csm} \mathbb{R}^{n+1}$. The cone K_g has a similar interpretation with respect to g, except for a reversal in the roles of last two components. Use the definition of conjugacy along with the relationships in 11.5 to analyze polarity. □

How does the duality between f and f^* affect situations where f represents a problem of optimization? Here are the central facts.

11.8 Theorem (dual properties in minimization). *The properties of a proper, lsc, convex function $f : \mathbb{R}^n \to \overline{\mathbb{R}}$ are paired with those of its conjugate function f^* in the following manner.*

(a) $\inf f = -f^*(0)$ *and* $\operatorname{argmin} f = \partial f^*(0)$.

(b) $\operatorname{argmin} f = \{\bar{x}\}$ *if and only f^* is differentiable at* 0 *with* $\nabla f^*(0) = \bar{x}$.

(c) f *is level-coercive (or level-bounded) if and only if* $0 \in \operatorname{int}(\operatorname{dom} f^*)$.

(d) f *is coercive if and only if* $\operatorname{dom} f^* = \mathbb{R}^n$.

Proof. The first property in (a) simply re-expresses the definition of $f^*(0)$, while the second comes from 11.3 and the fact that $\operatorname{argmin} f$ consists of the points x such that $0 \in \partial f(x)$; cf. 10.1. In (b) this is elaborated through the fact that $\partial f^*(0) = \{\bar{x}\}$ if and only if f^* is strictly differentiable at 0, this by 9.18 and the fact that because f^* is itself proper, lsc and convex, f is regular with $\partial^\infty f^*(0) = \partial f^*(0)^\infty$ (see 7.27 and 8.11). The regularity of f^* implies further that f^* is strictly differentiable wherever it's differentiable (cf. 9.20).

In (c) we recall that f is level-coercive if and only if $f^\infty(w) > 0$ for all $w \neq 0$ (see 3.26(a)), whereas the convex set $D = \operatorname{dom} f^*$ has $0 \in \operatorname{int} D$ if and only if $\sigma_D(w) > 0$ for all $w \neq 0$ (see 8.29(a)). The equivalence comes from 11.5, where $f^\infty(w)$ is identified with $\sigma_D(w)$. (Recall too that a convex function is level-bounded if and only if it is level-coercive; 3.27.) Similarly, in (d) we are seeing an instance of the fact that a convex set is the whole space if and only if it isn't contained in any closed half-space, i.e., its support function is $\delta_{\{0\}}$. □

The dualizations in 11.8 can be extended through elementary conjugacy relations like the ones in 11(3). Thus, one has

$$\inf_x\{f(x) - \langle a, x\rangle\} = -f^*(a), \quad \operatorname{argmin}_x\{f(x) - \langle a, x\rangle\} = \partial f^*(a), \quad 11(6)$$

the argmin being $\{b\}$ if and only if f^* is differentiable at a with $\nabla f^*(a) = b$. The function $f - \langle a, \cdot\rangle$ is level-coercive if and only if $a \in \operatorname{int}(\operatorname{dom} f^*)$.

So far, little has been said about how f^* can effectively be determined when f is given. Because of conjugacy's abstract uses in analysis and the dualization of properties for purposes of understanding them better, a formula for f beyond the defining one in 11(1) isn't always needed, but what about the times when it is? Ways of constructing f^* out of the conjugates of other functions that are part of the make up of f can be very helpful (and will be occupy our attention in due course), but somewhere along the line it's crucial to have a repertory of examples that can serve as building blocks, much as power functions, exponentials, logarithms and trigonometric expressions serve in classical differentiation and integration.

We've observed in 11.4(a) that f^* can sometimes be identified as a support function (here examples like 8.26 and 8.27 can be kept in mind), or in reverse as the indicator of a convex set defined by the system of linear constraints associated with a sublinear function (cf. 8.24) when f exhibits sublinearity. In this respect the results in 11.5 and 11.6 can be useful, and further also the subderivative-subgradient relations in Chapter 8 for functions that are subdifferentially regular. Then again, as in 11.4(b), f^* might be the indicator of a polar cone. For instance, the polar of $\mathbb{R}^n_+$ is $\mathbb{R}^n_-$, and the polar of a subspace M is $M^\perp$. The Farkas lemma in 6.45 and the relations between tangent cones and polar cones can provide assistance as well.

C. The Role of Differentiability

Beyond special cases such as these, there is the possibility of generating examples directly from the formula in 11(1) for f^* in terms of f. This may be intimidating, though, because it not only demands the calculation of a global supremum (the solution of a certain optimization problem), but requires this to be done parametrically—the supremum must be expressed as a function of the v element. For functions on $\mathbb{R}^1$, 'brute force' may succeed, but elsewhere some guidelines are needed. The next three examples will present important cases where f^* can be calculated from f by use of derivatives alone.

As long as f is convex and differentiable everywhere, one can hope to get the supremum in formula 11(1), and thereby the value of $f^*(v)$, by setting the gradient (with respect to x) of the expression $\langle v, x\rangle - f(x)$ equal to 0 and solving that equation for x. This is justified because the expression is concave with respect to x; the vanishing of its gradient corresponds therefore to the attainment of the global maximum. The equation in question is $v - \nabla f(x) = 0$, and its solutions are the vectors x, if any, belonging to $(\nabla f)^{-1}(v)$. An x identified in this manner can be substituted into $\langle v, x\rangle - f(x)$ to get $f^*(v)$. Pitfalls gape, however, in the fact that the range of the mapping ∇f might not be all of $\mathbb{R}^n$. For $v \notin \operatorname{rge} \nabla f$, the supremum would need to be determined through additional analysis. It might be ∞, with the meaning that $v \notin \operatorname{dom} f^*$, or it might be finite, yet not attained.

Putting such troubles aside to get a picture first of the nicest circumstances, one can ask what happens when ∇f is a one-to-one mapping from $\mathbb{R}^n$ onto $\mathbb{R}^n$, so that $(\nabla f)^{-1}$ is single-valued everywhere. Understandably, this is the historical case in which conjugate functions first attracted interest.

11.9 Example (classical Legendre transform). *Let f be a finite, coercive, convex function of class $\mathcal{C}^2$ (twice continuously differentiable) on $\mathbb{R}^n$ whose Hessian matrix $\nabla^2 f(x)$ is positive-definite for every x. Then the conjugate $g = f^*$ is likewise a finite, coercive, convex function of class $\mathcal{C}^2$ on $\mathbb{R}^n$ with $\nabla^2 g(v)$ positive-definite for every v and $g^* = f$. The gradient mapping ∇f is one-to-one from $\mathbb{R}^n$ onto $\mathbb{R}^n$, and its inverse is ∇g; one has*

$$\begin{aligned} g(v) &= \big\langle (\nabla f)^{-1}(v), v\big\rangle - f\big((\nabla f)^{-1}(v)\big), \\ f(x) &= \big\langle (\nabla g)^{-1}(x), x\big\rangle - g\big((\nabla g)^{-1}(x)\big). \end{aligned} \qquad 11(7)$$

Moreover the matrices $\nabla^2 f(x)$ and $\nabla^2 g(v)$ are inverse to each other when $v = \nabla f(x)$, or equivalently $x = \nabla g(v)$ (then x and v are conjugate points).

Detail. The assumption on second derivatives makes f strictly convex (see 2.14). Then for a fixed $a \in \mathbb{R}^n$ in 11(6) we not only have through coercivity the attainment of the infimum but its attainment at a *unique* point x (by 2.6). Then $\nabla f(x) - a = 0$ by Fermat's principle. This line of reasoning demonstrates that for each $v \in \mathbb{R}^n$ there is a unique x with $\nabla f(x) = v$, i.e., the mapping ∇f is invertible. The first equation in 11(7) is immediate, and the rest of the assertions can then be obtained from 11.8(d) and differentiation of g, using the standard inverse mapping theorem. □

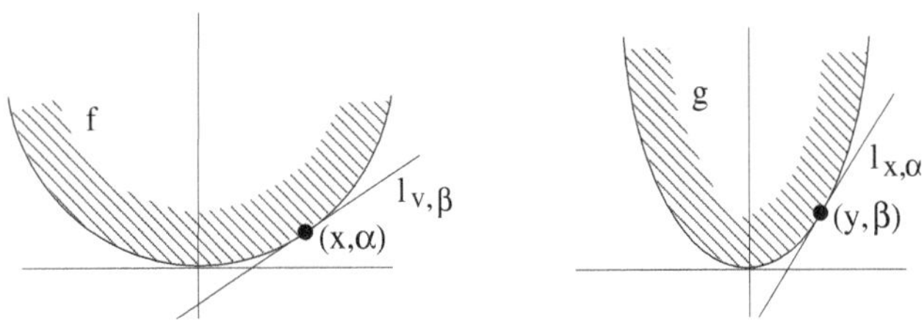

Fig. 11–3. Conjugate points in the classical setting.

The next example fits the pattern of the preceding one in part, but also illustrates how the approach to calculating f^* from the derivatives of f can be followed a bit more flexibly.

11.10 Example (linear-quadratic functions). *Suppose*

$$f(x) = \tfrac{1}{2}\langle x, Ax\rangle + \langle a, x\rangle + \alpha$$

with $A \in \mathbb{R}^{n\times n}$ symmetric and positive-semidefinite, so that f is convex. If A is nonsingular, the conjugate function is

$$f^*(v) = \tfrac{1}{2}\big\langle v - a, A^{-1}(v-a)\big\rangle - \alpha.$$

At the other extreme, if $A = 0$, so that f is merely affine, the conjugate function is given by $f^*(v) = \delta_{\{a\}}(v) - \alpha$.

In general, the column space of A (the range of $x \mapsto Ax$) is a linear subspace L, and there is a unique symmetric, positive-semidefinite matrix $A^\dagger$ (the pseudo-inverse of A) having $A^\dagger A = AA^\dagger =$ [orthogonal projector on L]. The conjugate function is given then by

$$f^*(v) = \begin{cases} \tfrac{1}{2}\big\langle v - a, A^\dagger(v-a)\big\rangle - \alpha & \text{when } v - a \in L,\\ \infty & \text{when } v - a \notin L.\end{cases}$$

Detail. The nonsingular case fits the pattern of 11.9, while the affine case is obvious on its own. The general case is made simple by reducing to $a = 0$ and $\alpha = 0$ through the relations 11(3) and invoking a change of coordinates that diagonalizes the matrix A. □

11.11 Example (self-conjugacy). *The function $f(x) = \frac{1}{2}|x|^2$ on $\mathbb{R}^n$ has $f^* = f$ and is the only function with this property.*

Detail. The self-conjugacy of this function is evident as the special case of 11.10 in which $A = I$, $a = 0$. Its uniqueness in this respect is seen as follows. If $f = f^*$, then $f^{**} = f$ by 11.1, and f is proper by 11.2. Formula 11(1) gives in this case $f(v) + f(x) \geq \langle v, x\rangle$ for all x and v, hence with $x = v$ that $f(x) \geq \frac{1}{2}|x|^2$ for all x. Passing to conjugates in this inequality, one sees on the other hand that $f^*(v) \leq \frac{1}{2}|v|^2$ for all v, hence from $f^* = f$ that $f(x) \leq \frac{1}{2}|x|^2$ for all x. Therefore, $f(x) = \frac{1}{2}|x|^2$ for all x. □

The formula in 11.6 for the support function of a level set can be illustrated through Example 11.10. For a convex set of the form

$$C = \left\{x \,\middle|\, \tfrac{1}{2}\langle x, Ax\rangle + \langle a, x\rangle + \alpha \leq 0\right\}$$

with A symmetric and positive-definite, and such that the inequality is satisfied strictly by at least one x, one necessarily has $\langle a, A^{-1}a\rangle - 2\alpha > 0$ (by 11.8(a) because this quantity is $2f^*(0)$), and thus the expression

$$\sigma_C(v) = \beta\sqrt{\langle v, A^{-1}v\rangle} - \langle b, v\rangle \quad \text{for} \quad b = A^{-1}a, \ \beta = \sqrt{\langle a, A^{-1}a\rangle - 2\alpha}.$$

In general, if one of the functions in a general conjugate pair is finite and coercive, so too must be the other function; this is clear from 11.8(d). Otherwise, at least one of the two functions in a conjugate pair must take on the value ∞ somewhere and thus have some convex set other than $\mathbb{R}^n$ itself as its effective domain. The support function relation in 11.5 shows in these cases how $\operatorname{dom} f^*$ relates to properties of f through the horizon function f^∞. For the same reason, since $f^{**} = f$ (when f is proper, lsc and convex), $\operatorname{dom} f$ relates to properties of f^* through the way it determines $f^{*\infty}$. Information about effective domains facilitates the calculation of conjugates in many cases.

The following example illustrates this principle as an extension of the method for calculating f^* from the derivatives of f.

11.12 Example (log-exponential function and entropy). *For $f(x) = \operatorname{logexp}(x)$, the conjugate f^* is the entropy function g defined for $v = (v_1, \ldots, v_n)$ by*

$$g(v) = \begin{cases} \sum_{j=1}^n v_j \log v_j & \text{when } v_j \geq 0,\ \sum_{j=1}^n v_j = 1, \\ \infty & \text{otherwise,} \end{cases}$$

with $0 \log 0 = 0$. The support function of the set $C = \{x \mid \operatorname{logexp}(x) \leq 0\}$ is the function $h : \mathbb{R}^n \to \overline{\mathbb{R}}$ defined under the same convention by

$$h(v) = \begin{cases} \sum_{j=1}^n v_j \log v_j - \left(\sum_{j=1}^n v_j\right) \log\left(\sum_{j=1}^n v_j\right) & \text{when } v_j \geq 0, \\ \infty & \text{otherwise.} \end{cases}$$

Detail. The horizon function of $f(x) = \operatorname{logexp}(x)$ is $f^\infty(x) = \operatorname{vecmax} x$ by 3(5), and this is the support function of the unit simplex C consisting of the vectors $v \geq 0$ with $v_1 + \cdots + v_n = 1$, as already noted in 8.26. Since f is a finite, convex function (cf. 2.16), it's in particular a proper, lsc, convex function (cf. 2.36). We may conclude from 11.5 that $\operatorname{dom} f^*$ has the same support function as C and therefore has $\operatorname{cl}(\operatorname{dom} f^*) = C$ (cf. 8.24). Hence in terms of relative interiors (cf. 2.40),

$$\operatorname{rint}(\operatorname{dom} f^*) = \operatorname{rint} C = \{v \mid v_j > 0, \sum_{j=1}^n v_j = 1\}.$$

From the formula $\nabla f(x) = \sigma(x)^{-1}(e^{x_1}, \ldots, e^{x_n})$ for $\sigma(x) := e^{x_1} + \cdots + e^{x_n}$ it is apparent that each $\bar{v} \in \operatorname{rint} C$ is of the form $\nabla f(\bar{x})$ for $\bar{x} = (\log \bar{v}_1, \ldots, \log \bar{v}_n)$. The inequality $f(x) \geq f(\bar{x}) + \langle \nabla f(\bar{x}), x - \bar{x} \rangle$ in 2.14(b) yields

$$\sup_x \{\langle x, \nabla f(\bar{x}) \rangle - f(x)\} = \langle \bar{x}, \nabla f(\bar{x}) \rangle - f(\bar{x}),$$

which is the same as

$$\sup_x \{\langle x, \bar{v} \rangle - f(x)\} = \sum_{j=1}^n (\log \bar{v}_j)\bar{v}_j - \log\left(\sum_{j=1}^n \bar{v}_j\right) = g(\bar{v}).$$

Thus $f^* = g$ on $\operatorname{rint} C$. The closure formula in 2.35 as translated to the context of relative interiors shows then that these functions agree on all of C, therefore on all of $\mathbb{R}^n$.

The function h is $\operatorname{pos} g$, where $g(0) = \infty$ and $\inf f = -\infty$; cf. 11.8(a). According to 11.6, h is then the support function of $\operatorname{lev}_{\leq 0} f$. □

With minor qualifications on the boundaries of domains, differentiability itself dualizes under the Legendre-Fenchel transform to strict convexity.

11.13 Theorem (strict convexity versus differentiability). *The following properties are equivalent for a proper, lsc, convex function $f : \mathbb{R}^n \to \overline{\mathbb{R}}$ and its conjugate function f^*:*

(a) *f is almost differentiable, in the sense that f is differentiable on the open, convex set $\operatorname{int}(\operatorname{dom} f)$, which is nonempty, but $\partial f(x) = \emptyset$ for all points $x \in \operatorname{dom} f \setminus \operatorname{int}(\operatorname{dom} f)$, if any;*

(b) *f^* is almost strictly convex, in the sense that f^* is strictly convex on every convex subset of $\operatorname{dom} \partial f^*$ (hence on $\operatorname{rint}(\operatorname{dom} f^*)$, in particular).*

Likewise, the function f^* is almost differentiable if and only if f is almost strictly convex.

Proof. Since $f = (f^*)^*$ under our assumptions (by 11.1), there's symmetry between the first equivalence asserted and the one claimed at the end. We can just as well work at verifying the latter. As seen from 11.8 (and its extension explained after the proof of that result), f^* is almost differentiable if and only if, for every $a \in \mathbb{R}^n$ such that the set $\operatorname{argmin}_x \{f(x) - \langle a, x \rangle\} = \partial f^*(a)$ is nonempty, it's actually a singleton. Our task is to show that this holds if and only if f is almost strictly convex.

Certainly if for some a this minimizing set, which is convex, contained two different points x_0 and x_1, it would contain $x_\tau := (1 - \tau)x_0 + \tau x_1$ for all $\tau \in (0, 1)$. Because $x_\tau \in \partial f^*(a)$ we would have $a \in \partial f(x_\tau)$ by 11.3, so the line segment joining x_0 and x_1 would lie in $\operatorname{dom} \partial f$. From the fact that

$\inf_x\{f(x) - \langle a, x\rangle\} = -f^*(a)$, we would have $f(x_\tau) - \langle a, x_\tau\rangle = -f^*(a)$ for $\tau \in (0,1)$. This implies $f(x_\tau) = (1-\tau)f(x_0) + \tau f(x_1)$ for $\tau \in (0,1)$, since $\langle a, x_\tau\rangle = (1-\tau)\langle a, x_0\rangle + \tau\langle a, x_1\rangle$. Then f isn't almost strictly convex.

Conversely, if f fails to be almost strictly convex there must exist $x_0 \neq x_1$ such that the points x_τ on the line segment joining them belong to $\operatorname{dom}\partial f$ and satisfy $f(x_\tau) = (1-\tau)f(x_0) + \tau f(x_1)$. Fix any $\bar\tau \in (0,1)$ and any $a \in \partial f(x_{\bar\tau})$. From 11.3 and formula 11(1) for f^* in terms of f, we have

$$f(x_\tau) \geq \langle a, x_\tau\rangle - f^*(a) \text{ for all } \tau \in (0,1), \text{ with equality for } \tau = \bar\tau.$$

The affine function $\varphi(\tau) := f(x_\tau)$ on $(0,1)$ thus attains its minimum at the intermediate point $\bar\tau$. But then φ has to be constant on $(0,1)$. In other words, for all $\tau \in (0,1)$ we must have $f(x_\tau) = \langle a, x_\tau\rangle - f^*(a)$, hence $x_\tau \in \partial f^*(a)$ by 11.3. In this event $\partial f^*(a)$ isn't a singleton. □

The property in 11.13(a) of being *almost* differentiable can be identified with the single-valuedness of the mapping ∂f relative to its domain (see 9.18, recalling from 7.27 that proper, lsc, convex functions are regular). It implies f is continuously differentiable—smooth—on $\operatorname{int}(\operatorname{dom} f)$; cf. 9.20.

D. Piecewise Linear-Quadratic Functions

Differentiability isn't the only tool available for understanding the nature of conjugate functions, of course. A major class of *nondifferentiable* functions with nice behavior under the Legendre-Fenchel transform consists of the convex functions that are piecewise linear (see 2.47) or more generally piecewise linear-quadratic (see 10.20).

11.14 Theorem (piecewise linear-quadratic functions in conjugacy). *Suppose that $f : \mathbb{R}^n \to \overline{\mathbb{R}}$ be proper, lsc and convex. Then*

(a) *f is piecewise linear if and only if f^* has this property;*

(b) *f is piecewise linear-quadratic if and only if f^* has this property.*

For proving part (b) we'll need a lemma, which is of some interest in itself. We take care of this first.

11.15 Lemma (linear-quadratic test on line segments). *In order that f be linear-quadratic relative to a convex set $C \subset \mathbb{R}^n$, in the sense of being expressible by a formula of type $f(x) = \frac{1}{2}\langle x, Ax\rangle + \langle a, x\rangle + \alpha$ for $x \in C$, it is necessary and sufficient that f be linear-quadratic relative to every line segment in C.*

Proof. The condition is trivially necessary, so the challenge is proving its sufficiency. Without loss of generality we can focus on the case where $\operatorname{int} C \neq \emptyset$ (cf. 2.40 and the discussion preceding it). It's enough actually to demonstrate that the condition implies f is linear-quadratic relative to $\operatorname{int} C$, because the formula obtained on $\operatorname{int} C$ must then extend to the rest of C through the fact that when a boundary point x of C is joined by a line segment to an interior point, all of the segment except x itself lies in $\operatorname{int} C$ (see 2.33).

We claim next that if f is linear-quadratic in some neighborhood of each point of $\operatorname{int} C$, then it's linear-quadratic relative to $\operatorname{int} C$. Consider any two points x_0 and x_1 of $\operatorname{int} C$. We'll show that the formula around x_0 must agree with the formula around x_1.

The line segment $[x_0, x_1]$ is a compact set, every point of which has an open ball relative to which f is linear-quadratic, and it can therefore be covered by a finite collection of such open balls, say O_k for $k = 1, \ldots, r$, each with a formula $f(x) = \frac{1}{2}\langle x, A_k x\rangle + \langle a_k, x\rangle + \alpha_k$. If two sets O_{k_1} and O_{k_2} overlap, their formulas have to agree on the intersection; this implies that $A_{k_1} = A_{k_2}$, $a_{k_1} = a_{k_2}$ and $\alpha_{k_1} = \alpha_{k_2}$. But as one moves along $[x_0, x_1]$ from x_0 to x_1, each transition out of one set O_k and into another passes through a region of overlap (again because of the line segment principle for convex sets, or more generally because line segments are connected sets). Thus, all the formulas for $k = 1, \ldots, r$ agree.

Having reduced the task to proving that f is linear-quadratic relative to a neighborhood of each point of $\operatorname{int} C$, we can take such neighborhoods to be cubes. The question then is whether, if f is linear-quadratic on every line segment in a certain cube, it must be linear-quadratic relative to the cube.

A cube in $\mathbb{R}^n$ is a product of n intervals, so an induction argument can be contemplated in which the product grows by one interval at a time until the cube is built up, and at each stage the linear-quadratic property of f relative to the partial product is verified. For a single interval, as the starter, the property holds by hypothesis.

To validate the induction argument we only have to show that *if U and V are convex neighborhoods of the origin in $\mathbb{R}^p$ and $\mathbb{R}^q$ respectively, and if a function f on $U \times V$ is such that $f(u, v)$ is linear-quadratic in u for fixed v, linear-quadratic in v for fixed u, and moreover f is linear-quadratic relative to all line segments in $U \times V$, then f is linear-quadratic relative to $U \times V$ as a whole.* We go about this by first using the property in u to express

$$f(u, v) = \tfrac{1}{2}\big\langle u, A(v)u\big\rangle + \big\langle a(v), u\big\rangle + \alpha(v) \quad \text{for } u \in U \text{ when } v \in V. \qquad 11(8)$$

We'll demonstrate next, invoking the linear-quadratic property of $f(u, v)$ in v, that $\alpha(v)$ and each component of the vector $a(v)$ and the matrix $A(v)$ must be linear-quadratic as a function of $v \in V$. For $\alpha(v)$ this is clear from having $\alpha(v) = f(0, v)$. In the case of $A(v)$ we observe that

$$\big\langle u', A(v)u\big\rangle = f(u + u', v) - f(u, v) - f(u', v) + \alpha(v)$$

for any u and u' in $\mathbb{R}^p$ small enough that they and $u + u'$ belong to U. Hence $\langle u', A(v)u\rangle$ is linear-quadratic in $v \in V$ for any such u and u'. Choosing u and u' among $\varepsilon e^1, \ldots, \varepsilon e^p$ for small $\varepsilon > 0$ and the canonical basis vectors e^k for $\mathbb{R}^p$ (where e^k has 1 in kth position and 0's elsewhere), we deduce that for every row i and column j the component $A_{ij}(v)$ is linear-quadratic in v. By writing

$$\langle a(v), u\rangle \;=\; f(u,v) - \tfrac{1}{2}\langle u, A(v)u\rangle - \alpha(v),$$

where the right side is now known to be linear-quadratic in $v \in V$, we see that $\langle a(v), u\rangle$ has this property for each u sufficiently near to 0. Again by choosing u from among $\varepsilon e^1, \ldots, \varepsilon e^p$ we are able to conclude that each component $a_j(v)$ of $a(v)$ is linear-quadratic in $v \in V$.

When such linear-quadratic expressions in v for $\alpha(v)$, $a(v)$ and $A(v)$ are introduced in 11(8), a polynomial formula for $f(u,v)$ is obtained in which there are no terms of degree higher than 4. We have to show that there really aren't any terms of degree higher than 2, i.e., that f is linear-quadratic relative to $U \times V$ as a whole. We already know that $\alpha(v)$ has no higher-order terms, so the issue concerns $a(v)$ and $A(v)$.

This is where we bring in the last of our assumptions, that f is linear-quadratic on all line segments in $U \times V$. We'll only need to look at line segments that join an arbitrary point $(\bar u, \bar v) \in U \times V$ to $(0,0)$. The assumption means then that $f(\theta\bar u, \theta\bar v)$ is linear-quadratic in $\theta \in [0,1]$. The argument just presented for reducing to individual components can be repeated by looking at $f(\theta[\bar u + \bar u'])$ with the vectors $\bar u$ and $\bar u'$ chosen from among $\varepsilon e^1, \ldots, \varepsilon e^p$, and so forth, to see that $\theta^2 A_{ij}(\theta\bar v)$ and $\theta a_j(\theta\bar v)$ are polynomials of at most degree 2 in θ for every choice of $\bar v \in V$. In the linear-quadratic expressions for $A_{ij}(v)$ and $a_j(v)$ as functions of v, it is obvious then that $A_{ij}(v)$ has to be constant in v, while $a_j(v)$ can at most have first-order terms in v. This finishes the proof. □

Proof of 11.14. The justification of (a) is relatively easy on the basis of earlier results. When the convex function f is piecewise linear, it can be expressed in the manner of 3.54: for some choice of vectors a_i and scalars c_i,

$$f(x) \;=\; \begin{cases} \text{infimum of } \;\; t_1c_1 + \cdots + t_mc_m + t_{m+1}c_{m+1} + \cdots + t_rc_r \\ \text{subject to } \;\; t_1a_1 + \cdots + t_ma_m + t_{m+1}a_{m+1} + \cdots + t_ra_r = x \\ \text{with } \;\; t_i \ge 0 \;\text{ for }\; i = 1,\ldots,r, \;\; \sum_{i=1}^m t_i = 1. \end{cases}$$

From $f^*(v) = \sup_x\{\langle v,x\rangle - f(x)\}$ we get $f^*(v) \;=\; \max_{i=1,\ldots,m}\{\langle v, a_i\rangle - c_i\} + \delta_C$ for the polyhedral set $C := \{v \,|\, \langle v, a_i\rangle \le c_i \text{ for } i = m+1,\ldots,r\}$. This signifies by 2.49 that f^* is piecewise linear. On the other hand, if f^* is piecewise linear, then so is f^{**} by this argument; but $f^{**} = f$.

For (b), suppose now that f is piecewise linear-quadratic: for $D := \operatorname{dom} f$ there are polyhedral sets C_k, $k = 1,\ldots,r$, such that $D = \bigcup_{k=1}^r C_k$ and

$$f(x) \;=\; \tfrac{1}{2}\langle x, A_kx\rangle + \langle a_k, x\rangle + \alpha_k \;\text{ when }\; x \in C_k. \qquad 11(9)$$

Our task is to show that f^* has a similar representation. We'll base our argument on the fact in 11.3 that

$$f^*(v) \;=\; \langle v, x\rangle - f(x) \;\text{ for any }\; x \;\text{ with }\; v \in \partial f(x). \qquad 11(10)$$

This requires careful investigation of the structure of the mapping ∂f.

Recall from 10.21 that the convex set $D = \operatorname{dom} f$, as the union of finitely many polyhedral sets C_k, is itself polyhedral. Any polyhedral set may be represented as the intersection of a finite collection of closed half-spaces, so we can contemplate a finite collection $\mathcal{H}$ of closed half-spaces in $\mathbb{R}^n$ such that (1) each of the sets $D, C_1, \ldots, C_r$ is the intersection of a subcollection of the half-spaces in $\mathcal{H}$, and (2) for every $H \in \mathcal{H}$ the opposite closed half-space H' (meeting H in a shared hyperplane) is likewise in $\mathcal{H}$.

Let $J_x = \{H \in \mathcal{H} \mid x \in H$ for each $x \in D$. Let $\mathcal{J}$ consist of all $J \subset \mathcal{H}$ such that $J = J_x$ for some $x \in D$, and for each $J \in \mathcal{J}$ let D_J be the intersection of the half-spaces $H \in J$. It is clear that each D_J is a nonempty polyhedral set contained in D; in fact, the half-spaces in $\mathcal{H}$ that intersect to form C_k belong to J_x if $x \in C_k$), so that $D_J \subset C_k$ when $J = J_x$ for any $x \in C_k$.

For each $J \in \mathcal{J}$, let $F_J = \operatorname{rint} D_J$, recalling that then $D_J = \operatorname{cl} F_J$. We claim that $J = J_x$ if and only if $x \in F_J$. For the half-spaces $H \in J_x$, there are only two possibilities: either $x \in \operatorname{int} H$ or x lies on the boundary of H, which corresponds to having both H and the opposite half-space H' belong to J_x. Thus, for $J = J_x$, D_J is the intersection of various hyperplanes along with some closed half-spaces having x in their associated open half-spaces. That intersection is the relatively open set F_J. Hence $x \in F_J$. On the other hand, for any x' in this set F_J, and in particular D_J, we have $J_{x'} \supset J = J_x$. If there were a half-space H in $J_{x'} \backslash J_x$, then x would have to lie outside of H, or more specifically, in the interior of the opposite half-space H' (likewise belonging to $\mathcal{H}$). In that case, however, $\operatorname{int} H'$ is one of the open half-spaces that includes F_J, and hence contains x', in contradiction to x' being in H. Thus, any $x' \in F_J$ must have $J_{x'} = J$. Indeed, we see from this that $\{F_J \mid J \in \mathcal{J}\}$ is a finite partition of D, comprised in effect of the equivalence classes under the relation that $x' \sim x$ when $J_{x'} = J_x$. Moreover, if any F_J touches a sets C_k, it must lie entirely in C_k, and the same is true then for its closure, namely D_J. In other words, the index set $K(x) = \{k \mid x \in C_k\}$ is the same set $K(J)$ for all $x \in F_J$.

It was shown in the proof of 10.21 that $df(x)$ is piecewise linear with $\operatorname{dom} df(\bar{x}) = T_D(x) = \bigcup_{k \in K(x)} T_{C_k}(x)$ and $df(x)(w) = \langle A_k x + a_k, w\rangle$ when $w \in T_{C_k}(x)$. Because f, being a proper, lsc, convex function, is regular (cf. 7.27), we know that $\partial f(x)$ consists of the vectors v such that $\langle v, w\rangle \leq df(x)(w)$ for all $w \in \mathbb{R}^n$ (see 8.30). Hence

$$\begin{aligned}\partial f(x) &= \bigcap\nolimits_{k \in K(x)} \left\{v \,\middle|\, \langle v - A_k x - a_k, w\rangle \leq 0 \text{ for all } w \in T_{C_k}(x)\right\}\\ &= \bigcap\nolimits_{k \in K(x)} \left\{v \,\middle|\, v - A_k x - a_k \in N_{C_k}(x)\right\}.\end{aligned}$$

In this we appeal to the polarity between $T_{C_k}(x)$ and $N_{C_k}(x)$, which results from C_k being convex (cf. 6.24). Observe next that the normal cone $N_{C_k}(x)$ (polyhedral) must be the same for all x in a given F_J. That's because having $v \in N_{C_k}(x)$ corresponds to the maximum of $\langle v, x'\rangle$ over $x' \in C_k$ being attained at x, and by virtue of F_J being relatively open, that can't happen unless this linear function is constant on F_J (and therefore attains its maximum at *every* point of F_J). This common normal cone can be denoted by $N_k(J)$, and in terms of the common index set $K(x) = K(J)$ for $x \in F_J$, we then have

$$\langle v, x' \rangle = \langle v, x \rangle \ \text{ for all } \ x, x' \in F_J \ \text{ when } \ v \in N_k(J), \ k \in K(J), \qquad 11(11)$$

along with $\partial f(x) = \{ v \mid v - a_k - A_k x \in N_k(J) \text{ for all } k \in K(J) \}$ when $x \in F_J$. Consider for each $J \in \mathcal{J}$ the polyhedral set

$$G_J = \{ (x, v) \mid x \in D_J \text{ and } v - a_k - A_k x \in N_k(J) \text{ for all } k \in K(J) \},$$

which is the closure of the analogous set with F_J in place of D_J. Because $\operatorname{gph} \partial f$ is closed (cf. 11.3), it follows now that $\operatorname{gph} \partial f = \bigcap_{J \in \mathcal{J}} G_J$.

For each $J \in \mathcal{J}$ let E_J be the image of G_J under $(x, v) \mapsto v$, which like G_J is polyhedral by 3.55(a), therefore closed. Since $\operatorname{dom} \partial f^* = \operatorname{rge} \partial f$ (by the inversion rule in 11.3), it follows that $\operatorname{dom} \partial f^* = \bigcup_{J \in \mathcal{J}} E_J$. Hence $\operatorname{dom} \partial f^*$ is closed, because the union of finitely many closed sets is closed. But since f^* is lsc and proper, $\operatorname{dom} \partial f^*$ is dense in $\operatorname{dom} f^*$ (see 8.10). The union of the polyhedral sets E_J is thus $\operatorname{dom} f^*$.

All that's left now is to show f^* is linear-quadratic relative to each set E_J. We'll appeal to Lemma 11.15. Consider any v_0 and v_1 in a given E_J, coming from G_J, and choose any x_0 and x_1 such that (x_0, v_0) and (x_1, v_1) belong to G_J. Then the pair $(x_\tau, v_\tau) := (1 - \tau)(x_0, v_0) + \tau(x_1, v_1)$ belongs to G_J too, so that $v_\tau \in \partial f(x_\tau)$. From 11(9) and 11(10) we get, for any $k \in K(J)$, that $f^*(v_\tau) = \langle v_\tau, x_\tau \rangle - f(x_\tau) = \langle v_\tau - A_k x_\tau - a_k, x_\tau \rangle - \alpha_k + \frac{1}{2} \langle x_\tau, A_k x_\tau \rangle = \langle v_\tau - A_k x_\tau - a_k, x_0 \rangle - \alpha_k + \frac{1}{2} \langle x_\tau, A_k x_\tau \rangle$, where the last equation is justified through the fact that $x_\tau = x_0 + \tau(x_1 - x_0)$ but $\langle v_\tau - A_k x_\tau - a_k, \, x_1 - x_0 \rangle = 0$ by 11(11). This expression for $f^*(v_\tau)$, being linear-quadratic in $\tau \in [0, 1]$, gives us what was required.

The fact that f^* is piecewise linear-quadratic *only* if f is piecewise linear-quadratic follows now by symmetry, because $f = f^{**}$. □

11.16 Corollary (minimum of a piecewise linear-quadratic function). *For any proper, convex, piecewise linear-quadratic function $f : \mathbb{R}^n \to \overline{\mathbb{R}}$, if $\inf f$ is finite, then $\operatorname{argmin} f$ is nonempty and polyhedral.*

Proof. We apply 11.8(a) with the knowledge that f^* is piecewise linear-quadratic like f, so that when $0 \in \operatorname{dom} f^*$ the set $\partial f^*(0)$ must be nonempty and polyhedral (cf. 10.21). □

11.17 Corollary (polyhedral sets in duality).

(a) *A closed, convex set C is polyhedral if and only if its support function σ_C is piecewise linear.*

(b) *A closed, convex cone K is polyhedral if and only if its polar cone K^* is polyhedral.*

Proof. This specializes 11.14(b) to the correspondences in 11.4. A convex indicator δ_C is piecewise linear if and only if C is polyhedral. The cone fact could also be deduced right from the Minkowski-Weyl theorem in 3.52. □

The preservation of the piecewise linear-quadratic property in passing to the conjugate of a given function, as in Theorem 11.14(b), is illustrated in Figure 11–4. As the figure suggests, this duality is closely tied to a property of the associated subgradient mappings through the inversion rule in 11.3. It will later be established in 12.30 that a convex function is piecewise linear-quadratic if and only if its subgradient mapping is piecewise polyhedral as defined in 9.57. The inverse of a piecewise polyhedral mapping is obviously still piecewise polyhedral.

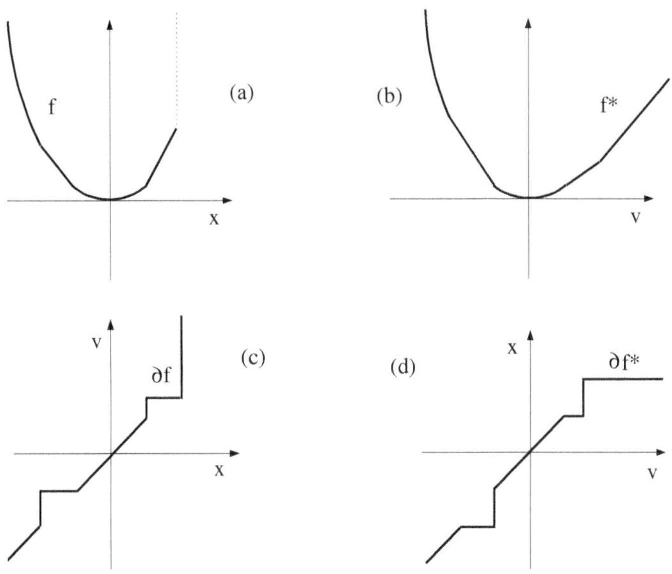

Fig. 11–4. Conjugate piecewise linear-quadratic functions.

An important application of the conjugacy in Theorem 11.14 comes up in the following class of functions $\theta : \mathbb{R}^m \to \overline{\mathbb{R}}$, which are useful in setting up 'penalty' expressions $\theta\big(f_1(x), \ldots, f_m(x)\big)$ in composite formats of optimization.

11.18 Example (piecewise linear-quadratic penalties). *For a nonempty polyhedral set $Y \subset \mathbb{R}^m$ and a symmetric positive-semidefinite matrix $B \in \mathbb{R}^{m \times m}$ (possibly $B = 0$), the function $\theta_{Y,B} : \mathbb{R}^n \to \overline{\mathbb{R}}$ defined by*

$$\theta_{Y,B}(u) := \sup_{y \in Y} \Big\{ \langle y, u \rangle - \tfrac{1}{2} \langle y, By \rangle \Big\}$$

is proper, convex and piecewise linear-quadratic. When $B = 0$, it is piecewise linear; $\theta_{Y,0} = \sigma_Y$ (support function). In general,

$$\operatorname{dom} \theta_{Y,B} = (Y^\infty \cap \ker B)^* =: D_{Y,B}, \quad \textit{where} \quad \ker B := \big\{ y \,\big|\, By = 0 \big\};$$

this is a polyhedral cone, and it is all of $\mathbb{R}^m$ *if and only if* $Y^\infty \cap \ker B = \{0\}$. *The subgradients of* $\theta_{Y,B}$ *are given by*

$$\begin{aligned}\partial\theta_{Y,B}(u) &= \operatorname*{argmax}_{y\in Y}\left\{\langle y,u\rangle - \tfrac{1}{2}\langle y,By\rangle\right\}\\ &= \left\{y \,\middle|\, u - By \in N_Y(y)\right\} = (N_Y + B)^{-1}(u),\\ \partial^\infty\theta_{Y,B}(u) &= \left\{y \in Y^\infty \cap \ker B \,\middle|\, \langle y,u\rangle = 0\right\} = N_{D_{Y,B}}(u),\end{aligned}$$

these sets being polyhedral as well, while the subderivative function $d\theta_{Y,B}$ *is piecewise linear and expressed by the formula*

$$d\theta_{Y,B}(u)(z) = \sup\left\{\langle y,z\rangle \,\middle|\, y \in (N_Y + B)^{-1}(u)\right\}.$$

Detail. We have $\theta_{Y,B} = (\delta_Y + j_B)^*$ for $j_B(y) := \frac{1}{2}\langle y,By\rangle$. The function $\delta_Y + j_B$ is proper, convex and piecewise linear-quadratic, and $\theta_{Y,B}$ therefore has these properties as well by 11.14. In particular, the effective domain of $\theta_{Y,B}$ is a polyhedral set, hence closed. The support function of this effective domain is $(\delta_Y + j_B)^\infty$ by 11.5, and

$$(\delta_Y + j_B)^\infty = \delta_Y^\infty + j_B^\infty = \delta_{Y^\infty} + \delta_{\ker B} = \delta_{Y^\infty \cap \ker B}.$$

Hence by 11.4, $\operatorname{dom}\theta_{Y,B}$ must be the polar cone $(Y^\infty \cap \ker B)^*$, which is $\mathbb{R}^m$ if and only if $Y^\infty \cap \ker B$ is the zero cone.

The argmax formula for $\partial\theta_{Y,B}(u)$ specializes the argmax part of 11.3. The maximum of the concave function $h(y) = \langle y,u\rangle - \frac{1}{2}\langle y,By\rangle$ over Y is attained at y if and only if the gradient $\nabla h(y) = u - By$ belongs to $N_Y(y)$; cf. 6.12. That yields the other expressions for $\partial\theta_{Y,B}(u)$. We know from the convexity of $\theta_{Y,B}$ that $\partial^\infty\theta_{Y,B}(u) = N_{D_{Y,B}}(u)$; cf. 8.12. Since the cones $D_{Y,B}$ and $Y^\infty \cap \ker B$ are polar to each other, we have from 11.4(b) that $y \in N_{D_{Y,B}}(u)$ if and only if $u \in D_{Y,B}$, $y \in Y^\infty \cap \ker B$, and $u \perp y$.

On the basis of 10.21, the sets $\partial\theta_{Y,B}(u)$ and $\partial^\infty\theta_{Y,B}(u)$ are polyhedral and the function $d\theta_{Y,B}(u)$ is piecewise linear. The formula for $d\theta_{Y,B}(u)(z)$ merely expresses the fact that this is the support function of $\partial\theta_{Y,B}(u)$. □

E. Polar Sets and Gauges

While most of the major duality correspondences, like convex sets versus sublinear functions, or polarity of convex cones, fit directly within the framework of conjugate convex functions as in 11.4, others, like polarity of convex sets that aren't necessarily cones but contain the origin, fit obliquely. In the next example we draw on the notion of the gauge γ_C of a set C in 3.50.

11.19 Example (general polarity of sets). *For any set* $C \subset \mathbb{R}^n$ *with* $0 \in C$, *the polar of* C *is the set*

$$C^\circ := \{v \mid \langle v, x\rangle \leq 1 \text{ for all } x \in C\},$$

which is closed and convex with $0 \in C^\circ$*; when* C *is a cone,* C° *agrees with the polar cone* C^**. The bipolar of* C*, which is the set*

$$C^{\circ\circ} := (C^\circ)^\circ = \{x \mid \langle v, x\rangle \leq 1 \text{ for all } v \in C^\circ\},$$

agrees always with $\operatorname{cl}(\operatorname{con} C)$*. Thus,* $C^{\circ\circ} = C$ *when* C *is a closed, convex set containing the origin, so the transformation* $C \mapsto C^\circ$ *maps that class of sets one-to-one onto itself. This correspondence is connected to conjugacy through the associated gauges, which obey the rule that*

$$\gamma_C = \sigma_{C^\circ} \underset{*}{\longleftrightarrow} \delta_{C^\circ}, \qquad \gamma_{C^\circ} = \sigma_C \underset{*}{\longleftrightarrow} \delta_C.$$

When two convex sets are polar to each other, one says that their gauges are polar to each other as well.

Detail. The facts about C° and $C^{\circ\circ}$ are evident from the envelope description of convex sets in 6.20. Clearly $C^{\circ\circ}$ is the intersection of all the closed half-spaces that include C and have the origin in their interior.

Because the gauge γ_C is proper, lsc and sublinear (cf. 3.50), we know from 8.24 that it's the support function of a certain nonempty, closed, convex set, namely the one consisting the vectors v such that $\langle v, x\rangle \leq \gamma_C(x)$ for all x. But in view of the definition of γ_C (in 3.50) this set is C°. Thus $\gamma_C = \sigma_{C^\circ}$, and since $C^{\circ\circ} = C$ also by symmetry $\gamma_{C^\circ} = \sigma_C$. These functions are conjugate to δ_{C° and δ_C respectively by 11.4(a). □

11.20 Exercise (dual properties of polar sets)**.** *Let* C *be a closed, convex subset of* $\mathbb{R}^n$ *containing the origin, and let* C° *be its polar as defined in 11.19.*

(a) *C is bounded if and only if* $0 \in \operatorname{int} C^\circ$*; likewise,* C° *is bounded if and only if* $0 \in \operatorname{int} C$.

(b) *C is polyhedral if and only if* C° *is polyhedral.*

(c) $C^\infty = (\operatorname{pos} C^\circ)^*$ *and* $(C^\circ)^\infty = (\operatorname{pos} C)^*$.

Guide. In (a) and (b), rely on the gauge interpretation of polarity in 11.19; apply 11.8(c) and 11.14. Argue the second equation in (c) from the intersection rule in 3.9 and the definition of C° as an intersection of half-spaces. Obtain the other equation in (c) then by symmetry. □

Polars of convex sets other than cones are employed most notably in the study of norms. Any closed, bounded, convex set $B \subset \mathbb{R}^n$ that's *symmetric* ($-B = B$) with nonempty interior (and hence has the origin in this interior) corresponds to a certain norm $\|\cdot\|$, given by its gauge γ_B, cf. 3.50. The polar set B° is likewise a closed, bounded, convex set that's symmetric with nonempty interior. Its gauge γ_{B° gives the norm $\|\cdot\|^\circ$ *polar* to $\|\cdot\|$. Of particular note is the famous rule for polarity in the family of l_p norms in 2.17, namely

$$\begin{aligned}&\|\cdot\|_p^\circ = \|\cdot\|_q \text{ when } 1 < p < \infty,\ 1 < q < \infty,\ p^{-1} + q^{-1} = 1,\\ &\|\cdot\|_1^\circ = \|\cdot\|_\infty, \quad \|\cdot\|_\infty^\circ = \|\cdot\|_1.\end{aligned} \qquad 11(12)$$

This can be derived from the next result, which furnishes additional examples of conjugate convex functions. The argument will be sketched after the proof.

11.21 Proposition (conjugate composite functions from polar gauges). *Consider the gauge γ_C of a closed, convex set $C \subset \mathbb{R}^n$ with $0 \in C$ and any lsc, convex function $\theta : \mathbb{R} \to \overline{\mathbb{R}}$ with $\theta(-r) = \theta(r)$. Under the convention $\theta(\infty) = \infty$ one has the conjugacy relation*

$$\theta\big(\gamma_C(x)\big) \;\overset{\longleftrightarrow}{*}\; \theta^*\big(\gamma_{C^\circ}(v)\big).$$

In particular, for any norm $\|\cdot\|$ and its polar norm $\|\cdot\|^\circ$, one has

$$\theta\big(\|x\|\big) \;\overset{\longleftrightarrow}{*}\; \theta^*\big(\|v\|^\circ\big).$$

Proof. Let $f(x) = \theta\big(\gamma_C(x)\big)$. The function θ has to be nondecreasing on $\mathbb{R}_+$ since for any $r > 0$ in $\operatorname{dom}\theta$ we have $\theta(-r) = \theta(r) < \infty$ and consequently

$$\theta\big((1-\tau)(-r) + \tau r\big) \;\leq\; (1-\tau)\theta(-r) + \tau\theta(r) = \theta(r) \;\text{ for }\; 0 < \tau < 1,$$

so that $\theta(r') \leq \theta(r)$ for all $r' \in (-r, r)$. This monotonicity ensures that f is convex and enables us to write $f(x) = \inf\big\{\theta(\lambda) \,\big|\, \lambda \geq \gamma_C(x)\big\}$. In calculating the conjugate we then have

$$\begin{aligned}
f^*(v) &= \sup\Big\{\langle v, x\rangle - \theta(\lambda) \,\Big|\, (x,\lambda) \in \operatorname{epi}\gamma_C\Big\} \\
&= \sup_{\substack{\lambda\geq 0\\ \lambda\in\operatorname{dom}\theta}} \;\sup\Big\{\langle v, x\rangle - \theta(\lambda) \,\Big|\, x \in \operatorname{lev}_{\leq\lambda}\gamma_C\Big\} \\
&= \sup_{\substack{\lambda\geq 0\\ \lambda\in\operatorname{dom}\theta}} \begin{Bmatrix} \lambda\sigma_C(v) - \theta(\lambda) & \text{for } \lambda > 0 \\ \delta_{C^{\infty *}}(v) & \text{for } \lambda = 0 \end{Bmatrix} \\
&= \sup_{\substack{\lambda\geq 0\\ \lambda\in\operatorname{dom}\theta}} \begin{Bmatrix} \lambda\gamma_{C^\circ}(v) - \theta(\lambda) & \text{for } \lambda > 0 \\ \delta_{\operatorname{cl}\left(\operatorname{dom}\gamma_{C^\circ}\right)}(v) & \text{for } \lambda = 0 \end{Bmatrix} = \theta^*\big(\gamma_{C^\circ}(v)\big),
\end{aligned}$$

relying here on $\operatorname{lev}_{\leq 0}\gamma_C = C^\infty$ and $C^{\infty *} = \operatorname{cl}(\operatorname{dom}\sigma_C)$; cf. the end of 8.24. □

An illustration of the possibilities in Proposition 11.21 is furnished by the case of composition with the dual functions

$$\begin{gathered}\theta_p(r) = \frac{1}{p}|r|^p \;\overset{\longleftrightarrow}{*}\; \theta_q^*(s) = \frac{1}{q}|s|^q \\ \text{when } 1 < p < \infty,\;\; 1 < q < \infty,\;\; p^{-1} + q^{-1} = 1.\end{gathered} \qquad 11(13)$$

This one-dimensional conjugacy can be used to deduce from Proposition 11.21 the polarity rule for the l_p-norms in 11(12). The argument proceeds as follows for $p \in (1,\infty)$. The function $f(x) := \theta_p(x_1) + \cdots + \theta_p(x_n)$ is convex, so the set $\mathbb{B}_p = \big\{x \,\big|\, f(x) \leq p^{-1}\big\}$ is convex; also, it's closed and symmetric about 0. The function $h := \|\cdot\|_p = (pf)^{1/p}$ is nonnegative, lsc and positively homogeneous with $\text{lev}_{\leq 1} h = \mathbb{B}_p$. This implies that $\|\cdot\|_p = \gamma_{\boldsymbol{B}_p}$ and also that $\|\cdot\|_p$ is convex (hence truly is a norm); furthermore $f = \theta_p \circ \gamma_{\boldsymbol{B}_p}$. In parallel, the function $g(v) := \theta_q(v_1) + \cdots \theta_q(v_n)$ agrees with $g = \theta_q \circ \gamma_{\boldsymbol{B}_q}$, with $\|\cdot\|_q = \gamma_{\boldsymbol{B}_q}$. But f and g are conjugate to each other, as seen directly through 11(13). It follows then from Proposition 11.21 that $\|\cdot\|_p$ and $\|\cdot\|_q$ must be polar to each other. (For $p = 1$ and $p = \infty$ the polarity in 11(12) can be deduced more simply from 11.19 and the fact that $\|\cdot\|_1$ is the support function of $\mathbb{B}_\infty = [-1, 1]^n$.)

F. Dual Operations

With a wealth of examples of conjugate convex functions now in hand, we turn to the question of how to dualize other functions generated from these by various operations. The effects of some elementary operations have already been compiled in 11(3), but we now take up the topic in earnest.

11.22 Proposition (conjugation in product spaces). *For proper functions f_i on $\mathbb{R}^{n_i}$, the function conjugate to $f(x_1, \ldots, x_m) = f_1(x_1) + \cdots + f_m(x_m)$ is $f^*(v_1, \ldots, v_m) = f_1^*(v_1) + \cdots + f_m^*(v_m)$.*

Proof. This is elementary from the definition of the transform. □

11.23 Theorem (dual operations).

(a) (addition/epi-addition). *For proper functions f_i, if $f = f_1 \# f_2$, then $f^* = f_1^* + f_2^*$. Dually, if $f = f_1 + f_2$ for proper, lsc, convex functions f_i such that* $\text{dom}\, f_1$ *meets* $\text{dom}\, f_2$, *then $f^* = \text{cl}(f_1^* \# f_2^*)$. Here the closure operation is superfluous when $0 \in \text{int}(\text{dom}\, f_1 - \text{dom}\, f_2)$, as is true in particular when* $\text{dom}\, f_1$ *meets* $\text{int}(\text{dom}\, f_2)$ *or when* $\text{dom}\, f_2$ *meets* $\text{int}(\text{dom}\, f_1)$.

(b) (composition/epi-composition). *If $g = Af$ for $f : \mathbb{R}^n \to \overline{\mathbb{R}}$ and $A \in \mathbb{R}^{m \times n}$, where $(Af)(u) := \inf\{f(x) \mid Ax = u\}$, then $g^* = f^* A^*$, where $(f^* A^*)(y) := f^*(A^* y)$ (with A^* the transpose of A). Dually, if $f = gA$ for a proper, lsc, convex function $g : \mathbb{R}^m \to \overline{\mathbb{R}}$ such that the subspace* $\text{rge}\, A$ *meets* $\text{dom}\, g$, *then $f^* = \text{cl}(A^* g^*)$. Here the closure operation is superfluous when $0 \in \text{int}(\text{dom}\, g - \text{rge}\, A)$, as is true in particular when* $\text{rge}\, A$ *meets* $\text{int}(\text{dom}\, g)$.

(c) (restriction/inf-projection). *If $p(u) = \inf_x f(x, u)$ for a proper function $f : \mathbb{R}^n \times \mathbb{R}^m \to \overline{\mathbb{R}}$, then $p^*(y) = f^*(0, y)$. Dually, if f is also convex and lsc, then $\varphi(x) = f(x, \bar{u})$ for some $\bar{u} \in U := \{u \mid \exists x,\ f(x, u) < \infty\}$, one has $\varphi^* = \text{cl}\, q$ for the function $q(v) = \inf_y \{f^*(v, y) - \langle y, \bar{u} \rangle\}$. Here the closure operation is superfluous when actually $\bar{u} \in \text{int}\, U$.*

(d) (pointwise sup/inf). *For a family of functions f_i, if $f = \inf_{i \in I} f_i$, then $f^* = \sup_{i \in I} f_i^*$. Dually, if $f = \sup_{i \in I} f_i$ with f_i proper, lsc and convex, and if f is proper, then $f^* = \text{cl con}(\inf_{i \in I} f_i^*)$.*

Proof. The first relation in (a) falls out of the definition of f^* and the formula for $f_1 \# f_2$ in 1(12). It implies that $(f_1^* \# f_2^*)^* = f_1^{**} + f_2^{**}$, the latter being the same as $f_1 + f_2$ when each f_i is proper, lsc and convex. When that is true and

$\operatorname{dom} f_1 \cap \operatorname{dom} f_2 \neq \emptyset$, so that $f = f_1 + f_2$ is proper, we get $f^* = (f_1^* \# f_2^*)^{**} = \operatorname{cl} \operatorname{con}(f_1^* \# f_2^*)$ by 11.1. The convex hull operation is superfluous because the convexity of f_i^* implies that of $f_1^* \# f_2^*$ (cf. 2.24). The closure operation can be omitted when $f_1^* \# f_2^*$ is lsc, which holds when the set $\operatorname{epi} f_1^* + \operatorname{epi} f_2^*$ is closed (cf. 1.28), a property guaranteed by the absence of any nonzero $(v, \beta) \in (\operatorname{epi} f_1^*)^\infty$ with $(-v, -\beta) \in (\operatorname{epi} f_2^*)^\infty$ (cf. 3.12).

Because $(\operatorname{epi} f_i^*)^\infty$ is the epigraph of $f_i^{*\infty}$, which we have identified in 11.5 with the support function of $D_i = \operatorname{dom} f_i$, this condition translates to the nonexistence of $v \neq 0$ such that $\sigma_{D_1}(v) \leq -\sigma_{D_2}(-v)$. But this is equivalent by 8.29(b) to having $0 \in \operatorname{int}(D_1 - D_2)$. Obviously $\operatorname{int}(D_1 - D_2)$ includes the sets $D_1 - (\operatorname{int} D_2)$ and $(\operatorname{int} D_1) - D_2$, since these are open.

For (b) the same pattern works. It's easily seen from the definitions that $(Af)^* = f^* A^*$. For the same reason, $(A^* g^*)^* = g^{**} A^{**} = gA$ when g is proper, lsc and convex. When $\operatorname{rge} A$ meets $\operatorname{dom} g$, so that gA is proper, we obtain from 11.1 that $(gA)^* = (A^* g^*)^{**} = \operatorname{cl} \operatorname{con} A^* g^*$. The convex hull operation can be omitted because the convexity of $A^* g^*$ accompanies that of g^* by 2.22(b). The closure operation can be omitted when $A^* g^*$ is lsc, which holds when the set $L(\operatorname{epi} g^*)$ is closed for the linear mapping $L(y, \alpha) = (A^* y, \alpha)$; cf. 1.31. This is implied by the absence of any nonzero (y, α) in $L^{-1}(0,0) \cap (\operatorname{epi} g^*)^\infty$, i.e., the absence of any $y \neq 0$ such that $A^* y = 0$, $g^{*\infty}(y) \leq 0$. Identifying $g^{*\infty}$ with the support function of $\operatorname{dom} g$ through 11.5, and identifying the indicator of the null space $\{y \mid A^* y = 0\}$ with the support function of the range space $\operatorname{rge} A$, we translate this condition into the absence of any $y \neq 0$ such that $\sigma_{\operatorname{dom} g}(y) \leq -\sigma_{\operatorname{rge} A}(-y)$. By 8.29(b), this means $0 \in \operatorname{int}(\operatorname{dom} g - \operatorname{rge} A)$. Here $\operatorname{int}(\operatorname{dom} g - \operatorname{rge} A)$ includes the open set $\operatorname{int}(\operatorname{dom} g) - \operatorname{rge} A$.

Likewise in (c), the definitions of p and p^* give

$$\begin{aligned} p^*(y) &= \sup_u \{ \langle y, u \rangle - \inf_x f(x, u) \} \\ &= \sup_{x,u} \{ \langle (0, y), (x, u) \rangle - f(x, u) \} = f^*(0, y). \end{aligned}$$

For parallel reasons, $q^*(x) = f^{**}(x, \bar{u})$. When f is proper, lsc and convex, we have $f^{**} = f$, so $q^* = \varphi$. But the convexity of f^* implies that of q by 2.22(a). Hence as long as $\bar{u} \in U$, so that φ is proper, we have $\varphi^* = \operatorname{cl} q$ by 11.1. To omit the closure operation as well, we can look to cases where q is known to be lsc. One such case, furnished by 3.31, is associated with having $f^{*\infty}(0, y) - \langle y, \bar{u} \rangle > 0$ when $y \neq 0$. But if f is proper and convex, $f^{*\infty}$ is the support function of $\operatorname{dom} f$ by 11.5, and $f^{*\infty}(0, \cdot)$ is then the support function of the image U of $\operatorname{dom} f$ under the projection $(x, u) \mapsto u$. The condition that $f^{*\infty}(0, y) - \langle y, \bar{u} \rangle > 0$ when $y \neq 0$ translates therefore to the condition that $\sigma_U(y) > 0$ when $y \neq 0$, which by 8.29(a) is equivalent to having $0 \in \operatorname{int} U$.

The first relation in (d) is immediate from the definition of f^*. It implies that $(\inf_{i \in I} f_i^*)^* = \sup_{i \in I} f_i^{**}$. When $f = \sup_{i \in I} f_i$ with f_i proper, lsc and convex, so that $f_i = f_i^{**}$ by 11.1, and f is proper (in addition to being convex by 2.9), we obtain from 11.1 that $f^* = (\inf_{i \in I} f_i^*)^{**} = \operatorname{cl} \operatorname{con}(\inf_{i \in I} f_i^*)$. □

In part (a) of Theorem 11.23, the condition $0 \in \operatorname{int}(\operatorname{dom} g - \operatorname{rge} A)$ is equivalent to the nonexistence of a 'separating hyperplane' for C_1 and C_2 (see 2.39). Likewise in part (b), the condition $0 \in \operatorname{int}(\operatorname{dom} g - \operatorname{rge} A)$ means that $\operatorname{dom} g$ can't be separated from $\operatorname{rge} A$, even improperly.

11.24 Corollary (rules for support functions).

(a) *If $D = \lambda C$ with $C \neq \emptyset$ and $\lambda > 0$, then $\sigma_D = \lambda\sigma_C$.*

(b) *If $C = C_1 + C_2$ with $C_i \neq \emptyset$, then $\sigma_C = \sigma_{C_1} + \sigma_{C_2}$.*

(c) *If $D = \{Ax \mid x \in C\}$ with $A \in I\!R^{m\times n}$, then $\sigma_D(y) = \sigma_C(A^*y)$.*

(d) *If $C = \{x \mid Ax \in D\}$ for a closed, convex set $D \subset I\!R^m$, and if $C \neq \emptyset$, then $\sigma_C = \operatorname{cl} A^*\sigma_D$, where $(A^*\sigma_D)(v) := \inf\{\sigma_D(y) \mid A^*y = v\}$. Here the closure operation is superfluous when $0 \in \operatorname{int}(D - \operatorname{rge} A)$, as is true in particular when the subspace $\operatorname{rge} A$ meets $\operatorname{int} D$.*

(e) *If $C = C_1 \cap C_2 \neq \emptyset$ with each set C_i convex and closed, then $\sigma_C = \operatorname{cl}(\sigma_{C_1} \# \sigma_{C_2})$. Here the closure operation is superfluous when $0 \in \operatorname{int}(C_1 - C_2)$, as is true in particular when C_1 meets $\operatorname{int} C_2$ or C_2 meets $\operatorname{int} C_1$.*

(f) *If $C = \bigcup_{i\in I} C_i$, then $\sigma_C = \sup_{i\in I} \sigma_{C_i}$.*

Proof. These six rules correspond to the cases where (a) $\delta_D = \lambda{\star}\delta_C$, (b) $\delta_C = \delta_{C_1} \# \delta_{C_2}$, (c) $\delta_D = A\delta_C$, (d) $\delta_C = \delta_D A$, (e) $\delta_C = \delta_{C_1} + \delta_{C_2}$, and (f) $\delta_C = \inf_{i\in I} \delta_{C_i}$. All except (a) are covered by Theorem 11.23 through 11.4(a), while (a) simply comes from 11(3)—but is best listed here. □

11.25 Corollary (rules for polar cones).

(a) *If $K = K_1 + K_2$ for cones K_i, then $K^* = K_1^* \cap K_2^*$. Likewise, if $K = \bigcup_{i\in I} K_i$ one has $K^* = \bigcap_{i\in I} K_i^*$.*

(b) *If $K = K_1 \cap K_2$ for closed, convex cones K_i, then $K^* = \operatorname{cl}(K_1^* + K_2^*)$. The closure operation is superfluous when $0 \in \operatorname{int}(K_1 - K_2)$.*

(c) *If $H = \{Ax \mid x \in K\}$ for $A \in I\!R^{m\times n}$ and a cone $K \subset I\!R^n$, then $H^* = \{y \mid A^*y \in K^*\}$.*

(d) *If $K = \{x \mid Ax \in H\}$ for $A \in I\!R^{m\times n}$ and a closed, convex cone $H \subset I\!R^m$, then $K^* = \operatorname{cl}\{A^*y \mid y \in H^*\}$. The closure operation is superfluous when $0 \in \operatorname{int}(H - \operatorname{rge} A)$.*

Proof. In (a) we apply 11.23(a) to $\delta_K = \delta_{K_1} \# \delta_{K_2}$, or 11.23(d) to $\delta_K = \inf_{i\in I} \delta_{K_i}$, whereas in (b) we apply 11.23(a) to $\delta_K = \delta_{K_1} + \delta_{K_2}$, each time making the observation in 11.4(b). In (c) and (d) it's the same story in applying 11.23(b) with $\delta_H = A\delta_K$ in the first case and $\delta_K = \delta_H A$ in the second. □

11.26 Example (distance functions, Moreau envelopes and proximal hulls).

(a) *For any nonempty, closed, convex set $C \subset I\!R^n$, the functions d_C and $\sigma_C + \delta_B$ are conjugate to each other.*

(b) *For any proper, lsc, convex function $f : I\!R^n \to \overline{I\!R}$ and any $\lambda > 0$, the functions $e_\lambda f$ and $f^* + \frac{\lambda}{2}|\cdot|^2$ are conjugate to each other. This entails*

$$e_\lambda f(x) + e_{\lambda^{-1}} f^*(\lambda^{-1}x) = \frac{1}{2\lambda}|x|^2 \text{ for all } x \in I\!R^n,\ \lambda > 0.$$

(c) *For any $f : \mathbb{R}^n \to \overline{\mathbb{R}}$, not necessarily convex, the Moreau envelope $e_\lambda f$ and proximal hull $h_\lambda f$ for each $\lambda > 0$ are expressed by*

$$\left.\begin{aligned} e_\lambda f(x) &= \lambda^{-1} j(x) - (f + \lambda^{-1} j)^*(\lambda^{-1} x) \\ h_\lambda f(x) &= (f + \lambda^{-1} j)^{**}(x) - \lambda^{-1} j(x) \end{aligned}\right\} \quad \text{for } j = \tfrac{1}{2}|\cdot|^2.$$

Here $(f + \lambda^{-1} j)^{**}$ can be replaced by $\operatorname{con}(f + \lambda^{-1} j)$ when f is lsc, proper and prox-bounded with threshold $\lambda_f > \lambda$. Then $\operatorname{dom} h_\lambda f = \operatorname{con}(\operatorname{dom} f)$, and on the interior of this convex set the function $h_\lambda f$ must be lower-$\mathcal{C}^2$.

(d) *A proper function $f : \mathbb{R}^n \to \overline{\mathbb{R}}$ is λ-proximal, in the sense that $h_\lambda f = f$, if and only if f is lsc and $f + \frac{1}{2\lambda}|\cdot|^2$ is convex.*

Detail. In (a) we have $d_C = \delta_C \# |\cdot|$ by 1(13), where the Euclidean norm $|\cdot|$ is the support function $\sigma_{\boldsymbol{B}}$, as recorded in 8.27, and $\sigma_{\boldsymbol{B}}^* = \delta_{\boldsymbol{B}}$ by 11.4(a). Then $d_C^* = \delta_C^* + \sigma_{\boldsymbol{B}}^*$ by 11.23(b), with $\delta_C^* = \sigma_C$ by 11.4(a). Because d_C is finite, hence continuous (cf. 2.36), we have $d_C^{**} = d_C$ (by 11.1).

In (b) the situation is similar. In terms of $j(x) = \frac{1}{2}|x|^2$ we have $e_\lambda f = f \# \lambda^{-1} j$ by 1(13), with $e_\lambda f$ finite and convex by 2.25. Here $\lambda^{-1} j$ is the same as $\lambda \star j$ and is conjugate to λj by 11.11 and the rules in 11(3). The conjugate function $(e_\lambda f)^*$ is calculated then from 11.23(a) to be $f^* + \lambda j$. But to say that $e_\lambda f$ is conjugate to $f^* + \lambda j$ is to say that for all x one has

$$\begin{aligned} e_\lambda f(x) &= \sup\nolimits_w \Big\{ \langle w, x\rangle - f^*(w) - \frac{\lambda}{2}|w|^2 \Big\} \\ &= \frac{1}{2\lambda}|x|^2 - \inf\nolimits_w \Big\{ f^*(w) + \frac{\lambda}{2}|w - \lambda^{-1} x|^2 \Big\} = \frac{1}{2\lambda}|x|^2 - e_{\lambda^{-1}} f^*(\lambda^{-1} x). \end{aligned}$$

This gives the identity claimed in (b).

The same calculation produces the first identity in (c), and the second then follows from the formula for $h_\lambda f$ in 1.44. Justification for replacing $(f + \lambda^{-1} j)^{**}$ by $\operatorname{con}(f + \lambda^{-1} j)$ comes from 3.28 and 3.47; $f + \lambda^{-1} j$ is coercive when $\lambda \in (0, \lambda_f)$. The domain assertion is then obvious. The claim about $h_\lambda f$ being lower-$\mathcal{C}^2$ on the interior is supported by 10.33. To get (d), we merely invoke the rule from (c) that $h_\lambda f = f$ if and only if $(f + \lambda^{-1} j)^{**} = (f + \lambda^{-1} j)$. □

11.27 Exercise (proximal mappings and projections as gradient mappings).

(a) *For any proper, lsc, convex function $f : \mathbb{R}^n \to \overline{\mathbb{R}}$ and any $\lambda > 0$ the proximal mapping $P_\lambda f$ is ∇g for $g = \lambda \star (e_{\lambda^{-1}} f^*)$.*

(b) *For any nonempty, closed, convex set $C \subset \mathbb{R}^n$, the projection mapping P_C is ∇g for $g = e_1 \sigma_C = \sigma_C \# \frac{1}{2}|\cdot|^2$.*

Guide. Derive the first expression from the formula for $\nabla e_\lambda f$ in 2.26 using the identity in 11.26(b). Then derive the second expression by specializing to $f = \delta_C$ and $\lambda = 1$; cf. 11.4. □

11.28 Example (piecewise linear-quadratic envelopes).

(a) *For $f : \mathbb{R}^n \to \overline{\mathbb{R}}$ proper, convex and piecewise linear-quadratic and for any $\lambda > 0$, the convex function $e_\lambda f$ is piecewise linear-quadratic.*

(b) *For $C \subset \mathbb{R}^n$ nonempty and polyhedral, the convex function d_C^2 is piecewise linear-quadratic.*

Detail. The assertion in (a) is justified by the conjugacy rules in 11.14(a) and 11.26(b). The one in (b) specializes this to $f = \delta_C$; then $e_\lambda f = \frac{1}{2\lambda} d_C^2$. □

The use of several dualizing rules during the course of a calculation is illustrated by the next example, which concerns adjoints of sublinear mappings as defined in 8(27) and 8(28). Relations are obtained between the outer and inner norms for such mappings that were introduced in 9(4) and 9(5).

11.29 Example (norm duality for sublinear mappings). *The outer norm $|H|^+$ and inner norm $|H|^-$ of a sublinear, osc mapping $H : \mathbb{R}^n \rightrightarrows \mathbb{R}^m$ are related to those of its upper adjoint H^{*+} and lower adjoint H^{*-} by*

$$|H|^+ = |H^{*+}|^- = |H^{*-}|^-, \qquad |H|^- = |H^{*+}|^+ = |H^{*-}|^+.$$

In addition, one has $d\big(0, H(w)\big) = \sigma_{H^{*+}(\mathbb{B})}(w)$ and $\sigma_{H(\mathbb{B})}(y) = d\big(0, H^{*-}(y)\big)$.

As a special case, if a mapping $S : \mathbb{R}^n \rightrightarrows \mathbb{R}^m$ is graphically regular at $\bar{x}$ for $\bar{u}$, one has

$$\big|D^*S(\bar{x}\,|\,\bar{u})\big|^+ = \big|DS(\bar{x}\,|\,\bar{u})\big|^-, \qquad \big|D^*S(\bar{x}\,|\,\bar{u})^{-1}\big|^+ = \big|DS(\bar{x}\,|\,\bar{u})^{-1}\big|^-.$$

Detail. This can be derived by using the support function rules in 11.24 along with the rules of cone polarity. From the definition of $|H|^+$ in 9(4) and the description of the Euclidean norm in 8.27 we have

$$|H|^+ = \sup_{z \in H(\mathbb{B})} |z| = \sup_{z \in H(\mathbb{B}),\, y \in \mathbb{B}} \langle y, z \rangle = \sup_{y \in \mathbb{B}} \sigma_{H(\mathbb{B})}(y).$$

To prove that this equals $|H^{*+}|^-$, hence also $|H^{*-}|^-$ (inasmuch as $H^{*-}(y) = -H^{*+}(-y)$), it's enough now to demonstrate that $\sigma_{H(\mathbb{B})}(y) = d\big(0, H^{*-}(y)\big)$. In terms of the projection $A : \mathbb{R}^n \times \mathbb{R}^m \to \mathbb{R}^m$ the set $H(\mathbb{B})$ has the representation AC for $C = C_1 \cap C_2$ with $C_1 = \mathbb{B} \times \mathbb{R}^m$ and $C_2 = \operatorname{gph} H$. From 11.24(c) we get $\sigma_{H(\mathbb{B})}(y) = \sigma_C(A^*y) = \sigma_C\big((0, y)\big)$. On the other hand we have $C_2 \cap \operatorname{int} C_1 \neq \emptyset$, so $\sigma_C\big((0, y)\big) = \min_{(w,u)}\big\{\sigma_{C_1}\big((0, y) - (w, u)\big) + \sigma_{C_2}\big((w, u)\big)\big\}$ by 11.24(e). Next we note that $\sigma_{C_1}\big((w, u)\big) = |w| + \delta_{\{0\}}(u)$ (cf. 8.27), whereas $\sigma_{C_2} = \delta_{(\operatorname{gph} H)^*}$ by 11.4(b), so that $\sigma_{C_2}\big((w, u)\big) = \delta_{\operatorname{gph} H^{*-}}\big((-w, u)\big)$ by the definition of H^{*-} in 8(28). Therefore,

$$\begin{aligned} \sigma_{H(\mathbb{B})}(y) &= \min_{(w,u)}\Big\{|0 - w| + \delta_{\{0\}}(y - u) + \delta_{\operatorname{gph} H^{*-}}\big((-w, u)\big)\Big\} \\ &= \min_{-w \in H^{*-}(y)} |w| = d\big(0, H^{*-}(y)\big). \end{aligned}$$

Thus, $|H|^+ = |H^{*+}|^-$ is confirmed. To get the remaining formulas, we merely have to apply the ones already obtained to $G = H^{*+}$, since $G^{*-} = H$.

The application at the end is based on having, in the presence of graphical regularity, $D^*S(\bar{x}\,|\,\bar{u}) = DS(\bar{x}\,|\,\bar{u})^{*-}$ and likewise $D^*S^{-1}(\bar{x}\,|\,\bar{u}) = DS^{-1}(\bar{x}\,|\,\bar{u})^{*-}$;

cf. 8.40. It's elementary that one always has $|D^*S^{-1}(\bar{x}\,|\,\bar{u})|^+ = |D^*S(\bar{x}\,|\,\bar{u})^{-1}|^+$ and $|DS^{-1}(\bar{x}\,|\,\bar{u})|^- = |DS(\bar{x}\,|\,\bar{u})^{-1}|^-$. □

11.30 Exercise (uniform boundedness of sublinear mappings). *If a family $\mathcal{H}$ of osc, sublinear mappings $H : \mathbb{R}^n \rightrightarrows \mathbb{R}^m$ has $\sup_{H\in\mathcal{H}} d\big(0, H(w)\big) < \infty$ for each $w \in \mathbb{R}^n$, it must actually have $\sup_{H\in\mathcal{H}} |H|^- < \infty$.*

Guide. Make use of 11.29, arguing that $h(w) = \sup_{H\in\mathcal{H}} d\big(0, H(w)\big)$ is the support function of the set $\bigcup_{H\in\mathcal{H}} H^{*+}(\mathbb{B})$. □

11.31 Exercise (duality in calculating adjoints of sublinear mappings).

(a) *For sublinear mappings $H : \mathbb{R}^n \rightrightarrows \mathbb{R}^m$ and $G : \mathbb{R}^m \rightrightarrows \mathbb{R}^p$ with $\operatorname{rint}(\operatorname{rge} H) \cap \operatorname{rint}(\operatorname{dom} G) \neq \emptyset$, one has*

$$(G\circ H)^{*+} = H^{*+}\circ G^{*+}, \qquad (G\circ H)^{*-} = H^{*-}\circ G^{*-}.$$

(b) *For sublinear mappings $H : \mathbb{R}^n \rightrightarrows \mathbb{R}^m$ and $G : \mathbb{R}^n \rightrightarrows \mathbb{R}^m$ with $\operatorname{rint}(\operatorname{dom} H) \cap \operatorname{rint}(\operatorname{dom} G) \neq \emptyset$, one has*

$$(H+G)^{*+} = H^{*+} + G^{*+}, \qquad (H+G)^{*-} = H^{*-} + G^{*-}.$$

(c) *For a sublinear mapping $H : \mathbb{R}^n \rightrightarrows \mathbb{R}^m$ and arbitrary $\lambda > 0$, one has*

$$(\lambda H)^{*+} = \lambda H^{*+}, \qquad (\lambda H)^{*-} = \lambda H^{*-}.$$

Guide. In (a), $\operatorname{gph}(G\circ H) = L(K)$ for $K = \big[\operatorname{gph} H \times \mathbb{R}^p\big] \cap \big[\mathbb{R}^n \times \operatorname{gph} G\big]$ and L the projection of $\mathbb{R}^n \times \mathbb{R}^m \times \mathbb{R}^p$ onto $\mathbb{R}^n \times \mathbb{R}^p$. Apply the rules in 11.25(b)(c), relating the polars $\operatorname{gph} H$, $\operatorname{gph} G$ and $\operatorname{gph}(G\circ H)$, to the graphs of the adjoint mappings by way of definitions 8(27) and 8(28).

In (b) use the representation $\operatorname{gph}(H+G) = L_2\big(L_1^{-1}(K)\big)$ for the cone $K = (\operatorname{gph} H) \times (\operatorname{gph} G)$, the linear mapping $L_1 : (x,y,z) \mapsto (x,y,x,z)$ and the linear mapping $L_2 : (x,y,z) \mapsto (x, y+z)$. Apply the rules in 11.25(c)(d). Get the elementary fact in (c) straight from the definitions of H^{*+} and H^{*-}. □

For the pairs of operations in Theorem 11.23 to be fully dual to each other, closures had to be taken, but a readily verifiable condition was provided under which this was superfluous. Another common case where it can be omitted comes up when the functions are piecewise linear-quadratic.

11.32 Proposition (operations on piecewise linear-quadratic functions).

(a) *If $f = f_1 \# f_2$ with f_i proper, convex and piecewise linear-quadratic, then f is proper, convex and piecewise linear-quadratic, unless f is improper with $\operatorname{dom} f$ a polyhedral set on which $f \equiv -\infty$. Either way, f is lsc.*

(b) *If $g(u) = (Af)(u) = \inf_x\big\{f(x) \,\big|\, Ax = u\big\}$ for $A \in \mathbb{R}^{m\times n}$ and f proper, convex and piecewise linear-quadratic on $\mathbb{R}^n$, then g is proper, convex and piecewise linear-quadratic on $\mathbb{R}^m$, unless g is improper with $\operatorname{dom} g$ a polyhedral set on which $g \equiv -\infty$. Either way, g is lsc.*

(c) *If $p(u) = \inf_x f(x,u)$ for a proper, convex, piecewise linear-quadratic function $f : \mathbb{R}^n \times \mathbb{R}^m \to \overline{\mathbb{R}}$, then p is proper, convex and piecewise linear-*

quadratic, unless p is improper with $\operatorname{dom} p$ *a polyhedral set on which* $p \equiv -\infty$. *Either way, p is lsc.*

Proof. Let's deal with (c) first; it will be leveraged into the other assertions. Because f is convex, its inf-projection p is convex; cf. 2.22(a). The set $\operatorname{dom} p$ is the image of $\operatorname{dom} f$ under the projection mapping $(x, u) \mapsto u$, and $\operatorname{dom} f$ is polyhedral. Hence $\operatorname{dom} p$ is itself polyhedral and in particular closed; cf. 3.55(a). The function conjugate to p is known from 11.23(c) to be $f^*(0, \cdot)$, which is not only convex but piecewise linear-quadratic by 10.22(c), due to the fact that f^* inherits being piecewise linear-quadratic from f by 11.14(b). As long as $f^*(0, \cdot)$ is proper, which is equivalent by 11.1 to p being proper, it follows further by 11.14(b) that p^{**}, as the function conjugate to $f^*(0, \cdot)$, is piecewise linear-quadratic too. The claim in (c) can be established therefore in showing that $p^{**} = p$ unless p is improper with no values other than $\pm\infty$.

Suppose p is finite at $\bar{u}$. The function $\varphi := f(\cdot, \bar{u})$, whose infimum over $\mathbb{R}^n$ is $p(\bar{u})$, is convex and piecewise linear-quadratic, hence its minimum is attained at some $\bar{x}$; cf. 11.16. Then $0 \in \partial\varphi(\bar{x})$. But $\partial\varphi(\bar{x}) = \{v \mid \exists\, y,\ (v, y) \in \partial f(\bar{x}, \bar{u})\}$ by 10.22(c). Hence there exists $\bar{y}$ with $(0, \bar{y}) \in \partial f(\bar{x}, \bar{u})$. Through convexity this subgradient condition can be written as

$$f(x, u) \geq f(\bar{x}, \bar{u}) + \langle (0, \bar{y}),\, (x, u) - (\bar{x}, \bar{u}) \rangle \quad \text{for all } (x, u)$$

(see 8.12), which gives us $p(u) \geq p(\bar{u}) + \langle \bar{y},\, u - \bar{u} \rangle$ for all u, implying that p is proper on $\mathbb{R}^m$ and lsc at $\bar{u}$. Thus, unless $p \equiv -\infty$ on $\operatorname{dom} p$, p is a proper, convex function which is lsc at every point of $\operatorname{dom} p$. Since $\operatorname{dom} p$ is closed, we conclude that p is lsc everywhere, so $p^{**} = p$. This proves (c).

We get (b) now as the case of (c) where $Af = p$ for $p(u) = \inf_x \bar{f}(x, u)$ with $\bar{f}(x, u) = f(x) + \delta_M(x, u)$ and $M = \{(x, u) \mid Ax = u\}$. Because δ_M, like f, is convex and piecewise linear-quadratic (the set M being affine, hence polyhedral; cf. 2.10), $\bar{f}$ is piecewise linear-quadratic by 10.22.

Finally, (a) specializes (b) to $f_1 \# f_2 = Ag$ with $g(x_1, x_2) = f_1(x_1) + f_2(x_2)$ on $\mathbb{R}^n \times \mathbb{R}^n$ and A the matrix of the linear mapping $(x_1, x_2) \mapsto x_1 + x_2$. □

11.33 Corollary (conjugate formulas in piecewise linear-quadratic case).

(a) *If $f = f_1 + f_2$ with f_i proper, convex and piecewise linear-quadratic, and if $f \not\equiv \infty$, then $f^* = f_1^* \# f_2^*$.*

(b) *If $f = gA$ with $A \in \mathbb{R}^m \times \mathbb{R}^n$ and g proper, convex and piecewise linear-quadratic, and if $f \not\equiv \infty$, then $f^* = A^* g^*$.*

(c) *If $\varphi(x) = f(x, \bar{u})$ with f proper, convex and piecewise linear-quadratic on $\mathbb{R}^n \times \mathbb{R}^m$, and if $\varphi \not\equiv \infty$, then $\varphi^*(v) = \inf_y \{f^*(v, y) - \langle y, \bar{u} \rangle\}$.*

Proof. We apply 11.23 in the light of the additional conditions furnished by 11.32 for the dual operations to produce a lower semicontinuous function, so that the closure operation can be omitted. □

G. Duality in Convergence

Switching to another topic, we look next at how the Legendre-Fenchel transform behaves with respect to the epi-convergence studied in Chapter 7.

11.34 Theorem (epi-continuity of the Legendre-Fenchel transform; Wijsman). *If the functions f^ν and f on $\mathbb{R}^n$ are proper, lsc, and convex, one has*

$$f^\nu \xrightarrow{\text{e}} f \quad\Longleftrightarrow\quad f^{\nu*} \xrightarrow{\text{e}} f^*.$$

More generally, as long as e-lim inf$_\nu f^\nu$ *nowhere takes on* $-\infty$ *and a bounded set B exists with* $\limsup_\nu[\inf_B f^\nu] < \infty$, *one has*

$$\begin{aligned} \text{e-}\liminf_\nu f^\nu \geq f \quad&\Longleftrightarrow\quad \text{e-}\limsup_\nu f^{\nu*} \leq f^*,\\ \text{e-}\limsup_\nu f^\nu \leq f \quad&\Longleftrightarrow\quad \text{e-}\liminf_\nu f^{\nu*} \geq f^*. \end{aligned}$$

Proof. For any $\lambda > 0$ we have through 7.37 that $f^\nu \xrightarrow{\text{e}} f$ if and only if $e_\lambda f^\nu \xrightarrow{\text{p}} e_\lambda f$. Likewise, $f^{\nu*} \xrightarrow{\text{e}} f$ if and only if $e_\lambda f^{\nu*} \xrightarrow{\text{p}} e_\lambda f^*$. In particular we can take $\lambda = 1$ in these conditions. But from 11.26 we have

$$e_1 f^\nu(x) + e_1 f^{\nu*}(x) = \tfrac{1}{2}|x|^2 = e_1 f(x) + e_1 f^*(x).$$

Thus, the two conditions are equivalent.

The assumption about B in the more general case means the existence of $\beta \in \mathbb{R}$ such that $\operatorname{epi} f^\nu \cap (B \times (-\infty, \beta]) \neq \emptyset$ for all ν in some index set in $\mathcal{N}_\infty$. This ensures that no subsequence of $\{f^\nu\}_{\nu\in\mathbb{N}}$ can escape epigraphically to the horizon; cf. 7.5. Then every subsequence has an epi-convergent subsequence by 7.6, but on the other hand, the epi-limit of such a sequence can't take on $-\infty$ because of the assumption about e-lim inf$_\nu f^\nu$. We are therefore in a setting where every subsequence of $\{f^\nu\}_{\nu\in\mathbb{N}}$ has a subsequence that epi-converges to some proper, lsc function g, which must of course be convex. Through the cluster description of outer limits in 4.19 as applied to epigraphs, we see that e-lim inf$_\nu f^\nu \geq f$ if and only if $g \geq f$ for every such function g; likewise in terms of inner limits, we have e-lim sup$_\nu f^\nu \leq f$ if and only if $g \geq f$ for every such g. It remains only to invoke for these epi-convergent sequences the continuity property established in the main part of the theorem. □

11.35 Corollary (convergence of support functions and polar cones).

(a) *For nonempty, closed, convex sets C^ν and C in $\mathbb{R}^n$, one has*

$$C^\nu \to C \quad\Longleftrightarrow\quad \sigma_{C^\nu} \xrightarrow{\text{e}} \sigma_C,$$

and if the sets are bounded this is equivalent to having $\sigma_{C^\nu}(v) \to \sigma_C(v)$ for each v. More generally, as long as $\limsup_\nu d(0, C^\nu) < \infty$ *one has*

$$\begin{aligned} \limsup_\nu C^\nu \subset C \quad&\Longleftrightarrow\quad \text{e-}\limsup_\nu \sigma_{C^\nu} \leq \sigma_C,\\ \liminf_\nu C^\nu \supset C \quad&\Longleftrightarrow\quad \text{e-}\liminf_\nu \sigma_{C^\nu} \geq \sigma_C. \end{aligned}$$

(b) *For closed, convex cones K^ν and K in $\mathbb{R}^n$, one has*

$$K^\nu \to K \iff K^{\nu *} \to K^*,$$

and more generally,

$$\begin{aligned} \limsup_\nu K^\nu \subset K &\iff \liminf_\nu K^{\nu *} \supset K^*, \\ \liminf_\nu K^\nu \supset K &\iff \limsup_\nu K^{\nu *} \subset K^*. \end{aligned}$$

Proof. We apply Theorem 11.34 in the setting of 11.4. The case of bounded C^ν, in which the support functions are finite-valued, appeals also to 7.18. □

The Legendre-Fenchel transform isn't just continuous; it's an *isometry* with respect to a suitable choice of metric on the space of proper, lsc, convex functions on $\mathbb{R}^n$. We'll establish that by developing isometric properties of the polarity correspondence in 6.24 and 11.4(b) and then appealing to the way that conjugate functions can be associated with polar cones, as in 11.7. In this geometric context, the set metric $d\!l$ in 4(12) will be apt.

11.36 Theorem (cone polarity as an isometry; Walkup-Wets). *For any cones $K_1, K_2 \subset \mathbb{R}^n$ that are convex, one has*

$$d\!l(K_1, K_2) = d\!l(K_1^*, K_2^*).$$

Proof. It can be assumed that K_1 and K_2 are closed, since $d\!l(K_1, K_2) = d\!l(\operatorname{cl} K_1, \operatorname{cl} K_2)$ with $\operatorname{cl} K_i$ convex and $[\operatorname{cl} K_i]^* = K_i^*$. Then $[K_i^*]^* = K_i$ by 11.4(b). We have $d\!l(K_1, K_2) = d\!l_1(K_1, K_2) = \hat{d\!l}_1(K_1, K_2)$ and $d\!l(K_1^*, K_2^*) = d\!l_1(K_1^*, K_2^*) = \hat{d\!l}_1(K_1^*, K_2^*)$ by 4.44. Thus, we need only demonstrate that $d\!l_1(K_1, K_2) \le \hat{d\!l}_1(K_1^*, K_2^*)$ and $d\!l_1(K_1^*, K_2^*) \le \hat{d\!l}_1(K_1, K_2)$; but the latter is just the former as applied to K_1^* and K_2^*, so the former suffices. Because of the way that K_1 and K_2 enter symmetrically in the definition of $d\!l_1(K_1, K_2)$ and $\hat{d\!l}_1(K_1^*, K_2^*)$ in 4(11), our task reduces simply to verifying for $\eta > 0$ that

$$d_{K_1} \le d_{K_2} + \eta \text{ on } \mathbb{B} \implies K_1^* \cap \mathbb{B} \subset K_2^* + \eta \mathbb{B}. \tag{11(14)}$$

The convex functions d_{K_1} and d_{K_2} are positively homogeneous, so the left side of 11(14) corresponds to the inequality $d_{K_1} \le d_{K_2} + \eta|\cdot|$ holding on $\mathbb{R}^n$, or equivalently $d_{K_1}^* \ge (d_{K_2} + \eta|\cdot|)^*$. Here $d_{K_1} = \delta_{K_1} \# |\cdot|$ and $d_{K_2} = \delta_{K_2} \# |\cdot|$ (cf. 1.20), while $|\cdot| = \sigma_{\boldsymbol{B}}$ (cf. 8.27) and $\eta|\cdot| = \sigma_{\eta \boldsymbol{B}}$, so the inequality in 11(14) dualizes to $\delta_{K_1}^* + \sigma_{\boldsymbol{B}}^* \ge [\delta_{K_2}^* + \sigma_{\boldsymbol{B}}^*] \# \sigma_{\eta \boldsymbol{B}}^*$ through the rules in 11.23(a). By 11.4, this is the same as $\delta_{K_1^*} + \delta_{\boldsymbol{B}} \ge [\delta_{K_2^*} + \delta_{\boldsymbol{B}}] \# \delta_{\eta \boldsymbol{B}}$, or in other words $K_1^* \cap \mathbb{B} \subset [K_2^* \cap \mathbb{B}] + \eta \mathbb{B}$, which implies the right side of 11(14). □

Theorem 11.36 can be applied to convex cones in the ray space model of cosmic space, most fruitfully to the cones in $\mathbb{R}^{n+2}$ that represent the epigraphs in $\mathbb{R}^{n+1}$ of convex functions on $\mathbb{R}^n$. Since the cosmic metric $d\!l_{\rm csm}$ for functions on $\mathbb{R}^n$, as defined in 7(28), is based on the set distance between such cones, the following result is obtained.

11.37 Corollary (Legendre-Fenchel transform as an isometry). *For functions $f_1, f_2 : \mathbb{R}^n \to \overline{\mathbb{R}}$ that are convex and proper, one has*

$$d\!l_{\rm csm}(f_1, f_2) = d\!l_{\rm csm}(f_1^*, f_2^*).$$

Proof. This comes out of the characterization of conjugate functions in terms of polar cones in 11.7. The cosmic epi-distances correspond to cone distances, and the isometry is apparent then from the one for cones in Theorem 11.36. □

It might be puzzling that this result is based on the cosmic epi-metric $d\!l_{\rm csm}$, which is known from 7.59 to characterize total epi-convergence, whereas the 'homeomorphism' in Theorem 11.34 refers to ordinary epi-convergence, which according to 7.58 is characterized by the ordinary epi-metric $d\!l$. The seeming discord vanishes when one recalls that, for sequences of convex functions, epi-convergence implies total epi-convergence (cf. 7.53). On the space of *convex* functions within lsc-fcns$_{\not\equiv\infty}(I\!R^n)$, the metrics $d\!l$ and $d\!l_{\rm csm}$ are actually equivalent topologically. Theorem 11.34 could equally well be stated in terms of total epi-convergence, but only $d\!l_{\rm csm}$ produces an isometry.

H. Dual Problems of Optimization

The rest of this chapter will be occupied with the important question of how *optimization problems* can be dualized. It will be shown that any optimization problem of convex type, when provided with a scheme of perturbation that respects convexity, is paired with a certain other optimization problem of convex type, which is provided in turn with a dual scheme of perturbation. The two problems are related to each other in remarkable ways. Even for problems that aren't of convex type, something analogous can be built up, although not as powerfully and not with full symmetry.

Hints of such duality are already present in the formulas we've been developing, and we'll work our way into the subject by looking there first. In principle, the value of the conjugate of a given function can be calculated at a given point by maximization in terms of the defining expression 11(1). But the formulas developed in 11.23, and more specially in 11.34, furnish an alternative approach to calculating the same value by minimization. This idea is captured most succinctly by the operation of inf-projection.

11.38 Lemma (dual calculations in parametric optimization). *For any function* $f : I\!R^n \times I\!R^m \to \overline{I\!R}$, *one has*

$$\left.\begin{array}{rcl} p(0) &=& \inf_x \varphi(x) \\ p^{**}(0) &=& \sup_y \psi(y) \end{array}\right\} \quad \textit{for} \quad \left\{\begin{array}{l} p(u) = \inf_x f(x,u) \\ \varphi(x) = f(x,0) \\ \psi(y) = -f^*(0,y). \end{array}\right.$$

Proof. This is immediate from 11.23(c). □

The circumstances under which $p(0) = p^{**}(0)$, and therefore $\inf \varphi = \sup \psi$ in this scheme, are of course governed in general by 11.1 and 11.2 and are rich in possibilities. The focus in 11.38 on $u = 0$ enhances symmetry and corresponds to interpreting u as a *perturbation* parameter. Much will be made of this perturbation idea as we proceed.

The next theorem develops out of 11.38 a symmetric framework in which some of the most distinguishing features of optimization problems of *convex* type find expression. Later we'll explore the extent to which dualization can be effective through 11.38 even for problems involving nonconvexities.

11.39 Theorem (dual problems of optimization). *Let $f : \mathbb{R}^n \times \mathbb{R}^m \to \overline{\mathbb{R}}$ be proper, lsc and convex, and consider the primal problem*

$$\text{minimize } \varphi \text{ on } \mathbb{R}^n, \qquad \varphi(x) := f(x, 0),$$

along with the dual problem

$$\text{maximize } \psi \text{ on } \mathbb{R}^m, \qquad \psi(y) := -f^*(0, y),$$

where φ is convex and lsc, but ψ is concave and usc. Let $p(u) = \inf_x f(x, u)$ and $U = \operatorname{dom} p$, while $q(v) = \inf_y f^*(v, y)$ and $V = \operatorname{dom} q$; these sets and functions are convex.

(a) *The inequality $\inf_x \varphi(x) \geq \sup_y \psi(y)$ holds always, and $\inf_x \varphi(x) < \infty$ if and only if $0 \in U$, whereas $\sup_y \psi(y) > -\infty$ if and only if $0 \in V$. Moreover,*

$$\inf_x \varphi(x) = \sup_y \psi(y) \quad \text{if either } 0 \in \operatorname{int} U \text{ or } 0 \in \operatorname{int} V.$$

(b) *The set $\operatorname{argmax}_y \psi(y)$ is nonempty and bounded if and only if $0 \in \operatorname{int} U$ and the value $\inf_x \varphi(x) = p(0)$ is finite, in which case $\operatorname{argmax}_y \psi(y) = \partial p(0)$.*

(c) *The set $\operatorname{argmin}_x \varphi(x)$ is nonempty and bounded if and only if $0 \in \operatorname{int} V$ and the value $\sup_y \psi(y) = -q(0)$ is finite, in which case $\operatorname{argmin}_x \varphi(x) = \partial q(0)$.*

(d) *Optimal solutions are characterized jointly through primal and dual forms of Fermat's rule:*

$$\left.\begin{aligned} \bar{x} &\in \operatorname{argmin}_x \varphi(x) \\ \bar{y} &\in \operatorname{argmax}_y \psi(y) \\ \inf_x \varphi(x) &= \sup_y \psi(y) \end{aligned}\right\} \iff (0, \bar{y}) \in \partial f(\bar{x}, 0) \iff (\bar{x}, 0) \in \partial f^*(0, \bar{y}).$$

$\varphi(x) := f(x, 0)$	$p(u) := \inf_x f(x, u)$	$U := \operatorname{dom} p$
$\uparrow *$	$\downarrow *$	
$q(v) := \inf_y f^*(v, y)$	$-\psi(y) := f^*(0, y)$	$V := \operatorname{dom} q$

Fig. 11–5. Notation for dual problems of convex type.

Proof. The convexity in the preamble is obvious from that of f and f^*. (The preservation of convexity under inf-projection is attested to by 2.22(a).)

We apply 11.23(c) in the context of 11.38, noting from 11.1 and 11.2 that $p(0) = p^{**}(0)$ in particular when $0 \in \operatorname{int} U$. The latter condition in combination with $p(0) > -\infty$ is equivalent by 11.8(c) to $-\psi$ being proper and level-bounded. But a proper, lsc, convex function is level-bounded if and only if its argmin set is nonempty and bounded; cf. 3.27 and 1.9. Then too, $\partial p(0) = \operatorname{argmin} p^* = \operatorname{argmax} \psi$ by 11.8(a).

Next we invoke the same facts with f replaced by f^*, using the relation $f^{**} = f$. This gives $-\sup \psi = q(0) \geq q^{**}(0) = -\inf \varphi$ with $\varphi = q^*$, where $q(0) = q^{**}(0)$ in particular when $0 \in \operatorname{int} V$. The latter condition in combination with $q(0) > -\infty$ corresponds by parallel argument to having $\operatorname{argmin} \varphi$ being nonempty and bounded. It also gives $\partial q(0) = \operatorname{argmin} q^* = \operatorname{argmin} \varphi$.

Turning to (d), we note that through 11.3 the relations $(0, \bar{y}) \in \partial f(\bar{x}, 0)$ and $(\bar{x}, 0) \in \partial f^*(0, \bar{y})$ are equivalent to each other and to having $\varphi(\bar{x}) = \psi(\bar{y})$. Since $\inf \varphi \geq \sup \psi$ in general by (a), they are equivalent further to having $\varphi(\bar{x}) = \inf \varphi = \sup \psi = \psi(\bar{y})$. □

11.40 Corollary (general best-case primal-dual relations). *In the context of Theorem 11.39, the following conditions are equivalent to each other and serve to guarantee that* $-\infty < \min \varphi = \max \psi < \infty$*:*

(a) $0 \in \operatorname{int} U$ *and* $0 \in \operatorname{int} V$*;*

(b) $0 \in \operatorname{int} U$*, and* $\operatorname{argmin} \varphi$ *is nonempty and bounded;*

(c) $0 \in \operatorname{int} V$*, and* $\operatorname{argmax} \psi$ *is nonempty and bounded;*

(d) $\operatorname{argmin} \varphi$ *and* $\operatorname{argmax} \psi$ *are nonempty and bounded.*

The relation $\operatorname{argmax} \psi = \partial p(0)$ in 11.39(b) shows the significance of optimal solutions $\bar{y}$ to the dual problem. In the situation in 11.39(b), the convex function p is finite on a neighborhood of 0, so the relation tells us that

$$dp(0)(w) = \max\{\langle \bar{y}, w \rangle \mid \bar{y} \in \operatorname{argmax} \psi\}$$

(cf. 8.30, 7.27) and further that a unique optimal solution $\bar{y}$ to the dual problem corresponds to having $\nabla p(0) = \bar{y}$ (cf. 9.18, 9.14). All this has to be seen in the light of $p(0)$ being the optimal value in the given 'primal' problem of minimizing φ, with $p(u)$ the optimal value obtained when this problem is perturbed by the amount u in the way prescribed by the chosen parametric representation.

This kind of interpretation of the dual elements $\bar{y}$ accompanying the primal elements $\bar{x}$ resembles one that was discussed in parametric minimization more generally (cf. 10.14 and 10.15), but without a dual *problem* being brought in for a supplementary description of the vectors $\bar{y}$ as optimal in their own right: $\bar{y} \in \operatorname{argmax} \psi$. In the present setup, fortified by convexity and the relation $\operatorname{argmin} \varphi = \partial q(0)$ in 11.39(c), the roles of $\bar{x}$ and $\bar{y}$ can be interchanged. Remarkably, the solutions $\bar{x} \in \operatorname{argmin} \varphi$ to the primal problem gain a parallel interpretation relative to perturbations to the dual problem, namely:

$$dq(0)(z) = \max\{\langle \bar{x}, z \rangle \mid \bar{x} \in \operatorname{argmin} \varphi\}.$$

Because $f^{**} = f$, everything is completely symmetric between primal and dual, apart from the sign convention in designating whether a problem should be viewed in maximization or minimization mode. The dualization scheme proceeds from a primal problem with perturbation vector u to a dual problem with perturbation vector v, and it does so in such a way that *the dual of the dual problem is identical to the primal problem.*

11.41 Example (Fenchel-type duality scheme). *Consider the two problems*

$$\begin{array}{ll} \text{minimize } \varphi \text{ on } I\!R^n, & \varphi(x) := \langle c, x\rangle + k(x) + h(b - Ax),\\ \text{maximize } \psi \text{ on } I\!R^m, & \psi(y) := \langle b, y\rangle - h^*(y) - k^*(A^*y - c), \end{array}$$

where $k : I\!R^n \to \overline{I\!R}$ and $h : I\!R^m \to \overline{I\!R}$ are proper, lsc and convex, and one has $A \in I\!R^{m\times n}$, $b \in I\!R^m$, and $c \in I\!R^n$. These problems fit the format of 11.39 with

$$\begin{aligned} f(x,u) &= \langle c, x\rangle + k(x) + h(b - Ax + u),\\ f^*(v,y) &= -\langle b, y\rangle + h^*(y) + k^*(A^*y - c + v). \end{aligned}$$

The assertions of 11.39 and 11.40 are valid then in the context of having

$$\begin{aligned} 0 \in \operatorname{int} U &\iff b \in \operatorname{int}\big(A \operatorname{dom} k + \operatorname{dom} h\big),\\ 0 \in \operatorname{int} V &\iff c \in \operatorname{int}\big(A^*\operatorname{dom} h^* - \operatorname{dom} k^*\big). \end{aligned}$$

Furthermore, optimal solutions are characterized by

$$\left.\begin{array}{c} \bar{x} \in \operatorname{argmin} \varphi\\ \bar{y} \in \operatorname{argmax} \psi\\ \inf \varphi = \sup \psi \end{array}\right\} \iff \left\{\begin{array}{c} \bar{y} \in \partial h(b - A\bar{x})\\ A^*\bar{y} - c \in \partial k(\bar{x}) \end{array}\right\} \iff \left\{\begin{array}{c} \bar{x} \in \partial k^*(A^*\bar{y} - c)\\ b - A\bar{x} \in \partial h^*(\bar{y}) \end{array}\right\}.$$

Detail. The function f is proper, lsc (by 1.39, 1.40) and convex (by 2.18, 2.20). The function p, defined by $p(u) := \inf_x f(x,u)$, has nonempty effective domain $U = A \operatorname{dom} k + \operatorname{dom} h - b$. Direct calculation of f^* yields

$$\begin{aligned} f^*(v,y) &= \sup_{x,u}\big\{\langle v, x\rangle + \langle y, u\rangle - \langle c, x\rangle - k(x) - h(b - Ax + u)\big\}\\ &= \sup_{x,w}\big\{\langle v, x\rangle + \langle y,\, w - b + Ax\rangle - \langle c, x\rangle - k(x) - h(w)\big\}\\ &= \sup_{x}\big\{\langle A^*y - c + v, x\rangle - k(x)\big\} + \sup_{w}\big\{\langle y, w\rangle - h(w)\big\} - \langle y, b\rangle\\ &= k^*(A^*y - c + v) + h^*(y) - \langle y, b\rangle \end{aligned}$$

as claimed. The effective domain of the function $q(v) = \inf_y f^*(v,y)$ is then $V = \operatorname{dom} k^* - A^* \operatorname{dom} h^* + c$.

To determine $\partial f(x,u)$ so as to verify the optimality condition claimed, we write $f = g\circ F$ for $g(x,w) = \langle c, x\rangle + k(x) + h(b+w)$ and $F(x,u) = (x, -Ax+u)$, noting that F is linear and nonsingular. We observe then that $\partial g(x,w) = \big\{(c+v, y) \,\big|\, v \in \partial k(x),\ y \in \partial h(b+w)\big\}$, and therefore through 10.7 that we have $\partial f(x,u) = \big\{(c + v - A^*y, y) \,\big|\, v \in \partial k(x),\ y \in \partial h(b - Ax + u)\big\}$. The condition $(0,\bar{y}) \in \partial f(\bar{x},0)$ in 11.39(d) reduces to having $A^*\bar{y} - c \in \partial k(\bar{x})$ for

$\bar{y} \in \partial h(b - A\bar{x})$. The alternate expression of optimality in terms of ∂k^* and ∂h^* comes immediately then from the inversion principle in 11.3. □

11.42 Theorem (piecewise linear-quadratic optimization). *For a proper, convex and piecewise linear-quadratic function $f : I\!R^n \times I\!R^m \to \overline{I\!R}$, consider the primal and dual problems*

$$\begin{aligned} &\text{minimize } \varphi \text{ on } I\!R^n, \quad \varphi(x) := f(x,0), \\ &\text{maximize } \psi \text{ on } I\!R^m, \quad \psi(y) := -f^*(0,y), \end{aligned}$$

along with the functions $p(u) = \inf_x f(x,u)$ and $q(v) = \inf_y f^*(v,y)$.

If either of the values $\inf \varphi$ or $\sup \psi$ is finite, then both are finite and both are attained. Moreover in that case one has $\inf \varphi = \sup \psi$ and

$$\begin{aligned} (\operatorname{argmin} \varphi) \times (\operatorname{argmax} \psi) &= \big\{ (\bar{x},\bar{y}) \,\big|\, (0,\bar{y}) \in \partial f(\bar{x},0) \big\} \\ &= \big\{ (\bar{x},\bar{y}) \,\big|\, (\bar{x},0) \in \partial f^*(0,\bar{y}) \big\} = \partial q(0) \times \partial p(0). \end{aligned}$$

Proof. This parallels Theorem 11.39, but in coming up with circumstances in which $p^{**}(0) = p(0)$, or $q^{**}(0) = q(0)$, it relies on the piecewise linear-quadratic nature of p and q in 11.32 instead of properties of general convex functions. □

11.43 Example (linear and extended linear-quadratic programming). *The problems in the Fenchel duality scheme in 11.41 fit the framework of piecewise linear-quadratic optimization in 11.42 when the convex functions k and h are piecewise linear-quadratic. A particular case of this, called* extended linear-quadratic programming, *concerns the primal and dual problems*

$$\begin{aligned} &\text{minimize } \langle c,x\rangle + \tfrac{1}{2}\langle x,Cx\rangle + \theta_{Y,B}(b - Ax) \text{ over } x \in X, \\ &\text{maximize } \langle b,y\rangle - \tfrac{1}{2}\langle y,By\rangle - \theta_{X,C}(A^*y - c) \text{ over } y \in Y, \end{aligned}$$

where the sets $X \subset I\!R^n$ and $Y \subset I\!R^m$ are nonempty and polyhedral, the matrices $C \in I\!R^{n\times n}$ and $B \in I\!R^{m\times m}$ are symmetric and positive-semidefinite, and the functions $\theta_{Y,B}$ and $\theta_{X,C}$ have expressions as in 11.18:

$$\theta_{Y,B}(u) = \sup_{y\in Y}\Big\{ \langle y,u\rangle - \tfrac{1}{2}\langle y,By\rangle \Big\}, \qquad \theta_{X,C}(v) = \sup_{x\in X}\Big\{ \langle v,x\rangle - \tfrac{1}{2}\langle x,Cx\rangle \Big\}.$$

When $C = 0$ and $B = 0$, while X and Y are cones, these primal and dual problems reduce to the linear programming problems

$$\begin{aligned} &\text{minimize } \langle c,x\rangle \text{ subject to } x \in X,\ b - Ax \in Y^*, \\ &\text{maximize } \langle b,y\rangle \text{ subject to } y \in Y,\ A^*y - c \in X^*. \end{aligned}$$

If in addition $X = I\!R^n_+$ and $Y = I\!R^m_+$, the linear programming problems have the form

$$\begin{aligned} &\text{minimize } \langle c,x\rangle \text{ subject to } x \geq 0,\ Ax \geq b, \\ &\text{maximize } \langle b,y\rangle \text{ subject to } y \geq 0,\ A^*y \leq c. \end{aligned}$$

More generally, if $X = I\!R^r_+ \times I\!R^{n-r}$ and $Y = I\!R^s_+ \times I\!R^{m-s}$ these linear programming problems have mixed inequality and equality constraints.

Detail. When k and h are piecewise linear-quadratic in the Fenchel scheme in 11.41, so too is f by the calculus in 10.22. Then Theorem 11.42 is applicable. In the special case described, we have $k = \delta_X + j_C$ for $j_C(x) := \frac{1}{2}\langle x, Cx\rangle$, whereas $h = (\delta_Y + j_B)^*$; cf. 11.18. □

Between linear programming and extended linear-quadratic programming in 11.43 is *quadratic programming*, where the primal problem takes the form

$$\text{minimize } \langle c, x\rangle + \tfrac{1}{2}\langle x, Cx\rangle \text{ subject to } x \in X,\ b - Ax \in Y^*,$$

for some choice of polyhedral cones $X \subset I\!R^n$, $Y \subset I\!R^m$ (such as $X = I\!R^n_+$, $Y = I\!R^m_+$) and a positive semidefinite matrix $C \in I\!R^{n\times n}$. This corresponds to the primal problem of extended linear-quadratic programming of 11.43 in the case of $B = 0$ and dualizes to

$$\text{maximize } \langle b, y\rangle - \theta_{X,C}(A^*y - c) \text{ over } y \in Y.$$

Thus, *the dual of a quadratic programming problem isn't another quadratic programming problem*, except in the linear programming subcase.

The general duality framework in Theorem 11.39 can be useful also in the derivation of conjugacy formulas. The next example shows this while demonstrating how the criteria in 11.39 can be verified in a particular case.

11.44 Example (dualized composition). *Let $g : I\!R^n \to \overline{I\!R}$ and $\theta : I\!R \to \overline{I\!R}$ be lsc, proper and convex with $\operatorname{dom}\theta \subset I\!R_+$ and $\lim_{\lambda\to\infty}\theta(\lambda)/\lambda > -g(0)$. With $g^\infty(z)$ replacing $\lambda g(\lambda^{-1}z)$ when $\lambda = 0$, one has for all $z \in I\!R^n$ that*

$$\inf_{\lambda\geq 0}\{\theta(\lambda) + \lambda g(\lambda^{-1}z)\} = (\theta^*\circ g^*)^*(z).$$

Detail. The assumption on $\operatorname{dom}\theta$ is equivalent to θ^* being nondecreasing; cf. 8.51. In particular, therefore, $\theta^*\circ g^*$ is another convex function; cf. 2.20(b).

Fixing any z, define $f(\lambda, u)$ to be $\theta(\lambda) + h(\lambda, z + u)$, with $h(\lambda, w) = \lambda g(\lambda^{-1}w)$ when $\lambda > 0$ but $g^\infty(w)$ when $\lambda = 0$; set $h(\lambda, w) = \infty$ when $\lambda < 0$. The left side of the claimed formula is then the optimal value in the problem of minimizing $\varphi(\lambda) = f(\lambda, 0)$ over $\lambda \in I\!R^1$.

The function f is lsc, proper and convex by 3.49(c), so we're in the territory of Theorem 11.39. To proceed, we have to determine f^*. Calculating from the definition (with μ as the variable dual to λ), we get

$$\begin{aligned} f^*(\mu, y) &= \sup_{\lambda,u}\Big\{\langle(\mu, y), (\lambda, u)\rangle - f(\lambda, u)\Big\} \\ &= \sup_{\lambda}\Big\{\mu\lambda - \theta(\lambda) + \sup_{w}\{\langle y, w - z\rangle - h(\lambda, w)\}\Big\}, \end{aligned} \qquad 11(15)$$

where the inner supremum only has to be analyzed for $\lambda \geq 0$, inasmuch as $\operatorname{dom}\theta \subset I\!R_+$. For $\lambda > 0$ it comes out as

$$\sup_w \{\langle y, w - z\rangle - (\lambda \star g)(w)\} = \lambda g^*(y) - \langle y, z\rangle,$$

cf. 11(3). For $\lambda = 0$, we use the fact in 11.5 that g^∞ is the support function of $D = \operatorname{dom} g^*$ in order to see that the inner supremum is

$$\sup_w \{\langle y, w - z\rangle - \sigma_D(w)\} = \delta_{\operatorname{cl} D}(y) - \langle y, z\rangle,$$

cf. 11.4(a). Substituting these into the last part of 11(15), we find that

$$f^*(\mu, y) = \theta^*\big(g^*(y) + \mu\big) - \langle y, z\rangle$$

(with $\theta^*(\infty)$ interpreted as ∞). The dual problem, which consists of maximizing $\psi(y) = -f^*(0, y)$ over $y \in I\!R^n$, therefore has optimal value $\sup \psi = (\theta^* \circ g^*)^*(z)$, which is the value on the right side of the claimed formula.

We can obtain the desired conclusion by verifying that $\inf \varphi = \sup \psi$. A criterion is provided for this purpose in 11.39(a): it suffices to know that $0 \in \operatorname{int}\{\mu \mid \exists y \text{ with } g^*(y) + \mu \in \operatorname{dom} \theta^*\}$. Because θ^* is nondecreasing, $\operatorname{dom} \theta^*$ is an interval that's unbounded below; its right endpoint (maybe ∞) is $\theta^\infty(1)$ by 11.5. What we need to have is $\inf g^* < \theta^\infty(1)$. But $\inf g^* = -g(0)$ by 11.8(a) as applied to g^*. The criterion thus means $-g(0) < \theta^\infty(1)$. Since $\theta^\infty(1) = \lim_{\lambda \nearrow \infty} \theta(\lambda)/\lambda$, we've reached our goal. □

I. Lagrangian Functions

Duality in the elegant style of Theorem 11.39 and its accompaniment is a feature of optimization problems of convex type only. In general, for an optimization problem represented as minimizing $\varphi = f(\cdot, 0)$ over $I\!R^n$ for a function $f : I\!R^n \times I\!R^m \to \overline{I\!R}$, regardless of convexity, we speak of the problem of maximizing $\psi = -f^*(0, \cdot)$ over $I\!R^n$ as the associated *dual* problem, in contrast to the given problem as the *primal* problem, but the relationships might not be as tight as the ones we've been seeing. The issue of whether $\inf \varphi = \sup \psi$ comes down to whether $p(0) = p^{**}(0)$ for $p(u) = \inf_x f(x, u)$, as observed in 11.38. The trouble is that in the absence of p being convex—which is hard to guarantee without simply assuming f is convex—there's no strong handle on whether $p(0) = p^{**}(0)$. We'll nonetheless eventually uncover some facts of considerable power about nonconvex duality and how it can be put to use.

An important step along the way is the study of general 'Lagrangian functions'. That has other motivations as well, most notably in the expression of optimality conditions and the development of methods for computing optimal solutions. Although tradition merely associates Lagrangian functions with constraint systems, their role can go far beyond just that.

The key idea is that a Lagrangian arises through *partial dualization* of a given problem in relation to a particular choice of perturbation parameters.

11.45 Definition (Lagrangians and dualizing parameterizations). *For a problem of minimizing φ on $\mathbb{R}^n$, a dualizing parameterization is a representation $\varphi = f(\cdot,0)$ in terms of a proper function $f : \mathbb{R}^n \times \mathbb{R}^m \to \overline{\mathbb{R}}$ such that $f(x,u)$ is lsc convex in u. The associated Lagrangian $l : \mathbb{R}^n \times \mathbb{R}^m \to \overline{\mathbb{R}}$ is given by*

$$l(x,y) := \inf_u\Big\{f(x,u) - \langle y,u\rangle\Big\}. \qquad 11(16)$$

This definition has its roots in the Legendre-Fenchel transform: for each $x \in \mathbb{R}^n$ the function $-l(x,\cdot)$ is conjugate to $f(x,\cdot)$ on $\mathbb{R}^m$. The conditions on f have the purpose of ensuring that $f(x,\cdot)$ is in turn conjugate to $-l(x,\cdot)$:

$$f(x,u) = \sup_y\Big\{l(x,y) + \langle y,u\rangle\Big\}. \qquad 11(17)$$

Indeed, they require $f(x,\cdot)$ to be proper, lsc and convex, unless $f(x,\cdot) \equiv \infty$; either way, $f(x,\cdot)$ then coincides with its biconjugate by 11.1 and 11.2.

The convexity of $f(x,u)$ in u is a vastly weaker requirement than convexity in (x,u), although it's certainly implied by the latter. The parametric representations in 11.39, 11.41, and 11.42 are dualizing parameterizations in particular. Before going any further with those cases, however, let's look at how Definition 11.45 captures the notion of a Lagrangian function as a vehicle for conditions about Lagrange multipliers. It does so with such generality that the multipliers aren't merely associated with constraints but equally well with penalties and other features of composite modeling structure in optimization.

11.46 Example (multiplier rule in Lagrangian form). *Consider the problem*

$$\text{minimize } f_0(x) + \theta\big(f_1(x),\ldots,f_m(x)\big) \text{ over } x \in X$$

for a nonempty, closed set $X \subset \mathbb{R}^n$, smooth functions $f_i : \mathbb{R}^n \to \mathbb{R}$, and a proper, lsc, convex function $\theta : \mathbb{R}^m \to \overline{\mathbb{R}}$. Taking $F(x) = \big(f_1(x),\ldots,f_m(x)\big)$, identify this with the problem of minimizing $\varphi(x) = f(x,0)$ over $x \in \mathbb{R}^n$ for

$$f(x,u) = \delta_X(x) + f_0(x) + \theta\big(F(x) + u\big).$$

This furnishes a dualizing parameterization for which the Lagrangian is

$$l(x,y) = \delta_X(x) + f_0(x) + \big\langle y, F(x)\big\rangle - \theta^*(y)$$

(with $\infty - \infty = \infty$ on the right). The optimality condition in the extended multiplier rule of 10.8, namely,

$$-\big[\nabla f_0(\bar{x}) + \nabla F(\bar{x})^*\bar{y}\big] \in N_X(\bar{x}) \text{ for some } \bar{y} \in \partial\theta\big(F(\bar{x})\big),$$

can then be written equivalently in the Lagrangian form

$$0 \in \partial_x l(\bar{x},\bar{y}), \qquad 0 \in \partial_y[-l](\bar{x},\bar{y}).$$

As a particular case, when $\theta = \theta_{Y,B}$ for a closed, convex set $Y \subset \mathbb{R}^m$ and a *symmetric, positive-semidefinite matrix* $B \in \mathbb{R}^{m\times m}$ *as in 11.18, possibly with* $B = 0$ *, the Lagrangian specializes to*

$$l(x,y) = \delta_X(x) + L(x,y) - \delta_Y(y) \qquad \text{for } L(x,y) = f_0(x) + \langle y, F(x)\rangle - \tfrac{1}{2}\langle y, By\rangle,$$

and then the Lagrangian form of the multiplier rule specializes to

$$-\nabla_x L(\bar{x},\bar{y}) \in N_X(\bar{x}), \qquad \nabla_y L(\bar{x},\bar{y}) \in N_Y(\bar{y}).$$

The case of $\theta = \delta_K$ *for a closed, convex cone corresponds to* $B = 0$ *and* $Y = K^*$.

Detail. It's elementary that Definition 11.45 yields the expression claimed for $l(x,y)$. (The convention about infinities reconciles what happens when both $x \notin X$ and $y \notin \operatorname{dom}\theta^*$; we have $f(x,\cdot) \equiv \infty$ when $x \notin X$, so the conjugate function $-l(x,\cdot)$ is then correctly the constant function $-\infty$, i.e., we have $l(x,\cdot) \equiv \infty$.) Subgradient inversion through 11.3 turns the multiplier rule into the condition on subgradients of l.

When $\theta = \theta_{Y,B}$ as in 11.18, we have $\theta = (\delta_Y + j_B)^*$ for $j_B(y) := \frac{1}{2}\langle y, By\rangle$, hence $\theta^* = \delta_Y + j_B$ by 11.1. The subgradient condition calculates out then to the one stated in terms of $\nabla_x L(\bar{x},\bar{y})$ and $\nabla_y L(\bar{x},\bar{y})$; cf. 8.8(c). □

The possibility of expressing conditions for optimality in Lagrangian form is useful for many purposes, such as the design of numerical methods. The Lagrangian brings out properties of the problem that otherwise might be obscured. This is seen in Example 11.46 when θ is a function of type $\theta_{Y,B}$, which may lack continuous derivatives through the incorporation of various penalty terms. From some perspectives, such nonsmoothness could be a handicap in contemplating how to minimize $f_0(x) + \theta\big(F(x)\big)$ directly. But in working with the Lagrangian $L(x,y)$, one has a finite, smooth function on a relatively simple product set $X \times Y$, and that may be more approachable.

Problems of extended linear-quadratic programming fit Example 11.46 with $f_0(x) = \langle c,x\rangle + \frac{1}{2}\langle x, Cx\rangle$, $F(x) = b - Ax$ and $\theta = \theta_{Y,B}$. Their Lagrangians can also be viewed as specializing the ones associated with problems in the Fenchel scheme.

11.47 Example (Lagrangians in the Fenchel scheme). *In the problem formulation of 11.41, the Lagrangian function is given by*

$$l(x,y) \;=\; \langle c,x\rangle + k(x) + \langle b,y\rangle - h^*(y) - \langle y, Ax\rangle$$

(with $\infty - \infty = \infty$ *on the right). The optimality conditions at the end of 11.41 can be written equivalently in the Lagrangian form*

$$0 \in \partial_x l(\bar{x},\bar{y}), \qquad 0 \in \partial_y[-l](\bar{x},\bar{y}).$$

For the special case of extended linear-quadratic programming described in 11.43, the Lagrangian reduces to

$$l(x,y) = \delta_X(x) + L(x,y) - \delta_Y(y)$$
$$\text{for } L(x,y) = \langle c,x\rangle + \tfrac{1}{2}\langle x,Cx\rangle + \langle b,y\rangle - \tfrac{1}{2}\langle y,By\rangle - \langle y,Ax\rangle,$$

and the optimality condition translates then to

$$-\nabla_x L(\bar{x},\bar{y}) \in N_X(\bar{x}), \qquad \nabla_y L(\bar{x},\bar{y}) \in N_Y(\bar{y}),$$

or in other words,

$$-c - C\bar{x} + A^*\bar{y} \in N_X(\bar{x}), \qquad b - A\bar{x} - B\bar{y} \in N_Y(\bar{y}).$$

Detail. The Lagrangian can be calculated right from Definition 11.45. The convention about $\infty - \infty$ comes in because the formula makes $l(x,y)$ have the value ∞ when $k(x) = \infty$, even if $h^*(y) = \infty$. The Lagrangian form of the optimality condition merely restates the conditions at the end of 11.41 in an alternative manner. □

Here $l(x,y)$ is convex in x and concave in y. This is characteristic of the convex duality framework in general.

11.48 Proposition (convexity properties of Lagrangians). *For any dualizing parameterization $\varphi = f(\cdot,0)$, the associated Lagrangian $l(x,y)$ is usc concave in y. It is convex in x besides if and only if $f(x,u)$ is convex in (x,u) rather than just with respect to u. In that case, one has*

$$(v,y) \in \partial f(x,u) \iff v \in \partial_x l(x,y),\ u \in \partial_y[-l](x,y),$$

the value $l(x,y)$ in these circumstances being finite and equal to $f(x,u) - \langle y,u\rangle$.

Proof. The fact that $l(x,y)$ is usc concave in y is obvious from $-l(x,\cdot)$ being conjugate to $f(x,\cdot)$. If $f(x,u)$ is convex in (x,u), so too for any $y \in \mathbb{R}^m$ is the function $f_y(x,u) := f(x,u) - \langle y,u\rangle$, and then $\inf_u f_y(x,u)$ is convex because the inf-projection of any convex function is convex; cf. 2.22. But $\inf_u f_y(x,u) = l(x,y)$ by definition. Thus, the convexity of $f(x,u)$ in (x,u) implies the convexity of $l(x,y)$ in x.

Conversely, if $l(x,y)$ is convex in x, the function $l_y(x,u) := l(x,y) + \langle y,u\rangle$ is convex in (x,u). But according to 11(17), f is the pointwise supremum of the collection of functions l_y as y ranges over $\mathbb{R}^m$. Since the pointwise supremum of a collection of convex functions is convex, we conclude that in this case $f(x,u)$ is convex in (x,u).

To check the equivalence in subgradient relations for particular x_0, u_0, v_0 and y_0, observe from 11.3 and the conjugacy between $f(x_0,\cdot)$ and $-l(x_0,\cdot)$ that the conditions $y_0 \in \partial_u f(x_0,u_0)$ and $u_0 \in \partial_y[-l](x_0,y_0)$ are equivalent to each other and to having $l(x_0,y_0) = f(x_0,u_0) - \langle y_0,u_0\rangle$ (finite). When f is convex, we can appeal to 8.12 to write the full condition $(v_0,y_0) \in \partial f(x_0,u_0)$ as the inequality

$$f(x,u) \geq f(x_0,u_0) + \langle v_0,\, x - x_0\rangle + \langle y_0,\, u - u_0\rangle \ \text{ for all } x,\ u,$$

cf. 8.12. From the case of $x = x_0$ we see that this entails $y_0 \in \partial_u f(x_0, u_0)$ and therefore is equivalent to having $u_0 \in \partial_y[-l](x_0, y_0)$ along with

$$\inf_u\{f(x,u) - \langle y_0, u\rangle\} \geq f(x_0,u_0) - \langle y_0, u_0\rangle + \langle v_0, x - x_0\rangle \text{ for all } x.$$

But the latter translates to $l(x, y_0) \geq l(x_0, y_0) + \langle v_0, x - x_0\rangle$ for all x, which by 8.12 and the convexity of $l(x, y_0)$ in x means that $v_0 \in \partial_x l(x_0, y_0)$. Thus, $(v_0, y_0) \in \partial f(x_0, u_0)$ if and only if $v_0 \in \partial_x l(x_0, y_0)$ and $u_0 \in \partial_y[-l](x_0, y_0)$. □

Whenever $l(x, y)$ is convex in x as well as concave in y, the Lagrangian condition in 11.46 and 11.47 can be interpreted through the following concept.

11.49 Definition (saddle points). *A vector pair $(\bar{x}, \bar{y})$ is said to be a saddle point of the function l on $I\!R^n \times I\!R^m$ (in the minimax sense, and under the convention of minimizing in the first argument and maximizing in the second) if $\inf_x l(x, \bar{y}) = l(\bar{x}, \bar{y}) = \sup_y l(\bar{x}, y)$, or in other words*

$$l(x,\bar{y}) \geq l(\bar{x},\bar{y}) \geq l(\bar{x},y) \text{ for all } x \text{ and } y. \tag{11(18)}$$

The set of all such saddle points $(\bar{x}, \bar{y})$ is denoted by argminimax l, *or in more detail by* $\operatorname{argminimax}_{x,y} l(x, y)$.

Likewise in the case of a function L on a product set $X \times Y$, a pair $(\bar{x}, \bar{y})$ is said to be a saddle point of L with respect to $X \times Y$ if

$$\begin{cases} \bar{x} \in X, \quad \bar{y} \in Y, \quad \text{and} \\ L(x,\bar{y}) \geq L(\bar{x},\bar{y}) \geq L(\bar{x},y) \text{ for all } x \in X,\ y \in Y. \end{cases} \tag{11(19)}$$

The notation for the saddle point set is then

$$\operatorname{argminimax}_{X,Y} L, \qquad \text{or } \operatorname*{argminimax}_{x \in X,\ y \in Y} L(x,y).$$

11.50 Theorem (minimax relations). *Let $l : I\!R^n \times I\!R^m \to \overline{I\!R}$ be the Lagrangian for a problem of minimizing φ on $I\!R^n$ with dualizing parameterization $\varphi = f(\cdot, 0)$, $f : I\!R^n \times I\!R^m \to \overline{I\!R}$. Let $\psi = -f^*(0, \cdot)$ on $I\!R^m$. Then*

$$\varphi(x) = \sup_y l(x,y), \qquad \psi(y) = \inf_x l(x,y), \tag{11(20)}$$

$$\inf_x \varphi(x) = \inf_x\big[\sup_y l(x,y)\big] \geq \sup_y\big[\inf_x l(x,y)\big] = \sup_y \psi(y), \tag{11(21)}$$

and furthermore

$$\left.\begin{array}{r} \bar{x} \in \operatorname{argmin}_x \varphi(x) \\ \bar{y} \in \operatorname{argmax}_y \psi(y) \\ \inf_x \varphi(x) = \sup_y \psi(y) \end{array}\right\} \iff (\bar{x},\bar{y}) \in \operatorname{argminimax}_{x,y} l(x,y) \tag{11(22)}$$
$$\iff \varphi(\bar{x}) = \psi(\bar{y}) = l(\bar{x},\bar{y}).$$

The saddle point condition $(\bar{x}, \bar{y}) \in \operatorname{argminimax}_{x,y} l(x, y)$ always entails the *subgradient condition*

$$0 \in \partial_x l(\bar{x}, \bar{y}), \qquad 0 \in \partial_y[-l](\bar{x}, \bar{y}), \tag{11(23)}$$

and is equivalent to it whenever $l(x, y)$ is convex in x and concave in y, in which case it is also the same as having $(0, \bar{y}) \in \partial f(\bar{x}, 0)$.

Proof. The expression for φ in 11(20) comes from 11(17) with $u = 0$, while the one for ψ is based on the observation that the formula

$$-f^*(v, y) \;=\; \inf_{x,u}\{f(x, u) - \langle v, x\rangle - \langle y, u\rangle\}$$

can be rewritten through 11(16) as

$$-f^*(v, y) \;=\; \inf_x\{l(x, y) - \langle v, x\rangle\}. \tag{11(24)}$$

From 11(20) we immediately get 11(21), since $\inf \varphi = p(0)$ and $\sup \psi = p^{**}(0)$ for $p(u) := \inf_x f(x, u)$; cf. 11.38 with $u = 0$. We then have 11(22), whose right side is now seen to be just another way of writing $\varphi(\bar{x}) = \psi(\bar{y})$. The final assertion about subgradients dualizes 11(23) through the relation in 11.48. □

An interesting consequence of Theorem 11.50 is the fact that the set of saddle points is always a product set:

$$\operatorname{argminimax} l \;=\; \begin{cases} (\operatorname{argmin} \varphi) \times (\operatorname{argmax} \psi) & \text{if } \inf \varphi = \sup \psi, \\ \emptyset & \text{if } \inf \varphi > \sup \psi. \end{cases} \tag{11(25)}$$

Moreover, l is constant on $\operatorname{argminimax} l$. This constant is called the *saddle value* of l and is denoted by $\operatorname{minimax} l$.

The crucial question of whether $\inf \varphi = \sup \psi$ in our Lagrangian setting is that of whether $p(0) = p^{**}(0)$ for $p(u) := \inf_x f(x, u)$, as we know already from 11.38. We'll return to this presently.

11.51 Corollary (saddle point conditions for convex problems)**.** *Consider a problem of minimizing $\varphi(x)$ over $x \in \mathbb{R}^n$ in the convex duality framework of 11.39, and let $l(x, y)$ be the corresponding Lagrangian. Suppose $0 \in \operatorname{int} U$, or merely that $0 \in U$ but f is piecewise linear-quadratic.*

Then for $\bar{x}$ to be an optimal solution it is necessary and sufficient that there exist a vector $\bar{y}$ such that $(\bar{x}, \bar{y})$ is a saddle point of the Lagrangian l on $\mathbb{R}^n \times \mathbb{R}^m$. Furthermore, $\bar{y}$ appears in this role with $\bar{x}$ if and only if $\bar{y}$ is an optimal solution to the dual problem of maximizing $\psi(y)$ over $y \in \mathbb{R}^m$.

Proof. From $0 \in \operatorname{int} U$ we have $\inf \varphi = \sup \psi < \infty$ by 11.39(a) along with the further fact in 11.39(b) that $\operatorname{argmax} \psi$ contains a vector $\bar{y}$ if $\inf \varphi > -\infty$. When $\bar{x} \in \operatorname{argmin} \varphi$, $\varphi(\bar{x})$ is finite (by the definition of 'argmin'). The claims follow then from the equivalences at the end of Theorem 11.50; cf. the preceding discussion. The piecewise linear-quadratic case substitutes the stronger results in 11.42 for those in 11.39. □

For instance, saddle points of the Lagrangian in 11.46 characterize optimality in the Fenchel scheme in 11.41 when $b \in \operatorname{int}(A \operatorname{dom} k + \operatorname{dom} h)$. When k and h are piecewise linear-quadratic, the sharper form of 11.51 is applicable—no constraint qualification is needed.

J* Minimax Problems

Problems of minimizing φ and maximizing ψ, in which φ and ψ are derived from some function l on $\mathbb{R}^n \times \mathbb{R}^m$ by the formulas in 11(18), are well known in game theory. It's interesting to see that the circumstances in which such problems also form a primal-dual pair arising out of a dualizing parameterization are completely described by the foregoing. All we need to know is that $l(x, y)$ is concave and usc in y, and that for each x either $l(x, \cdot) < \infty$ or $l(x, \cdot) \equiv \infty$. These conditions ensure that in defining f by 11(17) we get $\varphi = f(\cdot, 0)$ for a dualizing parameterization such that the associated Lagrangian is l.

11.52 Example (minimax problems). *Consider nonempty, closed, convex sets $X \subset \mathbb{R}^n$, $Y \subset \mathbb{R}^m$, and a continuous function $L : X \times Y \to \mathbb{R}$ with $L(x, y)$ convex in $x \in X$ for $y \in Y$ and concave in $y \in Y$ for $x \in X$. Let*

$$l(x,y) = \begin{cases} L(x,y) & \text{for } x \in X,\ y \in Y, \\ -\infty & \text{for } x \in X,\ y \notin Y, \\ \infty & \text{for } x \notin X, \end{cases}$$

$$\varphi(x) = \begin{cases} \sup_{y \in Y} L(x,y) & \text{for } x \in X, \\ \infty & \text{for } x \notin X, \end{cases}$$

$$\psi(x) = \begin{cases} \inf_{x \in X} L(x,y) & \text{for } y \in Y, \\ -\infty & \text{for } y \notin Y. \end{cases}$$

The saddle point set $\operatorname{argminimax}_{X,Y} L$, *which is closed and convex, coincides then with* $\operatorname{argminimax} l$ *and has the product structure in 11(25); on this set, L is constant, the saddle value* $\operatorname{minimax}_{X,Y} L$ *being* $\operatorname{minimax} l$.

Indeed, l is the Lagrangian for the problem of minimizing φ over $\mathbb{R}^n$ under the dualizing parameterization furnished by

$$f(x,u) = \begin{cases} \sup_{y \in Y} \{L(x,y) + \langle y, u \rangle\} & \text{for } x \in X, \\ \infty & \text{for } x \notin X. \end{cases}$$

The function f is lsc, proper, and convex with

$$-f^*(v,y) = \begin{cases} \inf_{x \in X} \{L(x,y) - \langle v, x \rangle\} & \text{for } y \in Y, \\ -\infty & \text{for } y \notin Y, \end{cases}$$

so the corresponding dual problem is that of maximizing ψ over $\mathbb{R}^m$.

If L is smooth on an open set that includes $X \times Y$, $\operatorname{argminimax}_{X,Y} L$ *consists of the pairs $(\bar{x}, \bar{y}) \in X \times Y$ satisfying*

$$-\nabla_x L(\bar{x},\bar{y}) \in N_X(\bar{x}), \qquad \nabla_y L(\bar{x},\bar{y}) \in N_Y(\bar{y}).$$

The equivalent conditions in 11.40 are necessary and sufficient for this saddle point set to be nonempty and bounded. In particular, the saddle point set is nonempty and bounded when X and Y are bounded.

Detail. Obviously $f(x,0) = \varphi(x)$. For $x \in X$, the function $-l(x,\cdot)$ is lsc, proper and convex with conjugate $f(x,\cdot)$, while for $x \notin X$ it is the constant function $-\infty$, whereas $f(x,\cdot)$ is the constant function ∞. Hence $-l(x,\cdot)$ is the conjugate of $f(x,\cdot)$ for all $x \in I\!R^n$. Thus, f furnishes a dualizing parameterization for the problem of minimizing φ on $I\!R^n$, and the associated Lagrangian is l. We have $l(x,y)$ convex in x and concave in y, so f is convex by 11.48. The continuity of L on $X \times Y$ ensures that $-l(x,\cdot)$ depends epi-continuously on $x \in X$, and the same then holds for $f(x,\cdot)$ by Theorem 11.34. It follows that f is lsc on $I\!R^n \times I\!R^m$. The rest is clear then from 11.50. When X and Y are bounded, the sets U and V in 11.40 are all of $I\!R^n$ and $I\!R^m$. □

A powerful rule about the behavior of optimal values in parameterized problems of convex type comes out of this principle.

11.53 Theorem (perturbation of saddle values; Golshtein). *Consider nonempty, closed, convex sets $X \subset I\!R^n$ and $Y \subset I\!R^m$, and an interval $[0,T] \subset I\!R$. Let $L : [0,T] \times X \times Y \to I\!R$ be continuous, with $L(t,x,y)$ convex in $x \in X$ and concave in $y \in Y$, and suppose that the saddle point set*

$$X_0 \times Y_0 = \operatorname*{argminimax}_{x\in X,\ y\in Y} L(0,x,y)$$

is nonempty and bounded. Then, relative to t in an interval $[0,\varepsilon]$ for some $\varepsilon > 0$, the saddle point set $\operatorname{argminimax}_{X,Y} L(t,\cdot,\cdot)$ is nonempty and bounded, the mapping $t \mapsto \operatorname{argminimax}_{X,Y} L(t,\cdot,\cdot)$ is osc and locally bounded at $t = 0$, and the saddle value

$$\lambda(t) := \operatorname*{minimax}_{x\in X,\ y\in Y} L(t,x,y)$$

converges to $\lambda(0)$ as $t \searrow 0$. If in addition the limit

$$L'_+(0,x,y) := \lim_{\substack{t \searrow 0\\ x' \underset{X}{\to} x\\ y' \underset{Y}{\to} y}} \frac{L(t,x',y') - L(0,x',y')}{t} \qquad 11(26)$$

exists for all $(x,y) \in X_0 \times Y_0$, then the value $L'_+(0,x,y)$ is continuous relative to $(x,y) \in X_0 \times Y_0$, convex in $x \in X_0$ for each $y \in Y_0$, and concave in $y \in Y_0$ for each $x \in X_0$. Indeed, the right derivative $\lambda'_+(0)$ exists and is given by

$$\lambda'_+(0) = \operatorname*{minimax}_{x\in X_0,\ y\in Y_0} L'_+(0,x,y).$$

Proof. We put ourselves in the picture of Example 11.52 and its detail, but with everything depending additionally on $t \in [0,T]$, in particular $f(t,x,u)$.

In extension of the earlier argument, the mapping $(t,x) \mapsto f(t,x,\cdot)$ is epi-continuous relative to $[0,T] \times X$. From this it follows that the mapping $t \mapsto f(t,\cdot,\cdot)$ is epi-continuous as well. The convex functions $p(t,\cdot)$ defined by $p(t,u) = \inf_x f(t,x,u)$ and the convex sets $U(t) = \operatorname{dom} p(t,\cdot)$ then have

$$p(0,\cdot) = \operatorname{e-lim}_{t \searrow 0} p(t,\cdot), \qquad U(0) \subset \liminf_{t \searrow 0} U(t),$$

by 7.57 and 7.4(h). Similarly, in using $f^*(t,\cdot,\cdot)$ to denote the convex function conjugate to $f(t,\cdot,\cdot)$, we have the epi-continuity of $t \mapsto f^*(t,\cdot,\cdot)$ by Theorem 11.34 and thus, for the convex functions $q(t,\cdot)$ defined by $q(t,v) = \inf_y f^*(t,v,y)$ and the convex sets $V(t) = \operatorname{dom} q(t,\cdot)$ that

$$q(0,\cdot) = \operatorname{e-lim}_{t \searrow 0} q(t,\cdot), \qquad V(0) \subset \liminf_{t \searrow 0} V(t).$$

Our assumption that $\operatorname{argminimax}_{X,Y} L(0,\cdot,\cdot)$ is nonempty and bounded corresponds by 11.40 to having $0 \in \operatorname{int} U(0)$ and $0 \in \operatorname{int} V(0)$. The inner limit inclusions for these sets imply then by 4.15 that also $0 \in \operatorname{int} U(t)$ and $0 \in \operatorname{int} V(t)$ for all t in some interval $[0,\varepsilon]$. Hence by 11.40 we have $\operatorname{argminimax} L(t,\cdot,\cdot)$ nonempty and bounded for t in this interval. We know further from Theorem 11.39, via 11.50 and 11.52, that

$$\operatorname{argminimax}_{X,Y} L(t,\cdot,\cdot) = \partial_u p(t,0) \times \partial_v q(t,0) \quad \text{for} \ \ t \in [0,\varepsilon].$$

The epi-convergence of $p(t,\cdot)$ to $p(0,\cdot)$ guarantees by the convexity of these functions that, as $t \searrow 0$, uniform convergence on a neighborhood of $u = 0$; cf. 7.17. Likewise, $q(t,\cdot)$ converges uniformly to $q(0,\cdot)$ around $v = 0$. The subgradient bounds in 9.14 imply that the mapping

$$t \mapsto \partial_u p(t,0) \times \partial_v q(t,0) \quad \text{on} \ \ [0,\varepsilon]$$

is osc and locally bounded at $t = 0$. And thus, the constant value $\lambda(t)$ that $L(t,\cdot,\cdot)$ has on $\partial_u p(t,0) \times \partial_v q(t,0)$ must converge as $t \searrow 0$ to the constant value $\lambda(0)$ that $L(0,\cdot,\cdot)$ has on $\partial_u p(0,0) \times \partial_v q(0,0)$.

The fact that the limit in 11(26) is taken over $x' \underset{X_0}{\to} x$ and $y' \underset{Y_0}{\to} y$ guarantees that $L'_+(0,x,y)$ is continuous relative to $(x,y) \in X_0 \times Y_0$. The convexity-concavity of $L'_+(0,\cdot,\cdot)$ on $X_0 \times Y_0$ follows immediately from that of $L(t,\cdot,\cdot)$ and the constancy of $L(0,\cdot,\cdot)$ on $X_0 \times Y_0$. Because the sets X_0 and Y_0 are closed and bounded, we know then from 11.52 that the saddle point set for $L'_+(0,\cdot,\cdot)$ on $X_0 \times Y_0$ is nonempty. Let $\mu = \operatorname{minimax}_{X_0,Y_0} L'_+(0,\cdot,\cdot)$. We have to demonstrate that

$$\lim_{t \searrow 0} \frac{1}{t}\big(\lambda(t) - \lambda(0)\big) = \mu.$$

Henceforth we can suppose for simplicity that $\varepsilon = T$ and write the set $\operatorname{argminimax}_{X,Y} L(t,\cdot,\cdot)$ as $X_t \times Y_t$; the mappings $t \mapsto X_t$ and $t \mapsto Y_t$ are nonempty-valued, and they are osc and locally bounded at $t = 0$. Consider any $(\bar{x},\bar{y}) \in X_0 \times Y_0$ and, for $t \in (0,T]$, pairs $(\bar{x}_t,\bar{y}_t) \in X_t \times Y_t$. As $t \searrow 0$, $(\bar{x}_t,\bar{y}_t)$

stays bounded, and any cluster point (x_0, y_0) belongs to $X_0 \times Y_0$. The saddle point conditions imply that

$$\begin{aligned} L(0,\bar{x}_t,\bar{y}) &\geq L(0,\bar{x},\bar{y}) \geq L(0,\bar{x},\bar{y}_t), \qquad & L(0,\bar{x},\bar{y}) = \lambda(0),\\ L(t,\bar{x},\bar{y}_t) &\geq L(t,\bar{x}_t,\bar{y}_t) \geq L(t,\bar{x}_t,\bar{y}), \qquad & L(t,\bar{x}_t,\bar{y}_t) = \lambda(t). \end{aligned}$$

Using these relations, we consider any sequence $t^\nu \searrow 0$ such that $\bar{y}_{t^\nu}$ converges to some y_0 and estimate that

$$\begin{aligned} \limsup_{\nu\to\infty} \frac{\lambda(t^\nu)-\lambda(0)}{t^\nu} &= \limsup_{\nu\to\infty} \frac{L(t^\nu,\bar{x}_{t^\nu},\bar{y}_{t^\nu})-L(0,\bar{x},\bar{y})}{t^\nu}\\ &\leq \limsup_{\nu\to\infty} \frac{L(t^\nu,\bar{x},\bar{y}_{t^\nu})-L(0,\bar{x},\bar{y}_{t^\nu})}{t^\nu}\\ &= L'_+(0,\bar{x},y_0) \;\leq\; \sup_{y\in Y_0} L'_+(0,\bar{x},y). \end{aligned}$$

Since $\bar{x}$ was an arbitrary point of X_0, this yields the bound

$$\limsup_{t\,\searrow\,0} \frac{\lambda(t)-\lambda(0)}{t} \leq \inf_{x\in X_0}\Big[\sup_{y\in Y_0} L'_+(0,x,y)\Big] = \mu.$$

A parallel argument with the roles of x and y switched shows also that

$$\liminf_{t\,\searrow\,0} \frac{\lambda(t)-\lambda(0)}{t} \geq \sup_{y\in Y_0}\Big[\inf_{x\in X_0} L'_+(0,x,y)\Big] = \mu.$$

Thus, $[\lambda(t)-\lambda(0)]/t$ does tend to μ as $t \searrow 0$. □

11.54 Example (perturbations in extended linear-quadratic programming). *In the format and assumptions of 11.43, but with dependence additionally on a parameter t, consider for each $t \in [0,T]$ the primal and dual problems*

$$\begin{aligned} &\text{minimize } \big\langle c(t),x\big\rangle + \tfrac{1}{2}\big\langle x,C(t)x\big\rangle + \theta_{Y,B(t)}\big(b(t)-A(t)x\big) \text{ over } x\in X,\\ &\text{maximize } \big\langle b(t),y\big\rangle - \tfrac{1}{2}\big\langle y,B(t)y\big\rangle - \theta_{X,C(t)}\big(A(t)^*y-c(t)\big) \text{ over } y\in Y, \end{aligned}$$

denoting their optimal solution sets by X_t and Y_t. Suppose that X_0 and Y_0 are nonempty and bounded.

If $c(t)$, $C(t)$, $b(t)$, $B(t)$, and $A(t)$ depend continuously on t, there exists $\varepsilon > 0$ such that, relative to $t \in [0,\varepsilon]$, the mappings $t \mapsto X_t$ and $t \mapsto Y_t$ are nonempty-valued and, at $t = 0$, are osc and locally bounded. Then too, the common optimal value in the two problems, denoted by $\lambda(t)$, behaves continuously at $t = 0$. If in addition the right derivatives

$$c_0 := c'_+(0), \quad C_0 := C'_+(0), \quad b_0 := b'_+(0), \quad B_0 := B'_+(0), \quad A_0 := A'_+(0),$$

exist, then the right derivative $\lambda'_+(0)$ exists and is the common optimal value in the problems

$$\text{minimize } \langle c_0, x\rangle + \tfrac{1}{2}\langle x, C_0 x\rangle + \theta_{Y_0,B_0}(b_0 - A_0 x) \text{ over } x \in X_0,$$
$$\text{maximize } \langle b_0, y\rangle - \tfrac{1}{2}\langle y, B_0 y\rangle - \theta_{X_0,C_0}(A_0^* y - c_0) \text{ over } y \in Y_0.$$

Detail. This specializes Theorem 11.53 to the Lagrangian setup for extended linear-quadratic programming in 11.47. □

Note that the matrices B_0 and C_0 in Example 11.54, although symmetric, might not be positive definite, so the subproblem giving the rate of change of the optimal value might not be one of extended linear-quadratic programming strictly as described in 11.43. But it's clear from the convexity-concavity of $L'_+(0,\cdot,\cdot)$ in Theorem 11.55 that the expression $\frac{1}{2}\langle x, C_0 x\rangle$ is convex *with respect to* $x \in X_0$, while $\frac{1}{2}\langle y, B_0 y\rangle$ is convex *with respect to* $y \in Y_0$. The primal and dual subproblems for $\lambda'_+(0)$ are thus of convex type nonetheless. (The notion of piecewise linear-quadratic programming can readily be refined in this direction. The special features of duality in Theorem 11.42 persist, and in 11.18 all that changes is the description of the effective domains of the θ functions.)

K*. Augmented Lagrangians and Nonconvex Duality

The question of the extent to which the duality relation $\inf \varphi = \sup \psi$ might hold for problems of nonconvex type, beyond the framework in 11.39, has its answer in a closer examination of the relationships in 11.38. The general situation is shown in Figure 11–6, where the notation is still that of $\varphi = f(\cdot, 0)$, $\psi = -f^*(0,\cdot)$, and $p(u) = \inf_x f(x,u)$. By the definition of p^{**}, the value $\sup \psi = p^{**}(0)$ is the supremum of the intercepts on the vertical axis that are achievable by affine functions majorized by p; the vectors $\bar{y} \in \operatorname{argmax} \psi$, if any, correspond to the affine functions that are 'highest' in this sense.

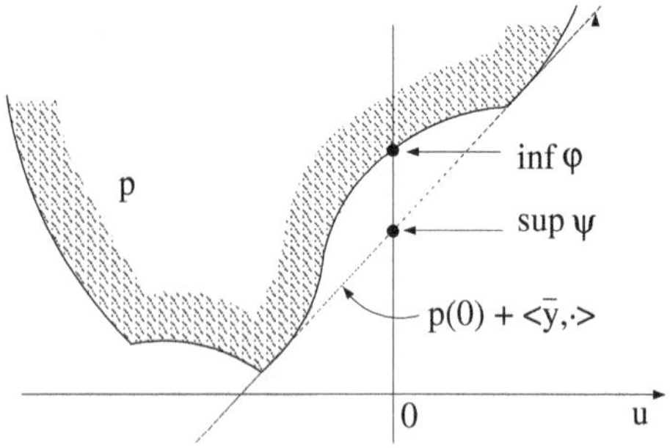

Fig. 11–6. Duality gap in minimization problems lacking adequate convexity.

A *duality gap* $\inf \varphi > \sup \psi$ arises precisely when the intercepts are prevented from getting as high as the value $\inf \varphi = p(0)$. A lack of convexity can evidently be the source of such a shortfall. (Of course, a duality gap can also occur even when p is convex if p fails to be lsc at 0 or if p takes on $-\infty$ but $0 \notin \mathrm{cl}(\mathrm{dom}\, p)$; cf. 11.2.) In particular, it's clear that

$$\left.\begin{array}{l}\inf \varphi = \sup \psi \\ \bar{y} \in \operatorname{argmax} \psi\end{array}\right\} \iff \left\{\begin{array}{l}p(u) \geq p(0) + \langle \bar{y}, u\rangle \text{ for all } u, \\ \text{with } p(0) \neq -\infty.\end{array}\right. \tag{11(27)}$$

This picture suggests that a duality gap might be avoided if the dual problem could be set up to correspond not to pushing *affine* functions up against epi p, but some other class of functions capable of penetrating possible 'dents'. This turns out to be attainable with only a little extra effort.

11.55 Definition (augmented Lagrangian functions). *For a primal problem of minimizing $\varphi(x)$ over $x \in \mathbb{R}^n$ and any dualizing parameterization $\varphi = f(\cdot, 0)$ for a choice of $f : \mathbb{R}^n \times \mathbb{R}^m \to \overline{\mathbb{R}}$, consider any augmenting function σ; by this is meant a proper, lsc, convex function*

$$\sigma : \mathbb{R}^m \to \overline{\mathbb{R}} \;\; \textit{with} \;\; \min \sigma = 0, \;\; \operatorname{argmin} \sigma = \{0\}.$$

The corresponding augmented Lagrangian with penalty parameter $r > 0$ is then the function $\bar{l} : \mathbb{R}^n \times \mathbb{R}^m \times (0, \infty) \to \overline{\mathbb{R}}$ defined by

$$\bar{l}(x, y, r) := \inf_u \Big\{ f(x, u) + r\sigma(u) - \langle y, u\rangle \Big\}.$$

The corresponding augmented dual problem consists of maximizing over all $(y, r) \in \mathbb{R}^m \times (0, \infty)$ the function

$$\bar{\psi}(y, r) := \inf_{x,u} \Big\{ f(x, u) + r\sigma(u) - \langle y, u\rangle \Big\}.$$

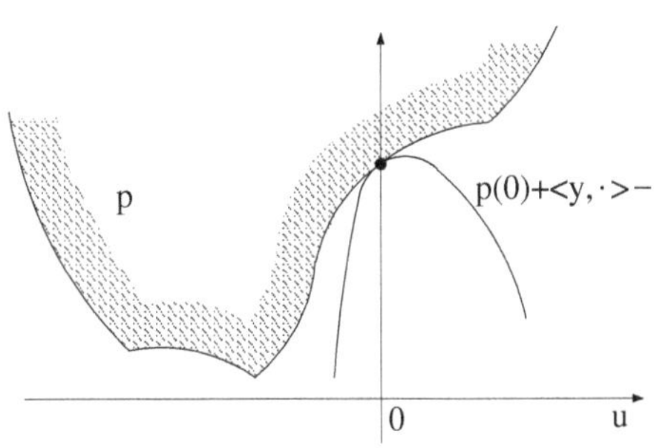

Fig. 11–7. Duality gap removed by an augmenting function.

The notion of the augmented Lagrangian grows from the idea of replacing the inequality in 11(27) by one of the form

$$p(u) \geq p(0) + \langle \bar{y}, u\rangle - r\sigma(u) \text{ for all } u$$

for some augmenting function σ (as just defined) and a parameter value $r > 0$ sufficiently high, as in Figure 11–7. What makes the approach successful in modifying the dual problem to get rid of the duality gap is that this inequality is identical to

$$p_{r\sigma}(u) \geq p_{r\sigma}(0) + \langle \bar{y}, u \rangle \text{ for all } u$$

where $p_{r\sigma}(u) := \inf_x f_{r\sigma}(x,u)$ for the function $f_{r\sigma}(x,u) := f(x,u) + r\sigma(u)$. Indeed, $p_{r\sigma}(u) = p(u) + r\sigma(u)$ and $p_{r\sigma}(0) = p(0)$, because $\sigma(0) = 0$.

We have $\varphi = f_{r\sigma}(\cdot, 0)$ as well as $\varphi = f(\cdot, 0)$. Moreover because σ is proper, lsc and convex, the representation of φ in terms of $f_{r\sigma}$, like that in terms of f, is a dualizing parameterization. The Lagrangian associated with $f_{r\sigma}$ is $l_{r\sigma}(x,y) = \bar{l}(x,y,r)$, as seen from the definition of $\bar{l}(x,y,r)$ above. The resulting dual problem, which consists of maximizing $\psi_{r\sigma} = -f^*_{r\sigma}(0,\cdot)$ over $y \in \mathbb{R}^m$, has $\psi_{r\sigma}(y) = \bar{\psi}(y,r)$. We can apply the theory already developed to this modified setting, where $f_{r\sigma}$ replaces f, and capture powerful new features.

Before translating this argument into a theorem about duality, we record some general consequences of Definition 11.55 for background, and we look at a couple of examples.

11.56 Exercise (properties of augmented Lagrangians). *For any dualizing parameterization f and augmenting function σ, the augmented Lagrangian $\bar{l}(x,y,r)$ is concave and usc in (y,r) and nondecreasing in r. It is convex in x if $f(x,u)$ is actually convex in (x,u). If σ is finite everywhere, the augmented Lagrangian is given in terms of the ordinary Lagrangian $l(x,y)$ by*

$$\bar{l}(x,y,r) = \sup_z \Big\{ l(x, y-z) - r\sigma^*(r^{-1}z) \Big\}. \tag{11(28)}$$

Likewise, the augmented dual expression $\bar{\psi}(y,r)$ is concave and usc in (y,r) and nondecreasing in r. In the case of $f(x,u)$ convex in (x,u) and σ finite everywhere, it is given in terms of the ordinary dual expression $\psi(y)$ by

$$\bar{\psi}(y,r) = \sup_z \Big\{ \psi(y-z) - r\sigma^*(r^{-1}z) \Big\}. \tag{11(29)}$$

Then in fact, $(\bar{y},\bar{r})$ maximizes $\bar{\psi}$ if and only if $\bar{y}$ maximizes ψ; the value of $\bar{r} > 0$ can be chosen arbitrarily.

Guide. Derive the properties of $\bar{l}$ straight from the formula in 11.55, using 2.22(a) for the convexity in x. Develop 11(28) out of the fact that $-\bar{l}(x,\cdot,r)$ is conjugate to $f(x,\cdot) + r\sigma$, taking note of 11.23(a). Handle $\bar{\psi}$ similarly. The final assertion about maximizing pairs $(\bar{y},\bar{r})$ falls out of 11(29). □

The monotonicity of $\bar{l}(x,y,r)$ and $\bar{\psi}(y,r)$ in r is consistent with r being a penalty parameter, an interpretation that will become clearer as we proceed. It lends special character to the augmented dual problem. In maximizing $\bar{\psi}(y,r)$ in y and r, the requirement that $r > 0$ doesn't act like a real constraint. There's no danger of having to move toward $r = 0$ to get higher values of $\bar{\psi}(y,r)$.

The fact at the end of 11.56, that, in the convex case, solutions to the augmented dual problem correspond simply to solutions to the ordinary dual problem, is reassuring. Augmentation doesn't destroy duality that may exist without it. This doesn't mean, though, that augmented Lagrangians have nothing to offer in the convex duality framework. Quite to the contrary, some of the main features of augmentation, for instance in achieving 'exact penalty representations' (as described below in 11.60), are most easily accessed in that special framework.

11.57 Example (proximal Lagrangians). *An augmented Lagrangian generated with the augmenting function $\sigma(u) = \frac{1}{2}|u|^2$ is a proximal Lagrangian. Then*

$$\begin{aligned}\bar{l}(x,y,r) &= \inf_u\Big\{f(x,u) + \frac{r}{2}|u|^2 - \langle y,u\rangle\Big\}\\ &= \sup_z\Big\{l(x,y-z) - \frac{1}{2r}|z|^2\Big\} = \sup_z\Big\{l(x,z) - \frac{1}{2r}|z-y|^2\Big\}.\end{aligned}$$

As an illustration, consider the case of Example 11.46 in which $\theta = \delta_D$ for a closed, convex, set $D \neq \emptyset$. This calculates out to

$$\bar{l}(x,y,r) = f_0(x) + \frac{r}{2}\Big[d_D\big(r^{-1}y + F(x)\big)^2 - |r^{-1}y|^2\Big],$$

which for $y = 0$ gives the standard quadratic penalty function for the constraint $F(x) \in D$. When $D = \{0\}$, one gets

$$\bar{l}(x,y,r) = f_0(x) + \big\langle y, F(x)\big\rangle + \frac{r}{2}\big|F(x)\big|^2.$$

Detail. These specializations are obtained from the first of the formulas for $\bar{l}(x,y,r)$ in writing $(r/2)|u|^2 - \langle y,u\rangle$ as $(r/2)\big(|u - r^{-1}y|^2 - |r^{-1}y|^2\big)$. □

11.58 Example (sharp Lagrangians). *An augmented Lagrangian generated with augmenting function $\sigma(u) = \|u\|$ (any norm $\|\cdot\|$) is a sharp Lagrangian. Then*

$$\begin{aligned}\bar{l}(x,y,r) &= \inf_u\Big\{f(x,u) + r\|u\| - \langle y,u\rangle\Big\}\\ &= \sup_z\Big\{l(x,y-z) - \delta_{rB^\circ}(z)\Big\} = \sup_{\|z-y\|^\circ \le r} l(x,z),\end{aligned}$$

where $\|\cdot\|^\circ$ is the polar norm and B° its unit ball. For Example 11.46 in the case of $\theta = \delta_D$ for a closed, convex set D, one gets

$$\bar{l}(x,y,r) = f_0(x) + \sup_{\|z-y\|^\circ \le r}\Big\{\big\langle z, F(x)\big\rangle - \sigma_D(z)\Big\},$$

which can be calculated out completely for instance when D is a box and $\|\cdot\| = \|\cdot\|_1$, so that $\|\cdot\|^\circ = \|\cdot\|_\infty$. Anyway, for $y = 0$ one has the standard linear penalty representation of the constraint $F(x) \in D$:

$$\bar{l}(x,0,r) = f_0(x) + r d_D\big(F(x)\big) \qquad \text{(distance in } \|\cdot\|\text{)}.$$

For general y but $D = \{0\}$ one obtains

$$\bar{l}(x, y, r) \;=\; f_0(x) + \langle y, F(x) \rangle + r\|F(x)\|.$$

Outfitted with these background facts and examples, we return now to the derivation of a duality theory in terms of augmented Lagrangians that is able even to cover certain nonconvex problems.

11.59 Theorem (duality without convexity). *For a problem of minimizing φ on $\mathbb{R}^n$, consider the augmented Lagrangian $\bar{l}(x, y, r)$ associated with a dualizing parameterization $\varphi = f(\cdot, 0)$, $f : \mathbb{R}^n \times \mathbb{R}^m \to \overline{\mathbb{R}}$, and some augmenting function $\sigma : \mathbb{R}^m \to \overline{\mathbb{R}}$. Suppose that $f(x, u)$ is level-bounded in x locally uniformly in u, and let $p(u) := \inf_x f(x, u)$. Suppose further that $\inf_x \bar{l}(x, y, r) > -\infty$ for at least one $(y, r) \in \mathbb{R}^m \times (0, \infty)$. Then*

$$\varphi(x) = \sup_{y,r} \bar{l}(x, y, r), \qquad \bar{\psi}(y, r) = \inf_x \bar{l}(x, y, r),$$

where actually $\varphi(x) = \sup_y \bar{l}(x, y, r)$ for every $r > 0$, and in fact

$$\begin{aligned} \inf_x \varphi(x) &= \inf_x \big[\sup_{y,r} \bar{l}(x, y, r) \big] \\ &= \sup_{y,r} \big[\inf_x \bar{l}(x, y, r) \big] \;=\; \sup_{y,r} \bar{\psi}(y, r). \end{aligned} \qquad 11(30)$$

Moreover, optimal solutions to the primal and augmented dual problems are characterized as saddle points of the augmented Lagrangian:

$$\begin{aligned} &\bar{x} \in \operatorname{argmin}_x \varphi(x) \;\text{ and }\; (\bar{y}, \bar{r}) \in \operatorname{argmax}_{y,r} \bar{\psi}(y, r) \\ &\iff \;\inf_x \bar{l}(x, \bar{y}, \bar{r}) \;=\; \bar{l}(\bar{x}, \bar{y}, \bar{r}) \;=\; \sup_{y,r} \bar{l}(\bar{x}, y, r), \end{aligned} \qquad 11(31)$$

the elements of $\operatorname{argmax}_{y,r} \bar{\psi}(y, r)$ being the pairs $(\bar{y}, \bar{r})$ with the property that

$$p(u) \;\geq\; p(0) + \langle \bar{y}, u \rangle - \bar{r}\,\sigma(u) \;\text{ for all }\; u. \qquad 11(32)$$

Proof. Most of this is evident from the monotonicity with respect to r in 11.56 and the explanation after Definition 11.55 of how 11(27) and Theorem 11.50 can be applied. The crucial new fact needing justification is the equality in the middle of 11(30), in place of merely the automatic '$\geq$'.

By hypothesis there's at least one pair $(\tilde{y}, \tilde{r})$ such that $\bar{\psi}(\tilde{y}, \tilde{r})$ is finite. To get the equality in question, it will suffice to demonstrate that $\bar{\psi}(\tilde{y}, r) \to p(0)$ as $r \to \infty$, since $p(0) = \inf_x \varphi(x)$. We can utilize along the way the fact that p is proper and lsc by virtue of the level boundedness assumption placed on f (cf. 1.17). From the definition of $\bar{\psi}$ in 11.55 we have for all $r > 0$ that

$$\bar{\psi}(\tilde{y}, r) \;=\; \inf_u \big\{ p(u) + r\sigma(u) - \langle \tilde{y}, u \rangle \big\}.$$

Let $\tilde{p}(u) := p(u) + \tilde{r}\sigma(u) - \langle \tilde{y}, u \rangle$, noting that $\tilde{p}(0) = p(0)$ because $\sigma(0) = 0$. This function is bounded below by $\bar{\psi}(\tilde{y}, \tilde{r})$, and like p it is proper and lsc because σ is proper and lsc. We can write

$$\bar{\psi}(\tilde{y}, \tilde{r} + s) \;=\; \inf_u \big\{ \tilde{p}(u) + s\sigma(u) \big\} \;\text{ for }\; s > 0$$

and concentrate on proving that the limit of this as $s \to \infty$ is $\tilde{p}(0)$. Because σ is convex with $\operatorname{argmin} \sigma = \{0\}$, it's level-coercive (by 3.27), and so too then is $\tilde{p} + s\sigma$, due to $\tilde{p}$ being bounded below. The positivity of σ away from 0 guarantees that $\tilde{p} + s\sigma$ increases pointwise as $s \to \infty$ to the function $\delta_{\{0\}} + \gamma$ for the constant $\gamma = \tilde{p}(0)$. In particular then $\tilde{p} + s\sigma$ epi-converges to $\delta_{\{0\}} + \gamma$ in the setting of Theorem 7.33 (see also 7.4(f)), and we are able to conclude that $\inf(\tilde{p} + s\sigma) \to \inf(\delta_{\{0\}} + \gamma) = \gamma$. This was what we needed. □

The importance of the solutions to the augmented dual problem is found in the following idea.

11.60 Definition (exact penalty representations). *In the augmented Lagrangian framework of 11.55, a vector $\bar{y}$ is said to support an exact penalty representation for the problem of minimizing φ on $\mathbb{R}^n$ if, for all $r > 0$ sufficiently large, this problem is equivalent to minimizing $\bar{l}(\cdot, \bar{y}, r)$ on $\mathbb{R}^n$ in the sense that*

$$\inf_x \varphi(x) = \inf_x \bar{l}(x, \bar{y}, r), \qquad \operatorname{argmin}_x \varphi(x) = \operatorname{argmin}_x \bar{l}(x, \bar{y}, r).$$

Specifically, a value $\bar{r} > 0$ is said to serve as an adequate penalty threshold in this respect if the property holds for all $r \in (\bar{r}, \infty)$.

11.61 Theorem (criterion for exact penalty representations). *In the notation and assumptions of Theorem 11.59, a vector $\bar{y}$ supports an exact penalty representation for the primal problem if and only if there exist $W \in \mathcal{N}(0)$ and $\hat{r} > 0$ such that*

$$p(u) \;\geq\; p(0) + \langle \bar{y}, u \rangle - \hat{r}\sigma(u) \;\; \text{for all} \;\; u \in W. \qquad 11(33)$$

This criterion is equivalent in turn to the existence of an $\bar{r} > 0$ with $(\bar{y}, \bar{r}) \in \operatorname{argmax}_{y,r} \bar{\psi}(y, r)$, and moreover such values $\bar{r}$ are the ones serving as adequate penalty thresholds for the exact penalty property with respect to $\bar{y}$.

Proof. As a starter, note that the assumptions of Theorem 11.59 guarantee that p is lsc and proper and hence that the condition in 11(33), and for that matter the one in 11(32), can't hold unless the value $p(0) = \inf \varphi$ is finite.

First we argue that the condition $(\bar{y}, \bar{r}) \in \operatorname{argmax}_{y,r} \bar{\psi}(y, r)$ is both necessary and sufficient for $\bar{y}$ to support an exact penalty representation with $\bar{r}$ as an adequate penalty threshold. For the necessity, note that the latter conditions imply through Definition 11.60 that $\bar{\psi}(\bar{y}, r) = \inf \varphi$ for all $r \in (\bar{r}, \infty)$ and hence, because $\bar{\psi}$ is usc (by 11.56), that $\bar{\psi}(\bar{y}, \bar{r}) \geq \inf \varphi$. Since $\sup \bar{\psi} = \inf \varphi$ by 11(31), it follows that $(\bar{y}, \bar{r})$ maximizes $\bar{\psi}$.

For the sufficiency, recall from the end of Theorem 11.59 that the condition $(\bar{y}, \bar{r}) \in \operatorname{argmax}_{y,r} \bar{\psi}(y, r)$ corresponds to the inequality 11(32) holding. When $\bar{r}$ is replaced in 11(32) by any higher value r, this inequality becomes strict for $u \neq 0$, because $\sigma(0) = 0$ but $\sigma(u) > 0$ for $u \neq 0$. Then

$$\operatorname{argmin}_u \Big\{ p(u) + r\sigma(u) - \langle \bar{y}, u \rangle \Big\} \;=\; \{0\}.$$

Fixing such $r > \bar{r}$, consider the function $g(x,u) := f(x,u) + r\sigma(u) - \langle \bar{y}, u\rangle$ and its associated inf-projections $h(u) := \inf_x g(x,u)$ and $k(x) := \inf_u g(x,u)$, noting that $h(u) = p(u) + r\sigma(u) - \langle \bar{y}, u\rangle$ whereas $k(x) = \bar{l}(x,\bar{y},r)$. According to the interchange rule in 1.35, one has

$$\left.\begin{array}{r}\bar{u} \in \operatorname{argmin}_u h(u)\\ \bar{x} \in \operatorname{argmin}_x g(x,\bar{u})\end{array}\right\} \iff \left\{\begin{array}{l}\bar{x} \in \operatorname{argmin}_x k(x)\\ \bar{u} \in \operatorname{argmin}_u g(\bar{x},u)\end{array}\right.$$

We've just seen that $\operatorname{argmin}_u h(u) = \{0\}$. The pairs $(\bar{x},\bar{u})$ on the left of this rule are thus the ones with $\bar{u} = 0$ and $\bar{x} \in \operatorname{argmin}_x g(x,0) = \operatorname{argmin}_x \varphi(x)$. These are then also the pairs on the right; in other words, in minimizing $\bar{l}(x,\bar{y},r)$ in x one obtains exactly the points $\bar{x} \in \operatorname{argmin}_x \varphi(x)$, and then in maximizing for any such $\bar{x}$ the expression $f(\bar{x},u) + r\sigma(u) - \langle \bar{y}, u\rangle$ in u one finds that the maximum is attained uniquely at 0. In particular, the sets $\operatorname{argmin}_x \varphi(x)$ and $\operatorname{argmin}_x \bar{l}(x,\bar{y},r)$ are the same.

To complete the proof of the theorem we need only show now that when 11(33) holds there must exist $\bar{r} \in (\hat{r},\infty)$ such that the stronger condition 11(32) holds. In assuming 11(33) there's no loss of generality in taking W to be a ball $\varepsilon I\!B$, $\varepsilon > 0$. Obviously 11(33) continues to hold when $\hat{r}$ is replaced by a higher value $\bar{r}$, so the question is whether, once $\bar{r}$ is high enough, we will have, in addition, the inequality in 11(32) holding for all $|u| > \varepsilon$. By hypothesis (inherited from Theorem 11.59) there exists $(\tilde{y},\tilde{r}) \in I\!R^m \times (0,\infty)$ such that $\bar{\psi}(\tilde{y},\tilde{r})$ is finite; then for $\alpha := \bar{\psi}(\tilde{y},\tilde{r})$ we have

$$p(u) \;\geq\; \alpha + \langle \tilde{y}, u\rangle - \tilde{r}\sigma(u) \;\text{ for all }\; u.$$

It will suffice to show that, when $\bar{r}$ is chosen high enough, one will have

$$\alpha + \langle \tilde{y}, u\rangle - \tilde{r}\sigma(u) > p(0) + \langle \bar{y}, u\rangle - \bar{r}\sigma(u) \;\text{ when }\; |u| > \varepsilon.$$

This amounts to showing that for high enough values of $\bar{r}$ one will have

$$\big\{u \,\big|\, (\bar{r} - \tilde{r})\sigma(u) \leq \langle \bar{y} - \tilde{y}, u\rangle + p(0) - \alpha\big\} \;\subset\; \varepsilon I\!B.$$

The issue can be simplified by working with $s = \bar{r} - \tilde{r} > 0$ and letting $\lambda := |\bar{y} - \tilde{y}|$ and $\mu := p(0) - \alpha$. Then we only need to check that the set

$$C(s) \;:=\; \big\{u \,\big|\, s\sigma(u) \leq \lambda|u| + \mu\big\}$$

lies in $\varepsilon I\!B$ when s is chosen high enough.

We know σ is level-coercive, because $\operatorname{argmin}\sigma = \{0\}$ (cf. 3.23, 3.27), hence there exist $\gamma > 0$ and β such that $\sigma(u) \geq \gamma|u| + \beta$ for all u (cf. 3.26). Let $s_0 > 0$ be high enough that $s_0\gamma - \lambda > 0$. For $u \in C(s_0)$ we have $s_0(\gamma|u| + \beta) \leq \lambda|u| + \mu$, hence $|u| \leq \rho := (\mu - s_0\beta)/(s_0\gamma - \lambda)$. But $C(s) \subset C(s_0)$ when $s > s_0$, inasmuch as $\sigma(u) \geq 0$ for all u. Therefore

$$C(s) \subset \big\{u \,\big|\, \sigma(u) \leq (\lambda\rho + \mu)/s\big\}$$

when $s > s_0$. On the other hand, because $\sigma(u) = 0$ only for $u = 0$, the level set $\{u \,|\, \sigma(u) \le \delta\}$ must lie in $\varepsilon I\!B$ for small $\delta > 0$. Taking such a δ and noting that $(\lambda\rho + \mu)/s \le \delta$ when s exceeds a certain s_1, we conclude that $C(s) \subset \varepsilon I\!B$ when $s > s_1$. □

11.62 Example (exactness of linear or quadratic penalties).

(a) *Consider the proximal Lagrangian of 11.57 under the assumption that* $\inf_x l(x, y, r) > -\infty$ *for at least one choice of* (y, r). *Then a necessary and sufficient condition for a vector* $\bar{y}$ *to support an exact penalty representation is that* $\bar{y}$ *be a proximal subgradient of the function* $p(u) = \inf_x f(x, u)$ *at* $u = 0$.

(b) *Consider the sharp Lagrangian of 11.58 under the assumption that* $\inf_x l(x, 0, r) > -\infty$ *for some* r. *Then a necessary and sufficient condition for the vector* $\bar{y} = 0$ *to support an exact penalty representation is that the function* $p(u) = \inf_x f(x, u)$ *be calm from below at* $u = 0$.

Detail. These results specialize Theorem 11.61. Relation 11(33) corresponds in (a) to the definition of a proximal subgradient (see 8.45), whereas in (b) it means calmness from below (see the material around 8.32). □

L.* Generalized Conjugacy

The notion of conjugate functions can be generalized in a number of ways, although none achieves the full power of the Legendre-Fenchel transform. Consider *any* nonempty sets X and Y and *any* function $\Phi : X \times Y \to \overline{I\!R}$. (The 'ordinary case', corresponding to what we have been involved with until now, has $X = I\!R^n$, $Y = I\!R^n$, and $\Phi(x, y) = \langle x, y \rangle$.) For any function $f : X \to \overline{I\!R}$, the *$\Phi$-conjugate* of f on Y is the function

$$f^{\Phi}(y) := \sup_{x \in X} \{\Phi(x, y) - f(x)\},$$

while the *Φ-biconjugate* of f back on X is the function

$$f^{\Phi\Phi}(x) := \sup_{y \in Y} \{\Phi(x, y) - f^{\Phi}(y)\}.$$

Likewise, for any function $g : Y \to \overline{I\!R}$, the Φ-conjugate of g on X is the function

$$g^{\Phi}(x) := \sup_{y \in Y} \{\Phi(x, y) - g(y)\},$$

while the Φ-biconjugate of g back on Y is the function

$$g^{\Phi\Phi}(y) := \sup_{x \in X} \{\Phi(x, y) - g^{\Phi}(x)\}.$$

Define the *Φ-envelope* functions on X to be the functions expressible as the pointwise supremum of some collection of functions of the form $\Phi(\cdot, y)$+constant for various choices of $y \in Y$, and define the Φ-envelope functions on Y analogously. (In the 'ordinary case', the proper Φ-envelope functions are the proper, lsc, convex functions.) Note that in circumstances where $X = Y$ but $\Phi(x,y)$ isn't symmetric in the arguments x and y, two different kinds of Φ-envelope functions might have to be distinguished on the same space.

11.63 Exercise (generalized conjugate functions). *In the scheme of Φ-conjugacy, for any function $f : X \to \overline{I\!R}$, f^Φ is a Φ-envelope function on Y, while $f^{\Phi\Phi}$ is the greatest Φ-envelope function on X majorized by f. Similarly, for any function $g : Y \to \overline{I\!R}$, g^Φ is a Φ-envelope function on X, while $g^{\Phi\Phi}$ is the greatest Φ-envelope function on Y majorized by g.*

Thus, Φ-conjugacy sets up a one-to-one correspondence between all the Φ-envelope functions f on X and all the Φ-envelope functions g on Y, with

$$g(y) = \sup_{x \in X}\{\Phi(x,y) - f(x)\}, \qquad f(x) = \sup_{y \in Y}\{\Phi(x,y) - g(y)\}.$$

Guide. Derive all this right from the definitions by elementary reasoning. □

11.64 Example (proximal transform). *For fixed $\lambda > 0$, pair $I\!R^n$ with itself under*

$$\Phi(x,y) = -\frac{1}{2\lambda}|x-y|^2 = \frac{1}{\lambda}\langle x,y\rangle - \frac{1}{2\lambda}|x|^2 - \frac{1}{2\lambda}|y|^2.$$

Then for any function $f : I\!R^n \to \overline{I\!R}$ and its Moreau envelope $e_\lambda f$ and λ-proximal hull $h_\lambda f$ one has

$$f^\Phi = -e_\lambda f, \qquad f^{\Phi\Phi}(x) = -e_\lambda[-e_\lambda f] = h_\lambda f.$$

In this case the Φ-envelope functions are the λ-proximal functions, defined as agreeing with their λ-proximal hull.

A one-to-one correspondence $f \leftrightarrow g$ in the collection of all proper functions of such type is obtained in which

$$g = -e_\lambda f, \qquad f = -e_\lambda g.$$

Detail. This follows from the definitions of f^Φ and $f^{\Phi\Phi}$ along with those of $e_\lambda f$ in 1.22 and $h_\lambda f$; see 1.44 and also 11.26(c). The symmetry of Φ in its two arguments yields the symmetry of the correspondence that is obtained. □

For the next example we recall from 2(13) the notation $I\!R^{n\times n}_{\rm sym}$ for the space of all symmetric real matrices of order n.

11.65 Example (full quadratic transform). *Pair $I\!R^n$ with $I\!R^n \times I\!R^{n\times n}_{\rm sym}$ under*

$$\Phi(x,y) = \langle v,x\rangle - j_Q(x) \text{ for } y = (v,Q), \text{ with } j_Q(x) = \frac{1}{2}\langle x, Qx\rangle.$$

In this case one has for any function $f : I\!R^n \to \overline{I\!R}$ that

$$f^\Phi(v,Q) = (f + j_Q)^*(v),$$

$$f^{\Phi\Phi}(x) = \begin{cases} (\operatorname{cl} f)(x) & \text{if } f \text{ is prox-bounded,} \\ -\infty & \text{otherwise.} \end{cases}$$

Here $f^\Phi \equiv -\infty$ if $f \equiv \infty$, whereas $f^\Phi \equiv \infty$ if f fails to be prox-bounded; otherwise f^Φ is a proper, lsc, convex function on the space $I\!R^n \times I\!R^{n\times n}_{\text{sym}}$.

Thus, Φ-conjugacy of this kind sets up a one-to-one correspondence $f \leftrightarrow g$ between the proper, lsc, prox-bounded functions f on $I\!R^n$ and certain proper, lsc, convex functions g on $I\!R^n \times I\!R^{n\times n}_{\text{sym}}$.

Detail. The formula for f^Φ is immediate from the definitions, and the one for $f^{\Phi\Phi}$ follows then from 1.44. The convexity of $f^{\Phi\Phi}$ comes from the linearity of $\Phi(x,y)$ with respect to the y argument for fixed x. □

11.66 Example (basic quadratic transform). *Pair $I\!R^n$ with $I\!R^n \times I\!R$ under*

$$\Phi(x,y) \;=\; \langle v,x\rangle - rj(x) \;\text{ for }\; y=(v,r), \;\text{ with }\; j(x) = \tfrac{1}{2}|x|^2.$$

In this case one has for any function $f : I\!R^n \to \overline{I\!R}$ that

$$f^\Phi(v,r) = (f + rj)^*(v), \qquad f^{\Phi\Phi}(x) = \begin{cases} (\operatorname{cl} f)(x) & \text{if } f \text{ is prox-bounded,} \\ -\infty & \text{otherwise.} \end{cases}$$

Here $f^\Phi \equiv -\infty$ if $f \equiv \infty$, whereas $f^\Phi \equiv \infty$ if f fails to be prox-bounded; aside from those extreme cases, f^Φ is a proper, lsc, convex function on $I\!R^n \times I\!R$. Thus, Φ-conjugacy of this kind sets up a one-to-one correspondence $f \leftrightarrow g$ between the proper, lsc, prox-bounded functions f on $I\!R^n$ and certain proper, lsc, convex functions g on $I\!R^n \times I\!R$.

Detail. This restricts the conjugate function in the preceding example to the subspace of $I\!R^{n\times n}_{\text{sym}}$ consisting of the matrices of form $Q = rI$. The biconjugate function is unaffected by this restriction because the derivation of its formula only relied on such special matrices, through 1.44. □

Yet another way that problems of optimization can be meaningfully be paired with one another is through the interchange rule for minimization in 1.35. In this framework the duality relations take the form of 'min = min' instead of 'min = max'. The idea is very simple. For comparison with the duality schemes in this chapter, it can be formulated as follows.

Given arbitrary nonempty sets X and Y and any function $l : X \times Y \mapsto \overline{I\!R}$, consider the problems,

$$\text{minimize } \varphi(x) \text{ over } x \in X, \text{ where } \varphi(x) := \inf\nolimits_y l(x,y),$$

$$\text{minimize } \psi(y) \text{ over } y \in Y, \text{ where } \psi(y) := \inf\nolimits_x l(x,y).$$

It's obvious that $\inf_X \varphi = \inf_Y \psi$; both values coincide with $\inf_{X\times Y} l$.

In contrast to the duality theory of convex optimization in 11.39, the two problems in this scheme aren't related to each other through perturbations, and the solutions to one problem can't be interpreted as 'multipliers' for the other. Rather, the two problems represent intermediate stages in an overall problem, namely that of minimizing $l(x,y)$ with respect to $(x,y) \in X \times Y$. Nonetheless, the scheme can be interesting in situations where a way exists for passing directly from one problem to the other without writing down the function l, especially if a correspondence between more than optimal values emerges. This is seen in the case of a Fenchel-type format resembling 11.41, where however the functions φ and ψ being minimized generally aren't convex.

11.67 Theorem (double-min duality in optimization). *Consider any primal problem of the form*

$$\text{minimize } \varphi(x) \text{ over } x \in I\!R^n, \quad \text{with } \varphi(x) = \langle c, x\rangle + k(x) - h(Ax - b),$$

for convex functions k on $I\!R^n$ and h on $I\!R^m$ such that k is proper, lsc, fully coercive and almost strictly convex, while h is finite and differentiable. Pair this with the dual problem

$$\text{minimize } \psi(y) \text{ over } y \in I\!R^m, \quad \text{with } \psi(y) = \langle b, y\rangle + h^*(y) - k^*(A^*y - c);$$

this similarly has h^* lsc, proper, fully coercive and almost strictly convex, while k^* is finite and differentiable. The objective functions φ and ψ in these two problems are amenable, and the two optimal values always agree: $\inf \varphi = \inf \psi$. Furthermore, in terms of

$$\begin{aligned} S :&= \{(x,y) \mid y = \nabla h(Ax - b),\ A^*y - c \in \partial k(x)\} \\ &= \{(x,y) \mid x = \nabla k^*(A^*y - c),\ Ax - b \in \partial h^*(y)\}, \end{aligned}$$

one has the following relations, which tie not only optimal solutions but also generalized stationary points of either problem to those of the other problem:

$$\begin{aligned} 0 \in \partial\varphi(\bar{x}) &\iff \exists\, \bar{y} \text{ with } (\bar{x},\bar{y}) \in S, \\ 0 \in \partial\psi(\bar{y}) &\iff \exists\, \bar{x} \text{ with } (\bar{x},\bar{y}) \in S, \\ (\bar{x},\bar{y}) \in S &\implies \varphi(\bar{x}) = \psi(\bar{y}). \end{aligned}$$

Proof. In the general format described before the theorem, these problems correspond to $l(x,y) = \langle c, x\rangle + k(x) + \langle b, y\rangle + h^*(y) - \langle y, Ax\rangle$.

The claim that h^* is fully coercive and almost strictly convex is justified by 11.5 and 3.27 for the coercivity and 11.13 for the strict convexity. The same results, in reverse implication, support the claim that k^* is finite and differentiable. Because convex functions that are finite and differentiable actually are $\mathcal{C}^1$ (by 9.20), the functions $h_0(x) = -h(Ax - b)$ and $k_0(y) = -k^*(A^*y - c)$ are $\mathcal{C}^1$, and consequently φ and ψ are amenable by 10.24(g). Then also

$$\begin{aligned} \partial\varphi(x) &= \partial k(x) + \nabla h_0(x) \text{ with } \nabla h_0(x) = -A^*\nabla h(Ax - b), \\ \partial\psi(y) &= \partial h^*(y) + \nabla k_0(x) \text{ with } \nabla k_0(x) = -A\nabla k^*(A^*y - c), \end{aligned}$$

from which the characterizations of $0 \in \partial\varphi(\bar{x})$ and $0 \in \partial\psi(\bar{y})$ are evident, the equivalence between the two expressions for S being a consequence of the subgradient inversion rule in 11.3. Pairs $(\bar{x},\bar{y}) \in S$ satisfy $h(A\bar{x}-b)+h^*(\bar{y}) = \langle A\bar{x}-b,\bar{y}\rangle$ and $k(\bar{x})+k^*(A^*\bar{y}-c) = \langle \bar{x}, A^*\bar{y}-c\rangle$ (again on the basis of 11.3), and these equations give $\varphi(\bar{x}) = \psi(\bar{y})$. □

A case worth noting in 11.67 is the one in which $k(x) = \delta_X(x) + \frac{1}{2}\langle x, Cx\rangle$ and $h(u) = \theta_{Y,B}(u)$ (cf. 11.18) for polyhedral sets X and Y and symmetric *positive-definite* matrices C and B. Then in minimizing φ one is minimizing the smooth piecewise linear-quadratic function

$$\varphi_0(x) = \langle c, x\rangle + \tfrac{1}{2}\langle x, Cx\rangle - \theta_{Y,B}(Ax-b)$$

subject to the linear constraints represented by $x \in X$, whereas in minimizing ψ one is minimizing the smooth piecewise linear-quadratic function

$$\psi_0(y) = \langle b, y\rangle + \tfrac{1}{2}\langle y, By\rangle - \theta_{X,C}(A^*y-c)$$

subject to the linear constraints represented by $y \in Y$. The functions φ_0 and ψ_0 needn't be convex, because the θ-expressions are subtracted rather than added as they were in 11.43, so this is a form of *nonconvex* extended linear-quadratic programming duality.

Commentary

A description of the 'Legendre transform' can be found in any text on the calculus of variations, since it's the means of generating the Hamiltonian functions and Hamiltonian equations that are crucial to that subject. The treatments in such texts often fall short in rigor, however. They revolve around inverting a gradient mapping ∇f as in 11.9, but typically without serious attention being paid to the mapping's domain and range, or to the conditions needed to ensure its *global* single-valued invertibility, such as (for most cases in practice) the strict convexity of f. The beauty of the Legendre-Fenchel transform, devised by Fenchel [1949], [1951], is that gradient inversion is replaced by an operation of maximization. Moreover, convexity properties are embraced from the start. In this way, a much more powerful tool is created which, interestingly, is perhaps the first in mathematical analysis to have relied on minimization/maximization in its very definition.

Mandelbrojt [1939] had earlier developed a limited case of conjugacy for functions of a single real variable. In the still more special context of nondecreasing convex functions on $\mathbb{R}_+$, similar notions were explored by Young [1912] and utilized in the theory of Banach spaces and beyond; cf. Birnbaum and Orlicz [1931] and Krasnosel'skii and Rutitskii [1961]. None of this captured the n-dimensional character of the Legendre transform, however, or allowed for the kind of focus on domains that's essential to handling *constraints* in the process of dualization.

Fenchel formulated the basic result in Theorem 11.1 in terms of pairs (C, f) consisting of a finite convex function f on a nonempty convex set C in $\mathbb{R}^n$. His transform was extended to infinite-dimensional spaces by Moreau [1962] and Brøndsted [1964] (publication of a dissertation written under Fenchel's supervision), with Moreau adopting the pattern of extended-real-valued functions f defined on all the whole space. The details of the Legendre case in 11.9 were worked out by Rockafellar [1967d]. All these authors were, in some way, aware of the facts in 11.2–11.4.

Rockafellar [1966b] discovered the support function meaning of horizon functions in 11.5 as well as the formula in 11.6 for the support function of a level set and the dualizations of coercivity and level-coercivity in 11.8(c)(d). The dualizations of differentiability in 11.8(b) and 11.13 were developed in Rockafellar [1970a], which is also the source for the connection in 11.7 between conjugate functions and cone polarity (probably known to Fenchel), and for the linear-quadratic examples in 11.10 and 11.11 and the log-exponential example in 11.12.

In the important case of convex functions on $\mathbb{R}^1$, special techniques can be used for determining the conjugate functions, for instance by 'integrating' right or left derivatives; for a thorough treatment see Chapter 8 of Rockafellar [1984b]. The one-dimensional version of Theorem 11.14, that f^* inherits from f the property of being piecewise linear, or piecewise linear-quadratic, can be found there as well.

In the full, n-dimensional version of Theorem 11.14, part (a) is new in its statement about the preservation of piecewise linearity when passing from a convex function to its conjugate, but this property corresponds through 2.49 to the fact that if $\operatorname{epi} f$ is polyhedral, the same is true also of $\operatorname{epi} f^*$. In that form, the observation goes back to Rockafellar [1963]. The assertion in 11.17(a) about the support functions of polyhedral sets being piecewise linear has similar status. The fact in 11.17(b), that the polar of a polyhedral cone is polyhedral, was recognized by Weyl [1935].

As for part (b) of Theorem 11.14, concerning the self-duality of the class of piecewise linear-quadratic functions, this was proved by Sun [1986], but with machinery from the theory of monotone mappings. Without the availability of that machinery in this chapter (it won't be set up until Chapter 12), we had to invent a different, more direct proof. The line segment test in 11.15, on which it rests, doesn't seem to have been formulated previously.

The consequent fact in 11.16, about piecewise linear-quadratic convex functions $f : \mathbb{R}^n \to \overline{\mathbb{R}}$ attaining their minimum value when that value is finite, can be compared to the result of Frank and Wolfe [1956] that a linear-quadratic convex function f_0 achieves its minimum relative to any polyhedral convex set C on which it's bounded from below. The latter can be identified with the case of 11.16 where $f = f_0 + \delta_C$.

The general polarity correspondence in 11.19, for sets that contain the origin but aren't necessarily cones, was developed first for bounded sets by Minkowski [1911], whose insights extended to the dualizations in 11.20. The formula for conjugate composite functions in 11.21 comes from Rockafellar [1970a].

The dual operations in 11.22 and 11.23 were investigated in various degrees by Fenchel [1951], Rockafellar [1963] and Moreau [1967]. The special cases in 11.24 for support functions and in 11.25 for cones were known much earlier. Moreau [1965] established the dual envelope formula in 11.26(b) (for $\lambda = 1$) and also the gradient interpretation in 11.27 for the proximal mappings associated with convex functions.

The results on the norms of sublinear mappings in 11.29 and 11.30 are new, but the adjoint duality for operations on such mappings in 11.31 was already disclosed in Rockafellar [1970a]. The special results in 11.32 and 11.33 on duality for operations on piecewise linear-quadratic functions are original as well.

The epi-continuity of the Legendre-Fenchel transform in Theorem 11.34 was discovered by Wijsman [1964], [1966], who also reported the corresponding behavior of support functions and polar cones in 11.35. The 'epi-semicontinuity' results in

Theorem 11.34, which provide inequalities for dualizing e-liminf and e-limsup, are actually new, and likewise for the support function version in 11.35(a). The inner and outer limit relations for cone polarity in 11.35(b) were noted by Walkup and Wets [1967], however. Those authors, in the same paper, established the polar cone isometry in Theorem 11.36, moreover not just for $\mathbb{R}^n$ but any reflexive Banach space. The proof of this isometry that we give here in terms of the duality between addition and epi-addition of convex functions is new. It readily generalizes to any norm and its polar, not only the Euclidean norm and unit ball, in setting up cone distances, and it too could therefore be followed in the Banach space setting.

The application in 11.37, showing that the Legendre-Fenchel transform is an isometry with respect to cosmic epi-distances, is new, but an isometry result with respect to a different metric, based instead on uniform convergence of Moreau envelopes on bounded sets, was obtained by Attouch and Wets [1986]. The technique of passing through cones and the continuity of the polar operation in 11.36 for purposes of investigating duality in the epi-convergence of convex functions (apart from questions of isometry), and thereby developing extensions of Wijsman's theorem, began with Wets [1980]. That approach has been followed more recently in infinite dimensions by Penot [1991] and Beer [1993]. Other epi-continuity results for the Legendre-Fenchel transform in spaces beyond $\mathbb{R}^n$ can be found in those works and in earlier papers of Mosco [1971], Joly [1973], Back [1986] and Beer [1990].

Many researchers have been fascinated by dual problems of optimization. The most significant early example was linear programming duality (cf. 11.43), which was laid out by Gale, Kuhn and Tucker [1951]. This duality grew out of the theory of minimax problems and two-person, zero-sum games that was initiated by von Neumann [1928]. Fenchel [1951], while visiting at Princeton where Gale and Kuhn were Ph.D. students of Tucker, sought to set up a parallel theory of dual problems in which the primal problem consisted of minimizing $h(x) + k(x)$, for what we now call lsc, proper, convex functions h and k on $\mathbb{R}^n$, while the dual problem consisted of maximizing $-h^*(y) - k^*(-y)$. Fenchel's duality theorem suffered from an error, however. Rockafellar [1963], [1964a], [1966c], [1967a], fixed the error and incorporated a linear mapping into the problem statement so as to obtain the scheme in 11.41. Perturbations played a role in that duality, but the scheme in 11.39 and 11.40, explicitly built around primal and dual perturbations, didn't emerge until Rockafellar [1970a].

Extended linear-quadratic programming was developed by Rockafellar and Wets [1986] along with the duality results in 11.43; further details were added by Rockafellar [1987]. It was in those papers that the dualizing penalty expressions $\theta_{Y,B}$ in 11.18 made their debut. The application to dualized composition in 11.44 is new.

The Lagrangian perturbation format in 11.45 came out in Rockafellar [1970a], [1974a], for the case of $f(x, u)$ convex in (x, u). The associated minimax theory in 11.50–11.52 was developed there also. The extension of Lagrangian dualization to the case of $f(x, u)$ convex merely in u started with Rockafellar [1993a], and that's where the multiplier rule in 11.46 was proved.

The formula in 11.48, relating the subgradients of $f(x, u)$ to those of $l(x, y)$ for convex f, can be interpreted as a generalization of the inversion rule in 11.3 through the fact that the functions $f(x, \cdot)$ and $-l(x, \cdot)$ are conjugate to each other. Instead of full inversion one has a partial inversion along with a change of sign in the residual argument. A similar partial inversion rule is known classically for smooth f with respect to taking the Legendre transform in the u argument, this being essential to the theory of Hamiltonian equations in the 'calculus of variations'. One could ask whether a broader formula of such type can be established for nonsmooth functions f that might only be convex in u, not necessarily in x and u jointly. Rockafellar [1993b], [1996], has answered this in the affirmative for particular function structures and more generally in terms of a partial convexification operation on subgradients.

The results on perturbing saddle values and saddle points in Theorem 11.53 are largely due to Golshtein [1972]. The linear programming case in Example 11.54 was treated earlier by Williams [1970]. The application here to linear-quadratic programming is new. For some nonconvex extensions of such perturbation theory that instead utilize augmented Lagrangians, see Rockafellar [1984a].

Augmented Lagrangians were introduced in connection with numerical methods for solving problems in nonlinear programming, and they have mainly been viewed in that special context; see Rockafellar [1974b], [1993a], Bertsekas [1982], and Golshtein and Tretyakov [1996]. They haven't been considered before with the degree of generality in 11.55–11.61. Exact penalty representations of the linear type in 11.62(b) have a separate history; see Burke [1991] for this background.

The generalized conjugacy in 11.64 was brought to light by Moreau [1967], [1970]. Something similar was noted by Weiss [1969], [1974], and also by Elster and Nehse [1974]. Such ideas were utilized by Balder [1977] in work related to augmented Lagrangians; this expanded on the strategy in Rockafellar [1974b], where the basic quadratic transform in 11.66 was implicitly utilized for this purpose. For related work see also Dolecki and Kurcyusz [1978]. The basic quadratic transform was fleshed out by Poliquin [1990], who put it to work in nonsmooth analysis; he demonstrated by this means, for instance, that any proper, lsc function $f : \mathbb{R}^n \to \overline{\mathbb{R}}$ that's bounded from below can be expressed as a composite function $g \circ F$ with g lsc convex and F of class $\mathcal{C}^\infty$. The full quadratic transform in 11.65 was set up earlier by Janin [1973]. He observed that its main properties could be derived as consequences of known features of the Legendre-Fenchel transform.

The earliest duality theory of the double-min variety as in 11.67 is due to Toland [1978], [1979]. He concentrated on the difference of two convex functions; there was no linear mapping, as associated here with the matrix A. The particular content of Theorem 11.67, in adding a linear transformation and drawing on facts in 11.8 (coercivity versus finiteness) and in 11.13 (strict convexity versus differentiability) hasn't been furnished before. The idea of relating 'extremal points' of one problem to those of another was carried forward on a broader front by Ekeland [1977], who, like Toland, was motivated by applications in the calculus of variations.

Also in the line of duality for nonconvex problems of optimization, the work of Aubin and Ekeland [1976] deserves special mentions. They constructed quantitative estimates of the 'lack of convexity' of a function and showed how these estimates can be utilized to get bounds on the size of the duality gap (between primal and dual optimal values) in a Fenchel-like format.

Although we haven't taken it up here, there is also a concept of conjugacy for convex-concave functions; see Rockafellar [1964b], [1970a]. A theory of dual minimax problems has been developed in such terms by McLinden [1973], [1974]. Epi-convergence isn't the right convergence for such functions and must be replaced by *epi-hypo-convergence*; see Attouch and Wets [1983a], Attouch, Azé and Wets [1988].

12. Monotone Mappings

A valuable tool in the study of gradient and subgradient mappings, solution mappings, and various other mappings of importance in variational analysis, both single-valued and set-valued, is the following concept of monotonicity.

12.1 Definition (monotonicity). *A mapping $T : \mathbb{R}^n \rightrightarrows \mathbb{R}^n$ is called monotone if it has the property that*

$$\langle v_1 - v_0, x_1 - x_0 \rangle \geq 0 \text{ whenever } v_0 \in T(x_0),\ v_1 \in T(x_1),$$

and strictly monotone if this inequality is strict when $x_0 \neq x_1$.

When T is single-valued, the monotonicity property takes the form of requiring that $\langle T(x_1) - T(x_0), x_1 - x_0 \rangle \geq 0$ for all x_0 and x_1.

A. Monotonicity Tests and Maximality

The monotonicity inequality holds for $T = \nabla f$ when f is a differentiable convex function, as observed in 2.14(a). We'll see in 12.17 that monotonicity is exhibited also by the subgradient mappings of nondifferentiable convex functions. Many other connections between monotonicity and convexity will come to light as well. First, though, we'll look at more general sources of monotonicity and a concept of 'maximal' monotone mappings.

12.2 Example (positive semidefiniteness). *An affine mapping $L(x) = Ax + a$ for a vector $a \in \mathbb{R}^n$ and a matrix $A \in \mathbb{R}^{n \times n}$, not necessarily symmetric, is monotone if and only if A is positive-semidefinite: $\langle x, Ax \rangle \geq 0$ for all x. It is strictly monotone if and only if A is positive-definite: $\langle x, Ax \rangle > 0$ for all $x \neq 0$. In particular, the identity mapping I is strictly monotone.*

Detail. The monotonicity inequality in 12.1 reduces to the requirement that $\langle A(x_1 - x_0), (x_1 - x_0) \rangle \geq 0$ for all x_0 and x_1, which obviously means that A is positive-semidefinite. The same goes for strict monotonicity versus positive definiteness. □

As a sidelight on the property in 12.2, recall that any matrix $A \in \mathbb{R}^{n \times n}$ can be written as

$$A = A_s + A_a \text{ with } A_s = \tfrac{1}{2}(A + A^*) \text{ and } A_a = \tfrac{1}{2}(A - A^*), \qquad 12(1)$$

where A_s is the *symmetric* part of A (satisfying $A_s^* = A_s$) and A_a the *antisymmetric* part of A (satisfying $A_a^* = -A_a$). The positive semidefiniteness of A is simply the positive semidefiniteness of its symmetric part A_s, because $\langle x, Ax\rangle = \langle x, A_s x\rangle$ always. The antisymmetric part A_a has no role in this. In particular, the linear mapping $x \mapsto Ax$ associated with any antisymmetric matrix A is monotone (since then $A_s = 0$).

12.3 Proposition (monotonicity of differentiable mappings). *A differentiable mapping $F : \mathbb{R}^n \to \mathbb{R}^n$ is monotone if and only if for each x the Jacobian matrix $\nabla F(x)$ (not necessarily symmetric) is positive-semidefinite. It is strictly monotone if the matrix $\nabla F(x)$ is positive-definite for each x (but this condition is only sufficient, not necessary).*

Proof. When F is monotone, one has

$$\big\langle F(x+\tau w) - F(x),\, (x+\tau w) - x\big\rangle \geq 0 \text{ for all } \tau > 0,\ x,\ w.$$

Dividing this by τ and taking the limit as $\tau \searrow 0$, we get $\langle w, \nabla F(x)w\rangle \geq 0$ for all choices of x and w. This is the claimed positive semidefiniteness of $\nabla F(x)$ for all x. (Strict monotonicity of F wouldn't yield positive definiteness of $\nabla F(x)$, though, because the strict inequality might not remain strict in the limit.)

Assuming conversely that $\nabla F(x)$ is positive-semidefinite for all x, consider arbitrary points x_0 and x_1 and the function $\varphi(\tau) := \langle x_1 - x_0,\, F(x_\tau) - F(x_0)\rangle$ with $x_\tau := (1-\tau)x_0 + \tau x_1$. We want to show that $\varphi(1) \geq 0$. We have $\varphi(0) = 0$ and $\varphi'(\tau) = \big\langle x_1 - x_0, \nabla F(x_\tau)(x_1 - x_0)\big\rangle$, so φ is nondecreasing. Hence $\varphi(1) \geq 0$ as desired. The same argument under the assumption that $\nabla F(x)$ is positive-definite everywhere yields $\varphi(1) > 0$ when $x_1 \neq x_0$ and gives the conclusion that F is strictly monotone. □

From the perspective of 12.3, monotonicity can be regarded as the natural generalization of positive semidefiniteness from linear to nonlinear and possibly even multivalued mappings. So far, only monotone mappings that are single-valued have been viewed, but for various theoretical reasons multivaluedness will be quite important. In fact, multivalued monotone mappings already arise from single-valued ones, as above, through various monotonicity-preserving operations like taking inverses.

12.4 Exercise (operations that preserve monotonicity).

(a) *If $T : \mathbb{R}^n \rightrightarrows \mathbb{R}^n$ is monotone, then so too is T^{-1}.*

(b) *If $T : \mathbb{R}^n \rightrightarrows \mathbb{R}^n$ is monotone, then so too is λT for any $\lambda > 0$. The same holds for strict monotonicity.*

(c) *If $T_1 : \mathbb{R}^n \rightrightarrows \mathbb{R}^n$ and $T_2 : \mathbb{R}^n \rightrightarrows \mathbb{R}^n$ are monotone, then $T_1 + T_2$ is monotone. If in addition either T_1 or T_2 is strictly monotone, then $T_1 + T_2$ is strictly monotone.*

(d) *If $S : \mathbb{R}^m \rightrightarrows \mathbb{R}^m$ is monotone, then for any matrix $A \in \mathbb{R}^{m\times n}$ and vector $a \in \mathbb{R}^m$ the mapping $T(x) = A^*S(Ax + a)$ is monotone. If in addition S is strictly monotone and A has rank n, then T is strictly monotone.*

Guide. Use Definition 12.1 directly. Note that in (d) there is no assumption of positive semidefiniteness on A. □

The most powerful features of monotonicity for mappings $T : \mathbb{R}^n \rightrightarrows \mathbb{R}^n$ only come forward in the presence of a companion property of 'maximality'. To make use of these features, a given mapping may need to be enlarged, and this is a process that can well lead to imbedding a single-valued mapping within a multivalued mapping, although the multivaluedness turns out to be of a very special kind.

12.5 Definition (maximal monotonicity). *A monotone mapping $T : \mathbb{R}^n \rightrightarrows \mathbb{R}^n$ is maximal monotone if no enlargement of its graph is possible in $\mathbb{R}^n \times \mathbb{R}^n$ without destroying monotonicity, or in other words, if for every pair $(\hat{x}, \hat{v}) \in (\mathbb{R}^n \times \mathbb{R}^n) \setminus \operatorname{gph} T$ there exists $(\tilde{x}, \tilde{v}) \in \operatorname{gph} T$ with $\langle \hat{v} - \tilde{v}, \hat{x} - \tilde{x} \rangle < 0$.*

More generally, T is maximal monotone locally around $(\bar{x}, \bar{v})$, a point of $\operatorname{gph} T$, if there is a neighborhood V of $(\bar{x}, \bar{v})$ such that for every pair $(\hat{x}, \hat{v}) \in V \setminus \operatorname{gph} T$ there exists $(\tilde{x}, \tilde{v}) \in V \cap \operatorname{gph} T$ with $\langle \hat{v} - \tilde{v}, \hat{x} - \tilde{x} \rangle < 0$.

The study of monotone mappings can largely be reduced to that of maximal monotone mappings on the following principle.

12.6 Proposition (existence of maximal extensions). *For any monotone mapping $T : \mathbb{R}^n \rightrightarrows \mathbb{R}^n$ there is a maximal monotone mapping $\overline{T} : \mathbb{R}^n \rightrightarrows \mathbb{R}^n$ (not necessarily unique) such that $\operatorname{gph} \overline{T} \supset \operatorname{gph} T$.*

Proof. Let $\mathcal{T}$ denote the collection of all monotone mappings $T' : \mathbb{R}^n \rightrightarrows \mathbb{R}^n$ such that $\operatorname{gph} T' \supset \operatorname{gph} T$, and consider $\mathcal{T}$ under the partial ordering induced by graph inclusion. According to Zorn's lemma, there's a maximal linearly ordered subset $\mathcal{T}_0$ of $\mathcal{T}$. Let $\overline{T}$ be the mapping whose graph is the union of the graphs of the mappings $T' \in \mathcal{T}_0$. It's evident that $\overline{T}$ is monotone, and that there can't be a monotone mapping T' with $\operatorname{gph} \overline{T} \subset \operatorname{gph} T'$, $\overline{T} \neq T'$. Hence $\overline{T}$ is maximal monotone. □

12.7 Example (maximality of continuous monotone mappings). *If a continuous mapping $F : \mathbb{R}^n \to \mathbb{R}^n$ is monotone, it is maximal monotone. In particular, every differentiable monotone mapping is maximal monotone, and so too is every affine monotone mapping.*

Detail. Suppose $(\hat{x}, \hat{v})$ has the property that $\langle \hat{v} - F(x), \hat{x} - x \rangle \geq 0$ for all x. Does this imply that $\hat{v} = F(\hat{x})$, as demanded by maximality? Taking $x = \hat{x} + \varepsilon u$ with $\varepsilon > 0$ and arbitrary $u \in \mathbb{R}^n$, we see that $\langle \hat{v} - F(\hat{x} + \varepsilon u), u \rangle \geq 0$. The assumed continuity of F guarantees that $F(\hat{x} + \varepsilon u) \to F(\hat{x})$ as $\varepsilon \searrow 0$, so that $\langle \hat{v} - F(\hat{x}), u \rangle \geq 0$. This can't hold for all $u \in \mathbb{R}^n$ unless $\hat{v} - F(\hat{x}) = 0$. □

Although maximal monotonicity is an automatic property for continuous monotone mappings, it isn't automatic for monotone mappings T of other kinds. Its presence is associated with powerful properties of a geometric nature, especially of the graph of T. The following list is merely a beginning.

12.8 Exercise (graphs and images under maximal monotonicity).

(a) *T is maximal monotone if and only if T^{-1} is maximal monotone.*

(b) *If T is maximal monotone, $\operatorname{gph} T$ is closed; thus, T is osc.*

(c) *If T is maximal monotone, both T and T^{-1} are closed-convex-valued.*

Guide. These facts are elementary consequences of Definition 12.5. For the convex-valuedness, it's enough to show that when $v_\tau = (1-\tau)v_0 + \tau v_1$ with $v_0 \in T(\bar{x})$ and $v_1 \in T(\bar{x})$, one has $\langle v - v_\tau, x - \bar{x}\rangle \geq 0$ for all $\tau \in (0,1)$ and all $(x,v) \in \operatorname{gph} T$; the maximality then forces $(\bar{x}, v_\tau) \in \operatorname{gph} T$. □

While monotonicity is readily perceived to persist under operations like the ones in 12.4(c)(d), maximal monotonicity need not persist unless certain 'constraint qualifications' are satisfied by the domains. Such matters will be taken up in 12.43–12.47.

The graph of a monotone mapping $T : \mathbb{R}^n \to \mathbb{R}^n$ lies in $\mathbb{R}^{2n}$ and therefore can't be pictured for $n > 1$. In the special case of $n = 1$, however, it lies in $\mathbb{R}^2$ and, when maximal, has the striking sort of appearance in Figure 12–1. In particular, the properties in 12.8 are confirmed.

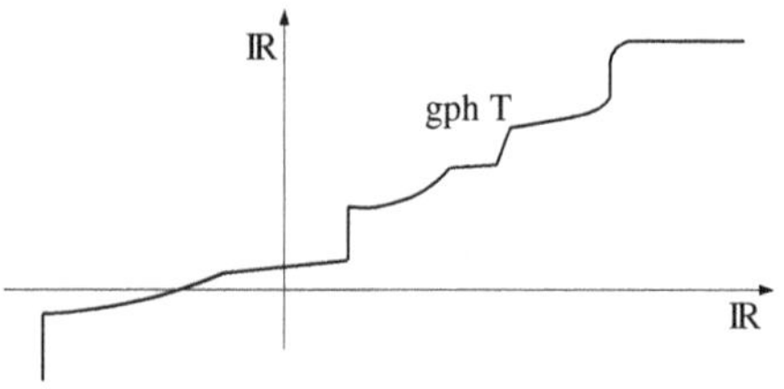

Fig. 12–1. A maximal monotone mapping in one dimension.

12.9 Exercise (maximal monotonicity in one dimension).

(a) *A mapping $T : \mathbb{R} \rightrightarrows \mathbb{R}$ is monotone if and only if $\operatorname{gph} T$ is totally ordered as a subset of $\mathbb{R}^2$ with respect to the partial ordering induced by the cone $\mathbb{R}^2_+$, or in other words, for every $(x_0, v_0) \in \operatorname{gph} T$ and $(x_1, v_1) \in \operatorname{gph} T$, either one has $x_0 \leq x_1$ and $v_0 \leq v_1$, or one has $x_1 \leq x_0$ and $v_1 \leq v_0$.*

(b) *A mapping $T : \mathbb{R} \rightrightarrows \mathbb{R}$ is maximal monotone if and only if there is a nondecreasing function $\varphi : \mathbb{R} \to \overline{\mathbb{R}}$ with $\varphi \not\equiv \infty$ and $\varphi \not\equiv -\infty$ such that*

$$T(x) = \{ v \in \mathbb{R} \mid \varphi(x_-) \leq v \leq \varphi(x_+) \} \;\; \textit{for}$$
$$\varphi(x_-) := \lim_{x' \nearrow x} \varphi(x'), \qquad \varphi(x_+) := \lim_{x' \searrow x} \varphi(x').$$

Guide. Verify (a) right from Definition 12.1, noting that maximal monotonicity corresponds then to $\operatorname{gph} T$ being a totally ordered subset of $\mathbb{R}^2$ to which no pair $(\bar{x}, \bar{v})$ can be added without destroying the total orderedness. In (b), start from the assumption that T has a representation of the kind described and argue that $\operatorname{gph} T$ has this property of maximal total orderedness. For the converse in (b), take $\operatorname{gph} T$ to be a maximal totally ordered subset of $\mathbb{R}^2$ and introduce the functions

$$\varphi_+(x) = \inf \{ v' \mid \exists x' > x \text{ with } v' \in T(x') \},$$
$$\varphi_-(x) = \sup \{ v' \mid \exists x' < x \text{ with } v' \in T(x') \}.$$

Show that $\varphi_-(x_0) \leq \varphi_+(x_0) \leq \varphi_-(x_1) \leq \varphi_+(x_1)$ when $x_0 < x_1$, and that $T(x)$ must include all finite values in the interval $[\varphi_-(x), \varphi_+(x)]$, if any. Construct φ by choosing the value $\varphi(x)$ (not necessarily finite) arbitrarily from this interval for each x, and demonstrate that a representation of T of the kind in (b) is thereby achieved, moreover with $\varphi_+(x) = \varphi(x_+)$ and $\varphi_-(x) = \varphi(x_-)$. □

For dimensions $n > 1$ there's no analog of the ordering property in 12.9(a), but much of the special geometry in Figure 12–1 will carry over nonetheless. For maximal monotone mappings, multivaluedness is very particular. Such mappings are remarkably 'function-like'. Our initial goal is the development of this aspect of monotonicity.

B. Minty Parameterization

The next result opens the technical terrain for establishing the main facts about the geometry of graphs of general maximal monotone mappings. Recall from 9.4 that a single-valued mapping $S : D \to \mathbb{R}^n$, with $D \subset \mathbb{R}^n$, is called nonexpansive if it is Lipschitz continuous with constant 1, or in other words, satisfies $|S(z_1) - S(z_0)| \leq |z_1 - z_0|$ for all z_0 and z_1 in D; it is contractive if the inequality is strict when $z_1 \neq z_0$. Nonexpansivity is closely connected with monotonicity in a certain way, but in order to see this clearly it will be helpful to think of a single-valued mapping $S : D \to \mathbb{R}^n$ as a special case of a set-valued mapping $S : \mathbb{R}^n \rightrightarrows \mathbb{R}^n$ with $\operatorname{dom} S = D$. From either perspective the graph of S, as a subset of $\mathbb{R}^n \times \mathbb{R}^n$, is the same.

We therefore now cast the definition in graphical terms by saying that a mapping $S : \mathbb{R}^n \rightrightarrows \mathbb{R}^n$ is *nonexpansive* if

$$|w_1 - w_0| \leq |z_1 - z_0| \text{ whenever } w_1 \in S(z_1),\ w_0 \in S(z_0),$$

and *contractive* if the inequality is strict when $z_1 \neq z_0$. Obviously this condition entails having $w_1 = w_0$ when $z_1 = z_0$. Thus, it entails the single-valuedness of S on $\operatorname{dom} S$ and amounts to the same thing as the condition previously used to express these concepts.

12.10 Exercise (operations that preserve nonexpansivity). *For nonexpansive mappings $S_0 : \mathbb{R}^n \rightrightarrows \mathbb{R}^n$ and $S_1 : \mathbb{R}^n \rightrightarrows \mathbb{R}^n$, one has that*

(a) *$S_1 \circ S_0$ is nonexpansive;*

(b) *$\lambda_0 S_0 + \lambda_1 S_1$ is nonexpansive if $|\lambda_0| + |\lambda_1| \leq 1$.*

12.11 Proposition (monotone versus nonexpansive mappings). *The one-to-one linear transformation $J : \mathbb{R}^n \times \mathbb{R}^n \to \mathbb{R}^n \times \mathbb{R}^n$ with $J(x, v) = (v + x,\ v - x)$*

induces a one-to-one correspondence between mappings $T : \mathbb{R}^n \rightrightarrows \mathbb{R}^n$ *and mappings* $S : \mathbb{R}^n \rightrightarrows \mathbb{R}^n$ *through*

$$\operatorname{gph} S = J(\operatorname{gph} T), \qquad \operatorname{gph} T = J^{-1}(\operatorname{gph} S),$$

in which S *is nonexpansive if and only if* T *is monotone, and on the other hand,* S *is contractive if and only if* T *is strictly monotone as well as single-valued on* $\operatorname{dom} T$*. Under this correspondence one has*

$$S = I - 2I{\circ}(I+T)^{-1}, \qquad T = (I-S)^{-1}{\circ}2I - I. \tag{12(2)}$$

Proof. We calculate for $(z_0, w_0) = J(x_0, v_0)$ and $(z_1, w_1) = J(x_1, v_1)$ that

$$\begin{aligned} |z_1 - z_0|^2 - |w_1 - w_0|^2 &= \big\langle (z_1 - z_0) + (w_1 - w_0),\, (z_1 - z_0) - (w_1 - w_0)\big\rangle \\ &= \big\langle (z_1 + w_1) - (z_0 + w_0),\, (z_1 - w_1) - (z_0 - w_0)\big\rangle \\ &= \langle 2v_1 - 2v_0,\, 2x_1 - 2x_0\rangle = 4\langle v_1 - v_0,\, x_1 - x_0\rangle \end{aligned}$$

and therefore

$$|w_1 - w_0| \le |z_1 - z_0| \iff \langle v_1 - v_0,\, x_1 - x_0\rangle \ge 0. \tag{12(3)}$$

Since this holds for all corresponding pairs $(z_i, w_i) \in \operatorname{gph} S$ and $(x_i, v_i) \in \operatorname{gph} T$, nonexpansivity of S is equivalent to monotonicity of T.

Further, we note that S is contractive if and only if the inequality on the left side of 12(3) is strict when $(z_0, w_0) \neq (z_1, w_1)$. At the same time, T is strictly monotone and also single-valued on $\operatorname{dom} T$ if and only if the inequality in the right side of 12(3) is strict when $(x_0, v_0) \neq (x_1, v_1)$.

To obtain the equivalence of the first formula in 12(2) with having $\operatorname{gph} S = J(\operatorname{gph} T)$, observe that the condition $(z, w) \in J(\operatorname{gph} T)$ means that for some x and some $v \in T(x)$ one has $z = v + x$ and $w = v - x$. This is the same as requiring the existence of x such that $z \in (I+T)(x)$ and $w = (z-x)-x = z-2x$. Thus, $w \in S(z)$ if and only if $w = z - 2x$ for some $x \in (I+T)^{-1}(z)$, which says that $S = I - 2I{\circ}(I+T)^{-1}$, as claimed.

Then we have $(I-S) = 2I{\circ}(I+T)^{-1}$, hence $(I+T) = (I-S)^{-1}{\circ}2I$, which is the second formula in 12(2). □

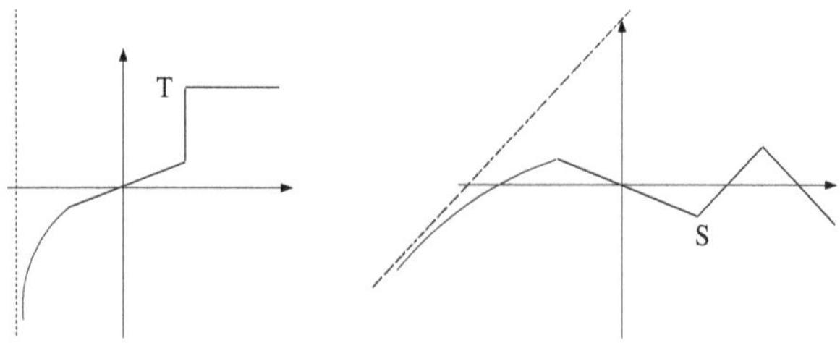

Fig. 12–2. Graph relationship between nonexpansive and monotone mappings.

The transformation J in Proposition 12.11 can be visualized in the case of $n = 1$ as a clockwise rotation of $I\!R \times I\!R$ by $\pi/4$ followed by an expansion (relative to the origin) by a factor of $\sqrt{2}$. The expansion isn't essential but simplifies the formulas in 12(2).

12.12 Theorem (maximality and single-valued resolvants). *Let $T : I\!R^n \rightrightarrows I\!R^n$ be monotone, and let $\lambda > 0$. Then $(I + \lambda T)^{-1}$ is monotone and nonexpansive.*

Moreover, T is maximal monotone if and only if $\operatorname{rge}(I + \lambda T) = I\!R^n$, or equivalently, $\operatorname{dom}(I + \lambda T)^{-1} = I\!R^n$. In that case $(I + \lambda T)^{-1}$ is maximal monotone too, and it is a single-valued mapping from all of $I\!R^n$ into itself.

Proof. Like T, the mapping λT is monotone, and it's maximal if and only if T is maximal. Without loss of generality we can replace T by λT, or what amounts to the same thing, assume that $\lambda = 1$. The monotonicity of T implies that of $I+T$ by 12.4(c) and the monotonicity of I in 12.1, hence that of $(I+T)^{-1}$ by 12.4(a). According to Proposition 12.11 we have $(I + T)^{-1} = \frac{1}{2}(I - S)$ for a certain nonexpansive mapping S. Since I is nonexpansive, so too is $\frac{1}{2}I - \frac{1}{2}S$ by 12.10(b). Thus, $(I + T)^{-1}$ is nonexpansive; on $D = \operatorname{dom}(I + T)^{-1}$ it's single-valued and Lipschitz continuous with constant 1.

In the framework of Proposition 12.11, T is a maximal monotone mapping if and only if S is a maximal nonexpansive mapping—its graph can't be enlarged without losing the nonexpansive property. But by Theorem 9.58 any nonexpansive mapping can be extended to all of $I\!R^n$. Thus, S is a maximal nonexpansive mapping if and only if $\operatorname{dom} S = I\!R^n$, which according to our formulas is the same as $\operatorname{dom}(I+T)^{-1} = I\!R^n$. Since $(I+T)^{-1}$ is single-valued and continuous on $\operatorname{dom}(I+T)^{-1}$ as well as monotone, it follows that T is maximal monotone; cf. 12.7. □

The mappings $(I + \lambda T)^{-1}$ for $\lambda > 0$ in Theorem 12.12 are called the *resolvants* of T. Their graphs are obtained from that of T by what amounts to just a linear change of coordinates in the graph space:

$$\operatorname{gph}(I + \lambda T)^{-1} = J_\lambda(\operatorname{gph} T) \quad \text{for} \quad J_\lambda : (x, v) \mapsto (x + \lambda v, x). \tag{12(4)}$$

This is true because $x = (I + \lambda T)^{-1}(z)$ if and only if $\lambda^{-1}(z - x) \in T(x)$.

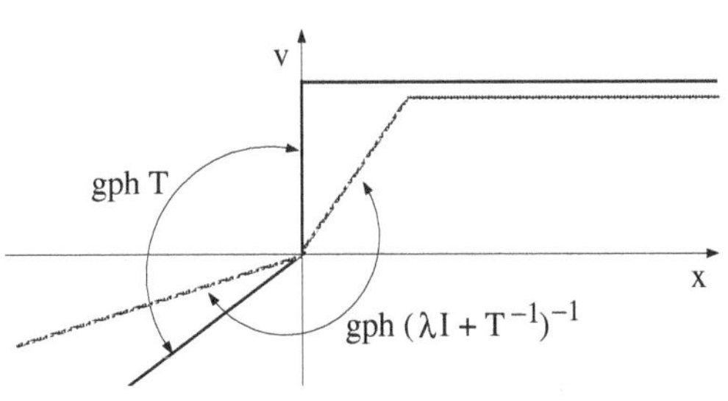

Fig. 12–3. Yosida λ-regularization.

12.13 Example (Yosida regularization). *For any mapping $T : \mathbb{R}^n \rightrightarrows \mathbb{R}^n$ and any $\lambda > 0$, the corresponding Yosida λ-regularization of T is the mapping*

$$(\lambda I + T^{-1})^{-1} = (I + \lambda^{-1}T^{-1})^{-1} \circ \lambda^{-1} I.$$

As $\lambda \searrow 0$, $(\lambda I + T^{-1})^{-1}$ converges graphically to $\operatorname{cl} T$, hence to T itself when T is osc, as when T is maximal monotone.

In the maximal monotone case, $(\lambda I + T^{-1})^{-1}$ is not only maximal monotone but single-valued and Lipschitz continuous globally with constant λ^{-1}.

Detail. It's elementary for any mapping S that $\lambda I + S \xrightarrow{g} \operatorname{cl} S$ as $\lambda \searrow 0$ (with $\operatorname{cl} S$ defined to be the mapping whose graph is $\operatorname{cl}(\operatorname{gph} S)$). Applying this to $S = T^{-1}$ and using the continuity of the inverse operation with respect to graph convergence, one sees that $(\lambda I + T^{-1})^{-1} \xrightarrow{g} \operatorname{cl} T$ as $\lambda \searrow 0$. The remaining assertions follow from Theorem 12.12 as applied to T^{-1} (cf. 12.8); the resolvant $(I + \lambda^{-1}T^{-1})^{-1}$ is maximal monotone and nonexpansive. □

Through the identity derived next, the resolvants of T will provide a parameterization of $\operatorname{gph} T$ that reveals how this graph must always resemble the graph of a Lipschitz continuous mapping.

12.14 Lemma (inverse-resolvant identity). *Every mapping $T : \mathbb{R}^n \rightrightarrows \mathbb{R}^n$ obeys the identity*

$$I - (I + T)^{-1} = (I + T^{-1})^{-1}.$$

Indeed, the Yosida regularizations of T are related to the resolvants of T by

$$(\lambda I + T^{-1})^{-1} = \lambda^{-1}\big[I - (I + \lambda T)^{-1}\big] \quad \textit{for any } \lambda > 0.$$

Proof. The first identity is the special case of the second where $\lambda = 1$, so we concentrate on the second identity. By direct manipulation we have

$$\begin{aligned} z \in \lambda^{-1}&\big[I - (I + \lambda T)^{-1}\big](w) \\ &\iff \lambda z \in w - (I + \lambda T)^{-1}(w) \\ &\iff w - \lambda z \in (I + \lambda T)^{-1}(w) \\ &\iff w \in (w - \lambda z) + \lambda T(w - \lambda z) \\ &\iff z \in T(w - \lambda z) \iff w - \lambda z \in T^{-1}(z) \\ &\iff w \in (\lambda I + T^{-1})(z) \iff z \in (\lambda I + T^{-1})^{-1}(w), \end{aligned}$$

and this is what was required. □

It's interesting to think of 12.13 and 12.14 as combining to give a sort of differentiation formula for the resolvants of T. Merely under the assumption that T is osc, one has $\operatorname{g-lim}_{\lambda \searrow 0} \lambda^{-1}\big[(I + \lambda T)^{-1} - I\big] = -T$.

12.15 Theorem (Minty parameterization of a graph). *Let $T : \mathbb{R}^n \rightrightarrows \mathbb{R}^n$ be maximal monotone. Then the mappings*

$$P = (I + T)^{-1}, \qquad Q = (I + T^{-1})^{-1},$$

are single-valued, in fact maximal monotone and nonexpansive, and the mapping $z \mapsto \big(P(z), Q(z)\big)$ *is one-to-one from* $\mathbb{R}^n$ *onto* $\operatorname{gph} T$. *This provides a parameterization of* $\operatorname{gph} T$ *that is Lipschitz continuous in both directions:*

$$\big(P(z), Q(z)\big) = (x, v) \iff z = x + v,\ (x, v) \in \operatorname{gph} T.$$

Proof. We know from 12.4(a) that T^{-1}, like T, is maximal monotone, and therefore from 12.12 that P and Q are single-valued, maximal monotone and nonexpansive, hence Lipschitz continuous. Furthermore, $P + Q = I$ by Lemma 12.14. When $\big(P(z), Q(z)\big) = (x, v)$ we consequently have $x + v = z$ and $z \in (I + T)(x)$, so that $z - x \in T(x)$, i.e., $(x, v) \in \operatorname{gph} T$.

Conversely, if $(x, v) \in \operatorname{gph} T$ and $x + v = z$ we have $z - x \in T(x)$, or $z \in (I + T)(x)$, so that $x = P(z)$. By symmetry, $v = Q(z)$. The mapping $z \mapsto \big(P(z), Q(z)\big)$ is Lipschitz continuous because P and Q are; cf. 9.8(d). The mapping $(x, v) \mapsto x + v$ is Lipschitz continuous because it's linear; cf. 9.3. □

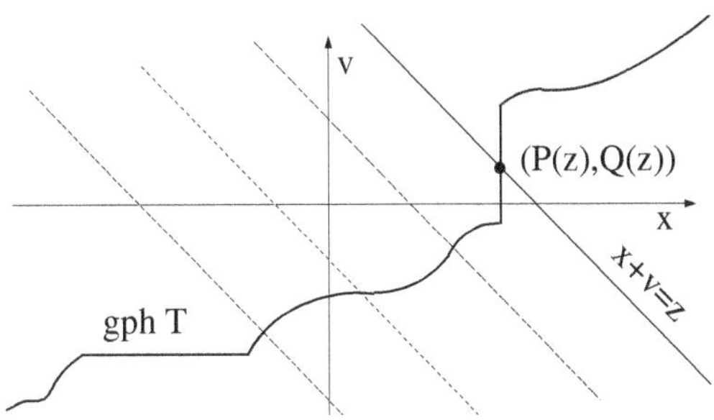

Fig. 12–4. Minty parameterization.

The special resolvants $(I + T)^{-1}$ and $(I + T^{-1})^{-1}$ are called the *Minty mappings* associated with T and T^{-1}. Other resolvants similarly yield Lipschitz continuous parameterizations of $\operatorname{gph} T$ when T is maximal monotone, although with formulas that lack symmetry. For instance, in terms of the λ-resolvant $P_\lambda = (I + \lambda T)^{-1}$ and the Yosida λ-regularization $Q_\lambda = (\lambda I + T^{-1})^{-1}$ one has

$$\big(P_\lambda(z), Q_\lambda(z)\big) = (x, v) \iff z = x + \lambda v,\ (x, v) \in \operatorname{gph} T.$$

This form of parameterization relies on the identity in Lemma 12.14, which says that $P_\lambda + \lambda Q_\lambda = I$. Otherwise the proof is virtually the same.

Along with the mappings $P = (I + T)^{-1}$ and $Q = (I + T^{-1})^{-1}$ being nonexpansive by Theorem 12.15, the mapping $S = Q - P$ has this property as well. Indeed, S is the nonexpansive mapping associated with T by Proposition 12.11, as seen from the first identity in Lemma 12.14.

12.16 Exercise (displacement mappings). *For any nonexpansive mapping* $F : \mathbb{R}^n \to \mathbb{R}^n$, *the displacement mapping* $I - F$, *giving the difference between* x *and its image under* F, *is maximal monotone.*

Guide. Argue by 12.11 that the mapping $T_0 := (I - F)^{-1} \circ 2I - I$ is maximal monotone. Express $I - F$ in terms of T_0 and apply 12.15 to T_0. □

The powerful insight gained through these parameterization results is that a maximal monotone mapping $T : \mathbb{R}^n \rightrightarrows \mathbb{R}^n$, despite its potential multivaluedness, has a graph that is a sort of full n-dimensional Lipschitz manifold within the $2n$-dimensional space $\mathbb{R}^{2n}$. This is clear from 12.15 and provides an extension of the picture in Figure 12–1 beyond the case of $n = 1$. What is more, T actually has the property of being *graphically Lipschitzian* in the sense defined in 9.66, namely that $\operatorname{gph} T$ can be transformed, under a change of coordinates in $\mathbb{R}^{2n}$, into the graph of a single-valued Lipschitz continuous mapping from $\mathbb{R}^n$ into $\mathbb{R}^n$. Indeed, the change of coordinates in 12.11 affords that perspective on the graphical geometry and does so globally, not just in some neighborhood of each point of $\operatorname{gph} T$.

C. Connections with Convex Functions

In the theorem coming next, the monotonicity of the gradient mappings associated with smooth convex functions (in 2.14) is generalized to the subgradient mappings associated with possibly nonsmooth, extended-real-valued convex functions. In the statement of this theorem it should be recalled from 11.13 that a convex function f is *almost strictly convex* if it is strictly convex along every line segment included in $\operatorname{dom} \partial f$. When thinking of ∂f as a set-valued mapping $\mathbb{R}^n \rightrightarrows \mathbb{R}^n$, we use the convention that

$$\partial f(x) := \emptyset \text{ for all } x \notin \operatorname{dom} f. \qquad 12(5)$$

12.17 Theorem (monotonicity versus convexity). *For any proper, convex function $f : \mathbb{R}^n \to \overline{\mathbb{R}}$, the mapping $\partial f : \mathbb{R}^n \rightrightarrows \mathbb{R}^n$ is monotone.*

Indeed, a proper, lsc function f is convex if and only if ∂f is monotone, in which case ∂f is maximal monotone. Such a function f is almost strictly convex if and only if ∂f is strictly monotone.

Proof. Suppose first that f is convex. If $v_0 \in \partial f(x_0)$ and $v_1 \in \partial f(x_1)$, we have from the subgradient inequality in 8.12 both $f(x_1) \geq f(x_0) + \langle v_0, x_1 - x_0 \rangle$, and $f(x_0) \geq f(x_1) + \langle v_1, x_0 - x_1 \rangle$, where the values $f(x_0)$ and $f(x_1)$ are finite. In adding these relations together and cancelling the term $f(x_0) + f(x_1)$ from both sides, we get $0 \geq \langle v_0, x_1 - x_0 \rangle + \langle v_1, x_0 - x_1 \rangle$, which is equivalent to the inequality in Definition 12.1. Thus, ∂f is monotone.

Assume now instead that f is proper and lsc with ∂f monotone. Add temporarily the assumption that f is prox-bounded, the threshold of prox-boundedness being $\lambda_f = \infty$. For any $\lambda \in (0, \lambda_f)$ we have $P_\lambda f \subset (I + \lambda \partial f)^{-1}$ by 10.2. Here $\operatorname{dom} P_\lambda f = \mathbb{R}^n$ by 1.25, and hence also $\operatorname{dom} (I + \lambda \partial f)^{-1} = \mathbb{R}^n$. Then ∂f is maximal monotone by Theorem 12.12, and $P_\lambda f$ is single-valued. Invoking the subgradient formula in 10.32 for the continuous function $-e_\lambda f$, we see that at every point x this function has a unique subgradient and therefore is differentiable everywhere (by 9.18). In fact we get

$$\begin{aligned}\nabla e_\lambda f(x) &= -\lambda^{-1}\big[P_\lambda f(x) - x\big]\\ &= \lambda^{-1}\big[I - (I+\lambda\partial f)^{-1}\big](x) = \big(\lambda I + \partial f^{-1}\big)^{-1}(x),\end{aligned}$$

with the final equation coming from Lemma 12.14. The monotonicity of ∂f implies that of ∂f^{-1}, hence that of $\lambda I + \partial f^{-1}$ and its inverse $(\lambda I + \partial f^{-1})^{-1}$; cf. 12.4(a)(b)(c). The gradient mapping $\nabla e_\lambda f$ is therefore monotone. But then $e_\lambda f$ must be convex; cf. 2.14(a). Because this is true for any $\lambda \in (0, \lambda_f)$, and $e_\lambda f(x) \nearrow f(x)$ as $\lambda \searrow 0$, we conclude that f itself is convex.

It remains to remove the additional assumption that f is prox-bounded. Suppose merely that f is lsc and proper. For each $\rho > 0$ let $f_\rho = f + g_\rho$ with $g_\rho(x) := (\rho^2 - |x|^2)^{-1}$ when $|x| < \rho$, but $g_\rho(x) = \infty$ when $|x| \geq \rho$. The function g_ρ is convex and smooth on its effective domain, which is the open ball $\rho \operatorname{int} I\!B$, so that (once ρ is large enough that this ball meets $\operatorname{dom} f$) the function f_ρ is proper and lsc with

$$\partial f_\rho(x) = \begin{cases} \partial f(x) + \nabla g_\rho(x) & \text{when } |x| < \rho,\\ \emptyset & \text{when } |x| \geq \rho,\end{cases}$$

The convexity of g_ρ makes ∇g_ρ be monotone, hence ∂f_ρ is monotone by 12.4(c). But f_ρ is bounded from below inasmuch as $\operatorname{dom} f_\rho \subset \rho I\!B$ (see 1.10), so f_ρ is prox-bounded (see the end of 1.24). Thus, the monotonicity of ∂f_ρ implies the convexity of f_ρ by the argument already given. We have $f_\rho(x) \searrow f(x)$ as $\rho \nearrow \infty$, so the convexity of f_ρ implies the convexity of f. Then f is prox-bounded after all (3.27, 3.28), and we conclude once more that ∂f is maximal.

For f lsc, proper and convex, if f isn't almost strictly convex there is a line segment $[x_0, x_1] \subset \operatorname{dom} \partial f$ on which f is affine. Then for any $\bar{x} \in \operatorname{rint}[x_0, x_1]$ and any $v \in \partial f(\bar{x})$ we must have $v \in \partial f(x_0) \cap \partial f(x_1)$, in which case $\langle v_1 - v_0, x_1 - x_0\rangle \geq 0$ for $v_0 = v_1 = v$. Hence ∂f isn't strictly monotone. Conversely, if $\langle v_1 - v_0, x_1 - x_0\rangle = 0$ for some $v_0 \in \partial f(x_0)$, $v_1 \in \partial f(x_1)$, $x_0 \neq x_1$, one sees from the inequality argument at the beginning of the present proof that $f(x_1) - f(x_0) = \langle v_0, x_1 - x_0\rangle$. Then by the convexity of f and the subgradient inequality satisfied by v_0 (cf. 8.12), f must be affine on the segment $[x_0, x_1]$, hence not almost strictly convex. □

12.18 Corollary (normal cone mappings). *For a closed, convex set $C \neq \emptyset$ in $I\!R^n$, the mapping $N_C : I\!R^n \rightrightarrows I\!R^n$ is maximal monotone; here*

$$N_C(x) := \emptyset \ \textit{for all}\ x \notin C. \tag{12(6)}$$

Proof. The function $f = \delta_C$ is proper, lsc, and convex with $\partial f = N_C$. □

12.19 Proposition (proximal mappings). *For a proper, lsc function $f : I\!R^n \to \overline{I\!R}$ and any $\lambda > 0$, the proximal mapping $P_\lambda f : I\!R^n \rightrightarrows I\!R^n$ is monotone.*

If f is also convex, then $P_\lambda f$ is maximal monotone and nonexpansive *(hence single-valued) with*

$$I - P_\lambda f = P_\lambda f^*, \quad \text{when } \lambda = 1$$

so that $\operatorname{gph} \partial f$ *has the Lipschitz continuous parameterization*

$$\bigl(P_1 f(z), P_1 f^*(z)\bigr) = (x, v) \iff z = x + v,\ v \in \partial f(x).$$

More generally in the possible absence of convexity, the following conditions on a given $\lambda > 0$ *are equivalent:*

(a) $P_\lambda f$ *is maximal monotone,*

(b) $f + \lambda^{-1} j$ *is convex for* $j = \frac{1}{2}|\cdot|^2$,

(c) $I + \lambda \partial f$ *is monotone.*

These conditions imply that $P_\lambda f = (I + \lambda \partial f)^{-1}$ *and that, for all* $\mu \in (0, \lambda)$, *the mapping* $P_\mu f$ *is Lipschitz continuous with constant* $\lambda/[\lambda - \mu]$.

Proof. The properties asserted for convex f can be obtained at once from 10.2, 12.15, and the maximal monotonicity of ∂f in 12.17. The connection with f^* comes from the fact that $\partial f^* = (\partial f)^{-1}$, cf. 11.3.

For the general situation, observe that vectors $\bar{x}$ and $\bar{w}$ satisfy $\bar{x} \in P_\lambda f(\bar{w})$ if and only if $f(\bar{x}) + (1/2\lambda)|\bar{x} - \bar{w}|^2 \le f(x) + (1/2\lambda)|x - \bar{w}|^2$ for all x, and that in terms of $g = j + \lambda f$ this inequality can be cast in the form $g(x) \ge g(\bar{x}) + \langle \bar{w}, x - \bar{x}\rangle$ for all x. The latter is equivalent in turn to having

$$\bar{w} \in \partial \bar{g}(\bar{x}) \text{ and } \bar{g}(\bar{x}) = g(\bar{x}) \text{ for } \bar{g} := \operatorname{cl}(\operatorname{con} g).$$

It follows that $(P_\lambda f)^{-1} \subset \partial \bar{g}$, where the mapping $\partial \bar{g}$ is maximal monotone by 12.17. In particular $(P_\lambda f)^{-1}$ is monotone, hence $P_\lambda f$ is monotone too. Furthermore, these mappings are maximal monotone if and only if $\bar{g}(x) = g(x)$ for all $x \in \operatorname{dom} \partial \bar{g}$. But $\operatorname{dom} \bar{g} \supset \operatorname{dom} \partial \bar{g} \supset \operatorname{rint} \operatorname{dom} \bar{g}$. This condition therefore requires g to agree with $\bar{g}$ on $\operatorname{rint}(\operatorname{dom} \bar{g})$. That's the same as g itself being convex, inasmuch as g is lsc while $\bar{g}$, being convex, lsc and proper, is completely determined by its values on $\operatorname{rint}(\operatorname{dom} \bar{g})$ (cf. 2.35). Thus, the maximal monotonicity in (a) corresponds to the convexity in (b). On the other hand, the convexity of g is equivalent by 12.17 to the monotonicity of ∂g, which is $I + \lambda \partial f$ (by the subgradient addition rule in 10.9), so the equivalence between (b) and (c) is correct too.

According to 10.2, it's always true that $P_\lambda f \subset (I + \lambda \partial f)^{-1}$, so the maximal monotonicity of $P_\lambda f$ in (a) and the monotonicity of $(I + \lambda \partial f)^{-1}$, implied by (c), necessitate $P_\lambda f = (I + \lambda \partial f)^{-1}$. For $\mu \in (0, \lambda)$ we then have the convexity of $f + \mu^{-1} j = f + \lambda^{-1} j + [\mu^{-1} - \lambda^{-1}] j$. This gives us

$$(P_\mu f)^{-1} = I + \mu \partial f = \mu \lambda^{-1} \bigl[(\mu^{-1}\lambda - 1) I + (I + \lambda \partial f)\bigr]$$

with $I + \lambda \partial f = (P_\lambda f)^{-1}$. Consequently for $\tau = \mu^{-1}\lambda - 1 > 0$ we obtain $P_\mu f = [\tau I + (P_\lambda f)^{-1}]^{-1} \circ \mu^{-1} \lambda I$. But $[\tau I + (P_\lambda f)^{-1}]^{-1}$ is the Yosida τ-regularization of the maximal monotone mapping $(P_\lambda f)^{-1}$, so it's Lipschitz continuous with constant τ^{-1} by 12.13. Then $P_\mu f$ is Lipschitz continuous with constant $\tau^{-1}\mu^{-1}\lambda = \lambda/(\lambda - \mu)$. □

12.20 Corollary (projection mappings). *For a nonempty, closed set $C \subset \mathbb{R}^n$, the mapping $P_C : \mathbb{R}^n \rightrightarrows C$ is monotone and the following are equivalent:*

(a) *C is convex;*

(b) *P_C is single-valued;*

(c) *P_C is nonexpansive;*

(d) *P_C is maximal monotone;*

(e) *$N_C + \rho I$ is monotone for some ρ.*

Proof. The function $f = \delta_C$ is proper and lsc with $\partial f = N_C$, and it's convex if and only if C is convex. For any $\lambda > 0$ we have $\lambda N_C = N_C$ (because $N_C(x)$ is a cone when nonempty, i.e., when $x \in C$), whereas $P_\lambda f = P_C$. Applying Proposition 12.19, we get the monotonicity of P_C. The equivalent conditions (a), (b) and (c) of 12.19 correspond respectively to conditions (d), (a) and (e) here. (Monotonicity of $N_C + \rho I$ for some $\rho \in (-\infty, \infty)$ implies its monotonicity for some $\rho \in (0, \infty)$, and that property is identical to the monotonicity of the mapping $I + \rho^{-1} N_C = I + N_C$.) Thus, the present (d), (a) and (e) are equivalent. We also know from the statement of 12.19 that (a) yields (c) here. Clearly (c) implies (b). But P_C is osc and locally bounded (cf. 7.44(b)), so (b) ensures that P_C is actually continuous (see 5.20). This yields (d) through 12.7 and confirms that the equivalence of (d), (a) and (e) extends to (c) and (b). □

The special parameterization of $\operatorname{gph} \partial f$ by proximal mappings in 12.19 works out in the context of 12.20 to a parameterization of $\operatorname{gph} N_C$ in terms of the projection mapping P_C. It was observed in 6.17 that for a convex set C one has $P_C = (I + N_C)^{-1}$. The Minty parameterization of Theorem 12.15 in the case of the graph of $T = N_C$ (a mapping which by 12.18 is maximal monotone when C is convex and closed) is effected therefore by

$$z \mapsto \big(P_C(z), Q_C(z)\big) \text{ with } Q_C = I - P_C = P_\lambda \sigma_C \text{ for } \lambda = 1.$$

12.21 Example (projections on subspaces). *For any linear subspace $M \subset \mathbb{R}^n$ and its orthogonal complement $M^\perp$, the mapping*

$$N_M(x) = \begin{cases} M^\perp & \text{when } x \in M, \\ \emptyset & \text{when } x \notin M, \end{cases}$$

is maximal monotone with $\operatorname{gph} N_M = M \times M^\perp \subset \mathbb{R}^n \times \mathbb{R}^n$ and inverse

$$N_M^{-1}(v) = \begin{cases} M & \text{when } v \in M^\perp, \\ \emptyset & \text{when } v \notin M^\perp. \end{cases}$$

The linear projection mapping onto M is $P_M = (I + N_M)^{-1}$, whereas the one onto $M^\perp$ is $P_{M^\perp} = (I + N_M^{-1})^{-1}$. These projection mappings are maximal monotone and nonexpansive.

Detail. This specializes 12.18 and 12.20 from a convex set to a subspace. □

12.22 Exercise (projections on a polar pair of cones). *With respect to a closed, convex cone $K \subset \mathbb{R}^n$ and its polar K^*, any point $z \in \mathbb{R}^n$ can be represented uniquely in the form*

$$z = x + v \quad \text{with} \quad x \in K, \quad v \in K^*, \quad x \perp v,$$

namely with $x = P_K(z)$ and $v = P_{K^*}(z)$, the projections of z on K and K^*. These projection mappings are maximal monotone and nonexpansive.

Guide. Take $C = K$ in 12.18 and 12.20. Use the fact that the mappings N_K and N_K^* are inverse to each other because δ_K and δ_{K^*} are conjugate to each other (see 11.3, 11.4). □

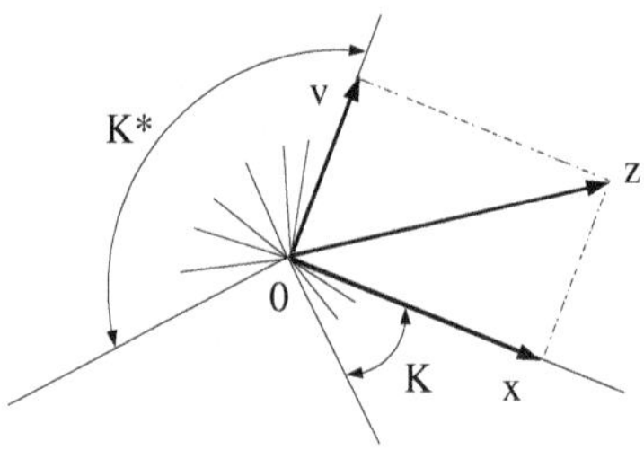

Fig. 12–5. Spatial decomposition with respect to polar cones.

12.23 Exercise (Moreau envelopes and Yosida regularizations). *For any proper, lsc, convex function $f : \mathbb{R}^n \to \mathbb{R}^n$, the Yosida regularization of the subgradient mapping ∂f for any $\lambda > 0$ is the gradient mapping $\nabla e_\lambda f$ associated with the Moreau envelope $e_\lambda f$: one has $\nabla e_\lambda f = \big(\lambda I + (\partial f)^{-1}\big)^{-1}$.*

This mapping is maximal monotone, single-valued and Lipschitz continuous globally on $\mathbb{R}^n$ with constant λ^{-1}. Consequently, $e_\lambda f$ is of class $\mathcal{C}^{1+}$.

Guide. Work with the facts in Example 12.13 and Theorem 12.17 using the gradient formula in 2.26. Apply the rule in Lemma 12.14. □

Theorem 12.17 characterizes the subgradient mappings associated with convex functions within the class of all subgradient mappings associated with proper, lsc functions $f : \mathbb{R}^n \to \overline{\mathbb{R}}$. It doesn't provide such a characterization within the class of all set-valued mappings $T : \mathbb{R}^n \rightrightarrows \overline{\mathbb{R}}$, however. Maximal monotonicity is identified as being necessary in order that $T = \partial f$ for a proper, lsc, convex function f, but it's certainly not sufficient. For example, a linear mapping $L(x) = Ax$ with A antisymmetric ($A^* = -A$) is maximal monotone by 12.2 and 12.7, but it can't be the subgradient mapping for any locally lsc function f unless $A = 0$. (If it were, f would have to be smooth with $\nabla f(x) = Ax$, cf. 9.13, but then $df(x)(w) = \langle w, Ax\rangle = 0$ for all x and w, in which case f must be constant, hence $A = 0$.)

For this reason we have to look further than maximal monotonicity if we wish to find a way of recognizing the mappings T that are the subgradient mappings of proper, lsc, convex functions. A stronger form of monotonicity turns out to be what is needed.

12.24 Definition (cyclical monotonicity). *A mapping $T : \mathbb{R}^n \rightrightarrows \mathbb{R}^n$ is cyclically monotone if for any choice of points $x_0, x_1, \ldots x_m$ (for arbitrary $m \geq 1$) and elements $v_i \in T(x_i)$, one has*

$$\langle v_0, x_1 - x_0 \rangle + \langle v_1, x_2 - x_1 \rangle + \cdots + \langle v_m, x_0 - x_m \rangle \leq 0. \qquad 12(7)$$

It is maximal cyclically monotone if it is cyclically monotone and its graph cannot be enlarged without destroying this property.

Note that the inequality in 12(7) reduces to the one in Definition 12.1 when $m = 1$. Every cyclically monotone mapping is thus monotone in particular. It doesn't immediately follow that every maximal cyclically monotone mapping is maximal monotone, but that's implied by the next theorem in conjunction with Theorem 12.17.

12.25 Theorem (characterization of convex subgradient mappings). *A mapping $T : \mathbb{R}^n \rightrightarrows \mathbb{R}^n$ has the form $T = \partial f$ for some proper, lsc, convex function $f : \mathbb{R}^n \to \overline{\mathbb{R}}$ if and only if T is maximal cyclically monotone. Then f is determined by T uniquely up to an additive constant.*

Proof. If f is a proper convex function and $v_i \in \partial f(x_i)$ for $i = 0, 1, \ldots, m$, we have $f(x_{i+1}) \geq f(x_i) + \langle v_i, x_{i+1} - x_i \rangle$ by the subgradient inequality for convex functions in 8.12, with x_{m+1} interpreted as x_0, hence

$$\sum_{i=0}^{m} \langle v_i, x_{i+1} - x_i \rangle \leq \sum_{i=0}^{m} \big[f(x_{i+1}) - f(x_i) \big] = 0.$$

Thus, ∂f is cyclically monotone. If f is also lsc, ∂f must be maximal cyclically monotone, because ∂f is maximal monotone by 12.17, and any cyclically monotone enlargement of ∂f would also be a monotone enlargement.

Conversely, if $T : \mathbb{R}^n \rightrightarrows \mathbb{R}^n$ is any cyclically monotone mapping and (x_0, v_0) is any vector pair in $\operatorname{gph} T$, define the function $f : \mathbb{R}^n \to \overline{\mathbb{R}}$ by

$$f(x) = \sup_{\substack{(x_i, v_i) \in \operatorname{gph} T \\ i=1,\ldots,m}} \Big\{ \langle v_m, x - x_m \rangle + \langle v_{m-1}, x_m - x_{m-1} \rangle + \cdots + \langle v_0, x_1 - x_0 \rangle \Big\},$$

where the supremum is over all choices of m as well as all choice of the pairs (x_i, v_i) for $i > 0$. This formula expresses f as the pointwise supremum of a collection of affine functions of x, so f is convex and lsc (cf. 1.26, 2.9). The cyclic monotonicity of T is equivalent to having $f(x_0) = 0$, so f is proper too, and hence by what we have just seen, the mapping ∂f is itself cyclically monotone. For any pair $(\bar{x}, \bar{v}) \in \operatorname{gph} T$ we have from the definition of f that

$$\begin{aligned} f(x) \geq & \sup_{\substack{(x_i,v_i)\in \operatorname{gph} T\\ i=1,\ldots,m}} \Big\{ \langle \bar{v}, x-\bar{x}\rangle + \langle v_m,\, \bar{x}-x_m\rangle \\ & \qquad + \langle v_{m-1},\, x_m - x_{m-1}\rangle + \cdots + \langle v_0,\, x_1 - x_0\rangle \Big\} \\ = & \ f(\bar{x}) + \langle \bar{v},\, x-\bar{x}\rangle, \end{aligned}$$

and this says that $\bar{v} \in \partial f(\bar{x})$ (again by 8.12). Thus, $\operatorname{gph} T \subset \operatorname{gph} \partial f$. If T is maximal cyclically monotone, it follows that $\operatorname{gph} T = \operatorname{gph} \partial f$, i.e., $T = \partial f$.

Finally, if f_1 and f_2 are proper, lsc, convex functions such that $\partial f_1 = \partial f_2$, then the mappings $P_\lambda f_1$ and $P_\lambda f_2$, which by 10.2 equal $(I + \lambda \partial f_1)^{-1}$ and $(I+\lambda \partial f_2)^{-1}$, must coincide for any $\lambda > 0$. But a proper, lsc, convex function is determined uniquely up to an additive constant by any of its proximal mappings (see 3.37). Hence f_1 and f_2 can differ only by a constant. □

12.26 Exercise (cyclical monotonicity in one dimension)**.** *Every maximal monotone mapping $T : \mathbb{R}^1 \rightrightarrows \mathbb{R}^1$ is maximal cyclically monotone. The maximal monotone mappings $T : \mathbb{R}^1 \rightrightarrows \mathbb{R}^1$ are precisely the subgradient mappings ∂f of the proper, lsc, convex functions $f : \mathbb{R} \to \overline{\mathbb{R}}$:*

$$T(x) = \begin{cases} \{ v \in \mathbb{R} \mid f'_-(x) \leq v \leq f'_+(x) \} & \text{for } x \in \operatorname{dom} f, \\ \emptyset & \text{for } x \notin \operatorname{dom} f. \end{cases}$$

Guide. Starting from 12.25, relate these claims to 12.9(b) and 8.52. □

Observe that the proximal mappings $P_\lambda f$ associated with proper, lsc, convex functions f and the projection mappings P_C associated with nonempty, closed, convex sets aren't just *maximal monotone*, in accordance with 12.19 and 12.20; they are *maximal cyclically monotone* as well. This follows from Theorem 12.25 through the fact that $(P_\lambda f)^{-1} = \partial g$ for $g = \frac{1}{2}|\cdot|^2 + \lambda f$ (see 12.19), whereas $P_C = P_\lambda f$ for $f = \delta_C$.

Although a monotone mapping can't actually be the subgradient mapping for some function unless that function is convex, an important class of monotone mappings arises in close association with the subgradients of certain nonconvex functions.

12.27 Example (monotone mappings from convex-concave functions)**.** *As an extended-real-valued function of $(x, y) \in \mathbb{R}^n \times \mathbb{R}^m$, let $l(x, y)$ be convex in x for fixed y, but concave in y for fixed x. Then l gives rise to a monotone mapping $T : \mathbb{R}^n \times \mathbb{R}^m \rightrightarrows \mathbb{R}^n \times \mathbb{R}^m$ through the formula*

$$T(x, y) \;=\; \partial_x l(x, y) \times \partial_y[-l](x, y).$$

In the Lagrangian case where $l(x, y) = \inf_u\{ f(x, u) - \langle y, u\rangle \}$ for a proper, convex function $f : \mathbb{R}^n \times \mathbb{R}^m$ that is lsc, T is maximal monotone.

Detail. Suppose $(v_i, u_i) \in T(x_i, y_i)$ for $i = 0, 1$. We have the subgradient inequalities

$$\begin{aligned} l(x_1, y_0) &\geq l(x_0, y_0) + \langle v_0,\, x_1 - x_0 \rangle, \\ l(x_0, y_1) &\geq l(x_1, y_1) + \langle v_1,\, x_0 - x_1 \rangle, \\ -l(x_0, y_1) &\geq -l(x_0, y_0) + \langle u_0,\, y_1 - y_0 \rangle, \\ -l(x_1, y_0) &\geq -l(x_1, y_1) + \langle u_1,\, y_0 - y_1 \rangle, \end{aligned}$$

where the function values have to be finite. Adding these up, we get

$$0 \geq \langle v_0,\, x_1 - x_0 \rangle + \langle v_1,\, x_0 - x_1 \rangle - \langle u_0,\, y_0 - y_1 \rangle - \langle u_1,\, y_1 - y_0 \rangle,$$

which works out to $0 \leq \big\langle (v_1, u_1) - (v_0, u_0),\, (x_1, y_1) - (x_0, y_0) \big\rangle$. This means that T is monotone.

When l comes from a convex function f as described, we have by 11.48 that $(v, u) \in T(x, y)$ if and only if $(v, y) \in \partial f(x, u)$. Then T, as the 'partial inverse' of ∂f, inherits the maximal monotonicity of the latter. Indeed, if T' were a monotone mapping with graph properly larger than that of T, its partial inverse $T'' : (x, u) \mapsto \big\{ (v, y) \,\big|\, (v, u) \in T'(x, y) \big\}$ would be a monotone mapping with graph properly larger than that of ∂f, which is impossible by 12.17. □

Another connection between monotonicity and the subgradient mappings of nonconvex functions is found in 'hypomonotonicity'.

12.28 Example (hypomonotone mappings). *A mapping $T : \mathbb{R}^n \rightrightarrows \mathbb{R}^n$ is hypomonotone at $\bar{x}$ if there exist $V \in \mathcal{N}(\bar{x})$ and $\rho \in \mathbb{R}_+$ such that $T + \rho I$ is monotone on V, or in other words,*

$$\big\langle v_1 - v_0,\, x_1 - x_0 \big\rangle \geq -\rho |x_1 - x_0|^2 \;\text{ when }\; x_i \in V,\; v_i \in T(x_i) \;\text{ for }\; i = 0, 1.$$

It is hypomonotone on O if it is hypomonotone at each $\bar{x} \in O$. Cyclical hypomonotonicity is defined analogously.

(a) *Any single-valued, strictly continuous mapping $F : O \to \mathbb{R}^n$ on an open set $O \subset \mathbb{R}^n$ is hypomonotone on O.*

(b) *A subgradient mapping ∂f for a proper, lsc function f is hypomonotone at $\bar{x}$ if and only if there exists $\rho \in \mathbb{R}_+$ such that $f + \frac{1}{2}\rho |\cdot|^2$ is convex on a neighborhood of $\bar{x}$, and then ∂f is cyclically hypomonotone at $\bar{x}$.*

(c) *A function f is lower-$\mathcal{C}^2$ on an open set O if and only if f is finite and lsc on O with ∂f nonempty-valued and hypomonotone on O.*

Detail. In (a), if κ is a constant of Lipschitz continuity for F on a neighborhood V of $\bar{x}$, then the mapping $G = F + \kappa I$ has on V the property that

$$\begin{aligned} \big\langle G(x_1) - G(x_0),\, x_1 - x_0 \big\rangle &= \big\langle F(x_1) - F(x_0),\, x_1 - x_0 \big\rangle + \kappa |x_1 - x_0|^2 \\ &\geq -|F(x_1) - F(x_0)||x_1 - x_0| + \kappa |x_1 - x_0|^2 \geq 0, \end{aligned}$$

so G is monotone on V. The facts in (b) are evident from 12.17 and 12.25; to apply these results locally, add to f the indicator of a closed, convex neighborhood of $\bar{x}$. The characterization in (c) falls out of (b) and 10.33, since a convex function has subgradients at every point of an open set where it is finite. □

Hypomonotonicity will be a key property in the second-order analysis of subgradient mappings ∂f in Chapter 13 (cf. 13.36).

Recall from 9.57 that a mapping is *piecewise polyhedral* if its graph is the union of a finite collection of polyhedral sets. This property, akin to piecewise linearity, has an interesting role in the present setting.

12.29 Proposition (monotonicity of piecewise polyhedral mappings). *Suppose that $T : \mathbb{R}^n \rightrightarrows \mathbb{R}^n$ is maximal monotone. Then T is piecewise polyhedral if and only if, for some $\lambda > 0$, the resolvant mapping $(I + \lambda T)^{-1}$ is piecewise linear, in which case this is true for every $\lambda > 0$.*

Proof. The graph of $(I + \lambda T)^{-1}$ corresponds to that of T under the invertible linear transformation J_λ in 12(4). Hence if one of these graphs is the union of finitely many polyhedral sets the other must have that form as well. But by Theorem 12.12, $(I + \lambda T)^{-1}$ is single-valued when T is maximal monotone. A single-valued mapping is piecewise polyhedral if and only if it's piecewise linear; see 2.48. □

12.30 Proposition (subgradients of piecewise linear-quadratic functions). *For a proper, lsc, convex function $f : \mathbb{R}^n \to \overline{\mathbb{R}}$ and any value $\lambda > 0$ the following properties are equivalent:*

(a) *the function f is piecewise linear-quadratic,*
(b) *the subgradient mapping ∂f is piecewise polyhedral,*
(c) *the envelope function $e_\lambda f$ is piecewise linear-quadratic,*
(d) *the proximal mapping $P_\lambda f$ is piecewise linear.*

Proof. The equivalence between (b) and (d) is evident from the preceding proposition, since $P_\lambda f = (I + \lambda \partial f)^{-1}$. The equivalence between (d) and (c) comes from the formula $\nabla e_\lambda f = \lambda^{-1}[I - P_\lambda f]$ in 2.26; the mapping $\lambda^{-1}[I - P_\lambda f]$ is piecewise linear if and only if $P_\lambda f$ is piecewise linear, whereas the smooth function $e_\lambda f$ is piecewise linear-quadratic if and only if $\nabla e_\lambda f$ is piecewise linear. According to 11.14, the property of a convex function being piecewise linear-quadratic is preserved under the Legendre-Fenchel transform. The function conjugate to $e_\lambda f$ under this transform is $f^* + \frac{1}{2}\lambda|\cdot|^2$ by 11.24(b), and the latter is piecewise linear-quadratic if and only if f^* itself is. Putting these facts together, we conclude that (c) is equivalent to (a). □

12.31 Example (piecewise linear projections). *For $C \subset \mathbb{R}^n$ nonempty, closed and convex, the following properties are equivalent:*

(a) *the set C is polyhedral,*
(b) *the normal cone mapping N_C is piecewise polyhedral,*
(c) *the function d_C^2 is piecewise linear-quadratic,*
(d) *the projection mapping P_C is piecewise linear.*

Detail. Apply Proposition 12.30 to $f = \delta_C$. □

D. Graphical Convergence

The connections between monotone mappings and nonexpansive mappings lead to a special theory of graphical convergence.

12.32 Theorem (graphical convergence of monotone mappings). *If a sequence of monotone mappings $T^\nu : I\!R^n \rightrightarrows I\!R^n$ converges graphically, the limit mapping T must be monotone. Moreover, if the mappings T^ν are maximal monotone, then T is maximal monotone.*

In general for maximal monotone mappings T^ν and T, one has for any choice of $\lambda > 0$ that

$$T^\nu \xrightarrow{\text{g}} T \quad \Longleftrightarrow \quad (I + \lambda T^\nu)^{-1} \xrightarrow{\text{p}} (I + \lambda T)^{-1}.$$

Then, for any sequence $\lambda^\nu \to \lambda$ with λ, $\lambda^\nu > 0$, the single-valued mappings $(I + \lambda^\nu T^\nu)^{-1}$ converge uniformly to $(I + \lambda T)^{-1}$ on all bounded sets.

Proof. Suppose $T = \text{g-lim}_\nu T^\nu$ with T^ν monotone, and consider $v_i \in T(x_i)$, $i = 0, 1$. Because $T = \text{g-lim inf}_\nu T^\nu$ in particular, there exist sequences $x_i^\nu \to x_i$ and $v_i^\nu \to v_i$ such that $v_i^\nu \in T^\nu(x_i^\nu)$. For each ν we have by assumption that $\langle v_1^\nu - v_0^\nu, x_1^\nu - x_0^\nu \rangle \geq 0$. Hence in the limit as $\nu \to \infty$ we get $\langle v_1 - v_0, x_1 - x_0 \rangle \geq 0$, which means that T is monotone.

Next take the mappings T^ν to be maximal monotone and, with λ fixed, let $P_\lambda^\nu := (I + \lambda T^\nu)^{-1}$. By 12.12, P_λ^ν is single-valued and nonexpansive. We have $x = P_\lambda^\nu(z)$ if and only if $\lambda^{-1}(z - x) \in T^\nu(x)$, so the graph of P_λ^ν is related to that of T^ν by $\text{gph}\, P_\lambda^\nu = J_\lambda(\text{gph}\, T^\nu)$ for the one-to-one linear transformation $J_\lambda : (x, v) \mapsto (x + \lambda v, x)$. Therefore, $T^\nu \xrightarrow{\text{g}} T$ if and only if $P_\lambda^\nu \xrightarrow{\text{g}} P_\lambda$, where $P_\lambda := (I + \lambda T)^{-1}$. But the mappings P_λ^ν are Lipschitz continuous with constant 1, hence equicontinuous on $I\!R^n$, so graphical convergence $P_\lambda^\nu \xrightarrow{\text{g}} P_\lambda$ is equivalent to pointwise convergence $P_\lambda^\nu \xrightarrow{\text{p}} P_\lambda$ (by 5.40) and entails uniform convergence on all bounded sets (cf. 5.45, 5.43). As a by-product, if $T^\nu \xrightarrow{\text{g}} T$ one must have $\text{dom}\, P_\lambda = I\!R^n$, making T be maximal monotone by 12.12.

For a sequence $\lambda^\nu \to \lambda$, this convergence picture can be extended to $P_{\lambda^\nu}^\nu = (I + \lambda^\nu T^\nu)^{-1}$ through the observation that $J_{\lambda^\nu} J_\lambda^{-1} \to I$. □

12.33 Corollary (convergent subsequences). *If a sequence of maximal monotone mappings $T^\nu : I\!R^n \rightrightarrows I\!R^n$ does not escape to the horizon, it has a subsequence that converges graphically to some maximal monotone mapping $T : I\!R^n \rightrightarrows I\!R^n$.*

Proof. This follows from Theorem 12.32 by way of the compactness property in Theorem 5.36. □

12.34 Exercise (maximal monotone limits).

(a) *If $\text{g-lim inf}\, T^\nu \supset T$ with T^ν monotone and T maximal monotone, then actually $T^\nu \xrightarrow{\text{g}} T$.*

(b) *If $T^\nu \xrightarrow{\text{p}} T$ with T^ν maximal monotone and the sequence $\{T^\nu\}_{\nu \in I\!N}$ is asymptotically equi-osc, then T is maximal monotone.*

Guide. For (a), use 12.33 and the characterization of inner limits in 4.19. To get (b), combine 12.32 with 5.40. □

Especially important is the connection between graphical convergence of subdifferential mappings and epi-convergence of the functions themselves, when the functions are convex.

12.35 Theorem (subgradient convergence for convex functions; Attouch). *For proper, lsc, convex functions f^ν and f, and any value $\lambda > 0$, the following properties are equivalent:*

(a) *the functions f^ν epi-converge to f;*

(b) *the mappings ∂f^ν converge graphically to ∂f, and for some choice of $\bar v^\nu \in \partial f^\nu(\bar x^\nu)$ and $\bar v \in \partial f(\bar x)$ with $(\bar x^\nu, \bar v^\nu) \to (\bar x, \bar v)$, one has $f^\nu(\bar x^\nu) \to f(\bar x)$;*

(c) *the proximal mappings $P_\lambda f^\nu$ converge pointwise to $P_\lambda f$, and for some choice of $\tilde x$ and a sequence $\tilde x^\nu \to \tilde x$ one has $e_\lambda f^\nu(\tilde x^\nu) \to e_\lambda f(\tilde x)$.*

Then, for any sequence $\lambda^\nu \to \lambda$ with $\lambda, \lambda^\nu > 0$, the single-valued mappings $P_{\lambda^\nu} f^\nu$ converge to $P_\lambda f$ uniformly on all bounded sets, and likewise the functions $e_{\lambda^\nu} f^\nu$ converge to $e_\lambda f$ uniformly on all bounded sets.

Proof. If (a) holds, the functions $e_\lambda f^\nu$ converge pointwise to $e_\lambda f$ by 7.37, in fact uniformly on all bounded sets. The argument for that result, based on 7.33, shows equally well that the mappings $P_\lambda f^\nu$ converge pointwise to $P_\lambda f$, although it wasn't mentioned at the time. Thus, (a) implies (c). But conversely, if (c) holds the mappings must converge uniformly on all bounded sets because they are equicontinuous; they are the nonexpansive mappings $(I + \lambda\partial f^\nu)^{-1}$; cf. 10.32 and 12.32 along with 12.19. Thus, (c) implies (a).

Because $P_\lambda f = (I + \lambda\partial f)^{-1}$, the convergence $P_\lambda f^\nu \xrightarrow{\text{p}} P_\lambda f$ in (c) is equivalent through 12.32 to the convergence $\partial f^\nu \xrightarrow{\text{g}} \partial f$ in (b). At the same time there is a correspondence between sequences $\tilde x^\nu \to \tilde x$ in (c) and sequences $(\bar x^\nu, \bar v^\nu) \to (\bar x, \bar v)$ as described in (b), which is given by

$$\tilde x^\nu = \bar x^\nu + \lambda \bar v^\nu \quad \longleftrightarrow \quad \begin{cases} \bar x^\nu = P_\lambda f^\nu(\tilde x^\nu), \\ \bar v^\nu = \lambda^{-1}\big[I - P_\lambda f^\nu\big](\tilde x^\nu), \end{cases}$$

and similarly for $\tilde x$, $\bar x$, and $\bar v$. Under this correspondence we have

$$e_\lambda f^\nu(\tilde x^\nu) = f^\nu(\bar x^\nu) + \frac{1}{2\lambda}|\bar x^\nu - \tilde x^\nu|^2, \qquad e_\lambda f(\tilde x) = f(\bar x) + \frac{1}{2\lambda}|\bar x - \tilde x|^2,$$

so that $e_\lambda f^\nu(\tilde x^\nu) \to e_\lambda f(\tilde x)$ if and only if $f^\nu(\bar x^\nu) \to f(\bar x)$. This shows that (c) is equivalent to (b). The additional claims about uniform convergence follow now via 12.32 and the gradient formula for envelope functions in 2.26. □

The assertion in 12.35(b) about convergence of function values is supplemented by the following fact.

12.36 Exercise (convergence of convex function values). *Let $f^\nu : \mathbb{R}^n \to \overline{\mathbb{R}}$ be proper, lsc and convex, and suppose $f^\nu \xrightarrow{\text{e}} f$ and $v^\nu \in \partial f^\nu(x^\nu)$ with $x^\nu \to \bar x$ and $v^\nu \to \bar v$. Then $f^\nu(x^\nu) \to f(\bar x)$ and $f^{\nu *}(v^\nu) \to f^*(\bar v)$.*

Guide. The crux of the matter is demonstrating that $\limsup_\nu f^\nu(x^\nu) \le f(\bar{x})$. Get this by applying the subgradient inequality for f^ν at x^ν while appealing to the existence of at least one sequence $\bar{x}^\nu \to \bar{x}$ having $f^\nu(\bar{x}^\nu) \to f(\bar{x})$ (as guaranteed by the meaning of epi-convergence). Then dualize by 11.3. □

E. Domains and Ranges

Returning to the study of maximal monotone mappings in general, we now use this convergence theory to gain deep insights about domains and ranges.

12.37 Theorem (normal vectors to domains). *Let $T : \mathbb{R}^n \rightrightarrows \mathbb{R}^n$ be maximal monotone, and let $D = \mathrm{cl}(\mathrm{dom}\, T)$. Then D is convex, and for all $x \in \mathrm{dom}\, T$ one has $T(x)^\infty = N_D(x)$. Furthermore*

$$\lambda T \xrightarrow{g} N_D \quad \text{as} \quad \lambda \searrow 0,$$

and as this happens the mappings $(I+\lambda T)^{-1}$ converge uniformly on all bounded sets to $(I + N_D)^{-1} = P_D$, the projection mapping onto D.

Proof. Let $T_+ :=$ g-$\limsup_{\lambda \searrow 0} \lambda T$ and $T_- :=$ g-$\liminf_{\lambda \searrow 0} \lambda T$. Observe that these mappings have $0 \in T_-(x) \subset T_+(x)$ for all $x \in D$, whereas $T_-(x) = T_+(x) = \emptyset$ for all $x \notin D$. In particular, no sequence $T^\nu = \lambda^\nu T$ with $\lambda^\nu \searrow 0$ escapes to the horizon. Furthermore, any mapping T_0 that is a cluster point of such a sequence (as must exist under graphical convergence by 12.33) has $T_0^{-1}(0) = D$. But such a limit mapping T_0 is maximal monotone by 12.32, and this implies that $T_0^{-1}(v)$ is convex for every v by 12.8(c). Therefore D is convex. Hence $(I + N_D)^{-1} = P_D$ by 6.17.

Next note that because of maximal monotonicity we have for any point $\bar{x}$ an expression for $T(\bar{x})$ as the set of solutions to a system of linear inequalities:

$$T(\bar{x}) = \big\{ \bar{v} \,\big|\, \langle \bar{v},\, x - \bar{x} \rangle \le \langle v,\, x - \bar{x} \rangle \;\text{ for all }\; (x,v) \in \mathrm{gph}\, T \big\}.$$

Then from the horizon cone formula in 3.24 we have, as long as $T(\bar{x}) \ne \emptyset$,

$$\begin{aligned} T(\bar{x})^\infty &= \big\{ \bar{v} \,\big|\, \langle \bar{v},\, x - \bar{x} \rangle \le 0 \;\text{ for all }\; (x,v) \in \mathrm{gph}\, T \big\} \\ &= \big\{ \bar{v} \,\big|\, \langle \bar{v},\, x - \bar{x} \rangle \le 0 \;\text{ for all }\; x \in D \big\} = N_D(\bar{x}) \end{aligned}$$

by the characterization of the normal cone of a convex set in 6.9.

Due to the convexity of $T(x)$ for all x (as noted in 12.8 but also seen in the constraint representation just constructed), this fact gives us $T(x) + N_D(x) = T(x)$ for all x, hence $\lambda T(x) + N_D(x) = \lambda T(x)$ for all x when $\lambda > 0$. It's obvious then that $T_-(\bar{x}) \supset N_D(\bar{x})$ for all $\bar{x}$. On the other hand, if $\bar{v} \in T_+(\bar{x})$ there exist $\lambda^\nu \searrow 0$, $x^\nu \to \bar{x}$ and $v^\nu \in T(x^\nu)$ with $\lambda^\nu v^\nu \to \bar{v}$. For every pair $(x,v) \in \mathrm{gph}\, T$ we have $\langle v^\nu - v,\, x^\nu - x \rangle \ge 0$, hence $\langle \lambda^\nu v^\nu - \lambda^\nu v,\, x^\nu - x \rangle \ge 0$ and in the limit $\langle \bar{v},\, \bar{x} - x \rangle \ge 0$. In other words, we have $\langle \bar{v},\, x - \bar{x} \rangle \le 0$ for all

$x \in \operatorname{dom} T$, hence for all $x \in D$, which means that $\bar{v} \in N_D(\bar{x})$. This establishes that $T_+(\bar{x}) \subset N_D(\bar{x})$ for all $\bar{x}$ and consequently that $T_- = T_+ = N_D$.

The last assertion of the theorem is justified then by 12.32. □

12.38 Corollary (local boundedness of monotone mappings). *A maximal monotone mapping $T : \mathbb{R}^n \rightrightarrows \mathbb{R}^n$ is locally bounded at $\bar{x}$ if and only if $\bar{x}$ is not a boundary point of* $\operatorname{cl}(\operatorname{dom} T)$. *Indeed, T is locally bounded at a point $\bar{x} \in \operatorname{dom} T$ if and only if $T(\bar{x})$ is a bounded set.*

Proof. The mapping T fails to be locally bounded at $\bar{x}$ if and only if there are sequences $x^\nu \to \bar{x}$, $v^\nu \in T(x^\nu)$ and $\lambda^\nu \searrow 0$, such that $\lambda^\nu v^\nu$ converges to some $\bar{v} \neq 0$. This is the same as saying that the mapping $T_+ := \limsup_{\lambda \searrow 0} \lambda T$ has a vector $\bar{v} \neq 0$ in $T_+(\bar{x})$. By 12.37, this is the case if and only if such a vector belongs to $N_D(\bar{x})$, where $D = \operatorname{cl}(\operatorname{dom} T)$. Normal cones are nontrivial precisely at boundary points; cf. 6.19. □

12.39 Corollary (full domains and ranges). *Let $T : \mathbb{R}^n \rightrightarrows \mathbb{R}^n$ be maximal monotone. Then*

(a) $\operatorname{dom} T = \mathbb{R}^n$ *if and only if T is locally bounded everywhere, as is true in particular when* $\operatorname{rge} T$ *is bounded;*

(b) $\operatorname{rge} T = \mathbb{R}^n$ *if and only if T^{-1} is locally bounded everywhere, as is true in particular if* $\operatorname{dom} T$ *is bounded.*

Proof. According to 12.38, T is locally bounded everywhere if and only if $\operatorname{cl}(\operatorname{dom} T) \setminus \operatorname{int}(\operatorname{dom} T) = \emptyset$. Since $\operatorname{dom} T \neq \emptyset$, this means that $\operatorname{dom} T = \mathbb{R}^n$. This gives (a), and then (b) follows by symmetry. □

For more on local boundedness of set-valued mappings and how it may be verified, see Chapter 5 (starting from 5.14).

12.40 Exercise (uniform local boundedness in convergence).

(a) *Let $\bar{x} \in \operatorname{int}(\operatorname{dom} T)$ and $T^\nu \xrightarrow{g} T$ with $T^\nu : \mathbb{R}^n \rightrightarrows \mathbb{R}^n$ maximal monotone. Then there exist $V \in \mathcal{N}(\bar{x})$, $N \in \mathcal{N}_\infty$ and bounded $B \supset T(V)$ such that $B \supset T^\nu(V)$ for all $\nu \in N$. If $T(\bar{x})$ is a singleton, then actually $T^\nu(x^\nu) \to T(\bar{x})$ as $x^\nu \to \bar{x}$.*

(b) *Let $\bar{x} \in \operatorname{int} \operatorname{dom} f$ with $f(\bar{x})$ finite and $f^\nu \xrightarrow{e} f$ with $f^\nu : \mathbb{R}^n \to \overline{\mathbb{R}}$ lsc, proper and convex. Then there exist $V \in \mathcal{N}(\bar{x})$, $N \in \mathcal{N}_\infty$ and bounded $B \supset \partial f(V)$ such that $B \supset \partial f^\nu(V)$ for all $\nu \in N$. If f is differentiable at $\bar{x}$, then actually $\partial f^\nu(x^\nu) \to \{\nabla f(\bar{x})\}$ as $x^\nu \to \bar{x}$.*

Guide. Derive (b) from (a) using 12.17 and 12.35 with 9.18 and 9.20. To get (a) via 12.32 and 12.38, consider a neighborhood $\operatorname{con}\{a_0, a_1, \ldots, a_n\}$ of $\bar{x}$ within $\operatorname{int} \operatorname{dom} T$ and argue (through graphical convergence and the simplex technology in 2.28) the existence of ρ_0, $\varepsilon > 0$ and $N \in \mathcal{N}_\infty$ such that, for any $\nu \in N$, one can find $a_i^\nu \in \rho_0 \mathbb{B}$ and $b_i \in T(a_i^\nu) \cap \rho_0 \mathbb{B}$ with $\mathbb{B}(\bar{x}, 2\varepsilon) \subset C^\nu := \operatorname{con}\{a_0^\nu, a_1^\nu, \ldots, a_n^\nu\}$. Estimate next that for any $x \in V := \mathbb{B}(\bar{x}, \varepsilon)$ and $v \in T(x)$ one has $\langle v, a_i^\nu - x \rangle \leq \langle b_i^\nu, a_i^\nu - x \rangle \leq \rho_0(\rho_0 + \varepsilon)$. Deduce then that the support function of $T^\nu(x)$ is bounded above by $\rho_0(\rho_0 + \varepsilon)$ on $C^\nu - x$ and hence on $\varepsilon \mathbb{B}$, and show by 12.8(c) that this implies $T^\nu(V) \subset \rho \mathbb{B}$ for $\rho := \rho_0(\rho_0 + \varepsilon)/\varepsilon$. □

12.41 Theorem (near convexity of domains and ranges). *For any maximal monotone mapping $T : \mathbb{R}^n \rightrightarrows \mathbb{R}^n$, the set* $\operatorname{dom} T$ *is nearly convex, in the sense that there is a convex set C such that $C \subset \operatorname{dom} T \subset \operatorname{cl} C$. The same applies to the set* $\operatorname{rge} T$.

Proof. Since $\operatorname{rge} T = \operatorname{dom} T^{-1}$ and T^{-1} is maximal monotone like T (cf. 12.8), we need only concern ourselves with the near convexity of domains. Let $D = \operatorname{cl}(\operatorname{dom} T)$, a set which we know from 12.37 to be convex as well as nonempty, and let $C = \operatorname{rint} D$. Then C is a convex set such that $\operatorname{cl} C = D$; cf. 2.40. Our task can therefore be accomplished by demonstrating that $C \subset \operatorname{dom} T$.

Consider first the case where D is n-dimensional, so that $C = \operatorname{int} D$, and any $\bar{x} \in C$. It must be shown that $\bar{x} \in \operatorname{dom} T$. We have T is locally bounded at $\bar{x}$ by 12.38. Then there is an open neighborhood $V \in \mathcal{N}(\bar{x})$ within C such that $T(V)$ is bounded, say $T(V) \subset B$ where B is compact. Because T is osc (cf. 12.8), $T^{-1}(B)$ is closed (cf. 5.25(b)). But $\operatorname{dom} T \supset T^{-1}(B) \supset V \cap \operatorname{dom} T$, whereas $\operatorname{cl}[V \cap \operatorname{dom} T] \supset V \cap D = V$. Therefore $\operatorname{dom} T \supset V$, hence $\bar{x} \in \operatorname{dom} T$.

For the general case we resort to a dimension-reducing argument. Let M be the affine hull of D. There is no loss of generality in supposing that $0 \in D$, so that M is a linear subspace of $\mathbb{R}^n$. Then $N_D(x) \supset M^\perp$ for all $x \in D$, so $T(x) + M^\perp = T(x)$ for all $x \in \operatorname{dom} T$. By an affine change of coordinates if necessary, we can reduce to the case where $\mathbb{R}^n = \mathbb{R}^{n_1} \times \mathbb{R}^{n_2}$ and M is identified with the factor $\mathbb{R}^{n_1}$. Then for $x = (x_1, x_2)$ the set $T(x)$ is empty when $x_2 \neq 0$, and otherwise it has the (still possibly empty) form $T_1(x_1) \times M^\perp$ for a certain mapping $T_1 : \mathbb{R}^n \to \mathbb{R}^{n_1}$. It's easy to see that T_1 inherits maximal monotonicity from T. We have $\operatorname{dom} T = \{(x_1, x_2) \mid x_1 \in \operatorname{dom} T_1,\ x_2 = 0\}$, so the near convexity of $\operatorname{dom} T$ comes down to that of $\operatorname{dom} T_1$. But the set $D_1 = \operatorname{dom} T_1$ is n_1-dimensional in $\mathbb{R}^{n_1}$ by our construction. The preceding argument can therefore be used to reach the desired conclusion. □

With the support of Theorem 12.41, we are allowed, in the case of any *maximal* monotone mapping T, to speak of the relative interiors

$$\operatorname{rint}(\operatorname{dom} T), \qquad \operatorname{rint}(\operatorname{rge} T),$$

these being nonempty convex sets that coincide with the relative interiors of the convex sets $\operatorname{cl}(\operatorname{dom} T)$ and $\operatorname{cl}(\operatorname{rge} T)$ and thus have those as their closures.

These results about domains and ranges apply through 12.17 in particular to the subgradient mapping $T = \partial f$ associated with a proper, lsc, convex function f. Even then, however, it's possible for $\operatorname{dom} T$ not actually to be convex, just nearly convex as described.

For instance, in terms of the convex function φ on $\mathbb{R}$ given by $\varphi(t) = 1 - \sqrt{1 - t^2}$ when $|t| \leq 1$ but $\varphi(t) = \infty$ when $|t| > 1$, define f on $\mathbb{R}^2$ by $f(x_1, x_2) = \max\{\varphi(x_1), |x_2|\}$. Then $\operatorname{dom} f$ is the closed strip $[-1, 1] \times (-\infty, \infty)$, while $\operatorname{rint}(\operatorname{dom} f) = \operatorname{int}(\operatorname{dom} f)$ is the open strip $(-1, 1) \times (-\infty, \infty)$, but $\operatorname{dom} \partial f$ is the nonconvex set consisting of the union of $(-1, 1) \times (-1, 1)$, $[-1, 1] \times [1, \infty)$ and $[-1, 1] \times (-\infty, -1]$. This union is nearly convex, though, because it lies between the open strip and the closed strip.

12.42 Corollary (ranges and growth). *Let $T : \mathbb{R}^n \rightrightarrows \mathbb{R}^n$ be maximal monotone and let $R = \operatorname{cl}(\operatorname{rge} T)$. For any choice of $\bar{x} \in \mathbb{R}^n$, one has*

$$\operatorname*{g-lim}_{\tau \nearrow \infty} T_\tau = \partial\sigma_R \quad \text{for} \quad T_\tau(w) := T(\bar{x} + \tau w),\ \ \tau > 0.$$

In particular, for any sequence $\tau^\nu \nearrow \infty$ one has $\bigcup_{\{w^\nu \to w\}} \limsup_\nu T(\bar{x}+\tau^\nu w^\nu) = \partial\sigma_R(w)$ for all w, where the union is taken over all sequences $\{w^\nu\}_{\nu\in\mathbb{N}}$ such that $w^\nu \to w$. Equivalently, this means that

$$\limsup_{x \to \operatorname{dir} w} T(x) = \operatorname*{argmax}_{v \in R} \langle v, w\rangle \quad \text{for all } w \neq 0.$$

Thus, whenever the linear function $\langle \cdot, w\rangle$ is unbounded from above on R, or is bounded from above on R without achieving its maximum, the set $T(x)$ must escape to the horizon as $x \to \operatorname{dir} w$.

Proof. Since $R = \operatorname{cl}(\operatorname{dom} T^{-1})$, the mappings $T_\tau^{-1} = \tau^{-1}(T^{-1} - \bar{x})$ converge graphically as $\tau \nearrow \infty$ to the normal cone mapping N_R; see 12.37. Hence the mappings T_τ converge graphically to N_R^{-1}. The set R is convex by 12.41, so $N_R^{-1}(w) = \partial\sigma_R(w) = \operatorname{argmax}_{v\in R}\langle v, w\rangle$; cf. 8.25, 11.3, 11.4. For any $w \neq 0$, the sequences $x^\nu \to \operatorname{dir} w$ are the ones that can be written in the form $x^\nu = \bar{x}+\tau^\nu w^\nu$ with $w^\nu \to w$ and $0 < \tau^\nu \nearrow \infty$. This yields all. □

F*. Preservation of Maximality

It was relatively easy to see that the sum of two monotone mappings is again monotone; cf. 12.4. But is the sum of two *maximal* monotone mappings again *maximal* monotone? That isn't automatic and requires closer analysis. Constraint qualifications involving the relative interiors of domains will be required. (Such relative interiors exist on the basis of the domains being nearly convex, as noted above after the proof of 12.41.)

It's expedient to begin with the question of maximality for mappings generated by composition in the sense of 12.4(d). Other operations can be construed as special instances of composition.

12.43 Theorem (maximal monotonicity under composition). *Suppose $T(x) = A^*S(Ax + a)$ for a maximal monotone mapping $S : \mathbb{R}^m \rightrightarrows \mathbb{R}^m$, a matrix $A \in \mathbb{R}^{m\times n}$, and a vector $a \in \mathbb{R}^m$. If $(\operatorname{rge} A + a) \cap \operatorname{rint}(\operatorname{dom} S) \neq \emptyset$, then T is maximal monotone.*

Proof. Without loss of generality we can take $a = 0$ and suppose that $0 \in \operatorname{rint}(\operatorname{dom} S)$, $0 \in S(0)$, since these properties can be arranged by translation. We can arrange also that the smallest subspace of $\mathbb{R}^m$ that includes the convex set $D := \operatorname{cl}(\operatorname{dom} S)$ along with the subspace $\operatorname{rge} A$ is $\mathbb{R}^m$ itself. (Otherwise we could replace $\mathbb{R}^m$ by a space of lower dimension.)

For $\lambda > 0$ let S_λ stand for the Yosida regularization $(\lambda I + S^{-1})^{-1}$. Note that $0 \in S_\lambda(0)$ because $0 \in S(0)$. According to 12.13, S_λ is monotone, single-valued and continuous. Define $T_\lambda(x) := A^* S_\lambda(Ax)$. The mapping T_λ is single-valued and continuous, and by 12.4(d) it inherits monotonicity from S_λ as well. Therefore it's maximal monotone by 12.7. Also, $0 \in T_\lambda(0)$.

For some sequence $\lambda^\nu \searrow 0$, the sequence of mappings T_{λ^ν} converges graphically to a maximal monotone mapping T_0; this is assured by 12.33, inasmuch as the sequence is precluded from escaping to the horizon by having $0 \in T_{\lambda^\nu}(0)$. By establishing that $\operatorname{gph} T_0 \subset \operatorname{gph} T$ we will be able to conclude that T itself must be maximal (and hence actually that $T_\lambda \xrightarrow{g} T$ as $\lambda \searrow 0$).

Suppose $\bar{v} \in T_0(\bar{x})$. There exist $x^\nu \to \bar{x}$ and $v^\nu \to \bar{v}$ with $v^\nu \in T_{\lambda^\nu}(x^\nu)$. The latter means that $v^\nu = A^* y^\nu$ for $y^\nu = S_{\lambda^\nu}(Ax^\nu)$. If the sequence of vectors y^ν has a cluster point $\bar{y}$, we have $\bar{v} = A^*\bar{y}$ but also $\bar{y} \in S(A\bar{x})$ by the graphical convergence of S_{λ^ν} to S, so that $\bar{v} \in T(\bar{x})$ as required. Otherwise we must have $|y^\nu| \to \infty$. We'll demonstrate that this case runs into a conflict with our relative interior assumption.

Even though $|y^\nu| \to \infty$, the sequence of vectors $\lambda^\nu y^\nu$ converges, because $\lambda^\nu y^\nu = \big[I - (I + \lambda^\nu S)^{-1}\big](Ax^\nu)$ by the formula in 12.14, with $Ax^\nu \to A\bar{x}$ and with the mappings $(I + \lambda^\nu S)^{-1}$ converging uniformly on bounded sets to the projection P_D on $D := \operatorname{cl}(\operatorname{dom} S)$; cf. 12.37. Using the fact that $S_{\lambda^\nu} = (\lambda^\nu I + S^{-1})^{-1}$ it's also possible to rewrite $y^\nu = S_{\lambda^\nu}(Ax^\nu)$ as meaning $y^\nu \in S(Ax^\nu - \lambda^\nu y^\nu)$, where we now recognize that the points $u^\nu := Ax^\nu - \lambda^\nu y^\nu$ must converge to a certain point $\bar{u}$.

Take $\mu^\nu = 1/|y^\nu|$, so that $\mu^\nu \searrow 0$. We have $\mu^\nu v^\nu \to 0$, but the sequence of vectors $\mu^\nu y^\nu$ has a cluster point $\tilde{y} \neq 0$. From $v^\nu = A^* y^\nu$ we get $0 = A^*\tilde{y}$, hence $\tilde{y} \perp \operatorname{rge} A$ (and in particular $\operatorname{rge} A \neq I\!R^m$), with

$$\langle \tilde{y}, \bar{u} \rangle = \lim_\nu \langle \mu^\nu y^\nu, Ax^\nu - \lambda^\nu y^\nu \rangle = \langle y, A\bar{x} \rangle - \lim_\nu \langle \mu^\nu y^\nu, \lambda^\nu y^\nu \rangle \leq 0.$$

We have $\mu^\nu y^\nu \in (\mu^\nu S)(u^\nu)$, but from 12.37 we know that $\mu^\nu S \xrightarrow{g} N_D$ for $D = \operatorname{cl}(\operatorname{dom} S)$. Therefore $\tilde{y} \in N_D(\bar{u})$. The hyperplane $H = \big\{ u \,\big|\, \langle \tilde{y}, u - \bar{u} \rangle = 0 \big\}$ then separates D from the subspace $\operatorname{rge} A$. This is impossible by 2.39 under our assumption that $\operatorname{rge} A$ meets the set $\operatorname{rint}(\operatorname{dom} S) = \operatorname{rint} D$, unless actually $D \subset \operatorname{rge} A$, which can't be the case because we've arranged that no proper subspace of $I\!R^n$ includes both D and $\operatorname{rge} A$. □

12.44 Corollary (maximal monotonicity under addition). *Let $T = T_1 + T_2$ for $T_i : I\!R^n \rightrightarrows I\!R^n$ maximal monotone. If $\operatorname{rint}(\operatorname{dom} T_1) \cap \operatorname{rint}(\operatorname{dom} T_2) \neq \emptyset$, then T is maximal monotone.*

Proof. We apply the theorem to $S(x_1, x_2) := T_1(x_1) \times T_2(x_2)$ and the linear mapping $x \mapsto (x, x)$. □

12.45 Example (truncation). *Let $T : I\!R^n \rightrightarrows I\!R^n$ be maximal monotone.*

(a) *For any compact, convex set $B \subset I\!R^n$ such that $\operatorname{int} B$ meets $\operatorname{dom} T$, the mapping $S = T + N_B$ is maximal monotone and agrees with T on $\operatorname{int} B$. Furthermore, $\operatorname{rge} S = I\!R^n$.*

(b) *For any compact, convex set $B \subset \mathbb{R}^n$ such that* $\operatorname{int} B$ *meets* $\operatorname{rge} T$, *the mapping* $S = (T^{-1} + N_B)^{-1}$ *is maximal monotone, and* S^{-1} *agrees with* T^{-1} *on* $\operatorname{int} B$. *Furthermore,* $\operatorname{dom} S = \mathbb{R}^n$.

Detail. We have N_B maximal monotone by 12.18, and $\operatorname{dom} N_B = B$. The condition that $\operatorname{dom} T \cap \operatorname{int} B \neq \emptyset$ is thus the same as $\operatorname{rint}(\operatorname{dom} T) \cap \operatorname{rint}(\operatorname{dom} N_B) \neq \emptyset$. We get (a) then from the maximality criterion in 12.44 and the observation about bounded domains in 12.39(b). Part (b) follows then by symmetry. □

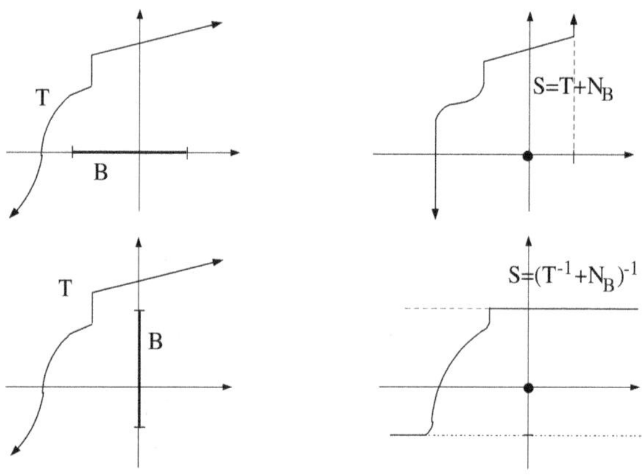

Fig. 12–6. Maximality-preserving truncations of a monotone mapping.

12.46 Exercise (restriction of arguments). *Let* $T : \mathbb{R}^{n_1} \times \mathbb{R}^{n_2} \rightrightarrows \mathbb{R}^{n_1} \times \mathbb{R}^{n_2}$ *be maximal monotone. Fix any* $\bar{x}_2 \in \mathbb{R}^{n_2}$ *and define* $T_1 : \mathbb{R}^{n_1} \rightrightarrows \mathbb{R}^{n_1}$ *by*

$$T_1(x_1) = \big\{ v_1 \,\big|\, \exists\, v_2 \text{ with } (v_1, v_2) \in T(x_1, \bar{x}_2) \big\}.$$

If $\bar{x}_2$ *is such that there exists* $\bar{x}_1 \in \mathbb{R}^{n_1}$ *with* $(\bar{x}_1, \bar{x}_2) \in \operatorname{rint}(\operatorname{dom} T)$, *then* T_1 *is maximal monotone.*

Guide. Deduce this from 12.43 with the affine mapping $x_1 \mapsto (x_1, \bar{x}_2)$. □

12.47 Example (monotonicity along lines). *Let* $T : \mathbb{R}^n \rightrightarrows \mathbb{R}^n$ *be maximal monotone, and consider any point* $x \in \operatorname{rint}(\operatorname{dom} T)$. *For any vector* $w \neq 0$ *define the mapping* $T_{x,w} : \mathbb{R}^1 \rightrightarrows \mathbb{R}^1$ *by* $T_{x,w}(\tau) = \big\{ \langle v, w \rangle \,\big|\, v \in T(x + \tau w) \big\}$. *Then* $T_{x,w}$ *is maximal monotone and thus has the form in 12.9(b) with respect to some nondecreasing function* $\varphi : \mathbb{R} \to \overline{\mathbb{R}}$.

Detail. This is the case of 12.43 in which T is composed with the affine mapping $\tau \mapsto x + \tau w$ from $\mathbb{R}^1$ into $\mathbb{R}^n$. □

G.* Monotone Variational Inequalities

Generalized equations, as described in Example 5.2, often involve monotone mappings, especially when they relate to optimality conditions. The fundamental problem of such type is this:

$$\text{given } T : \mathbb{R}^n \rightrightarrows \mathbb{R}^n, \text{ find } \bar{x} \text{ such that } T(\bar{x}) \ni 0. \qquad 12(8)$$

The solutions, if any, are obviously the elements of the set $T^{-1}(0)$.

In some situations it's useful to take a parametric stance and consider finding, for each choice of $\bar{v} \in \mathbb{R}^n$, one or all of the vectors $\bar{x}$, if any, such that $T(\bar{x}) \ni \bar{v}$. In this case the focus is on the entire mapping T^{-1}. The issue then is how to understand properties of T^{-1} well enough to be able to appreciate the existence and nature of solutions and the manner of their dependence on parameters such as $\bar{v}$, and to use this understanding in the design of numerical methods for computing such solutions.

Obviously, if T happens to be maximal monotone, a great deal of information is at our disposal, because T^{-1} is maximal monotone too, and its graph, domain and range have the geometric character revealed in the many results above. An immediate example is $T = \partial f$ with f proper, lsc, and convex. Then in solving $T(\bar{x}) \ni 0$ we are looking for the points $\bar{x} \in \operatorname{argmin} f$; cf. 10.1. More generally, the solution set to $T(\bar{x}) \ni \bar{v}$ in this case is $\operatorname{argmin}_x\{f(x) - \langle \bar{v}, x\rangle\} = \partial f^*(\bar{v})$; cf. 11.3, 11.8. Other examples relate to the first-order optimality conditions developed in 6.12 and the variational inequalities in 6.13.

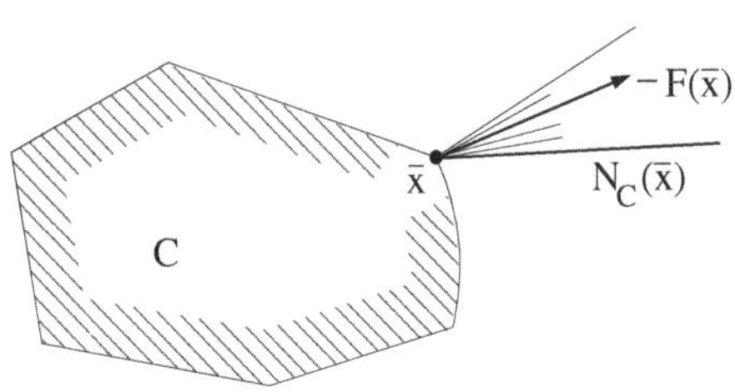

Fig. 12–7. The geometry of a variational inequality.

12.48 Example (monotonicity in variational inequalities). *For a closed, convex set $C \neq \emptyset$ in $\mathbb{R}^n$ and a mapping $F : C \to \mathbb{R}^n$, the associated variational inequality at a point $\bar{x}$ can be expressed as the condition $T(\bar{x}) \ni 0$ for the mapping $T : \mathbb{R}^n \rightrightarrows \mathbb{R}^n$ defined by*

$$T(x) = F(x) + N_C(x) \ \ (\text{with } T(x) = \emptyset \text{ when } x \notin C).$$

This mapping T is maximal monotone when F is continuous and monotone relative to C, i.e., has $\langle F(x_1) - F(x_0), x_1 - x_0\rangle \geq 0$ for all x_0 and x_1 in C, and in that case the (possibly empty) solution set $T^{-1}(0)$ is closed and convex.

Moreover, T is strictly monotone when F is strictly monotone relative to C. Then the solution set can have no more than one element.

Detail. The monotonicity of T follows from that of F as a special case of the addition rule in 12.4(b) (with F identified with a set-valued mapping $I\!R^n \rightrightarrows I\!R^n$ that happens to have C as its effective domain). Here we use the monotonicity of N_C in 12.18. For the rest, a reduction can be made (via the affine hull of C) to the case where $\operatorname{int} C \neq \emptyset$.

Let $\overline{T}$ be any maximal monotone extension of T, as exists by 12.6. Is $\overline{T} = T$, so that T is itself maximal? Obviously the convex set $D := \operatorname{cl}(\operatorname{dom} \overline{T})$ includes the convex set $\operatorname{dom} T = C$. For all $x \in C$ we have $N_C(x) = T(x)^\infty \subset \overline{T}(x)^\infty = N_D(x)$; cf. 12.37. But the latter implies that every boundary point of C is a boundary point of D (since boundary points x of C are distinguished precisely by having $N_C(x)$ be nontrivial). Therefore $D = C$. To verify that $\overline{T} = T$, it suffices therefore to fix any $\bar{x} \in C$, $\bar{v} \in \overline{T}(\bar{x})$, and demonstrate that $\bar{v} \in T(\bar{x})$, or equivalently, that $\bar{v} - F(\bar{x}) \in N_C(\bar{x})$.

Consider arbitrary $x \in C$ and let $x_\tau := \bar{x} + \tau(x - \bar{x})$ for $\tau \in (0,1)$. We have $x_\tau \in C$ and therefore $F(x_\tau) \in T(x_\tau) \subset \overline{T}(x_\tau)$ (because $0 \in N_C(x_\tau)$), so that $\langle F(x_\tau) - \bar{v},\, x_\tau - \bar{x}\rangle \geq 0$ by the monotonicity of $\overline{T}$. But $x_\tau - \bar{x} = \tau(x - \bar{x})$. It follows that $\langle F(x_\tau) - \bar{v},\, x - \bar{x}\rangle \geq 0$. This being true for any $\tau \in (0,1)$, we may take the limit as $\tau \searrow 0$ and conclude from the continuity of F that $\langle F(\bar{x}) - \bar{v},\, x - \bar{x}\rangle \geq 0$. The choice of $x \in C$ was arbitrary, so by invoking once more the characterization in 6.9 of normal vectors to convex sets, we find that $\bar{v} - F(\bar{x}) \in N_C(\bar{x})$, as required. □

Incidentally, the maximal monotonicity in this example generalizes that in 12.7, which can be regarded now as the case of 12.48 where $C = I\!R^n$.

12.49 Exercise (alternative expression of a variational inequality). *A point $\bar{x}$ satisfies the variational inequality for a closed, convex set $C \neq \emptyset$ and a continuous, monotone mapping F if and only if*

$$\bar{x} \in C, \qquad \big\langle F(x),\, \bar{x} - x\big\rangle \leq 0 \ \text{ for all } x \in C.$$

Guide. Show that this condition is equivalent to having $\big\langle 0 - v,\, \bar{x} - x\big\rangle \geq 0$ for all $(x, v) \in \operatorname{gph} T$ for the mapping T in Example 12.48. Appeal to the maximal monotonicity of T. □

It's noteworthy that 12.49 describes the set of solutions to the variational inequality for F and C as the set of solutions to a certain system of linear inequalities. When F is the mapping ∇f_0 from a $\mathcal{C}^1$ function f_0 on $I\!R^n$, this description reduces to the first-order necessary condition for f_0 to achieve its minimum over C at $\bar{x}$ (which is sufficient as well when f_0 is convex); cf. 6.12.

12.50 Example (variational inequalities for saddle points). *Let $X \subset I\!R^n$ and $Y \subset I\!R^m$ be nonempty, closed, convex sets, and let $L : I\!R^n \times I\!R^m \to I\!R$ be a $\mathcal{C}^1$ function such that $L(x, y)$ is convex in $x \in X$ for each $y \in Y$, but concave in $y \in Y$ for each $x \in X$. Then the saddle point condition*

$$\bar{x} \in X,\ \bar{y} \in Y, \qquad L(x, \bar{y}) \geq L(\bar{x}, \bar{y}) \geq L(\bar{x}, y) \ \textit{for all}\ x \in X,\ y \in Y,$$

is equivalent to the variational inequality on $(\bar{x}, \bar{y})$ *with respect to*

$$C = X \times Y, \qquad F : (x, y) \mapsto (\nabla_x L(x, y), -\nabla_y L(x, y)).$$

It corresponds to $T(\bar{x}, \bar{y}) \ni (0, 0)$ *for the maximal monotone mapping*

$$T(x, y) = \big(\nabla_x L(x, y) + N_X(x),\ -\nabla_y L(x, y) + N_Y(y)\big).$$

Detail. Here C is closed and convex with $N_C(x, y) = N_X(x) \times N_Y(y)$ (by 6.41), while F is continuous and monotone relative to C. (For the monotonicity, apply 12.27 to $l(x, y) + \delta_X(x) - \delta_Y(y)$ or appeal to an extension of the rule in 2.14: if a smooth function f is convex relative to a convex set D, then ∇f is monotone relative to D.) The indicated mapping T is thus $F + N_C$ as in 12.48. We have $T(\bar{x}, \bar{y}) \ni (0, 0)$ if and only if

$$\nabla_x L(\bar{x}, \bar{y}) + N_X(\bar{x}) \ni 0, \qquad -\nabla_y L(\bar{x}, \bar{y}) + N_Y(\bar{y}) \ni 0,$$

where the first condition is equivalent to $\bar{x} \in \operatorname{argmin}_{x \in X} L(x, \bar{y})$ and the second is equivalent to $\bar{y} \in \operatorname{argmax}_{y \in Y} L(\bar{x}, y)$ in accordance with the first-order optimality rules in 6.12. □

Example 12.50 recalls the minimax framework of 11.52 and ties in with the Lagrangian expressions of optimality in 11.46—11.51. It indicates how monotone variational inequalities arising from convex-concave functions and saddle point conditions characterize the primal-dual solution pairs in the duality schemes of Chapter 11.

Next, we look at a fundamental existence theorem.

12.51 Theorem (solutions to a monotone generalized equation). *For a maximal monotone mapping* $T : \mathbb{R}^n \rightrightarrows \mathbb{R}^n$, *consider the (closed, convex) solution set* $T^{-1}(0)$ *to the generalized equation* $T(\bar{x}) \ni 0$.

(a) $T^{-1}(0) \cap \rho\mathbb{B}$ *is nonempty if* $\langle v, x \rangle \geq 0$ *when* $v \in T(x)$, $|x| > \rho$.

(b) $T^{-1}(0)$ *is nonempty and bounded if and only if*

$$\begin{cases} \textit{for each nonzero}\ w \in (\operatorname{dom} T)^\infty \ \ \textit{(if any)} \\ \textit{there exists}\ v \in \operatorname{rge} T \ \textit{with}\ \langle v, w \rangle > 0. \end{cases}$$

Here $(\operatorname{dom} T)^\infty$ *consists, for any* $a \in \operatorname{rint}(\operatorname{dom} T)$, *of the vectors* w *such that* $a + \tau w \in \operatorname{dom} T$ *for all* $\tau > 0$.

Proof. (a) Since T is osc, the set $T(\rho\mathbb{B})$ is closed; cf. 5.25(a). We have to show that it contains 0 when the stated condition holds. For each $\nu \in \mathbb{N}$ we know from 12.12 that the mapping $(I + \nu\rho T)^{-1}$ is single-valued and has all of $\mathbb{R}^n$ as its domain. Let $x^\nu = (I + \nu\rho T)^{-1}(0)$, so that $0 \in x^\nu + \nu\rho T(x^\nu)$, or in other words, $v^\nu \in T(x^\nu)$ for $v^\nu := -x^\nu/\nu\rho$. If $|x^\nu| > \rho$, our condition would imply that $\langle v^\nu, x^\nu \rangle \geq 0$. But $\langle v^\nu, x^\nu \rangle = -|x^\nu|^2/\nu\rho < -\rho/\nu$, so this is impossible. Therefore $x^\nu \in \rho I\!B$, $v^\nu \in T(\rho I\!B)$ and $|v^\nu| \leq 1/\nu$. Then $v^\nu \to 0$, and it follows that $0 \in T(\rho I\!B)$, as needed.

(b) From applying 12.38 to T^{-1}, we see that $T^{-1}(0)$ is nonempty and bounded if and only if 0 belongs to the interior of the set $\operatorname{dom} T^{-1} = \operatorname{rge} T$. This set is nonempty and nearly convex in the sense given in 12.41, so its interior contains 0 if and only if its support function is positive away from the origin; cf. 8.29(a). Thus, $T^{-1}(0)$ is nonempty and bounded if and only if, for every $w \neq 0$, there exists $v \in \operatorname{rge} T$ such that $\langle v, w\rangle > 0$.

It remains only to demonstrate that the latter condition is sure always to be satisfied when $w \notin (\operatorname{dom} T)^\infty$. Consider any point $a \in \operatorname{rint}(\operatorname{dom} T)$. Through the property of $\operatorname{dom} T$ being almost convex, we know that for any $w \neq 0$ the half-line $\{a + \tau w \mid \tau \geq 0\}$ intersects $\operatorname{dom} T$ in a line segment; cf. 2.33, 2.41. Moreover, in consequence of 3.6, $(\operatorname{dom} T)^\infty$ consists of the vectors w such that this segment is unbounded, i.e., such that $a + \tau w \in \operatorname{dom} T$ for all $\tau \in [0, \infty)$. If $w \notin (\operatorname{dom} T)^\infty$, therefore, the segment is bounded. The maximal monotone mapping $T_{a,w} : I\!R^1 \rightrightarrows I\!R^1$ in 12.47 then has its effective domain bounded on the right. In view of the properties in 12.9, the values $\langle v, w\rangle$ realized by elements v and w with $v \in T(a + \tau w)$, $\tau \geq 0$, can't in that case even be bounded from above. □

12.52 Exercise (existence of solutions to variational inequalities). *For a closed, convex set $C \neq \emptyset$ in $I\!R^n$ and a continuous, monotone mapping $F : C \to I\!R^n$, let $X_{C,F}$ stand for the set of solutions $\bar{x}$ to the variational inequality for C and F. Let $a \in C$. Then*

(a) *$X_{C,F} \cap I\!B(a, \rho)$ is nonempty if $\langle F(x), x - a\rangle \geq 0$ for $x \in C$, $|x - a| > \rho$;*

(b) *$X_{C,F}$ is nonempty and bounded if and only if*

$$\begin{cases} \text{for each nonzero } w \in C^\infty \text{ (if any)}, \\ \text{there exists } x \in C \text{ with } \langle F(x), w\rangle > 0. \end{cases}$$

Guide. Apply Theorem 12.51 in the setting of 12.48, making a shift of domain in the case of part (a). Determine that when w belongs to the set $(\operatorname{dom} T)^\infty = C^\infty$, one has in particular that $w \in T_C(x)$ for all $x \in C$, and then $\langle v, w\rangle \leq 0$ for all $v \in N_C(x)$. □

The existence result in Theorem 12.51 is complemented by the fact that if T is strictly monotone the solution set $T^{-1}(0)$ can't contain more than one point, as is obvious from Definition 12.1. In the context of 12.52, the strict monotonicity of T comes down to the strict monotonicity of F relative to C.

H* Strong Monotonicity and Strong Convexity

For numerical purposes in solving generalized equations, strict monotonicity is significant in guaranteeing uniqueness of solutions, as just remarked. But the following concept is even more potent in numerical work, because it provides a simple test of existence and uniqueness simultaneously.

12.53 Definition (strong monotonicity). *A mapping $T : \mathbb{R}^n \rightrightarrows \mathbb{R}^n$ is strongly monotone if there exists $\sigma > 0$ such that $T - \sigma I$ is monotone, or equivalently:*

$$\langle v_1 - v_0,\, x_1 - x_0 \rangle \geq \sigma |x_1 - x_0|^2 \;\; \textit{whenever} \;\; v_0 \in T(x_0),\; v_1 \in T(x_1).$$

The equivalence in the definition is clear from writing the inequality in the form $\langle [v_1 - \sigma x_1] - [v_0 - \sigma x_0],\, x_1 - x_0 \rangle \geq 0$.

12.54 Proposition (consequences of strong monotonicity). *If a maximal monotone mapping $T : \mathbb{R}^n \rightrightarrows \mathbb{R}^n$ is strongly monotone with constant $\sigma > 0$, then T is strictly monotone and there is a unique point $\bar{x}$ such that $T(\bar{x}) \ni 0$. In fact T^{-1} is everywhere single-valued and is Lipschitz continuous globally with constant $\kappa = 1/\sigma$.*

Proof. The implication to strict monotonicity is obvious. The strong monotonicity of T with constant σ is equivalent to the monotonicity of $T_0 = T - \sigma I$, and we have $T^{-1} = (\sigma I + T_0)^{-1}$. When T is maximal monotone, T_0 must be maximal as well; for otherwise a proper monotone extension $\overline{T}_0$ of T_0 would yield a proper monotone extension $\overline{T}_0 + \sigma I$ of T. Then $(\sigma I + T_0)^{-1}$ is single-valued and Lipschitz continuous globally with constant $1/\sigma$ because it's the Yosida σ-regularization of T_0^{-1}; cf. 12.13. □

12.55 Corollary (inverse strong monotonicity). *If $T : \mathbb{R}^n \rightrightarrows \mathbb{R}^n$ is maximal monotone and T^{-1} is strongly monotone with constant $\sigma > 0$, i.e.,*

$$\langle v_1 - v_0,\, x_1 - x_0 \rangle \geq \sigma |v_1 - v_0|^2 \;\; \textit{whenever} \;\; v_0 \in T(x_0),\; v_1 \in T(x_1),$$

then T must be single-valued and Lipschitz continuous with constant $1/\sigma$.

Proof. Apply 12.54 to T^{-1}. □

Strong monotonicity is especially important in generating contractive mappings whose fixed points solve a given variational inequality.

12.56 Theorem (contractions from strongly monotone mappings). *If the mapping $T : \mathbb{R}^n \rightrightarrows \mathbb{R}^n$ is maximal monotone, then each of the mappings*

$$P_\tau := I - \tau(I + T^{-1})^{-1} = (1 - \tau)I + \tau(I + T)^{-1} \;\; \textit{for} \;\; \tau \in (0, 2]$$

is single-valued and nonexpansive, having $T^{-1}(0)$ as its set of fixed points.

In fact, if T is strongly monotone with constant $\sigma > 0$, the mappings P_τ for $\tau \in (0, 2)$ are contractive with Lipschitz modulus satisfying

$$\operatorname{lip} P_\tau \leq \begin{Bmatrix} 1 - \dfrac{\tau\sigma}{1+\sigma} & \textit{when } 0 < \tau \leq \bar{\tau} \\ \tau - 1 & \textit{when } \bar{\tau} \leq \tau < 2 \end{Bmatrix} \;\; \textit{for } \bar{\tau} = 1 + \frac{1}{1 + 2\sigma}, \tag{12(9)}$$

the lowest of these bounds being $\operatorname{lip} P_{\bar{\tau}} \leq 1/(1 + 2\sigma)$. Thus the mapping $P = (I + T)^{-1}$, which is P_τ for $\tau = 1$, is contractive with $\operatorname{lip} P \leq 1/(1 + \sigma)$.

Proof. The two forms of expression for P_τ agree by the identity in 12.14. The mappings $(I + T)^{-1}$ and $(I + T^{-1})^{-1}$ are single-valued by 12.12, so P_τ is single-valued everywhere. Let $w_i = P_\tau(z_i)$ for $i = 0, 1$, where $z_1 \neq z_0$. Then $w_i - (1-\tau)z_i = \tau(I+T)^{-1}(z_i)$, so that $z_i \in (I+T)\big(\tau^{-1}w_i + (1-\tau^{-1})z_i\big)$, or equivalently $\tau^{-1}(z_i - w_i) \in T\big(\tau^{-1}w_i + (1-\tau^{-1})z_i\big)$. The strong monotonicity of T gives us then the inequality

$$\Big\langle \frac{1}{\tau}(z_1 - w_1) - \frac{1}{\tau}(z_0 - w_0),\ \big(\frac{1}{\tau}w_1 + (1-\frac{1}{\tau})z_1\big) - \big(\frac{1}{\tau}w_0 + (1-\frac{1}{\tau})z_0\big)\Big\rangle$$
$$\geq \sigma\Big|\big(\frac{1}{\tau}w_1 + (1-\frac{1}{\tau})z_1\big) - \big(\frac{1}{\tau}w_0 + (1-\frac{1}{\tau})z_0\big)\Big|^2.$$

Multiplying through by τ^2 and rearranging, we obtain in terms of $\Delta w = w_1 - w_0$ and $\Delta z = z_1 - z_0$ that

$$\begin{aligned}
0 &\geq \Big\langle \sigma[\Delta w + (\tau-1)\Delta z] - \Delta z + \Delta w,\ [\Delta w + (\tau-1)\Delta z]\Big\rangle \\
&= \Big\langle [1+\sigma]\Delta w - [1+\sigma(1-\tau)]\Delta z,\ [\Delta w - (1-\tau)\Delta z]\Big\rangle \\
&= [1+\sigma]|\Delta w|^2 - [1+(1+2\sigma)(1-\tau)]\langle \Delta w, \Delta z\rangle \\
&\qquad + [1+\sigma(1-\tau)](1-\tau)|\Delta z|^2 \\
&\geq [1+\sigma]|\Delta w|^2 - |1+(1+2\sigma)(1-\tau)||\Delta w||\Delta z| \\
&\qquad + [1+\sigma(1-\tau)](1-\tau)|\Delta z|^2
\end{aligned}$$

and hence for the ratio $\rho = |\Delta w|/|\Delta z| = |w_1 - w_0|/|z_1 - z_0|$ that

$$0 \geq [1+\sigma]\rho^2 - |1+(1+2\sigma)(1-\tau)|\rho + [1+\sigma(1-\tau)](1-\tau).$$

The larger of the two roots of this quadratic polynomial in ρ gives an upper bound for ρ. This root is

$$\rho_{\max} = \frac{|1+(1+2\sigma)(1-\tau)| + \sqrt{\alpha}}{2(1+\sigma)},$$

where

$$\begin{aligned}
\alpha &= |1+(1+2\sigma)(1-\tau)|^2 - 4(1+\sigma)[1+\sigma(1-\tau)](1-\tau) \\
&= 1 + 2(1+2\sigma)(1-\tau) + (1+2\sigma)^2(1-\tau)^2 - 4(1+\sigma)\big[(1-\tau) + \sigma(1-\tau)^2\big] \\
&= 1 + \big[2(1+2\sigma) - 4(1+\sigma)\big](1-\tau) + \big[(1+2\sigma)^2 - 4\sigma(1+\sigma)\big](1-\tau)^2 \\
&= 1 - 2(1-\tau) + (1-\tau)^2 = \big[1-(1-\tau)\big]^2 = \tau^2.
\end{aligned}$$

Therefore

$$\rho \leq \frac{|1+(1+2\sigma)(1-\tau)| + \tau}{2(1+\sigma)}. \qquad 12(10)$$

The value $\bar{\tau}$ in 12(9) gives the threshold for the absolute value in the numerator:

$$|1+(1+2\sigma)(1-\tau)| = \begin{cases} 1+(1+2\sigma)(1-\tau) & \text{when } 0 < \tau \leq \bar{\tau}, \\ -1-(1+2\sigma)(1-\tau) & \text{when } \bar{\tau} \leq \tau < 2. \end{cases}$$

In the first case the numerator in 12(10) reduces to $2(1+\sigma) - 2\sigma\tau$ and the ratio comes out as $1-\tau\sigma(1+\sigma)^{-1}$, which verifies the first of the alternatives in 12(9). In the second case the numerator in 12(10) is $-[1+(1+2\sigma)(1-\tau)]+\tau = 2(1+\sigma)(\tau-1)$ and the ratio is simply $\tau - 1$. This yields the second of the alternatives in 12(9).

The case of $\tau = 1$, giving $P_\tau = P$, is covered by the first of the alternatives in 12(9) with the bound $1/(1+\sigma)$. In general, the right side of 12(9) is piecewise linear in τ. It decreases over $(0, \bar{\tau}]$ but increases over $[\bar{\tau}, 2)$, so its minimum is attained at $\bar{\tau}$, where its value is $\bar{\tau} - 1 = 1/(1+2\sigma)$. □

12.57 Example (contractions in solving variational inequalities). *Suppose $T = F+N_C$ for a nonempty, closed, convex set $C \subset \mathbb{R}^n$ and a continuous, monotone mapping $F : C \to \mathbb{R}^n$, as in 12.48. Then, for any $\tau \in (0,2)$, the mapping P_τ in 12.56 is nonexpansive, its fixed points being the solutions to the variational inequality for C and F. The iteration scheme $x^{\nu+1} = P_\tau(x^\nu)$ takes the form:*

$$\begin{cases} \text{calculate } \bar{x}^\nu \text{ by solving the variational inequality for } C \text{ and } F^\nu, \\ \text{where } F^\nu(x) := F(x) + x - x^\nu; \text{ then let } x^{\nu+1} := (1-\tau)x^\nu + \tau\bar{x}^\nu. \end{cases}$$

If F is strongly monotone relative to C with constant σ, the sequence $\{x^\nu\}_{\nu\in\mathbb{N}}$ generated from any starting point converges to $\bar{x}$, the unique solution to the variational inequality for C and F, and it does this at a linear rate:

$$\left|x^{\nu+1} - \bar{x}\right| \le \kappa_\tau\left|x^\nu - \bar{x}\right| \text{ for all } \nu \in \mathbb{N}, \text{ with } \kappa_\tau \in (0,1),$$

where κ_τ stands for the expression on the right side of the inequality in 12(9).

Detail. These facts are immediate from Theorem 12.56 in the context of Example 12.48. □

Next we explore connections between strong monotonicity and an enhanced form of convexity.

12.58 Definition (strong convexity). *A proper function $f : \mathbb{R}^n \to \overline{\mathbb{R}}$ is strongly convex if there is a constant $\sigma > 0$ such that*

$$f\big((1-\tau)x_0 + \tau x_1\big) \le (1-\tau)f(x_0) + \tau f(x_1) - \tfrac{1}{2}\sigma\tau(1-\tau)|x_0 - x_1|^2$$
$$\text{for all } x,\ x' \text{ when } \tau \in (0,1).$$

12.59 Exercise (strong monotonicity versus strong convexity). *For a function $f : \mathbb{R}^n \to \overline{\mathbb{R}}$ and a value $\sigma > 0$, the following properties are equivalent:*

(a) *∂f is strongly monotone with constant σ;*

(b) *f is strongly convex with constant σ;*

(c) *$f - \frac{1}{2}\sigma|\cdot|^2$ is convex.*

Guide. First verify the equivalence of (b) with (c) by applying the usual convexity inequality to $f - \frac{1}{2}\sigma|\cdot|^2$ and working out what it means for f. Then apply 12.17 to get the connection with (a). □

12.60 Proposition (dualization of strong convexity). *For a proper, lsc, convex function $f : \mathbb{R}^n \to \overline{\mathbb{R}}$ and a value $\sigma > 0$ the following properties are equivalent:*

(a) *f^* is strongly convex with constant σ;*

(b) *f is differentiable and ∇f is Lipschitz continuous with constant $1/\sigma$;*

(c) *$\langle v_1 - v_0, x_1 - x_0\rangle \geq \sigma|v_1 - v_0|^2$ whenever $v_0 \in \partial f(x_0)$ and $v_1 \in \partial f(x_1)$;*

(d) *$f = f_0 \# \frac{1}{2}\sigma^{-1}|\cdot|^2$ for some convex function f_0;*

(e) *f is differentiable and satisfies*

$$f(x') \leq f(x) + \langle \nabla f(x), x' - x\rangle + \frac{1}{2\sigma}|x' - x|^2 \text{ for all } x, x'.$$

Proof. Condition (a) is equivalent by 12.59 to having $f^* = g + (\sigma/2)|\cdot|^2$ for a proper, lsc, convex function g. The latter is equivalent in turn by 11.23(a) to having (d) for $f_0 = g^*$. Also, (a) means by 12.59 that ∂f^* is strongly monotone with constant σ, but $\partial f^* = (\partial f)^{-1}$ by 11.3, so this is the same as (c). Since ∂f is maximal monotone by 12.17, we see further through 12.55 that condition (c) implies (b) (where we recall that the single-valuedness of ∂f corresponds to the differentiability of f; cf. 9.18).

Next we observe that on the basis of (b) we get for any x and x' that the function $\varphi(t) = f\big(x + t(x' - x)\big)$ has $\varphi'(t) = \langle \nabla f\big(x + t(x' - x)\big), x' - x\rangle$ and $\varphi'(t) - \varphi'(0) \leq (t/\sigma)|x' - x|^2$. Then

$$\begin{aligned} f(x') - f(x) = \varphi(1) - \varphi(0) &= \int_0^1 \varphi'(t)dt \\ &\leq \int_0^1 \Big[\varphi'(0) + \frac{t}{\sigma}|x' - x|^2\Big]dt = \langle \nabla f(x), x' - x\rangle + \frac{1}{2\sigma}|x' - x|^2, \end{aligned}$$

so we obtain (e). Finally we verify that (e) implies (a). We have

$$\begin{aligned} f^*(v) &= \sup_{x'}\{\langle v, x'\rangle - f(x')\} \\ &= \sup_{x,x'}\Big\{\langle v, x'\rangle - f(x) - \langle \nabla f(x), x' - x\rangle - \frac{1}{2\sigma}|x' - x|^2\Big\} \\ &= \sup_{x,u}\Big\{\langle v, u + x\rangle - f(x) - \langle \nabla f(x), u\rangle - \frac{1}{2\sigma}|u|^2\Big\} \\ &= \sup_{x}\Big\{\langle v, x\rangle - f(x) + \sup_{u}\Big\{\langle v - \nabla f(x), u\rangle - \frac{1}{2\sigma}|u|^2\Big\}\Big\} \\ &= \sup_{x}\Big\{\langle v, x\rangle - f(x) + \frac{\sigma}{2}|v - \nabla f(x)|^2\Big\} = \frac{\sigma}{2}|v|^2 + g(v) \end{aligned}$$

for the function

$$g(v) := \sup_{x}\Big\{\langle v, x - \sigma\nabla f(x)\rangle + \frac{\sigma}{2}|\nabla f(x)|^2\Big\}.$$

This function is the pointwise supremum of a collection of affine functions of v, so it is lsc and convex. From having $f^*(v) - (\sigma/2)|v|^2 = g(v)$ we conclude that f^* is strongly convex with constant σ as stipulated by (a). □

Parallel in many ways to the theory of strongly convex functions is that of the λ-proximal functions in 1.44, which are 'hypoconvex' because of their characterization in 11.26(d). Here are some elementary properties.

12.61 Exercise (hypoconvexity properties of proximal functions). *For a proper, lsc function $f : \mathbb{R}^n \to \overline{\mathbb{R}}$ and a value $\lambda > 0$, the following are equivalent:*

(a) *f is λ-proximal, or in other words, $h_\lambda f = f$;*

(b) *$\partial f + \lambda^{-1} I$ is monotone;*

(c) $f\big((1-\tau)x_0 + \tau x_1\big) \leq (1-\tau)f(x_0) + \tau f(x_1) + \frac{1}{2}\lambda^{-1}\tau(1-\tau)|x_0 - x_1|^2$ *for all x, x' when $\tau \in (0,1)$.*

Guide. Derive these equivalences from 12.17 and 11.26(d). □

The Lasry-Lions double envelope functions in Example 1.46 fit into this pattern of 'hypoconvexity' as follows.

12.62 Proposition (hypoconvexity and subsmoothness of double envelopes). *For a proper, lsc function $f : \mathbb{R}^n \to \overline{\mathbb{R}}$ that is prox-bounded with threshold λ_f and any values $0 < \mu < \lambda < \lambda_f$, the double envelope $e_{\lambda,\mu} f$ is μ-proximal, while $-e_{\lambda,\mu} f$ is $[\lambda - \mu]$-proximal. Moreover, $e_{\lambda,\mu} f$ is of class $\mathcal{C}^{1+}$.*

Proof. We know from 1.46 that $e_{\lambda,\mu} f = h_\mu[e_{\lambda-\mu}]$, while $-e_{\lambda,\mu} f = -e_{\lambda-\mu}[h_\lambda f]$. The first equation tells us that $e_{\lambda,\mu}$ is μ-proximal (because $h_\mu[h_\mu g] = h_\mu g]$ for any function g), while the second says that $-e_{\lambda,\mu} f$ is $(\lambda-\mu)$-proximal (because $h_\alpha[-e_\alpha g] = -e_\alpha[-e_\alpha[-e_\alpha g]] = -e_\alpha[h_\alpha g] = -e_\alpha g$ by the rules in 1.44 and 1.46). Now let $g = -e_\lambda f$ and $g_0 = g + \frac{1}{2\lambda}|\cdot|^2$. Because g is λ-proximal (as just argued), g_0 is convex by 11.26(d). (This convexity can also be seen directly from the formula for $e_\lambda f$ in 1.22.) We have

$$-e_{\lambda-\mu} f(x) = \inf_w \Big\{ g_0(w) - \frac{1}{2\lambda}|w|^2 + \frac{1}{2\mu}|x-w|^2 \Big\}.$$

Taking $\eta = \lambda\mu/(\lambda-\mu)$, so that $\eta^{-1} = \mu^{-1} - \lambda^{-1}$, we can reorganize this as

$$\begin{aligned} -e_{\lambda-\mu} f(x) &= \inf_w \Big\{ g_0(w) + \frac{1}{2\eta}\Big|\frac{\eta}{\mu}x - w\Big|^2 \Big\} - \frac{\eta}{2\lambda\mu}|x|^2 \\ &= g_1\Big(\frac{\eta}{\mu}x\Big) - \frac{1}{2(\lambda-\mu)}|x|^2 \quad \text{for } g_1 = e_\eta g_0. \end{aligned}$$

The function g_1, as the η-envelope of a convex function, is of class $\mathcal{C}^{1+}$ by 12.23, so $-e_{\lambda,\mu}$ must belong to this class as well. □

I.* Continuity and Differentiability

Maximal monotone mappings have special properties of continuity and even differentiability. These are enjoyed by the mappings ∂f associated with proper, lsc, convex functions f in particular, since such mappings are maximal monotone by 12.17.

12.63 Theorem (continuity properties). *For any maximal monotone mapping $T : \mathbb{R}^n \rightrightarrows \mathbb{R}^n$, the following properties hold.*

(a) *If a sequence of points $x^\nu \in \operatorname{dom} T$ converges to a point $\bar{x}$ from some direction $\operatorname{dir} w$ and has $\limsup_\nu T(x^\nu) \neq \emptyset$, then $\bar{x} \in \operatorname{dom} T$ and*

$$\limsup_\nu T(x^\nu) \subset \operatorname*{argmax}_{v \in T(\bar{x})} \langle v, w \rangle = \{ v \in T(\bar{x}) \mid w \in N_{T(\bar{x})}(v) \}.$$

(b) *If a sequence of points $x^\nu \in \operatorname{dom} T$ converges to some $\bar{x} \in \operatorname{dom} T$ from a direction $\operatorname{dir} w$ with $w \in \operatorname{int} T_{\operatorname{dom} T}(\bar{x})$, then the sets $T(x^\nu)$ must eventually be nonempty and form a bounded sequence, so $\limsup_\nu T(x^\nu) \neq \emptyset$.*

(c) *T is continuous at a point $\bar{x} \in \operatorname{dom} T$ if and only if T is single-valued at $\bar{x}$, in which case necessarily $\bar{x} \in \operatorname{int}(\operatorname{dom} T)$.*

Proof. If x^ν converges to $\bar{x}$ from the direction $\operatorname{dir} w$, we can write $x^\nu = \bar{x} + \tau^\nu w^\nu$ with $\tau^\nu \searrow 0$ and $w^\nu \to w$. If, for ν in some index set $N \in \mathcal{N}_\infty^\#$, there exist $v^\nu \in T(x^\nu)$ with $v^\nu \underset{N}{\to} \bar{v}$, we have $\bar{v} \in T(\bar{x})$ because T is osc by its maximality. Then for any $v \in T(\bar{x})$ the monotonicity inequality gives us

$$0 \leq (1/\tau^\nu)\big\langle v^\nu - v, (\bar{x} + \tau^\nu w^\nu) - \bar{x} \big\rangle = \big\langle v^\nu - v, w^\nu \big\rangle \underset{N}{\to} \big\langle \bar{v} - v, w \big\rangle.$$

Hence $\langle \bar{v}, w \rangle \geq \langle v, w \rangle$ for all $v \in T(\bar{x})$. This proves the inclusion in (a). The subsequent equation comes then from the convexity of $T(\bar{x})$ in 12.8(c) and the characterization of normals to convex sets in 6.9.

For (b), suppose $\bar{x} \in \operatorname{dom} T$ and $w \in \operatorname{int} T_{\operatorname{dom} T}(\bar{x})$. Let $D = \operatorname{cl}(\operatorname{dom} T)$; then $T_{\operatorname{dom} T}(\bar{x}) = T_D(\bar{x})$, so that $w \in \operatorname{int} T_D(\bar{x})$. Since D is convex (by 12.37), this means that $\bar{x} + \varepsilon w \in \operatorname{int} D$ for some $\varepsilon > 0$; cf. the description of tangent cones to convex sets in 6.9. Hence there exists $N \in \mathcal{N}_\infty$ such that, for $\nu \in \mathcal{N}$, we have $\bar{x} + \varepsilon w^\nu \in \operatorname{int} D$ and $\tau^\nu < \varepsilon$. The point $x^\nu = \bar{x} + \tau^\nu w^\nu$ belongs then to $\operatorname{int} D$ as well (through the line segment principle in 2.33). But $\operatorname{int} D = \operatorname{int}(\operatorname{dom} T)$ because of the near convexity of $\operatorname{dom} T$ (cf. 12.41). Thus, for all $\nu \in N$, we have $x^\nu \in \operatorname{int}(\operatorname{dom} T)$, hence $T(x^\nu)$ nonempty and bounded by 12.38. We must show further now that in fact $\bigcup_{\nu \in N} T(x^\nu)$ is bounded.

If this weren't true, there would exist, for ν in an index set $N' \in \mathcal{N}_\infty^\#$ within N, vectors $v^\nu \in T(x^\nu)$ and scalars $\lambda^\nu \searrow 0$ such that $\lambda^\nu v^\nu \underset{N'}{\to} \tilde{v} \neq 0$. Then $\tilde{v} \in N_D(\bar{x})$, since $\lambda T \overset{g}{\to} N_D$ as $\lambda \searrow 0$; cf. 12.37. Because $w \in \operatorname{int} T_D(\bar{x})$ and $T_D(\bar{x})^* = N_D(\bar{x})$ (by the convexity of D), we must then have $\langle \tilde{v}, w \rangle < 0$ by the polarity rule in 6.22. But a contradiction to this is reached by considering any $v \in T(\bar{x})$ and calculating as above that

$$0 \leq (\lambda^\nu/\tau^\nu)\big\langle v^\nu - v, (\bar{x} + \tau^\nu w^\nu) - \bar{x} \big\rangle = \big\langle \lambda^\nu v^\nu - \lambda^\nu v, w^\nu \big\rangle \underset{N'}{\to} \big\langle \tilde{v}, w \big\rangle.$$

To get (c), we note first that T can't be continuous at a point $\bar{x} \in \operatorname{dom} T$ unless actually $\bar{x} \in \operatorname{int}(\operatorname{dom} T)$, in which case T is locally bounded at $\bar{x}$ by 12.38 and in particular $T(\bar{x})$ is compact. We then appeal to (a) and the fact that a nonempty, compact, convex set is a singleton if and only if it has the same argmax 'face' from all directions $\operatorname{dir} w$. □

Next we take note of the special nature of proto-differentiability in this setting of maximal monotonicity.

12.64 Exercise (proto-derivatives of monotone mappings). *If a maximal monotone mapping $T : \mathbb{R}^n \rightrightarrows \mathbb{R}^n$ is proto-differentiable at $\bar{x}$ for some $\bar{v} \in T(\bar{x})$, its graphical derivative mapping $DT(\bar{x} \,|\, \bar{v}) : \mathbb{R}^n \rightrightarrows \mathbb{R}^n$ is maximal monotone as well. The same holds for maximal cyclic monotonicity.*

Proto-differentiability of T at $\bar{x}$ for $\bar{v}$ is equivalent, for any $\lambda > 0$, to semidifferentiability of the resolvant $R_\lambda = (I + \lambda T)^{-1}$ at the point $\bar{z}_\lambda = \bar{x} + \lambda\bar{v}$. That corresponds in turn to single-valuedness of the mapping $DR_\lambda(\bar{z}_\lambda)$, which is given always by

$$DR_\lambda(\bar{z}_\lambda) = \big(I + \lambda DT(\bar{x} \,|\, \bar{v})\big)^{-1}.$$

In particular, R_λ is differentiable at $\bar{z}_\lambda$ if and only if $DT(\bar{x} \,|\, \bar{v})$ is a generalized linear mapping in the sense that its graph is a linear subspace of $\mathbb{R}^n \times \mathbb{R}^n$, the subspace then necessarily being n-dimensional.

Guide. Obtain the first fact from Theorem 12.32. Get the second by way of Theorems 12.25 and 12.35. For the resolvants, draw on the principle that proto-differentiability corresponds to geometric derivability of graphs and appeal to the linear transformation in $\mathbb{R}^{2n}$ that converts between $\operatorname{gph} T$ and $\operatorname{gph} R_\lambda$ and at the same time between $\operatorname{gph} DT(\bar{x} \,|\, \bar{v})$ and $\operatorname{gph} DR_\lambda(\bar{z}_\lambda)$. Make use then of the equivalence of proto-differentiability with semidifferentiability in the case of a Lipschitz continuous mapping like R_λ; cf. 9.50(b). In connection with the single-valuedness of $DR_\lambda(\bar{z}_\lambda)$, look to 9.25(a).

Finally, identify the differentiability of R_λ at $\bar{z}$ with the case where the graph of $DR_\lambda(\bar{z})$ is a linear subspace of $\mathbb{R}^n \times \mathbb{R}^n$; see also 9.25(b). □

Because a maximal monotone mapping T is graphically Lipschitzian everywhere, as observed after 12.16, its graph 'can be linearized almost everywhere' in consequence of Rademacher's theorem (in 9.60). In other words, at almost every point $(\bar{x}, \bar{v})$ of $\operatorname{gph} T$ in the n-dimensional sense of the Minty parameterization of this graph, T must be proto-differentiable with $DT(\bar{x} \,|\, \bar{v})$ 'generalized linear' as described in 12.64. The extent to which this property translates to actual differentiability of T will come out of the next theorem and its corollary.

Recall from the definition after 8.43 that T is *differentiable* at $\bar{x}$ if there exist $\bar{v} \in \mathbb{R}^n$, $A \in \mathbb{R}^{n \times n}$ and $V \in \mathcal{N}(\bar{x})$ such that

$$\emptyset \neq T(x) \subset \bar{v} + A[x - \bar{x}] + o(|x - \bar{x}|)\mathbb{B} \text{ when } x \in V.$$

Then $T(\bar{x}) = \{\bar{v}\}$, and the matrix A is denoted by $\nabla T(\bar{x})$. This reduces to the usual definition of differentiability when T is a single-valued mapping.

12.65 Theorem (differentiability and semidifferentiability). *Consider a maximal monotone mapping $T : \mathbb{R}^n \rightrightarrows \mathbb{R}^n$ and any $\bar{x} \in \operatorname{dom} T$ and $\bar{v} \in T(\bar{x})$. For any*

$\lambda > 0$, let $R_\lambda = (I + \lambda T)^{-1}$ and $\bar{z}_\lambda = \bar{x} + \lambda\bar{v}$. Then the following properties are equivalent and necessitate T being continuous at $\bar{x}$ with $T(\bar{x}) = \{\bar{v}\}$ while the mapping $DT(\bar{x}\,|\,\bar{v}) = DT(\bar{x})$, also maximal monotone, is everywhere continuous and such that $T(x) \subset T(\bar{x}) + DT(\bar{x})(x - \bar{x}) + o(|x - \bar{x}|)I\!B$:

(a) *T is semidifferentiable at $\bar{x}$ for $\bar{v}$;*

(b) *$DT(\bar{x}\,|\,\bar{v})$ is single-valued;*

(c) *R_λ is semidifferentiable at $\bar{z}_\lambda$ with $DR_\lambda(\bar{z}_\lambda)$ bijective.*

Indeed, the following, more special properties are equivalent as well:

(a′) *T is differentiable at $\bar{x}$;*

(b′) *$DT(\bar{x}\,|\,\bar{v})$ is a linear mapping;*

(c′) *R_λ is differentiable at $\bar{z}_\lambda$ with $\nabla R_\lambda(\bar{z}_\lambda)$ nonsingular.*

Proof. (a) $\Rightarrow$ (b). Semidifferentiability of T at $\bar{x}$ for $\bar{v}$ implies by 8.43 that $DT(\bar{x}\,|\,\bar{v})$ is a continuous mapping. But semidifferentiability also entails proto-differentiability, so it also implies by 12.64 that $DT(\bar{x}\,|\,\bar{v})$ is maximal monotone. Then by 12.63(c), $DT(\bar{x}\,|\,\bar{v})$ has to be single-valued at every point of $\operatorname{dom} DT(\bar{x}\,|\,\bar{v})$ and thus in particular at 0. In that case we have $0 \in \operatorname{int}(\operatorname{dom} DT(\bar{x}\,|\,\bar{v}))$ through 12.38, and because $DT(\bar{x}\,|\,\bar{v})$ is positively homogeneous we conclude that $\operatorname{dom} DT(\bar{x}\,|\,\bar{v})$ is all of $I\!R^n$. Hence $DT(\bar{x}\,|\,\bar{v})$ must be single-valued everywhere.

(b) $\Rightarrow$ (a). By definition, $DT(\bar{x}\,|\,\bar{v})$ is the outer graphical limit of the maximal monotone mappings $\Delta_\tau T(\bar{x}\,|\,\bar{v})$ as $\tau \searrow 0$. Thus, if a sequence $\tau^\nu \searrow 0$ is such that the corresponding sequence of mappings $\Delta_{\tau^\nu} T(\bar{x}\,|\,\bar{v})$ has a mapping H as a cluster point with respect to graphical convergence, then $H \subset DT(\bar{x}\,|\,\bar{v})$. In that situation, though, H too must be maximal monotone by 12.34(a), so this inclusion with $DT(\bar{x}\,|\,\bar{v})$ single-valued requires H to be locally bounded everywhere, hence by 12.38 nonempty-valued everywhere. Then $H = DT(\bar{x}\,|\,\bar{v})$. In particular, $DT(\bar{x}\,|\,\bar{v})$ must itself be maximal monotone and, by virtrue of its single-valuedness, be continuous everywhere; cf. 12.63(c).

On the other hand, since $0 \in \Delta_{\tau^\nu} T(\bar{x}\,|\,\bar{v})(0)$, no sequence $\{\Delta_{\tau^\nu} T(\bar{x}\,|\,\bar{v})\}_{\nu \in I\!N}$ can escape to the horizon. It follows from the compactness property in 12.33 that, no matter how the sequence $\tau^\nu \searrow 0$ is chosen, every subsequence of $\{\Delta_{\tau^\nu} T(\bar{x}\,|\,\bar{v})\}_{\nu \in I\!N}$ has a subsubsequence converging graphically to $DT(\bar{x}\,|\,\bar{v})$. This means that $\Delta_\tau T(\bar{x}\,|\,\bar{v}) \xrightarrow{g} DT(\bar{x}\,|\,\bar{v})$ as $\tau \searrow 0$, which is the property of T being proto-differentiable at $\bar{x}$ for $\bar{v}$.

Next we call upon the uniform local boundedness in 12.40(a): the graphical convergence of $\Delta_\tau T(\bar{x}\,|\,\bar{v})$ to $DT(\bar{x}\,|\,\bar{v})$, with $\operatorname{dom} DT(\bar{x}\,|\,\bar{v}) = I\!R^n$, implies the existence of $\varepsilon > 0$, $\delta > 0$ and $\rho > 0$ such that

$$\Delta_\tau T(\bar{x}\,|\,\bar{v})(\delta I\!B) \subset \rho I\!B \quad \text{when } \tau \in (0, \varepsilon).$$

That's the same as $T(\bar{x} + \tau w) \subset \bar{v} + \tau\rho I\!B$ when $|w| \leq \delta$ and $\tau \in (0, \varepsilon)$, and we therefore have

$$T(x) \subset \bar{v} + (\rho/\delta)|x - \bar{x}| I\!B \quad \text{when } |x - \bar{x}| < \varepsilon\delta,$$

as seen by setting $x = \bar{x} + \tau w$ and taking $\tau \in \big(|x - \bar{x}|/\delta, \varepsilon\big)$. This condition ensures also that $T(x) \neq \emptyset$ for all x in some neighborhood of $\bar{x}$, since it makes T be locally bounded at $\bar{x}$ and that requires $\bar{x} \in \operatorname{int}(\operatorname{dom} T)$ by 12.38.

Thus, we have deduced from (b) that T exhibits the properties in 8.43(e) at $\bar{x}$ for $\bar{v}$. According to the equivalences in 8.43, this not only guarantees the semidifferentiability of T at $\bar{x}$ for $\bar{v}$, hence (a), but also provides the expansion $T(x) \subset \bar{v} + DT(\bar{x} \,|\, \bar{v})(x - \bar{x}) + o(|x - \bar{x}|)\mathbb{B}$ that has been claimed here.

Having established that (a) $\Leftrightarrow$ (b), we pass now to (c). Utilizing the relation $DR_\lambda(\bar{z}_\lambda) = (I + \lambda DT(\bar{x} \,|\, \bar{v}))^{-1}$ in 12.64 in the equivalent form $DR_\lambda(\bar{z}_\lambda)^{-1} = I + \lambda DT(\bar{x} \,|\, \bar{v})$, we identify the single-valuedness of $DR_\lambda(\bar{z}_\lambda)^{-1}$ with that of $DT(\bar{x} \,|\, \bar{v})$. It's immediate then that (c) implies (b). But also, the combination of (a) and (b) yields (c) in turn by 12.64.

To get the equivalence of (a′), (b′) and (c′), we need only now specialize to the case of $DT(\bar{x} \,|\, \bar{v})$ being a generalized linear mapping as covered at the end of 12.64. Such a mapping $\mathbb{R}^n \rightrightarrows \mathbb{R}^n$, its graph a linear subspace of dimension n, is a linear mapping if and only if it is single-valued. □

12.66 Corollary (generic single-valuedness from monotonicity). *For a maximal monotone mapping $T : \mathbb{R}^n \rightrightarrows \mathbb{R}^n$, the following properties hold.*

(a) *The set of points of* $\operatorname{int}(\operatorname{dom} T)$ *where T fails to be differentiable is negligible. In particular, T is single-valued almost everywhere on* $\operatorname{int}(\operatorname{dom} T)$.

(b) *The set of points of* $\operatorname{int}(\operatorname{rge} T)$ *where T^{-1} fails to be differentiable is negligible. In particular, T^{-1} is single-valued almost everywhere on* $\operatorname{int}(\operatorname{rge} T)$.

Proof. Let E be the set of points where T is differentiable; at such points we have T single-valued in particular. Also, $E \subset \operatorname{int}(\operatorname{dom} T)$. Let $R = (I + T)^{-1}$. According to Theorem 12.65, $x \in E$ if and only if $x = R(z)$ for some z at which R is differentiable with $\nabla R(z)$ nonsingular. Since R is nonexpansive (and therefore Lipschitz continuous), the set of such points x has negligible complement within $\operatorname{rge} R = \operatorname{dom} T$ by 9.65. This gives (a). And (b) follows from the same argument applied to T^{-1}. □

12.67 Theorem (structure of maximal monotone mappings). *Let $T : \mathbb{R}^n \rightrightarrows \mathbb{R}^n$ be maximal monotone with* $\operatorname{int}(\operatorname{dom} T) \neq \emptyset$. *Let E be the subset of* $\operatorname{int}(\operatorname{dom} T)$ *on which T is single-valued, and let E_0 be any dense subset of E; in particular, E_0 can be taken to be E itself or the subset of E on which T is differentiable. Define the mapping $T_0 : \mathbb{R}^n \rightrightarrows \mathbb{R}^n$ by*

$$T_0(x) = \big\{ v \,\big|\, \exists\, x^\nu \underset{E_0}{\to} x \text{ with } T(x^\nu) \to v \big\}.$$

Then T_0 is single-valued on all of E and agrees there with T, and one has $\operatorname{dom} T = \operatorname{dom} T_0 \subset \operatorname{cl} E_0 = \operatorname{cl}(\operatorname{dom} T)$ *with*

$$T(x) = \operatorname{cl} \operatorname{con} T_0(x) + N_{\operatorname{cl} E_0}(x) \text{ for all } x \in \mathbb{R}^n.$$

Proof. We know from 12.63(c) that T is continuous on E and from 12.66 that the set of points where T is differentiable is a subset of E that is dense in

$\operatorname{dom} T$. Therefore T_0 is single-valued and continuous on E, coinciding on that set with T, and we have $\operatorname{cl} E_0 = \operatorname{cl} E = \operatorname{cl}(\operatorname{dom} T_0) = \operatorname{cl}(\operatorname{dom} T)$. Denote this set by D for simplicity and let $S(x) := \operatorname{cl}\operatorname{con} T_0(x)$. The formula to be verified can be written then as $T = S + N_D$.

Through its maximal monotonicity, T is closed-convex-valued and osc. Hence $S(x) \subset T(x)$ and $\operatorname{dom} S \subset \operatorname{dom} T \subset D$. According to 12.37, the closed set D is convex and, at every point $x \in \operatorname{dom} T$, we have $T(x)^\infty = N_D(x)$; then $S(x)^\infty \subset N_D(x)$ as well. Since $T(x) + T(x)^\infty = T(x)$ by 3.6, we get $S(x) + N_D(x) \subset T(x)$.

Now fix any $\bar{x} \in \operatorname{dom} T$ and let $C(\bar{x}) := S(\bar{x}) + N_D(\bar{x})$. We must show that $T(\bar{x}) \subset C(\bar{x})$. The set $C(\bar{x})$ is convex. Provided $C(\bar{x})$ is also closed, as will be established next, the inclusion $T(\bar{x}) \subset C(\bar{x})$ is equivalent to the support function inequality $\sigma_{T(\bar{x})} \leq \sigma_{C(\bar{x})}$, cf. 8.24.

We don't yet know that $C(\bar{x}) \neq \emptyset$, but if so, then not only is $C(\bar{x})$ closed but also $C(\bar{x})^\infty = N_D(\bar{x})$. This comes from the criterion in 3.12 and the fact that $S(\bar{x})^\infty \subset N_D(\bar{x}) = N_D(\bar{x})^\infty$, with the cone $N_D(x)$ being pointed since, by 12.41, $\operatorname{int} D = \operatorname{int}(\operatorname{dom} T) \neq \emptyset$. Under the assumption that $C(\bar{x}) \neq \emptyset$, we can apply the dualization rule in 11.5, as specialized to support functions and cones in the manner of 11.4, to see that polars of the convex cones $\operatorname{dom} \sigma_{C(\bar{x})}$ and $\operatorname{dom} \sigma_{T(\bar{x})}$ are $C(\bar{x})^\infty$ and $T(\bar{x})^\infty$, hence both equal to $N_D(\bar{x})$. Then $\operatorname{cl}[\operatorname{dom} \sigma_{T(\bar{x})}[= \operatorname{cl}[\operatorname{dom} \sigma_{C(\bar{x})}] = N_D(\bar{x})^* = \operatorname{cl} T_D(\bar{x})$, and it follows that

$$\begin{aligned} \operatorname{int}[\operatorname{dom} \sigma_{T(\bar{x})}] &= \operatorname{int}[\operatorname{dom} \sigma_{C(\bar{x})}] = \operatorname{int} T_D(\bar{x}) \\ &= \{ w \mid \exists \tau > 0 \text{ with } \bar{x} + \tau w \in \operatorname{int} D \}, \end{aligned} \qquad 12(11)$$

where the description of $\operatorname{int} T_D(\bar{x})$ comes from 6.9. A proper, lsc, convex function f is completely determined by its values on $\operatorname{int}(\operatorname{dom} f)$ when that's nonempty (cf. 2.36), so to confirm that $\sigma_{T(\bar{x})}(w) \leq \sigma_{C(\bar{x})}(w)$ for all w it's enough to check that this holds when w belongs to the open set in 12(11). Indeed, because a convex function f is continuous on $\operatorname{int}(\operatorname{dom} f)$, we need only deal with points w belonging to a dense subset of this open set. The dense subset that we focus on is

$$A := \{ w \mid \nabla \sigma_{T(\bar{x})}(w) \text{ exists } \}.$$

Its density is given by Rademacher's theorem (in 9.60), since a convex function is strictly continuous on an open set where it is finite (cf. 9.14).

We therefore fix now a vector $\bar{w} \in A$, bearing in mind that in this case $\bar{x} + \tau\bar{w} \in \operatorname{int} D = \operatorname{int}(\operatorname{dom} T)$ for all τ in some interval $(0, \varepsilon)$, $\varepsilon > 0$. In particular, $\bar{w} \in \operatorname{int} T_{\operatorname{dom} T}(\bar{x})$. Our goal is to demonstrate that

$$\sigma_{T(\bar{x})}(\bar{w}) \leq \sigma_{C(\bar{x})}(\bar{w}).$$

By accomplishing this without assuming outright that $C(\bar{x}) \neq \emptyset$, we will be able to *conclude* that $C(\bar{x}) \neq \emptyset$, since if $C(\bar{x})$ were empty we would have $\sigma_{C(\bar{x})}(\bar{w}) = -\infty$, whereas we know that $\sigma_{T(\bar{x})}(\bar{w}) > -\infty$ since $T(\bar{x}) \neq \emptyset$.

In general, the set $\partial\sigma_{T(\bar{x})}(\bar{w})$ equals $\operatorname{argmax}\{\langle v,\bar{w}\rangle \mid v\in T(\bar{x})\}$; cf. 8.25. Here $\partial\sigma_{T(\bar{x})}(\bar{w})$ is the singleton consisting of $\bar{v}:=\nabla\sigma_{T(\bar{x})}(\bar{w})$, so that

$$\operatorname{argmax}\{\langle v,\bar{w}\rangle \mid v\in T(\bar{x})\}=\{\bar{v}\},\qquad \sigma_{T(\bar{x})}(\bar{w})=\langle\bar{v},\bar{w}\rangle.$$

If we can establish that $\bar{v}\in T_0(\bar{x})$, we will have $\bar{v}\in C(\bar{x})$ and consequently $\sigma_{T(\bar{x})}(\bar{w})\le\sigma_{C(\bar{x})}(\bar{w})$ as required.

For this we can utilize the continuity properties in 12.63. Selecting $\tau^\nu\in(0,\varepsilon)$ arbitrarily with $\tau^\nu\searrow 0$, we obtain a sequence of points $\bar{x}^\nu=\bar{x}+\tau^\nu\bar{w}\in\operatorname{int}(\operatorname{dom}T)$ with $\bar{x}^\nu$ converging to $\bar{x}$ from the direction $\operatorname{dir}\bar{w}$. On the basis of the density fact in 12.66(a), we can approximate $\bar{x}^\nu$ by some $x^\nu\in E_0$ for each ν and do this in such a manner that x^ν likewise converges to $\bar{x}$ from the direction $\operatorname{dir}\bar{w}$. Then $\limsup_\nu T(x^\nu)\subset T_0(\bar{x})$, but also $\limsup_\nu T(x^\nu)\neq\emptyset$ by 12.63(b), inasmuch as $\bar{x}\in\operatorname{dom}T$ and $\bar{w}\in\operatorname{int}T_{\operatorname{dom}T}(\bar{x})$. At the same time, $\limsup_\nu T(x^\nu)\subset\operatorname{argmax}\{\langle v,\bar{w}\rangle \mid v\in T(\bar{x})\}$ by 12.63(a). The latter set is $\{\bar{v}\}$, so we must have $\bar{v}\in T_0(\bar{x})$. □

It deserves emphasis that all these properties of continuity and differentiability are valid in particular in the special case where T is the subgradient mapping ∂f associated with a proper, lsc, convex function f, since T is maximal monotone then by Theorem 12.17. On the 'negative' side, this means that such a mapping ∂f can't be continuous at a point x unless $x\in\operatorname{int}(\operatorname{dom}f)$ and f is strictly differentiable at $\bar{x}$, so that $\partial f(x)=\{\nabla f(x)\}$. That's a consequence of 12.63(c) and the characterization of strict differentiability in 9.18 (along with the strict continuity of finite convex functions in 9.14).

Remarkably, though, a relaxation from true subgradients to *ε-subgradients* produces continuity properties that are dramatically better. In comparison with the description of the set of true subgradients of f at $\bar{x}$ as

$$\begin{aligned}\partial f(x)&=\{v \mid f(x')\ge f(x)+\langle v,\,x'-x\rangle \text{ for all } x'\}\\&=\{v \mid f(x)+f^*(v)-\langle v,x\rangle\le 0\}=\operatorname{argmin}_v\{f^*(v)-\langle v,x\rangle\}\end{aligned}$$

through 11.3, the set of ε-subgradients of f at $\bar{x}$ is given for $\varepsilon>0$ by

$$\begin{aligned}\partial_\varepsilon f(x):&=\{v \mid f(x')\ge f(x)+\langle v,\,x'-x\rangle-\varepsilon \text{ for all } x'\}\\&=\{v \mid f(x)+f^*(v)-\langle v,x\rangle\le\varepsilon\}=\varepsilon\text{-}\operatorname{argmin}_v\{f^*(v)-\langle v,x\rangle\}\end{aligned}$$

where again the equivalences are seen through 11.3 and require f to be convex, proper and lsc. (Note: this concept of an ε-subgradient set is tuned to convex functions *only* and must not be confused with the ε-regular subgradient set $\widehat{\partial}_\varepsilon f(x)$ in 8(40), which is something different but does make good sense even for nonconvex functions.)

12.68 Proposition (strict continuity of ε-subgradient mappings). *For $\varepsilon>0$ and $f:\mathbb{R}^n\to\overline{\mathbb{R}}$ be proper, lsc and convex, define the mapping $\partial_\varepsilon f:\mathbb{R}^n\rightrightarrows\mathbb{R}^n$ as above. Consider $X\subset\operatorname{dom}f$ and suppose $\rho_0>0$ is such that $f(x)\le\rho_0$ and $d(0,\partial f(x))\le\rho_0$ when $x\in X$. Then for every $\rho\in(\rho_0+\varepsilon,\infty)$ one has*

$$\hat{d\!l}_\rho\big(\partial_\varepsilon f(x'), \partial_\varepsilon f(x)\big) \;\leq\; (2\rho/\varepsilon)(\rho+\varepsilon)\big|x'-x\big|$$
$$\text{when } x, x' \in X \text{ with } |x'-x| < \varepsilon/(\rho+\varepsilon).$$

Thus in particular, $\partial_\varepsilon f$ is strictly continuous relative to X.

Proof. Fix any points x_1, $x_2 \in X$ and let $g_i(v) := f^*(v) - \langle v, x_i\rangle$, so that $\partial_\varepsilon f(x_i) = \varepsilon\text{–}\operatorname{argmin} g_i$. The functions g_i are proper, lsc and convex, and we are able then to apply the estimate for approximate minimization in 7.69. Since $f(x_i) = -\inf g_i$ and $\partial f(x_i) = \operatorname{argmin} g_i$, our assumptions concerning ρ_0 translate to having $\inf g_i \geq -\rho_0$ and $\rho_0 I\!B \cap \operatorname{argmin} g_i \neq \emptyset$. Our choice of ρ ensures that $\varepsilon < \rho - \rho_0$. Theorem 7.69 tells us in this setting that

$$\hat{d\!l}^+_{\rho+\varepsilon}(g_1, g_2) < \varepsilon \;\Longrightarrow\; \hat{d\!l}_\rho\big(\partial_\varepsilon f(x'), \partial_\varepsilon f(x)\big) \;\leq\; \big(2\rho/\varepsilon\big)\hat{d\!l}^+_{\rho+\varepsilon}(g_1, g_2).$$

It will suffice now to demonstrate that $\hat{d\!l}^+_{\rho+\varepsilon}(g_1, g_2) \leq (\rho+\varepsilon)|x_1 - x_2|$. The desired inequality will thereby be confirmed, and the strict continuity of $\partial_\varepsilon f$ relative to X will fall out of Definition 9.28 and the estimates in 9(9).

By the definition of $\hat{d\!l}^+_{\rho+\varepsilon}(g_1, g_2)$, as specialized from 7.61 to our choice of g_1 and g_2 and the fact that $g_i(v) \geq -\rho_0 > -(\rho+\varepsilon)$ for all v, this distance is the infimum of the values $\eta > 0$ such that

$$\left.\begin{aligned} \min_{v' \in \boldsymbol{B}(v,\eta)} \big\{f^*(v') - \langle v', x_1\rangle\big\} \leq f^*(v) - \langle v, x_2\rangle + \eta \\ \min_{v' \in \boldsymbol{B}(v,\eta)} \big\{f^*(v') - \langle v', x_2\rangle\big\} \leq f^*(v) - \langle v, x_1\rangle + \eta \end{aligned}\right\} \text{ for all } v \in (\rho+\varepsilon)I\!B.$$

One of the candidates for v' in each case is v itself, so this condition on η certainly holds when both $-\langle v, x_1\rangle \leq -\langle v, x_2\rangle + \eta$ and $-\langle v, x_2\rangle \leq -\langle v, x_1\rangle + \eta$ for all $v \in (\rho+\varepsilon)I\!B$. But that's true when $\eta = (\rho+\varepsilon)|x_1 - x_2|$. Hence $\hat{d\!l}^+_{\rho+\varepsilon}(g_1, g_2) \leq (\rho+\varepsilon)|x_1 - x_2|$, as claimed. □

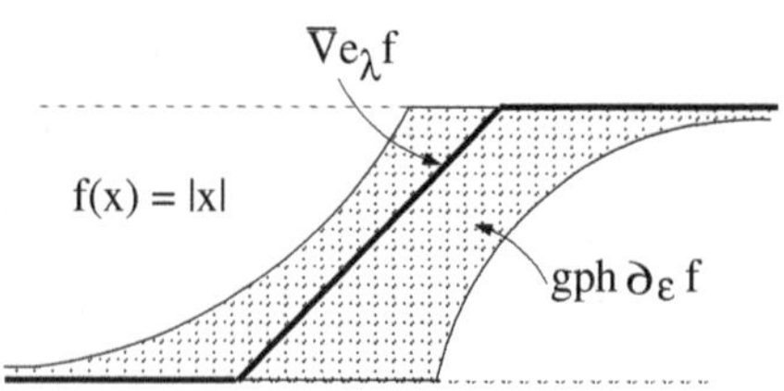

Fig. 12–8. A Lipschitz continuous selection of $\partial_\varepsilon f$.

It's interesting to note that when the convex function f in 12.68 is finite and globally Lipschitz continuous on $I\!R^n$, a single-valued Lipschitz continuous mapping $s : I\!R^n \to I\!R^n$ can be found with $s(x) \in \partial_\varepsilon f(x)$ for all x. Indeed, that can be accomplished by taking $s = \nabla e_\lambda f$ for any $\lambda \in (0, 2\varepsilon/\kappa^2)$, where κ is the Lipschitz constant for f, in which case the Lipschitz constant for s is $1/\lambda$; see

Figure 12–8. This can be seen from the conjugacy between $e_\lambda f$ and $f^* + \frac{\lambda}{2}|\cdot|^2$ in 11.26(b), which tells us through 11.3 that having $v = \nabla e_\lambda f(x)$ is equivalent to having $v \in \operatorname{argmin}_w \{f^*(w) + \frac{\lambda}{2}|w|^2 - \langle w, x\rangle\}$, i.e.,

$$f^*(v) + \frac{\lambda}{2}|v|^2 - \langle v, x\rangle \;\le\; f^*(w) + \frac{\lambda}{2}|w|^2 - \langle w, x\rangle \text{ for all } w \in \operatorname{dom} f^*.$$

The Lipschitz continuity of f with constant κ implies $\operatorname{dom} f^* \subset \kappa I\!B$ (e.g. by 11.5), so $f^*(v) - \langle v, x\rangle \le f^*(w) - \langle w, x\rangle + (\lambda\kappa^2/2)$ for all w. If $\lambda\kappa^2/2 \le \varepsilon$, then $v \in \partial_\varepsilon f(\bar{x})$.

Commentary

Like other basic concepts of analysis, 'monotonicity' in the sense of Definition 12.1 was molded by many hands. An obvious precedent is positive semidefiniteness of linear mappings. The earliest known condition of such sort for a nonlinear, single-valued mapping F is found in a paper of Golumb [1935], but in the form $\langle F(x_1) - F(x_0), x_1 - x_0\rangle \ge \sigma|x_1 - x_0|^2$ with a parameter σ that might be negative or positive as well as zero (so that hypomonotonicity and strong monotonicity are covered at the same time). The idea was rediscovered by Zarantonello [1960], who saw in it an useful adjunct to iterative methods for solving functional equations and obtained contraction results akin to cases of Theorem 12.56. Kachurovskii [1960] noted that *gradient* mappings of convex functions are monotone as in 2.14(a); he was the first to employ the term 'monotonicity' for this property. Really, though, the theory of monotone mappings began with papers of Minty [1961], [1962], where the concept was studied directly in its full scope and the significance of maximality was brought to light.

In those days, multivaluedness was in a temporary phase of being shunned by mathematicians, so Minty was obliged to speak of 'relations' described by subsets of $I\!R^n \times I\!R^n$, where we now think of such subsets as $\operatorname{gph} T$ for mappings $T : I\!R^n \rightrightarrows I\!R^n$. His discovery of maximal monotonicity as a powerful tool was one of the main impulses, however, along with the introduction of subgradients of convex functions, that led to the resurgence of multivalued mappings as acceptable objects of discourse, especially in variational analysis. The need for enlarging the graph of a monotone mapping in order to achieve maximal monotonicity, even if this meant that the graph would no longer be function-like, was clear to him from his previous work with optimization problems in networks, Minty [1960], which revolved around the one-dimensional case of this phenomenon; cf. 12.9.

Much of the early research on monotone mappings was centered on infinite-dimensional applications to integral equations and differential equations. The survey of Kachurovskii [1968] and the books of Brézis [1973] and Browder [1976] present this aspect well; more recently, the subject has been treated comprehensively by Zeidler [1990a], [1990b]. In such a setting, mappings are usually called 'operators'. But finite-dimensional applications to numerical optimization have also come to be widespread, particularly in schemes of decomposition; see, for example, the references and discussion of Chen and Rockafellar [1997].

Monotone linear mappings with possibly nonsymmetric matrix A (cf. 12.2) were called 'dissipative' by Dolph [1961] because of their tie to the physics of energy. This term might have carried over to nonlinear mappings through the Jacobian property in 12.3, derived by Minty [1962], were it not for the latter's preference for 'monotone' despite the ambiguities that arise with that word in spaces having a partial ordering. In a fundamental study of evolution equations that dissipate energy, Crandall and Pazy [1969] established a one-to-one correspondence between maximal monotone mappings and nonlinear semigroups of nonexpansive mappings. They too felt more comfortable with monotone relations in a product space than with multivaluedness, but their contribution was another big step on the road toward the eventual realization that major results in nonlinear mathematics couldn't satisfyingly be formulated within the confines of single-valued mappings.

The existence of maximal extensions of monotone mappings, and the correspondence between maximal monotone mappings $T : I\!R^n \rightrightarrows I\!R^n$ and nonexpansive mappings $S : I\!R^n \to I\!R^n$ through the transformation in 12.11, was proved by Minty [1962] (with the transformation $J/\sqrt{2}$ in place of J). In that landmark paper, Minty also established the maximality of continuous monotone mappings in 12.7 and the important facts about resolvants and parameterization in 12.12 and 12.15. The general inverse-resolvant identity in 12.14 was first stated by Poliquin and Rockafellar [1996a]. For the regularizations of the kind in 12.13 in the case of linear mappings, see Yosida [1971]; this type of approximation was subsequently utilized for other mappings by Brézis [1973].

The monotonicity of the subgradient mapping associated with a real-valued (but not necessarily differentiable) convex function f was shown by Minty [1964], but Moreau [1965] quickly pointed out that on the basis of his theory of proximal mappings one could deduce almost at once the *maximal* monotonicity of ∂f for *extended-real-valued* convex functions that are lsc and proper; cf. 12.17. His observation applied even to Hilbert spaces. Rockafellar [1966a], [1969a], [1970d] extended this result to general Banach spaces, also introducing cyclical monotonicity and obtaining the full characterization of the subgradient mappings of convex functions as in 12.25. (The argument in first of these papers had a gap which was fixed in the third; before this gap had surfaced, an alternative justification of the same facts was furnished in the second paper.)

For the subgradient mappings ∂f of proper, lsc functions f more generally, the valuable converse assertion in 12.17 that the monotonicity of ∂f implies the convexity of f, was first proved by Poliquin [1990a]; for the special case of strictly continuous functions (locally Lipschitz continuous), it was noted earlier by Clarke [1983]. Infinite-dimensional versions of this converse result have since been supplied by Correa, Jofré and Thibault [1992], [1994], [1995] and Clarke, Stern and Wolenski [1993]. Generalizations of the 'integration' result in Theorem 12.25 to the subgradient mappings of nonconvex functions have been developed by Poliquin [1991] and Thibault and Zagrodny [1993].

The maximal monotonicity of normal cone mappings (in 12.18) and projections (in 12.20, 12.21 and 12.22) was seen by Moreau and Rockafellar as an immediate consequence of their results about subgradients of extended-real-valued convex functions. The relations to envelopes and Yosida regularization (in 12.23) come from Moreau [1965]. It was Rockafellar [1970f] who uncovered how convex-concave functions too could give rise to maximal monotone mappings as in 12.27 and thereby showed how a variational inequality could correspond to solving an optimization problem even in cases where the underlying maximal monotone mapping in question wasn't itself a subgradient mapping; cf. 12.50. For a full description of the functions $l(x, y)$ that come up as Lagrangians in this setting, see Rockafellar [1964a] and [1970a].

The characterization of the subgradient mappings of lower-$\mathcal{C}^2$ functions by hypomonotonicity was developed by Rockafellar [1982b]. A characterization of the subgradient mappings of lower-$\mathcal{C}^1$ functions by 'submonotonicity', which we haven't treated here, was provided by Spingarn [1981]; for related work see Janin [1982].

The piecewise polyhedral characterization of the subgradient mappings of piecewise linear-quadratic convex functions in 12.30 was given in the dissertation of Sun [1986] but hasn't previously been published in full form.

The results in 12.32 on graphical convergence of monotone mappings are due to Attouch [1984]. Theorem 12.35 on the graphical convergence of subgradient mappings for convex functions comes mainly from Attouch [1977]; the extension to nonconvex functions was achieved by Poliquin [1992]. In infinite dimensions, a partial extension has been found by Levy, Poliquin and Thibault [1995]. The 'compactness' in 12.33 hasn't been noted before, nor have the convergence properties in 12.34(a) and 12.36; 12.34(b) is already in Bagh and Wets [1996].

The uniform local boundedness result in 12.40(a) is new. The version in 12.40(b) for subgradients of convex functions has precedents in Rockafellar [1970a] and Birge and Qi [1995], but in this full form is due to Bagh and Wets [1996].

The near-convexity result in 12.41 was obtained by Minty [1961] and extended to infinite-dimensional spaces by Rockafellar [1970e]. The characterization of local boundedness of maximal monotone mappings in 12.38, an early contribution of Rockafellar [1969b], helped to shape the subject through its consequences for domains and ranges in 12.39. (The sufficiency in the case of global boundedness was known already to Minty.) The proof of 12.38 by way of the convergence property in 12.37 is new, however. The fact in 12.37 that $\lambda T \xrightarrow{g} N_D$ as $\lambda \searrow 0$ hasn't been brought out until now, but it's equivalent, through inverse mappings and 12.41, to the graphical limit property in 12.42 (as shown in the proof of the latter). That limit property was established by P.-L. Lions [1978], although he didn't express it in terms of convergence to direction points.

A notion related to the graphical convergence in 12.42 is that of the limit of $\langle v, w\rangle$ for $v \in T(\bar{x} + \tau w)$ as $\tau \nearrow \infty$, which exists by monotonicity when the half-line $\{\bar{x} + \tau w \mid \tau \geq 0\}$ lies in $\operatorname{dom} T$ (as is true when $\bar{x} \in \operatorname{dom} T$ and $w \in (\operatorname{dom} T)^\infty$); cf. 12.47. It's evident that this limit can't be larger than $\sigma_R(w)$ for $R = \operatorname{cl}(\operatorname{rge} T)$. Strict inequality can hold, though, as for instance when $T(x) = Ax$ for a nonsingular, antisymmetric matrix A. Then the limit in question is always 0, but $\sigma_R(w) \equiv \infty$.

Such limits have played a role in 'coercivity' concepts for maximal monotone mappings and their substitutes, as in Browder [1965], Rockafellar [1970c], and Brézis and Nirenberg [1978]; see also Attouch, Chbani and Moudafi [1995]. They are reflected in the criteria for existence of solutions to variational inequalities and generalized equations $T(\bar{x}) \ni 0$ that are featured in 12.51 and 12.52. Parts (a) of these results come from Rockafellar [1970c], whereas parts (b) are new in the sharpness of their formulation.

For extensions of the results about ranges of maximal monotone mappings beyond the setting of reflexive Banach spaces, see Gossez [1971], Fitzpatrick and Phelps [1992], and Simons [1996]. For contributions to the study of local boundedness, see Borwein and Fitzpatrick [1988].

The question of whether maximality is preserved when a monotone mapping is constructed out of other such mappings, themselves maximal, has been perceived as basic to effective application of the general theory of monotonicity to individual cases.

The maximality criterion in 12.44 for sums, and its special version in 12.48 in the setting of a variational inequality, were established by Rockafellar [1970c] along with a case of the 'truncation' result in 12.45. The composition rule in 12.43 hasn't been looked at before, although here we make it the central fact from which the other rules are derived. It could in turn be derived from the addition rule in 12.44.

Strong monotonicity goes back to Zarantonello [1960]; strong convexity has a much longer history. Both concepts are especially useful for in developing the convergence rates of algorithms for finding solutions to variational inequalities and problems of optimization, as exemplified in 12.56 and 12.57. For more on this, and other aspects of the extensive numerical methodology that is based on maximal monotone mappings, see the articles of Rockafellar [1976b], [1976c], Lions and Mercier [1979], Spingarn [1983], Eckstein [1989], Eckstein and Bertsekas [1992], Güler [1991], Renaud [1993], and Chen and Rockafellar [1997].

The Lipschitz continuous gradient property in 12.62 for the double envelope functions of Example 1.46 is due to Lasry and Lions [1986].

The directional limit property in 12.63(a) was found by Rockafellar [1970a]. Closely related to this, and of particular significance for numerical methods, is 'semi smoothness' as defined by Mifflin [1989]. For more on how this operates, see Qi [1990], [1993]. The directional boundedness in 12.63(b) comes from Rockafellar [1969b].

The geometric relationship of proto-differentiability between T and its resolvants in 12.64 has heavily been used as a tool by Poliquin and Rockafellar [1996a], [1996b], for subgradient mappings. Some aspects of the differentiability rule Theorem 12.65 were established by those authors, but the results for general maximal monotone mappings in both 12.64 and 12.65 are new. The generic differentiability of monotone mappings in 12.66 was proved earlier by Mignot [1976].

The structure result in 12.67 is new also. It extends a theorem of Rockafellar [1970a] in the convex subgradient case.

The first to discover Lipschitzian properties of the ε-subgradient mappings $\partial_\varepsilon f$, with f proper, lsc and convex, were Nurminski [1978] and Hiriart-Urruty [1980b]; such mappings were introduced earlier by Rockafellar [1970a]. A stronger result was obtained by Attouch and Wets [1993b]. Proposition 12.68 improves on this in the assumptions and the sharpness of the estimate, furnishing also the connection with the theory of strict continuity of set-valued mappings.

The book of Phelps [1989] provides an exposition of the theory of monotone mappings in infinite-dimensional spaces and in particular addresses the geometry of differentiability in such spaces.

13. Second-Order Theory

For functions $f : \mathbb{R}^n \to \overline{\mathbb{R}}$, the notions of 'subderivative' and 'subgradient', along with semidifferentiability and epi-differentiability, have provided a broad and effective generalization of first-order differentiation. What can be said, though, on the level of generalized second-order differentiation? And what might the use of this be?

Classically, second derivatives carry forward the analysis of first derivatives and provide quadratic approximations of a given function, whereas first derivatives by themselves only provide linear approximations. They serve as an intermediate link in an endless chain of differentiation that proceeds to third derivatives, fourth derivatives, and so on. In optimization, derivatives of third order and higher are rarely of importance, but second derivatives help significantly in the understanding of optimality, especially the formulation of sufficient conditions for local optimality in the absence of convexity. Such conditions form the basis for numerical methodology and assist in studies of what happens to optimal solutions when the parameters on which a problem depends are perturbed.

In working toward a theory of second-order subdifferentiation that is effective for functions beyond the classical framework, such as have been treated successfully so far, we can aim at promoting these applications on a wider front as well as achieving new insights into results previously obtained by standard tools operating with difficulty in narrow confines. The technical and conceptual challenges are formidable, however. We have to make sense of some kinds of second derivatives for functions that need not have ordinary first derivatives and can be discontinuous and extended-real-valued. It's remarkable that despite such hurdles very much can be accomplished.

A. Second-Order Differentiability

Let's start by reviewing the classical notion of second-order differentiability and discussing a simple extension that can be made of it. The extension builds on Rademacher's theorem, the fact in 9.60 that when a function f is strictly continuous (hence locally Lipschitz continuous) on an open set O, the set of points where it fails to be differentiable is negligible in O, so that its complement is dense in O.

13.1 Definition (twice differentiable functions). *Let $f : \mathbb{R}^n \to \overline{\mathbb{R}}$ be a function that is finite at $\bar{x}$.*

(a) *f is twice differentiable at $\bar{x}$ in the classical sense if it is differentiable on a neighborhood of $\bar{x}$ and the gradient mapping ∇f on this neighborhood is differentiable at $\bar{x}$. The Jacobian matrix $\nabla(\nabla f)(\bar{x})$ (whose components are the second partial derivatives of f at $\bar{x}$) is then called the Hessian of f at $\bar{x}$ in the classical sense and is denoted by $\nabla^2 f(\bar{x})$.*

(b) *f is twice differentiable at $\bar{x}$ in the extended sense if it is differentiable at $\bar{x}$ and strictly continuous there, and ∇f is differentiable at $\bar{x}$ relative to D, its domain of existence (which has negligible complement relative to some neighborhood of $\bar{x}$): there is a matrix $A \in \mathbb{R}^{n\times n}$ such that*

$$\nabla f(x) \;=\; \nabla f(\bar{x}) + A[x - \bar{x}] + o\big(|x - \bar{x}|\big) \quad \text{for } x \in D.$$

This matrix A, necessarily unique, is then called the Hessian of f at $\bar{x}$ in the extended sense and is likewise denoted by $\nabla^2 f(\bar{x})$.

(c) *f has a quadratic expansion at $\bar{x}$ if it is differentiable at $\bar{x}$ and there is a matrix $A \in \mathbb{R}^{n\times n}$ such that*

$$f(x) \;=\; f(\bar{x}) + \big\langle \nabla f(\bar{x}), x - \bar{x}\big\rangle + \tfrac{1}{2}\big\langle x - \bar{x}, A[x - \bar{x}]\big\rangle + o\big(|x - \bar{x}|^2\big).$$

The extended definition in (b) is motivated by the desire to develop second-order differentiability at $\bar{x}$ without having to assume the existence of first partial derivatives at every point in some neighborhood of $\bar{x}$, which would be severely limiting. The components of A are regarded in this case as legitimate second partial derivatives, albeit from a broader perspective, hence the same notation $\nabla^2 f(\bar{x})$ for A. The uniqueness of A in (b) is apparent from the density of D in a neighborhood of $\bar{x}$; with A and A' both satisfying the condition, we get $0 = (A' - A)[x - \bar{x}] + o\big(|x - \bar{x}|\big)$ for $x \in D$, so $A' - A = 0$.

In speaking in (c) of a quadratic expansion, we could refer with seemingly greater generality to a right side in which the first two terms are written as $c + \langle v, x - \bar{x}\rangle$ for some $c \in \mathbb{R}$ and $v \in \mathbb{R}^n$, but then obviously $c = f(\bar{x})$ and $v = \nabla f(\bar{x})$ anyway, so nothing more would really be said. The elucidation of how the components of the matrix A in such a second-order approximation relate to second partial derivatives of f is one of the tasks we will face.

The quadratic term in (c) depends only on the symmetric part $\frac{1}{2}(A + A^*)$ of A and therefore determines only that part of A uniquely; the antisymmetric part $\frac{1}{2}(A - A^*)$ can be anything, unless symmetry of A is demanded. Of course, even for Hessian matrices in the classical sense symmetry can't be taken for granted. Symmetry in second partial derivatives is familiar when f belongs to $\mathcal{C}^2$ (this being the case in (a) where the Hessian exists at all x near $\bar{x}$ and depends continuously on x), but here we're pushing outside of that territory. Later, however, we'll find a substantial class of functions f beyond $\mathcal{C}^2$ for which the matrix $\nabla^2 f(\bar{x})$, when it exists in the manner of (a) or (b), must be symmetric. (See 13.42.)

In the theorem that follows, we utilize the notion of differentiability of a set-valued mapping that was introduced after 8.43. It requires the mapping to be single-valued at the point in question, but not necessarily elsewhere.

13.2 Theorem (extended second-order differentiability).

(a) *If $f : \mathbb{R}^n \to \overline{\mathbb{R}}$ is twice differentiable at $\bar{x}$ in the classical sense, it is also twice differentiable at $\bar{x}$ in the extended sense with the same Hessian.*

(b) *If $f : \mathbb{R}^n \to \overline{\mathbb{R}}$ is twice differentiable at $\bar{x}$ in the extended sense, it is strictly differentiable at $\bar{x}$ and its Hessian matrix furnishes a quadratic expansion for f at $\bar{x}$:*

$$f(\bar{x}+\tau w) = f(\bar{x}) + \tau\langle\nabla f(\bar{x}), w\rangle + \tfrac{1}{2}\tau^2\langle w, \nabla^2 f(\bar{x})w\rangle + o\big(\tau^2|w|^2\big).$$

(c) *In general, a function $f : \mathbb{R}^n \to \overline{\mathbb{R}}$ is twice differentiable at a point $\bar{x}$ in the extended sense if and only if f is finite and locally lsc at $\bar{x}$ and the subgradient mapping $\partial f : \mathbb{R}^n \rightrightarrows \mathbb{R}^n$ is differentiable at $\bar{x}$: there exist $v \in \mathbb{R}^n$ and $A \in \mathbb{R}^{n\times n}$ such that $\partial f(\bar{x}) = \{v\}$ and*

$$\emptyset \neq \partial f(x) \subset v + A[x-\bar{x}] + o\big(|x-\bar{x}|\big)\mathbb{B} \qquad 13(1)$$

with respect to x belonging to some neighborhood of $\bar{x}$, in which case one necessarily has $v = \nabla f(\bar{x})$ and $A = \nabla^2 f(\bar{x})$.

Proof. When f is twice differentiable at $\bar{x}$ in the classical sense, we have $\nabla f(x) = \nabla f(\bar{x}) + \nabla^2 f(\bar{x})[x-\bar{x}] + o\big(|x-\bar{x}|\big)$, which implies ∇f is locally bounded at $\bar{x}$. The existence of κ such that $|\nabla f(x)| \leq \kappa$ on a convex neighborhood V of $\bar{x}$ allows us to deduce that f is Lipschitz continuous on V with constant κ. Specifically, for any points $x_0, x_1 \in V$ the function $\varphi(\tau) = f(x_\tau)$ for $x_\tau := (1-\tau)x_0 + \tau x_1$ is differentiable and thus by the mean value theorem has $\varphi(1) - \varphi(0) = \varphi'(\tau)$ for some $\tau \in (0,1)$. But $\varphi'(\tau) = \langle\nabla f(x_\tau), x_1 - x_0\rangle$, so $|\varphi'(\tau)| \leq \kappa|x_1 - x_0|$ and we get $f(x_1) - f(x_0) \leq \kappa|x_1 - x_0|$.

The expansion for ∇f that expresses its differentiability at $\bar{x}$ then yields the one in 13.1(b) with $A = \nabla^2 f(\bar{x})$ and says that f is twice differentiable at $\bar{x}$ in the extended sense. This proves (a).

To obtain the quadratic expansion in (b), we observe that the differentiability of ∇f relative to D furnishes for any $\varepsilon > 0$ a $\delta > 0$ such that

$$\begin{aligned} &\big|\nabla f(\bar{x}+\tau w) - \nabla f(\bar{x}) - \tau A w\big| \leq \varepsilon\tau \\ &\qquad \text{for all } \tau \in [0,\delta] \text{ when } \bar{x}+\tau w \in D,\ |w| = 1. \end{aligned} \qquad 13(2)$$

Let V be an open neighborhood of $\bar{x}$ on which f is Lipschitz continuous. We can take δ small enough in 13(2) that $\mathbb{B}(\bar{x},\delta) \subset V$. The negligibility of $V \setminus D$ guarantees that in the unit sphere $S = \{w \mid |w| = 1\}$ there's a dense set of w's for which the intersection of the half-line $\{\bar{x}+\tau w \mid \tau \geq 0\}$ with $V \setminus D$ is negligible (in the one-dimensional sense). For such w the Lipschitz continuous function $\psi(\tau) = f(\bar{x}+\tau w)$ on $[0,\delta]$ has $\psi'(\tau) = \langle\nabla f(\bar{x}+\tau w), w\rangle$ except for a negligible set of τ values. Then $f(\bar{x}+\tau w) - f(\bar{x}) = \int_0^\tau \langle\nabla f(\bar{x}+tw), w\rangle dt$, and through 13(2) we get

$$\begin{aligned}
&\big|f(\bar{x}+\tau w)-f(\bar{x})-\tau\langle\nabla f(\bar{x}),w\rangle-\tfrac{1}{2}\tau^2\langle w,\nabla^2 f(\bar{x})w\rangle\big|\\
&\qquad=\Big|\int_0^\tau\big[\langle\nabla f(\bar{x}+tw),w\rangle-\langle\nabla f(\bar{x}),w\rangle-t\langle w,Aw\rangle\big]dt\Big|\\
&\qquad\le\int_0^\tau\big|\langle\nabla f(\bar{x}+tw),w\rangle-\langle\nabla f(\bar{x}),w\rangle-t\langle w,Aw\rangle\big|dt\\
&\qquad\le\int_0^\tau\big|\nabla f(\bar{x}+tw)-\nabla f(\bar{x})-tAw\big|\big|w\big|dt\le\varepsilon\tau^2\ \text{ when }\ \tau\in[0,\delta].
\end{aligned}$$

Due to the outer inequality in this chain holding for all w in a dense subset of the unit sphere S, and f being continuous on $I\!B(\bar{x},\delta)$, it must hold for all $w\in S$. This means that the targeted quadratic expansion is valid.

We turn now to the characterization in (c), from which we'll also derive the strict differentiability claimed in (b). Suppose first that f is twice differentiable at $\bar{x}$ in the extended sense, and again let V be an open neighborhood of $\bar{x}$ on which f Lipschitz continuous. We have $\emptyset\neq\partial f(x)\subset\operatorname{con}\overline{\nabla}f(x)$ at all points of V, where $\overline{\nabla}f(x)$ consists of all limits of sequences $\nabla f(x^\nu)$ as $x^\nu\to x$; see 9.61. The first-order expansion of ∇f relative to D then yields the proposed first-order expansion of ∂f with $v=\nabla f(\bar{x})$ and $A=\nabla^2 f(\bar{x})$.

Conversely, if f isn't assumed strictly continuous at $\bar{x}$ but just locally lsc, and ∂f is differentiable at $\bar{x}$ as in 13(1), we have ∂f locally bounded at $\bar{x}$, hence f strictly continuous at $\bar{x}$ after all by 9.13. Furthermore $\partial f(\bar{x})=\{v\}$, which implies by 9.18 that f is strictly differentiable at $\bar{x}$ with $\nabla f(\bar{x})=v$. The expansion in 13(1) then yields the one in Definition 13.1(b), inasmuch as $\nabla f(x)\in\partial f(x)$ whenever f is differentiable at x. □

Along with furnishing facts of direct interest, Theorem 13.2 lays out the patterns that will henceforth guide our efforts: the investigation of first-order approximations to ∂f at $\bar{x}$ and second-order approximations to f at $\bar{x}$, possibly nonlinear-nonquadratic and just one-sided in certain ways, and which are not based exclusively on pointwise convergence. We'll begin by exploring different kinds of limits of second-order difference quotient expressions.

B. Second Subderivatives

From an elementary standpoint, it's natural to think first of a simple generalization of the one-sided directional derivatives $f'(\bar{x};w)$ of 7(20), which are the right derivatives $\varphi'_+(0)$ of functions $\varphi(\tau)=f(\bar{x}+\tau w)$. Second right derivatives of such functions φ can be defined by

$$\varphi''_+(0)=\lim_{\tau\searrow 0}\frac{\varphi(\tau)-\varphi(0)-\tau\varphi'_+(0)}{\frac{1}{2}\tau^2},$$

and one can speak then of *one-sided second directional derivatives* of f at $\bar{x}$, a point where f is finite, as given by

$$f''(\bar{x};w) := \lim_{\tau \searrow 0} \frac{f(\bar{x}+\tau w) - f(\bar{x}) - \tau f'(\bar{x};w)}{\frac{1}{2}\tau^2} \tag{13(3)}$$

when this limit and the one for $f'(\bar{x};w)$ exist. Note that the $\frac{1}{2}$ in the denominator, cumbersome as it may be, is mandatory for keeping in step with classical conventions. In the case of $\varphi(\tau) = \tau^2$, for instance, one gets $\varphi'_+(\tau) = 2\tau$ and $\varphi''_+(\tau) = 2$, but this would fail without the factor $\frac{1}{2}$ being inserted.

This elementary notion of generalized derivatives is too weak to be satisfactory, however. We already know from experience in first-order theory that one-sided directional derivatives calculated just along half-lines emanating from $\bar{x}$ don't serve well for nonsmooth functions f. (See Chapter 7 for the discussion of first-order semidifferentiability, starting with 7.20.) Better results are promised by other types of limits that test the behavior of f as $\bar{x}$ is reached from the direction of a vector w but perhaps 'nonlinearly'.

Two forms of second-order difference quotients for f at $\bar{x}$ facilitate the study of such behavior. The first is

$$\Delta_\tau^2 f(\bar{x})(w) := \frac{f(\bar{x}+\tau w) - f(\bar{x}) - \tau df(\bar{x})(w)}{\frac{1}{2}\tau^2} \quad \text{for } \tau > 0, \tag{13(4)}$$

where $df(\bar{x})$ is the subderivative function of 8.1, so that in particular, $df(\bar{x})(w)$ comes out as the first-order semiderivative or the epi-derivative of f at $\bar{x}$ for w in the presence of the additional limit properties that define those terms. The finiteness of $f(\bar{x})$ is of course presupposed in this context, but to avoid trouble over $df(\bar{x})(w)$ also possibly having to be finite in order for the numerator in 13(4) even to make sense, we adopt the convention of inf-addition, where the sum of ∞ and $-\infty$ is interpreted as ∞. In other words, we take

$$\Delta_\tau^2 f(\bar{x})(w) := \infty \quad \text{if } f(\bar{x}+\tau w) = df(\bar{x})(w) = \infty \text{ or } -\infty.$$

The other form of second-order difference quotient, which also presupposes the finiteness of $f(\bar{x})$, obviates the need for special treatment of infinities by substituting a linear term $\langle v, w\rangle$ for $df(\bar{x})(w)$:

$$\Delta_\tau^2 f(\bar{x}\,|\,v)(w) := \frac{f(\bar{x}+\tau w) - f(\bar{x}) - \tau\langle v, w\rangle}{\frac{1}{2}\tau^2} \quad \text{for } \tau > 0. \tag{13(5)}$$

When f is differentiable at $\bar{x}$ and $v = \nabla f(\bar{x})$, this second form reduces to the first because $df(\bar{x})(w) = \langle v, w\rangle$.

It's instructive to observe that the elementary second directional derivative function $f''(\bar{x};\cdot)$ of 13(3), when everywhere defined, is the *pointwise* limit of the difference quotient functions $\Delta_\tau^2 f(\bar{x})$ as $\tau \searrow 0$. We can progress to more robust generalizations of second-order differentiation by studying other limits of $\Delta_\tau^2 f(\bar{x})$ and $\Delta_\tau^2 f(\bar{x}\,|\,v)$ as $\tau \searrow 0$ that instead look to uniform convergence, continuous convergence, or epi-convergence. The following lower limits are fundamental to the discussion of all such modes of convergence.

13.3 Definition (second subderivatives). *For $f : \mathbb{R}^n \to \overline{\mathbb{R}}$, any $\bar{x} \in \mathbb{R}^n$ with $f(\bar{x})$ finite and any $v \in \mathbb{R}^n$, the second subderivative at $\bar{x}$ for v and w is*

$$d^2 f(\bar{x}\,|\,v)(w) := \liminf_{\substack{\tau \searrow 0 \\ w' \to w}} \Delta_\tau^2 f(\bar{x}\,|\,v)(w'). \tag{13(6)}$$

The second subderivative at $\bar{x}$ for w (without mention of v) is

$$d^2 f(\bar{x})(w) := \liminf_{\substack{\tau \searrow 0 \\ w' \to w}} \Delta_\tau^2 f(\bar{x})(w'). \tag{13(7)}$$

Thus, the functions $d^2 f(\bar{x}\,|\,v) : \mathbb{R}^n \to \overline{\mathbb{R}}$ and $d^2 f(\bar{x}) : \mathbb{R}^n \to \overline{\mathbb{R}}$ are given by lower epi-limits:

$$d^2 f(\bar{x}\,|\,v) = \operatorname*{e-lim\,inf}_{\tau \searrow 0} \Delta_\tau^2 f(\bar{x}\,|\,v), \qquad d^2 f(\bar{x}) = \operatorname*{e-lim\,inf}_{\tau \searrow 0} \Delta_\tau^2 f(\bar{x}).$$

Distinctions in the behavior of the two kinds of second subderivative will be highlighted in what follows, but they agree when f is differentiable at $\bar{x}$:

$$d^2 f(\bar{x}\,|\,v) = d^2 f(\bar{x}) \;\text{ when }\; v = \nabla f(\bar{x}).$$

Unlike first-order subderivative functions, the second-order subderivative functions aren't positively homogeneous in the 'degree 1' sense used so far. Instead they have a higher-order property of such type.

13.4 Definition (positive homogeneity of degree p). *A function $h : \mathbb{R}^n \to \overline{\mathbb{R}}$ is called positively homogeneous of degree $p > 0$ if $h(\lambda w) = \lambda^p h(w)$ for all $\lambda > 0$ and w, and $h(0) < \infty$. (Then either $h(0) = 0$ or $h(0) = -\infty$.)*

The case of $p = 1$, important in first-order theory, takes the back stage now in favor of positive homogeneity of degree $p = 2$.

13.5 Proposition (properties of second subderivatives). *The functions $d^2 f(\bar{x})$ and $d^2 f(\bar{x}\,|\,v)$ are lsc and positively homogeneous of degree 2. In addition, $d^2 f(\bar{x}\,|\,v)(w)$ depends concavely on v and has*

$$\begin{cases} d^2 f(\bar{x}\,|\,v)(w) = \infty & \text{when } \langle v, w\rangle < df(\bar{x})(w), \\ d^2 f(\bar{x}\,|\,v)(w) = -\infty & \text{when } \langle v, w\rangle > df(\bar{x})(w), \end{cases}$$

so that finiteness of $d^2 f(\bar{x}\,|\,v)(w)$ necessitates having $\langle v, w\rangle = df(\bar{x})(w)$. Thus, $\operatorname{dom} d^2 f(\bar{x}\,|\,v) \subset \{w \,|\, df(\bar{x})(w) \le \langle v, w\rangle\}$ always, but properness of $d^2 f(\bar{x}\,|\,v)$ is not possible unless actually $df(\bar{x})(w) \ge \langle v, w\rangle$ for all w:

$$d^2 f(\bar{x}\,|\,v) \text{ proper} \quad\Longrightarrow\quad \begin{cases} v \in \widehat{\partial} f(\bar{x}), \\ \operatorname{dom} d^2 f(\bar{x}\,|\,v) \subset \{w \,|\, df(\bar{x})(w) = \langle v, w\rangle\} \\ \qquad\qquad\qquad\quad \subset N_{\widehat{\partial} f(\bar{x})}(v). \end{cases}$$

Proof. The lower semicontinuity is automatic from the expression of these functions as lower epi-limits; cf. 7.4(a). The positive homogeneity of degree 2 arises from the form of the second-order difference quotients in having $\Delta_\tau^2 f(\bar{x}\,|\,v)(\lambda w) = \lambda^2 \Delta_{\lambda\tau}^2 f(\bar{x}\,|\,v)(w)$ and $\Delta_\tau^2 f(\bar{x})(\lambda w) = \lambda^2 \Delta_{\lambda\tau}^2 f(\bar{x})(w)$ for $\lambda > 0$. The concavity of $d^2 f(\bar{x}\,|\,v)(w)$ with respect to v comes from

$$d^2 f(\bar{x}\,|\,v)(w) = \lim_{\delta \searrow 0} \Big[\inf_{\substack{w' \in \boldsymbol{B}(w,\delta) \\ \tau \in (0,\delta)}} \Delta_\tau^2 f(\bar{x}\,|\,v)(w') \Big]$$

and the fact that, for τ and w' such that $f(\bar{x} + \tau w')$ is finite, $\Delta_\tau^2 f(\bar{x}\,|\,v)(w')$ is affine with respect to v. (The pointwise infimum of a collection of affine functions is concave, and the pointwise limit of a sequence of concave functions is concave; cf. 2.9.) Next fix any vector v and note that

$$\Delta_\tau^2 f(\bar{x}\,|\,v)(w) = \frac{2}{\tau}\big[\Delta_\tau f(\bar{x})(w) - \langle v, w\rangle\big]. \qquad 13(8)$$

Consider sequences $w^\nu \to w$ and $\tau^\nu \searrow 0$ such that $\Delta_{\tau^\nu}^2 f(\bar{x}\,|\,v)(w^\nu)$ converges to some value $\alpha \in \overline{\mathbb{R}}$ while $\Delta_{\tau^\nu} f(\bar{x})(w^\nu)$ converges to some β. (If either sequence converges by itself, we can arrange, by passing to a subsequence if necessary that the other converges as well, so the stipulation of joint convergence entails no loss of generality.) The value $d^2 f(\bar{x}\,|\,v)(w)$ is by definition the lowest α attainable in such a scheme, whereas $df(\bar{x})(w)$ is the lowest β.

Here $\alpha = \lim_\nu (2/\tau^\nu)\big[\Delta_{\tau^\nu} f(\bar{x})(w^\nu) - \langle v, w^\nu\rangle\big]$ and $\Delta_{\tau^\nu} f(\bar{x})(w^\nu) - \langle v, w^\nu\rangle \to \beta - \langle v, w\rangle$. If $\beta - \langle v, w\rangle > 0$, then $\alpha = \infty$. Having $d^2 f(\bar{x}\,|\,v)(w) < \infty$ thus requires $\beta - \langle v, w\rangle \leq 0$, hence also $df(\bar{x})(w) - \langle v, w\rangle \leq 0$. On the other hand, if $\beta - \langle v, w\rangle < 0$, then $\alpha = -\infty$, so $d^2 f(\bar{x}\,|\,v)(w) = -\infty$. The latter must hold for sure when $df(\bar{x})(w) - \langle v, w\rangle < 0$, due to the fact that we can always choose the sequences such that $\beta = df(\bar{x})(w)$.

To have $d^2 f(\bar{x}\,|\,v)(w) > -\infty$ for all w, we therefore need $df(\bar{x})(w) \geq \langle v, w\rangle$ for all w. This means by 8.4 that $v \in \widehat{\partial} f(\bar{x})$. But $\langle v, w\rangle \leq df(\bar{x})(w)$ for every $v \in \widehat{\partial} f(\bar{x})$. It follows that any w with $d^2 f(\bar{x}\,|\,v)(w) < \infty$ must be such that $\langle v' - v, w\rangle \leq 0$ for all $v' \in \widehat{\partial} f(\bar{x})$, so $w \in N_{\widehat{\partial} f(\bar{x})}(v)$. □

The relationships that, according to Proposition 13.5, must hold if the second-order subderivative function $d^2 f(\bar{x}\,|\,v)$ is to be proper and have a given vector $w \neq 0$ in its effective domain are shown in Figure 13–1. Not only must v belong to the closed, convex set $\widehat{\partial} f(\bar{x})$, but also, w must be a normal vector to $\widehat{\partial} f(\bar{x})$ at v. Besides, $df(\bar{x})(w)$ must be finite, and the hyperplane $H = \big\{ v \,\big|\, \langle v, w\rangle = df(\bar{x})(w) \big\}$ must support $\widehat{\partial} f(\bar{x})$ at v. But of course the latter follows from the other relationships when f is regular at $\bar{x}$, since $df(\bar{x})$ is then the support function of $\widehat{\partial} f(\bar{x})$; cf. 8.30. Under regularity, $\widehat{\partial} f(\bar{x}) = \partial f(\bar{x})$.

Another thing to note about the situation described by this picture, in appreciating the content of Proposition 13.5, is that the concave function $v \mapsto d^2 f(\bar{x}\,|\,v)(w)$ must be $-\infty$ for v in one of the open half-spaces associated with H (the one toward $\operatorname{dir} w$), but ∞ for v in the other open half-space.

In particular, *the function $d^2 f(\bar{x}\,|\,v)$ can't be finite everywhere unless $df(\bar{x})$*

is the linear function $\langle v, \cdot\rangle$, *in which case* $\widehat{\partial} f(\bar{x})$ *must be the singleton* $\{v\}$. For functions f that are regular and strictly continuous, such as finite convex functions for instance, that isn't possible unless f is differentiable at $\bar{x}$ with $\nabla f(\bar{x}) = v$. At first this may seem disappointing, but the presence of ∞ values will later come out as a characteristic that's indispensable in capturing the key second-order properties of nonsmooth functions f.

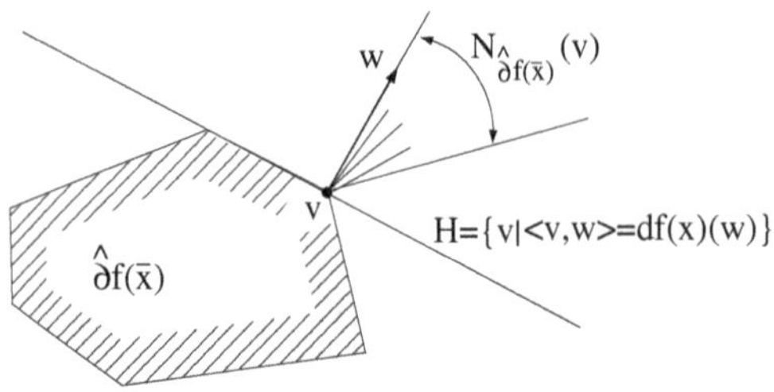

Fig. 13–1. Configuration necessary for properness of second subderivatives.

13.6 Definition (twice semidifferentiable and twice epi-differentiable functions). *Let* $\bar{x}$ *be a point where the function* $f : \mathbb{R}^n \to \overline{\mathbb{R}}$ *is finite.*

(a) f *is twice semidifferentiable at* $\bar{x}$ *if it is semidifferentiable at* $\bar{x}$ *and the functions* $\Delta_\tau^2 f(\bar{x})$ *converge continuously to* $d^2 f(\bar{x})$ *as* $\tau \searrow 0$.

(b) f *is twice epi-differentiable at* $\bar{x}$ *for* v *if the functions* $\Delta_\tau^2 f(\bar{x}\,|\,v)$ *epi-converge to* $d^2 f(\bar{x}\,|\,v)$ *as* $\tau \searrow 0$. *It is properly twice epi-differentiable at* $\bar{x}$ *for* v *if in addition the function* $d^2 f(\bar{x}\,|\,v)$ *is proper.*

The first-order semidifferentiability assumed in Definition 13.6(a) implies by 7.21 that $df(\bar{x})(w)$ is finite for all w, so the functions $\Delta_\tau^2 f(\bar{x})$ are well defined on $\mathbb{R}^n$ without any need for invoking the convention for handling conflicts of infinities. The requirement for second-order semidifferentiability is that the 'lim inf' giving $d^2 f(\bar{x})(w)$ in 13(6) should coincide with the 'lim sup', or in other words that for each w one should have the existence of

$$\lim_{\substack{\tau \searrow 0 \\ w' \to w}} \frac{f(\bar{x} + \tau w') - f(\bar{x}) - \tau df(\bar{x})(w')}{\frac{1}{2}\tau^2}.$$

Similarly in Definition 13.6(b), $d^2 f(\bar{x}\,|\,v)$ is already the e-lim inf of the functions $\Delta_\tau^2 f(\bar{x}\,|\,v)$ as $\tau \searrow 0$, so the requirement for second-order epi-differentiability is that $d^2 f(\bar{x}\,|\,v)$ should also be the corresponding e-lim sup. This means that for every $w \in \mathbb{R}^n$ and choice of $\tau^\nu \searrow 0$ there should exist $w^\nu \to w$ such that

$$\frac{f(\bar{x} + \tau^\nu w^\nu) - f(\bar{x}) - \tau^\nu \langle v, w^\nu\rangle}{\frac{1}{2}\tau^{\nu\,2}} \;\to\; d^2 f(\bar{x}\,|\,v)(w).$$

Second-order epi-differentiability doesn't assume first-order epi-differentiability, but it implies for each $w \in \operatorname{dom} d^2 f(\bar{x}\,|\,v)$ that the first-order epi-derivative of

f at $\bar{x}$ exists for w. This can be deduced from 13(8).

For still another notion of generalized differentiability, one might contemplate looking at *continuous* convergence of $\Delta_\tau^2 f(\bar{x}\,|\,v)$ to $d^2 f(\bar{x}\,|\,v)$ (with its implications of uniform convergence). But that would entail having $d^2 f(\bar{x}\,|\,v)(w) < \infty$ for all w (as seen from the case of $w = 0$) and consequently, when $d^2 f(\bar{x}\,|\,v)$ is proper, that $df(\bar{x})(w) = \langle v, w\rangle$ for all w (by Proposition 13.5). Then $\Delta_\tau^2 f(\bar{x}\,|\,v)$ and $d^2 f(\bar{x}\,|\,v)$ reduce to $\Delta_\tau^2 f(\bar{x})$ and $d^2 f(\bar{x})$, so nothing new would be obtained.

The virtue of second-order semidifferentiability is that it corresponds to f having a 'second-order expansion' at $\bar{x}$ in the following sense.

13.7 Exercise (characterizations of second-order semidifferentiability). *For any function $f : \mathbb{R}^n \to \overline{\mathbb{R}}$ and any point $\bar{x}$ at which f is semidifferentiable, the following conditions are equivalent:*

(a) *f is twice semidifferentiable at $\bar{x}$;*

(b) *as $\tau \searrow 0$, the difference quotient functions $\Delta_\tau^2 f(\bar{x})$ converge uniformly on all bounded subsets of $\mathbb{R}^n$ to a continuous function h;*

(c) *there is a finite, continuous, function h, which is positively homogeneous of degree 2 and such that*

$$f(x) \;=\; f(\bar{x}) + df(\bar{x})(x - \bar{x}) + \tfrac{1}{2}h(x - \bar{x}) + o\big(|x - \bar{x}|^2\big).$$

Under these conditions the function h must be $d^2 f(\bar{x})$, and therefore $d^2 f(\bar{x})(w)$ must everywhere be finite and depend continuously on w.

Guide. Mimic parts of the proof of Theorem 7.21. □

The two concepts of generalized second-order differentiability in 13.6 lead to the same thing when f is twice differentiable as in 13.1.

13.8 Example (connections with second-order differentiability). *Whenever a function $f : \mathbb{R}^n \to \overline{\mathbb{R}}$ is twice differentiable at $\bar{x}$ in the classical or the extended sense, it is both twice semidifferentiable at $\bar{x}$ and twice epi-differentiable at x for $v = \nabla f(\bar{x})$ and has*

$$d^2 f(\bar{x}\,|\,v)(w) = d^2 f(\bar{x})(w) = f''(\bar{x}; w) = \langle w, \nabla^2 f(\bar{x})w\rangle.$$

In general, f has a quadratic expansion at $\bar{x}$ if and only if it is differentiable at $\bar{x}$ and also twice semidifferentiable there with $d^2 f(\bar{x})$ quadratic.

Detail. This is immediate from the second-order expansion in Theorem 13.2(b) and the definition of quadratic expansion in 13.1(c). □

Beyond the classical setting it's possible for f to be twice semidifferentiable at $\bar{x}$ and at the same time properly twice epi-differentiable at $\bar{x}$ for every $v \in \partial f(\bar{x})$, yet with $d^2 f(\bar{x}\,|\,v)$ always different from $d^2 f(\bar{x})$. On the other hand, it's possible for f to be properly twice epi-differentiable at $\bar{x}$ for every $v \in \partial f(\bar{x})$ while failing to be twice semidifferentiable at $\bar{x}$. Both of these possibilities can be encountered even in working with functions f that are convex and piecewise linear-quadratic, as defined in 10.20. The analysis of this class of functions is important for such reasons but also as a stepping stone.

13.9 Proposition (piecewise linear-quadratic functions). *Suppose f is proper, convex and piecewise linear-quadratic on $\mathbb{R}^n$, and let $\bar{x} \in \operatorname{dom} f$. Then $d^2 f(\bar{x})$ is proper and piecewise linear-quadratic, but not necessarily convex, and the values of $d^2 f(\bar{x})$ are achievable by taking limits merely along half-lines:*

$$d^2 f(\bar{x})(w) \;=\; \lim_{\tau \searrow 0} \Delta_\tau^2 f(\bar{x})(w) \;= f''(\bar{x}; w) \geq 0.$$

For every $v \in \partial f(\bar{x})$, $d^2 f(\bar{x} \,|\, v)$ is proper, convex and piecewise linear-quadratic. With $K(\bar{x}, v) = \{ w \,|\, df(\bar{x})(w) = \langle v, w \rangle \}$, a polyhedral cone, one has

$$d^2 f(\bar{x} \,|\, v)(w) \;=\; \lim_{\tau \searrow 0} \Delta_\tau^2 f(\bar{x} \,|\, v)(w) \;=\; d^2 f(\bar{x})(w) + \delta_{K(\bar{x}, v)}(w).$$

Moreover f is twice epi-differentiable at $\bar{x}$ for every $v \in \partial f(\bar{x})$, whereas f is twice semidifferentiable at $\bar{x}$ if $\bar{x} \in \operatorname{int}(\operatorname{dom} f)$ but not if $\bar{x} \in \operatorname{bdry}(\operatorname{dom} f)$. Nonetheless there always exists $V \in \mathcal{N}(\bar{x})$ such that

$$f(x) = f(\bar{x}) + df(\bar{x})(x - \bar{x}) + \tfrac{1}{2} d^2 f(\bar{x})(x - \bar{x}) \;\text{ for }\; x \in V \cap \operatorname{dom} f. \qquad 13(9)$$

Further, there exists $\varepsilon > 0$ such that whenever $w \in T_{\operatorname{dom} f}(\bar{x}) \cap \mathbb{B}$ and $\tau \in (0, \varepsilon)$ one has, for every $z \in \mathbb{R}^n$, the additional expansion

$$df(\bar{x} + \tau w)(z) \;=\; dh_1(w)(z) + \tfrac{1}{2}\tau dh_2(w)(z) \;\text{ with }\; \begin{cases} h_1 = df(\bar{x}), \\ h_2 = d^2 f(\bar{x}). \end{cases} \qquad 13(10)$$

Proof. By the definition of 'piecewise linear-quadratic' in 10.20, the convex set $C := \operatorname{dom} f$ is the union of finitely many polyhedral sets C_k on which f coincides with a linear-quadratic function f_k. Fixing $\bar{x}$, we can take $f(\bar{x}) = 0$ and suppose that $\bar{x}$ belongs to every C_k (since this can be arranged by adding to f the indicator of a polyhedral neighborhood of $\bar{x}$, which won't affect the local properties of f at $\bar{x}$). Then

$$f_k(\bar{x} + \tau w) - f_k(\bar{x}) = \tau \langle a_k, w \rangle + \tfrac{1}{2}\tau^2 \langle w, A_k w \rangle \;\text{ when }\; \bar{x} + \tau w \in C_k$$

for vectors a_k and symmetric matrices A_k, and $T_C(\bar{x}) = \bigcup_{k=1}^r T_{C_k}(\bar{x})$, so

$$\Delta_\tau f(\bar{x})(w) = \begin{cases} \langle a_k, w \rangle + \frac{1}{2}\tau \langle w, A_k w \rangle & \text{when } w \in \tau^{-1}[C_k - \bar{x}], \\ \infty & \text{when } w \notin \tau^{-1}[C - \bar{x}], \end{cases}$$

$$df(\bar{x})(w) = \begin{cases} \langle a_k, w \rangle & \text{when } w \in T_{C_k}(\bar{x}), \\ \infty & \text{when } w \notin T_C(\bar{x}). \end{cases}$$

From this it's clear that f is epi-differentiable at $\bar{x}$, even semidifferentiable when $\bar{x} \in \operatorname{int} C$ (which corresponds to having $T_C(\bar{x}) = \mathbb{R}^n$). Further,

$$\Delta_\tau^2 f(\bar{x})(w) = \begin{cases} \langle w, A_k w \rangle & \text{when } w \in \tau^{-1}[C_k - \bar{x}], \\ \infty & \text{when } w \notin \tau^{-1}[C - \bar{x}], \end{cases}$$

$$d^2 f(\bar{x})(w) = \begin{cases} \langle w, A_k w \rangle & \text{when } w \in T_{C_k}(\bar{x}), \\ \infty & \text{when } w \notin T_C(\bar{x}), \end{cases}$$

with the lower limit that defines $d^2 f(\bar{x})(w)$ existing as the limit as $\tau \searrow 0$. (Note that here we use the rule of interpreting $\Delta_\tau^2 f(\bar{x})(w)$ as ∞ when $f(\bar{x}+\tau w) = df(\bar{x})(w) = \infty$.) It follows that f is twice semidifferentiable at $\bar{x}$ when $\bar{x} \in \operatorname{int} C$, but that the second-order expansion in 13(9) is valid locally around $\bar{x}$ even if $\bar{x} \in \operatorname{bdry} C$.

Consider next any $v \in \partial f(\bar{x})$. We have $\langle v, w\rangle \leq df(\bar{x})(w)$ for all w, and

$$\Delta_\tau^2 f(\bar{x}\,|\,v)(w) = \begin{cases} \langle w, A_k w\rangle + 2\tau^{-1}\langle a_k - v, w\rangle & \text{when } w \in \tau^{-1}[C_k - \bar{x}], \\ \infty & \text{when } w \notin \tau^{-1}[C - \bar{x}], \end{cases}$$

where $\langle a_k - v, w\rangle = df(\bar{x})(w) - \langle v, w\rangle \geq 0$ when $w \in \tau^{-1}[C_k - \bar{x}]$. Therefore

$$d^2 f(\bar{x}\,|\,v)(w) = \begin{cases} \langle w, A_k w\rangle & \text{when } w \in T_{C_k}(\bar{x}) \cap K(\bar{x}, v), \\ \infty & \text{when } w \notin T_C(\bar{x}) \cap K(\bar{x}, v). \end{cases}$$

The lower limit that defines $d^2 f(\bar{x}\,|\,v)(w)$ can be attained then as the limit as $\tau \searrow 0$, thus ensuring that $d^2 f(\bar{x}\,|\,v)$ is the epi-limit of $\Delta_\tau^2 f(\bar{x}\,|\,v)$ as $\tau \searrow 0$. Since $\Delta_\tau^2 f(\bar{x}\,|\,v)$ inherits convexity from f, and convexity is preserved under epi-convergence (cf. 7.17), $d^2 f(\bar{x}\,|\,v)$ must be convex.

To justify the formula for $df(\bar{x}+\tau w)$, recall first that because $\operatorname{dom} f$ is polyhedral we have for any $w \in T_{\operatorname{dom} f}(\bar{x})$ that $\bar{x} + \tau w \in \operatorname{dom} f$ for $\tau > 0$ sufficiently small. The function $df(\bar{x}+\tau w)$ is convex and piecewise linear (cf. 10.21), so for any $z \in I\!R^n$ we then have

$$df(\bar{x}+\tau w)(z) = \lim_{\sigma \searrow 0} \frac{f(\bar{x}+\tau w + \tau\sigma z) - f(\bar{x}+\tau w)}{\tau\sigma}.$$

The expansion in 13(9) gives $f(\bar{x}+\tau(w+\sigma z)) = f(\bar{x}) + \tau df(\bar{x})(w+\sigma z) + \frac{1}{2}\tau^2 df(\bar{x})(w+\sigma z)$ while $f(\bar{x}+\tau w) = f(\bar{x}) + \tau df(\bar{x})(w) + \frac{1}{2}\tau^2 df(\bar{x})(w)$. In substituting these expressions we get

$$df(\bar{x}+\tau w)(z) = \lim_{\sigma \searrow 0} \big\{\Delta_\sigma h_1(w)(z) + \tfrac{1}{2}\tau \Delta_\sigma h_2(w)(z)\big\},$$

hence $df(\bar{x}+\tau w)(z) = dh_1(w)(z) + \frac{1}{2}\tau dh_2(w)(z)$ because h_1 and h_2 are themselves piecewise linear-quadratic. □

The expansion in 13(10) is interesting alongside of the expansion in 13(9) because it generalizes 'cross partial derivatives'. In the simple case of $f(x) = \alpha + \langle a, x\rangle + \frac{1}{2}\langle x, Ax\rangle$ with A symmetric, so that $d^2 f(\bar{x})(w) = \langle w, Aw\rangle$, the term $dh_2(w)(z)$ in this formula would come out as $\langle z, Aw\rangle$. In general, the function $h_1 = df(\bar{x})$ here is the support function of the set $\partial f(\bar{x})$, which is polyhedral, and $dh_1(w)$ is in turn the support function of $\partial h_1(w)$, this being the face of $\partial f(\bar{x})$ where the linear function $v \mapsto \langle v, w\rangle$ attains its maximum (cf. 8.25).

The failure of $d^2 f(\bar{x})$ to exhibit convexity in the elementary context of 13.9, although the functions $d^2 f(\bar{x}\,|\,v)$ do inherit it from f, points to a troublesome shortcoming of $d^2 f(\bar{x})$ even in situations where semidifferentiability may be present. An instance of such behavior is seen for

$$f(x_1,x_2) = \tfrac{1}{2}x_1^2+\tfrac{1}{2}x_2^2+(1-x_1)|x_2|+\delta_D(x_1,x_2), \quad D=[-1,1]\times[-1,1].$$

This function on $\mathbb{R}^2$ is piecewise linear-quadratic with one formula on the half-square $[-1,1]\times[0,1]$ and another on $[-1,1]\times[-1,0]$. It's convex on each half-square separately, because the Hessians in the two quadratic formulas are positive-semidefinite, but it's actually convex also with respect to the whole of D. This can be verified by checking the convexity of f along each line segment in D that crosses the x_1-axis. For this we need only inspect segments of form $\{(\xi,0)+\tau(\omega,1) \mid -\varepsilon \le \tau \le \varepsilon\}$ for some $\xi \in (-1,1)$ and $\omega \in (-\infty,\infty)$, with $\varepsilon > 0$ small enough that the segment lies in D. Along such a segment we set $\varphi(\tau) = f(\xi+\tau\omega,\tau)$ and ask whether φ is convex on $[-\varepsilon,\varepsilon]$. This comes down to whether $\varphi'_-(0) \le \varphi'_+(0)$. It's easy to calculate that $\varphi'_+(0)-\varphi'_-(0) = 2(1-\xi) > 0$ when $\xi \in (-1,1)$, so f is convex as claimed. We note next that f is semidifferentiable at $\bar{x}=(0,0)$, inasmuch as $(0,0) \in \operatorname{int} D$, yet the function $d^2f(0,0)$ fails to be convex: we calculate that $df(0,0)(w_1,w_2) = |w_2|$ and consequently $d^2f(0,0)(w_1,w_2) = w_1^2+w_2^2-2w_1|w_2|$. For $w_1=1$ in particular, we get $d^2f(0,0)(1,w_2) = 1+w_2^2-2|w_2|$. This expression not only lacks convexity with respect to w_2 but has an 'upward kink'.

An example of a function f that is twice semidifferentiable at $\bar{x}$ without being properly twice epi-differentiable at $\bar{x}$ for any vector v is

$$f(x) = -|x| + \tfrac{1}{2}\max{}^2\{0,\langle a,x\rangle\} \text{ at } \bar{x}=0$$

for a vector $a \neq 0$. Here $df(\bar{x})(w) = -|w|$ and $d^2f(\bar{x})(w) = \max^2\{0,\langle a,w\rangle\}$. By the expansion criterion in 13.7(c), f is twice semidifferentiable at $\bar{x}$. But $\widehat{\partial} f(\bar{x}) = \emptyset$ by 8.4; it's impossible to find v such that $\langle v,w\rangle \le -|w|$ for all w. We know then from Proposition 13.5 that no vector v is eligible for f being properly twice epi-differentiable at $\bar{x}$ for v.

The property of second-order semidifferentiability, although appealing for its tie to second-order expansions, doesn't go far in treating nonsmoothness. Because it posits first-order semidifferentiability, it can't be present at $\bar{x}$ unless f is finite and continuous around $\bar{x}$. Thus, it can't be relevant to the analysis of points $\bar{x} \in \operatorname{bdry}(\operatorname{dom} f)$, so it can't lead anywhere in constrained optimization. But *even for functions that are finite and first-order semidifferentiable everywhere*, second-order semidifferentiability can be elusive. The class of functions given by the pointwise max of finitely many $\mathcal{C}^2$ functions reveals the nature of the difficulty.

13.10 Example (lack of second-order semidifferentiability of max functions). *Let $f = \max\{f_1,\ldots,f_m\}$ for functions f_i of class $\mathcal{C}^2$ on $\mathbb{R}^n$. At any $\bar{x}$, f is (once) semidifferentiable; one has for the index set $I(\bar{x}) := \{i \mid f_i(\bar{x}) = f(\bar{x})\}$ that*

$$df(\bar{x})(w) = \max_{i\in I(\bar{x})} df_i(\bar{x})(w) = \max_{i\in I(\bar{x})} \langle \nabla f_i(\bar{x}), w\rangle.$$

Whenever f is twice semidifferentiable at $\bar{x}$, one must further have for the index set $I'(\bar{x},w) := \{i \in I(\bar{x}) \mid df_i(\bar{x})(w) = df(\bar{x})(w)\}$ that

$$d^2 f(\bar{x})(w) = \max_{i \in I'(\bar{x},w)} d^2 f_i(\bar{x})(w) = \max_{i \in I'(\bar{x},w)} \langle w, \nabla^2 f_i(\bar{x}) w \rangle.$$

Thus, f can only be twice semidifferentiable at $\bar{x}$ in circumstances where the latter expression is continuous in w despite the dependence of $I'(\bar{x},w)$ on w. This fails in simple cases like $f(x) = \max\{|x+a|^2, 1\}$ at $\bar{x}=0$, with $|a|=1$.

Detail. The first-order fact is already known from 7.28 and 8.31. Around $\bar{x}$ one has $f(x) = \max_{i \in I(\bar{x})} f_i(x)$ and, for any w as long as τ is sufficiently small,

$$\begin{aligned} \Delta_\tau^2 f(\bar{x})(w) &= \frac{\big[\max_{i \in I(\bar{x})} f_i(\bar{x}+\tau w)\big] - f(\bar{x}) - \tau df(\bar{x})(w)}{\frac{1}{2}\tau^2} \\ &= \max_{i \in I(\bar{x})} \Big\{ \Delta_\tau^2 f_i(\bar{x})(w) - \frac{2}{\tau}\big[df(\bar{x})(w) - \langle \nabla f_i(\bar{x}), w \rangle\big] \Big\}. \end{aligned}$$

Here $df(\bar{x}) - \langle \nabla f_i(\bar{x}), w \rangle \geq 0$, and this inequality is strict whenever $i \notin I'(\bar{x},w)$. If f is twice semidifferentiable at $\bar{x}$, we must in particular have $d^2 f(\bar{x})(w) = \lim_{\tau \searrow 0} \Delta_\tau^2 f(\bar{x})(w)$. This limit is the maximum of $d^2 f_i(\bar{x})(w)$ over $i \in I'(\bar{x},w)$, as claimed. But by 13.7, $d^2 f(\bar{x})(w)$ has to be continuous relative to w when f is twice semidifferentiable at $\bar{x}$.

In the case described at the end, where $f = \max\{f_1, f_2\}$ for $f_1(x) = |x+a|^2$ and $f_2(x) \equiv 1$, one has $I(\bar{x}) = \{1,2\}$, $df(\bar{x})(w) = \max\{0, 2\langle a, w \rangle\}$, and

$$I'(\bar{x},w) = \begin{cases} \{1\} & \text{when } \langle a,w \rangle > 0, \\ \{1,2\} & \text{when } \langle a,w \rangle = 0, \\ \{2\} & \text{when } \langle a,w \rangle < 0. \end{cases}$$

Then $d^2 f(\bar{x})(w) = 2|w|^2$ when $\langle a,w \rangle > 0$, but $d^2 f(\bar{x})(w) = 0$ when $\langle a,w \rangle < 0$, so $d^2 f(\bar{x})$ isn't continuous at any $w \neq 0$ for which $\langle a,w \rangle = 0$. □

We'll see in 13.15, on the other hand, that max functions f of the type in this example are always twice *epi*-differentiable at $\bar{x}$ for every $v \in \partial f(\bar{x})$.

C. Calculus Rules

A key to establishing such facts will be the study of modes of variation around $\bar{x}$ that take into account not only the direction w from which $\bar{x}$ is approached, but also the way that w itself is approached. To this end we investigate 'arcs' $\xi : [0,\varepsilon] \to \mathbb{R}^n$ for which $\xi'_+(0)$ and $\xi''_+(0)$ both exist.

13.11 Definition (second-order tangent sets and parabolic derivability). *For $\bar{x} \in C \subset \mathbb{R}^n$, the second-order tangent set to C at $\bar{x}$ for a vector $w \in T_C(\bar{x})$ is*

$$T_C^2(\bar{x} \,|\, w) := \limsup_{\tau \searrow 0} \frac{C - \bar{x} - \tau w}{\frac{1}{2}\tau^2}.$$

The set C is said to be parabolically derivable at $\bar{x}$ for w if this outer limit is nonempty and achieved as a full limit, or in other words, if for each $z \in T_C^2(\bar{x} \,|\, w)$

there exists, for some $\varepsilon > 0$,

$$\xi : [0, \varepsilon] \mapsto C \;\; \textit{with} \;\; \xi(0) = \bar{x}, \; \xi'_+(0) = w, \; \xi''_+(0) = z. \tag{13(11)}$$

In general, the vectors z characterized by 13(11) are the ones belonging to the 'lim inf' of $[C - \bar{x} - \tau w]/\frac{1}{2}\tau^2$ as $\tau \searrow 0$ instead of the 'lim sup'. Without parabolic derivability there would be vectors $z \in T_C^2(\bar{x} \,|\, w)$ that aren't obtainable from an 'arc' ξ in 13(11) but merely from sequences $\tau^\nu \searrow 0$ and $z^\nu \to z$ with $\bar{x} + \tau^\nu w + \frac{1}{2}\tau^{\nu\,2} z^\nu \in C$.

The set $T_C^2(\bar{x} \,|\, w)$ needn't be a cone, and it might be empty. An example is $C = \{(x_1, x_2) \in I\!R^2 \,|\, x_1^4 - x_2^3 = 0\}$ at $\bar{x}$ for $w = (1, 0)$. Then there aren't any *bounded* sequences of vectors z^ν such that $\bar{x} + \tau^\nu w + \frac{1}{2}\tau^{\nu\,2} z^\nu \in C$ for some choice of $\tau^\nu \searrow 0$. In particular, C isn't parabolically derivable at $\bar{x}$ for w.

13.12 Proposition (properties of second-order tangents). *For any $\bar{x} \in C$ and $w \in T_C(\bar{x})$, the set $T_C^2(\bar{x} \,|\, w)$ is closed. If C is Clarke regular at $\bar{x}$ (as when C is closed and convex, in particular), one has*

$$T_C^2(\bar{x} \,|\, w) + u \subset T_C^2(\bar{x} \,|\, w) \quad \textit{for all} \;\; u \in T_C(\bar{x}), \tag{13(12)}$$

so that $T_C^2(\bar{x} \,|\, w)$, if nonempty, must be unbounded (apart from the degenerate case where $T_C(\bar{x}) = \{0\}$, i.e., $\bar{x}$ is an isolated point of C and $w = 0$).

If C is polyhedral, then C is parabolically derivable at $\bar{x}$ for every vector $w \in T_C(\bar{x})$, with

$$T_C^2(\bar{x} \,|\, w) = T_{T_C(\bar{x})}(w) = T_C(\bar{x}) + I\!R w. \tag{13(13)}$$

Proof. The closedness of $T_C^2(\bar{x} \,|\, w)$ is evident from its definition as an outer limit. The assumption of Clarke regularity ensures that $T_C(\bar{x})$ agrees with the regular tangent cone $\widehat{T}_C(\bar{x})$ (see 6.20). Our task for 13(12), therefore, is to show for any $z \in T_C^2(\bar{x} \,|\, w)$ and any $u \in \widehat{T}_C(\bar{x})$ that $z + u \in T_C^2(\bar{x} \,|\, w)$. From the definition of $\widehat{T}_C(\bar{x})$ in 6.25, there exists

$$\omega(x, \tau) \in [C - x]/\tau \;\; \text{with} \;\; \omega(x, \tau) \to u \;\; \text{as} \;\; \tau \searrow 0, \; x \xrightarrow[C]{} \bar{x}.$$

On the other hand, we have sequences $\tau^\nu \searrow 0$ and $z^\nu \to z$ such that the points $x^\nu = \bar{x} + \tau^\nu w + \frac{1}{2}\tau^{\nu 2} z^\nu$ lie in C. Let $u^\nu = \omega(x^\nu, \bar{\tau}^\nu)$ for $\bar{\tau}^\nu = \frac{1}{2}\tau^{\nu\,2}$. Then $x^\nu + \frac{1}{2}\tau^{\nu 2} u^\nu \in C$ with $u^\nu \to u$; in other words, $\bar{x} + \tau w + \frac{1}{2}\tau^{\nu 2}(z^\nu + u^\nu) \in C$ with $z^\nu + u^\nu \to z + u$. Hence $z + u \in T_C^2(\bar{x} \,|\, w)$ as claimed.

When C is polyhedral, it has a representation as the set of all x satisfying a system of finitely many linear inequalities, say $\langle a_i, x \rangle \le c_i$ for $i \in I$. Let $I(\bar{x})$ give the indices of the constraints that are active at $\bar{x}$. Then $T_C(\bar{x})$ consists of the vectors w such that $\langle a_i, w \rangle \le 0$ for $i \in I(\bar{x})$. Fixing w now, let $I(\bar{x}, w)$ consist of the indices such that $\langle a_i, w \rangle = 0$. Then for $K = T_C(\bar{x})$ the same reasoning expresses $T_K(w)$ as the set of all z such that $\langle a_i, z \rangle \le 0$ for $i \in I(\bar{x}, w)$. It's easy to see that any vector $z \in T_C^2(\bar{x} \,|\, w)$ must satisfy these inequalities, but also that, if they are satisfied, one has $\langle a_i, \, \bar{x} + \tau w + \frac{1}{2}\tau^2 z \rangle \le c_i$

for all $i \in I$ as long as $\tau > 0$ is sufficiently small. Thus, these inequalities on z describe $T_C^2(\bar{x} \,|\, w)$ too, and C is parabolically derivable at $\bar{x}$ for w. □

We'll need to analyze the images of arcs $\xi : [0, \varepsilon] \to I\!R^n$ with $\xi(0) = \bar{x}$ under mappings $F : I\!R^n \to I\!R^m$. If $\xi'_+(0)$ and $\xi''_+(0)$ exist and F is of class $\mathcal{C}^2$, it's elementary for $\omega(\tau) = F\big(\xi(\tau)\big)$ that $\omega'_+(0)$ and $\omega''_+(0)$ exist with

$$\omega'_+(0) = \nabla F(\bar{x})\xi'_+(0), \qquad \omega''_+(0) = \xi'_+(0)\nabla^2 F(\bar{x})\xi'_+(0) + \nabla F(\bar{x})\xi''_+(0).$$

Here we utilize the notation that

$$\begin{cases} w\nabla^2 F(\bar{x})w \ \text{ is the unique vector in } I\!R^m \text{ such that} \\ \big\langle w, \nabla^2(yF)(\bar{x})w \big\rangle = \big\langle y, w\nabla^2 F(\bar{x})w \big\rangle \ \text{ for all } \ w \in I\!R^n,\ y \in I\!R^m. \end{cases} \tag{13(14)}$$

In terms of $F = (f_1, \ldots, f_m)$, one has the expression

$$w\nabla^2 F(\bar{x})w = \big(\langle w, \nabla^2 f_1(\bar{x})w\rangle, \ldots, \langle w, \nabla^2 f_m(\bar{x})w\rangle\big).$$

13.13 Proposition (chain rule for second-order tangents). *Suppose that $C = \{x \,|\, F(x) \in D\}$ for a $\mathcal{C}^2$ mapping $F : I\!R^n \to I\!R^m$ and a closed set $D \subset I\!R^m$. Let $\bar{x} \in C$ satisfy the constraint qualification that the only vector $y \in N_D(F(\bar{x}))$ with $\nabla F(\bar{x})^* y = 0$ is $y = 0$. Then*

$$\left.\begin{array}{r} w \in T_C(\bar{x}) \\ z \in T_C^2(\bar{x} \,|\, w) \end{array}\right\} \iff \begin{cases} \nabla F(\bar{x})w \in T_D\big(F(\bar{x})\big) \\ \nabla F(\bar{x})z \in T_D^2\big(F(\bar{x}) \,|\, \nabla F(\bar{x})w\big) - w\nabla^2 F(\bar{x})w. \end{cases} \tag{13(15)}$$

Furthermore, C is parabolically derivable at $\bar{x}$ for w when D is Clarke regular at $F(\bar{x})$ and parabolically derivable at $F(\bar{x})$ for $\nabla F(\bar{x})w$. That is the case in particular when D is a polyhedral set, and then

$$T_D^2\big(F(\bar{x}) \,\big|\, \nabla F(\bar{x})w\big) = T_K\big(\nabla F(\bar{x})w\big) \ \text{ for } \ K = T_D(\bar{x}). \tag{13(16)}$$

Proof. We have $F(\bar{x}) \in D$ by assumption. The constraint qualification ensures that $w \in T_C(\bar{x})$ if and only if $\nabla F(\bar{x})w \in T_D(F(\bar{x}))$; cf. 6.31. Fix any such w and, for arbitrary $z' \in I\!R^n$, let

$$\Delta_\tau^2 F(\bar{x})(w \,|\, z') := \frac{F(\bar{x} + \tau w + \frac{1}{2}\tau^2 z') - F(\bar{x}) - \tau\nabla F(\bar{x})w}{\frac{1}{2}\tau^2},$$

noting that $\Delta_\tau^2 F(\bar{x})(w \,|\, z') \to w\nabla^2 F(\bar{x})w + \nabla F(\bar{x})z$ as $\tau \searrow 0$, $z' \to z$.

If $z \in T_C^2(\bar{x} \,|\, w)$, there exist $\tau^\nu \searrow 0$ and $z^\nu \to z$ with $\bar{x} + \tau^\nu w + \frac{1}{2}\tau^{\nu 2} z^\nu \in C$, i.e., $F(\bar{x} + \tau^\nu w + \frac{1}{2}\tau^{\nu 2} z^\nu) \in D$. The vectors $\Delta_{\tau^\nu}^2 F(\bar{x})(w \,|\, z^\nu)$ belong then to the set $[D - F(\bar{x}) - \tau^\nu \nabla F(\bar{x})w] / \frac{1}{2}\tau^{\nu 2}$, and in the limit we obtain

$$w\nabla^2 F(\bar{x})w + \nabla F(\bar{x})z \ \in \ T_D^2\big(F(\bar{x}) \,|\, \nabla F(\bar{x})w\big). \tag{13(17)}$$

This establishes the '⇒' half of the second-order equivalence in 13(15).

In order to establish the '⇐' half of the second-order equivalence in 13(15), we invoke the metric regularity property in Example 9.44: the constraint qualification provides the existence of $\kappa \in I\!R_+$ and $V \in \mathcal{N}(\bar{x})$ such that $d(x, C) \leq \kappa d(F(x), D)$ when $x \in V$. Applying this to $x = \bar{x} + \tau w + \frac{1}{2}\tau^2 z$ for any $z \in I\!R^n$, we see that

$$d\big(\bar{x} + \tau w + \tfrac{1}{2}\tau^2 z, C\big) \leq \kappa\, d\big(F\big(\bar{x} + \tau w + \tfrac{1}{2}\tau^2 z\big), D\big)$$

for $\tau > 0$ sufficiently small. This inequality can be written equivalently as

$$d\Big(z, \frac{C - \bar{x} - \tau w}{\frac{1}{2}\tau^2}\Big) \;\leq\; \kappa\, d\Big(\Delta_\tau^2 F(\bar{x})(w \,|\, z), \frac{D - F(\bar{x}) - \tau \nabla F(\bar{x}) w}{\frac{1}{2}\tau^2}\Big). \qquad 13(18)$$

Under the assumption that 13(17) holds, there's a sequence $\tau^\nu \searrow 0$ such that the right side of 13(18) goes to 0. Then the left side of 13(18) must go to 0 too, implying that $z \in T_C^2(\bar{x} \,|\, w)$.

The arguments provided so far are relevant not only to sequences. They show actually that the sets $[C - \bar{x} - \tau w]/\frac{1}{2}\tau^2$ converge to $T_C^2(\bar{x} \,|\, w)$ whenever the sets $[D - F(\bar{x}) - \tau \nabla F(\bar{x}) w]/\frac{1}{2}\tau^2$ converge to $T_D^2\big(F(\bar{x}) \,|\, \nabla F(\bar{x}) w\big)$. This property comes close to telling us that parabolic derivability of D at $F(\bar{x})$ for $\nabla F(\bar{x}) w$ implies parabolic derivability of C at $\bar{x}$ for w, but for that conclusion we must also verify that nonemptiness of $T_D^2\big(F(\bar{x}) \,|\, \nabla F(\bar{x}) w\big)$ implies nonemptiness of $T_C^2(\bar{x} \,|\, w)$, or in other words the existence of at least one vector z for which 13(17) holds.

For this we appeal once more to the constraint qualification, but write it in one of the equivalent dual forms in 6.39, as is possible under the additional assumption that D is Clarke regular at $F(\bar{x})$: $T_D(F(\bar{x})) + \nabla F(\bar{x}) I\!R^n = I\!R^m$. (Here we make use also of the fact that $T_D(F(\bar{x})) = \widehat{T}_D(F(\bar{x}))$ when D is regular at $F(\bar{x})$; cf. 6.29.) The nonemptiness of $T_D^2\big(F(\bar{x}) \,|\, \nabla F(\bar{x}) w\big)$ implies through relation 13(12) of Proposition 13.12 the existence of a vector a such that $T_D(F(\bar{x})) + a \subset T_D^2\big(F(\bar{x}) \,|\, \nabla F(\bar{x}) w\big)$. Then

$$T_D^2\big(F(\bar{x}) \,|\, \nabla F(\bar{x}) w\big) + \nabla F(\bar{x}) I\!R^n \;=\; I\!R^m.$$

The set on the left must therefore contain the vector $w \nabla^2 F(\bar{x}) w \in I\!R^m$, in particular. Hence 13(17) must hold for some z, as claimed.

The assertions for the special case of a polyhedral set D merely appeal to the result at the end of Proposition 13.12, applied to D. □

13.14 Theorem (chain rule for second subderivatives). *Let $f = g \circ F$ for a $\mathcal{C}^2$ mapping $F : I\!R^n \to I\!R^m$ and a proper, lsc function $g : I\!R^m \to \overline{I\!R}$. Let $\bar{x}$ be a point such that g is finite and regular at $F(\bar{x})$ and satisfies the constraint qualification*

$$y \in \partial^\infty g\big(F(\bar{x})\big),\;\; \nabla(yF)(\bar{x}) = 0 \quad \Longrightarrow \quad y = 0,$$

which is known to imply in particular that f is regular at $\bar{x}$ with

$$df(\bar{x})(w) = dg(F(\bar{x}))(\nabla F(\bar{x}) w), \quad \partial f(\bar{x}) = \big\{ \nabla(yF)(\bar{x}) \,\big|\, y \in \partial g(F(\bar{x})) \big\}.$$

Then for any $v \in \partial f(\bar{x})$ *the set*

$$Y(\bar{x}, v) := \{ y \in \partial g(F(\bar{x})) \mid \nabla(yF)(\bar{x}) = v \}$$

is compact as well as convex and nonempty, and for any $w \in I\!R^n$ *one has the inequality,*

$$d^2 f(\bar{x} \,|\, v)(w) \ge \sup_{y \in Y(\bar{x}, v)} \Big\{ d^2 g(F(\bar{x}) \,|\, y)(\nabla F(\bar{x}) w) + \big\langle w, \nabla^2 (yF)(\bar{x}) w \big\rangle \Big\}.$$

This holds with equality when g *is fully amenable at* $F(\bar{x})$*, and in that case* f *is properly twice epi-differentiable at* $\bar{x}$ *for every* $v \in \partial f(\bar{x})$*, with*

$$\operatorname{dom} d^2 f(\bar{x} \,|\, v) = \{ w \mid df(\bar{x})(w) = \langle v, w \rangle \} = N_{\partial f(\bar{x})}(v).$$

In particular these conclusions hold when g *is convex and piecewise linear-quadratic. Then more specifically*

$$d^2 f(\bar{x} \,|\, v)(w) = d^2 \bar{f}(\bar{x} \,|\, v)(w) + \max_{y \in Y(\bar{x}, v)} \big\langle w, \nabla^2 (yF)(\bar{x}) w \big\rangle \qquad 13(19)$$

for the function $\bar{f}(x) := g(F(\bar{x}) + \nabla F(\bar{x})[x - \bar{x}])$*. This function is convex and piecewise linear-quadratic with* $\bar{f}(\bar{x}) = f(\bar{x})$*,* $d\bar{f}(\bar{x}) = df(\bar{x})$*,* $\partial \bar{f}(\bar{x}) = \partial f(\bar{x})$*, and for* v *in the latter set one has*

$$\begin{aligned} d^2 \bar{f}(\bar{x} \,|\, v)(w) &= d^2 g(F(\bar{x}) \,|\, y)(\nabla F(\bar{x}) w) \quad \text{for any } y \in Y(\bar{x}, v) \\ &= \begin{cases} d^2 g(F(\bar{x}))(\nabla F(\bar{x}) w) & \text{if } dg(F(\bar{x}))(\nabla F(\bar{x}) w) = \langle v, w \rangle, \\ \infty & \text{otherwise.} \end{cases} \end{aligned}$$

Proof. The first-order facts in the preamble come from Theorem 10.6; recall that $\nabla(yF)(\bar{x})$ can be written as $\nabla F(\bar{x})^* y$. The convexity of $\partial g(F(\bar{x}))$ and therefore of $Y(\bar{x}, v)$ is due to the regularity of g at $F(\bar{x})$; cf. 8.6, 8.11. We have $Y(\bar{x}, v)^\infty \subset \{ y \in \partial^\infty g(F(\bar{x})) \mid \nabla F(\bar{x})^* y = 0 \} = \{0\}$, so $Y(\bar{x}, v)$ is compact. For any $y \in Y(\bar{x}, v)$ we can write

$$\begin{aligned} \Delta_\tau^2 f(\bar{x} \,|\, v)(w) &= \frac{g(F(\bar{x} + \tau w)) - g(F(\bar{x})) - \tau \langle v, w \rangle}{\tau^2/2} \\ &= \frac{g(F(\bar{x}) + \tau \Delta_\tau F(\bar{x})(w)) - g(F(\bar{x})) - \tau \langle \nabla F(\bar{x})^* y, w \rangle}{\tau^2/2} \\ &= \frac{g(F(\bar{x}) + \tau \Delta_\tau F(\bar{x})(w)) - g(F(\bar{x})) - \tau \langle y, \Delta_\tau F(\bar{x})(w) \rangle}{\tau^2/2} \\ &\qquad + \frac{\tau \langle y, \Delta_\tau F(\bar{x})(w) \rangle - \tau \langle y, \nabla F(\bar{x}) w \rangle}{\tau^2/2} \\ &= \Delta_\tau^2 g(F(\bar{x}) \,|\, y)(\Delta_\tau F(\bar{x})(w)) + \Delta_\tau^2 (yF)(w). \end{aligned}$$

Because $\Delta_\tau F(\bar{x})(w') \to \nabla F(\bar{x}) w$ when $\tau \searrow 0$ and $w' \to w$, while at the same time $\Delta_\tau^2 (yF)(\bar{x})(w') \to d^2 (yF)(w) = \langle w, \nabla^2 (yF)(\bar{x}) w \rangle$, it follows that

$$\begin{aligned} d^2 f(\bar{x}\,|\,v)(w) &= \liminf_{\substack{\tau \searrow 0 \\ w' \to w}} \Delta_\tau^2 f(\bar{x}\,|\,v)(w') \\ &= \liminf_{\substack{\tau \searrow 0 \\ w' \to w}} \left\{ \Delta_\tau^2 g(F(\bar{x})\,|\,y)\big(\Delta_\tau F(\bar{x})(w')\big) + \Delta_\tau^2 (yF)(w') \right\} \\ &\geq d^2 g\big(F(\bar{x})\,|\,y\big)\big(\nabla F(\bar{x})w\big) + \big\langle w, \nabla^2 (yF)(\bar{x})w \big\rangle. \end{aligned}$$

The general inequality in the theorem is thus established.

Next we suppose that g is convex and piecewise linear-quadratic. (The case of g just fully amenable will come last.) We know then from 13.9 that

$$\begin{aligned} &d^2 g\big(F(\bar{x})\,|\,y\big)\big(\nabla F(\bar{x})w\big) \\ &\qquad = \begin{cases} d^2 g\big(F(\bar{x})\big)\big(\nabla F(\bar{x})w\big) & \text{if } dg\big(F(\bar{x})\big)\big(\nabla F(\bar{x})w\big) = \langle y, \nabla F(\bar{x})w \rangle, \\ \infty & \text{otherwise.} \end{cases} \end{aligned}$$

But $\langle y, \nabla F(\bar{x})w \rangle = \langle v, w \rangle$ when $y \in Y(\bar{x}, v)$, whereas $dg(F(\bar{x}))(\nabla F(\bar{x})w) = df(\bar{x})(w)$. On the other hand, we have

$$d^2 g\big(F(\bar{x})\,|\,y\big)\big(\nabla F(\bar{x})w\big) = d^2 \bar{f}(\bar{x}\,|\,v)(w) \;\text{ for all }\; y \in Y(\bar{x}, v)$$

for $\bar{f}(x) := g(F(\bar{x}) + \nabla F(\bar{x})[x - \bar{x}])$. The function $\bar{f}$ is convex and piecewise linear-quadratic by 10.22(b), which also yields the claimed equality between the subgradients and first-order subderivatives of $\bar{f}$ and those of f. Then the characterization of $d^2 \bar{f}(\bar{x}\,|\,v)(w)$ at the end of the theorem follows from 13.9. In particular, $d^2 \bar{f}(\bar{x}\,|\,v)(w)$ is finite when $dg(F(\bar{x}))(\nabla F(\bar{x})w) = \langle v, w \rangle$.

We deduce from this and the general inequality for $d^2 f(\bar{x}\,|\,v)(w)$ already established that the '$\geq$' half of the equation in 13(19) is valid, with the right side being finite when $dg(F(\bar{x}))(\nabla F(\bar{x})w) = \langle v, w \rangle$ but ∞ otherwise. Here 'max' can replace 'sup' because of the compactness of $Y(\bar{x}, v)$, which follows from the constraint qualification since $Y(\bar{x}, v)^\infty = \big\{ y \in \partial g(F(\bar{x}))^\infty \,\big|\, \nabla F(\bar{x})^* y = 0 \big\}$ and $\partial g(F(\bar{x}))^\infty = \partial^\infty g(F(\bar{x}))$. We see also that equality holds automatically in 13(19) when $dg(F(\bar{x}))(\nabla F(\bar{x})w) \geq \langle v, w \rangle$ (both sides being ∞), and that the domain formula for $d^2 f(\bar{x}\,|\,v)$ will follow once 13(19) has been verified for other w as well. (The inclusion '$\subset$' in the domain formula is known from 13.5.)

Working toward the '$\leq$' half of 13(19) and the claim of second-order epi-differentiability, we can fix w satisfying $dg(F(\bar{x}))(\nabla F(\bar{x})w) = \langle v, w \rangle$ and view the 'max' term in 13(19) as the optimal value in the problem

$$\begin{aligned} &\text{maximize } \langle b, y \rangle \text{ subject to } y \in Y,\ A^* y - c = 0, \text{ where} \\ &b = w \nabla^2 F(\bar{x}) w,\ Y = \partial g(F(\bar{x})),\ A = -\nabla F(\bar{x}),\ c = -v. \end{aligned} \qquad 13(20)$$

(The notation $w \nabla^2 F(\bar{x}) w$ corresponds to 13(14).) This problem fits the framework of extended linear-quadratic programming in 11.43, inasmuch as the set $Y = \partial g\big(F(\bar{x})\big)$ is polyhedral (by virtue of g being piecewise linear-quadratic; cf. 10.21). It's dual within that framework to the problem

$$\text{minimize } \langle c, z \rangle + h(b - Az) \text{ over all } z \in I\!R^n$$

in which h is the support function $\sigma_Y = dg\big(F(\bar{x})\big)$. This comes out as:

$$\text{minimize } dg\big(F(\bar{x})\big)\big(w\nabla^2 F(\bar{x})w + \nabla F(\bar{x})z\big) - \langle v, z\rangle \text{ in } z. \qquad 13(21)$$

Theorem 11.42 says that the optimal value in 13(20) agrees with the one value in 13(21), which is attained. Hence there exist $\bar{z} \in \mathbb{R}^n$ and $\bar{y} \in Y(\bar{x}, v)$ with

$$dg\big(F(\bar{x})\big)\big(w\nabla^2 F(\bar{x})w + \nabla F(\bar{x})\bar{z}\big) - \langle v, \bar{z}\rangle \;=\; \big\langle w, \nabla^2(\bar{y}F)(\bar{x})w\big\rangle. \qquad 13(22)$$

Let $D = \operatorname{dom} g$ and $K = T_D\big(F(\bar{x})\big)$, so $K = \operatorname{dom} h$ in the notation $h = dg(F(\bar{x}))$, which we keep for simplicity. Because $h(\nabla F(\bar{x})w) = df(\bar{x})(w) = \langle v, w\rangle$, while the value $h\big(w\nabla^2 F(\bar{x})w - \nabla F(\bar{x})\bar{z}\big)$ is finite by 13(22), we have

$$\nabla F(\bar{x})w \in K, \qquad w\nabla^2 F(\bar{x})w + \nabla F(\bar{x})\bar{z} \in K.$$

Because K is a convex cone, we know that $K \subset T_K(u)$ for any vector $u \in K$, in particular $u = \nabla F(\bar{x})w$; hence $z \in T_C^2(\bar{x}\,|\,w)$ by 13.13. Through the parabolic derivability in 13.13, we obtain the existence of an arc

$$\xi : [0, \varepsilon] \mapsto \mathbb{R}^n \text{ with } \xi(0) = \bar{x},\ \xi'_+(0) = w,\ \xi''_+(0) = \bar{z},\ F(\xi(\tau)) \in D$$

(for some $\varepsilon > 0$). In working now with such an arc ξ, it will be useful to write

$$\xi(\tau) = \xi(0) + \tau w_\tau \text{ for } w_\tau := \frac{\xi(\tau) - \xi(0)}{\tau} \to w.$$

To reach our current goal (of validating for g convex and piecewise linear-quadratic the '$\leq$' part of 13(19) and the claim that f is twice epi-differentiable at $\bar{x}$ for v), it will suffice to demonstrate for our chosen w that

$$\begin{aligned}&\limsup_{\tau \searrow 0} \Delta_\tau^2 f(\bar{x}\,|\,v)(w_\tau)\\ &\qquad\leq\; d^2 g\big(F(\bar{x})\big)\big(\nabla F(\bar{x})w\big) + \big\langle w, \nabla^2(\bar{y}F)(\bar{x})w\big\rangle.\end{aligned} \qquad 13(23)$$

Let $\omega(\tau) = F\big(\xi(\tau)\big)$, so that $\omega(0) = F(\bar{x})$, $\omega'_+(0) = \nabla F(\bar{x})w$ and $\omega''_+(0) = w\nabla^2 F(\bar{x})w + \nabla F(\bar{x})\bar{z}$. Then

$$\omega(\tau) = \omega(0) + \tau\big[\omega'_+(0) + \tfrac{1}{2}\tau\zeta_\tau\big] \text{ for } \zeta_\tau := \frac{\omega(\tau) - \omega(0) - \tau\omega'_+(0)}{\frac{1}{2}\tau^2} \to \omega''_+(0).$$

In this scheme we can apply to g the expansion in 13(9), getting

$$\begin{aligned} g\big(\omega(\tau)\big) \;=\;& g\big(\omega(0)\big) + \tau dg\big(\omega(0)\big)\big(\omega'_+(0) + \tfrac{1}{2}\tau\zeta_\tau\big)\\ &+ \tfrac{1}{2}\tau^2 d^2 g\big(\omega(0)\big)\big(\omega'_+(0) + \tfrac{1}{2}\tau\zeta_\tau\big).\end{aligned}$$

Here it should be remembered that the function $dg\big(\omega(0)\big) = h$ is positively homogeneous of degree 1, while the function $d^2 g\big(\omega(0)\big)$ is positively homogeneous of degree 2; both are continuous on the cone $K = \operatorname{dom} h$. Furthermore, $h\big(\omega'_+(0)\big) = \langle v, w\rangle = \langle v, \xi'_+(0)\rangle$. We are enabled by this to calculate

$$\begin{aligned}\Delta_\tau^2 f(\bar{x}\,|\,v)(w_\tau) &= \frac{g\big(\omega(\tau)\big) - g\big(\omega(0)\big) - \tau\langle v, w_\tau\rangle}{\frac{1}{2}\tau^2}\\ &= d^2 g\big(\omega(0)\big)\big(\omega'_+(0) + \tfrac{1}{2}\tau\zeta_\tau\big) + \frac{h\big(\omega'_+(0) + \frac{1}{2}\tau\zeta_\tau\big) - h\big(\omega'_+(0)\big)}{\frac{1}{2}\tau}\\ &\qquad + \frac{\big\langle v, \xi'_+(0)\big\rangle - \big\langle v, \tau^{-1}[\xi(\tau) - \xi(0)]\big\rangle}{\frac{1}{2}\tau}\end{aligned}$$

and in consequence that

$$\begin{aligned}\lim_{\tau \searrow 0} \Delta_\tau^2 f(\bar{x}\,|\,v)(w_\tau) &= \lim_{\tau \searrow 0} d^2 g\big(\omega(0)\big)\big(\omega'_+(0) + \tfrac{1}{2}\tau\zeta_\tau\big)\\ &\qquad + \lim_{\tau \searrow 0} \frac{h\big(\omega'_+(0) + \frac{1}{2}\tau\zeta_\tau\big) - h\big(\omega'_+(0)\big)}{\frac{1}{2}\tau}\\ &\qquad - \Big\langle v,\ \lim_{\tau \searrow 0} \frac{\xi(\tau) - \xi(0) - \tau\xi'_+(0)}{\frac{1}{2}\tau^2}\Big\rangle\\ &= d^2 g\big(\omega(0)\big)\big(\omega'_+(0)\big) + dh\big(\omega'_+(0)\big)\big(\omega''_+(0)\big) - \big\langle v, \bar{z}\big\rangle.\end{aligned}$$

Here $d^2 g\big(\omega(0)\big)\big(\omega'_+(0)\big)$ is the term $d^2 g\big(F(\bar{x})\big)\big(\nabla F(\bar{x})w\big)$ sought in 13(23), while $\langle v, \bar{z}\rangle$ satisfies 13(22), where $w\nabla^2 F(\bar{x})w + \nabla F(\bar{x})\bar{z} = \omega''_+(0)$. Thus,

$$\begin{aligned}\lim_{\tau \searrow 0} \Delta_\tau^2 f(\bar{x}\,|\,v)(w_\tau) &= d^2 g\big(F(\bar{x})\big)\big(\nabla F(\bar{x})w\big) + \big\langle w, \nabla^2(\bar{y}F)(\bar{x})w\big\rangle\\ &\qquad + dh\big(\omega'_+(0)\big)\big(\omega''_+(0)\big) - h\big(\omega''_+(0)\big).\end{aligned}$$

All we need now, to get 13(23), is to check that $dh\big(\omega'_+(0)\big)\big(\omega''_+(0)\big) \le h\big(\omega''_+(0)\big)$. This rests on the fact that, because h is sublinear, one has $dh(u)(u') \le h(u')$ for any $u \in \operatorname{dom} h$, $u' \in \mathbb{R}^n$. Indeed, $\Delta_\tau h(u)(u') = \tau^{-1}[h(u+\tau u') - h(u)]$ with $h(u + \tau u') \le h(u) + \tau h(u')$, so that $\Delta_\tau h(u)(u') \le h(u')$ for all $\tau > 0$ and, in the limit as $\tau \searrow 0$, $dh(u)(u') \le h(u')$.

Finally, we take up the case where g isn't necessarily convex and piecewise linear-quadratic, but fully amenable at $F(\bar{x})$. This means (by Definition 10.23) that, at least locally around $F(\bar{x})$, we have $g = g_0 \circ G$ for a $\mathcal{C}^2$ mapping $G : \mathbb{R}^m \to \mathbb{R}^{m_0}$ and a convex, piecewise linear-quadratic function $g_0 : \mathbb{R}^{m_0} \to \overline{\mathbb{R}}$ satisfying for $\bar{u} := F(\bar{x})$ and $D_0 := \operatorname{dom} g_0$ the constraint qualification

$$y_0 \in N_{D_0}\big(G(\bar{u})\big),\ \nabla(y_0 G)(\bar{u}) = 0 \implies y_0 = 0. \qquad 13(24)$$

The convexity of g_0 makes $N_{D_0}\big(G(\bar{u})\big) = \partial^\infty g_0\big(G(\bar{u})\big)$ (cf. 8.12). Our result for the piecewise linear-quadratic case can thus be invoked for $g = g_0 \circ G$. We have $y \in \partial g(\bar{u})$ if and only if $\nabla(y_0 G)(\bar{u}) = y$ for some $y_0 \in \partial g_0\big(G(\bar{u})\big)$, and then

$$d^2 g(\bar{u}\,|\,y)(z) = \sup_{\substack{y_0 \in \partial g_0(G(\bar{u}))\\ \nabla(y_0 G)(\bar{u}) = y}} \Big\{ d^2 g_0(G(\bar{u})\,|\,y_0)(\nabla G(\bar{u})z) + \big\langle z, \nabla^2(y_0 G)(\bar{u})z\big\rangle \Big\}. \qquad 13(25)$$

Let $F_0 = G\circ F$, this being a $\mathcal{C}^2$ mapping such that $f = g_0\circ F_0$ and $F_0(\bar{x}) = G(\bar{u})$ as well as $\nabla F_0(\bar{x}) = \nabla G(\bar{u})\nabla F(\bar{x})$. Since $\nabla(y_0 F_0)(\bar{x}) = \nabla F_0(\bar{x})^* y_0 = \nabla F(\bar{x})^* \nabla G(\bar{u})^* y_0 = \nabla(yF)(\bar{x})$ for $y = \nabla(y_0 G)(\bar{u})$, we see from combining 13(24) with the original constraint qualification relative to g and F that

$$y_0 \in \partial^\infty g_0\big(F_0(\bar{x})\big),\ \ \nabla(y_0 F_0)(\bar{x}) = 0 \quad\Longrightarrow\quad y_0 = 0.$$

Our result for the piecewise linear-quadratic case can thus be applied to $f = g_0\circ F_0$ to get for $v \in \partial f(\bar{x})$ that f is twice epi-differentiable at $\bar{x}$ for v with

$$\begin{aligned} &d^2 f(\bar{x}\,|\,v)(w) = \\ &\qquad \sup_{\substack{y_0\in\partial g_0(F_0(\bar{x}))\\ \nabla(y_0F_0)(\bar{x})=v}} \Big\{ d^2 g_0(F_0(\bar{x})\,|\,y_0)(\nabla F_0(\bar{x})w) + \big\langle w, \nabla^2(y_0F_0)(\bar{x})w\big\rangle \Big\}. \end{aligned} \qquad 13(26)$$

In comparison, the targeted formula is

$$d^2 f(\bar{x}\,|\,v)(w) = \sup_{\substack{y\in\partial g(F(\bar{x}))\\ \nabla(yF)(\bar{x})=v}} \Big\{ d^2 g(F(\bar{x})\,|\,y)(\nabla F(\bar{x})w) + \big\langle w, \nabla^2(yF)(\bar{x})w\big\rangle \Big\}.$$

We'll demonstrate that this formula follows from 13(25), 13(26), and the underlying relationships. First, by specializing 13(25) to the case of $z = \nabla F(\bar{x})w$ we obtain

$$\begin{aligned} d^2 g_0(G(\bar{u})\,|\,y_0)\big(\nabla G(\bar{u})z\big) &= d^2 g_0\big(F_0(\bar{x})\,|\,y_0\big)\big(\nabla F_0(\bar{x})w\big), \\ \big\langle z, \nabla^2(y_0G)(\bar{u})z\big\rangle &= \big\langle \nabla F(\bar{x})w, \nabla^2(y_0G)(F(\bar{x}))\nabla F(\bar{x})w\big\rangle. \end{aligned}$$

On the other hand, the maximization in 13(26) can be realized in two stages:

$$\sup_{\substack{y_0\in\partial g_0(F_0(\bar{x}))\\ \nabla(y_0F_0)(\bar{x})=v}} \Big\{ \qquad \Big\} = \sup_{\substack{y\in\partial g(F(\bar{x}))\\ \nabla(yF)(\bar{x})=v}} \Big\{ \sup_{\substack{y_0\in\partial g_0(G(\bar{u}))\\ \nabla(y_0G)(\bar{u})=y}} \Big\{ \qquad \Big\}\Big\}.$$

It's apparent then that to get the formula we want we have only to prove that $\big\langle w, \nabla^2(y_0F_0)(\bar{x})w\big\rangle = \big\langle \nabla F(\bar{x})w, \nabla^2(y_0G)(F(\bar{x}))\nabla F(\bar{x})w\big\rangle + \big\langle w, \nabla^2(yF)(\bar{x})w\big\rangle$ when $\nabla(y_0G)(\bar{u}) = y$. We derive this equation by setting $\omega(\tau) = F(\bar{x}+\tau w)$ and computing

$$\begin{aligned} \big\langle w, \nabla^2(y_0F_0)(\bar{x})w\big\rangle &= \frac{d^2}{d\tau^2}(y_0F_0)(\bar{x}+\tau w)\Big|_{\tau=0} = \frac{d^2}{d\tau^2}(y_0G)\big(\omega(\tau)\big)\Big|_{\tau=0} \\ &= \big\langle \omega'(0), \nabla^2(y_0G)\big(\omega(0)\big)\omega'(0)\big\rangle + \big\langle \nabla(y_0G)\big(\omega(0)\big), \omega''(0)\big\rangle \\ &= \big\langle \nabla F(\bar{x})w, \nabla^2(y_0G)(\bar{u})\nabla F(\bar{x})w\big\rangle + \big\langle y, [w\nabla^2 F(\bar{x})w]\big\rangle. \end{aligned}$$

The last term is $\big\langle w, \nabla^2(yF)(\bar{x})w\big\rangle$ by definition, so we're done. □

13.15 Corollary (second-order epi-differentiability from full amenability). *When $f : \mathbb{R}^n \to \overline{\mathbb{R}}$ is fully amenable at $\bar{x}$, there exists $V \in \mathcal{N}(\bar{x})$ such that, for all $x \in V \cap \operatorname{dom} f$ and $v \in \partial f(x)$, f is properly twice epi-differentiable at x for v.*

Proof. At $\bar{x}$, the conclusion comes from the fact that fully amenable functions have, by definition (in 10.23), a representation $f = g \circ F$ for convex, piecewise linear-quadratic g meeting the constraint qualification in the theorem. Alternatively, this can be regarded as the case of the theorem in which $F = I$, $f = g$. Full amenability at $\bar{x}$ entails full amenability at all $x \in \operatorname{dom} f$ in some neighborhood of $\bar{x}$ (cf. 10.25(b)), and this justifies the broader claim. □

The class of fully amenable functions is rich, as observed in 10.24. The case of 'max' functions falls into this class in particular and highlights the differences between second-order epi-differentiability and second-order semidifferentiability through comparison with 13.10.

13.16 Example (second-order epi-differentiability of max functions). *Suppose $f = \max\{f_1, \ldots, f_m\}$ for functions f_i of class $\mathcal{C}^2$ on $I\!R^n$. Then f is properly twice epi-differentiable at every point $\bar{x}$ for every subgradient $v \in \partial f(\bar{x})$. Moreover in terms of the index sets*

$$I(\bar{x}) := \{i \mid f_i(\bar{x}) = f(\bar{x})\},$$
$$I'(\bar{x}, w) := \{i \in I(\bar{x}) \mid \langle \nabla f_i(\bar{x}), w \rangle = df(\bar{x})(w)\},$$

one has $v \in \partial f(\bar{x})$ if and only if $v \in \operatorname{con}\{\nabla f_i(\bar{x}) \mid i \in I(\bar{x})\}$, and then

$$d^2 f(\bar{x} \mid v)(w) = \delta_{K(\bar{x},v)}(w) + \max\Big\{ \sum_{i \in I(\bar{x})} y_i \langle w, \nabla^2 f_i(\bar{x}) w \rangle \;\Big|\; y_i \geq 0,\ \sum_{i \in I(\bar{x})} y_i = 1,\ \sum_{i \in I(\bar{x})} y_i \nabla f_i(\bar{x}) = v \Big\}$$

with $w \in K(\bar{x}, v)$ if and only if $v \in \operatorname{con}\{\nabla f_i(\bar{x}) \mid i \in I'(\bar{x}, w)\}$.

Detail. Let $F(x) = (f_1(x), \ldots, f_m(x))$. Then $f = g \circ F$ for $g = \operatorname{vecmax}$, which is convex and piecewise linear. The hypothesis of Theorem 13.14 is satisfied by this representation, and the epi-derivative formula there specializes to the one here by way of the subdifferential properties of g in 8.26. □

Example 13.16 gives occasion to note that in the final formula of Theorem 13.14, for g piecewise linear-quadratic, the term $d^2 g(F(\bar{x}))(\nabla F(\bar{x}) w)$ will drop out whenever g is just piecewise linear. This holds when g is the 'vecmax' function of 1.30, and also in the case described next.

13.17 Exercise (indicators and geometry). *Let $C = \{x \in X \mid F(x) \in D\}$ for a $\mathcal{C}^2$ mapping $F : I\!R^n \to I\!R^m$ and polyhedral sets $X \subset I\!R^n$ and $D \subset I\!R^n$. Let $\bar{x} \in C$ satisfy the constraint qualification:*

$$y \in N_D(F(\bar{x})),\ \nabla(yF)(\bar{x}) = 0 \implies y = 0.$$

Consider any vector $v \in N_C(\bar{x})$, such vectors being characterized by the existence of $y \in N_D(F(\bar{x}))$ with $v - \nabla(yF)(\bar{x}) \in N_X(\bar{x})$; denote the set of such y by $Y(\bar{x}, v)$. Then δ_C is properly twice epi-differentiable at $\bar{x}$ for v, and

$$d^2(\delta_C)(\bar{x}\,|\,v)(w) = \delta_{K(\bar{x},v)}(w) + \max_{y\in Y(\bar{x},v)} \langle w, \nabla^2(yF)(\bar{x})w\rangle$$

for $K(\bar{x},v) = \{w \in T_X(\bar{x}) \,|\, \nabla F(\bar{x})w \in T_D(F(\bar{x})),\ w \perp v\}$, *which is the normal cone to* $N_C(\bar{x})$ *at* v *and is polyhedral.*

Guide. When $X = I\!R^n$, one can get this by invoking the special formula at the end of Theorem 13.14 in the case of $g = \delta_D$, piecewise linear. For the general case, one can instead pass to the mapping $F_0 : x \mapsto (x, F(x))$ and use the representation $\delta_C = \delta_{D_0} \circ F_0$ with $D_0 = X \times D$. □

What's especially to be noted in 13.17 is that indicator functions δ_C can have nontrivial second epi-derivatives even though their first epi-derivatives are invariably 0, when finite: $d(\delta_C)(\bar{x})(w) = \delta_T(w)$ for $T = T_C(\bar{x})$. This phenomenon reflects the possible boundary curvature of C, which may be the source of second-order effects although not first-order. When C is polyhedral, the second epi-derivatives do have to be 0 when finite, as seen directly from 13.9 or as the case of 13.17 where F is affine and therefore $\nabla(yF)(\bar{x}) = 0$. Then one merely has $d^2(\delta_C)(\bar{x}\,|\,v) = \delta_{K(\bar{x},v)}$.

A concrete illustration of the situation in 13.17 may aid in understanding how the curvature of C is captured in the formula for the second epi-derivatives $d^2(\delta_C)(\bar{x})(w)$. Consider the half-disk

$$C = \{x = (x_1, x_2) \in I\!R^2 \,|\, x_1^2 + x_2^2 \le 2,\ x_1 - x_2 \le 0\}$$

and its corner point $\bar{x} = (1,1)$, as in Figure 13–2. Place this in the context of 13.17 by taking $X = I\!R^2$, $D = I\!R^2_-$ and $F = (f_1, f_2)$ for $f_1(x_1,x_2) = x_1^2 + x_2^2 - 2$ and $f_2(x_1,x_2) = x_1 - x_2$. (An alternative would be to define X by the linear constraint and use F only for the nonlinear constraint, in which case $D = I\!R^1_-$; the results for epi-derivatives of δ_C come out the same either way, of course.) We have $F(\bar{x}) = (0,0)$, $N_D(F(\bar{x})) = I\!R^2_+$, $T_D(F(\bar{x})) = I\!R^2_-$, and in terms of vectors $w = (w_1, w_2)$, $v = (v_1, v_2)$ and $y = (y_1, y_2)$ also

$$\begin{aligned}
\nabla F(\bar{x})w &= (2w_1 + 2w_2,\ w_1 - w_2),\\
T_C(\bar{x}) &= \{w \,|\, |w_2| \le -w_1\},\\
\nabla(yF)(\bar{x}) &= (2y_1 + y_2,\ 2y_1 - y_2) = y_1(2,2) + y_2(1,-1),\\
N_C(\bar{x}) &= \{v \,|\, |v_2| \le v_1\} = \{y_1(2,2) + y_2(1,-1) \,|\, y_1 \ge 0,\ y_2 \ge 0\},\\
Y(\bar{x},v) &= \{y \,|\, y_1 \ge 0,\ y_2 \ge 0,\ 2y_1 + y_2 = v_1,\ 2y_1 - y_2 = v_2\},\\
\nabla^2(yF)(\bar{x}) &= 2y_1 I, \quad \langle w, \nabla^2(yF)(\bar{x})w\rangle = 2y_1(w_1^2 + w_2^2).
\end{aligned}$$

Focusing on the particular normal vector $\bar{v} = (2,2) \in N_C(\bar{x})$, we find that

$$\begin{aligned}
Y(\bar{x},\bar{v}) &= \{\bar{y}\} \ \text{ for } \ \bar{y} = (1,0),\\
K(\bar{x},\bar{v}) &= \{w = (w_1,w_2) \,|\, -w_2 = w_1 \le 0\} = \{\lambda(-1,1) \,|\, \lambda \ge 0\},\\
d^2(\delta_C)(\bar{x}\,|\,\bar{v})(w) &= \begin{cases} 2\lambda^2 & \text{when } w = \lambda(-1,1) \text{ with } \lambda \ge 0,\\ \infty & \text{otherwise.}\end{cases}
\end{aligned}$$

Looking at the normal vector $\tilde{v} = (2, -2)$ instead, we get

$$\begin{aligned} Y(\bar{x}, \tilde{v}) &= \{\tilde{y}\} \text{ for } \tilde{y} = (0,1), \\ K(\bar{x}, \tilde{v}) &= \{w = (w_1, w_2) \mid w_2 = w_1 \leq 0\} = \{\lambda(-1,-1) \mid \lambda \geq 0\}, \\ d^2(\delta_C)(\bar{x} \,|\, \tilde{v})(w) &= \begin{cases} 0 & \text{when } w = \lambda(-1,-1) \text{ with } \lambda \geq 0, \\ \infty & \text{otherwise.} \end{cases} \end{aligned}$$

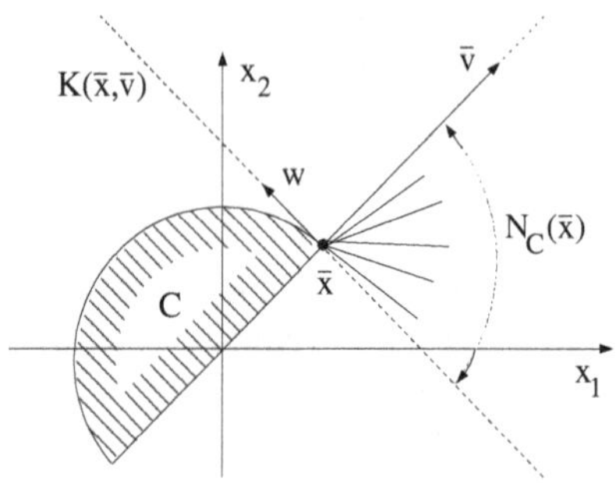

Fig. 13–2. Set curvature expressed through second epi-derivatives.

The difference between the two cases reflects the fact that the first detects a curved part of the boundary of C, while the second detects a straight part. A third normal vector worth inspecting is $\hat{v} = (2, 0)$, which yields

$$\begin{aligned} Y(\bar{x}, \hat{v}) &= \{\hat{y}\} \text{ for } \hat{y} = (1,1), \qquad K(\bar{x}, \hat{v}) = \{(0,0)\}, \\ d^2(\delta_C)(\bar{x} \,|\, \hat{v})(w) &= \begin{cases} 0 & \text{when } w = (0,0), \\ \infty & \text{when } w \neq (0,0). \end{cases} \end{aligned}$$

The second epi-derivatives trivialize in this case because $\hat{v}$ detects no part of the boundary of C beyond the corner point $\bar{x}$ itself.

Note that the uniqueness of the y vector in all three cases comes from the linear independence of the gradients of the constraints that are active at $\bar{x}$. To have an example of a set C, defined by finitely many inequality constraints, for which the maximum in the second epi-derivative formula is taken over more than just a singleton set of vectors y, we would need to pass to a situation in which the gradients aren't linearly dependent, although still such that the constraint qualification is satisfied.

Along with the chain rule in Theorem 13.14 there are other, simpler results that can assist in determining second subderivatives and checking whether a function is twice epi-differentiable, or for that matter twice semidifferentiable in a given situation.

13.18 Exercise (addition of a twice smooth function). *Suppose $f : \mathbb{R}^n \to \overline{\mathbb{R}}$ is properly twice epi-differentiable at $\bar{x}$ for v. Let $f_0 : \mathbb{R}^n \to \mathbb{R}$ be C^2. Then $f + f_0$ is properly twice epi-differentiable at $\bar{x}$ for $v + v_0$, where $v_0 := \nabla f_0(\bar{x})$,* *and one has*

$$d^2(f+f_0)(\bar{x}\,|\,v+v_0)(w) = d^2 f(\bar{x}\,|\,v)(w) + \langle w, \nabla^2 f_0(\bar{x})w\rangle.$$

Likewise, if f is twice semidifferentiable at $\bar{x}$, the same is true for $f+f_0$, and

$$d^2(f+f_0)(\bar{x})(w) = d^2 f(\bar{x})(w) + \langle w, \nabla^2 f_0(\bar{x})w\rangle.$$

Guide. Apply the definitions directly. □

13.19 Proposition (second subderivatives of a sum of functions). *Suppose that $f = \sum_{i=1}^m f_i$ for proper, lsc functions $f_i : \mathbb{R}^n \to \overline{\mathbb{R}}$. Let $\bar{x}$ be a point of $\operatorname{dom} f = \bigcap_{i=1}^m \operatorname{dom} f_i$ at which every f_i is regular and*

$$v_i \in \partial^\infty f_i(\bar{x}),\ \sum\nolimits_{i=1}^m v_i = 0 \implies v_i = 0 \text{ for all } i,$$

so that f is regular at $\bar{x}$ with $df(\bar{x}) = \sum_{i=1}^m df_i(\bar{x})$ and $\partial f(\bar{x}) = \sum_{i=1}^m \partial f_i(\bar{x})$. Then for any $v \in \partial f(\bar{x})$ one has

$$d^2 f(\bar{x}\,|\,v)(w) \geq \sup\Big\{ \sum\nolimits_{i=1}^m d^2 f_i(\bar{x}\,|\,v_i)(w) \,\Big|\, v_i \in \partial f_i(\bar{x}),\ \sum\nolimits_{i=1}^m v_i = v \Big\}.$$

This holds with equality when every f_i is fully amenable at $\bar{x}$, and then f, being itself fully amenable at $\bar{x}$, is twice epi-differentiable at $\bar{x}$ for v.

Proof. The first-order facts come from 10.9. We can think of f as $g \circ F$ for $g : (\mathbb{R}^n)^m \to \overline{\mathbb{R}}$ and $F : \mathbb{R}^n \to (\mathbb{R}^n)^m$ defined by

$$g(x_1,\ldots,x_m) = f_1(x_1) + \cdots + f_m(x_m), \qquad F(x) = (x,\ldots,x).$$

The linear mapping F has $\nabla F(x) \equiv [I,\cdots,I]^*$. It's elementary that

$$\Delta_\tau^2 g(x_1,\ldots,x_m\,|\,v_1,\ldots,v_m)(w_1,\ldots,w_m) = \sum\nolimits_{i=1}^m \Delta_\tau^2 f_i(x_i\,|\,v_i)(w_i),$$

and that the same relation then holds in the limit for second subderivatives. Also, g is fully amenable when every one of the functions f_i is fully amenable; this is covered as a special case of 10.26(a), which also provides then the full amenability of f. The results then follow from Theorem 13.14 as applied to this simple framework. □

D. Convex Functions and Duality

Second subderivatives of convex functions f have some special properties which can be viewed as extending the fact that, in the twice smooth case, the matrices $\nabla^2 f(x)$ are positive-semidefinite.

13.20 Proposition (second subderivatives of convex functions). *For any proper, convex function $f : \mathbb{R}^n \to \overline{\mathbb{R}}$, any $\bar{x} \in \operatorname{dom} f$ and any $\bar{v} \in \partial f(\bar{x})$, the following properties hold.*

(a) $d^2f(\bar{x}\,|\,v) \geq 0$ *everywhere, and when* f *is twice epi-differentiable at* $\bar{x}$ *for* $\bar{v}$*, the function* $d^2f(\bar{x}\,|\,v)$ *also has to be convex.*

(b) $d^2f(\bar{x}\,|\,v) = \gamma_C^2$ *for the gauge* γ_C *of a unique closed set* $C \ni 0$*, and when* f *is twice epi-differentiable at* $\bar{x}$ *for* $\bar{v}$*, the set* C *also has to be convex.*

(c) f *is twice semidifferentiable at* $\bar{x}$ *for* v *if and only if* f *is twice epi-differentiable at* $\bar{x}$ *for* v *and* $d^2f(\bar{x}\,|\,v)(w)$ *is finite for all* w*. Then in particular,* f *must be differentiable at* $\bar{x}$ *with* $v = \nabla f(\bar{x})$*.*

Proof. For $\tau > 0$ we have $f(\bar{x} + \tau w) - f(\bar{x}) - \tau\langle v, w\rangle \geq 0$ by the basic subgradient inequality for convex functions (in 8.12). Therefore $\Delta_\tau^2 f(\bar{x}\,|\,v)(w)$ is therefore nonnegative; obviously this expression is also convex with respect to w. It follows from the limit definition in 13.3 that $d^2f(\bar{x}\,|\,v) \geq 0$ everywhere. When f is twice epi-differentiable at $\bar{x}$ for v, so that $\Delta_\tau^2 f(\bar{x}\,|\,v) \xrightarrow{e} d^2f(\bar{x}\,|\,v)$, we deduce the convexity of $d^2f(\bar{x}\,|\,v)$ from the principle that epi-limits of convex functions are convex (see 7.17). This establishes (a).

In (b) the gauge relation clearly holds for $C = \{w \mid d^2f(\bar{x}\,|\,v)(w) \leq 1\}$, since $d^2f(\bar{x}\,|\,v)$ is not just nonnegative but lsc and positively homogeneous of degree 2 (by 13.5), hence vanishes at the origin. (For background on gauge functions, see 3.50.)

The equivalence in (c) is based on the observation that convex functions converge continuously to a finite function if and only if they epi-converge to that function; this follows from 7.14 and 7.17 in conjunction with the continuity of finite convex functions (in 2.36). We know from 13.5 that finiteness of $d^2f(\bar{x}\,|\,v)$ requires having $df(\bar{x}) = \langle v, \cdot\rangle$. But that corresponds for a convex function f to differentiability at $\bar{x}$ with $\nabla f(\bar{x}) = v$; cf. 9.18, 9.14. □

In the $\mathcal{C}^2$ case with $x = \nabla f(\bar{x})$ and $d^2f(\bar{x}\,|\,v)(w) = \langle w, \nabla^2 f(\bar{x})w\rangle$, the set C in 13.20(b) is the (possibly degenerate) 'ellipsoid' $\{w \mid \langle w, \nabla^2 f(\bar{x})w\rangle \leq 1\}$.

13.21 Theorem (second-order epi-differentiability in duality). *Let* $f : \mathbb{R}^n \to \overline{\mathbb{R}}$ *be lsc, proper and convex. Let* $\bar{x} \in \operatorname{dom} f$ *and* $\bar{v} \in \partial f(\bar{x})$*, so that the conjugate convex function* f^* *has* $\bar{v} \in \operatorname{dom} f^*$ *and* $\bar{x} \in \partial f^*(\bar{v})$*. Then* f *is properly twice epi-differentiable at* $\bar{x}$ *for* $\bar{v}$ *if and only if* f^* *is properly twice epi-differentiable at* $\bar{v}$ *for* $\bar{x}$*. In that case there is the conjugacy relation*

$$\tfrac{1}{2}d^2f(\bar{x}\,|\,\bar{v}) \quad \underset{*}{\longleftrightarrow} \quad \tfrac{1}{2}d^2f^*(\bar{v}\,|\,\bar{x}).$$

Proof. The equivalence of $\bar{v} \in \partial f(\bar{x})$ with $\bar{x} \in \partial f^*(\bar{v})$ is known from 11.3 with the fact that these relations entail $f(\bar{x}) + f^*(\bar{v}) = \langle \bar{v}, \bar{x}\rangle$. Let $\varphi_0 = \frac{1}{2}d^2f(\bar{x}\,|\,\bar{v})$ and $\psi_0 = \frac{1}{2}d^2f^*(\bar{v}\,|\,\bar{x})$, and for $\tau > 0$ define

$$\begin{aligned}\varphi_\tau(w) &= \tfrac{1}{2}\Delta_\tau^2 f(\bar{x}\,|\,\bar{v})(w) = \tau^{-2}\big[f(\bar{x} + \tau w) - f(\bar{x}) - \tau\langle \bar{v}, w\rangle\big],\\ \psi_\tau(z) &= \tfrac{1}{2}\Delta_\tau^2 f^*(\bar{v}\,|\,\bar{x})(z) = \tau^{-2}\big[f^*(\bar{v} + \tau z) - f^*(\bar{v}) - \tau\langle \bar{x}, z\rangle\big].\end{aligned}$$

Then φ_τ and ψ_τ are lsc, proper and convex, with $\varphi_0 = \text{e-lim inf}_{\tau \searrow 0} \varphi_\tau$ and $\psi_0 = \text{e-lim inf}_{\tau \searrow 0} \psi_\tau$. Applying the Legendre-Fenchel transform to φ_τ, we get

$$\begin{aligned}\varphi_\tau^*(z) &= \sup_w \{\langle z,w\rangle - \varphi_\tau(w)\}\\ &= \tau^{-2}\sup_w\{\langle \tau z,\tau w\rangle - f(\bar x+\tau w) + f(\bar x) + \tau\langle \bar v, w\rangle\}\\ &= \tau^{-2}\big[\sup_w\{\langle \tau z,\tau w\rangle + \langle \bar v,\, \bar x+\tau w\rangle - f(\bar x+\tau w)\} - f^*(\bar v)\big]\\ &= \tau^{-2}\big[\sup_u\{\langle \tau z,\, u-\bar x\rangle + \langle \bar v, u\rangle - f(u)\} - f^*(\bar v)\big]\\ &= \tau^{-2}\big[\sup_u\{\langle \bar v+\tau z, u\rangle - f(u)\} - \tau\langle z,\bar x\rangle - f^*(\bar v)\big]\\ &= \tau^{-2}\big[f^*(\bar v+\tau z) - \tau\langle z,\bar x\rangle - f^*(\bar v)\big] = \psi_\tau(z).\end{aligned}$$

By the same argument, using $f^{**}=f$, we get $\psi_\tau^* = \varphi_\tau$. Having f properly twice epi-differentiable at $\bar x$ for $\bar v$ corresponds to having $\varphi_\tau \overset{e}{\to} \varphi_0$ (proper) as $\tau \searrow 0$, whereas having f^* properly twice epi-differentiable at $\bar v$ for $\bar x$ corresponds to having $\psi_\tau \overset{e}{\to} \psi_0$ (proper) as $\tau\searrow 0$. These conditions are equivalent because epi-convergence is preserved under the Legendre-Fenchel transform (cf. 11.34), which pairs proper convex functions with proper convex functions. □

In the case of Theorem 13.21 corresponding to the classical Legendre transform as described in 11.9, where both f and f^* are $\mathcal{C}^2$ and one has $\bar v = \nabla f(\bar x)$ if and only if $\bar x = \nabla f^*(\bar v)$, the conjugacy between $\frac12 d^2 f(\bar x\,|\,\bar v)$ and $\frac12 d^2 f^*(\bar v\,|\,\bar x)$ says that

$$\nabla^2 f^*(\bar v) = \nabla^2 f(\bar x)^{-1}.$$

This is evident from the basic facts in 11.10 about conjugacy of linear-quadratic functions, but it's also clear immediately from the interpretation of $\nabla^2 f(\bar x)$ and $\nabla^2 f^*(\bar v)$ as Jacobian matrices associated with the smooth mappings ∇f and ∇f^* in light of having $\nabla f^* = (\nabla f)^{-1}$.

Another interesting feature of the duality in Theorem 13.21 is that if we represent $d^2 f(\bar x\,|\,\bar v) = \gamma_C^2$ in the manner of 13.20(b) using the gauge γ_C of a closed, convex set $C \ni 0$, we get $d^2 f^*(\bar v\,|\,\bar x) = \gamma_{C^\circ}^2$ with C° the polar of C. This follows from 11.21. In the Legendre case just described, the duality between second subderivative functions translates to the polarity between two ellipsoidal sets centered at the origin and expresses a reciprocity of eigenvalues in the associated quadratic forms.

Theorem 13.21 also provides examples of twice epi-differentiable functions which, in contrast to virtually all the examples so far, don't necessarily turn out to be fully amenable functions.

13.22 Corollary (conjugates of fully amenable functions). *Suppose f has the representation $f(x) = \sup_v\{\langle v,x\rangle - \varphi(v)\}$ for a function $\varphi : \mathbb{R}^n \to \overline{\mathbb{R}}$ that is lsc, proper and convex, in which case $\partial f(x) = \operatorname{argmax}_v\{\langle v,x\rangle - \varphi(v)\}$. If $\bar v \in \partial f(\bar x)$ and φ is fully amenable at $\bar v$, then f is twice epi-differentiable at $\bar x$ for $\bar v$ with*

$$d^2 f(\bar x\,|\,\bar v)(w) = \sup_z\{2\langle w,z\rangle - d^2\varphi(v\,|\,\bar x)(z)\}.$$

Proof. This combines 13.21 with 13.15, since $f=\varphi^*$ and $\varphi = f^*$ in the situation described. □

13.23 Example (piecewise linear-quadratic penalties). *For any nonempty polyhedral set $Y \subset \mathbb{R}^m$ and symmetric, positive-semidefinite matrix $B \in \mathbb{R}^{m\times m}$*

(possibly $B = 0$), the convex, piecewise linear-quadratic function

$$\theta_{Y,B}(u) := \sup_{y \in Y} \left\{ \langle y, u \rangle - \tfrac{1}{2} \langle y, By \rangle \right\}$$

on $\mathbb{R}^m$ is properly twice epi-differentiable at every point $\bar{u} \in \operatorname{dom} \theta_{Y,B}$ for every subgradient $\bar{y} \in \partial \theta_{Y,B}(\bar{u})$. Indeed, one has

$$d^2\theta_{Y,B}(\bar{u} \,|\, \bar{y})(z) = \sup_{w \in Y'(\bar{u},\bar{y})} \left\{ 2\langle w, z \rangle - \langle w, Bw \rangle \right\}$$

for the polyhedral cone $Y'(\bar{u}, \bar{y}) = \{ w \in T_Y(\bar{y}) \,|\, w \perp \bar{u} - B\bar{y} \}$, and consequently

$$\begin{aligned} d^2\theta_{Y,B}(\bar{u} \,|\, \bar{y}) &= 2\theta_{Y'(\bar{u},\bar{y}),B} \\ \operatorname{dom} d^2\theta_{Y,B}(\bar{u} \,|\, \bar{y}) &= \{ w \in T_Y(\bar{y}) \,|\, Bw = 0,\ w \perp \bar{u} \}^*. \end{aligned}$$

When $B = 0$, the function $d^2\theta_{Y,B}(\bar{u} \,|\, \bar{y})$ is the indicator of the cone $Y'(\bar{u}, \bar{y})^*$.

Detail. The function $\theta_{Y,B}$, already explored in 11.18, is φ^* for the convex, piecewise linear-quadratic function $\varphi := \delta_Y + j_B$, where $j_B(y) = \frac{1}{2}\langle y, By \rangle$. From 13.9 we have $d^2\varphi(\bar{y} \,|\, \bar{u})(w) = \langle w, Bw \rangle + \delta_K(w)$ for the convex cone $K = \{ w \,|\, d\varphi(\bar{y})(w) = \langle \bar{u}, w \rangle \}$. But $d\varphi(\bar{y})(w) = \delta_{T_Y(\bar{y})}(w) + \langle y, Bw \rangle$, so that $K = Y'(\bar{u}, \bar{y})$. Twice epi-differentiability and the expression for $d^2\theta_{Y,B}(\bar{u} \,|\, \bar{y})$ follow now from Corollary 13.22. The formula for $\operatorname{dom} d^2\theta_{Y,B}(\bar{u} \,|\, \bar{y})$ is the one for $\operatorname{dom} \theta_{Y'(\bar{u},\bar{y}),B}$ furnished by 11.18, namely the cone $[Y'(\bar{u}, \bar{y})^\infty \cap \ker B]^*$. We have $Y'(\bar{u}, \bar{y})^\infty = Y'(\bar{u}, \bar{u})$ because Y' is a cone. But $w \in Y'(\bar{u}, \bar{y}) \cap \ker B$ if and only if $w \in T_Y(\bar{y})$, $Bw = 0$ and $\langle w, \bar{u} - B\bar{y} \rangle = 0$. The latter is the same as $\langle w, \bar{u} \rangle = 0$ when $Bw = 0$, since $\langle w, B\bar{y} \rangle = \langle Bw, \bar{y} \rangle$. □

Functions of the kind in Example 13.23 are particularly useful in the role of penalty expressions in composite formats for problems of optimization (cf. 11.18, 11.43, 11.46).

E. Second-Order Optimality

One of the prime reasons for developing a second-order theory of subdifferentiation, of course, is the derivation of second-order conditions for optimality that go beyond the various versions of Fermat's rule and their applications. For this, the investments we have made can now pay off.

13.24 Theorem (second-order conditions for optimality). *For a proper function $f : \mathbb{R}^n \to \overline{\mathbb{R}}$, consider the problem of minimizing $f(x)$ over all $x \in \mathbb{R}^n$.*

(a) *If $\bar{x}$ is locally optimal, then $0 \in \partial f(\bar{x})$ and $d^2 f(\bar{x} \,|\, 0)(w) \geq 0$ for all w.*

(b) *If $0 \in \partial f(\bar{x})$ and $d^2 f(\bar{x} \,|\, 0)(w) > 0$ for $w \neq 0$, then $\bar{x}$ is locally optimal.*

(c) *Having $0 \in \partial f(\bar{x})$ and $d^2 f(\bar{x} \,|\, 0)(w) > 0$ for all $w \neq 0$ is equivalent to having the existence $\varepsilon > 0$ and $\delta > 0$ such that*

$$f(x) \geq f(\bar{x}) + \varepsilon|x - \bar{x}|^2 \text{ when } |x - \bar{x}| \leq \delta.$$

Proof. If $\bar{x}$ is locally optimal there exists $\delta > 0$ such that $f(x) \geq f(\bar{x})$ when $|x - \bar{x}| \leq \delta$, and then not only $0 \in \widehat{\partial} f(\bar{x}) \subset \partial f(\bar{x})$ but also $\Delta_\tau^2 f(\bar{x} \,|\, 0)(w) \geq 0$ when $|\tau w| \leq \delta$, i.e., $\Delta_\tau^2 f(\bar{x} \,|\, 0) \geq 0$ on $\delta\tau^{-1}\mathbb{B}$. From the fact that $d^2 f(\bar{x} \,|\, 0) = \text{e-lim inf}_{\tau \searrow 0} \Delta_\tau^2 f(\bar{x} \,|\, 0)$ we conclude that $d^2 f(\bar{x} \,|\, 0) \geq 0$. Thus, (a) is correct. To prove (b), it's obviously enough to verify (c).

In extension of the argument just given, the ε and δ condition in (c) yields $\Delta_\tau^2 f(\bar{x} \,|\, 0)(w) \geq 2\varepsilon|w|^2$ when $w \in \delta\tau^{-1}\mathbb{B}$, hence $d^2 f(\bar{x} \,|\, 0)(w) \geq 2\varepsilon|w|^2$ for all w. Conversely, if $\Delta_\tau^2 f(\bar{x} \,|\, 0)(w) > 0$ for all $w > 0$ choose ε by

$$3\varepsilon = \min_{w \in W} d^2 f(\bar{x} \,|\, 0)(w) \text{ for } W := \{w \mid |w| = 1\},$$

where the minimum exists and is positive because $d^2 f(\bar{x} \,|\, 0)$ is lsc and W is compact (cf. 13.5, 1.9). Because $d^2 f(\bar{x} \,|\, 0) = \text{e-lim inf}_{\tau \searrow 0} \Delta_\tau^2 f(\bar{x} \,|\, 0)$, there exists for each $w \in W$ a $\delta_w > 0$ and an open neighborhood $V_w \in \mathcal{N}(w)$ such that $\Delta_\tau^2 f(\bar{x} \,|\, 0)(w') \geq 2\varepsilon$ when $w' \in N_w$ and $\tau \in (0, \delta_w)$. The open sets V_w cover the compact set W, so a finite family of them cover it as well, say for $w_1, \ldots, w_r$ with corresponding $\delta_1, \ldots, \delta_r$. Let $\delta = \min\{\delta_1, \ldots, \delta_r\}$. Then $\delta > 0$ and we have $\Delta_\tau^2 f(\bar{x} \,|\, 0)(w') \geq 2\varepsilon$ for all $w' \in W$ when $\tau \in (0, \delta)$. Literally this means that $f(\bar{x} + \tau w') - f(\bar{x}) - \tau\langle 0, w' \rangle \geq \varepsilon\tau^2$ when $0 < \tau < \delta$ and $|w'| = 1$, and we deduce immediately that $f(x) \geq f(\bar{x}) + \varepsilon|x - \bar{x}|^2$ when $|x - \bar{x}| \leq \delta$. □

In bringing the conditions in Theorem 13.24 down to more detail, one can add particular structure to f and appeal to the formulas for second-order subderivatives that have been obtained. Observe that the basic inequalities for $d^2 f(\bar{x} \,|\, 0)$ that are available from the chain rule of 13.14 and the addition rule in 13.19, for instance, can operate effectively in getting second-order sufficient conditions for optimality, but they don't say anything about necessary conditions. For that one needs to focus on cases where the subderivative inequalities are manifested as equations. The results about twice epi-differentiability are then the main tool, and fully amenable functions come to the fore.

13.25 Example (second-order optimality with smooth constraints). *Consider the problem of minimizing f_0 over $C = \{x \in X \mid F(x) \in D\}$ for $f_0 : \mathbb{R}^n \to \mathbb{R}$ and $F : \mathbb{R}^n \to \mathbb{R}^m$ of class $\mathcal{C}^2$ and sets $X \subset \mathbb{R}^n$ and $D \subset \mathbb{R}^m$ that are polyhedral. Let $\bar{x} \in C$ satisfy the constraint qualification that*

$$y \in N_D(F(\bar{x})),\ -\nabla F(\bar{x})^* y \in N_X(\bar{x}) \implies y = 0.$$

Define

$$Y(\bar{x}) = \Big\{ y \in N_D(F(\bar{x})) \;\Big|\; -[\nabla f_0(\bar{x}) + \nabla F(\bar{x})^* y] \in N_X(\bar{x}) \Big\},$$
$$K(\bar{x}) = \Big\{ w \in T_X(\bar{x}) \;\Big|\; \nabla F(\bar{x}) w \in T_D(F(\bar{x})),\ w \perp \nabla f_0(\bar{x}) \Big\}.$$

Then $Y(\bar{x})$ is a polyhedral, compact set, $K(\bar{x})$ is a polyhedral cone, and the *possible optimality of $\bar{x}$ can be identified as follows.*

(a) (necessity) *If $\bar{x}$ is locally optimal, then $Y(\bar{x}) \neq \emptyset$ and*

$$\max_{y \in Y(\bar{x})} \langle w, \nabla^2_{xx}[f_0 + yF](\bar{x})w \rangle \geq 0 \quad \text{for all } w \in K(\bar{x}).$$

(b) (sufficiency) *Conversely, if these conditions hold and the inequality is strict when $w \neq 0$, then $\bar{x}$ is locally optimal; in fact there exist $\varepsilon > 0$ and $\delta > 0$ such that $f_0(x) \geq f_0(\bar{x}) + \varepsilon |x - \bar{x}|^2$ for all $x \in C \cap I\!B(\bar{x}, \delta)$.*

Detail. We apply Theorem 13.24 to $f = f_0 + \delta_C$, determining second subderivatives from 13.17 with the help of 13.18. The constraint qualification and the assumption that X and D are polyhedral make f be twice epi-differentiable. Note that the condition $Y(\bar{x}) \neq \emptyset$ refers to the existence of a vector y meeting the prescriptions of the Lagrange multiplier rule in 6.15. □

13.26 Exercise (second-order optimality with penalties). *Consider the problem*

$$\text{minimize } f_0(x) + \theta_{Y,B}\big(f_1(x), \ldots, f_m(x)\big) \text{ over all } x \in X$$

for $\mathcal{C}^2$ functions f_i on $I\!R^n$, a polyhedral set $X \subset I\!R^n$, and a function $\theta_{Y,B}$ of the kind described in 11.18 and 13.23. Viewing this as the problem of minimizing $f = \delta_X + f_0 + \theta_{Y,B} \circ F$ over all of $I\!R^n$, where $F = (f_1, \ldots, f_m)$, let $\bar{x} \in \operatorname{dom} f$ be a point at which X is fully amenable and

$$y \in Y^\infty, \ By = 0, \ -\nabla F(\bar{x})^* y \in N_X(\bar{x}) \implies y = 0.$$

Setting $L(x, y) := f_0(x) + \langle y, F(x) \rangle - \frac{1}{2}\langle y, By \rangle$ on $X \times Y$, define

$$Y(\bar{x}) := \Big\{ \bar{y} \Big| -\nabla_x L(\bar{x}, \bar{y}) \in N_X(\bar{x}), \ \nabla_y L(\bar{x}, \bar{y}) \in N_Y(\bar{y}) \Big\}.$$

Then $Y(\bar{x})$ is a polyhedral, compact set, and for $\hat{y} \in Y(\bar{x})$ the polyhedral cones

$$X'(\bar{x}) := T_X(\bar{x}) \cap \nabla_x L(\bar{x}, \hat{y})^\perp, \qquad Y'(\bar{x}) := T_Y(\hat{y}) \cap \nabla_y L(\bar{x}, \hat{y})^\perp,$$

are independent of the particular choice of $\hat{y}$. The possible optimality of $\bar{x}$ can be identified then as follows.

(a) (necessity) *If $\bar{x}$ is locally optimal, then $Y(\bar{x}) \neq \emptyset$ and*

$$\max_{y \in Y(\bar{x})} \langle w, \nabla^2_{xx} L(\bar{x}, y) w \rangle + 2\theta_{Y'(\bar{x}),B}\big(\nabla F(\bar{x}) w\big) \geq 0 \quad \text{for all } w \in X'(\bar{x}).$$

(b) (sufficiency) *Conversely, if these conditions hold and the inequality is strict when $w \neq 0$, then $\bar{x}$ is locally optimal; in fact there exist $\varepsilon > 0$ and $\delta > 0$ such that $f(x) \geq f(\bar{x}) + \varepsilon |x - \bar{x}|^2$ for all $x \in I\!B(\bar{x}, \delta)$.*

Guide. Apply 13.24 in the light of the calculus in 11.18, 13.14, 13.19, and the amenability facts in 10.24–10.26. See that $\delta_{X'(\bar{x})}(w) + 2\theta_{Y'(\bar{x}),B}\big(\nabla F(\bar{x}) w\big)$ is unaffected by the particular choice of $\hat{y}$ in the definition of $X'(\bar{x})$ and $Y'(\bar{x})$. □

An interesting thing about the second-order necessary condition in 13.26(a) is that it corresponds to $w = 0$ being an optimal solution to the subproblem

$$\text{minimize } g_0(w) + \theta_{Y'(\bar{x}),B}\big(g_1(w), \ldots, g_m(w)\big) \text{ over all } w \in X'(\bar{x}),$$

where $g_i(w) = \langle \nabla f_i(\bar{x}), w \rangle$ and $g_0(w) = \frac{1}{2}\max_{y \in Y(\bar{x})} \langle w, \nabla^2_{xx} L(\bar{x}, y) w \rangle$. The second-order sufficient condition in 13.26(b) corresponds to $w = 0$ being the only optimal solution. This subproblem has the same pattern as the original problem, except that g_0 need not be smooth. When the multiplier set $Y(\bar{x})$ is a singleton, however, g_0 is quadratic and the subproblem falls in the category of extended linear-quadratic programming that was described in 11.43.

A supplementary discussion of second-order optimality conditions in terms of 'parabolic subderivatives' (introduced in 13.59) will center on 13.66.

F. Prox-Regularity

It's time now for a closer look at the classical idea of obtaining second derivatives by differentiating first derivatives. How might this fit into the emerging picture of generalized second-order differentiation, focused on limits of second-order difference quotient functions? Theorem 13.2 points the way: the answer lies in the study of generalized first-order differentiation of subgradient mappings. The issue of when such mappings are proto-differentiable, say, is important not only in this respect but also in its own right.

A clue to the kind of connection between first and second derivatives that one might hope to develop can be found in the case of functions f that are twice differentiable in the extended sense; cf. Example 13.8. The function $h = d^2 f(\bar{x})$, the same as $d^2 f(\bar{x} \,|\, \bar{v})$ for $\bar{v} = \nabla f(\bar{x})$, is quadratic in this situation with matrix $\nabla^2 f(\bar{x})$. At the same time, the mapping ∇f is differentiable at $\bar{x}$ relative to its domain of definition, with associated derivative mapping $w \mapsto \nabla^2 f(\bar{x}) w$. The double role of the Hessian $\nabla^2 f(\bar{x})$ can be expressed by

$$\nabla[d^2 f(\bar{x})](w) = 2D[\nabla f](\bar{x})(w). \tag{13(27)}$$

This formula can be taken as a potential model for something much more general, especially in the light of Theorem 13.2, a result suggesting that ∇f on the right side might be replaceable by ∂f.

When f isn't twice differentiable at $\bar{x}$ or for that matter even once differentiable at $\bar{x}$, we do still have the functions $d^2 f(\bar{x} \,|\, v)$ corresponding to subgradients $v \in \partial f(\bar{x})$ and can contemplate substituting for the gradient $\nabla[d^2 f(\bar{x})](w)$ on the left side of 13(27) the subgradient set $\partial[d^2 f(\bar{x} \,|\, v)](w)$. Can the relation

$$\partial[d^2 f(\bar{x} \,|\, v)] = 2D[\partial f](\bar{x} \,|\, v),$$

with second-order epi-differentiability on the left and proto-differentiability on the right, possibly be realized as an appropriate generalization of 13(27) in major cases? The answer turns out to be yes.

The confirmation of this fact, ultimately in Theorem 13.40, will have the by-product of unveiling circumstances in which a subgradient mapping is proto-differentiable. This will illuminate what happens when optimality conditions are perturbed, since such conditions typically involve subgradient mappings. In the process, we'll make strong use of another kind of regularity property.

13.27 Definition (prox-regularity of functions). *A function $f : \mathbb{R}^n \to \overline{\mathbb{R}}$ is prox-regular at $\bar{x}$ for $\bar{v}$ if f is finite and locally lsc at $\bar{x}$ with $\bar{v} \in \partial f(\bar{x})$, and there exist $\varepsilon > 0$ and $\rho \geq 0$ such that*

$$f(x') \geq f(x) + \langle v, x' - x \rangle - \frac{\rho}{2}|x' - x|^2 \text{ for all } x' \in \mathbb{B}(\bar{x}, \varepsilon)$$
$$\text{when } v \in \partial f(x),\ |v - \bar{v}| < \varepsilon,\ |x - \bar{x}| < \varepsilon,\ f(x) < f(\bar{x}) + \varepsilon.$$

When this holds for all $\bar{v} \in \partial f(\bar{x})$, f is said to be prox-regular at $\bar{x}$.

Prox-regularity implies for all $(x, v) \in \operatorname{gph} \partial f$ near enough to $(\bar{x}, \bar{v})$, and with $f(x)$ near enough to $f(\bar{x})$, that v is a *proximal* subgradient of f at $\bar{x}$; cf. Definition 8.45. It goes beyond this, however, in requiring the constant ρ in that definition, and the neighborhood in it as well, to exhibit a local uniformity. Although proximal subgradients are regular subgradients in particular, prox-regularity doesn't necessarily imply subdifferential regularity at $\bar{x}$, since it may only involve some of the vectors $v \in \partial f(\bar{x})$ near to $\bar{v}$, not every $v \in \partial f(\bar{x})$.

The restriction to $f(x) < f(\bar{x}) + \varepsilon$ in Definition 13.27 makes sure that the subgradients under consideration correspond to normals to $\operatorname{epi} f$ at points $(x, f(x))$ sufficiently near to $(\bar{x}, f(\bar{x}))$ and thus truly reflect only the local geometry of $\operatorname{epi} f$ at $(\bar{x}, f(\bar{x}))$. This is superfluous when nearness of (x, v) to $(\bar{x}, \bar{v})$ in $\operatorname{gph} \partial f$ automatically entails nearness of $f(x)$ to $f(\bar{x})$, a property common to many types of functions, which we formalize as follows.

13.28 Definition (subdifferential continuity). *A function $f : \mathbb{R}^n \to \overline{\mathbb{R}}$ is called subdifferentially continuous at $\bar{x}$ for $\bar{v}$ if $\bar{v} \in \partial f(\bar{x})$ and, whenever $(x^\nu, v^\nu) \to (\bar{x}, \bar{v})$ with $v^\nu \in \partial f(x^\nu)$, one has $f(x^\nu) \to f(\bar{x})$. If this holds for all $\bar{v} \in \partial f(\bar{x})$, f is said to be subdifferentially continuous at $\bar{x}$.*

13.29 Exercise (characterization of subdifferential continuity). *For a function $f : \mathbb{R}^n \to \overline{\mathbb{R}}$ to be subdifferentially continuous at $\bar{x}$, it is necessary and sufficient that f be continuous at $\bar{x}$ relative to each set containing $\bar{x}$ that has the form $(\partial f)^{-1}(B)$ with B compact. Thus in particular, f is subdifferentially continuous at any point $\bar{x}$ where f is continuous relative to $\operatorname{dom} f$.*

An example of how a function f can fail to be subdifferentially continuous at a point $\bar{x} \in \operatorname{dom} f$ is displayed in Figure 13–3. Define $f : \mathbb{R} \to \mathbb{R}$ by $f(x) = 1$ for $x < 1$ but $f(x) = x - 2$ for $x \geq 1$. Obviously f is lsc everywhere. It's easy to see too that f is prox-regular everywhere. The graph of ∂f has a peculiarity at $(\bar{x}, \bar{v}) = (1, 0)$, though. The horizontal part of it shown as branching to the left from $(\bar{x}, \bar{v})$ comes from the constancy of f on $(-\infty, 1]$. It has no genesis

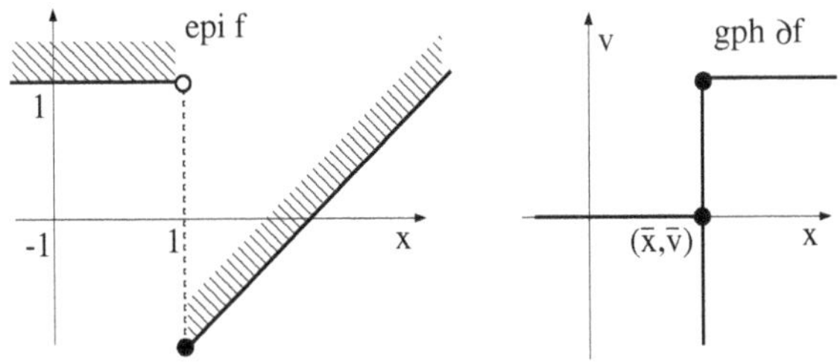

Fig. 13–3. A prox-regular function that lacks subdifferential continuity.

in the behavior of epi f close to $(\bar{x}, f(\bar{x})) = (1, -1)$. As $(x^\nu, v^\nu) \to (\bar{x}, \bar{v})$ in gph ∂f along this branch we have $f(x^\nu) \equiv 1$, so $f(x^\nu) \not\to f(\bar{x}) = -1$.

For terminology to help further with such distinctions, let's start from the idea that a *localization* of ∂f around $(\bar{x}, \bar{v})$ is a mapping whose graph is obtained by intersecting gph ∂f with some neighborhood of $(\bar{x}, \bar{v})$. Let's augment this by saying that an *f-attentive localization* of ∂f around $(\bar{x}, \bar{v})$ is a localization in which the pairs (x, v) are additionally restricted to have $f(x)$ within some neighborhood of $f(\bar{x})$. Of course, as long as f is locally lsc at $\bar{x}$, only upper bounds on $f(x) - f(\bar{x})$ need be of concern, since lower bounds are induced automatically. Anyway, ordinary localization intersects gph ∂f with a neighborhood of $(\bar{x}, \bar{v})$ in the ordinary topology on $\mathbb{R}^n \times \mathbb{R}^n$, while f-attentive localization takes the neighborhood instead in the sense of the product topology coming from the f-attentive topology on $\mathbb{R}^n$ in the x component but the ordinary topology on $\mathbb{R}^n$ in the v component.

In these terms, prox-regularity refers to a uniformity property possessed by some f-attentive localization of ∂f around $(\bar{x}, \bar{v})$. Subdifferential continuity describes the situation where f-attentive localization of ∂f amounts to the same thing as ordinary localization. In Figure 13–3, ordinary localization of ∂f around $(\bar{x}, \bar{v})$ would include part of the side branch, whereas f-attentive localization, when small enough, would eliminate the side branch.

13.30 Example (prox-regularity from convexity). *A proper, lsc, convex function f is prox-regular and subdifferentially continuous at every point of* dom f.

Detail. In this case the constant ρ can always be taken to be 0. In the circumstances of Definition 13.28 one has $f(\bar{x}) \geq f(x^\nu) + \langle v^\nu, \bar{x} - x^\nu \rangle$ with $(x^\nu, v^\nu) \to (\bar{x}, \bar{v})$, and this implies that $\limsup_\nu f(x^\nu) \leq f(\bar{x})$. Since $\liminf_\nu f(x^\nu) \geq f(\bar{x})$ because f is lsc, we have $f(x^\nu) \to f(\bar{x})$. □

13.31 Exercise (prox-regularity of sets). *For a set $C \subset \mathbb{R}^n$ and any point $\bar{x} \in C$, the indicator function δ_C is subdifferentially continuous at $\bar{x}$, and the following properties with respect to a vector $\bar{v}$ are equivalent:*

(a) *δ_C is prox-regular at $\bar{x}$ for $\bar{v}$;*

(b) *C is locally closed at $\bar{x}$ with $\bar{v} \in N_C(\bar{x})$, and there exist $\varepsilon > 0$ and $\rho \geq 0$ such that $\langle v, x' - x \rangle \leq \frac{1}{2}\rho |x' - x|^2$ for all $x' \in C \cap \mathbb{B}(\bar{x}, \varepsilon)$ when $v \in N_C(x)$, $|v - \bar{v}| < \varepsilon$ and $|x - \bar{x}| < \varepsilon$.*

A set C is called *prox-regular at $\bar{x}$ for $\bar{v}$* when the equivalent properties in 13.31 hold. It is called *prox-regular at $\bar{x}$* when this is true for all $\bar{v} \in N_C(\bar{x})$. Closed, convex sets C are everywhere prox-regular, in particular; this is apparent from combining 13.30 with 13.31. But prox-regularity prevails also for the much larger class of strongly amenable sets (cf. 10.23, 10.24) by virtue of the next proposition, which furnishes besides a prime source of examples of prox-regular functions.

Fig. 13–4. A nonconvex set that is everywhere prox-regular.

13.32 Proposition (prox-regularity from amenability). *Let $f : \mathbb{R}^n \to \overline{\mathbb{R}}$ be strongly amenable at $\bar{x}$. Then f is prox-regular and subdifferentially continuous at $\bar{x}$ and indeed at all points x in a neighborhood of $\bar{x}$ relative to* $\operatorname{dom} f$.

Proof. Consider a representation $f = g \circ F$ on a neighborhood V of $\bar{x}$ as provided by the definition of strong amenability in 10.23. This representation makes f be lsc relative to V, implying that f is locally lsc at $\bar{x}$ as required by Definition 13.27. Let $\bar{v} \in \partial f(\bar{x})$. For all x near $\bar{x}$ we have $\partial f(x) = \nabla F(x)^* \partial g\big(F(x)\big)$ (cf. 10.25). Thus, for $x \in V$ the vectors $v \in \partial f(x)$ are the ones of the form $v = \nabla F(x)^* y$ for some $y \in \partial g\big(F(x)\big)$. Moreover, there exists $\varepsilon > 0$ such that the set of all y having this property with respect to x and v satisfying $|x - \bar{x}| < \varepsilon$ and $|v - \bar{v}| < \varepsilon$ is bounded in norm, say by η (for otherwise a contradiction to the constraint qualification in the definition of amenability can be obtained); we suppose ε is small enough that $|x - \bar{x}| < \varepsilon$ implies $x \in V$. Because F is of class $\mathcal{C}^2$ in the stipulations for strong amenability, there exists $\rho > 0$ such that

$$\big\langle y, F(x') - F(x)\big\rangle \geq \big\langle \nabla F(x)^* y, x' - x\big\rangle - \frac{\rho}{2}|x' - x|^2$$
$$\text{when } |y| \leq \eta,\ |x - \bar{x}| < \varepsilon,\ |x' - \bar{x}| \leq \varepsilon.$$

Then, as long as $|x - \bar{x}| < \varepsilon$ and $v \in \partial f(x)$ with $|v - \bar{v}| < \varepsilon$, we have for any $y \in \partial g\big(F(x)\big)$ with $\nabla F(x)^* y = v$ and any point x' with $|x' - \bar{x}| \leq \varepsilon$ that

$$
\begin{aligned}
f(x')-f(x) &= g\big(F(x')\big) - g\big(F(x)\big) \geq \big\langle y,\, F(x') - F(x)\big\rangle \\
&\geq \big\langle \nabla F(x)^* y,\, x' - x\big\rangle - \frac{\rho}{2}|x' - x|^2 = \big\langle v,\, x' - x\big\rangle - \frac{\rho}{2}|x' - x|^2.
\end{aligned}
$$

This not only tells us that f is prox-regular at $\bar{x}$ but also yields the estimate $f(x) - f(x') \leq \eta|F(x') - F(x)|$. In the same way, if f has a subgradient v' at x' with $|v' - \bar{v}| < \varepsilon$, we have $f(x') - f(x) \leq \eta|F(x) - F(x')|$. Hence, whenever (x, v) and (x', v') are sufficiently near to $(\bar{x}, \bar{v})$ with $v \in \partial f(x)$ and $v' \in \partial f(x')$, we have $|f(x') - f(x)| \leq \eta|F(x') - F(x)|$. In particular, through the continuity of F we are able to conclude that f is subdifferentially continuous at $\bar{x}$ for $\bar{v}$.

The extension of the conclusions from $\bar{x}$ to all nearby x in $\operatorname{dom} f$ is validated by the fact that strong amenability persists at such points. □

13.33 Proposition (prox-regularity versus subsmoothness). *If f is lower-$\mathcal{C}^2$ on an open set $O \subset \mathbb{R}^n$, then f is prox-regular and subdifferentially continuous at every point of O as well as strictly continuous on O. Conversely, if f is prox-regular and strictly continuous on O, then f is lower-$\mathcal{C}^2$ on O.*

In particular, if f is prox-regular at a point $\bar{x}$ where it is strictly differentiable, it must be lower-$\mathcal{C}^2$ around $\bar{x}$.

Proof. Lower-$\mathcal{C}^2$ functions are strongly amenable by 10.36, so the initial assertion is a consequence of 13.32. Of course, lower-$\mathcal{C}^2$ functions, like all subsmooth functions, are strictly continuous in particular (see 10.31).

Suppose now that f is strictly continuous and prox-regular on O, and let $\bar{x} \in O$. The set $\partial f(\bar{x})$ is compact by 9.13. For each $\bar{v} \in \partial f(\bar{x})$ we can apply the definition of prox-regularity and get ρ and ε as described, but the compactness allows us to go further and, by selecting from the covering of $\partial f(\bar{x})$ by the various balls $\operatorname{int} \mathbb{B}(\bar{v}, \delta)$ a finite covering, we can obtain $\bar{\rho}$ and $\bar{\varepsilon}$ such that

$$
f(x') \geq f(x) + \big\langle v,\, x' - x\big\rangle - \tfrac{1}{2}\bar{\rho}|x' - x|^2 \;\text{ for }\; x, x' \in \mathbb{B}(\bar{x}, \bar{\varepsilon}),\; v \in \partial f(x).
$$

In terms of $g(x) := f(x) + \frac{1}{2}\bar{\rho}|x|^2$ this inequality can equally well be written as $g(x') \geq g(x) + \langle z,\, x' - x\rangle$ for all $x, x' \in \mathbb{B}(\bar{x}, \bar{\varepsilon})$ and $z \in \partial g(x)$. It implies that g is convex on $\mathbb{B}(\bar{x}, \bar{\varepsilon})$ (because g coincides with the pointwise supremum of the collection of affine functions appearing on the right sides). Therefore, on the open convex set $\operatorname{int} \mathbb{B}(\bar{x}, \bar{\varepsilon})$, f has the form $g - \frac{1}{2}\bar{\rho}|\cdot|^2$ for a finite convex function g. Then f is lower-$\mathcal{C}^2$ on this set by 10.33.

If f is strictly differentiable at $\bar{x}$, it's strictly continuous there by 9.18 with $\partial f(\bar{x}) = \{\nabla f(\bar{x})\}$. Prox-regularity at $\bar{x}$ for $\bar{v} = \nabla f(\bar{x})$ entails a neighborhood U of $(\bar{x}, \bar{v})$ such that, for all $(x, v) \in U \cap \operatorname{gph} \partial f$, f is prox-regular at x for v. Since f is strictly continuous, ∂f is osc and locally bounded at $\bar{x}$ (cf. 9.13), so by 5.19 there's an open neighborhood O of $\bar{x}$ such that $(x, v) \in U$ for all $v \in \partial f(x)$ when $x \in O$. Then f is prox-regular on O, hence lower-$\mathcal{C}^2$ on O. □

A finite function can be prox-regular and subdifferentially continuous everywhere without necessarily being lower-$\mathcal{C}^2$ or amenable. This is demonstrated in Figure 13–5 for a function $f : \mathbb{R} \to \mathbb{R}$, where the circumstances at $x = 0$ and $x = 1$ deserve particular attention.

Not to be overlooked as a special but important class of prox-regular functions are the $\mathcal{C}^{1+}$ functions, characterized as follows.

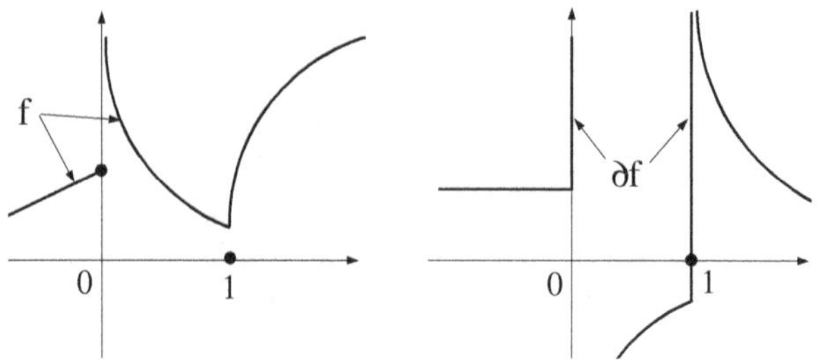

Fig. 13–5. Finite prox-regularity without subsmoothness or amenability.

13.34 Proposition (functions with strictly continuous gradient). *A function f is of class $\mathcal{C}^{1+}$ on an open set O (i.e., is differentiable with ∇f strictly continuous) if and only if f is simultaneously lower-$\mathcal{C}^2$ and upper-$\mathcal{C}^2$ on O. In particular, such a function f is prox-regular and subdifferentially continuous on O.*

Proof. If ∇f is strictly continuous it's hypomonotone by 12.28(a), and then f is lower-$\mathcal{C}^2$ by 12.28(c), hence prox-regular and subdifferentially continuous by 13.33. Applying this to $-f$, one sees that f is upper-$\mathcal{C}^2$ as well.

To verify the converse, suppose f is both lower-$\mathcal{C}^2$ and upper-$\mathcal{C}^2$ on O. In this case f is $\mathcal{C}^1$ on O by 10.30. We must demonstrate that ∇f isn't just continuous but strictly continuous a neighborhood of any point $\bar{x} \in O$. There's no loss of generality in assuming that $\bar{x} = 0$ and $\nabla f(\bar{x}) = 0$.

Let $j = \frac{1}{2}|\cdot|^2$. According to 10.33, there exist $\varepsilon > 0$ and $\rho > 0$ such that, on $2\varepsilon I\!B \subset O$, $f + \rho j$ is convex while $f - \rho j$ is concave. Increasing ρ if necessary, we can arrange that actually $f + \rho j$ is strictly convex on $2\varepsilon I\!B$. Let $g = f + \rho j + \delta_{2\varepsilon B}$. Then g is strictly convex, lsc and proper with $\operatorname{dom} g = 2\varepsilon I\!B$, and g is continuously differentiable on the interior of $2\varepsilon I\!B$ with $\nabla g(0) = 0$. Also, $g - 2\rho j$ is concave on $2\varepsilon I\!B$. It will suffice to show that ∇g is strictly continuous on some neighborhood of 0, since that will ensure the same property for ∇f.

The strict convexity of g implies through 11.13 that the conjugate function g^* is differentiable on $I\!R^n$, so $\partial g = (\nabla g^*)^{-1}$ (see 11.3). It follows that $\nabla g^* = (\nabla g)^{-1}$ on $\nabla g(\varepsilon I\!B)$, in particular. The concavity of $g - 2\rho j$ on $2\varepsilon I\!B$ gives us $(g - 2\rho j)(x') \leq (g - 2\rho j)(x) + \langle \nabla(g - 2\rho j)(x),\, x' - x \rangle$ when $x' - x \in \varepsilon I\!B$ and $x \in \varepsilon I\!B$, where the inequality can algebraically be rewritten as

$$g(x') \leq g(x) + \langle \nabla g(x),\, x' - x \rangle + \rho|x' - x|^2.$$

Define $\theta(u) = \rho|u|^2 + \delta_{\varepsilon B}(u)$, noting that θ is convex, lsc and proper. Then for each $x \in \varepsilon I\!B$ we can estimate

$$\begin{aligned} g^*(v) &= \sup_{x'}\{\langle v, x'\rangle - g(x')\} \\ &\geq \sup_{x'}\{\langle v, x'\rangle - g(x) - \langle \nabla g(x), x'-x\rangle - \theta(x'-x)\} \\ &= -g(x) + \langle v, x\rangle + \sup_u\{\langle v - \nabla g(x), u\rangle - \theta(u)\} \\ &= g^*(\nabla g(x)) + \langle v - \nabla g(x), x\rangle + \theta^*(v - \nabla g(x)), \end{aligned}$$

where at the end the relation has been used that $g^*(y) = \langle y, x\rangle - g(x)$ when $y \in \partial g(x)$; cf. 11.3. By invoking this estimate in the case of $x = x_0$ and $v = \nabla g(x_1)$ for any two points $x_0, x_1 \in \varepsilon I\!B$, we obtain

$$g^*(\nabla g(x_1)) \geq g^*(\nabla g(x_0)) + \langle \nabla g(x_1) - \nabla g(x_0), x_0\rangle + \theta^*(\nabla g(x_1) - \nabla g(x_0)),$$

but we can also get the corresponding inequality with the roles of x_0 and x_1 reversed. When added together, these two inequalities yield

$$\begin{aligned} \theta^*(\nabla g(x_1) - \nabla g(x_0)) &\leq \langle \nabla g(x_1) - \nabla g(x_0),\, x_1 - x_0\rangle \\ &\leq \big|\nabla g(x_1) - \nabla g(x_0)\big|\big|x_1 - x_0\big|. \end{aligned}$$

All that remains is the calculation that $\theta^* = (2\rho j + \delta_{\varepsilon B})^* = (2\rho j)^* \# \delta^*_{\varepsilon B} = (2\rho)^{-1} j \# \varepsilon|\cdot|$ (by rules in 11.23(a), 11.24, 11.11), from which one sees that $\theta^*(v) = (1/4\rho)|v|^2$ for v in a neighborhood of 0. This allows us to conclude that $|\nabla g(x_1) - \nabla g(x_0)| \leq 4\rho|x_1 - x_0|$ for x_0, x_1 in some neighborhood of 0, and hence that ∇g is strictly continuous in such a neighborhood. □

13.35 Exercise (perturbations of prox-regularity). *Let f be prox-regular at $\bar{x}$ for $\bar{v} \in \partial f(\bar{x})$. Then $f_0(x) = f(x) - \langle v, x\rangle$ is prox-regular at $\bar{x}$ for $0 \in \partial f_0(\bar{x})$.*

Indeed, for any function g that is $\mathcal{C}^2$ (or even just $\mathcal{C}^{1+}$) around $\bar{x}$, $f + g$ is prox-regular at $\bar{x}$ for the subgradient $\bar{v} + \nabla g(\bar{x}) \in \partial(f+g)(\bar{x})$. Subdifferential continuity, if present, is likewise preserved in these situations.

Guide. Develop these facts from the definitions of prox-regularity and subdifferential continuity, utilizing 8.8, 10.33, and 13.34. □

According to the next theorem, the subgradient mappings associated with prox-regular functions are distinguished by a 'hypomonotonicity' property akin to the one defined in 12.28, but requiring f-attentive localization.

13.36 Theorem (subdifferential characterization of prox-regularity). *Suppose $f : I\!R^n \to \overline{I\!R}$ is finite and locally lsc at $\bar{x}$, and let $\bar{v} \in \partial f(\bar{x})$ be a proximal subgradient. Then the following conditions are equivalent:*

(a) *f is prox-regular at $\bar{x}$ for $\bar{v}$;*

(b) *∂f has an f-attentive localization T around $(\bar{x}, \bar{v})$ such that $T + \rho I$ is monotone for some $\rho \in I\!R_+$.*

Proof. (a) $\Rightarrow$ (b). Taking ε and ρ as in the definition of prox-regularity in 13.27, let T be the f-attentive localization of ∂f around $(\bar{x}, \bar{v})$ whose graph consists of all $(x, v) \in \partial f$ with $|v - \bar{v}| < \varepsilon$, $|x - \bar{x}| < \varepsilon$, and $f(x) < f(\bar{x}) + \varepsilon$. As noted, the prox-regularity condition implies for every $(x, v) \in \operatorname{gph} T$ that v is a proximal subgradient of f at x, and this applies in particular to $(\bar{x}, \bar{v})$. Indeed, for any two pairs (x_0, v_0) and (x_1, v_1) in $\operatorname{gph} T$ we have

$$f(x_1) \geq f(x_0) + \langle v_0, x_1 - x_0 \rangle - \tfrac{1}{2}\rho|x_1 - x_0|^2,$$
$$f(x_0) \geq f(x_1) + \langle v_1, x_0 - x_1 \rangle - \tfrac{1}{2}\rho|x_0 - x_1|^2.$$

Adding these together, we get $0 \geq -\langle v_1 - v_0, x_1 - x_0 \rangle - \rho|x_1 - x_0|^2$. In the notation $\tilde{v}_0 = v_0 + \rho x_0 \in (T + \rho I)(x_0)$ and $\tilde{v}_1 = v_1 + \rho x_1 \in (T + \rho I)(x_1)$, the conclusion is that $\langle \tilde{v}_1 - \tilde{v}_0, x_1 - x_0 \rangle \geq 0$. Thus, $T + \rho I$ is monotone.

(b) $\Rightarrow$ (a). Both (a) and (b) refer to conditions dependent only on the nature of $\operatorname{epi} f$ near $(\bar{x}, f(\bar{x}))$. There's no loss of generality then in supposing f to be lsc on $\mathbb{R}^n$ with bounded domain, since that can be manufactured out of the local lsc property by adding some indicator function to f. On the same basis we can assume that the subgradient inequality satisfied by $\bar{v}$ at $\bar{x}$ holds globally (through 8.46(f)), and further that $\bar{v} = 0$ (cf. 13.35) and for that matter $\bar{x} = 0$. Then, by adding $\frac{1}{2}\bar{\rho}|\cdot|^2$ to f for $\bar{\rho}$ sufficiently large, which replaces ∂f by $\partial f + \bar{\rho} I$, we can pass (again with justification via 13.35) to the case where $\operatorname{argmin} f = \{0\}$ and an f-attentive localization of ∂f around $(0, 0)$ is known actually to be monotone. Let's denote this localization still by T.

Fix any $\lambda > 0$ and consider the proximal mapping $P_\lambda f : \mathbb{R}^n \rightrightarrows \mathbb{R}^n$, noting that $P_\lambda(0) = \{0\}$. By 1.25, $P_\lambda f$ is everywhere nonempty-valued and

$$x^\nu \in P_\lambda f(u^\nu),\ u^\nu \to 0 \quad \Longrightarrow \quad x^\nu \to 0,\ f(x^\nu) \to f(0), \qquad 13(28)$$

with the convergence of function values coming from the continuity of the envelope function $e_\lambda f$ and the relation $e_\lambda f(u^\nu) = f(x^\nu) + (1/2\lambda)|x^\nu - u^\nu|^2$.

From applying Fermat's rule 10.1 to the minimization problem having $P_\lambda f(u)$ as its optimal solution set, we know that whenever $x \in P_\lambda f(u)$ we have $0 \in \partial f(x) + (1/\lambda)(x - u)$. The f-attentiveness in 13(28) ensures, however, that $\partial f(x)$ can be replaced by $T(x)$ in this condition when u is sufficiently near to 0. Therefore, $P_\lambda f(u) \subset (I + \lambda T)^{-1}(u)$ for u around 0. In addition $(I + \lambda T)^{-1}(u)$ is nonempty for such u, yet because T is monotone it can't be more than a singleton; indeed, $(I + \lambda T)^{-1}$ is nonexpansive (cf. 12.12). Hence on some neighborhood U of 0 we have $P_\lambda f$ single-valued with $P_\lambda f = (I + \lambda T)^{-1}$. Choose $\varepsilon > 0$ small enough that

$$\left.\begin{array}{r} v \in \partial f(x),\ |v| < \varepsilon \\ |x| < \varepsilon,\ f(x) < f(0) + \varepsilon \end{array}\right\} \quad \Longrightarrow \quad v \in T(x),\ x + \lambda v \in U. \qquad 13(29)$$

Then for such (x, v) we have for $u = x + \lambda v$ that $x \in (I + \lambda T)^{-1}(u)$ and consequently $x = P_\lambda f(u)$. But this means that $f(x') + (1/2\lambda)|x' - u|^2 \geq f(x) + (1/2\lambda)|x - u|^2$ for all x'. On substituting $x + \lambda v$ for u we can rewrite this as the proximal subgradient inequality

$$f(x') \geq f(x) + \langle v, x' - x \rangle - \frac{1}{2\lambda}|x' - x|^2.$$

Because this is satisfied globally whenever x and v fulfill the conditions on the left side of 13(29), we may conclude that f is prox-regular at $\bar{x} = 0$ for $\bar{v} = 0$ with parameter value $\rho = \lambda^{-1}$. □

A close connection between prox-regularity and the theory of proximal mappings $P_\lambda f$ and Moreau envelopes $e_\lambda f$ is revealed by the preceding proof. Much more can be said about this.

13.37 Proposition (proximal mappings and Moreau envelopes). *Suppose that $f : \mathbb{R}^n \to \overline{\mathbb{R}}$ is prox-regular at $\bar{x}$ for $\bar{v} = 0$, and that f is prox-bounded. Then for all $\lambda > 0$ sufficiently small there is a neighborhood of $\bar{x}$ on which*

(a) *$P_\lambda f$ is monotone, single-valued and Lipschitz continuous; $P_\lambda f(\bar{x}) = \bar{x}$;*

(b) *$e_\lambda f$ is differentiable with $\nabla(e_\lambda f)(\bar{x}) = 0$, in fact of class $\mathcal{C}^{1+}$ with*

$$\nabla e_\lambda f \;=\; \lambda^{-1}[I - P_\lambda f] \;=\; [\lambda I + T^{-1}]^{-1}$$

for an f-attentive localization T of ∂f at $(\bar{x}, 0)$. Indeed, this localization can be chosen so that the set $U_\lambda := \operatorname{rge}(I + \lambda T)$ serves for all $\lambda > 0$ sufficiently small as a neighborhood of $\bar{x}$ on which these properties hold.

Proof. The justification proceeds along lines parallel to the proof of the converse in Theorem 13.36. We can take $\bar{x} = 0$. Because f is prox-bounded, the subgradient inequalities in the definition of prox-regularity in 13.27 can be taken to be global; cf. 8.46(f). Thus, there exist $\varepsilon > 0$ and $\lambda_0 > 0$ such that

$$\begin{aligned} f(x') > f(x) + \langle v, x' - x\rangle - \tfrac{1}{2}\lambda_0^{-1}|x' - x|^2 \;\text{ for all }\; x' \neq x \\ \text{when }\; v \in \partial f(x),\;\; |v| < \varepsilon,\;\; |x| < \varepsilon,\;\; f(x) < f(0) + \varepsilon. \end{aligned} \qquad 13(30)$$

Let T be the f-attentive localization of ∂f specified in 13(30). Let $\lambda \in (0, \lambda_0)$. The inequality in 13(30) persists with λ in place of λ_0 and can be written as

$$f(x') + \frac{1}{2\lambda}|x' - u|^2 > f(x) + \frac{1}{2\lambda}|x - u|^2 \;\text{ for }\; u = x + \lambda v.$$

Thus we have $P_\lambda(x + \lambda v) = \{x\}$ when $v \in T(x)$. On the other hand, as argued in the proof of 13.36, we have for any point u near enough to 0 and any $x \in P_\lambda f(u)$ that the vector $v = \lambda^{-1}[u - x]$ satisfies $v \in T(x)$. Therefore, the set $U_\lambda = \operatorname{rge}(I + \lambda T)$ is a neighborhood of 0 on which $P_\lambda f$ is single-valued and coincides with $(I + \lambda T)^{-1}$.

The mapping $T + \lambda_0^{-1} I$ is monotone as a consequence of 13(30) (by the argument employed in the first part of the proof of 13.36). Let $\delta = \lambda^{-1} - \lambda_0^{-1} > 0$, so that $T + \lambda^{-1} I = T + \lambda_0^{-1} I + \delta I$ and

$$(I + \lambda T)^{-1} = \big(\lambda\delta[I + \delta^{-1}(T + \lambda_0^{-1} I)]\big)^{-1} = (I + M)^{-1} \circ (\lambda\delta)^{-1} I$$

for the monotone mapping $M = \delta^{-1}(T + \lambda_0^{-1} I)$. We have $(I + M)^{-1}$ monotone and nonexpansive by 12.12, hence $(I + \lambda T)^{-1}$ monotone and Lipschitz continuous with constant $(\lambda\delta)^{-1}$. Since $P_\lambda f = (I + \lambda T)^{-1}$ on U_λ, it follows that $P_\lambda f$ has these properties on this neighborhood of 0.

Recalling from 10.32 the formula $\partial[-e_\lambda f](u) = \lambda^{-1}\big[\operatorname{con}(P_\lambda f)(u) - u\big]$, we obtain the single-valuedness of $\partial[-e_\lambda f]$ around 0 and deduce thereby through 9.18 that $-e_\lambda f$ is strictly differentiable at all points $u \in U_\lambda$ with $\nabla[-e_\lambda f](u) = \lambda^{-1}\big[P_\lambda f(u) - u\big]$. Then $e_\lambda f$ is strictly differentiable on this neighborhood with $\nabla[e_\lambda f] = \lambda^{-1}(I - P_\lambda)$. Because $P_{\lambda f}$ is Lipschitz continuous, $e_\lambda f$ is actually of class $\mathcal{C}^{1+}$ on U_λ. The relation $\lambda^{-1}(I - P_\lambda) = (\lambda I + T^{-1})^{-1}$ is equivalent to $P_\lambda f = (I + \lambda T)^{-1}$ by the identity in 12.14. □

The properties in Proposition 13.37, while implied by prox-regularity, aren't sufficient for it. This can be seen at $(\bar{x}, \bar{v}) = (0,0)$ for the function $f : I\!R \to I\!R$ defined by $f(0) = 0$ but $f(x) = |x|(1 + \sin x^{-1})$ for $x \neq 0$.

13.38 Exercise (projections on prox-regular sets). *For a closed set $C \subset I\!R^n$ and any point $\bar{x} \in C$, the following properties are equivalent:*

(a) *C is prox-regular at $\bar{x}$ for $\bar{v} = 0$,*

(b) *N_C has a hypomonotone localization around $(\bar{x}, 0)$.*

These properties, which hold in particular whenever C is fully amenable at $\bar{x}$, imply that

(c) *P_C is single-valued around $\bar{x}$,*

(d) *d_C is differentiable outside of C around $\bar{x}$.*

Indeed, there exists in this case a neighborhood V of $\bar{x}$ on which P_C is monotone and Lipschitz continuous with $P_C = (I + T)^{-1}$ for some localization T of N_C around $(\bar{x}, 0)$, while $\nabla d_C = [I - P_C]/d_C$ on $V \setminus C$.

Guide. Apply 13.36 and 13.37 to $f = \delta_C$. □

G. Subgradient Proto-Differentiability

We are ready to proceed with the analysis of how second-order epi-differentiability of f at $\bar{x}$ for $\bar{v}$ corresponds, at least for a large category of functions, to proto-differentiability of ∂f at $\bar{x}$ for $\bar{v}$ and even provides a rule for calculating the proto-derivatives.

13.39 Lemma (difference quotient relations). *For any function $f : I\!R^n \to \overline{I\!R}$, any point $\bar{x}$ where f is finite and any $\bar{v} \in \partial f(\bar{x})$, one has for all $\tau > 0$ that*

$$\partial\big[\tfrac{1}{2}\Delta_\tau^2 f(\bar{x}\,|\,\bar{v})\big] = [\Delta_\tau \partial f](\bar{x}\,|\,\bar{v}).$$

In the special case of $\bar{v} = 0$, $\bar{x} = 0$, and $f(\bar{x}) = 0 = \min f$, one further has for all $\tau > 0$ and $\lambda > 0$ that

$$e_\lambda\big[\tfrac{1}{2}\Delta_\tau^2 f(0\,|\,0)\big] = \big[\tfrac{1}{2}\Delta_\tau^2 e_\lambda f\big](0\,|\,0), \qquad P_\lambda\big[\tfrac{1}{2}\Delta_\tau^2 f(0\,|\,0)\big] = [\Delta_\tau P_\lambda f](0\,|\,0).$$

Proof. Fix $\tau > 0$. Let $\varphi(w) = \tau^{-2}[f(\bar{x} + \tau w) - f(\bar{x}) - \tau\langle\bar{v}, w\rangle]$, so that $\varphi = \frac{1}{2}\Delta_\tau^2 f(\bar{x}\,|\,\bar{v})$. Obviously $\partial\varphi(w) = \tau^{-2}[\tau\partial f(\bar{x} + \tau w) - \tau\bar{v}]$, hence $\partial\varphi = [\Delta_\tau \partial f](\bar{x}\,|\,\bar{v})$. This establishes the first of the formulas.

Now suppose $\bar{v}=0$, $\bar{x}=0$, $f(\bar{x})=0=\min f$. Then $\varphi(w)=\tau^{-2}f(\tau w)$. Calculating from the definition of a Moreau envelope, we get

$$\begin{aligned} e_\lambda\varphi(u) &= \min_w\big\{\tau^{-2}f(\tau w)+\tfrac{1}{2}\lambda^{-1}|w-u|^2\big\} \\ &= \tau^{-2}\min_z\big\{f(z)+\tfrac{1}{2}\lambda^{-1}|z-\tau u|^2\big\} = \tau^{-2}e_\lambda f(\tau u). \end{aligned}$$

But $e_\lambda f(0)=0$ because $\bar{x}\in\operatorname{argmin} f$, $\bar{x}=0$. Therefore, the final term equals $\frac{1}{2}\Delta_\tau^2[e_\lambda f](0\,|\,0)(u)$. In the footsteps of the same calculation we have for the corresponding proximal mappings that

$$\begin{aligned} P_\lambda\varphi(u) &= \operatorname{argmin}_w\big\{\tau^{-2}f(\tau w)+\tfrac{1}{2}\lambda^{-1}|w-u|^2\big\} \\ &= \tau^{-1}\operatorname{argmin}_z\big\{f(z)+\tfrac{1}{2}\lambda^{-1}|z-\tau u|^2\big\} = \tau^{-1}P_\lambda f(\tau u). \end{aligned}$$

Since $P_\lambda f(0)=0$ under our assumptions, the final term is $[\Delta_\tau P_\lambda f](0\,|\,0)(u)$. The other two formulas in the lemma are thus correct as well. □

13.40 Theorem (proto-differentiability of subgradient mappings). *Suppose that $f:\mathbb{R}^n\to\overline{\mathbb{R}}$ is prox-regular at $\bar{x}$ for $\bar{v}\in\partial f(\bar{x})$. As long as f is subdifferentially continuous at $\bar{x}$ for $\bar{v}$, the mapping ∂f is proto-differentiable at $\bar{x}$ for $\bar{v}$ if and only if f is twice epi-differentiable at $\bar{x}$ for $\bar{v}$, and then*

$$D(\partial f)(\bar{x}\,|\,\bar{v})=\partial h \quad \textit{for} \quad h=\tfrac{1}{2}d^2f(\bar{x}\,|\,\bar{v}) \ \textit{(proper)}. \qquad 13(31)$$

In the absence of subdifferential continuity, these conclusions hold with ∂f replaced by any of its sufficiently close f-attentive localizations at $(\bar{x},\bar{v})$.

Proof. We work directly with the case where f might not be subdifferentially continuous at $\bar{x}$ for $\bar{v}$, denoting by T an f-attentive localization of ∂f with the property that $T+\rho I$ is monotone for a certain $\rho\in\mathbb{R}_+$, as exists by Theorem 13.36. Everything refers to properties of $\operatorname{epi} f$ in the vicinity of $(\bar{x},f(\bar{x}))$, and $\operatorname{epi} f$ is locally closed there as part of the definition of prox-regularity. Without loss of generality, therefore, we can assume (as the result of adding some indicator function to f if necessary) that f is globally lsc with $\operatorname{dom} f$ bounded. Also, we can normalize to $\bar{v}=0$, $\bar{x}=0$, and $f(\bar{x})=0$ without any prejudice to the assertions. With the same impunity we can then add $\frac{1}{2}\sigma|\cdot|^2$ to f for any $\sigma>0$, and in this way, by taking σ sufficiently large, we can get T actually to be monotone and $\operatorname{argmin} f=\{0\}$, $\min f=0$, and such that

$$f(x')>f(x)+\langle v,\,x'-x\rangle \text{ for all } x'\neq x \text{ when } v\in T(x). \qquad 13(32)$$

This puts us in the framework of the special case in Lemma 13.39. By definition, f is twice epi-differentiable at $\bar{x}=0$ for $\bar{v}=0$ if and only if the functions $h_\tau:=\frac{1}{2}\Delta_\tau^2 f(0\,|\,0)$ epi-converge to something as $\tau\searrow 0$, the only candidate being of course $h=\frac{1}{2}d^2f(0\,|\,0)$. The functions h_τ and h, like f, are nonnegative everywhere, hence in particular prox-bounded uniformly. Therefore by Theorem 7.37, $h_\tau\xrightarrow{e}h$ if and only if $e_\lambda h_\tau\xrightarrow{p}e_\lambda h$ for all $\lambda>0$ sufficiently small. We seek next an understanding of what this convergence of envelopes entails.

We have $\partial h_\tau = \Delta_\tau(\partial f)(0\,|\,0)$ by Lemma 13.39, so $\operatorname{gph}\partial h_\tau = \tau^{-1}\operatorname{gph}\partial f$. Out of this it's clear that the monotone mapping $T_\tau := \Delta_\tau T(0\,|\,0)$, which has $\operatorname{gph} T_\tau = \tau^{-1}\operatorname{gph} T$, serves as an h_τ-attentive localization of ∂h_τ at $(0,0)$. Moreover $\operatorname{rge}(I+\lambda T_\tau) = \tau^{-1}U_\lambda$ for $U_\lambda := \operatorname{rge}(I+\lambda T)$. According to 13.37, U_λ is a neighborhood of 0 and we have $e_\lambda h_\tau$ differentiable with

$$\nabla(e_\lambda h_\tau) = (\lambda I + T_\tau^{-1})^{-1} \text{ on } \tau^{-1}U_\lambda.$$

Furthermore, because $(\lambda I + T_\tau^{-1})^{-1} = (I+\lambda^{-1}T_\tau^{-1})^{-1}\circ\lambda^{-1}I$ and the mapping $(I+\lambda^{-1}T_\tau^{-1})^{-1}$ is nonexpansive (by 12.12, since T_τ^{-1} like T_τ is monotone) we have $\nabla(e_\lambda h_\tau)$ Lipschitz continuous on $\tau^{-1}U_\lambda$ with constant λ^{-1}. For this reason, and the fact that $e_\lambda h_\tau(0) = 0$ and $\nabla(e_\lambda h_\tau)(0) = 0$ for all τ, pointwise convergence of $e_\lambda h_\tau$ to $e_\lambda h$ as $\tau \searrow 0$ (with λ fixed and sufficiently small) is equivalent to pointwise convergence of the gradient mappings $\nabla(e_\lambda h_\tau)$ eventually on any bounded set, in which case these mappings converge uniformly on such sets, hence graphically, and their limit must be $\nabla(e_\lambda h)$ (the function $e_\lambda h$ necessarily being differentiable). But graphical convergence of the mappings $\nabla(e_\lambda h_\tau)$ means such convergence for the mappings $(\lambda I + T_\tau^{-1})^{-1}$ and therefore that of the mappings $T_\tau = \Delta_\tau T(0,0)$, inasmuch as $\operatorname{gph}(\lambda I + T_\tau^{-1})^{-1}$ is simply the image of $\operatorname{gph} T_\tau$ under a certain nonsingular linear transformation from $\mathbb{R}^n \times \mathbb{R}^n$ onto itself that depends only on λ, not τ. Graphical convergence of $\Delta_\tau T(0\,|\,0)$ as $\tau \searrow 0$ is the property of proto-differentiability of T at 0 for 0; the limit has to be $DT(0\,|\,0)$.

Putting this together, we see that f is twice epi-differentiable at 0 for 0 if and only if T is proto-differentiable at 0 for 0, in which case $\nabla(e_\lambda h) = (\lambda I + DT(0\,|\,0)^{-1})^{-1}$. We have yet to verify that this implies $\partial h = DT(0\,|\,0)$.

For simplicity, let $T_0 = DT(0\,|\,0)$. As the graphical limit of the monotone mappings $T_\tau = \Delta_\tau T(0\,|\,0)$, we know that T_0 is monotone (by 12.32). The set $\operatorname{rge}(I+\lambda T_0)$ is the domain of $(\lambda I + T_0^{-1})^{-1} = \nabla(e_\lambda h)$, hence all of $\mathbb{R}^n$, so T_0 is maximal monotone (by 12.12). We claim now that T_0 is also cyclically monotone, which in combination with maximal monotonicity implies maximal cyclical monotonicity. This is verified as follows. Going back to 13(32), we recognize that T itself is cyclically monotone; the chain of inequalities used at the beginning of the proof of Theorem 12.25 gives this right away. It's elementary then that the difference quotient mappings T_τ are cyclically monotone, and in the graphical limit that T_0 too is cyclically monotone.

The maximal cyclical monotonicity of T_0 ensures the existence of a proper, lsc, convex function h_0 such that $\partial h_0 = T_0$. Since $0 \in T_0(0)$ we can fix the constant of integration by requiring $h_0(0) = 0$, and then h is nonnegative with 0 as its minimum value (inasmuch as $0 \in \partial h_0(0)$). Applying once more the theory of Moreau envelopes, this time to h_0, we determine that $e_\lambda h_0$ is differentiable with $\nabla(e_\lambda h_0) = (\lambda I + T_0^{-1})^{-1}$; cf. 13.37. Thus, $\nabla(e_\lambda h_0) = \nabla(e_\lambda h)$. Since also $h(0) = 0$, we must have $e_\lambda h_0 = e_\lambda h$. This holds on $\mathbb{R}^n$ for all $\lambda > 0$ sufficiently small. But as $\lambda \searrow 0$ we have for every $u \in \mathbb{R}^n$ that $e_\lambda h_0(u) \nearrow h_0(u)$ and $e_\lambda h(u) \nearrow h(u)$; cf. 1.25. Therefore $h_0 = h$. This tells us that $T_0 = \partial h$, which is what we needed. □

13.41 Corollary (subgradient proto-differentiability from amenability). *If f is fully amenable at $\bar{x}$, there exists a neighborhood V of $\bar{x}$ such that, for every $x \in V \cap \operatorname{dom} f$ and $v \in \partial f(x)$, ∂f is proto-differentiable at x for v.*

Proof. Full amenability persists at all the points x in some neighborhood V of $\bar{x}$. At such points we are sure that f is twice epi-differentiable, prox-regular and subdifferentially continuous at x for v; cf. 13.14 and 13.32. □

13.42 Corollary (second derivatives and quadratic expansions). *If a function $f : \mathbb{R}^n \to \overline{\mathbb{R}}$ is differentiable at $\bar{x}$ and prox-regular for $\bar{v} = \nabla f(\bar{x})$, the following properties are equivalent:*

(a) *f is twice differentiable at $\bar{x}$ in the extended sense;*

(b) *∂f is differentiable at $\bar{x}$;*

(c) *f has a quadratic expansion at $\bar{x}$.*

Moreover $\nabla^2 f(\bar{x})$ must then be symmetric, and f must be lower-$\mathcal{C}^2$ around $\bar{x}$.

Proof. This corresponds through 13.2 and 13.8 to the case of Theorem 13.40 where $d^2 f(\bar{x} \,|\, \bar{v})$ is quadratic or $D(\partial f)(\bar{x} \,|\, \bar{v})$ is linear. When $h(w) = \langle w, Aw \rangle$ we have $\partial h(w) = A_s w$ for $A_s = \frac{1}{2}[A + A^*]$, so the formula in the theorem implies that the Jacobian matrix at $\bar{x}$ for ∂f (or for ∇f on its domain of existence) is symmetric. The fact that f has to be lower-$\mathcal{C}^2$ locally falls out of 13.33 and the strict differentiability of f at $\bar{x}$ established in 13.2. □

13.43 Corollary (normal cone mappings and projections). *Let $C \subset \mathbb{R}^n$ be prox-regular at $\bar{x}$ for $\bar{v}$. Then the following are equivalent and hold in particular when C is fully amenable at $\bar{x}$:*

(a) *N_C is proto-differentiable at $\bar{x}$ for $\bar{v}$,*

(b) *P_C is proto-differentiable at $\bar{u} = \bar{x} + \bar{v}$ for $\bar{x}$,*

(c) *P_C is single-valued and semidifferentiable at $\bar{u} = \bar{x} + \bar{v}$,*

(d) *δ_C is twice epi-differentiable at $\bar{x}$ for $\bar{v}$,*

and then one has

$$DN_C(\bar{x} \,|\, \bar{v}) = \partial[\tfrac{1}{2} d^2 \delta_C(\bar{x} \,|\, \bar{v})], \qquad DP_C(\bar{u}) = [I + DN_C(\bar{x} \,|\, \bar{v})]^{-1}.$$

Proof. Apply Theorem 13.40 to $f = \delta_C$, invoking also 13.41 and the properties in 13.38. □

13.44 Example (polyhedral sets). *Let $C \subset \mathbb{R}^n$ be polyhedral. Let $\bar{x} \in C$ and $\bar{v} \in N_C(\bar{x})$ and define $K = \{ w \in T_C(\bar{x}) \,|\, w \perp \bar{v} \}$. Then*

(a) *N_C is proto-differentiable at $\bar{x}$ for $\bar{v}$ with $DN_C(\bar{x} \,|\, \bar{v}) = N_K$;*

(b) *P_C is semidifferentiable at $\bar{u} = \bar{x} + \bar{v}$ with $DP_C(\bar{u}) = P_K$.*

Detail. The function $f = \delta_C$ is fully amenable (as a special case of being convex, piecewise linear), hence covered by 13.42. From 13.9 we calculate $d^2 \delta_C = \delta_K$, and the conclusions then follow from 13.43. □

13.45 Exercise (semidifferentiability of proximal mappings). *Let $f : \mathbb{R}^n \to \overline{\mathbb{R}}$ be prox-regular at $\bar{x}$ for $\bar{v} = 0$ and prox-bounded. Then second-order epi-differentiability of f at $\bar{x}$ for 0 is equivalent to any of the following properties holding for some $\lambda > 0$ sufficiently small:*

(a) *$P_\lambda f$ is semidifferentiable at $\bar{x}$,*

(b) *$P_\lambda f$ is proto-differentiable at $\bar{x}$,*

(c) *$e_\lambda f$ is twice epi-differentiable at $\bar{x}$,*

(d) *$e_\lambda f$ is twice semidifferentiable at $\bar{x}$.*

All these properties hold then for all $\lambda > 0$ sufficiently small, and moreover one has the formulas

$$d^2\big[\tfrac{1}{2}e_\lambda f(\bar{x})\big] = e_\lambda\big[\tfrac{1}{2}d^2 f(\bar{x}\,|\,0)\big], \qquad D[P_\lambda f](\bar{x}) = P_\lambda\big[\tfrac{1}{2}d^2 f(\bar{x}\,|\,0)\big].$$

Guide. Apply Theorem 13.40 in the light of 13.37. □

The results about prox-regularity have revealed much about subgradient mappings and how the geometry of their graphs is related to second-order subdifferentiation. The notion in 9.66 of a mapping being graphically Lipschitzian helps to underscore the special nature of this geometry.

13.46 Proposition (Lipschitzian geometry of subgradient mappings). *When a function $f : \mathbb{R}^n \to \overline{\mathbb{R}}$ is prox-regular and subdifferentially continuous at $\bar{x}$ for $\bar{v}$, the mapping $\partial f : \mathbb{R}^n \rightrightarrows \mathbb{R}^n$ is graphically Lipschitzian of dimension n around $(\bar{x}, \bar{v})$. In the absence of subdifferential continuity this holds for any sufficiently close f-attentive localization of ∂f around $(\bar{x}, \bar{v})$.*

Proof. For simplicity we can normalize to $\bar{v} = 0$ (cf. 13.35); geometrically this just amounts to a translation of $\operatorname{gph} \partial f$ and its localizations. The formula in 13.37 then identifies $\operatorname{gph} T$ with the graph of the Lipschitz continuous mapping $\nabla e_\lambda f$ near $\bar{x}$ under a certain linear change of coordinates around $(\bar{x}, \bar{v})$. □

This geometry furnishes the insight that proto-differentiability of ∂f at $\bar{x}$ for $\bar{v}$ is essentially just semidifferentiability seen from a different angle. In graphical terms the two properties provide exactly the same kind of approximation to a graph that locally has the appearance of a 'Lipschitzian manifold of dimension n', but subdifferentiability is the more special way that the property can be described when the angle of view presents the graph as that of a Lipschitz continuous mapping, here $\nabla e_\lambda f$ around $\bar{x}$ in the case of $\bar{v} = 0$.

H. Subgradient Coderivatives and Perturbation

In the study of the subgradient mappings $\partial f : \mathbb{R}^n \rightrightarrows \mathbb{R}^n$ associated with functions $f : \mathbb{R}^n \to \overline{\mathbb{R}}$, one can look not only at the graphical derivative mappings

$$D(\partial f)(\bar{x}\,|\,\bar{v}) : \mathbb{R}^n \rightrightarrows \mathbb{R}^n,$$

as we have been doing so far, but also the coderivative mappings

$$D^*(\partial f)(\bar{x} \,|\, \bar{v}) : I\!R^n \rightrightarrows I\!R^n,$$

which offer yet another 'second derivative' concept. Significant conclusions can be obtained in a number of situations by applying to these mappings the various coderivative results in Chapters 8, 9 and 10.

Such a situation is found in sensitivity analysis of first-order conditions for optimality, with the parametric version of Fermat's rule in 10.12 serving for the basic format. This provides an intriguing application of Theorem 13.40 as well. For $f : I\!R^n \times I\!R^m \to \overline{I\!R}$ consider the problem

$$\mathcal{P}(u,v): \qquad \text{minimize } f(x,u) - \langle v, x\rangle \text{ over all } x \in I\!R^n,$$

where u and v are viewed as parameter elements. The constraint qualification

$$(0,y) \in \partial^\infty f(x,u) \implies y = 0$$

is shown by 10.12 to guarantee the existence of a vector y such that

$$(v,y) \in \partial f(x,u),$$

which can be interpreted as a generalized Lagrange multiplier element. (Although 10.12 dealt with $v = 0$, the extension to general v is immediate from applying that earlier result to $f_v(x,u) = f(x,u) - \langle v, x\rangle$.) We can study how the pair (x,y) depends on the pair (u,v). This leads to the discovery of relatively common circumstances in which the mapping $(u,v) \mapsto (x,y)$ is actually proto-differentiable, indeed with proto-derivatives that themselves can be computed by way of other optimality conditions.

13.47 Theorem (perturbation of optimality conditions). *For $f : I\!R^n \times I\!R^m \to \overline{I\!R}$ consider the mapping*

$$S : (u,v) \mapsto \big\{(x,y) \,\big|\, (v,y) \in \partial f(x,u)\big\}$$

along with particular elements $(\bar{x},\bar{y}) \in S(\bar{u},\bar{v})$. Suppose that f is prox-regular and subdifferentially continuous at $(\bar{x},\bar{u})$ for $(\bar{v},\bar{y})$. Then S is proto-differentiable at $(\bar{u},\bar{v})$ for $(\bar{x},\bar{y})$ if and only if f is twice epi-differentiable at $(\bar{x},\bar{u})$ for $(\bar{v},\bar{y})$, in which case the proto-derivatives have the formula

$$DS(\bar{u},\bar{v} \,|\, \bar{x},\bar{y})(u',v') = \big\{(x',y') \,\big|\, (v',y') \in \partial h(x',u')\big\}, \quad h = \tfrac{1}{2} d^2 f(\bar{x},\bar{u} \,|\, \bar{v},\bar{y}).$$

Moreover, S is graphically Lipschitzian of dimension $m+n$ around $(\bar{u},\bar{v};\bar{x},\bar{y})$, and it has the Aubin property at $(\bar{u},\bar{v})$ for $(\bar{x},\bar{y})$ if and only if

$$(0,y') \in D^*(\partial f)(\bar{x} \,|\, \bar{u})(x',0) \implies x' = 0,\ y' = 0.$$

Proof. The graph of S corresponds to that of ∂f under $(u,v,x,y) \leftrightarrow (x,u,v,y)$, which preserves proto-differentiability in particular. All one has to do is apply 13.46 and 13.40 in this framework. The result about the Aubin property invokes for S the Mordukhovich criterion in 9.40 with the observation that the graph of $D^*S(\bar{u},\bar{v}\,|\,\bar{x},\bar{y})$ is a copy of the graph of $D^*(\partial f)(\bar{x},\bar{u}\,|\,\bar{v},\bar{y})$ under the same affine correspondence. □

It can't escape attention that the proto-derivative formula in Theorem 13.47 concerns the first-order conditions for the auxiliary problem

$$\mathcal{P}'(\bar{x},\bar{u}\,|\,\bar{v},\bar{y})(u',v'): \quad \text{minimize } \tfrac{1}{2}d^2f(\bar{x},\bar{u}\,|\,\bar{v},\bar{y})(x',u')-\langle v',x'\rangle \text{ in } x'$$

in the same parameterized format as $\mathcal{P}(\bar{x},\bar{u})$. The perturbation vectors u' and v' associated with $\bar{u}$ and $\bar{v}$ are now the parameter elements, while the x' and y' that are so characterized give the corresponding perturbations of $\bar{x}$ and $\bar{v}$. Especially noteworthy is the case of convex optimization, where the first-order conditions are enough to guarantee optimality. The proto-derivatives of optimal solutions to a primal-dual pair of problems are obtained then by solving an auxiliary pair of such problems.

The result in Theorem 13.47 can be adapted to many situations where the formula for proto-derivatives can be fleshed out from the accompanying structure of f. Similar applications of second-order theory can also be made to other models of perturbation where subgradient mappings, such as normal cone mappings or projections, are involved. Variational inequalities, where the subgradient mapping is N_C for a closed, convex set C, serve as a major illustration. They are covered as a special case of the next result.

13.48 Theorem (perturbation of solutions to variational conditions). *For any function $f: I\!R^n \to \overline{I\!R}$ and $\mathcal{C}^1$ mapping $F: I\!R^d \times I\!R^n \to I\!R^n$, consider the generally set-valued mapping $S: I\!R^d \times I\!R^n \rightrightarrows I\!R^n$ defined by*

$$S(w,u) = \big\{x \,\big|\, F(w,x) + \partial f(x) \ni u\big\},$$

which in the special case of $f = \delta_C$, $\partial f = N_C$, would give the solutions to a parameterized variational condition. Let $\bar{x} \in S(\bar{w},\bar{u})$ and suppose f is fully amenable at $\bar{x}$. Then S is proto-differentiable at $(\bar{w},\bar{v})$ for $\bar{x}$ with

$$DS(\bar{w},\bar{u}\,|\,\bar{x})(w',u') = \big\{x' \,\big|\, \Phi(w',x') + \partial\varphi(x') \ni u'\big\},$$

where $\Phi(w',x') = \nabla_w F(\bar{w},\bar{x})w' + \nabla_x F(\bar{x},\bar{w})x'$ and $\varphi(x') = \frac{1}{2}d^2f(\bar{x}\,|\,\bar{v})(x')$ for $\bar{v} := \bar{u} - F(\bar{w},\bar{x})$. Further, S is graphically Lipschitzian of dimension $d+2n$ around $(\bar{w},\bar{u},\bar{x})$, and it has the Aubin property at $(\bar{w},\bar{u})$ for $\bar{x}$ if and only if

$$-\nabla_x F(\bar{w},\bar{x})^* z \in D^*(\partial f)(\bar{x}\,|\,\bar{v})(z) \implies z = 0.$$

Proof. Here $\operatorname{gph} S = \big\{(w,u,x) \,\big|\, \big(x,\, u - F(w,x)\big) \in \operatorname{gph}\partial f\big\}$. The smooth transformation $G: (w,u,x) \mapsto \big(x,\, u - F(w,x)\big)$ takes the point $(\bar{w},\bar{u},\bar{x}) \in \operatorname{gph} S$ to the point $(\bar{x},\bar{v}) \in \operatorname{gph}\partial f$, and its Jacobian has full rank:

$$\begin{aligned}\nabla G(\bar{w},\bar{u},\bar{x})(w',u',x') &= \big(x',\, u' - \nabla_w F(\bar{w},\bar{x})w' + \nabla_x F(\bar{w},\bar{x})x'\big),\\ \nabla G(\bar{w},\bar{u},\bar{x})^*(x'',v'') &= \big(-\nabla_w F(\bar{w},\bar{x})^* v'', v'',\, x'' - \nabla_x F(\bar{w},\bar{x})^* v''\big).\end{aligned}$$

Through the change-of-coordinates rule in 6.7 we have

$$T_{\operatorname{gph} S}(\bar{w}, \bar{u}, \bar{x}) = \Big\{ (w', u', x') \;\Big|\; \nabla G(\bar{w}, \bar{u}, \bar{x})(w', u', x') \in T_{\operatorname{gph} \partial f}(\bar{x}, \bar{v}) \Big\},$$
$$N_{\operatorname{gph} S}(\bar{w}, \bar{u}, \bar{x}) = \Big\{ \nabla G(\bar{w}, \bar{u}, \bar{x})^*(x'', v'') \;\Big|\; (x'', v'') \in N_{\operatorname{gph} \partial f}(\bar{x}, \bar{v}) \Big\}.$$

Proto-differentiability of ∂f corresponds to geometric derivability of $\operatorname{gph} \partial f$. It thus carries over to geometric derivability of $\operatorname{gph} S$ and furthermore to proto-differentiability of S. From the normal cone relation we have

$$\begin{aligned}(w', u') \in D^*S(\bar{w}, \bar{u} \,|\, \bar{x})(x') \iff& \; (w', u', -x') \in N_{\operatorname{gph} S}(\bar{w}, \bar{u}, \bar{x}) \\ \iff& \; \exists\, (x'', v'') \in N_{\operatorname{gph} \partial f}(\bar{x}, \bar{v}) \text{ with } \begin{cases} w' = -\nabla_w F(\bar{w}, \bar{x})^* v'', \\ u' = v'', \\ x' = x'' - \nabla_x F(\bar{w}, \bar{x})^* v'', \end{cases}\end{aligned}$$

where in the last description the normal cone description of (x'', v'') can equally be expressed by $x'' \in D^*(\partial f)(\bar{x} \,|\, \bar{v})(-v'')$. This tells us that

$$(w', u') \in D^*S(\bar{w}, \bar{u} \,|\, \bar{x})(x') \iff \begin{cases} x' + \nabla_x F(\bar{w}, \bar{x})^* u' \in D^*(\partial f)(\bar{x} \,|\, \bar{v})(-u'), \\ w' = -\nabla_w F(\bar{w}, \bar{x})^* u'. \end{cases}$$

The Mordukhovich criterion, necessary and sufficient for S to enjoy the Aubin property at $(\bar{w}, \bar{u})$ for $\bar{x}$, is $D^*S(\bar{w}, \bar{u} \,|\, \bar{x})(0) = \{(0,0)\}$. Our coderivative formula for S makes clear that this holds if and only if there's no $u' \neq 0$ with $\nabla_x F(\bar{w}, \bar{x})^* u' \in D^*(\partial f)(\bar{x} \,|\, \bar{v})(-u')$, which amounts to the stated condition. □

I.* Further Derivative Properties

The theory of prox-regularity and full amenability has additional consequences which we now proceed to lay out.

13.49 Proposition (prox-regularity and amenability of second subderivatives). *If f is prox-regular and properly twice epi-differentiable at $\bar{x}$ for $\bar{v}$, there exists $\rho \geq 0$ such that the function $d^2 f(\bar{x} \,|\, \bar{v}) + \rho|\cdot|^2$ is convex and the mapping $D(\partial f)(\bar{x} \,|\, \bar{v}) + \rho I$ is maximal monotone, implying that $d^2 f(\bar{x} \,|\, \bar{v})$ is prox-regular everywhere. In particular these properties hold whenever f is fully amenable at $\bar{x}$, and in that case $d^2 f(\bar{x} \,|\, \bar{v})$ is fully amenable everywhere.*

Proof. By Theorem 13.40, $\partial d^2 f(\bar{x} \,|\, \bar{v})$ has the form $2DT(\bar{x} \,|\, \bar{v})$ for a mapping T which, according to 13.36, has $T + \rho I$ is monotone for some $\rho \geq 0$. Then $2DT(\bar{x} \,|\, \bar{v}) + 2\rho I$ is monotone, but this is ∂k for $k = d^2 f(\bar{x} \,|\, \bar{v}) + \rho|\cdot|^2$, a function that's lsc because $d^2 f(\bar{x} \,|\, \bar{v})$ is lsc. This implies by 12.17 that k is convex and ∂k is maximal monotone.

In the fully amenable case we can invoke the facts about full amenability in 13.15 and 13.41, which plug this into the case just treated. Further, the formula in 13(19) can be interpreted as

$$d^2f(\bar{x}\,|\,\bar{v})(w) = \varphi(\Phi(w))$$

for the $\mathcal{C}^2$ mapping

$$\Phi : w \mapsto (w, w\nabla^2 F(\bar{x})w)$$

and the convex, piecewise linear-quadratic function

$$\varphi(w,z) = d^2\bar{f}(\bar{x}\,|\,\bar{v})(w) + \sigma(z),$$

with σ the support function of the polyhedral set $Y(\bar{x},\bar{v})$. The compactness of $Y(\bar{x},\bar{v})$ (cf. the proof of 13.14) causes σ to be finite on $I\!R^m$. This ensures the fulfillment of the constraint qualification required for this composition to meet the definition of amenability at w: no nonzero vector $(p,q) \in N_{\operatorname{dom}\varphi}(w, w\nabla^2 F(\bar{x})w)$ has $\nabla\Phi(w)^*(p,q) = 0$. Indeed, the normal cone in question is $N_D(w) \times \{0\}$ for $D = \operatorname{dom} d^2\bar{f}(\bar{x}\,|\,\bar{v})$, so it can't contain (p,q) unless $q = 0$. On the other hand, though, $\nabla\Phi(w)^*(p,0) = p$. □

The full amenability of the second epi-derivative functions in 13.49 doesn't extend to them actually being piecewise linear-quadratic—because of the stipulation in the definition of that property about the different 'pieces' of the domain being polyhedral. For an example, consider on $I\!R^3$ the fully amenable function $f(x_1,x_2,x_3) = |x_1^2 + x_2^2 - x_3^2|$ at $\bar{x} = 0$ for the vector $\bar{v} = \nabla f(\bar{x}) = 0$. It's evident that $d^2f(\bar{x}\,|\,\bar{v})(w) = 2f(w)$. The domain of this function is divided into two pieces on which different quadratic formulas have priority, but these pieces aren't polyhedral.

When f is fully amenable, the proto-derivatives in Theorem 13.40 (and Corollary 13.41) can be expressed in an alternative manner with the help of subgradient calculus. This expression too could be utilized in results on perturbation of solutions.

13.50 Exercise (subgradient derivative formula in full amenability). *Let $f : I\!R^n \to \overline{I\!R}$ be fully amenable at $\bar{x}$, and let $\bar{v} \in \partial f(\bar{x})$. Then, in the notation of 13.14, the proto-derivatives of ∂f at $\bar{x}$ for $\bar{v}$ have the expression*

$$\begin{aligned} D(\partial f)(\bar{x}\,|\,\bar{v})(w) &= D(\partial\bar{f})(\bar{x}\,|\,\bar{v})(w) + \big\{\nabla^2(yF)(\bar{x})w \,\big|\, y \in M(\bar{x},\bar{v},w)\big\}, \\ M(\bar{x},\bar{v},w) &= \operatorname*{argmax}_{y\in Y(\bar{x},\bar{v})} \big\langle w, \nabla^2(yF)(\bar{x})w\big\rangle. \end{aligned}$$

Guide. Write $d^2f(\bar{x}\,|\,\bar{v}) = \varphi\circ\Phi$ as in the proof of Proposition 13.49 and calculate the subgradients of this function using the basic chain rule in Theorem 10.6 and the formula for subgradients of support functions in Corollary 8.25. Then invoke Theorem 13.40. □

Next, let's note that the almost everywhere differentiability of maximal monotone mappings yields an analogous conclusion about second-order differentiability of functions whose subgradient mappings are hypomonotone.

13.51 Theorem (almost everywhere second-order differentiability). *Any lower-$\mathcal{C}^2$ function f on an open set O is twice differentiable in the extended sense at* *all but a negligible set of points $x \in O$, the matrices $\nabla^2 f(x)$ being symmetric where they exist. In particular this holds when f is finite and convex on O.*

Proof. For any $\bar{x} \in O$ there exists by 10.33 a compact neighborhood B of $\bar{x}$ within O along with $\rho \geq 0$ such that $f + \frac{1}{2}\rho|\cdot|^2$ is convex on B. Then the function $g := f + \frac{1}{2}\rho|\cdot|^2 + \delta_B$ is proper, lsc and convex on $I\!R^n$, and $\partial f = \partial g - \rho I$ on $\operatorname{int} B$. But ∂g is maximal monotone by 12.17, so ∂g is single-valued and differentiable except on a negligible subset of $\operatorname{int} B$ by 12.66. Then ∂f has this property, and it follows from 13.42 that twice differentiability of f prevails except on such a subset. Hessian symmetry follows then from 13.42. □

Theorem 13.51 covers as a special case all functions that are $\mathcal{C}^{1+}$ on O, inasmuch as they are lower-$\mathcal{C}^2$ (cf. 13.34). But for $\mathcal{C}^{1+}$ functions f the verification of almost everywhere twice differentiability in the extended sense doesn't need to pass through the theory of monotone mappings. It's immediate from Rademacher's theorem in 9.60 as applied to the strictly continuous mapping ∇f. The fact that the Hessian matrices are symmetric when they exist still requires an appeal to something further, though, the fact in 13.42.

Other interesting properties come up in the $\mathcal{C}^{1+}$ case as well.

13.52 Theorem (generalization of second-derivative symmetry). *Let $f : O \to \overline{I\!R}$ be of class $\mathcal{C}^{1+}$, with $O \subset I\!R^n$ open, and let $D \subset O$ consist of the points where f is twice differentiable (in the classical or extended sense). For $\bar{x} \in O$ define*

$$\overline{\nabla}^2 f(\bar{x}) := \left\{ A \in I\!R^{n\times n} \;\middle|\; \exists x^\nu \to \bar{x} \text{ with } x^\nu \in D,\ \nabla^2 f(x^\nu) \to A \right\}.$$

Then $\overline{\nabla}^2 f(\bar{x})$ is a nonempty, compact set of symmetric matrices, and for every $w \in I\!R^n$ one has

$$\operatorname{con} D_*(\nabla f)(\bar{x})(w) = \operatorname{con} D^*(\nabla f)(\bar{x})(w) = \operatorname{con}\{ Aw \mid A \in \overline{\nabla}^2 f(\bar{x}) \}, \qquad 13(33)$$

with $D_*(\nabla f)(\bar{x})$ denoting the strict derivative mapping of 9.53. Moreover for all $z, w \in I\!R^n$ one has

$$\begin{aligned} &\limsup_{\substack{\tau \searrow 0,\, \sigma \searrow 0 \\ x \to \bar{x}}} \frac{f(x+\tau w+\sigma z) - f(x+\tau w) - f(x+\sigma z) + f(x)}{\tau\sigma} \\ &\qquad = \limsup_{\substack{\tau \searrow 0 \\ x \to \bar{x}}} \frac{\langle z, \nabla f(x+\tau w)\rangle - \langle z, \nabla f(x)\rangle}{\tau} \\ &\qquad = \limsup_{\substack{\tau \searrow 0 \\ x \to \bar{x}}} \frac{\langle w, \nabla f(x+\tau z)\rangle - \langle w, \nabla f(x)\rangle}{\tau} \\ &\qquad = \max \langle z, D^*(\nabla f)(\bar{x})(w)\rangle = \max \langle w, D^*(\nabla f)(\bar{x})(z)\rangle \\ &\qquad = \max\{ \langle z, Aw\rangle \mid A \in \overline{\nabla}^2 f(\bar{x}) \}. \end{aligned} \qquad 13(34)$$

The upper limits in this equation would be unchanged if taken also over $w' \to w$ and $z' \to z$, or if σ were made the same as τ. Further, D could be replaced in *the formula for* $\overline{\nabla}^2 f(\bar{x})$ *by any set* $D' \subset D$ *such that* $D \setminus D'$ *is negligible.*

Proof. Everything except the first equation in 13(34) follows at once from applying 9.62 to $F = \nabla f$ and invoking the symmetry of $\nabla^2 f(x)$ when it exists; cf. 13.42. (The second and third expressions in 13(34) equal $\max \langle z, D_*(\nabla f)(\bar{x})(w)\rangle$ and $\max \langle w, D_*(\nabla f)(\bar{x})(z)\rangle$, respectively.)

Fix any w and z, and denote the first expression in 13(34) by $\Phi_{\tau,\sigma}(x)$ for $\tau > 0$ and $\sigma > 0$. We have $\Phi_{\tau,\sigma}(x) = [\theta_{x,\tau}(\sigma) - \theta_{x,\tau}(0)]/\sigma$ for $\theta_{x,\tau}(\sigma) := \Delta_\tau f(x+\sigma z)(w)$. By the mean-value theorem, $[\theta_{x,\tau}(\sigma) - \theta_{x,\tau}(0)]/\sigma$ equals $\theta'(\hat{\sigma})$ for some $\hat{\sigma} \in (0, \sigma)$. Because $\nabla_x\big[\Delta_\tau f(x)(w)\big] = \tau^{-1}\big[\nabla f(x+\tau w) - \nabla f(x)\big]$, we get $\theta'_{x,\tau}(\hat{\sigma}) = \tau^{-1}\langle z, \nabla f(x+\hat{\sigma} z + \tau w) - \nabla f(x+\hat{\sigma} z)\rangle$. Thus

$$\Phi_{\tau,\sigma}(x) = \frac{\langle z, \nabla f(x'+\tau w) - \langle z, \nabla f(x')\rangle}{\tau} \quad \text{for some } x' \in [x,\, x+\sigma z].$$

Clearly, then, the '$\leq$' half of the first equation in 13(34) is correct. To get the '$\geq$' half, we'll argue that the first expression in 13(34) can't be less than the final one, even when σ is restricted to being the same as τ.

Denote by α the max at the end of 13(34). For any $\varepsilon > 0$ we can find $x \in D \cap \mathbb{B}(\bar{x}, \varepsilon)$ with $\langle z, \nabla^2 f(x) w\rangle > \alpha - \varepsilon$. We have

$$\begin{aligned} f(x+\tau w+\tau z)) &= f(x) + \tau\langle \nabla f(x),\, w+z\rangle \\ &\qquad + \tfrac{1}{2}\tau^2\langle (w+z), \nabla^2 f(x)(w+z)\rangle + o(\tau^2|w+z|^2), \\ f(x+\tau w) &= f(x) + \tau\langle \nabla f(x), w\rangle + \tfrac{1}{2}\tau^2\langle w, \nabla^2 f(x) w\rangle + o(\tau^2|w|^2), \\ f(x+\tau z) &= f(x) + \tau\langle \nabla f(x), z\rangle + \tfrac{1}{2}\tau^2\langle z, \nabla^2 f(x) z\rangle + o(\tau^2|z|^2), \end{aligned}$$

and this gives us

$$\begin{aligned} f(x+\tau w{+}\tau z)) &- f(x+\tau w) - f(x+\tau z) + f(x) \\ &= \tau^2\big\langle z, \nabla^2 f(x) w\big\rangle + o(\tau^2|w+z|^2) - o(\tau^2|w|^2) - o(\tau^2|z|^2). \end{aligned}$$

It follows that $\Phi_{\tau,\tau}(x) \to \langle z, \nabla^2 f(x) w\rangle$ as $\tau \searrow 0$. Thus, by choosing τ small enough, we can make $\Phi_{\tau,\tau}(x) > \langle z, \nabla^2 f(x) w\rangle - \varepsilon$. Then $\Phi_{\tau,\tau}(x) > \alpha - 2\varepsilon$. This inequality being achievable for arbitrary $\varepsilon > 0$ by some $x \in D \cap \mathbb{B}(\bar{x}, \varepsilon)$ and $\tau \in (0, \varepsilon)$, we conclude, as needed, that the upper limit of $\Phi_{\tau,\tau}(x)$ as $x \to \bar{x}$ and $\tau \searrow 0$ is at least high as α.

The modification of the second and third expressions in 13(34) to upper limits with $w' \to w$ and $z' \to z$ would make no difference because of the Lipschitz continuity of ∇f. In this case the argument for the first equation in 13(34) could be made in essentially the same way. □

Is the convex hull operation in 13(33) really needed? Could it be true that the strict derivative $D_*(\nabla f)(\bar{x})(w)$ actually equals the coderivative $D^*(\nabla f)(\bar{x})(w)$? This can fail even in the one-dimensional case, unless, of course, ∇f happens to be strictly differentiable at $\bar{x}$. A simple counterexample on $\mathbb{R}^1$ is furnished by $f(x) = \frac{1}{2}x^2$ for $x \geq 0$, $f(x) = -\frac{1}{2}x^2$ for $x \leq 0$, which has $\nabla f(x) = f'(x) = |x|$. For this $\mathcal{C}^{1+}$ function at $\bar{x} = 0$,

$$D_*(\nabla f)(\bar{x})(w) = [-|w|, |w|] \text{ for all } w,$$
$$D^*(\nabla f)(\bar{x})(w) = \partial_x(w|x|)\big|_{x=0} = \begin{cases} \{-|w|, |w|\} & \text{for } w \geq 0, \\ [-|w|, |w|] & \text{for } w \leq 0. \end{cases}$$

The properties in Theorem 13.52 are applicable in particular to $e_\lambda f$ at all points $\bar{x}$ in a neighborhood of a point $\tilde{x}$ where $0 \in \partial f(\tilde{x})$, provided that f is prox-bounded as well as prox-regular at $\tilde{x}$ for $\tilde{v} = 0$, and $\lambda > 0$ is sufficiently small; this follows from 13.37. The relationship between $\nabla e_\lambda f$ and $P_\lambda f$ in 13.37 then gives us analogous properties for $P_\lambda f$ with respect to the matrix sets $\overline{\nabla}(P_\lambda f)(\bar{x})$, defined as in 9.62.

13.53 Corollary (symmetry in derivatives of proximal mappings). *Consider a prox-bounded function $f : \mathbb{R}^n \to \overline{\mathbb{R}}$ and a point $\tilde{x}$ with $0 \in \partial f(\tilde{x})$. Suppose f is prox-regular at $\tilde{x}$ for $\tilde{v} = 0$. Then there is an open set O containing $\tilde{x}$ such that, at every point $\bar{x} \in O$, the following properties hold for the mapping $P_\lambda f$, this being strictly continuous on O.*

The matrix set $\overline{\nabla}(P_\lambda f)(\bar{x})$ is nonempty and compact with all of its elements symmetric, and for every $w \in \mathbb{R}^n$ one has

$$\operatorname{con} D_*(P_\lambda f)(\bar{x})(w) = \operatorname{con} D^*(P_\lambda f)(\bar{x})(w) = \operatorname{con}\{Aw \mid A \in \overline{\nabla}(P_\lambda f)(\bar{x})\}.$$

Moreover for all $z, w \in \mathbb{R}^n$ one has

$$\begin{aligned} \max_{A \in \overline{\nabla}(P_\lambda f)(\bar{x})} \langle z, Aw \rangle &= \max \langle z, D^*(P_\lambda f)(\bar{x})(w) \rangle = \max \langle w, D^*(P_\lambda f)(\bar{x})(z) \rangle \\ &= \limsup_{\substack{\tau \searrow 0 \\ x \to \bar{x}}} \frac{\langle z, P_\lambda f(x + \tau w) \rangle - \langle z, P_\lambda f(x) \rangle}{\tau} \\ &= \limsup_{\substack{\tau \searrow 0 \\ x \to \bar{x}}} \frac{\langle w, P_\lambda f(x + \tau z) \rangle - \langle w, P_\lambda f(x) \rangle}{\tau}. \end{aligned}$$

Proof. This follows from 13.52 through 13.37, as noted above. □

13.54 Corollary (symmetry in derivatives of projections). *Let $C \subset \mathbb{R}^n$ be prox-regular at $\tilde{x}$ for $\tilde{v} = 0$. Then there is an open set O containing $\tilde{x}$ such that, at every point $\bar{x} \in O$, the following properties hold for the projection mapping P_C, this being strictly continuous on O.*

The matrix set $\overline{\nabla} P_C(\bar{x})$ is nonempty and compact with all of its elements symmetric, and for every $w \in \mathbb{R}^n$ one has

$$\operatorname{con} D_* P_C(\bar{x})(w) = \operatorname{con} D^* P_C(\bar{x})(w) = \operatorname{con}\{Aw \mid A \in \overline{\nabla} P_C(\bar{x})\}.$$

Moreover for all $z, w \in \mathbb{R}^n$ one has

$$\max_{A\in\overline{\nabla}P_C(\bar{x})} \langle z, Aw\rangle = \max\langle z, D^*P_C(\bar{x})(w)\rangle = \max\langle w, D^*P_C(\bar{x})(z)\rangle$$
$$= \limsup_{\substack{\tau\searrow 0\\ x\to\bar{x}}} \frac{\langle z, P_C(x+\tau w)\rangle - \langle z, P_C(x)\rangle}{\tau}$$
$$= \limsup_{\substack{\tau\searrow 0\\ x\to\bar{x}}} \frac{\langle w, P_C(x+\tau z)\rangle - \langle w, P_C(x)\rangle}{\tau}.$$

Proof. Here we specialize the preceding corollary to $f = \delta_C$. □

The expressions in the first lim sup in 13(34) are second-order difference quotients of yet another type. It might be wondered whether something can be made of them more generally, but their usefulness is limited to the class of $\mathcal{C}^{1+}$ functions because of the following fact.

13.55 Exercise (second-order Lipschitz continuity). *A function f is of class $\mathcal{C}^{1+}$ around $\bar{x}$ if and only if it is finite around $\bar{x}$ and for some $\varepsilon > 0$ there exists $\kappa \in \mathbb{R}_+$ such that*

$$\begin{aligned} f(x+w+z) - f(x+w) - f(x+z) + f(x) \ &\leq\ \kappa|w||z| \\ \text{when } |x-\bar{x}| \leq \varepsilon,\ |w| &\leq \varepsilon,\ |z| \leq \varepsilon. \end{aligned} \qquad 13(35)$$

The lowest limiting value of such constants κ as $\varepsilon \searrow 0$ is given then by

$$\limsup_{\substack{|w|\searrow 0,\ |z|\searrow 0\\ x\to\bar{x}}} \frac{f(x+w+z) - f(x+w) - f(x+z) + f(x)}{|w||z|} = \max_{A\in\overline{\nabla}^2 f(\bar{x})} |A|.$$

Guide. Get the necessity of the condition from Theorem 13.52, deducing the formula at the end by the same means. For the sufficiency, introduce $d_w(x) := [f(x+w) - f(x)]/|w|$ and argue that 13(35) means

$$|d_w(x') - d_w(x)| \leq \kappa|x'-x| \text{ when } |x-\bar{x}| \leq \varepsilon,\ |x'-x| \leq \varepsilon,\ 0 < |w| \leq \varepsilon. \quad 13(36)$$

Obtain from this the existence of some κ_0 such that $d_w(x) \leq \kappa_0$ when $|x-\bar{x}| \leq \varepsilon$ and $0 < |w| \leq \varepsilon$ and thereby the fact that f is Lipschitz continuous around $\bar{x}$ with constant κ_0. Next transform 13(36) into the condition that

$$|\Delta_\tau f(x')(w) - \Delta_\tau f(x)(w)| \leq \kappa|x'-x| \text{ when } |x-\bar{x}| \leq \varepsilon,\ \tau \in (0,\varepsilon],\ w \in \mathbb{B},$$

and by considering what happens when $\tau \searrow 0$ and invoking the existence of $\nabla f(x)$ almost everywhere around $\bar{x}$ (cf. 9.60), deduce that $\nabla f(x)$ must exist for all x near $\bar{x}$ and be strictly continuous with respect to x. □

This has a notable consequence for second-order differentiability.

13.56 Proposition (strict second-order differentiability). *The following properties of a function f at a point $\bar{x}$ are equivalent:*

(a) *f is differentiable on a neighborhood of $\bar{x}$, and the mapping ∇f is strictly differentiable at $\bar{x}$;*

(b) *f is of class $\mathcal{C}^{1+}$ on a neighborhood of $\bar{x}$ as well as twice differentiable at $\bar{x}$ (in the classical or extended sense), and the mapping $x \mapsto \nabla^2 f(x)$ is continuous at $\bar{x}$ relative to its domain of existence;*

(c) *f is finite on a neighborhood of $\bar{x}$, and for every choice of $z, w \in \mathbb{R}^n$ one has the existence of*

$$\lim_{\substack{\tau \searrow 0,\, \sigma \searrow 0 \\ x \to \bar{x}}} \frac{f(x+\tau w+\sigma z) - f(x+\tau w) - f(x+\sigma z) + f(x)}{\tau\sigma}.$$

In that case this limit equals $\langle z, \nabla^2 f(\bar{x}) w\rangle$.

Proof. In (a), the strict differentiability of ∇f at $\bar{x}$ entails its Lipschitz continuity there, so f must be $\mathcal{C}^{1+}$ around $\bar{x}$, as in (b)—and also in (c) by virtue of 13.55. All three cases thus fit the framework of Theorem 13.52. The strict differentiability of ∇f in (a) corresponds then to the existence of a matrix $\overline{A}$ such that the second and third expressions in 13(34) converge to $\langle z, \overline{A}w\rangle$ as $x \to \bar{x}$, $\tau \searrow 0$ and $\sigma \searrow 0$ (the lim sup equaling the lim inf), and this is therefore identical to having the quotient in (c) converge to $\langle z, \overline{A}w\rangle$, as well as to having

$$\max\{\langle z, Aw\rangle \mid A \in \overline{\nabla}^2 f(\bar{x})\} = \min\{\langle z, Aw\rangle \mid A \in \overline{\nabla}^2 f(\bar{x})\} = \langle z, \overline{A}w\rangle$$

for all z, w. Thus, $\overline{\nabla} f(\bar{x}) = \{\overline{A}\}$; in other words, $\nabla^2 f(x) \to \overline{A} = \nabla^2 f(\bar{x})$ as $x \to \bar{x}$ on the set where $\nabla^2 f(x)$ exists. This is (b). The max corresponds to the upper limit in (c) and the min to the lower limit of that quotient, so they can't coincide unless $\overline{\nabla}^2 f(\bar{x})$ is a singleton. Hence if the limit in (c) exists for all z, w, it must be of the form $\langle z, \overline{A}w\rangle$ for some $\overline{A}$. □

Graphical derivatives and coderivatives of ∂f are connected by still another result, which doesn't assume that the function f is $\mathcal{C}^{1+}$ or even finite around $\bar{x}$ but requires other properties instead.

13.57 Theorem (derivative-coderivative inclusion). *Suppose $\bar{v} \in \partial f(\bar{x})$ for a function $f : \mathbb{R}^n \rightrightarrows \overline{\mathbb{R}}$ that is prox-regular and subdifferentially continuous at $\bar{x}$ for $\bar{v}$ as well as twice epi-differentiable at $\bar{x}$ for $\bar{v}$. Then*

$$D(\partial f)(\bar{x} \,|\, \bar{v}) \subset D^*(\partial f)(\bar{x} \,|\, \bar{v}),$$

and in fact one has for all w that

$$\begin{aligned} &D(\partial f)(\bar{x} \,|\, \bar{v})(w) \cup [-D(\partial f)(\bar{x} \,|\, \bar{v})(-w)] \\ &\qquad \subset D^*(\partial f)(\bar{x} \,|\, \bar{v})(w) \cap [-D^*(\partial f)(\bar{x} \,|\, \bar{v})(-w)]. \end{aligned} \tag{13(37)}$$

Proof. Let $H = D(\partial f)(\bar{x} \,|\, \bar{v})$ and $(\bar{w}, \bar{z}) \in \operatorname{gph} H$. We have $\operatorname{gph} D^*H(\bar{w} \,|\, \bar{z}) \subset \operatorname{gph} D^*(\partial f)(\bar{x} \,|\, \bar{v})$ by 8.44(a). We wish to show that $\bar{z}$ belongs to the set $D^*(\partial f)(\bar{x} \,|\, \bar{v})(\bar{w}) \cap [-D^*(\partial f)(\bar{x} \,|\, \bar{v})(-\bar{w})]$, and for this it suffices to show

$$\bar{z} \in D^*H(\bar{w} \,|\, \bar{z})(\bar{w}) \quad \text{and} \quad -\bar{z} \in D^*H(\bar{w} \,|\, \bar{z})(-\bar{w}).$$

The theorem will thereby be proved, since, if we were to start instead by assuming $(-\bar{w},-\bar{z}) \in \operatorname{gph} H$, the identical reasoning would be applicable to $(\bar{w}',\bar{z}') := (-\bar{w},-\bar{z})$.

Because the graph of $D^*(\partial f)(\bar{x}\,|\,\bar{v})$ is closed, we don't really have to carry this out for arbitrary $(\bar{w},\bar{z})$, but merely for all pairs $(\bar{w},\bar{z})$ in some dense subset of $\operatorname{gph} H$. Indeed, density in some neighborhood of $(0,0)$ relative to $\operatorname{gph} H$ suffices, because $\operatorname{gph} H$ is a cone.

Through Theorem 13.40, our assumptions imply that $H = \partial h$ for a certain function h, which by 13.49 is globally prox-regular. For any $\lambda > 0$ sufficiently small we know from 13.37 that $(\lambda I + S^{-1})^{-1} = \nabla(e_\lambda h)$ on an open neighborhood V of 0, moreover with $e_\lambda h$ of class C^{1+} on V. In this relation the graphs of H and $F = \nabla(e_\lambda h)$ correspond to each other around $(0,0)$ under a linear change of variables in $I\!R^n \times I\!R^n$, namely

$$(w,z) \longleftrightarrow (u,z) \;\text{ with }\; u = w + \lambda z,\;\; w = u - \lambda z. \tag{13(38)}$$

Let D be a dense subset of V where F is differentiable, which exists by 13.51 as applied to $e_\lambda h$; for each $u \in D$ the matrix $\nabla F(u)$ is symmetric. The pairs $\big(u, F(u)\big)$ with $u \in D$ are locally dense in $\operatorname{gph} F$ around $(0,0)$, so the corresponding pairs $(w,z) = \big(u - \lambda F(u), F(u)\big)$ are dense in $\operatorname{gph} H$ around $(0,0)$. Our attention can therefore be focused on such pairs; we can assume that $(\bar{w},\bar{z}) = \big(\bar{u} - \lambda F(\bar{u}), F(\bar{u})\big)$ for a certain $\bar{u} \in D$, writing $A = \nabla F(\bar{u})$.

The tangent cone to $\operatorname{gph} F$ at $\big(\bar{u}, F(\bar{u})\big)$ is the subspace of $I\!R^n \times I\!R^n$ consisting of the pairs (u', Au') as u' ranges over $I\!R^n$. Under the linear change of variables in 13(38), the tangent cone $T_{\operatorname{gph} H}(\bar{w},\bar{z}) = \operatorname{gph} DH(\bar{w}\,|\,\bar{z})$ must therefore be the subspace consisting of the pairs (w',z') of form $(u' - \lambda Au', Au')$ for some $u' \in I\!R^n$. This subspace has to contain $(\bar{w},\bar{z})$ itself, inasmuch as $\operatorname{gph} H$ is a cone. Hence there exists $\bar{u}' \in I\!R^n$ such that $(\bar{w},\bar{z}) = (\bar{u}' - \lambda A\bar{u}', A\bar{u}')$.

Our goal of proving that $\bar{z} \in D^*H(\bar{w}\,|\,\bar{z})(\bar{w})$ and $-\bar{z} \in D^*H(\bar{w}\,|\,\bar{z})(-\bar{w})$ comes down to verifying that $(\bar{z},-\bar{w})$ and $(-\bar{z},\bar{w})$ lie in $N_{\operatorname{gph} H}(\bar{w},\bar{z})$. In fact $(\bar{z},-\bar{w})$ and $(-\bar{z},\bar{w})$ are regular normals to $\operatorname{gph} H$ at $(\bar{w},\bar{z})$; we'll get this by showing that $(\bar{z},-\bar{w})$ is orthogonal to the subspace $T_{\operatorname{gph} H}(\bar{w},\bar{z})$. For arbitrary $(w',z') = (u' - \lambda Au', Au')$ in this subspace, we have

$$\begin{aligned}\big\langle(\bar{z},-\bar{w}),(w',z')\big\rangle &= \big\langle(A\bar{u}', -\bar{u}' + \lambda A\bar{u}'),(u' - \lambda Au', Au')\big\rangle \\ &= \big\langle A\bar{u}', u' - \lambda Au'\big\rangle - \big\langle \bar{u}' - \lambda A\bar{u}', Au'\big\rangle \\ &= \big\langle A\bar{u}', u'\big\rangle - \big\langle \bar{u}', Au'\big\rangle = \big\langle \bar{u}', (A^* - A)u'\big\rangle = 0,\end{aligned}$$

because the matrix $A = \nabla F(\bar{u})$ is symmetric. This finishes the proof. □

13.58 Corollary (subgradient coderivatives of fully amenable functions). *If f is fully amenable at $\bar{x}$, one has for all $\bar{v} \in \partial f(\bar{x})$ that*

$$\mathop{\text{g-lim sup}}_{\substack{(x,v)\to(\bar{x},\bar{v})\\ v\in\partial f(x)}} D(\partial f)(x\,|\,v) \;\subset\; D^*(\partial f)(\bar{x}\,|\,\bar{v}).$$

Proof. The hypothesis entails f also being fully amenable at all $x \in \operatorname{dom} f$

near $\bar{x}$; cf. 10.25(b). At such x and for any $v \in \partial f(x)$ we have f prox-regular and subdifferentially continuous by 13.32 and twice epi-differentiable by 13.15. Then from 13.57 we get $D^*(\partial f)(x \,|\, v) \supset D(\partial f)(x \,|\, v)$. The fact in 8(18) that

$$D^*(\partial f)(\bar{x} \,|\, \bar{v}) \;=\; \operatorname*{g\text{-}lim\,sup}_{\substack{(x,v)\to(\bar{x},\bar{v})\\ v\in\partial f(x)}} D^*(\partial f)(x \,|\, v)$$

then produces the claimed inclusion with respect to $D^*(\partial f)(\bar{x} \,|\, \bar{v})$. □

The limit relation in Corollary 13.58 could also, of course, be elaborated further with symmetries in the pattern of 13(37).

J.* Parabolic Subderivatives

We look now at another kind of second-order difference quotient, which extends the one in 13(4) that was used in defining $d^2 f(\bar{x})(w)$:

$$\Delta_\tau^2 f(\bar{x})(w \,|\, z) := \frac{f(\bar{x} + \tau w + \frac{1}{2}\tau^2 z) - f(\bar{x}) - \tau df(\bar{x})(w)}{\frac{1}{2}\tau^2}.$$

This is called a *parabolic* difference quotient, because it tests the values of f along a parabola through $\bar{x}$ instead of a line.

13.59 Definition (parabolic subderivatives). *For a function $f : \mathbb{R}^n \to \overline{\mathbb{R}}$, any point $\bar{x}$ with $f(\bar{x})$ finite and any vector w with $df(\bar{x})(w)$ finite, the parabolic subderivative at $\bar{x}$ for w with respect to z is*

$$d^2 f(\bar{x})(w \,|\, z) := \liminf_{\substack{\tau \searrow 0\\ z' \to z}} \frac{f(\bar{x} + \tau w + \frac{1}{2}\tau^2 z') - f(\bar{x}) - \tau df(\bar{x})(w)}{\frac{1}{2}\tau^2}.$$

Thus, in the parabolic difference quotient notation, the function $d^2 f(\bar{x})(w \,|\, \cdot) : \mathbb{R}^n \to \overline{\mathbb{R}}$ is defined by

$$d^2 f(\bar{x})(w \,|\, \cdot) = \operatorname*{e\text{-}lim\,inf}_{\tau \searrow 0} \Delta_\tau^2 f(\bar{x})(w \,|\, \cdot).$$

If actually $d^2 f(\bar{x})(w \,|\, \cdot) = \text{e-lim}_{\tau \searrow 0} \Delta_\tau^2 f(\bar{x})(w \,|\, \cdot) \not\equiv \infty$, then f is said to be parabolically epi-differentiable at $\bar{x}$ for w.

Parabolic epi-differentiability at $\bar{x}$ for w thus refers to the case where, for any z, one can actually find $z_\tau \to z$ such that $\Delta_\tau^2 f(\bar{x})(w \,|\, z_\tau) \to d^2 f(\bar{x})(w \,|\, z)$ as $\tau \searrow 0$, or equivalently, to the existence of an arc $\xi : [0, \varepsilon] \to \mathbb{R}^n$ (for some $\varepsilon > 0$) with $\xi(0) = \bar{x}$, $\xi'_+(0) = w$, $\xi''_+(0) = z$, for which the function $\varphi(\tau) = f(\xi(\tau))$ has $\varphi''_+(0) = d^2 f(\bar{x})(w \,|\, z)$. The two perspectives correspond in terms of

$$\xi(\tau) = \bar{x} + \tau w + \tfrac{1}{2}\tau^2 z_\tau, \qquad z_\tau = [\xi(\tau) - \xi(0) - \tau \xi'(0)]/\tfrac{1}{2}\tau^2.$$

13.60 Example (parabolic aspects of twice smooth functions). *Let $f : \mathbb{R}^n \to \mathbb{R}$ be of class $\mathcal{C}^2$. Then at any point $\bar{x}$, f is parabolically epi-differentiable at $\bar{x}$ for every $w \in \mathbb{R}^n$, and*

$$d^2 f(\bar{x})(w \,|\, z) = \langle w, \nabla^2 f(\bar{x})w\rangle + \langle \nabla f(\bar{x}), z\rangle.$$

Detail. This follows from the quadratic expansion of f at $\bar{x}$. □

13.61 Exercise (parabolic aspects of piecewise linear-quadratic functions). *Let $f : \mathbb{R}^n \to \overline{\mathbb{R}}$ be convex and piecewise linear-quadratic, and let $\bar{x} \in \operatorname{dom} f$. Then f is parabolically epi-differentiable at $\bar{x}$ for every $w \in \operatorname{dom} df(\bar{x})$, and*

$$\begin{aligned} d^2 f(\bar{x})(w \,|\, z) &= d^2 f(\bar{x})(w) + \sigma_{\partial f(\bar{x}|w)}(z) \quad \text{with} \\ \partial f(\bar{x} \,|\, w) &:= \big\{ v \in \partial f(\bar{x}) \,\big|\, df(\bar{x})(w) = \langle v, w\rangle \big\} = \operatorname*{argmax}_{v \in \partial f(\bar{x})} \langle v, w\rangle, \end{aligned}$$

hence in particular,

$$d^2 f(\bar{x})(w \,|\, z) < \infty \iff z \in T_K(w) \ \text{ for } \ K = \operatorname{dom} df(\bar{x})(w).$$

Guide. Get this out of the facts in 13.9. □

Parabolic derivatives can be interpreted geometrically in terms of the second-order tangent sets in 13.11, as follows.

13.62 Example (parabolic subderivatives versus second-order tangents).

(a) *For a set $C \subset \mathbb{R}^n$ and any choice of $\bar{x} \in C$ and $w \in T_C(\bar{x})$, one has $d^2\delta_C(\bar{x})(w \,|\, z) = \delta_{T^2_C(\bar{x}|w)}(z)$. Parabolic epi-differentiability of the indicator function δ_C at $\bar{x}$ for w corresponds to parabolic derivability of C at $\bar{x}$ for w.*

(b) *Consider $f : \mathbb{R}^n \to \overline{\mathbb{R}}$, a point $\bar{x}$ with $f(\bar{x})$ finite and a vector w with $df(\bar{x})(w)$ finite; let $\alpha = f(\bar{x})$ and $\beta = df(\bar{x})(w)$. Then*

$$\operatorname{epi} d^2 f(\bar{x})(w \,|\, \cdot) = T^2_{\operatorname{epi} f}\big((\bar{x}, \alpha) \,\big|\, (w, \beta)\big).$$

Parabolic epi-differentiability of f at $\bar{x}$ for w is parabolic derivability of $\operatorname{epi} f$ at $(\bar{x}, \alpha)$ for (w, β).

Detail. These observations are immediate from the definitions. □

13.63 Exercise (chain rule for parabolic subderivatives). *Suppose $f(x) = g(F(x))$ for a $\mathcal{C}^2$ mapping $F : \mathbb{R}^n \to \mathbb{R}^m$ and a proper, lsc function $g : \mathbb{R}^m \to \overline{\mathbb{R}}$. Let $\bar{x} \in \operatorname{dom} f$ satisfy the constraint qualification that the only vector $y \in \partial^\infty g(F(\bar{x}))$ with $\nabla F(\bar{x})^* y = 0$ is $y = 0$. Then for any $w \in \operatorname{dom} df(\bar{x})$, i.e., with $\nabla F(\bar{x})w \in \operatorname{dom} dg(F(\bar{x}))$, one has*

$$d^2 f(\bar{x})(w \,|\, z) = d^2 g\big(F(\bar{x}) \,\big|\, w\nabla^2 F(\bar{x})w + \nabla F(\bar{x})z\big).$$

Further, f is parabolically epi-differentiable at $\bar{x}$ for w if g is subdifferentially regular at $F(\bar{x})$ and parabolically epi-differentiable at $F(\bar{x})$ for $\nabla F(\bar{x})w$.

Guide. Develop this by combining 13.13 with Example 13.62(b). □

13.64 Proposition (properties of parabolic subderivatives). *For $f : \mathbb{R}^n \to \overline{\mathbb{R}}$, any point $\bar{x}$ with $f(\bar{x})$ finite and any vector w with $df(\bar{x})(w)$ finite, the function $d^2 f(\bar{x})(w \,|\, \cdot) : \mathbb{R}^n \to \overline{\mathbb{R}}$ is lsc. For vectors v with $df(\bar{x})(w) = \langle v, w\rangle$ it satisfies*

$$\inf_z \Big\{ d^2 f(\bar{x})(w \,|\, z) - \langle v, z\rangle \Big\} \geq d^2 f(\bar{x} \,|\, v)(w), \qquad 13(39)$$

where the left side equals

$$\liminf_{\substack{\tau \searrow 0,\ w' \to w \\ [w'-w]/\tau \text{ bounded}}} \Delta_\tau^2 f(\bar{x} \,|\, v)(w')$$

in contrast to the right side, which has this form without the boundedness.

Proof. As a lower epi-limit, $d^2 f(\bar{x})(w \,|\, \cdot)$ is lsc. The left side of 13(39) is identical to the lowest limit attainable for $\lim_\nu \Delta_{\tau^\nu}^2(\bar{x})(w \,|\, z^\nu) - \langle v, z^\nu\rangle$ relative to $\tau^\nu \searrow 0$ and a bounded sequence of vectors z^ν (as seen from the cluster points z of such a sequence). In terms of $w^\nu = w + \frac{1}{2}\tau^\nu z^\nu$, which corresponds to $z^\nu = 2[w^\nu - w]/\tau^\nu$, we have, whenever $df(\bar{x})(w) = \langle v, w\rangle$, that

$$\begin{aligned} &\Delta_{\tau^\nu}^2 f(\bar{x})(w \,|\, z^\nu) - \langle v, z^\nu\rangle \\ &\quad = \frac{f(\bar{x} + \tau^\nu(w + \frac{1}{2}\tau^\nu z^\nu)) - f(\bar{x}) - \tau^\nu df(\bar{x})(w) - \langle v, \frac{1}{2}\tau^{\nu\,2} z^\nu\rangle}{\frac{1}{2}\tau^{\nu\,2}} \\ &\quad = \frac{f(\bar{x} + \tau^\nu(w + \frac{1}{2}\tau^\nu z^\nu) - f(\bar{x}) - \tau^\nu\langle v, w + \frac{1}{2}z^\nu\rangle}{\frac{1}{2}\tau^{\nu\,2}} = \Delta_{\tau^\nu}^2 f(\bar{x} \,|\, v)(w^\nu). \end{aligned}$$

It's clear then that the left side of 13(39) is the lowest value attainable as $\lim_\nu \Delta_{\tau^\nu} f(\bar{x} \,|\, v)(w^\nu)$ for sequences $\tau^\nu \searrow 0$ and $w^\nu \to w$ with $[w^\nu - w]/\tau^\nu$ bounded. Without the boundedness restriction we would get $d^2 f(\bar{x} \,|\, v)(w)$ this way, so the inequality in 13(39) is correct. □

13.65 Definition (parabolic regularity). *A function $f : \mathbb{R}^n \to \overline{\mathbb{R}}$ is parabolically regular at a point $\bar{x}$ for a vector v if $f(\bar{x})$ is finite and the inequality in 13(39) holds with equality for every w having $df(\bar{x})(w) = \langle v, w\rangle$, or in other words if for such w with $d^2 f(\bar{x} \,|\, v)(w) < \infty$ there exists, among the sequences $\tau^\nu \searrow 0$ and $w^\nu \to w$ with $\Delta_{\tau^\nu}^2 f(\bar{x} \,|\, v)(w^\nu) \to d^2 f(\bar{x} \,|\, v)(w)$, ones with the additional property that $\limsup_\nu |w^\nu - w|/\tau^\nu < \infty$.*

A major source of interest in parabolic subderivatives and parabolic regularity lies in second-order optimality conditions. A connection with the version of such conditions in terms of second subderivatives in Theorem 13.24 can be seen through the fact that, by 13.5, one has

$$\operatorname{dom} d^2 f(\bar{x} \,|\, 0) \subset \big\{ w \,\big|\, df(\bar{x})(w) \leq 0 \big\},$$

with $d^2 f(\bar{x} \,|\, 0)(w) = -\infty$ if $df(\bar{x})(w) < 0$. For comparison, 13.64 gives us

$$\inf_z d^2 f(\bar{x})(w \,|\, z) \geq d^2 f(\bar{x} \,|\, 0)(w) \ \text{ whenever } \ df(\bar{x})(w) = 0,$$

and by definition this holds with equality when f is parabolically regular at $\bar{x}$ for $v = 0$. These observations allow the optimality conditions in 13.24 to be translated into statements about parabolic subderivatives.

13.66 Theorem (parabolic version of second-order optimality). *Consider the problem of minimizing a proper function f over $\mathbb{R}^n$.*

(a) *If $\bar{x}$ is locally optimal, then $df(\bar{x})(w) \geq 0$ for all w, and in the case of $w \neq 0$ with $df(\bar{x})(w) = 0$, also $\inf_z d^2 f(\bar{x})(w \,|\, z) \geq 0$.*

(b) *If f is parabolically regular at $\bar{x}$ for $v = 0$ and has $df(\bar{x})(w) \geq 0$ for all w, and in the case of $w \neq 0$ with $df(\bar{x})(w) = 0$ also $\inf_z d^2 f(\bar{x})(w \,|\, z) > 0$, then $\bar{x}$ is locally optimal, and moreover there exist $\varepsilon > 0$ and $\delta > 0$ such that $f(x) \geq f(\bar{x}) + \varepsilon |x - \bar{x}|^2$ when $|x - \bar{x}| \leq \delta$.*

Proof. This is obvious from Theorem 13.24, the general inequality in 13(39) and the definition of parabolic regularity in 13.65(a), as applied to $v = 0$. □

The necessary condition in Theorem 13.66(a) is equivalent to the one in terms of second-order epi-derivatives in Theorem 13.24(a) as long as f is parabolically regular at $\bar{x}$ for $\bar{v} = 0$, although without that property it may be weaker. Parabolic regularity is absolutely essential, however, in obtaining the sufficient condition in 13.66(b).

An example of what can go wrong is furnished on $\mathbb{R}^2$ by the function $f(x_1, x_2) = |x_2 - x_1^{4/3}| - x_1^2$ at $\bar{x} = (0, 0)$. This function is amenable at $\bar{x}$ and semidifferentiable there with

$$df(\bar{x})(w_1, w_2) = |w_2|, \qquad \partial f(\bar{x}) = \{(v_1, v_2) \,|\, v_1 = 0,\ |v_2| \leq 1\},$$

hence $0 \in \partial f(\bar{x})$. It's twice epi-differentiable at $\bar{x}$ for $v = 0$ with

$$d^2 f(\bar{x} \,|\, 0)(w_1, w_2) = \begin{cases} -2w_1^2 & \text{if } w_2 = 0, \\ \infty & \text{if } w_2 \neq 0, \end{cases}$$

so f doesn't have a local minimum at $\bar{x}$ because the second-order necessary condition in 13.24(a) isn't satisfied. In the context of 13.66, however, we not only have $df(\bar{x})(w) \geq 0$ for all w, but, for vectors $w \neq 0$ with $df(\bar{x})(w) = 0$ (namely $w = (w_1, w_2)$ with $w_1 \neq 0$, $w_2 = 0$) also

$$d^2 f(\bar{x})(w \,|\, z) = \infty \ \text{ for all } \ z \in \mathbb{R}^n.$$

Hence $\inf_z d^2 f(\bar{x})(w \,|\, z) > 0$ for such w, and the second-order 'sufficient' condition in 13.66(b) is fulfilled despite the lack of a local minimum; the condition isn't really sufficient of course unless f is parabolically regular at $\bar{x}$ for $v = 0$, and that's false in this case.

The culprit in this example isn't infinite-valuedness of $d^2 f(\bar{x})(w \,|\, z)$ per se. If $f(x_1, x_2)$ were replaced by $f_0(x_1, x_2) = \min\{f(x_1, x_2), x_1^2\}$, the first-order optimality conditions would still be satisfied at $\bar{x} = (0, 0)$ and f_0 would be twice epi-differentiable there with $d^2 f_0(\bar{x} \,|\, 0) = d^2 f(\bar{x} \,|\, 0)$. But now $d^2 f(\bar{x})(w \,|\, z) = 2w_1^2$ for all $z \in \mathbb{R}^n$ when w is a vector with $df(\bar{x})(w) = 0$. Again, because of the absence of parabolic regularity, the second-order inequality in 13.66(b) is unable to detect that f doesn't have a local minimum at $\bar{x}$.

Fully amenable functions don't suffer from this sort of parabolically irregular behavior, and indeed they exhibit an especially tight relationship between parabolic subderivatives and second subderivatives.

13.67 Theorem (parabolic aspects of fully amenable functions). *Let $f : \mathbb{R}^n \to \overline{\mathbb{R}}$ be fully amenable at $\bar{x}$. Then f is parabolically epi-differentiable at $\bar{x}$ for every vector $w \in \operatorname{dom} df(\bar{x})$ and parabolically regular at $\bar{x}$ for every vector v included in $\partial f(\bar{x})$.*

Indeed, for every $w \in \operatorname{dom} df(\bar{x})$ the function $z \mapsto d^2 f(\bar{x})(w \,|\, z)$ is lower semicontinuous, proper, convex and piecewise linear, and it is conjugate to the function

$$v \mapsto \begin{cases} -d^2 f(\bar{x} \,|\, v)(w) & \text{if } v \in \partial f(\bar{x} \,|\, w) \\ \infty & \text{if } v \notin \partial f(\bar{x} \,|\, w) \end{cases} \quad \text{for } \partial f(\bar{x} \,|\, w) := \operatorname*{argmax}_{v \in \partial f(\bar{x})} \langle v, w \rangle,$$

which likewise is lsc, proper, convex and piecewise linear. With respect to any local representation $f = g \circ F$ around $\bar{x}$ as in the definition of full amenability (with g piecewise linear-quadratic), one has

$$\begin{aligned} d^2 f(\bar{x})(w \,|\, z) &= d^2 g(F(\bar{x}))\big(\nabla F(\bar{x})w \;\big|\; w\nabla^2 F(\bar{x})w + \nabla F(\bar{x})z\big) \\ &= d^2 g(F(\bar{x}))\big(\nabla F(\bar{x})w\big) \\ &\quad + \max\Big\{ \big\langle y,\, w\nabla^2 F(\bar{x})w + \nabla F(\bar{x})z \big\rangle \;\Big|\; y \in Y(\bar{x} \,|\, w) \Big\} \\ &\qquad \text{for } Y(\bar{x} \,|\, w) := \operatorname*{argmax}_{y \in \partial g(F(\bar{x}))} \langle y, \nabla F(\bar{x})w \rangle. \end{aligned} \qquad 13(40)$$

Proof. We'll proceed in terms of the representation $f = g \circ F$ in the amenability context. That immediately yields through 13.63 and 13.61 the parabolic epi-differentiability of f along with the formula in 13(40), so we concentrate on proving the duality relations.

Let $h(z) = d^2 f(\bar{x})(w \,|\, z)$ and denote by k the function of v that's claimed to be conjugate to h. Let $\gamma = d^2 g(F(\bar{x}))(\nabla F(\bar{x})w)$. From 13(40) we have

$$\begin{aligned} h(z) &= d^2 g(F(\bar{x}))\big(\nabla F(\bar{x})w \;\big|\; w\nabla^2 F(\bar{x})w + \nabla F(\bar{x})z\big) \\ &= \gamma + \sigma_{Y(\bar{x}|w)}\big(w\nabla^2 F(\bar{x})w + \nabla F(\bar{x})z\big) \end{aligned}.$$

The set $Y(\bar{x} \,|\, w)$ is not just convex, it is polyhedral, because $\partial g(F(\bar{x}))$ is polyhedral. On the other hand, Theorem 13.14 tells us that k is a proper convex function such that

$$-k(v) = \begin{cases} \gamma + \max\big\{ \langle y, w\nabla^2 F(\bar{x})w \rangle \;\big|\; y \in Y(\bar{x}, v) \big\} & \text{if } v \in \partial f(\bar{x} \,|\, w), \\ \infty & \text{otherwise,} \end{cases}$$

where

$$Y(\bar{x}, v) := \big\{ y \in \partial g(F(\bar{x})) \;\big|\; \nabla F(\bar{x})^* y = v \big\}.$$

The pairs (v,y) such that $v \in \partial f(\bar{x}\,|\,w)$ and $y \in Y(\bar{x},v)$ are the ones with $y \in Y(\bar{x}\,|\,w)$ and $\nabla F(\bar{x})^* y = v$. Thus, $k(v) = -\gamma + \min_y \psi(v,y)$ for the convex, piecewise linear function

$$\psi(v,y) = \begin{cases} -\langle y, w\nabla^2 F(\bar{x})w\rangle & \text{if } y \in Y(\bar{x}\,|\,w) \text{ and } \nabla F(\bar{x})^* y = v, \\ \infty & \text{otherwise.} \end{cases}$$

This description shows that k is piecewise linear (cf. 3.55(b)). We calculate

$$\begin{aligned} k^*(z) &= \sup_v \big\{\langle v,z\rangle - k(v)\big\} = \sup_{v,y}\big\{\langle v,z\rangle + \gamma - \psi(v,y)\big\} \\ &= \gamma + \sup_{y\in Y(\bar{x}|w)} \big\{\langle \nabla F(\bar{x})^* y, z\rangle + \langle y, w\nabla^2 F(\bar{x})w\rangle\big\} \\ &= \gamma + \sigma_{Y(\bar{x}|w)}\big(w\nabla^2 F(\bar{x})w + \nabla F(\bar{x})z\big) = h(z). \end{aligned}$$

Thus, $k^* = h$. It follows that h, like k, is proper, convex and piecewise linear (see 11.14), and $h^* = k$. □

13.68 Corollary (parabolic aspects of fully amenable sets). *If a set $C \subset \mathbb{R}^n$ is fully amenable at $\bar{x}$, then C is not only parabolically derivable at $\bar{x}$ for every $w \in T_C(\bar{x})$, but also parabolically regular at $\bar{x}$ for every $v \in N_C(\bar{x})$.*

Indeed, for every $w \in T_C(\bar{x})$ the second-order tangent set $T_C^2(\bar{x}\,|\,w)$ is nonempty and polyhedral, and its support function is given by

$$\sigma_{T_C^2(\bar{x}|w)}(v) = \begin{cases} -d^2(\delta_C)(\bar{x}\,|\,v)(w) & \textit{if } v \in N_C(\bar{x}) \textit{ and } v \perp w, \\ \infty & \textit{otherwise.} \end{cases}$$

Proof. Apply Theorem 13.67 to $f = \delta_C$. □

The parabolic derivability assertion in 13.68 reiterates, in different words, the result of 13.13 for the case of polyhedral D.

Commentary

The first major result about generalized second-order differentiability of functions was the theorem of Alexandrov [1939], according to which a finite, convex function on an open convex subset of $\mathbb{R}^n$ has a quadratic expansion at almost every point. A geometric counterpart to that result was independently obtained by Busemann and Feller [1936] in an investigation of convex hypersurfaces. Infinite-dimensional analogs have recently been explored by Borwein and Noll [1994].

Here we have found it advantageous not to interpret second-order differentiability beyond the classical setting as referring automatically just to the existence of a quadratic expansion, but to focus instead on the new notion in Definition 13.1(b). That notion of extended second-order differentiability fits better with our broader theme of 'subdifferentiability' while still supporting the derivation of results like Alexandrov's theorem. It's always sufficient for the existence of a quadratic expansion, as demonstrated in Theorem 13.2(b), and it's necessary too when the function f is 'prox-regular', according to 13.42. Equivalence thus holds for finite, convex functions in particular. Through the connection in 13.2(c) between extended second-order differentiability of f and first-order differentiability of ∂f, we are able to apply Rademacher's theorem (in 9.60) to the graph of ∂f and in that way get the generic existence of quadratic expansions to convex functions as a special case of 13.51.

The approach of using epigraphical instead of pointwise limits to define generalized second-derivatives originated with Rockafellar [1985b], [1988]. That work involved both of the subderivative expressions in Definition 13.3, but it concentrated on the situation where the lower epi-limit giving $d^2 f(\bar{x} \,|\, \bar{v})$ coincides with the corresponding upper epi-limit so as to yield the second-order epi-differentiability defined in 13.6(b). The goal was the establishment of the hitherto unsuspected fact, stated here in 13.14, that this property holds for fully amenable functions and thus prevails in the typical circumstances of finite-dimensional optimization, offering another tool for the design and analysis of numerical methods. Also in that early work of Rockafellar were most of the basic facts in 13.9 about piecewise linear-quadratic functions.

Second-order semidifferentiability, as a natural concept with its own attractions, hasn't previously been investigated. Its unfortunate shortcomings, powerfully illustrated in 13.10, make obvious the need a theory based on epi-convergence instead of uniform convergence of difference quotient functions.

Ioffe [1991] began the systematic study of second-order subderivatives defined by lower epi-limits rather than full epi-limits, as in 13.3. In that paper he obtained a chain rule inequality somewhat resembling (but weaker than) the new one in 13.14, as a complement to the chain rule equation of Rockafellar [1988] in the case of fully amenable functions. Cominetti [1991] identified the properties in composition that enable a chain rule equation for second-order subderivatives to be obtained more generally. For infinite-dimensional chain rules involving integral functionals, see Do [1992], Levy [1993], and Loewen and Zheng [1995].

For the formulas deduced from the chain rule and stated in 13.15, 13.16 and 13.18, see Poliquin and Rockafellar [1993], [1994], as well as Rockafellar [1989a], where optimality conditions were developed out of second-order epi-derivatives as well; cf. 13.24 and 13.25. The special optimality conditions in the penalty format of 13.26 are new but correspond to the same ideas in utilizing the duality for convex functions in 13.21, 13.22 and 13.23, which was established in Rockafellar [1990b] (cf. related work of Gorni [1991], Seeger [1992b], Penot [1994b]).

In the theory of second-order tangents, the parabolic derivability of fully amenable sets in 13.13 was proved by Rockafellar [1988]. There, as here, it was the platform for proving that fully amenable functions are twice epi-differentiable; cf. 13.14. The broader case of the chain rule in 13.13 in which D is convex but not necessarily polyhedral is due to Cominetti [1990], who worked with the tangent cone form of the constraint qualification. For nonconvex D, the rule is new.

Also developed in Rockafellar [1988] was the parabolic subderivative notion in 13.59, the focus being on parabolic epi-differentiability and the fact in 13.67 that fully amenable functions enjoy that property. The duality in 13.67 between parabolic subderivatives and second-order subderivatives of fully amenable functions was established in that paper as well. The support function formula in 13.68 for the second-order tangent sets associated with fully amenable sets, which furnishes parabolic regularity, wasn't made explicit there, but is simply the case where the function f is taken to be the indicator δ_C.

Parabolic 'directional derivatives', defined in analogy to traditional first-order directional derivatives but along parabolas instead of straight lines—or in other words by pointwise convergence of parabolic difference quotients instead of epi-convergence as with parabolic subderivatives—were featured earlier by Ben-Tal [1980] as a tool in the development of second-order optimality conditions; see also Ben-Tal and Zowe [1982], [1985]. This approach was continued by Cominetti [1990], who looked at second-order tangent sets in spaces of possibly infinite dimensions. Of course, because parabolic directional derivatives don't draw on epi-convergence, they don't correspond to second-order tangents to an epigraph in the way that parabolic subderivatives do in 13.62(b).

The approach to second-order optimality through second-order tangents was recently explored further by Bonnans, Cominetti and Shapiro [1996a], [1996b]. Those researchers formulated a property of 'second-order regularity' which is sufficient (although not necessary) for the parabolic regularity defined in 13.65, and they showed that this property holds for certain sets outside the fully amenable class, e.g. the cone of positive-definite, symmetric matrices. The necessary and sufficient conditions for optimality that they obtained fit the pattern of the parabolic conditions in 13.66 and thus can be construed as optimality conditions in terms of second-order subderivatives, echoing the ones in 13.24 (with some infinite-dimensional overtones).

For additional material on 'higher-order tangents', see Aubin and Frankowska [1990] and Penot [1992]. For other work on second-order optimality in the face of nonsmoothness, see Ioffe [1979], Chaney [1982a], [1982b], [1983], [1987], Hiriart-Urruty, Strodiot and Nguyen [1984], Burke [1987], Cominetti [1990], Cominetti and Correa [1990], Studniarski [1991], Penot [1994a], and Poliquin and Rockafellar [1997]. The precise relationship between Chaney's ideas and the ones in this chapter has yet to be clarified.

The bridge revealed in 13.20 between epi-derivatives of convex functions and the gauges of certain convex sets provides the way of relating the second-order theory presented here to an approach of Hiriart-Urruty [1982a], [1982c], [1984], [1986] in which those sets serve as the second-order objects and enjoy a calculus. Hiriart-Urruty's version of second-order differentiation has no apparent extension to nonconvex functions. For more on it, see also Hiriart-Urruty and Seeger [1989a], [1989b], Seeger [1992b], [1994], and Moussaoui and Seeger [1994].

The concept of prox-regularity in 13.27 was introduced along with the subdifferential continuity in 13.28 by Poliquin and Rockafellar [1996a], who also developed the connections with convexity, normality and amenability in 13.30, 13.31 and 13.32. A kindred notion of 'proximal smoothness' was formulated by Clarke, Stern and Wolenski [1995] in the geometric framework of projections on nonconvex sets, and in a global rather than local manner. They also got a result about projection mappings like 13.38. Their work encompassed subsets of general Hilbert spaces.

The fact in 13.34, that a function f belongs to class $\mathcal{C}^{1+}$ if and only if both f and $-f$ are lower-$\mathcal{C}^2$, is new.

Theorem 13.30, which uses prox-regularity to tie second-order epi-derivatives of f to proto-derivatives of ∂f, is a result of Poliquin and Rockafellar [1996a]. This pathway of analysis, invoking Attouch's theorem on graphical convergence of subgradient mappings (cf. 12.35), was opened up by Rockafellar [1985b], [1990b], in the case of convex functions and extended by Poliquin [1990b] to strongly amenable functions; for an infinite-dimensional version of the latter, see Penot [1995]. Other results of Poliquin and Rockafellar [1996a], [1996b], are seen in 13.45, 13.46 and 13.49, and in the symmetry assertion of 13.51. The formula in 13.50 follows Poliquin and Rockafellar [1993], [1994].

The proto-derivative model in 13.47 for the perturbation of first-order optimality conditions emerged in Rockafellar [1990a]. The one in 13.48 came up in Rockafellar [1989b] for the variational inequality case (where f is the indicator of a closed, convex set), but it has echoes in unpublished work of Robinson [1985] that wasn't focused on general proto-derivatives. It was extended in a number of ways by Levy and Rockafellar [1994], [1996] and Levy [1996]. Other precedents for the model 13.48 are found in the papers of Robinson [1977], [1980], [1982] on continuous dependence of solutions to generalized equations; more recently, Robinson [1991] has worked with semidifferentiability developed through Minty reparameterization.

For both perturbation models, the criteria for the Aubin property can be regarded as elaborations of ideas of Mordukhovich [1994]. Additional results on the model in 13.47 can be found in Poliquin and Rockafellar [1992] for an important special case. Lipschitzian properties (without proto-differentiability) in the general context of 13.48 have been investigated by Dontchev and Hager [1994]. In the case of 13.48 where $\partial f = N_C$ for a polyhedral set C, Dontchev and Rockafellar [1996] have shown that the Aubin property of S implies localized single-valuedness, hence semidifferentiability as well, and the criterion for the Aubin property can be stated as a 'critical face condition'.

The facts in Theorem 13.52 are largely attributable to Cominetti and Correa [1986], [1990], apart for the articulation furnished here in terms of the strict derivative mapping $D_*(\nabla f)(\bar{x})$. The consequences for proximal mappings and projections in 13.53 and 13.54 haven't previously been recorded. Also deserving credit for parts of the picture in 13.52 are Hiriart-Urruty [1982b] and Páles and Zeidan [1996]. The 1990 paper of Cominetti and Correa developed strict second-order differentiability as well, although not all of 13.56. The connection with second-order Lipschitz continuity in 13.55 was uncovered by Cominetti [1986]. It's known also, as a sort of partial converse to Theorem 13.52, that the finiteness of the double-difference quotient limit in 13(34) requires that f be C^{1+}; this was proved by Milosz [1988]. For other work on second-order properties of functions of class C^{1+}, see Yang and Jeyakumar [1992], [1995], Yang [1994], [1996], and Georgiev and Zlateva [1996].

The derivative-coderivative inclusions in 13.57 and 13.58 come from Rockafellar and Zagrodny [1997].

For another line of geometric development, aimed at generalizing the Dupin indicatrix of differential geometry, see Hiriart-Urruty [1989a] and Noll [1992]. Second-derivative ideas in terms of 'jets' have been explored by Ioffe and Penot [1995]. A different type of second subderivative has been defined by Michel and Penot [1994].

An application of second-order nonsmooth analysis to a fundamental question of statistics, as formulated with respect to the behavior of optimal solutions in problems that depend on random variables, can be seen in King and Rockafellar [1993].

14. Measurability

Problems involving 'integration' offer rich and challenging territory for variational analysis, and indeed it's especially around such problems, under the heading of the calculus of variations, that the subject has traditionally been organized. Models in which expressions have to be integrated with respect to time are central to the treatment of dynamical systems and their optimal control. When the systems are 'distributed', with states that, like density distributions, are conceived as elements of a function space, integration with respect to spatial variables or other parameters can enter the picture as well. In economics, it may be desirable to integrate over a space of infinitesimal agents. Applications in stochastic environments often concern expected values that are defined by integration with respect to a probability measure, or may demand a sturdy platform for working with concepts like that of a 'random set,' a 'random function', or even a 'random problem of optimization'.

Satisfactory handling of problem models in these categories usually requires an appeal to measure theory, and that inevitably raises questions about measurable dependence. This chapter is aimed at providing the technical machinery for answering such questions, so that analysis can go forward in full harmony with the ideas developed in the preceding chapters, where variable points are often replaced by variable sets, the geometry of graphs is replaced by that of epigraphs, and so forth.

Formally, 'measurability' can be regarded as analogous to 'continuity'. A need for results about measurability can be anticipated in just about every situation where results about continuity are worth knowing. Measurability has distinctive features, however. While continuous selections, as discussed at the end of Chapter 5, have a relatively limited role in the theory of continuity, measurable selections are the tool for resolving many of the issues that have to be faced in coping with measurability. Another feature is that a broader range of spaces has to be admitted. For continuity, we could conveniently focus our attention on mappings from one Euclidean space to another, thereby lessening the mathematical burden without undue sacrifice of generality, at least in an introductory phase of study. But for measurability it's important to proceed more abstractly, above all for the sake of stochastic applications. Probability spaces can be subsets of some $\mathbb{R}^d$, but commonly too they are discrete, or some mixture of continuous and discrete. A complication that must be accommodated as well is the presence of more than one probability measure, for instance a sequence of empirical measures that converges in some way to an underlying 'true' measure.

This obliges us to operate in the framework of a *measurable space* instead of a measure space. That space will be denoted by $(T, \mathcal{A})$; here T is any nonempty set and $\mathcal{A}$ is some σ-field of subsets of T, these being called the *measurable* subsets of T, or the $\mathcal{A}$*-measurable* subsets for emphasis. In many situations, T may have topological structure relevant to the designation of $\mathcal{A}$, but that isn't of immediate concern. Only occasionally will we be referring to some measure μ on $(T, \mathcal{A})$.

At the top of our agenda is the investigation of mappings $S : T \rightrightarrows I\!R^n$, what it means for them to be 'measurable', and what criteria can be used to check whether this property is at hand. We'll turn then toward the study of 'integrands' $f : T \times I\!R^n \to \overline{I\!R}$ as crucial elements in the expression of integral functionals of the form

$$I_f[x] = \int_T f\big(t, x(t)\big)\mu(dt) \qquad 14(1)$$

on spaces $\mathcal{X}$ of functions $x : T \to I\!R^n$. A hurdle that has to be surmounted in getting I_f to be well defined is that the standard assumptions on the integrand f that might be invoked to ensure the measurability of $t \mapsto f\big(t, x(t)\big)$ when $t \mapsto x(t)$ is measurable, such as measurability of f in the T argument in combination with continuity in the $I\!R^n$ argument, aren't adequate for the settings in variational analysis where $f(t, \cdot)$ is at best merely lsc on $I\!R^n$. The key to progress will be the class of integrands f that are epi-measurable, i.e., such that the set-valued mapping $t \mapsto \operatorname{epi} f(t, \cdot)$ is measurable.

A. Measurable Mappings and Selections

Measurability of set-valued mappings $S : T \rightrightarrows I\!R^n$ can be described from several different angles in analogy with the measurability of single-valued mappings, with which we suppose the reader is familiar. At the foundation is the measurability of various inverse images

$$S^{-1}(D) = \bigcup\nolimits_{x \in D} S^{-1}(x) = \big\{ t \in T \,\big|\, S(t) \cap D \neq \emptyset \big\}.$$

14.1 Definition (measurable mappings). *A set-valued mapping $S : T \rightrightarrows I\!R^n$ is measurable if for every open set $O \subset I\!R^n$ the set $S^{-1}(O) \subset T$ is measurable, i.e., $S^{-1}(O) \in \mathcal{A}$. In particular, the set $\operatorname{dom} S = S^{-1}(I\!R^n)$ must be measurable.*

When S happens to be single-valued, this definition of measurability reduces to the ordinary one for single-valued mappings (in one of its equivalent forms). An elementary class of examples beyond single-valuedness is furnished by the *constant* mappings, with $S(t) \equiv D$ for a set $D \subset I\!R^n$. On the heels of this is the class of 'step' mappings $S : T \rightrightarrows I\!R^n$ generated by partitioning T

into a collection of sets $T_j \in \mathcal{A}$ for $j \in J$, a countable (or finite) index set and, for some choice of sets $D_j \subset \mathbb{R}^n$, defining $S(t) \equiv D_j$ for $t \in T_j$.

Further examples on such a level are $S(t) = D + a(t)$ and $S(t) = \lambda(t)D$ for any set $D \subset \mathbb{R}^n$, as long as the vector $a(t) \in \mathbb{R}^n$ and scalar $\lambda(t) \in \mathbb{R}$ depend measurably on t. In the first case this is clear from having $S^{-1}(O) = a^{-1}(O - D)$, where $O - D$ is open when O is open. In the second case one has $S^{-1}(O) = \lambda^{-1}(U)$ for the open set $U = \{\tau \in \mathbb{R} \mid O \cap \tau D \neq \emptyset\}$.

To tap into the wealth of other examples, it's helpful to have a tool chest of criteria for measurability beyond the definition itself. Closed-valuedness of a mapping will need to be assumed in many cases, but that's not much of a handicap because closed-valued mappings S are typical in applications, and anyway the property is constructive in the following sense.

14.2 Proposition (preservation with closure of images). *For $S : T \rightrightarrows \mathbb{R}^n$,*

$$t \mapsto S(t) \text{ measurable} \quad \Longrightarrow \quad t \mapsto \operatorname{cl} S(t) \text{ measurable.}$$

Proof. For $O \subset \mathbb{R}^n$, clearly $O \cap S(t) \neq \emptyset$ if and only if $O \cap \operatorname{cl} S(t) \neq \emptyset$. □

Arguments about countability must frequently be made, since the σ-field $\mathcal{A}$ only allows for countable unions and intersections. It's no wonder then that we'll often have recourse to the rational numbers $\mathbb{Q}$ and the countable dense subset $\mathbb{Q}^n$ of $\mathbb{R}^n$. It will be convenient to speak of *rational closed balls* $\mathbb{B}(x, \rho)$ and *rational open balls* $\operatorname{int} \mathbb{B}(x, \rho)$ when $x \in \mathbb{Q}^n$ and $\rho \in \mathbb{Q}$, $\rho > 0$.

14.3 Theorem (measurability equivalences under closed-valuedness). *For closed-valued $S : T \rightrightarrows \mathbb{R}^n$, each of the following is equivalent to S being measurable:*

(a) *$S^{-1}(O) \in \mathcal{A}$ for all open sets $O \subset \mathbb{R}^n$ (the definition);*
(b) *$S^{-1}(C) \in \mathcal{A}$ for all closed sets $C \subset \mathbb{R}^n$;*
(c) *$S^{-1}(C) \in \mathcal{A}$ for all compact sets $C \subset \mathbb{R}^n$;*
(d) *$S^{-1}(B) \in \mathcal{A}$ for all closed balls $B \subset \mathbb{R}^n$;*
(e) *$S^{-1}(B) \in \mathcal{A}$ for all rational closed balls $B \subset \mathbb{R}^n$;*
(f) *$S^{-1}(O) \in \mathcal{A}$ for all open balls $O \subset \mathbb{R}^n$;*
(g) *$S^{-1}(O) \in \mathcal{A}$ for all rational open balls $O \subset \mathbb{R}^n$;*
(h) *$\{t \in T \mid S(t) \subset O\} \in \mathcal{A}$ for all open sets $O \subset \mathbb{R}^n$;*
(i) *$\{t \in T \mid S(t) \subset C\} \in \mathcal{A}$ for all closed sets $C \subset \mathbb{R}^n$;*
(j) *the function $t \mapsto d(x, S(t)) : T \to \mathbb{R}$ is measurable for each $x \in \mathbb{R}^n$.*

Proof. (a)⇒(c): For any nonempty, compact set $B \subset \mathbb{R}^n$ consider the sets $B^\nu := \{x \in \mathbb{R}^n \mid d(x, B) < 1/\nu\}$ for $\nu \in \mathbb{N}$. Since $S(t) \cap B \neq \emptyset$ if and only if $S(t) \cap B^\nu \neq \emptyset$ for all $\nu \in \mathbb{N}$, we have $S^{-1}(B) = \bigcap_\nu S^{-1}(B^\nu)$. Thus, $S(B)$ is the intersection of a countable collection of measurable sets and therefore is itself measurable.

(c)⇒(d)⇒(e): Closed balls are compact.

(e)⇒(a): Every open set O in $\mathbb{R}^n$ is the countable union of closed rational balls, say $O = \bigcup_{\nu \in \mathbb{N}} \mathbb{B}(x^\nu, \rho^\nu)$ with $x^\nu \in \mathbb{Q}^n$, $\rho^\nu \in \mathbb{Q}$, $\rho > 0$. Then $S^{-1}(O) = \bigcup_{\nu \in \mathbb{N}} S^{-1}(\mathbb{B}(x^\nu, \rho^\nu))$. As a countable union of sets in $\mathcal{A}$, $S^{-1}(O)$ is in $\mathcal{A}$.

(c)⇔(b): Noting that every compact set is closed and that every closed set $C = \bigcup_{\nu\in I\!N}(C \cap \nu I\!B)$ is the countable union of compact sets, one can draw on the argument used in proving that (e)⇒(a).

(a)⇒(f)⇒(g)⇒(a): Since every open ball is an open set, and every rational open ball is an open ball, it really suffices to check that (g)⇒(a). But that implication is immediate from the fact that every open subset of $I\!R^n$ can be expressed as the countable union of rational open balls.

(h)⇔(b) and (i)⇔(a): Here we simply use the fact that $\{t \mid S(t) \subset D\} = T \setminus S^{-1}(D')$ for $D' = I\!R^n \setminus D$.

(j)⇔(d): For any choice of $\alpha \in I\!R_+$, one has $d(x, S(t)) \leq \alpha$ if and only if $S(t) \cap (x + \alpha I\!B) \neq \emptyset$. (This utilizes the closedness of $S(t)$.) Therefore,

$$\{t \in T \mid d(x, S(t)) \leq \alpha\} = S^{-1}(x + \alpha I\!B).$$

Condition (j) corresponds to the measurability of the set on the left, while (d) corresponds to the measurability of the set on the right. □

The equivalences in Theorem 14.3 reassure us that, as long as we keep to closed-valued mappings, several properties that might be contemplated as alternative definitions of measurability are consistent with each other. The question of consistency with alternative interpretations also comes up in another way, however.

When a mapping $S : T \rightrightarrows I\!R^n$ is closed-valued, it can be identified with a single-valued mapping from the set $\operatorname{dom} S$ into the 'hyperspace' $\text{cl-sets}_{\neq\emptyset}(I\!R^n)$, which consists of all nonempty closed subsets of $I\!R^n$. Measurability of that single-valued mapping has a standard interpretation, namely that the inverse image of every Borel subset in the hyperspace is a measurable subset of T. The Borel field on $\text{cl-sets}_{\neq\emptyset}(I\!R^n)$ is the one generated by the topology of (Painlevé-Kuratowski) set convergence, or equivalently with respect to the hyperspace metric $d\!l$ developed at the end of Chapter 4. Does this interpretation clash with the definition of measurability in 14.1? No, according to the next theorem.

14.4 Theorem (measurability in hyperspace terms). *A closed-valued mapping $S : T \rightrightarrows I\!R^n$ is measurable if and only if it is measurable when viewed as a single-valued mapping from* $\operatorname{dom} S$, *a measurable subset of T, to the metric hyperspace* $(\text{cl-sets}_{\neq\emptyset}(I\!R^n), d\!l)$.

Proof. There's no loss of generality in taking $\operatorname{dom} S = T$. For clarity, let's denote by s the single-valued mapping from T to $(\text{cl-sets}_{\neq\emptyset}(I\!R^n), d\!l)$ that corresponds to S. The measurability of s means that $s^{-1}(B) \in \mathcal{A}$ for all $B \in \mathcal{B}$, where $\mathcal{B}$ is the Borel field on $(\text{cl-sets}_{\neq\emptyset}(I\!R^n), d\!l)$. The measurability of S, meaning that $S^{-1}(O) \in \mathcal{A}$ for every open set $O \subset I\!R^n$, can be expressed in terms of s through the fact that $S^{-1}(O) = \{t \in T \mid s(t) \in \mathcal{C}_O\} = s^{-1}(\mathcal{C}_O)$ for

$$\mathcal{C}_O := \{C \in \text{cl-sets}_{\neq\emptyset}(I\!R^n) \mid C \cap O \neq \emptyset\}.$$

The issue comes down then to whether the σ-field $\mathcal{E}$ on $\text{cl-sets}_{\neq\emptyset}(I\!R^n)$ that is generated by all sets of form $\mathcal{C}_O$ with $O \subset I\!R^n$ open (which is called the Effrös field) is the same as $\mathcal{B}$, the σ-field generated by all open sets in $(\text{cl-sets}_{\neq\emptyset}(\mathbb{R}^n), \mathbb{d})$.

Clearly $\mathcal{E} \subset \mathcal{B}$, because the sets $\mathcal{C}_O$ are open in $\text{cl-sets}_{\neq\emptyset}(\mathbb{R}^n)$, as follows immediately from Theorem 4.5(a). For the opposite inclusion, we'll show that the balls $\mathbb{B}_{\boldsymbol{d}}(C,\delta)$ for $C \in \text{cl-sets}_{\neq\emptyset}(\mathbb{R}^n)$, $\delta \in \mathbb{R}_+$, belong to $\mathcal{E}$. This is all that's needed, since such balls generate $\mathcal{B}$.

Let's start by observing that for each $x \in \mathbb{R}^n$ the function $d(x,\cdot)$: $\text{cl-sets}_{\neq\emptyset}(\mathbb{R}^n) \to \mathbb{R}_+$ is $\mathcal{E}$-measurable; for any $\gamma > 0$, we have

$$\{C \mid d(x,C) < \gamma\} = \mathcal{C}_O \quad \text{with} \quad O = \text{int}\, \mathbb{B}(x,\gamma).$$

Then for any $\widehat{C} \in \text{cl-sets}_{\neq\emptyset}(\mathbb{R}^n)$ and any $\rho \geq 0$ the function $C \mapsto \mathbb{d}_\rho(C,\widehat{C})$ is $\mathcal{E}$-measurable, inasmuch as

$$\mathbb{d}_\rho(C,\widehat{C}) = \sup_{x\in\rho\boldsymbol{B}} \left|d(x,C) - d(x,\widehat{C})\right| = \sup_{x\in\mathbb{Q}^n\cap\rho\boldsymbol{B}} \left|d(x,C) - d(x,\widehat{C})\right|,$$

the set $\mathbb{Q}^n \cap \rho\mathbb{B}$ being countable and dense in $\rho\mathbb{B}$. Next, the function $C \mapsto \mathbb{d}(C,\widehat{C}) := \int_0^\infty e^{-\rho}\mathbb{d}_\rho(C,\widehat{C})\,d\rho$ is $\mathcal{E}$-measurable on $\text{cl-sets}_{\neq\emptyset}(\mathbb{R}^n)$; by the definition of the integral it can be expressed as the limit of a countable collection of $\mathcal{E}$-measurable functions. Hence $\mathbb{B}_{\boldsymbol{d}}(\widehat{C},\delta) \in \mathcal{E}$. □

The next result gets to the heart of the relationship between the single-valued and set-valued versions of measurability for mappings into $\mathbb{R}^n$ itself.

14.5 Theorem (Castaing representations). *The measurability of a closed-valued mapping $S : T \rightrightarrows \mathbb{R}^n$ is equivalent to each of the following conditions (in combination with the measurability of* dom S*):*

(a) *S admits a Castaing representation: there is a countable family $\{x^\nu\}_{\nu\in\mathbb{N}}$ of measurable functions $x^\nu : \text{dom}\, S \to \mathbb{R}^n$ such that*

$$S(t) = \text{cl}\{x^\nu(t) \mid \nu \in \mathbb{N}\} \text{ for each } t \in T;$$

(b) *there is a countable family $\{y^\nu\}_{\nu\in\mathbb{N}}$ of measurable functions y^ν : $\text{dom}\, S \to \mathbb{R}^n$ such that*

(i) *for each $\nu \in \mathbb{N}$, the set $T^\nu = \{t \in T \mid y^\nu(t) \in S(t)\}$ is measurable,*

(ii) *for each $t \in T$, the set $S(t) \cap \{y^\nu(t) \mid \nu \in \mathbb{N}\}$ is dense in $S(t)$.*

Proof. We begin by showing that these two conditions are equivalent to each other. Clearly (a) implies (b). Now assume that (b) holds. Since $\text{dom}\, S = \bigcup_\nu T^\nu$, we have $\text{dom}\, S \in \mathcal{A}$. Define the function w recursively as follows: take $w = y^1$ on T^1, and for $\nu = 2, \ldots$ take $w = y^\nu$ on $T^\nu \setminus T^{\nu-1}$. Let

$$x^\nu(t) = \begin{cases} y^\nu(t) & \text{for } t \in T^\nu, \\ w(t) & \text{for } t \in \text{dom}\, S \setminus T^\nu. \end{cases}$$

Then x^ν is a measurable function, and $x^\nu(t) \in S(t)$ for all $t \in \text{dom}\, S$. Since $S(t) \cap \{y^\nu(t) \mid \nu \in \mathbb{N}\}$ is dense in $S(t)$, we have $S(t) = \text{cl}\,\{x^\nu(t) \mid \nu \in \mathbb{N}\}$. Thus, we have a Castaing representation for S.

Let's now show that if S admits a Castaing representation $\{x^\nu\}_{\nu\in\mathbb{N}}$, S must be a measurable mapping. Let O be any open subset of $\mathbb{R}^n$, and observe that $S^{-1}(O) = \bigcup_\nu \{t \in \operatorname{dom} S \mid x^\nu(t) \in O\}$. This says that $S^{-1}(O)$ is the countable union of measurable sets and consequently is itself measurable.

All that's left is verifying that if the mapping S is measurable, it must admit a Castaing representation. For any nonempty closed set $C \subset \mathbb{R}^n$, let $\pi_z C = C \cap \mathbb{B}\big(z, d(z,C)\big)$. Let I be the countable index set consisting of all $(n+1)$-tuples $i = (z_0, z_1, \ldots, z_n)$ of vectors $z_j \in \mathbb{Q}^n$ such that $z_0, z_1, \ldots, z_n$ are affinely independent. For $i \in I$ and $t \in \operatorname{dom} S$, let

$$x^i(t) := \pi_{z_n}\pi_{z_{n-1}}\cdots\pi_{z_1}\pi_{z_0}S(t);$$

this vector is well defined as the necessarily unique point in the nonempty intersection of a collection of $n+1$ different spheres in $\mathbb{R}^n$ whose centers are affinely independent. In particular, $x^i(t)$ is a point of $S(t)$ nearest to z_0. Since z_0 ranges over all of $\mathbb{Q}^n$, it's clear that $\operatorname{cl}\{x^i(t) \mid i \in I\} = S(t)$.

To check that every x^i is a measurable function, thereby confirming that the family $\{x^i\}_{i\in I}$ furnishes a Castaing representation for S, it suffices to show that, for any $z \in \mathbb{R}^n$, the mapping $t \mapsto \pi_z S(t) = P_{S(t)}(z)$ is measurable. (The general case will follow then by induction.) For each $\nu \in \mathbb{N}$, define $S^\nu : T \rightrightarrows \mathbb{R}^n$ by

$$S^\nu(t) := \big\{x \in \mathbb{R}^n \bigm| d(x, S(t)) < \nu^{-1},\ d(x,z) < d(z, S(t)) + \nu^{-1}\big\}.$$

Observe that $S^\nu(t)$ is open, and that it is nonempty if and only if $t \in \operatorname{dom} S$. For any open set O and any countable, dense subset D of O, say $D = O \cap \mathbb{Q}^n$, we have $O \cap \pi_z S(t) \neq \emptyset$ if and only if $D \cap S^\nu(t) \neq \emptyset$ for all $\nu \in \mathbb{N}$. Therefore,

$$(\pi_z S)^{-1}(O) = \bigcap_{\nu\in\mathbb{N}} \bigcup_{x\in D} (S^\nu)^{-1}(x).$$

The sets $(S^\nu)^{-1}(x) = \big\{t \in T \bigm| d(x, S(t)) < \nu^{-1},\ d(x,z) < d(z, S(t)) + \nu^{-1}\big\}$ belong to $\mathcal{A}$ by 14.2(h). Because the union and intersection operations in the formula for $(\pi_z S)^{-1}(O)$ are countable, it follows that $(\pi_z S)^{-1}(O) \in \mathcal{A}$. □

14.6 Corollary (measurable selections). *A closed-valued, measurable mapping $S : T \rightrightarrows \mathbb{R}^n$ always admits a measurable selection: there exists a measurable function $x : \operatorname{dom} S \to \mathbb{R}^n$ such that $x(t) \in S(t)$ for all $t \in \operatorname{dom} S$.*

Proof. Any of the functions taking part in a Castaing representation for S is a measurable selection for S. □

14.7 Example (convex-valued and solid-valued mappings). *A mapping $S : T \rightrightarrows \mathbb{R}^n$ is called solid-valued if $S(t) = \operatorname{cl}(\operatorname{int} S(t))$ for all $t \in T$, as is true in particular when S is closed-convex-valued with $\operatorname{int} S(t) \neq \emptyset$ for all $t \in T$.*

Such a mapping S is measurable if and only if it meets the simple test of having $S^{-1}(x) \in \mathcal{A}$ for every $x \in \mathbb{R}^n$.

Detail. When $S(t)$ is a closed, convex set with nonempty interior, we have

$S(t) = \text{cl}(\text{int}\, S(t))$ by 2.33. In general for a solid-valued mapping, the necessity of the proposed criterion for measurability follows from 14.3(c), because the singleton $\{x\}$ is a compact set.

For the sufficiency, consider for each point $a \in \mathbb{Q}^n$ the measurable function $y_a(t) \equiv a$. The countable family of such functions has $S(t) \cap \{y_a(t) \mid a \in \mathbb{Q}^n\} = S(t) \cap \mathbb{Q}^n$ is dense in $S(t)$, since $S(t) = \text{cl}(\text{int}\, S(t))$. Condition 14.5(b) is satisfied then, and we conclude from it that S is measurable. □

In the next theorem, recall that the σ-field $\mathcal{A}$ is said to be *complete* for a measure μ on $\mathcal{A}$ if it contains, for each $A \in \mathcal{A}$ with $\mu(A) = 0$, all subsets $A' \subset A$. The consequence of completeness that will be most important to us is the fact that it ensures the measurability of subsets in T obtained as the projections of certain measurable subsets G of $T \times \mathbb{R}^n$:

$$G \in \mathcal{A} \otimes \mathcal{B}(\mathbb{R}^n) \implies \{t \in T \mid \exists x \in \mathbb{R}^n \text{ with } (t,x) \in G\} \in \mathcal{A}. \qquad 14(2)$$

Here $\mathcal{B}(\mathbb{R}^n)$ is the Borel σ-field on $\mathbb{R}^n$ and $\mathcal{A} \otimes \mathcal{B}(\mathbb{R}^n)$ is the σ-field on $T \times \mathbb{R}^n$ generated by all sets $A \times D$ with $A \in \mathcal{A}$ and $D \in \mathcal{B}(\mathbb{R}^n)$. (For references, see the notes at the end of this chapter.)

14.8 Theorem (graph measurability). *Let $S : T \rightrightarrows \mathbb{R}^n$ be closed-valued. With respect to the Borel field $\mathcal{B}(\mathbb{R}^n)$, the implications (c) $\Rightarrow$ (a) $\Rightarrow$ (b) hold for the following properties:*

(a) *S is a measurable mapping;*

(b) *$\text{gph}\, S$ is an $\mathcal{A} \otimes \mathcal{B}(\mathbb{R}^n)$-measurable subset of $T \times \mathbb{R}^n$;*

(c) *$S^{-1}(D) \in A$ for all sets $D \in \mathcal{B}(\mathbb{R}^n)$.*

When the space $(T, \mathcal{A})$ is complete for some measure μ, these properties are equivalent. Then one has (b) $\Rightarrow$ (c) $\Rightarrow$ (a) even if S is not closed-valued.

Proof. It's evident that (c) implies (a), since $\mathcal{B}(\mathbb{R}^n)$ contains all the open subsets of $\mathbb{R}^n$. This implication doesn't require the closed-valuedness of S.

For the sake of establishing that (a) implies (b), note that, under the assumption that $S(t)$ is closed, a point x belongs to $S(t)$ if and only if, for every positive $\rho \in \mathbb{Q}$, there exists $a \in \mathbb{Q}^n$ such that $x \in \mathbb{B}(a,\rho)$ and $S(t) \cap \mathbb{B}(a,\rho) \neq \emptyset$. This means that

$$\text{gph}\, S = \bigcap_{0<\rho\in\mathbb{Q}} \bigcup_{a\in\mathbb{Q}^n} \left[S^{-1}(\mathbb{B}(a,\rho)) \times \mathbb{B}(a,\rho)\right].$$

By 14.3(d), each set $S^{-1}(\mathbb{B}(a,\rho)) \times \mathbb{B}(a,\rho)$ belongs to $\mathcal{A} \times \mathcal{B}(\mathbb{R}^n)$. The union and intersection are countable, so $\text{gph}\, S$ belongs to $\mathcal{A} \otimes \mathcal{B}(\mathbb{R}^n)$ by the definition of this product σ-field.

Assume now that (b) holds and that $\mathcal{A}$ is complete for some measure μ. It will be demonstrated that then (c) holds. Consider any set $D \in \mathcal{B}(\mathbb{R}^n)$. We have $T \times D \in \mathcal{A} \otimes \mathcal{B}(\mathbb{R}^n)$, and since $\text{gph}\, S \in \mathcal{A} \otimes \mathcal{B}(\mathbb{R}^n)$ by hypothesis, the set $G := \text{gph}\, S \cap [T \times D]$ is in $\mathcal{A} \otimes \mathcal{B}(\mathbb{R}^n)$. As recalled in 14(2), completeness ensures then that the set $\{t \in T \mid \exists x \text{ with } (t,x) \in G\}$ belongs to $\mathcal{A}$. But this set is $S^{-1}(D)$. Hence (c) is fulfilled. Closed-valuedness of S wasn't needed for this implication. □

The equivalence in Theorem 14.8 under completeness is a more subtle fact than might at first be appreciated. It's tempting in some situations to replace the Borel field $\mathcal{B}(\mathbb{R}^n)$ by the Lebesgue field $\mathcal{L}(\mathbb{R}^n)$, which consists of all sets $(D \setminus N) \cup (N \setminus D)$ with $D \in \mathcal{B}(\mathbb{R}^n)$ and N a negligible subset of $\mathbb{R}^n$ (as defined ahead of 9.60). The σ-field $\mathcal{L}(\mathbb{R}^n)$ is complete for the n-dimensional Lebesgue measure. But $\mathcal{A} \otimes \mathcal{B}(\mathbb{R}^n)$ can't be replaced by $\mathcal{A} \otimes \mathcal{L}(\mathbb{R}^n)$ in part (b) of the theorem without disrupting the validity of the implication from (b) to (a).

This must especially be kept in mind in the very common setting where T is a Lebesgue subset of some $\mathbb{R}^d$ and $\mathcal{A}$ is the restricted Lebesgue field

$$\mathcal{L}(T) := \{A \in \mathcal{L}(\mathbb{R}^d) \mid A \subset T\}. \tag{14(3)}$$

Hence, it might seem that $\mathcal{A} \otimes \mathcal{L}(\mathbb{R}^n) = \mathcal{L}(T) \otimes \mathcal{L}(\mathbb{R}^n)$ is a natural choice for consideration of graph measurability. But again, the $\mathcal{L}(T) \otimes \mathcal{L}(\mathbb{R}^n)$-measurability of the graph of a closed-valued mapping $S : T \rightrightarrows \mathbb{R}^n$ won't generally ensure that S is measurable. The stronger condition of $\mathcal{L}(T) \otimes \mathcal{B}(\mathbb{R}^n)$-measurability of the graph must be invoked. Moreover $\mathcal{L}(T) \otimes \mathcal{L}(\mathbb{R}^n)$ is only a subset of $\mathcal{L}(T \times \mathbb{R}^n)$ and thus not necessarily complete; equality would hold if T is a Borel set.

In the important case just mentioned, where T is a Borel subset of $\mathbb{R}^d$, it's instructive to pose questions about the relationship between measurability and continuity of mappings $S : T \rightrightarrows \mathbb{R}^n$. That can be done with $\mathcal{A} = \mathcal{L}(T)$ or with $\mathcal{A} = \mathcal{B}(T)$, the latter being the restricted Borel field:

$$\mathcal{B}(T) := \{A \in \mathcal{B}(\mathbb{R}^d) \mid A \subset T\}. \tag{14(4)}$$

The $\mathcal{A}$-measurability of S is termed *Borel measurability* when $\mathcal{A} = \mathcal{B}(T)$, in contrast to *Lebesgue measurability* when $\mathcal{A} = \mathcal{L}(T)$. Borel measurability always entails Lebesgue measurability, since $\mathcal{B}(T) \subset \mathcal{L}(T)$.

14.9 Exercise (measurability from semicontinuity). *Suppose T is a Borel subset of $\mathbb{R}^d$ (e.g., an open or closed subset), and let $S : T \rightrightarrows \mathbb{R}^n$ be closed-valued. If S is osc or isc relative to T, or in particular if S is continuous relative to T, then S is a measurable mapping with respect to $\mathcal{A} = \mathcal{B}(T)$ or $\mathcal{A} = \mathcal{L}(T)$.*

Guide. Use criterion 14.3(j) after observing that semicontinuity of the mapping S yields the semicontinuity of the function $t \mapsto d(x, S(t))$; cf. 5.11. □

As an illustration of the principle in 14.9 in the case where $T = \mathbb{R}^d$, one has for any mapping $S : \mathbb{R}^d \rightrightarrows \mathbb{R}^n$ that the horizon mapping $S^\infty : \mathbb{R}^d \rightrightarrows \mathbb{R}^n$ is Borel measurable (and therefore also Lebesgue measurable). This is observed from the fact that $\operatorname{gph} S^\infty = (\operatorname{gph} S)^\infty$, so that S^∞ is osc regardless of any assumptions, or lack of them, on S.

The next theorem generalizes Lusin's criterion for Lebesgue measurability to the context of set-valued mappings. By this criterion in the case of $T \in$

$\mathcal{B}(I\!R^d)$ and the Lebesgue measure μ on $\mathcal{A} = \mathcal{L}(T)$, a function $s : T \to I\!R^n$ is measurable if and only if, for each $\varepsilon > 0$, there is a closed set T_ε with $\mu(T \setminus T_\varepsilon) < \varepsilon$ such that s is continuous relative to T_ε.

14.10 Theorem (measurability versus continuity). *Suppose $T \in \mathcal{B}(I\!R^d)$ and let $S : T \rightrightarrows I\!R^n$ be closed-valued. Then, with respect to $\mathcal{A} = \mathcal{L}(T)$, and with μ the restriction of d-dimensional Lebesgue measure to this field, each of the following conditions is equivalent to S being measurable:*

(a) *for every $\varepsilon > 0$ there is a closed set $T_\varepsilon \subset T$ with $\mu(T \setminus T_\varepsilon) < \varepsilon$, such that S is continuous relative to T_ε;*

(b) *for every $\varepsilon > 0$ there is a closed set $T_\varepsilon \subset T$ with $\mu(T \setminus T_\varepsilon) < \varepsilon$, such that S is osc relative to T_ε;*

(c) *there is a mapping $R : T \rightrightarrows I\!R^n$ with $\operatorname{gph} R \in \mathcal{B}(T \times I\!R^n)$, such that R is closed-valued and agrees with S on a set $T' \subset T$ with $T' \in \mathcal{A}$, $\mu(T \setminus T') = 0$.*

(When $\mu(T) < \infty$, the sets T_ε in (a) and (b) can be taken to be compact.)

Proof. Obviously (a) implies (b). To prove that (b) implies (c), we proceed as follows. For $\varepsilon = 1/\nu$, $\nu \in I\!N$, let T^ν be a closed subset of T on which S is osc as postulated in (b). The set $G^\nu := [T^\nu \times I\!R^n] \cap \operatorname{gph} S$ is closed. Let $G = \bigcup_\nu G^\nu$ and $T' := \bigcup_\nu T^\nu$. Then $G \in \mathcal{B}(T \times I\!R^n)$, while T' is measurable with $\mu(T \setminus T') = 0$. The mapping R with $\operatorname{gph} R = G$ fills the demand in (c).

Next we verify that (c) suffices for the measurability of S. The mapping R in (c) is measurable by 14.8, due to the completeness of $\mathcal{A} = \mathcal{L}(T)$. In other words, for each open set $O \subset I\!R^n$, the set $R^{-1}(O) = \{t \mid R(t) \cap O \neq \emptyset\}$ belongs to $\mathcal{L}(T)$. But R agrees with S except on a negligible set, so $R^{-1}(O)$ can differ from $S^{-1}(O) = \{t \mid S(t) \cap O \neq \emptyset\}$ only by such a set. Hence $S^{-1}(O) \in \mathcal{L}(T)$ as well, and the measurability of S is confirmed.

For the remainder of the theorem, we assume that S is measurable and show this yields the property in (a). We begin by demonstrating that the task can be reduced to the case where $\mu(T) < \infty$. We express

$$T = \bigcup\nolimits_{\nu \in I\!N} T^\nu \ \text{ with } \ T^\nu := \{t \in T \mid \nu - 1 \leq |t| < \nu\}.$$

Every T^ν is a Borel set with $\mu(T^\nu) < \infty$. If the property in (a) is valid on sets of bounded measure, we can apply it with respect to each T^ν. Given any $\varepsilon > 0$, we can get a compact set $T^\nu_\varepsilon \subset T^\nu$ such that S is continuous relative to T^ν_ε and $\mu(T^\nu \setminus T^\nu_\varepsilon) < \varepsilon 2^{-\nu}$. No more than finitely many of the disjoint sets T^ν touch any bounded region. Therefore S is continuous relative to $T_\varepsilon := \bigcup_\nu T^\nu_\varepsilon$, which is a closed subset of T with $\mu(T \setminus T_\varepsilon) < \varepsilon$.

From now on, therefore, we can assume that $\mu(T) < \infty$. Restarting in this mode, fix any $\varepsilon > 0$. The measurability of S entails the measurability of the set $\operatorname{dom} S$, so there's a compact set $A_\varepsilon \subset T \setminus \operatorname{dom} S$ with $\mu([T \setminus \operatorname{dom} S] \setminus A_\varepsilon) < \varepsilon/2$. Relative to A_ε, of course, S is continuous by virtue of being empty-valued. Thus, we need only demonstrate that S is also continuous relative to some compact set $B_\varepsilon \subset \operatorname{dom} S$ with $\mu(\operatorname{dom} S \setminus B_\varepsilon) < \varepsilon/2)$, since then S will be continuous relative to the compact set $T_\varepsilon = A_\varepsilon \cup B_\varepsilon$ with $\mu(T \setminus T_\varepsilon) < \varepsilon$.

Consider any Castaing representation $\{x^\nu\}_{\nu\in I\!N}$ of S as provided in 14.5. By Lusin's theorem for measurable functions, there exists for each of the functions $x^\nu : \operatorname{dom} S \to I\!R^n$ a compact set $B^\nu_\varepsilon \subset \operatorname{dom} S$ with $\mu(\operatorname{dom} S \setminus B^\nu_\varepsilon) < \varepsilon 2^{-(\nu+1)}$, such that x^ν is continuous relative to B^ν_ε. Let $B_\varepsilon = \bigcap_\nu B^\nu_\varepsilon$. Then B_ε is compact with $\mu(\operatorname{dom} S \setminus B_\varepsilon) < \varepsilon/2$, and all the functions x^ν are continuous relative to B_ε. For any open set $O \subset I\!R^n$, we have

$$S^{-1}(O) \cap B_\varepsilon = \bigcup\nolimits_\nu [\,(x^\nu)^{-1}(O) \cap B_\varepsilon\,].$$

The set $(x^\nu)^{-1}(O) \cap B_\varepsilon$ is open relative to B_ε, since x^ν is continuous relative to B_ε, so that $S^{-1}(O) \cap B_\varepsilon$ is open relative to B_ε. In particular this means that S is isc relative to B_ε; cf. 5.7(c). We want continuity rather than just inner semicontinuity, but the task of obtaining this has greatly been reduced.

Instead of the continuity in (a), we now only have to produce the outer semicontinuity in (b) out of the assumptions that S is measurable and $\mu(T) < \infty$. Once again, it's best just to restart with this simpler aim. Fixing an arbitrary $\varepsilon > 0$, we need to construct a compact set $T_\varepsilon \subset T$ with $\mu(T \setminus T_\varepsilon) < \varepsilon$, such that the set $\{(t,x) \mid t \in T_\varepsilon,\ x \in S(t)\}$ is closed. (This will also justify the assertion at the end of the theorem.)

Enumerate by $\{C^\nu\}_{\nu\in I\!N}$ the countable family of sets of having the form $I\!R^n \setminus \operatorname{int} B(a,\rho)$ with $a \in \mathbb{Q}^n$ and $0 < \rho \in \mathbb{Q}$. For each $t \in T$, $S(t)$ is the intersection of the sets $C^\nu \supset S(t)$. Take $T^\nu := S^{-1}(I\!R^n \setminus C^\nu)$ and $\overline{T}^\nu := T \setminus T^\nu$. Since S is measurable, T^ν and $\overline{T}^\nu$ belong to $\mathcal{A}$. Moreover

$$\operatorname{gph} S = \bigcap\nolimits_\nu [\,(T^\nu \times I\!R^n) \cup (\overline{T}^\nu \times C^\nu)\,].$$

For each ν, there exist compact sets $T^\nu_\varepsilon \subset T^\nu$ and $\overline{T}^\nu_\varepsilon \subset \overline{T}^\nu$ such that $\mu(T \setminus (T^\nu_\varepsilon \cup \overline{T}^\nu_\varepsilon)) < \varepsilon 2^{-\nu}$. Define $T_\varepsilon := \bigcap_\nu (T^\nu_\varepsilon \cup \overline{T}^\nu_\varepsilon)$. Then T_ε is a compact set such that $\mu(T \setminus T_\varepsilon) \leq \varepsilon$, and we have

$$\{(t,x) \mid t \in T_\varepsilon, x \in S(t)\} = \bigcap\nolimits_\nu [\,(T^\nu_\varepsilon \times I\!R^n) \cup (\overline{T}^\nu_\varepsilon \times C^\nu)\,].$$

The set on the right is closed, so our goal has been reached. □

B. Preservation of Measurability

The results that come next provide tools for establishing the measurability of set-valued mappings as they arise through operations performed on other mappings, already known to be measurable.

14.11 Proposition (unions, intersections, sums and products). *For each $j \in J$, an index set, let $S_j : T \rightrightarrows I\!R^n$ be measurable. Then the following mappings are measurable:*

(a) $t \mapsto \bigcap_{j\in J} S_j(t)$, *if J is countable, S_j closed-valued;*

(b) $t \mapsto \bigcup_{j\in J} S_j(t)$, *if J is countable;*

(c) $t \mapsto \sum_{j\in J} \lambda_j S_j(t)$ *for arbitrary $\lambda_j \in I\!R$, if J is finite;*

and in the case of $S_j : T \to I\!R^{n_j}$ instead of $S_j : T \to I\!R^n$, also the set product

(d) $t \mapsto \big(S_1(t), \ldots, S_r(t)\big)$ *when $J = \{1, \ldots, r\}$.*

Proof. Beginning with (d), let's denote the mapping in question by R and recall that any open set $O \subset I\!R^{n_1} \times \cdots \times I\!R^{n_r}$ can be expressed as a union of a sequence of sets $O^\nu = O_1^\nu \times \cdots \times O_r^\nu$ with O_j^ν open in $I\!R^{n_j}$. (Open balls O_j^ν suffice.) Then $R^{-1}(O) = \bigcup_\nu \big[\bigcap_{j=1}^r S_j^{-1}(O_j^\nu)\big]$, where each set $S_j^{-1}(O_j^\nu)$ belongs to $\mathcal{A}$ because S_j is measurable. Countable unions and intersections of sets in $\mathcal{A}$ belong to $\mathcal{A}$, so we have $R^{-1}(O) \in \mathcal{A}$. Thus, R is measurable.

In (c) we may as well take $J = \{1, \ldots, r\}$. We can build on (d) by choosing R to be the mapping just treated, but in the special case of $n_j = n$ for all j. For any open set $O \subset I\!R^n$ consider the open set $O' \subset (I\!R^n)^r$ consisting of all $(x_1, \ldots, x_r)$ such that $\sum_{j=1}^r \lambda_j x_j \in O$. Then

$$\big\{t \,\big|\, (\textstyle\sum_{j=1}^r \lambda_j S_j)(t) \cap O \neq \emptyset\big\} = \big\{t \,\big|\, \big(S_1(t), \ldots, S_r(t)\big) \in O'\big\} = R^{-1}(O'),$$

where $R^{-1}(O') \in \mathcal{A}$ by (d). Hence $\big(\sum_{j=1}^r \lambda_j S_j\big)^{-1}(O) \in \mathcal{A}$, and we conclude that the mapping $\sum_{j=1}^r \lambda_j S_j$ is measurable.

The argument for (b) is elementary. For any open set $O \subset I\!R^n$ we have $(\bigcup_{j\in J} S_j)^{-1}(O) = \bigcup_{j\in J} S_j{}^{-1}(O) \in \mathcal{A}$ because $S_j{}^{-1}(O) \in \mathcal{A}$.

For the closed-valued mapping S in (a), we'll first treat the case where $J = \{1, 2\}$. Consider any compact set $C \subset I\!R^n$ and the truncated mappings $R_j(t) := S_j(t) \cap C$, $j = 1, 2$. These inherit the measurability of S_1 and S_2 by virtue of 14.3(c). We have

$$\begin{aligned} S^{-1}(C) &= \big\{t \,\big|\, S_1(t) \cap S_2(t) \cap C \neq \emptyset\big\} \\ &= \big\{t \,\big|\, 0 \in R_1(t) - R_2(t)\big\} = (R_1 - R_2)^{-1}(\{0\}). \end{aligned}$$

Here $R_1 - R_2$ is measurable via (c) and closed-valued because R_1 and R_2 are compact-valued (cf. 3.12). Hence $(R_1 - R_2)^{-1}(\{0\}) \in \mathcal{A}$ by 14.3(c) again, so $S^{-1}(C) \in \mathcal{A}$. Thus, S is measurable in this case, and it readily follows then by induction that S is measurable as long as the index set J is finite.

Suppose now that J is countably infinite; we can take $J = I\!N$. For each $\nu \in I\!N$ let $S^\nu(t) = \bigcap_{j=1}^\nu S_j(t)$. The mappings S^ν are measurable by the preceding. For any compact set $C \subset I\!R^n$ we have $S^{-1}(C) = \bigcap_\nu (S^\nu)^{-1}(C)$. Since $(S^\nu)^{-1}(C) \in \mathcal{A}$, it follows that $S^{-1}(C) \in \mathcal{A}$ as required. □

14.12 Exercise (hulls and polars). *If $S : T \rightrightarrows I\!R^n$ is measurable, so too are:*

(a) $t \mapsto \operatorname{con} S(t)$, *the convex hull of $S(t)$;*

(b) $t \mapsto \operatorname{pos} S(t)$, *the cone generated by $S(t)$;*

(c) $t \mapsto \operatorname{aff} S(t)$, *the affine hull of $S(t)$;*

(d) $t \mapsto \operatorname{lin} S(t)$, *the linear subspace generated by $S(t)$;*

(e) $t \mapsto S(t)^\circ = \big\{v \,\big|\, \langle v, x\rangle \leq 1 \text{ for all } x \in S(t)\big\}$;

(f) $t \mapsto S(t)^* = \{v \mid \langle v, x\rangle \leq 0 \text{ for all } x \in S(t)\}$.

Guide. For the mapping in (a), utilize Carathéodory's Theorem 2.26 to identify the inverse image of an open set $O \subset \mathbb{R}^n$ with $R^{-1}(O')$ for the mapping $R : t \mapsto S(t) \times \cdots \times S(t)$ [$n+1$ copies] and the open set $O' \subset (\mathbb{R}^n)^{n+1}$ consisting of all $(x_0, \ldots, x_n)$ such that $\sum_{i=0}^n \lambda_i x_i \in O$ for some choice of $\lambda_i \geq 0$ with $\sum_{i=0}^n \lambda_i = 1$. The mappings in (b), (c) and (d) can be handled similarly.

For the polar mapping $S^\circ : t \mapsto S(t)^\circ$ in (e), there's no harm in assuming that S is closed-valued; cf. 14.2. Argue that if C is a closed subset of $\mathbb{R}^n$ and C_0 is a dense subset of C, one has $C^\circ = C_0^\circ$; this can be applied to $S(t)$ and $S_0(t) = \{x^\nu(t) \mid \nu \in \mathbb{N}\}$ for a Castaing representation for S. Next, show that the measurability of the functions $x^\nu : \operatorname{dom} S \to \mathbb{R}^n$ ensures that of the closed-valued mappings $H^\nu : T \rightrightarrows \mathbb{R}^n$ defined by

$$H^\nu(t) = \begin{cases} \{v \mid \langle v, x^\nu(t)\rangle \leq 1\} & \text{when } t \in \operatorname{dom} S, \\ \emptyset & \text{otherwise.} \end{cases}$$

Finally, invoke the measurability in 14.11(a) for $t \mapsto \bigcap_\nu H^\nu(t)$. A parallel argument takes care of the mapping $S^* : t \mapsto S(t)^*$ in (f). □

Other operations of interest in connection with the preservation of measurability of a mapping $S : T \rightrightarrows \mathbb{R}^n$ fall in the category of constructing a new mapping by taking images of the sets $S(t)$ under various transformations.

14.13 Theorem (composition of mappings).

(a) *Let $S : T \rightrightarrows \mathbb{R}^n$ be measurable and let $M : \mathbb{R}^n \rightrightarrows \mathbb{R}^m$ be isc. Then the mapping $M \circ S : t \mapsto M(S(t))$ is measurable.*

(b) *Let $S : T \rightrightarrows \mathbb{R}^n$ be closed-valued and measurable, and for each $t \in T$ consider an osc mapping $M(t, \cdot) : \mathbb{R}^n \rightrightarrows \mathbb{R}^m$. Suppose the graphical mapping $t \mapsto \operatorname{gph} M(t, \cdot) \subset \mathbb{R}^n \times \mathbb{R}^m$ is measurable (as is true in particular when $M(t, \cdot)$ is the same for all t). Then the mapping $t \mapsto M(t, S(t))$ is measurable.*

Proof. To establish (a), consider any open set $O \subset \mathbb{R}^m$. We have $(M \circ S)^{-1}(O) = \{t \mid S(t) \cap M^{-1}(O) \neq \emptyset\} = S^{-1}(O')$ for the set $O' = M^{-1}(O)$, which is open because M is isc; cf. 5.7(c). Since S is measurable, we have $S^{-1}(O') \in \mathcal{A}$ and consequently $(M \circ S)^{-1}(O) \in \mathcal{A}$.

For the justification of (b), let $R(t) = M(t, S(t))$ and denote the closed-valued mapping $t \mapsto \operatorname{gph} M(t, \cdot)$ by G. Let P project from $\mathbb{R}^n \times \mathbb{R}^m$ to $\mathbb{R}^m$. We have $R = P \circ Q$ for $Q(t) = [S(t) \times \mathbb{R}^m] \cap G(t)$. The mapping $t \mapsto S(t) \times \mathbb{R}^m$ is closed-valued and measurable through 14.11(d), and its intersection Q with G has these properties through 14.11(a). On the other hand, P is single-valued and continuous. Then $P \circ Q$ is measurable by the result in part (a). □

14.14 Corollary (composition with measurable functions). *For each $t \in T$ consider an osc mapping $M(t, \cdot) : \mathbb{R}^n \rightrightarrows \mathbb{R}^m$, and suppose the graphical mapping $t \mapsto \operatorname{gph} M(t, \cdot) \subset \mathbb{R}^n \times \mathbb{R}^m$ is measurable (as is true in particular if $M(t, \cdot)$ is the same for all t). Then, whenever $t \mapsto x(t)$ is measurable, the mapping $t \mapsto M(t, x(t))$ is closed-valued and measurable.*

14.15 Example (images under Carathéodory mappings). *A single-valued mapping $F : T \times \mathbb{R}^n \to \mathbb{R}^m$ is called a Carathéodory mapping when $F(t,x)$ is measurable in t for each fixed x and continuous in x for each fixed t. In terms of such F, each of the following mappings is measurable as long as S is closed-valued and measurable:*

(a) $t \mapsto F\big(t, S(t)\big) \subset \mathbb{R}^m$ *with* $S : T \rightrightarrows \mathbb{R}^n$;

(b) $t \mapsto \big\{x \in \mathbb{R}^n \,\big|\, F(t,x) \in S(t)\big\} \subset \mathbb{R}^n$ *with* $S : T \rightrightarrows \mathbb{R}^m$.

Detail. Case (a) corresponds to the composition in Theorem 14.13(b) with $M(t,\cdot) = F(t,\cdot)$, whereas case (b) has $M(t,\cdot) = F(t,\cdot)^{-1}$. The task therefore is to verify that the sets $G(t) := \operatorname{gph} F(t,\cdot) \subset \mathbb{R}^n \times \mathbb{R}^m$, which are closed, depend measurably on t. For this we construct a Castaing representation, so that measurability can be drawn from 14.5. For each $q \in \mathbb{Q}^n$ define $z^q(t) = \big(q, F(t,q)\big)$. Through the Carathéodory properties, the functions z^q are measurable and such that $\operatorname{cl}\big\{z^q(t) \,\big|\, q \in \mathbb{Q}^n\big\} = G(t)$ for each t. The countable family $\{z^q\}_{q \in \mathbb{Q}^n}$ thus furnishes a Castaing representation for G. □

A variety of examples of measurable mappings can be seen under this umbrella. For instance, if $F(t,x) = A(t)x + a(t)$ for a matrix $A(t) \in \mathbb{R}^{m \times n}$ and a vector $a(t) \in \mathbb{R}^m$ whose elements depend measurably on a parameter $t \in T$, then F is a Carathéodory mapping. For any closed set $S(t) \subset \mathbb{R}^n$ depending measurably on t, the mapping $t \mapsto A(t)S(t) + a(t)$ is measurable by 14.15(a). The mapping $t \mapsto \operatorname{cl}[A(t)S(t) + a(t)]$ is both closed-valued and measurable, as seen then through the principle in 14.2. In particular one could take $S(t) \equiv C$ for a closed set $C \subset \mathbb{R}^n$.

On the other hand, for any closed set $D \subset \mathbb{R}^m$, the closed-valued mapping $t \mapsto \big\{x \,\big|\, A(t)x + a(t) \in D\big\}$ is measurable by 14.15(b). In this case we're looking at a 'feasible set' that depends measurably on $t \in T$. A wide range of nonlinear constraint systems can also be handled in this manner.

14.16 Theorem (implicit measurable functions). *Let $F : T \times \mathbb{R}^n \to \mathbb{R}^m$ be a Carathéodory mapping, and let $X(t) \subset \mathbb{R}^n$ and $D(t) \subset \mathbb{R}^m$ be closed sets that depend measurably on $t \in T$. Let*

$$E = \big\{t \in T \,\big|\, \exists x \in X(t) \text{ with } F(t,x) \in D(t)\big\}.$$

Then E is measurable, and there is a measurable function $x : E \to \mathbb{R}^n$ with

$$x(t) \in X(t) \;\text{ and }\; F\big(t, x(t)\big) \in D(t) \;\text{ for all } t \in E. \tag{14(5)}$$

Proof. Let $R(t) = \big\{x \in \mathbb{R}^n \,\big|\, F(t,x) \in D(t)\big\}$. The closed-valued mapping $R : T \rightrightarrows \mathbb{R}^n$ is measurable by 14.15(b). The mapping $S : t \mapsto X(t) \cap R(t)$ is closed-valued and measurable then by 14.11(a). Hence $\operatorname{dom} S$ is a measurable set relative to which S has a measurable selection; cf. 14.6. □

A result closely related to Theorem 14.16 will be provided in 14.35 in terms of a system of equations and inequalities, perhaps infinitely many.

14.17 Exercise (projection mappings). *For $S : T \rightrightarrows \mathbb{R}^n$ closed-valued and measurable, consider for each $t \in T$ the (osc) projection mapping $P_{S(t)} : \mathbb{R}^n \rightrightarrows \mathbb{R}^n$.*

(a) *The mapping $t \mapsto \operatorname{gph} P_{S(t)}$ is closed-valued and measurable.*

(b) *The mapping $t \mapsto P_{S(t)}\big(X(t)\big)$ is measurable if the mapping $t \mapsto X(t)$ is closed-valued and measurable; in particular, $X(t)$ could be a singleton $\{x(t)\}$.*

Guide. Get (a) from the foregoing observations by representing $\operatorname{gph} P_{S(t)}$ as the set of pairs $(x, y) \in C(t) = \mathbb{R}^n \times S(t)$ such that $|x - y| - d\big(x, S(t)\big) = 0$. Invoke 14.3(j) in the justification. Then get (b) from Theorem 14.13(b). □

14.18 Exercise (mixed operations). *For any closed-valued measurable mapping $S : T \rightrightarrows \mathbb{R}^d \times \mathbb{R}^n \times \mathbb{R}^m$ and any measurable function $u : T \to \mathbb{R}^m$, one has the measurability of the mapping*

$$t \mapsto \big\{x \,\big|\, \exists\, w \in \mathbb{R}^d \text{ with } \big(w, x, u(t)\big) \in S(t)\big\} \subset \mathbb{R}^n.$$

Guide. Show that $F(t, w, x) = \big(w, x, u(t)\big)$ is a Carathéodory mapping and apply 14.15(b). Then take images under $(w, x) \mapsto x$, utilizing 14.15(a). □

14.19 Example (variable scalar multiplication). *For measurable, closed-valued mappings $S_j : \mathbb{R}^n \rightrightarrows \mathbb{R}^m$, $j = 1, \ldots, r$, the mapping $S : t \mapsto \sum_{j=1}^r \lambda_j(t) S_j(t)$ is measurable as long as the scalars $\lambda_j(t) \in \mathbb{R}$ depend measurably on t.*

Detail. This goes beyond 14.11(c) in allowing non-constant multipliers, but in contrast to that earlier result imposes the assumption of closed-valuedness. The rule is justified through 14.15(a) in applying the mapping $F(t, x_1, \ldots, x_r) = \sum_{j=1}^r \lambda_j(t) x_j$ to the product mapping in 14.11(d). □

C. Limit Operations

Measurability of set-valued mappings is also preserved under taking various kinds of limits.

14.20 Theorem (graphical and pointwise limits). *For mappings $S^\nu : T \rightrightarrows \mathbb{R}^n$ that are measurable, the pointwise limit mappings*

$$\text{p-lim sup}_\nu\, S^\nu, \quad \text{p-lim inf}_\nu\, S^\nu, \quad \textit{and} \quad \text{p-lim}_\nu\, S^\nu \ \ \textit{(if it exists)},$$

are closed-valued and measurable. In the case where $T \in \mathcal{B}(\mathbb{R}^d)$ and $\mathcal{A} = \mathcal{B}(T)$ or $\mathcal{A} = \mathcal{L}(T)$ this conclusion applies also to the graphical limit mappings

$$\text{g-lim sup}_\nu\, S^\nu, \quad \text{g-lim inf}_\nu\, S^\nu, \quad \textit{and} \quad \text{g-lim}_\nu\, S^\nu \ \ \textit{(if it exists)}.$$

Proof. We have $\big(\text{p-lim sup}_\nu\, S^\nu\big)(t) = \bigcap_{\nu \in \mathbb{N}} \bar{S}^\nu(t)$ with $\bar{S}^\nu(t) = \operatorname{cl} \bigcup_{\kappa \geq \nu} S^\kappa(t)$. The mappings $\bar{S}^\nu$ are closed-valued and measurable by 14.11(b) and 14.2, hence p-lim inf$_\nu\, S^\nu$ has these properties by 14.11(a). The argument for p-lim inf$_\nu\, S^\nu$ relies on the formula

$$\left(\text{p-lim inf}_\nu S^\nu\right)(t) = \bigcap_{j=1}^{\infty} \text{cl}\left[\bigcup_{i=1}^{\infty}\bigcap_{\kappa=i}^{\infty} \text{cl}\left(S^\kappa(t) + j^{-1}\mathbb{B}\right)\right],$$

which comes out of 4(2). The mappings $t \mapsto S^\nu(t) + j^{-1}\mathbb{B}$ are measurable by 14.11(c), and the principles of intersection and union in 14.11(a) and 14.11(b) can again be invoked along with 14.2.

When $\text{p-lim}_\nu S^\nu$ exists, it coincides with $\text{p-lim inf}_\nu S^\nu$ and $\text{p-lim sup}_\nu S^\nu$, and thus is measurable. Graphical limit mappings are always osc, and thus through 14.9, closed-valued and measurable. □

14.21 Exercise (horizon cones and limits).

(a) *For any measurable mapping $S : T \rightrightarrows \mathbb{R}^n$, the horizon cone mapping $t \mapsto S(t)^\infty$ is (closed-valued and) measurable.*

(b) *For any sequence of measurable mappings $S^\nu : T \rightrightarrows \mathbb{R}^n$, one has the closed-valuedness and measurability of the horizon limit mappings*

$$t \mapsto \liminf_\nu^\infty S^\nu(t), \quad t \mapsto \limsup_\nu^\infty S^\nu(t), \quad t \mapsto \lim_\nu^\infty S^\nu(t) \ \ \textit{(if it exists).}$$

Guide. Handle (a) by introducing $S^\nu(t) = \{0\} \cup \{\lambda S(t) \mid 0 \le \lambda \le \nu^{-1}\}$. Argue the measurability of S^ν through a representation $S^\nu = F \circ R^\nu$ with $R^\nu(t) = [0, \nu^{-1}] \times [S(t) \cup \{0\}]$ and $F(\lambda, x) = \lambda x$, utilizing 14.15(a). Then apply to the mappings S^ν a fact in 14.20. Do similar constructions for (b). □

By a *simple* mapping, one means a mapping whose range consists of only finitely many points. Note that this condition means much more than just having image sets of finite cardinality.

14.22 Proposition (limits of simple mappings). *A closed-valued mapping $S : T \rightrightarrows \mathbb{R}^n$ is measurable if and only if it is the limit of a sequence of simple measurable mappings.*

Proof. In view of the Theorem 14.20, every pointwise limit of a sequence of simple measurable mappings is a closed-valued measurable mapping. For the converse, consider the sequence of mappings $S^\nu : T \rightrightarrows \mathbb{R}^n$ defined by

$$S^\nu(t) = \left(S(t) + \nu^{-1}\mathbb{B}\right) \cap \nu\mathbb{B} \cap \nu^{-1}\mathbb{Z}^n$$

where $\nu^{-1}\mathbb{Z}^n$ is the lattice of points in $\mathbb{R}^n$ whose coordinates are multiples of ν^{-1}. Each S^ν is a simple, measurable mapping, and it's evident that $\lim_\nu S^\nu(t) = S(t)$. □

For converging sequences of measurable mappings, one can always find convergent sequences of measurable selections.

14.23 Proposition (convergence of measurable selections). *Let $S = \text{p-lim inf}_\nu S^\nu$ for closed-valued measurable mappings $S^\nu : T \rightrightarrows \mathbb{R}^n$ with $\operatorname{dom} S^\nu = T$.*

(a) *If, for each ν, s^ν is a measurable selection of S^ν, and $\lim_\nu s^\nu(t) = s(t)$ for all $t \in T$, then s is a measurable selection of S. In the case where T is a* *closed subset of* $\mathbb{R}^d$ *and* $\mathcal{A} = \mathcal{B}(T)$, *so that the mapping* g-lim inf$_\nu S^\nu$ *makes sense, s also gives a measurable selection of that mapping.*

(b) *For any Castaing representation* $\{s^\kappa\}_{\kappa\in\mathbb{N}}$ *of* S, *there exist Castaing representations* $\{s^{\nu,\kappa}\}_{\kappa\in\mathbb{N}}$ *of the mappings* S^ν *such that, for each* κ, *one has* $\lim_\nu s^{\nu,\kappa}(t) = s^\kappa(t)$ *for all* $t \in T$. *In particular, for any measurable selection* s *of* S *there exist measurable selections* s^ν *of* S^ν *such that* $\lim_\nu s^\nu(t) = s(t)$ *for all* $t \in T$.

Proof. For (a), recall that a pointwise limit of measurable functions is measurable, and that $\liminf_\nu S^\nu(t) = (\text{p-}\liminf_\nu S^\nu)(t)$. Also, $(\text{p-}\liminf_\nu S^\nu)(t) \subset (\text{g-}\liminf_\nu S^\nu)(t)$ when the graphical limit mapping makes sense.

To prove (b), let's first demonstrate that for any measurable selection $s : T \to \mathbb{R}^n$ of S there are measurable selections $s^\nu : \operatorname{dom} S^\nu \to \mathbb{R}^n$ of the mappings S^ν such that $s^\nu(t) \to s(t)$ for all $t \in T$. For each $\nu \in \mathbb{N}$, the mapping $R^\nu : T \rightrightarrows \mathbb{R}^n$ defined by

$$R^\nu(t) := \big\{ x \in S^\nu(t) \,\big|\, |x - s(t)| \le d\big(s(t), S^\nu(t)\big) + \nu^{-1} \big\}$$

is closed-valued and measurable with $\operatorname{dom} R^\nu = T$. The measurability follows from that of $t \mapsto d\big(s(t), S^\nu(t)\big)$, which in turn is a consequence of 14.15 because $(t, x) \mapsto d\big(s(t), x\big)$ has the Carathéodory properties: it's continuous in x, and it's measurable in t as the composition of a continuous mapping with a measurable one. According to 14.6, each mapping R^ν admits a measurable selection, say $r^\nu : T \to \mathbb{R}^n$. Since

$$0 = d\big(s(t), S(t)\big) \ge \limsup_\nu d\big(s(t), S^\nu(t)\big) \ge 0$$

for all $t \in T$ by 4.7, it follows that $r^\nu(t) \to s(t)$.

Now for each member s^κ of the Castaing representation of S given in (b), let $r^{\nu,\kappa} : T \to \mathbb{R}^n$ be a sequence of measurable selections of the mappings S^ν that converges pointwise to s^κ, as just constructed. For each ν, let $\{x^{\nu,\kappa}\}_{\kappa\in\mathbb{N}}$, $x^{\nu,\kappa} : T \to \mathbb{R}^n$, be a Castaing representation of S^ν (as exists by 14.5), and for each $\kappa \in \mathbb{N}$ define

$$s^{\nu,\kappa}(t) = \begin{cases} r^{\nu,\kappa}(t) & \text{if } \kappa \le \nu; \\ x^{\nu,\kappa-\nu}(t) & \text{otherwise.} \end{cases}$$

Then $\{s^{\nu,\kappa}\}_{\kappa\in\mathbb{N}}$ furnishes a Castaing representation of S^ν, and for each κ one has $\lim_\nu s^{\nu,\kappa}(t) = s^\kappa(t)$ for all $t \in T$. □

Given μ a measure on $\mathcal{A}$, one can speak also of the *almost everywhere* (a.e.) convergence of a sequence of mappings $S^\nu : T \rightrightarrows \mathbb{R}^n$ to a mapping $S : T \rightrightarrows \mathbb{R}^n$. This means that, for all $t \in T$, except possibly for t in a set T_0 with $\mu(T_0) = 0$, one has $S^\nu(t) \to S(t)$; in other words, the mappings S^ν converge pointwise to S on $T \setminus T_0$. This relaxed property of pointwise convergence is indicated by

$$S = \text{p-}\lim_\nu S^\nu \ \text{ a.e.}, \quad \text{or } S^\nu \xrightarrow{\text{p}} S \ \text{ a.e.}$$

Our objective now, in addition to providing some characterizations of a.e. convergence, is to prove that, when $\mu(T) < \infty$, a.e. convergence implies *almost uniform* (a.u.) convergence, by which one means that for every $\delta > 0$ there is a measurable set $T_\delta \subset T$ with $\mu(T_\delta) < \delta$, such that, on $T \setminus T_\delta$, the mappings S^ν converge uniformly to S.

It will be helpful to use the following notation. In working with sets $A_1, A_2 \in \mathcal{A}$, we'll write

$$A_1 \subset_0 A_2 \text{ when } A_1 \subset A_2 \cup N \text{ for some } N \in \mathcal{A} \text{ with } \mu(N) = 0. \qquad 14(6)$$

Also, for any sequence $\{A^\nu\}_{\nu \in \mathbb{N}}$ of subsets of T, we'll write

$$*\liminf_\nu A^\nu = \bigcup_{\nu \in \mathcal{N}_\infty} \bigcap_{\nu \in N} A^\nu, \qquad *\limsup_\nu A^\nu = \bigcap_{\nu \in \mathcal{N}_\infty} \bigcup_{\nu \in N} A^\nu. \qquad 14(7)$$

These are the *set-theoretic* inner and outer limits of the sequence $\{A^\nu\}_{\nu \in \mathbb{N}}$. They can be interpreted as the inner and outer limits obtained in the framework of the set convergence theory of Chapter 4 when generalized from $\mathbb{R}^n$ to arbitrary topological spaces and then applied to T, equipped with the discrete topology. From that vantage point, the formulas in 4.2(b) are still applicable, but taking closures can be skipped since all sets are closed in the discrete topology, and $\bigcup_{\nu \in \mathcal{N}_\infty} \bigcap_{\nu \in N} A^\nu = \bigcap_{\nu \in \mathcal{N}_\infty^\#} \bigcup_{\nu \in N} A^\nu$.

14.24 Proposition (almost everywhere convergence). *The following conditions on closed-valued measurable mappings $S, S^\nu : T \rightrightarrows \mathbb{R}^n$, are equivalent with respect to a measure μ on $(T, \mathcal{A})$:*

(a) $S^\nu \xrightarrow{p} S$ *a.e.;*

(b) *for any compact set $B \subset \mathbb{R}^n$ and any open set $O \subset \mathbb{R}^n$, one has*

$$*\limsup_\nu (S^\nu)^{-1}(B) \subset_0 S^{-1}(B), \qquad S^{-1}(O) \subset_0 *\liminf_\nu (S^\nu)^{-1}(O);$$

(c) *for any $x \in \mathbb{R}^n$ and $\rho > 0$, one has*

$$\begin{aligned} S^{-1}\big(\operatorname{int} \mathbb{B}(x,\rho)\big) &\subset_0 *\liminf_\nu (S^\nu)^{-1}\big(\operatorname{int} \mathbb{B}(x,\rho)\big) \\ &\subset *\limsup_\nu (S^\nu)^{-1}\big(\mathbb{B}(x,\rho)\big) \subset_0 S^{-1}\big(\mathbb{B}(x,\rho)\big); \end{aligned}$$

(d) *for μ-almost all t, one has $d(x, S^\nu(t)) \to d(x, S(t))$ for all $x \in \mathbb{R}^n$;*

(e) $\lim_\nu \mu\big[\,(R_\varepsilon^\nu)^{-1}\big(\mathbb{B}(x,\rho)\big)\,\big] = 0$ *for all $x \in \mathbb{R}^n$, $\varepsilon > 0$ and $\rho > 0$, where*

$$R_\varepsilon^\nu(t) = \bigcup_{\kappa \geq \nu} \Big(\big[S^\kappa(t) \setminus (S(t) + \varepsilon \mathbb{B})\big] \cup \big[S(t) \setminus (S^\kappa(t) + \varepsilon \mathbb{B})\big] \Big).$$

Proof. The equivalence between (a), (b), (c) and (d) follows immediately from the definition of convergence a.e. and the various characterizations of set convergence featured in Chapter 4. More specifically, (b) comes from 4.5(a)(b), while (c) comes from 4.5(a′)(b′) and (d) from 4.7.

Let's now turn to (a)⇔(e). Without loss of generality we can choose

$x = 0$, and then the condition in (e) reads: for all $\varepsilon > 0$ and $\rho > 0$ one has $\lim_\nu \mu(W^\nu_{\varepsilon,\rho}) = 0$ for $W^\nu_{\varepsilon,\rho} := (R^\nu_\varepsilon)^{-1}(\rho I\!B)$. In view of 4.11(a)(b), we know that $S^\nu(t) \to S(t)$ if and only if $\lim_\nu R^\nu_\varepsilon(t) = \emptyset$ for all $\varepsilon > 0$; by the first part of that same result, this occurs if and only if, for all $\varepsilon > 0$ and $\rho > 0$, one can find $N \in \mathcal{N}_\infty$ such that $t \notin W^\nu_{\varepsilon,\rho}$ when $\nu \in N$.

Thus, $S^\nu \xrightarrow{\text{p}} S$ a.e. if and only if there exists T_0 with $\mu(T_0) = 0$ such that whenever $t \notin T_0$, then for all $\varepsilon > 0$, $\rho > 0$ there's an index set $N \in \mathcal{N}_\infty$ such that $t \notin W^\nu_{\varepsilon,\rho}$ for all $\nu \in N$. Or equivalently, because the sequence of sets $\{W^\nu_{\varepsilon,\rho}\}_{\nu \in I\!N}$ is nonincreasing: $S^\nu \xrightarrow{\text{p}} S$ a.e. if and only if there exists T_0 of μ-measure zero such that for all $\varepsilon > 0$ and $\rho > 0$, one has $W_{\varepsilon,\rho} := \bigcap_\nu W^\nu_{\varepsilon,\rho} \subset T_0$. And this certainly implies that $\mu(W^\nu_{\varepsilon,\rho}) \to 0$ since $\lim_\nu \mu(W^\nu_{\varepsilon,\rho}) = \mu(W_{\varepsilon,\rho})$. On the other hand, since $\mu(W^\nu_{\varepsilon,\rho}) \to 0$ for all $\varepsilon > 0$ and $\rho > 0$ if and only if the same holds for all ε and ρ belong to $\mathbb{Q}_+$, the sets

$$T_1 := \bigcup_{\varepsilon>0} \bigcup_{\rho>0} W_{\varepsilon,\rho}, \qquad T_2 := \bigcup_{\substack{\varepsilon>0\\ \varepsilon\in\mathbb{Q}}} \bigcup_{\substack{\rho>0\\ \rho\in\mathbb{Q}}} W_{\varepsilon,\rho}$$

are identical. Here $\mu(T_2) = 0$ when $\mu(W^\nu_{\varepsilon,\rho}) \to 0$, since T_2 is then the countable union of sets of measure zero; recall that $\lim_\nu \mu(W^\nu_{\varepsilon,\rho}) = \mu(W_{\varepsilon,\rho})$. Hence $\mu(T_1) = 0$, and because $S^\nu(t) \to S(t)$ for $t \in T \setminus T_1$, one has $S^\nu \xrightarrow{\text{p}} S$ a.e. □

14.25 Proposition (almost uniform convergence). *For closed-valued measurable mappings $S, S^\nu : T \rightrightarrows I\!R^n$ and a measure μ on $(T, \mathcal{A})$ with $\mu(T) < \infty$, one has*

$$S = \text{p-lim}_\nu S^\nu \text{ a.e.} \iff S = \text{p-lim}_\nu S^\nu \text{ a.u.}$$

Proof. Since $\mu(T) < \infty$, it's evident that convergence a.u. implies convergence a.e. For the converse, we rely on the metric characterization of uniform convergence provided by Proposition 5.49. For $\lambda \in I\!N$, let

$$A^\nu_\lambda = \big\{ t \in T \,\big|\, d\!\!l\big(S^\kappa(t), S(t)\big) > \lambda^{-1} \text{ for some } \kappa \geq \nu \big\}.$$

Due to convergence a.e., we have $\lim_\nu \mu(A^\nu_\lambda) = 0$. Choose ν_λ with $\mu(A^{\nu_\lambda}_\lambda) < \delta/2^\lambda$ and take $T_\delta := \bigcup_{\lambda=1}^\infty A^{\nu_\lambda}_\lambda$. Then $\mu(T_\delta) \leq \delta$ and

$$T \setminus T_\delta = \bigcap\nolimits_{\lambda=1}^\infty \big(T \setminus A^{\nu_\lambda}_\lambda\big) = \bigcap\nolimits_{\lambda=1}^\infty \big\{ t \,\big|\, d\!\!l\big(S^\kappa(t), S(t)\big) \leq \lambda^{-1},\ \forall \kappa \geq \nu_\lambda \big\};$$

in other words, if $t \in T \setminus T_\delta$, then for all $\varepsilon > 0$ one can choose $N = \{\nu \geq \nu_\lambda\} \in \mathcal{N}_\infty$ for some $\lambda \geq 1/\varepsilon$ and obtain $d\!\!l\big(S^\nu(t), S(t)\big) \leq \varepsilon$ for all $\nu \in N$. This says that the mappings S^ν converge a.u. to S. □

Questions about the measurability of mappings involving tangent and normal vectors can be answered from this platform of limits and other operations.

14.26 Theorem (tangent and normal cone mappings). *For any closed-valued measurable mapping $S : T \rightrightarrows I\!R^n$ and any measurable choice of $x(t) \in S(t)$, one has the closed-valuedness and measurability of the mappings*

$$t \mapsto T_{S(t)}\big(x(t)\big), \qquad t \mapsto \widehat{T}_{S(t)}\big(x(t)\big),$$
$$t \mapsto N_{S(t)}\big(x(t)\big), \qquad t \mapsto \widehat{N}_{S(t)}\big(x(t)\big).$$

Furthermore, the mapping $t \mapsto \operatorname{gph} N_{S(t)} = \big\{(x,v)\,\big|\, x \in S(t),\ v \in N_{S(t)}(x)\big\}$ *is closed-valued and measurable.*

Proof. The closed-valuedness of these mappings follows from their definitions; cf. 6.2, 6.5 and 6.26. To obtain the measurability of $G : t \mapsto \operatorname{gph} N_{S(t)}$, we'll appeal to the characterization of general normal vectors as limits of proximal normals in 6.18(a). This says that $G = \text{p-lim sup}_\nu G^\nu$ for the mapping $G^\nu : T \to I\!R^n \times I\!R^n$ defined by

$$\begin{aligned} G^\nu(t) &= \big\{(x,v)\,\big|\, x \in P_{S(t)}(x + \nu^{-1}v)\big\} \\ &= \big\{(x,v)\,\big|\, F^\nu(x,v) \in \operatorname{gph} P_{S(t)}\big\} \ \text{ for } \ F^\nu(x,v) = (x, x+\nu^{-1}v). \end{aligned}$$

The measurability of G^ν follows from 14.15(b) in the light of the measurability of $t \mapsto \operatorname{gph} P_{S(t)}$ in 14.17, and the measurability of G is clear then from 14.20.

The measurability of $t \mapsto N_{S(t)}\big(x(t)\big)$ is seen from the composition rule in 14.14, and the measurability of $t \mapsto \widehat{T}_{S(t)}\big(x(t)\big)$ then comes out of the polarity $\widehat{T}_{S(t)}\big(x(t)\big) = N_{S(t)}\big(x(t)\big)^*$ in 6.28(b) and the rule in 14.12(f).

Since $T_{S(t)}\big(x(t)\big) = \big\{v \in I\!R^n \,\big|\, \liminf_{\alpha \searrow 0} \alpha^{-1} d((t+\alpha v, S(t)) = 0\big\}$ and $(t,\alpha) \mapsto \liminf_{r \searrow 0} \alpha^{-1} d((t+\alpha v, S(t))$ is a Carathéodory mapping, applying 14.15(b) yields the measurability of $t \mapsto T_{S(t)}\big(x(t)\big)$. From the polarity rule in 14.12(f), we obtain the measurability of $t \mapsto \widehat{N}_{S(t)}\big(x(t)\big)$, since $\widehat{N}_{S(t)}\big(x(t)\big) = T_{S(t)}\big(x(t)\big)^*$ by 6.28(a). □

D. Normal Integrands

Up to now, we've been occupied with the issue of how a subset of $I\!R^n$ can depend measurably on a parameter t in T, but we're now going to shift the focus to how an extended-real-valued function on $I\!R^n$ can depend measurably on such a parameter. As in our treatment of continuity issues in Chapter 7, there are two ways of looking at such dependence, and both are important in concept and in practice.

On the one hand, we can think of a function-valued mapping that assigns to each $t \in T$ an element of the space $\operatorname{fcns}(I\!R^n)$. This element can be symbolized by $f(t,\cdot)$. Through that notation, however, we can regard a bivariate function $f : T \times I\!R^n \to \overline{I\!R}$ as furnishing the mathematical context instead. The bivariate approach, which emphasizes $f(t,x)$ in its dependence on both t and x, is simpler and conforms to the usual way of dealing with a parameter element t. But the function-valued approach is better at revealing, in our framework, the assumptions that guarantee essential properties such as the measurability of $t \mapsto f\big(t, x(t)\big)$ when $t \mapsto x(t)$ is measurable.

In straddling between these approaches, we'll study $f : T \times I\!R^n \to \overline{I\!R}$ from the perspective of two set-valued mappings associated with f, its *epigraphical mapping* S_f and its *domain mapping* D_f,

$$\begin{aligned} S_f(t) &:= \operatorname{epi} f(t,\cdot) = \big\{(x,\alpha) \in I\!R^n \times I\!R \,\big|\, f(t,x) \le \alpha\big\}, \\ D_f(t) &:= \operatorname{dom} f(t,\cdot) = \big\{x \in I\!R^n \,\big|\, f(t,x) < \infty\big\}. \end{aligned} \qquad 14(8)$$

That will enable us to apply very advantageously the theory of measurable set-valued mappings that has been built up.

14.27 Definition (normal integrands). *A function $f : T \times I\!R^n \to \overline{I\!R}$ will be called a normal integrand if its epigraphical mapping $S_f : T \to I\!R^n \times I\!R$ is closed-valued and measurable.*

14.28 Proposition (basic consequences of normality). *For any normal integrand $f : T \times I\!R^n \to \overline{I\!R}$, the domain mapping $D_f : T \rightrightarrows I\!R^n$ is measurable, and one has $f(t,x)$ lsc in x for each fixed t and measurable in t for each fixed x. Indeed, $f\big(t,x(t)\big)$ is measurable in t when $x(t)$ depends measurably on t.*

But not every function $f : T \times I\!R^n \to \overline{I\!R}$ with $f(t,x)$ lsc in x and measurable in t is a normal integrand, nor does such a function necessarily yield the measurability of $f\big(t,x(t)\big)$ in t when $x(t)$ depends measurably on t.

Proof. For any open set $O \subset I\!R^n$ we have $D_f^{-1}(O) = S_f^{-1}(O \times I\!R)$ and hence $D_f^{-1}(O) \in \mathcal{A}$ through the measurability of S_f. Thus, D_f is measurable. Next, for any measurable function $x(\cdot) : T \to I\!R^n$ and any $\alpha \in I\!R$ we have

$$\big\{t \,\big|\, f\big(t,x(t)\big) \le \alpha\big\} = \big\{t \,\big|\, S_f(t) \cap R(t) \ne \emptyset\big\} = \operatorname{dom}[S_f \cap R]$$

for the closed-valued mapping $R : t \mapsto \{x(t)\} \times (-\infty,\alpha]$, which is measurable by 14.11(d). The mapping $S_f \cap R$ is measurable by 14.11(a), so $\operatorname{dom}[S_f \cap R] \in \mathcal{A}$ by 14.5. Thus $\big\{t \,\big|\, f\big(t,x(t)\big) \le \alpha\big\} \in \mathcal{A}$, and the function $t \mapsto f\big(t,x(t)\big)$ is measurable. Taking $x(t)$ constant as a special case, we get the measurability of f with respect to its first argument. The lower semicontinuity with respect to its second argument is clear from the closedness of $S_f(t) = \operatorname{epi} f(t,\cdot)$.

For an elementary example of an integrand f that has $f(t,x)$ measurable in t and lsc in x, but isn't a normal integrand, let $n = 1$, $T = I\!R$, $\mathcal{A} = \mathcal{L}(I\!R)$. Take D to be any subset of $I\!R$ that isn't in $\mathcal{L}(I\!R)$ (it's well known in measure theory that such sets exist), and define $f(t,x) = 0$ when $t = x \in D$ but $f(t,x) = \infty$ otherwise. The measurability and lower semicontinuity properties hold trivially in this case. But the closed-valued mapping $S_f : I\!R \to I\!R^2$ isn't measurable, for if it were, we would have $\operatorname{gph} S_f \in \mathcal{L}(I\!R) \otimes \mathcal{B}(I\!R^2)$ by Theorem 14.8, and that's not true. The same example, through the choice $x(t) = t$, shows a lack of measurability of $t \mapsto f\big(t,x(t)\big)$. □

It will be convenient as well as suggestive to refer generally to a function $f : T \times I\!R^n \to \overline{I\!R}$ as an *integrand* when $f(t,x)$ is measurable with respect to t for every $x \in I\!R^n$, and to speak of it as being a *convex*, or *proper*, or *lsc* integrand, etc., when $f(t,x)$ exhibits such a property with respect to x for every $t \in T$. This terminology of 'integrands' f will serve as a reminder that the t argument in $f(t,x)$ has a distinctly different role from the x argument and ultimately relates to the possibility of some operation of integration taking place with respect to a measure μ on $(T,\mathcal{A})$. Such integration could be connected with an integral functional as in 14(1). But alternatively, when μ is a probability measure, say, it could be tied to the notion of $f(t,\cdot)$ being a random element of $\text{fcns}(\mathbb{R}^n)$, a function-valued random variable.

Proposition 14.28 tells us that every normal integrand is an lsc integrand, but not conversely, and that the class of lsc integrands wouldn't be able to uphold a theory of integral functionals, where the measurability of $t \mapsto f\big(t,x(t)\big)$ is crucial. The assumption that $f(t,x)$ is measurable with respect to t has to be replaced by the assumption of *epi-measurability*, i.e., the measurability of $t \mapsto \text{epi}\, f(t,\cdot)$, in situations where only the lower semicontinuity of the functions $f(t,\cdot)$ can be expected, as for instance when the domain sets $\text{dom}\, f(t,\cdot)$ carry the expression of constraints. When the functions $f(t,\cdot)$ are continuous, however, a bridge to classical theory is available.

14.29 Example (Carathéodory integrands). *Falling in the category of normal integrands are all Carathéodory integrands, i.e., the functions $f : T \times \mathbb{R}^n \to \mathbb{R}$ (finite-valued) such that $f(t,x)$ is measurable in t for each x and continuous in x for each t. Indeed, a function $f : T \times \mathbb{R} \to \overline{\mathbb{R}}$ is a Carathéodory integrand if and only if both f and $-f$ are proper normal integrands.*

Detail. When f is a Carathéodory integrand, its epigraphical mapping S_f has the constraint representation

$$S_f(t) = \big\{(x,\alpha) \in \mathbb{R}^{n+1} \,\big|\, F(t,x,\alpha) \leq 0\big\} \text{ for } F(t,x,\alpha) = f(t,x) - \alpha.$$

Here F is a Carathéodory mapping from $T \times \mathbb{R}^n$ to $\mathbb{R}^1$, and S_f is therefore closed-valued and measurable on the basis of $S_f(t)$ being the image of $\mathbb{R}_-$ under $F(t,\cdot)^{-1}$; cf. Example 14.15(b).

A function $f(t,\cdot)$ is finite and continuous on $\mathbb{R}^n$ if and only if both $f(t,\cdot)$ and $-f(t,\cdot)$ are proper and lsc. From 14.28, therefore, the Carathéodory conditions correspond to f and $-f$ being proper normal integrands. □

14.30 Example (autonomous integrands). *If $f : T \times \mathbb{R}^n \to \overline{\mathbb{R}}$ has $f(t,x) \equiv g(x)$ with $g : \mathbb{R}^n \to \overline{\mathbb{R}}$ lsc, then f is a normal integrand.*

Detail. Here the epigraphical mapping S_f is constant. □

14.31 Example (joint lower semicontinuity). *Suppose that T is a Borel subset of $\mathbb{R}^d$, and that $f : T \times \mathbb{R}^n \to \overline{\mathbb{R}}$ is lsc. Then f is a normal integrand.*

Detail. The epigraphical mapping S_f has closed graph; it's osc. Then S_f is measurable by 14.9. □

The measurability of the domain mapping D_f in Proposition 14.28 underscores the flexibility afforded by the concept of a normal integrand f, in departing from that of a Carathéodory integrand. When $f(t,x)$ is interpreted as a cost expression which, in taking on the value ∞, imposes an implicit constraint on x, the feasible set of x's that corresponds to t is $D_f(t)$. This set depends measurably on t, and the same is true then for $\operatorname{cl} D(t)$ by 14.2.

14.32 Example (indicator integrands). *The indicator $\delta_C : T \times \mathbb{R}^n \to \overline{\mathbb{R}}$ of a mapping $C : T \rightrightarrows \mathbb{R}^n$, with*

$$\delta_C(t,x) = \delta_{C(t)}(x) = \begin{cases} 0 & \text{if } x \in C(t), \\ \infty & \text{if } x \notin C(t), \end{cases}$$

is a normal integrand if and only if C is closed-valued and measurable. For any normal integrand $f_0 : T \times \mathbb{R}^n \to \mathbb{R}$, the function $f = f_0 + \delta_C$, with

$$f(t,x) = \begin{cases} f_0(t,x) & \text{if } x \in C(t), \\ \infty & \text{if } x \notin C(t), \end{cases}$$

is then a normal integrand having $D_f(t) = D_{f_0}(t) \cap C(t)$. When f_0 is a Carathéodory integrand, this holds with $D_f(t) = C(t)$.

Detail. For $f = f_0 + \delta_C$ we have $S_f(t) = S_{f_0}(t) \cap [C(t) \times \mathbb{R}]$ with $S_{f_0}(t)$ closed-valued and depending measurably on t, like $C(t)$. Then S_f is closed-valued and measurable by the rules in 14.11(d) and 14.11(a), hence f is a normal integrand. When f_0 is a Carathéory integrand, it's normal with $D_{f_0}(t) = \mathbb{R}^n$; cf. Example 14.29. The case of $f = \delta_C$ specializes to $f_0 \equiv 0$. □

The set $C(t)$ in this example could be specified by a system of equations or inequalities that depends on t, and normal integrands may be involved there as well.

14.33 Proposition (measurability of level-set mappings). *A bivariate function $f : T \times \mathbb{R}^n \to \overline{\mathbb{R}}$ is a normal integrand if and only if for all $\alpha \in \overline{\mathbb{R}}$, the level-set mapping*

$$t \mapsto \operatorname{lev}_{\leq \alpha} f(t,\cdot) = \{x \mid f(t,x) \leq \alpha\}. \tag{14(9)}$$

is closed-valued and measurable. Moreover, the mapping $t \mapsto \{x \mid f(t,x) \leq \alpha(t)\}$ is closed-valued and measurable as long as $t \mapsto \alpha(t)$ is measurable.

Proof. Let $L(t) = \{x \mid f(t,x) \leq \alpha(t)\}$ for measurable $t \mapsto \alpha(t) \in \overline{\mathbb{R}}$. The mapping $L : T \rightrightarrows \mathbb{R}^n$ is closed-valued because $f(t,x)$ is lsc in x, so we can test its measurability through the property in 14.3(b). The case of $\alpha(t)$ constant with respect to t is covered as a special case.

Fix any closed set $C \subset \mathbb{R}^n$ and consider the closed-valued mapping $R : t \mapsto C \times \{\beta \in \mathbb{R} \mid \beta \leq \alpha(t)\}$. This is measurable by 14.11(d), and we have

$$L^{-1}(C) = \{t \mid S_f(t) \cap R(t) \neq \emptyset\} = \operatorname{dom}[S_f \cap R].$$

The epigraphical mapping S_f being closed-valued and measurable by the definition of normality, we deduce from 14.11(a) that the mapping $S_f \cap R$ is measurable and hence that $\operatorname{dom}[S_f \cap R] \in \mathcal{A}$. Thus, $L^{-1}(C) \in A$, and the measurability of L has been confirmed.

For the converse: let $\{A^\nu\}_{\nu \in \mathbb{N}}$ be a collection of countable sets converging to $\mathbb{R}$. The measurability of the mappings $S^\nu(t) = \bigcup_{\alpha \in A^\nu} lev_{\leq \alpha} f(t, \cdot)$ comes from 14.11(b). Since, the epigraphical mappings S_f is the pointwise limit of these mappings it's closed-valued and measurable, cf. Theorem 14.20, i.e., f is a normal integrand. □

14.34 Corollary (joint measurability criterion). *If $f : T \times \mathbb{R}^n \to \overline{\mathbb{R}}$ is a normal integrand, then $f(t,x)$ is $\mathcal{A} \otimes \mathcal{B}(\mathbb{R}^n)$-measurable as a function of (t,x), in addition to being lsc as a function of x. Conversely, these properties ensure that f is a normal integrand when $(T, \mathcal{A})$ is complete with respect to some measure μ, as in particular when $T = \mathbb{R}^d$ and $\mathcal{A} = \mathcal{L}(\mathbb{R}^d)$.*

Proof. The question hinges on the $\mathcal{A} \otimes \mathcal{B}(\mathbb{R}^n)$-measurability of the sets $G_\alpha := \{(t,x) \mid f(t,x) \leq \alpha\}$, which are the graphs of the level-set mappings in 14(9). These mappings are closed-valued and measurable when f is normal, and their graphs G_α belong then to $\mathcal{A} \otimes \mathcal{B}(\mathbb{R}^n)$ by Theorem 14.8. That theorem justifies the converse claim as well when $(T, \mathcal{A})$ is complete. □

14.35 Exercise (closures of measurable integrands). *Suppose $f(t, \cdot) = \operatorname{cl} f_0(t, \cdot)$ for an $\mathcal{A} \otimes \mathcal{B}(\mathbb{R}^n)$-measurable function $f_0 : T \times \mathbb{R}^n \to \overline{\mathbb{R}}$. If $(T, \mathcal{A})$ is complete with respect to some measure μ, then f is a normal integrand.*

Guide. Verify that the mapping S_{f_0} has $\mathcal{A} \otimes \mathcal{B}(\mathbb{R}^n)$-measurable graph, and then apply 14.8 and 14.2. □

14.36 Theorem (measurability in constraint systems). *In terms of a closed set $X(t) \subset \mathbb{R}^n$ depending measurably on $t \in T$, define the set $C(t) \subset \mathbb{R}^n$ by*

$$x \in C(t) \iff x \in X(t) \text{ and } f_i(t,x) - \alpha_i(t) \begin{cases} \leq 0 & \text{for } i \in I_1, \\ = 0 & \text{for } i \in I_2, \end{cases} \qquad 14(10)$$

where f_i is a normal integrand for $i \in I_1$ and a Carathéodory integrand for $i \in I_2$, these index sets being (finite or) countable, and the functions α_i are measurable. Then $C(t)$ is closed and depends measurably on t.

In particular, therefore, the set $A = \{t \mid C(t) \neq \emptyset\} \subset T$ is measurable, and it is possible for each $t \in A$ to select a point $x(t) \in C(t)$ in such a manner that the function $t \mapsto x(t)$ is measurable.

Proof. Define $L_i(t)$ to consist of the points $x \in \mathbb{R}^n$ satisfying $f_i(t,x) \leq \alpha_i(t)$ for $i \in I_1$ and $i \in I_2$, and define $L_i'(t)$ by the opposite inequality for $i \in I_2$. The closed-valued mappings $L_i : t \mapsto L_i(t)$ and $L_i' : t \mapsto L_i'(t)$ are measurable by Theorem 14.32 (and the characterization of Carathéodory integrands in 14.28). The mapping $C : t \mapsto C(t)$ is given by their intersection with $X : t \mapsto X(t)$, so it's closed-valued measurable by 14.11(a). The set $A = \operatorname{dom} C$ is measurable then, and a measurable selection exists by 14.6. □

14.37 Theorem (measurability of optimal values and solutions). *For any normal integrand $f : T \times \mathbb{R}^n \to \overline{\mathbb{R}}$, let*

$$p(t) = \inf f(t,\cdot), \qquad P(t) = \operatorname{argmin} f(t,\cdot).$$

Then the function $p : T \to \overline{I\!R}$ *is measurable and the mapping* $P : T \rightrightarrows I\!R^n$ *is closed-valued and measurable.*

In particular, therefore, the set $A = \{t \mid \operatorname{argmin}_x f(t,x) \neq \emptyset\} \subset T$ *is measurable, and it is possible for each* $t \in A$ *to select a minimizing point* $x(t)$ *in such a manner that the function* $t \mapsto x(t)$ *is measurable.*

Proof. Consider the mapping $R : t \mapsto F\big(S_f(t)\big)$ obtained by taking F to be the projection from $I\!R^n \times I\!R$ to $I\!R$. This is measurable by 14.15(a), and the same is true then for $t \mapsto \operatorname{cl} R(t)$ by 14.2. But $\operatorname{cl} R(t) = \{\alpha \in I\!R \mid p(t) \leq \alpha\}$. Hence p is measurable. Then since $P(t) = \{x \mid f(t,x) \leq p(t)\}$ we have the measurability of P as well by Proposition 14.33. The assertion about the existence of a measurable selection for P is justified by 14.6. □

14.38 Exercise (Moreau envelopes and proximal mappings). *Consider a proper, normal integrand* $f : T \times I\!R^n \to \overline{I\!R}$ *such that, for each* $t \in T$, $f(t,\cdot)$ *is prox-bounded with threshold* $\lambda_f(t)$. *For a measurable function* $\lambda : T \rightrightarrows I\!R$ *with* $0 < \lambda(t) < \lambda_f(t)$ *for all* $t \in T$, *let*

$$e(t,x) := [e_{\lambda(t)} f_t](x), \qquad P(t,x) := [P_{\lambda(t)} f_t](x), \qquad \textit{where } f_t = f(t,\cdot).$$

Then e *is a Carathéodory integrand, while* $P(t,\cdot)$ *is an osc mapping with graph depending measurably on* t. *When* f *is a convex integrand, so that* $\lambda_f(t) = \infty$ *for all* t, P *is a (single-valued) Carathéodory mapping:* $T \times I\!R^n \to I\!R^n$.

Regardless of convexity, both $e\big(t, x(t)\big)$ *and* $P\big(t, x(t)\big)$ *depend measurably on* t *when* $x(t)$ *depends measurably on* t.

Guide. Work from the definition of Moreau envelopes and proximal mappings in 1.22 and the continuity properties in 1.25. First show for fixed $x \in I\!R^n$ that $e(t,x)$ is measurable with respect to t; do this by applying Theorem 14.37. Then get the measurability of $t \mapsto G(t) = \operatorname{gph} P(t,\cdot)$ by representing $G(t) := \{(x,w) \mid g(t,x,w) \leq e(t,x)\}$ for a certain normal integrand $g : T \times I\!R^n \times I\!R^n \to \overline{I\!R}$ and construing this to mean that $G(t)$ is the image under $(x,w,\alpha) \mapsto (x,w)$ of the set of (x,w,α) satisfying $g(t,x,w) \leq \alpha$ and $-e(t,x) \leq -\alpha$. (Rely on 14.33, 14.29 and 14.15(a).)

Make use of 2.26 in the convex case. In order to justify the final assertions, appeal to 14.28 and 14.14. □

Very similar to the case of Moreau envelopes in 14.38 is that of Pasch-Hausdorff envelopes, as considered in 9.11. From any normal integrand $f : T \times I\!R^n \rightrightarrows \overline{I\!R}$ and any $\kappa \in I\!R_+$ one can form

$$f_\kappa(t,x) = \inf_w \{f(t,w) + \kappa|w - x|\},$$

and as long as it doesn't take on $-\infty$, f_κ will be a Carathéodory integrand.

The normality of an integrand f corresponds through Theorem 14.5 to the existence of a Castaing representation for its epigraphical mapping S_f. When

f is a convex integrand (i.e., $f(t,x)$ is convex in x for each t), a comparable characterization can be given in terms of the domain mapping D_f instead, although that might not be closed-valued.

14.39 Proposition (normality test for convex integrands). *Let $f : T \times \mathbb{R}^n \to \overline{\mathbb{R}}$ be such that, for every $t \in T$, the function $f(t,\cdot)$ is lsc and convex. Then f is a normal integrand if and only if there is a countable family $\{x^\nu\}_{\nu \in \mathbb{N}}$ of measurable functions $x^\nu : T \mapsto \mathbb{R}^n$ such that*

(a) *for each $\nu \in \mathbb{N}$, $t \mapsto f(t, x^\nu(t))$ is measurable;*

(b) *for each $t \in T$, $\{x^\nu(t) \mid \nu \in \mathbb{N}\} \cap D_f(t)$ is dense in $D_f(t)$.*

In the case where $\operatorname{int} D_f(t) \neq \emptyset$ for all t with $D_f(t) \neq \emptyset$, there is a simpler test: f is a normal integrand if and only if $f(t,x)$ is measurable in t for each x. Thus, every lsc convex integrand is a normal integrand.

Proof. Suppose first that f is normal. Its epigraphical mapping S_f is then closed-valued and measurable and, by Theorem 14.5, admits a Castaing representation $\{(x^\nu, \alpha^\nu)\}_{\nu \in \mathbb{N}}$ with $(x^\nu, \alpha^\nu) : \operatorname{dom} S_f \to \mathbb{R}^n \times \mathbb{R}$ measurable. Extending each x^ν by taking $x^\nu(t) = 0$ (say) when $t \notin \operatorname{dom} S_f$, we get a family $\{x^\nu\}_{\nu \in \mathbb{N}}$ satisfying (b). We have (a) satisfied as well by virtue of 14.34.

For the converse, consider a family satisfying (a) and (b). The density in (b) and the lower semicontinuity and convexity of $f(t,x)$ with respect to x ensure that $f(t,\cdot) = \operatorname{cl} g(t,\cdot)$ for the function $g(t,\cdot) : T \to \overline{\mathbb{R}}$ defined by $g(t,x) = f(t, x^\nu(t))$ if $x = x^\nu(t)$ for some ν, but $g(t,x) = \infty$ otherwise; cf. Theorem 2.35 (which can be invoked for $\operatorname{rint} D_f(t)$ if $\operatorname{int} D_f(t) = \emptyset$). From this we see that the countable family $\{y^{\nu,\beta}\}_{(\nu,\beta) \in \mathbb{N} \times \mathbb{Q}}$ with $y^{\nu,\beta}(t) = (x^\nu(t), \beta)$ consists of measurable functions and has the property that, for each $t \in T$, the set $\{y^{\nu,\beta}(t) \mid (\nu,\beta) \in \mathbb{N} \times \mathbb{Q}\} \cap S_f(t)$ is dense in $S_f(t)$. We also have for each $(\nu,\beta) \in \mathbb{N} \times \mathbb{Q}$ that $\{t \in T \mid y^{\nu,\beta}(t) \in S_f(t)\} = \{t \in T \mid f(t, x^\nu(t)) \leq \beta\}$, and this set $T^{\nu,\beta}$ is measurable by (a). The set $\operatorname{dom} S_f = \bigcup_{\nu,\beta} T^{\nu,\beta}$ is measurable then as well. We have arrived at the pattern in Theorem 14.5(b) and thereby know that S_f is closed-valued and measurable.

The sufficiency of the test at the end of the proposition is justified by taking $x^\nu(t) \equiv q^\nu$ for an enumeration $\{q^\nu\}_{\nu \in \mathbb{N}}$ of $\mathbb{Q}^n$ and invoking the fact that any convex set with nonempty interior lies within the closure of that interior (see 2.33). The necessity of this condition in the special case described is apparent from 14.28 (last part) as applied to constant functions. □

Note that it's not possible to drop the convexity assumption from 14.39 without impairing the propositions' validity. The reason is that for a nonconvex function h on $\mathbb{R}^n$, even if h is lsc, one can't be sure of having $\operatorname{epi} h = \operatorname{cl}\{(x,\alpha) \in D \times \mathbb{R} \mid \alpha \geq h(x)\}$ when D is dense in $\operatorname{dom} h$. For example, let $h \equiv 1$ on $(0,1]$ and take $h(0) = 0$, but elsewhere $h \equiv \infty$. Then h is lsc. Let D be any countable dense subset of $(0,1)$. Then $(0,0)$ belongs to $\operatorname{epi} h$ but isn't a cluster point of any sequence in $[D \times \mathbb{R}] \cap \operatorname{epi} h$.

For lsc integrands, it's possible to describe epi-measurability, and thus normality, in terms of the measurability of certain families of extended real-valued functions generated by 'optimization'.

14.40 Proposition (scalarization of normal integrands). *Let $f : T \times \mathbb{R}^n \to \overline{\mathbb{R}}$ have $f(t,x)$ lsc in x for every t. For each $D \subset \mathbb{R}^n$, define*

$$p_D(t) := \inf_{x \in D} f(t,x).$$

Then f is a normal integrand if and only if all the functions in $\{p_D \mid D \in \mathcal{D}\}$ are measurable, where $\mathcal{D}$ is any of the following collections of subsets of $\mathbb{R}^n$:

(a) *all open sets;*
(b) *all open balls;*
(c) *all rational open balls;*
(d) *all closed sets;*
(e) *all compact sets;*
(f) *all closed balls;*
(g) *all rational closed balls.*

Proof. For (a), (b),..., let's denote the properties of measurability for the associated collections of functions p_D by (a′), (b′), etc. We have to demonstrate that these properties are equivalent to each other and to the normality of f. Obviously (a′) ⇒ (b′) ⇒ (c′) and (d′) ⇒ (e′) ⇒ (f′) ⇒ (g′). Furthermore, (g′) ⇒ (a′) on the basis that any open set is the union of countably many rational closed balls, and whenever $D = \bigcup_\nu D^\nu$ the function p_D is the pointwise infimum of the functions p_{D^ν} and thereby inherits their measurability.

Next, we observe that the normality of f implies (d′) through Theorem 14.37 as applied to f plus the indicator of D, which gives another normal integrand as seen in 14.32.

The task that remains is to demonstrate how (c′) implies the normality of f. This comes down to the measurability of the epigraphical mapping S_f, inasmuch as the closed-valuedness of S_f is assured by the lower semicontinuity assumption on f. In general, for a set $D \subset \mathbb{R}^n$ and arbitrary $\beta \in \mathbb{R}$ we have

$$\begin{aligned}\{t \in T \mid p_D(t) < \beta\} &= \{t \in T \mid S_f(t) \cap [D \times (-\infty, \beta)] \neq \emptyset\} \\ &= \{t \in T \mid S_f(t) \cap [D \times (\alpha, \beta)] \neq \emptyset\} \text{ for any } \alpha < \beta,\end{aligned}$$

with the second equality holding because $S_f(t)$ is an epigraph. We're operating now under the assumption that every such subset of T that arises from a rational open ball D belongs to $\mathcal{A}$. Every open set $O \subset \mathbb{R}^n \times \mathbb{R}$ can be expressed as the union of a sequence of sets $O^\nu = D^\nu \times (\alpha^\nu, \beta^\nu)$ with D^ν a rational open ball. Then $S_f^{-1}(O^\nu) \in \mathcal{A}$ and $S_f^{-1}(O) = \bigcup_\nu S_f^{-1}(O^\nu)$, so $S_f^{-1}(O) \in \mathcal{A}$. The measurability of S_f follows. □

Note that in the cases of Proposition 14.40 where the collection $\mathcal{D}$ consists of balls, one doesn't really need all balls of the type in question, but just a rich enough family to provide a neighborhood system of every point of $\mathbb{R}^n$. This is evident from the proof.

14.41 Corollary (alternative interpretation of normality). *For $f : T \times \mathbb{R}^n \to \overline{\mathbb{R}}$ to be a normal integrand, it is necessary and sufficient to have that*

(a) *$f(t,x)$ is lsc in x for each t, and*

(b) *for each x there exists $\varepsilon > 0$ such that, for all $\delta \in (0,\varepsilon)$, the function $t \mapsto \inf\{f(t,x') \mid x' \in \mathbb{B}(x,\delta)\}$ is measurable.*

Proof. This corresponds to (f) of 14.40 with the observation just made. □

Corollary 14.41 illuminates the difference between normal integrands and lsc integrands very clearly. To get normality out of an integrand that's lsc, the measurability of $t \mapsto f(t,x)$ has to be broadened to the property in 14.41(b). The measurability of $t \mapsto f(t,x)$, by itself, just corresponds to the extreme form of that property in which the ball reduces to a single point.

When the σ-field $\mathcal{A}$ is complete, with T a Borel subset of $\mathbb{R}^d$, normal integrands can be also characterized in terms of certain continuity properties in a manner reminiscent of Lusin's theorem.

14.42 Theorem (continuity properties of normal integrands). *Let $T \in \mathcal{B}(\mathbb{R}^d)$, and let $f : T \times \mathbb{R}^n \to \overline{\mathbb{R}}$ have $f(t,x)$ lsc in x for each t. Then, with $\mathcal{A} = \mathcal{L}(T)$ and μ the restriction of d-dimensional Lebesgue measure to this field, each of the following conditions is equivalent to f being a normal integrand:*

(a) *for every $\varepsilon > 0$ there is a closed set $T_\varepsilon \subset T$ with $\mu(T \setminus T_\varepsilon) < \varepsilon$, such that $t \mapsto f(t,\cdot)$ is epi-continuous relative to T_ε;*

(b) *for every $\varepsilon > 0$ there is a closed set $T_\varepsilon \subset T$ with $\mu(T \setminus T_\varepsilon) < \varepsilon$, such that f is lsc relative to $T_\varepsilon \times \mathbb{R}^n$;*

(c) *there exists $\mathcal{B}(T \times \mathbb{R}^n)$-measurable $g : T \times \mathbb{R}^n \to \overline{\mathbb{R}}$ with $g(t,\cdot)$ lsc, such that $f(t,\cdot) = g(t,\cdot)$ for all t in a set $T' \subset T$ with $T' \in \mathcal{A}$, $\mu(T \setminus T') = 0$.*

(When $\mu(T) < \infty$, the sets T_ε in (a) and (b) can be taken to be compact.)

Proof. Essentially this applies Theorem 14.10 to epigraphical mappings. With respect to S_f we have 14.10(a) equivalent to (a) through the definition of epi-continuity, and 14.10(b) equivalent to (b) because an epigraphical mapping is osc if and only if the underlying function is lsc. Furthermore, (c) yields 14.10(c): the graph of $R = S_g$ belongs to $\mathcal{B}(T \times \mathbb{R}^n \times \mathbb{R})$ because it consists of all (t,x,α) with $g(t,x) - \alpha \leq 0$, and the function $(t,x,\alpha) \mapsto g(t,x) - \alpha$ is Borel measurable because g is Borel measurable.

Through these implications and the equivalences in 14.10 itself, we can finish up by arguing that (b) ensures (c). For each $\nu \in \mathbb{N}$ let T^ν have the property in (b) with respect to $\varepsilon = 1/\nu$, and let $T' = \bigcup_\nu T^\nu$. Then $T' \in \mathcal{A}$, $\mu(T \setminus T') = 0$. Let g agree with f on $T' \times \mathbb{R}^n$ but have the value ∞ everywhere else. Then for each $\alpha \in \mathbb{R}$ we have

$$\{(t,x) \mid g(t,x) \leq \alpha\} = \bigcup_\nu \{(t,x) \mid t \in T^\nu,\ f(t,x) \leq \alpha\},$$

the sets in the union being closed, so g is $\mathcal{A} \otimes \mathcal{B}(\mathbb{R}^n)$-measurable. □

14.43 Corollary (Scorza-Dragoni property). *Let $T \in \mathcal{B}(I\!R^d)$ with $\mathcal{A} = \mathcal{L}(T)$ and μ the restriction of d-dimensional Lebesgue measure to this field. Then for any Carathéodory integrand $f : T \times I\!R^n \to \overline{I\!R}$ and any $\varepsilon > 0$ there is a closed set $T_\varepsilon \subset T$ with $\mu(T \setminus T_\varepsilon) < \varepsilon$, such f is continuous relative to $T_\varepsilon \times I\!R^n$.*

Proof. The theorem is applicable to both f and $-f$; cf. 14.29. □

E. Operations on Integrands

We now turn to the study of the operations that can be used in generating integrands and ascertaining the extent to which they preserve normality.

14.44 Proposition (addition and pointwise max and min). *For each $j \in J$, an index set, let $f_j : T \times I\!R^n \to \overline{I\!R}$ be a normal integrand. Then the following functions f are normal integrands:*

(a) *$f(t,x) = \sup_{j \in J} f_j(t,x)$ if J is countable;*

(b) *$f(t,x) = \inf_{j \in J} f_j(t,x)$ if J is countable and this expression is lsc in x (as when J is finite); or more generally the function obtained by applying cl_x (lsc regularization in the x argument) to this expression;*

(c) *$f(t,x) = \sum_{j \in J} \lambda_j f_j(t,x)$ for arbitrary $\lambda_j \in I\!R_+$, if J is finite;*

and in the case of $f_j : T \times I\!R^{n_j} \to \overline{I\!R}$ also

(d) *$f(t,x_1,\ldots,x_r) = f_1(t,x_1) + \cdots + f_r(t,x_r)$ when $J = \{1,\ldots,r\}$ and the functions $f_j(t,\cdot)$ are proper.*

Proof. We can immediately get (a) and (b) by applying the rules in 14.11(a) and 14.11(b) to the epigraphical mappings S_{f_j}. Looking next at (d), we introduce the mappings $E : t \mapsto S_{f_1}(t) \times \cdots \times S_{f_r}(t)$ and $S(t) = L\big(E(t)\big)$ with

$$L : (x_1,\alpha_1,\ldots,x_r,\alpha_r) \mapsto (x_1,\ldots,x_r,\alpha_1 + \cdots + \alpha_r).$$

We have E closed-valued in consequence of the closed-valuedness of each S_{f_j}, and E is measurable then by 14.11(d). Then S is measurable by the composition rule in 14.15(a). Clearly $S_f(t) = S(t)$, and since $S_f(t)$ is closed-valued (because, for proper functions, the addition formula preserves lower semicontinuity in x, cf. 1.39), we see that f is a normal integrand.

The argument for (c) is the same as the one for (d), except that L is replaced by the mapping

$$M : (x_1,\alpha_1,\ldots,x_r,\alpha_r) \mapsto \begin{cases} (x, \alpha_1 + \cdots + \alpha_r) & \text{if } x_1 = \cdots = x_r = x, \\ \emptyset & \text{otherwise.} \end{cases}$$

This mapping is osc and allows utilization of 14.13(b); the convention $0 \cdot [\pm\infty] = 0$ enters when $\lambda_j = 0$, □

14.45 Proposition (composition operations).

(a) *If* $f : T \times \mathbb{R}^n \to \overline{\mathbb{R}}$ *is given by* $f(t,x) = g\big(t, F(t,x)\big)$ *for a normal integrand* $g : T \times \mathbb{R}^m \to \overline{\mathbb{R}}$ *and a Carathéodory mapping* $F : T \times \mathbb{R}^n \to \mathbb{R}^m$, *then* f *is a normal integrand.*

(b) *If* $f : T \times \mathbb{R}^n \to \overline{\mathbb{R}}$ *is given by* $f(t,x) = \theta\big(t, g(t,x)\big)$ *for a normal integrand* $g : T \times \mathbb{R}^n \to \overline{\mathbb{R}}$ *and a normal integrand* $\theta : T \times \mathbb{R} \to \overline{\mathbb{R}}$, *with* $\theta(t,\alpha)$ *nondecreasing in* α *and the convention used that*

$$\theta(t,\infty) = \sup\big\{\theta(t,\alpha) \,\big|\, \alpha \in \mathbb{R}\big\}, \qquad \theta(t,-\infty) = \inf\big\{\theta(t,\alpha) \,\big|\, \alpha \in \mathbb{R}\big\},$$

then f *is a normal integrand.*

(c) *If* $f : T \times \mathbb{R}^n \to \overline{\mathbb{R}}$ *is given by* $f(t,x) = g\big(t, x, u(t)\big)$ *for a normal integrand* $g : T \times \mathbb{R}^n \times \mathbb{R}^m \to \overline{\mathbb{R}}$ *and a measurable function* $u : T \to \mathbb{R}^m$, *then* f *is a normal integrand.*

Proof. We have $f(t,x)$ lsc with respect to x in all cases (for (b) see 1.40), so only the measurability of S_f has to be checked. In (a), define the Carathéodory mapping $G : T \times \mathbb{R}^n \times \mathbb{R} \to \mathbb{R}^m \times \mathbb{R}$ by $G(t,x,\alpha) = \big(F(t,x),\alpha\big)$. We have

$$S_f(t) = \big\{(x,\alpha) \,\big|\, G(t,x,\alpha) \in S_g(t)\big\},$$

so S_f is measurable by 14.15(b). Then (c) follows immediately as well; think of $\big(x,u(t)\big)$ as $F(t,x)$. To obtain (b) we observe that

$$S_f(t) = M\big(t, S_g(t)\big) \quad \text{with} \quad M(t,x,\alpha) := \big\{(x,\beta) \,\big|\, \theta(t,\alpha) \le \beta\big\}.$$

Here $\operatorname{gph} M(t,\cdot,\cdot) = \big\{(x,\alpha,y,\beta) \,\big|\, (\alpha,\beta) \in S_\theta(t),\ x = y\big\}$, so this is a closed set depending measurably on t; cf. 14.11(d). The measurability of S_f is assured then by 14.13(b). □

14.46 Corollary (multiplication by nonnegative scalars). *If* $f : T \times \mathbb{R}^n \to \overline{\mathbb{R}}$ *is given by* $f(t,x) = \lambda(t)g(t,x)$ *for a normal integrand* $g : T \times \mathbb{R}^n \to \overline{\mathbb{R}}$ *and a measurable function* $\lambda : T \to \mathbb{R}_+$, *then* f *is a normal integrand.*

Proof. Take $\theta(t,\alpha) = \lambda(t)\alpha$ in part (b) of the preceding theorem. □

14.47 Proposition (inf-projection of integrands). *Let* $p(t,u) = \inf_x f(t,x,u)$ *for a normal integrand* $f : T \times \mathbb{R}^n \times \mathbb{R}^n \to \overline{\mathbb{R}}$. *If* $p(t,u)$ *is lsc in* u *(as holds in particular if, for each* t, $f(t,x,u)$ *is level-bounded in* x *locally uniformly in* u*), then* p *is a normal integrand.*

More generally, the function $\bar{p}(t,u) = \operatorname{cl}_u p(t,u)$ *(where* cl_u *refers to lsc regularization in the* u *argument) is always a normal integrand.*

Proof. This comes out of 14.13(a) and the representation $S_p = L \circ S_f$ with L the projection from (x,u,α) to (u,α). When lsc regularization is applied to ensure closed-valuedness of S_f, the closure operation in 14.2 is brought in. □

14.48 Example (epi-addition and epi-multiplication).

(a) *If* $f(t,\cdot) = \operatorname{cl}\big[f_1(t,\cdot) \# \cdots \# f_m(t,\cdot)\big]$ *with each* $f_i : T \times \mathbb{R}^n \to \overline{\mathbb{R}}$ *a normal integrand, then* f *is a normal integrand.*

(b) *If* $f(t,\cdot) = \lambda(t)\star g(t,\cdot)$ *for a normal integrand* $g : T \times \mathbb{R}^n \to \overline{\mathbb{R}}$ *and scalars* $\lambda(t) \geq 0$ *depending measurably on* t, *then* f *is a normal integrand.*

Detail. The situation in (a) can be placed in the parametric context of 14.47, but it's easier perhaps to note that $S_f = \text{cl}\left[S_{f_1} + \cdots + S_{f_m}\right]$; cf. 1.28. The normality of f corresponds to the closed-valuedness and measurability of S_f, and that's present because measurability of set-valued mappings is preserved when taking sums and closures, as seen in 14.11(c) and 14.2.

In case (b) we have $S_f(t) = \lambda(t)S_g(t)$ when $\lambda(t) > 0$, but $S_f(t) = \text{epi}\,\delta_{\{0\}}$ when $\lambda(t) = 0$; cf. the definition of epi-multiplication in 1(14). Let T_0 be the set of t values for which the latter alternative occurs, i.e., $T_0 = \lambda^{-1}(\{0\})$. The restriction of S_f to T_0 is measurable as a constant-valued mapping, whereas the restriction of S_f to $T \setminus T_0$ is measurable in consequence of 14.19. Hence for any open set $O \subset \mathbb{R}^n \times \mathbb{R}$ the set $\left\{t \in T_0 \,\middle|\, S_f(t) \cap O \neq \emptyset\right\}$ and the set $\left\{t \in T \setminus T_0 \,\middle|\, S_f(t) \cap O \neq \emptyset\right\}$ are both measurable. The union of these sets is $S_f^{-1}(O)$, so that's measurable as well. Thus, S_f is measurable, and since it's also closed-valued we conclude that f is a normal integrand. □

14.49 Exercise (convexification of normal integrands). *If* $f : T \times \mathbb{R}^n \to \overline{\mathbb{R}}$ *is a normal integrand and* $g(t,\cdot) = \text{cl}\,\text{con}\, f(t,\cdot)$, *then* g *is a normal integrand.*

Guide. Apply 14.12(a) and 14.2 to the epigraphical mapping S_f. □

Normality too is preserved under taking conjugates. This could be derived by invoking 14.12(f), but it's instructive and more convenient to rely instead on the properties of normal integrands that have been obtained so far. The conjugate f^* and biconjugate f^{**} of a normal integrand $f : T \times \mathbb{R}^n \to \overline{\mathbb{R}}$ are defined by applying the Legendre-Fenchel transform in the $\mathbb{R}^n$ argument with the T argument fixed:

$$
\begin{aligned}
f^*(t,v) &:= \sup_{x \in \mathbb{R}^n} \left\{\langle v, x\rangle - f(t,x)\right\},\\
f^{**}(t,x) &:= \sup_{v \in \mathbb{R}^n} \left\{\langle v, x\rangle - f^*(t,v)\right\}.
\end{aligned}
\qquad 14(11)
$$

Recall that $f^{**}(t,\cdot) \leq f(t,\cdot)$, and that equality holds when $f(t,\cdot)$ is proper, lsc and convex; cf. Theorem 11.1.

14.50 Theorem (conjugate integrands). *If* $f : T \times \mathbb{R}^n \to \overline{\mathbb{R}}$ *is a normal integrand, its conjugate* f^* *and biconjugate* f^{**} *are normal integrands.*

Proof. Let $T_0 = \left\{t \in T \,\middle|\, f(t,\cdot) \not\equiv \infty\right\} = \text{dom}\, S_f$ (measurable), and consider any countable family $\{(x^\nu, \alpha^\nu)\}_{\nu \in \mathbb{N}}$ of measurable functions $(x^\nu, \alpha^\nu) : T_0 \to \mathbb{R}^n \times \mathbb{R}$ that furnishes a Castaing representation of S_f (the existence being guaranteed by 14.5). For each $\nu \in \mathbb{N}$, define $g^\nu : T \times \mathbb{R}^n \to \overline{\mathbb{R}}$ by

$$
g^\nu(t,v) := \begin{cases} \langle x^\nu(t), v\rangle - \alpha^\nu(t) & \text{if } t \in T_0,\\ -\infty & \text{if } t \in T \setminus T_0. \end{cases}
$$

We have $f^* = \sup_\nu g^\nu$ by 14(11) and the density property of a Castaing representation (in 14.5(a)). The restriction of g^ν to $T_0 \times I\!R^n$ is a Carathéodory integrand and thus is a normal integrand (by 14.29), so the restriction of S_{g^ν} to T_0 is closed-valued and measurable. But also $S_{g^\nu}(t) \equiv I\!R^n \times I\!R$ for $t \in T \setminus T_0$, so the restriction of S_{g^ν} to $T \setminus T_0$ is closed-valued and measurable as well. Hence S_{g^ν} is closed-valued and measurable on T, and g^ν is a normal integrand on $T \times I\!R^n$. The fact that $f^* = \sup_\nu g^\nu$ ensures then through 14.43(a) that f^* is a normal integrand. The conjugate of f^* is f^{**}, so the normality of f^* in turn implies the normality of f^{**}. □

14.51 Example (support integrands). *Associate with $S : T \rightrightarrows I\!R^n$ the function $\sigma_S : T \times I\!R^n \to \overline{I\!R}$ defined by*

$$\sigma_S(t, v) := \sigma_{S(t)}(v) = \sup\{\langle v, x\rangle \,|\, x \in S(t)\}.$$

If S is measurable, then σ_S is a convex normal integrand. Conversely, if σ_S is a normal integrand and S is closed-convex-valued, then S is measurable.

In particular, when S is compact-convex-valued, S is measurable if and only if $\sigma_S(t, v)$ is measurable in t for each fixed v.

Detail. We get this by specializing 14.50 to the integrand δ_S in 14.32; cf. 11.4 and 11.1. For the last assertion, we appeal also to 14.39. □

14.52 Exercise (envelope representations through convexity).

(a) *A function $f : T \times I\!R^n \to \overline{I\!R}$ that nowhere takes on $-\infty$ is a convex normal integrand if and only if there is a collection $\{(v^\nu, \alpha^\nu)\}_{\nu \in I\!N}$ of measurable functions $(v^\nu, \alpha^\nu) : T \to I\!R^n \times I\!R$ such that*

$$f(t, x) = \sup_{\nu \in I\!N}\{\langle v^\nu(t), x\rangle - \alpha^\nu(t)\}.$$

(b) *A mapping $S : T \rightrightarrows I\!R^n$ is closed-convex-valued and measurable if and only if there is a collection $\{(v^\nu, \alpha^\nu)\}_{\nu \in I\!N}$ of measurable functions $(v^\nu, \alpha^\nu) : T \to I\!R^n \times I\!R$ such that*

$$S(t) = \{x \in I\!R^n \,\big|\, \langle v^\nu(t), x\rangle \le \alpha^\nu(t) \text{ for all } \nu \in I\!N\}.$$

Guide. Get the sufficiency in (a) from 2.9(b), 14.44 (and 14.39). Deduce the necessity in (a) from Proposition 14.49 by way of a Castaing representation (cf. 14.5) for the epigraphical mapping associated with the conjugate integrand. Derive (b) from (a) in the framework of the conjugacy between the indicator integrand δ_S (cf. 14.32) and the support integrand σ_S in 14.51. □

Next we take up the question of how normality of integrands is preserved under limit operations.

14.53 Proposition (limits of integrands). *Consider any sequence $\{f^\nu\}_{\nu \in I\!N}$ of normal integrands $f^\nu : T \times I\!R^n \to \overline{I\!R}$.*

(a) (epi-limits) *If $f : T \times I\!R^n \to \overline{I\!R}$ is given by $f(t, \cdot) = \text{e-lim}_\nu f^\nu(t, \cdot)$ for all $t \in T$, then f is a normal integrand. The same is true if $f(t, \cdot) =$*

e-lim sup$_\nu$ $f^\nu(t,\cdot)$ *for all* t, *or if* $f(t,\cdot) = \text{e-lim inf}_\nu f^\nu(t,\cdot)$ *for all* t.

(b) (pointwise limits) *If* $f : T \times \mathbb{R}^n \to \overline{\mathbb{R}}$ *has* $f(t,\cdot) = \text{cl}[\text{p-lim}_\nu f^\nu(t,\cdot)]$ *for all* $t \in T$, *then* f *is a normal integrand. The same is true if* $f(t,\cdot) = \text{cl}[\text{p-lim sup}_\nu f^\nu(t,\cdot)]$ *for all* t. *It holds if* $f(t,\cdot) = \text{cl}[\text{p-lim inf}_\nu f^\nu(t,\cdot)]$ *for all* t *under the assumption that* $(T,\mathcal{A})$ *is complete for some measure* μ.

Proof. To justify (a), we note that when $f(t,\cdot) = \text{e-lim}_\nu f^\nu(t,\cdot)$ for all t we have $S_f = \text{p-lim}_\nu S_{f^\nu}$. The closed-valuedness and measurability of S_f follows then from that of the mappings S_{f^ν}, cf. Theorem 14.20. The same argument takes care of e-lim sup and e-lim inf.

To justify (b), we look first at the case where $f(t,\cdot) = \text{cl}[\text{p-lim sup}_\nu f^\nu(t,\cdot)]$. This means $S_f(t) = \text{cl}\bigcup_\nu S^\nu(t)$ for $S^\nu(t) = \bigcap_{\kappa\geq\nu} S_{f^\kappa}(t)$. The mappings $S_{f^\nu}(t)$ are closed-valued and measurable, so S^ν has these properties by way of 14.11(a). Then S_f is closed-valued and measurable by 14.11(b) in combination with 14.2, so that f is a normal integrand. In the case of $f(t,\cdot) = \text{cl}[\text{p-lim}_\nu f^\nu(t,\cdot)]$ we of course have $f(t,\cdot) = \text{cl}[\text{p-lim sup}_\nu f^\nu(t,\cdot)]$ in particular.

The case of $f(t,\cdot) = \text{cl}[\text{p-lim inf}_\nu f^\nu(t,\cdot)]$ suffers from the fact that, although we can write $S_f(t) = \text{cl}\bigcap_\nu S^\nu(t)$ for $S^\nu(t) = \bigcup_{\kappa\geq\nu} S_{f^\kappa}(t)$, the mappings S^ν, while measurable by 14.11(b), might not be closed-valued, and the measurability of $S : t \mapsto \bigcap_\nu S^\nu(t)$ could then be in doubt because 14.11(a) wouldn't be applicable. When $(T,\mathcal{A})$ is complete for some measure μ, however, we can work instead with the joint measurability criterion in 14.34. We have $f(t,\cdot) = \text{cl}\, f_0(t,\cdot)$ for the function $f_0 = \text{p-lim inf}_\nu f^\nu$, which is $\mathcal{A}\otimes\mathcal{B}(\mathbb{R}^n)$ measurable because each f^ν is $\mathcal{A}\otimes\mathcal{B}(\mathbb{R}^n)$ measurable in consequence of normality. Then f is a normal integrand by 14.35. □

14.54 Exercise (horizon integrands and horizon limits).

(a) *For any normal integrand* $f : T\times\mathbb{R}^n \to \overline{\mathbb{R}}$, *the function* $h : T\times\mathbb{R}^n \to \overline{\mathbb{R}}$ *defined by* $h(t,\cdot) = f(t,\cdot)^\infty$ *is a normal integrand.*

(b) *For any sequence* $\{f^\nu\}_{\nu\in\mathbb{N}}$ *of normal integrands* $f^\nu : T \times \mathbb{R}^n \to \overline{\mathbb{R}}$, *the function* $h : T \times \mathbb{R} \to \overline{\mathbb{R}}$ *defined by* $h(t,\cdot) = \text{e-lim inf}^\infty_\nu f^\nu(t,\cdot)$ *is a normal integrand. The same is true for* $h(t,\cdot) = \text{e-lim sup}^\infty_\nu f^\nu(t,\cdot)$ *and for* $h(t,\cdot) = \text{e-lim}^\infty_\nu f^\nu(t,\cdot)$, *when that horizon epi-limit exists for all* $t \in T$.

Detail. Apply 14.21 to the epigraphical mappings. □

An integrand $f : T \times \mathbb{R}^n \to \overline{\mathbb{R}}$ is *simple* when $\text{rge}\, D_f$ and $\text{rge}\, f$ are finite sets. This means that, for each $x \in \mathbb{R}^n$, the function $t \mapsto f(t,x)$ is measurable with only finitely many values and in fact is identically ∞ unless x belongs to a certain finite $F \subset \mathbb{R}^n$.

14.55 Exercise (approximation by simple integrands). *Any simple integrand is a normal integrand, in particular. An integrand* $f : T \times \mathbb{R}^n$ *is normal if and only if there is a sequence of simple integrands* $f^\nu : T \times \mathbb{R}^n$ *such that* $f(t,\cdot) = \text{e-lim}_\nu f^\nu(t,\cdot)$ *for every* $t \in T$.

Guide. Establish the normality of a simple integrand from the definitions. Deduce the limit assertion from 14.53 and 14.22. □

14.56 Theorem (subderivatives and subgradient mappings). *For any proper normal integrand $f : T \times \mathbb{R}^n \to \overline{\mathbb{R}}$, and any choice of $x(t) \in D_f(t)$ depending measurably on $t \in T$, the subderivative functions*

$$(t, w) \mapsto \hat{d}f\big(t, x(t)\big)(w), \qquad (t, w) \mapsto df\big(t, x(t)\big)(w)$$

are normal integrands, and the subgradient mappings

$$t \mapsto \hat{\partial} f\big(t, x(t)\big), \qquad t \mapsto \partial f\big(t, x(t)\big), \qquad t \mapsto \partial^\infty f\big(t, x(t)\big),$$

are closed-valued and measurable. Also measurable are the mappings

$$t \mapsto \operatorname{gph} \partial f(t, \cdot), \qquad t \mapsto \operatorname{gph} \partial^\infty f(t, \cdot).$$

Proof. Let $\alpha(t) = f(t, x(t))$, noting that from 14.28 this depends measurably on t. In the case of the subderivative functions we have from 8.2 and 8.17 that

$$\operatorname{epi} \hat{d}f(t, x(t)) = \hat{T}_{S_f(t)}\big(x(t), \alpha(t)\big), \qquad \operatorname{epi} df(t, x(t)) = T_{S_f(t)}\big(x(t), \alpha(t)\big),$$

so the mappings $t \mapsto \operatorname{epi} \hat{d}f(t, x(t))$ and $t \mapsto \operatorname{epi} df(t, x(t))$ are closed-valued and measurable, as seen from 14.26. Similarly in the case of the subgradient mappings, we have from 8.9 that

$$\begin{aligned}
\hat{\partial} f(t, x(t)) &= \big\{ v \,\big|\, (v, -1) \in \hat{N}_{S_f(t)}(x(t), \alpha(t)) \big\}, \\
\partial f(t, x(t)) &= \big\{ v \,\big|\, (v, -1) \in N_{S_f(t)}(x(t), \alpha(t)) \big\}, \\
\partial^\infty f(t, x(t)) &= \big\{ v \,\big|\, (v, 0) \in N_{S_f(t)}(x(t), \alpha(t)) \big\},
\end{aligned}$$

and the closed-valuedness and measurability follows from the normal cone part of 14.26 in collaboration with 14.15(b). (For instance, we have $\partial f(t, x(t)) = \big\{ v \,\big|\, F(v) \in S(t) \big\}$ for $F(v) = (v, -1)$ and $S(t) = N_{S_f(t)}(x(t), \alpha(t))$.) The measurability of the graphical mappings likewise follows from that of the graphical mapping in 14.26 by this argument. $\square$

The graphical mappings at the end of Theorem 14.56, while measurable, aren't necessarily closed because the subgradient definitions in 8.3 only take f-attentive limits in x, not general limits. This restriction falls away in dealing with subdifferentially continuous functions like convex functions and amenable functions (cf. 13.28, 13.30, 13.32). Thus for instance, when f is a convex normal integrand, the mappings in question are closed-valued as well as measurable. Indeed, for such integrands there is also a converse property.

14.57 Proposition (subgradient characterization of convex normality). *Let $f : T \times \mathbb{R}^n \to \overline{\mathbb{R}}$ be such that $f(t, x)$ is a proper, lsc, convex function of $x \in \mathbb{R}^n$ for each $t \in T$. Then f is a normal integrand if and only if the following hold:*

(a) *the mapping $t \mapsto \operatorname{gph} \partial f(t, \cdot)$ is measurable;*

(b) *there is a measurable function $\bar{x} : T \to \mathbb{R}^n$ such that $\partial f(t, \bar{x}(t)) \neq \emptyset$ for all $t \in T$ and the function $t \mapsto f(t, \bar{x}(t))$ is measurable.*

Proof. The assumptions on f imply that the mapping $S : t \mapsto \operatorname{gph} \partial f(t, \cdot)$ is nonempty-closed-valued (since $\partial f(t, \cdot)$ is maximal monotone by 12.17). If f is a normal integrand we have (a) as a consequence of 14.56, as noted above, and we can choose $(x(t), v(t)) \in S(t)$ depending measurably on t; cf. 14.6. Then (b) holds through 14.28.

Conversely, if (a) and (b) hold, let $\{(x^\nu, v^\nu)\}_{\nu \in I\!N}$ give a Castaing representation for S; cf. 14.5(a). Select $\bar{v}(t) \in \partial f(t, \bar{x}(t))$ to depend measurably on t, as is possible through 14.6, because the mapping $t \mapsto \partial f(t, \bar{x}(t))$ is closed-valued and measurable by 14.56. By the central argument in the proof of Theorem 12.25, $f(t, x)$ is supremum of the expressions

$$\begin{aligned} f(t, \bar{x}(t)) + \langle \bar{v}(t),\, x_0 - \bar{x}(t) \rangle + \langle v_m,\, \bar{x}(t) - x_m \rangle \\ + \langle v_{m-1},\, x_m - x_{m-1} \rangle + \cdots + \langle v_0,\, x_1 - x_0 \rangle \end{aligned}$$

over all choices of $(x_k, v_k) \in S(t)$ (with m arbitrary). Because of the density of $\{(x^\nu(t), v^\nu(t)) \mid \nu \in I\!N\}$ in $S(t)$ that goes with having a Castaing representation, we still get $f(t, x)$ if we restrict to pairs $(x_k, v_k) = (x^{\nu_k}(t), v^{\nu_k}(t))$. In this way we can focus on a countable collection of expressions, each of which, as a function of t and x, is a Carathéodory integrand. Then by 14.44(a), f is a normal integrand. □

F. Integral Functionals

The theory of normal integrands is aimed especially at facilitating work with integral functionals I_f as expressed in 14(1). Although we won't, in this book, take up the many issues of infinite-dimensional variational analysis raised by such functionals, we'll develop now a key fact about the interchange of integration and minimization, which comes easily out of the results already obtained and demonstrates the usefulness of measurable selections.

To proceed with this we must first clear up a potential ambiguity in the what an integral might mean when an extended-real-valued function is integrated. Our approach to this is consistent with the extended arithmetic in Chapter 1, where we adopted the conventions that $0{\cdot}\infty = 0 = 0{\cdot}(-\infty)$ and $\infty + (-\infty) = (-\infty) + \infty = \infty$. Again we let ∞ dominate over $-\infty$ by taking

$$\int_T \alpha(t)\mu(dt) = \int_T \max\{\alpha(t), 0\}\mu(dt) + \int_T \min\{\alpha(t), 0\}\mu(dt)$$

for any measurable function $\alpha : T \to \overline{I\!R}$ and measure μ on $\mathcal{A}$, so that

$$\begin{aligned} \int_T \alpha(t)\mu(dt) < \infty \ \text{ when } \ \int_T \max\{\alpha(t), 0\}\mu(dt) < \infty, \\ \int_T \alpha(t)\mu(dt) = \infty \ \text{ when } \ \int_T \max\{\alpha(t), 0\}\mu(dt) = \infty. \end{aligned}$$

More specifically, $\int_T \alpha(t)\mu(dt)$ is called the *upper integral* of α with respect to μ under this interpretation, but in line with our handling of extended arithmetic, we take the "upper" for granted. Note that there's no ambiguity with $\int_T \max\{\alpha(t),0\}\mu(dt)$, which has a standard value, finite or ∞, or with $\int_T \min\{\alpha(t),0\}\mu(dt)$, which is finite or $-\infty$. Nor does 0 really have to be singled out; it's easy to see that the adopted convention conforms to taking

$$\int_T \alpha(t)\mu(dt) = \inf\left\{ \int_T \beta(t)\mu(dt) \,\middle|\, \beta \in \mathcal{L}^1(T,\mathcal{A},\mu),\ \beta \geq \alpha \right\}.$$

14.58 Proposition (integral functionals). *Under the extended interpretation of integration, the functional $I_f : \mathcal{X} \to \overline{I\!R}$ given by*

$$I_f[x] = \int_T f\big(t, x(t)\big)\mu(dt) \quad \textit{for } x \in \mathcal{X}$$

is well defined for any normal integrand $f : T \times I\!R^n \to \overline{I\!R}$ and any space $\mathcal{X}$ of measurable functions $x : T \to I\!R^n$. Furthermore,

$$I_f[x] < \infty \quad \Longrightarrow \quad x(t) \in D_f(t) = \operatorname{dom} f(t,\cdot) \quad \mu\textit{-almost everywhere.}$$

Proof. The measurable dependence of $x(t)$ on $t \in T$ ensures that of $\alpha(t) = f(t,x(t))$ by 14.28. Then $I_f[x]$ is well defined in the manner explained above, and one has $I_f[x] < \infty$ if and only if $\int_T \max\{f(t,x(t)),0\}\mu(dt) < \infty$. Then certainly $\mu\big(\{t \mid f(t,x(t)) = \infty\}\big) = 0$. □

The domain property at the end of this result shows how "pointwise" constraints on $x(t)$ can be passed through an integral functional to become constraints on the function $x \in \mathcal{X}$. When f has the form $f(t,\cdot) = f_0(t,\cdot) + \delta_{C(t)}$ in Example 14.32, for instance, we get $I_f[x] = I_{f_0}[x]$ when $x(t) \in C(t)$ for μ-almost every t, but $I_f[x] = \infty$ otherwise.

Typical spaces $\mathcal{X}$ of interest with respect to an integral functional I_f are the Lebesgue spaces $\mathcal{L}^p(T,\mathcal{A},\mu;I\!R^n)$ and the space of constant functions x. In the latter case, $\mathcal{X}$ can be identified with $I\!R^n$ itself, and I_f becomes, in this restriction, a function on $I\!R^n$ obtained by (potentially) "infinite addition" of nonnegative scalar multiples of the functions $f(t,\cdot)$ for $t \in T$. When T has topological structure, the space $\mathcal{C}(T;I\!R^n)$ of continuous functions $x : T \to I\!R^n$ is a candidate for $\mathcal{X}$ as well. Spaces of differentiable functions can also enter the scene along with Sobolev spaces, Orlicz spaces and more. A useful example that encompasses all the others is $\mathcal{M}(T,\mathcal{A};I\!R^n)$, the space of *all* measurable functions $x : T \to I\!R^n$.

In their role in the theory of integral functionals, these spaces fall into two very different categories, distinguished by the presence or absence of a certain property of decomposability.

14.59 Definition (decomposable spaces). *A space $\mathcal{X}$ of measurable functions $x : T \to I\!R^n$ is decomposable in association with a measure μ on $\mathcal{A}$ if for every function $x_0 \in \mathcal{X}$, every set $A \in \mathcal{A}$ with $\mu(A) < \infty$ and bounded, measurable* *function* $x_1 : A \to I\!R^n$, $\mathcal{X}$ *also contains the function* $x : T \to I\!R^n$ *defined by*

$$x(t) = \begin{cases} x_0(t) & \text{for } t \in T \setminus A, \\ x_1(t) & \text{for } t \in A. \end{cases}$$

In the preceding examples, the spaces $\mathcal{M}(T, \mathcal{A}; I\!R^n)$ and $\mathcal{L}^p(T, \mathcal{A}, \mu; I\!R^n)$ are decomposable, whereas $\mathcal{C}(T; I\!R^n)$ and the space of constant functions are not (except for some extreme choices of T). To be decomposable, a space $\mathcal{X}$ that's linear (and therefore contains the function 0) must in particular contain every bounded measurable function that vanishes outside some set of finite measure.

The main consequence of decomposability is the following result, which plays a fundamental role in many applications that involve a search for extremals. Basically, the result tells us when it's possible to replace optimality conditions in a functional space by pointwise conditions of optimality in $I\!R^n$.

14.60 Theorem (interchange of minimization and integration). *Let* $\mathcal{X}$ *be a space of measurable functions from* T *to* $I\!R^n$ *that is decomposable relative to* μ*, a* σ*-finite measure on* $\mathcal{A}$*. Let* $f : T \times I\!R^n \to \overline{I\!R}$ *be a normal integrand. Then the minimization of* I_f *over* $\mathcal{X}$ *can be reduced to pointwise minimization in the sense that, as long as* $I_f \not\equiv \infty$ *on* $\mathcal{X}$*, one has*

$$\inf_{x \in \mathcal{X}} \int_T f\big(t, x(t)\big)\mu(dt) = \int_T \Big[\inf_{x \in \boldsymbol{R}^n} f(t,x)\Big]\mu(dt).$$

Moreover, as long as this common value is not $-\infty$*, one has for* $\bar{x} \in \mathcal{X}$ *that*

$$\bar{x} \in \operatorname*{argmin}_{x \in \mathcal{X}} I_f[x] \iff \bar{x}(t) \in \operatorname*{argmin}_{x \in \boldsymbol{R}^n} f(t,x) \text{ for } \mu\text{-almost every } t \in T.$$

Proof. Let $p(t) = \inf_x f(t,x)$ and $P(t) = \operatorname{argmin}_x f(t,x)$, recalling from 14.37 that these depend measurably on t. For any $x \in \mathcal{X}$ we have $f\big(t, x(t)\big) \geq p(t)$ for all t; hence $\inf_{\mathcal{X}} I_f \geq \int_T p(t)\mu(dt)$ where we can assume that $\inf_{\mathcal{X}} I_f > -\infty$. To prove the inequality in the other direction, it suffices to show for finite $\beta > \int_T p(t)\mu(dt)$ that there exists $x \in \mathcal{X}$ with $I_f(x) < \beta$. We can build, in part, on the assumption that there exists $x_0 \in \mathcal{X}$ with $I_f[x_0] < \infty$.

Because $\int_T p(t)\mu(dt) < \infty$, we know that $\int_T \max\{p(t), 0\}\mu(dt) < \infty$ and therefore, in terms of $p_\varepsilon(t) = \max\{p(t), -\varepsilon^{-1}\}$, that $\int_T p_\varepsilon(t)\mu(dt) \to \int_T p(t)\mu(dt)$ as $\varepsilon \searrow 0$. Let $\psi : T \to I\!R$ be measurable with $\psi(t) > 0$ and $\int_T \psi(t)\mu(dt) < \infty$; the existence of such a function is guaranteed by the σ-finiteness of μ. The measurable function $\alpha_\varepsilon(t) := \varepsilon\psi(t) + p_\varepsilon(t)$ has

$$\int_T \alpha_\varepsilon(t)\mu(dt) = \varepsilon \int_T \psi(t)\mu(dt) + \int_T p_\varepsilon(t)\mu(dt) \to \int_T p(t)\mu(dt) < \beta,$$

but also $\alpha_\varepsilon(t) > p(t)$ for all t, so that the sets $S_\varepsilon(t) = \big\{x \in I\!R^n \,\big|\, f(t,x) \leq \alpha_\varepsilon(t)\big\}$ are nonempty. Fix any ε small enough that $\int_T \alpha_\varepsilon(t)\mu(dt) < \beta$. The mapping $S_\varepsilon : T \mapsto I\!R^n$ is measurable by 14.33, so it has a measurable selection: there exists by 14.6 a measurable function $x_1 : T \mapsto \mathbb{R}^n$ with $x_1(t) \in S_\varepsilon(t)$ for every $t \in T$. Then $\int_T f\big(t, x_1(t)\big)\mu(dt) < \beta$.

By σ-finiteness we can express T as the union of an increasing sequence of sets $T^\nu \in \mathcal{A}$ with $\mu(T^\nu) < \infty$. Let $A^\nu := \big\{t \in T^\nu \,\big|\, |x_1(t)| \le \kappa\big\}$. The sets A^ν likewise belong to $\mathcal{A}$ and increase to T with $\mu(A^\nu) < \infty$, and for x_1 and the earlier function $x_0 \in \mathcal{X}$ we have

$$\int_{T \setminus A^\nu} f(t, x_0(t)\mu(dt) \to 0, \quad \int_{A^\nu} f(t, x_1(t)\mu(dt) \to \int_T f(t, x_1(t)\mu(dt). \quad 14(12)$$

Define $x^\nu : T \to \mathbb{R}$ to agree with x_0 on $T \setminus A^\nu$ and with x_1 on A^ν. Then $x^\nu \in \mathcal{X}$ by our assumption that $\mathcal{X}$ is decomposable relative to μ. Furthermore,

$$I_f[x^\nu] = \int_{T \setminus A^\nu} f\big(t, x_0(t)\big)\mu(dt) + \int_{A^\nu} f\big(t, x_1(t)\big)\mu(dt),$$

so that $I_f[x^\nu] \to \int_T f\big(t, x_1(t)\big)\mu(dt) < \beta$ by 14(12) as $\nu \to \infty$, and consequently $I_f[x^\nu] < \beta$ for ν sufficiently large.

We pass now to the claim about a function $\bar{x} \in \mathcal{X}$ attaining the minimum of I_f on $\mathcal{X}$. Since $f\big(t, \bar{x}(t)\big) \ge \inf f(t, \cdot) = p(t)$ for all t, that's equivalent to having $\mu\big(\{t \mid f(t, \bar{x}(t)) > p(t)\}\big) = 0$ under our assumption that $\int_T p(t)\mu(dt)$ is finite. This is identical to the stated criterion, because $f(t, \bar{x}(t)) > p(t)$ means that $\bar{x}(t) \notin \operatorname{argmin} f(t, \cdot)$. □

14.61 Exercise (simplified interchange criterion). *Let $f : T \times \mathbb{R}^n \to \overline{\mathbb{R}}$ be a normal integrand, and let $\mathcal{M}_f$ be the collection of all $x \in \mathcal{M}(T, \mathcal{A}; \mathbb{R}^n)$ with $I_f[x] < \infty$, the measure μ in this functional being σ-finite. The conclusions of Theorem 14.60 then hold for any space $\mathcal{X}$ with $\mathcal{M}_f \subset \mathcal{X} \subset \mathcal{M}(T, \mathcal{A}; \mathbb{R}^n)$.*

Proof. Glean this as a simplification of the proof of previous theorem. □

An illustration of how the interchange of minimization and integration can work in practice is provided by the following example.

14.62 Example (reduced optimization). *For a σ-finite measure μ on $\mathcal{A}$ and a normal integrand $g : T \times (\mathbb{R}^n \times \mathbb{R}^m) \to \overline{\mathbb{R}}$, consider a problem of the form:*

$$(\mathcal{P}) \qquad \text{minimize } J[x] + I_g[x, u] \text{ over all } (x, u) \in \mathcal{X} \times \mathcal{U}$$

on spaces $\mathcal{X} \subset \mathcal{M}(T, \mathcal{A}; \mathbb{R}^n)$ and $\mathcal{U} \subset \mathcal{M}(T, \mathcal{A}; \mathbb{R}^m)$ with $\mathcal{U}$ decomposable; the functional $J : \mathcal{X} \to \overline{\mathbb{R}}$ is arbitrary. Letting

$$f(t, x) = \inf_{u \in \mathbb{R}^m} g(t, x, u)$$

and assuming the properties of g are such as to ensure that $f(t, \cdot)$ is lsc on $\mathbb{R}^n$ for each $t \in T$, consider also the problem

$$(\mathcal{P}') \qquad \text{minimize } J[x] + I_f[x] \text{ over all } x \in \mathcal{X}.$$

If $\inf(\mathcal{P}) < \infty$, one has $\inf(\mathcal{P}) = \inf(\mathcal{P}')$. If also $\inf(\mathcal{P}) > -\infty$, the pairs

$(\bar{x}, \bar{u}) \in \mathcal{X} \times \mathcal{U}$ furnishing an optimal solution to $(\mathcal{P})$ are the ones such that $\bar{x}$ is an optimal solution to (P') and

$$\bar{u}(t) \in \operatorname*{argmin}_{u \in \mathbb{R}^m} g\big(t, \bar{x}(t), u\big) \text{ for } \mu\text{-almost every } t \in T.$$

Thus in principle, $(\mathcal{P})$ can be solved by solving $(\mathcal{P}')$ for $\bar{x}$ and then taking $\bar{u}$ to be any measurable selection in $\mathcal{U}$ for the mapping $S : t \mapsto \operatorname{argmin} g\big(t, \bar{x}(t), \cdot\big)$.

Detail. In either problem $(\mathcal{P})$ or $(\mathcal{P}')$, the rules of extended arithmetic dictate that the expression being minimized is ∞ unless $J[x] < \infty$. We can therefore focus our attention on $x \in \mathcal{X}$ with $J[x] < \infty$, which necessarily exist when $\inf(\mathcal{P}) < \infty$ or $\inf(\mathcal{P}') < \infty$. Note that the assumption about the functions $f(t, \cdot)$ being lsc guarantees through 14.47 that f is a normal integrand on $T \times \mathbb{R}^n$, so that the integral functional I_f in $(\mathcal{P}')$ is well defined. (Refer to 1.28 for conditions on g that provide this lower semicontinuity property of f.)

Fixing any $x \in \mathcal{X}$ with $J[x] < \infty$, define $h : T \times \mathbb{R}^m \to \overline{\mathbb{R}}$ by $h(t, u) = g(t, x(t), u)$. According to 14.45, h is a normal integrand, and $\inf h(t, \cdot) = f\big(t, x(t)\big)$. In applying Theorem 14.60 to I_h, we see that

$$\inf_{u \in \mathcal{U}} I_g[x, u] = \inf_{u \in \mathcal{U}} I_h[u] = \int_T \Big[\inf_{u \in \mathbb{R}^d} h(t, u)\Big] \mu(dt) = \int_T f\big(t, x(t)\big) = I_f[x]$$

whenever $I_f[x] < \infty$. This fact and the others coming from Theorem 14.60 yield the result stated here. □

Commentary

Measurability of set-valued mappings has its roots in earlier studies of projections of classes of subsets of a product space, if these are viewed as graphs, but it mainly got going because of applications in control theory and, soon after, in economics. Pioneers like Filippov [1959] and Ważewski [1961] were motivated by technical questions arising in connection with differential inclusions, whereas Aumann [1965] and Debreu [1966] were interested in economic models involving 'infinitesimal agents'. The thesis of Castaing [1967] was instrumental in fostering a systematic development of the subject in terms of 'multivalued' mappings which could be added, intersected, composed with other mappings, and so forth, as here.

Debreu dealt with set-valued mappings as single-valued mappings from a measurable space into a metric hyperspace of compact sets. He saw measurability from that angle, which had the advantage of revealing how standard facts about measurable functions could be utilized. But he was also confined by the Pompeiu-Hausdorff metric and its unsuitability for handling unboundedness and nonemptiness.

That wasn't so with Castaing, who identified the measurability of a general set-valued mapping S with that of the sets $S^{-1}(C)$ for C closed. He showed that this was equivalent, for closed-valued S, to the measurability of $S^{-1}(O)$ for O open. Other researchers, notably Ioffe and Tikhomirov [1974], adopted the latter property as the definition of the measurability of S. We have followed that here, because it makes little difference in the long run, due to the closure rule in 14.2, and yet it simplifies constructions like those in 14.11(b)(c), 14.12(a)(b), 14.13(a) and 14.15(a), where the closure operation would otherwise have to be incorporated at every step.

Castaing demonstrated the equivalence of his approach to measurability with that of Debreu when nonempty-compact-valued mappings are involved. In Theorem 14.4 we have gone further by passing from an argument based on the Pompeiu-Hausdorff metric to one based on the integrated set metric, thereby removing the compactness restriction that was inherent in this setting. The proof is inspired by those of Salinetti and Wets [1986] and Hess [1983], [1986], who obtains the equivalence of the Effrös and Borel fields by endowing cl-sets$_{\neq\emptyset}(I\!R^n)$ with a uniform structure generated from uniform convergence of distance functions; cf. Theorem 4.35.

The basic tests in Theorem 14.3 for the measurability of a mapping S were established by Castaing [1967] and Rockafellar [1969c], with the latter concentrating on $I\!R^n$ as the range space and making explicit the simplifying features of that case; see also Rockafellar [1976a]. (*Note*: In this commentary we'll focus only on $I\!R^n$, although many of the cited works go beyond that.)

Debreu and Castaing further provided graphical criteria like those in 14.8 for some situations. Theorem 14.8 in its present form appeared in Rockafellar [1969c], but the underpinnings go back to some of the earliest work on measurable selections and to the theory of Suslin sets and their projections. See Sainte-Beuve [1974] for a broadened discussion of the projection fact quoted before the statement of this theorem. The Borel measurability of $\operatorname{gph} S$ (when T is a Borel subset of $I\!R^d$) was the property adopted as the measurability of S by Aumann [1965]; cf. 14.8(c).

The powerful characterization of measurability in terms of a Castaing representation in Theorem 14.5(a) comes from Castaing [1967] as well. Our proof, however, in taking advantage of $I\!R^n$, is that of Rockafellar [1976a], where also the alternative representation in 14.5(b) was developed together with its convenient application to convex-valued mappings in 14.7. The measurable selection result in 14.6, while depicted here as an immediate consequence of the existence of a Castaing representation, can mainly be attributed to Kuratowski and Ryll-Nardzewski [1965]. But there is a long history of precedents of varying generality, including an early proof by Rokhlin [1949] that was later found to be inadequate; see Wagner [1977] and Ioffe [1978a] for the full picture.

The set-valued extension of Lusin's theorem, obtained in equating the measurability of S with the property in 14.10(a), builds on Pliś [1961] and Castaing [1967] by allowing the sets $S(t)$ now to be unbounded, not just compact. Castaing made this Lusin mode of characterization his main tool in many arguments, for example in establishing the measurability of $t \mapsto \bigcap_{i\in I} S_i(t)$ when each S_i is measurable. But that technique requires T to have topological structure and for the discussion to revolve around a complete measure μ that's compatible with such structure; in his case the S_i's also had to be compact-valued. Those restrictions are avoided in our strategy for proving such results, which follows Rockafellar [1969c] for 14.11 and 14.12 and Rockafellar [1976a] for 14.13, 14.14 and 14.15. The latter article was the first to employ the measurability of the mapping $t \mapsto \operatorname{gph} M(t,\cdot)$ as in 14.13.

The relation $F(t,x(t)) \in D(t)$ in 14(5) reduces to the equation $F(t,x(t)) = c(t)$ when $D(t)$ is the singleton set $\{c(t)\}$. That case of the implicit measurable function theorem in 14.16, often referred to as Filippov's lemma, gave impetus to the study of measurable set-valued mappings and measurable selections because of a key application to differential equations with control parameters. The original result of Filippov [1959] required F to be continuous; Ważewski [1961] extended it to the Carathéodory conditions of continuity in x and measurability in t. Those conditions were developed much earlier by Carathéodory in his research on the existence of solutions to ordinary differential equations.

For results on the representation of a closed-valued measurable mapping S in the form $S(t) = F(t, C)$ for a Carathéodory mapping F, see Ioffe [1978b]. Such a representation may require allowing C to be a space more general than just a closed subset of $\mathbb{R}^d$ for some d.

Pointwise inner and outer limits of a sequence of measurable mappings, and the characterization in 14.22 of a measurable mapping as the limit of simple mappings, were already treated by Castaing [1967] in the compact-valued case. Our approach, however, follows that of Salinetti and Wets [1981], who drop the compact-valued restriction. The graphical limits in Theorem 14.20 are new, as are the horizon aspects in 14.21. The results about the convergence of measurable selections in 14.23 can be found in Salinetti and Wets [1981] and were further elaborated in Lucchetti, Papageorgiou and Patrone [1987]. The characterizations of almost everywhere and uniform convergence in 14.24 and 14.25 likewise go back to Salinetti and Wets [1981].

Aubin and Frankowska [1990] proved the measurability of the tangent cone mapping in Theorem 14.26. The measurability of the regular tangent cone mapping and the normal cone mappings in Theorem 14.26 hasn't been dealt with before now. But a subgradient result of Rockafellar [1969c], when applied to indicators, yields, as a special case, the measurability of the normal cone mapping when the mapping S is closed-convex-valued.

In problems involving integral functionals I_f such as in the 'calculus of variations', it was traditional to take the integrand $f(t, x)$ to be continuous in t and x jointly, or indeed differentiable to whatever extent was convenient. Later, the Carathéodory conditions of continuity in x and measurability in t came to the fore, especially in connection with the existence of optimal trajectories. With the emergence of modern control theory, however, inequality constraints became important, and this caused a shift in perspective once the idea of representing constraints through infinite penalties took hold.

That idea was a hallmark of the early work of Moreau and Rockafellar on the foundations of convex analysis in the mid 1960s, as explained in the notes to Chapter 1. It was natural to translate it to integral functionals by admitting integrands f that are extended-real-valued, so as to be able to represent 'pointwise' constraints of the sort described in 14.58. But when $f(t, x)$ can be ∞, there's little sense in asking $f(t, x)$ to depend continuously on x. The Carathéodory conditions are no longer suitable and have to be replaced by something else.

One could hope that it might be enough simply to replace the continuity of $f(t, x)$ in x by lower semicontinuity, while maintaining the measurability of $f(t, x)$ in t, but no. Counterexamples, of the kind on which the second part of 14.28 is based, show that although lower semicontinuity in x is certainly right, the assumption of measurability in t for each fixed x isn't adequate.

The way out of this impasse was found by Rockafellar [1968b] in the concept of a 'normal convex integrand', which in its initial formulation rested on the property in 14.39 and thus depended heavily on $f(t, x)$ being convex in x. That paper was actually submitted for publication in 1966. When Castaing's thesis came out in 1967, Rockafellar realized that his normality condition was equivalent to requiring the epigraphical mapping $t \mapsto \operatorname{epi} f(t, \cdot)$ to be closed-valued and measurable; see

Rockafellar [1969c].

Normal convex integrands and their associated convex integral functionals became a focus of work on 'fully convex' problems in optimal control and the calculus of variations, cf. Rockafellar [1970b], [1971a], [1972], 'convolution integrals', cf. Ioffe and Tikhomirov [1968], and infinite-dimensional convex analysis, cf. Rockafellar [1971b], [1971c], Ioffe and Levin [1972], Bismut [1973], and Castaing and Valadier [1977], as well as applications in stochastic programming, cf. Rockafellar and Wets [1976a], [1976b], [1976c], [1976d], [1977].

In the face of all this effort going into convex integral functionals, nonconvex integrands received comparatively little attention at first. Berliocchi and Lasry [1971], [1973], were the first to speak of normal integrands f beyond the case of $f(t,x)$ convex in x. They didn't take the closed-valuedness and measurability of $t \mapsto \operatorname{epi} f(t,\cdot)$ as the definition, however. Instead they essentially adopted the property in Theorem 14.42(c) for that purpose, which was possible because they were concerned only with spaces $T \in \mathcal{B}(\mathbb{R}^d)$. They showed that such normality agreed with that of Rockafellar [1968b] in the presence of convexity. Their definition was adopted by Ekeland and Temam [1974] in a book that broke new ground in treating nonconvex problems in the calculus of variations, including some problems related to partial differential equations. Around the same time, the book of Ioffe and Tikhomirov [1974] came out with other innovations in treating problems in the calculus of variations and optimal control. Those authors were the first to take the approach of epi-measurability directly in defining normal integrands without convexity.

The different views of the 'normality' of an integrand were reconciled in the lecture notes of Rockafellar [1976a], where Theorem 14.42 was first proved. The Berliocchi-Lasry property in 14.42(c) was translated in that work to a corresponding property of measurable set-valued mappings, which appears here as the characterization in Theorem 14.10(c). Many basic facts that had been established for convex normal integrands in Rockafellar [1969c] were extended at that time to the nonconvex case as well. Examples are the normality of sums and composites of normal integrands (in 14.44 and 14.45), the measurability of level-set mappings (in 14.33) and the inf function and argmin mapping (in 14.37), and the joint measurability criterion (in 14.34). The property of 14.43, generalizing that of Scorza-Dragoni [1948], was translated to normal integrands earlier by Ekeland and Temam [1974].

The new characterization of normality in 14.41 is based on the idea in 14.40 of 'scalarizing' normal integrands, which is due to Korf and Wets [1997]. The results in 14.38 (Moreau envelopes), 14.46 (scalar multiplication), 14.47 (inf-projections), 14.48 (epi-addition and epi-multiplication) and 14.49 (convexification) are newly presented here too. They add to the known catalogue of operations preserving normality.

Theorem 14.50, about the normality of the conjugate of a normal integrand, goes all the way back to Rockafellar [1968b], however, where it was seen as a key test of the concept. The consequences for support functions (in 14.51) and envelope representations (in 14.52) come from Rockafellar [1969c]; see Valadier [1974] for more on the support function case. The measurability properties of subderivatives and subgradients in Theorem 14.56 are new beyond the convex case of subgradients in Rockafellar [1969]. The subgradient characterization of convex normal integrands in 14.57 is due to Attouch [1975].

Proposition 14.53, about the limit of normal integrands, extends the results of Salinetti and Wets [1986]. The horizon developments in 14.54 are new, as are those in 14.55 about simple integrands.

Decomposable spaces were introduced by Rockafellar [1968b] for the purpose of identifying the circumstances in which the conjugate of a convex integral functional I_f on a linear space $\mathcal{X}$ would be the integral functional I_{f^*} associated with the conjugate integrand f^* on a space $\mathcal{Y}$ that is 'paired' with $\mathcal{X}$. The crucial part of the argument rested on interchanging minimization with integration. A need for similar patterns of reasoning came up later in connection with models of reduced optimization—various special cases of 14.62—which were developed in optimal control, cf. Rockafellar [1975], and in stochastic analysis, cf. Rockafellar and Wets [1976d]. The basic interchange rule in Theorem 14.60, however, wasn't distilled from the mix of ideas until Rockafellar [1976a].

For more on conjugate integral functionals and their properties, see also Rockafellar [1971b], [1971c]. Especially of interest in this theory are criteria for weakly compact level sets of integral functionals and the features that duality takes on when one is dealing with nonreflexive spaces, as for instance in calculating the conjugate of an integral functional on a space $\mathcal{C}(T;\mathbb{R}^n)$ of continuous functions as a functional on the dual space of measures.

References

Alart, P. and B. Lemaire [1991], Penalization in non-classical convex programming via variational convergence, Mathematical Programming, 51, 307–332.

Alexandrov, A. D. [1939], Almost everywhere existence of the second differential of a convex function and some properties of convex surfaces connected with it, Leningrad State University Annals [Uchenye Zapiski], Mathematical Series, 6, 3–35, Russian.

Artstein, Z. and S. Hart [1981], Law of large numbers for random sets and allocation processes, Mathematics of Operations Research, 6, 482–492.

Attouch, H. [1975], Mesurabilité et Monotonie, Ph.D. thesis, Paris-Orsay.

Attouch, H. [1977], Convergence de fonctions convexes, de sous-différentiels et semi-groupes, Comptes Rendus de l'Académie des Sciences de Paris, 284, 539–542.

Attouch, H. [1984], Variational Convergence for Functions and Operators, Applicable Mathematics Series, Pitman, London.

Attouch, H. and D. Azé [1993], Approximation and regularization of arbitrary functions in Hilbert spaces by the Lasry-Lions method, Annales de l'Institut H. Poincaré: Analyse Non Linéaire, 10, 289–312.

Attouch, H., D. Azé and R. J-B. Wets [1988], Convergence of convex-concave saddle functions: continuity properties of the Legendre-Fenchel transform and applications to convex programming, Annales de l'Institut H. Poincaré: Analyse Non Linéaire, 5, 537–572.

Attouch, H., Z. Chbani and A. Moudafi [1995], Recession operators and solvability of noncoercive optimization and variational problems, Advances in Mathematics for Applied Sciences, 18, World Scientific Publishers, Singapore.

Attouch, H., R. Lucchetti and R. J-B. Wets [1991], The topology of the ρ-Hausdorff distance, Annali di Matematica pura ed applicata, CLX, 303–320.

Attouch, H. and C. Sbordone [1980], Asymptotic limits for perturbed functionals of calculus of variations, Ricerche di Matematica, 29, 85–124.

Attouch, H. and R. J-B. Wets [1981], Approximation and convergence in nonlinear optimization, in Nonlinear Programming 4, edited by O. Mangasarian, R. Meyer and S. Robinson, pp. 367–394, Academic Press, New York.

Attouch, H. and R. J-B. Wets [1983a], A convergence theory for saddle functions, Transactions of the American Mathematical Society, 280, 1–41.

Attouch, H. and R. J-B. Wets [1983b], Convergence des points min/sup et de points fixes, Comptes Rendus de l'Académie des Sciences de Paris, 296, 657–660.

Attouch, H. and R. J-B. Wets [1986], Isometries for the Legendre-Fenchel transform, Transactions of the American Mathematical Society, 296, 33–60.

Attouch, H. and R. J-B. Wets [1989], Epigraphical analysis, in Analyse Non Linéaire, edited by H. Attouch, J.-P. Aubin, F. Clarke and I. Ekeland, pp. 73–100, Gauthier-Villars, Paris.

Attouch, H. and R. J-B. Wets [1991], Quantitative stability of variational systems: I. The epigraphical distance, Transactions American Mathematical Society, 328, 695–729.

Attouch, H. and R. J-B. Wets [1993a], Quantitative stability of variational systems: II. A framework for nonlinear conditioning, SIAM Journal on Optimization, 3, 359–381.

Attouch, H. and R. J-B. Wets [1993b], Quantitative stability of variational systems: III. ε-approximate solutions, Mathematical Programming, 61, 197–214.

Aubin, J.-P. [1981], Contingent derivatives of set-valued maps and existence of solutions to nonlinear inclusions and differential inclusions, in Mathematical Analysis and Applications, Part A, edited by L. Nachbin, Advances in Mathematics: Supplementary Studies, 7A, pp. 160–232, Academic Press, New York.

Aubin, J.-P. [1984], Lipschitz behavior of solutions to convex minimization problems, Mathematics of Operations Research, 9, 87–111.

Aubin, J.-P. [1991], Viability Theory, Birkhäuser, Boston.

Aubin, J.-P. and A. Cellina [1984], Differential Inclusions, Springer, Berlin.

Aubin, J.-P. and F. H. Clarke [1977], Monotone invariant solutions to differential inclusions, Journal of the London Mathematical Society, 16, 357–366.

Aubin, J.-P. and F. H. Clarke [1979], Shadow prices and duality for a class of optimal control problems, SIAM Journal on Control and Optimization, 17, 567–587.

Aubin, J.-P. and I. Ekeland [1976], Estimates of the duality gap in nonconvex optimization, Mathematics of Operations Research, 1, 225–245.

Aubin, J.-P. and I. Ekeland [1984], Applied Nonlinear Analysis, Wiley, New York.

Aubin, J.-P. and H. Frankowska [1987], On inverse function theorems for set-valued maps, Journal de Mathématiques Pures et Appliquées, 66, 71–89.

Aubin, J.-P. and H. Frankowska [1990], Set-Valued Analysis, Birkhäuser, Boston.

Aumann, R. J. [1965], Integrals of set-valued functions, Journal of Mathematical Analysis and Applications, 12, 1–12.

Auslender, A. [1996], Noncoercive optimization problems, Mathematics of Operations Research, 21, 769–782.

Averbukh, V. I. and O. G. Smolyanov [1967], The theory of differentiation in linear topological spaces, Russian Mathematical Surveys, 22, 201–258.

Averbukh, V. I. and O. G. Smolyanov [1968], The various definitions of the derivative in linear topological spaces, Russian Mathematical Surveys, 23, 67–113.

Azé, D. and J.-P. Penot [1990], Operations on convergent families of sets and functions, Optimization, 21, 521–534.

Back, K. [1983], Continuous and hypograph convergence of utilities, Northwestern University, Department of Economics.

Back, K. [1986], Continuity of the Fenchel transform of convex functions, Proceedings of the American Mathematical Society, 97, 661–667.

Bagh, A. and R. J-B. Wets [1996], Convergence of set-valued mappings: Equi-outer semicontinuity, Set-Valued Analysis, 4, 333–360.

Baiocchi, C., G. Buttazzo, F. Gastaldi and F. Tomarelli [1988], General existence theorems for unilateral problems in continuum mechanics, Archive for Rational Mechanics and Analysis, 100, 149–189.

Baire, R. L. [1905], Leçons sur les Fonctions Discontinues, Gauthier-Villars, Paris, (edited by A. Denjoy).

Baire, R. L. [1927], Sur l'origine de la notion de semi-continuité, Bulletin de la Sociéte Mathématique de France, 55, 141–142.

Balder, E. J. [1977], An extension of duality-stability relations to nonconvex optimization problems, SIAM Journal on Control and Optimization, 15, 329–343.

Bank, B., J. Guddat, D. Klatte, B. Kummer and K. Tammer [1982], Non-linear Parametric Optimization, Akademie-Verlag, Berlin.

Bazaraa, M., J. Gould and M. Nashed [1974], On the cones of tangents with application to mathematical programming, J. Optimization Theory and Applications, 13, 389–426.

Beer, G. [1983], Approximate selections for upper semicontinuous convex valued multifunctions, Journal of Approximation Theory 39 (1983), 39, 172–184.

Beer, G. [1989], Infima of convex functions, Proceedings of the American Mathematical Society, 315, 849–859.

Beer, G. [1990], Conjugate convex functions and the epi-distance topology, Proceedings of the American Mathematical Society, 108, 117–126.

Beer, G. [1993], Topologies on Closed and Closed Convex Sets, Kluwer Academic Publishers, Dordrecht, The Netherlands.

Beer, G. and P. Kenderov [1988], On the arg min multifuction for lower semicontinuous functions, Proceedings of the American Mathematical Society, 102, 107–113.

Beer, G. and V. Klee [1987], Limits of starshaped sets, Archives of Mathematics, 48, 241–248.

Beer, G. and R. Lucchetti [1989], Minima of quasi-convex functions, Optimization, 20, 581–596.

Beer, G. and R. Lucchetti [1991], Convex optimization and the epi-distance topology, Transactions of the American Mathematical Society, 327, 795–813.

Beer, G. and R. Lucchetti [1993], Weak topologies for the closed subsets of a metrizable space, Transactions of the American Mathematical Society, 335, 805–822.

Beer, G., R. T. Rockafellar and R. J-B. Wets [1992], A characterization of epi-convergence in terms of convergence of level sets, Proceedings of the American Mathematical Society, 116, 753–761.

Bellman, R. and K. Fan [1963], On systems of linear inequalities in Hermitian matrix variables, in Convexity, edited by V. L. Klee, Proceedings of Symposia in Pure Mathematics, 7, pp. 1–11, American Mathematical Society, Providence, Rhode Island.

Bellman, R. and W. Karush [1961], On a new functional transform in analysis: the maximum transform, Bulletin of the American Mathematical Society, 67, 501–503.

Bellman, R. and W. Karush [1962a], On the maximum transform and semigroups of transformations, Bulletin of the American Mathematical Society, 68, 516–518.

Bellman, R. and W. Karush [1962b], Mathematical programming and the maximum transform, SIAM Journal, 10, 550–567.

Bellman, R. and W. Karush [1963], On the maximum transform, Journal of Mathematical Analysis and Applications, 6, 67–74.

Benoist, J. [1992], Approximation and regularization of arbitrary sets in finite dimension case, Séminaire d'Analyse Numérique, Univeristé Paul Sabatier Toulouse, 92.

Benoist, J. [1992], Convergence de la dérivée régularisée Lasry-Lions, Comptes Rendus de l'Académie des Sciences de Paris, 315, 941–944.

Benoist, J. [1994], Approximation and regularization of arbritary sets in finite dimension, Set-Valued Analysis, 94, 95–115.

Ben-Tal, A. [1980], Second order and related extremality conditions in nonlinear programming, Journal of Optimization Theory and Applications, 31, 143–165.

Ben-Tal, A. and J. Zowe [1982], Necessary and sufficient conditions for a class of nonsmooth minimization problems, Mathematical Programming, 24, 70–91.

Ben-Tal, A. and J. Zowe [1985], Directional derivatives in nonsmooth optimization, Journal of Optimization Theory and Applications, 47, 483–490.

Berge, C. [1959], Espaces Topologiques et Fonctions Multivoques, Dunod, Paris.

Berliocchi, H. and J. M. Lasry [1971], Intégrandes normales et mesures paramétrées en calcul des variations, Comptes Rendus de l'Académie des Sciences de Paris, 274, 839–842.

Berliocchi, H. and J. M. Lasry [1973], Intégrandes normales et mesures paramétrées en calcul des variations, Bulletin de la Société Mathématique de France, 10, 129–184.

Bertsekas, D. P. [1982], Constrained Optimization and Lagrange Multiplier Methods, Academic Press, New York.

Birge, J. and L. Qi [1995], Subdifferential convergence in stochastic programming, SIAM Journal on Optimization, 5, 436–453.

Birnbaum, Z. and W. Orlicz [1931], Über die Verallgemeinerung des Begriffes der zueinander konjugierten Potenzen, Studia Mathematica, 3, 1–67.

Bismut, J. -M. [1973], Intégrales convexes et probabilités, Journal of Mathematical Analysis and Applications, 42, 639–773.

Blanc, E. [1933], Sur la structure de certaines lois générales régissant des correspondances multiformes, Comptes Rendus de l'Académie des Sciences de Paris, 196, 1769–1771.

Blaschke, W. [1914], Kreis und Kugel, Jahresbericht der Deutschen Mathematische Verein, 24, 195–207.

Bonnans, J. F., R. Cominetti and A. Shapiro [1996a], Second order necessary and sufficient optimality conditions under abstract constraints, Rapport de Recherche INRIA 2952, Roquencourt, France.

Bonnans, J. F., R. Cominetti and A. Shapiro [1996b], Sensitivity analysis of optimization problems under second order regular constraints, Rapport de Recherche INRIA 2989, Roquencourt, France.

Bonnesen, T. and W. Fenchel [1934], Theorie der konvexen Körper, Ergebnisse der Mathematik, 3, Springer, Berlin.

Borwein, D., J. M. Borwein and X. Wang [1996], Approximate subgradients and coderivatives in R^n, Set-Valued Analysis, 4, 375–398.

Borwein, J. M. and S. Fitzpatrick [1988], Local boundedness of monotone operators under minimal hypotheses, Bulletin of the Australian Mathematical Society, 39 RP 439-441.

Borwein, J. M. and A. D. Ioffe [1996], Proximal analysis in smooth spaces, Set-Valued Analysis, 4, 1–24.

Borwein, J. M. and A. Jofré [1996], A nonconvex separation property on Banach spaces and applications, Report MA-97B437, Ingenieria Matematica, Universidad de Chile.

Borwein, J. M. and D. Noll [1994], Second order differentiability of convex functions in Banach spaces, Transactions of the American Mathematical Society, 342, 43–81.

Borwein, J. M. and D. Preiss [1987], A smooth variational principle with applications to subdifferentiability and to differentiability of convex functions, Transactions of the American Mathematical Society, 303, 517–527.

Borwein, J. M. and H. Strojwas [1985], Tangential approximations, Nonlinear Analysis: Theory, Methods & Applications, 9, 1347–1366.

Borwein, J. M. and J. D. Vanderwerff [1995], Convergence of Lipschitz regularizations of convex functions, Journal of Functional Analysis, 128, 139–162.

Borwein, J. M. and D. M. Zhuang [1988], Verifiable necessary and sufficient conditions for openness and regularity of set-valued and single-valued maps, Journal of Mathematical Analysis and Applications, 134, 441–459.

Bouligand, G. [1930], Sur les surfaces dépourvues de points hyperlimites, Annales de la Société Polonaise de Mathématique, 9, 32–41.

Bouligand, G. [1932a], Sur la semi-continuité d'inclusions et quelques sujets connexes, Enseignement Mathématique, 31, 14–22.

Bouligand, G. [1932b], Introduction à la géométrie infinitésimale directe, Gauthier-Villars, Paris.

Bouligand, G. [1933], Propriétés générales des correspondances multiformes, Comptes Rendus de l'Académie des Sciences de Paris, 196, 1767–1769.

Bouligand, G. [1935], Géométrie infinitésimale directe et physique mathématique classique, Mémorial des Sciences Mathématiques, 71, Gauthier-Villars, Paris.

Bourbaki, N. [1967], Variétés différentielles et analytiques, Eléments de Mathématiques, 33, Hermann, Paris.

Boyd, S. and L. Vandenberghe [1996], Semidefinite programming, SIAM Review, 38, 49–95.

Brezis, H. [1968], Equations et inéquations non linéaires dans les espaces vectoriels en dualité, Annales de l'Institut Fourier, 18, 115–175.

Brezis, H. [1972], Problèmes unilatéraux, Journal de Mathématiques Pures et Appliquées, 51, 1–168.

Brezis, H. [1973], Opérateurs Maximaux Monotones et Semi-groupes de Contractions dans les Espaces de Hilbert, North Holland, Amsterdam.

Brezis, H. and L. Nirenberg [1978], Characterizations of ranges of some nonlinear operators and applications to boundary value problems, Annali Scuola Normale Superiore Pisa, Classe di Scienze (Serie IV), 5, 225–325.

Brøndsted, A. [1964], Conjugate convex functions in topological vector spaces, Fys. Medd. Dans. Vid. Selsk., 34, 1–26.

Brøndsted, A. and R. T. Rockafellar [1965], On the subdifferentiability of convex functions, Proceedings of the American Mathematical Society, 16, 605–611.

Browder, F. E. [1968a], Nonlinear maximal monotone operators in Banach space, Mathematische Annalen, 175, 89–113.

Browder, F. E. [1968b], The fixed-point theory of multi-valued problems, Mathematische Annalen, 177, 283–301.

Browder, F. E. [1968c], Multi-valued monotone nonlinear mappings and duality mappings in Banach spaces, Transactions of the American Mathematical Society, 118, 89–113.

Browder, F. E. [1976], Nonlinear Operators and Nonlinear Equations of Evolution in Banach Spaces, Proceedings of Symposia in Pure Mathematics, 18 (no.2), American Mathematical Society, Providence, Rhode Island.

Burke, J. V. [1987], Second order necessary and sufficient conditions for convex composite NDO, Mathematical Programming, 38, 287–302.

Burke, J. V. [1991], An exact penalization viewpoint of constrained optimization, SIAM Journal on Control and Optimization, 29, 968–998.

Burke, J. V. and M. L. Overton [1994], Differential properties of the spectral abscissa and the spectral radius for analytic matrix-valued mappings, Nonlinear Analysis: Theory, Methods & Applications, 23, 467–488.

Burke, J. V. and P. Tseng [1996], A unified analysis of Hoffman's bound via Fenchel duality, SIAM Journal on Optimization, 6, 265–282.

Busemann, H. and W. Feller [1936], Krümmungseigenschaften konvexer Flächen, Acta Mathematica, 66, 1–47.

Buttazzo, G. [1977], Su una definizione generale dei Γ-limiti, Bollettino dell' Unione Ma tematica Italiana, 14-B, 722–744.

Castaing, C. [1967], Sur les multi-applications mesurables, Revue d'Informatique et de Recherche Opérationelle, 1, 91–167.

Castaing, C. and M. Valadier [1977], Convex Analysis and Measurable Multifunctions, Lecture Notes in Mathematics, 580, Springer, Berlin.

Cellina, A. [1969a], Approximation of set-valued functions and fixed point theorems, Annali di Matematica pura ed applicata, 82, 17–24.

Cellina, A. [1969b], A theorem on the approximation of compact multivalued mappings, Rendiconti Atti della Accademia Nazionale dei Licei, (8) 47, 429–433.

Chaney, R. W. [1982a], On sufficient conditions in nonsmooth optimization, Mathematics of Operations Research, 7, 463–475.

Chaney, R. W. [1982b], Second-order sufficient conditions for nondifferentiable programming problems, SIAM Journal on Control and Optimization, 20, 20–33.

Chaney, R. W. [1983], A general sufficiency theorem for nonsmooth nonlinear programming, Transactions of the American Mathematical Society, 276, 235–245.

Chaney, R. W. [1987], Second-order directional derivatives for nonsmooth functions, Journal of Mathematical Analysis and Applications, 128, 495–511.

Chen, G. H.-G. and R. T. Rockafellar [1997], Convergence rates in forward-backward splitting, SIAM Journal on Optimization, 7, forthcoming.

Chernoff, H. [1954], On the distribution of the likelihood ratio, Annals of Mathematical Statistics, 25, 573–578.

Choquet, G. [1947], Convergence, Annales de l'Université de Grenoble, 23, 55–112.

Choquet, G. [1962], Ensembles et cônes convexes faiblement complets, Comptes Rendus de l'Académie des Sciences de Paris, 254, 1908–1910.

Choquet, G. [1969], Outils topologiques et métriques de l'analyse mathématique, Centre de documentation universitaire et SEDES, Paris.

Christensen, J.P. [1974], Topology and Borel Structure, Mathematics Studies, 10, North-Holland, Amsterdam.

Clarke, F. H. [1973], Necessary conditions for Nonsmooth Problems in Optimal Control and the Calculus of Variations, Ph.D. dissertation, University of Washington, Seattle.

Clarke, F. H. [1975], Generalized gradients and applications, Transactions of the American Mathematical Society, 205, 247–262.

Clarke, F. H. [1976a], A new approach to Lagrange multipliers, Mathematics of Operations Research, 1, 165–174.

Clarke, F. H. [1976b], On the inverse function theorem, Pacific Journal of Mathematics, 64, 97–102.

Clarke, F. H. [1983], Optimization and Nonsmooth Analysis, Wiley, New York.

Clarke, F. H. [1989], Methods of Dynamic and Nonsmooth Optimization, CBMS-NSF Regional Conference Series in Applied Mathematics, 57, SIAM Publications, Philadelphia.

Clarke, F. H. and R. M. Redheffer [1993], The proximal subgradient and constancy, Canadian Mathematics Bulletin, 36, 30–32.

Clarke, F. H., R. J. Stern and P. R. Wolenski [1993], Subgradient criteria for monotonicity, the Lipschitz condition, and convexity, Canadian Journal of Mathematics, 45, 1167–1183.

Clarke, F. H., R. J. Stern and P. R. Wolenski [1995], Proximal smoothness and the lower-C^2 property, Journal of Convex Analysis, 1-2, 117–144.

Cominetti, R. [1986], Contribuciones al análisis no-diferenciable: derivadas generalizadas de primer y segundo orden, Ph.D. thesis, Universidad de Chile, Santiago.

Cominetti, R. [1990], Metric regularity, tangent cones, and second-order optimality conditions, Applied Mathematics and Optimization, 21, 265–287.

Cominetti, R. [1991], On pseudo-differentiability, Transactions of the American Mathematical Society, 324, 843–865.

Cominetti, R. and R. Correa [1986], Sur une dérivée du second ordre en analyse non différentiable, Comptes Rendus de l'Académie des Sciences de Paris, 303, 761–764.

Cominetti, R. and R. Correa [1990], A generalized second-order derivative in nonsmooth optimization, SIAM Journal on Control and Optimization, 28, 789–809.

Correa, R., A. Jofré and L. Thibault [1992], Characterization of lower semicontinuous convex functions, Proceedings of the American Mathematical Society, 116, 67–72.

Correa, R., A. Jofré and L. Thibault [1994], Subdifferential monotonicity as characterization of convex functions, Numerical Functional Analysis and Optimization, 15, 531–535.

Correa, R., A. Jofré and L. Thibault [1995], Subdifferential characterization of convexity, in Recent Advances in Nonsmooth Optimization, edited by D. Du, L. Qi and R. Womersley, pp. 18–23, World Scientific Publishing, Singapore.

Costantini, C., S. Levi and J. Zieminska [1993], Metrics that generate the same hyperspace convergence, Set-Valued Analysis, 1, 141–158.

Cottle, R. W., F. Giannessi and J.-L. Lions [1980], Variational Inequalities and Complementarity Problems, Wiley, New York.

Cottle, R. W., J.-S. Pang and R. E. Stone [1992], The Linear Complementarity Problem, Academic Press, Boston.

Crandall, M. G., L. C. Evans and P.-L. Lions [1984], Some properties of viscosity solutions of the Hamilton-Jacobi equation, Transactions of the American Mathematical Society, 282, 487–502.

Crandall, M. G. and P.-L. Lions [1981], Conditions d'unicité sur les solutions généralisées des equations d'Hamilton-Jacobi, Comptes Rendus de l'Académie des Sciences de Paris, 292, 183–186.

Crandall, M. G. and P.-L. Lions [1983], Viscosity solutions of Hamilton-Jacobi equations, Transactions of the American Mathematical Society, 277, 1–42.

Crandall, M. G. and A. Pazy [1969], Semi-groups of nonlinear contractions and dissipative sets, Journal of Functional Analysis, 3, 376–418.

Dal Maso, G. [1993], An introduction to Γ-Convergence, Birkhäuser, Basel.

Danskin, J. M. [1967], The Theory of Max-Min and its Applications to Weapons Allocations Problems, Springer, New York.

Dantzig, G. B., J. Folkman and N. Shapiro [1967], On the continuity of the minimum set of a continuous function, Journal of Mathematical Analysis and Applications, 17, 519–548.

Davis, C. [1957], All convex invariant functions of hermitian matrices, Archiv der Mathematik, 8, 276–278.

De Giorgi, E. [1977], Γ-convergenza e G-convergenza, Bollettino dell' Unione Matematica Italiana, 14-A, 213–220.

De Giorgi, E. and G. Dal Maso [1981], Γ-convergence and calculus of variations, in Mathematical Theories of Optimization, edited by P. Cecconi and T. Zolezzi, Lecture Notes in Mathematics, 979, pp. 121–143, Springer, Berlin.

De Giorgi, E. and T. Franzoni [1975], Su un tipo di convergenza variazionale, Atti Accademia Nazionale de Lincei, Rendiconti della Classe di Scienze Fisiche, Matematiche e Naturali, 58, 842–850.

De Giorgi, E. and T. Franzoni [1979], Su un tipo di convergenza variazionale, Rendiconti del Seminario di Brescia, 3, 63–101.

Debreu, G. [1959], Theory of Value, Wiley, New York.

Debreu, G. [1967], Integration of correspondences, in Proceedings of the Fifth Berkeley Symposium on Mathematical Statistics and Probability II, edited by L. LeCam, J. Neyman and E. Scott, pp. 351–372, University of California Press, Berkeley, California.

Del Prete, I., S. Dolecki and M. B. Lignola [1986], Continuous convergence and preservation of convergence of sets, Journal of Mathematical Analysis and Applications, 120, 454–471.

Del Prete, I. and M. B. Lignola [1983], On convergence of closed-valued multifunctions, Bollettino dell'Unione Matematica Italiana, B(6), 819–834.

Demyanov, V. F. and V. N. Malozemov [1971], The theory of nonlinear minimal problems, Russian Mathematical Surveys, 26, 55–115.

Deutsch, F. and P. Kenderov [1983], Continuous selections and approximate selection for set-valued mappings and applications to metric projections, SIAM Journal on Mathematical Analysis, 14, 185–194.

Deville, R., G. Godefroy and V. Zissler [1993], Smoothness and Renormings in Banach Spaces, Longman Scientific & Technical (Wiley), New York.

Dieudonné, J. [1966], Sur la séparation des ensembles convexes, Mathematische Annalen, 163, 1–3.

Dini, U. [1878], Fondamenti per la Teoria delle Funzioni di Variabili Reali, Pisa, Italy.

Dmitruk, A., A. A. Miliutin and N. Osmolovskii [1981], Liusternik's theorem and the theory of extrema, Russian Mathematical Surveys, 35, 11–51.

Do, C. [1992], Generalized second-order derivatives of convex functions in reflexive Banach spaces, Transactions of the American Mathematical Society, 334, 281–301.

Dolecki, S. [1982], Tangency and differentiation: Some applications of convergence theory, Annali di Matematica pura ed applicata, 130, 223–255.

Dolecki, S. and S. Kurcyusz [1978], On Φ-convexity in extremal problems, SIAM Journal on Control and Optimization, 16, 277–300.

Dolecki, S., G. Salinetti and R. J-B. Wets [1983], Convergence of functions: Equi-semiconti nuity, Transactions of the American Mathematical Society, 276, 409–429.

Dontchev, A. L. [1996], The Graves theorem revisited, Journal of Convex Analysis, 3, 45–54.

Dontchev, A. L. and W. W. Hager [1994], Implicit functions, Lipschitz maps and stability in optimization, Mathematics of Operations Research, 19, 753–768.

Dontchev, A. L. and R. T. Rockafellar [1996], Characterizations of strong regularity for variational inequalities over polyhedral convex sets, SIAM Journal on Optimization, 6, 1087–1105.

Dontchev, A.L. and R.T. Rockafellar [2009], Implicit Functions and Solution Mappings, Springer, New York.

Dontchev, A. L. and T. Zolezzi [1993], Well-Posed Optimization Problems, Lecture Notes in Mathematics, 1543, Springer, New York.

Dubovitzkii, A. Y. and A. A. Miliutin [1965], Extremum problems in the presence of restrictions, U.S.S.R. Computational Mathematics and Mathematical Physics, 5, 1–80.

Duffin, R. J. [1956], Infinite programs, in Linear Inequalities and Related Systems, edited by H. W. Kuhn and A. W. Tucker, Annals of Mathematics Studies, 38, pp. 157–170, Princeton University Press, Princeton, New Jersey.

Duffin, R. J., E. L. Peterson and C. Zener [1967], Geometric Programming — Theory and Application, Wiley, New York.

Eckstein, J. [1989], Splitting Methods for Monotone Operators with Applications to Parallel Optimization, Ph.D. thesis, Massachusetts Institute of Technology, Cambridge, Massachusetts.

Eckstein, J. and D. P. Bertsekas [1992], On the Douglas-Rachford splitting method and the proximal point algorithm for maximal monotone mappings, Mathematical Programming, 55, 293–318.

Ekeland, I. [1974], On the variational principle, Journal of Mathematical Analysis and Applications, 47, 324–353.

Ekeland, I. [1977], Legendre duality in nonconvex optimization and calculus of variations, SIAM Journal on Control and Optimization, 15, 905–934.

Ekeland, I. and G. Lebourg [1976], Generic Fréchet-differentiability and perturbed optimization problems in Banach spaces, Transactions of the American Mathematical Society, 224, 193–216.

Ekeland, I. and R. Temam [1974], Analyse Convexe et Problèmes Variationnels, Dunod, Paris.

Ekeland, P. and B. Valadier [1971], Represensation of set valued functions, Journal of Mathematical Analysis and Applications, 35, 621–629.

Elster, K. H. and R. Nehse [1974], Zur Theorie der Polarfunktionale, Mathematische Operationsforschung und Statistik, 5, 3–21.

Ermoliev, Y. M., V. I. Norkin and R. J-B. Wets [1995], The minimization of semicontinuous functions: Mollifier subgradients, SIAM Journal on Control and Optimization, 33, 149–167.

Evans, J. P. and F. J. Gould [1970], Stability in nonlinear programming, Operations Research, 18, 107–118.

Evans, L. C. and R. F. Gariepy [1992], Measure Theory and Fine Properties of Functions, CRC Press, Boca Raton, Florida.

Federer, H. [1959], Curvature measures, Transactions of the American Mathematical Society, 93, 418–491.

Fell, J. M. [1962], A Hausdorff topology for the closed subsets of a locally compact non-Hausdorff space, Proceedings of the American Mathematical Society, 13, 472–474.

Fenchel, W. [1949], On conjugate convex functions, Canadian Journal of Mathematics, 1, 73–77.

Fenchel, W. [1951], Convex Cones, Sets and Functions, Lecture Notes, Princeton University, Princeton, New Jersey.

Fiacco, A. V. and Y. Ishizuka [1990], Sensitivity and stability analysis for nonlinear programming, Annals of Operations Reasearch, 27, 215–236.

Fiacco, A. V. and G. P. McCormick [1968], Nonlinear Programming, Wiley, New York.

Filippov, A. F. [1959], On a certain question in the theory of optimal control, Vestnik Mos kovskovo Universiteta, Ser. Mat. Mech. Astr., 2, 22–32, Russian, English translation: SIAM Journal on Control 1(1962), 76-84.

Filippov, A. [1967], Classical solutions of differential equations with multivalued right-hand side, SIAM Journal on Control, 5, 609–621.

Fisher, B. [1981], Common fixed points of mappings and set-valued mappings, Rostock Mathematische Kolloquium, 18, 69–77.

Fitzpatrick, S. and R. R. Phelps [1992], Bounded approximants to monotone operators on Banach spaces, Annales de l'Institut H. Poincaré: Analyse Non Linéaire, 9, 573–595.

Fort, M. K. [1951], Points of continuity of semicontinuous functions, Public. Mathem. Debrecem, 2, 100–102.

Fougères, A. and A. Truffert [1988], Régularisation s.c.i. et Γ-convergence: approximations inf-convolutives associées à un référentiel, Annali di Matematica pura ed applicata, 152, 21–51.

Frank, M. and P. Wolfe [1956], An algorithm for quadratic programming, Naval Research Logistics Quarterly, 3, 95–110.

Frankowska, H. [1984], The first order necessary conditions for nonsmooth variational and control problems, SIAM Journal on Control and Optimization, 22, 1–12.

Frankowska, H. [1986], Théorème d'application ouverte pour les correspondences, Comptes Rendus de l'Académie des Sciences de Paris, 302, 559–562.

Frankowska, H. [1987a], Théorèmes d'application ouverte et de fonction inverse, Comptes Rendus de l'Académie des Sciences de Paris, 305, 773–776.

Frankowska, H. [1987b], An open mapping principle for set-valued maps, Journal of Mathematical Analysis and Applications, 127, 172–180.

Frankowska, H. [1989], High order inverse function theorems, in Analyse Non Linéaire, edited by H. Attouch, J.-P. Aubin, F. Clarke and I. Ekeland, pp. 283–304, Gauthier-Villars, Paris.

Frankowska, H. [1990], Some inverse mapping theorems, Annales de l'Institut H. Poincaré: Analyse Nonlinéaire, 7, 183–234.

Gajek, L. and D. Zagrodny [1995], Geometric Mean Value Theorems for the Dini Derivative, Journal of Mathematical Analysis and Applications, 191, 56–76.

Gale, D., H. W. Kuhn and A. W. Tucker [1951], Linear programming and the theory of games, in Activity Analysis of Production and Allocation, edited by T. C. Koopmans, Wiley, New York.

Georgiev, P. G. and N. P. Zlateva Second-order subdifferential of $C^{1,1}$ functions and optimality conditions, Set-Valued Analysis, 4, 101–117.

Ginsburg, B. and A. D. Ioffe [1996], The maximum principle in optimal control theory of systems governed by semilinear equations, in Nonlinear Analysis and Geometric Methods in Deterministic Optimal Control, edited by B. Mordukhovich and H. Sussman, IMA Volumes in Mathematics and its Applications, 78, pp. 81–110, Springer, New York.

Golshtein, E. G. [1972], Theory of Convex Programming, Translations of Mathematical Monographs, 26, American Mathematical Society, Providence, Rhode Island.

Golshtein, E. G. and N. V. Tretyakov [1996], Modified Lagrangians and Monotone Maps in Optimization, Wiley, New York.

Golumb, M. [1935], Zur Theorie der nichtlinearen Integralgleichungen, Integralgleichungs systeme und allgemeinen Functionalgleichungen, Mathematische Zeitschrift, 39, 45–75.

Gorni, G. [1991], Conjugation and second-order properties of convex functions, Journal of Mathematical Analysis and Applications, 158, 293–315.

Gossez, J.-P. [1971], Opérateurs monotones non linéaires dans les espaces de Banach non réflexifs, Journal of Mathematical Analysis and Applications, 34, 371–395.

Gould, J. F. and J. W. Tolle [1971], A necessary and sufficient qualification for constrained optimization, SIAM Journal on Applied Mathematics, 20, 164–172.

Gowda, M. S. and R. Sznajder [1996], On the Lipschitzian properties of polyhedral multifunctions, Mathematical Programming, 74, 267–278.

Graves, L. M. [1950], Some mapping theorems, Duke Mathematical Journal, 17, 111–114.

Güler, O. [1991], On the convergence of the proximal point algorithm for convex minimization, SIAM Journal on Control and Optimization, 29, 403–419.

Hahn, H. [1932], Reelle Funktionen I, Akademische Verlagsgesellschaft, Leipzig.

Halkin, H. [1976a], Interior mapping theorem with set-valued derivatives, Journal d'Analyse Mathématique, 30, 200–207.

Halkin, H. [1976b], Mathematical programming without differentiability, in Calculus of Variations and Control Theory, edited by D. Russell, pp. 279–287, Academic Press, New York.

Hanner, O. and H. Rådström [1951], A generalization of a theorem of Fenchel, Proceedings of the American Mathematical Society, 2, 589–593.

Hausdorff, F. [1919], Über halbstetige Funktionen und deren Verallgemeinerung, Mathematische Zeitschrift, 5, 292–309.

Hausdorff, F. [1927], Mengenlehre, Walter de Gruyter & Co., Berlin.

Henrion, R. [1997], The Approximate Subdifferential and Parametric Optimization, Habilitation Thesis, Humboldt University, Berlin.

Hess, C. [1983], Loi de probabilité des ensembles aléatoires à valeurs fermées dans un espace métrique séparable, Comptes Rendus de l'Académie des Sciences de Paris, 296, 883–886.

Hess, C. [1986], Contribution à l'étude de la mesurabilité, de la loi de probabilité et de la convergence des multifonctions, Thèse, Université des Sciences et Techniques du Languedoc, Montpellier, France.

Hestenes, M. R. [1966], Calculus of Variations and Optimal Control Theory, Wiley, New York.

Hestenes, M. R. [1975], Optimization Theory, Wiley, New York.

Higle, J. L. and S. Sen [1992], On the convergence of algorithms with implications for stochastic and nondifferentiable optimization, Mathematics of Operations Research, 17, 112–131.

Higle, J. L. and S. Sen [1995], Epigraphical nesting: A unifying theory for the convergence of algorithms, Journal of Optimization Theory and Applications, 84, 339–360.

Hildenbrand, W. [1974], Core and Equilibria of a Large Economy, Princeton University Press, Princeton, New Jersey.

Hiriart-Urruty, J.-B. [1979], Tangent cones, generalized gradients and mathematical programming in Banach spaces, Mathematics of Operations Research, 4, 79–97.

Hiriart-Urruty, J.-B. [1980a], Extension of Lipschitz functions, Journal of Mathematical Analysis and Applications, 77, 539–544.

Hiriart-Urruty, J.-B. [1980b], Lipschitz r-continuity of the approximate subdifferential of a convex function, Matematica Scandinavia, 47, 123–134.

Hiriart-Urruty, J.-B. [1982a], Approximating a second order directional derivative for nonsmooth convex functions, SIAM Journal on Control and Optimization, 10, 783–807.

Hiriart-Urruty, J.-B. [1982b], Characterization of the plenary hull of the generalized Jacobian matrix, Mathematical Programming Studies, 17, 1–12.

Hiriart-Urruty, J.-B. [1982c], Limiting behavior of the approximate first and second order directional derivatives for a convex function, Nonlinear Analysis: Theory, Methods & Applications, 6, 1309–1326.

Hiriart-Urruty, J.-B. [1984], Calculus rules on the approximate second order directional derivatives for a convex function, SIAM Journal on Control and Optimization, 22, 381–404.

Hiriart-Urruty, J.-B. [1986], A new set-valued second-order derivative for convex functions, in Mathematics for Optimization, edited by J.-B. Hiriart-Urruty, Mathematics Study Series, 129, pp. 157–182, North-Holland, Amsterdam.

Hiriart-Urruty, J.-B. and C. Lemaréchal [1993], Convex Analysis and Minimization Algorithms I & II, Springer, New York.

Hiriart-Urruty, J.-B., M. Moussaoui, A. Seeger and M. Volle [1995], Subdifferential calculus without qualification conditions, using approximate subdifferentials, Nonlinear Analysis: Theory, Methods & Applications, 24, 1727–1754.

Hiriart-Urruty, J.-B. and A. Seeger [1989a], The second order subdifferential and the Dupin indicatrices of a non-differentiable convex function, Proceedings of the London Mathematical Society, 58, 351–365.

Hiriart-Urruty, J.-B. and A. Seeger [1989b], Calculus rules on a new set-valued second-order derivative for convex functions, Nonlinear Analysis: Theory, Methods & Applications, 13, 721–738.

Hiriart-Urruty, J.-B., J.-J. Strodiot and V. H. Nguyen [1984], Generalized Hessian matrix and second-order optimality conditions for problems with $C^{1,1}$ data, Applied Mathematics and Optimization, 11, 43–56.

Hoffman, A. J. [1952], On approximate solutions of systems of linear inequalities, Journal of the National Bureau of Standards, 49, 263–265.

Hogan, W. W. [1973], Point-to-set maps in mathematical programming, SIAM Review, 15, 591–603.

Holmes, R. B. [1966], Approximating best approximations, Nieuw Archief voor Wiskunde, 14, 106–113.

Hörmander, L. [1954], Sur la fonction d'appui des ensembles convexes, Arkiv för Matematik, 3, 181–186.

Ioffe, A. D. [1978a], Survey of measurable selection theorems: Russian literature supplement, SIAM Journal on control and optimization, 16, 728–723.

Ioffe, A. D. [1978b], Representation theorems for multifunctions and analytic sets, Bulletin of the American Mathematical Society, 84, 142–144.

Ioffe, A. D. [1979a], Regular points of Lipschitz functions, Transactions of the American Mathematical Society, 251, 61–69.

Ioffe, A. D. [1979b], Necessary and sufficient conditions for a local minimum, III: Second order conditions and augmented duality, SIAM Journal on Control and Optimization, 17, 266–288.

Ioffe, A. D. [1981a], Approximate subdifferentials of nonconvex functions, CEREMADE Publication 8120, Universiteé de Paris IX "Dauphine".

Ioffe, A. D. [1981b], Nonsmooth analysis: differential calculus of non-differentiable mappings, Transactions of the American Mathematical Society, 266, 1–56.

Ioffe, A. D. [1981c], Sous-différentielles approchées de fonctions numériques, Comptes Rendus de l'Académie des Sciences de Paris, 292, 675–678.

Ioffe, A. D. [1984a], Calculus of Dini subdifferentials of functions and contingent coderivatives of set-valued maps, Nonlinear Analysis: Theory, Methods & Applications, 8, 517–39.

Ioffe, A. D. [1984b], Approximate subdifferentials and applications, I: the finite dimensional theory, Transactions of the American Mathematical Society, 281, 389–416.

Ioffe, A. D. [1984c], Necessary conditions in nonsmooth optimization, Mathematics of Operations Research, 9, 159–188.

Ioffe, A. D. [1986], Approximate subdifferentials and applications, II: functions on locally convex spaces, Mathematika, 33, 111–128.

Ioffe, A. D. [1987], On the local surjection property, Nonlinear Analysis: Theory, Methods & Applications, 11, 565–592.

Ioffe, A. D. [1989], Approximate subdifferentials and applications, III: the metric theory, Mathematika, 71, 1–38.

Ioffe, A. D. [1991], Variational analysis of a composite function: A formula for the lower second order epi-derivative, Journal of Mathematical Analysis and Applications, 160, 379–405.

Ioffe, A. D. [1997], Codirectional compactness, metric regularity and subdifferential calculus, paper presented at Serdica Mathematics Journal, forthcoming.

Ioffe, A. D. and V. L. Levin [1972], Subdifferentials of convex functions, Trudy Moskov. Mat. Ob., 26, 3–73, Russian.

Ioffe, A. D. and J.-P. Penot [1995], Limiting subjets and their calculus, Comptes Rendus de l'Académie des Sciences de Paris, 321, 799–804.

Ioffe, A. D. and J.-P. Penot [1996], Subdifferentials of performance functions and calculus of coderivatives of set-valued mappings, Serdica Mathemathics Journal, 22, 359–384.

Ioffe, A. D. and V. M. Tikhomirov [1968], The duality of convex functions and extremal problems, Uspekhi Matematicheskikh Nauk, 23, 51–116, Russian.

Ioffe, A. D. and V. M. Tikhomirov [1974], Theory of Extremal Problems, Nauka, Moscow, Russian.

Janin, R. [1973], Sur une classe de fonctions sous-linéarisables, Comptes Rendus de l'Académie des Sciences de Paris, 277, 265–267.

Janin, R. [1982], Sur des multiapplications qui sont des gradients généralisés, Comptes Rendus de l'Académie des Sciences de Paris, 294, 115–117.

Jiménez, B. and V. Novo [2006], Characterization of the cone of attainable directions, J. of Optimization Theory and Applications, 131, 493–499.

Jofré, A. [1989], Image of critical point sets for nonsmooth mappings and applications, in Contributions à l'analyse non différentiable avec applications à l'optimisation et à l'analyse non linéaire, Thèse, Pau, France.

Jofré, A. and J. Rivera [1995a], A nonconvex separation property and applications, University of Chile, Santiago, (Mathematical Programming, submitted).

Jofré, A. and J. Rivera [1995b], The second welfare theorem in nonconvex, nontransitive economies, University of Chile, Santiago, (Journal of Mathematical Economics, submitted).

John, F. [1948], Extremum problems with inequalities as side conditions, in Studies and Essays: Courant Anniversary Volume, edited by K. O. Fredrichs, O. E. Neugebauer and J. J. Stoker, pp. 187–204, Wiley-Interscience, New York.

Joly, J.-L. [1973], Une famille de topologies sur l'ensemble des fonctions convexes pour lesquelles la polarité est bicontinue, Journal de Mathématiques Pures et Appliquées, 52, 421–441.

Jourani, A. and L. Thibault [1993], Approximate subdifferentials of composite functions, Bulletin of the Australian Mathematical Society, 47, 443–455.

Jourani, A. and L. Thibault [1995], Verifiable conditions for openness and regularity of multivalued mappings, Transactions of the American Mathematical Society, 347, 1255–1268.

Jourani, A. and L. Thibault [1997], Chain rules for coderivatives of multivalued mappings in Banach spaces, Université du Languedoc, Montpellier, (forthcoming: Proceedings of the American Mathematical Society).

Kachurovskii, R. I. [1960], On monotone operators and convex functionals, Uspekhi Matematicheskikh Nauk, 15, 213–214.

Kachurovskii, R. I. [1968], Nonlinear monotone operators in Banach spaces, Uspekhi Matematicheskikh Nauk, 23, 121–168.

Kakutani, S. [1941], A generalization of Brouwer's fixed point theorem, Duke Mathematical Journal, 8, 457–459.

Kall, P. [1986], Approximation to optimization problems: an elementary review, Mathematics of Operations Research, 11, 9–18.

Karush, W. [1939], Minima of functions of several variables with inequalities as side conditions, Master's thesis, University of Chicago, Chicago.

Kato, T. [1966], Perturbation Theory for Linear Operators, Springer, Berlin.

King, A. J. and R. T. Rockafellar [1993], Asymptotic theory for solutions in generalized M-estimation and stochastic programming, Mathematics of Operations Research, 18, 148–162.

Klatte, D. and G. Thiere [1995], Error bounds for solutions of linear equations and inequalities, Zeitschrift für Operations Research, 41, 191–214.

Klee, V. L. [1964], Mathematical Reviews, 28, # 514.

Klee, V. L. [1968], Maximal separation theorems for convex sets, Transactions of the American Mathematical Society, 134, 133–148.

Klein, E. and A. C. Thompson [1984], Theory of Correspondences, Including Applications to Mathematical Economics, Wiley, New York.

Korf, L. and R. J-B. Wets [1997], Ergodic limit laws for stochastic optimization problems, Technical Report, University of California, Davis.

Kowalczyk, S. [1994], Topological convergence of multivalued maps and topological convergence of graphs, Demonstratio Mathematica, XXVII, 79–87.

Krasnoselskii, M. A. and Y. B. Rutitskii [1961], Convex Functions and Orlicz Spaces, Noordhoff, Groningen, The Netherlands.

Kruger, A. Y. [1981], Generalized Differentials of Nonsmooth Functions and Necessary Conditions for an Extremum, doctoral dissertation, University of Minsk, Russian.

Kruger, A. Y. [1985], Properties of generalized differentials, Siberian Mathematical Journal, 26, 822–832.

Kruger, A. Y. [1988], A covering theorem for set-valued mappings, Optimization, 19, 763–780.

Kruger, A. Y. and B. S. Mordukhovich [1980], Extremal points and the Euler equation in nonsmooth problems of optimization, Doklady Akademia Nauk BSSR (Belorussian Academy of Sciences), 24, 684–687, Russian.

Kudo, H. [1954], Dependent experiments and sufficient statistics, Nat. Sci. Rep. Ochanomizu Univ., 4, 151–163.

Kuhn, H. W. [1976], Nonlinear programming: a historical view, in Nonlinear Programming, edited by R. W. Cottle and C. E. Lemke, SIAM-AMS Proceedings, 9, pp. 1–26, American Mathematical Society, Providence, Rhode Island.

Kuhn, H. W. and A. W. Tucker [1951], Nonlinear programming, in Proceedings of the Second Berkeley Symposium on Mathematical Statistics and Probability, edited by J. Neyman, pp. 481–492, University of California Press, Berkeley, California.

Kummer, B. [1991], Lipschitzian inverse functions, directional derivatives and application in $\mathcal{C}^{1,1}$ optimization, Journal of Optimization Theory and Applications, 158, 35–46.

Kuratowski, K. [1932], Les fonctions semi-continues dans l'espace des ensembles fermés, Fundamenta Mathematicae, 18, 148–159.

Kuratowski, K. [1933], Topologie, I & II, Panstowowe Wyd Nauk, Warsaw, (Academic Press, New York, 1966).

Kuratowski, K. and C. Ryll-Nardzewski [1965], A general theorem on selectors, Bulletin de l'Académie Polonaise de Sciences, Série Mathématiques, Astronomie et Physique, 13, 397–411.

Kushner, H. [1984], Robustness and approximation of escape times and large deviation estimates for systems with small noise effects, SIAM Journal on Applied Mathematics, 44, 160–182.

Lasry, J.-M. and P.-L. Lions [1986], A remark on regularization in Hilbert spaces, Israel Journal of Mathematics, 55, 257–266.

Laurent, P.G. [1972], Approximation et Optimisation, Herman, Paris.

Lax, P. D. [1957], Hyperbolic systems of conservation laws, Communications in Pure and Applied Mathematics, 10, 537–566.

Leach, E. B. [1961], A note on inverse function theorems, Proceedings of the American Mathematical Society, 12, 694–697.

Leach, E. B. [1963], On a related function theorem, Proceedings of the American Mathematical Society, 14, 687–689.

Lebourg, G. [1975], Valeur moyenne pour gradient généralisé, Comptes Rendus de l'Académie des Sciences de Paris, 281, 795–797.

Lemaire, B. [1986], Méthode d'optimisation et convergence variationelle, Séminaire d'Analy se Convexe, Université du Languedoc, 17, 9.1–9.18.

Levy, A. B. [1993], Second-order epi-derivatives of integral functionals, Set-Valued Analysis, 1, 379–392.

Levy, A. B. [1996], Implicit multifunction theorems for the sensitivity analysis of variational conditions, Mathematical Programming, 74, 333–350.

Levy, A. B., R. A. Poliquin and L. Thibault [1995], A partial extension of Attouch's theorem and its applications to second-order epi-differentiation, Transactions of the American Mathematical Society, 347, 1269–1294.

Levy, A. B. and R. T. Rockafellar [1994], Sensitivity analysis of solutions to generalized equations, Transactions of the American Mathematical Society, 345, 661–671.

Levy, A. B. and R. T. Rockafellar [1995], Sensitivity of solutions in nonlinear programming problems with nonunique multipliers, in Recent Advances in Optimization, edited by D. Du, L. Qi and R. A. Womersley, pp. 215–223, World Scientific Publishers, Singapore.

Levy, A. B. and R. T. Rockafellar [1996], Variational conditions and the proto-differentiation of partial subgradient mappings, Nonlinear Analysis: Theory, Methods & Applications, 26, 1951–1964.

Lewis, A. S. [1996], Nonsmooth analysis of eigenvalues, Combinatorics and Optimization Department, University of Waterloo, Research Report CORR 196-15, Waterloo, Ontario.

Lewis, A. S. and M. L. Overton [1996], Eigenvalue optimization, paper presented at Acta Numerica.

Lions, P.-L. [1978], Two remarks on the convergence of convex functions and monotone operators, Nonlinear Analysis: Theory, Methods & Applications, 2, 553–562.

Lions, P.-L. and B. Mercier [1979], Splitting algorithms for the sum of two nonlinear operators, SIAM Journal on Numerical Analysis, 16, 964–979.

Lions, J.-L. and G. Stampacchia [1965], Inéquations variationelles non coercives, Comptes Rendus de l'Académie des Sciences de Paris, 261, 25–27.

Lions, J.-L. and G. Stampacchia [1967], Variational inequalities, Communications in Pure and Applied Mathematics, 1967, 493–519.

Lipschitz, R. [1877], Lehrbuch der Analysis, Cohen & Sohn, Bonn.

Liusternik, L. A. [1934], On the conditional extrema of functionals, Matematicheskii Sbornik, 41, 390–401.

Loewen, P. D. [1992], Limits of Fréchet normals in nonsmooth analysis, in Optimization and Nonlinear Analysis, edited by A. D. Ioffe, M. Marcus and S. Reich, Pitman Research Notes in Mathematics, 244, pp. 178–188, Longman Scientific & Technical (Wiley), New York.

Loewen, P. D. [1993], Optimal Control via Nonsmooth Analysis, CRM Proceedings & Lecture Notes, American Mathematical Society, Providence, Rhode Island.

Loewen, P. D. [1995], A mean value theorem for Fréchet subgradients, Nonlinear Analysis: Theory, Methods & Applications, 23, 1365–1381.

Loewen, P. D. and R. T. Rockafellar [1994], Optimal control of unbounded differential inclusions, SIAM Journal on Control and Optimization, 32, 442–470.

Loewen, P. D. and H. Zheng [1995], Epi-differentiability of integral functionals with applications, Transactions of the American Mathematical Society, 347, 443–459.

Lubben, R. G. [1928], Concerning limiting sets in abstract spaces, Transactions of the American Mathematical Society, 29, 668–685.

Lucchetti, R., N. S. Papageorgiou and F. Patrone [1987], Convergence and approximation results for measurable multifunctions, Proceedings of the American Mathematical Society, 100, 551–556.

Lucchetti, R., G. Salinetti and R. J-B. Wets [1994], Uniform convergence of probability measures: Topological criteria, Journal of Multivariate Analysis, 51, 252–264.

Lucchetti, R. and A. Torre [1994], Classical set convergences and topologies, Set-Valued Analysis, 36, 219–240.

Lucchetti, R., A. Torre and R. J-B. Wets [1993], A topology for the solid subsets of a topological space, Canadian Mathematical Bulletin, 36, 197–208.

Luo, Z.-Q., J.-S. Pang and D. Ralph [1996], Mathematical Programs with Equilibrium Constraints, Cambridge University Press, Cambridge, UK.

Luo, Z.-Q. and P. Tseng [1994], Perturbation analysis of a condition number for linear systems, SIAM Journal on Matrix Analysis and Applications, 15, 636–660.

Mandelbrojt, S. [1939], Sur les fonctions convexes, Comptes Rendus de l'Académie des Sciences de Paris, 209, 977–978.

Mangasarian, O. L. and S. Fromovitz [1967], The Fritz John conditions in the presence of equality and inequality constraints, Journal of Mathematical Analysis and Applications, 17, 73–74.

Marino, A. and M. Tosques [1990], Some variational problems with lack of convexity and some partial differential inequalities, in Methods of Nonconvex Analysis, edited by A. Cellina, Lecture Notes in Mathematics, 1446, pp. 58–83.

McLinden, L. [1973], Dual Operations on saddle functions, Transactions of the American Mathematical Society, 179, 363–381.

McLinden, L. [1974], An extension of Fenchel's duality theorem to saddle functions and dual minimax problems, Pacific Journal of Mathematics, 50, 135–150.

McLinden, L. [1982], Successive approximation and linear stability involving convergent sequences of optimization problems, Journal of Approximation Theory, 35, 311–354.

McLinden, L. and R. Bergstrom [1981], Preservation of convergence of sets and functions in finite dimensions, Transactions of the American Mathematical Society, 268, 127–142.

McShane, E. J. [1973], The Lagrange multiplier rule, American Mathematics Monthly, 80.

Michael, E. [1951], Topologies on spaces of subsets, Transactions of the American Mathematical Society, 71, 152–182.

Michael, E. [1956], Continuous selections I, Annals of Mathematics, 63, 361–382.

Michael, E. [1959], Dense families of continuous selections, Fundamenta Mathematicae, 67, 173–178.

Michel, P. and J.-P. Penot [1984], Calcul sous-différentiel pour des fonctions lipschitziennes et non-lipschitziennes, Comptes Rendus de l'Académie des Sciences de Paris, 298, 684–687.

Michel, P. and J.-P. Penot [1992], A generalized derivative for calm and stable functions, Differential and Integral Equations, 5, 189–196.

Michel, P. and J.-P. Penot [1994], Second-order moderate derivatives, Nonlinear Analysis: Theory, Methods & Applications, 22, 809–822.

Mickle, E. J. [1949], On the extension of a transformation, Bulletin of the American Mathematical Society, 55, 160–164.

Mifflin, R. [1977], Semismooth and semiconvex functions in constrained optimization, Mathematics of Operations Research, 2, 191–207.

Mignot, F. [1976], Contrôle dans les inéquations variationelles élliptiques, Journal of Functional Analysis, 22, 130–185.

Milosz, T. [1988], Necessary and sufficient second order conditions for nonsmooth extremal problems, Ph.D. thesis, Institute of Mathematics, Polish Academy of Sciences, Warsaw.

Minkowski, H. [1910], Geometrie der Zahlen, Teubner, Leipzig.

Minkowski, H. [1911], Theorie der konvexen Körper, insbesondere Begründung ihres Ober flächenbegriffs, Gesammelte Abhandlungen, II, Teubner, Leipzig.

Minty, G. J. [1960], Monotone Networks, Proceedings of the Royal Society of London (Series A), 257, 194–212.

Minty, G. J. [1961], On the maximal domain of a 'monotone' function, Michigan Mathematical Journal, 8, 135–137.

Minty, G. J. [1962], Monotone (nonlinear) operators in Hilbert space, Duke Mathematical Journal, 29, 341–346.

Minty, G. J. [1964], On the monotonicity of the gradient of a convex function, Pacific Journal of Mathematics, 14, 243–247.

Minty, G. J. [1967], A theorem on monotone sets in Hilbert spaces, Journal of Mathematical Analysis and Applications, 11, 434–439.

Mordukhovich, B. S. [1976], Maximum principle in the problem of time optimal response with nonsmooth constraints, Journal of Applied Mathematics and Mechanics, 40, 960–969.

Mordukhovich, B. S. [1980], Metric approximations and necessary optimality conditions for general classes of nonsmooth extremal problems, Soviet Mathematical Doklady, 22, 526–530.

Mordukhovich, B. S. [1984], Nonsmooth analysis with nonconvex generalized differentials and adjoint mappings, Doklady Akademia Nauk BSSR (Belorussian Academy of Sciences), 28, 976–979, Russian.

Mordukhovich, B. S. [1988], Approximation Methods in Problems of Optimization and Control, Nauka, Moscow, Russian.

Mordukhovich, B. S. [1992], Sensitivity analysis in nonsmooth optimization, in Theoretical Aspects of Industrial Design, edited by D. A. Field and V. Komkov, SIAM Proceedings in Applied Mathematics, 58, pp. 32–46.

Mordukhovich, B. S. [1993], Complete characterization of openness, metric regularity, and Lipschitzian properties of multifunctions, Transactions of the American Mathematical Society, 340, 1–36.

Mordukhovich, B. S. [1994a], Stability theory for parametric generalized equations and variational inequalities via nonsmooth analysis, Transactions of the American Mathematical Society, 343, 609–657.

Mordukhovich, B. S. [1994b], Sensitivity analysis for constraint and variational systems by means of set-valued differentiation, Optimization, 31, 13–46.

Mordukhovich, B. S. [1994c], Lipschitzian stability of constraint systems and generalized equations, Nonlinear Analysis: Theory, Methods & Applications, 22, 173–206.

Mordukhovich, B. S. [1994d], Generalized differential calculus for nonsooth and set-valued mappings, Journal of Mathematical Analysis and Applications, 183, 250–288.

Mordukhovich, B. S. [1996], Generalized differentiation and stability of solution maps to parametric variational systems, in World Congress of Nonlinear Analysts '92, edited by V. Lakshmikantham, pp. 2311–2323, Walter de Gruyter, Berlin.

Mordukhovich, B. S. and Y. Shao [1995a], Differential characterizations of covering, metric regularity, and Lipschitzian properties of multifunctions between Banach spaces, Nonlinear Analysis: Theory, Methods & Applications, 25, 1401–1424.

Mordukhovich, B. S. and Y. Shao [1996a], Nonsmooth sequential analysis in Asplund spaces, Transactions of the American Mathematical Society, 348, 1235–1280.

Mordukhovich, B. S. and Y. Shao [1996b], Nonconvex differential calculus for infinite dimensional multifunctions, Set-Valued Analysis, 4, 205–256.

Mordukhovich, B. S. and Y. Shao [1997a], Fuzzy calculus for coderivatives of multifunctions, Nonlinear Analysis: Theory, Methods & Applications, 29.

Mordukhovich, B. S. and Y. Shao [1997b], Stability of set-valued mappings in infinite dimensions: point criteria and applications, SIAM Journal on Control and Optimization, 35, 285–314.

Moreau, J.-J. [1962], Fonctions convexes duales et points proximaux dans un espace hilbertien, Comptes Rendus de l'Académie des Sciences de Paris, 255, 2897–2899.

Moreau, J.-J. [1963a], Propriétés des applications 'prox', Comptes Rendus de l'Académie des Sciences de Paris, 256, 1069–1071.

Moreau, J.-J. [1963b], Inf-convolution des fonctions numériques sur un espace vectoriel, Comp tes Rendus de l'Académie des Sciences de Paris, 256, 5047–5049.

Moreau, J.-J. [1963c], Remarques sur les fonctions à valeurs dans $[-\infty, \infty]$ définies sur un demi-groupe, Comptes Rendus de l'Académie des Sciences de Paris, 257, 3107–3109.

Moreau, J.-J. [1963d], Fonctionelles sous-différentiables, Comptes Rendus de l'Académie des Sciences de Paris, 257, 4117–4119.

Moreau, J.-J. [1965], Proximité et dualité dans un espace hilbertien, Bulletin de la Société Mathématique de France, 93, 273–299.

Moreau, J.-J. [1967], Fonctionelles convexes, College de France, lecture notes.

Moreau, J.-J. [1970], Inf-convolution, sous-additivité, convexité des fonctions numerique, Journal de Mathématiques Pures et Appliquées, 47, 33–41.

Moreau, J.-J. [1978], Approximation en graphe d'une évolution discontinue, RAIRO, 12, 75–84.

Morrey, C. B. [1952], Quasiconvexity and the lower semicontinuity of multiple integrals, Pacific Journal of Mathematics, 2, 25–53.

Morse, M. [1976], Global Variational Analysis: Weierstrass Integrals on Riemannian Manifolds, Mathematical Notes, 16, Princeton University Press, Princeton, New Jersey.

Mosco, U. [1969], Convergence of convex sets and of solutions of variational inequalities, Advances in Mathematics, 3, 510–585.

Mosco, U. [1971], On the continuity of the Young-Fenchel transform, Journal of Mathematical Analysis and Applications, 35, 518–535.

Mouallif, K. and P. Tossings [1988], Variational metric and exponential penalization, Rapport de Recherche, Ecole Hassania des Travaux Publics, Casablanca.

Moussaoui, M. and A. Seeger [1994], Analyse du second ordre de fonctionnelles intégrales: le cas convexe non différentiable, Comptes Rendus de l'Académie des Sciences de Paris, 318, 613–618.

Nadler, S. B. [1978], Hyperspaces of Sets, Dekker, New York.

Nijenhuis, A. [1974], Strong derivatives and inverse mappings, American Mathematical Monthly, 81, 969–980.

Noll, D. [1992], Generalized second fundamental form for Lipschitzian hypersurfaces by way of second epi-derivatives, Canadian Mathematical Bulletin, 35, 523–536.

Noll, D. [1993], Second order differentiability of integral functionals on Sobolev spaces and L^2-spaces, Journal für die Reine und Angewandte Mathematik, 436, 1–17.

Nurminskii, E. [1978], Continuity of ε-subgradient mappings, Kibernetika (Kiev), 5, 148–149.

Olech, C. [1968], Approximation of set-valued functions by continuous functions, Colloquia Mathematica, 19, 285–293.

Páles, Z. and V. Zeidan [1996], Generalized Hessian for C^{1+} functions in infinite-dimensional normed spaces, Mathematical Programming, 74, 59–78.

Papageorgiou, N. S. and D. A. Kandilakis [1987], Convergence in approximation and nonsmooth analysis, Journal of Approximation Theory, 49, 41–54.

Pennanen, T., M. Théra and R. T. Rockafellar [2002], Convergence of the sum of monotone mappings, Proceedings of the American Mathematical Society, 130, 2261–2269.

Penot, J.-P. [1974], Sous-différentielles de fonctions numériques non-convexes, Comptes Rendus de l'Académie des Sciences de Paris, 278, 1553–1555.

Penot, J.-P. [1978], Calcul sous-différentiel et optimisation, Journal of Functional Analysis, 27, 248–276.

Penot, J.-P. [1981], A characterization of tangential regularity, Nonlinear Analysis: Theory, Methods & Applications, 5, 625–643.

Penot, J.-P. [1982], On regularity conditions in mathematical programming, Mathematical Programming Studies, 19, 167–199.

Penot, J.-P. [1984], Differentiability of relations and differential stability of perturbed optimization problems, SIAM Journal on Control and Optimization, 22, 529–551.

Penot, J.-P. [1989], Metric regularity, openness and Lipschitzean behavior of multifunctions, Nonlinear Analysis: Theory, Methods & Applications, 13, 629–643.

Penot, J.-P. [1991], The cosmic Hausdorff topology, the bounded Hausdorff topology, and the continuity of polarity, Proceedings of the American Mathematical Society, 113, 275–286.

Penot, J.-P. [1992], Second-order generalized derivatives: Comparisons of two types of epiderivatives, in Advances in Optimization, Proceedings, edited by W. Oettli and D. Pallaschke, Lecture Notes in Economics and Mathematical Systems, 382, pp. 52–76.

Penot, J.-P. [1994a], Optimality conditions in mathematical programming and composite optimization, Mathematical Programming, 67, 225–245.

Penot, J.-P. [1994b], Sub-Hessians, super-Hessians and conjugation, Nonlinear Analysis: Theory, Methods & Applications, 23, 689–702.

Penot, J.-P. [1995], On the interchange of subdifferentiation and epi-convergence, Journal of Mathematical Analysis and Applications, 196, 676–698.

Penot, J.-P. [1996], Favorable classes of mappings and multimappings in nonlinear analysis and optimization, Journal of Convex Analysis, 3, 97–116.

Phelps, R. [1978], Gaussian sets and differentiability of Lipschitz functions on Banach spaces, Pacific Journal of Mathematics, 77, 523–531.

Phelps, R. R. [1989], Convex Functions, Monotone Operators, and Differentiability, Lecture Notes in Mathematics, 1364, Springer, New York.

Pliś, A. [1961], Remark on measurable set valued functions, Bulletin de l'Académie Polonaise des Sciences, Série Mathématiques, Astronomie et Physique, 12, 857–859.

Polak, E. [1993], On the use of consistent approximations in the solution of semi-infinite optimization problems, Mathematical Programming, 62, 385–414.

Polak, E. [1997], Optimization: Algorithms and Consistent Approximations, Springer, New York.

Poliquin, R. A. [1990a], Subgradient monotonicity and convex functions, Nonlinear Analysis: Theory, Methods & Applications, 14, 385–398.

Poliquin, R. A. [1990b], Proto-differentiation of subgradient set-valued mappings, Canadian Journal of Mathematics, 42, 520–532.

Poliquin, R. A. [1991], Integration of subdifferentials of nonconvex functions, Nonlinear Analysis: Theory, Methods & Applications, 17, 385–398.

Poliquin, R. A. [1992], An extension of Attouch's theorem and its application to second-order epi-differentiation of convexly composite functions, Transactions of the American Mathematical Society, 332, 861–874.

Poliquin, R. A. and R. T. Rockafellar [1992], Amenable functions in optimization, in Nonsmooth Optimization Methods and Applications, edited by F. Giannessi, pp. 338–353, Gordon and Breach, Philadelphia.

Poliquin, R. A. and R. T. Rockafellar [1993], A calculus of epi-derivatives applicable to optimization, Canadian Journal of Mathematics, 45, 879–896.

Poliquin, R. A. and R. T. Rockafellar [1994], Proto-derivative formulas for basic subgradient mappings in mathematical programming, Set-Valued Analysis, 2, 275–290.

Poliquin, R. A. and R. T. Rockafellar [1996a], Prox-regular functions in variational analysis, Transactions of the American Mathematical Society, 348, 1805–1838.

Poliquin, R. A. and R. T. Rockafellar [1996b], Generalized Hessian properties of regularized nonsmooth functions, SIAM Journal on Optimization, 6, 1121–1137.

Poliquin, R. A. and R. T. Rockafellar [1997], Tilt stability of a local minimum, SIAM Journal on Optimization, 7.

Pompeiu, D. [1905], Fonctions de variables complexes, Annales de la Faculté des Sciences de l'Université de Toulouse, 7, 265–315.

Preiss, D. [1990], Differentiability of Lipschitz functions on Banach spaces, Journal of Functional Analysis, 91, 312–345.

Pshenichnyi, B. [1971], Necessary Conditions for an Extremum, Nauka, Moscow, Russian.

Qi, L. [1990], Semismoothness and decomposition of maximal normal operators, Journal of Mathematical Analysis and Applications, 146, 271–279.

Qi, L. [1993], Convergence Analysis of some algorithms for solving nonsmooth equations, Mathematics of Operations Research, 18, 227–244.

Rådström, H. [1952], An embedding theorem for spaces of convex sets, Proceedings of the American Mathematical Society, 3, 165–169.

Renaud, A. [1993], Algorithmes de Régularisation et Décomposition pour les Problèmes Variationnels Monotones, doctoral thesis, L'École Nationale Supérieure des Mines, Paris.

Richter, H. [1963], Verallgemeinerung einer in der Statistik benötigten Satzes der Masstheorie, Mathematische Annalen, 150, 85–90, 440-441.

Robert, R. [1974], Convergence de fonctionnelles convexes, Journal of Mathematical Analysis and Applications, 45, 533–555.

Robinson, S. M. [1972], Normed convex processes, Transactions of the American Mathematical Society, 174, 127–140.

Robinson, S. M. [1973a], Bounds for error in the solution set of a perturbed linear program, Linear Algebra and its Applications, 6, 69–81.

Robinson, S. M. [1973b], An inverse-function theorem for a class of multivalued functions, Proceedings of the American Mathematical Society, 41, 211–218.

Robinson, S. M. [1975], Stability theory for systems of inequalities, part I: linear systems, SIAM Journal on Numerical Analysis, 12, 754–769.

Robinson, S. M. [1976a], Regularity and stability for convex multivalued functions, Mathematics of Operations Research, 1, 130–143.

Robinson, S. M. [1976b], Stability theory for systems of inequalities, part II: differentiable nonlinear systems, SIAM Journal on Numerical Analysis, 13, 497–513.

Robinson, S. M. [1977], Generalized equations and their solutions, part I: basic theory, Mathematical Programming Studies, 10, 128–141.

Robinson, S. M. [1980], Strongly regular generalized equations, Mathematics of Operations Research, 5, 43–62.

Robinson, S. M. [1981], Some continuity properties of polyhedral multifunctions, Mathematical Programming Studies, 14, 206–214.

Robinson, S. M. [1982], Generalized equations and their solutions, part II: basic theory, Mathematical Programming Studies, 19, 200–221.

Robinson, S. M. [1983], Local structure of feasible sets in nonlinear programming, part I: regularity, in Numerical Methods, edited by V. Pereyra and A. Reinoza, Lecture Notes in Mathematics, 1005, Springer, Berlin.

Robinson, S. M. [1985], Implicit B-differentiability in generalized equations, Mathematics Research Center, University of Wisconsin, Technical Report #2854.

Robinson, S. M. [1987a], Local structure of feasible sets in nonlinear programming, part III: stability and sensitivity, Mathematical Programming Studies, 30, 45–66.

Robinson, S. M. [1987b], Local epi-continuity and local optimization, Mathematical Programming, 37, 208–222.

Robinson, S. M. [1991], An implicit-function theorem for a class of nonsmooth functions, Mathematics of Operations Research, 16, 292–309.

Rockafellar, R. T. [1963], Convex Functions and Dual Extremum Problems, Ph.D. dissertation, Harvard University, Cambridge, Massachusetts.

Rockafellar, R. T. [1964a], Duality theorems for convex functions, Transactions of the American Mathematical Society, 123, 46–63.

Rockafellar, R. T. [1964b], Minimax theorems and conjugate saddle functions, Mathematica Scandinavica, 14, 151–173.

Rockafellar, R. T. [1966a], Characterization of the subdifferentials of convex functions, Pacific Journal of Mathematics, 17, 497–510.

Rockafellar, R. T. [1966b], Level sets and continuity of conjugate convex functions, Transactions of the American Mathematical Society, 123, 46–63.

Rockafellar, R. T. [1966c], Extensions of Fenchel's duality theorem, Duke Mathematics Journal, 33, 81–90.

Rockafellar, R. T. [1967a], Duality and stability in extremum problems involving convex functions, Pacific Journal of Mathematics, 21, 167–187.

Rockafellar, R. T. [1967b], A monotone convex analog of linear algebra, in Proceedings of the Colloquium on Convexity, Copenhagen, 1965, edited by W. Fenchel, pp. 261–276, Matematisk Institut, Copenhagen.

Rockafellar, R. T. [1967c], Monotone Processes of Convex and Concave Type, Memoirs of the American Mathematical Society, 77, American Mathematical Society, Providence, Rhode Island.

Rockafellar, R. T. [1967d], Conjugates and Legendre transforms of convex functions, Canadian Journal of Mathematics, 19, 200–205.

Rockafellar, R. T. [1968a], Duality in nonlinear programming, in Mathematics of the Decision Sciences, edited by G. B. Dantzig and A. F. Veinott, AMS Lectures in Applied Mathematics, II, pp. 401–422.

Rockafellar, R. T. [1968b], Integrals which are convex functionals, Pacific Journal of Mathematics, 24, 252–540.

Rockafellar, R. T. [1969a], Convex functions, monotone operators and variational inequalities, in Theory and Applications of Monotone Operators, edited by A. Ghizzetti, pp. 34–65, Tipografia Oderisi Editrice, Gubbio, Italy.

Rockafellar, R. T. [1969b], Local boundedness of nonlinear monotone operators, Michigan Mathematics Journal, 16, 397–407.

Rockafellar, R. T. [1969c], Measurable dependence of convex sets and functions on parameters, Journal of Mathematical Analysis and Applications, 28, 4–25.

Rockafellar, R. T. [1970a], Convex Analysis, Princeton University Press, Princeton, New Jersey.

Rockafellar, R. T. [1970b], Conjugate convex functions in optimal control and the calculus of variations, Journal of Mathematical Analysis and Applications, 32, 174–222.

Rockafellar, R. T. [1970c], On the maximality of sums of nonlinear monotone operators, Transactions of the American Mathematical Society, 149, 75–88.

Rockafellar, R. T. [1970d], On the maximal monotonicity of subdifferential mappings, Pacific Journal of Mathematics, 33, 209–216.

Rockafellar, R. T. [1970e], On the virtual convexity of the domain and range of a nonlinear maximal monotone operator, Mathematische Annalen, 185, 81–90.

Rockafellar, R. T. [1970f], Monotone operators associated with saddle-functions and minimax problems, in Nonlinear Functional Analysis, edited by F. Browder, Proceedings of Symposia in Pure Mathematics, 18(1), pp. 241–250, American Mathematical Society, Providence, Rhode Island.

Rockafellar, R. T. [1971a], Existence and duality theorems for convex problems of Bolza, Transactions of the American Mathematical Society, 159, 1–40.

Rockafellar, R. T. [1971b], Integrals which are convex functionals, II, Pacific Journal of Mathematics, 39, 439–469.

Rockafellar, R. T. [1971c], Convex integral functionals and duality, in Contributions to Nonlinear Functional Analysis, edited by E. Zarantonello, pp. 215–236, Academic Press.

Rockafellar, R. T. [1972], State constraints in convex problems of Bolza, SIAM Journal on Control, 10, 691–715.

Rockafellar, R. T. [1974a], Conjugate Duality and Optimization, Conference Board of Mathematical Sciences Series, 16, SIAM Publications, Philadelphia.

Rockafellar, R. T. [1974b], Augmented Lagrange multiplier functions and duality in nonconvex programming, SIAM Journal on Control and Optimization, 12, 268–285.

Rockafellar, R. T. [1975], Existence theorems for general control problems of Bolza and Lagrange, Advances in Mathematics, 15, 315–333.

Rockafellar, R. T. [1976a], Integral functionals, normal integrands and measurable selections, in Nonlinear Operators and the Calculus of Variations; Bruxelles 1975, Lecture Notes in Mathematics, 543, pp. 157–207, Springer, Berlin.

Rockafellar, R. T. [1976b], Monotone operators and the proximal point algorithm, SIAM Journal on Control and Optimization, 14, 877–898.

Rockafellar, R. T. [1976c], Augmented Lagrangians and applications of the proximal point algorithm in convex programming, Mathematics of Operations Research, 1, 97–116.

Rockafellar, R. T. [1979a], Directionally Lipschitzian functions and subdifferential calculus, Proceedings of the London Mathematical Society, 39, 331–355.

Rockafellar, R. T. [1979b], Clarke's tangent cones and the boundaries of closed sets in R^n, Nonlinear Analysis: Theory, Methods & Applications, 3, 145–154.

Rockafellar, R. T. [1980], Generalized directional derivatives and subgradients of nonconvex functions, Canadian Journal of Mathematics, 32, 157–180.

Rockafellar, R. T. [1981a], The Theory of Subgradients and its Applications to Problems of Optimization: Convex and Nonconvex Functions, Helderman Verlag, Berlin.

Rockafellar, R. T. [1981b], Proximal subgradients, marginal values and augmented Lagrangians in nonconvex optimization, Mathematics of Operations Research, 6, 424–436.

Rockafellar, R. T. [1982a], Lagrange multipliers and subderivatives of optimal value functions in nonlinear programming, Mathematical Programming Studies, 17, 28–66.

Rockafellar, R. T. [1982b], Favorable classes of Lipschitz continuous functions in subgradient optimization, in Progress in Nondifferentiable Optimization, edited by E. Nurminski, pp. 125–143, IIASA, Laxenburg, Austria.

Rockafellar, R. T. [1983], Generalized subgradients in mathematical programming, in Mathematical Programming, Bonn 1982, edited by A. Bachem, M. Groetschel and B. Korte, pp. 368–380, Springer, New York.

Rockafellar, R. T. [1984a], Directional differentiability of the optimal value function in a nonlinear programming problem, Mathematical Programming Studies, 21, 213–226.

Rockafellar, R. T. [1984b], Network Flows and Monotropic Optimization, Wiley, New York.

Rockafellar, R. T. [1985a], Extensions of subgradient calculus with applications to optimization, Nonlinear Analysis: Theory, Methods & Applications, 9, 665–698.

Rockafellar, R. T. [1985b], Maximal monotone relations and the second derivatives of nonsmooth functions, Annales de l'Institut H. Poincaré: Analyse Non Linéaire, 2, 167–184.

Rockafellar, R. T. [1985c], Lipschitzian properties of multifunctions, Nonlinear Analysis: Theory, Methods & Applications, 9, 867–885.

Rockafellar, R. T. [1987], Linear-quadratic programming and optimal control, SIAM Journal on Control and Optimization, 25, 781–814.

Rockafellar, R. T. [1988], First and second-order epi-differentiability in nonlinear programming, Transactions of the American Mathematical Society, 307, 75–108.

Rockafellar, R. T. [1989a], Second-order optimality conditions in nonlinear programming obtained by way of epi-derivatives, Mathematics of Operations Research, 14, 462–484.

Rockafellar, R. T. [1989b], Proto-differentiability of set-valued mappings and its applications in optimization, in Analyse Non Linéaire, edited by H. Attouch, J.-P. Aubin, F. Clarke and I. Ekeland, pp. 449–482, Gauthier-Villars, Paris.

Rockafellar, R. T. [1990a], Nonsmooth analysis and parametric optimization, in Methods of Nonconvex Analysis, edited by A. Cellina, Lecture Notes in Mathematics, 1446, pp. 137–151, Springer .

Rockafellar, R. T. [1990b], Generalized second derivatives of convex functions and saddle functions, Transactions of the American Mathematical Society, 322, 51–77.

Rockafellar, R. T. [1993a], Lagrange multipliers and optimality, SIAM Review, 35, 183–238.

Rockafellar, R. T. [1993b], Dualization of subgradient conditions for optimality, Nonlinear Analysis: Theory, Methods & Applications, 20, 627–646.

Rockafellar, R. T. [1996], Equivalent subgradient versions of Hamiltonian and Euler-Lagrange equations in variational analysis, SIAM Journal on Control and Optimization, 34, 1300–1315.

Rockafellar, R. T. and R. J-B. Wets [1976a], Stochastic convex programming: basic duality, Pacific Journal of Mathematics, 62, 173–195.

Rockafellar, R. T. and R. J-B. Wets [1976b], Stochastic convex programming: singular multipliers and extended duality, Pacific Journal of Mathematics, 62, 507–522.

Rockafellar, R. T. and R. J-B. Wets [1976c], Stochastic convex programming: relatively complete recourse and induced feasibility, SIAM Journal on Control and Optimization, 14, 547–589.

Rockafellar, R. T. and R. J-B. Wets [1976d], Nonanticipativity and L^∞-martingales in stochastic optimization problems, Mathematical Programming Studies, 6, 170–187.

Rockafellar, R. T. and R. J-B. Wets [1977], Stochastic convex programming: Kuhn-Tucker conditions, Journal of Mathematical Economics, 2, 349–370.

Rockafellar, R. T. and R. J-B. Wets [1984], Variational systems, an introduction, in Multifunctions and Integrands: Stochastic Analysis, Approximation and Optimization, edited by G. Salinetti, Lecture Notes in Mathematics, 1091, pp. 1–54, Springer Verlag , Berlin.

Rockafellar, R. T. and R. J-B. Wets [1986], A Lagrangian finite generation technique for solving linear-quadratic problems in stochastic programming, Mathematical Programming Studies, 28, 55–73.

Rockafellar, R. T. and R. J-B. Wets [1992], Cosmic convergence, in Optimization and Nonlinear Analysis, edited by A. Ioffe, M. Marcus and S. Reich, Pitman Research Notes in Mathematics, 244, pp. 249–272, Longman Scientific and Technical, Harlow, UK.

Rockafellar, R. T. and D. Zagrodny [1997], A derivative coderivative inclusion in second-order nonsmooth analysis, Set-Valued Analysis, 5, 1–17.

Rokhlin, V. A. [1949], Selected topics from the metric theory of dynamical systems, Uspekhi Matematicheskii Nauk, 4, 57–128, Russian.

Sainte-Beuve, M. F. [1974], Sur la généralisation d'un théorème de section mesurable de von Neumann-Aumann, Journal of Mathematical Analysis and Applications, 17, 112–129.

Salinetti, G. [1984], Multifunctions and Integrands: Stochastic Analysis, Approximation and Optimization, Lecture Notes in Mathematics, 1091, Springer, Berlin.

Salinetti, G. and R. J-B. Wets [1977], On the relation between two types of convergence for convex functions, Journal of Mathematical Analysis and Applications, 60, 211–226.

Salinetti, G. and R. J-B. Wets [1979], On the convergence of sequences of convex sets in finite dimensions, SIAM Review, 21, 16–33.

Salinetti, G. and R. J-B. Wets [1981], On the convergence of closed-valued measurable multifunctions, Transactions of the American Mathematical Society, 266, 275–289.

Salinetti, G. and R. J-B. Wets [1986], On the convergence in distribution of measurable multifunctions (random sets), normal integrands, stochastic processes and stochastic infima, Mathematics of Operations Research, 11, 385–419.

Sbordone, C. [1974], Su alcune applicazioni di un tipo di convergenza variazionale, Annali Scuola Normale Superiore-Pisa, 2, 617–638.

Schwartz, L. [1966], Théorie des Distributions, Hermann, Paris.

Scorza-Dragoni, G. [1948], Un teorema sulla funzioni continue rispetto ad una e misurabili rispetto ad un' altra variabile, Rend. Sem. Math. Univ. Padova, 17, 219–224.

Seeger, A. [1988], Second-order directional derivatives in parametric optimization problems, Mathematics of Operations Research, 13, 124–139.

Seeger, A. [1992a], Limiting behavior of the approximate second-order subdifferential of a convex function, Journal of Optimization Theory and Applications, 74, 527–544.

Seeger, A. [1992b], Second derivatives of a convex function and of its Legendre-Fenchel transformate, SIAM Journal on Optimization, 2, 405–424.

Seeger, A. [1994], Second-order normal vectors to a convex epigraph, Bulletin of the Australian Mathematical Society, 50, 123–134.

Sendov, B. [1990], Hausdorff Approximations, Kluwer, Dordrecht, Holland.

Severi, F. [1930], Le curve intuitive, Rendicoti Circolo Matematica Palermo, 54, 213–228.

Severi, F. [1930], Su alcune questioni de topologia infinitesimale, Annals de la Société Polonnaise de Mathématique, 9, 340–348.

Severi, F. [1934], Sull differenziabilitá totale delle funzioni di piú variabili reali, Annali di Matematica, 4, 1–35.

Shapiro, A. [1990], On concepts of directional differentiability, Journal of Mathematical Analysis and Applications, 66, 477–487.

Shunmugaraj, P. and D. V. Pai [1991], On stability of approximate solutions of minimization problems, Numerical Functional Analysis and Optimization, 12, 559–610.

Simons, S. [1996], The range of a monotone operator, Journal of Mathematical Analysis and Applications, 199, 176–201.

Slater, M. [1950], Lagrange multipliers revisited: a contribution to non-linear programming, Cowles Commission Discussion Paper, Math. 403.

Sobolev, S. L. [1963], Applications of Functional Analysis in Mathematical Physics, Translations of Mathematical Monographs, 7, American Mathematical Society, Providence.

Sonntag, Y. [1976], Interprétation géométrique de la convergence d'une suite de convexes, Comptes Rendus de l'Académie des Sciences de Paris, 282, 1099–1100.

Sonntag, Y. and C. Zălinescu [1993], Set convergence, an attempt of classification, Transactions of the American Mathematical Society, 340, 199–226.

Spagnolo, S. [1976], Convergence in energy for elliptic operators, in Proceedings of the Third Symposium on the Numerical Solutions of Partial Differential Equations, pp. 469–498, Academic Press, New York.

Spanier, E. H. [1966], Algebraic Topology, McGraw-Hill, New York.

Spingarn, J. E. [1981], Submonotone subdifferentials of Lipschitz functions, Transactions of the American Mathematical Society, 264, 77–89.

Spingarn, J. E. [1983], Partial inverse of a monotone mapping, Applied Mathematics and Optimization, 10, 247–265.

Stampacchia, G. [1964], Formes bilinéaires coercitives sur les ensemble convexes, Comptes Rendus de l'Académie des Sciences de Paris, 258, 4413–4416.

Steinitz, E. [1913-16], Bedingt konvergente Reihen und konvexe Systeme, I, II, III, Journal der Reine und Angewandte Mathematik, 143-144-146, 128–175;1-40;1-52.

Steklov, V. A. [1907], On the asymptotic expressions of certain functions defined by differential equations of the second order and their application to the development in series of arbitrary functions, Communication of the Charkov Mathematical Society, Series 2, 10, 97–199, Russian.

Stoker, J. J. [1940], Unbounded convex sets, American Journal of Mathematics, 62, 165–179.

Strömberg, T. [1994], A Study of the Operation of Infimal Convolution, doctoral thesis, Luleå University of Technology, Luleå, Sweden.

Strömberg, T. [1996], On regularization in Banach spaces, Arkiv för Matematik, 34, 383–406.

Studniarski, M. [1991], Second order necessary conditions for optimality in nonsmooth nonlinear programming, Journal of Mathematical Analysis and Applications, 154, 303–317.

Sun, J. [1986], On monotropic piecewise linear-quadratic programming, Ph.D. dissertation, University of Washington, Seattle.

Teboulle, M. [1992], Entropic proximal mappings with applications to nonlinear programming, Mathematics of Operations Research, 17, 670–690.

Thibault, L. [1978], Sous-différentiels de fonctions vectorielles compactement lipschitziennes, Comptes Rendus de l'Académie des Sciences de Paris, 286, 995–998.

Thibault, L. [1980], Subdifferentials of compactly Lipschitzian vector-valued functions, Annali di Matematica pura ed applicata, CXXV, 157–192.

Thibault, L. [1982], On generalized differentials and subdifferentials of Lipschitz vector-valued functions, Nonlinear Analysis, 6, 1037–1053.

Thibault, L. [1983], Tangent cones and quasi-interiorly tangent cones to multifunctions, Transactions of the American Mathematical Society, 277, 601–621.

Thibault, L., D. Zagrodny [1995], Integration of subdifferentials of lower semicontinuous functions, Journal of Mathematical Analysis and Applications, 189, 33–58.

Toland, J. F. [1978], Duality in non-convex optimization, Journal of Mathematical Analysis and Applications, 66, 399–415.

Toland, J. F. [1979], A duality principle for non-convex optimization and the calculus of variations, Archive of Rational Mechanics and Analysis, 71, 41–61.

Treiman, J. S. [1983a], Characterization of Clarke's tangent and normal cones in finite and infinite dimensions, Nonlinear Analysis: Theory, Methods & Applications, 7, 771–783.

Treiman, J. S. [1983b], A New Characterization of Clarke's Tangent Cone and its Applications to Subgradient Analysis and Optimization, Ph.D. dissertation, University of Washington, Seattle.

Treiman, J. S. [1986], Clarke's gradients and ε-subgradients in Banach spaces, Transactions of the American Mathematical Society, 294, 65–78.

Treiman, J. S. [1988], Shrinking generalized gradients, Nonlinear Analysis: Theory, Methods & Applications, 12, 1429–1450.

Tsukada, M. [1984], Convergence of best approximations in Banach spaces, Journal of Approximation Theory, 40, 304–309.

Ursescu, C. [1975], Multifunctions with closed convex graph, Czechoslovak Mathematical Journal, 25, 438–441.

Valadier, M. [1974], Sur le théorème de Strassen, Comptes Rendus de l'Académie des Sciences de Paris, 278, 1021–1024.

Vasilesco, F. [1925], Essai sur les fonctions multiformes de variables réelles, Thèse, Université de Paris, Gauthier-Villars & Co., Paris.

Vervaat, W. [1988], Random upper semicontinuous functions and extremal processes, Report # MS-8801, Centrum voor Wiskunde en Informatica, Amsterdam.

Vietoris, L. [1921], Berichte zweite Ordnung, Monatshefte für Mathematik und Physik, 31, 173–204.

Volle, M. [1984], Convergence en niveaux et en épigraphes, Comptes Rendus de l'Académie des Sciences de Paris, 299, 295–298.

Von Neumann, J. [1928], Zur Theorie der Gesellschaftsspiele, Mathematische Annalen, 100, 295–320.

Wagner, D. H. [1977], Survey of measurable selection theorems, SIAM Journal on Control and Optimization, 15, 859–903.

Walkup, D. W. and R. J-B. Wets [1967], Continuity of some convex-cone-valued mappings, Proceedings of the American Mathematical Society, 18, 229–235.

Walkup, D. W. and R. J-B. Wets [1968], A note on decision rules for stochastic programs, J. Computer Science, 2, 305–311.

Walkup, D. W. and R. J-B. Wets [1969a], A Lipschitzian characterization of convex polyhedra, Proceedings of the American Mathematical Society, 23, 167–173.

Walkup, D. W. and R. J-B. Wets [1969b], Some practical regularity conditions for nonlinear programs, SIAM Journal on Control, 7, 430–436.

Ward, D. E. [1991], Chain rules for nonsmooth functions, Journal of Mathematical Analysis and Applications, 158, 519–538.

Ward, D. E. and J. M. Borwein [1987], Nonsmooth calculus in finite dimensions, SIAM Journal on Control and Optimization, 25, 1312–1340.

Warga, J. [1975], Necessary conditions without differentiability assumptions in optimal control, Journal of Differential Equations, 15, 41–61.

Warga, J. [1976], Derivate containers, inverse functions, and controllability, in Calculus of Variations and Control Theory, edited by D. L. Russell, Academic Press, New York.

Warga, J. [1981], Fat homeomorphisms and unbounded derivate containers, Journal of Mathematical Analysis and Applications, 81, 545–560.

Ważewski, T. [1961a], Systèmes de commandes et équations au contingent, Bulletin de l'Académie Polonaise des Sciences, Serie Mathématiques, Astronomie et Physique, 9, 151–155.

Ważewski, T. [1961b], Sur une condition d'existence des fonctions implicites mesurables, Bulletin de l'Académie Polonaise de Sciences, Serie Mathématiques, Astronomie et Physique, 9, 861–863.

Ważewski, T. [1962], Sur une généralisation de la notion des solutions d'une équation au contingent, Bulletin de l'Académie Polonaise des Sciences, Serie Mathématiques, Astronomie et Physique, 10, 11–15.

Weiss, E. A. [1969], Konjugierte Funktionen, Archiv der Mathematik, 20, 538–545.

Weiss, E. A. [1974], Verallgemeinerte konvexe Funktionen, Mathematika Scandinavica, 35, 129–144.

Wets, R. J-B. [1974], On inf-compact mathematical programs, in Fifth Conference on Optimization Techniques, edited by R. Conti and A. Ruberti, Lecture Notes in Computer Science, 5, pp. 426–436, Springer, New York.

Wets, R. J-B. [1980], Convergence of convex functions, variational inequalities and convex optimization problems, in Variational Inequalities and Complementarity Problems, edited by R. Cottle, F. Giannessi and J.-L. Lions, pp. 375–403, Wiley, NewYork.

Wets, R. J-B. [1981], A formula for the level sets of epi-limits and some applications, in Mathematical Theories of Optimization, edited by P. Cecconi and T. Zolezzi, Lecture Notes in Mathematics, 979, pp. 258–268, Springer, Berlin.

Weyl, H. [1935], Elementare Theorie der konvexen Polyeder, Commentarii Mathematici Helvetici, 7, 290–306.

Wicks, K. R. [1994], Topologies on sets of packings and tilings, Mathematics Report no. PM-94-04, University of Bristol.

Wijsman, R. A. [1964], Convergence of sequences of convex sets, cones and functions, Bulletin of the American Mathematical Society, 70, 186–188.

Wijsman, R. A. [1966], Convergence of sequences of convex sets, cones and functions. II, Transactions of the American Mathematical Society, 123, 32–45.

Williams, A. C. [1970], Nonlinear activity analysis and duality, Management Science, 17, 127–137.

Yang, X. Q. [1994], Generalized second-order derivatives and optimality conditions, Nonlinear Analysis: Theory, Methods & Applications, 23, 767–784.

Yang, X. Q. On second-order derivatives, Nonlinear Analysis: Theory, Methods & Applications, 26, 55–66.

Yang, X. Q. and V. Jeyakumar [1992], Generalized second-order directional derivatives and optimization with $C^{1,1}$ functions, Optimization, 26, 165–185.

Yang, X. Q. and V. Jeyakumar [1995], Convex composite minimization with $C^{1,1}$ functions, Journal of Optimization Theory and Applications, 86, 631–648.

Yen, N. D. [1992], Hölder continuity of solutions to a parametric variational inequality, Technical Report 3.188(706), Universitá di Pisa.

Yosida, K. [1971], Functional Analysis, Springer, New York, 3rd edition.

Young, W. H. [1912], On classes of summable functions and their Fourier series, Proceedings of the Royal Society (A), 87, 225–229.

Yuan, G.X. [1999], KKM Theory and Applications in Nonlinear Analysis, Marcel Dekker, New York.

Zagrodny, D. [1988], Approximate mean value theorem for upper subderivatives, Nonlinear Analysis: Theory, Methods & Applications, 12, 1413–1428.

Zagrodny, D. [1992], Recent mean value theorems in nonsmooth analysis, in Nonsmooth Optimization Methods and Applications, edited by F. Giannessi, pp. 421–425, Gordon and Breach, Philadelphia.

Zagrodny, D. [1994], The cancellation law for inf-convolution of convex functions, Studia Mathematica, 110, 271–282.

Zălinescu, C. [1989], Stability for a class of nonlinear optimization problems and applications, in Nonsmooth Optimization and Related Topics, edited by F. H. Clarke, V. F. Demyanov and F. Gianessi, pp. 437–458, Plenum Press, New York.

Zarankiewicz, C. [1928], Über eine topologische Eigenschaft der Ebene, Fundamenta Mathematicae, 11, 19–26.

Zarantonello, E. [1960], Solving functional equations by contractive averaging, University of Wisconsin, MRC Report 160, Madison, Wisconsin.

Zeidler, E. [1990a], Linear Monotone Operators, Nonlinear Functional Analysis and its Applications, II/A, Springer, New York.

Zeidler, E. [1990b], Nonlinear Monotone Operators, Nonlinear Functional Analysis and its Applications, II/B, Springer, New York.

Zolezzi, T. [1985], Continuity of generalized gradients and multipliers under perturbations, Mathematics of Operations Research, 10, 664–673.

Zoretti, L. [1909], Un théorème de la théorie des ensembles, Bulletin de la Société Mathéma tique de France, 37, 116–119.

Index of Statements

1.1 Example (equality and inequality constraints).
1.2 Example (box constraints).
1.3 Example (penalties and barriers).
1.4 Example (principle of abstract minimization).
1.5 Definition (lower limits and lower semicontinuity).
1.6 Theorem (characterization of lower semicontinuity).
1.7 Lemma (characterization of lower limits).
1.8 Definition (level boundedness).
1.9 Theorem (attainment of a minimum).
1.10 Corollary (lower bounds).
1.11 Example (level boundedness relative to constraints).
1.12 Exercise (continuity of functions).
1.13 Exercise (closures and interiors of epigraphs).
1.14 Exercise (growth properties).
1.15 Exercise (lower and upper limits of sequences).
1.16 Definition (uniform level boundedness).
1.17 Theorem (parametric minimization).
1.18 Proposition (epigraphical projection).
1.19 Exercise (constraints in parametric minimization).
1.20 Example (distance functions and projections).
1.21 Example (convergence of penalty methods).
1.22 Definition (Moreau envelopes and proximal mappings).
1.23 Definition (prox-boundedness).
1.24 Exercise (characteristics of prox-boundedness).
1.25 Theorem (proximal behavior).
1.26 Proposition (semicontinuity under pointwise max and min).
1.27 Proposition (properties of epi-addition).
1.28 Exercise (sums and multiples of epigraphs).
1.29 Exercise (calculus of indicators, distances, and envelopes).
1.30 Example (vector-max and log-exponential functions).
1.31 Exercise (images of epigraphs).
1.32 Proposition (properties of epi-composition).
1.33 Definition (local semicontinuity).
1.34 Exercise (consequences of local semicontinuity).
1.35 Proposition (interchange of order of minimization).
1.36 Exercise (max and min of a sum).
1.37 Exercise (scalar multiples versus max and min).
1.38 Proposition (lower limits of sums).
1.39 Proposition (semicontinuity of sums, scalar multiples).
1.40 Exercise (semicontinuity under composition).
1.41 Exercise (level-boundedness of sums and epi-sums).
1.42 Theorem (minimizing sequences).
1.43 Proposition (variational principle; Ekeland).
1.44 Example (proximal hulls).
1.45 Exercise (proximal inequalities).
1.46 Example (double envelopes).

2.1 Definition (convex sets and convex functions).
2.2 Theorem (convex combinations and Jensen's inequality).
2.3 Exercise (effective domains of convex functions).
2.4 Proposition (convexity of epigraphs).
2.5 Exercise (improper convex functions).
2.6 Theorem (characteristics of convex optimization).
2.7 Proposition (convexity of level sets).
2.8 Example (affine functions, half-spaces and hyperplanes).
2.9 Proposition (intersection, pointwise supremum and pointwise limits).
2.10 Example (polyhedral sets and affine sets).
2.11 Exercise (characterization of affine sets).
2.12 Lemma (slope inequality).
2.13 Theorem (one-dimensional derivative tests).
2.14 Theorem (higher-dimensional derivative tests).
2.15 Example (linear-quadratic functions).
2.16 Example (convexity of vector-max and log-exponential).
2.17 Example (norms).
2.18 Exercise (addition and scalar multiplication).
2.19 Exercise (set products and separable functions).
2.20 Exercise (convexity in composition).
2.21 Proposition (images under linear mappings).
2.22 Proposition (convexity in inf-projection and epi-composition).
2.23 Proposition (convexity in set algebra).
2.24 Exercise (epi-sums and epi-multiples).
2.25 Example (distance functions and projections).
2.26 Theorem (proximal mappings and envelopes under convexity).
2.27 Theorem (convex hulls from convex combinations).
2.28 Exercise (simplex technology).
2.29 Theorem (convex hulls from simplices; Carathéodory).
2.30 Corollary (compactness of convex hulls).
2.31 Proposition (convexification of a function).
2.32 Proposition (convexity and properness of closures).
2.33 Theorem (line segment principle).
2.34 Proposition (interiors of epigraphs and level sets).
2.35 Theorem (continuity properties of convex functions).
2.36 Corollary (finite convex functions).
2.37 Corollary (convex functions of a single real variable).
2.38 Example (discontinuity and unboundedness).
2.39 Theorem (separation).
2.40 Proposition (relative interiors of convex sets).
2.41 Exercise (relative interior criterion).
2.42 Proposition (relative interiors of intersections).
2.43 Proposition (relative interiors in product spaces).
2.44 Proposition (relative interiors of set images).
2.45 Exercise (relative interiors in set algebra).
2.46 Exercise (closures of functions).
2.47 Definition (piecewise linearity).
2.48 Exercise (graphs of piecewise linear mappings).
2.49 Theorem (convex piecewise linear functions).
2.50 Lemma (polyhedral convexity of unions).
2.51 Exercise (functions with convex level sets).
2.52 Example (posynomials).
2.53 Example (weighted geometric means).
2.54 Exercise (eigenvalue functions).
2.55 Exercise (matrix inversion as a gradient mapping).
2.56 Proposition (convexity property of matrix inversion).
2.57 Exercise (convex functions of positive-definite matrices).

3.1 Definition (cosmic convergence to direction points).
3.2 Theorem (compactness of cosmic space).
3.3 Definition (horizon cones).
3.4 Exercise (cosmic closedness).
3.5 Theorem (horizon criterion for boundedness).
3.6 Theorem (horizon cones of convex sets).
3.7 Exercise (convex cones).
3.8 Proposition (convex cones versus subspaces).
3.9 Proposition (intersections and unions).
3.10 Theorem (images under linear mappings).
3.11 Exercise (products of sets).
3.12 Exercise (sums of sets).
3.13 Definition (pointed cones).
3.14 Proposition (pointedness of convex cones).
3.15 Theorem (convex hulls of cones).
3.16 Exercise (images under nonlinear mappings).
3.17 Definition (horizon functions).
3.18 Definition (positive homogeneity and sublinearity).
3.19 Exercise (epigraphs of positively homogeneous functions).
3.20 Exercise (sublinearity versus linearity).
3.21 Theorem (properties of horizon functions).
3.22 Corollary (monotonicity on lines).
3.23 Proposition (horizon cone of a level set).
3.24 Exercise (horizon cones from constraints).
3.25 Definition (coercivity properties).
3.26 Theorem (horizon functions and coercivity).
3.27 Corollary (coercivity and convexity).
3.28 Example (coercivity and prox-boundedness).
3.29 Exercise (horizon functions in addition).
3.30 Proposition (horizon functions in pointwise max and min).
3.31 Theorem (coercivity in parametric minimization).
3.32 Corollary (boundedness in convex parametric minimization).
3.33 Corollary (coercivity in epi-addition).
3.34 Theorem (cancellation in epi-addition).
3.35 Corollary (cancellation in set addition).
3.36 Corollary (functions determined by their Moreau envelopes).
3.37 Corollary (functions determined by their proximal mappings).
3.38 Proposition (vector inequalities).
3.39 Example (matrix inequalities).
3.40 Exercise (operations on cones and positively homogeneous functions).
3.41 Definition (convexity in cosmic space).
3.42 Exercise (cone characterization of cosmic convexity).
3.43 Exercise (extended line segment principle).
3.44 Exercise (cosmic convex hulls).
3.45 Proposition (extended expression of convex hulls).
3.46 Corollary (closures of convex hulls).
3.47 Corollary (convex hulls of coercive functions).
3.48 Exercise (closures of positive hulls).
3.49 Exercise (convexity of positive hulls).
3.50 Example (gauge functions).
3.51 Exercise (entropy functions).
3.52 Theorem (generation of polyhedral cones; Minkowski-Weyl).
3.53 Corollary (polyhedral sets as convex hulls).
3.54 Exercise (generation of convex, piecewise linear functions).
3.55 Proposition (polyhedral and piecewise linear operations).

4.1 Definition (inner and outer limits).
4.2 Exercise (characterizations of set limits).
4.3 Exercise (limits of monotone and sandwiched sequences).

4.4 Proposition (closedness of limits).
4.5 Theorem (hit-and-miss criteria).
4.6 Exercise (index criterion for convergence).
4.7 Corollary (pointwise convergence of distance functions).
4.8 Exercise (equality in distance limits).
4.9 Proposition (set convergence through projections).
4.10 Theorem (uniformity of approximation in set convergence).
4.11 Corollary (escape to the horizon).
4.12 Corollary (limits of connected sets).
4.13 Example (Pompeiu-Hausdorff distance).
4.14 Exercise (limits of cones).
4.15 Proposition (limits of convex sets).
4.16 Exercise (convergence of convex sets through truncations).
4.17 Exercise (limits of star-shaped sets).
4.18 Theorem (extraction of convergent subsequences).
4.19 Proposition (cluster description of inner and outer limits).
4.20 Exercise (cosmic limits through horizon limits).
4.21 Exercise (properties of horizon limits).
4.22 Example (eventually bounded sequences).
4.23 Definition (total set convergence).
4.24 Proposition (horizon criterion for total convergence).
4.25 Theorem (automatic cases of total convergence).
4.26 Theorem (convergence of images).
4.27 Theorem (total convergence of linear images).
4.28 Example (convergence of projections of convex sets).
4.29 Exercise (convergence of products and sums).
4.30 Proposition (convergence of convex hulls).
4.31 Exercise (convergence of unions).
4.32 Theorem (convergence of solutions to convex systems).
4.33 Exercise (convergence of convex intersections).
4.34 Lemma (distance function relations).
4.35 Theorem (uniformity in convergence of distance functions).
4.36 Theorem (quantification of set convergence).
4.37 Proposition (distance estimates).
4.38 Corollary (Pompeiu-Hausdorff distance as a limit).
4.39 Proposition (distance between convex truncations).
4.40 Exercise (properties of Pompeiu-Hausdorff distance).
4.41 Lemma (estimates for the integrated set distance).
4.42 Theorem (metric description of set convergence).
4.43 Corollary (local compactness in metric spaces of sets).
4.44 Example (distances between cones).
4.45 Proposition (separability and approximation by finite sets).
4.46 Theorem (metric description of cosmic set convergence).
4.47 Corollary (metric description of total set convergence).
4.48 Exercise (cosmic metric properties).

5.1 Example (constraint systems).
5.2 Example (generalized equations and implicit mappings).
5.3 Example (algorithmic mappings and fixed points).
5.4 Definition (continuity and semicontinuity).
5.5 Example (profile mappings).
5.6 Exercise (criteria for semicontinuity at a point).
5.7 Theorem (characterizations of semicontinuity).
5.8 Example (feasible-set mappings).
5.9 Theorem (inner semicontinuity from convexity).
5.10 Example (parameterized convex constraints).
5.11 Proposition (continuity of distances).
5.12 Proposition (uniformity of approximation in semicontinuity).
5.13 Exercise (uniform continuity).

5.14 Definition (local boundedness).
5.15 Proposition (boundedness of images).
5.16 Exercise (local boundedness of inverses).
5.17 Example (level boundedness as local boundedness).
5.18 Theorem (horizon criterion for local boundedness).
5.19 Theorem (outer semicontinuity under local boundedness).
5.20 Corollary (continuity of single-valued mappings).
5.21 Corollary (continuity of locally bounded mappings).
5.22 Example (optimal-set mappings).
5.23 Example (proximal mappings and projections).
5.24 Exercise (continuity of perturbed mappings).
5.25 Theorem (closedness of images).
5.26 Exercise (horizon criterion for a closed image).
5.27 Proposition (cosmic semicontinuity).
5.28 Definition (total continuity).
5.29 Proposition (criteria for total continuity).
5.30 Exercise (images of converging sets).
5.31 Definition (pointwise limits of mappings).
5.32 Definition (graphical limits of mappings).
5.33 Proposition (graphical limit formulas at a point).
5.34 Exercise (uniformity in graphical convergence).
5.35 Example (graphical convergence of projection mappings).
5.36 Theorem (extraction of graphically convergent subsequences).
5.37 Theorem (approximation of generalized equations).
5.38 Definition (equicontinuity properties).
5.39 Exercise (equicontinuity of single-valued mappings).
5.40 Theorem (graphical versus pointwise convergence).
5.41 Definition (continuous and uniform limits of mappings).
5.42 Exercise (distance function descriptions of convergence).
5.43 Theorem (continuous versus uniform convergence).
5.44 Theorem (graphical versus continuous convergence).
5.45 Corollary (graphical convergence of single-valued mappings).
5.46 Proposition (graphical convergence from uniform convergence).
5.47 Theorem (Arzelà-Ascoli, set-valued version).
5.48 Theorem (uniform convergence of monotone sequences).
5.49 Proposition (metric version of continuous and uniform convergence).
5.50 Theorem (quantification of graphical convergence).
5.51 Proposition (addition of set-valued mappings).
5.52 Proposition (composition of set-valued mappings).
5.53 Theorem (images of converging mappings).
5.54 Exercise (convergence of positive hulls).
5.55 Theorem (generic continuity from semicontinuity).
5.56 Corollary (generic continuity of extended-real-valued functions).
5.57 Example (projections as continuous selections).
5.58 Theorem (Michael representations of isc mappings).
5.59 Corollary (extensions of continuous selections).

6.1 Definition (tangent vectors and geometric derivability).
6.2 Proposition (tangent cone properties).
6.3 Definition (normal vectors).
6.4 Definition (Clarke regularity of sets).
6.5 Proposition (normal cone properties).
6.6 Proposition (limits of normal vectors).
6.7 Exercise (change of coordinates).
6.8 Example (tangents and normals to smooth manifolds).
6.9 Theorem (tangents and normals to convex sets).
6.10 Example (tangents and normals to boxes).
6.11 Theorem (gradient characterization of regular normals).
6.12 Theorem (basic first-order conditions for optimality).

6.13 Example (variational inequalities and complementarity).
6.14 Theorem (normal cones to sets with constraint structure).
6.15 Corollary (Lagrange multipliers).
6.16 Example (proximal normals).
6.17 Proposition (proximality of normals to convex sets).
6.18 Exercise (approximation of normals).
6.19 Exercise (characterization of boundary points).
6.20 Theorem (envelope representation of convex sets).
6.21 Corollary (polarity correspondence).
6.22 Exercise (pointedness and polarity).
6.23 Example (orthogonal subspaces).
6.24 Example (polarity of normals and tangents to convex sets).
6.25 Definition (regular tangent vectors).
6.26 Theorem (regular tangent cone properties).
6.27 Proposition (normals to tangent cones).
6.28 Theorem (tangent-normal polarity).
6.29 Corollary (characterizations of Clarke regularity).
6.30 Corollary (consequences of Clarke regularity).
6.31 Theorem (tangent cones to sets with constraint structure).
6.32 Exercise (regular tangents under a change of coordinates).
6.33 Definition (recession vectors).
6.34 Exercise (recession cones and convexity).
6.35 Proposition (recession vectors from tangents and normals).
6.36 Theorem (interior tangents and recession).
6.37 Corollary (internal approximation in tangent limits).
6.38 Exercise (convexified normal and tangent cones).
6.39 Exercise (constraint qualification in tangent cone form).
6.40 Example (Mangasarian-Fromovitz constraint qualification).
6.41 Proposition (tangents and normals to product sets).
6.42 Theorem (tangents and normals to intersections).
6.43 Theorem (tangents and normals to image sets).
6.44 Exercise (tangents and normals under set addition).
6.45 Lemma (polars of polyhedral cones; Farkas).
6.46 Theorem (tangents and normals to polyhedral sets).
6.47 Exercise (exactness of tangent approximations).
6.48 Exercise (separation of convex cones).
6.49 Proposition (generic continuity of normal-cone mappings).

7.1 Definition (lower and upper epi-limits).
7.2 Proposition (characterization of epi-limits).
7.3 Exercise (alternative epi-limit formulas).
7.4 Proposition (properties of epi-limits).
7.5 Exercise (epigraphical escape to the horizon).
7.6 Theorem (extraction of epi-convergent subsequences).
7.7 Proposition (level sets of epi-convergent functions).
7.8 Exercise (convergence-preserving operations).
7.9 Exercise (equi-semicontinuity properties of sequences).
7.10 Theorem (epi-convergence versus pointwise convergence).
7.11 Theorem (epi-convergence versus continuous convergence).
7.12 Definition (uniform convergence, extended).
7.13 Proposition (characterizations of uniform convergence).
7.14 Theorem (continuous versus uniform convergence).
7.15 Proposition (epi-convergence from uniform convergence).
7.16 Exercise (Arzelà-Ascoli theorem, extended).
7.17 Theorem (epi-limits of convex functions).
7.18 Corollary (uniform convergence of finite convex functions).
7.19 Example (functions mollified by averaging).
7.20 Definition (semiderivatives).
7.21 Theorem (characterizations of semidifferentiability).

7.22 Corollary (differentiability versus semidifferentiability).
7.23 Definition (epi-derivatives).
7.24 Proposition (geometry of epi-differentiability).
7.25 Definition (subdifferential regularity).
7.26 Theorem (derivative consequences of regularity).
7.27 Example (regularity of convex functions).
7.28 Example (regularity of max functions).
7.29 Proposition (characterization of epi-convergence via minimization).
7.30 Proposition (epigraphical nesting).
7.31 Theorem (inf and argmin).
7.32 Exercise (eventual level boundedness).
7.33 Theorem (convergence in minimization).
7.34 Example (bounds on converging convex functions).
7.35 Definition (eventually prox-bounded sequences).
7.36 Exercise (characterization of eventual prox-boundedness).
7.37 Theorem (epi-convergence through Moreau envelopes).
7.38 Exercise (graphical convergence of proximal mappings).
7.39 Definition (epi-continuity properties).
7.40 Exercise (epi-continuity versus bivariate continuity).
7.41 Theorem (optimal-set mappings).
7.42 Corollary (fixed constraints).
7.43 Corollary (solution mappings in convex minimization).
7.44 Example (projections and proximal mappings).
7.45 Exercise (local minimization).
7.46 Theorem (epi-convergence under addition).
7.47 Exercise (epi-convergence of sums of convex functions).
7.48 Proposition (epi-convergence of max functions).
7.49 Definition (horizon limits of functions).
7.50 Exercise (horizon limit formulas).
7.51 Definition (total epi-convergence).
7.52 Proposition (properties of horizon limits).
7.53 Theorem (automatic cases of total epi-convergence).
7.54 Theorem (horizon criterion for eventual level-boundedness).
7.55 Corollary (continuity of the minimization operation).
7.56 Proposition (convergence of epi-sums).
7.57 Exercise (epi-convergence under inf-projection).
7.58 Theorem (quantification of epi-convergence).
7.59 Theorem (quantification of total epi-convergence).
7.60 Exercise (epi-distance estimates).
7.61 Proposition (auxiliary epi-distance estimate).
7.62 Example (functions satisfying a Lipschitz condition).
7.63 Example (conditioning functions in solution stability).
7.64 Theorem (approximations of optimality).
7.65 Corollary (linear and quadratic growth at solution set).
7.66 Example (proximal estimates).
7.67 Exercise (projection estimates).
7.68 Proposition (distances between convex level sets).
7.69 Theorem (estimates for approximately optimal solutions).

8.1 Definition (subderivatives).
8.2 Theorem (subderivatives versus tangents).
8.3 Definition (subgradients).
8.4 Exercise (regular subgradients from subderivatives).
8.5 Proposition (variational description of regular subgradients).
8.6 Theorem (subgradient relationships).
8.7 Proposition (subgradient semicontinuity).
8.8 Exercise (subgradients versus gradients).
8.9 Theorem (subgradients from epigraphical normals).
8.10 Corollary (existence of subgradients).

8.11 Corollary (subgradient criterion for subdifferential regularity).
8.12 Proposition (subgradients of convex functions).
8.13 Theorem (envelope representation of convex functions).
8.14 Exercise (normal vectors as indicator subgradients).
8.15 Theorem (optimality relative to a set).
8.16 Definition (regular subderivatives).
8.17 Theorem (regular subderivatives versus regular tangents).
8.18 Theorem (subderivative relationships).
8.19 Corollary (subderivative criterion for subdifferential regularity).
8.20 Exercise (subdifferential regularity from smoothness).
8.21 Proposition (subderivatives of convex functions).
8.22 Theorem (interior formula for regular subderivatives).
8.23 Exercise (regular subderivatives from subgradients).
8.24 Theorem (sublinear functions as support functions).
8.25 Corollary (subgradients of sublinear functions).
8.26 Example (vector-max and the canonical simplex).
8.27 Exercise (subgradients of the Euclidean norm).
8.28 Example (support functions of cones).
8.29 Proposition (properties expressed by support functions).
8.30 Theorem (subderivative-subgradient duality for regular functions).
8.31 Exercise (elementary max functions).
8.32 Proposition (calmness from below).
8.33 Definition (graphical derivatives and coderivatives).
8.34 Example (subdifferentiation of smooth mappings).
8.35 Example (subgradients and subderivatives via profile mappings).
8.36 Proposition (positively homogeneous mappings).
8.37 Proposition (basic derivative and coderivative properties).
8.38 Definition (graphical regularity of mappings).
8.39 Example (graphical regularity from graph-convexity).
8.40 Theorem (characterizations of graphically regular mappings).
8.41 Proposition (proto-differentiability).
8.42 Exercise (proto-differentiability of feasible-set mappings).
8.43 Exercise (semidifferentiability of set-valued mappings).
8.44 Exercise (subdifferentiation of derivatives).
8.45 Definition (proximal subgradients).
8.46 Proposition (characterizations of proximal subgradients).
8.47 Corollary (approximation of subgradients).
8.48 Exercise (approximation of coderivatives).
8.49 Theorem (Clarke subgradients).
8.50 Proposition (recession properties of epigraphs).
8.51 Example (nonincreasing functions).
8.52 Example (convex functions of a single real variable).
8.53 Example (subdifferentiation of distance functions).
8.54 Exercise (generic continuity of subgradient mappings).

9.1 Definition (Lipschitz continuity and strict continuity).
9.2 Theorem (Lipschitz continuity from bounded modulus).
9.3 Example (linear and affine mappings).
9.4 Example (nonexpansive and contractive mappings).
9.5 Exercise (rate of convergence to a fixed point).
9.6 Example (distance functions).
9.7 Theorem (strict continuity from smoothness).
9.8 Exercise (calculus of the modulus).
9.9 Exercise (scalarization of strict continuity).
9.10 Proposition (strict continuity in pointwise max and min).
9.11 Example (Pasch-Hausdorff envelopes).
9.12 Exercise (extensions of Lipschitz continuity).
9.13 Theorem (subdifferential characterization of strict continuity).
9.14 Theorem (Lipschitz continuity from convexity).

9.15 Exercise (subderivatives under strict continuity).
9.16 Theorem (subdifferential regularity under strict continuity).
9.17 Definition (strict differentiability).
9.18 Theorem (subdifferential characterization of strict differentiability).
9.19 Corollary (smoothness).
9.20 Corollary (continuity of gradient mappings).
9.21 Corollary (upper subgradient property).
9.22 Exercise (criterion for constancy).
9.23 Proposition (local boundedness of positively homogeneous mappings).
9.24 Proposition (derivatives and coderivatives of single-valued mappings).
9.25 Exercise (differentiability properties of single-valued mappings).
9.26 Definition (Lipschitz continuity of set-valued mappings).
9.27 Definition (sub-Lipschitz continuity).
9.28 Definition (strict continuity of set-valued mappings).
9.29 Proposition (alternative descriptions of strict continuity).
9.30 Theorem (Lipschitz versus sub-Lipschitz continuity).
9.31 Theorem (sub-Lipschitz continuity via strict continuity).
9.32 Example (transformations acting on a set).
9.33 Theorem (truncations of convex-valued mappings).
9.34 Corollary (strict continuity from graph-convexity).
9.35 Example (polyhedral graph-convex mappings).
9.36 Definition (Aubin property and graphical modulus).
9.37 Exercise (distance characterization of the Aubin property).
9.38 Theorem (graphical localization of strict continuity).
9.39 Lemma (extended formulation of the Aubin property).
9.40 Theorem (Mordukhovich criterion).
9.41 Theorem (reinterpretation of normal vectors and subgradients).
9.42 Exercise (epi-Lipschitzian sets and directionally Lipschitzian functions).
9.43 Theorem (metric regularity and openness).
9.44 Example (metric regularity in constraint systems).
9.45 Example (openness from graph-convexity).
9.46 Proposition (inverse properties).
9.47 Example (global metric regularity from polyhedral graph-convexity).
9.48 Theorem (metric regularity from graph-convexity; Robinson-Ursescu).
9.49 Exercise (Lipschitzian properties of derivative mappings).
9.50 Proposition (semidifferentiability from proto-differentiability).
9.51 Example (semidifferentiability of feasible-set mappings).
9.52 Example (semidifferentiability from graph-convexity).
9.53 Definition (strict graphical derivatives).
9.54 Theorem (single-valued localizations).
9.55 Corollary (single-valued Lipschitzian invertibility).
9.56 Theorem (implicit mappings).
9.57 Example (calmness of piecewise polyhedral mappings).
9.58 Theorem (Lipschitz continuous extensions of single-valued mappings).
9.59 Proposition (Lipschitzian properties in limits).
9.60 Theorem (almost everywhere differentiability; Rademacher).
9.61 Theorem (Clarke subgradients of strictly continuous functions).
9.62 Theorem (coderivative duality for strictly continuous mappings).
9.63 Corollary (coderivative criterion for Lipschitzian invertibility).
9.64 Exercise (strict differentiability and derivative continuity).
9.65 Theorem (singularities of Lipschitzian equations; Mignot).
9.66 Definition (graphically Lipschitzian mappings).
9.67 Theorem (subgradients via mollifiers).
9.68 Proposition (constraint relaxation).
9.69 Proposition (extremal principles).

10.1 Theorem (Fermat's rule generalized).
10.2 Example (subgradients in proximal mappings).
10.3 Proposition (normals to level sets).

10.4 Example (level sets with epi-Lipschitzian boundary).
10.5 Proposition (separable functions).
10.6 Theorem (basic chain rule).
10.7 Exercise (change of coordinates).
10.8 Example (composite Lagrange multiplier rule).
10.9 Corollary (addition of functions).
10.10 Exercise (subgradients of Lipschitzian sums).
10.11 Corollary (partial subgradients and subderivatives).
10.12 Example (parametric version of Fermat's rule).
10.13 Theorem (subdifferentiation in parametric minimization).
10.14 Corollary (parametric Lipschitz continuity and differentiability).
10.15 Example (subdifferential interpretation of Lagrange multipliers).
10.16 Proposition (Lipschitzian meaning of constraint qualification).
10.17 Exercise (geometry of sets depending on parameters).
10.18 Exercise (epi-addition).
10.19 Proposition (nonlinear rescaling).
10.20 Definition (piecewise linear-quadratic functions).
10.21 Proposition (properties of piecewise linear-quadratic functions).
10.22 Exercise (piecewise linear-quadratic calculus).
10.23 Definition (amenable sets and functions).
10.24 Example (prevalence of amenability).
10.25 Exercise (consequences of amenability).
10.26 Exercise (calculus of amenability).
10.27 Exercise (calculus of semidifferentiability).
10.28 Example (semidifferentiability of eigenvalue functions).
10.29 Definition (subsmooth functions).
10.30 Proposition (smoothness versus subsmoothness).
10.31 Theorem (consequences of subsmoothness).
10.32 Example (subsmoothness of Moreau envelopes).
10.33 Theorem (subsmoothness and convexity).
10.34 Corollary (higher-order subsmoothness).
10.35 Exercise (calculus of subsmoothness).
10.36 Exercise (subsmoothness and amenability).
10.37 Theorem (coderivative chain rule).
10.38 Corollary (Lipschitzian properties under composition).
10.39 Exercise (outer composition with a single-valued mapping).
10.40 Theorem (inner composition with a single-valued mapping).
10.41 Theorem (coderivatives of sums of mappings).
10.42 Corollary (Lipschitzian properties under addition).
10.43 Exercise (addition of a single-valued mapping).
10.44 Proposition (variational principle for subgradients).
10.45 Exercise (variational principle for regular subgradients).
10.46 Proposition (ε-regular subgradients).
10.47 Proposition (calmness as a constraint qualification).
10.48 Theorem (mean-value, extended).
10.49 Theorem (extended chain rule for subgradients).
10.50 Corollary (normals for Lipschitzian constraints).
10.51 Corollary (pointwise max of strictly continuous functions).
10.52 Exercise (nonsmooth Lagrange multiplier rule).
10.53 Proposition (representations of subsmoothness).
10.54 Lemma (smooth bounds on semicontinuous functions).
10.55 Theorem (properties under epi-addition).
10.56 Example (subsmoothness of distance functions).
10.57 Theorem (subsmoothness in parametric optimization).
10.58 Proposition (minimization estimate for a convex function).

11.1 Theorem (Legendre-Fenchel transform).
11.2 Exercise (improper conjugates).
11.3 Proposition (inversion rule for subgradient relations).

11.4 Example (support functions and cone polarity).
11.5 Theorem (horizon functions as support functions).
11.6 Exercise (support functions of level sets).
11.7 Exercise (conjugacy as cone polarity).
11.8 Theorem (dual properties in minimization).
11.9 Example (classical Legendre transform).
11.10 Example (linear-quadratic functions).
11.11 Example (self-conjugacy).
11.12 Example (log-exponential function and entropy).
11.13 Theorem (strict convexity versus differentiability).
11.14 Theorem (piecewise linear-quadratic functions in conjugacy).
11.15 Lemma (linear-quadratic test on line segments).
11.16 Corollary (minimum of a piecewise linear-quadratic function).
11.17 Corollary (polyhedral sets in duality).
11.18 Example (piecewise linear-quadratic penalties).
11.19 Example (general polarity of sets).
11.20 Exercise (dual properties of polar sets).
11.21 Proposition (conjugate composite functions from polar gauges).
11.22 Proposition (conjugation in product spaces).
11.23 Theorem (dual operations).
11.24 Corollary (rules for support functions).
11.25 Corollary (rules for polar cones).
11.26 Example (distance functions, Moreau envelopes and proximal hulls).
11.27 Exercise (proximal mappings and projections as gradient mappings).
11.28 Example (piecewise linear-quadratic envelopes).
11.29 Example (norm duality for sublinear mappings).
11.30 Exercise (uniform boundedness of sublinear mappings).
11.31 Exercise (duality in calculating adjoints of sublinear mappings).
11.32 Proposition (operations on piecewise linear-quadratic functions).
11.33 Corollary (conjugate formulas in piecewise linear-quadratic case).
11.34 Theorem (epi-continuity of the Legendre-Fenchel transform; Wijsman).
11.35 Corollary (convergence of support functions and polar cones).
11.36 Theorem (cone polarity as an isometry; Walkup-Wets).
11.37 Corollary (Legendre-Fenchel transform as an isometry).
11.38 Lemma (dual calculations in parametric optimization).
11.39 Theorem (dual problems of optimization).
11.40 Corollary (general best-case primal-dual relations).
11.41 Example (Fenchel-type duality scheme).
11.42 Theorem (piecewise linear-quadratic optimization).
11.43 Example (linear and extended linear-quadratic programming).
11.44 Example (dualized composition).
11.45 Definition (Lagrangians and dualizing parameterizations).
11.46 Example (multiplier rule in Lagrangian form).
11.47 Example (Lagrangians in the Fenchel scheme).
11.48 Proposition (convexity properties of Lagrangians).
11.49 Definition (saddle points).
11.50 Theorem (minimax relations).
11.51 Corollary (saddle point conditions for convex problems).
11.52 Example (minimax problems).
11.53 Theorem (perturbation of saddle values; Golshtein).
11.54 Example (perturbations in extended linear-quadratic programming).
11.55 Definition (augmented Lagrangian functions).
11.56 Exercise (properties of augmented Lagrangians).
11.57 Example (proximal Lagrangians).
11.58 Example (sharp Lagrangians).
11.59 Theorem (duality without convexity).
11.60 Definition (exact penalty representations).
11.61 Theorem (criterion for exact penalty representations).
11.62 Example (exactness of linear or quadratic penalties).

11.63 Exercise (generalized conjugate functions).
11.64 Example (proximal transform).
11.65 Example (full quadratic transform).
11.66 Example (basic quadratic transform).
11.67 Theorem (double-min duality in optimization).

12.1 Definition (monotonicity).
12.2 Example (positive semidefiniteness).
12.3 Proposition (monotonicity of differentiable mappings).
12.4 Exercise (operations that preserve monotonicity).
12.5 Definition (maximal monotonicity).
12.6 Proposition (existence of maximal extensions).
12.7 Example (maximality of continuous monotone mappings).
12.8 Exercise (graphs and images under maximal monotonicity).
12.9 Exercise (maximal monotonicity in one dimension).
12.10 Exercise (operations that preserve nonexpansivity).
12.11 Proposition (monotone versus nonexpansive mappings).
12.12 Theorem (maximality and single-valued resolvants).
12.13 Example (Yosida regularization).
12.14 Lemma (inverse-resolvant identity).
12.15 Theorem (Minty parameterization of a graph).
12.16 Exercise (displacement mappings).
12.17 Theorem (monotonicity versus convexity).
12.18 Corollary (normal cone mappings).
12.19 Proposition (proximal mappings).
12.20 Corollary (projection mappings).
12.21 Example (projections on subspaces).
12.22 Exercise (projections on a polar pair of cones).
12.23 Exercise (Moreau envelopes and Yosida regularizations).
12.24 Definition (cyclical monotonicity).
12.25 Theorem (characterization of convex subgradient mappings).
12.26 Exercise (cyclical monotonicity in one dimension).
12.27 Example (monotone mappings from convex-concave functions).
12.28 Example (hypomonotone mappings).
12.29 Proposition (monotonicity of piecewise polyhedral mappings).
12.30 Proposition (subgradients of piecewise linear-quadratic functions).
12.31 Example (piecewise linear projections).
12.32 Theorem (graphical convergence of monotone mappings).
12.33 Corollary (convergent subsequences).
12.34 Exercise (maximal monotone limits).
12.35 Theorem (subgradient convergence for convex functions; Attouch).
12.36 Exercise (convergence of convex function values).
12.37 Theorem (normal vectors to domains).
12.38 Corollary (local boundedness of monotone mappings).
12.39 Corollary (full domains and ranges).
12.40 Exercise (uniform local boundedness in convergence).
12.41 Theorem (near convexity of domains and ranges).
12.42 Corollary (ranges and growth).
12.43 Theorem (maximal monotonicity under composition).
12.44 Corollary (maximal monotonicity under addition).
12.45 Example (truncation).
12.46 Exercise (restriction of arguments).
12.47 Example (monotonicity along lines).
12.48 Example (monotonicity in variational inequalities).
12.49 Exercise (alternative expression of a variational inequality).
12.50 Example (variational inequalities for saddle points).
12.51 Theorem (solutions to a monotone generalized equation).
12.52 Exercise (existence of solutions to variational inequalities).
12.53 Definition (strong monotonicity).

12.54 Proposition (consequences of strong monotonicity).
12.55 Corollary (inverse strong monotonicity).
12.56 Theorem (contractions from strongly monotone mappings).
12.57 Example (contractions in solving variational inequalities).
12.58 Definition (strong convexity).
12.59 Exercise (strong monotonicity versus strong convexity).
12.60 Proposition (dualization of strong convexity).
12.61 Exercise (hypoconvexity properties of proximal functions).
12.62 Proposition (hypoconvexity and subsmoothness of double envelopes).
12.63 Theorem (continuity properties).
12.64 Exercise (proto-derivatives of monotone mappings).
12.65 Theorem (differentiability and semidifferentiability).
12.66 Corollary (generic single-valuedness from monotonicity).
12.67 Theorem (structure of maximal monotone mappings).
12.68 Proposition (strict continuity of ε-subgradient mappings).

13.1 Definition (twice differentiable functions).
13.2 Theorem (extended second-order differentiability).
13.3 Definition (second subderivatives).
13.4 Definition (positive homogeneity of degree p).
13.5 Proposition (properties of second subderivatives).
13.6 Definition (twice semidifferentiable and twice epi-differentiable functions).
13.7 Exercise (characterizations of second-order semidifferentiability).
13.8 Example (connections with second-order differentiability).
13.9 Proposition (piecewise linear-quadratic functions).
13.10 Example (lack of second-order semidifferentiability of max functions).
13.11 Definition (second-order tangent sets and parabolic derivability).
13.12 Proposition (properties of second-order tangents).
13.13 Proposition (chain rule for second-order tangents).
13.14 Theorem (chain rule for second subderivatives).
13.15 Corollary (second-order epi-differentiability from full amenability).
13.16 Example (second-order epi-differentiability of max functions).
13.17 Exercise (indicators and geometry).
13.18 Exercise (addition of a twice smooth function).
13.19 Proposition (second subderivatives of a sum of functions).
13.20 Proposition (second subderivatives of convex functions).
13.21 Theorem (second-order epi-differentiability in duality).
13.22 Corollary (conjugates of fully amenable functions).
13.23 Example (piecewise linear-quadratic penalties).
13.24 Theorem (second-order conditions for optimality).
13.25 Example (second-order optimality with smooth constraints).
13.26 Exercise (second-order optimality with penalties).
13.27 Definition (prox-regularity of functions).
13.28 Definition (subdifferential continuity).
13.29 Exercise (characterization of subdifferential continuity).
13.30 Example (prox-regularity from convexity).
13.31 Exercise (prox-regularity of sets).
13.32 Proposition (prox-regularity from amenability).
13.33 Proposition (prox-regularity versus subsmoothness).
13.34 Proposition (functions with strictly continuous gradient).
13.35 Exercise (perturbations of prox-regularity).
13.36 Theorem (subdifferential characterization of prox-regularity).
13.37 Proposition (proximal mappings and Moreau envelopes).
13.38 Exercise (projections on prox-regular sets).
13.39 Lemma (difference quotient relations).
13.40 Theorem (proto-differentiability of subgradient mappings).
13.41 Corollary (subgradient proto-differentiability from amenability).
13.42 Corollary (second derivatives and quadratic expansions).
13.43 Corollary (normal cone mappings and projections).

13.44 Example (polyhedral sets).
13.45 Exercise (semidifferentiability of proximal mappings).
13.46 Proposition (Lipschitzian geometry of subgradient mappings).
13.47 Theorem (perturbation of optimality conditions).
13.48 Theorem (perturbation of solutions to variational conditions).
13.49 Proposition (prox-regularity and amenability of second subderivatives).
13.50 Exercise (subgradient derivative formula in full amenability).
13.51 Theorem (almost everywhere second-order differentiability).
13.52 Theorem (generalization of second-derivative symmetry).
13.53 Corollary (symmetry in derivatives of proximal mappings).
13.54 Corollary (symmetry in derivatives of projections).
13.55 Exercise (second-order Lipschitz continuity).
13.56 Proposition (strict second-order differentiability).
13.57 Theorem (derivative-coderivative inclusion).
13.58 Corollary (subgradient coderivatives of fully amenable functions).
13.59 Definition (parabolic subderivatives).
13.60 Example (parabolic aspects of twice smooth functions).
13.61 Exercise (parabolic aspects of piecewise linear-quadratic functions).
13.62 Example (parabolic subderivatives versus second-order tangents).
13.63 Exercise (chain rule for parabolic subderivatives).
13.64 Proposition (properties of parabolic subderivatives).
13.65 Definition (parabolic regularity).
13.66 Theorem (parabolic version of second-order optimality).
13.67 Theorem (parabolic aspects of fully amenable functions).
13.68 Corollary (parabolic aspects of fully amenable sets).

14.1 Definition (measurable mappings).
14.2 Proposition (preservation with closure of images).
14.3 Theorem (measurability equivalences under closed-valuedness).
14.4 Theorem (measurability in hyperspace terms).
14.5 Theorem (Castaing representations).
14.6 Corollary (measurable selections).
14.7 Example (convex-valued and solid-valued mappings).
14.8 Theorem (graph measurability).
14.9 Exercise (measurability from semicontinuity).
14.10 Theorem (measurability versus continuity).
14.11 Proposition (unions, intersections, sums and products).
14.12 Exercise (hulls and polars).
14.13 Theorem (composition of mappings).
14.14 Corollary (composition with measurable functions).
14.15 Example (images under Carathéodory mappings).
14.16 Theorem (implicit measurable functions).
14.17 Exercise (projection mappings).
14.18 Exercise (mixed operations).
14.19 Example (variable scalar multiplication).
14.20 Theorem (graphical and pointwise limits).
14.21 Exercise (horizon cones and limits).
14.22 Proposition (limits of simple mappings).
14.23 Proposition (convergence of measurable selections).
14.24 Proposition (almost everywhere convergence).
14.25 Proposition (almost uniform convergence).
14.26 Theorem (tangent and normal cone mappings).
14.27 Definition (normal integrands).
14.28 Proposition (basic consequences of normality).
14.29 Example (Carathéodory integrands).
14.30 Example (autonomous integrands).
14.31 Example (joint lower semicontinuity).
14.32 Example (indicator integrands).
14.33 Proposition (measurability of level-set mappings).

14.34 Corollary (joint measurability criterion).
14.35 Exercise (closures of measurable integrands).
14.36 Theorem (measurability in constraint systems).
14.37 Theorem (measurability of optimal values and solutions).
14.38 Exercise (Moreau envelopes and proximal mappings).
14.39 Proposition (normality test for convex integrands).
14.40 Proposition (scalarization of normal integrands).
14.41 Corollary (alternative interpretation of normality).
14.42 Theorem (continuity properties of normal integrands).
14.43 Corollary (Scorza-Dragoni property).
14.44 Proposition (addition and pointwise max and min).
14.45 Proposition (composition operations).
14.46 Corollary (multiplication by nonnegative scalars).
14.47 Proposition (inf-projection of integrands).
14.48 Example (epi-addition and epi-multiplication).
14.49 Exercise (convexification of normal integrands).
14.50 Theorem (conjugate integrands).
14.51 Example (support integrands).
14.52 Exercise (envelope representations through convexity).
14.53 Proposition (limits of integrands).
14.54 Exercise (horizon integrands and horizon limits).
14.55 Exercise (approximation by simple integrands).
14.56 Theorem (subderivatives and subgradient mappings).
14.57 Proposition (subgradient characterization of convex normality).
14.58 Proposition (integral functionals).
14.59 Definition (decomposable spaces).
14.60 Theorem (interchange of minimization and integration).
14.61 Exercise (simplified interchange criterion).
14.62 Example (reduced optimization).

Index of Notation

$\mathbb{R}$: the real numbers
$\overline{\mathbb{R}}$: the extended real numbers
$\mathbb{N}$: the natural numbers
$\mathbb{Q}$: the rational numbers
$\mathbb{B}$: closed unit ball
$|x|$: Euclidean norm
$\langle x, y \rangle$: canonical inner product
$\mathcal{C}^{k+}$: strictly continuous derivatives
$C \setminus D$: relative complement
$\operatorname{bdry} C$: boundary
$\operatorname{cl} C$: closure
$\operatorname{int} C$: interior
$\operatorname{rint} C$: relative interior
$\operatorname{csm} C$: cosmic closure
$\operatorname{hzn} C$: horizon
$\operatorname{con} C$: convex hull of set C
$\operatorname{con} f$: convex hull of function f
$\operatorname{pos} C$, $\operatorname{pos} f$: positive hull
$\operatorname{dom} f$: effective domain
$\operatorname{epi} f$: epigraph
$\operatorname{hypo} f$: hypograph
$\operatorname{dir} w$: the direction of w
$\operatorname{gph} S$: graph of S
$\operatorname{cl} f$: lsc regularization
$\operatorname{cl} S$: graphical closure
$\operatorname{lip} f$: Lipschitz modulus of function f
$\operatorname{lip} F$, $\operatorname{lip} S$: Lipschitz modulus of mappings
C^∞: horizon cone
f^∞: horizon function
S^∞: horizon mapping
$\operatorname{lev}_{\leq \alpha} f$: lower level set
$\mathrm{d}f(\bar{x})$: subderivative function
$\widehat{\mathrm{d}}f(\bar{x})$: regular subderivative function
$\partial f(\bar{x})$: subgradient set
$\widehat{\partial} f(\bar{x})$: regular subgradient set
$\partial^\infty f(\bar{x})$: horizon subgradient set
$\bar{\partial} f(\bar{x})$: convexified subgradient set
$\mathrm{d}^2 f(\bar{x})$, $\mathrm{d}^2 f(\bar{x} \mid v)$: second subderivatives
$e_\lambda f$: Moreau envelope
$P_\lambda f$: proximal mapping
P_C: projection mapping
$T_C(\bar{x})$: tangent cone
$\widehat{T}_C(\bar{x})$: regular tangent cone
$N_C(\bar{x})$: normal cone
$\widehat{N}_C(\bar{x})$: regular normal cone
$R_C(\bar{x})$: local recession cone
K^*: polar cone
C°: polar set
f^*, f^{**}: conjugate, biconjugate
δ_C: indicator function
σ_C: support function
γ_C: gauge function
$d_C(x)$, $d(x, C)$: distance from C
logexp: log-exponential function
vecmax: vector-max function
$DS(\bar{x} \mid \bar{v})$, $DS(\bar{x})$: graphical derivative
$\widehat{D}S(\bar{x} \mid \bar{v})$, $\widehat{D}S(\bar{x})$: regular derivative
$D^*S(\bar{x} \mid \bar{v})$, $D^*S(\bar{x})$: coderivative
$\widehat{D}^*S(\bar{x} \mid \bar{v})$, $\widehat{D}^*S(\bar{x})$: regular coderivative
$D_*S(\bar{x} \mid \bar{v})$, $D_*S(\bar{x})$: strict derivative
$f \# g$: epi-addition
$\lambda \star f$: epi-multiplication
$\mathcal{N}(\bar{x})$: neighborhood collection
$\mathcal{N}_\infty$: subsets of $\mathbb{N}$ containing all ν sufficiently large
$\mathcal{N}_\infty^\#$: infinite subsets of $\mathbb{N}$
g-lim: graphical limits
p-lim: pointwise limits
e-lim: epigraphical limits
h-lim: hypographical limits
$\lim^\infty$, $\liminf^\infty$, $\limsup^\infty$: horizon limits
$\xrightarrow{p}$: pointwise convergence
$\xrightarrow{e}$: epigraphical convergence
$\xrightarrow{h}$: hypographical convergence
$\xrightarrow{c}$: cosmic convergence
$\xrightarrow{t}$: total convergence
$\xrightarrow{g}$: graphical convergence
$\underset{C}{\to}$: convergence within C
$\underset{f}{\to}$: f-attentive convergence
$\underset{N}{\to}$: convergence indexed by N
sets, cl-sets: spaces of sets
fcns, lsc-fcns: spaces of functions
maps, osc-maps: spaces of mappings
$\hat{d\!l}_\rho$: estimates of set and epi-distances
$d\!l$, $d\!l_\rho$: set distances, epi-distances

Index of Topics

addition rules, 85, 93, 230, 445, 471, 493
  for coderivatives, 456+
  for graphical derivatives, 457
  in epi-convergence, 247, 275
  in monotonicity, 534, 557
  in subdifferentiation, 430+, 441
  in subsmoothness, 452
  second-order, 602+
  set-valued, 151, 184
adjoint mappings, 329, 348, 498+
  norms of, 497, 530
affine
  functions, 42, 474+
  hulls, 64, 652
  independence, 54
  mappings, 352
  sets, 43+
  supports, 308+
Alexandrov's theorem, 638
algorithmic mappings, 151
almost differentiable functions, 483
almost strictly convex functions, 483, 542
amenable functions, 442+, 471, 595, 599, 612, 621+, 625+, 632, 637+, 639+
  in subsmoothness, 452
amenable sets, 442+, 621, 638, 639
argmin, argmax, 2, 8, 11
Arzelà-Ascoli theorem, 180, 192, 195, 252, 294
asymptotic cone, 106
asymptotic equicontinuity, 173+, 248+
Attouch's theorem, 552, 640
Aubin property, 376+, 393, 417
  calculus, 453, 455, 457
  in convergence, 402
  inverse form, 386
  of derivative mappings, 391
augmented Lagrangians, 518+
augmenting functions, 519+
autonomous integrands, 662

balls, 25, 48, 111
  rational, 644
barrier functions, 4, 20, 35+
barycentric coordinates, 54
B-differentiability, 294
biconjugate functions, 474

bifunction, 270
Borel σ-fields, 648
boundary characterization, 214
boxes, 3, 204, 211

calmness, 322, 348, 351, 391, 399, 420
  as a constraint qualification, 348, 459, 471
cancellation rules, 94+, 107
Castaing representations, 646, 657, 680
Carathéodory integrands, 662+, 681
Carathéodory mappings, 654+, 680
Carathéodory's theorem, 55, 75
celestial model, 78
chain rules, 427+, 438, 445, 462+, 470
  for coderivatives, 452+
  geometric, 202, 208+, 221+, 236
  in conjugacy, 493
  second-order, 593+, 639
Clarke regularity: see regularity
closedness criteria, 84+
closures, 14
  in convexity, 57+
  of epigraphs, 14+
  of functions, 14, 59, 67, 88, 100
  of graphs, 154+
cluster points, 9
  in set limits, 121, 145
coderivatives, 324+, 348, 631+
  approximation, 336
  calculus, 452+, 470
  duality with graphical derivatives, 330, 405
  nonsingularity condition, 387
  of single-valued mappings, 366
  regular, 324+
coercivity, 91+, 99+, 530, 557
  convex case, 92
  dualization, 479
  in epi-convergence, 266, 280+
  in minimization, 93
co-finiteness, 107
compactification, 79, 81, 105
compactness principles, 82
  in epi-convergence, 252, 283, 293
  in graphical convergence, 170, 551
  in set convergence, 120+, 140, 142, 145, 147

σ-compactness, 189
complementarity, 208, 236
complete measurable spaces, 648
composite optimization, 428+, 464, 509, 608+
composition, 30, 50, 151, 184
  calculus: see chain rules
  in epi-convergence, 247, 276
  in monotonicity, 556
  in subsmoothness, 452
concave functions, 39, 42
conditioning functions, 286+, 297
cones, 80, 232+
  as epigraphs, 87+
  as representing directions, 80, 106
  convergence, 119
  convex, 83+, 86, 215
  derivable, 198
  horizon, 80+
  operations, 97
  normal: see normals
  pointed, 86, 99, 107, 385
  polyhedral, 103+
  set distances, 138
  tangent: see tangents
conjugate functions, 473+, 604+
  generalized, 525+, 532
  in one dimension, 530
  self-conjugacy, 482
  via cone polarity, 478, 530
conjugate points, 480
connected sets, 55, 75, 117, 145
constancy criterion, 364, 417
constraint qualifications, 209, 236, 310, 348, 429, 432, 435+, 442, 459, 463+, 470+
  in tangent form, 225
  Lipschitzian interpretation, 436, 471
  Mangasarian-Fromovitz, 226+, 236+
  Slater, 237
constraint relaxation, 4, 413, 420
constraints, 2+, 12, 150, 208+, 221, 231, 272
  implicit, 2
  parameterized, 18, 150
contingent cone, 232
contingent derivatives, 345, 294
continuous convergence, 175+, 195
  of difference quotients, 256, 332
  metric description, 181
  relation to epigraphical, 250
  relation to graphical, 178+
  relation to uniform, 177, 252
continuity, 13, 152+, 192+
  cosmic, 164+, 194
  from semicontinuity, 152, 160
  Lipschitz: see Lipschitz continuity
  of gradient mappings, 363, 406, 448
  of monotone mappings, 567+
  of operations, 125+, 145+, 275+, 292, 296
  of subgradient mappings, 573
  strict: see strict continuity
  subdifferential, 610+
  total, 164+, 184+, 194
  uniform, 157
  with local boundedness, 161
contractive mappings, 353, 563+
control theory, 192, 679, 682
convergence
  almost everywhere, 657+
  almost uniform, 658+
  attentive to a function, 271+, 301
  continuous: see continuous convergence
  cosmic, 122+, 145, 301, 416
  directional, 197
  epigraphical: see epi-convergence
  graphical: see graphical convergence
  of coderivatives, 402
  of convex hulls, 128
  of difference quotients, 197+, 256+, 326, 331
  of images, 125+, 129+
  of intersections, 131
  of matrices, 352
  of measurable mappings, 655+
  of monotone mappings, 551+
  of positive hulls, 186
  of projections, 127
  of subgradient mappings, 402, 552
  of unions, 128
  pointwise, 166+, 170+, 194, 239+, 252
  quantification, metrics, 131+
  rough, 145
  to a direction point, 79
  total, 124+, 146, 279+
  uniform: see uniform convergence
  variational: see epi-convergence
convex analysis, 74+
convex combinations, 39+, 53+
convex-concave functions, 511+, 514+, 532, 548
convex functions, 38, 74+
  continuity properties, 59+, 359
  derivative tests, 45+
  differentiability properties, 360+, 364, 403+, 603+
  in subsmoothness, 450
  operations, 49+
  subdifferential regularity, 261
convex hulls, 53+, 99+, 107+, 155, 652
  compactness, 56
  convergence, 128
  cosmic, 99
  of functions, 57, 99, 474, 671
convex integrands, 666, 674
convex optimization, 41+, 237
convex sets, 38+, 74+
  algebra, 51

cosmic version, 98, 107
operations, 49+
polyhedral: see polyhedral sets
convex-valued mappings, 148, 155+, 184, 189+, 195, 375+
cosmic closure, 77, 80+
cosmic convexity, 97+
cosmic limits, 79, 122, 145
cosmic metric, 141+, 164
cosmic space, 77, 104
counter-coercivity, 91+
cyclical monotonicity, 547+

decomposable spaces, 676+, 682
derivability, 197+, 201, 217, 221, 233
epigraphical, 260
parabolic, 591+
derivable cone, 198, 217, 233
derivatives
directional, 257, 582+
epigraphical: see epi-derivatives
graphical: see graphical derivatives
subdifferentiation of, 333, 348
determinants, 73
diagrams of relationships,
in graphical differentiation, 329
in subdifferentiation, 320
in variational geometry, 221
difference quotients, 197, 255+, 299, 236
second-order, 583+, 633
differentiability 255+, 258, 367
almost everywhere, 344, 403+, 448, 469, 571, 626+
duality with strict convexity, 483
of subgradient mappings, 581, 621
strict, 362, 368, 396, 406, 416, 457, 464
second-order, 580+, 587, 609, 621, 626, 630, 638, 641
Dini derivatives, 345
Dini normals, 236
Dini subgradients: see subgradients, regular
directional derivatives, 257+, 582+
directionally Lipschitzian functions, 385, 417
direction of a vector, 77
direction points, 77+, 105, 233
displacement mappings, 541
ρ-distance, 134+, 146+
distance functions, 6, 19, 24, 26, 550
dualization, 495
in constraint relaxation, 413
in semicontinuity, 156+
in set convergence, 110, 113, 132+, 176
in uniform convergence, 176
Lipschitz continuity of, 353
subdifferentiation, 340, 344, 348, 415
subsmoothness, 468
under prox-regularity, 618
domain mappings, 661
domains, 5, 40, 149+
duality gap, 518+, 532
dualizing parameterizations, 508+, 519
dual operations, 493+, 530
dual problems, 502+, 518, 531
double-min type, 527+, 532
nonconvex, 519+, 521, 532

effective domains, 5, 40
eigenvalue functions, 73, 76, 446, 471
Ekeland's principle, 31, 472
entropy functions, 102, 107, 296, 482
envelopes
double, Lasry-Lions, 33, 37, 244, 450, 567
in generalized conjugacy, 525+
Moreau: see Moreau envelopes
Pasch-Hausdorff, 296, 357, 420, 665
representing convex functions, 309, 344, 473+, 672
representing convex sets, 214
representing sublinear functions, 309
epi-addition, 23+, 36, 437, 467, 493, 670
cancellation rule, 94
convexity, 51
role of coercivity, 94
epi-composition, 27+, 37, 51, 493, 498
epi-continuity, 270, 296
of Legendre-Fenchel transform, 500, 530+
epi-convergence, 240+, 292+
convex case, 252, 267, 276, 280
cosmic, 279, 283
dualization, 499
in minimization, 262+, 281, 295
in second-order theory, 639
of indicators, 244
operations, 247, 275+, 292, 296+
quantification, metrics, 282+, 296+
relation to continuous, 250, 275
relation to pointwise, 247+
relation to uniform, 250+
total, 279+, 283, 297, 502
via level sets, 246, 293
epi-derivatives, 259+, 294, 295
parabolic, 633+
second-order, 586+, 595, 599+, 625, 639+
epi-distances, 531, 282+, 297
epigraphical projection, 18+, 35
epigraphs, 7, 34+
conical, 88
convex, 40
polyhedral, 104+
epigraphical mappings, 270, 661
epigraphical nesting, 264, 296
epi-hypo-convergence, 532
epi-limits, 240+
epi-Lipschitzian sets, 385+, 417, 425, 470
epi-measurability, 662, 681+

epi-multiplication, 24+, 37, 100, 475, 478, 670
epi-sums, 23
equations, 150
  generalized, 150, 172, 561
  Lipschitzian, 406
equi-coercive sequences, 280
equicontinuity, 173+, 195, 248
equilibrium conditions, 208+
equi-lsc, 248+, 293
  aymptotic 248+
escape to the horizon, 116, 170, 245
Euclidean norm, 6, 318
eventually bounded sequences, 123, 126, 249
expansions, 580+, 587+, 621, 638
extended arithmetic, 15+, 37, 354
extended-real-valued functions, 1
extended linear-quadratic programming, 506
extremal principles, 414, 420, 458, 528, 532

fans, 418
Farkas' lemma, 230
feasible-set mappings, 154, 154, 156, 194, 331, 392, 419, 664
feasible solutions, 6
Fenchel duality scheme, 505+, 510, 531
Fermat's rule, 421+, 432, 459, 469+
Filippov's lemma, 680
filters, 109
fixed points, 151, 353
Fréchet normals, 236
full cone, 81
function spaces, 270, 281+, 676
function-valued mappings, 270

gap distance, 113, 144
gauge functions, 101+, 490+
geometric derivability: see derivability
geometric means, 72
generalized linear mappings, 329
generic continuity, 187+, 195+, 232, 237, 343, 348
  almost everywhere sense, 570
generic differentiability: see differentiability, a.e.
generic single-valuedness, 571, 573
Golshtein's theorem, 515, 532
gradient mappings, 363, 406, 448, 580, 609, 614+, 627
  inversion 480
gradients, 47
graph distances, 182, 195
graph-convex mappings, 155, 159, 348, 376, 390, 393, 417+, 454, 456+
graphical convergence, 166+, 194
  compactness property, 170
  of difference quotients, 331
  of measurable mappings, 655, 681
  of monotone mappings, 551+
  of projection mappings, 169
  of subgradient mappings, 552+, 640
  quantification, metrics, 182
  relation to pointwise, 170+
  relation to uniform, 178+
  total, 185+
  uniformity in the large, 168
graphical derivatives, 324+, 347+, 631+
  duality with coderivatives, 330, 348, 631+
  norms of, 366
  of single-valued mappings, 366
  regular, 324+
  strict, 393+, 419, 627+
graphically Lipschitzian mappings, 234, 408, 420, 622
graphical modulus, 377+
graphs of mappings, 148
grill, 109
growth properties, 14, 91+, 106, 288+, 479, 556

half-spaces, 42+, 204, 214+
Hahn-Banach theorem, 75
Hausdorff metric: see Pompeiu-Hausdorff distance
hemispherical model, 79
Hessian matrices, 47, 480, 580+, 621, 627+
hit-and-miss criteria, 112, 144+
horizon, 77, 80+, 116
horizon cones, 80+, 106+, 163, 237
  convex case, 82
  epigraphical, 87+, 106
  graphical, 159
  of level sets, 90+
horizon functions, 87+, 673
  as support functions, 477, 530
  convex case, 88+
  of indicators, 89
  of quadratic functions, 89
  of translates, 89
horizon limits, 122+, 145, 184, 278+, 296, 301, 656, 673
horizon mappings, 159, 163, 185+, 194
hyperplanes, 43+, 62
hyperspaces of sets, 134+, 144, 146, 182, 645, 679
  metrics, 138+, 146
hypo-convergence, 242+, 252
hypoconvexity, 567
hypographs, 13, 41
hypomonotone mappings, 549, 575, 577, 618

image sets, 149
  boundedness, 158
  closedness, 84, 163
  convergence, 125+, 165+, 186
  convexity, 50

horizon cones, 84, 87
variational geometry, 228+
implicit mappings, 150, 397, 419, 654
improper functions, 5, 41, 475
indicator functions, 6+, 26, 35
as integrands, 663
convex, 40
epi-convergence, 244
second-order properties, 600, 621
subderivatives of, 300, 600
subgradients of, 310
indicator mappings, 270
infimum, 1
inf-addition, 15
inf-compactness, 12, 35, 37, 266
inf-convolution, 23
inf-projection, 16+, 51, 93, 101, 281, 433+, 493, 498
inner semicontinuity, 144, 152+, 190, 193+
from convexity, 155
integral functionals, 76, 643, 675+, 681+
conjugate, 682+
integration, 192, 643
of subgradient mappings, 547+
interchange rules, 28, 677, 682
interior, 14
in convexity, 58
of a level set, 59
of an epigraph, 14, 59
relative, 64+, 75
inverse images: see image sets
inverse of a mapping, 149
inverse-resolvant identity, 540, 576
invertibility, single-valued, 396, 406, 419
isc mappings: see inner semicontinuity

Jacobian, 202
generalized, 419
Jensen's inequality, 39
jets, 641

Lagrange multipliers, 208, 211, 236, 428+, 464, 471, 509, 608+
interpretation, 435+
Lagrangian functions, 508, 531, 548, 608
augmented, 519
proximal, 520
Lebesgue σ-fields, 649
Legendre-Fenchel transform, 473+, 480, 529
as an isometry, 501, 530
epi-continuity, 500+
level boundedness, 11+, 30, 35, 159, 266+
eventual, 266
uniform, 16+, 159
level coercivity, 91+, 530
dualization, 479
level sets, 8+, 424, 470, 663
convexity, 42+, 71
distance estimates, 291
in epi-convergence, 246
support functions of, 478
limiting subgradients, 345
limits, 108+
continuous, 175+
cosmic, 79, 122+, 145
epigraphical, 240+
graphical, 167+
horizon, 122+, 145, 278+, 297
inner, outer, 109+, 144, 167+
lower, upper, 8, 13+, 15, 22, 29+
pointwise, 166+
uniform, 175+
linear functions, mappings, 329, 352, 481
linear programming, 236, 506, 531
linear-quadratic functions, 48
line segment principle, 58, 98
line segments, 38
Lipschitz continuity, 349+, 415
constants, 350+, 370+
directional, 385
extensions, 358, 400, 420
from convexity, 359
in derivatives, 295, 391
in epi-distances, 285
in limits, 401+
local, 350, 373
modulus, 350+, 416
of derivative mappings, 391+
of set-valued mappings, 368+, 453+, 457
relation to strict continuity, 349+, 373, 416
second-order, 630, 641
Lipschitzian invertibility, 396, 416
local boundedness, 157+, 160+, 184+, 19r3 194, 454
of monotone mappings, 554
relative, 162
localizations of mappings, 394, 419, 615
locally closed sets, 28
local minimization, 273
local semicontinuity, 28
log-determinant function, 73, 76
log-exponential function, 27, 48, 72, 76, 89, 482
lower bounds, 12
lower-$\mathcal{C}^k$ functions, 447+, 452, 465, 471, 496, 613+
lower limits, 8+, 29+
lower semicontinuity, 8+, 16, 22, 30, 35+
local, 28+
of mappings, 193+
lsc functions: see lower semicontinuity
lsc regularization, 14
Lusin's criterion, 649+, 680

majorize, 56

mapping terminology, 148+
marginal functions, 470
matrix functions, 73
matrix inversion, 73
matrix orderings, 73, 97
max functions, 356, 431, 443, 446+, 471, 590, 600
  epi-convergence, 277
  subdifferentiation, 261, 321, 446+, 463
maximal monotonicity, 535+
maximum, 1
meager sets, 187
mean-value theorem, 461, 472
measurable mappings, 643+, 679+
  measurability-preserving operations, 651+
measurable sets, spaces, 643+
metric regularity, 386+, 417+
  global, 389
  in constraint systems, 388
Michael representations, 190, 192, 195
minimax problems, 512, 514+, 531
minimax relations, 512
minimization estimates, 287+, 469
minimizing sequences, 30+
minimum, 1,5, 6, 11+
  convergence, 262+
Minkowski addition, 24, 37, 151
Minkowski-Weyl theorem, 76, 103, 107
minorize, 56
Minty parameterization, 540+, 545
mollifiers, 254, 294, 409, 420
monotone mappings, 192, 533+, 575+
  as relations in $I\!R^2$, 343, 536, 547+, 575
  continuity, differentiability, 567+
  convergence, 551, 554
  cyclical, 547
  domains, ranges, 553+
  local boundedness, 554
  maximal, 535+, 575
  maximality rules, 556+
  nonexpansive, 537, 543+
  strict, 559
  strong, 562+
monotone sequences, 111
Mordukhovich criterion, 379+, 418
Moreau envelopes, 20+, 24, 26, 31, 450, 468, 472, 526, 617+, 665
  convex case, 52, 75, 95, 268
  in duality, 495+
  in epi-convergence, 244, 268, 296, 552
  in Yosida regularization, 546
  piecewise linear-quadratic, 550
  subsmoothness, 450
multivalued mappings, 148, 575

near convexity, 555
neighborhoods, 6, 54
nondecreasing, nonincreasing functions, 89, 339, 348, 386
nonexpansive mappings, 353, 537+, 543
nonnegative, nonpositive orthant, 3
nonsmooth manifolds, 234
nonsmoothness, 23+, 26, 196
normal cone mappings, 202, 550, 621, 624, 641
  measurability, 659, 681
  monotonicity, 543, 545, 618
  via projections, 213
normal cones: see normals
normal integrands, 660+, 681+
  continuity properties, 668
  convergence, 672
  operations on, 669, 682
normals, 199+, 232+
  approximation, 214, 237
  calculus rules, 202, 208, 227+
  Clarke convexified, 225+, 234, 415
  convex case, 203+, 477
  duality with tangents, 200, 216, 219+, 225
  epigraphical, 304
  graphical, 324
  limiting, 234+
  Lipschitzian interpretation, 383
  proximal, 212+, 234+
  regular, 199+, 205, 213, 235
  to boxes, 204+
  to cones, 219, 237, 477
norms, 6, 49, 101
  matrix, 352
  positively homogeneous mappings, 364+, 417, 530
  polar, 491
  subgradients of, 318
nowhere dense sets, 187

openness with linear rate, 386+, 417+
optimality, 205+, 233, 236, 310, 422, 428+, 432+, 459
  local, global, 6, 41
  second-order, 606+, 635+
optimal-set mappings, 16, 162, 272+, 281
optimal solutions, 6, 16+, 479, 488, 503+
  ε-optimality, 30, 291
  measurable parameterization, 662
optimal values, 6, 11, 16
  dualization, 479
  measurable parameterization, 664
orderings, 73, 94+, 107
orthant, 3
orthogonality, 216
osc hull, 155
osc mappings: see outer semicontinuity
outer semicontinuity, 144, 152+, 184+, 193+
  cosmic, 165+
  in local boundedness, 160

relative, 162
total, 165

Pasch-Hausdorff envelopes, 296, 357+, 420, 665
Painlevé-Kuratowski convergence, 111+, 144+
parabolic derivability, 591+, 634, 639
parabolic regularity, 635+, 639
parabolic subderivatives, 633+
parametric minimization, 16+, 35+, 93, 105, 107, 272+, 283, 296, 433, 459, 468, 502+
paratingent derivatives, 419
partial subdifferentiation, 431, 442
penalty functions, 4, 18, 35+, 414, 489, 506, 509, 529, 605+, 608, 639
epi-convergence, 245
exact, 523+, 532
penalty methods, 18, 35, 293
perturbations, 150
in duality, 502+, 517, 518+, 531
in minimax format, 515+, 532
of continuity, 162
of solutions, 623+, 641+
piecewise linearity, 67+, 76, 104+, 107, 399, 420, 471, 550+
conjugacy, 484
operations, 105
piecewise linear-quadratic functions, 440+, 496, 550, 588, 595, 605+, 637, 639
attainment of minimum, 488, 530
conjugacy, 484, 530
operations on, 498+, 530
subdifferentiation, 440
subgradient characterization, 550
piecewise linear-quadratic optimization, 506, 513
piecewise polyhedral mappings, 399, 418, 420, 550+
piecewise polyhedral sets, 76
pointedness of cones, 85, 99, 107, 386
pointwise infimum, supremum, 22+
polar cones, 215+, 477+, 495, 546
convergence, 500+
isometry, 501, 531
polyhedral, 230, 489, 530
polar gauges, norms, 491+
polar sets, 490+
polyhedral functions, 104+
polyhedral sets, 43, 76, 491, 496, 592, 621
as convex hulls, 104+, 107
as epigraphs, 76, 104+
operations, 105
unions, 69
variational geometry, 231, 237

Pompeiu-Hausdorff distance, 111, 117+, 120, 124, 137+, 144, 146, 161, 192, 368+, 417
positive-definiteness, 47+, 533+
positive-definite programming, 107
positively homogeneous
functions, 87, 97+, 258+, 280
higher order, 584
mappings, 328, 365, 497
positive hulls, 100+, 103+, 186, 652
posynomials, 71, 76
profile mappings, 152, 248, 250+, 294, 325
projection mappings, 18, 162, 273, 340, 618, 621, 629, 655
as selections, 188
as gradient mappings, 496
convex case, 51+, 213
estimates, 290
graphical convergence, 169
in set convergence, 114, 145, 169
in variational geometry, 212+
maximal monotonicity, 545+
onto domains, 553+
piecewise linear, 550
proper functions, 5
proper integrands, 661
proto-derivatives, 331+, 348, 391, 395 417, 419, 640+
of monotone mappings, 569
of projection mappings, 621
of subgradient mappings, 619+, 640
prox-boundedness, 21+, 92, 296
eventual, 267+
proximal hulls, 31+, 495+, 526
proximal mappings, 20+, 37, 162, 273, 422, 617+, 621, 629+, 665
as gradient mappings, 496
convex case, 52, 75, 96
estimates, 289
graphical convergence, 269
in epi-convergence, 269, 552
maximal monotonicity, 543
piecewise linear, 550
proximal normals: see normals
proximal smoothness, 640
proximal subgradients, 333+, 345
proximal threshold, 21+
proximal transform, 526
prox-regularity, 610+, 619, 625, 638, 640
subdifferential characterization, 615
pseudo-inverse of a matrix, 481
pseudo-Lipschitz: see Aubin property
pseudo-metrics, 134, 135

quadratic expansions, 580+, 587+, 621, 638
quadratic functions, 48, 481, 484
quadratic programming, 236, 507, 530
quadratic transform, 526+, 532

quasi-convexity, 76
quasi-differentiable functions, 346

Rademacher's theorem, 403+, 415, 420, 639
range of a mapping, 149
rays, 77
ray space model, 78, 98, 307
recession vectors, 222+, 237
  convex case, 106, 222
  epigraphical, 338+, 348
reduced optimization, 678, 682
regularity
  Clarke, 199+, 209, 220+, 232, 235, 368, 424, 444
  graphical, 348, 368, 380, 387, 392, 455, 464, 497
  metric: see metric regularity
  parabolic, 635+, 639+
  prox-: see prox-regularity
  second-order, 640
  subdifferential, 260+, 295, 307, 313+, 344, 409, 422, 444, 448
relative interiors, 64+, 75
  in monotonicity, 555
resolvants, 539+, 550, 553, 569+
Robinson-Ursescu theorem, 390, 418
rough convergence, 145

saddle points, 512+, 560+
Sard's theorem, 420
scalar multiples, 438
Scorza-Dragoni property, 668
second-order theory, 234, 579+
selections
  continuous, 188+, 192, 195, 574
  measurable, 647, 656, 680
self-conjugacy, 482
semiderivatives, 256+, 295, 345, 360, 367, 417, 419, 455, 457, 471
  calculus, 446, 457
  from regularity, 261
  in perturbations, 641+
  in subsmoothness, 448+
  of mappings, 332, 348, 367, 392+, 419, 455, 457, 570
  of projections, proximal mappings, 621+
  of subgradient mappings, 619+
  second-order, 586+, 604, 639
separable functions, 49, 426, 493
separable spaces, 141, 147, 183
separation of sets, 62+, 75, 129, 231, 234
  nonconvex, 414+, 420
sequences, 108+
  minimizing, 30+
set algebra, 25
set convergence, 108+, 144+
  compactness property, 120+
  convex case, 119+
  cosmic, 122+, 141+
  projection characterization, 169
  quantification, metrics, 131+
  total, 124+, 142
  uniformity, 114
set distances, 117, 135+, 139, 146+, 176
  between cones, 140
set-valued mappings, 148+
simple mappings, 656
simplex, 54
  canonical, 318
single-valuedness, 148
singularity, 406
solid sets, 647
solution estimates, 286+, 469
smooth manifolds, 203
smoothness, 23, 304, 314, 325, 363, 447, 634
star-shaped sets, 120, 145
stationary points, 470+, 528
stereographic metric, 146
stochastic programming, 641, 681, 682
strict continuity 349+, 416
  calculus of, 356+, 453+
  differentiability consequences, 403
  from convexity, 359
  graphical localization, 378
  in optimization, 435
  in subsmoothness, 448
  of convex-valued mappings, 375+
  of derivatives, classes $\mathcal{C}^{k+}$, 355, 627, 630
  of set-valued mappings, 370+
  of ε-subgradient mappings, 574
  modulus, 350, 371+
  relation to Lipschitz, 349+, 373, 416
  scalarization, 356, 366
  subdifferential characterization, 358
strict convexity, 38, 483
strong convexity, 565+
strict derivatives, 361, 368, 396, 406, 419, 457, 464
  graphical, 393+, 627+
  second-order, 630
strict monotonicity, 533, 559
strong monotonicity, 562+
subderivatives, 299+, 325, 345, 348
  Clarke type, 416
  convex case, 314, 340
  distance functions, 340
  duality with subgradients, 301, 316, 320, 344
  of integrands, 673
  parabolic, 633+, 639
  regular, 311+, 344
  second-order, 584+, 594+
  under strict continuity, 360+
subdifferential continuity, 610+
subgradient mappings, 366, 542, 554, 573, 673+

coderivatives, 622+, 627+
continuity properties, 302, 573
convergence, 292, 335, 348, 402, 552, 554
derivative properties, 573, 581, 619, 625+
inversion, 476, 480, 511, 531
monotonicity properties, 542+, 547+, 549, 575+, 615+
piecewise polyhedral, 550
subgradients, 301+, 325
'approximate', 347
approximation, 335
Clarke convexified, 336, 344, 346, 403, 415, 472
convergence, 335
convex case, 308, 340, 343, 363, 422, 547
duality with subderivatives, 301, 316, 320, 344
existence, 307
horizon, 301+, 336
in infinite-dimensional spaces, 347, 415
in viscosity theory, 347
of distance functions, 340
of norms, support functions, 318, 477
Lipschitzian interpretation, 383, 423
mollifier, 409
proximal, 333+, 345
regular, 301
ε-regular, 346, 458, 472
singular, 345+
upper, 302, 363
variational principles, 458
sublinear functions, 87, 101, 309, 317+
subgradients of, 318
sublinear mappings, 328, 348, 497+, 530
sub-Lipschitz continuity, 370+, 417
inverse form, 389
relation to Lipschitz continuity, 373
subsequence notation, 108+
subsmooth functions, 447+, 466, 468+, 471
from amenability, 452
prox-regularity of, 613
relation to convexity, 450, 549
rules for preserving, 452
second derivatives, 626+, 630+
subspaces, 83, 216, 477, 545
sup-addition, 15
support functions, 317+, 477, 495, 672
convergence, 500
entropy case, 482
of domains, 477, 556
of level sets, 478, 482
of polyhedral sets, 489
supporting half-spaces, 204
supremum, 1
symmetry in second derivatives, 621, 627

tangent cone mappings, 220
measurability, 659, 581
tangent cones: see tangents
tangents, 197, 232+
calculus rules, 202, 221+, 227+, 236+
convex case, 203+
convexified, 225
derivable, 197, 217
duality with normals, 200, 203+, 216, 219+, 225
epigraphical, 300, 312
graphical, 324
higher-order, 591+, 634, 639+
interior, 224+, 237
regular, 217, 220+, 385
relation to recession, 224+
to boxes, 204
threshold of prox-boundedness, 21+, 267
translates of sets, 25
triangle inequalities, 135, 137
truncations, 137, 161
in set convergence, 145, 147
inverse, 162
of convex-valued mappings, 375
of monotone mappings, 557

uniform convergence, 175+, 250+
extended-real-valued, 251+, 294
metric description, 181
of convex functions, 253+
of difference quotients, 257+
of monotone sequences, 181
relation to continuous, 177, 252
relation to epigraphical, 252, 294
relation to graphical, 178
uniform set approximation, 114, 157, 168
uniqueness in optimality, 41
upper-$\mathcal{C}^k$ functions, 447, 467+, 471, 614+
upper integrals, 675
upper limits, 13+
upper Lipschitzian, 420
upper semicontinuity, 13+, 36+
of mappings, 193+
usc functions: see upper semicontinuity
usc regularization, 14

variational inequalities, conditions, 208, 236, 292, 559+, 565, 624
variational principles, 31, 37, 414, 458
vector inequalities, 96+, 107
vector-max function, 27, 48, 89, 318
Vietoris topology, 145, 194

Walkup-Wets theorem, 501, 531
well-posedness, 36
Wijsman's theorem, 500, 530

Yosida regularization, 38, 540+, 546

zero cone, 80

Grundlehren der mathematischen Wissenschaften

A Series of Comprehensive Studies in Mathematics

A Selection

247. Suzuki: Group Theory I
248. Suzuki: Group Theory II
249. Chung: Lectures from Markov Processes to Brownian Motion
250. Arnold: Geometrical Methods in the Theory of Ordinary Differential Equations
251. Chow/Hale: Methods of Bifurcation Theory
252. Aubin: Nonlinear Analysis on Manifolds. Monge-Ampère Equations
253. Dwork: Lectures on ρ-adic Differential Equations
254. Freitag: Siegelsche Modulfunktionen
255. Lang: Complex Multiplication
256. Hörmander: The Analysis of Linear Partial Differential Operators I
257. Hörmander: The Analysis of Linear Partial Differential Operators II
258. Smoller: Shock Waves and Reaction-Diffusion Equations
259. Duren: Univalent Functions
260. Freidlin/Wentzell: Random Perturbations of Dynamical Systems
261. Bosch/Güntzer/Remmert: Non Archimedian Analysis – A System Approach to Rigid Analytic Geometry
262. Doob: Classical Potential Theory and Its Probabilistic Counterpart
263. Krasnosel'skiĭ/Zabreĭko: Geometrical Methods of Nonlinear Analysis
264. Aubin/Cellina: Differential Inclusions
265. Grauert/Remmert: Coherent Analytic Sheaves
266. de Rham: Differentiable Manifolds
267. Arbarello/Cornalba/Griffiths/Harris: Geometry of Algebraic Curves, Vol. I
268. Arbarello/Cornalba/Griffiths/Harris: Geometry of Algebraic Curves, Vol. II
269. Schapira: Microdifferential Systems in the Complex Domain
270. Scharlau: Quadratic and Hermitian Forms
271. Ellis: Entropy, Large Deviations, and Statistical Mechanics
272. Elliott: Arithmetic Functions and Integer Products
273. Nikol'skiĭ: Treatise on the shift Operator
274. Hörmander: The Analysis of Linear Partial Differential Operators III
275. Hörmander: The Analysis of Linear Partial Differential Operators IV
276. Liggett: Interacting Particle Systems
277. Fulton/Lang: Riemann-Roch Algebra
278. Barr/Wells: Toposes, Triples and Theories
279. Bishop/Bridges: Constructive Analysis
280. Neukirch: Class Field Theory
281. Chandrasekharan: Elliptic Functions
282. Lelong/Gruman: Entire Functions of Several Complex Variables
283. Kodaira: Complex Manifolds and Deformation of Complex Structures
284. Finn: Equilibrium Capillary Surfaces
285. Burago/Zalgaller: Geometric Inequalities
286. Andrianaov: Quadratic Forms and Hecke Operators
287. Maskit: Kleinian Groups
288. Jacod/Shiryaev: Limit Theorems for Stochastic Processes
289. Manin: Gauge Field Theory and Complex Geometry

290. Conway/Sloane: Sphere Packings, Lattices and Groups
291. Hahn/O'Meara: The Classical Groups and K-Theory
292. Kashiwara/Schapira: Sheaves on Manifolds
293. Revuz/Yor: Continuous Martingales and Brownian Motion
294. Knus: Quadratic and Hermitian Forms over Rings
295. Dierkes/Hildebrandt/Küster/Wohlrab: Minimal Surfaces I
296. Dierkes/Hildebrandt/Küster/Wohlrab: Minimal Surfaces II
297. Pastur/Figotin: Spectra of Random and Almost-Periodic Operators
298. Berline/Getzler/Vergne: Heat Kernels and Dirac Operators
299. Pommerenke: Boundary Behaviour of Conformal Maps
300. Orlik/Terao: Arrangements of Hyperplanes
301. Loday: Cyclic Homology
302. Lange/Birkenhake: Complex Abelian Varieties
303. DeVore/Lorentz: Constructive Approximation
304. Lorentz/v. Golitschek/Makovoz: Construcitve Approximation. Advanced Problems
305. Hiriart-Urruty/Lemaréchal: Convex Analysis and Minimization Algorithms I. Fundamentals
306. Hiriart-Urruty/Lemaréchal: Convex Analysis and Minimization Algorithms II. Advanced Theory and Bundle Methods
307. Schwarz: Quantum Field Theory and Topology
308. Schwarz: Topology for Physicists
309. Adem/Milgram: Cohomology of Finite Groups
310. Giaquinta/Hildebrandt: Calculus of Variations I: The Lagrangian Formalism
311. Giaquinta/Hildebrandt: Calculus of Variations II: The Hamiltonian Formalism
312. Chung/Zhao: From Brownian Motion to Schrödinger's Equation
313. Malliavin: Stochastic Analysis
314. Adams/Hedberg: Function spaces and Potential Theory
315. Bürgisser/Clausen/Shokrollahi: Algebraic Complexity Theory
316. Saff/Totik: Logarithmic Potentials with External Fields
317. Rockafellar/Wets: Variational Analysis
318. Kobayashi: Hyperbolic Complex Spaces
319. Bridson/Haefliger: Metric Spaces of Non-Positive Curvature
320. Kipnis/Landim: Scaling Limits of Interacting Particle Systems
321. Grimmett: Percolation
322. Neukirch: Algebraic Number Theory
323. Neukirch/Schmidt/Wingberg: Cohomology of Number Fields
324. Liggett: Stochastic Interacting Systems: Contact, Voter and Exclusion Processes
325. Dafermos: Hyperbolic Conservation Laws in Continuum Physics
326. Waldschmidt: Diophantine Approximation on Linear Algebraic Groups
327. Martinet: Perfect Lattices in Euclidean Spaces
328. Van der Put/Singer: Galois Theory of Linear Differential Equations
329. Korevaar: Tauberian Theory. A Century of Developments
330. Mordukhovich: Variational Analysis and Generalized Differentiation I: Basic Theory
331. Mordukhovich: Variational Analysis and Generalized Differentiation II: Applications
332. Kashiwara/Schapira: Categories and Sheaves. An Introduction to Ind-Objects and Derived Categories
333. Grimmett: The Random-Cluster Model
334. Sernesi: Deformations of Algebraic Schemes
335. Bushnell/Henniart: The Local Langlands Conjecture for GL(2)
336. Gruber: Convex and Discrete Geometry
337. Maz'ya/Shaposhnikova: Theory of Sobolev Multipliers. With Applications to Differential and Integral Operators
338. Villani: Optimal Transport: Old and New